Henderson's Dictionary of

BIOLOGICAL TERMS

Henderson's Dictionary of

BIOLOGICAL TERMS

Ninth Edition

Sandra Holmes

Longman
Scientific &
Technical

Longman Scientific & Technical
Longman Group UK Limited
Longman House, Burnt Mill, Harlow,
Essex CM20 2JE, England
and Associated Companies throughout the world

Originally published by Oliver and Boyd under the title
A Dictionary of Scientific Terms
First edition 1920
Seventh edition 1960
First published under the title
A Dictionary of Biological Terms
Eighth edition 1963
Reprinted 1967, 1968, 1972
Reprinted by Longman Group Limited 1975, 1976
Ninth edition 1979
Reprinted 1982
First published in paperback 1985
Reprinted 1986, 1987

British Library Cataloguing in Publication Data
Henderson, I.F.
Henderson's dictionary of biological terms.—
9th ed.
1. Biology—Dictionaries
I. Title   II. Henderson, W.D.   III. Holmes,
Sandra   IV. Henderson, I.F. Dictionary of
biological terms
574'.03'21   QH302.5

ISBN 0-582-44759-3

Produced by Longman Group (FE) Ltd
Printed in Hong Kong

CONTENTS

PREFACE

In the sixteen years that have elapsed since the publication of the Eighth Edition of Henderson's *Dictionary of Biological Terms* there has been a marked expansion in research in the biological sciences. Indeed, it is no exaggeration to say that certain important branches of biology have actually come into being during this time; others have certainly come to maturity. As a result, the revision of a book such as this has had to be extensive, with the incorporation of many new terms and the redefinition of existing ones. The number of headwords has been increased from about 16,500 to 22,500. Taxonomic terms have also been included down to the order level. Some obsolete terms in the eighth edition have been deleted, but those considered to be of historic interest are retained.

The headwords are now arranged in strict alphabetical order and acronyms are incorporated in the main body of the dictionary. Most headwords are entered in the singular, but the reader is advised that some are in the plural. Definitions under a headword are separated by a semi-colon, except where there are two parts of speech which are then separated by a full-stop. Material at the end of a headword, such as a synonym or comparison, is separated by a semi-colon if it refers to the whole headword, while material referring to just one definition of a headword is separated by a comma.

An important departure from previous editions is the inclusion of tables of classification of the Plant and Animal Kingdoms. These can be found as appendices to the dictionary proper. It is hoped that the users of the book will find them useful.

I would like to thank the Science Publisher and Science Editor and their staff at Longman House for their continued help and encouragement throughout this project.

Finally, comments concerning errors or omissions in this edition will be greatly appreciated, so that they may be rectified in future reprints or editions.

Sandra Holmes
Windsor 1979

ABBREVIATIONS

a. adjective
adv. adverb
alt. alternative name (synonym)
appl. applies or applied to
Ar. Arabic
A.S. Anglo-Saxon
ca. (*circa*) approximately
cf. (*confer*) compare
Dan. Danish
dim. diminutive
Dut. Dutch
EC Enzyme Commission number
  (1979)
e.g. (*exempli gratia*) for example
et al. (*et alii*) and others
etc. (*et cetera*) and so forth
F. French
Ger. German
Gk. Greek
ib., ibid. (*ibidem*) in the same place
Icel. Icelandic
i.e. (*id est*) that is
i.q. (*idem quod*) the same as
It. Italian
L. Latin
L. Lower (*appl.* geological periods)
L.L. Late Latin

M. Mid (*appl.* geological periods)
Mal. Malaysian
M.E. Middle English
n haploid number of chromosomes
$2n$ diploid number of chromosomes
n. noun
O.E. Old English
O.F. Old French
pert. pertaining to
π (pi) 3.14159265 (22/7)
plu. plural
p.p.m. parts per million
q.v. (*quod vide*) which see
r.n. Enzyme Commission
  recommended name
Russ. Russian
sing. singular
s.l. (*sensu lato*) in the broad sense
Sp. Spanish
sp. species (*sing.*)
spp. species (*plu.*)
s.s. (*sensu stricto*) in the narrow sense
Sw. Swedish
U. Upper (*appl.* geological periods)
v. verb
x haploid generation
$2x$ diploid generation

UNITS AND CONVERSIONS

acre	4046.86 m² (4840 sq. yd)
angstrom unit*, Å	10^{-10} m
atmosphere, standard, atm	101325 Pa (14.72 p.s.i)
bar	10^5 Pa
British thermal unit, Btu	1.055 kJ
British thermal unit/hour, Btu/h	0.293 W
bushel, bu	0.0364 m³
bushel (US), bu	0.0352 m³
calorie, thermochemical	4.184 J
centimetre, cm	10^{-2} m (0.394 in)
cubic foot, ft³	0.0283 m³
cubic inch, in³	16.387 cm³
cubic yard, yd³	0.7645 m³

* The use of this unit is being discouraged. It is being replaced by the nanometre
  (nm, 10^{-9} m).

degree Celsius (centigrade), °C†	(9/5) °F
degree Fahrenheit, °F†	(5/9) °C
dram (avoirdupois), dr	1.772 g
fathom	1.829 m (6 ft)
fluid ounce, fl oz	28.413 cm³
foot, ft	0.3048 m
gallon, gal	4.546 dm³
gallon (US), US gal	3.785 dm³
grain, gr	64.799 mg
hectare, ha	10^4 m² (2.471 acres)
hour, h	3600 s
hundredweight, cwt	50.802 kg
inch, in	25.4 mm
joule, J (SI unit of energy)	kg m² s^{-2} (0.239 cal)
kilocalorie/hour, kcal/h	1.163 W
kilogram, kg (SI unit of mass)	2.20 lb
litre, l	dm³ (1.76 pt)
metre, m (SI unit of length)	39.37 in
micron ‡, μm	10^{-6} m
mile	1.6093 km
millibar, mbar	100 Pa
millimetre, mm	10^{-3} m (0.039 in)
millimetre of mercury, mmHg	133.322 Pa
millimetre of water	9.807 Pa
minute (time), min	60 s
molar, mol/l, m	1 mol dm^{-3}
nanometre, nm	10^{-9} m (10Å)
ounce (avoirdupois), oz	0.0283 kg
pascal, Pa (SI unit of pressure)	kg m^{-1} s^{-2}
pint, pt	0.568 dm³
pound (avoirdupois), lb	0.4536 kg
square foot, ft²	0.0929 m²
square inch, in²	645.16 mm²
square mile, sq mile	2.590 km² (640 acres)
square yard, yd²	0.836 m²
ton (long)	1016.05 kg (2240 lb)
tonne (metric ton), t	1 Mg (0.984 tons)
watt, W (SI unit of power)	kg m² s^{-3}
yard, yd	0.9144 m

† To convert temperature in °C to °F multiply by 9/5 and add 32; to convert °F to °C subtract 32, then multiply by 5/9.

‡ The term 'micron' is no longer recommended; 'micrometre' is preferred. The symbols μ and mμ should not be used; they should give way to μm (micrometre) and nm (nanometre) respectively. See p. xi for a full list of SI prefixes.

SI PREFIXES

The following prefixes may be used to construct decimal multiples of units.

Multiple	Prefix	Symbol	Multiple	Prefix	Symbol
10^{-1}	deci	d	10	deca	da
10^{-2}	centi	c	10^2	hecto	h
10^{-3}	milli	m	10^3	kilo	k
10^{-6}	micro	μ	10^6	mega	M
10^{-9}	nano	n	10^9	giga	G
10^{-12}	pico	p	10^{12}	tera	T
10^{-15}	femto	f			
10^{-18}	atto	a			

GREEK ALPHABET

Greek characters are often used in biology as symbols or in nomenclature. They are therefore listed below for ease of reference.

Name	Greek letter	English equivalent	Name	Greek letter	English equivalent
alpha	A α	a	nu	N ν	n
beta	B β	b	xi	Ξ ξ	x
gamma	Γ γ	g	omicron	O o	o
delta	Δ δ	d	pi	Π π	p
epsilon	E ε	e	rho	P ρ	r
zeta	Z ζ	z	sigma	Σ σ	s
eta	H η	e	tau	T τ	t
theta	Θ θ	th	upsilon	Y ν	u
iota	I ι	i	phi	Φ ϕ	ph
kappa	K κ	k	chi	X χ	ch
lambda	Λ λ	l	psi	Ψ ψ	ps
mu	M μ	m	omega	Ω ω	o

COMMON LATIN AND GREEK NOUN ENDINGS

sing.	plu.
-a	-ae (L.)
-a	-ata (Gk.)
-is	-es (L.)
-on	-a (Gk.)
-um	-a (L.)
-us	-i (L.)

THE DICTIONARY

A

ABA abscisic acid, *q.v.*

abactinal *a.* [L. *ab*, from; Gk. *aktis*, ray] not situated on the ambulacral area, *appl.* area of echinoderm body without tube-feet and in which madreporite is usually included; *alt.* abambulacral, antambulacral, anti-ambulacral.

abambulacral *a.* [L. *ab*, from; *ambulare*, to walk] abactinal, *q.v.*

A-band the anisotropic or doubly refracting band of a sarcomere, which appears dark and is made of both actin and myosin filaments; *alt.* A-disc, Q-disc, sarcous disc.

abapical *a.* [L. *ab*, from; *apex*, summit] *pert.* or situated at lower pole; away from the apex.

abaxial *a.* [L. *ab*, from; *axis*, axle] *pert.* that surface of any structure which is remote or turned away from the axis; *cf.* adaxial.

abaxile *a.* [L. *ab*, from; *axis*, axle] *appl.* embryo whose axis has not the same direction as axis of seed.

abbreviated *a.* [L. *ad*, to; *brevis*, short] shortened; curtailed.

abcauline *a.* [L. *ab*, from; *caulis*, stalk] outwards from or not close to the stem; *cf.* adcauline.

abdomen *n.* [L. *abdomen*, belly] in vertebrates, the part of the body cavity containing the digestive organs, in mammals separated from the thorax by the diaphragm; in arthropods and certain polychaetes, the posterior part of the body; in tunicates, the section of the body containing stomach and intestines.

abdominal *a.* [L. *abdomen*, belly] *pert.* abdomen; *appl.* structures, organs, or parts of organs situated in, on, or closely related to, the abdomen.

abdominal pores single or paired openings leading from coelom to exterior, in cyclostomes and certain fishes.

abdominal reflex contraction of abdominal wall muscles when skin over side of abdomen is stimulated.

abdominal regions 9 areas into which the abdomen is divided by 2 horizontal and 2 vertical imaginary lines; hypochondriac (2), lumbar (2), inguinal (2), epigastric, umbilical, hypogastric.

abdominal ribs ossifications occurring in fibrous tissue in abdominal region between skin and muscles in certain reptiles.

abdominal ring one of 2 openings in fasciae of abdominal muscles through which passes spermatic cord in male, round ligament in female; *alt.* inguinal ring.

abducens *n.* [L. *abducere*, to lead away] the 6th cranial nerve, supplying the rectus externus muscle of the eyeball.

abduction *n.* [L. *abducere*, to lead away] movement away from the median axis; *cf.* adduction.

abductor *n.* [L. *abductus*, led away] a muscle that draws a limb or part outwards.

aberrant *a.* [L. *aberrare*, to stray] with characteristics not in accordance with type, *appl.* species, etc.

abhymenial *a.* [L. *ab*, from; Gk. *hymēn*, membrane] on or *pert.* the side of the lamella opposite that of the hymenium in agarics.

abience *n.* [L. *abire*, to depart] retraction from stimulus; avoiding reaction; *cf.* adience.

abient *a.* [L. *abire*, to depart] avoiding the source of stimulation; *cf.* adient.

abiocoen *n.* [Gk. *a*, not; *bios*, life; *koinōs*, in common] the abiotic parts of the environment, in total.

abiogenesis *n.* [Gk. *a*, not; *bios*, life; *genesis*, birth] the production of living from non-living matter, considered either in terms of the origin of life on earth, or as spontaneous generation, a theory held up to the 19th century which stated that micro-organisms or higher organisms could arise from non-living material; *alt.* archebiosis, archegenesis, archigenesis, autogenesis, heterogenesis; *cf.* biogenesis.

abiology *n.* [Gk. *a*, not; *bios*, life; *logos*, discourse] the study of non-living things; *alt.* anorganology.

abioseston *n.* [Gk. *a*, not; *bios*, life; *sēsis*, sifting] tripton, *q.v.*

abiosis *n.* [Gk. *a*, not; *biosis*, living] apparent suspension of life.

abiotic *a.* [Gk. *a*, not; *biōtikos*, *pert.* life] non-living.

abiotic environment the non-living part of the environment consisting of topography, climatic factors, and inorganic nutrients.

abiotrophy *n.* [Gk. *a*, not; *bios*, life; *trophē*, maintenance] differential vitality or longevity of cells or tissues.

abjection *n.* [L. *abjicere*, to cast away] the shedding of spores, as from sporophores, usually with some force.

abjunction *n.* [L. *abjungere*, to unyoke] the delimitation of spores by septa at tip of hypha; *v.* abjoint.

ablactation *n.* [L. *ab*, from; *lactare*, to give milk] cessation of milk secretion; weaning.

abomasum *n.* [L. *ab*, from; *omasum*, paunch] in ruminants, the 4th chamber of the stomach, being the true digestive stomach; *alt.* read, reed, rennet-stomach.

aboospore *n.* [L. *ab*, from; Gk. *ōon*, egg; *sporos*, seed] a spore developed from an unfertilized female gamete; *alt.* azygospore, parthenospore.

aboral *a.* [L. *ab*, from; *os, oris*, mouth] away from, or opposite to, the mouth; *cf.* oral.

abortion *n.* [L. *abortus*, premature birth] premature birth; arrest of development of an organ.

abranchiate *a.* [Gk. *a*, without; *brangchia*, gills] without gills.

abrupt *a.* [L. *abrumpere*, to break off] appearing as if broken, or cut off, at extremity; *alt.* truncate.

abruptly-acuminate having a broad extremity from which a point arises; *appl.* leaf.

abruptly-pinnate having the main axis of epipodium not winged, but bearing a number of secondary axes which are winged.

abrupt speciation the formation of a species as the result of a sudden change in chromosome number or constitution.

abcise *v.* [L. *abscidere*, to cut off] to become separated; to fall off, as leaves, fruit, etc.

abscisic acid (ABA) *n.* [L. *abscidere*, to cut off] a hormone found in many higher plants which promotes senescence, leaf fall, dormancy in buds, and antagonizes the effect of the growth-promoting hormones; *alt.* abscisin, dormin.

abscisin *n.* [L. *abscidere*, to cut off] abscisic acid, *q.v.*, especially abscisin II, abscisin I being a less effective and chemically unrelated substance.

absciss, abscissile, abscission layer the layer of cells in the abscission zone, the breakdown of which separates a leaf, fruit, flower, etc., from the plant; *alt.* separation layer.

abscission *n.* [L. *abscindere*, to cut off] the separation of parts.

abscission zone the region at the base of a leaf, flower, fruit, or other part of the plant consisting of the abscission layer of weak cells and the protective layer of corky cells which protect the wound when the part falls.

absorption *n.* [L. *absorbere*, to suck in] intussusception of fluid by living cells or tissues; passage of nutritive material through living cells; of light when neither reflected nor transmitted.

abstriction *n.* [L. *abstringere*, to cut off] the process of detaching spores or conidia by rounding off of tips of sporophores; abjunction and abscission.

abterminal *a.* [L. *ab*, from; *terminus*, limit] going from the end inwards.

abyssal *a.* [Gk. *abyssos*, unfathomed] *pert.* depths of ocean beyond the continental shelf, *appl.* organisms or material usually found there; *alt.* abysmal.

abyssobenthic *a.* [Gk. *abyssos*, unfathomed; *benthos*, depths of sea] *pert.* or found on the ocean floor at the depths of the ocean, in the abyssal zone.

abyssopelagic *a.* [Gk. *abyssos*, unfathomed; *pelagos*, sea] *pert.* or inhabiting the ocean depths of the abyssal zone, but floating, not on the ocean floor.

Acalephae *n.* [Gk. *akalēphē*, stinging nettle] in some classifications, a group of coelenterates including the hydroids, jellyfish, and related forms.

acanaceous *a.* [Gk. *akanos*, thistle] prickly; bearing prickles, as leaves.

acanth- *alt.* akanth-.

acantha *n.* [Gk. *akantha*, thorn] prickle; spinous process.

acanthaceous *a.* [Gk. *akantha*, thorn] bearing thorns or prickles.

Acantharia *n.* [Gk. *akantha*, thorn] an order of Sarcodina, formerly included in the Radiolaria, whose spicules are made of acanthin.

acanthin *n.* [Gk. *akantha*, thorn] the substance forming the skeleton of Acantharia, now thought to be strontium sulphate.

acanthion *n.* [Gk. *akanthion*, small thorn] the most prominent point on the anterior nasal spine.

Acanthobdellae *n.* [Gk. *akantha*, thorn; *bdella*, leech] an order of Hirudinea somewhat intermediate between Hirudinea and Oligochaeta.

acanthocarpous *a.* [Gk. *akantha*, thorn; *karpos*, fruit] having fruit covered with spines or prickles.

Acanthocephala *n.* [Gk. *akantha*, thorn; *kephalē*, head] a phylum of pseudocoelomate animals, commonly called thorny-headed worms, that as adults are intestinal parasites of vertebrates and as larvae have an arthropod host.

acanthocephalous *a.* [Gk. *akantha*, thorn; *kephalē*, head] with hooked proboscis.

acanthocladous *a.* [Gk. *akantha*, thorn; *klados*, branch] having spiny branches.

Acanthocotyloidea *n.* [Gk. *akantha*, thorn; *kotylē*, cup; *eidos*, form] an order of Monopisthocotylea having one or more testes and a small opisthaptor with 16 spines.

acanthocyst *n.* [Gk. *akantha*, thorn; *kystis*, bladder] a sac containing lateral or reserve stylets in Nemertini.

Acanthodiformes *n.* [Gk. *akantha*, thorn; L. *forma*, shape] an order of acanthodians having only one dorsal fin.

Acanthodii, acanthodians *n., n.plu.* [Gk. *akanthōdēs*, thorny] a group of fish, at present considered a subclass of Osteichthyes but formerly placed with the sharks or placoderms, which existed from Silurian to Permian times.

acanthodion *n.* [Gk. *akanthōdēs*, thorny] a tarsal seta containing extension of a sensory basal cell, in Acarina; *plu.* acanthodia.

acanthoid *a.* [Gk. *akantha*, thorn; *eidos*, shape] resembling a spine or prickle; *alt.* spiniform.

acanthophore *n.* [Gk. *akantha*, thorn; *pherein*, to bear] a conical mass, the basis of median stylet in Nemertini.

acanthopore *n.* [Gk. *akantha*, thorn; *poros*, passage] a tubular spine in certain Polyzoa.

Acanthopterygii, acanthopterygians *n., n.plu.* [Gk. *akantha*, thorn; *pterygion*, fin] a large advanced group of teleosts, having spiny fins and spiny scales, existing from upper Cretaceous to the present day and including perch, mackerel, and plaice.

acanthosphenote *a.* [Gk. *akantha*, thorn; *sphēn*, wedge] *appl.* echinoid spine made of solid wedges separated by porous tissue.

acanthozooid *n.* [Gk. *akantha*, thorn; *zōon*, animal; *eidos*, form] tail part of proscolex of cestodes; *cf.* cystozooid.

acapnia *n.* [Gk. *akapnos*, without smoke] condition of low carbon dioxide content in blood.

Acari, Acarina *n.* [Gk. *akarēs*, tiny; L.L. *acarus*, mite] a very large and varied order of arachnids, commonly called mites and ticks, usually having a rounded body.

acarocecidium n. [Gk. akarēs, tiny; kēkis, gall] a gall caused by gall mites (Eriophytidae).

acarology n. [Gk. akarēs, tiny; L.L. acarus, mite; Gk. logos, discourse] the study of mites and ticks.

acarophily n. [L.L. acarus, mite; philein, to love] symbiosis of plants and Acarina; alt. acarophytism; a. acarophilous.

acarpous a. [Gk. a, not; karpos, fruit] not fruiting.

acaryote akaryote, q.v.

acaudate ecaudate, q.v.

acaulescent a. [Gk. a, without; kaulos, stalk] having a shortened stem.

acauline, acaulous a. [Gk. a, without; kaulos, stalk] having no stem or stipe.

accelerator n. [L. accelerare, to hasten] appl. muscle or nerve which increases rate of action; cf. augmentor.

acceptor n. [L. accipere, to accept] a substance which receives and unites with another substance, as in oxidation–reduction processes where oxygen acceptor is the substance oxidized, hydrogen acceptor the substance reduced.

acceptor RNA transfer RNA, q.v.

accessorius n. [L. accedere, to support] a muscle aiding in action of another; spinal accessory or 11th cranial nerve.

accessory bodies minute argyrophil particles originating from Golgi body in spermatocytes; alt. chromatoid bodies.

accessory bud an additional axillary bud; a bud formed on a leaf.

accessory cells auxiliary cells, q.v.

accessory chromosomes supernumerary chromosomes, q.v.; sex chromosomes, q.v.

accessory disc N-disc, q.v.

accessory food factor vitamin, q.v.

accessory glands detached portions of glands; glands in relation with genital ducts.

accessory nerve one of the 11th pair of cranial nerves which is a motor nerve, arising partly from the medulla and partly from the spinal cord.

accessory pulsatory organs sac-like structures of insects, variously situated, pulsating independently of the heart; alt. accessory hearts.

acclimatation acclimation, acclimatization, q.q.v.

acclimation n. [L. ad, to; Gk. klima, climate] the habituation of an organism to a different climate or environment; alt. acclimatation, acclimatization.

acclimatization n. [L. ad, to; Gk. klima, climate] acclimation, q.v.; acclimation to an environment under human management, such as a zoo; alt. acclimatation.

accommodation n. [L. ad, to; commodus, fitting] adjustment of eye to receive clear images of objects at different distances, by changing the focal length of the lens; adaptation of receptors to a different stimulus; trend towards absence of sensation as a result of continuous stimulation; capacity of a plant to adapt to new conditions if these are introduced gradually.

accrescence n. [L. accrescere, to increase] growth through addition of similar tissues; continued growth after flowering; a. accrescent.

accrete a. [L. accrescere, to increase] grown or joined together; formed by accretion.

accretion n. [L. accrescere, to increase] growth by external addition of new matter; cf. intussusception.

accumbent a. [L. accumbere, to lie on] appl. embryo having cotyledons with edges turned towards radicle, as in dicots of the family Cruciferae.

accumulators n.plu. [L. ad, to; cumulus, heap] plants with a relatively high concentration of certain chemical elements in tissues.

A-cells alpha cells of islets of Langerhans.

acellular a. [L. a, without; cellula, small room] not containing cells; not considered as cells but as complete organisms, i.e. Protista.

acelomate acoelomate, q.v.

acelous acoelous, q.v.

acentric a. [Gk. a, without; kentron, centre] having no centromere, appl. chromosomes and chromosome segments; alt. akinetic.

acentrous a. [L. a. without; centrum, centre] with no vertebral centra, but persistent notochord, as certain fishes.

acephalocyst n. [Gk. a, without; kephalē, head; kystis, bladder] hydatid stage of certain tapeworms.

acephalous a. [Gk. a, not; kephalē, head] having no structure comparable to head, appl. some molluscs, appl. larvae of certain Diptera, appl. ovary without terminal stigma.

acerate a. [L. acer, sharp] needle-shaped; pointed at one end, appl. monaxon or oxeote spicules.

acerose a. [L. acer, sharp] narrow and slender, with sharp point, as leaf of pine.

acerous a. [Gk. a, without; keras, horn] hornless; without antennae; without tentacles.

acervate a. [L. acervare, to amass] heaped together; clustered.

acervuline a. [L.L. dim. of acervus, heap] irregularly heaped together, appl. shape of foraminiferal tests.

acervulus n. [L.L. dim. of acervus, heap] a small heap or cluster, especially of sporogenous mycelium; a. acervulate.

acervulus cerebri brain sand, q.v.

acetabulum n. [L. acetabulum, vinegar cup] the cotyloid cavity or socket in pelvic girdle for head of femur; in insects, cavity of thorax in which leg is inserted; socket of coxa in Arachnida; cavity in proximal end of spine, for articulation with mamelon, in echinoids; sucker in trematodes and cestodes; large posterior sucker in leeches; sucker on arm of cephalopod; one of the cotyledons of placenta in ruminants; alt. cotyle; a. acetabular.

acetylcholine (ACh) a neurotransmitter which is an acetyl ester of choline, secreted at the synapses of cholinergic nerves, e.g. parasympathetic nerve fibres, and is broken down by the enzyme acetylcholinesterase (cholinesterase); cf. noradrenaline.

acetylcoenzyme A, acetylcoA an acetyl thioester of coenzyme A which is widely used as a donor of acetyl groups, e.g. in the formation of citric acid from oxaloacetic acid in the Krebs' cycle.

acetylmuramic acid an amino sugar which is a component of bacterial cell walls and is derived from glucosamine and lactic acid.

ACh acetylcholine, q.v.

achaenocarp n. [Gk. a, not; chainein, to gape; karpos, fruit] achene, q.v.

achaetous a. [Gk. a, without; chaitē, hair] without chaetae.

acheilary *a*. [Gk. *a*, without; *cheilos*, lip] having labellum undeveloped, as some orchids.

achelate *a*. [Gk. *a*, not; *chēlē*, claw] without claws or chelae; not cheliform.

achene *n*. [Gk. *a*, not; *chainein*, to gape] a 1-seeded dry, indehiscent fruit formed from one carpel, usually with one seed not fused to the fruit wall; *alt*. achenium, achaenocarp, akene.

achenial *appl*. 1-seeded, dry, indehiscent fruits, as achene, cypsela, caryopsis, samara, and nut.

achiasmatic *a*. [Gk. *a*, without; *chiasma*, cross] lacking chiasmata in meiosis, as some Diptera.

Achillis tendo, Achilles tendon *n*. [Gk. *Achilles*; L. *tendo*, tendon] the tendon of the heel, the united strong tendon of gastrocnemius and solaeus muscles; *alt*. tendo calcaneus.

achlamydate *a*. [Gk. *a*, without; *chlamys*, cloak] not having a mantle, as certain gastropods.

achlamydeous *a*. [Gk. *a*, without; *chlamvs*, cloak] having neither calyx nor corolla; *alt*. gymnanthous.

Acholeplasmas *n.plu*. [Gk. *a*, without; *cholē*, bile; *plasma*, form] a group of Mollicutes recently considered to be distinct from mycoplasmas.

achondroplasia *n*. [Gk. *a*, without; *chondros*, cartilage; *plasis*, a moulding] heritable dwarfism due to disturbance of ossification in the long bones of the limbs and of certain facial bones during development; *cf*. ateleosis.

achroacyte *n*. [Gk. *a*, not; *chrōs*, colour; *kytos*, hollow] colourless cell; lymphocyte, *q.v*.

achroglobin *n*. [Gk. *a*, not; *chrōs*, colour; L. *globus*, sphere] a colourless respiratory pigment of some tunicates and molluscs.

achroic achrous, *q.v*.

achromasie *n*. [Gk. *a*, not; *chrōma*, colour] emission of chromatin from nucleus; *cf*. chromasie.

achromatic *a*. [Gk. *a*, without; *chrōma*, colour] *appl*. threshold, the minimal stimulus inducing sensation of luminosity or brightness; *appl*. neutral colours; achromatinic, *q.v*.

achromatin *n*. [Gk. *a*, without; *chrōma*, colour] the non-staining ground substance and linin of the nucleus; *alt*. nuclear sap.

achromatinic *a*. [Gk. *a*, without; *chrōma*, colour] *pert*. achromatin, or resembling achromatin in properties; *alt*. achromatic.

achromic *a*. [Gk. *a*, without; *chrōma*, colour] unpigmented; colourless; *alt*. achromatous.

achromite centromere, *q.v*.

A-chromosomes the normal chromosomes of a diploid chromosome set, as opposed to the B-chromosomes which differ from them structurally and functionally.

achrous *a*. [Gk. *a*, without; *chrōs*, complexion] unpigmented; colourless; *alt*. achroic.

acicle *n*. [L. *acicula*, small needle] a thorn-shaped scaphocerite, as in Paguridae (hermit crabs); acicula, *q.v*.

acicula *n*. [L. *acicula*, small needle] a small needle-like bristle, spine or crystal, *alt*. acicle, *plu*. aciculae; *plu*. of aciculum.

acicular like a needle in shape; sharp-pointed; *alt*. aciculiform.

aciculate *a*. [L. *acicula*, small needle] having acicles or aciculae.

aciculiform acicular, *q.v*.

aciculum *n*. [L. *acicula*, small needle] a stiff basal seta in parapodium of Chaetopoda; *plu*. acicula.

acid–base balance the maintenance of the correct ratio of acids to bases in the blood in order to maintain the most suitable pH.

acid-fast remaining stained with aniline dyes on treatment with acids.

acid-gland acid-secreting gland of Hymenoptera; oxyntic cells, *q.v*.

acidic *a*. [L. *acidus*, sour] having the properties of an acid; *appl*. stains whose colour determinant plays the part of an acid, acting on protoplasm, *cf*. basic.

acidophil *a*. [L. *acidus*, sour; Gk. *philein*, to love] oxyphil, *q.v*.; growing in acid media, *alt*. aciduric.

acid tide transient increase in acidity of body fluids which follows the alkaline tide.

aciduric *a*. [L. *acidus*, sour; *durus*, hardy] tolerating acid media; *alt*. acidophil.

aciform *a*. [L. *acus*, needle; *forma*, shape] belonoid, *q.v*.

acinaciform *a*. [L. *acinaces*, short sword; *forma*, shape] shaped like a sabre or scimitar, *appl*. leaf.

acinar *a*. [L. *acinus*, berry] *pert*. acinus; *appl*. cells of pancreas having sac-like terminations.

acinarious *a*. [L. *acinarius*, *pert*. grapes] having globose vesicles, as some algae.

Acinetae *n*. [Gk. *akinētos*, motionless] Suctoria, *q.v*.

aciniform *a*. [L. *acinus*, berry; *forma*, shape] grape- or berry-shaped; *appl*. a type of silk gland in spiders.

acinus *n*. [L. *acinus*, berry] drupel, *q.v*.; a cluster of cells forming the inner secretory region of a gland, usually a branched or compound gland, *alt*. alveolus: *plu*. acini.

Acipenseriformes *n*. [*Acipenser*, generic name of sturgeon; L. *forma*, shape] a group of degenerate palaeoniscids including the sturgeon, having a mainly cartilaginous skeleton with little bone and a skin naked or with a few bony scales; *alt*. Acipenseroidea.

acleidian *a*. [Gk. *a*, without; *kleis*, collar bone] with clavicles vestigial or absent.

acme *n*. [Gk. *akmē*, prime] the highest point attained, or prime, in phylogeny and ontogeny; *cf*. epacme, paracme.

Acnidosporidia *n*. [Gk. *a*, not; *knidē*, nettle; *sporos*, seed; *idion*, dim.] an order of Neosporidia having spores without polar capsules.

Acochlidiacea *n*. [Gk. *a*, not; *kochlias*, snail] an order of very small opisthobranchs which live as interstitial fauna in sand.

Acoela *n*. [Gk. *a*, not; *koilos*, hollow] an order of small turbellarians with no gut and having simple gonads and ducts; an order of almost symmetrical opisthobranchs which have no shell or mantle cavity, *alt*. Nudibranchia.

acoelomate *a*. [Gk. *a*, without; *koilōma*, hollow] *appl*. animals not having a true coelom; *alt*. acoelomatous, acelomate, acoelous.

acoelous *a*. [Gk. *a*, without; *koilos*, hollow] *appl*. vertebrae with flattened centra; acoelomate, *q.v*.; *alt*. acelous.

acondylous *a*. [Gk. *a*, without; *kondylos*, knuckle] without nodes or joints.

acone *a*. [Gk. *a*, without; *kōnos*, cone] *appl*. insect compound eye without crystalline or liquid secretion in cone cells.

aconitase *n*. [*Aconitum*, generic name of monkshood] the enzyme which converts citric acid to *cis*-aconitic acid and also converts isocitric acid to *cis*-aconitic acid; EC 4.2.1.3; *r.n.* aconitate dehydratase.

aconitic acid [*Aconitum*, generic name of monkshood] a colourless acid obtained from monkshood, horsetail, and some other plants which takes part in the Krebs' cycle in the form of *cis*-aconitic acid.

acont akont, *q.v.*

acont- *alt.* akont-.

Aconta *n.* [Gk. *a,* without; *kontos,* punting pole] a group of eukaryotic algae that never produce flagella, i.e. the red algae; *cf.* Contophora.

Acontae *n.* [Gk. *a,* without; *kontos,* punting pole] Zygnematales, *q.v.*

acontia *n.plu.* [Gk. *akontion,* small javelin] threadlike processes of mesenteric filaments armed with nematocysts in some actinians.

acotyledon *n.* [Gk. *a,* without; *kotylēdōn,* a cupshaped hollow] a plant without a cotyledon.

acoustic *a.* [Gk. *akouein,* to hear] *pert.* organs or sense of hearing, *appl.* meatus, nerve, etc.; *pert.* science of sound.

acoustico-lateralis system a system of receptors in fish and amphibians, which detects slow vibrations and consists of neuromasts that are either scattered or arranged in a lateral-line system.

ACP acyl carrier protein, *q.v.*

acquired behaviour behaviour brought about by conditioning and learning.

acquired character a modification or permanent structural or functional change effected during the lifetime of the individual organism and induced by use or disuse of a particular organ, by disease, trauma, or other functional or environmental influences.

acral *a.* [Gk. *akros,* tip] *pert.* extremities.

acrandry *n.* [Gk. *akros,* tip; *anēr,* male] the condition of having antheridia borne at the tips, in bryophytes; *a.* acrandrous.

Acrania *n.* [Gk. *a,* not; *kranion,* skull] a group of chordates including all groups except Craniata, i.e. including the Urochordata and Cephalochordata, sometimes considered a subphylum with the urochordates and cephalochordates as classes; *alt.* Protochordata.

acranthous *a.* [Gk. *akros,* tip; *anthos,* flower] having the inflorescence borne on the tip of the main axis; *cf.* pleuranthous.

Acrasiales, Acrasieae *n.* [Gk. *akrasia,* bad mixture] a group of cellular slime moulds whose simple amoeboid cells aggregate into a pseudoplasmodium.

acrasin *n.* [Gk. *akrasia,* bad mixture] a chemotactic substance produced by certain slime moulds which causes the aggregation of cells and has now been shown to be cyclic AMP.

Acraspeda *n.* [Gk. *akraspedos,* without fringes] in some classifications, a group of coelenterates having a medusa without a velum.

acraspedote *a.* [Gk. *a,* without; *kraspedon,* border] having no velum.

acroblast *n.* [Gk. *akros,* tip; *blastos,* bud] a body in spermatid which gives rise to acrosome, *alt.* idiosphaerotheca; outer layer of mesoblast.

acrobryous *a.* [Gk. *akros,* tip; *bryein,* to swell] growing at the tip only.

Acrocarpi *n.* [Gk. *akros,* tip; *karpos,* fruit] a group of mosses comprising acrocarpous forms; *cf.* Pleurocarpi.

acrocarpic, acrocarpous *a.* [Gk. *akros,* tip; *karpos,* fruit] with terminal fructifications; *appl.* mosses bearing archegonia and therefore capsules at the tips of the stem or main branches; *cf.* pleurocarpous.

acrocentric *a.* [Gk. *akros,* tip; *kentron,* centre] with centromere at end, *appl.* chromosome. *n.* A rod-shaped chromosome. *Cf.* metacentric, telocentric.

acrochordal *a.* [Gk. *akros,* tip; *chorde,* cord] *appl.* a chondrocranial unpaired frontal cartilage in birds.

acrochroic *a.* [Gk. *akros,* tip; *chrōs,* colour] with coloured tips, as of hyphae.

acrocoracoid *n.* [Gk. *akros,* tip; *korax,* crow; *eidos,* form] a process at dorsal end of coracoid in birds.

acrocyst *n.* [Gk. *akros,* tip; *kystis,* bladder] the spherical gelatinous cyst formed by gonophores at maturation of generative cells.

acrodont *a.* [Gk. *akros,* tip; *odous,* tooth] *appl.* teeth attached to the summit of a parapet of bone, as in lizards.

acrodrome, acrodromous *a.* [Gk. *akros,* tip; *dramein,* to run] *appl.* leaf with veins converging at its point; *alt.* campylodrome.

acrogenous *a.* [Gk. *akros,* tip; *-genēs,* producing] increasing in growth at summit or apex.

acrogenous *a.* [Gk. *akros,* tip; *-genēs,* producing] ferns and mosses, collectively.

acrogynous *a.* [Gk. *akros,* tip; *gynē,* female] with archegonia arising from apical cell, *appl.* certain liverworts; *cf.* anacrogynous.

acromegaly *n.* [Gk. *akros,* tip; *megalon,* great] gigantism due to excessive activity of part of pituitary gland.

acromial *a.* [Gk. *akros,* summit; *ōmos,* shoulder] *pert.* acromion, *appl.* artery, process, ligament, etc.

acromio-clavicular *a.* [Gk. *akrōmion,* shoulder summit; L. *clavicula,* dim. of *clavis,* key] *appl.* ligaments covering joint between acromion and clavicle.

acromion *n.* [Gk. *akros,* summit; *ōmos,* shoulder] ventral prolongation of scapular spine.

acron *n.* [Gk. *akron,* top] preoral region of insects; anterior, unsegmented part of young trilobite.

acronematic *a.* [Gk. *akros,* tip; *nēma,* thread] *appl.* flagella which are smooth and whip-like.

acroneme *n.* [Gk. *akros,* tip; *nēma,* thread] the slender part of acronematic flagella.

acropetal *a.* [Gk. *akros,* summit; L. *petere,* to seek] ascending; *appl.* leaves, flowers, or roots, developing successively from an axis so that youngest arise at apex; *alt.* basifugal; *cf.* basipetal.

acrophyte *n.* [Gk. *akron,* peak; *phyton,* plant] a plant growing at a high altitude; alpine plant.

acroplasm *n.* [Gk. *akros,* tip; *plasma,* form] cytoplasm of the apex of an ascus.

acropodium *n.* [Gk. *akros,* tip; *pous,* foot] digits,—fingers or toes; *cf.* metapodium.

acrorhagus *n.* [Gk. *akros,* summit; *rhax,* grape] a tubercle near the margin of certain Actiniaria, containing specialized nematocysts.

acrosarc *n.* [Gk. *akros,* summit; *sarx,* flesh] a pulpy berry resulting from union of ovary and calyx.

acroscopic *a.* [Gk. *akros,* tip; *skopein,* to view] facing towards the apex; *cf.* basiscopic.

Acrosiphonales

Acrosiphonales n. [Gk. *akros*, tip; *siphōn*, tube] an order of green algae which have an unusual type of cellulose in their cell walls, but are otherwise mainly similar to the Cladophorales.

acrosome n. [Gk. *akros*, tip; *sōma*, body] organelle at apex of spermatozoon which digests the coatings around the egg so that the sperm can enter; *alt.* apical body, head-cap, idiosphaerosome, perforatorium.

acrospire n. [Gk. *akros*, tip; *speira*, something twisted] the 1st shoot or sprout, being spiral, at end of germinating seed.

acrospore n. [Gk. *akros*, tip; *sporos*, seed] the spore at the apex of a sporophore or hypha.

acrostical a. [Gk. *akros*, tip; *stichos*, row] *appl.* hairs: small bristles along dorsal surface of thorax of some Diptera such as *Drosophila*.

acrostichoid a. [Gk. *akros*, tip; *stichos*, row; *eidos*, form] *appl.* fern sporangia produced all over the surface, not in sori over a vein.

acrosyndesis n. [Gk. *akros*, tip; *syndēsai*, to bind together] telosyndesis, *q.v.*

acroteric a. [Gk. *akrōtēria*, extremities] *pert.* outermost points, as tips of digits, nose, ears, tail.

Acrothoracica n. [Gk. *akros*, tip; *thōrax*, chest] an order of barnacles having a reduced number of thoracic limbs.

acrotonic, acrotonous a. [Gk. *akros*, tip; *tonos*, brace] having anther united at its apex with rostellum; *cf.* basitonic.

acrotroch n. [Gk. *akros*, tip; *trochos*, hoop] a circlet of cilia anterior to prototroch of trochophore, in certain polychaetes.

acrotrophic a. [Gk. *akros*, tip; *trophē*, nourishment] *appl.* ovariole having nutritive cells at apex which are joined to oocytes by nutritive cords; *alt.* telotrophic; *cf.* meroistic, polytrophic.

ACTH adrenocorticotrophic hormone, *q.v.*

actin n. [Gk. *aktis*, ray] a protein occurring in muscle where it combines with myosin to form actomyosin, in striped muscle found alone in the I-disc, and also found in flagella and other contractile systems.

actinal a. [Gk. *aktis*, ray] *appl.* area of echinoderm body with tube-feet, *alt.* ambulacral; *appl.* oral area with tentacles in Actiniaria; starshaped.

actine n. [Gk. *aktis*, ray] a star-shaped spicule.

actinenchyma n. [Gk. *aktis*, ray; *en*, in; *cheein*, to pour] cellular tissue having a stellate appearance.

Actiniaria n. [Gk. *aktis*, ray] an order of Zoantharia, commonly called sea anemones, which are solitary without a skeleton, and have complete and incomplete septa in multiples of 6.

actinic a. [Gk. *aktis*, ray] *appl.* or *pert.* rays with wavelengths between those of visible violet and of X-rays, and having certain chemical effects, e.g. on ergosterol, *q.v.*

actiniform actinoid, *q.v.*

Actinistia n. [Gk. *aktis*, ray] Coelacanthina, *q.v.*

actinobiology n. [Gk. *aktis*, ray; *bios*, life; *logos*, discourse] the study of the effects of radiation upon living organisms.

actinoblast n. [Gk. *aktis*, ray; *blastos*, bud] the mother cell from which a spicule is developed, as in Porifera.

actinocarpous a. [Gk. *aktis*, ray; *karpos*, fruit] *appl.* plants with flowers and fruit radially arranged; *alt.* actinocarpic.

actinochitin n. [Gk. *aktis*, ray; *chitōn*, tunic] anisotropic or birefringent chitin; *cf.* isotropic chitin.

actinodrome a. [Gk. *aktis*, ray; *dromos*, course] veined palmately; *alt.* actinodromous.

actinogonidial a. [Gk. *aktis*, ray; *gonos*, offspring] having radiately arranged genital organs.

actinoid a. [Gk. *aktis*, ray; *eidos*, shape] rayed; star-shaped, stellate; *alt.* actiniform.

actinology n. [Gk. *aktis*, ray; *logos*, discourse] the study of the action of radiation; study of radially symmetrical animals; homology of successive regions or parts radiating from a common central region.

actinomere n. [Gk. *aktis*, ray; *meros*, part] a radial segment; *alt.* antimere.

actinomorphic a. [Gk. *aktis*, ray; *morphē*, form] radially symmetrical; *alt.* actinomorphous, regular; *cf.* zygomorphic.

Actinomycetales, actinomycetes n., n.plu. [Gk. *aktis*, ray; *mykēs*, fungus] an order of mainly rod-shaped non-motile bacteria which often branch in culture, and including many members that produce important antibiotics; sometimes considered to be a separate class of Schizophyta.

actinomycin D n. [Gk. *aktis*, ray; *mykēs*, fungus] an antibiotic produced by the actinomycete *Streptomyces chrysomallus* which prevents the formation of mRNA.

actinopharynx n. [Gk. *aktis*, ray; *pharynx*, gullet] the gullet of a sea anemone.

Actinopoda n. [Gk. *aktis*, ray; *pous*, foot] a subclass of Sarcodina, in some classifications.

Actinopterygii, actinopterygians n., n.plu. [Gk. *aktis*, ray; *pterygion*, fin] a subclass of bony fishes, often called ray-finned fishes, consisting of many extinct groups and most living bony fish, having ganoid scales and usually paired fins with broad bases.

actinospore n. [Gk. *aktis*, ray; *sporos*, seed] a spore of actinomycetes.

actinost n. [Gk. *aktis*, ray; *osteon*, bone] basal bone of fin-rays in teleosts.

actinostele n. [Gk. *aktis*, ray; *stēlē*, pillar] a stele with no pith, having xylem and phloem in alternating or radial groups, so the xylem is starshaped in cross-section.

actinostome n. [Gk. *aktis*, ray; *stoma*, mouth] the mouth of a sea anemone; 5-rayed oral aperture of starfish.

actinotrichia n.plu. [Gk. *aktis*, ray; *thrix*, hair] unjointed horny rays at edge of fins in many fishes.

actinotrocha n. [Gk. *aktis*, ray; *trochos*, wheel] free-swimming larval form of Phoronida.

Actinozoa n. [Gk. *aktis*, ray; *zōon*, animal] Anthozoa, *q.v.*

actinula n. [Gk. *aktis*, ray] a larval stage in some hydrozoans.

Actinulida n. [Gk. *aktis*, ray] a small order of hydrozoans, being very small individuals living as meiofauna and retaining the appearance and cilia of actinula larvae.

action potential a potential difference produced in a nerve or muscle when it is stimulated, reversing the resting potential from about −70 millivolts to about +30 millivolts, and being an easily observed manifestation of a nerve impulse; *cf.* resting potential.

action system the pattern of behaviour in an organism.

activating enzymes a group of enzymes involved in protein synthesis, that catalyse the reaction of ATP with an amino acid and the formation of amino acyl-tRNA; also known as amino acyl synthetases; EC sub-subgroup 6.1.1.

activator n. [L. activus, active] a substance which promotes or protects enzyme action; a substance which stimulates development of any particular embryonic tissue or organ.

active centre or site the region of an enzyme molecule which interacts with the substrate molecules and where activation and reaction take place.

active transport the movement of materials into cells other than by diffusion, usually involving energy expenditure by the cell and often against concentration gradients.

actomyosin n. [Gk. aktis, ray; mys, muscle] a protein in muscle formed from the linkage of the 2 proteins actin and myosin in myofilaments, which shortens when stimulated and so causes muscles to contract.

aculeate a. [L. aculeus, prickle] having prickles, sharp points, or a sting.

aculeiform a. [L. aculeus, prickle; forma, shape] formed like a prickle or thorn.

aculeus n. [L. aculeus, prickle] a prickle growing from bark, as in rose; a sting; a hair-like projection; a microtrichium, q.v.

acumen n. [L. acumen, point] the point of an acuminate leaf.

acuminate a. [L. acumen, point] drawn out into long point; tapering; pointed.

acuminiferous a. [L. acumen, point; ferre, to carry] having pointed tubercles.

acuminulate a. [L. acuminulus, dim. of acumen, point] having a very sharp tapering point.

acute a. [L. acutus, sharpened] ending in a sharp point; temporarily severe, not chronic.

acyclic a. [Gk. a, without; kyklos, circle] appl. flowers with floral leaves arranged in a spiral; cf. cyclic.

acyl carrier protein (ACP) a small protein which carries acyl groups in the metabolic cycles concerned with fat synthesis.

adamantoblast n. [Gk. adamas, diamond; blastos, bud] enamel cell; ameloblast, q.v.

adambulacral a. [L. ad, to; ambulare, to walk] appl. structures adjacent to ambulacral areas in echinoderms.

Adam's apple laryngeal prominence, q.v.

Adapedonta n. [L. adaperire, to open fully; dens, tooth] an order of burrowing lamellibranchs with eulamellibranch gills and gaping shells; cf. Eulamellibranchiata.

adaptation n. [L. adaptare, to fit to] the process by which an organism becomes fitted to its environment; a structure or habit fitted for some special environment; the fitting of sensations to a point when discomfort ceases; adjustment of disturbance of nervous system without involving higher coordinating centres.

adaptive a. [L. adaptare, to fit to] capable of fitting different conditions; adjustable; inducible, appl. enzymes formed when their specific substrates are available, cf. constitutive enzymes; appl. radiation: evolution from a common ancestry of a number of morphologically and ecologically different types adjusted to different environments.

adaptogens n.plu. [L. adaptare, to fit to; Gk. genos, birth] a hypothetical group of drugs which return body processes to normal in the presence of stress, but are ineffective when there is no stress.

adaptor RNA transfer RNA, q.v.

adaxial a. [L. ad, to; axis, axle] turned towards the axis; cf. abaxial.

adcauline a. [L. ad, to; caulis, stalk] towards or nearest the stem; cf. abcauline.

ad-digital n. [L. ad, to; digitus, finger] a primary wing quill connected with phalanx of 3rd digit.

adduction n. [L. ad, to; ducere, to lead] movement towards the median axis; cf. abduction.

adductor n. [L. ad, to; ducere, to lead] a muscle which brings one part towards another.

adeciduate a. [L. a, away from; decidere, to fall down] not falling, or coming away, appl. evergreens, appl placenta; alt. indeciduate.

adecticous a. [Gk. a, without; dēktikos, biting] without functional mandibles to escape from puparium or cocoon, appl. pupa of some insects; cf. decticous.

adelocodonic a. [Gk. adēlos, concealed; kōdōn, bell] appl. undetached medusa of certain Gymnoblastea, which degenerates after discharging ripe sexual cells; cf. phanerocodonic.

adelomorphic a. [Gk. adēlos, concealed; morphē, shape] indefinite in form; appl. central cells of peptic glands; alt. adelomorphous.

adelomycete n. [Gk. adēlos, concealed; mykēs, fungus] a fungus lacking the sexual spore stage, an imperfect fungus; see Fungi Imperfecti.

adelophycean a. [Gk. adēlos, concealed; phykion, seaweed] appl. stage or generation of many seaweeds when they appear as prostrate microthalli.

adelphogamy n. [Gk. adelphos, brother; gamos, marriage] brother–sister mating, as in certain ants; union of mother cell and one of the daughter cells formed from it by mitosis.

adelphous a. [Gk. adelphos, brother] joined together in bundles, as filaments of stamens; cf. monadelphous, diadelphous.

adenase n. [Gk. adēn, gland; -ase] a hydrolysing enzyme which catalyses the deamination of adenine with the formation of hypoxanthine and ammonia; EC 3.5.4.2; r.n. adenine deaminase.

adendritic, adendric a. [Gk. a, not; dendron, tree] without dendrites or branches, appl. cells.

adendroglia n. [Gk. a, not; dendron, tree; gloia, glue] a type of neuroglia lacking processes.

adenine n. [Gk. adēn, gland] a purine base which is part of the genetic code of DNA where it pairs with thymine, and of RNA where it pairs with uracil, and is part of the molecule of NAD, NADP, FAD, and adenosine.

adenoblast n. [Gk. adēn, gland; blastos, bud] embryonic glandular cell.

adenocheiri n.plu. [Gk. adēn, gland; cheir, hand] elaborate accessory copulatory organs, outgrowths of atrial walls in Turbellaria; alt. adenodactyli.

adenocyte n. [Gk. adēn, gland; kytos, hollow] secretory cell of a gland.

adenodactyli n.plu. [Gk. adēn, gland; daktylos, finger] adenocheiri, q.v.

adenohypophysis n. [Gk. adēn, gland; hypo, under; physis, growth] the glandular lobe or portions of the pituitary body, derived from

Rathke's pouch, consisting of the pars distalis, pars tuberalis, and pars intermedia; *cf.* neurohypophysis.

adenoid *a.* [Gk. *adēn*, gland; *eidos*, shape] *pert.* or resembling a gland or lymphoid tissue. *n.* Nasopharyngeal tonsil.

Adenophora *n.* [Gk. *adēn*, gland; *phora*, producing] Aphasmidia, *q.v.*

adenophore *n.* [Gk. *adēn*, gland, *pherein*, to carry] the stalk of a gland, especially a nectar gland.

adenophyllous *a.* [Gk. *adēn*, gland; *phyllon*, leaf] bearing glands on leaves.

adenopodous *a.* [Gk. *adēn*, gland; *pous*, foot] bearing glands on peduncles or petioles.

adenose *a.* [Gk. *adēn*, gland] glandular.

adenosine *n.* [Gk. *adēn*, gland] a nucleoside with adenine as its base.

adenosine diphosphate (ADP) a nucleotide cofactor made of adenosine and 2 phosphate groups, involved in energy transfers; *cf.* adenosine triphosphate.

adenosine monophosphate (AMP) a nucleotide made of adenosine and one phosphate group; *alt.* adenylic acid; *cf.* adenosine triphosphate, cyclic AMP.

adenosine triphosphate (ATP) a nucleotide cofactor made up of adenosine and 3 phosphate groups, important in many biological reactions where energy is transferred, and which can be converted to adenosine diphosphate and monophosphate with the release of energy and phosphates, or made from these compounds with the uptake of energy in phosphate bonds.

adenostemonous *a.* [Gk. *adēn*, gland; *stēmōn*, spun thread] having glands on stamens.

adenoviruses *n.plu.* [Gk. *adēn*, gland; L. *virus*, poison] a group of viruses containing doublestranded RNA, infecting various mammals including those causing repiratory diseases in man.

adenylic acid adenosine monophosphate, *q.v.*; formerly called vitamin B_8 when found in striated muscle.

adequate *appl.* stimulus which normally acts on a given receptor, and induces the appropriate sensation.

adermin pyridoxine, *q.v.*; broadly, vitamin B_6.

adesmic *a.* [Gk. *adesmos*, unfettered] *appl.* cyclomorial scales made up of separate lepidomorial units; *cf.* monodesmic, polydesmic.

adesmy *n.* [Gk. *adesmos*, unfettered] a break or division in an organ usually entire.

adetopneustic *a.* [Gk. *adetos*, free; *pnein*, to breathe] having dermal gills occurring beyond abactinal surface, as in certain stelleroids.

adfrontal *a.* [L. *ad*, to; *frons*, forehead] *appl.* oblique plates beside frons of certain insect larvae.

ADH antidiuretic hormone, *q.v.*

adherent *a.* [L. *ad*, to; *haerere*, to stick] exhibiting adhesion, *q.v.*; attached to substratum, *appl.* zooecia of polyzoan colony; *cf.* coherent.

adhesion *n.* [L. *ad*, to; *haerere*, to stick] condition of touching without growing together of parts normally separate, as between members of different series of floral leaves; *cf.* cohesion.

adhesive cells various glandular or specialized cells for purposes of attachment, as on tentacles of Ctenophora, on epidermis of Turbellaria, on pedal disc of hydra.

adiabatic *a.* [Gk. *a*, not; *diabatos*, passable] with-

out losing or gaining heat; incapable of translocation.

adience *n.* [L. *adire*, to approach] urge, or advance, towards stimulus; approaching reaction; *cf.* abience.

adient *a.* [L. *adire*, to approach] approaching the source of stimulation; *cf.* abient.

adipocellulose *n.* [L. *adeps*, fat; *cellula*, small room] cellulose with a large amount of suberin, as in cork tissue.

adipocyte *n.* [L. *adeps*, fat; Gk. *kytos*, hollow] one of the cells forming the fat-body in insects.

adipoleucocyte *n.* [L. *adeps*, fat; Gk. *leukos*, white; *kytos*, hollow] leucocyte containing fat droplets or wax, in insects.

adipolysis *n.* [L. *adeps*, fat; Gk. *lysis*, loosing] lipolysis, *q.v.*

adipose *a.* [L. *adeps*, fat] *pert.* animal fat; fatty.

adipose body fat-body, *q.v.*

adipose fin modified rayless posterior dorsal fin, as in Salmoniformes, Characiformes, Suluriformes.

adipose tissue a type of connective tissue whose cells are filled with fat.

A-disc A-band, *q.v.*

aditus *n.* [L. *aditus*, entrance] anatomical structure forming approach or entrance to a part, e.g. to antrum, larynx, etc.

adjustor *n.* [L.L. *adjustare*, to adjust, from L. *ad*, to; *justus*, just] a muscle connecting stalk and valve in Brachiopoda; ganglionic part of a reflex arc, connecting receptor and effector.

adjuvant *n.* [L. *ad*, to; *juvenalis*, youthful] a substance that increases the production and lifetime of antibodies when a body is injected with antigen.

adlacrimal *n.* [L. *ad*, to; *lacrima*, tear] lacrimal bone of reptiles, not homologous to that of mammals.

admedial *a.* [L. *ad*, towards; *medius*, middle] near the middle, *alt.* mediad; near the median plane, *alt.* admedian.

adminiculum *n.* [L. *adminiculum*, support] a locomotory spine of certain pupae; posterior fibres of linea alba attached to os pubis.

adnasal *n.* [L. *ad*, to; *nasus*, nose] a small bone in front of each nasal in certain fishes.

adnate *a.* [L. *ad*, to; *gnatus*, born] joined to another organ of a different kind; *pert.* or designating the condition of being closely attached to side of petiole or stalk, as stipules or leaves; designating condition of anther with back attached throughout its length to filament, or to its continuation the connective; *appl.* gills of an agaric which are fused with the stem for the whole of their width; *alt.* conjoined.

adnephrin(e) adrenaline, *q.v.*

adnexa *n.plu.* [L. *ad*, to; *nectere*, to bind] structures or parts closely related to an organ; extraembryonic structures, as foetal membranes, placenta.

adnexed *a.* [L. *ad*, to; *nectere*, to bind] reaching to the stem only; *appl.* gills of an agaric which are fused to the stem for only part of their width.

adolescaria *n.* [L. *adolescere*, to grow up] metacercaria, *q.v.*

adoral *a.* [L. *ad*, to; *os*, mouth] near or *pert.* mouth.

ADP adenosine diphosphate, *q.v.*

adpressed *a.* [L. *ad*, to; *pressus*, pressed] appressed, *q.v.*

adradius *n.* [L. *ad*, to; *radius*, radius] in coelenterates, the radius midway between perradius and interradius, a radius of 3rd order.

adrectal *a.* [L. *ad*, to; *rectum*, rectum] near to or closely connected with rectum.

adrenal *a.* [L. *ad*, to; *rectum*, rectum] situated near kidneys; *alt.* suprarenal, surrenal.

adrenal body one of a pair of bodies adjacent to the kidneys in mammals consisting of the suprarenal and interrenal (adrenal) glands, the suprarenals forming the central portion (medulla) secreting adrenaline and noradrenaline, and the interrenals forming the outer portion (cortex) secreting the adrenocortical hormones, in some vetebrates the 2 glands being separate; *alt.* adrenal gland, suprarenal gland, paranephros.

adrenalergic adrenergic, *q.v.*

adrenalin(e) *n.* [L. *ad*, to; *renes*, kidneys] a hormone secreted by the suprarenal medulla of the adrenal bodies and by nerve endings of the sympathetic nervous system, which prepares the animal for fight or flight reactions and maintains muscle tone, may act as a neurotransmitter, and is also found in some invertebrates; *alt.* adrenine, adnephrine, epinephrine, suprarenin.

adrenergic *a.* [L. *ad*, to; *renes*, kidneys; Gk. *ergon*, work] *appl.* sympathetic nerves, which liberate adrenaline or noradrenaline from their terminations; *alt.* adrenalergic; *cf.* cholinergic.

adrenin(e) adrenaline, *q.v.*

adrenocortical *a.* [L. *ad*, to; *renes*, kidneys; *cortex*, bark] *pert.* or secreted in the adrenal cortex, *appl.* various hormones including sex hormones, glucocorticoids, mineralocorticoids.

adrenocorticotrop(h)ic *a.* [L. *ad*. to; *renes*, kidneys; *cortex*, bark; Gk. *trophē*, nourishment] *appl.* polypeptide hormone secreted by the adenohypophysis which controls the growth and activity of the adrenal cortex, *alt.* ACTH, corticotrophin; *alt.* corticotrophic.

adrenotropic, adrenotrophic *a.* [L. *ad*, to; *renes*, kidneys; Gk. *tropē*, turn; *trophē*, nourishment] *appl.* a pituitary hormone acting on the adrenal medulla.

adrostral *a.* [L. *ad*, to; *rostrum*, beak] near to or closely connected with beak or rostrum.

adsorption *n.* [L. *ad*, to; *sorbere*, to suck in] the adhesion of molecules to solid bodies; formation of unimolecular surface layer; taking up of a substance at a surface.

adtidal *a.* [L. *ad*, to; A.S. *tid*, time] *appl.* organisms living just below low-tide mark.

adultoid *appl.* nymph having imaginal characters differentiated further than in normal nymph.

aduncate *a.* [L. *aduncus*, hooked] crooked; bent in the form of a hook.

adust(ous) *a.* [L. *adustus*, sunburnt] browned; appearing as if scorched.

advehent *a.* [L. *advehere*, to carry to] afferent; carrying to an organ.

adventitia *n.* [L. *adventitius*, extraordinary] external connective tissue layer of blood vessels; *alt.* tunica adventitia.

adventitious *a.* [L. *adventitius*, extraordinary] accidental; found in an unusual place; *appl.* tissues and organs arising in abnormal positions; secondary, *appl.* dentine.

adventive *a.* [L. *advenire*, to arrive] not native. *n.* An organism in a new habitat but not completely established there.

aecia *plu.* of aecium.

aecial aecidial, *q.v.*

aecidia *plu.* of aecidium.

aecidial *a.* [L. *aecidium*, cup] *pert.* aecidia, or aecidium; *alt.* aecial.

aecidiosorus *n.* [L. *aecidium*, cup; Gk *sōros*, heap] a cluster or row of aecidiospores.

aecidiospores *n.plu.* [L. *aecidium*, cup; Gk. *sporos*, seed] the spores produced in an aecidium; *alt.* aeciospores.

aecidium *n.* [L. *aecidium*, cup] in rust fungi, a cup-shaped structure containing chains of aecidiospores; *alt.* cluster cup, aecium; *plu.* aecidia.

aeciospores aecidiospores, *q.v.*

aecium aecidium, *q.v.*

aedeagus *n.* [Gk. *aidoia*, genitals] the male intromittent organ of insects; *alt.* aedoeagus, edeagus.

aegithognathous *a.* [Gk. *aigithos*, hedge sparrow; *gnathos*, jaw] with maxillopalatines separate, vomers forming a wedge in front and diverging behind, *appl.* a type of palate found in Passeres.

aeolian *a.* [L. *Aeolus*, god of the winds] windborne, *appl.* deposits.

Aepyornithiformes *n.* [*Aepyornis*, generic name; L. *forma*, shape] an order of very large Pleistocene birds of the subclass Neornithes from Madagascar, known as elephant birds.

aerenchyma *n.* [Gk. *aēr*, air; *engchyma*, infusion] parenchyma tissue with large intercellular spaces; air-storing tissue in cortex of various aquatic plants; tissue between spore mass and capsule wall in mosses.

aerial *a.* [L. *aer*, air] inhabiting the air; *appl.* roots growing above ground, e.g. from stems of ivy, for purposes of climbing; *appl.* small bulbs appearing in leaf axils, *alt.* bulbils.

aero-aquatic *a.* [L. *aer*, air; *aqua*, water] *appl.* or *pert.* fungi growing in water and liberating spores in the air.

aerobe *n.* [Gk. *aēr*, air; *bios*, life] an organism capable of living in the presence of oxygen; *alt.* aerobiont; *a.* aerobic; *cf.* anaerobe.

aerobic respiration respiration occurring in the presence of oxygen.

aerobiology *n.* [Gk. *aēr*, air; *bios*, life; *logos*, discourse] the study of air-borne organisms and their distribution.

aerobiont *n.* [Gk. *aēr*, air; *bion*, living] aerobe, *q.v.*; an organism living mainly in the air.

aerobiosis *n.* [Gk. *aēr*, air; *biōsis*, manner of life] existence in presence of oxygen.

aerobiotic *a.* [Gk. *aēr*, air; *biōtikos*, *pert.* life] living mainly in the air.

aerocyst *n.* [Gk. *aēr*, air; *kystis*, bladder] an air vesicle of algae.

aerogenic *a.* [Gk. *aēr*, air; *gennaein*, to produce] gas-producing, *appl.* certain bacteria.

aerolae *n.plu.* [L. *aer*, air] in the walls of diatoms, large depressed box-like structures.

aeromorphosis *n.* [Gk. *aēr*, air; *morphōsis*, form] modification of form or structure owing to exposure to air or wind.

aerophora *n.* [Gk. *aēr*, air; *pherein*, to bear] aerating outgrowth or pneumatophore in certain ferns.

aerophyte *n.* [Gk. *aēr*, air; *phyton*, plant] an epiphyte attached to the aerial portion of another plant.

aeroplankton *n.* [Gk. *aēr*, air; *plangktos*, wandering] living particles drifting in the air, as spores, pollen, bacteria, etc.; also applied to non-living particles; *alt.* anemoplankton.

aerostat *n.* [L. *aer*, air; *stare*, to stand] an air sac in insect body or in bird bone.

aerostatic *a.* [L. *aër*, air; *stare*, to stand] containing air spaces; *alt.* pneumatic.

aerotaxis *n.* [Gk. *aër*, air; *taxis*, arrangement] the arrangement of micro-organisms or motile gametes towards or away from oxygen; *a.* aerotactic.

aerotropism *n.* [Gk. *aër*, air; *tropē*, turn] reaction to gases, generally to oxygen, particularly the growth curvature of roots or other parts of plants to changes in oxygen tension; *a.* aerotropic.

aesth- *alt.* esth-.

aesthacyte *n.* [Gk. *aisthēsis*, sensation; *kytos*, hollow] a sensory cell of primitive animals such as sponges.

aesthesis *n.* [Gk. *aisthēsis*, sensation] sensibility; sense-perception; *alt.* aesthesia.

aesthetasc *n.* [Gk. *aisthētēs*, perceiver; *askein*, to exercise] an olfactory receptor on antennule of some crustaceans such as *Daphnia*.

aesthetes *n.plu.* [Gk. *aisthētēs*, perceiver] sense organs.

aestival *a.* [L. *aestivus*, of summer] produced in, or *pert.* summer; *pert.* early summer, *cf.* serotinal; *alt.* estival.

aestivation *n.* [L. *aestivus*, of summer] the mode in which different parts of flower are disposed in flower bud, *cf.* prefloration; torpor during heat and drought during summer in some animals, *cf.* hibernation; *alt.* estivation; *v.* aestivate.

aethalium *n.* [Gk. *aithalos*, soot] an aggregation of plasmodia or sporangia to form a compound fruit body in slime moulds; *cf.* pseudoaethalium.

aethogametism *n.* [Gk. *aēthēs*, unaccustomed; *gametēs*, spouse] asynethogametism, *q.v.*; *alt.* aethogamety.

aetiolation etiolation, *q.v.*

aetiolin etiolin, *q.v.*

aetiology *n.* [Gk. *aitia*, cause; *logos*, discourse] the science of causation; origin of causes; *alt.* etiology.

affectional *appl.* behaviour concerned with social relationships as in monkeys, important in development and maintenance of social cohesion and organization.

afferent *a.* [L. *afferre*, to bring] bringing towards; *appl.* nerves carrying impulses to nervous centres, *alt.* centripetal; *appl.* blood vessels carrying blood to an organ or set of organs; *cf.* efferent.

afferent neurone sensory neurone, *q.v.*

aflagellar *a.* [Gk. *a*, without; L. *flagellum*, whip] akont, *q.v.*

aflatoxin a mycotoxin produced by *aspergillus* which is thought to cause liver damage.

AFP alphafetoprotein, *q.v.*

afterbirth *n.* [A.S. *aefter*, behind; *beran*, to bring forth] placenta and foetal membranes expelled after offspring's birth; *alt.* decidua, secundines.

after-brain myelencephalon, *q.v.*

after-ripening the period after a seed has been dispersed when it cannot germinate, even if conditions are favourable, and during which physiological changes occur so that it can germinate.

after-sensation persistent sensation, due to continued activity in sense receptor, after cessation of external stimulation.

aftershalf *n.* [A.S. *aefter*, farther away; *sceaft,*

shaft] a small tuft of down near superior umbilicus of a feather; *alt.* hypoptilum.

agameon *n.* [Gk. *a*, without; *gamos*, marriage; *on*, being] a species comprising only apomictic individuals.

agamete *n.* [Gk. *a*, without; *gametēs*, spouse] a young form, gamete, or amoebula, which develops directly without syngamy into an adult, in Sporozoa being a merozoite or schizozoite.

agametoblast *n.* [Gk. *a*, not; *gametēs*, spouse; *blastos*, bud] *see* cytomere; *alt.* schizontoblast.

agamic *a.* [Gk. *a*, without; *gamos*, marriage] asexual; parthenogenetic; *alt.* agamous.

agamic complexes a group of apomictic plants that are usually allopolyploids and consist of many different biotypes forming a taxonomically difficult group.

agamobium *n.* [Gk. *a*, without; *gamos*, marriage; *bios*, life] the asexual generation in alternation of generations, i.e. the sporophyte; *cf.* gamobium.

agamodeme *n.* [Gk. *a*, not; *gamos*, marriage; *dēmos*, people] a deme consisting predominantly of apomictic plants or asexual organisms.

agamogenesis *n.* [Gk. *a*, without; *gamos*, marriage; *genesis*, descent] any reproduction without the male gamete such as parthenogenesis; asexual reproduction.

agamogenetic *a.* [Gk. *a*, without; *gamos*, marriage; *genesis*, descent] asexual; produced asexually.

agamogony *n.* [Gk. *a*, without; *gamos*, marriage; *gonos*, generation] schizogony, *q.v.*; any reproduction without the sexual process.

agamohermaphrodite *a.* [Gk. *a*, without; *gamos*, marriage; *hermaphroditos*, combining both sexes] with neuter and hermaphrodite flowers on the same plant, usually in the same inflorescence.

agamont *n.* [Gk. *a*, without; *gamos*, marriage; *on*, being] a schizont; that stage which gives rise to agametes.

agamospecies *n.* [Gk. *a*, without; *gamos*, marriage; L. *species*, particular kind] a species that reproduces only non-sexually.

agamospermy *n.* [Gk. *a*, without; *gamos*, marriage; *sperma*, seed] any form of apomixis in which embryos and seeds are produced asexually.

agamotropic *a.* [Gk. *a*, not; *gamos*, marriage; *tropē*, turn] *appl.* flowers which once having opened, remain so without closing.

agamous agamic, *q.v.*

agar *n.* [Mal. *agar-agar*, a seaweed] a medium for bacterial and other cultures, prepared from agaragar, a gelatinous substance, yielded by red algae.

Agaricales, agarics *n., n.plu.* [L. *agaricum*, fungus] an order of Basidiomycetes having basidia developed on the surface of gills, or in some classifications also on pores or occasionally on a smooth surface.

age and area hypothesis of Willis that older species occur in a more extensive area than that occupied by more recent species.

agenesia, agenesis *n.* [Gk. *a*, not; *genesis*, origin] failure to develop.

agennesis *n.* [Gk. *a*, without; *gennēsis*, an engendering] sterility.

ageotropism *n.* [Gk. *a*, not; *gē*, earth; *tropē*, turn] not responding to gravity; negative geotropism, *alt* apogeotropism.

agglomerate *a.* [L. *ad*, to; *glomus*, ball] clustered, as a head of flowers; *appl.* adhering mass of Protozoa, as in agglomeration of trypanosomes.

agglutinate *v.* [L. *agglutinare*, to glue on] to cause or to undergo agglutination. *n.* The mass formed by agglutination. *a.* Stuck together; obtect, *q.v.*

agglutination *n.* [L. *ad*, to; *glutinare*, to glue] the formation of clumps or floccules by pollen, bacteria, erythrocytes, spermatozoa, and some protozoans.

agglutinin *n.* [L. *ad*, to; *glutinare*, to glue] a substance or specific antibody which causes agglutination; *alt.* heteroagglutinin.

agglutinogen *n.* [L. *ad*, to; *glutinare*, to glue; Gk. *gennaein* to produce] substance or antigen that produces agglutinin.

aggregate *a.* [L. *ad*, to; *gregare*, to collect into a flock] formed in a cluster; *appl.* fruit formed from apocarpous gynaecium of a single flower, as raspberry, *alt.* etaerio; *appl.* fruit formed from several flowers, as pineapple; *appl.* certain medullary rays; *appl.* a type of silk gland in certain spiders; *appl.* soil particles cemented together by humus, inorganic salts, and mucilage to form clumps of various sizes and shapes, *alt.* ped.

aggregation *n.* [L. *ad*, to; *gregare*, to collect] a grouping or crowding of separate organisms; movement of protoplasm in tentacle or tendril cells of sensitive plants, which causes tentacle or tendril to bend towards the point stimulated.

aggressin *n.* [L. *aggressus*, attacked] toxic substance produced by pathogenic organisms, inhibiting defensive reactions of host; *alt.* virulin.

aggression animal behaviour involving threats or attacks on other animals or sometimes inanimate objects.

aglomerular *a.* [Gk. *a*, without; L. *glomerare*, to form into a ball] devoid of glomeruli, as kidney in certain fishes.

aglossate *n.* [Gk. *a*, without; *glōssa*, tongue] having no tongue.

aglutones *n.plu.* [Gk. *a*, not; *glykys*, sweet] the non-sugar residue produced, together with glucose or some other sugar, on hydrolysis of glycosides.

agminated *a.* [L. *agmen*, a crowd] clustered; *appl.* glands: Peyer's patches, *q.v.*

Agnatha *n.* [Gk. *a*, without; *gnathos*, jaw] a class or superclass of primitive jawless vertebrates, including the lampreys, hagfish, and their extinct relatives; *see also* Cyclostomata.

agnathostomatous, agnathous *a.* [Gk. *a*, without; *gnathos*, jaw; *stoma*, mouth] having mouth unfurnished with jaws, as lamprey.

agon *n.* [Gk. *agōn*, contest] the active principle of an enzyme; *cf.* pheron, symplex.

agonist *n.* [Gk. *agōnistēs*, champion] a prime mover or muscle directly responsible for change in position of a part.

agonistic *a.* [Gk. *agōnistēs*, champion] *appl.* behaviour involving contest, combat, escape, attack, or appeasement.

agranular *a.* [L. *a*, away; *granulum*, small grain] without granules; without a conspicuous layer of granular cells, *appl.* cortex of brain: the motor areas.

agranulocyte *n.* [Gk *a*, without; L. *granulum*, small grain; Gk. *kytos*, hollow] a non-granular or lymphoid leucocyte.

agrestal *a.* [L. *agrestis*, rural] *appl.* uncultivated plants growing on arable land.

agriotype *n.* [Gk. *agrios*, wild; *typos*, image] wild or ancestral type.

agrostology *n.* [Gk. *agrōstis*, grass; *logos*, discourse] that part of botany dealing with grasses; *alt.* graminology.

aheliotropism apheliotropism, *q.v.*

A-horizon the upper, or leached, soil layers; *alt.* eluvial layer.

air bladder *n.* [L. *aer*; A.S. *blædre*, bladder] the swim bladder in fishes; hollow dilatation of thallus in bladderwrack.

air cells thin-walled cavities in ethmoidal labyrinth; numerous cavities in mastoid; alveoli of lungs; air spaces in plant tissue.

air chamber gas-filled compartment of *Nautilus* shell, previously occupied by the animal; accessory respiratory organ or respiratory sac in certain air-breathing teleosts.

air duct duct connecting the swim bladder and gut of certain fishes.

air pore stoma, *q.v.*, of plants; lenticel, *q.v.*

air saccules small terminal sacs of alveolar ducts of bronchioles.

air sacs spaces filled with air and connected with lungs in birds; dilatations of tracheae in many insects; sacs representing tracheal system and having hydrostatic function in certain insect larvae; *alt.* aerostats.

air sinuses cavities in frontal, ethmoid, sphenoid, and maxillary bones, with passages to nasal cavities.

Aistopoda *n.* [Gk. *aistos*, unseen; *pous*, foot] an extinct order of limbless lepospondyl amphibians existing from Carboniferous to Permian times.

aitiogenic *a.* [Gk. *aitios*, causing; *gennaein*, to generate] resulting from causation; *appl.* reaction, as movement induced by an external agent.

aitionastic *a.* [Gk. *aitios*, causing; *nastos*, close-pressed] *appl.* curvature of part of a plant, induced by a diffuse stimulus.

akanth- acanth-, *q.v.*

akaryocyte akaryote, *q.v.*; erythrocyte, *q.v.*

akaryote *n.* [Gk. *a*, without; *karyon*, nut] a cell in which nucleoplasm has not collected together to form a nucleus; a non-nucleated cell; *all.* akaryocyte. *a.* Non-nucleated; *alt.* acaryote.

akene achene, *q.v.*

akinesis *n.* [Gk. *a*, not; *kinēsis*, movement] absence or arrest of motion.

akinete *n.* [Gk. *a*, not; *kinein*, to move] a resting cell in certain algae, which will later reproduce, corresponding to chlamydospore of fungi.

akinetic *a.* [Gk. *a*, not; *kinein*, to move] acentric, *q.v.*

akont *a.* [Gk. *a*, without; *kontos*, punting pole] without flagella; *alt.* aflagellar, acont, atrichous.

akont- *also* acont-, *q.v.*

ala alanine, *q.v.*

ala *n.* [L. *ala*, wing] any wing-like projection or structure; lateral petal of papilionaceous flower; membranous expansion on some fruits or seeds for wind dispersal; basal lobe of moss leaves; outgrowth from petiole of a decurrent leaf; a wing-like projection on bone; *plu.* alae.

alanine *n.* [Gk. *alanin*, irregular from *aldehyd*, aldehyde] a non-essential amino acid, amino-propionic acid; abbreviated to ala.

alar wing-like; *pert.* wings or alae; axillary.

alarm pheromone a pheromone released into the environment which induces a fright response in other members of the species.

alary *a.* [L. *ala*, wing] wing-like, *alt.* aliform; *pert.* wings.

ala spuria bastard wing, *q.v.*

alate *a.* [L. *alatus*, winged] having a wing-like expansion, as of petiole or stem; broad-lipped, *appl.* shells; *appl.* a spicular system in Calcarea which is sagittal because of inequality of angles; winged; *alt.* pterote.

ala temporalis alisphenoid, *q.v.*

albedo *n.* [L. *albus*, white] diffused reflection, the ratio of the amount of light reflected by a surface to the amount of incident light; mesocarp, white tissue of rind, of hesperidium; *cf.* flavedo.

albescent *a.* [L. *albescere*, to grow white] growing whitish.

albicant *a.* [L. *albicare*, to be white] tending to become white.

albinism *n.* [L. *albus*, white] absence of pigmentation in animals normally pigmented; state of having colourless chromatophores.

albino *n.* [Sp. *albino*, white, from L. *albus*] any animal with congenital deficiency of pigment in skin, hair, eyes, etc.; a plant with colourless chromatophores, due to absence of chloroplasts or undeveloped chromoplasts.

albomaculus *a.* [L. *albus*, white; *macula*, spot] *appl.* variegation in plants consisting of an irregular distribution of green and white patches on leaves due to mitotic segregation of genes or chloroplasts.

albuginea *n.* [L. *albus*, white; *gignere*, to beget] white, dense connective tissue surrounding testis, ovary, corpora cavernosa, spleen, or eye; *alt.* tunica albuginea, perididymis.

albumen *n.* [L. *albumen*, white of egg] white of egg containing several proteins including ovalbumin; in higher animals the nutritive material around the yolk; endosperm, *q.v.*

albumin *n.* [L. *albumen*, white of egg] one of a group of heat-coagulable, water-soluble proteins, such as ovalbumin in egg white, serum albumin in blood, lactalbumin in milk, and leucosin in wheat.

albuminoids *n.plu.* [L. *albumen*, white of egg; Gk. *eidos*, form] scleroproteins, *q.v.*

albuminous *a.* [L. *albumen*, white of egg] *pert.*, containing, or of nature of, albumen or an albumin.

albuminous cells parenchyma cells asssociated with sieve cells, as in pteridophytes and gymnosperms.

alburnum *n.* [L. *albus*, white] the young wood of dicotyledons, containing functional xylem, often white in colour and next to the bark; *alt.* sapwood, splintwood; *cf.* duramen.

Alcyonacea *n.* [*Alcyonium*, generic name] an order of Octocorallia, commonly called soft corals, in which the lower parts of the polyps fuse to form a soft mass.

Alcyonaria *n.* [Alcyonium, generic name] Octocorallia, *q.v.*

alder flies *see* Neuroptera.

aldohexoses aldehydic sugars with 6 carbon atoms.

aldolase *n.* [*Alde*hyde] an enzyme which cleaves fructose-1,6-bisphosphate (formerly diphosphate) to form 2 types of triose phosphates; formerly used for any of a group of enzymes which in general catalyse the formation of aldehydes from ketose phosphates; EC 4.1.2.13; *r.n.* fructose-bis-phosphate aldolase.

aldose *n.* [*Alde*hyde] a group of monosaccharides containing an aldehyde group; *cf.* ketose.

aldosterone *n.* [*Alde*hyde, Gk. *stear*, suet] a hormone of the adrenal cortex, promoting retention of sodium ions, excretion of potassium, and influencing carbohydrate metabolism.

alecithal *a.* [Gk. *a*, without; *lekithos*, yolk] with little or no yolk, *appl.* ova; *alt.* alecithic.

alepidote *a.* [Gk. *a*, not; *lepidōtos*, scaly] without scales; *alt.* elepidote.

aletocyte *n.* [Gk. *alētēs*, wanderer; *kytos*, hollow] wandering cell, *q.v.*

aleurispore, aleuriospore, aleurium aleurospore, *q.v.*

aleuron(e) *n.* [Gk. *aleuron*, flour] protein grains in general cytoplasm and used as a reserve food material. *n. Appl.* layer of endosperm containing protein, as in cereals.

aleuroplast *n.* [Gk. *aleuron*, flour; *plastos*, formed] a plastid storing protein.

aleurospore *n.* [Gk. *aleuron*, flour; *sporos*, seed] a lateral conidium of certain fungal parasites of skin; spore or tip early separated from hypha by a septum, or by contraction of protoplasm; *alt.* aleuriospore, aleurispore, aleurium, microconidium.

alexin(e) *n.* [Gk. *alexein*, to ward off] a substance in blood serum which combines with an amboceptor to produce lysis; *alt.* complement.

algae *n.* [L. *alga*, seaweed] a major division of the plant kingdom consisting of simple non-vascular photosynthetic plants with a unicellular, colonial, filamentous, or thalloid body, and being aquatic in marine or fresh water or found in damp habitats on land; *alt.* Phycophyta, algae.

algesis *n.* [Gk. *algēsis*, sense of pain] the sense of pain.

algicolous *a.* [L. *alga*, seaweed; *colere*, to inhabit] living on algae.

algin *n.* [L. *alga*, seaweed] a gel-like polysaccharide, the salt of alginic acid, found in cell walls of brown algae.

alginic acid an acid occurring in the middle lamellae of cells of many brown algae.

algoid *a.* [L. *alga*, seaweed; Gk. *eidos*, shape] *pert.*, resembling, or of the nature of an alga.

algology *n.* [L. *alga*, seaweed; Gk. *logos*, discourse] the study of algae; *alt.* phycology.

Algonkian *a.* [*Algonquian* tribe of Indians] *pert.* late Proterozoic era.

alien a plant thought to have been introduced by man but now more or less naturalized.

aliform *a.* [L. *ala*, wing; *forma*, shape] wing-shaped; *appl.* muscles as in insects; *appl.* some wood parenchyma with wing-like extensions; *alt.* alary.

alima *n.* [Gk. *halios*, *pert.* sea] a larval stage of certain Crustacea.

alimentary *a.* [L. *alimentarius*, *pert.* sustenance] *pert.* nutritive functions.

alimentary canal the tube from mouth to anus which ingests and digests foodstuffs and from which they are absorbed into the body; *alt.* alimentary tract, digestive tract, gut.

alimentary system the alimentary canal and its associated glands.

alimentation n. [L. *alimentum*, nourishment] the process of nourishing or being nourished.

Alismatales n. [*Alisma*, generic name] an order of monocots, placed in the Alismatidae or Helobiae and used in slightly different ways by different authorities.

Alismatidae n. [*Alisma*, generic name] a subclass of monocots having many primitive characteristics such as an apocarpous gynaecium with spirally arranged carpels, and being aquatic or semi-aquatic herbs; *alt.* Helobiae.

alisphenoid n. [L. *ala*, wing; Gk. *sphēn*, wedge; *eidos*, form] wing-like portion of sphenoid forming part of cranium; *alt.* ala temporalis.

alitrunk n. [L. *ala*, wing; *truncus*, trunk] thorax of insect when fused with 1st segment of abdomen.

alkaline gland Dufour's gland, *q.v.*

alkaline tide transient decrease in acidity of body fluids after taking food.

alkaloid n. [Ar. *al*, the; *qali*, ash; Gk. *eidos*, form] any of a group of nitrogenous organic bases found in plants, having poisonous or medicinal properties, such as caffeine, morphine, nicotine, strychnine, etc.; *alt.* vegetable base.

alkanes a group of saturated hydrocarbons thought to be chemical fossils indicating life, which have been found in Pre-Cambrian geological strata.

allaesthetic a. [Gk. *allos*, other; *aisthētēs*, perceiver] *appl.* characters effective when perceived by other organisms; *alt.* allesthetic.

allantochorion n. [Gk. *allas*, sausage; *chorion*, skin] foetal membrane formed of outer wall of allantois and the primitive chorion, being the true chorion.

allantoic a. [Gk. *allas*, sausage] *pert.* allantois.

allantoic acid an oxidation product of allantoin, formed by the action of the enzyme allantoinase, which can be broken down to urea and glyoxylic acid before excretion.

allantoicase an enzyme catalysing the hydrolytic breakdown of allantoic acid to glyoxylic acid and urea, which is found in amphibia, certain fishes, and invertebrates; EC 3.5.3.4; *r.n.* allantoicase.

allantoid a. [Gk. *allas*, sausage; *eidos*, form] botuliform, *q.v.*

allantoin n. [Gk. *allas*, sausage] the end-product of purine and pyrimidine metabolism occurring in allantoic fluid and urine of certain mammals, gastropods, and insects.

allantoinase the enzyme which catalyses the conversion of allantoin to allantoic acid; EC 3.5.2.5; *r.n.* allantoinase.

allantois n. [Gk. *allas*, sausage] an embryonic organ, a membranous sac arising from posterior part of alimentary canal in higher vertebrates, and acting as an organ of respiration and/or nutrition and/or excretion; *a.* allantoic.

allassotonic a. [Gk. *allassein*, to change; *tonos*, strain] induced by stimulus, *appl.* movements of grown plants; *cf.* auxotonic.

allatectomy n. [L. *allatum*, aided; Gk. *ektomē*, a cutting out] excision or removal of corpora allata.

allatum hormone juvenile hormone, *q.v.*

allele n. [Gk. *allēlōn*, one another] one of a pair or more of alternative hereditary characters; a gene which can occupy the same locus as another gene in a particular chromosome; *alt.* allelomorph; *a.* allelic; *cf.* multiple alleles.

allelism n. [Gk. *allēlōn*, one another] the relationship between alleles and their inheritance; *alt.* allelomorphism, alternative inheritance.

allelocatalysis n. [Gk. *allēlōn*, one another; *katalysis*, dissolution] mutually accelerating or retarding effect of cells by their secretion of growth-accelerating or -retarding substances, as in certain Protozoa which accelerate their rate of fission due to this, with increase in number of individuals present.

allelomimetic a. [Gk. *allēlōn*, one another; *mimētikos*, imitative] *appl.* animal behaviour involving imitating another animal usually of the same species.

allelomorph n. [Gk. *allēlōn*, one another; *morphē*, form] allele, *q.v.*; *a.* allelomorphic.

allelomorphism allelism, *q.v.*

allelopathy n. [Gk. *allēlōn*, one another; *pathos*, suffering] the influence or effect of one living plant upon another.

allelotype n. [Gk. *allēlōn*, one another; *typos*, pattern] the frequency of alleles in a population.

Allen's rule the idea that the protruding parts of an animal body such as tails, ears, limbs are shorter in animals in colder regions than those of the same species in warmer regions.

allergen n. [Gk. *allos*, other; *ergon*, activity; *-genēs*, producing] a substance, usually a protein, which induces allergy; *alt.* atopen.

allergy n. [Gk. *allos*, other; *ergon*, activity] changed reactivity on second or subsequent infection or poisoning; exaggerated or unusual susceptibility, *alt.* anaphylaxis; atopy, *q.v.*

allesthetic allaesthetic, *q.v.*

alliaceous a. [L. *allium*, garlic] *pert.* or like garlic or onion.

alloantigen n. [Gk. *allos*, other; *anti*, against; *genos*, birth] isoantigen, *q.v.*

allobiosis n. [Gk. *allos*, other; *biōsis*, manner of life] changed reactivity of an organism in a changed internal or external environment.

allocarpy n. [Gk. *allos*, other; *karpos*, fruit] the production of fruit after cross-fertilization.

allocheiral a. [Gk. *allos*, other; *cheir*, hand] having right and left sides reversed; *pert.* reversed symmetry.

allochroic a. [Gk. *allos*, other; *chrōs*, colour] able to change colour; with colour variation.

allochronic a. [Gk. *allos*, other; *chronos*, time] not contemporary, *appl.* species, etc.; *alt.* allogenic; *cf.* synchronic.

allochthonous a. [Gk. *allos*, other; *chthōn*, the ground] exotic; not aboriginal; acquired; *cf.* autochthonous.

allocortex n. [Gk. *allos*, other; L. *cortex*, bark] the primitive cortical areas or cortex of olfactory brain; *cf.* isocortex.

allocycly n. [Gk. *allos*, other; *kyklos*, circle] the differences in coiling behaviour in chromosomes or regions of a chromosome, as seen in some sex chromosomes, the nucleolar organizer, and centromere.

Alloeocoela, Alloiocoela n. [Gk. *alloios*, different; *koilos*, hollow] an order of turbellarians having a pharynx, and an intestine with short diverticula.

allogamy n. [Gk. *allos*, other; *gamos*, marriage] cross-fertilization, *q.v.*; *cf.* autogamy; *a.* allogamous.

allogene

allogene *n.* [Gk. *allos*, other; *genos*, descent] a recessive allele; *cf.* protogene.

allogenic *a.* [Gk. *allos*, other; *genos*, descent] caused by external factors, *appl.* plant successions; *pert.* allogenes; of different genetic constitution, *alt.* allogeneic; derived from elsewhere, *alt.* allogenetic; allogenous, *q.v.*; exogenous, *q.v.*; allochronic, *q.v.*; *cf.* autogenic.

allogenous *a.* [Gk. *allos*, other; *genos*, descent] *appl.* floras persisting from an earlier environment; *alt.* allogenic.

allograft *n.* [Gk. *allos*, other; O.F. *graffe*, graft] homograft, *q.v.*

alloheteroploid *n.* [Gk. *allos*, other; *heteros*, other; *aploos*, onefold; *eidos*, form] heteroploid derived from specifically distinct genomes; *cf.* autoheteroploid.

alloiogenesis *n.* [Gk. *alloios*, different; *genesis*, descent] alternation of generations, *q.v.*

alloiometron *n.* [Gk. *alloios*, different; *metron*, measure] measurable change of proportion or intensity of development within species or races, e.g. head, limb, tooth, etc., proportions.

allokinesis *n.* [Gk. *allos*, other; *kinēsis*, movement] reflex, or passive, movement; involuntary movement.

allokinetic *a.* [Gk. *allos*, other; *kinētikos*, putting in motion] moving passively; drifting, as plankton; *cf.* autokinetic.

allometric *a.* [Gk. *allos*, other; *metron*, measure] differing in growth rate; *pert.* allometry; *alt.* heterogonic.

allometry *n.* [Gk. *allos*, other; *metron*, measure] study of relative growth; change of proportions with increase of size; growth rate of a part differing from a standard growth rate or from the growth rate of the whole; *alt.* heterogony, dysharmony.

allomixis *n.* [Gk. *allos*, other; *mixis*, mingling] cross-fertilization, *q.v.*

allomone *n.* [Gk. *allos*, other; hor*mone*] a chemical secreted by an individual which causes another organism of a different species to react favourably to it, such as scent given out by flowers attracting pollinating insects.

allomorphosis *n.* [Gk *allos*, other; *morphōsis*, shaping] evolution with a rapid increase of specialization; *cf.* aromorphosis.

alloparalectotype *n.* [Gk. *allos*, other; *para*, beside; *lektos*, chosen; *typos*, pattern] specimen, from the original collection, of the sex opposite to that of the holotype, and described subsequently.

allopatric *a.* [Gk. *allos*, other; *patra*, native land] having separate and mutually exclusive areas of geographical distribution; *cf.* sympatric.

allopelagic *a.* [Gk. *allos*, other; *pelagos*, sea] *pert.* organisms found at any depth of the sea.

allophene *n.* [Gk. *allos*, other; *phainein*, to appear] a phenotype not due to a mutation in the actual cells showing the characteristic but due to other cells of the host, and which will show a normal phenotype if transplanted to a normal host; *cf.* autophene.

allophore *n.* [Gk. *allos*, other; *pherein*, to bear] a cell or chromatophore containing red pigment, in skin of fishes, amphibians, and reptiles.

allophytoid *n.* [Gk. *allos*, other; *phytos*, growing; *eidos*, form] a propagative bud, differing from a vegetative bud; a bulbil, as of some lilies.

alloplasm *n.* [Gk. *allos*, other; *plasma*, mould] the differentiated portion of protoplasm such as myofibrils or cilia, not forming independent organelles; protoplasmic derivatives such as intercellular substance and cell walls.

alloplast *n.* [Gk. *allos*, other; *plastos*, formed] a morphological cell unit of more than one kind of tissue; *cf.* homoplast.

allopolyploid, alloploid *n.* [Gk. *allos*, other; *polys*, many; *aploos*, onefold; *eidos*, form] a polyploid produced from a hybrid between 2 or more species and therefore possessing 2 or more unlike sets of chromosomes; *alt.* multiple diploid, amphiploid; *cf.* autopolyploid.

allorhizal *a.* [Gk. *allos*, other; *rhiza*, root] having opposed root and shoot poles; *cf.* homorhizal.

all-or-none principle that response to a stimulus is either completely effected or is absent, first observed in heart muscle; *alt.* Bowditch's law.

alloscutum *n.* [Gk. *allos*, another; L. *scutum*, shield] dorsal area or sclerite behind scutum in larval ticks; *cf.* conscutum.

allosematic *a.* [Gk. *allos*, other; *sēma*, sign] having markings or coloration imitating warning signs in other, usually dangerous species; *cf.* aposematic.

allosomal *a.* [Gk. *allos*, other; *sōma*, body] *pert.* allosome; *appl.* inheritance of characters controlled by genes located in an allosome.

allosome *n.* [Gk. *allos*, other; *sōma*, body] a chromosome other than a typical one, such as a sex chromosome; *alt.* heterochromosome; *cf* autosome.

allosteric *a.* [Gk. *allos*, other; *stereos*, solid] *appl.* effect of binding of certain small molecules on to a protein, such as an enzyme, at a site or sites distant from the active site, which change the properties of the active site.

allostoses *n.plu.* [Gk. *allos*, other; *osteon*, bone] bones formed by direct ossification of areolar connective tissue without passing through a cartilage stage; *alt.* membrane bone, investing bone; *cf.* autostoses.

allosynapis, allosyndesis *n.* [Gk. *allos*, other; *synapsis*, union; *syndesis*, a binding together] pairing of homologous chromosomes from opposite parents, in a polyploid; *cf.* autosyndesis.

allotetraploid *n.* [Gk. *allos*, other; *tetraplē*, fourfold] an allopolyploid produced when a hybrid between 2 species doubles its chromsome number; *alt.* amphidiploid.

Allotheria *n.* [Gk. *allos*, other; *thērion*, small animal] a subclass of Jurassic to Eocene herbivorous mammals, including the single order Multituberculata, which may be primitive and are sometimes considered as a group of therians.

allotherm *n.* [Gk. *allos*, other; *thermē*, heat] an organism with body temperature dependent on environmental temperature.

allotopotype *n.* [Gk. *allos*, other; *topos*, place; *typos*, model] an allotype obtained from the type locality.

Allotriognathi *n.* [Gk. *allotrios*, strange; *gnathos*, jaw] Lampridiformes, *q.v.*

allotriploid *n.* [Gk. *allos*, other; *triploos*, threefold; *eidos*, form] an organism whose somatic cells contain 3 sets of chromosomes, 1 of which differs from the other 2.

allotrophic *a.* [Gk. *allos*, other; *trophē*, nourishment] heterotrophic, *q.v.*; with a changed, usually lowered, nutritive value, *appl.* foods.

allotropism *n.* [Gk. *allos*, other; *trepein*, to turn] tendency of certain cells or structures to approach each other; mutual tropism, as between gametes; *alt.* allotropy; *a.* allotropic.

allotropous *a.* [Gk. *allos*, any other; *tropos*, direction] *appl.* insects not limited to or adapted to visiting special kinds of flowers; *appl.* flowers whose nectar is available to all kinds of insects; *cf.* eutropous; *n.* allotropy.

allotropy allotropism, *q.v.*; allotropous condition.

allotype *n.* [Gk. *allos*, other; *typos*, pattern] paratype of the sex opposite to that of the holotype.

allotypy *n.* [Gk. *allos*, other; *typos*, type] the property in proteins of existing in antigenically distinguishable forms that it has not so far been possible to distinguish by chemical means.

allozygote *n.* [Gk. *allos*, other; *zygōtos*, yoked] a homozygote having recessive characters, *cf.* protozygote; a homozygote in which the 2 homologous genes are thought to have an independent origin, *cf.* autozygote.

alluvial *a.* [L. *alluere*, to wash to] *pert.* deposits formed by finely divided material laid down by running water; *appl.* A-horizon of soil.

alpha (α) cells oxyphilic cells in pars glandularis of pituitary gland; cells which secrete glucagon in islets of Langerhans, *alt.* A-cells.

alphafetoprotein (AFP) a blood protein, the levels of which can indicate whether a pregnant women is carrying a foetus with spina bifida or anencephaly.

alpha (α) globulin one of the constituents of blood plasma, some of which are hormone-transporters.

alpha (α) granules metachromatic granules in central region of protoplast, as in blue–green algae.

alpha (α) helix a right-handed helix which forms the most stable structure in any polypeptide chains; *cf.* beta pleated sheet.

alpha (α) rhythm spontaneous rhythmic fluctuations of electric potential of cerebral cortex during mental inactivity.

alpha (α) tocopherol vitamin E, *q.v.*

alphitomorphous *a.* [Gk. *alphiton*, pearl barley; *morphē*, form] having the appearance of peeled (pearl) barley, *appl.* certain fungi.

alsinaceous *a.* [Gk. *alsinē*, chickweed] *appl.* polypetalous corolla where intervals occur between petals, as in chickweed.

alteration theory explains electromotive forces of nerve and muscle by alterations in chemical compostion of tissue at cross-section.

alternate *a.* [L. *alternus*, one after another] not opposite; *appl.* leaves, branches, etc., occurring at different levels successively on opposite sides of stem; every other; taking turns.

alternating cleavage spiral clevage, *q.v.*

alternation of generations the occurrence in one life history of 2 or more different forms differently produced, usually an alternation of a sexual with an asexual form; *alt.* alloiogenesis, metagenesis, digenesis, heterogamy, heterogenesis, heterogony.

alternation of parts general rule that leaves of different whorls alternate in position with each other, sepals with petals, stamens with petals.

alternative inheritance allelism, *q.v.*

alterne *n.* [L. *alternus*, one after another] vegetation exhibiting disturbed zonation due to abrupt change in environment, or to interference with normal plant succession.

alternipinnate *a.* [L. *alternus*, one after another; *pinna*, wing] *appl.* leaflets or pinnae arising alternately on each side of midrib.

Altmann's granules [*R. Altmann*, German histologist] hypothetical units, *q.v.*; mitochondria, *q.v.*

altrices *n.plu.* [L. *altrix*, nourisher] birds whose young are hatched in a very immature condition; *cf.* praecoces.

altricial *a.* [L. *altrix*, nourisher] requiring care or nursing after hatching or birth.

altruistic adaptation an adaptation favouring the survival of a group at the expense of an individual member of that group.

alula *n.* [L. *alula*, *dim.* of *ala*, wing] a small lobe separated off from wing base on its posterior edge in certain insects; lower tegula or squama thoracicalis of Diptera; bastard wing, *q.v.*

alutaceous *a.* [L. *aluta*, alum-dressed leather] tancoloured; leathery; having appearance of minute cracks, *appl.* markings on elytra of certain beetles.

alveola *n.* [L. *alveolus*, small cavity] a pit on the surface of an organ, *alt.* faveolus; alveolus, *q.v.*

alveolar *a.* [L. *alveolus*, small pit] *pert.* an alveolus; *pert.* tooth socket; *appl.* artery, nerve, process, canal, in connection with the jaw bone; *appl.* small cavities in lungs, glands, etc.; *appl.* pores connecting adjacent to air cells or pulmonary alveoli.

alveolar theory a 19th-century theory of the structure of cytoplasm, that protoplasm was made of a series of bubbles called 'alveolar spheres' scattered in the ground substance; *alt.* foam theory.

alveolate *a.* [L. *alveolatus*, pitted] deeply pitted or honeycombed; *alt.* faveolate.

alveolation *n.* [L. *alveolatus*, pitted] the formation of alveoli; alveolate appearance.

alveolus *n.* [L. *alveolus*, small pit] a small pit or depression; tooth socket, *alt.* odontobothrion; pyramidal ossicle supporting tooth in Aristotle's lantern of sea urchin; air cavity of lungs; a cavity in glands, *alt.* acinus; cavity in tarsus of spiders; receptacle for haematodoca; pit for articulation of macrotrichia; a subdivision of a vacuole; *alt.* alveola.

alveus *n.* [L. *alveus*, cavity] a white layer of fibres on ventricular surface of hippocampus; utricle of ear; dilatation of thoracic duct.

amacrine *a.* [Gk. *a*, not; *makros*, long; *is*, fibre] having no conspicuous axon; *appl.* cells in inner nuclear layer of retina, with dendrites in inner plexiform layer.

amb *n.* [L. *ambulare*, to walk] ambulacral area.

ambiens *n.* [L. *ambire*, to go round] a thigh muscle in certain birds, the action of which causes the toes to maintain grasp on perch.

ambient *a.* [L. *ambire*, to go round] surrounding; *appl.* vein, the costal nervure when encircling insect wing.

ambilateral *a.* [L. *ambo*, both; *latus*, side] *pert.* both sides.

ambiparous *a.* [L. *ambo*, both; *parere*, to produce] containing the beginnings of both flowers and leaves, *appl.* buds.

ambisexual *a.* [L. *ambo*, both; *sexus*, sex] *pert.* both sexes, *alt.* ambosexual; monoecious, *q.v.*

ambisporangiate *a.* [L. *ambo*, both; Gk. *sporos*, seed; *anggeion*, vessel] amphisporangiate, *q.v.*

ambital *a.* [L. *ambire*, to go round] *appl.* inter-ambulacral and antambulacral plates of asteroids; *appl.* outer skeleton of ophiuroid arm.

ambitus *n.* [L. *ambitus*, going around] the outer edge or margin; outline of echinoid shell viewed from apical pole.

amblychromatic *a.* [Gk. *amblys*, dull; *chrōma*, colour] staining or stained slightly; *cf.* trachychromatic.

amblypoda, amblypods *n., n.plu.* [Gk. *amblys*, dull; *pous*, foot] an order of Palaeocene to Eocene placental mammals of the New World and Asia, that were archaic ungulates probably from several unrelated groups.

Amblypygi *n.* [Gk. *amblys*, dull; *pygē*, rump] an order of arachnids that have a long slender 1st pair of walking legs used as antennae.

amboceptor *n.* [L. *ambo*, both; *capere*, to take] a specific antibody or immune body necessary for action of complement on a toxin or a red blood corpuscle; lysin, *q.v.*; *alt.* desmone.

ambon *n.* [Gk. *ambon*, raised platform] fibrocartilaginous ring surrounding an articular socket, as around acetabulum; circumferential fibrocartilage; *alt.* labrum.

ambosexual *a.* [L. *ambo*, both; *sexus*, sex] common to, or *pert.*, both sexes, *alt.* ambisexual, monoecious; activated by both male and female hormones.

abrosial *a.* [Gk. *ambrosia*, food of the gods] *appl.* a class of odours including those typified by ambergris and musk.

ambulacra *n.plu.* [L. *ambulare*, to walk] region containing tube-feet of echinoderms; the bands of tube-feet themselves; *sing.* ambulacrum.

ambulacral *pert.* or used for walking, *appl.* limbs of arthropods; *pert.* ambulacra; *n.* ambulacralium.

ambulacralia *n.plu.* [L. *ambulare*, to walk] ambulacral plates, i.e. plates through which tube-feet protrude; *alt.* ambulacrals; *sing.* ambulacralium.

ambulacriform *a.* [L. *ambulare*, to walk; *forma*, shape] having the form or appearance of ambulacra.

ameba amoeba, *q.v.*

ameiosis *n.* [Gk. *a*, without; *meiōsis*, diminution] occurrence of only one division in meiosis instead of 2; the absence of pairing of chromosomes in meiosis.

ameiotic *a.* [Gk. *a*, without; *meiōn*, smaller] *appl.* parthenogenesis in which meiosis is suppressed.

amelification *n.* [M.E. *amell*, enamel; L. *facere*, to make] formation of tooth enamel.

ameloblast *n.* [M.E. *amell*, enamel; Gk. *blastos*, bud] a columnar or hexagonal cell of internal epithelium which secretes enamel and is part of the enamel organ; *alt.* enamel cell, adamantoblast, ganoblast.

amensalism *n.* [Gk. *a*, not; L. *mensa*, table] a form of antagonism or competition between 2 species or organisms in which one is inhibited and the other is not; *alt.* amentalism.

ament amentum, *q.v.*

amenta *plu.* of amentum.

amentaceous, amentiferous *a.* [L. *amentum*, thong; *ferre*, to carry] *appl.* plants bearing catkins.

amentalism amensalism, *q.v.*

Amentiferae *n.* [L. *amentum*, thong; *ferre*, to carry] trees that bear catkins, not a taxonomic group; *alt.* Amentiflorae.

Amentiflorae *n.* [L. *amentum*, thong; *flos*, flower] Hamamelididae, *q.v.*; Amentiferae, *q.v.*

amentum *n.* [L. *amentum*, thong] catkin, *q.v.*; *alt.* ament.

ameristic *a.* [Gk. *a*, without; *meristos*, divided] not divided into parts; unsegmented; undifferentiated or undeveloped.

Ametabola *n.* [Gk. *a*, without; *metabolē*, change] Apterygota, *q.v.*

ametabolic, ametabolous *a.* [Gk. *a*, without; *metabolē*, change] not changing form, *appl.* ciliates; *appl.* insects that do not pass through marked metamorphosis.

ametoecious *a.* [Gk. *a*, without; *meta*, after; *oikos*, house] parasitic on one host during one life cycle; *alt.* autoecious, autoxenous; *cf.* metoecious.

amicron *n.* [Gk. *a*, without; *mikros*, small] a particle smaller than 1 nm, so that the ultramicroscope can only indicate it as a diffuse illumination in the track of the beam; *cf.* submicron.

amicronucleate *a.* [Gk. *a*, without; *mikros*, small; L. *nucleus*, kernel] *appl.* fragments of certain Protozoa in which there is no micronucleus.

amictic *a.* [Gk. *a*, not; *miktos*, mixed] *appl.* eggs that cannot be fertilized and which develop parthenogenetically into females; *appl.* females producing such eggs.

amidase an enzyme catalysing the hydrolysis of a monocarboxylic acid amide to a monocarboxylic acid and ammonia; EC 3.5.1.4.

amidases a group of enzymes hydrolysing non-peptide C–N linkages of amides including urease, asparaginase, glutaminase.

amides *n.plu.* [Gk. *ammoniakon*, gum] derivatives of carboxylic acids in which the hydroxyl (–OH) group has been replaced by an amino group ($-NH_2$).

amidinases a group of enzymes hydrolysing non-peptide C–N linkages of amidines; EC sub-sub-groups 3.5.3 and 3.5.4.

amidines simple organic compounds of general formula $RCNH.NH_2$.

Amiiformes *n.* [*Amia*, generic name of bowfin; L. *forma*, shape] an order of holosteans including the modern bowfins and their fossil relatives from Triassic times.

aminases hydrolysing enzymes which attack the C–NH and $C-NH_2$ linkages in non-protein compounds and are important in purine metabolism and the production of urea.

amines *n.plu.* [Gk. *ammoniakon*, resinous gum] simple organic compounds of nitrogen that are derivatives of ammonia in which the hydrogen atoms may be replaced by organic groups, formed by plants and by bacterial action on amino acids, and of general formula $RCONH_2$.

amino acid racemization the conversion of a 'left-handed' to a 'right-handed' amino acid after the death of the organism containing it, which can be used as a method of dating fossils over a timespan of 15,000 to 100,000 years.

amino acids compounds containing amino ($-NH_2$) and carboxyl (–COOH) groups, α-amino acids being constituents of proteins, synthesized in autotrophic organisms, produced from proteins by hydrolysis in heterotrophic organisms.

amino acyl RNA an amino acid attached to its molecule of tRNA by a high-energy bond.

amino-peptidase any of various proteolytic enzymes which attack peptides containing a free amino group.

aminopherases transaminases, *q.v.*

amino sugars substances such as glucosamine and galactosamine in which amino groups have been substituted for various hydroxyl groups of sugars.

amitosis n. [Gk. *a*, without; *mitos*, thread] division of the nucleus by constriction without the formation of chromosomes or a spindle and without the breakdown of the nuclear membrane; *alt.* holoschisis, direct nuclear division, fragmentation; *cf.* mitosis.

amixia n. [Gk. *a*, not; *mixis*, a mixing] cross-sterility between members of the same species as a result of morphological, geographical, or physiological isolating mechanisms.

amixis n. [Gk. *a*, not; *mixis*, a mixing] absence of fertilization; sometimes used for absence of gonads; apomixis in haploid organisms.

ammochaeta n. [Gk. *ammos*, sand; *chaitē*, hair] bristle on head of desert ants, arranged in groups and used for removal of sand from forelegs.

ammocoetes n.plu. [Gk. *ammos*, sand; *koitē*, bed] the larval form of lampreys.

ammonia n. [Gk. *ammoniakon*, resinous gum] a colourless pungent alkaline gas formed by the decomposition of protein, nitrogenous bases, and urea.

ammonification n. [Gk. *ammoniakos*, resinous gum; L. *facere*, to make] the production of ammonium ions by heterotrophic soil organisms from organic matter in the soil; *alt.* ammonization.

ammonifiers ammonia-releasing organisms, especially bacteria.

ammoniotelic a. [*Ammonia*; Gk. *telos*, end] excreting nitrogen mainly as ammonia as various invertebrates, teleosts, tadpoles, etc.

ammonitiferous a. [Gk. *Ammōn*, Jupiter; L. *ferre*, to carry] containing fossil remains of ammonites.

ammonization ammonification, *q.v.*

Ammonoidea, ammonites n., n.plu. [Gk. *Ammōn*, Jupiter; *eidos*, form] an extinct subclass of tetrabranch cephalopods that were similar to the nautiloids but probably had a calcareous protoconch.

amniomucoids n.plu. [Gk. *amnion*, foetal membrane; L. *mucus*, mucus; Gk. *eidos*, form] a class of hexosamine-rich glycoproteins isolated from amniotic fluid.

amnion n. [Gk. *amnion*, foetal membrane] a foetal membrane of reptiles, birds, and mammals; inner embryonic membrane of insects; a membrane like a amnion found in other invertebrates; viscous envelope of certain ovules; *alt.* caul.

amnionic amniotic, *q.v.*

Amniota, amniotes n., n.plu. [Gk. *amnion*, foetal membrane] a group of Craniata, usually considered a superclass, including reptiles, birds, and mammals, which have an amnion around the embryo.

amniotic a. [Gk. *amnion*, foetal membrane] pert.

amnion; *appl.* folds, sac, cavity, fluid; *alt.* amnionic.

amoeba n. [Gk. *amoibē*, change] a protozoon in which the shape is subject to constant alterations due to formation and retraction of pseudopodia; *alt.* ameba.

amoebadiastase n. [Gk. *amoibē*, change; *dia*, through; *histanai*, to set] amylase secreted by amoeba; any of various digestive enzymes secreted by amoeba.

amoebic a. [Gk. *amoibē*, change] pert., or caused by, amoebae.

Amoebidiales n. [Gk. *amoibē*, change] an order of Trichomycetes reproducing by amoeboid cells.

amoebiform a. [Gk. *amoibē*, change; L. *forma*, shape] amoeboid, *q.v.*

Amoebina n. [Gk. *amoibē*, change] an order of Sarcodina including amoeba and having the body of naked cytoplasm and blunt pseudopodia.

amoebism n. [Gk. *amoibē*, change] amoeboid form or behaviour, as of leucocytes.

amoebocyte n. [Gk. *amoibē*, change; *kytos*, hollow] any cell having the shape or properties of an amoeba; one of certain cells in coelom of echinoderms; leucocyte, *q.v.*

amoeboid a. [Gk. *amoibē*, change; *eidos*, shape] resembling an amoeba in shape, in properties, or in locomotion; *alt.* amoebiform.

amoebula n. [Gk. *amoibē*, change] the amoeboid swarm spore of Protista furnished with pseudopodia, *alt.* pseudopodiospore; any protozoon having amoeboid movement.

amorph n. [Gk. *a*, without; *morphē*, form] a recessive allele which does not influence phenotype.

amorphous a. [Gk. *a*, without; *morphē*, shape] of indeterminate or irregular form; with no visible differentiation in structure.

AMP adenosine monophosphate, *q.v.*; *see also* cyclic AMP.

ampheclexis n. [Gk. *amphi*, both; *eklexis*, choice] sexual selection.

ampherotoky amphitoky, *q.v.*

amphetamine n. [*Alpha*; *methyl*; *phenyl*; *ethyl*; *amine*] a sympathomimetic drug sold under the trade name of Benzedrine that is chemically related to adrenaline and is used as a powerful stimulant of the CNS.

amphiapomict n. [Gk. *amphi*, both; *apo*, away; *miktos*, mixed] a biotype reproduced from facultative sexual forms.

amphiarthrosis n. [Gk. *amphi*, both; *arthron*, joint] a slightly movable articulation, as a symphysis or a syndesmosis.

amphiaster n. [Gk. *amphi*, both; *astēr*, star] the 2 asters connected by the achromatic spindle in mitosis and meiosis; a sponge spicule star-shaped at both ends; *a.* amphiastral.

Amphibia, amphibians n., n.plu. [Gk. *amphi*, both; *bios*, life] a class of anamniote vertebrates which have gills during the larval stage, but adults typically have lungs and are tetrapods with a skin usually without scales.

amphibian a. [Gk. *amphi*, both; *bios*, life] adapted for life either on land or in water; emersed, *q.v.*; *alt.* amphibious.

amphibiotic a. [Gk. *amphi*, both; *biōtikos*, pert. life] living in water as a larva, on land in the adult stage; *alt.* amphibious.

amphibious amphibian, *q.v.*; amphibiotic, *q.v.*

amphibivalent n. [Gk. amphi, both; L. bis, twice; valere, to be strong] a ring of chromosomes arising in the metaphase and anaphase of the 1st meiotic division as a result of the reciprocal translocation of chromosome segments betwen 2 chromosomes.

amphiblastic a. [Gk. amphi, both; blastos, bud] appl. telolecithal ova with complete but unequal segmentation.

amphiblastula n. [Gk. amphi, both; blastos, bud] larval stage in development of certain sponges in which posterior end is composed of granular archaeocytes, and anterior end is made of flagellate cells.

amphibolic a. [Gk. amphi, both; bolē, throw] capable of turning backwards or forwards, as outer toe of certain birds.

amphicarpous a. [Gk. amphi, both; karpos, fruit] producing fruit of 2 kinds; alt. amphicarpic.

amphicaryon amphikaryon, q.v.

amphichromatism n. [Gk. amphi, both; chrōmatismos, colouring] production of flowers of 2 different colours on the same plant.

amphicoelous, amphicelous a. [Gk. amphi, both; koilos, hollow] concave on both surfaces, appl. biconcave vertebral centra.

amphicondylous a. [Gk. amphi, both; kondylos, knuckle] having 2 occipital condyles.

amphicone n. [Gk. amphi, both; kōnos, cone] cusp of molar of extinct mammals, believed to have evolved into metacone and paracone.

amphicribral a. [Gk. amphi, both; L. cribrum, sieve] with the phloem surrounding the xylem, appl. some concentric vascular bundles; alt. amphiphloic; cf. amphivasal.

amphicytes n.plu. [Gk. amphi, both; kytos, hollow] endothelial cells surrounding, or forming, capsules of cells of a dorsal root ganglion; alt. capsule cells.

amphidelphic a. [Gk. amphi, both; delphys, womb] having a paired uterus, as in certain nematodes; alt. didelphic.

amphidetic a. [Gk. amphi, both; detos, bound] extending behind and in front of umbo, appl. hinge ligaments of some bivalve shells; cf. opisthodetic, parivincular.

amphidial a. [Gk. amphi, both] pert. amphids, appl. a unicellular gland in nematodes.

amphidiploid a. [Gk. amphi, both; diploos, double] allotetraploid, q.v.

amphidisc n. [Gk. amphi, both; diskos, round plate] a grapnel-shaped spicule of some freshwater sponges.

Amphidiscophora n. [Gk. amphi, both; diskos, round plate; phora, producing] an order of Hexactinellida in which the microscleres are all amphidiscs.

amphids n.plu [Gk. amphi, both] two anterior lateral chemoreceptive organs in nematodes.

amphigastria n.plu. [Gk. amphi, both; gaster, stomach] rudimentary leaves, or scales, on under surface of leafy liverworts.

amphigenesis n. [Gk. amphi, both; genesis, descent] amphigony, q.v.

amphigenous a. [Gk. amphi, both; -genes, producing] borne or growing on both sides of a structure as of a leaf; borne or growing on all sides of an organism or structure, such as oogonium of certain fungi growing through antheridium, alt.

perigenous; being sexually attractive to members of the same or opposite sex.

amphigonic a. [Gk. amphi, both; gonē, seed] producing male and female gametes in separate gones in different individuals, alt. bisexual; pert. amphigony; cf. digonic, syngonic.

amphigony n. [Gk. amphi, both; gonos, offspring] sexual reproduction involving 2 individuals; alt. amphigenesis.

amphigynous a. [Gk. amphi, both; gynē, female] appl. antheridium surrounding the base of the oogonium, as in some Peronosporales.

amphihaploid n. [Gk. amphi, both; haploos, simple; eidos, form] a haploid arising from an amphidiploid species.

amphikaryon n. [Gk. amphi, both; karyon, nut] amphinucleus, q.v.; nucleus with 2 haploid sets of chromosomes; alt. amphicaryon; cf. diplokaryon.

Amphilinoidea n. [Amphilina, generic name; Gk. eidos, form] an order of Cestodaria having a protrusible proboscis and glands at the anterior end.

amphimict n. [Gk. amphi, both; miktos, mixed] a biotype resulting from sexual reproduction; an obligate sexual organism.

amphimixis n. [Gk. amphi, both; mixis, mingling] reproducing by seed produced by normal sexual fusion; a. amphimictic; cf. apomixis.

Amphineura n. [Gk. amphi, both; neuron, nerve] a class of molluscs, commonly called chitons, that have an elongated body and a mantle bearing calcareous plates.

Amphinomorpha n. [Amphinoe, generic name; Gk. morphē, form] an order of polychaetes having a small prostomium and well-developed parapodia.

amphinucleolus n. [Gk. amphi, both; L. nucleolus, a small kernel] a double nucleolus comprising basiphil and oxyphil components.

amphinucleus n. [Gk. amphi, both; L. nucleus, kernel] a nucleus with large karyosome, in reference to supposed encapsuling of kinetic nucleus by trophic nucleus; alt. amphikaryon.

amphiodont a. [Gk. amphi, both; odous, tooth] appl. an intermediate state of mandible development in stag beetles.

amphiont n. [Gk. amphi, both; on, being] zygote or sporont formed by coming together of 2 individuals.

amphiphloic a. [Gk. amphi, both; phloios, inner bark] with phloem both external and internal to xylem; alt. amphicribral.

amphiplatyan a. [Gk. amphi, both; platys, flat] flat on both ends, appl. vertebral centra.

amphiploid n. [Gk. amphi, both; aploos, onefold; eidos, form] allopolyploid, q.v.

amphipneustic a. [Gk. amphi, both; pnein, to breathe] having both gills and lungs throughout life history; with only anterior and posterior pairs of spiracles functioning, as in most dipterous larvae; alt. amphipneustous; cf. peripneustic.

Amphipoda, amphipods n., n.plu. [Gk. amphi, both; pous, foot] an order of freshwater and marine malacostracans, having a laterally compressed body, elongated abdomen, and no carapace.

amphipodous a. [Gk. amphi, both; pous, foot] having feet for walking and feet for swimming.

amphipyrenin n. [Gk. amphi, both; pyrēn, fruit stone] substance of which nuclear membrane is composed.

amphirhinal a. [Gk. amphi, both; rhines, nostrils] having, or pert., 2 nostrils.

amphisarca n. [Gk. amphi, both; sarx, flesh] a superior indehiscent many-seeded fruit with pulpy interior and woody exterior.

amphispermous a. [Gk. amphi, both; sperma, seed] having seed closely surrounded by pericarp.

amphisporangiate a. [Gk. amphi, both; sporos, seed; anggeion, vessel] having sporophylls bearing both megasporangia and microsporangia; hermaphrodite, appl. flowers; alt. ambisporangiate, bisexual.

amphispore n. [Gk. amphi, both; sporos, seed] a reproductive spore which functions as a resting spore in certain algae, alt. mesospore; a uredospore modified to withstand dry environment.

amphisternous a. [Gk. amphi, both; sternon, breast bone] appl. type of sternum structure in Atelostomata.

amphistomatic a. [Gk. amphi, both; stoma, mouth] having stomata on both surfaces, appl. certain types of leaves.

amphistomous a. [Gk. amphi, both; stoma, mouth] having a sucker at each end of body, as leeches.

amphistylic a. [Gk. amphi, both; stylos, pillar] having jaw arch connected with skull by both hyoid and quadrate, or by both hyoid and palatoquadrate; n. amphistyly.

amphitelic a. [Gk. amphi, both; telos, fulfilment] appl. orientation of chromosomes on the equator of the spindle at metaphase in mitosis, with the centromere exactly on the equator and equidistant from both poles.

amphitene a. [Gk. amphi, both; tainia, band] zygotene, q.v.

amphithecium n. [Gk. amphi, both; thēkē, box] in bryophytes, peripheral layer of cells in sporogonium, formed from the outer region of the embryo; cf. endothecium.

amphitoky n. [Gk. amphi, both; tokos, birth] parthenogenetic reproduction of both males and females; alt. ampherotoky, amphoterotoky.

amphitriaene n. [Gk. amphi, both; triaina, trident] a double trident-shaped spicule.

amphitrichous a. [Gk. amphi, both; thrix, hair] with a flagellum at each pole, appl. bacteria; alt. amphitrichate, amphitrichic.

amphitrocha n. [Gk. amphi, both; trochos, wheel] a free-swimming annelid larva with 2 rings of cilia.

amphitrophs n.plu. [Gk. amphi, both; trophē, nourishment] normally autotrophic organisms which adapt themselves to heterotrophic nutrition if placed in the dark for long periods.

amphitropous a. [Gk. amphi, both; tropē, turning] having the ovule inverted, with hilum in middle of one side.

amphivasal a. [Gk. amphi, both; L. vas, vessel] with the xylem surrounding the phloem, appl. some concentric vascular bundles; alt. amphixylic; cf. amphicribral.

amphixylic amphivasal, q.v.

amphochromatophil amphophil, q.v.

amphocyte n. [Gk. ampho, both of two; kytos, hollow] an amphophil cell.

amphogenic a. [Gk. ampho, both of two; -genēs, producing] producing offspring consisting of both males and females.

amphophil a. [Gk. ampho, both of two; philein, to love] appl. cells staining with basic and acid dyes; alt. amphochromatophil; cf. neutrophil; n. amphocyte.

amphosome n. [Gk. ampho, both of two; sōma, body] a vestigial structure found in algae of the class Cryptophyceae, which may be a vestigial eye-spot or pyrenoid.

amphoteric a. [Gk. amphoterē, in both ways] possessing both acidic and basic properties; with opposite characters.

amphoterotoky amphitoky, q.v.

amplectant a. [L. amplecti, to embrace] clasping or winding tightly round some support, as tendrils.

amplexicaul a. [L. amplecti, to embrace; caulis, stem] clasping or surrounding the stem, as base of sessile leaf.

amplexus n. [L. amplexus, embrace] in frogs and toads, the mating embrace when eggs are shed into the water and fertilized.

ampliate a. [L. ampliatus, made wider] having outer edge of wing prominent, as in certain insects.

amplification n. [L. amplificatio, enlargement] changes towards increased structural or functional complexity in ontogeny or phylogeny; cf. reduction

ampulla n. [L. ampulla, flask] a membranous vesicle; dilatation of a lactiferous tubule beneath areola; dilated portion at one end of each semicircular canal of ear; dilatation of united common bile duct and pancreatic duct; part of oviduct between infundibulum and isthmus; dilated portion of vas deferens at fundus of urinary bladder; terminal dilatation of rectum; pit in skeleton of Hydrocorallina, for medusa; internal reservoir on ring canal of water vascular system in echinoderms; terminal vesicle of sensory canals of elasmobranchs; submerged bladder of Utricularia, the bladderwort.

ampullaceal, ampullaceous a. [L. ampulla, flask] flask-shaped; appl. arachnid spinning glands which furnish silk for foundations, lines, and radii; appl. sensillae.

ampullary a. [L. ampulla, flask] pert. or resembling an ampulla.

ampullula n. [Dim. of L. ampulla, flask] a small ampulla, as of some lymphatic vessels.

ampyx n. [Gk. ampyx, fillet] a transverse bar connecting the rostralia of Palaeospondylus.

amyelinic, amyelinate a. [Gk. a, without; myelos, marrow] non-medullated, q.v.

amygdala n. [L. from Gk. amygdalē, almond] almond, or almond-shaped structure; one of palatal tonsils; rounded lobe at side of vallecula of cerebellum.

amygdalin n. [Gk. amygdalē, almond] a cyanophoric glycoside found in fruit kernels of bitter almonds and other Rosaceae such as peach and cherry, producing hydrocyanic acid, glucose, and benzaldehyde upon hydrolysis, and also known as laetrile, q.v.

amylase n. [L. amylum, starch] the general name for several enzymes including diastase, ptyalin, and amylopsin which hydrolyse starch to maltose, alt. amylolytic enzyme; now known to be a mixture of 2 enzymes, α-amylase which catalyses the hydrolysis of α-1-4 linkages of polysaccharides made of glucose units, alt. ptyalin, diastase, EC

3.2.1.1, and β-amylase which catalyses the hydrolysis of polysaccharides from the ends, splitting off maltose units, EC 3.2.1.2, many enzymes known as amylase being a mixture of the two.

amyliferous *a*. [L. *amylum*, starch; *ferre*, to carry] containing or producing starch.

amyloclastic amylolytic, *q.v.*

amylogenesis *n*. [L. *amylum*, starch; Gk. *genesis*, descent] starch formation.

amyloid *a*. [Gk. *amylon*, starch; *eidos*, form] starch-like. *n*. Starch-like substance.

amyloid bodies corpora amylacea, *q.v.*

amylolytic *a*. [Gk. *amylon*, starch; *lysis*, loosing] starch-digesting, *appl.* enzymes; *alt.* amyloclastic.

amylom(e) *n*. [Gk. *amylon*, starch] starch-containing wood parenchyma; layer of starch-containing cells between central cylinder and leptoids of certain moss rhizoids.

amylopectin *n*. [Gk. *amylon*, starch; *pēktos*, congealed] a branch-chained polysaccharide found in starch with a structure similar to glycogen, having chains with α-1-4 glycosidic links, cross-linked by α-1-6 glycosidic links.

amyloplast(id) *n*. [Gk. *amylon*, starch; *plastos*, formed] a leucoplast or colourless starch-forming granule in plants.

amylopsin *n*. [Gk. *amylon*, starch; *opson*, seasoning] pancreatic amylase.

amylose *n*. [L. *amylum*, starch] a straight chain polysaccharide found in starch, containing 100 to 1000 glucose residues joined by α-1-4 glycosidic linkages.

amylostatolith *n*. [Gk. *amylon*, starch; *statos*, stationary; *lithos*, stone] a starch grain which moves under the influence of gravity in a statocyte; *cf.* statolith, chlorostatolith.

amylum *n*. [L. *amylum*, starch] starch, *q.v.*, specifically vegetable starch; *cf.* glycogen.

anabiosis *n*. [Gk. *anabiōsis*, recovery of life] a condition of suspended animation or apparent death produced in certain organisms, e.g. by desiccation and from which they can be revived to normal metabolism, e.g. by return of moisture.

anabolism *n*. [Gk. *ana*, up; *bolē*, throw] the constructive chemical processes in living organisms involving the formation of complex molecules from simpler ones, and the taking up and storing of energy; *alt.* assimilation; *cf.* catabolism.

anabolite *n*. [Gk. *ana*, up; *bolē*, throw] a substance participating in anabolism; *alt.* anastate; *cf.* catabolite.

Anacanthini *n*. [Gk. *an*, not; *akantha*, prickle] Gadiformes, *q.v.*

anacanthous *a*. [Gk. *an*, not; *akantha*, prickle] without spines or thorns.

anachoresis *n*. [Gk. *ana*, up; *chōrein*, to spread] the phenomenon of living in holes or crevices.

anacrogynous *a*. [Gk. *an*, not; *akros*, at the top; *gynē*, female] with archegonia not arising at or near the apex of the shoot, *appl.* certain liverworts; *cf.* acrogynous.

anacromyoidian *a*. [Gk. *ana*, up; *akros*, apex; *mys*, muscle; *eidos*, form] with syringeal muscles attached at dorsal ends of bronchial semirings.

anadromous *a*. [Gk. *ana*, up; *dramein*, to run] *appl.* fishes which migrate from salt to fresh water annually; *cf.* catadromous.

anaerobe *n*. [Gk. *an*, without; *aēr*, air; *bios*, life] an organism which can respire in the absence of oxygen; *alt.* anaerobiont; *cf.* aerobe; *a.* anaerobic, anoxybiotic.

anaerobic respiration respiration occurring in the absence of oxygen; *alt.* anaerobiosis.

anaerobiont *n*. [Gk. *an*, without; *aēr*, air; *bion*, living] anaerobe, *q.v.*

anaerobiosis *n*. [Gk. *an*, without; *aēr*, air; *biōsis*, manner of life] existence in absence of free oxygen; anaerobic respiration, *q.v.*

anagen *n*. [Gk. *anagennaein*, to regenerate] active stage in the hair growth cycle.

anagenesis *n*. [Gk. *ana*, again; *genesis*, origin] regeneration of tissues; progressive evolution; speciation brought about by the transformation of species in time, *cf.* cladogenesis.

anahaemin *n*. [Gk. *ana*, again; *haima*, blood] a proteid substance of liver, acting in regeneration of erythrocytes; *alt.* haemopoietic principle.

anakinetic *a*. [Gk. *ana*, up; *kinein*, to move] *appl.* processes which restore energy with the formation of energy-rich compounds; *cf.* katakinetic.

anakinetomeres *n.plu.* [Gk. *ana*, up; *kinein*, to move; *meros*, part] energy-rich atoms, molecules, or protoplasm; *cf.* katakinetomeres.

anal *a*. [L. *anus*, anus] *pert.*, or situated at or near the anus; *appl.* posterior median ventral fin of fishes, margin and vein of insect wing, posterior ventral scute of reptiles, etc.; *appl.* columns: rectal columns, *q.v.*

analgesic *a*. [Gk. *analgēsia*, painlessness] pain killing. *n*. A pain killer.

analogues *n.plu.* [Gk. *analogia*, proportion] organs of different plants or animals with like function but of unlike origin.

analogy *n*. [Gk. *analogia*, proportion] resemblance in function though not in structure or development.

anamestic *a*. [Gk. *ana*, up; *mestos*, filled] *appl.* small variable bones filling spaces between larger bones of more fixed position, as in fish skulls.

Anamniota, anamniotes *n*., *n.plu.* [Gk. *a*, not; *amnion*, foetal membrane] a group of Craniata, usually considered a superclass, including the fish and amphibians, having no amnion around the embryo.

anamorpha *n.plu.* [Gk. *ana*, backwards; *morphē*, form] larvae hatched with incomplete number of segments; *cf.* epimorpha.

Anamorpha *n*. [Gk. *ana*, backwards; *morphē*, form] a subclass of centipedes in which the eggs are not laid in a group and are not tended by the female.

anamorphosis *n*. [Gk. *ana*, throughout; *morphōsis*, shaping] evolution from one type to another through a series of gradual changes; excessive or abnormal formation of a plant organ.

anandrous *a*. [Gk. *a*, without; *anēr*, male] ananatherous, *q.v.*

anangian *a*. [Gk. *a*, without; *anggeion*, vessel] without a vascular system.

anantherous *a*. [Gk. *a*, without; *anthēros*, flowering] without anthers; *alt.* anandrous, inantherate.

ananthous *a*. [Gk. *a*, without; *anthos*, flower] not flowering; without inflorescence.

anaphase *n*. [Gk. *ana*, up; *phasis*, appearance] the phase of mitosis and meiosis that follows metaphase and in which half chromosomes or homologous chromosomes separate and move towards opposite poles of the spindle; sometimes used for all the stages of mitosis up to the division of chromatin into chromosomes, cf. kataphase.

anaphylaxis n. [Gk. *ana*, up; *phylax*, guard] condition of being hypersensitive to a serum or foreign protein, caused by first or sensitizing dose; *alt.* allergy.

anaphysis n. [Gk. *ana*, up; *phyein*, to grow] an outgrowth; sterigma-like filament in apothecium of certain lichens.

anaphyte n. [Gk. *ana*, up; *phyton*, plant] transverse segment of a shoot, an internode.

anaplasia n. [Gk. *ana*, again; *plassein*, to form] undifferentiation; reversion to a less differentiated structure.

anaplasis n. [Gk. *ana*, up; *plassein*, to form] progressive stage in development of an individual preceding the mature phase or metaplasis.

anaplast(id) n. [Gk. *ana*, up; *plastos*, formed] leucoplast, *q.v.*

anapleurite a. [Gk. *ana*, up; *pleura*, side] upper thoracic pleurite, as in certain Thysanura.

anapophysis n. [Gk. *ana*, up; *apo*, from; *physis*, origin] a small dorsal projection rising near transverse process in lumbar vertebrae.

anapsid a. [Gk. *ana*, up; *apsis*, arch] with skull wholly imperforate or completely roofed over, the only gaps on the dorsal surface being the nares, orbits, and parietal foramen; *alt.* stegocrotaphic.

Anapsida, anapsids n., n.plu. [Gk. *ana*, up; *apsis*, arch] a subclass of reptiles having a sprawling gait and a skull with no temporal opening.

anaptychus n. [Gk. *ana*, throughout; *ptychē*, plate] aptychus or operculum consisting of a single plate, as in certain ammonites; cf. synaptychus.

anarthrous a. [Gk. *a*, without; *arthron*, joint] having no distinct joints.

anaschistic a. [Gk. *ana*, up to; *schistos*, split] *appl.* type of tetrads which divide twice longitudinally in meiosis; *alt.* eumitotic; cf. diaschistic.

Anaspida, anaspids n., n.plu. [Gk. *a*, without; *aspis*, shield] an order of U. Silurian and Devonian ostracoderms (Osteostraci) having a head shield made of small scales.

Anaspidacea n. [Gk. *a*, without; *aspis*, shield] an order of Syncarida having the 1st thoracic segment fused with the head.

Anaspidea n. [Gk. *a*, without; *aspis*, shield] an order of opisthobranchs having a reduced internal shell and no head shield; *alt.* Aplysiomorpha.

anastate n. [Gk. *ana*, up to; *statos*, standing] anabolite, *q.v.*; cf. catastate.

anastomosis n. [Gk. *ana*, up to; *stoma*, mouth] union of ramifications of leaf veins; union of blood vessels arising from a common trunk; union of nerves; fine threads joining chromonemata in resting nucleus; formation of a network or anastomotic meshwork.

anastral a. [Gk. *an*, not; *astēr*, star] *appl.* type of mitosis without aster formation.

anatomy n. [Gk. *ana*, up; *tomē*, cutting] the science which treats of the structure of plants and of animals, as determined by dissection; cf. anthropotomy, phytotomy, zootomy.

anatoxin toxoid, *q.v.*

anatrepsis n. [Gk. *anatrepein*, to turn over] stage of increasing movement in blastokinesis; cf. katatrepsis.

anatriaene n. [Gk. *ana*, up; *triaina*, trident] triaene with backwardly directed branches.

anatropal, anatropous a. [Gk. *anatropē*, overturning] inverted; *appl.* ovule bent over against the funicle so that hilum and micropyle are close together and chalaza is at the other end; *appl.* sporangium as in Equisetales.

anautogenous a. [Gk. *a*, without; *autos*, self; *-genēs*, producing] *appl.* adult female insect that must feed if her eggs are to mature; *alt.* non-autogenous; cf. autogenous.

anaxial a. [Gk. *an*, without; L. *axis*, axle] having no distinct axis; asymmetrical.

anaxon(e) n. [Gk. *an*, without; *axōn*, axis] a nerve cell having no evident axon.

ancestrula n. [L. *antecedere*, to go before] first zooecium of polyzoan colony.

anchor n. [L. *ancora*, anchor] anchor-shaped spicule found in skin of Holothuria.

anchylosis n. [Gk. *angcheein*, to press tight] union of 2 or more bones or hard parts to form one part, e.g. of bone to bone, or tooth to bone; *alt.* ankylosis, synosteosis.

ancipital a. [L. *anceps*, double] flattened and having 2 edges.

ancistroid ankistroid, *q.v.*

anconaeus anconeus, *q.v.*

anconeal a. [Gk. *angkōn*, elbow] *pert.* the elbow.

anconeus n. [Gk. *angkōn*, elbow] small extensor muscle situated over elbow; *alt.* anconaeus.

Ancylistales n. [*Ancylistes*, generic name] a former name for the Lagenidiales, *q.v.*

andrangium n. [Gk. *anēr*, male; *anggeion*, vessel] a gonangium in which male gametes are produced; spermangium, *q.v.*

andrase n. [Gk. *anēr*, male] a male-determining factor in form of an enzyme or hormone.

Andreaeales, Andreacales n. [*Andreaea*, generic name] an order or in some classifications superorder of bryophytes of the subclass Andreaeidae, *q.v.*

Andreaeidae n. [*Andreaea*, generic name] a subclass of mosses with one order, Andreaeales, having blackish gametophores growing on rocks, and capsules opening by longitudinal slits.

andric a. [Gk. *andrikos*, masculine] male; cf. gynic.

andrin n. [Gk. *anēr*, male] the testicular androgens.

androchory n. [Gk. *anēr*, male; *chōrein*, to spread] anthropochory, *q.v.*

androclinium clinandrium, *q.v.*

androconia n.plu. [Gk. *anēr*, male; *konia*, dust] modified wing scales producing a sexually attractive scent in certain male butterflies; *alt.* plumules, scent scales.

androcyte n. [Gk. *anēr*, male; *kytos*, hollow] a cell arising by growth from an androgonium and giving rise to antherozoid; *alt.* antherozoid mother cell.

androdioecious a. [Gk. *anēr*, male; *dis*, two; *oikos*, house] having male and hermaphrodite flowers on different plants; cf. andromonoecious.

androecium n. [Gk. *anēr*, male; *oikos*, house] male reproductive organs of a plant; stamens taken collectively.

androgametangium n. [Gk. *anēr*, male; *gametēs*, spouse; *anggeion*, vessel] antheridium, *q.v.*

androgamone n. [Gk. *anēr*, male; *gamos*, marriage] a gamone of male gametes; cf. antifertilizin.

androgen n. [Gk. *anēr*, male; *genos*, descent] a male sex hormone concerned with production and maintenance of secondary sexual characteristics.

androgenesis n. [Gk. anēr, male; genesis, descent] development in which the embryo contains paternal chromosomes only, due to the failure of the nucleus of the female gamete to participate in fertilization; development from a male gamete, i.e. male parthenogenesis; a. androgenetic.

androgenic a. [Gk. anēr, male; gennaein, to produce] stimulating male characters, masculinizing, appl. hormones; appl. tissue capable of elaborating an androgenic hormone; androgenous, q.v.

androgenous a. [Gk. anēr, male; genos, descent] producing only male offspring; alt. androgenic.

androgonidia n.plu. [Gk. anēr, male; gonos, offspring; idion, dim.] male sexual elements formed after repeated divisions of parthenogonidia of the protistan, Volvox.

androgonium n. [Gk. anēr, male; gonos, offspring] a cell in the antheridium which gives rise to androcytes; cf. jacket cell.

androgynal a. [Gk. anēr, male; gynē, female] hermaphrodite; bearing both staminate and pistillate flowers in the same inflorescence; with antheridium and oogonium on the same hypha; alt. androgynous.

androgynary a. [Gk. anēr, male; gynē, female] having flowers with stamens and pistils developed into petals.

androgyne a., n. [Gk. anēr, male; gynē, female] hermaphrodite, q.v.

androgynism n. [Gk. anēr, male; gynē, female] the condition of bearing both stamens and pistils; alt. hermaphroditism.

androgynous androgynal, q.v.

andromerogony n. [Gk. anēr, male; meros, part; gonē, generation] the development of an egg fragment with only paternal chromosomes.

andromonoecious a. [Gk. anēr, male; monos, alone; oikos, house] having male and hermaphrodite flowers on the same plant; cf. androdioecious.

andropetalous a. [Gk. anēr, male; petalon, leaf] having petaloid stamens.

androphore n. [Gk. anēr, male; phoros, carrying] a hyphal branch bearing antheridia; stalk supporting androecium or stamens; stalk carrying male gonophores in Siphonophora.

androphyll n. [Gk. anēr, make; phyllon, leaf] microsporophyll, q.v.

androsome n. [Gk. anēr, male; sōma, body] a male-limited chromosome.

androsporangium n. [Gk. anēr, male; sporos, seed; anggeion, vessel] a sporangium containing androspores.

androspore n. [Gk. anēr, male; sporos, seed] an asexual zoospore which gives rise to a male dwarf plant; a male spore such as a microspore or pollen grain.

androsterone n. [Gk. anēr, male; stear, suet] an androgen that is less active than testosterone and is found especially in human male urine.

androtype n. [Gk. anēr, male; typos, pattern] type specimen of the male of a species.

anebous a. [Gk. anēbos, before manhood] immature; before puberty, alt. prepubertal.

anelectrotonus n. [Gk. ana, up; elektron, amber; tonos, tension] decrease in irritability of a nerve under influence of an electric current.

anellus n. [L. anellus, little ring] a small ring-shaped or triangular plate supported by valves and vinculum, in Lepidoptera.

anelytrous a. [Gk. an, not; elytron, sheath] without elytra.

anemochorous a. [Gk. anemos, wind; chōrein, to spread] dispersed by wind; with seeds so dispersed; alt. anemochoric.

anemochory dispersal by wind.

anemophily n. [Gk. anemos, wind; philein, to love] wind pollination; any other kind of plant fertilization brought about by wind; a. anemophilous.

anemoplankton n. [Gk. anemos, wind; plangktos, wandering] wind-borne organisms and living particles; aeroplankton, q.v.

anemosporic a. [Gk. anemos, wind; sporos, seed] having spores or seeds disseminated by air currents.

anemotaxis n. [Gk. anemos, wind; taxis, arrangement] a taxis in response to air currents.

anemotropism n. [Gk. anemos, wind; tropē, turn] orientation of body, or plant curvature, in response to air currents.

anencephaly n. [Gk. an, not; engkephalon, brain] condition of having no brain.

anenterous a. [Gk. an, without; enteron, gut] having no alimentary tract; alt. anenteric.

aner n. [Gk. anēr, male] the male of insects, especially of ants.

anestrum anoestrus, q.v.

aneucentric a. [Gk. a, without; eu, well; kentron, centre] acentric and dicentric resulting from translocation involving centromere of a chromosome.

aneuploid a. [Gk. a, without; eu, well; aploos, onefold] having fewer or more chromosomes than an exact multiple of the haploid number; alt. dysploid, heteroploid; n. aneuploidy; cf. euploid.

aneurin(e) n. [Gk. a, without; neuron, nerve] thiamin(e), q.v.

aneuronic a. [Gk. a, without; neuron, nerve] without innervation, appl. chromatophores controlled by hormones.

aneusomic a. [Gk. a, without; eu, well; sōma, body] appl. organisms whose cells have varying numbers of chromosomes.

anfractuose a. [L. anfractus, bending] wavy, sinuous.

angienchyma n. [Gk. anggeion, vessel; engcheein, to pour] vascular tissue, q.v.

angioblast n. [Gk. anggeion, vessel; blastos, bud] one of cells from which lining of blood vessels is derived; alt. vasifactive cell.

angiocarpic, angiocarpous a. [Gk. anggeion, vessel; karpos, fruit] having fruit invested in a masking covering; having closed apothecia, as in certain lichens; having spores enclosed in some kind of receptacle; cf. gymnocarpic.

angiocarpy n. [Gk. anggeion, vessel; karpos, fruit] the angiocarpic condition; alt. endocarpy, hermetism.

angiology n. [Gk. anggeion, vessel; logos, discourse] anatomy of blood and lymph vascular systems.

Angiospermae, angiosperms n., n.plu. [Gk. anggeion, vessel; sperma, seed] a major division of the plant kingdom, in some classifications considered a subdivision of the Spermatophyta, in others of the Pteropsida, commonly called flowering plants as their reproductive organs are in flowers, having seeds which develop in a closed ovary made of carpels, a very reduced gametophyte, and endosperm developed from a triple fusion nucleus; in some classifications, also called Magnoliophytina; formerly also called metasperms; *alt.* Angiospermophyta, Anthophyta.

angiosporous *a.* [Gk. *anggeion*, vessel; *sporos*, seed] having spores contained in a theca or spore capsule.

angiostomatous *a.* [Gk. *anggeion*, vessel; *stoma*, mouth] narrow-mouthed, *appl.* molluscs and snakes with non-distensible mouth.

angiotensin, angiotonin *n.* [Gk. *anggeion*, vessel; *tonos*, tension] polypeptide in blood, formed by reaction between hypertensinogen elaborated in the liver, and renin, causing constriction of arterioles; *alt.* hypertensin.

Ångström *n.* [*A. J. Ångström*, Swedish physicist] a unit of microscopic measurement and of wavelength of some electromagnetic radiation, being 10^{-10} metres, one ten-millionth mm, one-tenth nm; also called Ångström unit; symbol Å or ÅU.

Anguilliformes *n.* [*Anguilla*, generic name of eel; L. *forma*, shape] Apodes, *q.v.*

angular *n.* [L. *angulus*, corner] a membrane bone of lower jaw in most vertebrates, *alt.* angulare. *a.* Having, or *pert.*, an angle; *appl.* leaf originating at forking of stem, as in many ferns; *appl.* collenchyma with cell walls thickened in the angles of the cells; *appl.* line: collarette, *q.v.*

angulosplenial *n.* [L. *angulus*, corner; *splenium*, patch] bone forming most of lower and inner part of mandible in Amphibia.

angulus *n.* [L. *angulus*, angle] an angle, as that formed by junction of manubrium and body of sternum; *alt.* angle of Louis, angulus Ludovici, sternal angle.

angustifoliate *a.* [L. *angustus*, narrow; *folium*, leaf] with narrow leaves; *cf.* latifoliate.

angustirostral *a.* [L. *angustus*, narrow; *rostrum*, beak] with narrow beak or snout; *alt.* angustirostrate; *cf.* latirostral.

angustiseptate *a.* [L. *angustus*, narrow; *septum*, partition] having a silicula laterally compressed with a narrow septum; *cf.* latiseptate.

anholocyclic *a.* [Gk. *an*, not; *holos*, whole; *kyklos*, circle] *pert.* alternation of generations with suppression of sexual part of cycle; permanently parthenogenetic.

anidian *a.* [Gk. *an*, not; *eidos*, form] formless, *appl.* blastoderm without apparent embryonic axis.

animal cellulose tunicine, *q.v.*

animal pole the side of a blastula at which the micromeres collect; the upper, more rapidly segmenting portion of a telolecithal egg containing little yolk; *cf.* vegetal pole.

animal starch glycogen, *q.v.*

anion *n.* [Gk. *ana*, up; *ienai*, to go] a negatively charged ion which moves towards the anode, the positive electrode; *cf.* cation.

anisocarpous *a.* [Gk. *anisos*, unequal; *karpos*, fruit] having number of carpels less than that of other floral whorls.

anisocercal *a.* [Gk. *anisos*, unequal; *kerkos*, tail] with lobes of tail fin unequal.

anisochela *n.* [Gk. *anisos*, unequal; *chēlē*, claw] a chela with the 2 parts unequally developed.

Anisochytridiales *n.* [Gk. *anisos*, unequal; *chytridion*, little pot] Hyphochytriales, *q.v.*

anisocytic *a.* [Gk. *anisos*, unequal; *kytos*, hollow] *appl.* stomata of a type in which 3 subsidiary cells, 1 distinctly smaller than the other 2, surround the stoma; formerly called cruciferous; *cf.* anomocytic, diacytic, paracytic.

anisodactylous *a.* [Gk. *anisos*, unequal; *daktylos*, finger] having unequal toes, 3 toes forward, 1 backward.

anisodont *a.* [Gk. *anisos*, unequal; *odous*, tooth] heterodont, *q.v.*; *cf.* isodont.

anisogamete *n.* [Gk. *anisos*, unequal; *gametēs*, spouse] one of 2 conjugating gametes differing in form or size; *alt.* heterogamete.

anisogametism, anisogamety the production of anisogametes, as of macrogametes and microgametes.

anisogamy *n.* [Gk. *anisos*, unequal; *gametēs*, spouse] the union of unlike gametes, which in its extreme form becomes oogamy; *alt.* heterogamy, anisomerogamy; *cf.* isogamy; *a.* anisogamous.

anisognathous *a.* [Gk. *anisos*, unequal; *gnathos*, jaw] with jaws of unequal width; having teeth in upper and lower jaws unlike.

anisomeres *n.plu.* [Gk. *anisos*, unequal; *meros*, part] homologous parts of polyisomeres when differing amongst themselves; *cf.* polyanisomere.

anisomerogamy anisogamy, *q.v.*

anisomerous *a.* [Gk. *anisos*, unequal; *meros*, part] having unequal numbers of parts in floral whorls; *n.* anisomery.

anisomorphic *a.* [Gk. *anisos*, unequal; *morphē*, form] differing in shape, size, or structure.

Anisomyaria *n.* [Gk. *anisos*, unequal; *mys*, muscle] an order of lamellibranchs with filibranch gills and which are usually attached to the substratum by byssus threads or cement.

anisophylly *n.* [Gk. *anisos*, unequal; *phyllon*, leaf] heterophylly, *q.v.*

anisopleural *a.* [Gk. *anisos*, unequal; *pleura*, side] asymmetrical bilaterally.

anisoploid *a.* [Gk. *anisos*, unequal; *aploos*, onefold; *eidos*, form] with an odd number of chromosome sets in somatic cells. *n.* An anisoploid individual.

anisopogonous *a.* [Gk. *anisos*, unequal; *pōgōn*, beard] unequally webbed, *appl.* feathers.

anisopterous *a.* [Gk. *anisos*, unequal; *pteron*, wing] unequally winged, *appl.* seeds.

anisospore *n.* [Gk. *anisos*, unequal; *sporos*, seed] a spore with sexual dimorphism; *cf.* isospore.

anisostemonous *a.* [Gk. *anisos*, unequal; *stēmōn*, spun thread] having the number of stamens unequal to the number of parts in other floral whorls; having stamens of unequal size.

anisotropic, anisotropous *a.* [Gk. *anisos*, unequal; *tropē*, turn] *appl.* eggs with predetermined axis or axes; doubly refracting; *appl.* dark bands of voluntary muscle fibre; *cf.* isotropic; *n.* anisotropy.

ankistroid *a.* [Gk. *agkistron*, fish-hook; *eidos*, form] like a barb; barbed; *alt.* ancistroid.

ankyloblastic *a.* [Gk. *agkylos*, crooked; *blastos*, bud] with a curved germ band; *cf.* orthoblastic.

ankylosis anchylosis, *q.v.*

ankyroid *a.* [Gk. *agkyra*, hook; *eidos*, form] hook-shaped.

anlage *n.* [Ger. *Anlage*, predisposition] the 1st structure or cell group indicating development of a part or organ; *alt.* inception, primordium, lébauche.

annectent *a.* [L. *annectere*, to bind together] linking, *appl.* intermediate species or genera.

Annelida, annelids *n., n.plu.* [L. *annulus*, ring; Gk. *eidos*, form] a phylum of segmented coelomate worms, commonly called ringed worms, having a thin, flexible cuticle and a preoral prostomium and postanal pygidium.

annidation *n.* [Gk. *a*, without; L. *nidus*, nest] the situation where a mutant organism survives in a population because an ecological niche exists which the normal organism cannot use.

annotinous *a.* [L. *annus*, year] a year old; *appl.* growth during the previous year.

annual *a.* [L. *annus*, year] *appl.* structures or features that are marked off or completed yearly; living for a year only; completing life cycle in one year from germination. *n.* An annual plant or therophyte.

annual ring one of the rings, seen in transverse sections of dicotyledons, indicating the secondary growth during a year; a ring indicating growth in bivalve shells; *alt.* growth ring.

annular *a.* [L. *annulus*, ring] ring-like; *appl.* certain ligaments of wrist and ankle; *appl.* orbicular ligament encircling head of radius and attached to radial notch of ulna; *appl.* certain lamina or sternal plates in ants; *appl.* certain vessels in xylem, owing to ring-like thickenings in their interior; *appl.* bands formed on inner surface of cell wall.

Annulata, annulates *n., n.plu* [L. *annulus*, ring] a group of invertebrates, including the annelids, arthropods, and some related forms, having bilateral symmetry and true metameric segmentation.

annulate *a.* [L. *annulus*, ring] ring-shaped; composed of ring-like segments; with ring-like constrictions; having colour arranged in ring-like bands or annuli.

annulus *n.* [L. *annulus*, ring] any ring-like structure; special ring in fern sporangium, by action of which sporangium bursts, *alt.* gyroma; remains of veil in mushrooms; ring of cells in moss capsule whose rupture causes opening; circular groove for transverse flagellum in Dinoflagellata; ring of annelid; growth ring of fish scale; 4th digit of hand.

anococcygeal *a.* [L. *anus*, anus; Gk. *kokkyx*, cuckoo] *pert.* region between coccyx and anus, *appl.* body of fibrous and muscular tissue, nerves, etc.

anoestrus *n.* [Gk. *an*, not; *oistros*, gadfly] the non-breeding period; period of absence of sexual urge; *alt.* anoestrum, anestrum; *cf.* dioestrus.

Anomalodesmata *n.* [Gk. *anōmalos*, uneven; *desma*, bond] an order of lamellibranch molluscs in which the gills are eulamellibranch and the mantle edges almost entirely fused together; *cf.* Eulamellibranchiata.

anomaly *n.* [Gk. *anōmalos*, uneven] any departure from type characteristics.

anomocytic *a.* [Gk. *anomos*, lawless; *kytos*, hollow] *appl.* stomata of a type in which no subsidiary cells are associated with the guard cells; formerly called ranunculaceous; *cf.* anisocytic.

anomophyllous *a.* [Gk. *anomos*, lawless; *phyllon*, leaf] with irregularly placed leaves.

Anopla *n.* [Gk. *an*, without; *hoplon*, weapon] a class of proboscis worms having the mouth posterior to the brain and an unarmed proboscis.

Anoplura *n.* [Gk. *an*, without; *hoplon*, weapon; *oura*, tail] an order of insects, commonly called sucking lice or body lice, which are ectoparasitic on mammals, sometimes included with the Mallophaga in the order Phthiraptera; *alt.* Siphunculata.

anorganology *n.* [Gk. *an*, not; *organon*, instrument; *logos*, discourse] abiology, *q.v.*

anorthogenesis *n.* [Gk. *an*, not; *orthos*, straight; *genesis*, descent] evolution manifesting changes in direction of adaptations owing to preadaptation, 'zigzag' evolution.

anorthospiral *a.* [Gk. *an*, not; *orthos*, straight; *speira*, coil] relationally coiled, spirals not interlocking; *alt.* paranemic; *cf.* orthospiral, plectonemic.

anosmatic, anosmic *a.* [Gk. *a*, without; *osmē*, smell] having no sense of smell; *n.* anosmia.

Anostraca *n.* [Gk. *an*, without; *ostrakon*, shell] an order of Branchiopoda with no carapace and having stalked eyes.

anoxybiotic *a.* [Gk. *a*, not; *oxys*, sharp; *biōtos*, to be lived] capable of living in absence of oxygen; *alt.* anaerobic; *n.* anoxybiosis.

ANS autonomic nervous system, *q.v.*

ansa *n.* [L. *ansa*, handle] loop, as of certain nerves.

Anseriformes *n.* [L. *anser*, goose; *forma*, shape] an order of birds of the subclass Neornithes, including the swans, geese, and ducks; *a.* anseriform.

anserine *n.* [L. *anser*, goose] a dipeptide, methylcarnosine, found in the muscle tissue of birds, reptiles, and fishes, and in mammalian urine; *pert.* a goose.

ansiform *a.* [L. *ansa*, handle; *forma*, shape] loop-shaped, or looped, *appl.* outer cytoplasm in cerebrospinal ganglia.

antagonism *n.* [Gk. *antagōnistēs*, adversary] the ability of one toxic substance to reduce or eliminate the effect of another; the inhibitory action of one species on the other, such as the action of a substance given out by a plant into the soil which inhibits other plants nearby.

antagonist *n.* [Gk. *antagōnistēs*, adversary] a muscle acting in opposition to the action produced by a prime mover or agonist; antihormone, *q.v.*

antambulacral *a.* [Gk. *anti*, against; L. *ambulare*, to walk] abactinal, *q.v.*

antapex *n.* [Gk. *anti*, opposite; L. *apex*, tip] tip of hypocone in Dinoflagellata.

antapical *a.* [Gk. *anti*, opposite; L. *apex*, tip] at or *pert.* antapex; *pert.* region opposite apex.

ante- *see also* anti-.

anteaters *see* Edentata, Pholidota, Monotremata.

anteclypeus *n.* [L. *ante*, before; *clypeus*, shield] anterior portion of clypeus when differentiated by suture; *cf.* postclypeus.

antecosta *n.* [L. *ante*, before; *costa*, rib] internal ridge of tergum for attachment of intersegmental muscles in insects, extended to phragma in alar segments.

antecubital *a.* [L. *ante*, before; *cubitus*, elbow] in front of the elbow, *appl.* fossa.

antedorsal *a.* [L. *ante*, before; *dorsum*, back] situated in front of dorsal fin in fishes.

antefrons n. [L. *ante*, before; *frons*, forehead] the portion of frons anterior to antennary base line in certain insects.

antefurca n. [L. *ante*, in front; *furca*, fork] forked process or sternal apodeme of anterior thoracic segment in insects.

antelabrum n. [L. *ante*, before; *labrum*, lip] the anterior portion of insect labrum when differentiated.

antemarginal a. [L. *ante*, before; *margo*, edge] *appl.* sori of ferns when they lie within margin of frond.

antenna n. [L. *antenna*, sailyard] a jointed feeler on head of various Arthropoda; feeler of Rotifera; in some fish, a modified flap on dorsal fin which attracts prey; *a.* antennal, antennary.

antennifer n. [L. *antenna*, sailyard; *ferre*, to carry] socket of antenna in arthropods; projection on rim of antennal socket, acting as a pivot, in myriapods; *alt.* torulus.

antennule n. [*Dim.* of L. *antenna*] a small antenna or feeler, specifically the 1st pair of antennae in Crustacea.

antephase n. [L. *ante*, before; Gk. *phasis*, appearance] a phase presumed to occur immediately before the beginning of prophase and during which DNA is synthesized.

anteposition n. [L. *ante*, before; *ponere*, to place] a position opposite rather than alternate to another structure in a flower.

anterior a. [L. *anterior*, former] nearer head end; ventral in human anatomy; facing outwards from axis; previous.

anterior lobe of pituitary gland the part of the pituitary gland including the pars distalis and pars tuberalis.

anterior vena cava precava, *q.v.*

anterolateral ventrolateral, *q.v.*

antesternite n. [L. *ante*, before; *sternum*, breast bone] anterior sternal sclerite of insects, *alt.* basisternum, eusternum.

anthela n. [Gk. *anthein*, to bloom] the cymose inflorescence of the rush family, Juncaceae.

anthelix antihelix, *q.v.*

anther n. [Gk. *antheros*, flowering] the part of a stamen which produces pollen.

antheraxanthin n. [Gk. *anthos*, flower; *xanthos*, yellow] a xanthophyll carotinoid pigment found in the Euglenophyceae.

antherid antheridium, *q.v.*

antheridia *plu.* of antheridium.

antheridial cell the larger of 2 cells derived from a microspore and giving rise to an antheridium, as in Lycopodiales, or to a cell representing an antheridium, as in Gymnospermae.

antheridiophore n. [Gk. *anthos*, flower; *idion, dim.*; *pherein*, to bear] a branch of a bryophyte or of a fern prothallus bearing antheridia; *cf.* archegoniophore.

antheridium n. [Gk. *anthos*, flower; *idion, dim.*] an organ or receptacle in which male gametes are produced in many cryptogams, i.e. a male gametangium; cluster of microgametes, as in certain Flagellata; *alt.* antherid, androgametangium; *plu.* antheridia; *cf.* archegonium, pollinodium.

antheridogen n. [Gk. *anthos*, flower; *idion, dim.*; *genos*, birth] a hypothetical hormone which promotes development of antheridia on gametophytes.

antherophore n. [Gk. *antheros*, flowering; *pherein*, to bear] the stalk of a stamen bearing several anthers, in male cone of certain gymnosperms.

antherozoid mother cell androcyte, *q.v.*

antherozoids, antherozooids *n.pl.* [Gk. *anthos*, flower; *zoon*, animal; *eidos*, form] motile male gametes produced in antheridia; *alt.* spermatozoids.

anthesis n. [Gk. *anthe*, flower] stage or period at which flower bud opens; flowering, *alt.* florescence; period of flowering.

anthoblast n. [Gk. *anthos*, flower; *blastos*, bud] in Madreporaria, a young sessile polyp producing anthocyathus.

anthocarp n. [Gk. *anthos*, flower; *karpos*, fruit] a collective, composite or aggregated fruit formed from an entire inflorescence, as sorosis, syconus; *alt.* infructescence, inflorescence fruit, multiple fruit; *a.* anthocarpous.

anthocaulis n. [Gk. *anthos*, flower; L. *caulis*, stem] the pedicle of a late trophozooid stage of madrepore development.

Anthoceropsida Anthocerotopsida, *q.v.*

Anthocerotae Anthocerotopsida, *q.v.*

Anthocerotales n. [*Anthoceros*, generic name] an order of liverworts, commonly called hornworts, in some classifications referred to the class Anthocerotopsida because of striking differences from other liverworts, or to the subclass Anthocerotidae, and having a thalloid gametophyte with only one chloroplast in each cell.

Anthocerotidae n. [*Anthoceros*, generic name] subclass of liverworts, sometimes raised to class status, Anthocerotopsida, or reduced to order status, Anthocerotales, *q.v.*; *cf.* Hepaticidae.

Anthocerotopsida n. [*Anthorceros*, generic name; Gk. *opsis*, appearance] in some classifications, a class of bryophytes given equal status with the Hepaticopsida, in other classifications referred to the order Anthocerotales, *q.v.*; *alt.* Anthoceropsida, Anthocerotae.

anthochlore n. [Gk. *anthos*, flower; *chloros*, yellow] a yellow pigment dissolved in cell sap of corolla, as of primrose.

anthocodia n. [Gk. *anthos*, flower; *kodeia*, head] the distal portion of a zooid bearing mouth and tentacles, in Alcyonaria.

anthocyan n. Gk. *anthos*, flower; *kyanos*, dark blue] any of various water-soluble pigments found in the vacuoles of higher plants, comprising the anthocyanidins and anthocyanins.

anthocyanidins *n.plu.* [Gk. *anthos*, flower; *kyanos*, dark blue] flavonoid plant pigments whose glucosides are anthocyanins.

anthocyanins n. *plu.* Gk. *anthos*, flower; *kyanos*, dark blue] important plant pigments found in flowers, leaves, fruits, and stems, which are sap-soluble glucosides giving scarlet, purple, and blue colours, also found in some insects, absorbed with plant food.

anthocyathus n. [Gk. *anthos*, flower; *kyathos*, cup] the discoid crown of trophozooid stage in madrepore development.

anthodium n. [Gk. *anthos*, flower; *eidos*, form] capitulum of florets, as of Compositae; *alt.* cephalanthium.

anthogenesis n. [Gk. *anthos*, flower; *genesis*, descent] in certain aphids, production of both males and females by asexual forms.

Anthomedusae

Anthomedusae *n.* [Gk. *anthos*, flower; *Medousa*, Medusa] Athecata, *q.v.*

anthophilous *a.* [Gk. *anthos*, flower; *philein*, to love] attracted by flowers; feeding on flowers.

anthophore *n.* [Gk. *anthos*, flower; *pherein*, to bear] elongation of receptacle between calyx and corolla.

Anthophyta, anthophytes *n.*, *n.plu.* [Gk. *anthos*, flower; *phyton*, plant] Phanerogamia, *q.v.*; Angiospermae, *q.v.*

anthostrobilus *n.* [Gk. *anthos*, flower; *strobilos*, fir cone] strobilus of certain cycads.

anthotaxis, anthotaxy *n.* [Gk. *anthos*, flower; *taxis*, arrangement] arrangement of flowers on an axis.

anthoxanthins *n.plu.* [Gk. *anthos*, flower; *xanthos*, yellow] sap-soluble flavone flower pigments giving colours from ivory to deep yellow, also found in insects, having been absorbed from the plant on which the insect feeds.

Anthozoa, anthozoans *n.*, *n.plu.* [Gk. *anthos*, flower; *zōon*, animal] a class of cnidarian coelenterates including sea anemones, sea fans, sea pens, and corals, having a dominant polyp phase and usually no medusa; *alt.* Actinozoa.

anthracobiontic *a.* [Gk. *anthrax*, charcoal; *biōnai*, to live] growing on burned-over soil or scorched material.

Anthracosauria, anthracosaurs *n.*, *n.plu.* [Gk. *anthrax*, charcoal; *sauros*, lizard] an order of Carboniferous to Permian labyrinthodonts, which may include the ancestors of reptiles; *alt.* Batrachosauria.

anthraquinones a group of orange or red pigments important in lichens, fungi, higher plants, and insects such as the cochineal and lac insect.

anthropeic *a.* [Gk. *anthrōpeios*, by human means] due to influence of man.

anthropochory *n.* [Gk. *anthrōpos*, man; *chōrein*, to spread] dispersal of diaspores by man; *alt.* androchory.

anthropogenesis *n.* [Gk. *anthrōpos*, man; *genesis*, descent] the ontogenesis and phylogenesis of man, the descent of man; *a.* anthropogenetic.

anthropogenic *a.* [Gk. *anthrōpos*, man; *genēs*, produced] produced or caused by man.

anthropoid *a.* [Gk. *anthrōpos*, man; *eidos*, form] resembling man, *appl.* tailless apes.

anthropology *n.* [Gk. *anthrōpos*, man; *logos*, discourse] the natural history of man.

anthropometry *n.* [Gk. *anthrōpos*, man; *metron*, measure] that part of biology dealing with proportional measurements of parts of the human body.

anthropomorphism *n.* [Gk. *anthrōpos*, man; *morphē*, shape] ascribing human emotions to animals.

anthropomorphous *a.* [Gk. *anthrōpos*, man; *morphē*, shape] resembling man.

anthropotomy *n.* [Gk. *anthrōpos*, man; *temnein*, to cut] human anatomy.

Anthropozoic Psychozoic, *q.v.*

anti- *see also* ante-.

antiae *n.plu.* [L. *antiae*, forelock] feathers at base of bill-ridge of some birds.

anti-alopecia factor inositol, *q.v.*

anti-ambulacral abactinal, *q.v.*

anti-apex lower end of axis, as in rootless plants.

Antiarchi *n.* [Gk. *anti*, against; *archi*, first] an order of Devonian placoderms having much thoracic armour, and pectoral fins replaced by long jointed spines; *alt.* Antiarchiformes.

antiauxin *n.* [Gk. *anti*, against; *auxein*, to grow] any organic compound which regulates or inhibits growth stimulation by auxins.

antiavidin biotin, *q.v.*; *cf.* avidin.

antibiont *n.* [Gk. *anti*, against; *biōnai*, to live; *onta*, beings] any antibiotic organism.

antibiosis *n.* [Gk. *anti*, against; *biōsis*, way of life] antagonistic association of organisms in which one produces compounds, antibiotics, harmful to the other.

antibiotic *n.* [Gk. *anti*, against; *biōtikos*, pert. life] a substance produced by a micro-organism which inhibits the growth of other microorganisms. *a.* Inhibiting or destroying life as of parasitic organisms; *pert.* antibiosis.

antibiotin avidin, *q.v.*

antiblastic *a.* [Gk. *anti*, against; *blastos*, bud] *appl.* immunity due to factors which inhibit growth of invading organism.

antibody *n.* [Gk. *anti*, against; A.S. *bodig*, body] a globulin protein made by a vertebrate in response to an antigen, which binds to the antigen; *alt.* antitrope, antitropin, immunoprotein.

antibrachium *n.* [L. *ante*, before; *brachium*, arm] the forearm, or corresponding portion of a forelimb; *a.* antibrachial.

antical *n.* [L. *ante*, before] *appl.* the upper or front surface of a thallus, leaf, or stem, especially in liverworts; *cf.* postical.

anticipation *n.* [L. *ante*, before; *capere*, to take] the manifestation of a condition or disease at a progressively earlier age in successive generations.

anticlinal *a.* [Gk. *anti*, against; *klinein*, to slope] *appl.* line of division of cells at right angles to surface of apex of a growing point; in quadrupeds, *appl.* one of lower thoracic vertebrae with upright spine towards which those on either side incline.

anticoagulin *n.* [Gk. *anti*, against; L. *coagulum*, rennet] a substance which prevents coagulation of drawn blood, as hirudin; *alt.* anticoagulant.

anticodon sequence of bases of transfer RNA which codes for an amino acid and matches a codon of messenger RNA, consists of 3 bases (triplet) and is designated by 3 letters.

anticryptic *a.* [Gk. *anti*, against; *kryptos*, hidden] *appl.* protective coloration facilitating attack.

antidiuretic *a.* [Gk. *anti*, against; *dia*, through; *ouron*, urine] reducing the volume of urine; *appl.* hormone (ADH) which controls water reabsorption by kidney tubules: vasopressin, *q.v.*

antidiuretin *n.* [Gk. *anti*, against; *dia*, through; *ouron*, urine] pituitrin, *q.v.*

antidromic, antidromous *a.* [Gk. *anti*, opposite; *dromos*, running] contrary to normal direction; *appl.* conduction of impulse along axon towards body of nerve cell; *appl.* stipules with fused outer margins; *cf.* orthodromic.

antidromy *n.* [Gk. *anti*, against; *dromos*, running] condition of spiral phyllotaxis with genetic spiral changing direction after each cycle.

anti-enzyme *n.* [Gk. *anti*, against; *en*, within; *zymē*, leaven] a substance which retards or stops enzyme activity; *cf.* inhibitor.

antifertilizin *n.* [Gk. *anti*, against; L. *fertilis*, fertile] an acid protein, of varying composition according to species, in cytoplasm of spermatozoa, and reacting with fertilizin; *alt.* androgamone II.

antigen *n.* [Gk. *anti*, against; *genos*, birth] substance which causes a series of physiologicochemical changes resulting in formation of antibodies.

antigeny sexual dimorphism, *q.v.*; causing antibody production.

antihaemorrhagic *a.* [Gk. *anti*, against; *haimorrhagēs*, bleeding] *appl.* vitamin: vitamin K, *q.v.*

antihelix *n.* [Gk. *anti*, opposite; *helix*, a convolution] the curved prominence in front of helix of ear; *alt.* anthelix.

antihormones *n.plu.* [Gk. *anti*, against; *hormaein*, to excite] substances which prevent the effect of hormones; *alt.* chalones, antagonists.

antilobium tragus, *q.v.*

antilysin *n.* [Gk. *anti*, against; *lyein*, to dissolve] a substance which counteracts a lysin or lysis.

antimeres *n.plu.* [Gk. *anti*, opposite; *meros*, part] corresponding parts, as left and right limbs, of a bilaterally symmetrical animal; a series of equal radial parts or actinomeres of a radially symmetrical animal.

antimetabolite *n.* [Gk. *anti*, against; *metabolē*, change] a compound which inhibits metabolism.

antimitotic *a.* [Gk. *anti*, against; *mitos*, thread] inhibiting mitosis.

antimorph *n.* [Gk. *anti*, against; *morphē*, form] a mutant gene acting in the opposite way to the normal allele.

antimutagen *n.* [Gk. *anti*, against; L. *mutare*, to change; Gk. *gennaein*, to produce] any substance or other agent which slows down the mutation rate or reverses the action of a mutagen.

antineuritic *a.* [Gk. *anti*, against; *neuron*, nerve] *appl.* vitamin: thiamine, vitamin B$_1$, lack of which causes polyneuritis.

Antipatharia *n.* [*Antipathes*, generic name] an order of Ceriantipatharia, commonly called black corals, which are colonial and have a black or brown skeleton.

antipepsin *n.* [Gk. *anti*, against; *pepsis*, digestion] a stomach secretion which prevents action of pepsin on tissue proteins.

antiperistalsis *n.* [Gk. *anti*, against; *peri*, around; *stalsis*, contraction] reversed peristalsis, i.e. peristaltic action in postero-anterior direction.

anti-pernicious anaemia factor vitamin B$_{12}$, *see* cobalamine.

antiperosis factor biotin, *q.v.*

antipetalous *a.* [Gk. *anti*, opposite; *petalon*, leaf] inserted opposite the insertion of the petals.

antiphyte *n.* [Gk. *anti*, opposite; *phyton*, plant] the sporophyte in the antithetic alternation of generations; *cf.* protophyte.

antipodal *a.* [Gk. *anti*, against; *pous*, foot] *appl.* group of 3 cells at chalazal end of embryo sac, the end away from the micropyle; *appl.* cone of astral rays opposite spindle fibres.

antiprostate *n.* [Gk. *anti*, opposite; *prostatēs*, one who stands before] bulbo-urethral gland, *q.v.*

antiprothrombin *n.* [Gk. *anti*, against; *pro*, before; *thrombos*, clot] heparin, *q.v.*

antipygidial *a.* [Gk. *anti*, against; *pygidion*, narrow rump] *appl.* bristles of 7th abdominal segment which extend to pygidium, in fleas.

antipyretic *a.* [Gk. *anti*, against; *pyretos*, fever]

appl. a substance which lowers the body temperature.

antirachitic *a.* [Gk. *anti*, against; *rhachis*, spine] preventing or counteracting rickets, *appl.* vitamin: vitamin D; *see* calciferol.

antiscorbutic *a.* [Gk. *anti*, against; L.L. *scorbutus*, scurvy] preventing or counteracting scurvy, *appl.* vitamin: vitamin C, *q.v.*

antisepalous *a.* [Gk. *anti*, opposite; F. *sépale*, from L. *separare*, to separate] inserted opposite the insertion of the sepals.

antiseptic *a.* [Gk. *anti*, against; *sepsis*, putrefaction] preventing putrefaction. *n.* A substance which destroys harmful micro-organisms.

antiserum *n.* [Gk. *anti*, against; L. *serum*, whey] a serum containing antibodies.

antispadix *n.* [Gk. *anti*, against; *spadix*, palm branch] a group of 4 modified tentacles in internal lateral lobes of Tetrabranchiata such as *Nautilus.*

antisquama *n.* [Gk. *anti*, against; L. *squama*, scale] basal lobe next to squama of insect wing; *alt.* squama alaris, antitegula.

antisterility factor vitamin E, *q.v.*

antistiffness factor stigmasterol, *q.v.*

antistyle *n.* [Gk. *anti*, against; *stylos*, pillar] basal projection of stylifer in certain insects.

antitegula *n.* [Gk. *anti*, against; L. *tegula*, tile] upper tegula or antisquama, *q.v.*

antithetic *a.* [Gk. *antithesis*, opposition] *appl.* a theory concerning, or the generations in, alternation of generations in plants where the sporophyte and gametophyte are thought to be fundamentally distinct, the gametophyte being the ancestral phase and the sporophyte arising from it by delay of meiosis after sexual fusion; *cf.* homologous.

antithrombin *n.* [Gk. *anti*, against; *thrombos*, clot] heparin, *q.v.*

antitoxin *n.* [Gk. *anti*, against; *toxikon*, poison] a substance which neutralizes a toxin by combining with it.

antitragus *n.* [Gk. *anti*, opposite; *tragos*, goat] prominence opposite tragus of external ear.

antitrochanter *n.* [Gk. *anti*, against; *trochanter*, a runner] in birds, an articular surface on ilium against which trochanter of femur plays.

antitropal antitropous, *q.v.*

antitrope *n.* [Gk. *anti*, opposite; *tropē*, turn] any structure which forms a bilaterally symmetrical pair with another; antibody, *q.v.*

antitropic *a.* [Gk. *anti*, against; *tropē*, turn] turned or arranged in opposite directions; arranged to form bilaterally symmetric pairs, as ribs of opposite sides; *cf.* syntropic.

antitropin antibody, *q.v.*

antitropous *a.* [Gk. *anti*, against; *tropē*, turn] inverted; *appl.* embryos with radicle directed away from hilum; *alt.* antitropal.

antitype *n.* [Gk. *anti*, equal to; *typos*, pattern] a specimen of the same type as that chosen for designation of a species, and gathered at the same time and place.

antivitamins *n.plu.* [Gk. *anti*, against; L. *vita*, life; *ammoniacum*, a gum] chemical compounds which displace, split, or combine with, vitamins.

antlia *n.* [L. *antlia*, pump] the spiral suctorial proboscis of Lepidoptera.

ant lions *see* Neuroptera.

antorbital *a.* [L. *ante*, before; *orbis*, orbit]

situated in front of orbit, *appl.* bone, cartilage, process.

antrorse *a., adv.* [L. *ante*, before; *vertere*, to turn] directed forwards or upwards; *cf.* retrorse.

antrum *n.* [L. *antrum*, cavity] a cavity or sinus. e.g. maxillary sinus, cavity of pylorus.

anucleate *a.* [Gk. *a*, without; L. *nucleus*, kernel] without a nucleus.

anucleolate *a.* [Gk. *a*, without; L. *nucleolus*, little kernel] without a nucleolus.

anura *n.* [Gk. *a*, without; *oura*, tail] an order of amphibians, placed in the subclass Apsidospondyli in some classifications or in the Lissamphibia in others, including the living frogs and toads; *alt.* Batrachia.

anural, anurous *a.* [Gk. *a*, without; *oura*, tail] tailless.

anus *n.* [L. *anus*, anus] the opening of the alimentary canal (usually posterior) through which undigested food is voided; *a.* anal.

aorta *n.* [Gk. *aortē*, the great artery] the great trunk artery which carries pure blood to the body through arteries and their branches.

aortic *a.* [Gk. *aortē*, the great artery] *pert.* aorta, *appl.* arch, hiatus, isthmus, lymph glands, semilunar valves, etc.

aortic bodies two small masses of chromaffin cells in a capillary plexus, one on each side of foetal abdominal aorta, being part of system for controlling oxygen content and acidity of blood; *alt.* Zuckerkandl's bodies.

apandrous *a.* [Gk. *apo*, away; *anēr*, male] without functional male sex organs, especially antheridia; parthenogenetic, as oospores in certain Oomycetes; *n.* apandry.

apatetic *a.* [Gk. *apatētikos*, fallacious] *appl.* misleading coloration.

aperispermic *a.* [Gk. *a*, without; *peri*, around; *sperma*, seed] *appl.* seeds without nutritive tissue.

apertura piriformis anterior nasal aperture of skull.

apes *see* Primates.

Apetalae *n.* [Gk. *a*, without; *petalon*, leaf] Monochlamydeae, *q.v.*

apetalous *a.* [Gk. *a*, without; *petalon*, leaf] without petals; *alt.* monochlamydeous; *cf.* apopetalous, petaliferous; *n.* apetaly.

apex *n.* [L. *apex*, summit] tip or sumit, as of lungs, heart, root, stem; styloid process of fibula; tip of epicone in Dinoflagellata; wing tip in insects; *plu.* apices.

Aphaniptera *n.* [Gk. *aphanes*, unseen; *pteron*, wing] Siphonaptera, *q.v.*

aphanipterous *a.* [Gk. *aphanes*, unseen; *pteron*, wing] apparently without wings; *appl.* a member of the Aphaniptera.

aphantobiont *n.* [Gk. *aphantēs*, invisible; *biōnai*, to live] an ultramiscroscopic organism, a virus.

Aphasmidia *n.* [Gk. *a*, without; *phasma*, apparition; *dim.*] a class of nematodes which have no phasmids and whose amphids open on the posterior part of the head capsule; *alt.* Adenophora.

apheliotropism *n.* [Gk. *apo*, away; *hēlios*, sun; *tropē*, turn] tendency to turn away from light, strictly from the sun; *alt.* aheliotropism; *cf.* aphototropism.

Aphetohyoidea *n.* [Gk. *aphetos*, let loose; *hyoeidēs*, Υ-shaped] Placodermi, *q.v.*

aphins *n.plu.* [*Aphis*, generic name] red and yellow fat-soluble pigments extracted from various

aphids (greenfly), probably arising after death from protaphin.

aphlebia *n.* [Gk. *a*, without; *phleps*, vein] lateral outgrowth from base of frond stalk in certain ferns.

aphodus *n.* [Gk. *aphodos*, departure] the short tube leading from flagellate chamber to excurrent canal in a type of canal system in sponges; *a.* aphodal.

aphotic *a.* [Gk. *a*, without; *phōs*, light] *pert.* absence of light; *appl.* zone of deep sea where daylight fails to penetrate; *cf.* photic.

aphototropism *n.* [Gk. *a*, not; *phōs*, light; *tropē*, turn] tendency to turn away from light; *cf.* apheliotropism.

Aphyllophorales *n.* [Gk. *a*, without; *phyllon*, leaf; *phora*, producing] Poriales, *q.v.*

aphyllous *a.* [Gk. *a*, without; *phyllon*, leaf] without foliage leaves.

aphylly *n.* [Gk. *a*, without; *phyllon*, leaf] suppression or absence of leaves.

aphytic *a.* [Gk. *a*, without; *phyton*, plant] without seaweeds, *appl.* zone of coastal waters below approximately 100 m.

apical *a.* [L. *apex*, summit] at tip of summit; *pert.* distal end; *appl.* cell at tip of growing point; *appl.* style arising from summit of ovary; *appl.* dominance, of terminal bud; *appl.* aboral plates of echinoderms; *appl.* neural plate of trochophore and tornaria.

apical meristem the dividing tissue at the tips of the plant, which gives rise to the primary plant body; *alt.* vegetative cone.

apical placentation placentation where the ovule is at the apex of the ovary.

apices *plu.* of apex.

apicotransverse *adv.* [L. *apex*, summit; *transversus*, crosswise] situated across, at, or near the tip, *appl.* mitotic spindle.

apiculate *a.* [*Dim.* of L. *apex*, summit] forming abruptly to a small tip, as leaf.

apiculus *n.* [*Dim.* of L. *apex*, summit] a small apical termination, as in some Protozoa, or hilar appendix of certain spores; reflected portion of antennal club in some Lepidoptera.

apilary *a.* [Gk. *a*, not; *pilos*, felt cap] having upper lip missing or suppressed in corolla.

apileate *a.* [L. *a*, away; *pileatus*, wearing a cap] without a pileus.

apitoxin *n.* [L. *apis*, bee; Gk. *toxikon*, poison] main toxic fraction of bee venom.

apituitarism *n.* [L. *a*, away; *pituita*, phlegm] absence or deficiency of pituitary gland secretion; *cf.* hypopituitarism.

aplacental *a.* [L. *a*, away; *placenta*, flat cake] having no placenta, as monotremes; *alt.* implacental.

Aplacophora *n.* [Gk. *a*, without; *plax*, plate; *pherein*, to bear] a subclass, or in some classifications an order, of aberrant chitons which are long and worm-like and have no shell.

aplanetic *a.* [Gk. *a*, not; *planētēs*, wanderer] not motile, *appl.* spores.

aplanetism *n.* [Gk. *a*, not; *planētēs*, wanderer] absence of motile spores.

aplanogametangium *n.* [Gk. *a*, not; *planos*, wandering; *gametēs*, spouse; *anggeion*, vessel] cell in which aplanogametes are formed.

aplanogamete *n.* [Gk. *a*, not; *planos*, wandering; *gametēs*, spouse] a non-motile gamete.

aplanoplastid aplanospore, *q.v.*

aplanosporangium *n.* [Gk. *a*, not; *planos*, wandering; *sporos*, seed; *anggeion*, vessel] a sporangium producing aplanospores.

aplanospore *n.* [Gk. *a*, not; *planos*, wandering; *sporos*, seed] a non-motile resting spore; *alt.* aplanoplastid; *cf.* planospore.

aplasia *n.* [Gk. *a*, without; *plassein*, to mould] arrested development; non-development; defective development.

aplastic *a.* [Gk. *a*, without; *plastos*, formed] *pert.* aplasia; without change in development or structure.

aplerotic *a.* [Gk. *a*, not; *plēroun*, to fill] not entirely filling a space; *appl.* oospore not extended to oogonial wall; *cf.* plerotic.

aploperistomatous *a.* [Gk. *aploos*, single; *peri*, around; *stoma*, mouth] having a peristome with one row of teeth, as mosses.

aplostemonous *a.* [Gk. *aploos*, single; *stēmōn*, spun thread] with a single row of stamens.

Aplysiomorpha *n.* [*Aplysia*, generic name; Gk. *morphē*, form] Anaspidea, *q.v.*

apneustic *a.* [Gk. *apneustos*, breathless] with spiracles closed or absent, *appl.* aquatic larvae of certain insects.

apobasidium *n.* [Gk. *apo*, sprung from; *basis*, base; *idion*, *dim.*] protobasidium, *q.v.*; a basidium having sterigmata with terminal spores; *cf.* autobasidium.

apobiotic *a.* [Gk. *apo*, away; *biōtikos*, *pert.* life] causing or *pert.* death of certain cells or tissues rather than death of the entire body.

apocarp *n.* [Gk. *apo*, away; *karpos*, fruit] the individual carpel of an apocarpous fruit.

apocarpous *a.* [Gk. *apo*, away; *karpos*, fruit] having separate or partially united carpels; *alt.* dialycarpic; *cf.* syncarpous.

apocarpy apocarpous condition; *cf.* syncarpy.

apocentric *a.* [Gk. *apo*, away; *kentron*, centre] diverging or differing from the original type; *cf.* archecentric.

apochlorosis *n.* [Gk. *apo*, away; *chlōros*, grass green] the absence of chlorophyll in Flagellata.

apocrine *a.* [Gk. *apo*, away; *krinein*, to separate] *appl.* glands whose secretion accumulates below the free surface and is released by breaking away of the distal part of the cell, involving cytoplasmic loss, as in mammary glands; *cf.* holocrine, merocrine.

apocyte, apocytium *n.* [Gk. *apo*, away; *kytos*, hollow] a multinucleate cell; a plurinucleate mass of protoplasm without cell walls.

Apoda *n.* [Gk. *a*, without; *pous*, foot] an order of parasitic barnacles having no thoracic limbs; an order of burrowing sea cucumbers having no podia or respiratory trees, and a greatly reduced water vascular system, *alt.* Paractinopoda, Synaptida; an order of limbless burrowing amphibians placed in the Lepospondyli in some classifications and in the Lissamphibia in others, having a reduced or absent larval stage and minute calcified scales in the skin, *alt.* Gymnophiona.

apodal *a.* [Gk. *a*, without; *pous*, foot] having no feet or foot-like locomotory appendages; having no ventral fin; stemless; *alt.* apodous.

apodema, apodeme *n.* [Gk. *apo*, away; *demas*, body] an internal skeletal projection in Arthropoda.

apoderma *n.* [Gk. *apo*, later; *derma*, skin] enveloping membrane secreted during resting stage between instars by certain Acarina.

Apodes *n.* [Gk. *a*, without; *pous*, foot] an order of elongated teleosts including the eels, and sometimes subdivided into several orders; *alt.* Anguilliformes.

Apodiformes *n.* [Gk. *a*, without; *pous*, foot; L. *forma*, shape] an order of birds of the subclass Neornithes, including the swifts and hummingbirds; *alt.* Micropodiformes.

apodous apodal, *q.v.*

apoenzyme *n.* [Gk. *apo*, away; *en*, in; *zymē*, leaven] the protein part of an enzyme which requires a coenzyme to become a holoenzyme.

apogamy *n.* [Gk. *apo*, away; *gamos*, marriage] a type of apomixis in which the embryo is produced from the unfertilized female gamete or from an associated cell.

apogeotropism *n.* [Gk. *apo*, away; *gaia*, earth; *tropē*, turn] growing away from gravity; *alt.* negative geotropism, ageotropism; *a.* apogeotropic.

apolegamic *a.* [Gk. *apolegein*, to choose; *gamos*, marriage] *appl.* mating associated with sexual selection.

apomeiosis *n.* [Gk. *apo*, away; *meion*, smaller] sporogenesis without meiosis.

apomict *n.* [Gk. *apo*, away; *miktos*, mixed] an organism reproducing by apomixis.

apomixis *n.* [Gk. *apo*, away; *mixis*, a mixing] a reproductive process without fertilization in plants, akin to parthenogenesis but including development from cells other than ovules, as apogamy and apospory, *q.v.*; *cf.* amphimixis.

aponeurosis *n.* [Gk. *apo*, from; *neuron*, sinew] the flattened tendon for insertion of, or membrane investing, certain muscles.

aponeurosis epicranialis galea aponeurotica, *q.v.*

apopetalous *a.* [Gk. *apo*, away; *petalon*, leaf] polypetalous, *q.v.*; *cf.* apetalous.

apophyllous *a.* [Gk. *apo*, away; *phyllon*, leaf] having free perianth leaves.

apophysis *n.* [Gk. *apo*, away; *phyein*, to grow] process from a bone, usually for muscle attachment; endosternite or sternal apodeme; swelling beneath reproductive structure on fungal hypha; photosynthetic region forming swelling at base of capsule in some mosses; small protuberance at apex of ovuliferous scale in pine.

apoplasmodial *a.* [Gk. *apo*, away; *plasma*, something moulded] not forming a typical plasmodium.

apoplast *n.* [Gk. *apo*, away; *plastos*, formed] the total of adjacent cell walls throughout the plant.

apoplastid *n.* [Gk. *apo*, away; *plastos*, formed; *idion*, *dim.*] a plastid having no chromatophores.

apopyle *n.* [Gk. *apo*, away; *pylē*, gate] exhalent pore of sponge.

aporogamy *n.* [Gk. *a*, without; *poros*, channel; *gamos*, marriage] entry of pollen tube into ovule by some method other than through the micropyle; *cf.* chalazogamy, porogamy.

aporrhysa *n.plu.* [Gk. *aporrheein*, to flow away] exhalent canals in sponges; *cf.* epirrhysa.

aposematic *a.* [Gk. *apo*, away; *sēma*, signal] warning colours or markings which signal to a predator that the organism is harmful, such as the colours of some stinging insects; *alt.* warning coloration; *cf.* allosematic, sematic.

aposporogony *n.* [Gk. *apo*, away; *sporos*, seed; *gonos*, birth] absence of sporogony.

apospory *n.* [Gk. *apo*, away; *sporos*, seed] a type of apomixis in which a diploid gametophyte is produced from the sporophyte without spore formation.

apostasis *n.* [Gk. *apo*, away; *stasis*, standing] condition of abnormal growth of axis which thereby causes separation of perianth whorls from one another.

apostaxis *n.* [Gk. *apostaxis*, a dribbling] excessive or abnormal exudation.

apostrophe *n.* [Gk. *apo*, away; *strophē*, turn] arrangement of chloroplasts along lateral walls of leaf cells as in bright light.

apothecium *n.* [Gk. *apothēkē*, store] a cup-shaped ascocarp in Discomycetes; ascocarp in lichens of the Gymnocarpeae; *alt.* discocarp.

apothelium *n.* [Gk. *apo*, away; *thēlē*, nipple] a secondary tissue derived from a primary epithelium.

apotome *n.* [Gk. *apo*, away; *tomē*, a cutting] a part appearing as if cut off, as from episternum, trochanter, etc., in Arthropoda.

apotracheal *a.* [Gk. *apo*, away; L. *trachia*, windpipe] with xylem parenchyma independent of vessels, or dispersed, *appl.* wood.

apotropous *a.* [Gk. *apo*, away; *tropē*, turn] *appl.* anatropous ovule with a ventrally situated raphe.

apotype hypotype, *q.v.*

apotypic *a.* [Gk. *apo*, away; *typos*, pattern] diverging from a type.

apparato reticolare *see* Golgi complex.

appeasement *appl.* behaviour which ends the attack of one animal on another of the same species by the loser adopting a submissive posture or gesture.

appendage *n.* [L. *ad*, to; *pendere*, to hang] an organ or part attached to a trunk, as a limb, branch, etc.; a hyphal or rigid structure for attachment or detachment of perithecium to or from mycelium, varying in structure and function in different Ascomycetes; *a.* appendicular.

appendical *a.* [L. *appendix*, appendage] *pert.* an appendix; *pert.* vermiform appendix.

appendices *plu.* of appendix.

appendicular *a.* [L. *ad*, to; *pendere*, to hang] *pert.* appendages; *appl.* skeleton of limbs, *cf.* axial skeleton; *pert.* vermiform appendix.

Appendicularia *n.* [L. *appendicula*, small appendage] Larvacea, *q.v.*

appendiculate *a.* [L. *ad*, to; *pendere*, to hang] having a small appendage, as a stamen or filament; having an appendiculum.

appendiculum *n.* [L. *appendicula*, small appendage] remains of the partial veil on rim of pileus.

appendifer *n.* [L. *ad*, to; *pendere*, to hang; *ferre*, to carry] a ventral projection in a thoracic segment, for attachment of limb muscles, in trilobites.

appendix *n.* [L. *ad*, to; *pendere*, to hang] an outgrowth especially the vermiform appendix; *plu.* appendices; *a.* appendical; *cf.* epityphlon.

appendix colli *n.* [L. *ad*, to; *pendere*, to hang; *collum*, neck] exterior throat appendage of goat, sheep, pig, etc; *alt.* tassel, wattle.

appetite juice gastric juice secreted in response to the sight or smell of food.

appetitive *appl.* behaviour at the beginning of an instinctive behaviour pattern which can be very variable from unoriented wanderings to apparently purposeful behaviour.

applanate *a.* [L. *ad*, to; *planatus*, flattened] flattened.

apposition *n.* [L. *ad*, to; *ponere*, to place] the formation of successive layers in growth of a cell wall; *cf.* intussusception.

appressed *a.*, *adv.* [L. *ad*, to; *pressare*, to press] pressed together without being united; *alt.* adpressed.

appressorium *n.* [L. *ad*, to; *pressare*, to press] adhesive disc, as of haustorium or sucker; modified hyphal tip which may form haustorium or penetrate substrate, as of parasitic fungi; *alt.* holdfast.

aproterodont *a.* [Gk. *a*, without; *proteros*, fore; *odous*, tooth] having no premaxillary teeth.

Apsidospondyli *n.* [Gk. *apsis*, arch; *sphondylos*, vertebra] a group of amphibians including the Labyrinthodontia and Salientia, having vertebral centra primitively formed from pleurocentra and intercentra.

apteria *n.plu.* [Gk. *a*, without; *pteron*, feather] naked or down-covered surfaces between pterylae.

apterous *a.* [Gk. *a*, without; *pteron*, wing] wingless; having no wing-like expansions on stems or petioles, *alt.* exalate.

apterygial *a.* [Gk. *a*, without; *pterygion*, little wing or fin] wingless; without fins.

Apterygiformes *n.* [Gk. *a*, without; *pterygion*, little wing; L. *forma*, shape] an order of birds of the subclass Neornithes, including the kiwis.

Apterygota *n.* [Gk. *a*, without; *pterygion*, little wing] a subclass of insects, thought to be primitive, that have no wings and little or no metamorphosis and have abdominal appendages in the adult; *alt.* Ametabola; *a.* apterygotous.

aptychus *n.* [Gk. *a*, together; *ptychē*, plate] a horny or calcareous structure, possibly an operculum, of ammonites.

apyrene *a.* [Gk. *a*, not; *pyrēn*, fruit stone] *appl.* spermatozoa lacking nucleus, *cf.* eupyrene, oligopyrene; seedless, *appl.* certain cultivated fruits.

aqueduct *n.* [L. *aqua*, water; *ducere*, to lead] a channel or passage, as that of cochlea, and of vestibule of ear; *alt.* aquaeductus, iter.

aqueduct of Sylvius [*F. de Boë* or *Sylvius*, Flemish anatomist] cerebral aqueduct, aqueduct of the mid-brain, or iter, connecting 3rd and 4th ventricle; *alt.* mesocoel.

aqueous *a.* [L. *aqua*, water] watery, *appl.* humour, fluid occupying space between lens and cornea; *appl.* tissue consisting of thin-walled watery parenchymatous cells.

arabinan *n.* [Gum *arabic*] a pentosan occurring in gums and mucilage which on hydrolysis gives arabinose; *alt.* araban.

arabinose *n.* [Gum *arabic*] an aldose pentose sugar found in gums, pectins, and mucilages.

arachidonic acid [*Arachis*, generic name of peanut] a fatty acid essential for growth of mammals, but which can be synthesized from linoleic acid.

arachnactis *n.* [Gk. *arachnē*, spider; *aktis*, ray] larval stage of cerianthid Zoantharia.

Arachnida, arachnids *n.*, *n.plu.* [Gk. *arachnē*, spider] a class of mainly terrestrial carnivorous chelicerates, e.g. spiders and scorpions, having 4 pairs of walking legs and chelicerae and pedipalps on the prosoma; *alt.* Arachnoidea.

arachnidium *n.* [Gk. *arachnē*, spider; *idion*, *dim.*] the spinning apparatus of a spider, including spinning glands and spinnerets.

arachniform arachnoid, *q.v.*; stellate, *q.v.*, *appl.* cells.

arachnoid *a.* [Gk. *arachnē*, spider, cobweb; *eidos*, form] *pert.* or resembling a spider; like a cobweb, *alt.* histioid; consisting of fine entangled hairs, *alt.* araneose; *appl.* the thin membrane between dura and pia mater, *alt.* arachniform. *n.* The arachnoid membrane.

Arachnoidea *n.* [Gk. *arachnē*, spider; *eidos*, form] Arachnida, *q.v.*

arachnoideal *a.* [Gk. *arachnē*, cobweb; *eidos*, form] *pert.* the arachnoid; *appl.* granulations: Pacchionian bodies, *q.v.*

Araeoscelidia *n.* [Gk. *araios*, thin; *-skelis*, leg] an order of Permian to Triassic, lizard-like, primitive euryapsid reptiles.

Arales *n.* [*Arum*, generic name] an order of monocots (Arecidae), which are usually herbaceous perennials or climbers and including the families Araceae (aroids) and Lemnaceae (duckweeds); *alt.* Spathiflorae.

Araliales *n.* [*Aralia*, generic name] an order of woody or herbaceous dicots (Rosidae) with compound or lobed leaves and including the Araliaceae (e.g. ivy) and the umbellifers.

Araneida *n.* [*Aranea*, generic name] an order of arachnids, commonly called spiders, having spinning glands on opisthosoma and poison glands on the chelicerae; *alt.* Araneae.

araneose, araneous *a.* [L. *araneosus*, cobwebby] covered with, or consisting of fine entangled filaments; *alt.* arachnoid.

arbacioid *see* diadematoid.

Arbacoida *n.* [*Arbacia*, generic name; Gk. *eidos*, form] an order of Echinacea having unperforated tubercles on the spines.

arboreal *a.* [L. *arbor*, tree] of the nature of a tree; *pert.* trees.

arborescence arborization, *q.v.*

arborescent *a.* [L. *arborescens*, growing like a tree] branched like a tree.

arborization *n.* [L. *arbor*, tree] tree-like branching, as of nerve cell processes; *alt.* arborescence.

arboroid *a.* [L. *arbor*, tree; Gk. *eidos*, like] tree-like, designating general structure of a protozoan colony; *alt.* dendroid.

arborvirus *n.* [arthropod-*borne virus*] a virus which replicates in an arthropod as its intermediate host and in a vertebrate as its definitive host; *alt.* arbovirus.

arbor vitae *n.* [L. *arbor*, tree; *vita*, life] the tree of life, *appl.* arborescent appearance of cerebellum in section.

arbovirus arborvirus, *q.v.*

arbuscle *n.* [L. *arbuscula*, shrub] a tree-like small shrub or a dwarf tree; a branched haustorium as in certain fungi, such as the complex branched hyphal systems in root cells of vesicular–arbuscular mycorrhizae; *alt.* arbuscula; *a.* arbuscular.

arbutoid *a.* [*Arbutus*, generic name of strawberry tree; Gk. *eidos*, form] *appl.* mycorrhiza constituting a link between ecto- and endotrophic mycorrhizae.

arcade *n.* [L. *arcus*, arch] an arched channel or passage; a bony arch, as supra- and infratemporal

arches in skull; transverse canal connecting lateral canals, in the nematode *Ascaris*.

arch- *see also* arche-, archi-.

Archaean *a.* [Gk. *archaios*, ancient] *pert.* or belonging to or containing the group of rocks of the earlier part of the Pre-Cambrian; sometimes considered equivalent to the whole of the Pre-Cambrian.

archaeo- *alt.* archeo-.

Archaeoceti *n.* [Gk. *archaios*, primitive; *kētos*, whale] an order of Eocene to Miocene whale-like marine placental mammals.

archaeocytes *n.plu.* [Gk. *archaios*, primitive; *kytos*, hollow] cells arising from undifferentiated blastomeres and ultimately giving rise to gametes.

Archaeogastropoda *n.* [Gk. *archaios*, primitive; *gastēr*, stomach; *pous*, foot] an order of marine prosobranchs having 1 or 2 bipectinate ctenidia, some families lacking ctenidia or having secondary gills.

Archaeopteridales *n.* [Gk. *archaios*, primitive; *pteris*, fern] in some classifications, an order of Primofilices, *q.v.*

Archaeopteryx **and** *Archaeornis* genera of fossil birds from the Jurassic, which show many reptilian features and provide a missing link between reptiles and birds.

Archaeornithes *n.* [Gk. *archaios*, primitive; *ornis*, bird] a subclass of primitive birds comprising the order Archaeornithiformes which contains the fossils *Archaeopteryx* and *Archaeornis*.

archaeostomatous *a.* [Gk. *archaios*, primitive; *stoma*, mouth] having the blastopore persistent and forming mouth.

Archaeozoic *a.* [Gk. *archaios*, ancient; *zōē*, life] *pert.* earliest geological era, the lower division of the Pre-Cambrian, the time of Archaean rocks and of unicellular life; *alt.* Palaeolaurentian; *cf.* Proterozoic.

arch-centra *n.plu.* [L. *arcus*, bow; *centrum*, centre] centra formed by fusion of basal growths of primary arcualia external to chordal sheath; *cf.* chordacentra; *a.* archecentrous, archicentrous, archocentrous.

arche- *see also* archi-, arch-.

archebiosis *n.* [Gk. *urchē*, beginning; *biōsis*, living] abiogenesis, *q.v.*

archecentric *a.* [Gk. *archē*, beginning; *kentron*, centre] conforming more or less with the original type; *cf.* apocentric.

archedictyon *n.* [Gk. *archē*, beginning; *diktyon*, net] an intervein network in wings of some primitive insects; *alt.* archidictyon.

archegenesis *n.* [Gk. *archē*, beginning; *genesis*, descent] abiogenesis, *q.v.*

archegonia *plu.* of archegonium.

Archegoniatae, archegoniates *n.*, *n.plu.* [Gk. *archē*, beginning; *gonos*, offspring] a division of the plant kingdom consisting of the Bryophyta and Pteridophyta, being plants having archegonia as the female sex organ and having clear alternation of generations.

archegoniophore *n.* [Gk. *archē*, beginning; *gonos*, offspring; *pherein*, to bear] branches of bryophytes, or parts of fern prothalli, bearing archegonia; *alt.* archegonial receptacle; *cf.* antheridiophore.

archegonium *n.* [Gk. *archē*, beginning; *gonos*, offspring] the female sex organ in bryophytes, pteridophytes, and in a simplified form in some spermatophytes, consisting of a neck and a venter which contains the ovum (oosphere) and in which the young sporophyte begins development; *alt.* ooangium, pistillidium; *cf.* antheridium; *plu.* archegonia.

archencephalon *n.* [Gk. *archē*, beginning; *engkephalos*, brain] the primitive fore-brain or cerebrum.

archenteron *n.* [Gk. *archē*, beginning; *enteron*, gut] the cavity of gastrula which forms primitive gut of embryo.

archeo- archaeo-, *q.v.*

archespore *n.* [Gk. *archē*, beginning; *sporos*, seed] the tetrahedral or meristematic cell of a sporangium; cell of an archesporium; *alt.* sporoblast.

archesporium *n.* [Gk. *archē*, beginning; *sporos*, seed] a cell or mass of cells, dividing to form spore mother cells; in liverworts, spore mother cells, and elater-forming cells.

archetype architype, *q.v.*

archi- *see also* arche-, arch-.

Archiacanthocephala *n.* [Gk. *archi*, first; *akantha*, thorn; *kephalē*, head] an order of Acanthocephala which are parasites of terrestrial hosts and have concentric circles of proboscis spines.

archiamphiaster *n.* [Gk. *archi*, first; *amphi*, on both sides; *astēr*, star] the amphiaster forming 1st or 2nd polar body in maturation of ovum.

Archiannelida *n.* [Gk. *archi*, first; L. *annulus*, ring; Gk. *eidos*, form] an order of polychaetes with reduced parapodia and moving by cilia, formerly considered a separate class of annelids.

archibenthic *a.* [Gk. *archi*, first; *benthos*, depths of sea] *pert.* bottom of sea from edge of continental shelf to upper limit of abyssobenthic zone, at depths of *ca.* 200 to 1000 m.

archiblast *n.* [Gk. *archi*, first; *blastos*, bud] egg protoplasm.

archiblastic *a.* [Gk. *archi*, first; *blastos*, bud] *pert.* archiblast; *pert.* archiblastula, having total and equal segmentation.

archiblastula *n.* [Gk. *archi*, first; *blastos*, bud] typical hollow ball of cells derived from an egg with total and equal segmentation.

archicarp *n.* [Gk. *archi*, first; *karpos*, fruit] spirally coiled region of thallus or stalk bearing female sex organ of certain fungi; a cell which gives rise to a fruit body.

archicerebrum *n.* [Gk. *archi*, first; L. *cerebrum*, brain] the primitive brain, as the supraoesophageal ganglia of higher invertebrates; primary brain of arthropods.

Archichlamydeae *n.* [Gk. *archi*, first; *chlamys*, cloak] a subdivision of dicotyledons which are achlamydeous, haplochlamydeous, or dichlamydeous with a corolla of a number of unjoined petals; *alt.* Choripetalae.

archichlamydeous *a.* [Gk. *archi*, first; *chlamys*, cloak] having no petals, or having petals entirely separate from one another.

archicoel *n.* [Gk. *archi*, first; *koilos*, hollow] the primary body cavity or space between alimentary canal and ectoderm in development of various animals; *alt.* blastocoel.

Archidiales *n.* [*Archidium*, generic name] order of mosses of the subclass Bryidae having an unstalked capsule with no columella, lid, or peristome.

archidictyon archedictyon, *q.v.*

archigenesis *n.* [Gk. *archi*, first; *genesis*, descent] abiogenesis, *q.v.*

archigony *n.* [Gk. *archi*, first; *gonos*, begetting] the first origin of life.

Archimycetes, Archimycetales *n.* [Gk. *archi*, first; *mykēs*, fungus] a group of primitive fungi including the Chytridiales and some related forms.

archinephric *a.* [Gk. *archi*, first; *nephros*, kidney] *appl.* duct into which pronephric tubules open; *pert.* archinephros; *pert.* archinephridium.

archinephridium *n.* [Gk. *archi*, first; *nephros*, kidney; *idion*, dim.] excretory organ of certain larval invertebrates, usually a solenocyte.

archinephros *n.* [Gk. *archi*, first; *nephros*, kidney] the primitive kidney.

archipallium *n.* [Gk. *archi*, first; L. *pallium*, mantle] the olfactory region of cerebral hemispheres, comprising olfactory bulbs and tubercles, pyriform lobes, hippocampus, and fornix; *cf.* neopallium.

archiplasm *n.* [Gk. *archi*, first; *plasma*, mould] the substance of attraction sphere, astral rays, and spindle fibres, *alt.* archoplasm, ergastoplasm, kinoplasm; idiosome, *q.v.*

archipterygium *n.* [Gk. *archi*, first; *pterygion*, little wing or fin] type of fin in which skeleton consists of elongated segmented central axis and 2 rows of jointed rays.

archisternum *n.* [Gk. *archi*, first; L.L. *sternum*, breast bone] cartilaginous elements in myocommata of ventral region of thorax, as in tailed amphibians.

Archistia *n.* [Gk. *archi*, first; *histion*, sail] an order of primitive Palaeozoic fish with ganoid scales and superficially resembling herrings.

architomy *n.* [Gk. *archi*, first; *tomē*, cutting] reproduction by fission with subsequent regeneration, in certain annelids; *cf.* paratomy.

architype *n.* [Gk. *archi*, first; *typos*, type] an original type from which others may be derived; *alt.* archetype.

archoplasm archiplasm, *q.v.*

Archosauria, archosaurs *n., n.plu.* [Gk. *archon*, ruler; *sauros*, lizard] a subclass of reptiles, the ruling reptiles, mostly extinct but including the crocodiles, having specializations in skeleton associated with trends towards bipedalism.

aricentrous arocentrous, *q.v.*

arciferous *a.* [L. *arcus*, bow; *ferre*, to carry] *appl.* pectoral arch of toads, etc., where precoracoid and coracoid are separated and connected by arched epicoracoid.

arciform *a.* [L. *arcus*, bow; *forma*, shape] shaped like an arch or bow; *alt.* arcuate.

arcocentrous *a.* [L. *arcus*, bow; *centrum*, centre] *appl.* vertebral column with inconspicuous chordal sheath and centra mainly derived from arch tissue; *alt.* aricentrous.

arcocentrum *n.* [L. *arcus*, bow; *centrum*, centre] a centrum formed from parts of neural and haemal arches.

Arctogaea *n.* [Gk. *Arktos*, Great Bear; *gaia*, earth] zoogeographical area comprising Holarctic, Ethiopian, and Oriental regions.

arcualia *n.plu.* [L. *arcus*, bow] small cartilaginous pieces, dorsal and ventral, fused or free, on vertebral column of fishes.

arcuate *a.* [L. *arcuatus*, curved] arciform, *q.v.*

arculus *n.* [*Dim.* of L. *arcus*, bow] arc formed by 2 wing veins of certain insects.

ardellae *n.plu.* [Gk. *ardein*, to sprinkle] small apothecia of certain lichens, having appearance of dust.

area *n.* [L. *area*, ground space] a surface, as area opaca, area pellucida, area vasculosa, etc.; part enclosed by a raised ridge, as in Polyzoa; a region.

Arecales *n.* [*Areca*, generic name] an order of monocots (Arecidae) including only the family Arecaceae (Palmae), the palms; *alt.* Principes.

Arecidae *n.* [*Areca*, generic name] a subclass of monocots having numerous inconspicuous flowers borne on a club-shaped inflorescence called a spadix subtended by a conspicuous bract called the spathe; *alt.* Spadiciflorae.

arenaceous *a.* [L. *arena*, sand] having properties or appearance of sand; sandy; arenicolous, *q.v.*

arenicolous *a.* [L. *arena*, sand; *colere*, to inhabit] living or growing in sand, *alt.* arenaceous; psammophilous, *q.v.*

areola *n.* [L. *areola*, dim. of *area*, space] a small coloured circle round a nipple; part of iris bordering pupil of eye; one of small spaces or interstices in some connective tissue; area defined by cracks on surface of lichens; poroids when surrounded by thickened margins; scrobicula, *q.v.*; a small pit; *alt.* areole.

areolar *a.* [L. *areola*, small space] of or like an areola; *pert.* an areola; pitted.

areolar glands sebaceous glands on areola of mammary papilla.

areolar tissue a type of connective tissue consisting of cells (macrophages, fibroblasts, and mast cells) separated by a mucin matrix in which are embedded yellow and white fibres.

areolate *a.* [L. *areola*, small space] divided into small areas defined by cracks or other margins.

areolation *n.* [L. *areola*, small space] areolar pattern or network appearance, as of cell margins in tissue.

areole *n.* [L. *areola*, small space] areola, *q.v.*; space occupied by a group of hairs or spines, as in cacti; small area of mesophyll in leaf delimited by intersecting veins.

arescent *a.* [L. *arescere*, to dry up] becoming dry.

arg arginine, *q.v.*

argentaffin *a.* [L. *argentum*, silver; *affinis*, related] staining with silver salts, *appl.* cells; *alt.* argyrophil, argentophil.

argentaffin(e) cells certain cells in the gut wall in gastric glands and crypts of Lieberkühn which secrete digestive enzymes.

argenteal *a.* [L. *argenteus*, silvern] *appl.* layer of eye as in fish, having a silvery appearance.

argenteous *a.* [L. *argenteus*, silvern] like silver.

argenteum *n.* [L. *argenteus*, silvern] a dermal silvery reflecting tissue layer of iridocytes, without chromatophores, in fishes.

argentophil argyrophil, *q.v.*

arginase the enzyme which catalyses the hydrolysis of arginine to ornithine and urea, and is important in the urea cycle; EC 3.5.3.1; *r.n.* arginase.

arginine an essential amino acid, formed in the liver and kidney and hydrolysed by the enzyme arginase, which appears to be a constituent of all proteins and is used in the production of creatine: abbreviated to arg.

arginine phosphate phosphagen, *q.v.*, in invertebrates.

arginine–urea cycle urea cycle, *q.v.*

argyrophil *a.* [Gk. *argyros*, silver; *philos*, loving] staining with silver salts, *appl.* fibres of reticular tissue, *appl.* , blepharoplasts or basal bodies; *alt.* argentaffin, argentophil.

aril *n.* [F. *arille*, Sp. *arillo*, a small hoop] an additional integument formed on some seeds after fertilization, and which may be spongy, fleshy, or a tuft of hairs; *alt.* arillus.

arillate possessing an aril.

arillode *n.* [F. *arille*, hoop; Gk. *eidos*, like] a false aril arising from region of micropyle as an expansion of exostome.

arillus *n.* [L.L. *arillus*, aril] aril, *q.v.*

arista *n.* [L. *arista*, awn] awn: long-pointed process as in many grasses; a bristle borne by antenna of many brachycerous Diptera; *alt.* style.

aristate *a.* [L. *arista*, awn] provided with awns, or with a well-developed bristle.

arthrogenous *a.* [Gk. *arthron*, joint; *genos*, descent] an evolutionary theory, now largely discredited, that evolution occurs due to a creative principle or potentiality in living matter which manifests itself in response to external stimuli so that an adaptation anticipates a need.

aristogenic eugenic, *q.v.*

Aristolochiales [*Aristolochia*, generic name] an order of dicotyledons of the Magnoliidae or Archichlamydeae, the term being used in different ways by different authorities.

Aristotle's lantern masticating apparatus of sea urchin.

aristulate *a.* [*Dim.* of L. *arista*, awn] having a short awn or bristle.

arkyochrome *a.* [Gk. *arkys*, net; *chrōma*, colour] with Nissl granules arranged like network, *appl.* certain neurones.

armature *n.* [L. *armatura*, armour] anything which serves to defend, as hairs, prickles, thorns, spines, stings, etc.

armilla *n.* [L. *armilla*, armlet] a bracelet-like fringe; superior annulus or manchette of certain fungi.

armillate fringed around; having an armilla.

arm-palisade palisade tissue in which the chloroplast-bearing surface is enlarged by infolding of cell walls beneath the epidermis.

arolium *n.* [Gk. *arole*, protection] median lobe or pad on pretarsus of many insects.

aromorph *n.* [Gk. *airein*, to raise; *morphē*, form] a character or structure resulting from aromorphosis.

aromorphosis *n.* [Gk. *airein*, to raise; *morphōsis*, shaping] evolution with an increase in the degree of organization without much increased specialization; *cf.* allomorphosis.

arousal a level of responsiveness in an animal due to nerve impulses from the reticular activating system.

array *n.* [O.F. *arroi*, order] arrangement in order of magnitude.

arrect *a.* [L. *arrectus*, set upright] upright; erect.

arrectores pilorum bundles of non-striped muscular fibres associated with hair follicles,—contraction causing hair to stand on end; *sing.* arrector pili.

arrhenogenic *a.* [Gk. *arrhēn*, male; *genos*, offspring] producing offspring preponderantly or entirely male; *n.* arrhenogeny, *cf.* monogeny, thelygeny.

arrhenoid *a.* [Gk. *arrhēn*, male; *eidos*, form] exhibiting male characteristics, as genetically female animals undergoing sex reversal. *n.* Sperm aster during fertilization of ovum.

arrhenoplasm *n.* [Gk. *arrhēn*, male; *plasma*, mould] male plasm, in reference to theory that all protoplasm consists of arrhenoplasm and thelyplasm.

arrhenotoky *n.* [Gk. *arrhēn*, male; *tokos*, birth] parthenogenetic production of males; *cf.* deuterotoky, thelyotoky.

arrhizal *a.* [Gk. *arrhizos*, not rooted] without true roots, as some parasitic plants; *alt.* arrhizous.

arrhostia *n.* [Gk. *arrhōstia*, ill health] a normal condition or trend in development or evolution, which resembles a diseased condition, e.g. extreme size in certain extinct vertebrates resembling over-action of pituitary gland.

arrow worms the common name for the Chaetognatha, *q.v.*

artefact *n.* [L. *ars*, art; *factus*, made] an apparent structure due to method of preparation and not part of the specimen; a man-made object, especially a stone implement; *alt.* artifact.

artenkreis *n.* [Ger. *Art*, species; *Kreis*, circle] complex of species which replace one another geographically.

arterenol noradrenalin(e), *q.v.*

arterial *a.* [L. *arteria*, artery] *pert.* an artery, or system of channels by which blood issues to body from heart.

arterial circle circulus arteriosus, *q.v.*

arteriolar–venular *pert.* arterioles and venules, *appl.* anastomosis.

arteriole *n.* [L. *arteriola*, small artery] a small artery.

artery *n.* [L. *arteria*, artery] a vessel which conveys blood from heart to body.

arthral, arthritic *a.* [Gk. *arthron*, joint] *pert.* or at joints.

arthrobranchiae *n.plu.* [Gk. *arthron*, joint; *brangchia*, gills] joint gills, arising at junction of thoracic appendage with trunk, of Arthropoda.

arthrodia *n.* [Gk. *arthrōdēs*, well-jointed] a joint admitting of only gliding movements.

arthrodial *appl.* articular membranes connecting thoracic appendages with trunk, as in arthropods.

Arthrodira, arthrodires *n., n.plu.* [Gk. *arthron*, joint; *deirē*, neck] an order of Devonian to Carboniferous placoderms, called 'jointed neck' fishes, having a joint between the skull and the thoracic armour; *alt.* Arthrodiriformes.

arthrogenous *a.* [Gk. *arthron*, joint; *genos*, descent] formed as a separate joint,. as spores; developed from separated portions of a plant.

arthromere *n.* [Gk. *arthron*, joint; *meros*, part] an arthropod body segment or somite.

Arthrophyta, arthrophytes *n., n.plu.* [Gk. *arthron*, joint; *phyton*, plant] Sphenopsida, *q.v.*

arthropleure *n.* [Gk. *arthron*, joint; *pleura*, side] the lateral part of an arthropod body segment.

Arthropoda, arthropods *n., n.plu.* [Gk. *arthron*, joint; *pous*, foot] a phylum of metamerically seg-

mented animals with jointed legs and a thickened chitinous cuticle forming an exoskeleton, and having a haemocoel, head, and sometimes a telson.

arthropodins *n.plu.* [Gk. *arthron*, joint; *pous*, foot] proteins found in the arthropod skeleton associated with chitin.

arthropterous *a.* [Gk. *arthron*, joint; *pteron*, wing or fin] having jointed fin-rays, as fishes.

arthrosis *n.* [Gk. *arthron*, joint] articulation, *q.v.*

arthrospore *n.* [Gk. *arthron*, joint; *sporos*, seed] in some blue–green algae, a thick-walled resting cell formed by segmentation of the filament; a cell formed by segmentation of hyphae, as oidia in fungi; in bacterial cultures, a refractive body which is not an endospore and is thought to be a resistant vegetative cell, a degenerate cell, or concerned with the sexual cycle.

arthrosterigmata *n.plu.* [Gk. *arthron*, joint; *stērigma*, support] jointed sterigmata.

arthrostracous *a.* [Gk. *arthron*, joint; *ostrakon*, shell] having a segmented shell.

arthrotergal *a.* [Gk. *arthron*, joint; L. *tergum*, back] *appl.* median dorsal flexor of opisthosoma in the king crab, *Limulus*.

arthrous *a.* [Gk. *arthron*, joint] articulate, *q.v.*

Articulae *n.* [L. *articulus*, joint] Sphenopsida, *q.v.*; *alt.* Articulatae.

articulamentum *n.* [L. *articulus*, joint; *mentum*, chin] in chitons, the lower part of each of the body plates; *cf.* tegmentum.

articular *a.* [L. *articulus*, joint] *pert.* or situated at a joint, *appl.* cartilage, lamellae, surface, capsule, etc. *n.* Bone articulating with quadrate to constitute suspensorium, *alt.* articulare.

articularis genus subcrureus, *q.v.*

Articulata *n.* [L. *articulus*, joint] a class of brachiopods having the valves joined by a tooth and socket at the posterior edge; an order of crinoids with living members having a flexible cup to the theca.

Articulatae *n.* [L. *articulus*, joint] Articulae, *q.v.*

articulate *a.* [L. *articulus*, joint] jointed, *alt.* articulated, arthrous; separating easily at certain points. *v.* To form a joint.

articulation *n.* [L. *articulus*, joint] a joint between bones or segments, or between segments of a stem or fruit; *alt.* arthrosis.

artifact artefact, *q.v.*

artificial classification a classification that groups organisms or objects together on the basis of only one or a very few specifically chosen characters; *cf.* natural classification.

artificial insemination the introduction, by human manipulation, of sperm into the female reproductive tract in mammals; *alt.* telegenesis.

artiodactyl *a.* [Gk. *artios*, even; *daktylos*, finger] having an even number of digits; *cf.* perissodactyl.

Artiodactyla, artiodactyls *n., n.plu.* [Gk. *artios*, even; *daktylos*, finger] an order of placental mammals known from the Eocene, commonly called even-toed ungulates, and including pigs, cattle, sheep, and camels, which have a complex stomach for dealing with plant food, and the 3rd and 4th digit forming a cloven hoof; *cf.* Perissodactyla.

arytaenoid *a.* [Gk. *arytaina*, ladle; *eidos*, form] pitcher-like, *appl.* 2 cartilages at back of larynx, also glands, muscles, etc.

asc ascus, *q.v.*

Ascarida(ta) *n.* [*Ascaris*, generic name] in some classifications, an order of Phasmidia or a subdivision of Rhabditida, having the mouth with 3 to 6 papillae.

ascending curving or sloping upwards.

ascending aestivation aestivation where each petal overlaps the edge of the one posterior to it; *cf.* descending aestivation.

Aschelminthes *n.* [Gk. *askos*, bag; *helmins*, worm] a phylum including the groups Rotifera, Gastrotricha, Kinorhyncha, Nematoda, and Nematomorpha, which are sometimes raised to phylum status.

asci *plu.* of ascus.

Ascidiacea *n.* [Gk. *askidion*, little bag] a class of urochordates, commonly called sea squirts, which are often colonial and usually fixed to a substratum; *alt.* ascidians.

ascidial *a.* [Gk. *askidion*, dim. of *askos*, bag] saclike; *appl.* certain specialized, or abnormal, floral and foliage leaves; *pert.* ascidium.

ascidian *n.* [Gk. *askidion*, little bag] a member of the Ascidacea, *q.v. a.* Like an ascidian.

ascidiozooid *n.* [Gk. *askidion*, little bag; *zōon*, animal; *eidos*, like] in Pyrosomida, 1 of a chain of 4 blastozooids (buds) arising on the cyathozooid.

ascidium *n.* [Gk. *askidion*, little bag] a pitcher-leaf or part of a leaf as in *Nepenthes* and other pitcher plants.

asciferous, ascigerous *a.* [Gk. *askos*, bag; L. *ferre, gerere,* to bear] bearing asci, as certain hyphae in fungi.

ascocarp *n.* [Gk. *askos*, bag; *karpos*, fruit] the fruit body of Ascomycetes containing asci surrounded by their protective covering, and which may be an apothecium, cleistocarp, or perithecium; *alt.* sporangiocarp, sporocarp.

ascogenous *n.* [Gk. *askos*, bag; *-genēs*, producing] producing asci, *appl.* hypae, cells.

ascogonium *n.* [Gk. *askos*, bag; *gonos*, offspring] a specialized hyphal branch which gives rise to ascogenous hyphae or an ascus; female sex organ of Ascomycetes; *alt.* carpogonium.

Ascohymeniales *n.* [Gk. *askos*, bag; *hymēn*, skin] a subdivision of Ascomycetes having perithecia or apothecia with true walls.

Ascolichenes, ascolichens *n., n.plu.* [Gk. *askos*, bag; *leichēn*, lichen] a group of lichens in which the fungus is an ascomycete.

ascoma *n.* [Gk. *askōma*, leather padding] disc-shaped ascocarp in certain fungi; *plu.* ascomata.

Ascomycetes, Ascomycetae, Ascomycotina *n.* [Gk. *askos*, bag; *mykēs*, fungus] a major class of fungi which are characterized by the production of ascospores inside asci.

ascophore *n.* [Gk. *askos*, bag; *pherein*, to bear] a hypha producing asci in an ascocarp.

ascoplasm *n.* [Gk. *askos*, bag; *plasma*, mould] cytoplasm of an ascus involved in spore formation; *cf.* epiplasm.

ascorbic acid vitamin C, *q.v.*

Ascorhynchomorpha *n.* [Gk. *askos*, bag; *rhynchos*, snout; *morphē*, form] an order of Pycnogonida having a very long proboscis.

Ascoseirales *n.* [Gk. *askos*, bag; *seira*, chain] an order of brown algae found in the Antarctic.

ascospore *n.* [Gk. *askos*, bag; *sporos*, seed] one of the spores produced in an ascus; *alt.* theca-spore.

ascostome *n.* [Gk. *askos*, bag; *stoma*, mouth] apical pore of an ascus; *alt.* ascuspore.

ascostroma *n.* [Gk. *askos*, bag; *strōma*, bedding] a structure which produces asci in some fungi (Loculoascomycetidae) where the stroma forms the wall of loculi in which asci are formed; *plu.* ascostromata.

Ascothoracica *n.* [Gk. *askos*, bag; *thorāx*, chest] an order of parasitic barnacles having reduced prehensile thoracic limbs.

ascus *n.* [Gk. *askos*, bag] a cell inside which spores (ascospores), usually 8, are formed in Ascomycetes; *plu.* asci; *alt.* theca, asc.

ascuspore ascostome, *q.v.*

ascyphous *a.* [Gk. *a*, without; *skyphos*, cup] without a cup-shaped expansion of the podetium, as some lichens.

-ase suffix denoting an enzyme, and joined to a root naming the substance acted on or the type of reaction.

Asellariales *n.* [Gk. *a*, not; L. *sella*, saddle] order of Trichomycetes reproducing by arthrospores.

asemic *a.* [Gk. *asēmos*, without sign] without markings.

aseptate *a.* [L. *a*, not; *septum*, partition] without any septum; *alt.* coenocytic.

asexual *a.* [Gk. *a*, without; L. *sexus*, sex] *appl.* reproduction which does not involve fusion of gametes, including spore production, budding, fission, vegetative propagation, and parthenogenesis.

asiphonate *a.* [L. *a*, not; *sipho*, tube] without siphons; *appl.* larvae whose respiratory tubes open directly to exterior.

asp aspartic acid, *q.v.*

asparaginase a hydrolysing enzyme widely distributed in plants, and also found in animal tissue, especially liver, which hydrolyses asparagine, with the production of aspartic acid and ammonia; EC 3.5.1.1; *r.n.* asparaginase.

asparagine *n.* [Gk. *asparagos*, asparagus] an amino acid, the monoamide of aspartic acid, which was first detected in asparagus, found also in young leaves of other plants and in leguminous and other seeds, and which is important in nitrogen metabolism in plants; abbreviated to asp-NH_2.

aspartase an enzyme catalysing hydrolysis of aspartic acid from which ammonia is removed to yield fumaric acid, and which is found in some bacteria and in higher plants; EC 4.3.1.1; *r.n.* aspartate ammonia-lyase.

aspartic acid an amino acid, amino succinic acid, which takes part in transamination reactions; abbreviated to asp.

aspect *n.* [L. *aspicere*, to look toward] direction facing part of a surface; appearance or look; seasonal appearance.

aspection *n.* [L. *aspicere*, to look toward] seasonal succession of phytological and zoological phenomena.

asperate *a.* [L. *asperare*, to roughen] having a rough surface; *alt.* asperous.

Aspergillales *n.* [*Aspergillus*, generic name] Plectascales, *q.v.*

aspergilliform *a.* [*Aspergillus*, generic name; L. *forma*, shape] tufted like a brush.

asperity *n.* [L. *asperitas*, roughness] roughness, as on a leaf.

asperous asperate, *q.v.*

asperulate *a.* [*Dim.* of L. *asperare*, to roughen] minutely rough.

Aspidobothria *n.* [Gk. *aspidēs*, broad; *bothrion*, small pit] Aspidogastrea, *q.v.*

aspidobranch *a.* [Gk. *aspidēs*, broad; *brangchia*, gills] of mollusc gills, bipectinate, *q.v.*

Aspidochirota(e) *n.* [Gk. *aspidēs*, broad; *cheir*, hand] an order of sea cucumbers having podia on trunk, and respiratory trees.

Aspidogastrea *n.* [Gk. *aspidēs*, broad; *gastēr*, stomach] a subclass or order of flukes with a simple life cycle and having no anterior adhesive organ but very large ventral suckers; *alt.* Aspidobothria.

aspirin acetylsalicylic acid, used as an analgesic and antipyretic.

asplanchnic *a.* [Gk. *a*, without; *splangchna*, viscera] without alimentary canal.

asp-NH₂ asparagine, *q.v.*

asporocystid *a.* [Gk. *a*, not; *sporos*, seed; *kystis*, bladder; *idion*, *dim.*] *appl.* oocyst of Sporozoa when zygote divides into sporozoites without sporocyst formation.

asporogenic, asporogenous *a.* [Gk. *a*, without; *sporos*, seed; *gennaein*, to produce] not originating from spores; not producing spores.

asporous *a.* [Gk. *a*, without; *sporos*, seed] having no spores.

assimilation *n.* [L. *ad*, to; *similis*, like] in autotrophic organisms, the process by which organic molecules are built up from inorganic ones from the environment; in heterotrophic organisms, the conversion of digested food materials into the complex molecules of the body of the organism; *alt.* anabolism.

assimilative *a.* [L. *assimilare*, to make like] *pert.* or used for assimilation, *appl.* hyphae, *appl.* growth preceding reproduction.

association *n.* [L. *ad*, to; *socius*, fellow] a plant community forming a division of a formation or larger unit of vegetation, as of tundra, grassland, forest, and characterized by dominant species; adherence of gregarines without fusion of nuclei; a form of learning in which a stimulus produces an increased positive response, *cf.* habituation; a group of animals in a certain habitat. *a. Appl.* fibres connecting white matter of interior of brain with cortex; *appl.* neurone: internuncial neurone, *q.v.*

associes *n.* [L. *ad*, to; *socius*, fellow] an association representing a stage in the process of succession.

assortive mating or breeding a kind of nonrandom mating where males and females mate with an individual like themselves.

astacene, astacin *n.* [L. *astacus*, crayfish] carotenoid pigment of certain crustaceans, echinoderms, and fishes.

astaxanthin *n.* [L. *astacus*, crayfish; Gk. *xanthos*, yellow] a xanthophyll carotenoid pigment found in the Chlorophyceae and Euglenophyceae algae and other plants, ingested by animals and found in certain insects and crustaceans in association with proteins.

astely *n.* [Gk. *a*, without; *stēlē*, pillar] absence of a central cylinder, axis, or stele; *alt.* schizostely; *a.* astelic.

aster *n.* [Gk. *astēr*, star] the star-shaped system of microtubules radiating from the centriole, seen in many cells during cell division, at either end of

the spindle, but not found in higher plants, *alt.* cytaster; a star-shaped arrangement of chromosomes during cell division, *alt.* karyaster.

Asterales *n.* [*Aster*, generic name] a very advanced order of dicots (Asteridae) having inflorescences organized into heads (capitula) with a calyx-like involucre, and including the families Asteraceae (Compositae–Tubuliforae) and Cichoriaceae (Compositae–Liguliforae); also called Compositae, especially when the group is reduced to family status.

Asteridae *n.* [*Aster*, generic name] a subclass of dicots comprising the most advanced sympetalous orders.

asterigmate *a.* [Gk. *a*, without; *stērigma*, support] not borne on sterigmata, *appl.* spores.

asterion *n.* [Gk. *astēr*, star] the region of posterolateral fontanelle where lambdoid, parieto-mastoid, and occipito-mastoid sutures meet.

asteriscus *n.* [Gk. *asteriskos*, *dim.* of *astēr*, star] a small otolith in rudimentary cochlea of teleosts.

asternal *a.* [L. *a*, from; L.L. *sternum*, breast bone] *appl.* ribs: false ribs, *q.v.*

asteroid *a.* [Gk. *astēr*, star; *eidos*, form] stellate, *q.v.*; *pert.* starfish. *n.* A member of the Asteroidea.

Asteroidea *n.* [Gk. *astēr*, star; *eidos*, form] a class of echinoderms of the subphylum Eleutherozoa, commonly called starfish or sea stars, having a star-shaped body with 5 radiating arms not sharply marked off from the central disc; *alt.* asteroids.

asterophysis *n.* [Gk. *astēr*, star; *physis*, constitution] a rayed cystidium-like structure or seta in an hymenium; *alt.* asteroseta.

asteroseta asterophysis, *q.v.*

asterospondylous *a.* [Gk. *astēr*, star; *sphondylos*, vertebra] having centrum with radiating calcified cartilage; *alt.* asterospondylic.

aster phase metaphase, *q.v.*

asthenic *a.* [Gk. *asthenēs*, feeble] weak; tall and slender, *appl.* physical constitutional type, *alt.* linear; leptosome, *q.v.*

asthenobiosis *n.* [Gk. *asthenēs*, feeble; *biōsis*, manner of life] life during a phase of lessened metabolic activity.

astichous *a.* [Gk. *a*, without; *stichos*, row] not set in a row or in rows.

astigmatous *a.* [Gk. *a*, without; *stigma*, mark] without stigmata; without spiracles.

astipulate exstipulate, *q.v.*

astogeny *n.* [Gk. *astos*, citizen; *genos*, descent] the development of a colony by budding.

Astomatida, Astomata *n.* [Gk. *a*, without; *stoma*, mouth] an order or suborder of Holotricha that are endozoic and mouthless.

astomatous *a.* [Gk. *a*, without; *stoma*, mouth] not having a mouth; without a cytostome; without stomata.

astomous *a.* [Gk. *a*, without; *stoma*, mouth] without a stomium or line of dehiscence; bursting irregularly.

astragalus *n.* [Gk. *astragalos*, ankle bone] second largest tarsal bone in man, *alt.* talus; a tarsal bone in vertebrates.

astral rays polar rays, *q.v.*

astral sphere astrosphere, *q.v.*

Astrapotheria, astrapotheres *n.*, *n.plu.* [Gk. *astrapē*, lightning; *thērion*, small animal] an extinct order of South American placental mammals of the Eocene to Miocene that in some ways were similar to perissodactyls.

astroblast n. [Gk. aster, star; blastos, bud] a cell giving rise to protoplasmic or to fibrillar astrocytes.

astrocentre n. [L. aster, star; centrum, centre] centrosome, q.v.

astrocyte n. [Gk. aster, star; kytos, hollow] certain neuroglia cells forming packing tissue in the CNS; alt. astroglia, macroglia; see also fibrous astrocytes, protoplasmic astrocytes.

astroglia n. [Gk. aster, star; glia, glue] astrocyte, q.v.

astropodia n.plu. [Gk. aster, star; pous, foot] fine unbranched radiating pseudopodia, as in Heliozoa and some Radiolaria.

astropyle n. [Gk. aster, star; pyle, gate] chief aperture of central capsule, in certain Radiolaria.

astrosclereid n. [Gk. aster, star; skleros, hard; eidos, form] a multiradiate sclereid or stone cell; a spiculate or ophiuroid cell.

astrosphere n. [Gk. aster, star; sphaira, ball] central mass of aster without rays; aster exclusive of centrosome; alt. astral sphere.

asymmetric(al) a. [Gk. asymmetros, disproportionate] pert. want of symmetry; having 2 sides unlike or disproportionate; appl. structures or organs which cannot be divided into similar halves by any plane.

asynapsis n. [Gk. a, not; synapsis, union] absence of pairing of chromosomes in meiosis; alt. asyndesis.

asyndesis asynapsis, q.v.

asynethogametism n. [Gk. a, not; synethes, well-suited; gametes, spouse] incapability of 2 apparently suitable gametes to unite, owing to presence of an inhibiting factor; gametal incompatibility; alt. aethogametism; cf. synethogametism.

atactostele n. [Gk. ataktos, irregular; stele, post] a complex stele having bundles scattered in the ground tissue, as in monocotyledons.

atavism n. [L. atavus, ancestor] occurrence of an ancestral characteristic not observed in more immediate progenitors; alt. reversion.

atavistic a. [L. atavus, ancestor] pert., marked by, or tending to atavism; alt. recapitulatory.

ateleosis n. [Gk. ateles, imperfect] dwarfism where individual is a miniature adult; cf. achondroplasia.

atelia n. [Gk. ateles, ineffectual] the apparent uselessness of a character of unknown biological significance; incomplete development.

atelomitic a. [Gk. a, not; telos, end; mitos, thread] appl. other than terminal attachment of chromosome to spindle.

Atelostomata n. [Gk. a, not; telos, end; stoma, mouth] a superorder of irregular sea urchins of the subclass Euechinoidea having a rigid test but no Aristotle's lantern.

athalamous a. [Gk. a, without; thalamos, inner room] lacking a thalamus.

Athecanephria n. [Gk. a, without; theke, case; nephros, kidney] an order of Pogonophora having free tentacles and diverging nephric ducts.

Athecata n. [Gk. a, without; theke, case] an order of hydrozoans having deeply bell-shaped medusae with no statocysts, and hydranths with skeletal material only around the stalk; alt. Anthomedusae, Gymnoblastea.

Atherinomorpha n. [Atherina, generic name of sand smelt; Gk. morphe, form] a group of teleosts existing from the Eocene to the present day, including flying fish and garfish and having both primitive and advanced features; alt. Atheriniformes.

athrocyte n. [Gk. athroos, collective; kytos, hollow] a large resorptive cell of nephridium in Bryozoa, alt. paranephrocyte; a type of coelomocyte in nematodes.

athrocytosis n. [Gk. athroos, collective; kytos, hollow] the capacity of cells to selectively absorb and retain solid particles in suspension, as dyes.

atlanto-occipital occipito-atlantal, q.v.

atlas n. [Gk. Atlas, a Titan] the 1st cervical vertebra.

atokous a. [Gk. atokos, childless] without offspring.

atoll n. [Mal. atoll] a coral reef surrounding a central lagoon.

atopy n. [Gk. atopia, unusual nature] idiosyncrasy; genetic sensitivity to poisonous effects of particular antigens or atopens, as of certain proteins, pollen, etc.; alt. allergy.

ATP adenosine triphosphate, q.v.

atractoid a. [Gk. atraktos, spindle; eidos, form] spindle-shaped; alt. fusiform.

atractosome n. [Gk. atraktos, spindle; soma, body] a spindle-shaped particle in mucus-secreting cells.

Atremata n. [Gk. a, without; trema, hole] an order of inarticulate brachiopods having both valves notched to allow the stalk to pass.

atretic a. [Gk. a, not; tretos, perforated] having no opening; imperforate; appl. vesicles resulting from degeneration of Graafian follicles, spurious corpora lutea; n. atresia.

atrial a. [L. atrium, central room] pert. atrium, appl. cavity, pore, canal, siphon, lobes, etc.

atrichic, atrichous a. [Gk. a, not; thrix, hair] having no flagella or cilia; alt. aflagellar, akont.

atriocoelomic a. [L. atrium, central room; Gk. koilōma, a hollow] connecting atrium and coelom, appl. funnels, of uncertain functions, in Cephalochordata, alt. brown funnels.

atriopore n. [L. atrium, central room; porus, passage] the opening from atrial cavity to exterior in Cephalochordata; spiracle in tadpole.

atrioventricular a. [L. atrium, chamber; ventriculus, ventricle] pert. atrium and ventricle of heart, appl. bundle: His' bundle, q.v., appl. groove, openings; appl. node: a mass of tissue in the wall of the right auricle structurally and functionally related to the sinu-auricular node, alt. Tawara's node.

atrium n. [L. atrium, chamber] the main chamber of the auricle of the heart; the embryonic structure that will give rise to the auricle of the heart; the whole auricle; the tympanic cavity; a division of the vestibule at the end of bronchiole; chamber surrounding pharynx in Tunicata and Cephalochordata; vestibule in insects from which tracheae extend into body; a chamber or cavity in various invertebrates.

atrochal a. [Gk. a, without; trochos, wheel] without preoral circlet of cilia; appl. trochophore when preoral circlet is absent and surface is uniformly ciliated.

atropal atropous, q.v.

atrophy *n.* [Gk. *a*, without; *trophē*, nourishment] emaciation; diminution in size and function.

atropine *n.* [*Atropa bella-donna*, deadly night-shade] an alkaloid obtained originally from *Atropa bella-donna*, and now from a wide range of plants of the family Solanaceae, used medicinally.

atropous *a.* [Gk. *a*, without; *tropē*, turn] *appl.* ovule which is not inverted; *alt.* atropal, orthotropous.

attachment the spindle attachment; a lasting fusion of 2 chromosomes.

attachment constriction centromere, *q.v.*

attenuate(d) *a.* [L. *attenuare*, to thin] thinned; gradually tapering to a point; reduced in density, strength, or pathogenic activity.

atterminal *a.* [L. *ad*, to; *terminus*, end] towards a terminal; *appl.* current directed toward thermal cross-section.

attic *n.* [Gk. *attikos*, Athenian] the epitympanic recess.

attraction cone fertilization cone, *q.v.*

attraction particle centriole, *q.v.*

attraction sphere centrosphere, *q.v.*

auditory *a.* [L. *audire*, to hear] *pert.* hearing apparatus, *appl.* organ, nucleus, ossicle, capsule, canal, meatus, nerve, vesicle, etc.; *pert.* sense of hearing; *cf.* aural.

auditory teeth of Huschke, projections on upper part of limbus of osseous spiral lamina of cochlea.

Auerbach's plexus [L. *Auerbach*, German anatomist] a gangliated plexus of non-medullated nerve fibres, found between the circular and longitudinal layers of muscular coat of small intestine; *alt.* plexus myentericus, myenteric plexus.

augmentation *n.* [L. *augere*, to increase] increase in number of whorls; *cf.* chorisis.

augmentor *a.* [L. *augere*, to increase] *appl.* nerves rising from sympathetic system and acting on heart, with antagonistic relation to vagi; *alt.* accelerator.

aulophyte *n.* [Gk. *aulōn*, hollow way; *phyton*, plant] a non-parasitic plant growing in a hollow of another.

aulostomatous *a.* [Gk. *aulos*, tube; *stoma*, mouth] having a tubular mouth or snout.

Auluroidea *n.* [Gk. *aulos*, tube; *eidos*, form] in former classifications an extinct class of echinoderms of the subphylum Eleutherozoa, existing from Ordovician to Mississippian times.

aural *a.* [L. *auris*, ear] *pert.* ear or hearing; *cf.* auditory.

auricle *n.* [L. *auricula*, small ear] any ear-like lobed appendage; the external ear; atrium or anterior chamber of heart, or the auricular appendage of atrium; lateral chemical receptor in Turbellaria; lateral outgrowth on 2nd abdominal tergum in dragonflies of the group Anisoptera; small ear-like projections at base of a leaf, especially in grasses; *alt.* auricula.

auricula *n.* [L. *auricula*, small ear] auricle, *q.v.*

auricular *n.* [L. *auricula*, small ear] eat covert of birds. *a.* Pert. an auricle, *appl.* artery, nerve, tubercle, vein, appendage of atrium, etc.

auricularia *n.* [L. *auricula*, small ear] a type of larva found among Holothuria.

Auriculariales *n.* [L. *auricula*, a small ear] an order of jelly fungi which have either the hypobasidium or epibasidium divided into 4 cells.

auricularis *n.* [L. *auricula*, small ear] superior,

anterior, posterior, extrinsic muscles of the external ear.

auriculate *a.* [L. *auricula*, small ear] eared, having auricles; *appl.* leaf with expanded bases surrounding stem; *appl.* leaf with lobes separate from rest of blade; *alt.* hastate-auricled.

auriculotemporal *a.* [L. *auricula*, small ear; *tempora*, temples] *pert.* external ear and temporal region; *appl.* nerve: a branch of the mandibular nerve.

auriculoventricular *a.* [L. *auricula*, small ear; *ventriculus*, ventricle] *pert.* or connecting auricle and ventricle of heart, *appl.* bundle: His' bundle, *q.v.*, *appl.* valve.

auriform *a.* [L. *auris*, ear; *forma*, shape] resembling the external ear in shape, as shell of the gastropod *Haliotis*, the ormer.

aurones *n.plu.* [L. *aurum*, gold] a group of yellow flavonoid plant pigments.

aurophore *n.* [L. *auris*, ear; Gk. *pherein*, to bear] an organ projecting from base of pneumatophore of certain Siphonophora.

austral *a.* [L. *australis*, southern] *appl.* or *pert.* southern biogeographical region, or restricted to North American between transitional and tropical zones.

Australian *a.* [L. *australis*, southern] *appl.* or *pert.* a zoogeographical region including Papua, Australia, New Zealand, and Pacific islands; *alt.* Australasian.

australopithecine *a.* [L. *australis*, southern; Gk. *pithēkos*, ape] *appl.* a group of man-like apes of the early Pleistocene, whose fossils have been found in southern Africa, including the genus *Australopithecus* and some related genera.

Australopithecus *n.* [L. *australis*, southern; Gk. *pithēkos*, ape] a genus of fossil tool-using hominids found in various parts of Africa including the Taung skull from Botswana named *A. africanus*, and more robust skeletons called *A. robustus*, formerly *Paranthropus robustus*, from southern Africa, and *A. boisei*, formerly called *Zinjanthropus boisei*, from Olduvai gorge; *cf. Homo.*

Austro-Columbian Neotropical, *q.v.*

autacoid *n.* [Gk. *autos*, self; *akos*, remedy; *eidos*, form] internal secretion, a hormone or a chalone; *alt.* autocoid, incretion.

autarchic *a.* [Gk. *autarchos*, autocratic] *appl.* genes in mosaic organisms which are not inhibited from expressing their effect by the presence of neighbouring genes of different genotype; *cf.* hyparchic.

autarticular(e) *n.* [Gk. *autos*, self; L. *articulus*, joint] goniale, *q.v.*

autecic, autecious autoecious, *q.v.*

autecology *n.* [Gk. *autos*, self; *oikos*, household; *logos*, discourse] the biological relations between a single species and its environment; ecology of an individual organism; *alt.* auto-ecology; *cf.* synecology.

autoagglutination *n.* [Gk. *autos*, self; L. *agglutinare*, to cement to] the clumping of an individual organism's cells by its own serum.

autoantibiosis *n.* [Gk. *autos*, self; *anti*, against; *biōsis*, a living] retardation or inhibition of growth in a medium made stale by the same organism.

Autobasidiomycetes *n.* [Gk. *autos*, self; *basis*, base; *idion*, *dim.*; *mykēs*, fungus] Homobasidiomycetes, *q.v.*

autobasidium n. [Gk. *autos*, self; *basis*, base; *idion*, *dim*.] a basidium having sterigmata bearing spores laterally, *cf.* apobasidium; a non-septate basidium, *alt.* holobasidium.

autobiology idiobiology, *q.v.*

autobiotic n. [Gk. *autos*, self; *bios*, life] one of a group of substances produced by cells whose behaviour they then participate in controlling.

autoblast n. [Gk. *autos*, self; *blastos*, bud] an independent micro-organism or cell.

autocarp n. [Gk. *autos*, self; *karpos*, fruit] fruit resulting from self-fertilization.

autocatalysis n. [Gk. *autos*, self; *kata*, down; *lysis*, loosing] dissolution or reaction of a cell or substance due to influence of a product or secretion of its own; catalysis of a reaction by one of its own products.

autochory n. [Gk. *autos*, self; *chōrein*, to spread] self-dispersal of diaspores by an explosive mechanism.

autochthon n. [Gk. *autochthōn*, aboriginal] an indigenous species.

authochthonous a. [G. *autos*, self; *chthōn*, ground] aboriginal; indigenous; inherited or hereditary, native, *appl.* characteristics; originating within an organ, as pulsation of excised heart; formed where found; *cf.* allochthonous.

autocoid autacoid, *q.v.*

autocyst n. [Gk. *autos*, self; *kystis*, bladder] a thick membrane formed by Neosporidia separating them from host tissues.

autodeliquescent a. [Gk. *autos*, self; L. *deliquescere*, to become liquid] becoming liquid as a result of self-digestion, as in pileus and gills of *Coprinus* fungi.

autodeme n. [Gk. *autos*, self; *dēmos*, people] members of a taxon which practise self-fertilization.

autodermalia n.plu. [Gk. *autos*, self; *derma*, skin] dermal spicules with axial cross, within dermal membrane.

autodont a. [Gk. *autos*, self; *odous*, tooth] designating or *pert.* teeth not directly attached to jaws, as in cartilaginous fishes.

autoecious a. [Gk. *autos*, self; *oikos*, house] passing different stages of life history in the same host; *alt.* autecic, autecious, autoxenous, ametoecious.

auto-ecology autecology, *q.v.*

autogamy n. [Gk. *autos*, self; *gamos*, marriage] self-fertilization, *cf.* allogamy; conjugation of nuclei within a single cell; conjugation of 2 protozoans originating from division of the same individual; *alt.* orthogamy; a. autogamous.

autogenesis n. [Gk. *autos*, self; *genesis*, origin] abiogenesis, *q.v.*; origin, production or reproduction within the same organism; *alt.* autogeny, autogony.

autogenetic a. [Gk. *autos*, self; *genesis*, birth] reproducing spontaneously, as body cells.

autogenic a. [Gk. *autos*, self; *gennaein*, to produce] caused by reactions of organisms themselves, *appl.* plant successions, *cf.* allogenic; autonomic or spontaneous, *appl.* movements, *cf.* ectogenic.

autogenous a. [Gk. *autos*, self; *-genēs*, producing] produced in the same organism, *appl.* enzymes; *appl.* graft reimplanted in same animal; *appl.* vaccine injected into same animal; *appl.* variations due to changes within chromosomes; *appl.* adult female insect that does not need to feed for her eggs to mature, *cf.* anautogenous.

autogeny autogenesis, *q.v.*

autogony n. [Gk. *autos*, self; *gonos*, offspring] autogenesis, *q.v.*

autograft n. [Gk. *autos*, self; O.F. *graffe*, graft] the grafting of a piece of tissue from one part of the body to another part of the same organism.

autoheteroploid n. [Gk. *autos*, self; *heteros*, other; *aploos*, onefold; *eidos*, form] heteroploid derived from a single genome or multiplication of some of its chromosomes; *cf.* alloheteroploid.

autoicous a. [Gk. *autos*, self; *oikos*, house] monoecious, *q.v.*

autoinfection n. [Gk. *autos*, self; L. *inficere*, to taint] reinfection from host's own parasites.

autointoxication n. [Gk. *autos*, self; L. *in*, in; Gk. *toxikon*, poison] reabsorption of toxic substances produced by the body.

autokinetic a. [Gk. *autos*, self; *kinein*, to move] moving by its own action; *cf.* allokinetic.

autologous a. [Gk. *autos*, self; *logos*, discourse] *appl.* graft from one part to another of the same organism.

autolysin n. [Gk. *autos*, self; *lysis*, loosing] any special autolytic enzyme.

autolysis n. [Gk. *autos*, self; *lysis*, loosing] self-digestion, i.e. cell or tissue disintegration by action of autogenous enzymes; *cf.* heterolysis.

autolytic a. [Gk. *autos*, self; *lysis*, loosing] causing or *pert.* autolysis, *appl.* enzymes; *cf.* heterolytic.

automimicry n. [Gk. *autos*, self; *mimikos*, imitating] intraspecific mimicry as when some members of a species are unpalatable, and palatable members of the same species mimic them.

automixis n. [Gk. *autos*, self; *mixis*, mingling] the union, in a cell, of chromatin derived from common parentage, i.e. self-fertilization.

automutagen n. [Gk. *autos*, self; L. *mutare*, to change; Gk. *gennaein*, to generate] a mutagen produced as a metabolite in an organism.

autonarcosis n. [Gk. *autos*, self; *narkē*, numbness] state of being poisoned, rendered dormant, or arrested in growth, owing to self-produced carbon dioxide.

autonomic a. [Gk. *autos*, self; *nomos*, law] autonomous; self-governing, spontaneous, *appl.* the involuntary nervous system as a whole, comprising parasympathetic and sympathetic systems; induced by internal stimuli, as movements of development, growth, unfolding, etc., *cf.* paratonic; internal, *appl.* environment, *cf.* choronomic.

autonomic nervous system (ANS) the system of motor (efferent) nerve fibres supplying smooth and cardiac muscles and glands, which is divided into the sympathetic and parasympathetic nervous system.

autopalatine n. [Gk. *autos*, self; L. *palatum*, palate] in a few teleosts, an ossification at anterior end of pterygoquadrate.

autoparasite n. [Gk. *autos*, self; *parasitos*, parasite] a parasite subsisting on another parasite.

autoparthenogenesis n. [Gk. *autos*, self; *parthenos*, virgin; *genesis*, descent] development from unfertilized eggs activated by a chemical or physical stimulus.

autophagous a. [Gk. *autos*, self; *phagein*, to eat] *appl.* birds capable of running about and securing food for themselves when newly hatched.

autophagy *n.* [Gk. *autos*, self; *phagein*, to eat] subsistence by self-absorption of products of metabolism, as consumption of their own glycogen by yeasts.

autophene *n.* [Gk. *autos*, self; *phainein*, to appear] a phenotype due to a mutation in the actual cells showing it, and which will not show the normal phenotype if transplanted to a normal host; *cf.* allophene.

autophilous *a.* [Gk. *autos*, self; *philein*, to love] self-pollinating.

autophya *n.plu.* [Gk. *autos*, self; *phyein*, to produce] elements in formation of shell or skeleton secreted by animal itself; *cf.* xenophya.

autophyllogeny *n.* [Gk. *autos*, self; *phyllon*, leaf; *genos*, birth] growth of one leaf upon or out of another.

autophyte *n.* [Gk. *autos*, self; *phyton*, plant] an autotrophic plant; *cf.* saprophyte, heterophyte; *a.* autophytic.

autoplasma *n.* [Gk. *autos*, self; *plasma*, mould] plasma from same animal used as medium for tissue culture; *cf.* homoplasma, heteroplasma.

autoplast chloroplast, *q.v.*

autoplastic *a.* [Gk. *autos*, self; *plastos*, formed] *appl.* graft to another position in the same individual.

autoploid autopolyploid, *q.v.*

autopodium *n.* [Gk. *autos*, self; *pous*, foot] the hand or foot.

autopolyploid *n.* [Gk. *autos*, self; *polys*, many; *aploos*, onefold; *eidos*, form] an organism having more than 2 sets of homologous chromosomes; a polyploid in which chromosome sets are all derived from a single species; *alt.* autoploid, duplicational polyploid; *cf.* allopolyploid.

autopotamic *a.* [Gk. *autos*, self; *potamos*, river] thriving in a stream, not in its backwaters, *appl.* potamoplankton.

autoradiography *n.* [Gk. *autos*, self; L. *radius*, ray; Gk. *graphein*, to write] method of demonstrating the presence of specific chemical substances by first making them radioactive, then recording on a photographic film their distribution in the body, organs, or tissues.

autoskeleton *n.* [Gk. *autos*, self; *skeletos*, dried] a true skeleton formed from autophya within the animal.

autosome *n.* [Gk. *autos*, self; *sōma*, body] a typical chromosome rather than a sex chromosome; *alt.* euchromosome; *cf.* allosome.

autospasy *n.* [Gk. *autos*, self; *spaein*, to pluck off] self-amputation as in autotilly and autotomy.

autospore *n.* [Gk. *autos*, self; *sporos*, seed] an aplanospore which resembles the parent cell; protoplast resulting from longitudinal division of a diatom, and forming new valves.

autostoses *n.plu* [Gk. *autos*, self; *osteon*, bone] bones formed by the ossification of cartilage; *alt.* cartilage bones; *cf.* allostoses.

autostyly *n.* [Gk. *autos*, self; *stylos*, pillar] the condition of having the mandibular arch self-supporting, articulating directly with skull; *a.* autostylic; *cf.* holostyly, hyostyly.

autosynapsis *n.* [Gk. *autos*, self; *synapsis*, union] autosyndesis, *q.v.*

autosyndesis *n.* [Gk. *autos*, self; *syndesis*, a binding together] pairing of chromosomes from the same parent, in an autopolyploid or allopolyploid; pairing of homogenetic chromosomes; *alt.* autosynapsis; *cf.* allosyndesis.

autotetraploid *n.* [Gk. *autos*, self; *tetraplē*, fourfold] a tetraploid whose nuclei contain 4 sets of chromosomes of the same origin, produced by autopolyploidy.

autotheca *n.* [Gk. *autos*, self; *thēkē*, case] a theca budded from a stolotheca, and surrounding the female polyp in graptolites.

autotilly *n.* [Gk. *autos*, self; *tillesthai*, to pluck] autotomy, as in certain spiders; *alt.* autospasy.

autotomy *n.* [Gk. *autos*, self; *tomē*, cutting] self-amputation of a part, as in certain worms, arthropods, and lizards; *alt.* autospasy.

autotransplantation transplantation of tissue or organ to another part of same organism; *cf.* homoiotransplantation.

autotrephones *n.plu.* [Gk. *autos*, self; *trephein*, to nourish] intracellular substances essential for maintenance of metabolism in the same cell.

autotriploid *n.* [Gk. *autos*, self; *triploos*, threefold] an individual with 3 homologous sets of chromosomes in the body cells.

autotroph *n.* [Gk. *autos*, self; *trophē*, nourishment] an organism using inorganic carbon as carbon dioxide as the principle carbon source, such as photoautotrophs and chemoautotrophs; often used synonymously with photoautotroph; *alt.* producer, primary producer, lithotroph; *cf.* heterotroph.

autotrophic *a.* [Gk. *autos*, self; *trephein*, to nourish] obtaining food as an autotroph; *alt.* holophytic, phytophytic, autophytic.

autotropism *n.* [Gk. *autos*, self; *tropē*, turn] tendency to grow in a straight line, *appl.* plants unaffected by external influence; tendency of organs to resume original form, after bending or straightening due to external factors; *alt.* rectipetality.

autoxenous *a.* [Gk. *autos*, self; *xenos*, host] autoecious, *q.v.*

autozoid, autozooid *n.* [Gk. *autos*, self; *zōon*, animal; *eidos*, form] a fully formed alcyonarian zooid or individual with tentacles, as distinct from a siphonozooid.

autozygote *n.* [Gk. *autos*, self; *zygōtos*, yoked] a homozygote in which the 2 homologous genes have a common origin; *cf.* allozygote.

auxenolonic acid auxin b, *q.v.*

auxentriolic acid auxin a, *q.v.*

auxesis *n.* [Gk. *auxēsis*, growth] growth; increase in size owing to increase in cell size; induction of cell division; *cf.* merisis.

auxetic *n.* [Gk. *auxein*, to increase] any agent which induces cell division. *a.* Stimulating cell proliferation; *appl.* growth of multicellular organisms due to enlargement of cells, rather than multiplicative growth by increase in number of cells.

auxilia *n.plu.* [L. *auxilium*, assistance] two small sclerites between unguitractor and claws in insects.

auxiliary cells two or more modified epidermal cells adjoining guard cells, or surrounding stomata; *alt.* accessory or subsidiary cells.

auximone *n.* [Gk. *auximos*, promoting growth] a substance thought necessary in small quantities for vigorous growth of plants.

auxins *n.plu.* [Gk. *auxein*, to increase] growth-promoting hormones that cause cell elongation and are responsible for many developmental responses including phototropism; natural auxin is indole-acetic acid and various synthetic auxins have been made including indole-butyric acid; formerly grouped as auxin a (auxentriolic acid) isolated from growing tips of oat seedlings and human urine, and auxin b (auxenolonic acid) from various plants and urine, which accelerated mycelium growth; formerly also called correlators.

auxoautotroph *n.* [Gk. *auxein*, to increase; *autos*, self; *trophē*, nourishment] an organism which can synthesize all the growth substances needed for its development.

auxocyte *n.* [Gk. *auxein*, to increase; *kytos*, hollow] androcyte, sporocyte, oocyte, or spermatocyte at growth period; any cell in which meiosis occurs; *alt.* gonotokont, meiocyte, cyte.

auxoheterotroph *n.* [Gk. *auxein*, to increase; *heteros*, other; *trophē*, nourishment] an organism that cannot itself synthesize all the growth substances needed for its development.

auxospireme *n.* [Gk. *auxein*, to increase; *speirēma*, coil] spireme formed after syndesis.

auxospore *n.* [Gk. *auxein*, to increase; *sporos*, seed] zygote of diatoms, formed by union of 2 individuals at limit of decrease in size resulting in a resting spore; *alt.* auxozygote.

auxotonic *a.* [Gk. *auxein*, to increase; *tonos*, strain] induced by growth, *appl.* movements of immature plants, *cf.* allassotonic; *appl.* contraction against an increasing resistance in muscle.

auxotroph *n.* [Gk. *auxein*, to increase; *trophē*, nourishment] a mutant micro-organism lacking the capacity of forming an enzyme present in the parental strain, and therefore requiring a supplementary substance for growth.

auxozygote auxospore, *q.v.*

aversion zone barrage, *q.v.*

aversive *appl.* stimuli which decrease the strength of a response if applied several times, and evoke fear and avoidance behaviour patterns.

Aves *n.* [L. *avis*, bird] a class of vertebrates, the birds, derived from archosaur reptiles, having the body clothed in feathers.

avicularium *n.* [L. *avicula*, *dim.* of *avis*, bird] in Polyzoa, a modified zooecium with muscular movable attachments resembling a bird's beak.

avidin *n.* [L. *avis*, bird] a protein in the white of eggs which combines with antiavidin (biotin) in the digestive tract and inactivates it, thereby producing biotin deficiency; *alt.* antibiotin.

avifauna *n.* [L. *avis*, bird; *Faunus*, rural deity] all the bird species or birds of a region or period; *alt.* ornis.

avitaminosis *n.* [L. *a*, from; *vita*, life; *ammoniacum*, resinous gum] a condition or disease resulting from vitamin deficiency.

avoidance reaction a movement away from a stimulus, a negative taxis or tropism.

awn *n.* [Icel. *ögen*, chaff] a stiff bristle-like projection from the tip or back of the lemma or glumes in grasses, or from a fruit (usually the hard style), or from the tip of a leaf; *alt.* arista, beard.

axenic *a.* [Gk. *axenos*, inhospitable] without, or deprived of, any commensals, symbionts, or parasites; not contaminated, *appl.* cultures; *alt.* sterile.

axerophthol *n.* [Gk. *a*, not; *xēros*, dry; *ophthalmos*, eye] *see* vitamin A.

axial *a.* [L. *axis*, axle] *pert.* axis or stem.

axial cell a single elongated cell constituting the axis of a rhombogen, or 1 of 2 or 3 cells arranged end to end in a nematogen, of Dicyemidae.

axial filament central filament, as of a stiff radiating pseudopodium or of a flagellum.

axial gradient gradation along an axis, e.g. of metabolic rate.

axial sinus a nearly vertical canal in echinoderms, opening into internal division of oral ring sinus, and communicating with stone canal.

axial skeleton skeleton of head and trunk; *cf.* appendicular skeleton.

axiate pattern arrangement of parts with reference to a definite axis.

axil *n.* [L. *axilla*, armpit] the angle between leaf or branch and axis from which it springs; *alt.* axilla.

axile *a.* [L. *axis*, axle] *pert.*, situated in, or belonging to the axis; forming an axis, *appl.* columella in sporangium or gleba.

axilemma *n.* [L. *axis*, axle; Gk. *lemma*, skin] in medullated nerve fibres, the sheath surrounding the axis cylinder.

axile placentation placentation in which ovules are situated in the middle of the ovary in the angles formed by the meeting of the septa; *cf.* free central placentation.

axilla *n.* [L. *axilla*, armpit] the armpit; axil, *q.v.*

axillary *a.* [L. *axilla*, armpit] *pert.* axil; growing in axil, as buds; *pert.* armpit; *appl.* 7th longitudinal or anal vein of insect wing; *appl.* sclerites: pteralia, *q.v. n.* Any of the pteralia, *q.v.*

axinost axonost, *q.v.*

axipetal *a.* [L. *axis*, axle; *petere*, to seek] passing towards attachment of axon, *appl.* nerve impulses; *alt.* axopetal.

axis *n.* [L. *axis*, axle] the main stem or central cylinder; the fundamentally central line of a structure; rachis of trilobites; structure at base of insect wing; the 2nd cervical vertebra.

axis cylinder the axon only of a medullated nerve fibre; *alt.* neuraxon, neurite.

axoblast *n.* [Gk. *axōn*, axle; *blastos*, bud] a germ cell or agamete of Dicyemidae.

axodendritic *a.* [Gk. *axōn*, axle; *dendron*, tree] *appl.* synapse in which end-brush of axon is in contact with dendritic processes.

axon(e) *n.* [Gk. *axōn*, axle] the fibre of a neurone which carries impulses away from the cell body, being a non-medullated nerve fibre, or the centre part (axis cylinder) of a medullated nerve fibre; *alt.* neuraxon, neuraxis, neurite.

axonal *pert.* an axon or axones.

axoneme *n.* [Gk. *axōn*, axle; *nēma*, thread] the central fibrils of a flagellum or cilium; genoneme of a chromosome.

axon hill or hillock cone of origin, *q.v.*

axonost *n.* [Gk. *axōn*, axle; *osteon*, bone] the basal portion of rods supporting dermotrichia of fin-rays; *alt.* axinost, interspinal.

axopetal axipetal, *q.v.*

axoplasm *n.* [Gk. *axōn*, axle; *plasma*, form] plasma surrounding the neurofibrils within the axis cylinder; *alt.* perifibrillar substance.

axoplast *n.* [Gk. *axōn*, axle; *plastos*, formed] a filament extending from kinetoplast to end of body in some trypanosomes.

axopodium *n.* [Gk. *axōn*, axle; *pous*, foot] a pseudopodium with a strengthening axial filament.

axosomatic *a.* [Gk. *axōn*, axle; *sōma*, body] *appl.*

synapse in which end-brush of axon terminates about nerve cell body.

axospermous *a.* [Gk. *axōn*, axle; *sperma*, seed] with axile placentation.

axostyle *n.* [Gk. *axon*, axle; *stylos*, pillar] a slender flexible rod of organic substance forming a supporting rod for the body or organ of locomotion of many Flagellata.

azoic *a.* [Gk. *a*, without; *zōikos*, *pert.* life] uninhabited; without remains of organisms or of their products, *appl.* Pre-Cambrian era or rocks; *cf.* zoic.

azonal *a.* [Gk. *a*, without; *zōnē*, girdle] not zoned; *appl.* soils without definite horizons.

azonic not restricted to a zone.

azurophil *a.* [F. *azur*, from Ar. *al azurd*, lapis lazuli; Gk. *philein*, to love] staining readily with blue aniline dyes.

azygobranchiate *a.* [Gk. *a*, without; *zygon*, yoke; *brangchia*, gills] having gills or ctenidia not developed on one side.

azygoid *a.* [Gk. *a*, without; *zygon*, yoke; *eidos*, form] not zygoid; haploid, *appl.* parthenogenesis.

azygomatous *a.* [Gk. *a*, without; *zygōma*, a bar] without a zygoma or cheek bone arch.

azygomelous *a.* [Gk. *a*, without; *zygon*, yoke; *melos*, limb] having unpaired appendages, *appl.* fin of Acrania and Cyclostomata; *cf.* zygomelous.

azygos *n.* [Gk. *a*, without; *zygon*, yoke] an unpaired structure such as a muscle, artery, vein, process.

azygosperm azygospore, *q.v.*

azygospore *n.* [Gk. *a*, without; *zygon*, yoke; *sporos*, seed] a spore developed directly from a gamete without conjugation, but resembling a zygospore; *alt.* aboospore, azygosperm, parthenospore, pseudozygospore.

azygote *n.* [Gk. *a*, without; *zygon*, yoke] an organism resulting from haploid parthenogenesis.

azygous *a.* [Gk. *a*, without; *zygon*, yoke] unpaired; *alt.* impar.

azymic *a.* [Gk. *a*, without; *zymē*, leaven] not fermented; devoid of enzymes; not resulting from fermentation.

B

Babes-Ernst bodies [*V. Babes,* Romanian bacteriologist; *H. C. Ernst,* American bacteriologist] metachromatic or volutin granules, in bacteria.

bacca *n.* [L. *bacca*, berry] berry, *q.v.*; berry formed from an inferior ovary, *cf.* nuculanium.

baccate *a.* [L. *bacca*, berry] pulpy, fleshy, as a berry; bearing berries, *alt.* bacciferous; resembling a berry, *alt.* bacciform.

bacciferous *a.* [L. *bacca*, berry; *ferre*, to bear] berry-producing, or -bearing; *alt.* baccate.

bacciform *a.* [L. *bacca*, berry; *forma*, shape] berry-shaped; *alt.* baccate.

Bacillariales *n.* [*Bacillaria*, generic name] the Bacillariophyceae, *q.v.*, when reduced to order status in the Chrysophyceae; *alt.* Diatomales.

Bacillariophyceae *n.* [*Bacillaria*, generic name: Gk. *phykos*, seaweed] a class of unicelluar algae, commonly called diatoms, which have a silicified wall of 2 halves, chlorophyll and carotinoid pigments, and store oil and leucosin instead of starch; in some classifications raised to division status as Bacillariophyta, in others placed in the Chrysophyta or sometimes reduced to order status (Bacillariales) in the Chrysophyceae.

Bacillariophyta *n.* [*Bacillaria*, generic name; Gk. *phyton*, plant] in some classifications, the Bacillariophyceae, *q.v.*, raised to division status.

bacillary *a.* [L. *bacillum*, small staff] rod-like; *appl.* layer of rods and cones of retina, *alt.* Jacob's membrane; *pert.* bacilli.

bacillus *n.* [L. *bacillum*, small staff] a rod-shaped bacterium, particularly one of the genus *Bacillus.*

backbone vertebral column, *q.v.*

back cross the mating of a heterozygous individual to one of its parents, usually the double recessive parent; *cf.* test cross.

back mutation reversion by mutation of a mutant gene to its original state; *alt.* reverse mutation, reversion.

Bacteria *n.* [Gk. *baktērion*, small rod] a group of unicellular microscopic prokaryotic plants, sometimes aggregated into filaments, which are found in soil, water, or parasitic or saprophytic on plants and animals, and have a cell wall and reproduce by fission, produce asexual spores and also have sexual processes; *alt.* Schizomycetes, fission fungi, bacteria.

bactericidal *a.* [Gk. *baktērion*, small rod; L. *caedere*, to kill] causing death of bacteria; *alt.* bacteriocidal.

bactericidin a substance that kills bacteria without causing lysis.

bacteriochlorin *n.* [Gk. *baktērion*, small rod; *chlōros*, green] bacteriochlorophyll, *q.v.*

bacteriochlorophyll *n.* [Gk. *baktērion*, small rod; *chlōros*, green; *phyllon*, leaf] a photosynthetic pigment in photosynthetic bacteria related to but not identical with chlorophyll of higher plants; *alt.* bacteriochlorin.

bacteriocidal bactericidal, *q.v.*

bacteriocins *n.plu.* [Gk. *baktērion*, small rod; L. *caedere*, to kill] proteins produced by certain bacteria which are toxic to sensitive strains of bacteria.

bacterioid bacteroid, *q.v.*

bacteriology *n.* [Gk. *baktērion*, small rod; *logos*, discourse] the science dealing with bacteria.

bacteriolysin *n.* [Gk. *baktērion*, small rod; *lysis*, loosing] a substance which causes dissolution of bacteria.

bacteriolysis *n.* [Gk. *baktērion*, small rod; *lysis*, loosing] the disintegration and dissolution of bacteria.

bacteriophage *n.* [Gk. *baktērion*, small rod; *phagein*, to devour] a virus which infects bacteria; *alt.* phage.

bacteriopurpurin *n.* [Gk. *baktērion*, small rod; L. *purpura*, purple] a complex of photosynthetic pigments causing the red, purple, or violet appearance of certain bacteria.

bacteriorhodopsin *n.* [Gk. *baktērion*, small rod; *rhodon*, rose; *opsis*, sight] a purple protein found in the purple membrane of extreme halobacteria (salt-loving bacteria), which acts as a light-driven trans-membrane protein pump.

bacteriorrhizae *n.plu.* [Gk. *baktērion*, small rod; *rhiza*, root] root nodules, *q.v.*

bacteriostatic *a.* [Gk. *baktērion*, small rod; *statikos*, causing to stand] inhibiting development of bacteria.

bacteriotropin n. [Gk. *baktērion*, small rod; *tropē*, turn] an ingredient of blood serum which renders bacteria more readily phagocytable such as opsonin.

bacteroid n. [Gk. *baktērion*, small rod; *eidos*, form] an irregularly shaped or branched form of certain bacteria such as those formed in root cells during nodule formation; *alt.* bacterioid.

bacteroidal *appl.* cells containing rod-shaped uric acid particles, in certain annelids.

baculiform a. [L. *baculum*, rod; *forma*, shape] rod-shaped, *appl.* chromosomes, *appl.* ascospores.

baculum n. [L. *baculum*, rod] the penis bone; *alt.* os priapi.

bailer scaphognathite, q.v.

Baillarger's line [*J. F. G. Baillarger*, French neurologist] outer and inner layer of white fibres parallel to surface of cerebral cortex.

balanced lethals heterozygotes in which different lethal genes are in such close proximity on a pair of homologous chromosomes that there is usually no crossing-over.

balanced polymorphism the coexistence of 2 or more distinct types of individuals in the same breeding population, both maintained actively all the time by selection; *cf.* transient polymorphism.

balancers n.plu. [L. *bilanx*, having two scales] halteres, q.v.; paired larval head appendages functioning as props until forelegs are developed in certain salamanders.

balanic a. [Gk. *balanos*, acorn] *pert.* glans penis; *pert.* glans clitoridis.

balanoid a. [Gk. *balanos*, acorn; *eidos*, like] acornshaped; *pert.* barnacles.

Balanopsidales n. [*Balanopsis*, generic name] an order of woody dicots (Archichlamydeae) which have male flowers in spikes and solitary female flowers with an ovary of 2 carpels which develops into a drupe.

balanus n. [L. *balanus*, acorn] glans penis, q.v.; a genus of barnacles.

balausta n. [Gk. *balaustion*, blossom] a manycelled, many-seeded, indehiscent fruit with tough pericarp such as fruit of pomegranate.

Balbiani's body or nucleus yolk nucleus, q.v.

Balbiani ring puff, q.v., especially in the gnat *Chironomus* where they were first discovered by E. G. Balbiani in 1881.

baleen n. [L. *balaena*, whale] whalebone, horny plates attached to upper jaw of true whales.

baler scaphognathite, q.v.

ball-and-socket *appl.* joint in which the hemispherical end of one bone fits into the socket of another allowing movement in several planes, as in shoulder and hip joints, *alt.* enarthrosis; *appl.* ocelli on secondary wing feathers of male Argus pheasant, apparent convexity being due to shading.

ballast n. [Sw. *barlast*] *appl.* elements present in plants and which are not apparently essential for growth, such as aluminium or silicon.

ball centrosome the centrosome found in spermatozoa, from which the fibres of the tail arise.

ballistic a. [Gk. *ballein*, to throw] *appl.* fruits with explosive dehiscence and discharge of seeds.

ballistospores n.plu. [Gk. *ballein*, to throw; *sporos*, seed] asexual spores, formed on sterigmata and suddenly discharged with excretion of droplet, as in fungi of the Sporobolomycetes; *alt.* ballospores.

balsam n. [L. *balsamum*, balsam] any of various substances of high molecular weight found in mixtures in plants, consisting of resin acids, esters, and terpenes, which are often exuded from wounds and may protect plants from insects and fungi; *cf.* resin.

balsamic a. [Gk. *balsamon*, resinous oil of *Balsamodendron*] *appl.* a class of odours typified by those of vanilla, coumarin, heliotrope, resins, etc., and due to a benzene nucleus with various lateral chains.

balsamiferous a. [L. *balsamum*, balsam; *ferre*, to bear] producing balsam.

Bangiales n. [*Bangia*, generic name] an order of red algae of the subclass Bangiophycidae.

Bangioideae n. [*Bangia*, generic name; Gk. *eidos*, form] Bangiophycidae, q.v.

Bangiophycidae n. [*Bangia*, generic name; Gk. *phykos*, seaweed] a subclass of red algae having each cell with a single stellate chloroplast and never having a parenchymatous structure; *alt.* Bangioideae.

banner vexillum, q.v., of papilionaceous flower; muscle banner, q.v., of Anthozoa.

baraesthesia n. [Gk. *baros*, weight; *aisthēsis*, sensation] the sensation of pressure; *alt.* baresthesia, baryaesthesia.

barb n. [L. *barba*, beard] one of delicate threadlike structures extending obliquely from a feather rachis, and forming the vane; a hooked hair-like bristle.

barbate a. [L. *barbatus*, bearded] bearded; having hair tufts.

barbel n. [L.L. *barbellus*, barbel] a tactile process arising from the head of various fishes; *alt.* barbule, wattle.

barbellate a. [L. *barba*, beard] with stiff hooked hair-like bristles, *appl.* pappus.

barbicel n. [L. *barba*, beard] small process on a feather barbule.

barbula n. [L. *barbula*, dim. of *barba*, beard] row of teeth in peristome of certain mosses; *alt.* barbule.

barbule n. [L. *barbula*, dim. of *barba*, beard] one of small hooked processes fringing barbs of feather; appendage of lower jaw in some teleosts; barbel, q.v.; barbula, q.v.

bare name nomen nudum, q.v.

baresthesia baraesthesia, q.v.

bark n. [Dan. *bark*, bark] the tissues external to the vascular cambium collectively, being the secondary phloem, cortex, and periderm; the periderm alone, also known as outer bark, *alt.* rhytidome, epiphloem; inner bark: secondary phloem, q.v.

bark lice a common name for some members of the Psocoptera, q.v.

barnacles the common name for the Cirripedia, q.v.

baroceptor n. [Gk. *baros*, pressure; L. *capere*, to take] a receptor sensitive to changes in pressure, such as those in the wall of blood vessels which react to changes in blood pressure; *alt.* baroreceptor.

bar of Sanio crassula, q.v.

barognosis n. [Gk. *baros*, weight; *gnōsis*, recognition] the ability to detect changes of pressure.

baroreceptor baroceptor, q.v.

barotaxis n. [Gk. *baros*, weight; *taxis*, arrangement] a taxis in response to a pressure stimulus.

barrage n. [F. *barrage*, dam] zone of inhibition

between certain bacterial or fungal colonies, not between others; *alt.* aversion zone.

Barr body in female mammals, the condensed X-chromosome of sex chromatin seen in somatic cell nuclei.

Bartholin's duct [*C. Bartholin, jr.,* Danish anatomist] the larger duct of the sublingual gland.

Bartholin's glands the greater vestibular glands on each side of vagina, homologues of male bulbo-urethral glands.

baryaesthesia baraesthesia, *q.v.*

basad *adv.* [L. *basis*, base; *ad* to] towards the base.

basal *a.* [L. *basis*, base] *pert.*, at, or near the base.

basalar *a.* [L. *basis*, base; *ala*, wing] *appl.* sclerites below wing base in insects.

basal body basal granule, *q.v.*

basal bone os basale, basale, *q.v.*

basal cell uninucleate cell which supports the dome and tip cells of a hyphal crozier, *alt.* stalk cell; a myoepithelial cell, as in coelenterates.

basal disc in corals, the area of ectoderm which secretes the calcareous skeleton; in the hydrozoan, hydra, the lower end of the body by which it attaches to the substratum; in spirochaetes, basal granule, *q.v.*

basale *n.* [L. *basis*, base] a bone of variable structure arising from fusion of pterygiophores and supporting fish fins; os basale, the fused basioccipital and parasphenoid in Gymnophiona; *alt.* basal bone; *plu.* basalia.

basal ganglia ganglia connecting cerebrum with other centres.

basal granule the granule usually found at the base of cilia and flagella; *alt.* basal body, blepharoplast.

basal knobs swellings or granules at points of emergence of cilia in ciliated epithelial cells.

basal leaf one of the leaves produced near base of stem, a radical leaf.

basal metabolic rate rate of metabolism of a resting organism, expressed as percentage of normal heat production per hour per square metre surface area.

basal metabolism standard metabolism, tissue activity or physicochemical changes of a resting organism.

basal placentation placentation where ovules are situated at the base of the ovary.

basal plates certain plates in echinoderms, situated at or near top of stalk in crinoids, in echinoids forming part of apical disc; fused parachordal plates in skull development; of placentae, outer wall of intervillous space.

basal ridge cingulum of a tooth.

basal wall the 1st plane of division of zygotes of pteridophytes and bryophytes.

basapophysis *n.* [Gk. *basis*, base; *apo*, away; *phyein*, to grow] a transverse process arising from ventrolateral side of a vertebra.

base exchange capacity the extent to which cations can be exchanged in the soil; *alt.* cation exchange capacity.

basement membrane a membrane of modified connective tissue beneath epithelial tissue, as of a gland containing acini or special secreting portions; *alt.* membrana propria, basilemma.

baseost *n.* [Gk. *basis*, base; *osteon*, bone] distal element of pterygiophore of teleosts.

base pair two bases, a purine and a pyrimidine, which pair together in the genetic code of DNA and RNA, cytosine pairing with guanine in both, and adenine pairing with thymine in DNA and with uracil in RNA.

base ratio the ratio between the amount of adenine and thymine bases (A + T) and the amount of guanine and cytosine bases (C + G) in a DNA sample.

base-rich *appl.* soils containing a relatively large amount of free basic ions such as calcium or magnesium.

basialveolar *a.* [L. *basis*, base; *alveolus*, small pit] extending from basion to centre of aveolar arch.

basibranchial *n.* [Gk. *basis*, base; *brangchia*, gills] median ventral or basal skeletal portion of branchial arch.

basibranchiostegal *n.* [Gk. *basis*, base; *brangchia*, gills; *stege*, roof] urohyal, *q.v.*

basic *a.* [Gk. *basis*, base] *appl.* stains which act in general on nuclear contents of cell, *cf.* acidic; *appl.* number: the minimum haploid chromosome number occurring in a series of euploid species of a genus; chromosome number in gametes of diploid ancestor of a polyploid organism.

basichromatin *n.* [Gk. *basis*, base; *chroma*, colour] chromatin which stains deeply with basic dyes.

basiconic *a.* [Gk. *basis*, base; *konos*, cone] having, or consisting of, a conical process above general surface, *appl.* sensillae.

basicoxite *n.* [L. *basis*, base; *coxa*, hip] basal ring of coxa.

basicranial *a.* [Gk. *basis*, base; *kranion*, skull] situated at or relating to base of skull.

basicyte *n.* [Gk. *basis*, base; *kytos*, hollow] mast cell, *q.v.*

basidia *plu.* of basidium.

basidial *a.* [Gk. *basis*, base; *idion*, dim.] *pert.* basidia or a basidium.

basidiocarp *n.* [Gk. *basis*, base; *idion*, dim.; *karpos*, fruit] the fruit body of Basidiomycetes.

Basidiolichenes, basidiolichens *n., n.plu.* [Gk. *basis*, base; *idion*, dim.; *leichen*, lichen] a group of lichens in which the fungus is a basidiomycete; *alt.* Hymenolichenes.

basidiolum *n.* [L.L. *dim.* of Gk. *basidion*, small pedestal] an undeveloped basidium, *alt.* pseudo-paraphysis; formerly: paraphysis.

Basidiomycetes, Basidiomycetae, Basidiomycotina [Gk. *basis*, base; *idion*, dim.; *mykes*, fungus] a major class of fungi which are characterized by the production of basidiospores on the outside of basidia.

basidiophore *n.* [Gk. *basis*, base; *idion*, dim.; *pherein*, to bear] a sporophore which carries basidia.

basidiospore *n.* [Gk. *basis*, base; *idion*, dim.; *sporos*, seed] a spore formed by meiosis on a basidium, in Basidiomycetes.

basidium *n.* [Gk. *basis*, base; *idion*, dim.] in Basidiomycetes, a cell of the hymenium, on the surface of which basidiospores, usually 4 in number, are formed by meiosis; *plu.* basidia.

basidorsal *a.* [L. *basis*, base; *dorsum*, back] *appl.* small cartilaginous neural plate.

basifemur *n.* [L. *basis*, base; *femur*, thigh] proximal segment of femur, between trochanter and telofemur, in certain Acarina.

basifixed *a.* [L. *basis*, base; *figere*, to make fast] attached by base; having filament attached to anther base, *alt.* innate.

basifugal *a.* [L. *basis*, base; *fugere*, to flee] growing away from base; acropetal, *q.v.*

basifuge *n.* [L. *basis*, base; *fugere*, to flee] a plant unable to tolerate basic soils, *alt.* calcifuge. *a.* Oxyphilous, *q.v.*

basigamic, basigamous *a.* [Gk. *basis*, base; *gamos*, marriage] having ovum and synergids at the far end of the embryo sac, away from the micropyle.

basigynium *n.* [Gk. *basis*, base; *gynē*, woman] podogynium, *q.v.*

basihyal *n.* [Gk. *basis*, base; *hyoeidēs*, Υ-shaped] broad median plate, the basal or median ventral portion of hyoid arch.

basikaryotype *n.* [Gk. *basis*, base; *karyon*, nucleus; *typos*, pattern] the haploid karyotype.

basilabium *n.* [L. *basis*, base; *labium*, lip] sclerite formed by fusion of labiostipites in insects.

basilar *a.* [L. *basis*, base] *pert.*, near, or growing from base: as artery, crest, membrane, plexus, plate, process, style, etc.

basilemma *n.* [Gk. *basis*, base; *lemma*, skin] basement membrane, *q.v.*

basilic *a.* [Gk. *basilikos*, royal] *appl.* a large vein on inner side of biceps of arm.

basilingual *a.* [L. *basis*, base; *lingua*, tongue] *appl.* a broad cartilaginous plate, the body of the hyoid, in crocodiles, turtles, and amphibians.

basimandibula *n.* [L. *basis*, base; *mandibulum*, lower jaw] a small sclerite, on insect head, at base of mandible.

basimaxilla *n.* [L. *basis*, base; *maxilla*, upper jaw] a sclerite at base of maxilla in some insects.

basinym *n.* [Gk. *basis*, base; *onyma*, name] the name upon which new names of species, etc., have been based; *cf.* isonym.

basioccipital *n.* [L. *basis*, base; *occiput*, back of head] the median basilar bone or element in occipital region of skull.

basion *n.* [Gk. *basis*, base] the middle of anterior margin of foramen magnum.

basiophthalmite *n.* [Gk. *basis*, base; *ophthalmos*, eye] the proximal joint of eye-stalk in crustaceans.

basiotic *a.* [Gk. *basis*, base; *ous*, ear] mesotic, *q.v.*

basipetal *a.* [L. *basis*, base; *petere*, to seek] descending; developing from apex to base, *appl.* leaves, roots, or flowers; *cf.* acropetal.

basipharynx *n.* [Gk. *basis*, base; *pharyngx*, gullet] in insects, epipharynx and hypopharynx united.

basiphil *a.* [Gk. *basis*, base; *philein*, to love] basophil, *q.v. n.* A basiphil cell; a mast cell, *q.v.*

basipodite *n.* [Gk. *basis*, base; *pous*, foot] the 2nd or distal joint of the protopodite of certain limbs of Crustacea, *alt.* basium; trochanter of spiders; *alt.* basopodite.

basipodium *n.* [Gk. *basis*, base; *pous*, foot] wrist or ankle.

basiproboscis *n.* [Gk. *basis*, base; *proboskis*, trunk] membranous portion of proboscis of some insects, consisting of mentum, submentum, and maxillary cardines and stipites.

basipterygium *n.* [Gk. *basis*, base; *pterygion*, little wing or fin] a large flat triangular bone in pelvic fin of teleosts, and a bone or cartilage in other fishes.

basipterygoid *n.* [Gk. *basis*, base; *pteryx*, wing; *eidos*, form] a process of the basisphenoid in some birds.

basirostral *a.* [L. *basis*, base; *rostrum*, bill] situated at, or *pert.*, the base of a beak or rostrum.

basiscopic *a.* [Gk. *basis*, base; *skopein*, to view] facing towards the base; *cf.* acroscopic.

basisphenoid *n.* [Gk. *basis*, base; *sphēn*, wedge; *eidos*, form] cranial bone between basioccipital and presphenoid.

basisternum *n.* [L. *basis*, base; *sternum*, breast bone] antesternite, *q.v.*

basistyle *n.* [Gk. *basis*, base; *stylos*, pillar] proximal part of coxite of gonostyle in mosquitoes; *cf.* dististyle.

basitarsus *n.* [Gk. *basis*, base; *tarsos*, sole of foot] proximal tarsomere or 'metatarsus' of spiders; *cf.* telotarsus.

basitemporal *n.* [L. *basis*, base; *tempora*, temples] a broad membrane bone covering basisphenoidal region of skull.

basitonic *a.* [Gk. *basis*, base; *tonos*, brace] having anther united at its base with rostellum; *alt.* basitonous; *cf.* acrotonic.

basium *see* basipodite.

basivertebral *a.* [L. *basis*, base; *vertebra*, vertebra] *appl.* veins within bodies of vertebrae and communicating with vertebral plexuses.

basket cells myoepithelial cells surrounding glandular cells; cerebellar cortical cells with axon branches surrounding Purkinje cells; *cf.* tendril fibres.

basocyte *n.* [Gk. *basis*, base; *kytos*, hollow] a basophil cell; a basophil leucocyte.

Basommatophora *n.* [Gk. *basis*, base; *omma*, eye; *pherein*, to bear] an order of mainly aquatic pulmonates with eyes at the base of the tentacles.

basophil *a.* [Gk. *basis*, base; *philein*, to love] having a strong affinity for basic stains, *alt.* basiphil, basiphilic, basophile, basophilic, basophilous. *n.* A cell which stains with basic dyes such as a white blood cell or a secretory cell of the anterior lobe of the pituitary gland, *alt.* basocyte.

basoplasm *n.* [Gk. *basis*, base; *plasma*, anything moulded] cytoplasm which stains readily with basic dyes.

basopodite basipodite, *q.v.*

bass *see* bast.

bassorin *n.* [*Bassora* (Basra) gum] a carbohydrate food store in orchids, used to make saloop.

bast *n.* (A.S. *baest*, bast] the inner fibrous bark of certain trees, *alt.* bass, liber; any phloem fibres; *alt.* leptome, leptomestome. *a. Appl.* any non-xylem fibres.

bastard merogony activation of an enucleated egg fragment by spermatozoon of a different species.

bastard wing a group of 3 quill feathers borne on the 1st digit of bird's wing; *alt.* alula, ala spuria.

Batesian mimicry [*H. W. Bates*, English naturalist] the resemblance of one animal (the mimic) to another (the model) to the benefit of the mimic, such as when the model is a dangerous or unpalatable species; *cf.* Müllerian mimicry.

bathmotropic *a.* [Gk. *bathmos*, degree; *tropikos*, turning] affecting the excitability of tissue, as of muscular tissue; *n.* bathmotropism.

bathyaesthesia *n.* [Gk. *bathys*, deep; *aisthēsis*, perception] sensation of stimuli within the body; deep sensibility.

bathyal *a.* [Gk. *bathys*, deep] *appl.* or *pert.* zone of continental slope.

bathylimnetic *a.* [Gk. *bathys*, deep; *limnētēs*, living in marshes] living or growing in the depths of lakes or marshes.

bathymetric *a.* [Gk. *bathys*, deep; *metron*, measure] *pert.* vertical distribution of organisms in space.

Bathynellacea *n.* [*Bathynella*, generic name] an order of Syncarida which do not have the 1st thoracic segment fused with the head.

bathypelagic *a.* [Gk. *bathys*, deep; *pelagos*, sea] *pert.*, or inhabiting, the deep sea; *cf.* abyssopelagic.

bathysmal *a.* [Gk. *bathys*, deep] *pert.* deepest depths of the sea.

Batidales *n.* [*Batis*, generic name] an order of dicots (Archichlamydeae) which are coastal shrubs with opposite fleshy leaves and panicles of spikes with unisexual flowers.

Batoidea *n.* [*Batis*, generic name; Gk. *eidos*, form] an order of Jurassic to modern elasmobranch fish including the skates and rays, having pectoral fins enlarged as organs of locomotion and the tail and other fins reduced.

batonnet(te) *n.* [F. *bâtonnet*, small stick] an element of the Golgi apparatus, *q.v.*; a rod-like assembly or fusion of fragments of chromatoid bodies.

Batrachia, batrachians *n.*, *n.plu.* [Gk. *batrachos*, frog] Anura, *q.v.*; *a.* batrachian.

Batrachoidiformes *n.* [Gk. *batrachos*, frog; *eidos*, form; L. *forma*, shape] an order of teleosts including the toadfishes; *alt.* Haplodoci.

Batrachosauria, batrachosaurs *n.*, *n.plu.* [Gk. *batrachos*, frog; *sauros*, lizard] Anthracosauria, *q.v.*

bats *see* Chiroptera.

B-cells beta cells of islets of Langerhans.

B-chromosome supernumerary chromosome, *q.v.*; *cf.* A-chromosome.

B-complex vitamin B complex, *q.v.*

bdelloid *a.* [Gk. *bdella*, leech; *eidos*, form] having the appearance of a leech.

Bdelloidea *n.* [Gk. *bdella*, leech; *eidos*, form] an order of freshwater rotifers whose corona has 2 ciliary discs.

Bdellomorpha, Bdellonemertini, bdellonemertines *n.*, *n.plu.* [Gk. *bdella*, leech; *morphē*, form; *Nēmertēs*, a nereid] an order of parasitic Enopla having an unarmed proboscis.

beak *n.* (O.F. *bec*, beak) the mandibles, jaw or bill of birds; a beak-like structure or projection, as avicularium, rostellum, rostrum.

beak-cushion a fold of skin at proximal angle of beak in nestlings.

beard *n.* [A.S. *beard*, beard] any of the arrangements of hairs which resemble a man's beard on heads of animals; barbed or bristly hair-like outgrowths on grain; awn, *q.v.*

bedeguar *n.* [From Persian through F. *bédeguar*] a mossy gall produced on rosebushes by gall wasps.

beetles the common name for the Coleoptera, *q.v.*

beet sugar sucrose, *q.v.*, especially that derived from sugar beet, but identical chemically with that from sugar cane.

Beggiatoales *n.* [*Beggiatoa*, generic name] an order of bacteria having rigid cell walls, with large cells often arranged in trichomes.

Begoniales *n.* [*Begonia*, generic name] an order of dicots (Dilleniidae), the begonias.

behaviorism *n.* [A.S. *behabban*, to hold in] theory that the manner in which animals act may be explained in terms of conditioned neuromotor and glandular reactions.

belemnoid *a.* [Gk. *belemnon*, javelin; *eidos*, form] shaped like a dart, *appl.* styloid process.

Belemnoidea *n.* [Gk. *belemnon*, javelin; *eidos*, form] an extinct group of Mesozoic dibranchiate cephalopods, which had a cigar-shaped internal shell; *alt.* belemnites.

bell *n.* [A.S. *belle*, bell] a bell-shaped structure; nectocalyx, *q.v.*; umbrella, *q.v.*; campanulate corolla.

Bellini's ducts [L. *Bellini*, Italian anatomist] tubes opening at apex of kidney papilla, and formed by union of smaller straight or collecting tubules.

bell-nucleus a solid mass of cells, derived from ectoderm and lying between ordinary ectoderm and mesogloea at apex of medusoid bud.

Beloniformes *n.* [Gk. *belonē*, needle; L. *forma*, shape] an order of mainly marine teleosts including the garfish and flying fish; *alt.* Synentognathi.

belonoid *a.* [Gk. *belonē*, needle; *eidos*, form] shaped like a needle; *alt.* aciform.

Beltian or Belt's bodies glandular structures on the swollen thorn acacia which provide food for ants that live in the tree and drive off herbivores trying to feed on it; *alt.* food bodies.

Bennettitales *n.* [*Bennettites*, generic name] an order of fossil Cycadopsida which have typically massive stems, pinnate leaves, and are usually monoecious with sporophylls arranged in a flower-like structure, and have been considered as ancestors of the angiosperms; *alt.* Cycadeoideales.

Bennettitatae *n.* [*Bennettites*, generic name] in some classifications, a group of Cycadophytina, equivalent to the Bennettitales, or including both the Bennettitales and Pentoxylales.

Benson–Calvin–Bassham cycle Calvin cycle, *q.v.*

benthal, benthic *a.* [Gk. *benthos*, depths of sea] *pert.*, or living on, sea bottom.

benthon *n.* [Gk. *benthos*, depths of sea] the flora and fauna of the sea or lake bottom; *cf.* benthos.

benthophyte *n.* [Gk. *benthos*, depths; *phyton*, plant] a bottom-living plant.

benthopotamous *a.* [Gk. *benthos*, depths; *potamos*, river] *pert.*, growing, or living, on bed of a river or stream.

benthos *n.* [Gk. *benthos*, depths of sea] the sea bottom; benthon, *q.v.*

Benzedrine the trade name for amphetamine, *q.v.*

Bergmann's rule the idea that a warm-blooded animal species has a smaller body size in the warmer parts of its range than in the colder.

Berlese's organ [A. *Berlese*, Italian zoologist] a glandular organ in haemocoel on right side of female abdomen in *Cimex*, a hemipteran, secreting during passage of spermatozoa to spermatheca.

Beroida *n.* [*Beroe*, generic name; Gk. *eidos*, form] an order of conical Ctenophora of the class Nuda.

berry *n.* [A.S. *berie*, berry] a several-seeded indehiscent fruit with a fleshy pericarp and without a stony layer surrounding the seeds, *see also* bacca, nuculanium; egg of some decapod crustaceans such as lobster or crayfish; dark knob-like structure on bill of swan.

Bertin's columns [*E. J. Bertin*, French anatomist] renal columns, *q.v.*

Beryciformes *n.* [*Beryx*, generic name; L. *forma*, shape] an order of teleosts similar to the Perciformes but having an orbitosphenoid; *alt.* Berycomorphi.

Berycomorphi *n.* [*Beryx*, generic name; Gk. *morphē*, form] Beryciformes, *q.v.*

beta (β) cells basophil cells in pars glandularis of pituitary gland; cells elaborating insulin, in islets of Langerhans, *alt.* B-cells.

betacyanins, betaxanthins, betalains pigments found in flowers of the order Caryophyllales.

beta-galactosidase an enzyme which catalyses the hydrolysis of lactose to glucose and galactose; EC 3.2.1.21; *r.n.* ß-D-glucosidase.

beta (β) globulin a globulin in blood plasma.

beta (β) granules cyanophycin, *q.v.*

betaine *n.* [L. *beta*, beet] a derivative of glycine, used as a donor of methyl groups during choline synthesis, found in beet juice and some other plants and in animals; *alt.* oxyneurine.

beta (β) oxidation fatty acid oxidation, *q.v.*

beta (β) pleated sheet a regular folding of polypeptide chains found in some proteins; *cf.* alpha helix.

beta (β) rhythm spontaneous rhythmic fluctuations of electric potential of cerebral cortex during mental activity.

between-brain diencephalon, *q.v.*

Betz cells [*V. A. Bets*, Russian histologist] giant pyramidal cells in motor area of cerebral cortex.

B-horizon illuvial layer, *q.v.*

biacuminate *a.* [L. *bis*, twice; *acumen*, point] having 2 tapering points.

biarticulate *a.* [L. *bis*, twice; *articulus*, joint] two-jointed; *alt.* diarthric.

biaxial *a.* [L. *bis*, twice; *axis*, axle] with 2 axes; allowing movement in 2 planes, as condyloid and ellipsoid joints.

bicapsular *a.* [L. *bis*, twice; *capsula*, little box] having 2 capsules; having a biloculate capsule.

bicarinate *a.* [L. *bis*, twice; *carina*, keel] with 2 keel-like processes.

bicarpellate *a.* [L. *bis*, twice; Gk. *karpos*, fruit] with 2 carpels; *alt.* bicarpellary.

bicaudate *a.* [L. *bis*, twice; *cauda*, tail] possessing 2 tail-like processes; *alt.* bicaudal.

bicellular *a.* [L. *bis*, twice; *cellula*, little cell] composed of 2 cells.

bicentric *a.* [L. *bis*, twice; *centrum*, centre] *pert.* 2 centres; *appl.* distribution of species, etc., discontinuous owing to alteration in the intervening area.

biceps *n.* [L. *bis*, twice; *caput*, head] a muscle with 2 heads or origins, as biceps brachii and femoris.

biciliate *a.* [L. *bis*, twice; *cilium*, eyelash] furnished with 2 cilia.

bicipital *a.* [L. *bis*, twice; *caput*, head] *pert.* biceps; *appl.* fascia, an aponeurosis of distil tendon of the biceps brachii, *alt.* lacertus fibrosus; a groove, the intertubercular sulcus, on upper part of humerus; ridges, the crests of the greater and lesser tubercles of the humerus; *appl.* a rib with dorsal tuberculum and ventral capitulum; divided into 2 parts at one end.

bicollateral *a.* [L. *bis*, twice; *con*, together; *latus*, side] having the 2 sides similar; *appl.* a vascular bundle with phloem on both sides of xylem, as in the dicot families Cucurbitaceae and Solanaceae.

bicolligate *a.* [L. *bis*, twice; *cum*, together; *ligare*, to bind] with 2 stretches of webbing on the foot.

biconjugate *a.* [L. *bis*, twice; *cum*, with; *jugum*, yoke] with 2 similar sets of pairs.

Bicornes *n.* [L. *bis*, twice; *corneus*, horny] Ericales, *q.v.*

bicornute *a.* [L. *bis*, twice; *cornutus*, horned] with 2 horn-like processes.

bicuspid *a.* [L. *bis*, twice; *cupis*, point] having 2 longitudinal ridges or ribs, as a leaf.

bicrenate *a.* [L. *bis*, twice; *crena*, notch] doubly crenate, as crenate leaves with notched toothed margins; having 2 rounded teeth.

biscuspid *a.* [L. *bis*, twice; *cuspis*, point] having 2 cusps or points, *alt.* bicuspidate; *appl.* valve consisting of anterior and posterior cusps attached to circumference of left atrioventricular orifice, *alt.* mitral valve; *appl.* tooth: premolar, *q.v.*, *alt.* bitubercular.

bicyclic *a.* [L. *bis*, twice; Gk. *kyklos*, circle] arranged in 2 whorls.

Bidder's ganglia [*F. H. Bidder*, Estonian anatomist] a collection of nerve cells in region of the auriculoventricular groove in some anurans.

Bidder's organ a rudimentary ovary attached to anterior end of generative organs in some toads.

bidental *a.* [L. *bis*, twice; *dens*, tooth] having 2 teeth, or tooth-like processes; *alt.* bidentate.

bidenticulate *a.* [L. *bis*, twice; *dim.* of *dens*, tooth] with 2 small teeth or tooth-like processes, as some scales.

bidiscoidal *a.* [L. *bis*, twice; Gk. *diskos*, round plate; *eidos*, form] consisting of 2 disc-shaped parts, *appl.* a placental type.

biennial *a.* [L. *bis*, twice; *annus*, year] *appl.* plant living for 2 years and fruiting only in the second; *n.* a biennial plant.

bifacial *a.* [L. *bis*, twice; *facies*, face] dorsiventral, *q.v.*

bifarious *a.* [L. *bis*, twice; *fariam*, in rows] arranged in 2 rows, one on each side of axis.

bifid *a.* [L. *bis*, twice; *findere*, to split] forked; opening with a median cleft; divided nearly to middle line.

biflabellate *a.* [L. *bis*, twice; *flabellum*, fan] doubly flabellate, each side of antennal joints sending out flabellate processes.

biflagellate *a.* [L. *bis*, twice; *flagellum*, whip] having 2 flagella; *alt.* dicont, dikont, dimastigote.

biflex *a.* [L. *bis*, twice; *flectere*, to bend] twice curved.

biflorate *a.* [L. *bis*, twice; *flos*, flower] bearing 2 flowers; *alt.* biflorous.

bifoliar *a.* [L. *bis*, twice; *folium*, leaf] having 2 leaves.

bifoliate *a.* [L. *bis*, twice; *folium*, leaf] *appl.* palmate compound leaf with 2 leaflets; diphyllous, *q.v.*; *appl.* polyzoan colony with 2 opposed layers of zooecia.

biforate *a.* [L. *biforis*, having double doors] having 2 foramina or pores; *alt.* biforous.

biforin *n.* [L. *bis*, twice; *foris*, door] an oblong raphidian cell opening at each end.

biforous

biforous *a.* [L. *biforis*, with two openings] biforate, *q.v.*, *appl.* spiracles in larvae of certain beetles.

bifurcate *a.* [L. *bis*, twice; *furca*, fork] forked; having 2 prongs; having 2 joints, the distal V-shaped and attached by its middle to the proximal.

bigeminal *a.* [L. *bis*, twice; *geminus*, double] with structures arranged in double pairs; *appl.* arrangement of pore-pairs in 2 rows, in ambulacra of some echinoids; *pert.* corpora bigemina.

bigeminate *a.* [L. *bis*, twice; *germinus*, double] doubly-paired; twin-forked.

bigeminum one of the corpora bigemina.

bigener *n.* [L. *bis*, twice; *genus*, race] a bigeneric hybrid.

bigeneric *a.* [L. *bis*, twice; *genus*, race] *appl.* hybrids between 2 distinct genera.

bijugate *a.* [L. *bis*, twice; *jugare*, to join] with 2 pairs of leaflets.

bilabiate *a.* [L. *bis*, twice; *labium*, lip] two-lipped, *appl.* calyx, corolla; *appl.* dehiscence by a transverse split at the top.

bilamellar *a.* [L. *bis*, twice; *lamella*, plate] formed of 2 plates; having 2 lamellae.

bilaminar *a.* [L. *bis*, twice; *lamina*, thin plate] having 2 plate-like layers; diploblastic, *q.v.*; *alt.* bilaminate.

bilateral *a.* [L. *bis*, twice; *latus*, side] *pert.* or having 2 sides.

bilateral symmetry having 2 sides symmetrical only about one median axis; in flowers also called zygomorphic or irregular; *cf.* radial symmetry.

bile *n.* [L. *bilis*, bile] a secretion of the liver cells which collects in the gall bladder and passes via the bile duct to the duodenum and contains the bile salts, bile pigments, cholesterol, lecithin, and some other substances; *alt.* gall.

bile cyst gall bladder, *q.v.*

bile pigments pigments produced by breakdown of haemoglobin and including bilirubin and biliverdin.

bile salts the salts sodium glycocholate and sodium taurocholate found in bile which aid the emulsification of fats during digestion.

biliary *a.* [L. *bilis*, bile] conveying or *pert.* bile.

biliation the secretion of bile.

bilicyanin *n.* [L. *bilis*, bile; Gk. *kyanos*, dark blue] a blue pigment resulting from the oxidation of biliverdin and bilirubin; *alt.* cholecyanin.

bilifulvin bilirubin, *q.v.*

bilineurine choline, *q.v.*

biliphaein bilirubin, *q.v.*

bilipurpurin *n.* [L. *bilis*, bile; *purpura*, purple] phylloerythrin, *q.v.*

bilirubin *n.* [L. *bilis*, bile; *ruber*, red] a reddish bile pigment, though to be produced as a breakdown product of haemoglobin; *alt.* bilifulvin, biliphaein, cholochrome, cholephaein, cholepyrrhin, cholerythrin.

biliverdin *n.* [L. *bilis*, bile; F. *vert*, green] a green bile pigment, an oxidation product of bilirubin.

bilobate, bilobed *a.* [L. *bis*, twice; L.L. *lobus*, from Gk. *lobos*, rounded flap] having 2 lobes.

bilobular *a.* [L. *bis*, twice; L.L. *lobulus*, dim. of *lobus*, lobe] having 2 lobules.

bilocellate *a.* [L. *bis*, twice; *locellus*, dim. of *locus*, place] divided into 2 compartments; having 2 locelli.

bilocular, biloculine *a.* [L. *bis*, twice; *locus*, place] having 2 cavities or loculi.

bilophodont *a.* [L. *bis*, twice; Gk. *lophos*, ridge; *odous*, tooth] *appl.* molar teeth of tapir, which have ridges joining the 2 anterior and posterior cusps.

biloproteins *n.plu.* [L. *bilis*, bile; Gk. *prōteion*, first] chromoproteins found in some groups of algae, especially the Cyanophyceae, Cryptophyceae, and Rhodophyceae, including phycoerythrin and phycocyanin; *alt.* phycobilins.

bimaculate *a.* [L. *bis*, twice; *macula*, spot] marked with 2 spots or stains.

bimanous *a.* [L. *bis*, twice; *manus*, hand] having 2 hands, *appl.* certain Primates.

bimastism *n.* [L. *bis*, twice; Gk. *mastos*, breast] condition of having 2 mammae; *a.* bimastic.

bimuscular *a.* [L. *bis*, twice; *musculus*, muscle] having 2 muscles.

binary *a.* [L. *binarius*, from *bini*, pair] composed of 2 units; *appl.* e.g. compounds of only 2 chemical elements.

binary fission division of a cell into 2 by an apparently simple division of nucleus and cytoplasm.

binary nomenclature binominal nomenclature, *q.v.*

binate *a.* [L. *bini*, two by two] growing in pairs; *appl.* leaf composed of 2 leaflets; *alt.* geminate.

binaural *a.* [L. *bini*, pair; *auris*, ear] *pert.* both ears; *alt.* binotic.

bineme *n.* [L. *bis*, twice; Gk. *nēma*, thread] *appl.* suggested chromatid structure consisting of 2-stranded DNA; *cf.* mononeme, polyneme.

binervate *a.* [L. *bis*, twice; *nervus*, nerve] having 2 nervures or veins, *appl.* leaf, *appl.* insect's wing.

binocular *a.* [L. *bini*, pair; *oculus*, eye] having or *pert.* 2 eyes; stereoscopic, *appl.* vision.

binodal *a.* [L. *bis*, twice; *nodus*, knob] having 2 nodes, as stem of plant.

binomial *a.* [L. *bis*, twice; *nomen*, name] consisting of 2 names, *appl.* nomenclature, the system of double names given to plants and animals,—first generic name, then specific, as *Felis* (genus) *tigris* (species); *alt.* binominal, binary nomenclature; *cf.* monomial, multinomial.

binomialism the system of binomial nomenclature.

binominal binomial, *q.v.*

binotic binaural, *q.v.*

binovular *a.* [L. *bini*, pair; *ovum*, egg] *pert.* 2 ova, as in twinning; *alt.* biovular, dizygotic.

binuclear, binucleate *a.* [L. *bis*, twice; *nucleus*, small nut] having 2 nuclei.

bioblast *n.* [Gk. *bios*, life; *blastos*, bud] hypothetical unit, *q.v.*; cytoblast, *q.v.*; mitochondrion, *q.v.*

biocatalyst *n.* [Gk. *bios*, life; *katalysis*, dissolving] enzyme, *q.v.*; ferment, *q.v.*

biocellate *a.* [L. *bis*, twice; *ocellus*, dim. of *oculus*, eye] having 2 ocelli.

biocenosis biocoenosis, *q.v.*

Biochemical Oxygen Demand (BOD) a measurement of the amount of organic pollution in water, measured as the amount of oxygen taken up from a sample containing a known amount of oxygen kept at 20°C for 5 days, a low BOD meaning little pollution and a high BOD meaning much pollution.

biochemistry *n.* [Gk. *bios*, life; *chēmeia*, transmutation] the chemistry of living organisms.

biochore n. [Gk. *bios*, life; *chōris*, separate] boundary of a floral or faunal region; climatic boundary of a floral region; a group of similar biotopes; *cf*. chore.

biochrome n. [Gk. *bios*, life; *chrōma*, colour] any natural colouring matter of plants and animals, a biological pigment.

biocoen n. [Gk. *bios*, life; *koinōs*, in common] the biotic parts of the environment in total.

biocoenosis n. [Gk. *bios*, life; *koinōs*, in common] a community of organisms inhabiting a biotope; *alt*. biocenosis.

biocontrol biological control, *q.v.*

biocycle n. [Gk. *bios*, life; *kyklos*, place of assembly] one of the 3 main divisions of the biosphere: marine, or freshwater, or terrestrial habitat.

biodegradable appl. organic compounds that can be broken down by micro-organisms; *cf*. recalcitrant.

biodemography n. [Gk. *bios*, life; *dēmos*, people; *graphein*, to write] science dealing with the integration of ecology and genetics of populations.

biodynamics n. [Gk. *bios*, life; *dynamis*, power] the science of the active vital phenomena of organisms.

bioecology n. [Gk. *bios*, life; *oikos*, household; *logos*, discourse] ecology of plants and animals.

bioelectric a. [Gk. *bios*, life; *elektron*, amber] appl. currents produced in living organisms.

bioenergetics n. [Gk. *bios*, life; *energeia*, action] the energy flow in an ecosystem, the study of energy transformations in living organisms.

bio-engineering the use of artificial replacements for body organs; the use of technology in the biosynthesis of economically important products; the application of the physical sciences and technology to the study of the human body; *alt*. biological engineering.

bioflavonoids n.plu. [Gk. *bios*, life; L. *flavus*, yellow; Gk. *eidos*, form] a group of flavonoids occurring in citrus and other fruits such as paprika, which have biological activity due to their reducing and chelating properties, and include vitamin P.

biogen(e) n. [Gk. *bios*, life; *genos*, descent] a hypothetical unit, *q.v.*; a large living molecule: precursor of bios, *q.v.*

biogenesis n. [Gk. *bios*, life; *genesis*, descent] the theory of the descent of living matter from living matter—*omne vivum e vivo*; *alt*. germ theory; *cf*. abiogenesis.

biogenetic law recapitulation theory, *q.v.*

biogenic a. [Gk. *bios*, life; *genos*, descent] originating from living organisms, appl. deposits such as coal, oil, chalk.

biogenous a. [Gk. *bios*, life; *genos*, offspring] inhabiting living organisms, as parasites.

biogeny n. [Gk. *bios*, life; *genesis*, descent] the science of the evolution of organisms, comprising ontogeny and phylogeny.

biogeochemistry n. [Gk. *bios*, life; *gē*, earth; *chēmeia*, transmutation] the study of the distribution and migration of chemical elements present in living organisms and in interaction with their geographical environment.

biogeocoenosis n. [Gk. *bios*, life; *gē*, earth; *koinōs*, shared in common] a community of organisms in relation to its special habitat.

biogeography n. [G. *bios*, life; *gē*, earth; *graphein*, to write] the part of biology dealing with the geographical distribution of plants (phytogeography) and animals (zoogeography); *alt*. chorology, geonemy.

biological a. [Gk. *bios*, life; *logos*, discourse] relating to the science of life.

biological clock a hypothetical mechanism in an organism which enables it to 'tell the time' in that it can perform metabolic or behavioural endogenous rhythms.

biological control control of pests or weeds by other living organisms; *alt*. biocontrol.

biological races strains of a species which are alike morphologically but differ in some physiological way, such as a parasite or saprophyte with particular host requirement or a free-living organism with a food or habitat preference.

biological species biospecies, *q.v.*

biological spectrum the list of percentages of the different life forms of plants represented in the total flora of a region, a term coined by Raunkiaer.

biology n. [Gk. *bios*, life; *logos*, discourse] the science of life and living.

bioluminescence n. [G. *bios*, life; L. *luminescere*, to grow light] the production of light by living organisms; *alt*. biophotogenesis, photogenesis; *cf*. phosphorescence.

biolysis n. [Gk. *bios*, life; *lysis*, loosing] the decomposition of organic matter resulting from activity of living organisms; disintegration of life; a. biolytic.

biomass n. [Gk. *bios*, life; *massein*, to squeeze] total weight of organisms per unit area.

biome n. [Gk. *bios*, life] a major community of living organisms; a complex of climax communities of plants and animals in a major region, as tundra, forest, grassland, desert, mountain; *alt*. major life zone, biotic formation.

biometeorology n. [Gk. *bios*, life; *meteōrologia*, treatise on the heavenly bodies] the study of the effects of atmospheric conditions upon plants and animals.

biometrics, biometry n. [Gk. *bios*, life; *metron*, measure] the statistical study of living organisms and their variations.

bion n. [Gk. *bion*, living] an independent living organism; an individual organism; *alt*. biont.

bionergy n. [Gk. *bios*, life; *energeia*, action] vital force, *q.v.*

bionomics, bionomy n. [Gk. *bios*, life; *nomos*, law] ecology, *q.v.*

biont bion, *q.v.*

biophage n. [Gk. *bios*, life; *phagein*, to eat] an heterotrophic organism feeding upon other living organisms; *cf*. saprophage.

biophore n. [Gk. *bios*, life; *pherein*, to carry] a hypothetical unit, *q.v.*

biophotogenesis n. [Gk. *bios*, life; *phōs*, light; *genesis*, origin] bioluminescence, *q.v.*

biophysics n. [Gk. *bios*, life; *physis*, nature] study of biological phenomena interpreted in terms of physical principles; physics as applicable to biology.

biophyte n. [Gk. *bios*, life; *phyton*, plant] a plant which is a parasite.

bioplasm n. [Gk. *bios*, life; *plasma*, mould] protoplasm, *q.v.*

bioplast

bioplast *n.* [Gk. *bios*, life; *plastos*, formed] a minute quantity of living protoplasm capable of reproducing itself.

biopoiesis *n.* [Gk. *bios*, life; *poiēsis*, making] the origination of living organisms from non-living replicating molecules.

biopsy *n.* [Gk. *bios*, life; *opsis*, sight] examination of living organisms, organs, or tissues.

biopterin(e) *n.* [Gk. *bios*, life; *pteron*, wing] a pteridine precursor of the drosopterins, *q.v.*, which is also found in the royal jelly fed to the queen bee.

biorgan *n.* [Gk. *bios*, life; *organon*, instrument] an organ in the physiological sense, not necessarily a morphological unit.

bios *n.* [Gk. *bios*, life] organic life, plant or animal; vitamin B complex, *q.v.*; a mixture of certain of the vitamins of the B-complex, including biotin, inositol, and pantothenic acids extracted from some yeasts and necessary for their growth.

bioseries *n.* [Gk. *bios*, life; L. *series*, row] a succession of changes of any single heritable character.

bioses disaccharides, *q.v.*

biosis *n.* [Gk. *biōsis*, a living] mode of living; vitality.

biosomes *n.plu.* [Gk. *bios*, life; *sōma*, body] structural and functional units in cytoplasm, as mitochondria, chromidia, and plastids.

biospecies *n.* [Gk. *bios*, life; L. *species*, particular kind] a population of individuals which can breed together, i.e. a true species; *alt.* biological species.

biospeleology *n.* [Gk. *bios*, life; *spēlaion*, cave; *logos*, discourse] the biology of cave-dwelling organisms.

biosphere *n.* [Gk. *bios*, life; *sphaira*, globe] the part of the globe containing living organisms; *alt.* ecumene.

biostasis *n.* [Gk. *bios*, life; *stasis*, standing] the ability of organisms to withstand environmental changes without being changed themselves.

biostatics *n.* [Gk. *bios*, life; *statos*, stationary] the science of structure in relation to function of organisms.

biosynthesis *n.* [Gk. *bios*, life; *synthesis*, composition] the formation of organic compounds by living organisms.

biosystem ecosystem, *q.v.*

biosystematics genonomy, *q.v.*; taxonomy, *q.v.*

biota *n.* [Gk. *bios*, life] the fauna and flora of a region.

biotic *a.* [Gk. *biōtikos*, pert. life] *pert.* life; vital, alive.

biotic climax a plant association that is maintained at climax status by a biotic factor such as grazing.

biotic community a community of plants and animals as a whole.

biotic environment the part of an organism's environment produced by its interaction with other organisms.

biotic formation biome, *q.v.*

biotic potential highest possible rate of population increase, resulting from maximum natality and minimum mortality.

biotin *n.* [Gk. *bios*, life] a vitamin of the B-complex, found in liver, yeast, egg yolk, royal jelly, and pollen, the absence of which causes a type of dermatitis and paralysis induced in many animals by e.g. eating large amounts of raw egg white, by combining with the protein avidin in egg white so preventing the absorption of biotin, and which is involved in carbon dioxide fixation and in fat and protein metabolism; *alt.* vitamin B_4, anti-avidin, antiperosis factor, coenzyme R, vitamin H.

biotomy *n.* [Gk. *bios*, life; *tomē*, cutting] the dissection of living organisms, vivisection.

biotonus *n.* [Gk. *bios*, life; *tonos*, tension] the ratio between assimilation and dissimilation of biogens.

biotope *n.* [Gk. *bios*, life; *topos*, place] an area in which the main environmental conditions and biotypes adapted to them are uniform; a place where organisms can survive; microhabitat, *q.v.*; *cf.* biochore, chore.

biotype *n.* [Gk. *bios*, life; *typos*, pattern] all the individuals of identical genotype; in bacteria, a taxon distinguished by a certain physiological feature; type specimen of an organism.

biovular binovular, *q.v.*

biovulate *a.* [L. *bis*, twice; *ovum*, egg] containing 2 ovules.

bipaleolate *a.* [L. *bis*, twice; *palea*, chaff] furnished with 2 small paleae.

bipalmate *a.* [L. *bis*, twice; *palma*, palm of hand] lobed with the lobes again lobed.

biparietal *a.* [L. *bis*, twice; *paries*, wall] connected with the 2 parietal eminences.

biparous *a.* [L. *bis*, twice; *parere*, to bear] having 2 young at a time; dichotomous, *appl.* branching; dichasial, *appl.* cyme.

bipectinate *a.* [L. *bis*, twice; *pecten*, comb] having the 2 margins furnished with teeth like a comb; *alt.* aspidobranch.

biped *n.* [L. *bis*, twice; *pes*, foot] a 2-footed animal; *a.* bipedal.

bipeltate *a.* [L. *bis*, twice; *pelta*, shield] having, or consisting of, 2 shield-like structures.

bipennate *a.* [L. *bis*, twice; *penna*, feather] bipenniform, *q.v.*; *appl.* muscles in which the tendon of insertion extends through the middle.

bipenniform *a.* [L. *bis*, twice; *penna*, feather; *forma*, shape] feather-shaped, with sides of vane of equal size; *alt.* bipennate.

bipetalous *a.* [L. *bis*, twice; Gk. *petalon*, leaf] with 2 petals.

bipinnaria *n.* [L. *bis*, twice; *pinna*, feather] in Asteroidea, a larva with 2 bands of cilia.

bipinnate *a.* [L. *bis*, twice; *pinna*, feather] *appl.* compound pinnate leaf in which leaflets grow in pairs on paired stems.

bipinnatifid *a.* [L. *bis*, twice; *pinna*, feather; *findere*, to cleave] *appl.* pinnatifid (segmented) leaf whose segments are again divided.

bipinnatipartite *a.* [L. *bis*, twice; *pinna*, feather; *partiri*, to divide] bipinnatifid, but with divisions extending nearly to midrib.

bipinnatisect *a.* [L. *bis*, twice; *pinna*, feather; *secare*, to cut] bipinnatifid, but with division extending completely to midrib.

biplicate *a.* [L. *bis*, twice; *plicare*, to fold] having 2 folds; having 2 distinct wavelengths, *appl.* flagellar movement in certain bacteria.

bipocillus *n.* [L. *bis*, twice; *pocillum*, little cup] a microsclere with curved shaft and cup-shaped expansion at each end.

bipolar *a.* [L. *bis*, twice; *polus*, pole] having located at, or *pert.* 2 ends or poles; *appl.* nerve cells having a process at each end; *appl.* allied species occurring towards Arctic and Antarctic regions; *alt.* dipolar.

bipolarity *n.* [L. *bis*, twice; *polus*, pole] the condition of having 2 polar processes; condition of having 2 distinct poles, as vegetative and animal poles in an egg; bipolar distribution, as of species.

biradial *a.* [L. *bis*, twice; *radius*, ray] symmetrical both radially and bilaterally, as some coelenterates; *alt.* disymmetrical.

biradiate two-rayed; *cf.* diactinal, diaxon.

biramous *a.* [L. *bis*, twice; *ramus*, branch] divided into 2 branches; *alt.* biramose.

bird lice a common name for the Mallophaga, *q.v.*

birds the common name for the Aves, *q.v.*

bird's nest fungi the common name for the Nidulariales, *q.v.*

birostrate *a.* [L. *bis*, twice; *rostrum*, beak] furnished with 2 beak-like processes.

birth pore uterine pore of trematodes and cestodes; birth-opening of redia of trematodes.

biscoctiform *a.* [L. *bis*, twice; *coctus*, baked; *forma*, shape] biscuit-shaped, *appl.* spores.

bisect a stratum transect chart with root system as well as shoot included.

bisegmental *a.* [L. *bis*, twice; *segmentum*, a slice] *pert.*, or involving, 2 segments.

biseptate *a.* [L. *bis*, twice; *septum*, fence] with 2 partitions.

biserial, biseriate *a.* [L. *bis*, twice; *series*, row] arranged in 2 rows or series.

biserrate *a.* [L. *bis*, twice; *serra*, saw] having marginal teeth which are themselves notched.

bisexual *a.* [L. *bis*, twice; *sexus*, sex] having both male and female reproductive organs, *alt.* hermaphrodite; amphisporangiate, *q.v.*

bisporangiate *a.* [L. *bis*, twice; Gk. *sporos*, seed; *anggeion*, vessel] having both micro- and megasporangia; *appl.* strobilus consisting of both micro- and megasporophylls.

bispore *n.* [L. *bis*, twice; Gk. *sporos*, seed] a paired spore, as of certain Rhodophyceae.

bisporic *a.* [L. *bis*, twice; Gk. *sporos*, seed] disporous, *q.v.*

bistephanic *a.* [L. *bis*, twice; Gk. *stephanos*, crown] joining 2 points where coronal suture crosses superior temporal ridges.

bistipulate *a.* [L. *bis*, twice; *stipula*, stem] provided with 2 stipules.

bistrate *a.* [L. *bis*, twice; *stratum*, layer] having 2 layers.

bistratose *a.* [L. *bis*, twice; *stratum*, layer] with cells arranged in 2 layers.

bisulcate *a.* [L. *bis*, twice; *sulcus*, groove] having 2 grooves; cloven-hoofed.

bitemporal *a.* [L. *bis*, twice; *tempora*, temples] *appl.* 2 temporal bones; *appl.* line joining posterior ends of 2 zygomatic processes.

biternate *a.* [L. *bis*, twice; *terni*, three by three] ternate with each division itself again ternate.

bitheca *n.* [L. *bis*, twice; *theca*, case] a theca budded from a stolotheca, and surrounding the male polyp in graptolites.

biting lice a common name for the Mallophaga, *q.v.*

bitubercular *a.* [L. *bis*, twice; *tuberculum*, small swelling] with 2 tubercles or cusps; biscuspid, *appl.* teeth.

bivalent *n.* [L. *bis*, twice; *valere*, to be strong] a pair of homologous chromosomes at meiosis.

bivalve *a.* [L. *bis*, twice; *valvae*, folding doors] consisting of 2 plates or valves, as shell of brachiopods and bivalves; *appl.* a seed capsule of similar structure.

Bivalvia, bivalves *n.*, *n.plu.* [L. *bis*, twice; *valvae*, folding doors] a class of bilaterally symmetrical molluscs which are laterally compressed and have a shell made of 2 hinged valves, a reduced head and large ciliated ctenidia; in some classifications the class is subdivided into the subclasses Protobranchia and Lamellibranchia, although the latter term is often regarded as a synonym for Bivalvia; *alt.* Pelecypoda.

biventer cervicis *n.* [L. *bis*, twice; *venter*, belly; *cervix*, neck] the spinalis capitis, or medial part of semispinalis, a muscle of neck, consisting of 2 fleshy ends with narrow tendinous portion in middle.

biventral *a.* [L. *bis*, twice; *venter*, belly] *appl.* muscles of the biventer type; digastric, *q.v.*

biverticillate *a.* [L. *bis*, twice; *verticillus*, small whorl] having 2 verticils or whorls; *alt.* diverticillate.

bivittate *a.* [L. *bis*, twice; *vitta*, band] with 2 oil receptacles; with 2 stripes.

bivium *n.* [L. *bis*, twice; *via*, way] generally the posterior pair of ambulacral areas in certain Echinoidea; the 2 rays between which the madreporite lies; *cf.* trivium.

bivoltine *a.* [L. *bis*, twice; It. *volta*, time] having 2 broods in a year.

black corals the common name for the Antipatharia, *q.v.*

black earth chernozem, *q.v.*

bladder *n.* [A.S. *blaedre*, bag] a membranous sac filled with air or fluid; a cyst; *alt.* vesica.

bladder cell a globular modified hyphal cell in integument or carpophore; in tunicates, a large vacuolated cell in outer layer of tunic.

bladderworm cysticercus, *q.v.*, in tapeworms.

blade *n.* [A.S. *blaed*, leaf] the flat part of leaf; *alt.* lamina.

Blandin's glands [*P.-F. Blandin*, French surgeon] anterior lingual glands; *alt.* glands of Nuhn.

blanket bog bog developed in a very wet climate and covering great stretches of country.

blastaea *n.* [Gk. *blastos*, bud] a ciliated planula, a hypothetical stage in evolution; *alt.* planaea.

blastelasma *n.* [Gk. *blastos*, bud; *elasma*, plate] any germ layer formed after formation of epiblast and hypoblast.

blastema *n.* [Gk. *blastēma*, bud] the protoplasm of an egg, as distinct from the yolk; a group of undifferentiated cells which develop into an organ; thallus of a lichen; a bud or cell group which develop into an organism by vegetative reproduction, as distinct from the single cell (zygote) in sexual reproduction.

blastic *a.* [Gk. *blastos*, bud] *pert.* or stimulating enlargement by cell division.

blastocarpous *a.* [Gk. *blastos*, bud; *karpos*, fruit] developing while still surrounded by pericarp.

blastocele blastocoel, *q.v.*

blastocheme *n.* [Gk. *blastos*, bud; *ochēma*, vessel] a reproductive individual in some medusae.

blastocholines

blastocholines *n.plu.* [Gk. *blastos*, bud; *chōlos*, halting] various substances, present in sporangia, seeds, and fruits, which prevent premature germination, being germination inhibitors.

blastochyle *n.* [Gk. *blastos*, bud; *chylos*, juice] the fluid in a blastocoel or segmentation cavity.

Blastocladiales *n.* [Gk. *blastos*, bud; *klados*, sprout] an order of Phycomycetes having a more-or-less extensive true mycelium, asexual reproduction by posteriorly uniflagellate zoospores produced in sporangia, and isogamous or anisogamous sexual reproduction.

blastocoel(e) *n.* [Gk. *blastos*, bud; *koilos*, hollow] segmentation cavity, *q.v.*; archicoel, *q.v.*; *alt.* blastocele.

blastocolla *n.* [Gk. *blastos*, bud; *kolla*, glue] a gummy substance coating certain buds.

blastocone *n.* [Gk. *blastos*, bud; *kōnos*, cone] an outer larger cell of 1st circumferential division, in segmentation of certain eggs.

blastocyst *n.* [Gk. *blastos*, bud; *kystis*, bladder] the mammalian embryo, a modified blastula, at the stage resulting from the cleavage of the zygote at the time of implantation in the uterine wall; *alt.* blastodermic vesicle, gastrocystis.

blastocyte *n.* [Gk. *blastos*, bud; *kytos*, hollow] any undifferentiated embryonic cell.

blastoderm *n.* [Gk. *blastos*, bud; *derma*, skin] a blastodisc after cleavage and the formation of the blastocoel; the layer of cells around the blastocoel, *alt.* primary epithelium; a corresponding part of the embryo in various invertebrates.

blastodermic vesicle blastula, *q.v.*; blastocyst, *q.v.*

blastodisc *n.* [Gk. *blastos*, bud; *diskos*, disc] in large yolky eggs as of birds or reptiles, a disc-shaped superficial layer of cells formed by cleavage, and which will form the embryo; *alt.* germinal disc; *cf.* blastoderm.

blastogene plasmagene, *q.v.*

blastogenesis *n.* [Gk. *blastos*, bud; *genesis*, descent] gemmation or reproduction by budding; transmission of inherited characters by means of germ plasm only; *cf.* embryogenesis.

blastogenic *a.* [Gk. *blastos*, bud; *genos*, offspring] *appl.* inactive idioplasm unalterable till time and place of activity are reached; arising from changes in germ cells; *appl.* characteristics of germinal constitution; *appl.* reproduction by budding; *cf.* cytogenic.

Blastoidea, blastoids *n., n.plu.* [Gk. *blastos*, bud; *eidos*, form] a subclass of ovoid cystoid echinoderms.

blastokinesis *n.* [Gk. *blastos*, bud; *kinēsis*, movement] movement of embryo in the egg, as in certain insects and cephalopods.

blastomere *n.* [Gk. *blastos*, bud; *meros*, part] one of the cells formed during primary divisions of an egg; *alt.* cleavage cell, mere.

blastoneuropore *n.* [Gk. *blastos*, bud; *neuron*, nerve; *poros*, passage] a temporary passage connecting blastopore and neuropore.

blastophore *n.* [Gk. *blastos*, bud; *pherein*, to bear] embryonic origin of plumule; the reproductive body in Alcyonaria; central part of spermocyte mass which remains unchanged through spermatogenesis in Annelida, *alt.* cytophore.

blastophthoria *n.* [Gk. *blastos*, bud; *phthora*, corruption] any injurious effect on germ cells or on germ plasm.

blastopore *n.* [Gk. *blastos*, bud; *poros*, passage] the original mouth of a gastrula, the channel leading into archenteron; *alt.* protostoma.

blastosphere *n.* [Gk. *blastos*, bud; *sphaira*, globe] blastula, *q.v.*; a hollow ball of cells.

blastospore *n.* [Gk. *blastos*, bud; *sporos*, seed] an attached thallospore developed by budding and itself capable of budding, as of yeast cells.

blastostyle *n.* [Gk. *blastos*, bud; *stylos*, pillar] in Hydrozoa, a columniform zooid with or without mouth and tentacles, bearing gonophores; *alt.* gonoblastid, gonoblastidium.

blastozoite *n.* [Gk. *blastos*, bud; *zōē*, life] blastozooid, *q.v.*

blastozooid *n.* [Gk. *blastos*, bud; *zōon*, animal; *eidos*, form] an individual or zooid formed by budding; *alt.* blastozoite; *cf.* oozooid.

blastula *n.* [L. *dim.* of Gk. *blastos*, bud] a hollow ball of cells with the wall usually one layer thick, formed by cleavage of the zygote; *alt.* blastosphere, blastodermic vesicle, mesembryo; *plu.* blastulae.

blastulation *n.* [L. *blastula*, little bud] formation of blastulae.

bleeding of plants, exudation of watery sap from vessels at a cut surface, due to root pressure.

bleeding pressure root pressure, *q.v.*

blematogen *n.* [Gk. *blēma*, coverlet; *gennaein*, to produce] primordial covering of a carpophore; undeveloped universal veil in agarics; primordial cuticle.

blended inheritance mingling or non-segregation of parental characteristics, so that offspring appear a blend of the 2; mixed race or descent.

blendling *n.* [A.S. *blandan*, to mix] a racial hybrid.

blennogenous *a.* [Gk. *blennos*, mucus; *gennaein*, to produce] mucus-producing.

blennoid *a.* [Gk. *blennos*, mucus; *eidos*, form] resembling mucus.

blephara *n.* [Gk. *blepharis*, eyelash] peristome tooth in mosses.

blepharal *a.* [Gk. *blepharon*, eyelid] *pert.* eyelids.

blepharoplast *n.* [Gk. *blepharis*, eyelash; *plastos*, formed] a basal granule especially in relation to an organelle of locomotion such as a flagellum of Flagellata; *alt.* blepharoblast, mastigosome.

blight *n.* [A.S. *blaecan*, to grow pale] an insect or fungus producing a plant disease; the disease itself.

blind pit a cell wall pit which is not backed by a complementary pit.

blind spot region of retina devoid of rods and cones and where optic nerve enters; *alt.* optic disc.

blocky *appl.* soil crumbs where the vertical and horizontal axes are approximately of the same size and rather angular.

blood *n.* [A.S. *blód*, blood] the fluid circulating in the vascular system of animals, distributing food material, oxygen, etc., and collecting waste products.

blood anlage blood islands, *q.v.*

blood cells cells derived by mitosis from ordinary mesoderm cells forming haematoblast; blood corpuscles, *q.v.*

blood coagulation factor formerly, vitamin K, *q.v.*

blood corpuscle a cell in suspension in blood plasma, as red blood corpuscle (erythrocyte) and white blood corpuscle (leucocyte); *alt.* blood cell.

blood crystals crystals of haemoglobin, haemin, or haematoidin, which form when blood is shaken up with chloroform or ether.

blood dust fine droplets of neutral fats present in the blood stream; haemoconia, *q.v.*

blood gills delicate blood-filled sacs functioning in uptake of salts, in certain insects.

blood groups classification of types of blood depending on presence or absence of 2 agglutinogens (A and B) in the red corpuscles and 2 agglutinins (α or anti-A, and β or anti-B) in serum or plasma: A cells agglutinate with B type serum, B with A type, AB with A and B type, and O cells not agglutinating with A and B types, *cf.* universal donor, universal recipient; there are other classifications, such as Rh and MN types.

blood islands isolated reddish mesodermal cell groups in which primitive erythroblasts are enclosed by the peripheral cells which develop into endothelium; *alt.* blood anlage, haemangioblast, insula.

blood platelets colourless bodies in the blood, about one-third the size of red corpuscles, which are thought to release thrombokinase when they disintegrate at a wound and which are formed from megakaryocytes and agglutinate in shed blood; *alt.* thrombocytes, thromboplastids.

blood plates minute amoeboid protoplasmic bodies found in blood.

blood serum fluid or plasma left after removal of corpuscles and fibrin.

blood shadow the colourless stroma of red blood corpuscles.

blood sugar glucose, *q.v.*

blood vessel any vessel or space in which blood circulates; strictly used only in regard to special vessls with well-defined walls.

bloom *n.* [A.S. *blōwan*, to bloom] a layer of wax particles on external surface of certain fruits, as grapes, peaches; blossom or flower; seasonal dense phytoplankton.

blubber *n.* [M.E. *blober*, a bubble] fat of whales, seals, etc., lying between outer skin and muscle layer.

blue coral the common name for the Coenothecalia, *q.v.*

blue–green algae the common name for the Cyanophyceae, *q.v.*

Bobasatraniiformes *n.* [*Bobasatrania*, generic name; L. *forma*, shape] an order of Triassic Chondrostei having a laterally compressed body.

BOD Biochemical Oxygen Demand, *q.v.*

body cavity the internal cavity in many animals in which various organs are suspended and which is bounded by the body wall; *alt.* perigastrium.

body cell a somatic cell as distinct from a germ cell; an antheridial cell; cell in pollen grain from which male nuclei arise.

body lice a common name for the Anoplura, *q.v.*

body stalk a band of mesoderm connecting caudal end of embryo with chorion.

Boettcher's cells granular cells between Claudius' cells and basilar membrane in organ of Corti.

bog *n.* [Celtic *bog*, soft] a community on wet, very acid peat; *cf.* fen.

Bohr effect the influence of pH on the ability of haemoglobin to carry oxygen, an increase in carbonic acid from respiration resulting in a decreased capacity for the haemoglobin to bind oxygen.

Bojanus, organ of [*L. H. Bojanus*, Alsation zoologist] excretory organ in lamellibranchs.

Boletales *n.* [*Boletus*, generic name] an order of Basidiomycetes, now usually included as a family (Boletaceae) in the Agaricales, and having the hymenium lining pores instead of gills.

boletiform *a.* [L. *boletus*, a mushroom; *forma*, shape] shaped like a somewhat elliptic spindle, *appl.* spore of some Boletales; *alt.* subfusiform.

boll *n.* [A.S. *bolla*; L. *bulla*, knob] a capsule or globular pericarp as in cotton plant.

bolting in plants, the elongation of internodes in a normally rosette form or dwarf plant.

bolus *n.* [L. *bolus*, from Gk. *bolos*, lump] a rounded mass; lump of chewed food.

bone *n.* [A.S. *ban*, bone] connective tissue in which the ground substance contains calcium salts, mainly phosphate and carbonate.

bone-beds deposits formed largely by remains of bones of fishes and reptiles, as Liassic bonebeds.

bone cell or corpuscle osteoblast, *q.v.*; osteocyte, *q.v.*

bone enzyme *see* phosphatase.

bones of Bertin [*E. J. Bertin*, French anatomist] thin anterior coverings of sphenoidal sinuses.

bonitation *n.* [L. *bonitas*, goodness] the evaluation of the numerical distribution of a species in a particular locality or season, in relation to agricultural, veterinary, or medical implications.

bony fishes the common name for the Osteichthyes, *q.v.*

book gill gill book, *q.v.*

book lice a common name for some members of the Psocoptera, *q.v.*

book lung lung book, *q.v.*

booted *a.* [O.F. *bote*, boot] equipped with raised horny plates of skin, as feet of some birds; caligate, *q.v.*

bordered pit a form of pit, developed on walls of tracheids and wood vessels, with overarching border of secondary cell wall.

boreal *a.* [L. *boreas*, north wind] *appl.* or *pert.* northern biogeographical region; Holarctic except Sonoran, or restricted to Nearctic; *pert.* postglacial age with continental type of climate.

bossed bosselated, *q.v.*; umbonate, *q.v.*

bosselated *a.* [O.F. *boce*, knob] covered with knobs; *alt.* bossed.

bosset *n.* [O.F. *boce*, knob] the beginning of horn formation in deer in the 1st year.

bostryx *n.* [Gk. *bostrychos*, curl] helicoid cyme, *q.v.*

Botallo's duct [*L. Botallo*, Italian surgeon] ductus arteriosus, a small blood vessel representing 6th gill arch and connecting pulmonary with systemic arch.

botany *n.* [Gk. *botanē*, pasture] the branch of biology dealing with plants; *alt.* phytology.

bothrenchyma *n.* [Gk. *bothros*, pit; *engchyma*, infusion] a plant tissue formed of pitted ducts.

bothridium *n.* [Gk. *bothros*, trench; *idion*, *dim.*] a muscular cup-shaped outgrowth from scolex of certain tapeworms, used for attachment to host; *alt.* phyllidium.

bothrionic *a.* [Gk. *bothros*, pit] *appl.* seta arising from the bottom of a pit in the integument.

bothrium

bothrium *n*. [Gk. *bothros*, trench] a sucker; a sucking groove in scolex of certain tapeworms.

botryoid(al) *a*. [Gk. *botrys*, bunch of grapes; *eidos* form] in the form of a bunch of grapes; *appl.* tissue of branched canals surrounding enteric canal in leeches; *alt.* botryose.

botryology *n*. [Gk. *botrys*, cluster; *logos*, discourse] the science of arranging things or concepts in groups and clusters.

botryose *a*. [Gk. *botrys*, bunch of grapes] racemose, *q.v.*; botryoidal, *q.v.*

botuliform *a*. [L. *botulus*, sausage; *forma*, form] sausage-shaped; *alt.* allantoid.

bouquet *n*. [F. *bouquet*, nosegay] arrangement of chromosomes in loops with their ends near one side of nuclear wall during zygotene and pachytene in some organisms; bunch of muscles and ligaments connected with the styloid process of the temporal bone.

bourrelet *n*. [F. *bourrelet*, circular pad] poison gland associated with sting in ants; in some echinoids, a swelling on the interambulacral plates where they meet the peristome.

bouton *n*. [F. *bouton*, bud] terminal bulb of arborization of an axon, *alt.* end-bulb; labellum in Hymenoptera.

Bowditch's law [*H. P. Bowditch*, American physiologist] all-or-none principle, *q.v.*

Bowman's capsule [*Sir W. Bowman*, English histologist] the dilated vesicle at the end of a nephron; *alt.* capsula glomeruli.

Bowman's glands serous glands in corium of olfactory mucous membrane.

Bowman's membrane anterior elastic membrane of cornea.

braccate *a*. [L. *braccae*, breeches] having additional feathers on legs or feet, *appl.* birds.

brachelytrous *a*. [Gk. *brachys*, short; *elytron*, sheath] having short elytra.

brachia *n.plu.* [L. *brachium*, arm] the arms; 2 spirally coiled structures, one at each side of mouth, in Brachiopoda; cerebellar peduncles; white lateral bands of colliculi of corpora quadrigemina; *sing.* brachium, *q.v.*

brachial *pert.* arm; arm-like.

brachialis *n*. [L. *brachialis*, *pert.* arm] a flexor muscle of the forearm, from lower half of front of humerus to coronoid process of ulna; *alt.* brachialis anticus.

brachiate *a*. [L. *brachium*, arm] branched, *alt.* brachiferous; having opposite, widely spread, paired branches on alternate sides; having arms. *v.* To move along by swinging by the arms from one hold to another as in the gibbon.

brachiation *n*. [L. *brachium*, arm] abduction or movement of forelimbs away from the median longitudinal plane; the act of brachiating.

brachidia *n.plu.* [Gk. *brachiōn*, arm; *idion*, dim.] calcareous skeleton supporting brachia in certain Brachiopoda.

brachiferous, brachigerous *a*. [L. *brachium*, arm; *ferre, gerere*, to carry] branched; *alt.* brachiate.

brachiocephalic *a*. [Gk. *brachiōn*, arm; *kephalē*, head] *pert.* arm and head, *appl.* artery, veins; *alt.* innominate.

brachiocubital *a*. [L. *brachium*, arm; *cubitum*, forearm] *pert.* arm and forearm.

brachiolaria *n*. [L. *brachiolum*, small arm] a larval stage in some Asteroidea.

brachiole *n*. [L. *brachiolum*, small arm] a pinnule-like structure on ambulacral margin in Blastoidea; 1 of the 3 arms or outgrowths containing an extension of the coelom during development of a bipinnaria into a brachiolaria.

Brachiopoda *n*. [Gk. *brachiōn*, arm; *pous*, foot] a phylum of coelomate animals, commonly called lamp shells, and widely represented as fossils, that have a bivalved shell and a lophophore on the anterior part of the body; *alt.* brachiopods.

Brachiopterygii *n*. [Gk. *brachiōn*, arm; *pterygion*, fin] Polypterini, *q.v.*

brachiorachidian *a*. [Gk. *brachiōn*, arm; *rhachis*, spine] *pert.* arm and spine.

brachioradialis *n*. [L. *brachium*, arm; *radius*, ray] the supinator longus muscle of forearm.

brachium *n*. [L. *brachium*, arm] arm or branching structure; the forelimb of vertebrates; a bundle of fibres connecting cerebellum to cerebrum or to pons; *plu.* brachia, *q.v.*

brachyblast brachyplast, *q.v.*

brachyblastic *a*. [Gk. *brachys*, short; *blastos*, bud] with a short germ band; *cf.* tanyblastic.

brachycephalic *a*. [Gk. *brachys*, short; *kephalē*, head] short-headed; with cephalic index of over 80; *cf.* dolichocephalic.

brachycerous *a*. [Gk. *brachys*, short; *keras*, horn] short-horned; with short antennae.

brachycnemic *a*. [Gk. *brachys*, short; *knēmē*, tibia] *appl.* arrangement of mesenteries of Zoantharia where the 6th protocneme is imperfect.

brachydactyly *n*. [Gk. *brachys*, short; *daktylos*, digit] the condition of having digits abnormally short; *a.* brachydactylous.

brachydont *a*. [Gk. *brachys*, short; *odous*, tooth] *appl.* molar teeth with low crowns; *alt.* brachyodont; *cf.* hypsodont.

brachymeiosis *n*. [Gk. *brachys*, short; *meion*, smaller] in the asci of some fungi, a further reduction division thought to occur after normal meiosis; the whole of meiosis in certain fungi involving 2 reduction divisions which may restore the haploid condition where double fertilization has produced a tetraploid; meiosis involving only one division.

brachyodont brachydont, *q.v.*

brachyourous brachyural, *q.v.*

brachyplast *n*. [Gk. *brachys*, short; *plastos*, formed] a short branch bearing leaf tufts, occurring with normal branches on the same plant; *alt.* brachyblast, short shoot, spur; *cf.* foliar.

brachypleural *a*. [Gk. *brachys*, short; *pleuron*, side] with short pleura or side plates.

brachypodous *a*. [Gk. *brachys*, short; *pous*, foot] with short legs, or stalk.

brachypterous *a*. [Gk. *brachys*, short; *pteron*, wing] with short wings, *appl.* insects; *n.* brachypterism; *cf.* macropterous.

brachysclereid(e) *n*. [Gk. *brachys*, short; *sklēros*, hard; *eidos*, form] stone cell, *q.v.*

brachysm *n*. [Gk. *brachys*, short] dwarfism in plants caused by shortening of internodes.

brachystomatous *a*. [Gk. *brachys*, short; *stoma*, mouth] with short proboscis, *appl.* certain insects.

brachytic *a*. [Gk. *brachytēs*, shortness] dwarfish, *appl.* plants; exhibiting or *pert.* brachysm.

brachytmema *n*. [Gk. *brachys*, short; *tmēma*, segment, from *tmēgein*, to cut] truncated condition or appearance; a cell which ruptures, releasing a gemma, as in bryophytes.

brachyural, brachyurous a. [Gk. brachys, short; oura, tail] having short abdomen usually tucked in below thorax, appl. certain crabs; alt. brachyourous.

brachyuric a. [Gk. brachys, short; oura, tail] short-tailed.

bract n. [L. bractea, thin plate of metal] a modified leaf in whose axil an inflorescence or a flower arises, cf. bracteole; a floral leaf; one of the leaves or leaf-like structures associated with sporangia or antheridia and archegonia in bryophytes and pteridophytes; hydrophyllium in Siphonophora; distal exite of 6th appendage in some Branchiopoda such as Apus.

bracteal a. [L. bractea, thin metal plate] pert. a bract.

bracteate a. [L. bractea, thin metal plate] having bracts.

bracteiform a. [L. bracteola, thin metal leaf; forma, form] like a bract.

bracteolate a. [L. bracteola, thin metal leaf] appl. flowers with bracteoles.

bracteole n. [L. bracteola, thin metal leaf] secondary bract at the base of an individual flower; alt. bractlet.

bracteose a. [L. bractea, thin metal plate] with many bracts.

bractlet bracteole, q.v.

bract scales small scales developed directly on axis of cones of conifers; alt. cover scales; cf. ovuliferous scales.

bradyauxesis n. [Gk. bradys, slow; auxēsis, growth] relatively slow growth; growth of a part at a slower rate than that of the whole; cf. isauxesis, tachyauxesis.

bradygenesis n. [Gk. bradys, slow; genesis, descent] retarded development in phylogeny; cf. tachygenesis.

bradykinins n.plu. [Gk. bradys, slow; kinein, to move] a group of polypeptides found in blood and causing dilatation of blood vessels and contraction of smooth muscle; alt. kinins.

Bradyodonti n. [Gk. bradys, slow; odous, tooth] Holocephali, q.v.

bradytelic a. [Gk. bradys, slow; telos, fulfilment] evolving at a rate slower than the standard rate; cf. tachytelic, horotelic; n. bradytely.

brain n. [A.S. braegen, brain] centre of nervous system; mass of nervous matter in vertebrates at anterior end of spinal cord, lying in cranium; in invertebrates, supraoesophageal or suprapharyngeal ganglia; alt. encephalon.

brain hormone a substance secreted by the insect brain which activates the prothoracic glands.

brain sand granular bodies of calcium, ammonium, and magnesium phosphates, occurring in pineal gland and pia mater; alt. corpora arenacea, acervulus cerebri.

brain stem the mid-brain, pons, and medulla oblongata.

brain vesicles three dilatations at anterior end of the embryonic neural tube, which give rise to fore-brain, mid-brain, and hind-brain.

branch a taxonomic group used in different ways by different specialists, but usually referring to a level between subphylum and class.

branch gaps gaps in the vascular cylinder of a main stem, subtending branch traces.

branchia sing. of branchiae.

branchiac branchial, q.v.

branchiae n.plu. [L. branchiae, gills] gills, of aquatic animals.

branchial a. [Gk. brangchia, gills] pert. gills; alt. branchiac; see also gill.

branchial arch gill arch, q.v.

branchial cleft gill cleft, q.v.

branchial grooves outer pharyngeal grooves or visceral clefts, q.v.

branchial siphon incurrent siphon of molluscs.

branchiate a. [L. branchiae, gills] having gills; alt. branchiferous.

branchicolous a. [L. branchiae, gills; colere, to inhabit] parasitic on fish gills.

branchiferous branchiate, q.v.

branchiform a. [L. branchiae, gills; forma, shape] gill-like.

branchihyal n. [Gk. brangchia, gills; hyoeidēs, Υ-shaped] an element of a branchial arch.

branchiocardiac a. [Gk. brangchia, gills; kardia, heart] pert. gills and heart; appl. vessel given off ventrally from ascidian heart; appl. vessels conveying blood from gills to pericardial sinus in certain crustaceans.

branchiomere n. [Gk. brangchia, gills; meros, part] a segment of an animal bearing gills.

branchiomeric pert. branchiomeres; appl. muscles derived from gill arches.

branchiopallial a. [Gk. brangchia, gills; L. pallium, mantle] pert. gill and mantle of molluscs.

branchiopneustic a. [Gk. brangchia, gills; pneustikos, disposed to blow] appl. insects having spiracles replaced functionally by gills.

Branchiopoda, branchiopods n., n.plu. [Gk. brangchia, gills; pous, foot] a subclass of crustaceans whose carapace, if present, forms a dorsal shield or bivalved shell, and which have broad lobed trunk appendages fringed with hairs.

branchiostegal a. [Gk. brangchia, gills; stegē, roof] with or pert. a gill cover, appl. membrane, rays.

branchiostege n. [Gk. brangchia, gills; stegē, roof] the branchiostegal membrane.

branchiostegite n. [Gk. brangchia, gills; stegē, roof] expanded lateral portion of carapace forming gill cover in certain Crustacea.

branchireme n. [L. branchiae, gills; remus, oar] a branchiate limb; locomotory and respiratory limb of Branchiopoda.

Branchiura n. [Gk. brangchia, gills; oura, tail] a subclass of crustaceans, commonly called fish lice, that are ectoparasites on fish and amphibia.

branch traces the vascular bundles connecting those of a main stem to those of a branch.

brand n. [A.S. beornan, to burn] a burnt appearance on leaves, caused by rust and smut fungi.

brand fungi a common name for the Ustilaginales, q.v.

brand spore a thick-walled spore of Ustilaginales; uredospore of Uredinales; teleutospore, q.v.

breakage and reunion the generally accepted mechanism of crossing-over in meiosis, in which chromatids break and rejoin in a way that differs from the original, resulting in a visible chiasma; cf. copy choice.

bregma n. [Gk. bregma, top of the head] that part of skull where frontals and parietals meet; intersection of sagittal and coronal sutures.

brephic

brephic *a.* [Gk. *brephikos*, newborn] *appl.* a larval phase preceding that of adult form; *alt.* neanic.

brephoplastic *a.* [Gk. *brephikos*, newborn; *plastos*, formed] *appl.* graft in which embryonic tissue is transplanted to an adult.

brevicaudate *a.* [L. *brevis*, short; *cauda*, tail] with a short tail.

brevifoliate *a.* [L. *brevis*, short; *folium*, leaf] having short leaves.

brevilingual *a.* [L. *brevis*, short; *lingua*, tongue] with short tongue.

breviped *a.* [L. *brevis*, short; *pes*, foot] having short legs, *appl.* certain birds.

brevipennate *a.* [L. *brevis*, short; *penna*, feather] with short wings.

brevirostrate *a.* [L. *brevis*, short; *rostrum*, beak] with short beak or rostrum.

brevissimus oculi obliquus inferior, shortest muscle of eye.

bridge a chromosomal arrangement seen at anaphase of meiosis which is produced from a dicentric strand.

bridge corpuscle desmosome, *q.v.*

brigade a taxonomic group used in different ways by different authors.

bristletails a common name for many members of the Thysanura, *q.v.*, and sometimes for the Diplura, *q.v.*

bristle worms the common name for the Polychaeta, *q.v.*

brittle stars the common name for the Ophiuroidea, *q.v.*

Broca's area [*P. Broca*, French surgeon] parolfactory area of brain; Broca's gyrus, *q.v.*

Broca's gyrus left inferior frontal gyrus, speech centre in cerebral cortex; *alt.* Broca's area.

brochidodrome *a.* [Gk. *brochos*, loop; *dramein*, to run] *appl.* veins in leaves when they form loops within the blade.

brochonema *n.* [Gk. *brochos*, loop; *nēma*, thread] the spireme in loops equivalent to the number of chromosome pairs to be formed.

bromatium *n.* [Gk. *brōma*, food] a hyphal swelling on a fungus cultivated by ants, and serving as food.

Bromeliales *n.* [*Bromelia*, generic name] an order of monocots (Liliidae) comprising the single family Bromeliaceae, the bromeliads, which are usually rosette plants with narrow leaves, and include the pineapple.

bromelin *n.* [*Bromelia*, generic name] a proteolytic and milk clotting enzyme found in pineapples; EC 3.4.22.4; *r.n.* bromelain.

bronchi *n.plu.* [Gk. *brongchos*, windpipe] tubes connecting trachea with lungs; *sing.* bronchus.

bronchia *n.plu.* [Gk. *brongchos*, windpipe] the subdivisions or branches of each bronchus.

bronchial *pert.* bronchi.

bronchiole *n.* [Gk. *brongchos*, windpipe] a small terminal branch of bronchi.

bronchopulmonary *a.* [Gk. *brongchos*, windpipe; L. *pulmo*, lung] *pert.* bronchi and lungs.

bronchotracheal *a.* [Gk. *brongchos*, windpipe; L. *trachia*, trachea] *pert.* bronchi and trachea.

bronchovesicular *a.* [Gk. *brongchos*, windpipe; L. *vesicula*, little sac] *pert.* bronchial tubes and lung cells.

bronchus *sing.* of bronchi.

brood bud a spore of certain types of sporangia; soredium, *q.v.*; bulbil, *q.v.*

brood cells gonidia, *q.v.*

brood pouch a sac-like cavity in which eggs or embryos are placed, such as a space formed by overlapping plates attached to bases of thoracic limbs in certain Crustacea; *alt.* ovisac.

brown algae the common name for the Phaeophyceae, *q.v.*

brown body a brown, rounded mass of compacted degenerate organs in some Polyzoa; brown cell, *q.v.*

brown cell or body nephrocyte in ascidians.

brown earths, brown forest soils soils associated with areas of the world's land surface, originally covered with deciduous forest.

brown funnels a single pair of organs on dorsal aspect of posterior end of pharynx in *Amphioxus*, the lancelet; *alt.* atriocoelomic funnels, brown canals.

Brownian movements [*R. Brown*, Scottish botanist] movement of small particles such as pollen grains when suspended in a colloidal solution, due to their bombardment by colloidal molecules.

Bruch's membrane [*C. W. L. Bruch*, German anatomist] a thin basal membrane forming the inner layer of the choroid; *alt.* lamina basalis, lamina vitrea.

Brunner's glands [*J. C. Brunner*, Swiss anatomist] racemose glands in the submucosa of the duodenum which open into the crypts of Lieberkühn and with the crypts secrete succus entericus; *alt.* duodenal glands.

brush border a dense arrangement of very minute cylindrical processes, microvilli, with intervening canals, on surface of epithelial cells bordering lumen of intestine, and on renal epithelial cells.

brush cell echinidium, *q.v.*

Bryales *n.* [Gk. *bryon*, moss] an order or in some classifications superorder of mosses of subclass Bryidae, *q.v.*

Bryidae *n.* [Gk. *bryon*, moss] a subclass of mosses having the sporophyte consisting of a seta and a capsule opening by a peristome, and a filamentous protonema.

bryology *n.* [Gk. *bryon*, moss; *logos*, discourse] the science dealing with mosses and liverworts; *cf.* muscology, hepaticology.

Bryophyta, bryophytes *n.*, *n.plu.* [Gk. *bryon*, moss; *phyton*, plant] a major division of the plant kingdom including the classes Musci (mosses), Hepaticae (liverworts), and in some classifications the Anthocerotae (hornworts), having well-marked alternation of generations, with an independent photosynthetic gametophyte and a sporophyte completely or partly depending on it.

Bryopsida *n.* [Gk. *bryon*, moss] Musci, *q.v.*

Bryozoa, bryozoans *n.*, *n.plu.* [Gk. *bryon*, moss; *zoön*, animal] a phylum of small marine colonial animals which superficially resemble seaweeds and are known as moss animals, the individuals being called zooids and bearing a crown of ciliated tentacles; consists of 2 very distinct groups, the Endoprocta and Ectoprocta which are now usually considered as separate phyla; *alt.* Polyzoa.

bryozoon *n.* [Gk. *bryon*, moss; *zöon*, animal] a member of the Bryozoa, *q.v.*

B-substance intermedin, *q.v.*

buccae *n.plu.* [L. *bucca*, cheek] the cheeks.

buccal *a.* [L. *bucca*, cheek] *pert.* the cheek or mouth.

buccal cavity the part of the alimentary tract between the mouth and pharynx.

buccinator *n.* [L. *buccinator*, trumpeter] a broad thin muscle of the cheek.

buccolabial *a.* [L. *bucca*, cheek; *labium*, lip] *pert.* mouth cavity and lips.

buccolingual *a.* [L. *bucca*, cheek; *lingua*, tongue] *pert.* cheeks and tongue.

bucconasal *a.* [L. *bucca*, cheek; *nasus*, nose] *pert.* cheek and nose; *appl.* membrane closing posterior end of olfactory pit.

buccopharyngeal *a.* [L. *bucca*, cheek; Gk. *pharyngx*, throat] *pert.* cheeks and pharynx, *appl.* membrane and fascia.

bucculae *n.plu.* [L. *buccula*, little cheek] in Heteroptera, ridges under the head on either side of rostrum.

buckle clamp connection, *q.v.*

bud *n.* [M.E. *budde*, bud] a rudimentary shoot or flower; gemma, *q.v.*; an incipient outgrowth, as of limbs, etc.

budding *n.* [M.E. *budde*, bud] the production of buds; reproduction by development of one or more outgrowths or buds which may or may not be set free, in plants and many primitive animals; artificial propagation by insertion of a bud within the bark of another plant.

buffer *n.* [O.F. *buffe*, blow] *appl.* salt solution which minimizes changes in pH when an acid or alkali is added; *appl.* cells: conidia formed in a chain, as in certain Phycomycetes.

buffer(ing) gene polygene, *q.v.*

bufotoxins *n.plu.* [L. *bufo*, toad; Gk. *toxikon*, poison] toad venoms, as bufotoxin, $C_{34}H_{46}O_{10}$, and bufonin, $C_{34}H_{54}O_2$.

bugs the common name for the Hemiptera, *q.v.*

bulb *n.* [L. *bulbus*, globular root] a specialized underground perennating and reproductive organ consisting of a short stem bearing a number of swollen fleshy leaf bases or scale leaves, the whole enclosing next year's bud; a part resembling a bulb; a bulb-like dilatation; basal part of intromittent organ in spiders; medulla oblongata, *q.v.*; *alt.* bulbus.

bulbar *a.* [L. *bulbus*, globular root] *pert.* a bulb or bulb-like part; *pert.* medulla oblongata.

bulbet a small bulb.

bulbiferous *a.* [L. *bulbus*, bulb; *ferre*, to carry] bearing bulbs or bulbils.

bulbil *n.* [L. *bulbus*, bulb] a fleshy axillary bud which may fall and produce a new plant, as in some lilies, *alt.* aerial bulb; any small bulb-shaped structure or dilatation; *alt.* brood bud, tubercle.

bulbocavernosus *n.* [L. *bulbus*, bulb; *cavernosus*, cavernous] a muscle of perinaeum ejaculator urinae in the male; sphincter of vagina.

bulbonuclear *a.* [L. *bulbus*, bulb; *nucleus*, kernel] *pert.* medulla oblongata and nuclei of cranial nerves.

bulbo-urethral *a.* [L. *bulbus*, bulb; Gk. *ourethra*, urethra] *appl.* 2 racemose glands opening into bulb of male urethra, *alt.* antiprostate, Cowper's or Mery's glands; *appl.* the greater vestibular glands in the female, *alt.* Bartholin's glands.

bulbous *a.* [L. *bulbus*, bulb] like a bulb; developing from a bulb; having bulbs.

bulbus *n.* [L. *bulbus*, bulb] bulb, *q.v.*, swollen base of stipe in agarics; the knob-like part found in

connection with various nerves; a dilatation of base of aorta.

bulla *n.* [L. *bulla*, bubble] *appl.* rounded prominence formed by bones of ear, tympanic bulla; *appl.* prominence of middle ethmoidal air cells; *appl.* structure in head of certain parasitic copepods, becoming extruded and attached to gill filament of fish.

bullate *a.* [L. *bulla*, bubble] blistered-like; puckered like a savoy cabbage leaf.

bulliform *a.* [L. *bulla*, bubble; *forma*, shape] bubble-shaped; *appl.* thin-walled cells which cause rolling, folding, or opening of leaves by turgor changes, *alt.* motor cell.

Bullomorpha *n.* [*Bulla*, generic name; Gk. *morphē*, form] Cephalaspidea, *q.v.*

bundle sheath one or more layers of large parenchyma or sclerenchyma cells surrounding a vascular bundle.

bunodont *a.* [Gk. *bounos*, mound; *odous*, tooth] having molar teeth with low conical cusps.

bunoid *a.* [Gk. *bounos*, mound; *eidos*, form] *appl.* cusps of cheek teeth, low and conical.

bunolophodont *a.* [Gk. *bounos*, mound; *lophos*, crest; *odous*, tooth] between bunodont and lophodont in structure, *appl.* cheek teeth.

bunoselenodont *a.* [Gk. *bounos*, mound; *selēnē*, moon; *odous*, tooth] having internal cusps bunoid, external selenoid, *appl.* cheek teeth.

bunt fungi a common name for the Ustilaginales, *q.v.*

bursa *n.* [L. *bursa*, purse] a sac-like cavity; a sac with viscid fluid to prevent friction at joints.

bursa copulatrix a genital pouch of various animals.

bursa entiana the short duodenum in Chondropterygii.

bursa Fabricii a sac opening into dorsal part of posterior region of cloaca in birds, and usually degenerating during adolescence.

bursa seminalis fertilization chamber of female genital ducts, as in Turbellaria.

bursicle *n.* [L. *dim.* of *bursa*, purse] in orchid flowers, a flap- or purse-like structure surrounding a viscidium and containing a sticky liquid to prevent the disc of the viscidium from drying up.

bursicon *n.* [L. *dim.* of *bursa*, purse] a hormone found in the blood of insects after moulting, needed for tanning and hardening the new cuticle.

bursicule *n.* [L. *dim.* of *bursa*, purse] a small sac.

butterflies the common name for many members of the Lepidoptera, which have clubbed antennae.

butterfly bone sphenoid, *q.v.*

buttress roots branch roots given off above ground, arching away from stem before entering soil, forming additional props; *alt.* prop roots, stilt roots, strut roots, drop roots.

butyrinase *n.* [L. *butyrum*, butter] an enzyme occurring mainly in blood serum, which hydrolyses butyrin.

butyrin(e) *n.* [L. *butyrum*, butter] one of the 3 glycerides of butyric acid, especially tributyrin.

Buxbaumiales *n.* [*Buxbaumia*, generic name] an order, or in some classifications, superorder of mosses of subclass Bryidae, having an asymmetrical capsule.

byssaceous byssoid, *q.v.*

byssal *a.* [Gk. *byssos*, fine flax] *pert.* the byssus.

byssogenous *a.* [Gk. *byssos*, fine flax; *genos*, birth] byssus-forming, *appl.* glands.

byssoid *a.* [Gk. *byssos*, fine flax; *eidos*, shape] resembling a byssus; formed of fine threads; *alt.* byssaceous.

byssus *n.* [Gk. *byssos*, fine flax] the tuft of strong filaments secreted by a gland of certain bivalve molluscs, by which they become attached; the stalk of certain fungi.

C

caco- *alt.* kako-.

cacogenesis *n.* [Gk. *kakos*, bad; *genesis*, descent] inability to hybridize.

cacogenic *a.* [Gk. *kakos*, bad; *genos*, birth] dysgenic, *q.v.*

Cactales, Cactiflorae *n.* [*Cactus*, generic name; L. *flos*, flower] Opuntiales, *q.v.*

cacuminous *a.* [L. *cacumen*, peak] with a pointed top, *appl.* trees.

cadavericole *n.* [L. *cadere*, to fall; *colere*, to dwell] an organism feeding on carrion.

cadaverine *n.* [L. *cadere*, to fall] a toxic diamine occurring in putrefied foods, especially meat, formed by the action of bacterial decarboxylases on lysine.

caddis flies the common name for the Trichoptera, *q.v.*

cadophore *n.* [Gk. *kados*, cask; *pherein*, to bear] a dorsal bud-bearing outgrowth in certain tunicates.

caducibranchiate *a.* [L. *caducus*, falling; *branchiae*, gills] with temporary gills.

caducous *a.* [L. *caducus*, falling] *pert.* parts that fall off early, e.g. calyx, stipules; *alt.* fugacious; *cf.* deciduous.

caeca *plu.* of caecum.

caecal *a.* [L. *caecus*, blind] ending without outlet, *appl.* stomach with cardiac part prolonged into blind sac; *pert.* caecum; *alt.* cecal.

caecum *n.* [L. *caecus*, blind] a blind diverticulum or pouch from some part of alimentary canal; *alt.* cecum; *plu.* caeca.

caecum cupulare the closed apical end of the cochlear canal.

caecum vestibulare the closed lower end of the cochlear duct.

caeno- *alt.* caino-, ceno-, kaino-.

Caenogaea *n.* [Gk. *kainos*, recent; *gaia*, earth] a zoogeographical region which includes the Nearctic, Palearctic, and Oriental regions; *alt.* Cainogea, Kainogaea; *cf.* Eogaea.

caenogenesis *n.* [Gk. *kainos*, recent; *genesis*, origin] the non-phylogenetic processes in development of an individual; development of transitory adaptations in early stages of an individual.

caenogenetic *a.* [Gk. *kainos*, recent; *genesis*, origin] of recent origin.

Caenozoic *a.* [Gk. *kainos*, recent; *zōē*, life] the geological era following the Mesozoic, subdivided into the Tertiary and Quaternary, and known as the age of mammals.

caeruloplasmin ceruloplasmin, *q.v.*

caespitose *a.* [L. *caespes*, turf] *pert.* turf; having low, closely matted stems; growing densely in tufts; *alt.* caespitulose, cespitose.

caffeine *n.* [Gk. *kaffein*, coffee] a bitter compound found in coffee, tea, maté, and kola nuts, which is a stimulant of the CNS and a diuretic.

caino- *alt.* caeno-, ceno-, kaino-.

caisson *n.* [F. *caisson*, coffer] box-like arrangement of longitudinal muscle fibres in Lumbricidae, earthworms.

calamistrum *n.* [L. *calamistrum*, curling-iron] a comb-like structure on metatarsus of certain spiders.

Calamitales *n.* [*Calamites*, generic name] an order of fossil Sphenopsida having woody, often tree-like sporophytes and small, often forked leaves.

Calamophyta *n.* [*Calamites*, generic name; Gk. *phyton*, plant] Sphenopsida, *q.v.*

calamus *n.* [L. *calamus*, reed] a hollow reed-like stem without nodes; the quill of a feather; calamus scriptorius: the tip of posterior part of floor of 4th ventricle.

Calanoida *n.* [*Calanus*, generic name; Gk. *eidos*, form] an order of free-living, usually planktonic copepods.

calcaneus, calcaneum *n.* [L. *calx*, heel] the heel; large bone of tarsus which forms the heel, *alt.* os calcis; process on metatarsus of birds, *alt.* hypotarsus; *alt.* calx.

calcar *n.* [L. *calcar*, spur] a hollow prolongation or tube at base of sepal or petal; spur-like process on leg or wing of birds, *alt.* spica; tibial spine in insects; process of calcaneus which supports web between leg and tail in bats; prehallux of amphibians; internal bony plate strengthening neck of femur.

calcarate *a.* [L. *calcar*, spur] spurred, *appl.* petal, corolla.

calcar avis eminence in posterior part of lateral ventricle; *alt.* hippocampus minor, unguis.

Calcarea *n.* [L. *calcarius*, limy] a class of sponges of the subphylum Gelatinosa, commonly called calcareous sponges, having a calcareous skeleton of 1-, 3- or 4-rayed spicules; the orders are variable; using adult features, *see* Homocoela, Heterocoela; using cytological and embryological features, *see* Calcinea, Calcaronea.

calcareous *a.* [L. *calcarius*, limy] limy; growing on soil derived from decomposition of calcareous rocks; *pert.* limestone.

calcariform *a.* [L. *calcar*, spur; *forma*, shape] spur-like.

calcarine *a.* [L. *calcar*, spur] *pert.* calcar avis; *appl.* fissure extending to hippocampal gyrus, on medial surface of cerebral hemisphere.

Calcaronea *n.* [L. *calcarius*, limy] a subclass of Calcarea having an amphiblastula larva, the adults with the flagellum of choanocytes arising from an apical nucleus.

calceiform calceolate, *q.v.*

calceolate *a.* [L. *calceolus*, small shoe] slipper-shaped, *appl.* corolla; *alt.* calceiform.

calcicole *n.* [L. *calx*, lime; *colere*, to dwell] a plant which thrives in soil rich in calcium salts; *alt.* calcipete, calciphile, calciphyte, gypsophyte; *a.* calcicolous; *cf.* calcifuge.

calciferol an unsaturated alcohol, vitamin D_2, which can be made by ultraviolet irradiation of ergosterol, occurs in fish liver oils, egg yolk, milk, etc., controls levels of calcium and phosphorus in the body, and is used in the treatment of rickets, *alt.* ergocalciferol; a number of chemically related compounds which prevent or cure rickets; *alt.* antirhachitic vitamin; *cf.* viosterol.

calciferous *a.* [L. *calx*, lime; *ferre*, to carry] containing or producing lime salts; *alt.* calcigerous.

calcific *a.* [L. *calx*, lime; *facere*, to make] producing lime salts; *appl.* part of oviduct forming eggshell in reptiles and birds.

calcification *n.* [L. *calx*, lime; *facere*, to make] the deposition of lime salts in tissue; the process of accumulation of lime salts in soil development; *a.* calcified.

calcifuge *n.* [L. *calx*, lime; *fugere*, to flee] a plant which thrives only in soils poor in calcium carbonate and usually acid; *alt.* basifuge, calciphobe, oxyphyte, oxylophyte; *cf.* calcicole, silicole.

calcigerous calciferous, *q.v.*

Calcinea *n.* [L. *calx*, lime] a subclass of Calcarea with a parenchymula-like larva, the adults having the flagellum of choanocytes arising independently of the nucleus.

calcipete *n.* [L. *calx*, lime; *petere*, to go towards] calcicole, *q.v.*

calciphile *n.* [L. *calx*, lime; Gk. *philein*, to love] calciphyte, *q.v.*

calciphobe *n.* [L. *calx*, lime; Gk. *phobos*, fear] calcifuge, *q.v.*

calciphyte *n.* [L. *calx*, lime; Gk. *phyton*, plant] a plant which thrives only on calcareous soils: *alt.* calcicole, calcipete, calciphile, gypsophyte.

calcitonin *n.* [L. *calx*, lime; *tonos*, tension] a polypeptide hormone secreted by the thyroid and/or parathyroid gland in mammals and by the ultimobranchial bodies in other vertebrates, which lowers the level of calcium in the blood (opposing the activity of parathyroid hormone) by reducing release of calcium from bone; *alt.* thyrocalcitonin.

calcivorous *a.* [L. *calx*, lime; *vorare*, to devour] *appl.* plants which live on limestone.

calcospherites *n.plu.* [L. *calx*, lime; *sphaera*, globe] concentrically laminated granules of calcium carbonate in Malpighian tubes of some insects, in cells associated with fat-body in certain larval Diptera.

calice calyx, *q.v.*; calycle, *q.v.*

Caliciales *n.* [*Calicium*, generic name] an order of ascolichens in which the asci disintegrate so that the spores form a loose mass.

calicle calycle, *q.v.*

caligate *a.* [L. *caliga*, boot] sheathed; veiled; peronate, *q.v.*; laminiplantar, *q.v.*; *alt.* booted.

Caligoida *n.* [*Caligus*, generic name; Gk. *eidos*, form] an order of copepods that are mostly ectoparasites on fish, and attach to the host by modified 2nd antennae.

calines *n.plu.* [Gk. *kalein*, to summon] plant hormones influencing growth of specific parts, as of root, stem, or leaf.

caliology *n.* [Gk. *kalia*, cabin; *logos*, discourse] the study of homes or shelters made by animals, as of burrows, nests, hives, etc.

callosal *a.* [L. *callosus*, hard] *pert.* corpus callosum.

callose *n.* [L. *callum*, hard skin] an amorphous polysaccharide which gives glucose on hydrolysis, usually found on sieve plates, but also on parenchyma cells after injury. *a.* Having callosities.

callosity *n.* [L. *callositas*, hardness] hardened and thickened area on skin, or on bark; *alt.* callus.

callosum corpus callosum, *q.v.*

callow *n.* [A.S. *calu*, bald] a newly hatched worker ant. *a.* Unfledged.

callus *n.* [L. *callum*, hard skin] tissue that forms over cut or damaged plant surface; deposit of callose on sieve plates; small hard outgrowth at base of spikelet in some grasses; a growth of shell-like material within umbilicus of shell; a mesonotal swelling in some insects; callosity, *q.v.*

Calobryales *n.* [*Calobryum*, generic name] an order of liverworts, by some authorities included in the Jungermanniales, *q.v.*

Caloplacales *n.* [*Caloplaca*, generic name] an order of ascolichens having thick-walled, usually 2-celled spores.

caloricity *n.* [L. *calere*, to be warm] in animals, the power of developing and maintaining a certain degree of heat.

calorie *n.* [L. *calere*, to be warm] amount of heat required to raise temperature of one gramme of water 1°C (small calorie); one large Calorie (kilocalorie) equals 1000 small calories.

calorifacient, calorigenic *a.* [L. *calor*, heat; *facere*, to make; *genere*, to beget] promoting oxygen consumption and heat production.

calotte *n.* [F. *calotte*, skull cap] an outer cell group or polar cap in Dicyemidae, for adhesion to kidney of Cephalopoda; a retractile disc with sensory cilia, in larval Polyzoa; lid of an ascus.

caltrop *n.* [A.S. *coltraeppe*, thistle] a sponge spicule with 4 rays so disposed that any 3 being on the ground the 4th projects vertically upwards; *alt.* calthrop.

calvaria *n.* [L. *calvaria*, skull] the dome of the skull.

Calvin cycle [*Melvin Calvin*, American biophysicist] a cycle of reactions resulting in the fixation of carbon dioxide into C_3-carbohydrate in photosynthesis in plants; *alt.* Benson–Calvin–Bassham cycle.

calx *n.* [L. *calx*, lime, heel] lime; calcaneus, *q.v.*

calycanthemy *n.* [Gk. *kalyx*, calyx; *anthemon*, flower] abnormal development of parts of calyx into petals.

calyces *plu.* of calyx.

calyciflorous *a.* [L. *calyx*, calyx; *flos*, flower] *appl.* flowers in which stamens and petals are adnate to the calyx.

calyciform *a.* [L. *calyx*, calyx; *forma*, shape] calyx-like in shape; *alt.* calycoid.

calycine *a.* [L. *calyx*, calyx] *pert.* a calyx; cup-like.

calycle *n.* [L. *calyculus*, little calyx] epicalyx, *q.v.*; a cup-shaped cavity in a coral; a theca in a hydroid; calyx, *q.v.*; *alt.* calice, calicle, calyculus.

calycoid *a.* [Gk. *kalyx*, calyx; *eidos*, form] calyciform, *q.v.*

calyculus *n.* [L. *calyculus*, little calyx] cup-shaped or bud-shaped structure; calycle, *q.v.*

calyculus gustatorius taste bud, *q.v.*; *alt.* gustatory calyculus.

calyculus ophthalmicus optic cup, formed by invagination of the optic bulb and developing into the retina.

calymma kalymma, *q.v.*

calymmocytes kalymmocytes, *q.v.*

calypter *n.* [Gk. *kalyptos*, hidden] antitegula or modified alula covering haltere in certain Diptera; *alt.* calyptron.

Calyptoblastea n. [Gk. *kalyptos*, hidden; *blastos*, bud] Thecata, *q.v.*

calyptoblastic a. [Gk. *kalyptos*, hidden; *blastos*, bud] *pert.* hydroids in which gonophore is enclosed in a gonotheca; *cf.* gymnoblastic.

calyptobranchiate a. [Gk. *kalyptos*, hidden; *brangchia*, gills] with gills not visible from exterior.

calyptopsis n. [Gk. *kalyptos*, hidden; *opsis*, sight] a larva of some crustaceans, being a modified zoea.

calyptra n. [Gk. *kalyptra*, covering] tissue enclosing developing sporogonium in liverworts; remains of archegonium which surround apex of capsule in mosses; neck of archegonium in prothallus of some pteridophytes; root cap, *q.v.*; *alt.* veil; *cf.* calyptrogen.

calyptrate a. [Gk. *kalyptra*, covering] *appl.* caducous calyx separating from its lower portion or from thalamus; operculate, *q.v.*; *appl.* Diptera with halteres hidden by squamae.

calyptrogen n. [Gk. *kalyptra*, covering; *gennaein*, to produce] the meristem giving rise to the root cap independently of other initials in an apical meristem; *cf.* calyptra.

calyptron n. [Gk. *kalyptra*, covering] calypter, *q.v.*

calyx n. [Gk. *kalyx*, calyx] the sepals collectively, forming the outer whorl of the flower; cup-like portion of pelvis of kidney; theca of certain hydroids; cup-like body of crinoids; cup or head of pedunculate bodies in insects; a cup- or pouch-like structure; *plu.* calyces; *alt.* calice, calycle.

CAM Crassulacean Acid Metabolism, *q.v.*

Camallanida, **Camallanata** n. [*Camallanus*, generic name] an order of Phasmidia having adults parasitic in vertebrate connective tissue and larvae in copepods.

camarodont a. [Gk. *kamara*, vault; *odous*, tooth] *appl.* Aristotle's lantern having the 2 epiphyses of a pair of pyramids fused together.

Camaroidea n. [Gk. *kamara*, vault; *eidos*, form] an order of graptolites having bithecae and autothecae, the latter specialized in shape.

cambial a. [L. *cambium*, change] *pert.* cambium.

cambiform a. [L. *cambium*, change; *forma*, shape] similar to cambium cells.

cambiogenetic a. [L. *cambium*, change; Gk. *genesis*, origin] *appl.* cells which produce cambium.

cambium n. [L. *cambium*, change] the meristematic tissue from which secondary growth arises in stems and roots.

Cambrian a. [L. *Cambria*, Wales] *pert.* earliest period, or system of rocks, of Palaeozoic era.

cameration n. [L. *cameratio*, vaulting] division into a large number of separate chambers.

camerostome n. [L. *camera*, chamber; Gk. *stoma*, mouth] hollow in anterior part of podosoma, for reception of gnathostoma in Acarina.

cAMP cyclic AMP, *q.v.*

Campanales n. [*Campanula*, generic name] an order of dicots (Gamopetalae) of the Bentham and Hooker system, including the Campanulaceae, bellflowers.

campaniform a. [L.L. *campana*, bell; *forma*, shape] bell- or dome-shaped; *appl.* sensilla of certain insects.

campanula Halleri [Dim. of L.L. *campana*, bell; *A. von Haller*, Swiss anatomist] expansion of falciform process at lens in many fishes.

Campanulales n. [*Campanula*, generic name] an order of dicots of the Asteridae or Sympetalae, used in different ways by different authorities, and including the Campanulaceae, bellflowers.

campanulate a. [Dim. of L.L. *campana*, bell] bell-shaped, *appl.* corolla.

campodeiform a. [Gk. *kampē*, caterpillar; *eidos*, form; L. *forma*, shape] having a larva resembling the thysanuran *Campodea* having well-developed legs, antennae, cerci, and mouth-parts, as in many Coleoptera; *alt.* thysanuriform.

camptodrome a. [Gk. *kamptos*, flexible; *dromos*, course] *pert.* leaf venation in which secondary veins bend forward and anastomose before reaching margin.

camptotrichia n.plu. [Gk. *kamptos*, flexible; *thrix*, hair] jointed dermal fin-rays in certain primitive fishes.

camptotrophism n. [Gk. *kamptein*, to bend; *trophē*, maintenance] the effects of bending strains upon plant structures and functions.

campylodrome a. [Gk. *kampylos*, curved; *dromos*, course] acrodrome, *q.v.*

campylospermous a. [Gk. *kampylos*, curved; *sperma*, seed] *appl.* seeds with groove along inner face.

campylotropous a. [Gk. *kampylos*, curved; *tropē*, turning] *appl.* ovules bent so that the funicle appears attached to the side halfway between the chalaza and micropyle; *alt.* notorhizal.

canalicular a. [L. *canaliculus*, small channel] *pert.* canals, or canaliculi.

canalicular apparatus formerly the Golgi bodies, regarded as a system of canals.

canaliculus n. [L. *canaliculus*, small channel] one of the small canals containing cell processes of bone corpuscles and connecting lacunae in Haversian system; small channel for passage of nerves through various bones; one of the minute acid-containing channels in cytoplasm of oxyntic cells; a small channel.

canaliform a. [L. *canalis*, channel; *forma*, shape] canal-like.

canavanine n. [*Canavalia*, genus of tropical twining plants including jack bean] an amino acid found in the jack bean.

cancellous a. [L. *cancellosus*, latticed] consisting of slender fibres and lamellae, which join to form a reticular structure, *alt.* cancellated; *appl.* inner, more spongy, portion of bony tissue; *appl.* anterior portion of cuttlebone.

cancrisocial a. [L. *cancer*, crab; *socius*, ally] *appl.* commensals with crabs.

cane sugar sucrose, *q.v.*, occurring both in sugar cane and sugar beet, and in some other plants.

canine n. [L. *caninus*, *pert.* dog] the tooth next to incisors. a. *Pert.* canine tooth; *pert.* a fossa and eminence on anterior surface of maxilla; *pert.* a dog.

caninus n. [L. *caninus*, canine] muscle from canine fossa to angle of mouth; *alt.* levator anguli oris.

cannabis n. [*Cannabis*, generic name] Indian hemp, *Cannabis sativa*, the plant from which the drug cannabis is extracted, also cultivated for its stem fibres that are used to make ropes; the hallucinogenic drug obtained from the plant as a resin from the flowering tops of the female flowers (hashish, charas), or as dried leaves or resin from the leaves, *alt.* marijuana, marihuana.

cannon bone bone supporting limb from hock to fetlock, enlarged and fused metacarpals or metatarsals; in birds, the tarsometatarsus, *q.v.*

canopy *n.* [Gk. *kōnōpeion*, curtained bed] topmost layer of leaves, twigs, and branches of forest trees, or of other woody plants.

canthal *a.* [Gk. *kanthos*, corner of eye] *pert.* canthus; *appl.* a scale in certain reptiles.

cantharidin *n.* [*Cantharidae*, blister beetles, from Gk. *kantharos*] a toxic irritant and blister-causing terpene derivative produced by the Spanish fly and other beetles of the family Meloidae, blister beetles.

canthus *n.* [Gk. *kanthos*, corner of eye] the angle where upper and lower eyelids meet; *alt.* commissura palpebrarum.

capillary *a.* [L. *capillus*, hair] hair-like; *appl.* moisture held between and around particles of soil. *n.* One of the minute thin-walled vessels which form networks in various parts of body, e.g. blood, lymph, or biliary capillaries.

capillitium *n.* [L. *capillus*, hair] a structure in the sporangium of slime moulds and in the fruit body of some Gastromycetes, made of tangled strands in which spores are entangled.

capitate *a.* [L. *caput*, head] enlarged or swollen at tip; gathered into a mass at apex, as compound stigma, some inflorescences; *appl.* a bone, os capitatum.

capitatum *n.* [L. *caput*, head] the 3rd carpale; *alt.* os magnum, os capitatum.

capitellum *n.* [*Dim.* of L. *caput*, head] a capitulum or articulatory protuberance at end of a bone; tentacle-bearing structure of a polyp.

capitular *a.* [L. *capitulum*, small head] *pert.* a capitellum or capitulum.

capitulum *n.* [L. *capitulum*, small head] a knoblike swelling at end of a bone, e.g. on humerus for articulation with radius; part of cirripede body enclosed in mantle, *cf.* pedicle; swollen end of hair or tentacle; enlarged end of insect proboscis, or antenna; exsert part of head in ticks; part of column above parapet in sea anemones; spherical apothecium containing a powdery mass of spores, in certain lichens; spherical cell at inner end of manubrium in Charales; an inflorescence forming a head of sessile flowers or florets crowded together on a receptacle and usually surrounded by an involucre; *cf.* anthodium; *plu.* capitula.

Capparales *n.* [*Capparis*, generic name] an order of dicots (Dilleniidae) having hermaphrodite, often tetramerous flowers, formerly included with the Papaverales in the Rhoeadales.

capreolate *a.* [L. *capreolus*, tendril] supplied with tendrils; tendril-shaped.

caprification *n.* [L. *caprificus*, wild fig tree] pollination of flowers of fig tree by chalcid wasps; the process of hanging caprifig branches in female fig trees to cause pollination.

Caprimulgiformes *n.* [*Caprimulgus*, generic name of nightjar; L. *forma*, shape] an order of birds of the subclass Neornithes, including the nightjars.

Capsaloidea *n.* [*Capsala*, generic name; Gk. *eidos*, form] an order of Monopisthocotylea that are oviparous and have testes anterior to the ovary.

capsid *n.* [L. *capsa*, box] the protein capsule around the nucleic acid of a virus. [*Capsis*, generic name] the common name for a bug of the family Capsidae.

capsomere *n.* [L. *capsa*, box; Gk. *meros*, part] one of the protein units of which a capsid is made.

capsula glomeruli Bowman's capsule, *q.v.*

capsula lentis *see* phacocyst.

capsular *a.* [L. *capsula*, little box] like or *pert.* a capsule; *appl.* dry, dehiscent, many-seeded fruits, as capsule, follicle, legume, silicula, siliqua.

capsule *n.* [L. *capsula*, little box] a sac-like membrane enclosing an organ; thickened slime layer surrounding certain bacteria; any closed box-like vessel containing spores, seeds, or fruits; in bryophytes, the portion of the sporogonium containing the spores; a superior, one or more celled, many-seeded, dehiscent fruit; membrane surrounding nerve cells of sympathetic ganglia; *appl.* cells: amphicytes, *q.v.*

capsuliferous, capsuligerous, capsulogenous *a.* [L. *capsula*; little box; *ferre, gerere*, to carry] with, or forming, a capsule.

captacula *n.plu.* [L. *captare*, to lie in wait for] exsertile filamentous tactile organs with sucker-like ends near mouth of Scaphopoda.

Captorhinomorpha, captorhinomorphs *n.*, *n.plu.* [*Captorhinus*, generic name; Gk. *morphē*, form] one of the orders into which the cotylosaurs are sometimes divided, having a small pineal opening and a lightly built skull.

caput *n.* [L. *caput*, head] head; knob-like swelling at apex; peridium of certain fungi; a globule of coherent conidia at tip of a sterigma or phialide.

caput caecum coli former name of caecum.

carapace *n.* [Sp. *carapacho*, covering] a chitinous or bony shield covering whole or part of back of certain animals.

carbamide urea, *q.v.*

carbohydrases hydrolysing enzymes which catalyse various stages in the hydrolysis of higher carbohydrates to simple sugars.

carbohydrates *n.plu.* [L. *carbo*, coal; Gk. *hydor*, water] compounds of carbon, hydrogen, and oxygen, the latter 2 elements usually in the proportion of 2 to 1, and including monosaccharides, disaccharides, oligosaccharides, and polysaccharides.

carbon assimilation photosynthesis resulting in the incorporation of carbon dioxide into organic carbon.

carbon cycle a biological cycle involving the uptake of carbon dioxide into plants and its fixation in photosynthesis, and the various ways in which it is returned to the atmosphere.

carbon dioxide fixation the pathway incorporating carbon dioxide into carbohydrates which occurs in the stroma of the chloroplasts; *alt.* dark reaction.

carbonic anhydrase an enzyme which catalyses the decomposition of carbonic acid to carbon dioxide and water, found in red blood corpuscles, other parts of animals, and in plants; EC 4.2.1.1; *r.n.* carbonate dehydratase.

carbonicole, carbonicolous *a.* [L. *carbo*, charcoal; *colere*, to inhabit] living on burnt soil or burnt wood.

Carboniferous *a.* [L. *carbo*, coal; *ferre*, to carry] *pert.* period of late Palaeozoic era including formation of coal measures.

carboxyhaemoglobin a compound of carbon monoxide and haemoglobin formed in the blood following carbon monoxide poisoning.

carboxylases a group of enzymes, mostly biotin-containing proteins, which form carbon-carbon bonds using carbon dioxide and ATP; EC subgroup 6.4; *r.n.* carboxylases.

carboxypeptidases a group of proteolytic enzymes which hydrolyse the C-terminal amino acid from a peptide.

carcerule, carcerulus *n.* [L. *carcer*, prison] a superior, dry, many-celled fruit, with indehiscent 1- or few-seeded carpels cohering by united styles to a central axis.

carcinology *n.* [Gk. *karkinos*, crab; *logos*, discourse] the study of Crustacea.

cardenolides chemicals, cardiac glycosides, found in milkweeds which cause vertebrates feeding on the plant to vomit, and hence avoid it.

cardia *n.* [Gk. *kardia*, stomach, heart] the opening between oesophagus and stomach; in sucking insects, the enlarged anterior part of the proventriculus.

cardiac *a.* [Gk. *kardiakos*, pert. heart, stomach] *pert.*, near, or supplying heart, *appl.* cycle, etc.; *pert.* anterior part of stomach.

cardiac impulse motion caused by rapid increase in tension of ventricle.

cardiac jelly gelatinous substance between endocardium and myocardium of embryonic tubular heart, replaced by endocardial cells.

cardiac muscle the special type of muscle of which the vertebrate heart is made.

cardiac sphincter the thick ring of muscle around the opening between stomach and oesophagus.

cardiac valve in insects, a valve at the junction of the fore-gut and mid-gut, probably serving to prevent or reduce regurgitation of food from the mid-gut.

cardinal *a.* [L. *cardo*, hinge] *pert.* that upon which something depends or hinges; *pert.* hinge of bivalve shell; *pert.* cardo of insects; *appl.* points for plant growth: minimum, optimum, and maximum temperatures or temperature ranges.

cardinal sinuses and veins veins uniting in Cuvier's duct, persistent in most fishes, embryonic in other vertebrates.

cardines *plu.* of cardo.

cardioblast *n.* [Gk. *kardia*, heart; *blastos*, bud] one of embryonic cells destined to form walls of heart.

cardiobranchial *a.* [Gk. *kardia*, heart; *brangchia*, gills] *appl.* enlarged posterior basibranchial cartilage ventral to heart in elasmobranchs.

cardiogenic *a.* [Gk. *kardia*, heart; *gennaein*, to produce] arising in the heart.

cardiovascular *a.* [Gk. *kardia*, heart; L. *vasculum*, a small vessel] *pert.* heart and blood vessels.

cardo *n.* [L. *cardo*, hinge] the hinge of a bivalve shell; basal sclerite of maxilla in insects, itself divided into eucardo and paracardo; *plu.* cardines.

carina *n.* [L. *carina*, keel] a keel-like ridge on certain bones, as of breast bone of birds; median dorsal plate of a barnacle; the 2 coherent lower petals of a leguminous flower; ridge on bracts of certain grasses.

carinal like, or *pert.*, a keel or ridge; *appl.* median strand of xylem passing from stem to leaf; *appl.* canals in protoxylem beneath ridges of stem in Equisetales; *appl.* dots or puncta on keel of diatom valves; *appl.* cartilage at the bifurcation of the trachea.

Carinatae *n.* [L. *carina*, keel] Neognathae, *q.v.*

carinate *a.* [L. *carina*, keel] having a ridge or keel; *cf.* ratite.

cariniform *a.* [L. *carina*, keel; *forma*, shape] keel-shaped; *alt.* tropeic.

carnassial *a.* [L. *caro*, flesh] *pert.* cutting teeth of Carnivora, 4th premolar above and 1st molar below,—in upper the protocone is reduced, in lower the metaconid.

carneous *a.* [L. *caro*, flesh] flesh-coloured; carnose, *q.v.*

carnitine *n.* [L. *caro*, flesh] a substance found in muscle that transports fatty acid across membranes of mitochondria during fatty acid oxidation.

Carnivora, carnivores *n.* [L. *carno*, flesh; *vorare*, to devour] an order of placental mammals existing from the Palaeocene to the present day, including the dogs, cats, bears, and seals, having strongly developed canine teeth and eating meat or fish, but otherwise showing wide variation.

carnivorous *a.* [L. *caro*, flesh; *vorare*, to devour] flesh-eating, especially as in Carnivora, as secondary or tertiary consumer; *appl.* certain plants which feed on trapped insects or other animals.

Carnosa *n.* [L. *carnosus*, fleshy] an order of Tetractinellida having megascleres and microscleres not clearly different.

carnose *a.* [L. *carnosus*, fleshy] fleshy or pulpy, *appl.* mushrooms, or fruit; *alt.* carneous.

carnosine *n.* [L. *carnosus*, fleshy] a dipeptide found in muscle which has a histamine-like action and produces a fall in blood pressure; *alt.* β-alanyl-L-histidine.

carotenase *n.* [L. *carota*, carrot] a liver enzyme which activates vitamin A formation from carotenes; EC 1.13.11.12; *r.n.* lipoxygenase.

carotene *n.* [L. *carota*, carrot] any of a series of hydrocarbons with the formula $C_{40}H_{56}$, widely distributed often as a yellow pigment in plants and animals, possibly important in plants in phototropism and in animals important as a precursor for the synthesis of vitamin A, so it is also known as provitamin A; *alt.* carotin.

carotenoids *n.plu.* [L. *carota*, carrot; Gk. *eidos*, form] fat-soluble pigments widely distributed in plants and animals and including carotenes and xanthophylls; *alt.* carotinoids.

carotenophore *n.* [L. *carota*, carrot; Gk. *phoros*, bearing] a pigmented stigma or eye-spot.

carotid *a.* [Gk. *karos*, heavy sleep] *pert.* chief arteries in the neck, *appl.* arch, ganglion, nerve, etc.

carotid bodies two small masses of chromaffin cells associated with carotid sinus, and being part of system for controlling oxygen content and acidity of blood; *alt.* glomera carotica.

carotiform *a.* [L. *carota*, carrot; *forma*, shape] shaped like a carrot, *appl.* certain cystidia.

carotin carotene, *q.v.*

carotinoids carotenoids, *q.v.*

carpal *n.* [L. *carpus*, wrist] a wrist bone, *alt.* carpale. *a.* Pert. wrist, carpus.

carpel *n.* [Gk. *karpos*, fruit] a megasporophyll in angiosperms, containing one or more ovules, the carpels together making up the gynaecium; *alt.* carpophyll.

carpellary *a.* [Gk. *karpos*, fruit] *pert.* carpels; containing a carpel or carpels.

carpellate having carpels.

carpellody *n.* [Gk. *karpos*, fruit; *eidos*, form] the changing of a part of a flower into a carpel.

carpocerite *n.* [L. *carpus*, wrist; Gk. *keras*, horn] fifth antennal joint in certain Crustacea.

carpogenic *a.* [Gk. *karpos*, fruit; *gennaein*, to produce] *appl.* those cells in red algae which form the carpogonium; *appl.* cell: oogonium of archicarp; *alt.* carpogenous.

carpogenous *a.* [Gk. *karpos*, fruit; *gennaein*, to produce] growing on or in fruit, *appl.* fungi; carpogenic, *q.v.*

carpogonium *n.* [Gk. *karpos*, fruit; *gonos*, birth] lower portion of procarp, which contains female nucleus, in some thallophytes such as red algae; female gametangium in red algae; ascogonium, *q.v.*; *a.* carpogonial.

Carpoidea *n.* [Gk. *karpos*, fruit; *eidos*, form] an extinct class of stalked Cambrian to Devonian echinoderms of the subphylum Pelmatozoa.

carpolith *n.* [Gk. *karpos*, fruit; *lithos*, stone] a fossil fruit; *alt.* lithocarp.

carpometacarpus *n.* [Gk. *karpos*, wrist; *meta*, after] portion of wing skeleton formed by fusion of carpal and metacarpal bones, in birds.

carpomycetous *a.* [Gk. *karpos*, fruit; *mykēs*, fungus] producing fruit bodies, *appl.* higher fungi.

carpophagous *a.* [Gk. *karpos*, fruit; *phagein*, to eat] feeding on fruit.

carpophore *n.* [Gk. *karpos*, fruit; *pherein*, to bear] part of flower axis to which carpels are attached; stalk of sporocarp.

carpophyll *n.* [Gk. *karpos*, fruit; *phyllon*, leaf] carpel, *q.v.*; megasporophyll, *q.v.*

carpophyte *n.* [Gk. *karpos*, fruit; *phyton*, plant] a thallophyte which forms sporocarps.

carpopodite *n.* [Gk. *karpos*, wrist; *pous*, foot] the 3rd joint of endopodite in certain Crustacea; patella in spiders.

carposoma *n.* [Gk. *karpos*, fruit; *sōma*, body] nonreproductive part of a carpophore; an immature carpophore.

carposperm *n.* [Gk. *karpos*, fruit; *sperma*, seed] the fertilized ovum of any alga; the ovum of a red alga.

carposporangium *n.* [Gk. *karpos*, fruit; *sporos*, seed; *anggeion*, vessel] the terminal cells of filaments developed from fertilized carpogonium in red algae.

carpospore *n.* [Gk. *karpos*, fruit; *sporos*, seed] a spore formed at end of filaments of cystocarp, and developed from carpogonium in red algae; *alt.* cystospore.

carposporophyte *n.* [Gk. *karpos*, fruit; *sporos*, seed; *phyton*, plant] the diploid generation of red algae, which consists of filaments forming carpospores at their apices.

carpostome *n.* [Gk. *karpos*, fruit; *stoma*, mouth] opening for emission of spores from the cystocarp of red algae.

carpostrote *n.* [Gk. *karpos*, fruit; *strotos*, spread] a plant which is dispersed by the whole fruit.

carpotetrasporangium *n.* [Gk. *karpos*, fruit; *tetras*, four; *sporos*, seed; *anggeion*, vessel] a type of sporangium found in some red algae and containing 4 spores.

carpotetraspore *n.* [Gk. *karpos*, fruit; *tetras*,

four; *sporos*, seed] any of the 4 spores contained in a carpotetrasporangium.

carpotropism *n.* [Gk. *karpos*, fruit; *tropē*, turn] the movements of fruit stalk especially after fertilization to place the fruit in a good position for ripening and dispersal.

carpus *n.* [L. *carpus*, wrist] the wrist; region of forelimb between forearm and metacarpus.

carr *n.* [M.E. *ker*, underbrush] fen woodland, usually dominated by alder.

cartilage *n.* [L. *cartilago*, cartilage] a firm, elastic skeletal connective tissue in which the cells, chondrocytes, are embedded in a matrix of chondrin, usually bluish-white and translucent in appearance; formerly called gristle; *see* hyaline cartilage, fibrocartilage.

cartilage bones autostoses, *q.v.*

cartilaginous *a.* [L. *cartilagineus*, gristly] gristly, consisting of or *pert.* cartilage; resembling consistency of cartilage, as cortex of certain fungi.

cartilaginous fishes the common name for the Chondrichthyes, *q.v.*

caruncle *n.* [L. *caruncula*, small piece of flesh] a naked fleshy excrescence; small conical body at inner junction of upper and lower eyelids, caruncula lacrimalis; one of the carunculae hymenales, rounded vestiges of ruptured hymen; a fleshy outgrowth on head of certain birds, and on certain caterpillars; a little horny elevation at end of beak of embryo chicks; piston-like structure within acetabulum of dibranchiate Cephalopoda; suckingdisc on tarsi of certain mites; strophiole, *q.v.*

caryo- *also* karyo-, *q.v.*

caryolite *n.* [Gk. *karyon*, nut; *lytikos*, loosing] a nucleated muscle fragment undergoing phagocytosis in development of insects.

Caryophanales *n.* [*Caryophanon*, generic name] an order of bacteria having rigid cell walls and cells arranged in trichomes.

caryophyllaceous *a.* [*Caryophyllus*, generic name] diacytic, *q.v.*; *pert.* dicotyledon flower family, Caryophyllaceae; *appl.* flowers having long clawed petals.

Caryophyllales *n.* [*Caryophyllus*, generic name] an order of dicots of the Archichlamydeae or Caryophyllidae, used in different ways by different authorities; *alt.* Centrospermae.

Caryophyllidae *n.* [*Caryophyllus*, generic name] a subclass of mainly herbaceous dicots, having actinomorphic, mostly trimerous or pentamerous flowers and simple leaves.

caryopsis *n.* [Gk. *karyon*, nut; *opsis*, appearance] an achene with pericarp and testa inseparably fused, as in grasses; *alt.* grain.

casein *n.* [L. *caseus*, cheese] a mixture of soluble phosphoproteins found in milk, which are precipitated by acidification or the action of rennin, to form an insoluble curd known as paracasein; in some terminologies casein is known as caseinogen and paracasein as casein.

caseinogen *n.* [L. *caseus*, cheese; Gk. *genos*, birth] *see* casein.

Casparian band [R. *Caspary*, German botanist] a cork- or wood-like strip encircling radial walls of endodermis cells; *alt.* Casparian strip.

casque *n.* [Sp. *casco*, helmet] a helmet-like structure in animals, as head-scutes of certain extinct fishes; horny outgrowth of beak in hornbill; frontal extension of beak in coot.

cassideous *a.* [L. *cassis*, helmet] helmet-like.

Cassiduloida n. [*Cassidulus*, generic name; Gk. *eidos*, form] an order of mostly extinct sea urchins of the superorder Atelostomata.

caste n. [L. *castus*, pure] one of the distinct forms found among certain social insects.

castrate a. [L. *castrare*, to castrate] *pert.* flowers from which androecium has been removed, *alt.* emasculated. n. An animal deprived of functional gonads. v. To deprive of testes; to gonadectomize; to inhibit development of gonads; to emasculate.

casual a plant which has been introduced but has not become established as a wild plant, though it does occur uncultivated.

Casuariiformes n. [*Casuarius*, generic name of cassowary; L. *forma*, shape] an order of birds of the subclass Neornithes, including the cassowaries and emus.

Casuarinales, Casuariniflorae n. [*Casuarina*, generic name; L. *flos*, flower] an order of dicots of the Archichlamydeae or Hamamelididae, comprising one family the Casuarinaceae, having the single genus *Casuarina*, she oak; *alt.* Verticillatae, Unisexales.

cat- *also* kat-, *q.v.*

catabolism n. [Gk. *kata*, down; *bolē*, throw] the breaking down of complex organic molecules by living organisms with release of energy; *alt.* dissimilation; *cf.* anabolism.

catabolite n. [Gk. *kata*, down; *bolē*, throw] any product of catabolism; *alt.* catastate; *cf.* anabolite.

catacorolla n. [Gk. *kata*, against; L. *corolla*, little wreath] a secondary corolla.

catadromous a. [Gk. *kata*, down; *dramein*, to run] tending downward; having branches arising from lower side of pinnae, in ferns; having 1st set of nerves in a frond segment given off on basal side of midrib; *appl.* fishes which migrate from fresh to salt water annually, *cf.* anadromous.

catagen n. [Gk. *kata*, down; *gennaein*, to produce] intermediate stage in hair growth cycle.

catalase an enzyme present in all aerobic tissues which catalyses the breakdown of hydrogen peroxide into water and molecular oxygen, and does not depend on an oxygen acceptor; EC 1.11.1.6; *r.n.* catalase; *cf.* peroxidase.

catalepsis n. [Gk. *katalēpsis*, seizure] a so-called shamming-dead reflex, as in spiders; *cf.* kataplexy.

catallact homoplast, *q.v.*

catalysis n. [Gk. *katalysis*, dissolution] acceleration or retardation of reaction due to presence of a catalyst.

catalysor catalyst, *q.v.*

catalyst n. [Gk. *katalysis*, dissolving] an agent, e.g. an enzyme, which can accelerate or retard a reaction and apparently remains unchanged; *alt.* catalysor.

catamenia n. [Gk. *kata*, according to; *mēn*, month] menses, *q.v.*

catapetalous a. [Gk. *kata*, over; *petalon*, leaf] having petals united with the base of monadelphous stamens.

cataphoresis n. [Gk. *katapherein*, to carry down] electrophoresis, *q.v.*

cataphyll n. [Gk. *kata*, down; *phyllon*, leaf] simple form of leaf on lower part of plant, as cotyledon, bud scale, bulb scale, scale leaf; *alt.* cataphyllary leaf; *cf.* hypsophyll.

cataphyllary a. [Gk. *kata*, down; *phyllon*, leaf] *appl.* rudimentary or scale-like leaves which act as covering of buds.

cataplasis n. [Gk. *kata*, downward; *plasis*, moulding] regression or decline following the mature period or metaplasis.

cataplasmic a. [Gk. *kataplassein*, to spread over] *appl.* irregular galls caused by parasites or other factors.

catapleurite n. [Gk. *kata*, down; *pleura*, side] thoracic pleurite between anapleurite and trochantin, as in certain Thysanura; *alt.* coxopleurite.

cataplexis kataplexy, *q.v.*

catastate n. [Gk. *kata*, down; *stasis*, to stand] catabolite, *q.v.*; *cf.* anastate.

catechins n.plu. [Gk. *katechu*, catechu] a group of colourless flavonoids.

catecholamines n.plu. [Gk. *katechu*, catechu; *ammoniakon*, gum] a group of chemicals acting as neurotransmitters or hormones, which are amine derivatives of catechol (2-hydroxyphenol) and include noradrenaline, adrenaline, and dopamine.

catelectrotonus n. [Gk. *kata*, down; *elektron*, amber; *tonos*, tension] increase in irritability of a nerve under influence of non-polarizing electric current.

catena n. [L. *catena*, chain] a sequence of soil types which is repeated in a corresponding sequence of topographical sites, as between ridges and valleys of a region; a bast fibre in the tree *Heliocarpus*; chain behaviour, *q.v.*

catenation n. [L. *catenatus*, chained] end-to-end arrangement of chromosomes; ring formation of alternating paternally and maternally derived chromosomes; a chain, as of diatom frustules.

catenoid a. [L. *catena*, chain; Gk. *eidos*, form] chain-like; *appl.* certain protozoan colonies; *alt.* catenuliform.

catenular a. [L. *catenula*, little chain] chain-like; *appl.* colonies of bacteria, colour-markings on butterfly wings, shells, etc.; *alt.* catenuliform.

catenulate forming a chain-like series.

catenuliform catenoid, *q.v.*; catenular, *q.v.*

caterpillar n. [L.L. *cattus*, cat; L. *pilosus*, hairy] a fleshy thin-skinned larva, found especially in Lepidoptera, having true legs and often also having prolegs on abdomen, and no cerci; *alt.* eruca.

cathammal a. [Gk. *kathamma*, anything tied] *appl.* plates forming endoderm lamella in some Coelenterata.

cathepsins n. [Gk. *kathepso*, I digest] a group of proteolytic enzymes found in many animal tissues such as liver, spleen, kidney, which are thought to be concerned in autolysis in some diseased conditions and after death.

cation n. [Gk. *kata*, down; *ienai*, to go] a positively charged ion which moves towards cathode or negative pole; *cf.* anion.

cation exchange capacity base exchange capacity, *q.v.*

catkin n. [A.S. *catkin*, little cat] an inflorescence consisting of a pendulous spike with bracted axis bearing unisexual flowers as in poplar and willow; *alt.* amentum, ament.

cauda n. [L. *cauda*, tail] a tail, or tail-like appendage; posterior part of an organ, e.g. cauda equina, cauda epididymis; a tube at posterior end of abdomen of certain insects, suggesting presence of a further segment.

caudad *adv.* [L. *cauda*, tail; *ad*, toward] towards tail region or posterior end; *cf.* rostrad.

caudal *a.* [L. *cauda*, tail] of or *pert.* a tail, e.g. caudal fin.

Caudata *n.* [L. *cauda*, tail] in some classifications, an order of lepospondyl amphibians including the newts and salamanders.

caudate *a.* [L. *cauda*, tail] having a tail; *appl.* nucleus: intraventricular portion of corpus striatum; *appl.* a lobe of the liver.

caudatolenticular *a.* [L. *cauda*, tail; *lens*, lentil] *appl.* caudate and lenticular (lentiform) nuclei of corpus striatum.

caudex *n.* [L. *caudex*, tree trunk] the axis or stem of a woody plant, as of tree ferns, palms, etc.

caudicle *n.* [*Dim.* of L. *cauda*, tail] stalk of pollinium in orchids.

caudihaemal *a.* [L. *cauda*, tail; Gk. *haima*, blood] *appl.* posterior lower portion of a sclerotome.

caudineural *a.* [L. *cauda*, tail; Gk. *neuron*, nerve] *appl.* posterior upper portion of a sclerotome.

caudostyle *n.* [L. *cauda*, tail; Gk. *stylos*, column] a terminal structure in certain parasitic amoebae.

caudotibialis *n.* [L. *cauda*, tail; *tibia*, shin] a muscle connecting caudal vertebrae and tibia, as in Phocidae (hair seals).

caul *n.* [M.E. *calle*, covering] an enclosing membrane; amnion, *q.v.*; omentum, *q.v.*; greater omentum, *q.v.*

Caulerpales *n.* [*Caulerpa*, generic name] an order of marine green algae whose vegetative plant is diploid and has cells with discoid chloroplasts; *alt.* Siphonales.

caulescent *a.* [L. *caulis*, stalk] with leaf-bearing stem above ground.

caulicle *n.* [L. *cauliculus*, small stalk] a small or rudimentary stem; axis of a young seedling.

caulicolous *a.* [L. *caulis*, stalk; *colere*, to inhabit] growing on the stem of another plant, usually *appl.* fungi.

cauliflory *n.* [L. *caulis*, stalk; *flos*, flower] condition of having flowers arising from axillary buds on the main stem or older branches; *alt.* cauliflorous habitus; *cf.* ramiflory.

cauliform *a.* [L. *caulis*, stalk; *forma*, shape] stemlike.

cauligenous *a.* [Gk. *kaulos*, stem; *genos*, birth] borne on the stem.

cauline *a.* [L. *caulis*, stalk] *pert.* stem; *appl.* leaves growing on upper portion of a stem; *appl.* vascular bundles not passing into leaves.

caulis *n.* [L. *caulis*, stalk] the stem especially in herbaceous plants.

caulocaline *n.* [Gk. *kaulos*, stem; *kalein*, to summon] a hormone, not an auxin, which may play a part in development of stems; *cf.* rhizocaline.

caulocarpous *a.* [Gk. *kaulos*, stem; *karpos*, fruit] with fruit-bearing stem; fruiting repeatedly, *alt.* polytokous.

caulocystidium *n.* [Gk. *kaulos*, stalk; *kystis*, bag; *idion*, *dim.*] one of the cystidium-like structures on stipe of certain Basidiomycetes; caulotrichome, *q.v.*

caulome *n.* [Gk. *kaulos*, stem] the shoot structure of a plant as a whole.

caulomer *n.* [Gk. *kaulos*, stem; *meros*, part] a secondary axis in a sympodium.

caulonema *n.* [Gk. *kaulos*, stem; *nēma*, thread] profusely branched portion of a protonema with relatively few chloroplasts found in some genera of mosses; *cf.* chloronema.

caulotaxis *n.* [Gk. *kaulos*, stem; *taxis*, arrangement] the arrangement of branches on a stem; *alt.* caulotaxy.

caulotrichome *n.* [Gk. *kaulos*, stem; *trichōma*, growth of hair] hair-like or filamentous outgrowths on a stem; *alt.* caulocystidia.

caveolae *n.plu.* [L. *cavea*, excavated place; *dim.*] minute vesicles or elongated invaginations in plasma membranes, facilitating passage of molecules and droplets; *alt.* caveolae intracellulares.

cavernicolous *a.* [L. *caverna*, cavern; *colere*, to dwell] cave-inhabiting.

cavernosus *a.* [L. *cavernosus*, chambered] full of cavities; hollow, or resembling a hollow or honeycomb, *appl.* tissue, nerve, arteries.

cavicorn *a.* [L. *cavus*, hollow; *cornu*, horn] hollow-horned, *appl.* certain ruminants.

cavitation *n.* [L. *cavitas*, from *cavus*, hollow] formation of a cavity by a parting of cell groups, or within a cell mass.

cavum *n.* [L. *cavus*, hollow] the lower division of concha caused by origin of helix; cavity of mouth, larynx, long bones, etc.; any hollow or chamber.

Caytoniales *n.* [*Caytonia*, generic name] an order of fossil Cycadopsida having palmate leaves and sporangia on pinnate sporophylls with the megasporangia in carpel-like structures; in some classifications, considered a group of pteridosperms.

C-cells cells with non-granular cytoplasm in islets of Langerhans, possibly giving rise to A-cells; a group of cells which secrete calcitonin, in lower vertebrates forming the ultimobranchial bodies, and in higher vertebrates incorporated into the thyroid and parathyroid glands; *alt.* parafollicular cells.

^{12}C/^{13}C ratio a ratio of the proportion of ^{12}C to ^{13}C in a geological deposit which determines whether or not it is of living matter.

C_4 dicarboxylic acid pathway Hatch–Slack pathway, *q.v.*

CDP cytidine diphosphate, *q.v.*

cecal caecal, *q.v.*

cecidium *n.* [Gk. *kēkis*, gall; *idion*, *dim.*] *see* gall.

cecidogen *n.* [Gk. *kēkis*, gall; *idion*, *dim.*; *genos*, birth] any substance that causes galls to form.

cecum caecum, *q.v.*

Celastrales *n.* [*Celastrus*, generic name] an order of polypetalous dicots of the Polypetalae or Rosidae, used in different ways by different authorities.

celiac coeliac, *q.v.*

cell *n.* [L. *cella*, compartment] one of the units consisting of a nucleus and cytoplasm surrounded by a cell membrane which make up the building blocks of plant and animal bodies, and in plants are surrounded by a non-living cell wall; a small cavity or hollow; loculus, *q.v.*; space between veins of insect wing.

cell body in neurone, the part of the cell containing nucleus and some cytoplasm; *alt.* neurocyton, perikaryon.

cell cycle the changes that occur in a cell from the beginning of one mitosis to the beginning of the next, composed of phases known as G_1, S, M, G_2, and G_0 phases.

cell division the splitting of a cell into 2 by mitosis or amitosis, or into 4 by meiosis.

cell family coenobium, *q.v.*

cellifugal *a.* [L. *cella*, cell; *fugere*, to flee] moving away from a cell.

cellipetal *a*. [L. *cella*, cell; *petere*, to seek] moving towards a cell.

cell lineage the derivation of a tissue or part from a definite blastomere of embryo.

cell membrane plasma membrane, *q.v.*; ectoplast and tonoplast of a plant cell.

cellobiase the enzyme which hydrolyses cellobiose to glucose; EC 3.2.1.21; *r.n. β-D-glucosidase*.

cellobiose *n*. [L. *cella*, cell; *bis*, twice] a disaccharide which does not occur in an unbound form, but is obtained when cellulose is hydrolysed by dilute acids or by the enzyme cellobiase.

cell organ a part of a cell having a special function, such as a centrosome; *alt*. organoid, organelle.

cell plate equatorial thickening of spindle fibres from which partition wall arises during division of plant cells; mid-body of animal cells.

cell sap the fluid in vacuoles of plant cells, being a mixture of organic solutes in water; sometimes used for the more fluid part of the cytoplasm, especially in Protozoa, *alt*. enchylema.

cell theory the idea that plant and animal bodies are made up of cells, first proposed by Schleiden for plants and Schwann for animals in 1838–40.

cellular *a*. [L. *cellula*, small cell] *pert*. or consisting of cells.

cellulase *n*. [L. *cellula*, small cell] the enzyme which hydrolyses cellulose to cellobiose, sometimes also called cytase; EC 3.2.1.4; *r.n.* cellulase.

cellulin *n*. [L. *cellula*, small cell] a carbohydrate found in hyphae.

cellulose *n*. [L. *cellula*, small cell] a polysaccharide that is a structural material in green plants, forming young cell walls, which has large molecular weight and consists of ß-1, 4 linked glucose units.

cell wall the non-living rigid structure surrounding the cell membrane of most plant cells, usually made of cellulose in green plants, sometimes with other materials such as lignin or cutin, made of chitin in some fungi, and of various materials in prokaryotes; *alt*. cytoderm.

celo- coelo-, *q.v.*

cement *n*. [L. *caementum*, mortar] a substance chemically and physically allied to bone, investing parts of teeth, *alt*. crusta petrosa, substantia ossea; a uniting substance, as between cells of plants or animals, or between parts and substrate.

cementocytes cells resembling osteocytes, in lacunae of cement of teeth.

cenanthy kenanthy, *q.v.*

cenchrus *n*. [Gk. *kengchros*, millet] a pale-coloured area on mesothorax of saw flies.

cenenchyma coenenchyma, *q.v.*

cenesthesis coenaesthesia, *q.v.*

ceno- *see* caeno-, coeno-.

censer mechanism method of seed distribution by which seeds are jerked out from fruit by wind.

centipedes the common name for the Chilopoda, *q.v.*

centradenia *n*. [Gk. *kentron*, centre; *adēn*, gland] a type of siphonophore colony.

central *a*. [L. *centrum*, centre] situated in, or *pert*., the centre; *pert*. a vertebral centrum. *n*. A bone, os centrale, in wrist or ankle, situated between proximal and distal rows.

central apparatus cytocentrum, *q.v.*

central body centrosome, *q.v.*; in blue–green algae and some bacteria, the colourless inner portion of protoplasm, thought to contain the nuclear material.

central cylinder stele, *q.v.*

central dogma the idea that genetic information is only transferred in one direction, from DNA to RNA to protein, the DNA acting as a template for its own replication and for the transcription of RNA which then undergoes translation into a protein; some RNA viruses seem to be able to direct the synthesis of DNA from RNA; *see* reverse transcriptase.

Centrales *n*. [L. *centrum*, centre] an order of diatoms having radial symmetry; *cf*. Pennales.

central nervous system (CNS) the part of the nervous system containing a large number of cell bodies and synapses, and consisting of a brain and cerebral ganglia, and a nerve cord which may be dorsal or ventral, single or double; *cf*. peripheral nervous system.

central sulcus lateral middle fissure of cerebral hemisphere, formerly called fissure of Rolando.

centrarch *a*. [L. *centrum*, centre; *archē*, beginning] *appl*. protostele with central protoxylem.

centric *a*. [L. *centrum*, centre] *appl*. leaves which are cylindrical or terete; having a centromere; *appl*. diatoms that are members of the Centrales.

centrifugal *a*. [L. *centrum*, centre; *fugere*, to flee] turning or turned away from centre or axis; *appl*. radicle; *appl*. compact cymose inflorescences having youngest flowers towards outside; *appl*. xylem differentiating from centre towards edge of stem or root; *appl*. thickening of cell wall when material is deposited on outside of cell wall, as in pollen grains; *appl*. nerves transmitting impressions from nerve centre to parts supplied by nerve, *alt*. efferent; *cf*. centripetal.

centriole *n*. [L. *centrum*, centre] a minute rod or granule forming the central part of the centrosome, *alt*. attraction particle, division centre, microcentrosome; sometimes, the whole centrosome.

centripetal *a*. [L. *centrum*, centre; *petere*, to seek] turning or turned towards centre or axis; *appl*. radicle; *appl*. racemose inflorescences having youngest flowers at apex; *appl*. xylem differentiating from the edge towards the centre of a stem or root; *appl*. thickening of cell wall where material is deposited on the inside of the cell wall, as most cells; *appl*. nerves transmitting impressions from peripheral extremities to nerve centres, *alt*. afferent; *cf*. centrifugal.

centripetal canals blind canals growing from circular canal backwards towards apex of bell in certain Trachymedusae.

centro-acinar *a*. [L. *centrum*, centre; *acinus*, berry] *pert*. centre of an alveolus, as in pancreas.

centrodesm(us), centrodesmose *n*. [Gk. *kentron*, centre; *desmos*, bond] the fibril or system of fibrils temporarily connecting 2 centrosomes.

centrodorsal *a*. [L. *centrum*, centre; *dorsum*, back] *appl*. plate in middle of aboral surface of unstalked crinoids.

centrogenous *a*. [Gk. *kentron*, centre; *gennaein*, to produce] *appl*. a skeleton of spicules which meet in a common centre and grow outwards.

centrolecithal *a*. [Gk. *kentron*, centre; *lekithos*, yolk] with yolk aggregated in the centre, *appl*. ovum; *cf*. mesolecithal.

centromere *n.* [Gk. *kentron*, centre; *meros*, part] the part of the chromosome located at the point lying on the equator of the spindle at metaphase and dividing at anaphase, controlling chromosome activity; *alt.* spindle-attachment region, achromite, kinetochore, kinomere, kinetic constriction, attachment constriction, polar granule.

centron *n.* [Gk. *kentron*, centre] neurocyton, *q.v.*

centrophormium *n.* [Gk. *kentron*, centre; *phormis*, small basket] the Golgi bodies when in round basket-like form.

centroplasm *n.* [Gk. *kentron*, centre; *plasma*, mould] substance of centrosphere; a more-or-less definite concentric zone round the aster in mitosis; in blue–green algae, the inner colourless region of the cell, *cf.* chromatoplasm.

centroplast *n.* [Gk. *kentron*, centre; *plastos*, formed] an extranuclear spherical body forming division centre of mitosis, as in some Radiolaria.

centrosome *n.* [Gk. *kentron*, centre; *sōma*, body] an organelle in many animal and some plant cells situated near the nucleus and dividing during nuclear division, thought to be the centre of dynamic activity in nuclear division, and consisting of 1 or 2 centrioles and the attraction sphere; *alt.* astrocentre, central body, cytocentrum, microcentrum, spindle-fibre locus, centrum, centrosphere, periplast, polar corpuscle.

Centrospermae *n.* [Gk. *kentron*, centre; *sperma*, seed] Caryophyllales, *q.v.*

centrosphere *n.* [Gk. *kentron*, centre; *sphaira*, ball] the layer of differentiated cytoplasm in the centrosome surrounding the centriole; the whole centrosome; *alt.* astrosphere, attraction sphere, microsphere, directive sphere, centrum, macrocentrosome.

centrotaxis *n.* [Gk. *kentron*, centre; *taxis*, arrangement] orientation of chromatin thread towards cytocentrum during leptotene stage.

centrotheca *n.* [Gk. *kentron*, centre; *thēkē*, case] *see* idiozome.

centrum *n.* [L. *centrum*, centre] the main body of a vertebra from which neural and haemal arches arise; centrosome, *q.v.*; centrosphere, *q.v.*

ceph-, cephal- *alt.* keph-, kephal-.

cephalad *adv.* [Gk. *kephalē*, head; L. *ad*, towards] towards head region or anterior end.

cephalanthium *n.* [Gk. *kephalē*, head; *anthos*, flower] anthodium, *q.v.*

Cephalaspida, cephalaspids *n.*, *n.plu.* [Gk. *kephalē*, head; *aspis*, shield] an order of U. Silurian and Devonian ostracoderms (Osteostraci) having a solid head shield with electric organs.

Cephalaspidea *n.* [Gk. *kephalē*, head; *aspis*, shield] an order of opisthobranchs having a fairly large shell and a shield-like head for burrowing; *alt.* Bullomorpha.

cephaletron *n.* [Gk. *kephalē*, head; *ētron*, belly] the anterior region of Xiphosura.

cephalic *a.* [Gk. *kephalē*, head] *pert.* head; in head region.

cephalic index one-hundred times maximum breadth divided by maximum length of skull.

cephalin *n.* [Gk. *kephalē*, head] phospholipid which on hydrolysis yields glycerol, fatty acids, phosphoric acid, and the nitrogenous base colamine, found in brain, nerve fibres, and egg yolk, *alt.* phosphatidylethanolamine; an epimerite bearing trophozoites.

cephalis *n.* [Gk. *kephalis*, little bulb] the uppermost chamber of monaxonic shells of Radiolaria.

cephalization *n.* [Gk. *kephalē*, head] increasing differentiation and importance of anterior end in animal development.

Cephalobaenida *n.* [*Cephalobaena*, generic name] an order of Pentastomida having a ventral nerve cord with 4 separate ganglia.

Cephalocarida *n.* [Gk. *kephalē*, head; L. *caris*, a crab] a subclass of minute marine crustaceans living in sand and having a horseshoe-shaped carapace; sometimes considered an order of Branchiopoda.

Cephalochorda(ta), cephalochordates *n.*, *n.plu.* [Gk. *kephalē*, head; *chordē*, string] a subphylum of chordates commonly called lancelets and including *Amphioxus*, which are metamerically segmented with a notochord persisting in the adult and a large sac-like pharynx with gill slits for food collection and respiration, sometimes considered a group of protochordates.

cephalodia *n.plu.* [Gk. *kephalē*, head] gall-like outgrowths, usually brown, developing on, or occasionally within, certain lichens, and containing a different alga from that characteristic of the lichen; *sing.* cephalodium.

Cephalodiscida *n.* [Gk. *kephalē*, head; *diskos*, disc] an order of hemichordates of the class Pterobranchia that form aggregations but do not have organic union between members.

cephalogenesis *n.* [Gk. *kephalē*, head; *genesis*, origin] development of the head region, embryonic stage after notogenesis.

cephalon *n.* [Gk. *kephalē*, head] the head of some arthropods; head shield of trilobites.

cephalont *n.* [Gk. *kephalē*, head] a sporozoan about to proceed to spore formation.

Cephalopoda, cephalopods *n.*, *n.plu.* [Gk. *kephalē*, head; *pous*, foot] a class of molluscs including octopus, squid, nautiloids and ammonites, having a well-developed head surrounded by prehensile tentacles, and a large mantle cavity communicating with the exterior by a siphon; *alt.* Siphonopoda.

cephalopodium *n.* [Gk. *kephalē*, head; *pous*, foot] the head and arms constituting the head region in cephalopods.

cephalopsin *n.* [Gk. *kephalē*, head; *opsis*, sight] a photopigment resembling visual purple, in eyes of cephalopods and some other invertebrates.

cephalosporinase penicillinase, *q.v.*

cephalosporium *n.* [Gk. *kephalē*, head; *sporos*, seed] a globular mucilaginous mass of spores; *alt.* spore ball.

cephalostegite *n.* [Gk. *kephalē*, head; *stegē*, roof] anterior part of cephalothoracic shield.

cephalostyle *n.* [Gk. *kephalē*, head; *stylos*, pillar] anterior end of notochord enclosed in sheath, in animals with a chondrocranium.

cephalotheca *n.* [Gk. *kephalē*, head; *thēkē*, case] head integument in insect pupa.

cephalothorax *n.* [Gk. *kephalē*, head; *thōrax*, chest] the body region formed by fusion of head and thorax in Crustacea; in Arachnida, prosoma, *q.v.*

cephalotrocha *n.* [Gk. *kephalē*, head; *trochos*, wheel] Müller's larva, *q.v.*

Cephaloxeniformes *n.* [Gk. *kephalē*, head; *xenos*, strange; L. *forma*, shape] an order of

Triassic Chondrostei with thick head bones and crushing teeth.

cephalula *n.* [Gk. *kephalē*, head] free-swimming embryonic stage in certain brachiopods.

ceptor receptor, *q.v.*

cer- also ker-, *q.v.*

ceraceous *a.* [L. *cera*, wax] waxy, wax-like; *alt.* cereous.

ceral *a.* [L. *cera*, wax] *pert.* wax; *pert.* the cere of birds.

Ceramiales *n.* [*Ceramium*, generic name] an order of red algae having the auxiliary cell produced after fertilization.

ceramide an N-acylated derivative of sphingosine, widely distributed in plant and animal tissues.

cerata *n.plu.* [Gk. *keras*, horn] lobes or leaf-like processes acting as gills on back of nudibranch molluscs; *sing.* ceras.

Ceratiomyxomycetidae *n.* [*Ceratiomyxa*, generic name; Gk. *mykēs*, fungus] a subclass of slime moulds with spores borne externally and containing one order, the Ceratiomyxales.

ceratium *n.* [Gk. *keration*, little horn] a siliqua without the replum.

ceratobranchial *n.* [Gk. *keras*, horn; *brangchia*, gills] an element of branchial arch.

ceratohyal *n.* [Gk. *keras*, horn; *hyoeidēs*, Υ-shaped] the component of hyoid arch next below epihyal.

ceratoid *a.* [Gk. *keras*, horn; *eidos*, form] keratoid, *q.v.*

ceratotheca *n.* [Gk. *keras*, horn; *thēkē*, case] the part of the casing of an insect pupa which protects the antennae.

ceratotrichia *n.plu.* [Gk. *keras*, horn; *thrix*, hair] horny and non-cellular actinotrichia of elasmobranchs; *cf.* lepidotrichia.

cercal *a.* [Gk. *kerkos*, tail] *pert.* the tail; *pert.* cerci, *appl.* hairs, nerve.

cercaria *n.* [Gk. *kerkos*, tail] a heart-shaped trematode larva with tail.

cerci *plu.* of cercus.

cercid *n.* [Gk. *kerkis*, shuttle] one of minute wandering cells produced by division of archaeocytes in certain sponges.

cercoid *n.* [Gk. *kerkos*, tail; *eidos*, shape] one of paired appendages on 9th, or 10th, abdominal segment of certain insect larvae.

cercopod cercus, *q.v.*

cercus *n.* [Gk. *kerkos*, tail] a jointed appendage at end of abdomen in many arthropods; appendage bearing acoustic hairs in some insects; *plu.* cerci; *alt.* cercopod.

cere *n.* [L. *cera*, wax] a swollen fleshy patch at proximal end of bill in birds; *alt.* ceroma.

cerebellar *a.* [L. *cerebrum*, brain] *pert.* the cerebellum or hind-brain.

cerebellum *n.* [L. *cerebrum*, brain] the 4th division of brain, arising from differentiation of anterior part of 3rd primary vesicle; *alt.* epencephalon.

cerebral *a.* [L. *cerebrum*, brain] *pert.* the brain; *pert.* hemispheres or cerebrum.

cerebral aqueduct passage in mid-brain, connecting 3rd and 4th ventricles; *alt.* mesocoel, aqueduct of Sylvius.

cerebral organs chemical sense organs, paired ciliated tubes associated with dorsal ganglion and opening to exterior, in nemertines.

cerebral peduncles crura cerebri, *q.v.*

cerebriform cerebrose, *q.v.*

cerebrifugal *a.* [L. *cerebrum*, brain; *fugere*, to flee] *appl.* nerve fibres which pass from brain to spinal cord.

cerebroganglion *n.* [L. *cerebrum*, brain; Gk. *ganglion*, swelling] the supraoesophageal ganglia of invertebrates.

cerebroid cerebrose, *q.v.*

cerebropedal *a.* [L. *cerebrum*, brain; *pes*, foot] *appl.* nerve strands connecting cerebral and pedal ganglia in molluscs.

cerebrose *a.* [L. *cerebrum*, brain] resembling convolutions of the brain, *appl.* surface of spores, of pileus, etc.; *alt.* cerebriform, cerebroid, convolute.

cerebrosides *n.plu.* [L. *cerebrum*, brain] glycolipids, *q.v.*, in which the sugar is sometimes glucose, but usually galactose (galactosides, galactolipids), e.g. phrenosin and kerasin.

cerebrospinal *a.* [L. *cerebrum*, brain; *spina*, spine] *pert.* brain and spinal cord; *alt.* encephalospinal.

cerebrospinal fluid (CSF) in vertebrates, a fluid filling the cavity in the brain and spinal cord and the pia-arachnoid space, secreted by the choroid plexuses and reabsorbed by veins on the brain surface; *alt.* neurolymph.

cerebrovisceral *a.* [L. *cerebrum*, brain; *viscera*, viscera] *appl.* connective joining cerebral and visceral ganglia in molluscs.

cerebrum *n.* [L. *cerebrum*, brain] the fore-brain arising from differentiation of 1st primary vesicle; *alt.* cerebral hemispheres.

cereous *a.* [L. *cereus*, waxen] ceraceous, *q.v.*

Ceriantharia *n.* [*Cerianthus*, generic name] an order of long and solitary Ceriantipatharia with 2 rings of tentacles and a single dorsal siphonoglyph.

Ceriantipatharia *n.* [*Cerianthus*, *Antipathes*, generic names] a subclass of anthozoans which are primitive with a simple arrangement of tentacles and septa, formerly included in the Zoantharia.

ceriferous *a.* [L. *cera*, wax; *ferre*, to carry] wax-producing.

cernuous *a.* [L. *cernuus*, with face turned downwards] drooping; pendulous.

ceroma *n.* [Gk. *kērōma*, waxed surface] cere, *q.v.*

cerous *a.* [L. *cera*, wax] *appl.* structure resembling a cere.

certation *n.* [L. *certatio*, contest] competition in growth rate of pollen tubes of genetically different types.

ceruloplasmin *n.* [L. *caeruleus*, dark blue; Gk. *plasma*, form] a protein containing copper found in plasma and blue in colour; *alt.* caeruloplasmin.

cerumen *n.* [L. *cera*, wax] wax-like secretion from ceruminous glands of ear; wax secreted by scale insects; wax of nest of certain bees.

ceruminous *a.* [L. *cera*, wax] *appl.* glands of the external auditory meatus of the ear, which are modified sweat glands and secrete cerumen.

cervical *a.* [L. *cervix*, neck] *appl.* or *pert.* structures connected with neck, as nerves, bones, blood vessels; *appl.* cervix or neck of an organ; *appl.* groove across dorsal surface of cephalothorax in certain crustaceans.

cervicum *n.* [L. *cervix*, neck] the neck region of Arthropoda; the neck of vertebrates.

cervix *n.* [L. *cervix*, neck] the neck or narrow mouth of an organ; cervix uteri: the neck of the uterus just above vagina.

cespitose caespitose, *q.v.*

Cestida *n.* [*Cestum*, generic name] an order of Ctenophora of the class Tentaculata having a flattened body and 2 rows of tentacles.

Cestoda *n.* [L. *cestus*, girdle] a class of endoparasitic segmented flatworms, commonly called tapeworms, lacking a gut or mouth, and having a complex life cycle with 2 or more hosts.

Cestodaria *n.* [L. *cestus*, girdle] a class of flatworms that are endoparasites in fish or occasionally chelonians and have an unsegmented leaflike body; sometimes considered a subclass or order of Cestoda.

Cetacea, cetaceans *n.*, *n.plu.* [L. *cetus*, whale] an order of placental mammals known from the Eocene and including the whales and dolphins, having the body highly adapted for swimming, with the forelimbs modified as flippers and the hindlimbs invisible externally.

cetolith *n.* [Gk. *kētos*, whale; *lithos*, stone] the fused tympanic and petrosal of whales, found in deep-sea dredging.

cetology *n.* [L. *cetus*, whale; Gk. *logos*, discourse] the study of cetaceans.

cevitamic acid vitamin C, *q.v.*

chaeta *n.* [Gk. *chaitē*, hair] chitinous bristle, as of certain annelids, embedded in the body wall; seta, *q.v.*; *alt.* cheta.

chaetic *a.* [Gk. *chaitē*, hair] bristle-like, *appl.* a type of tactile sensilla in insects.

chaetiferous, chaetigerous *a.* [Gk. *chaitē*, hair; L. *ferre, gerere*, to bear] bristle-bearing; *alt.* setiferous, setigerous.

Chaetodermomorpha *n.* [Gk. *chaitē*, hair; *derma*, skin; *morphē*, form] an order of chitons of the subclass Aplacophora which live in deep water and have no vestigial foot or groove.

Chaetognatha *n.* [Gk. *chaitē*, hair; *gnathos*, jaw] a phylum of pelagic coelomate animals found in swarms in plankton, commonly called arrow worms, having an elongated transparent body with a head, trunk, and tail.

Chaetonotoidea *n.* [*Chaetonotus*, generic name; Gk. *eidos*, form] an order of mostly freshwater Gastrotricha that have a fusiform body and are mostly parthenogenetic females.

Chaetophorales *n.* [Gk. *chaitē*, hair; *pherein*, to bear] an order of green algae having a branching filamentous plant body with uninucleate cells.

chaetophorous *a.* [Gk. *chaitē*, hair; *pherein*, to bear] bristle-bearing, *appl.* some annelids and certain insects.

Chaetopoda *n.* [Gk. *chaitē*, hair; *pous*, foot] the group of annelids that have chaetae, the Polychaeta and Oligochaeta.

chaetosema *n.* [Gk. *chaitē*, hair; *sēma*, sign] one of 2 small sensory organs located on head of certain Lepidoptera, and provided with bristles and sensory cells connected by a sheathed nerve to brain; *alt.* Jordan's organ.

chaetotaxy *n.* [Gk. *chaitē*, hair; *taxis*, arrangement] bristle pattern or arrangement.

chain behaviour a series of actions, each being induced by the antecedent action and being an integral part of a unified performance; *alt.* reflex chain, catena.

chalaza *n.* [Gk. *chalaza*, hail] one of 2 spiral bands attaching yolk to membrane of a bird's egg; base of nucellus of ovule, from which integuments arise.

chalaziferous *a.* [Gk. *chalaza*, hail; L. *ferre*, to bear] *appl.* layer of albumen surrounding yolk and continuous with chalazae.

chalazogam *n.* [Gk. *chalaza*, hail; *gamos*, union] a plant whose pollen grain enters the ovule by the chalaza; *cf.* porogram.

chalazogamy *n.* [Gk. *chalaza*, hail; *gamos*, marriage] fertilization in which the pollen tube pierces chalaza of ovule; *cf.* porogamy, aporogamy.

chalcones *n.plu.* [Gk. *chalkos*, copper] a group of yellow flavonoid plant pigments.

chalice *n.* [L. *calix*, goblet] *appl.* simple gland cells or goblet cells; a modified columnar epithelial gland cell; arms and disc of a crinoid.

chalones *n.plu.* [Gk. *chalinos*, curb] internal secretions which depress or inhibit cellular proliferation; *alt.* antihormones; *cf.* hormones.

chalonic *a.* [Gk. *chalinos*, curb] depressor, inhibitory, or restraining, *appl.* internal secretions; *cf.* hormonic.

chamaephyte *n.* [Gk. *chamai*, on the ground; *phyton*, plant] a plant with shoots that bear dormant buds lying on or near the ground.

Chamaesiphonales *n.* [*Chamaesiphon*, generic name] an order of blue–green algae which are unicellular or colonial and produce endospores.

chambered organ in sea lilies, an aboral coelomic structure with 5 compartments, which occupies the body region enclosed by the thecal plates and sends branches to the cirri.

Channiformes *n.* [*Channus*, generic name; L. *forma*, shape] an order of modern advanced teleosts including the snakeheads.

Chantransia stage a filament produced from carpospores by some red algae and resembling the alga *Chantransia*.

Characiformes *n.* [*Charax*, generic name; L. *forma*, shape] an order of freshwater teleosts of Africa and tropical America which have strong jaws and a scaly body.

Charadriiformes *n.* [*Charadrius*, generic name; L. *forma*, shape] an order of birds of the subclass Neornithes, including the gulls, terns, auks, and waders.

Charales *n.* [*Chara*, generic name] an order of algae, in many classifications placed in the division Charophyta, *q.v.*, and in the class Charophyceae, or sometimes placed in the Chlorophyceae.

charas hashish, *q.v.*

Charophyceae *n.* [*Chara*, generic name; Gk. *phykos*, seaweed] a class of algae, in many classifications placed in the division Charophyta, *q.v.*, or sometimes in the division Chlorophyta.

Charophyta *n.* [*Chara*, generic name; Gk. *phyton*, plant] a division of algae that are encrusted with calcium carbonate so are commonly called stoneworts, and have a filamentous or thalloid plant body bearing lateral branches in whorls; in some classifications reduced to a class (Charophyceae) or order (Charales).

chartaceous *a.* [L. *charta*, paper] like paper; *alt.* papyraceous.

chasmatoplasm *n.* [Gk. *chasma*, expanse; *plasma*, mould] an expanded form of plasson.

chasmochomophyte *n.* [Gk. *chasma*, opening; *chōma*, mound; *phyton*, plant] a plant growing on detritus in rock crevices.

chasmocleistogamy the condition of having both chasmogamous and cleistogamous flowers.

chasmogamy *n.* [Gk. *chasma*, opening; *gamos*,

marriage] opening of a mature flower in the normal way to ensure fertilization; *a.* chasmogamic, chasmogamous; *cf.* cleistogamy.

chasmophyte, chasmophil *n.* [Gk. *chasma*, opening; *phyton*, plant; *philein*, to love] a plant which grows in crevices of rocks; *a.* chasmophilous.

cheek *n.* [A.S. *céace*, cheek] the fleshy wall of mouth in mammals; side of face; in invertebrates the lateral portions of head, as fixed and free cheeks of trilobites.

cheilocystidium *n.* [Gk. *cheilos*, edge; *kystis*, bag; *idion*, dim.] a cystidium in hymenium at edge of gill lamella; *alt.* cheilotrichome; *cf.* pleurocystidium.

Cheilostomata *n.* [Gk. *cheilos*, edge; *stoma*, mouth] an order of Gymnolaemata (Ectoprocta) in which the case enclosing the zooids is membranous or calcified.

cheilotrichome cheilocystidium, *q.v.*

cheiro- *also* chiro-, *q.v.*

Cheiroptera Chiroptera, *q.v.*

cheiropterygium *n.* [Gk. *cheir*, hand; *pterygion*, little wing] the pentadactyl limb of higher vertebrates; *alt.* chiropterygium; *cf.* ichthyopterygium.

chela *n.* [Gk. *chēlē*, claw] the large claw borne on certain limbs of some arthropods; a short sponge spicule with talon-like projections at one or each end; *plu.* chelae.

chelate *a.* [Gk. *chēlē*, claw] claw-like or pincer-like, *alt.* cheliform; cheliferous, *q.v.*

chelating agent a chemical which can react with a metal ion and form a stable water-soluble complex with it.

chelation *n.* [Gk. *chēlē*, claw] the structural combination of organic compounds and metal atoms, as in chlorophyll, cytochromes, haemoglobins, etc.

Cheleutoptera *n.* [Gk. *cheleutos*, plaited; *pteron*, wing] Phasmida, *q.v.*

chelicerae *n.plu.* [Gk. *chēlē*, claw; *keras*, horn] first pair of chelate or subchelate prehensile preoral appendages in arachnids and other chelicerates; *alt.* cheliceres; *cf.* falces.

Chelicerata, chelicerates *n.*, *n.plu.* [Gk. *chēlē*, claw; *keras*, horn] a subphylum of arthropods, including the classes Merostomata and Arachnida, having the body divided into 2 regions, the prosoma and opisthosoma, and a single preoral pair of appendages modified as chelicerae.

cheliferous *a.* [Gk. *chēlē*, claw; L. *ferre*, to bear] supplied with chelae or claws; *alt.* chelate.

cheliform *a.* [Gk. *chēlē*, claw; L. *forma*, shape] *see* chelate.

cheliped *n.* [Gk. *chēlē*, claw; L. *pes*, foot] a claw-bearing appendage; forceps of decapod crustaceans.

Chelonia, Chelonea *n.* [*Chelonia*, generic name] an order of anapsid reptiles including the turtles and tortoises having a short broad trunk protected by a dorsal shield (carapace) and ventral shield (plastron) composed of bony plates overlain by epidermal plates of tortoiseshell; *alt.* chelonians, Testudinata.

chelophores *n.plu.* [Gk. *chēlē*, claw; *pherein*, to bear] first pair of appendages in Pycnogonida.

chemical fossils organic molecules such as alkanes and porphyrins found in geological strata such as Pre-Cambrian, that are thought to indicate that living matter existed before the formation of true fossils.

chemiluminescence *n.* [Gk. *chēmeia*, transmutation; L. *luminescere*, to grow light] light production at ordinary temperature during a chemical reaction, as bioluminescence, *q.v.*

chemiotaxis chemotaxis, *q.v.*

chemoautotrophs *n.plu.* [Gk. *chēmeia*, transmutation; *autos*, self; *trophē*, nourishment] organisms deriving energy from the oxidation of inorganic compounds, and using carbon dioxide as the principal carbon source, including several groups of specialized bacteria such as the nitrifying bacteria and thiobacilli; *alt.* chemotrophs, chemolithotrophs.

chemoceptor chemoreceptor, *q.v.*

chemodifferentiation *n.* [Gk. *chēmeia*, transmutation; L. *differentia*, difference] the chemical change in cells which precedes their visible differentiation in embryonic development.

chemodinesis *n.* [Gk. *chēmeia*, transmutation; *dinēeis*, eddying] protoplasmic streaming induced by chemical agents.

chemoheterotrophs *n.plu.* [Gk. *chēmeia*, transmutation; *heteros*, other; *trophē*, nourishment] organisms using organic compounds as both energy and carbon source, including animals, fungi, and many bacteria; *alt.* chemoorganotrophs.

chemokinesis *n.* [Gk. *chēmeia*, transmutation; *kinēsis*, movement] a kinesis in response to the intensity of chemical stimuli, including those of scent.

chemolithotrophs *n.plu.* [Gk. *chēmeia*, transmutation; *lithos*, stone; *trophē*, nourishment] chemoautotrophs, *q.v.*

chemonasty *n.* [Gk. *chēmeia*, transmutation; *nastos*, close pressed] a nastic movement in response to diffuse or indirect chemical stimuli.

chemoorganotrophs *n.plu.* [Gk. *chēmeia*, transmutation; *organon*, instrument; *trophē*, nourishment] chemoheterotrophs, *q.v.*

chemoreceptor *n.* [Gk. *chēmeia*, transmutation; L. *recipere*, to receive] a terminal sense organ or cell receiving chemical stimuli; *alt.* chemoceptor.

chemoreflex *n.* [Gk. *chēmeia*, transmutation; L. *reflectere*, to bend back] a reflex caused by chemical stimulus.

chemosensory *a.* [Gk. *chēmeia*, transmutation; L. *sensus*, sense] sensitive to chemical stimuli, *appl.* certain hairs in insects and to other chemoreceptors.

chemostat *n.* [Gk. *chēmeia*, transmutation; *statos*, standing] any organ concerned in maintaining constancy of chemical conditions, as of hydrogen ion concentration in blood.

chemosynthesis *n.* [Gk. *chēmeia*, transmutation; *synthesis*, composition] the building up of organic chemical compounds in organisms; the process of obtaining energy by the oxidation of inorganic compounds, rather than from light, to make organic compounds as in some bacteria, *cf.* photosynthesis; *a.* chemosynthetic.

chemotaxis *n.* [Gk. *chēmeia*, transmutation; *taxis*, arrangement] the reaction of cells or freely motile organisms to chemical stimuli; *alt.* chemiotaxis.

chemotroph *n.* [Gk. *chēmeia*, transmutation; *trophē*, nourishment] an organism using chemical sources of energy as chemoautotrophs and chemoheterotrophs; an organism using inorganic chemical sources of energy only, as chemoautotrophs; *cf.* phototroph; *a.* chemotrophic.

chemotropism *n.* [Gk. *chēmeia*, transmutation;

trope, turn] curvature of a plant or plant organ in response to chemical stimuli.

chernozem *n*. [Russ. *cherni*, black; *zemlya*, soil] black soil, characteristic of steppe and grassland and formed under continental climatic conditions; *alt*. black earth.

chersophilous *a*. [Gk. *chersos*, dry land; *philein*, to love] thriving on dry waste land.

chersophyte *n*. [Gk. *chersa*, waste places; *phyton*, plant] a plant which grows on waste land or on shallow soil.

chestnut soils dark-brown soils of semiarid steppelands, fertile under adequate rainfall or when irrigated.

chet- chaet-, *q.v.*

cheta chaeta, *q.v.*

chevron *a*. [F. *chevron*, rafter, from L. *caper*, goat] *appl*. V-shaped bones articulating with ventral surface of spinal column in caudal region of many vertebrates.

chewing the cud rumination, *q.v.*

chiasma *n*. [Gk. *chiasma*, cross] a decussation of fibres as optic chiasma; a visible cross formation seen in homologous chromosomes during prophase of meiosis, and thought to be due to crossing-over; *plu*. chiasmata.

chiasmatypy *n*. [Gk. *chiasma*, cross; *typos*, character] a form of recombination of chromosome material in synapsis; chiasmatype, *appl*. theory that chiasmata and crossing-over are causally correlated.

chiastic *a*. [Gk. *chiastos*, diagonally arranged] decussating; crossing; obliquely or at right angles to axis; *pert*. chiasmata.

chiastobasidium *n*. [Gk. *chiastos*, diagonally arranged; *basis*, base; *idion*, dim.] a club-shaped basidium having nuclear spindles at right angles to axis.

chiastoneural *a*. [Gk. *chiastos*, diagonally arranged; *neuron*, nerve] *appl*. certain gastropods in which visceral nerve cords cross and form a figure 8.

chief cells peptic cells, *q.v.*

chilaria *n.plu.* [Gk. *cheilos*, lip] pair of processes between 6th pair of appendages in the king crab, *Limulus*.

chilidium *n*. [Gk. *cheilos*, lip; *idion*, dim.] a shelly plate covering deltidial fissure in dorsal valve of certain Brachiopoda.

Chilognatha *n*. [Gk. *cheilos*, lip; *gnathos*, jaw] a subclass of millipedes with calcification and tanning of the exoskeleton, and body hairs, if present, not barbed.

Chilopoda *n*. [Gk. *cheilos*, lip; *pous*, foot] a group of arthropods, commonly called centipedes, having numerous and similar body segments each with one pair of walking legs, except the 1st segment which bears a pair of poison claws; the group may be considered an order or subclass of the Myriapoda, or raised to class status.

chim(a)era *n*. [L. *chimaera*, monster] a single organism developing from different individuals, or composed of tissues of 2 different genotypes, *alt*. mosaic; graft chimaera: graft hybrid, *q.v.*

Chimaericoloidea *n*. [*Chimaericola*, generic name; Gk. *eidos*, form] an order of Polyopisthocotylea whose posterior end is drawn out into a long stalk bearing the opisthaptor which has a pincer on the end.

Chimaeriformes *n*. [*Chimaera*, generic name; L.

forma, shape] in some classifications, an order of Holocephali including the chimaeras.

chimochlorous *a*. [Gk. *cheima*, winter; *chloros*, pale green] retaining green leaves in winter.

chimonophilous *a*. [Gk. *cheimon*, winter; *philein*, to love] thriving or growing during winter.

chimopelagic *a*. [Gk. *cheima*, winter; *pelagos*, sea] *appl*. or *pert*. certain deep-sea organisms which inhabit surface-water only in winter.

chionophyte *n*. [Gk. *chion*, snow; *phyton*, plant] snow-loving plant.

chiro- also cheiro-, *q.v.*

Chiroptera *n*. [Gk. *cheir*, hand; *pteron*, wing] an order of placental mammals, including the bats and flying foxes, having the forelimbs modified for flight and the hindlimbs reduced; *alt*. Cheiroptera.

chiropterophilous *a*. [Gk. *cheir*, hand; *pteron*, wing; *philos*, loving] pollinated by agency of bats; *n*. chiropterophily.

chiropterygium cheiropterygium, *q.v.*

chirotype *n*. [Gk. *cheir*, hand; *typos*, pattern] the specimen of a species designated by a manuscript name or chironym, ratified on publication as being the type specimen.

chisel teeth chisel-shaped, scalpriform, incisors of rodents.

chitin *n*. [Gk. *chiton*, tunic] a linear array of β-linked N-acetyl-glucosamine units, a mucopolysaccharide, found in annelid cuticle and arthropod exoskeleton and in some plants, especially fungi, which on hydrolysis yields acetic acid and glucosamine; *alt*. isotropic chitin; *cf*. actinochitin.

chitinase *n*. [Gk. *chiton*, tunic] an enzyme which hydrolyses chitin; EC 3.2.1.14; *r.n.* chitinase.

Chitonida *n*. [Gk. *chiton*, tunic] an order of advanced chitons of the subclass Polyplacophora.

chitons *n.plu.* [Gk. *chiton*, tunic] the common name for the Amphineura, *q.v.*

chlamydate *a*. [Gk. *chlamys*, cloak] supplied with a mantle.

chlamydeous *a*. [Gk. *chlamys*, cloak] with a perianth.

Chlamydobacteriales *n*. [Gk. *chlamys*, cloak; *bakterion*, small rod] an order of filamentous bacteria which are mainly free living and aquatic and are enclosed in a sheath.

chlamydocyst *n*. [Gk. *chlamys*, cloak; *kystis*, bladder] an encysted zoosporangium, as in certain saprobic fungi.

Chlamydospermae *n*. [Gk. *chlamys*, cloak; *sperma*, seed] Gnetatae or Gnetopsida, *q.v.*

chlamydospore *n*. [Gk. *chlamys*, cloak; *sporos*, seed] a thick-walled resting spore of certain fungi and protozoans; *alt*. gemma, paulospore.

chloragen chloragogen, *q.v.*

chloragocyte *n*. [Gk. *chloros*, sandy yellow; *kytos*, hollow] a chloragogen cell.

chloragogen *a*. [Gk. *chloros*, sandy yellow; *genos*, descent] *appl*. yellow cells found in connection with alimentary canal of annelids; *alt*. chloragen.

chloragosomes *n.plu.* [Gk. *chloros*, sandy yellow; *soma*, body] yellow or brownish globules formed in chloragogen cells.

Chloramoebales *n*. [*Chloramoeba*, generic name] Heterochloridales, *q.v.*

chloramphenicol *n*. [*Chlorine*; *amido*; *phenyl*; *nitr*; glycol] an antibiotic produced by the actinomycete *Streptomyces venezuelae* which inhibits protein synthesis in prokaryotes by attaching to a

50-S subunit of the ribosome and preventing the addition of an amino acid to the polypeptide chain; *alt.* chloromycetin.

chloranthy *n.* [Gk. *chlōros*, grass green; *anthos*, flower] reversion of floral leaves to ordinary green leaves.

chlorenchyma *n.* [Gk. *chlōros*, grass green; *engchyma*, infusion] tissue containing chlorophyll; formerly used for tissues collectively.

chloride cell the columnar cell of gill filament, specialized for excretion of chlorides, in certain fishes.

chloride shift the complex of processes occuring in and around red blood cells, resulting in chloride ions moving into the cells and bicarbonate ions into the plasma.

Chlorococcales *n.* [*Chlorococcum*, generic name] an order of green algae having non-motile unicellular and colonial vegetative stages, the colonial ones with no vegetative cell division; *alt.* Protococcales.

chlorocruorin *n.* [Gk. *chlōros*, grass green; L. *cruor*, blood] a green haemochrome respiratory pigment found in the blood of certain polychaete annelids; *cf.* oxychlorocruorin.

chlorofucin *n.* [Gk. *chlōros*, green; L. *fucus*, seaweed] a former name for chlorophyll c, in diatoms and brown algae; chlorophyll γ.

chlorolabe *n.* [Gk. *chlōros*, pale green; *labē*, grasping] the green-sensitive pigment of the normal human eye.

chloroleucite chloroplast, *q.v.*

Chloromonadina *n.* [Gk. *chlōros*, green; *monas*, unit] an order of minute Phytomastigina having 2 unequal flagella and food reserves of oil.

Chloromonadophyceae *n.* [Gk. *chlōros*, green; *monas*, unit; *phykos*, seaweed] a small class of motile unicellular algae having no cell wall but a soft periplast allowing change of shape, sometimes raised to phylum (division) status and then known as Chloromonadophyta.

chloromycetin chloramphenicol, *q.v.*

chloronema *n.* [Gk. *chlōros*, pale green; *nēma*, thread] in mosses, a type of protonemal branch which grows along the surface of the substrate or into the air for a short distance, and contains many conspicuous chloroplasts; *cf.* caulonema.

chlorophane *n.* [Gk. *chlōros*, grass green; *phainein*, to appear] a green chromophane.

chlorophore *n.* [Gk. *chlōros*, grass green; *phora*, carrying] a chlorophyll granule in Protista.

Chlorophyceae *n.* [Gk. *chlōros*, grass green; *phykos*, seaweed] a large class of algae, commonly called green algae, which have chlorophylls and carotinoids similar to those of higher plants and appear green, store food as starch and have cellulose cell walls; in some classifications raised to division status called Chlorophyta.

chlorophyll *n.* [Gk. *chlōros*, grass green; *phyllon*, leaf] any of several green pigments found in plants, which absorb mainly red and violet–blue light energy for photosynthesis, being a porphyrin with magnesium at centre and characteristically having one acid side chain, an esterified form of the alcohol, phytol, different chlorophylls having different side chains.

chlorophyllose cells elongated very narrow living cells containing chloroplasts, separated from each other by large empty cells with cuticu-

larized ribs, in *Sphagnum* moss leaves; *alt.* green cells.

Chlorophyta *n.* [Gk. *chlōros*, grass green; *phyton*, plant] in some modern classifications, the Chlorophyceae, *q.v.*, raised to division (phylum) status, and sometimes also including the Charophyceae.

chloroplast(id) *n.* [Gk. *chlōros*, grass green; *plastos*, moulded] a plastid containing chlorophyll, sometimes with other pigments, found in the cytoplasm of plant cells, usually discoid in higher plants but variously shaped in lower ones; formerly also called autoplast, chloroleucite.

chloroplastin *n.* [Gk. *chlōros*, grass green; *plastos*, formed] a hypothetical chromoprotein of which chlorophyll is the prosthetic group.

chloroplast pigments chlorophylls, carotene, and xanthophyll.

chlorosis *n.* [Gk. *chlōros*, pallid] abnormal condition characterized by absence of green pigments in plants, owing to lack of light, or to magnesium- or iron-deficiency, or to genetic factors inhibiting chlorophyll synthesis; green-sickness, a rare form of anaemia in humans.

chlorostatolith *n.* [Gk. *chlōros*, grass green; *statos*, stationary; *lithos*, stone] a chloroplast which moves under the influence of gravity in a statocyte; *cf.* statolith, amyloplastolith.

chlorotic *a.* [Gk. *chlōros*, pallid] *pert.* or affected by chlorosis.

choana *n.* [Gk. *choanē*, funnel] a funnel-shaped opening; posterior naris, *q.v.*

Choanichthyes *n.* [Gk. *choanē*, funnel; *ichthys*, fish] Sarcopterygii, *q.v.*

choanocyte *n.* [Gk. *choanē*, funnel; *kytos*, hollow] a cell with funnel-shaped rim or collar round the base of a flagellum, as in Choanoflagellata and Parazoa; *alt.* collar cell.

choanoderm *n.* [Gk. *choanē*, funnel; *derma*, skin] choanosome, *q.v.*

Choanoflagellata, choanoflagellates *n., n.plu.* [Gk. *choanē*, funnel; *flagellum*, whip] a group of flagellates, sometimes considered an order, which have a protoplasmic collar around the flagellum; *cf.* Lissoflagellata.

choanoid *a.* [Gk. *choanē*, funnel; *eidos*, like] funnel-shaped, *appl.* eye muscle, retractor bulbi, absent in snakes, birds, and higher primates; *alt.* infundibular.

choanosome *n.* [Gk. *choanē*, funnel; *sōma*, body] in sponges, the inner layer with flagellate cells; *alt.* choanoderm; *cf.* pinacoderm.

cholangioles *n.plu.* [Gk. *cholē*, bile; *anggeion*, vessel] terminal or interlobular biliary ducts; *alt.* bile capillaries.

cholecalciferol vitamin D₃, occurring in fish liver oils and formed in the skin of animals on exposure to ultraviolet light.

cholecyanin *n.* [Gk. *cholē*, bile; *kyanos*, dark blue] bilicyanin, *q.v.*

cholecyst *n.* [Gk. *cholē*, bile; *kystis*, bladder] gall bladder, *q.v.*

cholecystokinin *n.* [Gk. *cholē*, bile; *kystis*, bladder; *kinein*, to move] a duodenal hormone which induces contraction of gall bladder and relaxation of Oddi's sphincter.

choledoch *a.* [Gk. *cholē*, bile; *dochos*, containing] *appl.* common bile duct.

choleh(a)ematin cholohaematin, *q.v.*

choleic *a.* [Gk. *cholē*, bile] *appl.* a bile acid: deoxycholic or choleinic acid.

cholephaein, cholepyrrhin, cholerythrin bilirubin, *q.v.*

cholesterol *n.* [Gk. *cholē*, bile; *stereos*, solid] a sterol which is the parent compound for many steroids, found in animals in many tissues including nerve, bile, and yolk, and in some plants, especially red algae.

cholic *a.* [Gk. *cholē*, bile] *pert.*, present in, or derived from, bile.

cholic acid the product of hydrolysis of various bile acids, forming glycocholic acid with glycine and taurocholic acid with taurine.

choline *n.* [Gk. *cholē*, bile] a nitrogenous base and lipotropic factor present in bile, egg yolk, and brain, found in lecithin and other lipines, which is concerned with regulating the decomposition of fat in the liver, and as its acetyl ester, acetylcholine, with nerve activity, and can be obtained from neurine; *alt.* bilineurine.

cholinergic *a.* [Gk. *cholē*, bile; *ergon*, work] *appl.* parasympathetic nerve fibres which liberate acetylcholine from their terminations; *cf.* adrenergic.

cholinesterase an enzyme which hydrolyses acetylcholine into choline and acetic acid, and also other acylcholines to choline and carboxylic acids; EC 3.1.1.8; *r.n.* cholinesterase.

cholochrome *n.* [Gk. *cholos*, bile; *chrōma*, colour] bilirubin, *q.v.*

cholohaematin *n.* [Gk. *cholos*, bile; *haima*, blood] phylloerythrin, *q.v.*; *alt.* cholehaematin, cholohematin.

cholophaein *n.* [Gk. *cholos*, bile; *phaios*, dusky] bilirubin, *q.v.*

chomophyte *n.* [Gk. *chōma*, mound; *phyton*, plant] a plant growing in detritus on rocks.

chondral *n.* [Gk. *chondros*, cartilage] *pert.* cartilage.

Chondrenchelyiformes *n.* [Gk. *chondros*, cartilage; *enchelys*, eel; L. *forma*, shape] an order of L. Carboniferous Holocephali found in Scotland.

chondric gristly, cartilaginous.

Chondrichthyes *n.* [Gk. *chondros*, cartilage; *ichthys*, fish] a class of fishes from the Devonian period to the present day, commonly called cartilaginous fishes, having a cartilaginous endoskeleton, a spiral valve in the gut, and no lungs or air bladder.

chondrification *n.* [Gk. *chondros*, cartilage; L. *facere*, to make] the formation of chondrin and thus the production of cartilage; *alt.* chondrogenesis.

chondrigen *n.* [Gk. *chondros*, cartilage; *gennaein*, to produce] the base matrix of all cartilaginous substance, a collagen; *alt.* chondrogen.

chondrillasterol a sterol found in algae of the classes Bacillariophyceae and some Chlorophyceae, and possibly occurring in some other classes.

chondrin *n.* [Gk. *chondros*, cartilage] a gelatinous bluish-white substance which forms the ground substance of cartilage, having a firm, elastic consistency.

chondrioclast chondroclast, *q.v.*

chondriocont *n.* [Gk. *chondros*, grain; *kontos*, punting pole] a rod-like or fibrillar type of mitochondrion; *alt.* plastocont, plastokont.

chondriodieresis *n.* [Gk. *chondros*, grain; *dieresein*, to swing about] changes in mitochondria during cell division.

chondriokinesis *n.* [Gk. *chondros*, grain; *kinēsis*,

movement] division of mitochondria in mitosis and meiosis.

chondriolysis *n.* [Gk. *chondros*, grain; *lysis*, loosing] the disintegration of mitochondria.

chondrioma, chondriome *n.* [Gk. *chondros*, grain] the mitochondria content of a cell.

chondriomere *n.* [Gk. *chondros*, grain; *meros*, part] plastomere, *q.v.*; cytomere, *q.v.*

chondriomite *n.* [Gk. *chondros*, grain; *mitos*, thread] a linear type of mitochondrion.

chondrioplasm *n.* [Gk. *chondros*, grain; *plasma*, form] the apparently almost structureless ground substance enclosed between outer and inner membranes of a mitochondrion.

chondrioplast *n.* [Gk. *chondros*, grain; *plastos*, formed] a term not now much used, but originally coined for any rod-like organelle such as a part of the Golgi body or a mitochondrion.

chondriosomes *n.plu.* [Gk. *chondros*, grain; *sōma*, body] mitochondria, *q.v.*, especially granular or globular ones.

chondriosphere *n.* [Gk. *chondros*, grain; *sphaira*, globe] a spherical type of mitochondrion; mitochondria which have coalesced.

chondroblast *n.* [Gk. *chondros*, cartilage; *blastos*, bud] a cartilage-producing cell which secretes the chondrin matrix.

chondroclast *n.* [Gk. *chondros*, cartilage; *klastos*, broken down] a large multinucleate cell which destroys cartilage matrix; *alt.* chondrioclast.

chondrocranium *n.* [Gk. *chondros*, cartilage; *kranion*, skull] the skull when in a cartilaginous condition, either temporarily as in embryos, or permanently as in some fishes.

chondrocyte *n.* [Gk. *chondros*, cartilage; *kytos*, hollow] a cartilage cell formed from a chondroblast.

chondrogen chondrigen, *q.v.*

chondrogenesis *n.* [Gk. *chondros*, cartilage; *genesis*, descent] chondrification, *q.v.*

chondroglossus *n.* [Gk. *chondros*, cartilage; *glōssa*, tongue] an extrinsic muscle of the tongue, arising from hyoid bone, between genioglossus and hyoglossus.

chondroid *a.* [Gk. *chondros*, cartilage; *eidos*, shape] cartilage-like, *appl.* tissue, undeveloped cartilage or pseudo-cartilage, serving as support in certain invertebrates and lower vertebrates, *appl.* vesicular supporting tissue of notochord; *alt.* fibrohyaline.

chondroitin *n.* [Gk. *chondros*, cartilage] a mucopolysaccharide with N-acetylchondrosine, which occurs in cartilage as chondroitinsulphuric acid.

chondromucoid *n.* [Gk. *chondros*, cartilage; L. *mucus*, mucus; Gk. *eidos*, form] a substance found in the matrix of cartilage which on hydrolysis yields a protein and chondroitinsulphuric acid (sulphate); *alt.* chondromucin.

Chondrophora *n.* [Gk. *chondros*, cartilage; *pherein*, to bear] an order of pelagic hydrozoans with polymorphic colonies of polyps only, and a float sometimes with a vertical sail; now often included in the Athecata, they were formerly placed in the Siphonophora.

chondrophore *n.* [Gk. *chondros*, cartilage; *pherein*, to bear] a structure which supports the inner hinge cartilage in a bivalve shell.

chondroproteins *n.plu.* [Gk. *chondros*, cartilage; *prōteion*, first] a group of glycoproteins which on hydrolysis yield a protein and chondroitinsulphuric acid (sulphate).

Chondropterygii *n.* [Gk. *chondros*, cartilage; *pterygion*, fin] a group of fishes which includes the lampreys, sturgeons, and elasmobranchs.

chondrosamine [Gk. *chondros*, cartilage; *ammoniakon*, resinous gum] an amino sugar containing galactose, which is the basis of chondroitin, *q.v.*; *alt.* galactosamine.

chondroseptum *n.* [Gk. *chondros*, cartilage; L. *septum*, partition] the cartilaginous part of the septum of the nose.

chondroskeleton *n.* [Gk. *chondros*, cartilage; *skeleton*, dried body] a cartilaginous skeleton.

Chondrostei *n.* [Gk. *chondros*, cartilage; *osteon*, bone] a group of primitive actinopterygians including the sturgeons and having a heterocercal or anomalous tail, usually a spiral valve in the gut, and retaining the spiracle.

Chondrosteiformes *n.* [Gk. *chondros*, cartilage; *osteon*, bone; L. *forma*, shape] an order of marine Triassic to Jurassic Chondrostei from which the sturgeons probably arose, some being suctorial feeders.

chondrosteosis *n.* [Gk. *chondros*, cartilage; *osteon*, bone] the conversion of cartilage to bone.

chondrosteous *a.* [Gk. *chondros*, cartilage; *osteon*, bone] having a cartilaginous skeleton.

chondrosternal *a.* [Gk. *chondros*, cartilage; *sternon*, chest] *pert.* rib cartilages and sternum.

chone *n.* [Gk. *chōnē*, funnel] a passage through cortex of sponges, with one or more external openings, and one internal opening.

Chonotricha *n.* [Gk. *chōnē*, funnel; *thrix*, hair] an order of Ciliata having only one type of nucleus and being permanently fixed to the body of crustaceans.

chorda *n.* [Gk. *chordē*, string] any cord-like structure; chorda dosalis: notochord, *q.v.*; chorda tympani: a branch of the facial nerve; chorda umbilicus: umbilical cord; chorda vocalis: vocal chord; *plu.* chordae.

chordacentra *n.plu.* [Gk. *chordē*, string; L. *centrum*, centre] centra formed by conversion of chordal sheath into a number of rings; *cf.* archcentra.

chordae tendineae tendinous cords connecting papillary muscles with valves of heart.

chordae willisii fibrous bands crossing superior sagittal sinus of dura mater.

chordal *pert.* a chorda or chordae; *pert.* the notochord.

chordamesoderm *n.* [Gk. *chordē*, string; *mesos*, middle; *derma*, skin] an undifferentiated group or layer of embryonic cells which give rise to notochordal and mesodermal cells in vertebrates.

Chordariales *n.* [*Chordaria*, generic name] an order of brown algae having a branched filamentous sporophyte.

Chordata, chordates *n.*, *n.plu* [Gk. *chordē*, string] a phylum of coelomate animals having a notochord and gill clefts in the pharynx at some point in their life history and a dorsal hollow nerve cord with the anterior end usually dilated to form a brain.

Chordeumida *n.* [*Chordeuma*, generic name] an order of Helminthomorpha having silk glands.

chordotonal *a.* [Gk. *chordē*, string; *tonos*, tone] *appl.* sensilla: rod-like or bristle-like receptor for mechanical or sound vibrations in various parts of body of insects; *alt.* scolophore, scolopidium.

chore *n.* [Gk. *chōrē*, place] an area manifesting a unity of geographical or environmental conditions; *cf.* biochore, biotope.

choreiathetose *a.* [Gk. *choreia*, dance; *athetōs*, lawlessly] arhythmic and uncoordinated, *appl.* foetal movements.

chorioallantoic *a.* [Gk. *chorion*, skin; *allas*, sausage] *appl.* placenta when chorion is lined by allantois, allantoic vessels conveying blood to embryo, as in certain marsupials and all placental mammals.

choriocapillaris *n.* [Gk. *chorion*, skin; L. *capillaris*, capillary] the innermost vascular layer of choroid.

chorioid choroid, *q.v.*

chorion *n.* [Gk. *chorion*, skin] an embryonic membrane external to and enclosing the amnion; allantochorion, *q.v.*; a hardened shell covering egg of insects; outer membrane of seed.

chorion frondosum villous placental part of chorion.

chorionic *a.* [Gk. *chorion*, skin] *pert.* the chorion; *appl.* hormone, see gonadotrophins.

chorion laeve smooth non-placental part of chorion.

chorioretinal *a.* [Gk. *chorion*, skin; L. *retina*, retina] *pert.* choroid and retina.

choriovitelline *a.* [Gk. *chorion*, skin; L. *vitellus*, yolk] *appl.* placenta when part of chorion is lined with yolk sac, vitelline blood vessels being connected with uterine wall, as in certain marsupials.

Choripetalae *n.* [Gk. *chōris*, separate; *petalon*, leaf] Archichlamydeae, *q.v.*

choripetalous *a.* [Gk. *chōris*, separate; *petalon*, leaf] polypetalous, *q.v.*

choriphyllous *a.* [Gk. *chōris*, separate; *phyllon*, leaf] having perianth parts distinct.

chorisepalous *a.* [Gk. *chōris*, separate; F. *sépale*, sepal] polysepalous, *q.v.*

chorisis *n.* [Gk. *chōris*, separate] increase in parts of floral whorl due to division of its primary members; *alt.* deduplication, duplication, dédoublement; *cf.* augmentation.

Choristida *n.* [Gk. *chōris*, separate] an order of Tetractinellida having clearly different megascleres and microscleres.

choroid *a.* [Gk. *chorion*, skin; *eidos*, form] *appl.* delicate and highly vascular membranes. *n.* Layer of eye between retina and sclera; *a.* choroidal. *Alt.* chorioid.

chorology *n.* [Gk. *chōros*, place; *logos*, discourse] biogeography, *q.v.*; distribution of organs.

choronomic *a.* [Gk. *chōros*, place; *nomos*, law] external, *appl.* influences of geographical or regional environment; *cf.* autonomic.

chorotypes *n.plu.* [Gk. *chōros*, place; *typos*, pattern] local types.

chresard *n.* [Gk. *chrēsis*, use; *ardo*, I water] soil water available for plant growth; *cf.* echard, holard.

chroma *n.* [Gk. *chrōma*, colour] the hue and saturation of a colour.

Chromadorida *n.* [*Chromadora*, generic name; Gk. *eidos*, form] an order of Aphasmidia having spiral amphids and an annulated cuticle.

chromaffin(e) *a.* [Gk. *chrōma*, colour; L. *affinis*, related] *appl.* a tissue made of modified neural cells that secrete catecholamines and are stained by chromic acid or its salts, and which is found in all vertebrates especially in the adrenal glands, and known as the suprarenal medulla, and in certain invertebrates such as annelids; *alt.* chromaphil, chromophil, phaeochrome.

chromaphil *a.* [Gk. *chrōma*, colour; *philein*, to love] chromaffin, *q.v.*

chromaphobe *a.* [Gk. *chrōma*, colour; *phobos*, fear] chromophobe, *q.v.*

chromasie *n.* [Gk. *chrōma*, colour] increase of chromatin in nucleus and formation of nucleolus; *cf.* achromasie.

chromatic *a.* [Gk. *chrōma*, colour] colourable by means of staining reagents; *pert.* colour; having hue and saturation; having chromatophores.

chromatic body chromatoid body, *q.v.*

chromatic sphere the sphere formed by coalescence of chromosomes after anaphase in mitosis.

chromatic threshold the minimal stimulus, varying with wavelength of light, which induces a colour sensation.

chromatid *n.* [Gk. *chrōma*, colour] a half chromosome resulting from the longitudinal duplication of a chromosome, observable during prophase and metaphase, and becoming a daughter chromosome at anaphase in mitosis and in the 2nd division of meiosis.

chromatid bridge a chromatid joining 2 centromeres during anaphase, in paracentric inversions.

chromatin *n.* [Gk. *chrōma*, colour] a substance in the nucleus which stains with basic dyes and is made of nucleoprotein; *alt.* idioplasm, karyotin.

chromatocyte *n.* [Gk. *chrōma*, colour; *kytos*, hollow] any cell containing a pigment; *alt.* chromocyte, pigment cell.

chromatogen organ a brownish lobed body, the axial organ of certain echinoderms.

chromatoid body a body consisting mainly of RNA, in cytoplasm during certain stages of spermatogenesis, *alt.* accessory body; limosphere of mosses; *alt.* chromatic body.

chromatoid grains grains in cell protoplasm, which stain similarly to chromatin.

chromatolysis *n.* [Gk. *chrōma*, colour; *lysis*, loosing] the breaking up of chromophil substances in a cell, such as chromatin or Nissl granules; *cf.* tigrolysis.

chromatophil(ous) *a.* [Gk. *chrōma*, colour; *philein*, to love] chromophilous, *q.v.*

chromatophore *n.* [Gk. *chrōma*, colour; *pherein*, to bear] a coloured plastid of plants and animals; a colourless body in cytoplasm and developing into a leucoplast, chloroplast, or chromoplast; a pigment cell, or group of cells, which under control of the sympathetic nervous system can be altered in shape to produce a colour change.

chromatophoric *a.* [Gk. *chrōma*, colour; *pherein*, to bear] containing pigment; *pert.* chromatophores.

chromatophorotropic *a.* [Gk. *chrōma*, colour; *pherein*, to bear; *tropē*, turn] *appl.* hormone: intermedin, *q.v.*

chromatophyll *n.* [Gk. *chrōma*, colour; *phyllon*, leaf] the colouring matter of plant-like flagellates; *alt.* chromophyll.

chromatoplasm *n.* [Gk. *chrōma*, colour; *plasma*, mould] in blue–green algae, the peripheral contents of the cell, containing the photosynthetic pigments, *cf.* centroplasm; formerly used for any pigment matter in cells.

chromatosome chromosome, *q.v.*

chromatospherite *n.* [Gk. *chrōma*, colour; *sphaira*, globe] nucleolus, *q.v.*

chromatotropism *n.* [Gk. *chrōma*, colour; *tropos*, direction] orientation in response to stimulation by a particular colour.

chromidia *n.plu.* [Gk. *chrōma*, colour; *idion*, dim.] extranuclear particles of chromatin, which may replace or be re-formed into nuclei; gonidia, *q.v.*

chromidial substance minute basophil granules containing iron, occurring in cytoplasm as chromophil or tigroid bodies.

chromidiogamy *n.* [Gk. *chrōma*, colour; *idion*, dim.; *gamos*, marriage] the union of chromidia from 2 conjugants.

chromidiosomes *n.plu.* [Gk. *chrōma*, colour; *idion*, dim.; *sōma*, body] the smallest chromatin particles of which the chromidial mass is composed.

chromiole *n.* [Gk. *chrōma*, colour] one of the minute granules of which a chromomere is composed.

chromo-argentaffin *a.* [Gk. *chrōma*, colour; L. *argentum*, silver; *affinis*, related] staining with bichromates and silver nitrate; *appl.* flask-shaped cells in epithelium of crypts of Lieberkühn, *alt.* yellow cells.

chromoblast *n.* [Gk. *chrōma*, colour; *blastos*, bud] an embryonic cell giving rise to a pigment cell.

chromocentre *n.* [Gk. *chrōma*, colour; *kentron*, centre] a granule of heterochromatin, many of which are found in interphase nuclei.

chromocyte *n.* [Gk. *chrōma*, colour; *kytos*, hollow] chromatocyte, *q.v.*

chromogen *n.* [Gk. *chrōma*, colour; *genos*, birth] the substance which is converted into a pigment, e.g. by oxidation; a chromogenic organism.

chromogenesis *n.* [Gk. *chrōma*, colour; *genesis*, origin] the production of colour or pigment.

chromogenic *a.* [Gk. *chrōma*, colour; *genos*, birth] colour-producing.

chromoleucite chromoplast, *q.v.*

chromolipid(e)s *n.plu.* [Gk. *chrōma*, colour; *lipos*, fat] the carotenoids and related pigments.

chromomere *n.* [Gk. *chrōma*, colour; *meros*, part] one of the chromatin granules of which a chromosome is formed and which may correspond to a gene, formerly called id; *cf.* nucleosome; granular part of blood platelet, *cf.* hyalomere.

chromomorphosis *n.* [Gk. *chrōma*, colour; *morphōsis*, form] change of form in response to light of different or particular wavelengths; *cf.* photomorphosis.

chromonema *n.* [Gk. *chrōma*, colour; *nēma*, thread] chromosome threads during interphase or prophase when they are extended and dispersed in the nucleus; genonema, *q.v.*; *plu.* chromonemata.

chromoparous *a.* [Gk. *chrōma*, colour; L. *parere*, to bring forth] having coloured excreta, *appl.* bacteria.

chromophanes *n.plu.* [Gk. *chrōma*, colour; *phainein*, to show] red (rhodophane), yellow (xanthophane), and green (chlorophane) lipochrome oil globules found in retina of birds, reptiles, fishes, marsupials; any retinal pigments.

chromophil(ic) *a.* [Gk. *chrōma*, colour; *philein*, to love] chromaffin, *q.v.*; chromophilous, *q.v.*

chromophil bodies Nissl granules, *q.v.*

chromophilous *a.* [Gk. *chrōma*, colour; *philos*, loving] staining readily; *alt.* chromatophil, chromophil, chromophilic.

chromophobe *a.* [Gk. *chrōma*, colour; *phobos*, fear] non-stainable or staining slightly, *appl.* certain cells of pituitary gland; *alt.* chromaphobe.

chromophore *n.* [Gk. *chrōma*, colour; *pherein*, to bear] any substance to whose presence colour in a compound is due.

chromophyll chromatophyll, *q.v.*

Chromophyta *n.* [Gk. *chrōma*, colour; *phyton*, plant] in some classifications, the groups of Contophora that contain pigments additional to chlorophyll.

chromoplast(id) *n.* [Gk. *chrōma*, colour; *plastos*, moulded] in plant cells, a plastid containing pigment; *alt.* chromoleucite; *cf.* leucoplast.

chromoproteins *n.plu.* [Gk. *chrōma*, colour; *prōteion*, first] proteins combined with a pigment, including the respiratory pigments haemoglobin, chlorocruorin, and haemocyanin (the haemochromes), the algal pigments phycocyanin and phycoerythrin, the cytochromes and flavoproteins.

chromosomal aberration any difference from the normal in chromosome structure or number.

chromosomal vesicle karyomere, *q.v.*

chromosome *n.* [Gk. *chrōma*, colour; *sōma*, body] one of the deeply staining bodies, the number of which is constant for the somatic cells of a species, into which chromatin resolves itself during mitosis and meiosis, composed mainly of DNA and basic protein, and bearing the genes in a linear order; formerly called chromatosome, idant, karyomite, karyosome; *cf.* chromatid.

chromosome complement karyotype, *q.v.*

chromosome map a plan which shows the position of genes on a chromosome; *alt.* linkage map, genetic map.

chromosome mutation a mutation involving visible changes in chromosome structure or number in a cell, including polyploidy, aneuploidy, duplications, deletions, and inversions.

chromosome-races races differing in number of chromosomes or of chromosome sets.

chromosomin *n.* [Gk. *chrōma*, colour; *sōma*, body] one of the protein constituents of chromosomes.

chromospire *n.* [Gk. *chrōma*, colour; *speira*, coil] a spireme-like thread formed from nuclear granules in haplomitosis.

chromotropic *a.* [Gk. *chrōma*, colour; *tropikos*, turning] controlling pigmentation, *appl.* hormone of pars intermedia of pituitary gland; *cf.* intermedin.

Chromulinales *n.* [*Chromulina*, generic name] an order of Chrysophyta having unicellular and colonial motile forms.

chronaxia, chronaxie, chronaxy *n.* [Gk. *chronos*, time; *axia*, value] latent period between electrical stimulus and muscular response; minimal excitation time required with a current of an intensity twice the threshold necessary for excitation when the duration of the stimulus is prolonged; *alt.* excitation time.

chronotropic *a.* [Gk. *chronos*, time; *tropē*, turning] affecting the rate of action, as accelerator and inhibitory cardiac nerves.

Chroococcales *n.* [*Chroococcus*, generic name] an order of blue-green algae which are unicellular

or colonial and never have trichomes or endospores.

chrysalis *n.* [Gk. *chrysallis*, gold, golden thing] the pupa of insects with complete metamorphosis, enclosed in a protective case; the case itself.

Chrysocapsales *n.* [Gk. *chrysos*, gold; L. *capsula*, little box] an order of Chrysophyta having palmelloid forms.

chrysocarpous *a.* [Gk. *chrysos*, gold; *karpos*, fruit] with golden-yellow fruit.

chrysolaminarin *n.* [Gk. *chrysos*, gold; *Laminaria*, genus of brown algae] leucosin, *q.v.*

Chrysomonadales *n.* [Gk. *chrysos*, gold; *monas*, unit] an order of Chrysophyta having motile vegetative phases.

Chrysomonadina *n.* [Gk. *chrysos*, gold; *monas*, unit] an order of Phytomastigina having 1 or 2 flagella, a fatty food reserve not starch, a cyst wall containing silica, being yellowish brown in colour and including the marine silicoflagellates and coccoliths.

chrysophanic *a.* [Gk. *chrysos*, gold; *phainein*, to show] having a golden or bright orange colour, *appl.* an acid formed in certain lichens and in leaves.

Chrysophyceae *n.* [Gk. *chrysos*, gold; *phykos*, seaweed] a class of Chrysophyta having a pantonematic flagellum, sometimes an acronematic flagellum, and no haptonema.

chrysophyll *n.* [Gk. *chrysos*, gold; *phyllon*, leaf] a yellow colouring matter in plants, a decomposition product of chlorophyll.

Chrysophyta *n.* [Gk. *chrysos*, gold; *phyton*, plant] a division of algae which have a large amount of carotinoid pigments, reserve products of oil and chrysolaminarin, no cellulose wall but the membrane tends to become silicified; it contains the classes Chrysophyceae and Haptophyceae, and in some classifications also the Xanthophyceae.

chrysopsin *n.* [Gk. *chrysos*, gold; *opsis*, sight] a photolabile retinal pigment in certain deep-sea fishes.

Chrysosphaer(i)ales *n.* *Chrysosphaera*, generic name] an order of Chrysophyta which have coccoid non-motile bodies and may be unicellular or colonial.

Chrysotrichales *n.* [*Chrysothrix*, generic name] an order of Chrysophyceae having simple or branched filaments; *alt.* Phaeothamniales.

chylaceous *a.* [Gk. *chylos*, juice] of the nature of chyle.

chyle *n.* [Gk. *chylos*, juice] lymph containing globules of emulsified fat, found in the lacteals during digestion.

chylifaction *n.* [Gk. *chylos*, juice; L. *facere*, to make] formation of chyle; *alt.* chylification, chylopoiesis.

chyliferous *a.* [Gk. *chylos*, juice; L. *ferre*, to carry] chyle-conducting, *appl.* tubes or vessels; *alt.* chylophoric.

chylific *a.* [Gk. *chylos*, juice; L. *facere*, to make] chyle-producing, *appl.* ventricle or true stomach of insects.

chylification chylifaction, *q.v.*

chylocaulous *a.* [Gk. *chylos*, juice; *kaulos*, stem] with fleshy stems.

chylocyst *n.* [Gk. *chylos*, juice; *kystis*, bladder] receptaculum chyli, *q.v.*

chylomicrons *n.plu.* [Gk. *chylos*, juice; *mikros*,

small] minute fatty particles in plasma, plentiful during fat digestion.

chylophagy *n*. [Gk. *chylos*, juice; *phagein*, to consume] the exchange of substances in solution by haustorial hyphae.

chylophoric chyliferous, *q.v.*

chylophyllous *a*. [Gk. *chylos*, juice; *phyllon*, leaf] with fleshy leaves, *appl.* certain xeromorphic plants.

chylopoiesis *n*. [Gk. *chylos*, juice; *poiēsis*, a making] chylifaction, *q.v.*

chymase rennin, *q.v.*

chyme *n*. [Gk. *chymos*, juice] the partially digested food after leaving the stomach.

chymification *n*. [Gk. *chymos*, juice; L. *facere*, to make] the process of converting food into chyme.

chymosin the EC recommended name for rennin, *q.v.*, being a carboxyl peptidase with a specificity resembling that of pepsin.

chymotrypsin *n*. [Gk. *chymos*, juice; *tripsai*, to rub down; *pepsis*, digestion] an enzyme which, in the small intestine, splits the various protein products of the action of pepsin and trypsin; EC 3.4.21.1; *r.n.* chymotrypsis.

chymotrypsinogen *n*. [Gk. *chymos*, juice; *tripsai*, to rub down; *pepsis*, digestion; *-genēs*, producing] a pancreatic protein which is activated by trypsin and converted into chymotrypsin.

Chytridiales *n*. [Gk. *chytridion*, little pot] a group of Phycomycetes having simple hyphae which never form a true mycelium, asexual reproduction by flagellate zoospores, and isogamous sexual reproduction with flagellate gametes; in some classifications, raised to class status as Chytridiomycetes.

Chytridiomycetes *n*. [Gk. *chytridion*, little pot; *mykēs*, fungus] the Chytridiales, *q.v.*, raised to class status in some classifications, sometimes containing the orders Chytridiales, Blastocladiales, and Monoblepharidales.

chytridium *n*. [Gk. *chytridion*, little pot] the spore vessel of certain fungi.

cibarium *n*. [L. *cibaria*, victuals] the part of the buccal cavity anterior to pharynx, in insects.

cicatricial tissue newly-formed fibrillar connective tissue which closes and draws together wounds.

cicatrice, cicatricle, cicatricula, cicatrix *n*. [L. *cicatrix*, scar] the blastoderm in bird and reptile eggs; a small scar in place of previous attachment of an organ; a scar; the mark left after healing of a wound in plants.

cicinnal *a*. [Gk. *kikinnos*, curled lock] *appl.* uniparous cymose branching in which daughter axes are developed right and left alternately; *alt.* cincinnal.

Ciconiiformes *n*. [*Ciconia*, generic name of stork; L. *forma*, shape] an order of birds of the subclass Neornithes, including the herons, bitterns, and storks.

Cidaroidea *n*. [*Cidaris*, generic name; Gk. *eidos*, form] an order of Perischoechinoidea with living members.

cilia *n.plu.* [L. *cilium*, eyelid] hair-like vibratile outgrowths of ectoderm; processes of cell surface, consisting of a closed tube of plasma membrane enclosing filaments on a basal granule; barbicels of a feather; eyelashes; in lichens, hair-like thalline-appendages which are decolorized or carbonized strands of hyphae along lobe margins; *sing.* cilium.

ciliaris *n*. [L. *cilium*, eyelid] unstriped muscle forming a ring outside anterior part of choroid and, attached to ciliary processes, acting on convexity of lens.

ciliary *a*. [L. *cilium*, eyelid] *pert.* cilia; *pert.* eyelashes; *appl.* sudoriferous glands; *appl.* certain structures in the eyeball, as arteries, body, processes, muscle; *appl.* branches of nasocilary nerve and to ganglion.

Ciliata *n*. [L. *cilium*, eyelid] a subclass of Ciliophora, commonly called ciliates, having cilia as adults and not possessing suctorial tentacles.

ciliate(d) *a*. [L. *cilium*, eyelid] having cilia; having regularly arranged hairs projecting from edge.

ciliated epithelium an epithelium found lining various passages, usually with columnar cells provided with cilia on the free surface.

ciliates *n.plu.* [L. *cilium*, eyelid] the common name for the Ciliata or Ciliophora, formerly sometimes called infusorians.

ciliograde *a*. [L. *cilium*, eyelid; *gradus*, step] progressing by movement of cilia.

ciliolum *n*. [*Dim.* of L. *cilium*, eyelid] a minute cilium.

Ciliophora *n*. [L. *cilium*, eyelid; *pherein*, to bear] a class of Protozoa, commonly called ciliates, having cilia at least when young, usually having 2 nuclei, and never being amoeboid.

ciliospore *n*. [L. *cilium*, eyelid; Gk. *sporos*, seed] a ciliated protozoan swarm spore.

cilium *n*. [L. *cilium*, eyelid] *sing.* of cilia.

cinchonine *n*. [After Countess *de Chinchón*] alkaloid found in various dicots of the family Rubiaceae.

cincinnal cicinnal, *q.v.*

cincinnus *n*. [L. *cincinnus*, curl] a scorpioid cyme, *q.v.*

cinclides *n.plu.* [Gk. *kingklis*, latticed gate] perforations, in body wall of certain Anthozoa, for extrusion of acontia; *sing.* cinclis.

cinerea *n*. [L. *cinereus*, ashen] the grey matter of the nervous system.

cinereous ashy-grey; *alt.* tephrous.

cingula *plu.* of cingulum. *n. sing.* Ring formed by hyphal proliferation around upper part of stipe, uniting with incurved edge of pileus; *plu.* cingulae.

cingulate *a*. [L. *cingulum*, girdle] having a girdle or cingulum; shaped like a girdle; *appl.* a gyrus and sulcus above corpus callosum.

cingulum *n*. [L. *cingulum*, girdle] any structure which is like a girdle; part of plant between root and stem; part of diatom frustule uniting valves; a ridge round base of crown of a tooth, *alt.* basal ridge; a tract of fibres connecting callosal and hippocampal convolutions of brain; outer ciliary zone on disc of rotifers, *cf.* trochus; clitellum, *q.v.*

cion scion, *q.v.*

circa-annual rhythm a rhythm or cycle of behaviour of approximately one year.

circadian rhythm a metabolic or behavioural rhythm with a cycle of about 24 hours, which may be endogenous or exogenous; *alt.* diurnal rhythm.

circinate *a*. [L. *circinatus*, made round] rolled on the axis, so that apex is centre.

circulation *n*. [L. *circulatio*, act of circulating] the regular movement of any fluid within definite channels in the body; streaming movement of protoplasm of plant cells.

circulus *n*. [L. *circulus*, circle] any ring-like arrangement, as of blood vessels caused by branching or connection with one another, as circulus major of iris, or as of markings of fish scales.

circulus arteriosus a vascular ring at base of brain; *alt*. circle of Willis, arterial circle.

circumduction *n*. [L. *circum*, around; *ductus*, led] the form of motion exhibited by a bone describing a conical space with the articular cavity as apex.

circumesophageal circumoesophageal, *q.v.*

circumferential *a*. [L. *circum*, around; *ferre*, to bear] *appl*. canal in medusoids; *appl*. cartilages which surround certain articulatory fossae; *appl*. primary lamellae parallel to circumference of bone.

circumfila *n.plu*. [L. *circum*, around; *filum*, thread] looped or wreathed filaments on antennal segments, as in gall midges.

circumflex *a*. [L. *circum*, around; *flectere*, to bend] bending round, *appl*. certain arteries, veins; *appl*. nerve: the axillary nerve.

circumfluence *n*. [L. *circum*, around; *fluens*, flowing] in Protozoa, ingestion by cytoplasm flowing towards food and surrounding it after contact.

circumgenital *a*. [L. *circum*, around; *gignere*, to beget] surrounding the genital pore; *appl*. glands secreting waxy powder in oviparous species of Coccidae bugs.

circumnutation *n*. [L. *circum*, around; *nutare*, to nod] the irregular elliptical or spiral movement exhibited by apex of a growing stem, shoot, or tendril.

circumoesophageal *a*. [L. *circum*, around; Gk. *oisophagos*, gullet] *appl*. structures or organs surrounding or passing along the gullet; *alt*. circumesophageal.

circumoral *a*. [L. *circum*, around; *os*, mouth] encircling a mouth, *appl*. cilia, nerve ring, perihaemal canal, etc.

circumorbital *a*. [L. *circum*, around; *orbis*, eye socket] surrounding the orbit, *appl*. bones of skull.

circumpolar *a*. [L. *circum*, around; *polus*, end of axle] *appl*. flora and fauna of Polar regions.

circumpulpar *a*. [L. *circum*, around; *pulpa*, fruit pulp] *appl*. dentine forming layer around pulp cavity of teeth, as in fishes.

circumscissile *a*. [L. *circum*, around; *scindere*, to cut] splitting along a circular line, *appl*. dehiscence exhibited by a pyxidium.

circumscript *a*. [L. *circumscribere*, to draw line around] *appl*. marginal sphincter when sharply defined, in sea anemones.

circumvallate *a*. [L. *circum*, around; *vallum*, rampart] encircled by a wall, as of tissue; vallate, *appl*. certain tongue papillae.

circumvallation *n*. [L. *circum*, around; *vallare*, to wall] ingestion of food by extruded pseudopodia, as in Protozoa or in phagocytes.

circumvascular *a*. [L. *circum*, around; *vasculum*, small vessel] *appl*. dentine lining vascular canals in pulp cavity of teeth, as in fishes.

cirral *a*. [L. *cirrus*, curl] *pert*. cirri or a cirrus. *n*. Any of the hollow ossicles in cirri of crinoids.

cirrate *a*. [L. *cirratus*, having curls] having cirri; *alt*. cirriferous.

cirrhi, cirrhus cirri, *q.v.*; cirrus, *q.v.*

cirri *n.plu*. [L. *cirrus*, curl] tendrils or tendril-like structures; appendages of barnacles; jointed filaments of axis or of aboral surface of crinoids; barbels of fishes; respiratory and tactile appendages of annelids; organs of copulation in some molluscs and trematodes; hair-like structures on appendages of insects; *alt*. cirrhi; *sing*. cirrus.

cirriferous *a*. [L. *cirrus*, curl; *ferre*, to bear] cirrate, *q.v.*

Cirripedia, cirripedes *n.*, *n.plu*. [L. *cirrus*, curl; *pes*, foot] a subclass of crustaceans, commonly called barnacles, which as adults are stalked or sessile sedentary animals with the head and abdomen reduced and the body enclosed in a carapace.

cirrose *a*. [L. *cirrus*, curl] with cirri or tendrils; *alt*. cirrous.

cirrous cirrose, *q.v.*; *appl*. leaf with prolongation of midrib forming a tendril.

cirrus *n*. [L. *cirrus*, curl] *sing*. of cirri, *q.v.*; coherent spores discharged through an ostiole; *alt*. cirrhus.

cirrus sac the sac containing the seminal vesicle and retracted copulatory organ in trematodes.

cis-configuration the situation where 2 mutants at different sites in the same cistron are on the same chromosome; *cf*. *trans*-configuration.

cisterna *n*. [L. *cisterna*, cistern] closed space containing fluid, as any of the subarachnoid spaces; cisterna chyli: receptaculum chyli, *q.v.*; expanded flattened sacs at end of vesicles of Golgi body or endoplasmic reticulum; any of various other flattened membrane-enclosed fluid-filled vesicles.

cis–trans effect the situation where 2 mutants at different sites in the same cistron may act as a single gene in the *cis*-configuration or as a pair of alleles in the *trans*-configuration.

cistron the portion of a chromosome within which a number of mutational entities or loci are integrated for one function; the section of a DNA molecule that specifies a polypeptide chain, i.e. a structural gene.

citric acid an acid of the Krebs' cycle which is converted to isocitric acid by the enzyme aconitase, and which is the acid giving tartness to oranges and grapefruits.

citric acid cycle Krebs' cycle, *q.v.*

citrin *n*. [L.L. *citrus*, lemon] a mixture of flavonoid pigments first isolated from lemon peel, of which the active constituent is hesperidin, and which is also known as vitamin P or the permeability factor.

citrulline *n*. [L. *citrullus*, water melon] an amino acid, a derivative of valeric acid, first obtained from water melon, also occurring as an intermediate product in the formation of urea from ornithine in the urea cycle.

cladanthous *a*. [Gk. *klados*, sprout; *anthos*, flower] having terminal archegonia on short lateral branches; *alt*. cladocarpous, cladogenous.

cladautoicous *a*. [Gk. *klados*, sprout; *autos*, self; *oikos*, house] with antheridia on a special stalk, as in mosses.

Cladistia *n*. [Gk. *klados*, branch; *histion*, sail] an order of primitive teleosts including the bichir and reedfish.

cladistic *a*. [Gk. *klados*, sprout] in plants, *pert*. similarity due to recent origin from a common ancestor; *cf*. patristic.

clado- *alt*. klado-.

cladocarpous cladanthous, *q.v.*

Cladocera n. [Gk. klados, sprout; keras, horn] an order of Branchiopoda including the water fleas; sometimes considered as a suborder and included with the Conchostraca in the Diplostraca.

Cladocopa n. [Gk. klados, sprout; ōdēs, like] an order of marine ostracods having biramous 2nd antennae and no trunk limbs.

cladode n. [Gk. klados, sprout] a green flattened lateral shoot, arising from the axil of a leaf, and resembling a foliage leaf, as in Ruscus, butcher's broom; alt. cladophyll, cladophyllum; cf. phylloclade.

cladodont a. [Gk. klados, sprout; odous, tooth] having or appl. teeth with prominent central and small lateral cusps.

Cladodontiformes n. [Gk. klados, sprout; odous, tooth; L. forma, shape] an order of Selachii existing from Devonian to Permian times.

cladogenesis n. [Gk. klados, sprout; genesis, descent] branching of evolutionary lineages so as to produce new types; speciation brought about by the multiplication of species at any one time; cf. anagenesis.

cladogenous a. [Gk. klados, sprout; gennaein, to produce] stem-borne, appl. certain roots; borne on branches; cladanthous, q.v.

cladogram n. [Gk. klados, sprout; graphein, to write] a tree-like diagram illustrating the evolutionary descent of any group of organisms.

cladome n. [Gk. klados, sprout] the group of superficially situated rays in a triaene.

Cladophorales n. [Cladophora, generic name] an order of green algae having multinucleate cylindrical cells united end to end in simple or branched filaments.

cladophyll(um) n. [Gk. klados, sprout; phyllon, leaf] cladode, q.v.

cladoptosis n. [Gk. klados, sprout; ptōsis, falling] annual or other shedding of twigs.

cladose a. [Gk. klados, sprout] branched.

Cladoselachii n. [Cladoselache, generic name] an extinct order of Devonian to Permian elasmobranch fish having primitive broad-based fins and no claspers; alt. Cladoselachiformes.

cladosiphonic a. [Gk. klados, sprout; siphōn, tube] with insertion of leaf trace on periphery of the axial stele; cf. phyllosiphonic.

cladotyle n. [Gk. klados, sprout; tylos, knob] a rhabdus with one actine branched, the other tylote.

Cladoxylales n. [Cladoxylon, generic name] an order of Primofilices, q.v.

cladus n. [Gk. klados, branch] a branch, as of a branched spicule.

clamp connections swellings on certain dikaryotic hyphae, for passage of daughter nuclei to cell below, with subsequent septum formation; also occurring in whorls, for distribution of nuclei to hyphal branches; alt. buckles.

clan phratry, q.v.

clandestine a. [L. clandestinus, from clam, secretly] appl. evolution which is not apparent in adult forms; or of adult characters from ancestral embryonic characters.

clasmatoblast n. [Gk. klasma, fragment; blastos, bud] mast cell, q.v.

clasmatocyte n. [Gk. klasma, fragment; kytos, hollow] a variable basiphil phagocyte or macrophage in areolar tissue; histiocyte, q.v.

claspers n.plu. [A.S. clyppan, to embrace] rod-like processes on pelvic fins of certain male elasmobranchs; outer gonapophyses of insects; valves or harpes of male Lepidoptera; any modification of an organ or part to enable the 2 sexes to clasp one another; tendrils or climbing shoots.

claspettes harpagones, q.v.

class n. [L. classis, division] a taxonomic group into which a phylum or division is divided, and which is itself divided into orders.

classical conditioning Pavlovian conditioning, q.v.

classification the arrangement of plants and animals in groups, based on some or all of their similarities and differences.

clathrate, clathroid a. [Gk. klēthra, lattice] lattice-like.

Clathrinida n. [Clathrina, generic name] an order of Calcinea having no true dermal membrane or cortex and the central cavity always lined with choanocytes.

clathrovirus n. [Gk. klēthra, lattice; L. virus, poison] a virus containing RNA, and usually with a latticed substructure.

Claudius' cells outer columnar or cuboid cells adjoining Hensen's cells in organ of Corti.

claustrum n. [L. claustrum, bar] in cerebral hemispheres, a thin layer of grey matter lateral to external capsule; one of the Weberian ossicles in Ostariophysi.

clava n. [L. clava, club] a club-shaped spore-bearing structure of certain fungi; the knob-like end of antenna of certain insects; hypostoma of ticks; swelling at end of fasciculus gracilis of medulla oblongata; a club-shaped structure.

clavate a. [L. clava, club] club-shaped, thickened at one end; alt. claviform.

Claviceptales, Clavicipitales n. [Claviceps, generic name] an order of Pyrenomycetes having a fairly pseudoparenchymatous mycelium like a cushion, and in which asci are formed in perithecium-like cavities; sometimes included in the Sphaeriales or Hypocreales.

clavicle n. [L. clavicula, small key] collar bone, forming anterior or ventral portion of the shoulder girdle; alt. proscapula.

clavicular a. [L. clavicula, small key] pert. clavicle.

clavicularium n. [L. clavicula, small key] the epiplastron of Chelonia, probably corresponding to clavicles of other forms.

claviform n. [L. clava, club; forma, form] clavate, q.v.

clavola n. [L. clava, club] the flagellar portion, or terminal joints, of insect antenna.

clavula n. [L. clava, club] a monactinal modification of triaxon spicule; a minute ciliated spine on fasciole of Spatangoida; a clavate sporophore of certain fungi.

clavus n. [L. clavus, nail] the part of an hemelytron lying next to the scutellum in Hemiptera; a projection or crotchet from scape of spiders; ergot disease in grasses.

claw n. [A.S. clawu, claw] the unguis or stalk of a petal; a sharp curved nail on finger or toe; forceps of certain crustaceans; curved process on limb of insect.

clay a soil having most particles below 0·002 mm, made of hydrated aluminosilicates, and which is poorly drained and aerated.

clearing foot filamentous process of exopodite of 2nd maxilla in Phyllocarida.

cleavage *n.* [A.S. *cleofan,* to cut] the series of mitotic divisions which change the zygote into a multicellular embryo.

cleavage cell blastomere, *q.v.*

cleavage nucleus nucleus of fertilized egg or zygote produced by union of male and female pronuclei; the egg nucleus of parthenogenetic eggs.

cleidoic *a.* [Gk. *kleis,* bar; *ōon,* egg] having or *pert.* eggs enclosed within a shell or membrane.

cleisto- *also,* clisto-, kleisto-, *q.v.*

cleistocarp *n.* [Gk. *kleistos,* closed; *karpos,* fruit] cleistothecium, *q.v.*

cleistocarpous *a.* [Gk. *kleistos,* closed; *karpos,* fruit] having closed ascocarps; with non-operculate capsules, *appl.* mosses; *alt.* cleistocarpic; *cf.* stegocarpous.

cleistogamy *n.* [Gk. *kleistos,* closed; *gamos,* marriage] the condition of having flowers which never open and are self-pollinated, and are often small and inconspicuous; *a.* cleistogamic, cleistogamous; *cf.* chasmogamy.

cleistogene *n.* [Gk. *kleistos,* closed; *genos,* descent] a plant with cleistogamous flowers.

cleistothecium *n.* [Gk. *kleistos,* closed; *thēkē,* box] a closed ascocarp whose spores are produced internally and which releases the spores by breakdown of the wall; *alt.* cleistocarp.

cleithrum *n.* [Gk. *kleithron,* bar] the pair of additional clavicles in Stegocephalia; clavicular element of some fishes.

climacteric *n.* [Gk. *klimaktēr,* step of staircase] a critical phase or period of changes, in living organisms; *appl.* change associated with menopause, or with recession of male function; *appl.* phase of increased respiratory activity at ripening of fruit.

Climatiiformes *n.* [*Climatius,* generic name; L. *forma,* shape] an order of acanthodians existing in Silurian to Carboniferous times, having 2 dorsal fins, and spines between pelvic and pectoral fins.

climatype *n.* [Gk. *klima,* climate; *typos,* image] a biotype resulting from selection in a particular climate, a climatic ecotype.

climax *n.* [Gk. *klimax,* ladder] the mature or stabilized stage in a successional series of communities, when dominant species are completely adapted to environmental conditions; completion of development, *appl.* leaves.

clin- *also* klin-, *q.v.*

clinandrium *n.* [Gk. *klinē,* bed; *anēr,* man] a cavity in the column between anthers in orchids; *alt.* androclinium.

clinanthium *n.* [Gk. *klinē,* bed; *anthos,* flower] a dilated floral receptacle, as in capitulum of Compositae.

cline *n.* [Gk. *klinein,* to slant] a series of form changes; gradient of biotypes; character-gradient.

clinging fibres tendril fibres, *q.v.*

clinidium *n.* [Gk. *klinidion,* small couch] a filament in a pycnidium, which produces spores.

clinoid *a.* [Gk. *klinē,* couch; *eidos,* form] *appl.* processes of sella turcica.

clinology *n.* [Gk. *klinein,* to decline; *logos,* discourse] the study of the decline of organisms after maturity, or after their prime in groups or in phylogeny.

clinosporangium *n.* [Gk. *klinē,* bed; *spora,* seed; *anggeion,* vessel] pycnidium, *q.v.*

clinospore *n.* [Gk. *klinē,* bed; *spora,* seed] a spore abjointed from a clinidium; conidium, *q.v.*

clisere *n.* [*climate; sere*] succession of communities which results from a changing climate.

clisto- cleisto-, *q.v.*

Clitellata *n.* [L. *clitellae,* pack saddle] a class of annelids having few chaetae and no parapodia, and a clitellum all or some of the time; the group includes the Oligochaeta and Hirudinea but is not recognized in all classifications.

clitellum *n.* [L. *clitellae,* pack saddle] the swollen glandular portion of skin of certain annelids, such as earthworm, which secretes the coccoon and nutrient material for embryo; *alt.* saddle, cingulum.

clitoris *n.* [Gk. *kleiein,* to enclose] an erectile organ, homologous with penis, at upper part of vulva.

clivus *n.* [L. *clivus,* slope] a shallow depression in sphenoid, behind dorsum sellae; posterior sloped part of the monticulus.

cloaca *n.* [L. *cloaca,* sewer] the common chamber into which intestinal, genital, and urinary canals open, in vertebrates except most mammals; posterior end of intestinal tract in certain invertebrates.

cloacal *pert.* cloaca, *appl.* gland, *appl.* excurrent siphon of molluscs; *appl.* membrane of ectoderm and endoderm temporarily separating cloaca and proctodaeum during embryonic development.

clone *n.* [Gk. *klōn,* twig] a group of individuals propagated by mitosis from a single ancestor; an apomict strain; *alt.* klon.

clonotype *n.* [Gk. *klōn,* twig; *typos,* pattern] a specimen of an asexually propagated part of a type specimen or holotype.

clonus *n.* [Gk. *klonos,* violent motion] a series of muscular contractions when individual contractions are discernible, incomplete tetanus.

club hair a hair forming a keratinized club-shaped bulb, becoming detached from papilla, and eventually shed.

club moss the common name for a member of the Lycopsida or sometimes restricted to the orders Lycopodiales and/or Selaginellales.

clunes *n.plu.* [L. *clunes,* buttocks] buttocks; *alt.* nates.

Clupeiformes *n.* [*Clupea,* generic name of herring; L. *forma,* shape] Clupeomorpha, *q.v.*; sometimes considered to be an order of Clupeomorpha, where the Clupeomorpha are considered as a superorder.

clupeine *n.* [*Clupea,* generic name of herring] a protamine obtained from herring sperm which forms arginine on hydrolysis.

Clupeomorpha *n.* [*Clupea,* feneric name of herring; Gk. *morphē,* form] a group of primitive teleosts existing from Jurassic or Cretaceous to present day and including herring and salmon; *alt.* Clupeiformes, Isospondyli.

cluster-crystals globular aggregates of calcium oxalate crystals in plant cells; *alt.* sphaeraphides.

cluster-cup aecidium, *q.v.*

clypeal *a.* [L. *clypeus,* shield] *pert.* clypeus of insects.

Clypeasteroida *n.* [*Clypeaster,* generic name; Gk. *eidos,* form] an order of sea urchins of the superorder Gnathostomata, commonly called sand dollars and having a flattened test.

clypeate *a.* [L. *clypeus,* shield] round or shield-like, *alt.* clypeiform; having a clypeus.

clypeola, clypeole n. [L. clypeus, shield] a sporophyll in the cone of a horsetail.

clypeo-labral a. [L. clypeus, shield; labrum, lip] appl suture between clypeus and labrum.

clypeus n. [L. clypeus, shield] a sclerite on anteromedian part of insect head; the strip of cephalothorax between eyes and cheliceral bases in spiders; a band of tissue round mouth of perithecium of certain fungi; alt. shield.

CMP cytidine monophosphate, q.v.

cnemial a. [Gk. knēmē, tibia] pert. tibia; appl. ridge along dorsal margin of tibia.

cnemidium n. [Gk. knēmis, legging; idion, dim.] lower part of bird's leg devoid of feathers, generally scaly.

cnemis n. [Gk. knēmis, legging] shin or tibia.

cnida n. [Gk. knidē, nettle] cnidoblast, q.v.

Cnidaria, cnidarians n., n.plu. [Gk. knidē, nettle] a phylum or subphylum of coelenterates having nematocysts on tentacles, and hydroid (polyp) and/or medusa forms.

cnidoblast n. [Gk. knidē, nettle; blastos, bud] strictly, nematoblast, q.v., but sometimes also used for nematocyst, q.v.; alt. cnida.

cnidocil n. [Gk. knidē, nettle; L. cilium, eyelid] a minute process projecting externally from a cnidoblast, whose stimulation causes discharge of a nematocyst.

cnidophore n. [Gk. knidē, nettle; pherein, to bear] a modified zooid which bears nematocysts.

cnidopod n. [Gk. knidē, nettle; pous, foot] drawnout basal part of a nematocyst, embedded in mesogloea.

cnidosac n. [Gk. knidē, nettle; sakkos, bag] a kidney-shaped swelling or battery, often protected by a hood, found on dactylozooids of Siphonophora.

Cnidosporidia n. [Gk. knidē, nettle; sporos, seed; idion, dim.] an order of Neosporidia having spores possessing polar capsules.

CNS central nervous system, q.v.

CoA coenzyme A, q.v.

coacervate n. [L. coacervare, to heap up] a collection of emulsoid particles into liquid droplets before flocculation, thought to be a possible method by which the prebiotic soup was concentrated at the origin of life. v: To mass in a small heap.

coactate a. [L. coacta, felt] closely matted but smooth, appl. surface.

coaction n. [L. cum, with; actio, action] the reciprocal activity of organisms within a community.

coactor any organism participating in coaction.

coadaptation n. [L. cum, with ad, to; aptare, to fit] the correlated variation and adaption in 2 mutually dependent organs.

coagulation n. [L. cum, together; agere, to drive] curdling or clotting; the changing from a liquid to a viscous or solid state by chemical reaction; appl. factor: vitamin K, q.v.

coagulin n. [L. coagulum, rennet] any agent capable of coagulating proteins.

coagulocyte n. [L. cum, together; agere, to drive; Gk. kytos, hollow] a granular haemocyte (blood cell) in insects; alt. cystocyte.

coagulum n. [L. coagulum, rennet] any coagulated mass; clot; curd.

coal ball a petrified more-or-less globular aggregation of plant structures found in certain coal measures.

coaptation n. [L. cum, together; aptare, to fit]

mutual adjustment of parts; dependence of function upon the presence of an organic structure or character.

coarctate a. [L. coarctare, to press together] compressed; closely connected; with abdomen separated from thorax by a constriction.

coarctate larva or pupa a larval stage of certain Diptera in which the larval skin is retained as a protective puparium; alt. pseudopupa, semipupa.

cobalamin(e) a member of the vitamin B complex, vitamin B_{12}, an organic compound containing cobalt and consisting of a nucleotide and a porphyrin, that is found in liver, kidney, and seafood but is produced commercially by bacteria and is necessary for the formation of red blood cells, growth and reproduction in man, and by many animals, and is used to treat pernicious and other anaemias; it exists in 3 forms, vitamin B_{12a}: cyanocobalamin(e) the first isolated and sometimes also called vitamin B_{12}, vitamin B_{12b}: hydroxy-cobalamin(e), vitamin B_{12c}: nitrocobalamin(e); alt. anti-pernicious anaemia factor, liver factor, erythrocyte-maturing factor, erythrotin.

cocaine n. [Gk. kokain, cocaine] an alkaloid obtained from the leaves of the coca plant, Erythroxylon coca, and having stimulating, narcotic, and locally anaesthetic effects.

cocarboxylase a coenzyme necessary for a carboxylase to work, e.g. thiamine pyrophosphate.

cocci n.plu. [Gk. kokkos, berry] septicidal carpels; spore mother cells of certain hepatics; rounded cells, as certain bacteria; sing. coccus.

Coccidiomorpha n. [Gk. kokkos, berry; morphē, form] an order of Telosporidia having adult phase intracellular and female gametes hologamous.

coccogone n. [Gk. kokkos, berry; gonos, begetting] a reproductive cell in certain algae.

coccoid a. [Gk. kokkos, berry; eidos, form] like or pert. a coccus; spherical or globose.

coccolith n. [Gk. kokkos, berry; lithos, stone] a calcareous plate found in coccolithophorids.

coccolithophorids n.plu. [Gk. kokkos, berry; lithos, stone; pherein, to bear] a group of unicellular organisms possessing coccoliths, considered as flagellates of the order Chrysomonadina (Coccolithophoridae) or as algae of the Chrysophyta, possibly Haptophyceae (Coccolithophoridaceae).

coccospheres n.plu. [Gk. kokkos, berry; sphaira, globe] remains of hard parts of certain algae and radiolarians.

coccus n. [Gk. kokkos, berry] sing. of cocci, q.v.

coccygeal a. [Gk. kokkyx, cuckoo] pert. or in region of coccyx.

coccygeomesenteric a. [Gk. kokkyx, cuckoo; mesos, middle; enteron, gut] appl. a branch of the caudal vein, as in birds.

coccyges plu. of coccyx.

coccyx n. [Gk. kokkyx, cuckoo] the terminal part of the vertebral column beyond the sacrum; plu. coccyges.

cochlea n. [Gk. kochlias, snail] the part of the inner ear concerned with hearing, which is spirally coiled like a snail's shell; a coiled legume.

cochlear appl. aestivation when wholly internal leaf is next but one to wholly external leaf; pert. the cochlea.

cochleariform a. [Gk. kochlias, snail; L. forma,

cochleate

shape] screw- or spoon-shaped; *pert.* thin plate or process of bone separating tensor tympani canal from Eustachian tube.

cochleate *a.* [Gk. *kochlias*, snail] screw-like; spiral; like a snail shell.

cockroaches the common name for many members of the Dictyoptera, *q.v.*

cocoon *n.* [F. *cocon*, cocoon] the protective case of many larval forms before they become pupae; silky or other covering formed by many animals for their eggs.

code *see* genetic code.

codehydrogenases I and II coenzymes I and II, *q.q.v.*

codeine *n.* [Gk. *kōdeia*, poppy head] an alkaloid found with morphine in opium and having effects similar to but weaker than morphine.

codominant two species being equally dominant in climax vegetation; 2 genes being equally dominant, so that when present together the organism shows a different characteristic from when each is present separately, *alt.* semidominant.

codon the sequence of 3 bases in DNA or mRNA, that code for a particular amino acid, and are matched by an anticodon on tRNA.

Coelacanthina *n.* [Gk. *koilos*, hollow; *akantha*, thorn] the order of crossopterygians that includes the coelacanth and have a reduced air bladder and no choanae; *alt.* Actinistia.

coelarium coelomic epithelium; *alt.* mesothelium.

Coelenterata, coelenterates *n.*, *n.plu.* [Gk. *koilos*, hollow; *enteron*, intestine] a group of diploblastic metazoan animals having radial symmetry, a single body cavity (coelenteron), body wall of ectoderm and endoderm separated by a usually non-cellular mesogloea; the group is usually considered a phylum containing 2 subphyla the Cnidaria and Ctenophora, but it may be raised to superphylum level and these groups considered phyla.

coelenteron *n.* [Gk. koilos, hollow; *enteron*, intestine] body cavity of coelenterates; *alt.* enteron, gastrovascular cavity.

coeliac *a.* [Gk. *koilia*, belly] *pert.* the abdominal cavity, *appl.* arteries, veins, nerves; *appl.* plexus: solar plexus, *q.v.*; *alt.* celiac.

coelo- *alt.* celo-.

coeloblast *n.* [Gk. *koilos*, hollow; *blastos*, bud] a division of the embryonic hypoblast; the whole hypoblast.

coeloconic *a.* [Gk. *koilos*, hollow; *kōnos*, cone] having, or consisting of, a conical process situated in a pit, *appl.* sensillae.

coelogastrula *n.* [Gk. *koilos*, hollow; *gaster*, stomach] a gastrula developed from a blastula with a segmentation cavity.

Coelolepida, coelolepids *n.*, *n.plu.* [Gk. *koilos*, hollow; *lepis*, scale] an order of Silurian and Devonian ostracoderms (Heterostraci) having small scales covering the whole body; *alt.* Thelodonti.

coelom *n.* [Gk. *koilōma*, hollow] a secondary body cavity lying completely within the mesoderm; *alt.* perigastrium, deuterocoele.

Coelomata *n.* [Gk. *koilōma*, hollow] in some classifications, a taxonomic group comprising those metazoans that have a coelom at some stage in their life cycle.

coelomate *a.* [Gk. *koilōma*, hollow] having a coelom.

coelomesoblast *n.* [Gk. *koilos*, hollow; *mesos*, middle; *blastos*, bud] in segmentation, the mesoblastic bands destined to form wall of coelom and outgrowths.

coelomic *a.* [Gk. *koilōma*, hollow] *pert.* a coelom.

coelomocytes *n.plu.* [Gk. *koilōma*, hollow; *kytos*, hollow vessel] coelomic corpuscles, including amoebocytes and elaeocytes, in annelids; mesenchymatous cells in body cavity of nematodes; cells in coelomic fluid and in water vascular and haemal systems, including morula-shaped cells, spindle-shaped cells, phagocytes, and crystal cells, in echinoderms.

coelomoduct *n.* [Gk. *koilōma*, hollow; L. *ducere*, to lead] a channel leading from body cavity to exterior.

coelomopores *n.plu.* [Gk. *koilōma*, hollow; *poros*, passage] ducts leading directly from pericardial cavity to exterior, peculiar to the nautiloid, *Nautilus*.

coelomostome *n.* [Gk. *koilōma*, hollow; *stoma*, mouth] the external opening of a coelomoduct.

coelomyarian *n.* [Gk. *koilos*, hollow; *mys*, muscle] having a longitudinal row of muscle cells bulging into the pseudocoel, *appl.* some nematodes.

coelosperm *n.* [Gk. *koilos*, hollow; *sperma*, seed] a carpel, hollow on its inner surface.

coelozoic *a.* [Gk. *koilos*, hollow; *zōon*, animal] *appl.* a trophozoite when situated in some cavity of the body.

coen- *alt.* cen-.

coenaesthesia *n.* [Gk. *koinos*, common; *aisthēsis*, sensation] the undifferentiated sensation caused by the body as a whole; common sensibility; *alt.* cenesthesia.

coenangium *n.* [Gk. *koinos*, common; *anggeion*, vessel] a coenocytic sporangium.

coenanthium *n.* [Gk. *koinos*, common; *anthos*, flower] inflorescence with a nearly flat receptacle having upcurved margins.

coenenchyma, coenenchyme *n.* [Gk. *koinos*, common; *engchyma*, infusion] common tissue which connects the polyps or zooids of a compound coral; *alt.* cenenchyma.

coeno- *alt.* ceno-, coino-, koino-.

coenobium *n.* [Gk. *koinos*, common; *bios*, life] a colony of unicellular organisms having a definite form and organization, which behave as an individual and reproduce to give daughter coenobia; *alt.* cell family, catallact, homoplast.

coenoblast *n.* [Gk. *koinos*, common; *blastos*, bud] a germ layer which gives origin to endoderm and mesoderm.

coenocentre, coenocentrum *n.* [Gk. *koinos*, common; *kentron*, centre] a deeply-staining body accompanying the ovum in certain fungi; *alt.* central body.

coenocyte, coenocytia *n.* [Gk. *koinos*, common; *kytos*, hollow] a plant tissue in which constituent protoplasts are not separated by cell walls; syncytium, *q.v.*; *alt.* symplast; *a.* coenocytic, *alt.* aseptate.

coenoecium *n.* [Gk. *koinos*, common; *oikos*, house] the common groundwork of a polyzoan colony.

coenogametangium *n.* [Gk. *koinos*, common; *gametēs*, spouse, *anggeion*, vessel] a coenocytic gametangium, as in Zygomycetes.

coenogamete *n.* [Gk. *koinos*, common; *gametēs*, spouse] a multinucleate gamete.

coenogamodeme *n.* [Gk. *koinos*, common;

gamos, marriage; *dēmos*, people] a unit made up of the hologamodemes that can, under certain specified conditions, exchange genes.

coenogamy n. [Gk. *koinos*, common; *gamos*, marriage] the union of coenogametangia.

coenogenesis n. [Gk. *koinos*, common; *genesis*, descent] common descent from the same ancestry; *alt*. syngenesis.

coenogony n. [Gk. *koinos*, common; *gonē*, generation] reproduction by means of coenocytes.

Coenopteridales n. [Gk. *koinos*, common; *pteris*, fern] an order of Primofilices, *q.v.*

coenosarc n. [Gk. *koinos*, common; *sarx*, flesh] the common tissue uniting the polyps in a compound colony; *cf.* perisarc.

coenosite n. [Gk. *koinos*, common; *sitos*, food] an organism habitually sharing food with another; commensal, *q.v.*

coenosorus n. [Gk. *koinos*, common; *sōros*, heap] in some ferns, a sorus formed by merging of separate sori.

coenospecies n. [Gk. *koinos*, common; L. *species*, particular kind] a group of taxonomic units such as species, ecospecies, or varieties, which can intercross to form hybrids that are sometimes fertile, the group being equivalent to a subgenus or superspecies.

coenosteum n. (Gk. *koinos*, common; *osteon*, bone] the common colonial skeleton in corals and polyzoans.

Coenothecalia n. [Gk. *koinos*, common; *thēkē*, case] an order of Octocorallia, commonly called blue corals, and containing only one genus, *Heliopora*, having a solid calcareous skeleton with vertical tubular cavities containing polyps; sometimes included in the Alcyonacea.

coenotrope n. [Gk. *koinos*, common; *tropē*, turning] behaviour common to a group of organisms or to a species.

coenozygote n. [Gk. *koinos*, common; *zygōtos*, yoked] a zygote formed by coenogametes.

coenurus n. [Gk. *koinos*, common; *oura*, tail] a metacestode with large bladder, from whose walls many daughter cysts arise, each with one scolex.

coenzyme n. [L. *cum*, with; Gk. *en*, in; *zymē*, leaven] an accessory substance, not a protein, necessary to the protein part of an enzyme (apoenzyme) to work, and which is only weakly bound to a protein; *cf.* cofactor, prosthetic group.

coenzyme A (CoA) a nucleotide built up from phosphoadenosine diphosphate and pantetheine, which acts as a transporter of acyl groups, forming acyl CoA, from one of many donors, such as pyruvic acid, and transfers it to one of many acceptors, such as oxaloacetic acid in the Krebs' cycle.

coenzyme I NAD, *q.v.*; *alt*. codehydrogenase I.

coenzyme II NADP, *q.v.*; *alt*. codehydrogenase II.

coenzyme Q ubiquinone, *q.v.*

coenzyme R biotin, *q.v.*

coevolution n. [L. *cum*, with; *evolvere*, to unroll] evolution of 2 species in relation to each other, such as predator-prey.

cofactor any accessory substance that is not an enzyme, but is necessary for metabolic pathways, such as ATP, a metallic ion, or a coenzyme.

cog-tooth spur or projection of incudal facet of malleus.

coherent a. [L. *cohaerere*, to stick together] with similar parts united but capable of separating with a little tearing; *cf.* adherent.

cohesion n. [L. *cohaerere*, to stick together] condition of union of separate parts of floral whorl; *cf.* adhesion.

Cohnheim's areas a group of myofibrils in a muscle fibre separated by sarcoplasm, in transverse section of the fibre giving the appearance of dotted areas.

cohort n. [L. *cohors*, enclosure] an indefinite taxonomic group used in different ways by different authorities, such as a group in rank above a superorder, a group between class and order, or a group of related families.

coincidence the ratio of observed double cross-overs to expected double cross-overs calculated on the basis of independent occurrence, and used as a measure of interference in crossing-over.

coino- coeno-, *q.v.*

coition, coitus n. [L. *coire*, to go together] sexual intercourse, copulation.

colamine n. [Alcohol, *amine*] a nitrogenous base found in cephalins.

colchicine n. [L. *colchicum*, meadow saffron, from *Colchis*, ancient Mingrelia] an alkaloid obtained from *Colchicum autumnale*, autumn crocus or meadow saffron, which is used to suppress the spindle and induce polyploidy in plants.

cold-blooded poikilothermal, *q.v.*

coleogen n. [Gk. *koleos*, sheath; *gennaein*, to produce] meristematic layer giving rise to endodermis.

Coleoidea n. [Gk. *koleos*, sheath; *eidos*, form] a subclass of cephalopods in which the shell is internal and the head has 8 sucker-like arms and often a pair of long retractile arms as well.

Coleoptera n. [Gk. *koleos*, sheath; *pteron*, wing] a very large order of insects, commonly called beetles, having complete metamorphosis, with the fore-wings modified as elytra and covering membranous hind-wings which may be reduced or absent.

coleoptile n. [Gk. *koleos*, sheath; *ptilon*, feather] a protective sheath surrounding the plumule of some monocotyledons, especially grasses; *alt*. plumule sheath.

coleorhiza n. [Gk. *koleos*, sheath; *rhiza*, root] a protective sheath surrounding the radicle of some flowering plants such as grasses; *alt*. radicle sheath, root sheath.

colic a. [Gk. *kolon*, colon] *pert*. the colon.

colicinogenic producing colicins.

colicins n.plu. [Gk. *kolon*, colon; L. *caedere*, to kill] substances produced by certain bacteria, especially some strains of *E.coli*, from their surfaces, which kill or inhibit other bacteria.

coliform a. [L. *colum*, strainer; *forma*, shape] sieve-like *alt*. cribriform. [Gk. *kolon*, colon] resembling colon bacilli.

Coliiformes n. [*Colius*, generic name of mousebird; L. *forma*, shape] an order of Neornithes including the colies or mousebirds.

coliphage n. [Gk. *kolon*, colon; *phagein*, to eat] a bacteriophage of *E.coli*.

collagen n. [Gk. *kolla*, glue; *genos*, descent] one of a family of proteins having a high content of hydroxyproline, containing hydroxylysine but lacking cysteine, cystine, and tryptophan; scleroprotein, occurring as chief constituent of white connective tissue fibres and organic part of bone, where it is called ossein, also of some fish scales; *alt.* collogon; *cf.* tropocollagen.

collaplankton *n.* [Gk. *kolla*, glue; *plangktos*, wandering] the plankton organisms rendered buoyant by a mucilaginous or gelatinous envelope.

collar *n.* [M.E. *coler*, collar] the choana of a collared cell; the reflexed peristomium in certain Sabellidae, polychaetes; a prominent fold behind the proboscis in Hemichordata; the fleshy rim projecting beyond the edge of a snail shell; any structure comparable with a collar; collum, *q.v.*; junction between root and stem; region of junction between blade and leaf sheath of grasses; collet, *q.v.*

collar cell choanocyte, *q.v.*

collarette line of junction between pupillary and ciliary zones of anterior surface of iris; *alt.* iris frill, angular line.

collateral *a.* [L. *cum*, with; *latera*, sides] side by side; *appl.* ovules; *appl.* bundles with xylem and phloem on the same radius, the phloem to the outside of xylem; *appl.* bud at side of axillary bud; *appl.* fine lateral branches from the axon of a nerve cell; *appl.* prevertebral ganglia of sympathetic system; *appl.* inheritance of character from a common ancestor in individuals not lineally related; *appl.* circulation established through anastomosis with other parts when the chief vein is obstructed.

collecting hair collector, *q.v.*

collecting tubules more-or-less straight ducts which convey urine from cortical to pelvic region of kidney, terminating in Bellini's ducts to renal papillae.

collective fruit anthocarp, *q.v.*

collector *n.* [L. *colligere*, to collect] one of the pollen-retaining hairs on stigma or style of certain flowers; *alt.* collecting hair.

Collembola *n.* [Gk. *kolla*, glue; *embolos*, wedge] an order of wingless insects, commonly called springtails, having a forked springing organ on the 4th abdominal segment.

collenchyma *n.* [Gk. *kolla*, glue; *engchyma*, infusion] parenchymatous peripheral supporting tissue with cells more-or-less elongated and thickened, either at the angles (angular c.), or on walls adjoining intercellular spaces (lacunar c.), or tangentially (lamellar c.); the middle layer of sponges; *alt.* collenchyme.

collencyte *n.* [Gk. *kolla*, glue; *en*, in; *kytos*, hollow] a clear cell with thread-like pseudopodia found in sponges making up collenchyma.

collet *n.* [F. *collet*, collar] root zone of hypocotyl, where cuticle is absent; *alt.* collar.

colleterium *n.* [Gk. *kolla*, glue] a mucus-secreting gland in female reproductive system of insects, for production of egg case or for cementing eggs to substrate; *alt.* colleterial gland, sebific gland.

colleters *n.plu.* [Gk. *kollētos*, glued] the hairs, usually secreting a gluey substance, which cover many resting buds; *alt.* multicellular glandular trichomes.

colletocystophore *n.* [Gk. *kollētos*, glued; *kystis*, bladder; *pherein*, to bear] the statorhabd of the scyphozoan, *Haliclystus*.

colliculate, colliculose *a.* [L. *dim.* of *collis*, hill] having small elevations.

colliculus *n.* [L. *colliculus*, little hill] a prominence of corpora quadrigemina; a rounded elevation near apex of anterolateral surface of arytae-noid cartilages; slight elevation formed by optic nerve at entrance to retina; one of the rounded prominences surrounding the embryonic external auditory meatus; elevation of urethral crest, with openings of ejaculatory ducts and prostatic utricle.

colligation *n.* [L. *colligare*, to bind together] the combination of persistently discrete units.

colloblast *n.* [Gk. *kolla*, glue; *blastos*, bud] a cell on tentacles and pinnae of ctenophores, which carries little globules of adhesive substance; *alt.* lasso cell.

collogen collagen, *q.v.*

colloid *n.* [Gk. *kolla*, glue; *eidos*, form] a substance of high molecular weight which does not readily diffuse through a semipermeable membrane, *cf.* crystalloid; a substance composed of 2 homogeneous parts or phases, one of which is dispersed in the other.

collophore *n.* [Gk. *kolla*, glue; *pherein*, to bear] the ventral tube of 1st abdominal segment in Collembola.

collum *n.* [L. *collum*, neck] neck; collar, *q.v.*; any collar-like structure; dorsal plate of 1st body segment in Diplopoda; basal portion of sporogonium in mosses.

Colobognatha *n.* [Gk. *kolobos*, curtailed; *gnathos*, jaw] a superorder of Chilognatha having highly modified or reduced mouth-parts, sometimes forming a piercing and sucking beak.

colon *n.* [Gk. *kolon*, colon] the 2nd portion of intestine of insects; part of the large intestine of vertebrates.

colony *n.* [L. *colonia*, farm] any collection of organisms living together, *appl.* ants, bees; a group of animals or plants living together and sexually isolated, or established in a new area; coenobium, *q.v.*; a group of bacteria or of other micro-organisms in a culture; organisms with organic connection between them.

Colossendeomorpha *n.* [*Colossendeis*, generic name; Gk. *morphē*, form] an order of Pycnogonida having small or absent chelicerae.

colostrum *n.* [L. *colostrum*, first milk] a clear fluid secreted at the end of pregnancy and differing from the milk secreted later; *alt.* fore-milk.

colpate *a.* [Gk. *kolpos*, fold] furrowed, *appl.* pollen grains.

colulus *n.* [*Dim.* of L. *colus*, distaff] a small conical structure between anterior spinnerets of spiders.

Columbiformes *n.* [*Columba*, generic name of pigeon; L. *forma*, shape] an order of birds of the subclass Neornithes, including the doves and pigeons.

columella *n.* [L. *columella*, small column] the column of sterile cells in the centre of the capsule of mosses; central core in root cap; central pillar in skeleton of some corals; the central pillar in gasteropod shells; epipterygoid, *q.v.*; the rod, partly bony, partly cartilaginous, connecting tympanum with inner ear in birds, reptiles, and amphibians; the axis of cochlea; lower part of nasal septum; *a.* columellar.

column(a) *n.* [L. *columna*, pillar] any structure like a column, as spinal column; cylindrical body of hydroids up to the tentacles; stalk of a crinoid; longitudinal bundle of nerve fibres in white matter of spinal cord; nasal septum edge; thick muscular strands found in ventricle; stamens in mallows; united stamens and style in orchids.

columnals n.plu. [L. columna, pillar] stem ossicles in crinoids.

columnar a. [L. columna, pillar] pert., or like, a column or columna; appl. cells longer than broad; appl. epithelium of columnar cells.

Columniferae n. [L. columna, pillar; ferre, to bear] Malvales, q.v.

Colymbiformes n. [Colymbus, generic name; L. forma; shape] an order of small water birds without webbed feet including the grebes; alt. Podicipitiformes.

colyone kolyone, q.v.

coma n. [Gk. kome, hair] a terminal cluster of bracts, as in pineapple; hair-tufts on certain seeds; a compact head of clustered leaves or branches in certain mosses such as Sphagnum.

Comanchean a. [Comanche County, Texas] Lower Cretaceous in North America.

comb n. [A.S. comb, comb] a comb-like structure, as swimming plate, ctenidium, ctene, pecten, strigilis, honeycomb, fleshy crest, mushroom gills.

combination colours those produced by a structural feature in conjunction with a layer of pigment.

comb jellies a common name for the Ctenophora, q.v.

comb-rib swimming plate, q.v.

comes n. [L. comes, companion] a blood vessel that runs alongside a nerve; plu. comites.

comitalia n.plu. [L. comitari, to accompany] small di- or tri-actine spicules in sponges.

comma n. [Gk. komma, short clause] sarcomere, q.v.; appl. tract: certain nerve fibres in dorsal or posterior column of spinal cord; appl. bacillus: the spirillum causing cholera.

Commelinales n. [Commelina, generic name] an order of monocots (Liliidae) including the family Commelinaceae, e.g. Tradescantia.

commensal n. [L. cum, with; mensa, table] an organism living with another and sharing the food, both species as a rule benefiting by the association, or one benefiting and the other not being harmed; alt. coenosite.

commensalism the association between commensals, or between a commensal and its host.

comminator a. [L. cum, with; minari, to threaten] appl. muscles which connect adjacent jaws in Aristotle's lantern.

commissure n. [L. commissura, seam] the union-line between 2 parts; inner side of mericarp; carpellary cohesion plane; a connecting band of nerve tissue.

community a well-defined assemblage of plants and/or animals, clearly distinguishable from other such assemblages.

community biomass total weight per unit area of the organisms in a community.

comose a. [L. comosus, hairy] hairy; having a tuft of hairs.

companion cell a narrow cell, retaining its nucleus, derived from a cell giving rise also to a sieve tube element, in phloem of angiosperms.

compass n. [L. cum, together; passus, pace] a curved bifid ossicle, part of Aristotle's lantern.

compass plants certain plants with permanent north and south direction of their leaf edges.

compensation point point of balance between respiration and photosynthesis, as determined by intensity of light at a given temperature, alt. compensation intensity; limit of sea or lake depth below which plants lose more by respiration than they gain by photosynthesis, alt. compensation depth or level; cf. extinction point.

competence n. [L. competere, to suit] reactive state permitting directional development and differentiation in response to a stimulus, as of part of an embryo in response to an evocator or organizer stimulus.

competition active demand by 2 or more organisms for a material or condition, so that both are inhibited by the demand, e.g. plants competing for light, water, etc.; cf. amensalism.

complement n. [L. complere, to fill up] alexine. q.v.; a group composed of 1, 2, or more genomes or chromosome sets derived from a single nucleus.

complemental air volume of air which can be taken in addition to that drawn in during normal breathing.

complemental male a purely male form, usually small, found living in close proximity to the ordinary hermaphrodite form in certain animals, as in certain polychaetes and cirripedes; alt. pigmy male.

complementary n. [L. complere, to fill up] the coronoid bone, a. Appl. non-suberized cells loosely arranged in cork tissue and forming air passages in lenticels, alt. filling tissue; appl. genes or factors producing a similar effect when inherited separately but a different effect together.

complementation the situation where 2 mutant alleles brought together in a cell produce a wild type phenotype.

complete metamorphosis insect metamorphosis in which the young are usually different from the adult and are called larvae, and go through a resting stage, the pupa, before reaching the adult stage, imago, the wings being developed inside the body during the pupal stage; cf. incomplete metamorphosis.

complexus n. [L. complexus, embrace] an aggregate formed by a complicated interweaving of parts; appl. a muscle, the semispinalis capitis.

complicant a. [L. cum, together; plicare, to fold] folding over one another, appl. wings of certain insects.

complicate a. [L. cum, together; plicare, to fold] folded, appl. leaves folded longitudinally so that right and left halves are in contact, appl. insect wings, alt. conduplicate; compound appl. fruit body composed of pileoli with stipes joining to form a somewhat central stipe, as in some Hymenomycetes.

Compositae, composites n., n.plu. [L. cum, together; ponere, to place] Asterales, q.v., especially when the group is reduced to family level, when it is the largest family of dicots, having an inflorescence called a capitulum made up of florets, and including dandelion, daisy, thistles, etc.

composite a. [L. cum, together; ponere, to place] closely packed, as a capitulum, appl. fruits, as sorosis, syconus, strobilus; appl. a member of the plant family Compositae, q.v.

compound a. [L. cum, together; ponere, to place] made up of several elements, appl. inflorescences, pistils, leaves, medullary rays, eyes, etc.; appl. starch grains with 2 or more hila.

compound loci sites in close proximity on a chromosome where genetic units are separable by crossing-over; alt. compound allelomorphs.

compound spore sporodesm, q.v.

compression wood in conifers, reaction wood formed on the lower sides of crooked stems or branches, and having a dense structure and much lignification.

compressor n. [L. *cum*, together; *premere*, to press] something that serves to compress, *appl.* muscles, as compressor naris.

Compsopogonales n. [Gk. *compos*, elegant; *pōgōn*, beard] an order of red algae having branched filaments.

conalbumin n. [L. *cum*, with; *albumen*, white of egg] an iron-building glucoprotein from white of egg.

conarium n. [Gk. *kōarion*, little cone] transparent deep-sea larva of the coelenterate *Velella*; pineal gland, *q.v.*

concatenate a. [L. *cum*, together; *catenatus*, chained] forming a chain, as spores; linked at their bases, as processes.

concavoconvex heterocoelous, *q.v.*, *appl.* centra.

concentric a. [L. *cum*, together; *centrum*, centre] having a common centre; *appl.* vascular bundles with one kind of tissue surrounding another; *appl.* corpuscles: Hassall's corpuscles, *q.v.*

conceptacle n. [L. *concipere*, to conceive] a depression in thallus of certain algae in which gametangia are borne.

conceptive a. [L. *concipere*, to conceive] capable of being fertilized and producing an embryo.

concha n. [Gk. *kongchē*, shell] the cavity of the external ear, which opens into the external acoustic meatus; a superior, middle, and inferior projection from lateral wall of nasal cavity; turbinate body; one of 2 curved plates of sphenoidal bone; a marine shell.

conchiform a. [L. *concha*, shell; *forma*, shape] shaped like a concha; shell-shaped; *alt.* conchoid.

conchiolin n. [Gk. *kongchē*, shell] the organic component of ligament and periostracum of shells of molluscs.

conchoid conchiform, *q.v.*

conchology n. [Gk. *kongchē*, shell; *logos*, discourse] the branch of zoology dealing with molluscs or their shells.

Conchostraca n. [Gk. *kongchē*, shell; *ostrakon*, shell] an order of Branchiopoda having the head and body enclosed in a carapace, and 10 to 27 pairs of trunk limbs; sometimes considered a suborder and included with the Cladocera in the Diplostraca.

conchula n. [L. *concha*, shell] the conspicuous protuberant lip of the modified sulcus in the zoantharian, *Peachia*.

concolorate, concolorous a. [L. *concolor*, of the same colour] similarly coloured on both sides or throughout; of the same colour as a specified structure.

concrescence n. [L. *concrescere*, to grow together] the growing together of parts.

concrete a. [L. *concretus*, grown together] grown together to form a single structure.

condensation n. [L. *condensatio*; from *cum*, together; *densare*, to make thick] process of making or becoming thick; contraction, thickening, and spiralization of chromatids during prophase; the formation of a larger molecule by union of smaller ones with the elimination of a simpler group such as water, *cf.* hydrolysis.

condensed a [L. *condensare*, to press close together] *appl.* inflorescence with short-stalked or sessile flowers closely crowded.

condensing enzyme formerly an enzyme causing condensation, specifically the one catalysing the reaction between acetylcoenzyme A and oxaloacetic acid to form citric acid.

conditional *appl.* dominance owing to influence of modifying genes.

conditioned *appl.* reflex depending on new functional connections in central nervous system as a result of Pavlovian conditioning; *appl.* stimulus (CS) inducing a conditioned reflex in Pavlovian conditioning; *appl.* response (CR) given by an animal after it has undergone conditioning.

conditioning producing a response to a new stimulus by associating the new stimulus with an old one; *cf.* Pavlovan conditioning, trial-and-error conditioning.

conducting a. [L. *conducere*, to lead together] conveying, *appl.* tissues, bundles, etc.

conduction n. [L. *conducere*, to lead together] the transference of soluble matter from one part of a plant to another; the transmission of an excitation, function of nervous system.

conductivity n. [L. *conducere*, to lead together] power of transmitting an impulse.

conductor n. [L. *conducere*, to lead together] that which can transmit; a projection at base of embolus in spiders.

conduplicate a. [L. *conduplicare*, to double] *appl.* cotyledons folded to embrace the radicle; *appl.* vernation when one half of the leaf is folded upon the other; *alt.* complicate.

Condylarthra, condylarths n., n.plu. [Gk. *kondylos*, knuckle; *arthron*, joint] an order of extinct placental mammals of Cretaceous to Miocene times that were herbivores and probably ancestral ungulates.

condyle n. (Gk. *kondylos*, knuckle] the antheridium of stoneworts; a process on a bone for purposes of articulation; a rounded structure adapted to fit into a socket; a. condylar.

condyloid n. (Gk. *kondylos*, knuckle; *eidos*, form] shaped like, or situated near, a condyle.

cone n. (Gk. *kōnos*, cone] strobile, *q.v.*; a conical elevation on an egg just before fertilization; a conical or flask-shaped light-sensitive cell of the retina, responsible for colour vision and vision in good light.

cone bipolars bipolar cells whose inner ends ramify in contact with dendrites of ganglionic cells.

cone of origin small clear area of nerve cell at point of exit of axon, where there are no Nissl granules; *alt.* implantation cone, axon hill.

cone of Wulzen [R. Wulzen, American physiologist] a structure projecting forwards from pars intermedia into hypophysial cavity in pituitary region of ox and pig.

conferted a. [L. *confertus*, crowded] closely assembled or packed.

confluence n. [L. *confluere*, to flow together] angle of union of superior sagittal and transverse sinuses at occipital bone; *alt.* confluens sinuum, torcular Herophili.

congeneric a. [L. *congener*, of same race] belonging to the same genus.

congenetic a. [L. *cum*, with; Gk. *genesis*, descent] having the same origin.

congenital *a.* [L. *cum*, with; *gignere*, to beget] present at birth.

congestin *n.* [L. *congestus*, heaped up] a toxin of sea anemone tentacles.

conglobate *a.* [L. *conglobatus*, formed into a ball] ball-shaped; *appl.* gland on lower side of ductus ejaculatorius in insects.

conglomerate *a.* [L. *cum*, together; *glomerare*, to wind] bunched or crowded together.

congression *n.* [L. *congressio*, meeting] chromosome movement to equatorial plane of spindle at metaphase.

coni *n.plu.* [L. *conus*, cone] cones; coni vasculosi: lobules forming head of epididymis; *sing.* conus.

conidia *plu.* of conidium.

conidial *a.* [Gk. *konis*, dust; *idion*, *dim.*] *pert.* a conidium.

conidiiferous *a.* [Gk. *konis*, dust; *idion*, *dim.*; L. *ferre*, to bear] bearing conidia.

conidiocarp *n.* [Gk. *konis*, dust; *idion*, *dim.*; *karpos*, fruit] a collection of conidiophores enclosed in a covering; pycnidium, *q.v.*

conidiogenous *a.* [Gk. *konis*, dust; *idion*, *dim.*; *gennaein*, to produce] leading to or stimulating the formation of conidia, *appl.* compounds, media, etc.

conidiole *n.* [*Dim.* of *conidium*] a small or a secondary conidium.

conidiophore *n.* [Gk. *konis*, dust; *idion*, *dim.*; *pherein*, to bear] a hypha with sterigmata which bear conidia.

conidiosporangium *n.* [Gk. *konis*, dust; *idion*, *dim.*; *sporos*, seed; *anggeion*, vessel] conidium, especially in Phycomycetes, which may produce zoospores as well as germinating directly.

conidiospore *n.* [Gk. *konis*, dust; *idion*, *dim.*; *sporos*, seed] conidium, *q.v.*

conidium *n.* [Gk. *konis*, dust; *idion*, *dim.*] a fungal spore asexually produced by constriction of sterigma or part of a hypha and not enclosed in a sporangium; formerly called gonidium; *alt.* clinospore, conidiospore, stylogonidium, exospore, exosporium; *plu.* conidia.

Coniferales, Coniferae, conifers *n.*, *n.plu.* [L. *conus*, cone; *ferre*, to bear] an order of Coniferopsida, being branching woody plants often with long and short shoots, and with reproductive organs in unisexual cones.

Coniferophytina *n.* [L. *conus*, cone; *ferre*, to bear; Gk. *phyton*, plant] in some classifications, a major division of Spermatophyta, comprising the fork- and needle-leaved gymnosperms; *alt.* Pinicae.

Coniferopsida *n.* [L. *conus*, cone; *ferre*, to bear, *opsis*, sight] A class of gymnosperms having simple often needle-like leaves, or sometimes broad or long flat ones, branched stems, and megasporangia usually in compound strobili; *alt.* Coniferophyta.

coniferous *a.* [L. *conus*, cone; *ferre*, to bear] cone-bearing; *pert.* to conifers or to a cone-bearing plant.

coniform cone-shaped; conoid, *q.v.*

conjoined *a.* [L. *cum*, with; *jungere*, to join] adnate, *q.v.*

Conjugales, Conjugatae *n.* [L. *conjugare*, to join together] Zygnematales, *q.v.*

conjugate *v.* [L. *conjugare*, to join together] to unite, as Protozoa; to undergo conjugation. *a.* United in pairs; *appl.* pores united by a groove;

appl. division in pairs of monoploid nuclei; co-ordinated, *appl.* movements of the 2 eyes.

conjugated united; *appl.* protein, when molecule is united to non-protein molecule.

conjugation *n.* [L. *cum*, together; *jugare*, to yoke] the temporary union or complete fusion of 2 gametes; the pairing of chromosomes; in unicellular organisms a process of sexual reproduction which may involve the exchange of genetic material as in the ciliate *Paramecium*, or a one-way transfer of genetic material as in the bacterium *Escherichia coli* or the green alga, *Spirogyra*.

conjugation canal tube formed in fused outgrowths from opposite cells of parallel filaments, for passage of male gametes to the other filament, as in scalariform conjugation, *q.v.*, in the green alga, *Spirogyra*.

conjugon in bacteria, a genetic element necessary for conjugation.

conjunctiva *n.* [L. *cum*, together; *jungere*, to join] mucous membrane of eye, lining eyelids and reflected over fore-part of sclera and constituting corneal epithelium.

conjunctive *a.* [L. *cum*, together; *jungere*, to join] *appl.* tissue: ground tissue, *q.v.*; connective, *q.v.*; *appl.* symbiosis in which the partners are organically connected, *cf.* disjunctive symbiosis.

connate *a.* [L. *cum*, together; *gnatus*, born] firmly joined together from birth; *appl.* organs growing together and becoming joined, but separate at birth.

connate-perfoliate joined together at base so as to surround stem, *appl.* opposite sessile leaves.

connective *n.* [L. *connectere*, to bind together] a connecting band of nerve tissue between 2 ganglia; tissues separating 2 lobes of anther; the structure and zone between successive conidia, *alt.* disjunctor, bridge; vestige of middle layer of cortex of rootlets, connecting inner and outer layers, as in the fossil pteridophyte, *Stigmaria*; *alt.* conjunctive.

connective tissue animal tissue developed from the mesoderm, having a large amount of non-living intercellular matrix, often having fibres connecting and supporting other tissues; includes areolar tissue, bone, cartilage, and sometimes blood.

connector neurone internuncial neurone, *q.v.*

connexivum *n.* [L. *connectere*, to fasten together] flattened lateral margin of abdomen in bugs.

connivent *a.* [L. *connivere*, to close the eyes] converging; arching over so as to meet.

conodonts *n.plu.* [Gk. *kōnos*, cone; *odous*, tooth] a group of Palaeozoic tooth-like fossils of obscure affinities.

conoid *a.* [Gk. *kōnos*, cone; *eidos*, form] cone-like, but not quite conical; *alt.* coniform.

conoid ligament one of the fasciculi of the coraco-clavicular ligament.

conoid tubercle coracoid tuberosity, a small rough eminence on posterior border of clavicle, serving for attachment of conoid ligament.

conopodium *n.* [Gk. *kōnos*, cone; *pous*, foot] a conical receptacle or thalamus of a flower.

conotheca *n.* [Gk. *kōnos*, cone; *thēkē*, case] thin integument of phragmocone.

conscutum *n.* [L. *cum*, together with; *scutum*, shield] dorsal shield formed by united scutum and alloscutum in certain ticks.

consensual *a.* [L. *consensus*, agreement] *appl.*

involuntary action correlated with voluntary action; relating to excitation of a corresponding organ; *appl.* contraction of both pupils when only one retina is directly stimulated.

consimilar *a.* [L. *consimilis*, entirely similar] similar in all respects; with both sides alike, as some diatoms.

consociation *n.* [L. *consociatio*, partnership] a unit of a plant association, characterized by a single dominant species; a group formed by consocies.

consocies *n.* [L. *cum*, together; *socius*, fellow] a consociation representing a stage in the process of succession; a group of mores.

consortes *n.plu.* [L. *consortes*, partners] associate organisms other than symbionts, commensals, or hosts and parasites; *sing.* consors.

consortism *n.* [L. *consortium*, partnership] a kind of symbiosis in which the relationship between the partners is a consortium; *cf.* helotism.

consortium *n.* [L. *consortium*, partnership] the mutual relationship of the alga and fungus in a lichen in which both gain benefit; the compound lichen thallus itself.

conspecific [L. *cum*, together; *species*, particular kind; *facere*, to make] belonging to the same species.

consperse *a.* [L. *conspersus*, besprinkled] densely scattered, *appl.* dot-like markings, pores, etc.

constitutional type habitus, *q.v.*

constitutive *a.* [L. *constituere*, to establish] naturally present in an organism, *appl.* enzymes, *cf.* adaptive or inducible enzymes.

constricted *a.* [L. *constrictus*, drawn together] narrowed; compressed at regular intervals.

constriction *n.* [L. *constrictus*, drawn together] a constricted part or place, as a node of Ranvier; non-spiralizing chromosome segments at metaphase, either associated with the centromere, or acentric, or controlled by the nucleolus.

constrictor *n.* [L. *constrictus*, drawn together] a muscle which compresses or constricts, e.g. constrictor pharyngis, c. urethrae.

consumer heterotroph, *q.v.*; *see also* primary consumer, secondary consumer.

consummatory act an action at the end of an instinctive behaviour pattern such as eating.

consute *a.* [L. *consuere*, to sew together] with stitch-like markings, as elytra of certain beetles.

contabescence *n.* [L. *contabescere*, to waste away] abortion or atrophy of stamens.

contact receptor a receptor in epidermis or in dermis.

context *n.* [L. *cum*, together; *texere*, to weave] the layers developed between hymenium and true mycelium in certain fungi.

contiguous touching each other at the edges but not actually united.

continental drift the gradual movement of continents across the surface of the earth.

continuity *n.* [L. *continuus*, continuous] succession without a break, especially continuity of germ plasm.

continuous variation variations between individuals of a population in which differences are slight and grade into each other; *alt.* quantitative inheritance; *cf.* discontinuous variation.

continuum a form of vegetation in which one type passes almost imperceptibly into another and no 2 types are repeated exactly.

Contophora *n.* [Gk. *kontos*, punting pole; *pherein*, to bear] the eukaryotic algae that do produce flagella, i.e. all except red algae; *cf.* Aconta.

Contortae *n.* [L. *contortus*, twisted together] an order of sympetalous dicots, the term being used in different way by different authorities; in some classifications, Gentianales, *q.v.*

contorted *a.* [L. *contortus*, twisted together] twisted; *appl.* aestivation in which one leaf overlaps the next with one margin, and is overlapped by the previous on the other.

contortuplicate *a.* [L. *cum*, with; *torquere*, to twist; *plicare*, to fold] *appl.* bud with contorted and plicate leaves.

contour *n.* [F. *contour*, circuit] outline of a figure or body; *appl.* outermost feathers that cover the body of a bird, *alt.* plumae.

contractile *a.* [L. *cum*, together; *trahere*, to draw] capable of contracting.

contractile cell any cell in a sporangium or anther wall which by hygroscopic contraction helps to open the organ.

contractile fibre cells elongated, spindle-shaped, more-or-less polyhedral, nucleated muscle cells, containing a central bundle of fibrillae.

contractile vacuole a small spherical vesicle found in cytoplasm of many freshwater Protozoa which expels surplus water; *alt.* pulsating vacuole.

contractility *n.* [L. *cum*, together; *trahere*, to draw] the power by which muscle fibres are enabled to contract; the capacity to change shape.

contractin presumable neurohumor inducing contraction of chromatophores in crustaceans; *cf.* expantin.

contracture *n.* [L. *contractus*, drawn together] contraction of muscles persisting after stimulus has been removed.

contra-deciduate *a.* [L. *contra*, opposite to; *decidere*, to fall off] *appl.* foetal placenta and distal part of allantois which are absorbed by maternal leucocytes at birth.

contralateral *a.* [L. *contra*, opposite to; *latus*, side] *pert.* or situated on the opposite side; *cf.* ipsilateral.

contranatant *a.* [L. *contra*, against; *natare*, to swim] swimming or migrating against the current; *cf.* denatant.

conuli *n.plu.* [*Dim.* of L. *conus*, cone] tent-like projecions on surface of certain sponges caused by principal skeletal elements.

conus *n.* [L. *conus*, cone] any cone-shaped structure; conus arteriosus: a structure between ventricle and aorta in fishes and amphibians; diverticulum of right ventricle from which pulmonary artery arises; conus medullaris: the tapering end of spinal cord; *plu.* coni, *q.v.*

convergence *n.* [L. *convergere*, to incline together] homoplasty, *q.v.*; coordinated movement of eyes when focusing a near point.

convolute *a.* [L. *cum*, together; *volvere*, to wind] rolled together, *appl.* leaves and cotyledons; *appl.* shells in which outer whorls overlap inner; coiled, *appl.* parts of renal tubule, *alt.* convoluted; cerebrose, *q.v.*

convolution *n.* [L. *cum*, together; *volvere*, to wind] a coiling or twisting, as of brain, intestine.

coordinated growth sympathetic growth, *q.v.*

copal *n.* [Sp. *copalli*, resin] a resin obtained from certain fossil plants and exuding from various tropical trees and hardening to colourless, yellow, red, or brown masses.

Copelata n. [Gk. kōpēlatēs, rower] the only order of urochordates of the class Larvacea, q.v.

Copepoda, copepods n., n.plu. [Gk. kōpē, oar; pous, foot] a subclass of free-living or parasitic crustaceans that have no carapace, and have one median eye in the adult.

Copodontiformes n. [Gk. kōpē, oar; odous, tooth; L. forma, shape] an order of Devonian to Carboniferous Holocephali, known only from their teeth, and of uncertain taxonomic position.

coprobiont n. [Gk. kopros, dung; bionai, to live] a coprophytic or coprozoic organism; coprophage, q.v.

coprodaeum n. [Gk. kopros, dung; hodos, way] the division of cloaca which receives rectum.

coprolite, coprolith n. [Gk. kopros, dung; lithos, stone] petrified faeces.

coprophage n. [Gk. copros, dung; phagein, to eat] an organism feeding on dung; alt. coprobiont.

coprophagous a. [Gk. kopros, dung; phagein, to eat] feeding on dung; alt. scatophagous.

coprophagy n. [Gk. kopros, dung; pahgein, to eat] habitual feeding on dung; refection or re-ingestion of faeces.

coprophil, -ic, -ous a. [Gk. kopros, dung; philos, loving] growing in or on dung, appl. dung bacteria and flagellates; alt. coprophytic.

coprophyte n. [Gk. kopros, dung; phyton, plant] a dung-inhabiting plant; a. coprophytic.

coprosterol n. [Gk. kopros, dung; stereos, solid; alcohol] a sterol produced by bacterial reduction of cholesterol, which is a constituent of the faeces.

coprozoic a. [Gk. kopros, dung; zōon, animal] inhabiting faeces, as some Protozoa.

coprozoite n. [Gk. kopros, dung; zōon, animal] a dung-inhabiting or coprozoic animal

copula n. [L. copula, bond] a ridge in development of the tongue, formed by union of ventral ends of 2nd and 3rd arches; basihyal or os interglossum in certain reptiles; fused basihyal and basibranchial in birds; any bridging or connecting structure.

copulant n. [L. copulare, to couple] a unit in conjugation with another, as nuclei, cells, hyphae, thalli, etc.

copularium n. [L. copula, bond] a cyst formed around 2 associated gametocytes, in gregarines.

copulation n. [L. copula, bond] sexual union, alt. coition; in Protozoa, complete fusion of 2 individuals; conjugation, as in yeasts.

copy choise a mechanism of crossing-over in meiosis in which crossing-over occurs at the time of DNA replication; cf. breakage and reunion.

coracidium n. [Gk. korax, crow; idion, dim.] ciliated embryo of certain cestodes, developing into a procercoid within 1st intermediate host.

Coraciiformes n. [Coracias, generic name of roller; L. forma, shape] an order of birds of the subclass Neornithes including the kingfishers, rollers, and hornbills.

coracoid a. [Gk. korax, crow; eidos, form] appl. or pert. bone or part of the pectoral girdle between scapula and sternum; appl. ligament which stretches over the suprascapular notch.

coracoid process the rudimentary coracoid element fused to the scapula in most mammals.

coralliferous a. [Gk. korallion, coral; L. ferre, to bear] coral-forming; containing coral.

coralliform coralloid, q.v.

coralligenous a. [Gk. korallion, coral; gennaein, to produce] coral-forming.

Corallimorpharia n. [Gk. korallion, coral; morphē, form] an order of Zoantharia which resemble true corals but have no skeletons.

coralline a. [Gk. korallion, coral] resembling a coral, appl. Hydrozoa and Polyzoa; composed of coral; appl. a Pliocene crag or deposit containing fossil Polyzoa and Mollusca. [Corallina, a genus of Rhodophyceae] appl. zone of coastal waters at about 30–100 m between laminarian and aphytic zones.

corallite n. (Gk. korallion, coral] cup of a single polyp of coral.

coralloid a. [Gk. korallion, coral; eidos, from] resembling, or branching like a coral, appl. gleba, roots, etc.; alt. coralliform.

corallum n. [Gk. korallion, coral] skeleton of compound coral.

corals see true corals, black corals, soft corals, blue corals, horny corals.

corbicula n. [L. dim. of corbis, basket] basket-like arrangement of a teleutosorus or telium; plu. of corbiculum, q.v.; plu. corbiculae.

corbiculum n. [L. dim. of corbis, basket] fringe of hair on insect tibia; the pollen basket of a bee; plu. corbicula.

corbula n. [L. corbula, little basket] a phylactocarp in which alternate branches rise upwards and form a pod-like structure.

cord n. [L. chordē, cord] any cord-like structure, as spinal cord, spermatic cord, etc.

Cordaitales n. [Cordaites, generic name] an order of fossil Coniferopsida being mostly tall trees with slender trunks and a crown of branches, with spirally arranged simple grass-like or paddle-like leaves and with mega- and microsporangia in compound strobili.

cordate, cordiform a. [L. cor, heart] heart-shaped.

cordiform tendon the central aponeurosis of the diaphragm.

cordylus n. [Gk. kordylē, swelling] an intertentacular exumbral structure with core of vacuolated cells and flattened ectoderm.

coremata n.plu. [Gk. korēma, broom] paired sacs bearing hairs, on membrane between 7th and 8th abdominal segments, accessory copulatory organ in moths; sing. corema.

coremiform a. [Gk. korēma, broom; L. forma, shape] formed like a broom or sheaf.

coremiospore n. [Gk. korēma, broom; sporos, seed] one of a series of spores in the top of a coremium.

coremium n. [Gk. korēma, broom] a sheaf-like aggregation of conidiophores, or of hyphae; alt. synnema.

coriaceous, corious a. [L. corium, leather] leathery, appl. leaves.

corium n. [L. corium, leather] the middle division of an elytron; deeper-seated layer of the skin, consisting of a vascular connective tissue ; alt. cutis vera, dermis.

cork n. [Sp. alcorque, cork] a tissue derived from the cork cambium and made of dead suberized cells, forming a seal impermeable to water; alt. phellum, suber.

cork cambium phellogen, *q.v.*

corm *n.* [Gk. *kormos*, trunk] an enlarged solid subterranean stem, rounded in shape, composed of 2 or more internodes, and covered externally by a few thin membranous scales or cataphyllary leaves; cormus, *q.v.*; protopodite and endopodite when constituting an axis.

cormel *n.* (Gk. *kormos*, trunk) a secondary corm produced by an old corm.

cormidium *n.* [Gk. *kormos*, trunk; *idion, dim.*] an aggregation of individuals in a siphonophore, borne on the coenosarc and capable of liberation therefrom.

Cormobionta *n.* [Gk. *kormos*, trunk; *bios*, life] Cormphyta, *q.v.*

cormoid *a.* [Gk. *kormos*, trunk; *eidos*, form] like a corm.

cormophylogeny *n.* [Gk. *kormos*, trunk; *phylē*, tribe; *genos*, offspring] development of families or races.

Cormophyta, cormophytes *n., n.plu.* [Gk. *kormos*, trunk; *phyton*, plant] in some classifications, plants which are differentiated into roots, stems, and leaves and are well-adapted to life on land, comprising the Pteridophyta and Spermatophyta; *alt.* Cormobionta; *cf.* Thallophyta.

cormous *a.* [Gk. *kormos*, trree trunk] corm-producing.

cormus *n.* [Gk. *kormos*, tree trunk] corm, *q.v.*; body of a plant which is developed into root and shoot systems, *cf.* thallus; body or colony of a compound animal.

Cornales *n.* [*Cornus*, generic name] an order of usually woody dicots (Rosidae) including the families Cornaceae, e.g. dogwood, and Garryaceae.

cornea *n.* [L. *corneus*, horny] the transparent covering of anterior surface of eyeball; outer transparent part of each elememt of a compound eye.

corneagen *a.* [L. *cornu*, horn; Gk. *gennaein*, to produce] cornea-producing, *appl.* cells immediately below cuticle, which secrete cuticular lens and are renewed on ecdysis.

corneal *a.* [L. *corneus*, horny] *pert.* the cornea.

corneosclerotic *a.* [L. *corneus*, horny; Gk. *sklēros*, hard] *pert.* cornea and sclera; *alt.* sclerocorneal.

corneoscute *n.* [L. *corneus*, horny; *scutum*, shield] an epidermal scale.

corneous *a.* [L. *corneus*, horny] horny, *appl.* sheath covering bill of birds.

cornicle *n.* [L. *corniculum*, little horn] a wax-secreting organ of aphids; corniculum, *q.v.*

corniculate *a.* [L. *corniculum*, little horn] having small horns.

corniculate cartilages two small, conical, elastic cartilages articulating with apices of arytaenoids; *alt.* Santorini's cartilages, cornicula laryngis.

corniculum *n.* [L. *dim.* of *cornu*, horn] a small horn or horn-like process; *alt.* cornicle.

cornification *n.* [L. *cornu*, horn; *facere*, to make] formation of outer horny layer of epidermis; *alt.* keratinization.

cornua *n.plu.* [L. *cornu*, horn] horns; horn-like prolongations, as of bones, nerve tissues, cavities, etc; the dorsal, lateral, and ventral columns of grey matter in spinal cord; *sing.* cornu.

cornucopia *n.* [L. *cornu*, horn; *copia*, plenty] part of taeniae of 4th ventricle, covering chorioid plexus.

cornule *n.* [L. *cornulum, dim.* of *cornu*, horn] a small horn-like process; one of the horny jaw plates of *Ornithorhynchus*, the platypus.

cornute *a.* [L. *cornutus*, horned] with horn-like processes.

corolla *n.* [L. *corolla*, small crown] the petals of a flower as a whole.

corollaceous *a.* [L. *corolla*, crown] *pert.* a corolla.

corolliferous *a.* [L. *corolla*, small crown; *ferre*, to bear] having a corolla.

corona *n.* [L. *corona*, crown] a cup-shaped body formed by union of scales on perianth leaves, as in daffodil; theca and arms of a crinoid; echinoid test excepting apical and antapical plates; ciliated disc or circular band of certain animals such as rotifers; head or upper portion of any structure; *alt.* crown.

coronal *a.* [L. *corona*, crown] *pert.* corona; *appl.* suture between frontal and parietal bones; situated in the coronal sutural plane; *appl.* later roots of grasses, *cf.* seminal root.

corona radiata layer of cells surrounding mammalian egg; fibres of internal capsule of brain.

coronary *a.* [L. *corona*, crown] crown-shaped or crown-like; encircling; *appl.* arteries, bones, sinus, ligaments, plexus, veins.

coronary arteries arteries supplying tissue of heart; *alt.* labial arteries.

coronary bone a small conical bone in mandible of reptiles; small pastern bone of horse.

coronary sinus channel receiving most cardiac veins and opening into right auricle.

Coronatae *n.* [L. *corona*, crown] an order of scyphozoans having free-swimming medusae with the edge of the bell scalloped and separated from the rest by a groove.

coronate *a.* [L. *corona*, crown] having a corona; having a row of tubercles encircling a structure, or mounted on whorls of spiral shells.

coronaviruses *n.plu.* [L. *corona*, crown; *virus*, poison] suggested name for a group of viruses that have a fringe of round or petal-shaped projections as well as other common properties, and including the mouse hepatitis virus.

coronet *n.* [L. *corona*, crown] a small terminal ring of hairs, spines, etc.; corona of certain flowers; burr or knob at base of an antler.

coronoid *a.* [Gk. *korōnis*, crook-beaked; *eidos*, form] shaped like a beak, *appl.* processes. [L. *corona*, crown] *n.* Coronary bone of reptiles.

Coronophorales *n.* [L. *corona*, crown; Gk. *phora*, producing] an order of Ascomycetes having asci in ascostromata with round or irregular openings.

coronula *n.* [*Dim.* of L. *corona*, crown] a group of cells forming a crown on the oosphere, as in Charophyta; circle of pointed processes around frustule of certain diatoms.

corpora *n.plu.* [L. *corpus*, body] bodies; *see also* corpus.

corpora albicantia white bodies or scars formed in ovarian follicle after disintegration of luteal cells; corpora mammillaria, *q.v.*

corpora allata endocrine glands in insects situated just behind the brain, which secrete juvenile hormone, and may be paired ovoid whitish structures, or fuse during development to form a single median structure.

corpora amylacea spherical bodies of nucleic

acid and protein, more numerous with age, in alveoli of prostate gland; *alt.* amyloid bodies.

corpora arenacea brain sand, *q.v.*

corpora bigemina the optic lobes of vertebrate brain, corresponding to the superior colliculi of corpora quadrigemina of mammals.

corpora cardiaca neuroglandular bodies between cerebral ganglia and corpora allata, in some insects.

corpora cavernosa erectile masses of tissue, forming anterior part of body of penis; erectile tissue of clitoris.

corpora mamillaria two white bodies enclosing grey matter in hypothalamus, beneath floor of 3rd ventricle; *alt.* corpora albicantia, mamillary bodies.

corpora pedunculata groups of association cells with axons forming bundles in protocerebrum of insects, especially highly developed in Hymenoptera; *alt.* mushroom bodies, pedunculate bodies.

corpora quadrigemina four rounded eminences or colliculi which form dorsal part of mesencephalon; *alt.* quadrigeminal bodies.

corpus *n.* [L. *corpus*, body] body; any fairly homogeneous structure which forms part of an organ; core of apical meristem within the tunica; *plu.* corpora, *q.v.*

corpus adiposum fat-body, *q.v.*

corpus callosum the broad transverse band of white matter connecting the cerebral hemispheres; *alt.* callosum, trabs cerebri.

corpuscle *n.* [L. *corpusculum*, small body] a cell, floating freely in a fluid, or embedded in a matrix; any minute particle, as in a cell; any of various small multicellular structures, as Malpighian corpuscle, tactile corpuscle, etc.

corpuscular *a.* [L. *corpusculum*, small body] like or *pert.* a corpuscle or small particle; compact or globular.

corpusculum lamellosum Pacinian bodies, *q.v.*

corpus fibrosum fibrous tissue remaining after disintegration of corpus luteum.

corpus geniculatum geniculate body, *q.v.*

corpus haemorrhagicum body developed from ruptured Graafian follicle around blood clot, and later developing into corpus luteum.

corpus highmoreanum mediastinum testis, *q.v.*

corpus luteum the glandular body developed from a Graafian follicle after extrusion of ovum; *alt.* yellow body.

corpus spongiosum a mass of erectile tissue forming posterior wall of penis; *alt.* corpus cavernosum urethrae.

corpus sterni sternebrae fused into a single mesosternal bone; *alt.* mesosternum, gladiolus.

corpus striatum a mass of grey matter containing white nerve fibres and consisting of the caudate nucleus which projects into the lateral ventricle and of the lenticular nucleus; *alt.* striatum.

corpus tentorii tentorium, *q.v.*

correlation *n.* [L.L. *correlatio*, relationship] mutual relationship; proportional growth; interdependence of characters, particularly of quantitive characters, measured by correlation coefficient which is plus or minus one if characters are exactly interrelated, and zero is entirely unrelated; combination of nervous impulses in sensory centres, resulting in adaptive reactions; determina-

tion of the relation of homotaxis to geological time; *alt.* covariation.

correlator *n.* [L.L. *correlatio*, relationship] a diffusible substance correlating activities of colcoptile tip and hypocotyl; a former name for auxin.

Corrodentia *n.* [L. *corrodere*, to gnaw to pieces] Psocoptera, *q.v.*

corrugator *a.* [L. *corrugare*, to wrinkle] wrinkled or wrinkling, *appl.* muscles.

cortex *n.* [L. *cortex*, bark] the extrastelar fundamental tissue of the sporophyte; outer or more superficial part of an organ; envelope of a bacterial spore, between cytoplasm, or spore wall, and the spore coat; *plu.* cortices.

cortical *a.* [L. *cortex*, bark] *pert.* the cortex.

corticate *a.* [L. *cortex*, bark] having a special outer covering.

cortices *plu.* of cortex.

corticicolous corticolous, *q.v.*

corticiferous *n.* [L. *cortex*, bark; *ferre*, to carry] forming or having bark-like cortex.

corticoids *n.plu.* [L. *cortex*, bark; ster*oid*] steroids secreted by the cortex of the adrenal bodies, some being hormones, and including the glucocorticoids and mineralocorticoids; *alt.* corticosteroids.

corticolous *a.* [L. *cortex*, bark; *colere*, to inhabit] inhabiting, or growing on, bark; *alt.* corticicolous.

corticospinal *a.* [L. *cortex*, bark; *spina*, spine] *pert.* or connecting cerebral cortex and spinal cord.

corticosteroids corticoids, *q.v.*

corticosterone a steroid hormone which can be extracted from the cortex of the adrenal gland and made synthetically.

corticostriate *a.* [L. *cortex*, bark; *stria*, channel] *appl.* fibres which join corpus striatum to cerebral cortex.

corticotrop(h)ic adrenocorticotrophic, *q.v.*

corticotrop(h)in *n.* [L. *cortex*, bark; Gk. *tropē*, turn; *trophē*, nutrition] adrenocorticotrophic hormone, *q.v.*

corticotrophin-releasing factor (CFR) a hormone secreted by the hypothalamus which regulates ACTH secretion.

cortin *n.* [L. *cortex*, bark] adrenal cortex extract, containing cortical hormones.

cortina *n.* [L. *cortina*, vault] the velum in some agarics.

cortinate *a.* [L. *cortina*, vault] craspedote, *q.v.*; of a cobweb-like texture.

Corti's membrane [A. *Corti*, Italian histologist] tectorial membrane covering spiral organ of Corti.

cortisol hydrocortisone, *q.v.*

cortisone *n.* [L. *cortex*, bark] a glucocorticoid hormone produced by the adrenal body which has many complex effects including healing and reduction of inflammation, and promoting carbohydrate formation from fat and protein.

Corti's organ the auditory organ on the inner portion of membrane basilaris of ear; *alt.* organ of Corti, organon spirale, spiral organ.

Corti's rods double row of arching rods based on basilar membrane and forming the spiral tunnel of Corti.

coruscation *n.* [L. *coruscatio*, flash] twinkle, rapid fluctuation in a flash or oscillation in light emission, as of fireflies.

corymb *n.* [Gk. *korymbos*, cluster of flowers] a

raceme with lower pedicels elongated so that the top is nearly flat; *a.* corymbose, corymbous.

corymbose cyme a flat-topped cyme which therefore resembles a corymb in appearance but not in development.

corynebacteria *n.* [*Corynebacterium*, generic name] a group of bacteria having club-shaped cells, and including pathogenic species.

Coryneliales *n.* [Gk. *korynē*, club] an order of Ascomycetes, having asci in ascostromata with funnel-shaped openings.

coscinoid *a.* [Gk. *koskinon*, sieve; *eidos*, form] sieve-like.

cosmine *n.* [Gk. *kosmios*, regular] a kind of dentine found in cosmoid scales in which the tiny canals penetrating it branch, sometimes considered homologous with the dentine of ganoid scales.

cosmoid scale [Gk. *kosmios*, regular; *eidos*, form] the type of scale found in typical crossopterygians and early lungfish having an outer single layer of enamel, then a layer of cosmine, then bone, growth in thickness being by addition of inner layers only; *cf.* ganoid scale.

cosmopolitan *a.* [Gk. *kosmos*, world; *politēs*, citizen] world-wide in distribution; *alt.* pandemic.

cosmopolite a cosmopolitan species.

costa *n.* [L. *costa*, rib] a rib; anything rib-like in shape, as a ridge on shell, coral, etc.; anterior vein, or margin, of insect wing; swimming plate of Ctenophora; structure at base of undulating membrane in protozoans of the Trichomonadidae; *plu.* costae.

costaeform *a.* [L. *costa*, rib; *forma*, shape] rib-like; *appl.* unbranched parallel leaf veins.

costal *a.* [L. *costa*, rib] *pert.* ribs or rib-like structures; *appl.* bony shields of Chelonia; *pert.* costa of insect wing; *pert.* primary branchial series in crinoids; *pert.* a main rib.

costalia *n.plu.* [L. *costa*, rib] the supporting plates in theca.

costa spuria false rib, *q.v.*

costate *a.* [L. *costa*, rib] with one or more longitudinal ribs; with ridges or costae.

coterminous *a.* [L. *cum*, with; *terminus*, end] of similar distribution; bordering on; having a common boundary.

cotransduction the transduction of 2 characters in a single event.

cotyle *n.* [Gk. *kotylē*, cup] acetabulum, *q.v.*

cotyledon *n.* [Gk. *kotylēdōn*, cup] the 1st leaf or leaves of a seed plant, found in the embryo of the seed, which may form the 1st photosynthetic leaves or may remain below ground; a patch of villi on mammalian placenta; *a.* cotyledonary.

cotyliform, cotyloid *a.* [Gk. *kotylē*, cup; *eidos*, form; L. *forma*, form] cup-shaped; *pert.* acetabulum.

cotylophorous *a.* [Gk. *kotylē*, cup; *pherein*, to bear] with a cotyledonary placenta.

Cotylosauria, cotylosaurs *n.*, *n.plu.* [Gk. *kotylē*, cup; *sauros*, lizard] an order of primitive anapsid reptiles of the U. Carboniferous to Triassic, the 'stem reptiles' from which many later forms developed, sometimes subdivided into 2 orders, the Procolophonia and Captorhinomorpha.

cotype *n.* [L. *cum*, with; *typus*, image] an additional type specimen, frequently collected in same place at same time, or a specimen from a description of which, along with others, the type is defined; usually not now used, formerly used for syntype, isotype, or paratype.

coumarin [F. *coumarine*, from coumaron, the tonka bean tree, *Dipteryx*] a substance found in many plants, especially clover and tonka (tonquin) beans, having an odour of new-mown hay and used in perfumery, soaps, and making dicoumarin; *alt.* cumarin.

counterflow the flow of 2 fluids in opposite directions.

countershading the condition of an animal being dark dorsally and pale ventrally, so that when the lighting is from above, the ventral shadow is obscured and the animal appears evenly coloured and inconspicuous.

coupling factors proteins in the membrane of mitochondria which control oxidative phosphorylation.

courtship the behaviour pattern in animals prior to copulation, which ensures copulation occurs between members of the same species, of different sexes, and of animals in a receptive condition.

COV cross-over value, *q.v.*

covariation *n.* [L. *con*, with; *varius*, diverse] correlation, *q.v.*

cover scales bract scales, *q.v.*

coverts *n.plu.* [F. *couvrir*, to cover] feathers covering bases of quills in birds; *alt.* tectoria.

cowled *a.* [L. *cucullus*, hood] furnished with or shaped like a hood; *alt.* cucullate.

Cowper's glands [*W. Cowper*, English surgeon] bulbo-urethral glands, *q.v.*

coxa *n.* [L. *coxa*, hip] proximal joint of leg of an insect, arachnid, and some other arthropods; the hip.

coxal *pert.* the coxa, *appl.* glands; *pert.* the hip.

coxite *n.* [L. *coxa*, hip] one of paired lateral plates in contiguity with insect sternum; limb base bearing stylus in Thysanura.

coxocerite *n.* [L. *coxa*, hip; Gk. *keras*, horn] the proximal or basal joint of insect antenna.

coxopleurite catapleurite, *q.v.*

coxopodite *n.* [L. *coxa*, hip; Gk. *pous*, foot] the proximal part of protopodite of crustacean limb; coxa of arachnids.

coxosternum *n.* [L. *coxa*, hip; *sternum*, breast bone] plate formed by fusion of coxites and sternum; vinculum in Lepidoptera.

C₃ plants plants which, when they fix carbon in photosynthesis, first make 3-carbon (C_3) compounds such as phosphoglyceric acid.

C₄ plants plants which use the Hatch–Slack pathway of photosynthesis, producing C_4 dicarboxylic acids.

CR conditioned response, *q.v.*

crampon *n.* [F. *crampon*, adventive root] an aerial root, as in ivy.

craniad *adv.* [Gk. *kranion*, skull; L. *ad.* towards] towards the head; cephalad in Craniata.

cranial *a.* [Gk. *kranion*, skull] *pert.* skull, or that part which encloses the brain; *appl.* bones, fossae, nerves, muscles, blood vessels, etc.

Craniata, craniates *n.*, *n.plu.* [Gk. *kranion*, skull] a subphylum of metamerically segmented chordates, commonly called vertebrates, that have a high degree of cephalization of the CNS to give a brain, enclosed in a skull (cranium), and an endoskeleton of cartilage and/or bone, including a backbone; *alt.* Vertebrata.

cranidium n. [Gk. *kranion*, skull; *idion*, dim.] glabella together with fixed genae, in trilobites.

cranihaemal a. [Gk. *kranion*, skull; *haima*, blood] appl. anterior lower portion of a sclerotome.

cranineural a. [Gk. *kranion*, skull; *neuron*, nerve] appl. anterior upper portion of a sclerotome.

craniology n. [Gk. *kranion*, skull; *logos*, discourse] the study of the skull.

craniometry n. [Gk. *kranion*, skull; *metron*, measure] the science of the measurement of skulls.

craniosacral a. [Gk. *kranion*, skull; L. *sacer*, sacred] pert. skull and sacrum, appl. nerves, the parasympathetic system.

cranium n. [Gk. *kranion*; L. *cranium*, skull] the skull of any craniate animal, or more particularly, that part enclosing the brain.

craspedodromous a. [Gk. *kraspedon*, edge; *dramein*, to run] with nerves running directly from midrib to margin.

craspedote a. [Gk. *kraspedon*, edge] having a velum; alt. cortinate.

craspedum n. [Gk. *kraspedon*, edge] a mesenteric filament of sea anemones; alt. craspedon.

crassula n. [L. *crassus*, thick] thickened bar on middle lamella between 2 bordered pits in tracheids of wood of conifers; alt. bar of Sanio; plu. crassulae.

Crassulacean Acid Metabolism (CAM) fixation of carbon dioxide in the dark into organic acids which are used in photosynthesis by day, found in certain succulent plants especially in the dicot family, Crassulaceae.

crateriform a. [L. *crater*, bowl; *forma*, shape] bowl-shaped, appl. receptacle, calyx, etc.

craticular a. [L. *craticula*, gridiron] crate-like; appl. stage in life history of a diatom where new valves are formed before the old are lost.

creatine n. [Gk. *kreas*, flesh] a substance present in all vertebrate muscle where it combines with phosphate to form creatine phosphate (phosphagen), which is also found in the blood, and is used to phosphorylate ADP to ATP in muscle under anaerobic conditions.

creatine phosphate see phosphagen.

creatinine n. [Gk. *kreas*, flesh] a substance produced from creatine by dehydration, found in muscles, blood, and urine.

cremaster n. [Gk. *kremastos*, hung] a thin muscle along the spermatic cord; a stout terminal abdominal spine in subterranean insect pupae; the anal hooks for suspension of pupae.

cremocarp n. [Gk. *kremasai*, to hang down; *karpos*, fruit] a bicarpellary, bilocular fruit, as in the Umbelliferae, in which the 2 carpels separate into 1-seeded indehiscent mericarps, which remain attached to the central supporting axis (carpophore) before dispersal.

crena n. [L. *crena*, notch] notch in a crenate margin, as of leaf; cleft, as anal cleft; deep groove, as longitudinal sulcus of heart.

crenate a. [L. *crena*, notch] with scalloped margin; alt. gimped.

crenation n. [L. *crenatus*, notched] a scalloped margin, or rounded tooth, as of leaf, alt. crenature; notched or wrinkled appearance, as of erythrocytes exposed to hypertonic solutions.

crenulate a. [Dim. of L. *crena*, notch] with margins minutely crenate; alt. crenellated, crenulated.

Creodonta, creodonts n., n.plu. [Gk. *kreas*, flesh; *odous*, tooth] an order of extinct placental mammals of the Cretaceous to Pliocene periods which were probably archaic carnivores.

crepis n. [Gk. *krēpis*, foundation] the fundamental spicule by deposition of silica upon which a desma is formed.

crepitaculum n. [L. *crepitaculum*, rattle] a stridulating organ, as in some Orthoptera; rattle in rattlesnake's tail.

crepitation n. [L. *crepitare*, to crackle] in insects, the discharge of fluid with an explosive sound for defence.

crepuscular a. [L. *crepusculum*, dusk] pert. dusk; flying before sunrise or in twilight; alt. vespertine.

crescent n. [L. *crescere*, to grow] a crescent-shaped structure.

crescentic, crescentiform a. [L. *crescere*, to grow; *forma*, shape] crescent-shaped.

crescents of Gianuzzi demilunes, q.v.

crest n. [L. *crista*, crest] a ridge on a bone; a fleshy longitudinal ridge, as in newts; crown or feather tuft on head of birds; a ridge in certain seeds; a ridge-like structure.

Cretaceous a. [L. *creta*, chalk] appl. last period of Mesozoic era, during which chalk was being laid down, occurring after the Jurassic and before the Tertiary; U. Cretaceous in the system used in North America.

CRF corticotrophin-releasing factor, q.v.

cribell-, cribi- see cribell-, cribri-.

cribrellate a. [L. dim. of *cribrum*, sieve] having many pores as certain spores.

cribrellum n. [L. dim. of *cribum*, sieve] a plate perforated by openings of silk ducts in certain spiders; a perforated chitinous plate in some insects; alt. cribellum.

cribriform a. [L. *cribrum*, sieve; *forma*, shape] sieve-like; alt. coliform.

cribriform organ folded membrane carrying papillae in interradial angles of certain starfishes.

cribriform plate the portion of ethmoid, or of mesethmoid, perforated by many foramina for exit of olfactory nerves; alt. lamina cribrosa.

cribrose a. [L. *cribrum*, sieve] having sieve-like pitted markings.

crickets the common name for many members of the Orthoptera, q.v.

cricoid a. [Gk. *krikos*, ring; *eidos*, form] ring-like; appl. cartilage in larynx, articulating with thyroid and arytaenoid cartilages; appl. placenta lacking villi on central part of disc, as in certain Edentata.

crico-thyroid a. [Gk. *krikos*, ring; *thyreos*, shield; *eidos*, form] pert. cricoid and thyroid cartilages; appl. tensor muscle of vocal cord.

crinite a. [L. *crinitus*, having locks of hair] with hairy or hair-like structures or tufts. n. A fossil crinoid.

Crinoidea, crinoids n., n.plu. [Gk. *krinon*, lily; *eidos*, form] a class of echinoderms of the subphylum Pelmatozoa, commonly called sea lilies, having a cup-shaped body attached to the substratum by a stalk.

crinome n. [L. *crinis*, hair] network formed in cytoplasm by basophil substances reacting to vital staining.

crinose a. [L. *crinis*, hair] hairy; with long hairs.

criocone a. [Gk. *krios*, ram; *kōnos*, cone] with uncoiled spiral shaped like ram's horn, appl. shell of certain ammonites.

crisped, crispate curled or frizzled.

crissal *a.* [L. *crissare*, to move haunches] *pert.* the crissum.

criss-cross *appl.* inheritance when offspring resemble the parent of the opposite sex.

crissum *n.* [L. *crissare*, to move haunches] the circumcloacal region of a bird; vent feathers or lower tail coverts.

crista *n.* [L. *crista*, crest] a crest or ridge; projection from ectoloph into median valley in lophodont molars; a fine membrane attached to body of certain spirochaetes; ligule of palm leaves; a longitudinal membranous ridge on certain bacteria; *sing.* of cristae, *q.v.*

crista acustica thickening, covered with neuroepithelium, of membrane lining ampullae of semicircular canals; a chordotonal structure in Orthoptera.

cristae *plu.* of crista; folds of the inner membrane of a mitochondrion.

crista galli anterior median process of cribriform plate.

cristate *a.* [L. *cristatus*, crested] crested; shaped like a crest, *alt.* cristiform.

crista urethralis verumontanum, *q.v.*

critical frequency the maximum frequency of successive stimuli at which they can produce separate sensations, minimum frequency for a continuous sensation.

critical group a taxonomic group containing organisms which cannot be subdivided into smaller groups, as in apomictic species.

crithidial *a.* [Gk. *krithidion*, *dim.* of *krithē*, barleycorn] trypanomonad, *q.v.*

crochet *n.* [F. *crochet*, small hook] the projection of the protoloph in lophodont molars; a balancer in larval salamanders; a larval locomotory hook in insects; distal joint or unguis of chelicerae in arachnids.

Crocodilia, crocodiles *n.*, *plu.* [L. *crocodilus*, crocodile] an order of archosaurs occurring from Triassic to present day, which are often armoured and have front limbs shorter than hind and a body elongated for swimming; *alt.* crocodilians.

Cro-Magnon man an early type of modern man, *Homo sapiens*, whose fossils were first found at Cro-Magnon in the Dordogne in France.

crop *n.* [M.E. *croppe*, craw] sac-like dilatation of gullet of a bird; a similar structure in alimentary canal of insect or annelid; *alt.* ingluvies.

crop-milk secretion of epithelium of crop in pigeons, stimulated by prolactin, for nourishment of nestlings.

crosier crozier, *q.v.*

cross *n.* [O.F. *crois*, cross] an organism produced by mating parents of different breeds. *v.* To hybridize.

cross-fertilization the fusion of male and female gametes from different individuals, especially of different genotypes; *alt.* allogamy, allomixis, xenogamy.

crossing cross-pollination or cross-fertilization.

crossing-over interchange of corresponding chromatid segments by homologous pairs of chromosomes during prophase of meiosis, resulting in a visible chiasma.

Crossopterygii, crossopterygians *n.*, *n.plu.* [Gk. *krossoi*, tassels; *pterygion*, fin] a group of bony fishes sometimes placed in the subclass Sarcopterygii, commonly called tassel-finned or lobe-finned fishes, which are known mainly as fossils from Devonian times, probably ancestral to land vertebrates, but also including the coelacanth.

cross-over a chromatid formed as a result of crossing-over.

cross-over unit a 1% frequency of crossing over between 2 linked genes.

cross-over value (COV) the frequency of crossing-over between 2 genes on a chromosome, which is used to make a chromosome map.

cross-pollination the transfer of pollen from the anther of one flower to the stigma of another flower; the term is sometimes restricted to the transfer of pollen from an anther to the stigma of a flower of a different plant, especially of a different genotype.

cross-reflex reaction of an effector on one side of the body to stimulation of a receptor on the other side.

cross-striation in striped and cardiac muscle fibres, dark lines running across a fibre perpendicular to the myofibrils and representing lines of A-bands.

crotaphite *n.* [Gk. *krotaphos*, side of forehead] the temporal fossa.

crotchet *n.* [F. *crochet*, small hook] a curved chaeta, notched at the end, *alt.* uncinus; clavus in spiders; crochet of larval insects.

crown *n.* [L. *corona*, crown] the exposed part of a tooth, especially the grinding surface; distal part of anther; crest; head; cup and arms of a crinoid; corona, *q.v.*; leafy upper part of a tree; a short rootstock with leaves.

crozier *n.* [L.L. *crocia*, crook] circinate young frond of fern; hook formed by terminal cells of ascogenous hyphae; flat spiral shell, as of the cephalopod *Spirula*; *alt.* crosier.

crucial, cruciate *a.* [L. *crux*, cross] cruciform, *q.v.*; with leaves or petals in form of a cross; X-shaped or +-shaped, *appl.* muscles, ligaments.

cruciferous *a.* [L. *crux*, cross; *ferre*, to bear] anisocytic, *q.v.*; *appl.* dicotyledon flower family Cruciferae, having petals in the form of a cross.

cruciform *a.* [L. *crux*, cross; *forma*, shape] arranged like the points of a cross; *appl.* division: promitosis in Plasmodiophorales; *alt.* crucial, cruciate.

crumb a soil aggregate, especially in the form of a granule.

crumena *n.* [L. *crumena*, purse] a sheath for retracted stylets, as in Hemiptera.

cruor *n.* [L. *cruor*, blood] the clots in coagulated blood.

cruorin *n.* [L. *cruor*, blood] haemoglobin, *q.v.*

crura *n.plu.* [L. *crura*, legs] the shanks; leg-like or columnar structures; lumbar part of diaphragm muscle fibres; proximal process of corpora cavernosa penis; branches of incus and stapes; pillars of subcutaneous inguinal ring; posterior pillars of fornix; crura cerebri, *q.v.*; *sing.* crus, *q.v.*

crura cerebri the cerebral peduncles, 2 cylindrical masses forming the ventrolateral portion of mid-brain.

crural *a.* [L. *crus*, leg] *pert.* the thigh; *alt.* femoral.

crureus *n.* [L. *crus*, leg] vastus intermedius muscles of thigh.

crus *n.* [L. *crus*, leg] the shank or zeugopodium of hindlimb of vertebrates; any leg-like organ;

common duct of superior and posterior semicircular canals; anterior end of helix of external ear; *plu.* crura, *q.v.*

crusta *n.* [L. *crusta* shell] ventral part or base or pes of cerebral peduncles; cement layer of teeth, *alt.* crusta petrosa; any of various other hard coverings.

Crustacea, crustaceans *n., n.plu.* [L. *crusta*, shell] a class of mainly aquatic gill-breathing arthropods of the subphylum Mandibulata usually having 2 pairs of antennae and a carapace.

crustaceous *a.* [L. *crusta*, shell] with crustacean characteristics; crustose, *q.v.*; thin or brittle.

crustose *a.* [L. *crusta*, shell] forming crusts on substratum, *appl.* certain lichens; *alt.* crustaceous.

crymophil *a.* [Gk. *krymos*, frost; *philein*, to love] cryophil, *q.v.*; psychrophil, *q.v.*

cryophil(ic) *a.* [Gk. *kryos*, frost; *philein*, to love] thriving at a low temperature; *alt.* crymophil.

cryophylactic *a.* [Gk. *kryos*, frost; *phylaktikos*, preservative] resistant to low temperatures, *appl.* bacteria.

cryophytes *n.plu.* [Gk. *kryos*, frost; *phyton*, plant] algae, bacteria, and fungi on snow and ice.

cryoplankton *n.* [Gk. *kryos*, frost; *plangktos*, wandering] glacial and polar plankton.

crypsis *n.* [Gk. *kryptos*, hidden] the phenomenon of being cryptic especially in the sense of camouflaged to resemble part of the environment.

crypt *n.* [Gk. *kryptos*, hidden] a simple glandular tube or cavity; pit of stoma; depression in uterine mucous membrane.

cryptic *a.* [Gk. *kryptos*, hidden] *appl.* protective coloration facilitating concealment; *appl.* polymorphism due to presence of recessive genes; *appl.* species extremely similar as to external appearance but which do not normally interbreed; *n.* crypsis.

cryptocarp *n.* [Gk. *kryptos*, hidden; *karpos*, fruit] cystocarp, *q.v.*

cryptochiasmatic *a.* [Gk. *kryptos*, hidden; *chiasma*, cross] *appl.* meiosis where chiasmata are present but are seen for only a very short time just before 1st anaphase.

Cryptococcales *n.* [*Cryptococcus*, generic name] an order of Cryptophyceae whose cells are nonmotile and arranged in very short filaments.

cryptocyst *n.* [Gk. *kryptos*, hidden; *kystis*, bladder] in Cheilostomata, a calcareous plate which serves as a hydrostatic organ in the colony.

cryptofauna *n.* [Gk. *kryptos*, hidden; L. *faunus*, god of woods] organisms living in concealment in protected situations, such as in crevices in coral reefs; *alt.* phytalfauna.

Cryptogamia, cryptogams *n., n.plu.* [Gk. *krytos*, hidden; *gamos*, union] in former systems of classification, plants reproducing by spores, often also used for plants without flowers, or without true stems, roots, and leaves; *cf.* Phanerogamia.

cryptogene *a.* [Gk. *kryptos*, hidden; *genos*, descent] of unknown descent; having an indeterminate phylogeny.

cryptogram *n.* [Gk. *kryptos*, hidden; *graphein*, to write] a method of expressing in a standard form a collection of data used in classification.

cryptohaplomitosis *n.* [Gk. *kryptos*, hidden;

haploos, simple; *mitos*, thread] type of cell division in some flagellates where chromatin divides into 2 masses which pass to opposite poles without spireme formation.

cryptomere *n.* [Gk. *kryptos*, hidden; *meros*, part] a 'hidden,' i.e. recessive, gene.

cryptomitosis *n.* [Gk. *kryptos*, hidden; *mitos*, thread] division of unicellular organisms, in which chromatin assembles in the equatorial region without apparent chromosome formation.

Cryptomonadales *n.* [Gk. *kryptos*, hidden; *monas*, unit] an order of Cryptophyceae, many of which are the symbiotic algae called zooxanthellae; *alt.* cryptomonads.

Cryptomonadina *n.* [Gk. *kryptos*, hidden; *monas*, unit] an order of Phytomastigina, that are ovoid and flattened with a gullet from which arise 2 unequal flagella: *alt.* cryptomonads.

cryptomonads the common name for the Crytomonadina or Cryptomonadales, many of which are considered as both protozoans and algae.

cryptonema *n.* [Gk. *kryptos*, hidden; *nēma*, thread] a filamentous outgrowth or paraphysis in a cryptostoma.

Cryptonemiales *n.* [Gk. *kryptos*, hidden; *nēma*, thread] an order of red algae whose auxiliary cells are borne on special branches.

cryptonephridial *a.* [Gk. *kryptos*, hidden; *nephros*, kidney; *idion*, *dim.*] with the distal ends of the Malpighian tubules closely applied to the hind-gut, *appl.* insects; *alt.* cryptonephric; *n.* cryptonephry.

cryptoneurous *a.* [Gk. *kryptos*, hidden; *neuron*, nerve] with no definite or distinct nervous system.

Cryptophycales *n.* [Gk. *kryptos*, hidden; *phykos*, seaweed] an order of algae, in some classifications considered to be Pyrrophyceae, which include zooxanthellae that live symbiotically in marine animals.

Cryptophyceae *n.* [Gk. *kryptos*, hidden; *phykos*, seaweed] a class of algae, in some classifications of the Pyrrophyta, in other of the Cryptophyta, having flattened cells, usually with a tubular gullet, and photosynthetic pigments including unusual biloproteins.

Cryptophyta *n.* [Gk. *kryptos*, hidden; *phyton*, plant] in some classifications, the Cryptophyceae, *q.v.*, raised to division status.

cryptophyte *n.* [Gk. *kryptos*, hidden; *phyton*, plant] a plant perennating by means of rhizomes, corms, or bulbs underground, or by under water buds; *cf.* geophyte.

cryptoplasm *n.* [Gk. *kryptos*, hidden; *plasma*, form] the non-granular portion of cytoplasm.

cryptoptile *n.* [Gk. *kryptos*, hidden; *ptilon*, feather] a feather filament, developed from a papilla.

cryptorchid *a.* [Gk. *kryptos*, hidden; *orchis*, testis] having testes abdominal in position.

cryptorhetic *a.* [Gk. *kryptos*, hidden; *rheein*, to flow] secreting internally; *alt.* endocrine.

cryptosolenial *a.* [Gk. *kryptos*, hidden; *sōlēn*, channel] *appl.* region of attachment of Malpighian vessels to hind-gut in certain Coleoptera.

cryptosphere *n.* [Gk. *kryptos*, hidden; *sphaira*, globe] the habitat of the cryptozoa.

cryptostomata *n.plu.* [Gk. *kryptos*, hidden; *stoma*, mouth] non-sexual conceptacles in some large brown seaweeds, as family Fucaceae; *sing.* cryptostoma; *alt.* hair pits.

cryptozoa *n.* [Gk. *kryptos*, hidden; *zōon*, animal] an ecological grouping, being small terrestrial animals living on the ground but above the soil in leaf litter and twigs, among and under pieces of bark, stones, etc.; a. cryptozoic.

cryptozoite n. [Gk. kryptos, hidden; zōon, animal] stage of sporozoite when living in tissues before entering blood.

cryptozone n. [Gk. kryptos, hidden; zōnē, girdle] not having clearly distinct marginals, appl. starfish.

cryptozoology n. [Gk. kryptos, hidden; zōon, animal; logos, discourse] the study of the cryptozoa.

crypts of Lieberkühn [J. N. Lieberkühn, German anatomist] simple tubular glands in the intestines which secrete succus entericus; in the duodenum Brunner's glands open into crypts, and the secretion of the 2 together make up succus entericus; alt. Lieberkühn's crypts.

crystal cells coelomocytes containing rhomboid crystals, in echinoderms.

crystallin n. [Gk. krystallos, ice] a globulin which is the principal constituent of lens of eye.

crystalline a. [Gk. krystallinos, crystalline] transparent, appl. various structures.

crystalline style a proteinaceous hyaline rod concerned with carbohydrate digestion in the alimentary canal of some molluscs.

crystalloid n. [Gk. krystallos, ice; eidos, form] a substance which in solution readily diffuses through a semi-permeable membrane, cf. colloid; a protein crystal found in certain plant cells.

crystal sand a deposit of minute crystals of calcium oxalate, as in certain plant cells.

crystal spore an isospore containing a crystal, of Radiolaria.

CS conditioned stimulus, q.v.

CSF cerebrospinal fluid, q.v.

cteinophyte n. [Gk. ktanai, to kill; phyton, plant] a parasitic plant, e.g. fungus, which destroys its host.

cteinotrophic a. [Gk. ktanai, to kill; trophē, nourishment] parasitic and destroying the host, as cteinophytes.

ctene n. [Gk. kteis, comb] swimming plate, q.v. of Ctenophora; alt. costa, comb-rib.

ctenidium n. [Gk. kteis, comb; idion, dim.] a feathery or comb-like structure, especially respiratory apparatus in molluscs, cf. gill plume; rows of fused cilia forming comb plates or ribs of Ctenophora; a row of spines forming a comb in some insects.

cteniform a. [Gk. kteis, comb; L. forma, shape] comb-shaped.

ctenocyst n. [Gk. kteis, comb; kystis, bladder] aboral sense organ of Ctenophora.

ctenoid a. [Gk. kteis, comb; eidos, form] comb-like; with comb-like margin, appl. scales.

Ctenophora, ctenophores n., n.plu. [Gk. kteis, comb; pherein, to bear] a subphylum of marine coelenterates (sometimes considered a phylum), commonly called sea gooseberries, sea combs, or comb jellies, which are free swimming and biradially symmetrical with 8 meridional rows of ciliary comb-ribs.

ctenose a. [Gk. kteis, comb] comb-like, appl. type of seta.

Ctenostomata n. [Gk. kteis, comb; stoma, mouth] an order of Gymnolaemata (Ectoprocta) in which the case enclosing the zooid is membranous and the pore terminal or subterminal.

Ctenothrissiformes n. [Ctenothrissus, generic name; L. forma, shape] an order of Cretaceous marine fishes of uncertain position.

ctetology n. [Gk. ktetos, acquired; logos, discourse] aspect of biology concerned with acquired characters.

ctetosome n. [Gk. ktetos, acquired; soma, body] a supernumerary chromosome associated with a sex chromosome during meiosis.

CTP cytidine triphosphate, q.v.

cubical a. [L. cubus, cube] appl. cells as long as broad.

cubital a. [L. cubitalis, of elbow] pert. the elbow, appl. joint including the humeroulnar, humeroradial, and proximal radioulnar articulations; pert. the ulna or cubitus. n. A secondary wing quill, connected with the ulna.

cubitus n. [L. cubitum, elbow] the ulna, forearm; primary vein in an insect wing.

cuboid a. [Gk. kyboeidēs, cube-like] nearly cubic in shape. n. Outermost of distal tarsal bones.

cuboidal a. [Gk. kyboeidēs, cube-like] pert. the cuboid.

Cubomedusae n. [L. cubus, cube; Gk. Medousa, Medusa] an order of scyphozoans having free-swimming cuboidal medusae with tentacles at each corner.

Cuculiformes n. [Cuculus, generic name of cuckoo; L. forma, shape] an order of birds of the subclass Neornithes, including the cuckoos.

cucullate a. [L. cucullus, hood] cowled, q.v.; with hood-like sepals or petals; with prothorax hood-shaped.

cucullus n. [L. cucullus, hood] a hood-shaped structure; upper part of harpe, in Lepidoptera.

Cucurbitales n. [Cucurbita, generic name] an order of dicots (Dilleniidae) formerly placed in the Sympetalae, having sympetalous flowers and shoot tendrils, comprising one family, Cucurbitaceae, the marrows, cucumbers, and gourds.

cuiller n. [F. cuiller, spoon] spoon-like terminal portion of male insect clasper.

cuirass n. [F. cuirasse, leathern jacket] bony plates or scales arranged like a cuirass; lorica, q.v.

culm n. [L. culmus, stalk] the stem of grasses and sedges.

culmen n. [L. culmen, summit] median longitudinal ridge of a bird's beak; part of superior vermis, continuous laterally with quadrangular lobules of anterior lobe of cerebellum.

culmicole, culmicolous a. [L. culmus, stalk; colere, to inhabit] living on stems, as of grasses.

cultellus n. [L. cultellus, little knife] a sharp knife-like organ, one of mouth-parts of certain blood-sucking flies.

cultivar n. [Cultural variety] a plant variety obtained in agriculture or horticulture.

culture n. [L. cultura, cultivation; colere, to till] the cultivation of micro-organisms or tissues in prepared media.

Cumacea n. [Cuma, generic name] an order or suborder of marine Peracarida having a small carapace with 4 or 5 of the thoracic segments exposed.

cumarin coumarin, q.v.

cumulose a. [L. cumulus, heap] appl. deposits consisting chiefly of plant remains, e.g. peat.

cumulus *n*. [L. *cumulus*, heap] the mass of epithelial cells bulging into cavity of an ovarian follicle and in which ovum is embedded; cumulus oophorus: discus proligerus, *q.v.*

cuneate *a*. [L. *cuneatus*, wedge-shaped] wedge-shaped; *appl.* leaves with broad abruptly pointed apex and tapering to the base; *appl.* a fasciculus and tubercle formed by a grey nucleus at posterior end of rhomboid fossa of medulla oblongata; *cf.* cuneiform, sphenoid.

cuneiform *a*. [L. *cuneus*, wedge; *forma*, shape] wedge-shaped; *appl.* distal tarsal bones; *appl.* a carpal bone, os triquetrum; *appl.* 2 small cartilages of larynx; *cf.* cuneate, sphenoid, sphenoidal.

cuneus *n*. [L. *cuneus*, wedge] division of elytron of certain insects; a wedge-shaped area of the occipital lobe between calcarine fissure and medial part of parieto-occipital fissure.

cup *n*. [A.S. *cuppe*, cup] any structure resembling a cup, e.g. apothecium, acetabulum, cotyle, cupule, cyathus, scyphus.

cup fungi the common name for the Discomycetes. *q.v.*

cupula *n*. [L. *cupula*, little tub] the bony apex of cochlea; the part of pleura over the apex of lung; cupule, *q.v.*

cupulate *a*. [L. *cupula*, little tub] cup-shaped, *appl.* certain aecidia; having a cup-shaped structure or a cupule.

cupule *n*. [L. *cupula*, little tub] the involucre around the fruit of some trees, e.g. acorns; the gemma cup of some liverworts; the outer seed coat in pteridosperms; a small sucker of various animals; *alt.* cupula.

curviserial *a*. [L. *curvus*, curved; *series*, row] *appl.* phyllotaxis in which divergence is such that orthostichies themselves are slightly twisted spirally.

cushion *n*. [O.F. *coissin*, cushion] the central thick region in prothallus of fern; *appl.* habitus of many plants, as in certain alpine species; torus tubarius, prominence behind pharyngeal opening of Eustachian tube; tubercle or elevation of laryngeal surface of epiglottis; embryonic endocardial thickening of wall of atrial canal; pulvillus, *q.v.*

cusp *n*. [L. *cuspis*, point] a prominence, as on teeth; a sharp point; *alt.* tubercle.

cuspidate *a*. [L. *cuspidare*, to make pointed] terminating in a point, *appl.* cystidium, pileus, leaves, teeth.

cutaneous *a*. [L. *cutis*, skin] *pert.* the skin.

cuticle *n*. [L. *cuticula*, thin skin] an outer skin or pellicle; the epidermis, especially when impermeable to water; a layer of waxy material, cutin, on the outer wall of epidermal cell walls in plants, making them fairly impermeable to water; a layer of material laid down over the epidermis in animals; *alt.* cuticula; *a.* cuticular.

cuticularization *n*. [L. *cuticula*, thin skin] cutinization in external layers of epidermal cells; the formation of the cuticle.

cuticulin *n*. [L. *cuticula*, thin skin] a protein united with a fatty compound, lipoprotein, secreted by epidermal cells and forming the epicuticle of insects.

cutin *n*. [L. *cutis*, skin] a wax-like mixture of fatty substances impregnated into epidermal walls of plant cells and also forming a separate layer, the cuticle, on the outer surface of the epidermis, making the surface impermeable to water; *alt.* cutose; *cf.* suberin.

cutinization *n*. [L. *cutis*, skin] the deposition of cutin in cell wall, thereby forming a cuticle; the formation of cutin.

cutis *n*. [L. *cutis*, skin] the corium, or deeper layer of the skin, *alt.* cutis vera, dermis; layer investing pileus and stipe.

cutis-lamella dermatome, *q.v.*

Cutleriales *n*. [*Cutleria*, generic name] an order of brown algae having partly trichothallic growth.

cutocellulose *n*. [L. *cutis*, skin; *cellula*, small cell] cellulose with cutin, as in plant epidermis.

cutose *n*. [L. *cutis*, skin] cutin, *q.v.* .

Cuvier, ducts of [*G.L.C.F.D. Cuvier*, French comparative anatomist] short veins opening into sinus venosus, and formed by union of anterior and posterior cardinal veins, in lower vertebrates.

Cuvierian organs glandular tubes extending from cloaca of holothurians; *alt.* slime tubes.

Cyamoidea *n*. [Gk. *kyamos*, pebble; *eidos*, form] an extinct class of small M. Cambrian echinoderms of the subphylum Pelmatozoa, having a theca of non-prismatic material.

cyanic *a*. [Gk. *kyanos*, dark blue] blue, bluish, *appl.* flowers, birds' eggs.

cyanin, cyanidin *n*. [Gk. *kyanos*, cornflower] a common blue–violet anthocyanin pigment found in flowers, e.g. cornflower.

cyano- *alt.* kyano-.

cyanocobalamin(e) vitamin B_{12a}, *see* cobalamine; *alt.* erythrocyte-maturing factor.

cyanogenesis *n*. [Gk. *kyanos*, blue; *genesis*, origin] the elaboration of hydrocyanic acid, as in certain plants; *a.* cyanogenic.

cyanolabe *n*. [Gk. *kyanos*, blue; *labē*, grasping] the blue-sensitive pigment of the normal human eye.

cyanophil *a*. [Gk. *kyanos*, blue; *philein*, to love] with special affinity for blue or green stains.

Cyanophilales *n*. [Gk. *kyanos*, blue; *philein*, to love] an order of ascolichens with blue–green algae.

cyanophoric *a*. [Gk. *kyanos*, blue; *pherein*, to bear] *appl.* glycosides which on hydrolysis yield glucose, hydrocyanic acid, and benzaldehyde, such as amygdalin.

Cyanophyceae *n*. [Gk. *kyanos*, blue; *phykos*, seaweed] a class of prokaryotic marine and freshwater unicellular, colonial, or filamentous algae, commonly called blue–green algae, that have the pigments chlorophyll, unique carotinoids called myxoxanthin and myxoxanthophyll, phycocyanin and phycoerythrin and often appear blue; in some classifications the group is raised to division status called Cyanophyta; *alt.* Myxophyceae, Schizophyceae.

cyanophycin *n*. [Gk. *kyanos*, blue; *phykos*, seaweed] protein reserve forming granules in peripheral region of cells in blue–green algae; *alt.* β granules.

cyanophyll *n*. [Gk. *kyanos*, blue; *phyllon*, leaf] a bluish–green colouring matter in plants.

Cyanophyta *n*. [Gk. *kyanos*, blue; *phyton*, plant] in some classifications, the Cyanophyceae, *q.v.*, raised to division (phylum) status; *alt.* Myxophyta.

cyathiform *a*. [L. *cyathus*, cup; *forma*, form] cup-shaped.

cyathium *n*. [Gk. *kyathos*, cup] the peculiar inflorescence in some members of the Euphorbiaceae, such as the spurges, *Euphorbia*, consisting of a cup-shaped involucre of bracts surrounding

staminate flowers each consisting of a single stamen, with a central pistillate flower consisting of
a stalked tricarpellary ovary.

cyathozooid *n.* [Gk. *kyathos*, cup; *zōon*, animal;
eidos, shape] the primary zooid in certain tunicates.

cyathus *n.* [L. *cyathus*, cup] a small cup-shaped
organ; gemma cup, *q.v.*

cybernetics *n.* [Gk. *kybernētikos*, skilled in gover-
ning] science of communication and control, as
by nervous system and brain; *cf.* kybernetics.

Cycadales, cyads *n., n.plu.* [*Cycas*, generic
name] an order of Cycadopsida known both
as fossils and living plants, having a palm-like
appearance with massive stems which may be
short or tree-like, pinnate leaves, sporophylls
in cones with male and female cones usually
similar.

Cycadatae *n.* [*Cycas*, generic name] in some classifications, a group of Cycadophytina, including
the Cycadales and Nilssoniales.

Cycadeoideales *n.* [*Cycas*, generic name; Gk.
eidos, form] Bennettitales, *q.v.*

Cycadicae *n.* [*Cycas*, generic name] Cycadophytina, *q.v.*

Cycadofilicales *n.* [*Cycas*, generic name; L. *filix*,
fern] Lyginopteridales, *q.v.*

Cycadophyta *n.* [*Cycas*, generic name; Gk.
phyton, plant] Cycadopsida, *q.v.*

Cycadophytina *n.* [*Cycas*, generic name; Gk.
phyton, plant] in some classifications, a major
division of Spermatophyta, comprising the
pinnate-leaved gymnosperms; *alt.* Cycadicae.

Cycadopsida *n.* [*Cydas*, generic name; Gk.
opsis, sight] a class of gymnosperms including living
and fossil orders, having little-branched stems,
large pinnate leaves and sporophylls resembling
leaves or grouped into cones, and motile male
gametes; *alt.* Cycadophyta.

Cyclanthales *n.* [Gk. *kyklos*, cycle; *anthos*, flow-
er] an order of monocots (Arecidae) including
the family Cyclanthaceae, which is somewhat
intermediate between the palms and Arales.

cycle *n.* [Gk. *kyklos*, circle] the circulation of a
fluid through a definite series of vessels; recurrent
series of phenomena, as life cycle, ovarian cycle,
etc.

cyclic *a.* [Gk. *kyklos*, circle] having parts of
flower arranged in whorls, *alt.* cyclical, *cf.* acyclic;
periodic.

cyclic AMP a form of AMP in which the phosphate group is bonded internally to form a cyclic
molecule that is concerned with the regulation of
metabolism and gene expression, and acts as an
acrasin in slime moulds; *alt.* cAMP, adenosine
3′,5′-phosphate.

cyclic photophosphorylation a type of photophosphorylation in which light energy is converted to ATP but no $NADPH_2$ is generated; *cf.*
non-cyclic.

cyclocoelic *a.* [Gk. *kyklos*, circle; *koilia*, intes-
tines] with the intestine coiled in one or more distinct spirals.

Cycloganoidei *n.* [Gk. *kyklos*, circle; *ganos*, sheen;
eidos, form] in some classifications, a group of
Ganoidei with cycloid scales.

cyclogenous *a.* [Gk. *kyklos*, circle; *gennaein*, to
produce] *appl.* a stem growing in concentric
circles.

cyclogeny *n.* [Gk. *kykos*, circle; *genos*, genera-

tion] production of a succession of different morphological types in a life cycle.

cyclohexamide *n.* [Gk. *kykos*, circle; *hex*, six;
ammoniakon, gum] an antibiotic produced by the
actinomycete *Streptomyces griseus*, which inhibits
protein synthesis by inhibiting translation.

cycloid *a.* [Gk. *kyklos*, circle; *eidos*, form] *appl.*
scales with evenly curved free border.

Cycloidea, cycloids *n., n.plu.* [Gk. *kyklos*; circle;
eidos, form] an extinct class of small M. Cambrian
echinoderms of the subphylum Pelmatozoa having
a theca of prismatic material.

cyclomorial *a.* [Gk. *kyklos*, circle; *morion*, constituent part] *appl.* scales, growing in area by apposition of marginal zones, as in Palaeozoic elasmobranchs.

cyclomorphosis *n.* [Gk. *kyklos*, circle; *morphōsis*,
form] a cycle of changes in form, usually seasonal, found especially in marine zooplankton,
possibly in response to changes in salinity.

Cyclomyaria *n.* [Gk. *kyklos*, circle; *mys*, muscle]
Doliolida, *q.v.*

cyclopean, cyclopic *a.* [Gk. *kyklos*, circle; *ōps*,
eye] *appl.* single median eye developed under certain artificial conditions, or as a mutation, instead
of the normal pair.

Cyclophyllidea *n.* [Gk. *kyklos*, circle; *phyllidion*,
little leaf] an order of segmented tapeworms
having a scolex with apical hooks and 4 strong
suckers (acetabula); *alt.* Taeniidea.

cyclophysis *n.* [Gk. *kyklos*, circle; *physis*, consti-
tution] differences in parts at different times in
the growth cycle of a plant.

cyclopoid *a.* [Gk. *kyklos*, circle; *ōps*, eye; *eidos*,
shape] naupliiform, *q.v.*

Cyclopoida *n.* [Gk. *kyklos*, circle; *ōps*, eye; *eidos*,
shape] an order of copepods with free-living and
parasitic forms that have uniramous 2nd antennae.

cyclops *n.* [Gk. *kyklos*, circle; *ōps*, eye] a larval
stage in some copepods, formed after a metanauplius and having many characteristics of the
adult.

cyclosis *n.* [Gk. *kyklosis*, whirling around] circulation, as of protoplasm within a cell.

cyclospermous *a.* [Gk. *kyklos*, circle; *sperma*,
seed] with embryo coiled in a circle or spiral.

cyclospondylic *a.* [Gk. *kyklos*, circle; *sphondylos*,
vertebra] *appl.* vertebral centra in which the calcified material forms a single cylinder surrounding the notochord; *alt.* cyclospondylous.

Cyclosporae *n.* [Gk. *kyklos*, circle; *sporos*,
seed] a group of brown algae having a life cycle
with no alternation of generations and a diploid
thallus.

Cyclostomata, cyclostomes *n., n.plu.* [Gk.
kyklos, circle; *stoma*, mouth] a subclass of
Agnatha, consisting of the living lampreys and
hagfish, and sometimes used as an alternative
name to Agnatha, which have lost their bone and
have no paired fins, *alt.* Marsipobranchia; an
order of Gymnolaemata (Ectoprocta) in which
the zooids are fused and the case enclosing them
is completely calcified, and the pore has no lid.

cydippid *n.* [Gk. *Kydippe*, mythological girl] a
larva of Ctenophora.

Cydippida *n.* [Gk. *Kydippe*, mythological girl] an
order of Ctenophora of the class Tentaculata
having a round or oval shape.

cyesis *n.* [Gk. *kyesis*, conception] pregnancy.

cygneous *a.* [Gk. *kyknos*, swan] shaped like a swan's neck.

cylindrical *a.* [Gk. *kylindros*, cylinder] *Appl.* leaves rolled on themselves, or to solid cylinder-like leaves; tubuliform, *q.v.*

cymasin *n.* [Gk. *kyma*, wave] a saponin glycoside obtained from Indian hemp.

cymba *n.* [L. *cymba*, boat] upper part of concha of ear; a boat-shaped sponge spicule.

cymbiform *a.* [L. *cymba*, boat; *forma*, shape] boat-shaped; *alt.* navicular, scaphoid.

cymbium *n.* [Gk. *kymbion*, small boat] boat-shaped tarsus of pedipalpus in certain spiders.

cyme *n.* [L. *cyma*, young sprout] any determinate inflorescence in which each growing point ends in a flower.

cymose *a.* [L. *cyma*, young sprout] sympodially branched, *appl.* inflorescence.

cymotrichous *a.* [Gk. *kyma*, wave; *thrix*, hair] having wavy hair.

cynarrhodium, cynarrhodon *n.* [Gk. *kyōn*, dog; *rhodon*, rose] an etaerio with achenes placed on concave thalamus; *alt.* hip.

cynopodous *a.* [Gk. *kyōn*, dog; *pous*, foot] with non-retractile claws as dogs.

Cyperales *n.* [*Cyperus*, generic name] an order of monocots (Liliidae) comprising the Cyperaceae, sedges.

cyphella *n.* [Gk. *kyphella*, hollow of ear] small cavity on thallus of certain lichens thought to be concerned with aeration of the thallus.

cyphonautes *n.* [Gk. *kyphos*, bent; *nautēs*, sailor] young free-swimming larva of certain Polyzoa.

Cypriniformes *n.* [*Cyprinus*, generic name; L. *forma*, shape] Ostariophysi, *q.v.*, *alt.* Cyprinoidea, Cyprini, Cyprinidae; an order of Ostariophysi, when this is considered a superorder.

cyprinine *n.* [Gk. *kyprinos*, carp] a protamine present in carp sperm.

Cyprinodontiformes *n.* [*Cyprinodon*, generic name; L. *forma*, shape] an order of small, mainly tropical, and viviparous teleosts including the toothed carps; *alt.* Microcyprini.

cypris *n.* [L. L. *Cypris*, Venus] a larval stage in some Cirripedia formed after a nauplius and showing many features of the adult.

cypsela *n.* [Gk. *kypselē*, hollow vessel] an inferior bicarpellary achene, as in Compositae.

cys cysteine, *q.v.*

(cys)² cystine, *q.v.*

cyst *n.* [Gk. *kystis*, bladder] the enclosing membrane round a resting cell or apocyte; a bladder or air vesicle in certain seaweeds; abnormal sac containing fluid; a bladder-like structure.

cysteine *n.* [Gk. *kystis*, bladder] a sulphur-containing amino acid, which is a constituent of many proteins and can be oxidized to cystine; abbreviated to cys.

cystenchyma *n.* [Gk. *kystis*, bladder; *engchyma*, infusion] a parenchymula in sponges with large vesicular cell structure.

cystencytes *n.plu.* [Gk. *kystis*, bladder; *en*, in; *kytos*, hollow] in sponges, collencytes which have acquired a vesicular structure; *alt.* cystocytes.

cystic *a.* [Gk. *kystis*, bladder] *pert.* cyst; *pert.* gall bladder or to urinary bladder.

cysticercoid *a.* [Gk. *kystis*, bladder; *kerkos*, tail; *eidos*, from] *appl.* larval tapeworm; *appl.* bladderworm in some tapeworms with small bladder and tail-like appendages.

cysticercus *n.* [Gk. *kystis*, bladder; *kerkos*, tail] in tapeworms, a larval stage consisting of fluid-filled sac containing a proscolex; *alt.* bladderworm, metacestode; the whole larval stage may be called a proscolex.

cysticolous *a.* [Gk. *kystis*, bladder; L. *colere*, to inhabit] living in a cyst.

cystid *n.* [Gk. *kystis*, bladder; *idion*, *dim.*] a fossil cystoid; cystidium, *q.v.*

Cystidea *n.* [Gk. *kystis*, bladder; *idion*, *dim.*] Cystoidea, *q.v.*

cystidia *plu.* of cystidium.

cystidiform *a.* [Gk. *kystis*, bladder; *idion*, *dim.*; L. *forma*, form] *appl.* clavate cells on gill margin in agarics.

cystidiole *n.* [Gk. *kystis*, bladder; *dim.*; L. *dim.*] a projecting sterile hymenial cell in Hymenomycetes.

cystidium *n.* [Gk. *kystis*, bladder; *idion*, *dim.*] a hair-like inflated cell in the hymenial layer of some fungi; *alt.* cystid; *plu.* cystidia.

cystine *n.* [Gk. *kystis*, bladder] a sulphur-containing amino acid, which is a constituent of many proteins and can be reduced to 2 molecules of cysteine; abbreviated to (cys)₂.

cystiphragm *n.* [Gk. *kystis*, bladder; *phragma*, fence] a calcareous plate curving from wall towards base of zooecium in Polyzoa.

cystoarian *a.* [Gk. *kystis*, bladder; *ōarion*, small egg] *appl.* gonads when enclosed in coelomic sacs, as in most teleosts; *cf.* gymnoarian.

cystocarp *n.* [Gk. *kystis*, bladder; *karpos*, fruit] a cyst arising from carpogonial branch and containing spores, in certain Rhodophyceae; *alt.* cryptocarp, gonimocarp.

cystochroic *a.* [Gk. *kystis*, bladder; *chrōs*, complexion] having pigment in cell vacuoles.

cystocyte *n.* [Gk. *kystis*, bladder; *kytos*, hollow] cystencyte, *q.v.*; coagulocyte, *q.v.*

cystogenous *a.* [Gk. *kystis*, bladder; *-genēs*, producing] cyst-forming; *appl.* large nucleated cells which secrete the cyst, in cercaria.

Cystoidea, cystoids *n.,* *n.plu.* [Gk. *kystis*, bladder; *idion*, *dim.*] an extinct class of vase-shaped echinoderms of the subphylum Pelmatozoa which existed from the Ordovician to Permian times; *alt.* Cystidea.

cystolith *n.* [Gk. *kystis*, bladder; *lithos*, stone] a mass of calcium carbonate, occasionally of silica, formed on ingrowths of epidermal cell walls in some plants.

cyston *n.* [Gk. *kystis*, bladder] a dactylozooid modified for excretory purposes, in Siphonophora.

cystosorus *n.* [Gk. *kystis*, bladder; *sōros*, heap] a cluster of cystospores, as in certain Phycomycetes.

cystospore *n.* [Gk. *kystis*, bladder; *sporos*, seed] carpospore, *q.v.*; a cyst containing a zoospore.

cystozooid *n.* [Gk. *kystis*, bladder; *zōon*, animal; *eidos*, form] the body portion of a metacestode; *cf.* acanthozooid.

cytase *n.* [Gk. *kytos*, hollow] *see* cellulase, protopectinase.

cytaster *n.* [Gk. *kytos*, hollow; *astēr*, star] the aster, *q.v.*, found in the cytoplasm rather than in the nucleus.

cytes *n.plu.* [Gk. *kytos*, hollow] spermatocyte and oocyte stages of germ cell formation; *alt.* auxocytes.

cytidine a nucleoside with cytosine as its base.

cytidine diphosphate (CDP)

cytidine diphosphate (CDP) a nucleotide cofactor analogous to ADP.

cytidine monophosphate (CMP) a nucleotide cofactor formed from cytidine and phosphoric acid, analogous to AMP; *alt.* cytidylic acid.

cytidine triphosphate (CTP) a nucleotide cofactor analogous to ATP and essential to protein synthesis.

cytidylic acid cytidine monophosphate, *q.v.*

cyto- *alt.* kyto-.

cytobiotaxis cytoclesis, *q.v.*; cytotaxis, *q.v.*

cytoblast *n.* [Gk. *kytos*, hollow; *blastos*, bud] one of the granules in the cytoplasm seen by Altmann in the 19th century, which he regarded as elementary organisms, and which were actually mitochondria, *alt.* bioblast; hypothetical unit, *alt.* bioblast; the cell nucleus.

cytoblastema *n.* [Gk. *kytos*, hollow; *blastēma*, growth] the formative material from which cells were supposed to arise.

cytocentrum *n.* [Gk. *kytos*, hollow; *kentron*, centre] centrosome, *q.v.*; centrosome surrounded by an area of differentiated cytoplasm; *alt.* central apparatus, idiosome.

cytochemistry *n.* [Gk. *kytos*, hollow; *chēmeia*, transmutation] the study of the chemical composition and chemical processes of cells.

cytochimaera *n.* [Gk. *kytos*, hollow; L. *chimaera*, monster] a combination of tissues in which cells in the same organism have different chromosome numbers.

cytochroic *a.* [Gk. *kytos*, hollow; *chrōs*, complexion] with pigmented cytoplasm.

cytochrome oxidase a terminal oxidase enzyme which catalyses the transfer of electrons in the reduction of oxygen to water, at the end of electron transfer chains; *alt.* Warburg's factor, Warburg's respiratory enzyme; EC 1.9.3.1; *r.n.* cytochrome c oxidase.

cytochromes *n.plu.* [Gk. *kytos*, hollow; *chrōma*, colour] chromoproteins with a haem (iron-porphyrin) group that are important in electron and/or hydrogen transfer because of a reversible valency change of their haem iron; *cf.* histohaematins, myohaematin.

cytochylema *n.* [Gk. *kytos*, hollow; *chylos*, juice] cytolymph, *q.v.*

cytocidal *a.* [Gk. *kytos*, hollow; L. *caedere*, to kill] cell-destroying.

cytoclesis *n.* [Gk. *kytos*, hollow; *klēsis*, summons] the influence of a cell group or placode upon development or differentiation of neighbouring cells; *alt.* cytobiotaxis; *cf.* organizer.

cytococcus *n.* [Gk. *kytos*, hollow; *kokkos*, kernel] the nucleus of a fertilized egg.

cytocyst *n.* [Gk. *kytos*, hollow; *kystis*, bladder] the envelope formed by remains of host cell within which a protozoan parasite multiplies.

cytode *n.* [Gk. *kytos*, hollow; *eidos*, form] a nonnucleated protoplasmic mass.

cytodeme *n.* [Gk. *kytos*, hollow; *dēmos*, people] a unit in a taxon differing cytologically, usually in chromosome number, from the rest of the taxon.

cytoderm *n.* [Gk. *kytos*, hollow; *derma*, skin] cell wall, *q.v.*

cytodiaeresis *n.* [Gk. *kytos*, hollow; *diairesis*, division] mitosis, *q.v.*

cytoflav(in) *n.* [Gk. *kytos*, hollow; L *flavus*, yellow] a water-soluble yellow pigment in certain cells, resembling, or identical with, riboflavin; *alt.* yellow enzyme.

cytogamy *n.* [Gk. *kytos*, hollow; *gamos*, marriage] cell conjugation.

cytogene plasmagene, *q.v.*

cytogenesis *n.* [Gk. *kytos*, hollow; *genesis*, descent] development or formation of cells.

cytogenetic *a.* [Gk. *kytos*, hollow; *genesis*, descent] *pert.* cytogenesis; *pert.* cytogenetics; *appl.* map showing location of genes within a chromosome.

cytogenetics *n.* [Gk. *kytos*, hollow; *genesis*, descent] genetics in relation to cytology, the cytological aspect of genetics.

cytogenic *a.* [Gk. *kytos*, hollow; *genos*, offspring] *appl.* reproduction by cell division, as in a clone; *cf* blastogenic.

cytogenous *a.* [Gk. *kytos*, hollow; *genos*, offspring] producing cells, *appl.* lymphatic tissue.

cytoglobin *n.* [Gk. *kytos*, hollow; L. *globus*, globe] a protein which retards coagulation of blood.

cytohistology *n.* [Gk. *kytos*, hollow; *histos*, tissue; *logos*, discourse] the study of cells and tissues.

cytohyaloplasma *n.* [Gk. *kytos*, hollow; *hyalos*, glass; *plasma*, mould] the substance of the cytomitome; *alt.* hyaloplasm.

cytokinesis *n.* [Gk. *kytos*, hollow; *kinēsis*, movement] the changes occurring in the general cytoplasm during nuclear division; the separation of the cytoplasm of the parent cell into daughter cells after nuclear division.

cytokinins *n.plu.* [Gk. *kytos*, hollow; *kinein*, to move] plant growth hormones which promote cell division and are concerned with various other developmental processes and include kinetin and zeatin; *alt.* kinins, phytokinins.

cytolemma *n.* [Gk. *kytos*, hollow; *lemma*, skin] plasma membrane, *q.v.*

cytology *n.* [Gk. *kytos*, hollow vessel; *logos*, discourse] the science dealing with structure, functions, and life history of cells.

cytolosome cytolysosome, *q.v.*

cytolymph *n.* [Gk. *kytos*, hollow; L. *lympha*, water] cell sap; the fluid part of cytoplasm; *alt.* cytochylema.

cytolysin *n.* [Gk. *kytos*, hollow; *lysis,* loosing] a substance inducing cytolysis.

cytolysis *n.* [Gk. *kytos*, hollow; *lysis*, loosing] cell dissolution; cell degeneration.

cytolysosome *n.* [Gk. *kytos*, hollow; *lyein*, to loose; *sōma*, body] an enlarged lysosome functioning in cell autolysis; *alt.* cytosegresome, cytolosome.

cytome *n.* [Gk. *kytos*, hollow] the microsome, mitochondrial or cytosome system of a cell.

cytomegalovirus *n.* [Gk. *kytos*, hollow; *megas*, big; L. *virus*, poison] a virus associated with the presence of very large cells, particularly in glands of vertebrates.

cytomeres *n.plu.* [Gk. *kytos*, hollow; *meros*, part] cells formed by division of schizont and giving rise to merozoites, *alt.* agametoblasts; nonnuclear portion of sperms, when containing mitochondria also called chondriomeres, *alt.* plastomeres.

cytometry *n.* [Gk. *kytos*, hollow; *metreo*, to compute] count of cells; blood count.

cytomicrosome *n.* [Gk. *kytos*, hollow; *mikros*,

small; *sōma*, body] a microsome of cytoplasm; *cf*.
karyomicrosome.

cytomitome *n*. [Gk. *kytos*, hollow; *mitos*, thread]
cytoreticulum, *q.v.*

cytomorphosis *n*. [Gk. *kytos*, hollow; *morphōsis*,
shaping] the life history of cells; the series of
structural modifications of cells or successive
generations of cells; cellular change, as in senescence.

cyton *n*. [Gk. *kytos*, hollow] neurocyton, q.v.

cytopempsis *n*. [Gk. *kytos*, hollow; *pempsis*, a
sending] the engulfment by, passage through,
and discharge from, a cell of a droplet or particle;
alt. podocytosis; *cf*. pinocytosis, phagocytosis.

cytophan *n*. [Gk. *kytos*, hollow; *phaneros*, visible]
ovoid matrix surrounding karyophans in spironeme
and axoneme fibres in stalk of some ciliates.

cytopharynx *n*. [Gk. *kytos*, hollow; *pharyngx*,
gullet] a tube-like structure leading from mouth
into endoplasm in certain Protozoa; *alt*. gullet.

cytophil *a*. [Gk. *kytos*, hollow; *philein*, to love]
pert. haptophorous groups; having an affinity for
cells.

cytophilic antibody a globulin component of
immune serum which becomes attached *in vitro* to
certain cells in such a way that those cells are subsequently capable of specifically absorbing antigen.

cytophore *n*. [Gk. *kytos*, hollow; *phora*, bur-
den] a cell regarded as bearer of parasitic
Sporozoa; residual cytoplasm associated with
groups of spermatozoa in certain invertebrates,
also called blastophore in annelids.

cytoplasm *n*. [Gk. *kytos*, hollow; *plasma*,
mould] protoplasm of cell body other than the
nucleus; *cf*. nucleoplasm.

cytoplasmic inheritance inheritance by particles in the cytoplasm instead of by genes on
chromosomes.

cytoproct, cytopyge *n*. [Gk. *kytos*, hollow;
prōktos, anus; *pygē*, rump] a cell-anus.

cytoreticulum *n*. [Gk. *kytos*, hollow; L. *reticulum*, little net] the cytoplasmic threadwork; *alt*.
cytomitome, spongioplasm.

cytosegresome *n*. [Gk. *kytos*, hollow; L. *segregare*, to separate; Gk. *sōma*, body] a body with a
single or double outer membrane, and containing
cytoplasmic organelles; cytolysosome, *q.v.*

cytosine *n*. [Gk. *kytos*, hollow] a pyrimidine base
present in all living tissues, as part of the genetic
code of DNA and RNA pairing with guanine and
also forming part of the molecule of cytidine.

cytoskeleton a suggested cytoplasmic skeleton.

cytosol *n*. [Gk. *kytos*, hollow; L. *solutus*, dis-
solved] the ground matrix of the cytoplasm when
organelles or other particles have been removed.

cytosome *n*. [Gk. *kytos*, hollow; *sōma*, body] the
cytoplasmic part of a cell; microsome, *q.v.*

cytostatic *a*. [Gk. *kytos*, hollow; *statikos*, causing
to stand] appl. any substance suppressing cell
growth and multiplication.

cytostome *n*. [Gk. *kytos*, hollow; *stoma*, mouth] a
cell-mouth.

cytotaxis *n*. [Gk. *kytos*, hollow; *taxis*, arrange-
ment] rearrangement of cells on stimulation; *alt*.
cytobiotaxis.

cytotaxonomy n. [Gk. *kytos*, hollow; *taxis*,
arrangement; *nomos*, law] classification of organisms based on characteristics of chromosome
structure and number.

cytothesis *n*. [Gk *kytos*, hollow; *thesis*, arrang-
ing] regenerative tendency of a cell.

cytotoxin *n*. [Gk. *kytos*, hollow; *toxikon*, pois-
on] a cell-poisoning substance formed in blood
serum, such as cytolysin; *alt*. enzymoid.

cytotrophoblast *n*. [Gk. *kytos*, hollow; *trophē*,
nourishment; *blastos*, bud] inner layer of trophoblast; *alt*. layer of Langhans.

cytotropism *n*. [Gk. *kytos*, hollow; *tropē*, turn-
ing] the mutual attraction of 2 or more cells.

cytozoic *a*. [Gk. *kytos*, hollow; *zōon*, animal] living within a cell, *appl*. sporozoan trophozoite.

cytozyme *a*. [Gk. *kytos*, hollow; *zymē*, leaven]
thrombokinase, *q.v.*

cytula *n*. [Gk. *kytos*, hollow] the fertilized ovum
or parent cell.

D

Dacromycetales, Dacrymycetales *n*. [Gk. *dakryon*, tear; *mykēs*, fungus] a group of jelly fungi
having basidia in which the distal end of the
hypobasidium bears 2 divergent epibasidia.

dacryocyst n. [Gk. *dakryon*, tear; *kystis*, bladder]
a pouch in the angle of the eye between upper
and lower eyelids, which receives tears from the
lacrimal ducts; *alt*. lacrimal sac, saccus lacrimalis,
tear pit, larmier.

dacryoid *a*. [Gk. *dakryon*, tear; *eidos*, shape] tear-
shaped; *alt*. lacrimiform.

dacryon *n*. [Gk. *dakryon*, tear] point of junction
of anterior border of lacrimal with frontal bone
and frontal process of maxilla.

dactyl *n*. [Gk. *daktylos*, finger] a digit; finger, or
toe; terminal ventral projection of praetarsus in
scorpions, *alt*. dactylus.

dactylar *a*. [Gk. *daktylos*, finger] *pert*. finger or
digit.

dactyline dactyloid, *q.v.*

dactylognathite *n*. [Gk. *daktylos*, finger;
gnathos, jaw] terminal segment of a maxillipede.

dactyloid *a*. [Gk. *daktylos*, finger; *eidos*, form]
like a finger or fingers; *alt*. dactyline.

dactylopatagium *n*. [Gk. *daktylos*, finger; L.
patagium, border] ectopatagium, *q.v.*

dactylopodite *n*. [Gk. *daktylos*, finger; *pous*,
foot] distal joint in certain limbs of Crustacea;
metatarsus and tarsus of spiders.

dactylopore *n*. [Gk. *daktylos*, finger; *poros*, chan-
nel] opening in skeleton of Milleporina, for protrusion of a dactylozooid.

Dactylopteriforms *n*. [*Dactylopterus*, generic
name; L. *forma*, shape] an order of teleosts including the flying gurnards.

dactylopterous *a*. [Gk. *daktylos*, finger; *pteron*,
wing] with anterior rays of pectoral fins more or
less free.

dactylozooid *n*. [Gk. *daktylos*, finger; *zōon*,
animal; *eidos*, form] in colonial hydrozoans, a
hydroid modified for catching prey and defence,
being long and slender with tentacles or short
knobs and usually without a mouth; *alt*. hydrocyst, palpon; *cf*. gastrozooid.

dactylus *n*. [Gk. *daktylos*, finger] part of tarsus of
an insect; dactyl of scorpions.

dahlia starch inulin, *q.v.*

dammar n. [Mal. *damar*, dammar] a resin obtained from several Malayan trees.

dance of bees a series of movements performed by honey bees on returning to the hive after finding a food source, which informs the other bees in the hive of the location of the food; *see also* round dance, sickle dance, waggle dance.

Danielli membrane a theoretical structure of the cell membrane put forward by Danielli, consisting of a double layer of phospholipid, with protein arranged along each outer edge; *alt.* Danielli–Davson membrane.

daphnid any of various small water fleas especially of the genus *Daphnia*.

dark reaction in photosynthesis, the reactions occurring in the stroma of the chloroplasts, being those of carbon dioxide fixation; *cf.* light reaction.

Darlington's rule the fertility of an allopolyploid is inversely proportional to the fertility of the original hybrid.

dart n. [O.F. *dart*, dagger] any structure resembling a dart; a crystalline structure in molluscs, used in copulation, *alt.* spiculum; in nematodes, a pointed structure used to penetrate host.

dartoid a. [Gk. *dartos*, flayed] *pert.* the dartos.

dartos n. [Gk. *dartos*, flayed] a thin layer of non-striped muscle united to skin of scrotum or of labia majora; *alt.* tunica dartos.

dart sac a small sac, containing a limy dart, attached to vagina near its orifice in some gastropods.

Darwinian tubercle the slight prominence on helix, of external ear, near the point where it bends downwards; *alt.* Woolner's tubercle.

Darwinism n. [C. *Darwin*, English naturalist] the theory of origin of species by natural selection working on slight variations that occur, thereby selecting those best adapted to survive.

Dasycladales n. [Gk. *dasys*, hairy; *klados*, sprout] an order of green algae having a uninucleate juvenile vegetative stage and a multinucleate reproductive stage consisting of an erect axis with whorls of branches near the apex.

dasypaedes n.plu. [Gk. *dasys*, hairy; *pais*, child] birds whose young are downy at hatching; a. dasypaedic, *alt.* hesthogenous.

dasyphyllous a. [Gk. *dasys*, hairy; *phyllon*, leaf] with thickly haired leaves.

dauermodification n. [Ger. *Dauer*, duration; L. *modificatio*, modification] a change induced by environmental factors and persisting for several generations but not permanently, the organism eventually reverting to type.

daughter n. [A.S. *dohtor*, daughter] offspring of 1st generation with no reference to sex, as daughter cell, daughter nucleus, etc.; daughter chromosome: a chromatid during anaphase.

day-neutral *appl.* plants in which flowering can be induced by either a long or a short photoperiod or by neither; *cf.* long-day, short-day.

dealation n. [L. *de*, away; *alatus*, winged] the removal of wings, as by female ants after fertilization, or by termites.

deamination n. [L. *de*, down; Gk. *ammōniakon*, resinous gum] the removal of amino ($-NH_2$) groups from an amine, e.g. during the breakdown of an amino acid.

death n. [A.S. *déath*, death] complete and permanent cessation of vital functions in an organism.

death point temperature, or other environmental variable, above or below which organisms cannot exist.

Débove's membrane [*M.G. Débove*, French histologist] layer between tunica propria and epithelium of tracheal, bronchial, and intestinal mucous membranes; *alt.* subepithelial endothelium.

decagynous a. [Gk. *deka*, ten; *gynē*, female] having 10 pistils.

decalcify v. [L. *de*, away; *calx*, lime; *facere*, to make] to deprive of lime salts; to treat with acids for removal of calcareous part.

decamerous a. [Gk. *deka*, ten; *meros*, part] with the various parts arranged in 10s.

decandrous a. [Gk. *deka*, ten; *anēr*, male] having 10 stamens.

decaploid a. [Gk. *deka*, ten; *aploos*, onefold; *eidos*, form] having 10 times the haploid number of chromosomes.

Decapoda n. [Gk. *deka*, ten; *pous*, foot] an order of freshwater, marine, and terrestrial malacostracans, including the crabs and lobsters, having 5 pairs of legs on the thorax, and the carapace completely covering the thorax; an order of cephalopods of the subclass Coleoidea, including squids and cuttlefish, having 2 retractile arms as well as 8 normal arms.

decapodiform a. [Gk. *deka*, ten; *pous*, foot; L. *forma*, shape] resembling a decapod, *appl.* certain insect larvae.

decarboxylases a group of enzymes that catalyse decarboxylation, especially of amino acids; EC sub-subgroup 4.1.1; *r.n.* carboxylyases.

decarboxylation removal of a carboxyl group, usually as carbon dioxide, from an organic compound.

decemfid a. [L. *decem*, ten; *findere*, to cleave] cut into 10 segments.

decemfoliate a. [L. *decem*, ten; *folium*, leaf] tenleaved.

decemjugate a. [L. *decem*, ten; *jugare*, to join] with 10 pairs of leaflets.

decempartite a. [L. *decem*, ten; *partiri*, to divide] having 10 lobes.

decidua n. [L. *decidere*, to fall off] the mucous membrane lining the pregnant uterus, cast off after parturition; *alt.* afterbirth.

decidua basalis decidua placentalis, *q.v.*

decidua capsularis portion of the decidua over the ovum.

decidual a. [L. *decidere*, to fall off] *pert.* decidua.

decidua parietalis the decidua vera lining the body of the uterus.

decidua placentalis portion of the decidua between myometrium and ovum; *alt.* decidua basalis.

deciduate a. [L. *decidere*, to fall off] characterized by having a decidua; partly formed by the decidua.

deciduous a. [L. *decidere*, to fall down] falling at end of growth period or at maturity, *cf.* caducous; *appl.* teeth: milk teeth, *q.v.*

declinate a. [L. *de*, away; *clinare*, to bend] bending aside in a curve, as anther filament in horse chestnut.

declivis n. [L. *declivis*, sloping] part of superior vermis, continuous laterally with lobulus simplex of cerebellar hemispheres.

decollated *a.* [L. *de*, away from; *collum*, neck] with apex of spire wanting.

decomposed *a.* [L. *de*, away; *cum*, with; *pausare*, to rest] not in contact; not adhering, said of barbs of feather when separate; decayed; rather shapeless and gelatinous, *appl.* cortical hyphae in lichens.

decomposers organisms that feed on dead plants and animals, thus breaking them down physically and chemically and recycling organic and inorganic materials to the environment.

decomposite decompound, *q.v.*

decompound *a.* [L. *de*, away; *cum*, with; *ponere*, to place] *appl.* compound leaf whose leaflets are also compound; *alt.* decomposite.

deconjugation *n.* [L. *de*, away from; *conjugare*, to join together] separation of paired chromosomes, as before end of meiotic prophase.

decorticate *v.* [L. *decorticare*, to peel] to remove bark or cortex. *a.* With bark or cortex removed, *alt.* excorticate.

decticous *n.* [Gk. *dēktikos*, biting] having functional mandibles for opening puparium or cocoon, *appl.* pupa of some insects; *cf.* adecticous.

decumbent *a.* [L. *decumbere*, to lie down] lying on the ground but rising at apex, *appl.* stem, stipe, etc.

decurrent *a.* [L. *decurrere*, to run downwards] having leaf base prolonged down stem as a winged expansion or rib; prolonged down stipe, as gills of agaric; *cf.* surcurrent.

decurved curved downwards.

decussate *a.* [L. *decussare*, to cross] crossed; having paired leaves, succeeding pairs crossing at right angles.

decussation *n.* [L. *decussare*, to cross] decussate condition of leaves; crossing of nerves with interchange of fibres, as in optic and pyramidal tracts.

dedifferentiation *n.* [L. *de*, away from; *differentia*, difference] the losing of characteristics of specialized cells and regression to a more simple state.

dediploidization *n.* [L. *de*, away from; Gk. *diploos*, double; *eidos*, form] in Basidiomycetes and Ascomycetes, the production of haploid cells or hyphae by a dikaryotic diploid cell or mycelium; *alt.* dekaryotization.

dédoublement *n.* [F. *dédoublement*, dividing into two] chorisis, *q.v.*

deduplication *n.* [L. *de*, by reason of; *duplicare*, to double] chorisis, *q.v.*

defaecation, defecation *n.* [L. *defaecatio*, voiding of excrement] the expulsion of faeces.

defaunation *n.* [L. *de*, away from; *faunus*, god of woods] removal of animal life, especially of the symbiotic protozoan fauna from an insect.

deferent *a.* [L. *deferre*, to carry away] conveying away; *appl.* ducts: vasa deferentia, *q.v.*

deferred *a.* [L. *deferre*, to carry away] *appl.* shoots arising from dormant buds.

deficiency inactivation or absence of a chromosomal segment or gene.

deficiency diseases pathological conditions in plants and animals, due to lack of certain necessary nutritive substances, e.g. crown rot in sugar beet due to boron deficiency; diseases in mammals due to vitamin deficiency, absence of vitamin A causing poor growth and xerophthalmia,—

of B_1, beri-beri,—of B_2, retardation of growth,— of C, scurvy,—of D, rickets,—of E, infertility and paralysis,—of K, bleeding,—etc.

definite *a.* [L. *definire*, to limit] fixed, constant; cymose, *appl.* inflorescences with primary axis, terminating early in a flower; *appl.* stamens limited to 20 in number.

definitive *a.* [L. *definire*, to limit] defining or limiting; complete, fully developed; final, *appl.* host of adult parasite.

deflex *v.* [L. *deflectere*, to turn aside] to bend or turn downwards or aside.

deflorate *a.* [L. *deflorere*, to shed blossoms] after the flowering stage.

defoliate *a.* [L.L. *defoliare*, to strip of leaves] bared at the annual fall. *v.* To deprive of leaves.

degenerate code a genetic code in which several triplet codes code for the same amino acid.

degeneration *n.* [L. *degenerare*, to degenerate] change to a less specialized or functionally less active form; retrogressive evolution, i.e. changing from a complex to a simpler state.

deglutition *n.* [L. *de*, down; *glutire*, to swallow] the process of swallowing.

degrowth *n.* [L. *de*, down from; A.S. *grówan*, to grow] decrease in mass of living matter.

dehiscence *n.* [L. *dehiscere*, to gape] the spontaneous opening of an organ or structure along certain lines or in a definite direction.

dehydrogenases formerly, enzymes which catalyse the reaction involving the transfer of hydrogen from a substrate to a hydrogen acceptor, now all classified as oxidoreductases; EC group 1.

dehydrotheelin oestradiol, *q.v.*

deimatic behaviour frightening behaviour consisting of the adoption of a posture by one animal to intimidate another one; *alt.* dymantic behaviour.

deinopore *n.* [Gk. *deinos*, urn; *poros*, channel] a cell bridge.

deirids *n.plu.* [Gk. *deiras*, chain of hills; *idion*, *dim.*] cervical papillae in Nematoda.

Deiters' cells [*O.F.C. Deiters*, German anatomist] supporting cells between rows of outer hair cells in organ of Corti; *alt.* outer phalangeal cells.

Deiters' nucleus lateral nucleus of vestibular nerve.

dekaryotization dediploidization, *q.v.*

delamination *n.* [L. *de*, down; *lamina*, layer] the dividing off of cells to form new layers; splitting of a layer.

deletion *n.* [L. *delere*, to efface] a deficiency of an acentric part of chromosome; absence of a chromosome segment and of the genes involved; absence of a base in a nucleic acid leading to a phase change in the reading of the triplets following the deletion.

deliquescent *a.* [L. *deliquescere*, to become fluid] having lateral buds the more vigorously developed, so that the main stem seems to divide into a number of irregular branches; becoming fluid.

delitescence *n.* [L. *delitescere*, to lie hidden] the latent period of a poison; incubation period of a pathogenic organism.

delomorphic *a.* [Gk. *dēlos*, visible; *morphē*, shape] with definite form, *appl.* oxyntic cells of the gastric glands; *alt.* delomorphous.

delphinidin

delphinidin a deep blue anthocyanin pigment found in delphinium and other flowers.

delphinology n. [Gk. *delphis*, dolphin; *logos*, discourse] the study of dolphins.

delthyrium n. [Gk. *delos*, visible; *thyrion*, little door] the opening, between hinge and beak, for peduncle exit in many Brachiopoda.

deltidium n. [Gk. Δ, delta, *idion*, dim.] a plate covering the delthyrium.

deltoid a. [Gk. Δ, delta; *eidos*, form] more or less triangular in shape, *appl.* muscles, etc., *appl.* oral plates on calyx of Blastoidea, *appl.* leaf.

demanian a. [*J.G. de Man*, French zoologist] *appl.* a complex system of paired efferent tubes connecting with intestine and uteri in Nematoda, and associated with gelatinous secretion for protection of eggs.

deme n. [Gk. *dēmos*, people] assemblage of taxonomically closely related individuals; aggregate of single cells; a small breeding group.

demersal a. [L. *demergere*, to plunge into] living on or near bottom of sea or lake; sunk.

demersed a. [L. *demergere*, to plunge into] growing under water, *appl.* parts of plants.

demibranch hemibranch, *q.v.*

demifacet n. [L. *dimidius*, half; *facies*, face] part of parapophysis facet when divided between centra of 2 adjacent vertebrae.

demilunes crescentic cells; crescentic bodies of mucous alveoli of salivary glands, *alt.* crescents of Gianuzzi, demilunes of Heidenhain.

demiplate n. [L. *dimidius*, half; F. *plate*, flat] plate cut off by fusion of adjoining plates behind it from central suture line of ambulacral area in echinoderms.

demisheath n. [L. *dimidius*, half; A.S. *sceath*, sheath] one of paired protecting covers of insect ovipositor.

demography n. [Gk. *dēmos*, people; *graphein*, to write] the study of numbers of organisms in a population and their variation with time.

demoid a. [Gk. *dēmōdēs*, common] abundant.

Demospongea, Demospongiae n. [Gk. *desmos*, bond; L. *spongia*, sponge] a class of sponges, commonly called horny sponges, of the subphylum Gelatinosa having a skeleton made of silica and/or spongin spicules which are usually differentiated into micro- and megascleres and are not 6-rayed.

Demour's membrane Descemet's membrane, *q.v.*

denatant a. [L. *de*, down from; *natare*, to swim] swimming, drifting, or migrating with the current; *cf.* contranatant.

denaturation an alteration in the structural properties of a protein, in enzymes or hormones resulting in change of activity, brought about by heating, change in pH, ultraviolet radiation, etc.

dendraxon n. [Gk. *dendron*, tree; *axōn*, axis] a nerve cell with axis cylinder branching close to the cell body; *cf.* inaxon.

dendriform a. [Gk. *dendron*, tree; L. *forma*, shape] dendroid, *q.v.*

dendrite n. [Gk. *dendron*, tree] a fine branch of a dendron; sometimes used as a synonym for dendron; *alt.* neurodendron.

dendritic a. [Gk. *dendron*, tree] dendroid, *appl.* tree-like structures or markings; like, *pert.*, or having, dendrites or dendrons.

Dendrochirota(e) n. [Gk. *dendron*, tree; *cheir*, hand] an order of sea cucumbers with many podia, tree-like oral tentacles, and respiratory trees.

dendrochronology n. [Gk. *dendron*, tree; *chronos*, time; *logos*, discourse] determination of age of trees or timber; dating by comparative study of tree rings; science of tree-ring analysis and its implications.

Dendrogaea n. [Gk. *dendron*, tree; *gaia*, earth] a biogeographical region including all the neotropical region except temperate South America.

dendrogram n. [Gk. *dendron*, tree; *graphein*, to write] any branching, tree-like diagram illustrating the relationship between organisms or objects.

dendroid a. [Gk. *dendron*, tree; *eidos*, form] tree-like; much branched; *alt.* dendritic, dendriform, arboroid.

Dendroidea n. [Gk. *dendron*, tree; *eidos*, form] an order of graptolites having thecae arranged regularly on one side of the stipe only.

dendrology n. [Gk. *dendron*, tree; *logos*, discourse] the study of trees.

dendron n. [Gk. *dendron*, tree] a protoplasmic process of the nerve cell, which conducts impulses towards the cell body.

denitrification n. [L. *de*, away; Gk. *nitron*, soda; L. *facere*, to make] reduction of nitrates, to nitrites and ammonia, as in plant tissues, or to molecular nitrogen, as by certain soil bacteria.

denitrifiers denitrifying bacteria.

denitrifying *appl.* bacteria causing denitrification, *alt.* denitrifiers, nitrate-reducing bacteria.

dens n. [L. *dens*, tooth] tooth or tooth-like process; *alt.* odontoid process; *plu.* dentes, *q.v.*

dens serotinus the 3rd molar or wisdom tooth.

dental a. [L. *dens*, tooth] *pert.* teeth, *appl.* nerves, blood vessels, canals, furrows, papillae, sac, tissue, etc.

dental formula a method of expressing the numbers of each type of teeth in a mammal, consisting of a series of fractions indicating the number of teeth of each type in one half of the upper jaw, divided by those of the lower jaw.

dentary a. [L. *dens*, tooth] *pert.* dentaries, membrane bones in lower jaw of many vertebrates. n. Dentary bone, os dentale.

dentate a. [L. *dens*, tooth] toothed; with large saw-like teeth on the margin.

dentate-ciliate with teeth and hairs on the margins, *appl.* leaves.

dentate-crenate with marginal teeth somewhat rounded.

dentes *plu.* of dens; prongs of the furcula borne on manubrium in Collembola.

denticidal a. [L. *dens*, tooth; *caedere*, to cut] dehiscent with tooth-like formation at top of capsule, as in the dicot family Caryophyllaceae.

denticles *n.plu.* [L. *denticulus*, little tooth] small tooth-like processes; the paragnaths of certain Polychaeta, *alt.* grater; the teeth within the secondary orifice in Polyzoa; placoid scales, *q.v.*

denticulate a. [L. *denticulus*, little tooth] having denticles; with minute marginal teeth.

dentin dentine, *q.v.*

dentinal a. [L. *dens*, tooth] *pert.* dentine; *appl.* tubules, canaliculi dentales.

dentine n. [L. *dens*, tooth] a hard, elastic substance, chemically resembling bone, composing the greater part of teeth and denticles; *alt.* dentin, substantia eburnea.

dentition *n.* [L. *dens*, tooth] the number, arrangement, and kind of teeth; teething.

deoxycorticosterone an active constituent of adrenal cortical extract; *alt.* desoxycorticosterone.

deoxynucleoside a nucleoside consisting of a purine or pyrimidine base attached to deoxyribose.

deoxyribo- *alt.* desoxyribo-.

deoxyribonuclease DNase, *q.v.*

deoxyribonucleic acid (DNA) the self-replicating molecule forming the genetic hereditary material as chromosomes in the nuclei of eukaryotes and as strands in prokaryotes, consisting of deoxyribose sugar, phosphate, the purines adenine and guanine, and the pyrimidines thymine and cytosine; *alt.* desoxyribonucleic acid; *see also* Watson–Crick DNA model; *cf.* ribonucleic acid.

deoxyribonucleotide a nucleotide consisting of a purine or pyrimidine attached to deoxyribose which is attached to a phosphate group.

deoxyribose the sugar occurring in DNA which is similar to ribose but lacks one oxygen atom.

deoxyvirus a virus containing DNA; *alt.* DNA virus.

deperulation *n.* [L. *de*, away; *dim.* of *pera*, wallet] the pushing apart or throwing off, of bud scales.

depigmentation *n.* [L. *de*, away; *pingere*, to paint] the destruction of colour in a cell, by natural or experimental physiological processes.

depilation *n.* [L. *de*, away; *pilus*, hair] loss of hairy covering, as of plants when maturing; removal of hair.

deplanate *a.* [L. *deplanare*, to level] levelled, flattened.

deplasmolysis *n.* [L. *de*, away from; Gk. *plasma*, form; *lysis*, loosing] re-entrance of water into a plant cell after plasmolysis, and reversal of shrinkage.

deplumation *n.* [L. *de*, away from; *pluma*, feather] moulting in birds.

depolarization a reduction of polarization (i.e. of potential difference) across a membrane.

depressant *n.* [L. *deprimere*, to keep down] anything that lowers vital activity.

depressomotor *n.* [L. *deprimere*, to keep down; *movere*, to move] any nerve which lowers muscular activity, a depressor nerve.

depressor *n.* [L. *deprimere*, to keep down] any muscle which lowers or depresses any structure, *cf.* levator; *appl.* a nerve which lowers the activity of an organ; *appl.* compounds which slow down metabolic rate.

depula *n.* [Gk. *depas*, goblet] invaginated blastula preceding gastrula stage in development of embryo.

Derbesiales *n.* [*Derbesia*, generic name] an order of Chlorophyceae whose gametophytes may be oval vesicles and whose sporophytes are siphonaceous and branched.

deric *a.* [Gk. *deros*, skin] dermic *appl.* epithelium: epidermis, *q.v.*

derm derma, *q.v.*; dermis, *q.v.*

derma *n.* [Gk. *derma*, skin] the layers of integument below the epidermis, *alt.* dermis; epiderma and hypoderma of fungi; *alt.* derm.

dermal *pert.* derma; *pert.* skin; *alt.* dermatic, dermic.

dermalia *n.plu.* [Gk. *derma*, skin] microscleres in the dermal membrane of sponges.

Dermaptera *n.* [Gk. *derma*, skin; *pteron*, wing] an order of insects, commonly called earwigs, having slight incomplete metamorphosis, short leathery fore-wings and membranous hind-wings.

dermarticular(e) *n.* [Gk. *derma*, skin; L. *articulus*, joint] the goniale, *q.v.*

dermatic dermal, *q.v.*

dermato- *see also* dermo-.

dermatocystidium pileocystidium, *q.v.*

dermatogen *n.* [Gk. *derma*, skin; *genea*, birth] the young epidermis in plants; the histogen giving rise to the epidermis in plants; antigen of skin disease.

dermatoglyphics *n.* [Gk. *derma*, skin; *glyphein*, to carve] skin, palm, finger, sole, toe prints; print formulae.

dermatoid *a.* [Gk. *derma*, skin; *eidos*, form] resembling a skin; functioning as a skin.

dermatomes *n.plu.* [Gk. *derma*, skin; *tomē*, cutting] lateral parts of segmental mesoderm, which develop into connective tissue of corium or dermis, *alt.* dermatomeres; skin areas supplied by individual spinal nerves; *alt.* cutis-lamellae.

dermatophyte *n.* [Gk. *derma*, skin; *phyton*, plant] any fungous parasite of skin; *alt.* dermatophyton, dermophyte, epidermophyte.

dermatopsy *n.* [Gk. *derma*, skin; *opsis*, sight] condition of seeing with the skin, i.e. with a skin sensitive to light.

dermatoskeleton exoskeleton, *q.v.*

dermatosome *n.* [Gk. *derma*, skin; *sōma*, body] in one theory of the structure of the plant cell wall, one of the units joined to another by cytoplasm; a unit of cellulose of the plant cell wall.

dermatozoon *n.* [Gk. *derma*, skin; *zōon*, animal] any animal parasite of the skin; *alt.* dermozoon.

dermentoglossum *n.* [Gk. *derma*, skin; *entos*, within; *glōssa*, tongue] a bone arising by fusion of dentinal bases, covering entoglossum, in some fishes.

dermethmoid *n.* [Gk. *derma*, skin; *ēthmos*, sieve; *eidos*, form] supraethmoid, *q.v.*

dermic *a.* [Gk. *derma*, skin] *pert.*, or derived from, skin; *alt.* dermal.

dermis *n.* [Gk. *derma*, skin] derma, *q.v.*; corium, *q.v.*

dermo- *see also* dermato-.

dermoblast *n.* [Gk. *derma*, skin; *blastos*, bud] the layer of mesoblast which gives rise to the derma.

dermoccipitals *n.plu.* [Gk. *derma*, skin; L. *occiput*, back of head] two bones taking the place of interparietal in some lower vertebrates and in embryonic development in higher vertebrates.

dermomyotome *n.* [Gk. *derma*, skin; *mys*, muscle; *tomē*, cutting] the dorsilateral part of mesodermal somites.

dermo-ossification *n.* [Gk. *derma*, skin; L. *os*, bone; *fieri*, to become] a bone formed in the skin.

dermopharyngeal *n.* [Gk. *derma*, skin; *pharyngx*, gullet] superior or inferior plate of membrane bone supporting pharyngeal teeth in some fishes.

dermophyte dermatophyte, *q.v.*

Dermoptera *n.* [Gk. *derma*, skin; *pteron*, wing] an order of placental mammals known from Palaeocene times to the present day, including the flying lemurs, having the fore- and hindlimbs connected by a skin-fold forming a parachute.

dermosclerites

dermosclerites *n.plu.* [Gk. *derma*, skin; *sklēros*, hard] masses of spicules found in tissues of some Alcyonaria.

dermoskeleton *n.* [Gk. *derma*, skin; *skeletos*, dried] exoskeleton, *q.v.*

dermosphenotic *n.* [Gk. *derma*, skin; *sphēn*, wedge] a circumorbital bone, between supraorbitals and suborbitals, as in teleosts; *alt.* intertemporal.

dermotrichia *n.plu.* [Gk. *derma*, skin; *thrix*, hair] dermal fin-rays.

dermozoon dermatozoon, *q.v.*

Derocheilocarida *n.* [Gk. *deros*, skin; *cheilos*, edge; L. *caris*, kind of crab] the only order of the Mystacocarida, *q.v.*

derotreme *n.* [Gk. *deros*, skin; *trēma*, aperture] skin forming an operculum as in *Megalobatrachus*, the giant salamander.

dertrotheca *n.* [Gk. *dertron*, caul; *thēkē*, box] the horny casing of bird maxilla; *alt.* unguicorn.

dertrum *n.* [Gk. *dertron*, caul] any modification of the casing of maxilla in birds.

Descemet's membrane [*J. Descemet*, French anatomist] the posterior elastic layer of cornea; *alt.* Demour's membrane.

descending *a.* [L. *de*, down; *scandere*, to climb] directed downwards, or towards caudal region, *appl.* blood vessels, nerves, etc.; in plants, growing or hanging downwards; *appl.* aestivation where each petal overlaps the one in front of it.

desegmentation *n.* [L. *de*, from; *segmentum*, piece cut off] fusion of segments originally separate.

deserticolous *a.* [L. *desertus*, waste; *colere*, to inhabit] desert-inhabiting.

desma *n.* [Gk. *desma*, bond] megasclere which forms characteristic skeletal network of Lithistida.

desmactinic *a.* [Gk. *desma*, bond; *aktis*, ray] with podia continued upwards to apical plate, *appl.* Stelleroidea; *cf.* lysactinic.

Desmarestiales *n.* [*Desmarestia*, generic name] an order of brown algae having thalli of a single filament at each growing apex behind which is a pseudoparenchymatous thallus.

desmergate *n.* [Gk. *desma*, bond; *ergatēs*, worker] a type of ant intermediate between worker and soldier.

desmids *n.plu.* [Gk. *desmos*, band, chain] unicellular or colonial green algae of the order Zygnematales, especially of the family Desmidiaceae (placoderm or true desmids) and sometimes also of the family Mesotaeniaceae (saccoderm desmids).

desmin *n.* [Gk. *desmos*, bond] a protein found in muscle which seems to help to hold muscle filaments in position, and may be important in other cells.

Desmocontae Desmokontae, *q.v.*

desmocyte *n.* [Gk. *desmos*, bond; *kytos*, hollow] fibroblast, *q.v.*

desmoenzymes *n.plu.* [Gk. *desmos*; bond; *en*, in; *zymē*, leaven] intracellular enzymes bound to the protoplasm and which cannot therefore be extracted by the methods at present available.

desmogen *n.* [Gk. *desmos*, bond; *genos*, descent] merismatic or growing tissue.

desmognathous *a.* [Gk. *desmos*, bond; *gnathos*, jaw] having maxillopalatines fused in middle line

owing to other peculiarities in skull, *appl.* certain birds.

desmoid *a.* [Gk. *desmos*, bond; *eidos*, form] bandlike; forming a chain or ribbon; resembling desmids.

Desmokontae *n.* [Gk. *desmos*, bond; *kontos*, punting pole] formerly a subclass of Dinophyceae, now considered to be a separate class Desmophyceae, *q.v.*; in some classifications reduced to an order of Pyrrophyceae, known as Desmokontales (Desmocontales); *alt.* Desmocontae.

desmolases *n.plu.* [Gk. *desmos*, bond] formerly a general term for enzymes which catalyse the splitting of C − C linkages, including carboxylases and decarboxylases.

desmology *n.* [Gk. *desmos*, bond; *logos*, discourse] the anatomy of ligaments; *cf.* syndesmology.

desmoneme *n.* [Gk. *desmos*, bond; *nēma*, thread] a nematocyst in which the distal end of the thread or closed tube, when discharged, coils round prey; *alt.* volvent.

desmones *n.plu.* [Gk. *desmos*, bond] chemical substances exchanged by way of protoplasmic bridges between cells; amboceptors, *q.v.*

Desmophycaea *n.* [Gk. *desmos*, bond; *phykos*, seaweed] a class of Pyrrophyta having the cell wall divided into 2 valves and plate-like or lobed chloroplasts with pyrenoids.

desmose *n.* [Gk. *desmos*, bond] a strand connecting blepharoplasts at mitosis.

desmosome *n.* [Gk. *desmos*, bond; *sōma*, body] one of the strong points anchoring cells to each other, and consisting of a thickening of cell membranes running parallel to each other, separated by an enlarged intercellular space, and bundles of fine cytoplasmic filaments extending to the cytoplasm; *alt.* bridge corpuscle, macula adherans.

Desmostylia *n.* [Gk. *desmos*, bond; *stylos*, pillar] an order of Miocene to Pliocene placental mammals from the north Pacific with a sirenianlike skull and with legs.

desoxycorticosterone deoxycorticosterone, *q.v.*

desoxyribo- deoxyribo-, *q.v.*

desoxyribonucleic acid deoxyribonucleic acid, *q.v.*

desquamation *n.* [L. *de*, away; *squama*, scale] shedding of cuticle or epidermis in flakes.

desynapsis, desyndesis *n.* [L. *de*, away from; Gk. *synapsis*, union] failure of synapsis, caused by disjunction of homologous chromosomes.

determinant *n.* [L. *determinare*, to limit] a hypothetical unit, being an aggregation of biophores determining the development of a cell or of an independently viable group of cells; hereditary factor, gene.

determinate *a.* [L. *determinare*, to limit] with certain limits; *appl.* inflorescence with primary axis terminated early with a flower bud, *alt.* cyme, cymose; *appl.* cleavage in which the part of the body to be formed by each blastomere is fixed from the beginning; *appl.* evolution: orthogenesis, *q.v.*; with a well-marked edge.

determination *n.* [L. *determinatio*, boundary] the process adjusting regional development according to relative location of region and organization centre.

determinator *n.* [L. *determinare*, to determine] a gene that controls the male or female character of haploid mycelium at the site of formation of a fruit body.

determined *appl.* embryonic cells once their fate has been irrevocably established.

detorsion *n.* [L. *de*, away; *torquere*, to twist] torsion in an opposite direction to that of original, resulting in a more-or-less posterior position of anus and circumanal complex.

detoxication *n.* [L. *de*, away; Gk. *toxikon*, poison] the removal of toxic materials from the body, by converting them to relatively harmless substances which are eliminated in excretion or by antibodies.

detritivore *n.* [L. *detritus*, rubbed off; *vorare*, to devour] an organism feeding on detritus.

detritus *n.* [L. *detritus*, rubbed off] aggregate of fragments of a structure, as of detached or broken-down tissues; small pieces of dead and decomposing plants and animals.

detrusor *n.* [L. *detrudere*, to thrust from] the outer of 3 layers of the muscular coat of the urinary bladder; physiologically, all 3 layers; *alt.* detrusor urinae, detrusor vesicae.

detumescence *n.* [L. *de*, down from; *tumescere*, to swell] subsidence of swelling; *cf.* intumescence.

deuter cell eurycyst, *q.v.*

deuter- *see also* deuto-.

deuterocerebrum *n.* [Gk. *deuteros*, second; L. *cerebrum*, brain] that portion of crustacean brain from which antennular nerves arise; *alt.* mesocerebrum; *cf.* deutocerebrum.

deuterocoele *n.* [Gk. *deuteros*, second; *koilos*, hollow] the coelom, *q.v.*

deuterocone *n.* [Gk. *deuteros*, second; *kōnos*, cone] mammalian premolar cusp corresponding to molar protocone.

deuteroconidium *n.* [Gk. *deuteros*, second; *konis*, dust; *idion*. dim.] one of the conidia produced by division of a hemispore or protoconidium in dermatophytes.

deuterogamy *n.* [Gk. *deuteros*, second; *gamos*, marriage] secondary fertilization; pairing substituting for the union of gametes as in fungi.

deuterogenesis *n.* [Gk. *deuteros*, second; *genesis*, origin] second phase of embryonic development, involving growth in length and consequent bilateral symmetry; *cf.* protogenesis.

Deuteromycetes, Deuteromycetae, Deuteromycotine *n.* [Gk. *deuteros*, second; *mykes*, fungus] Fungi Imperfecti, *q.v.*

deuteroplasm deutoplasm, *q.v.*

deuteropolydesmic *a.* [Gk. *deuteros*, second; *polys*, many; *desmos*, bond] *appl.* cyclomorial scales composed mainly of synpolydesmic scales.

deuteroproteose *n.* [Gk. *deuteros*, second; *prôteion*, first] a secondary product from digestion of proteins.

deuterostoma *n.* [Gk. *deuteros*, second; *stoma*, mouth] a mouth formed secondarily, as distinct from gastrula mouth.

deuterostome *n.* [Gk. *deuteros*, second; *stoma*, mouth] *appl.* coelomates having a true coelom, with radial cleavage of the egg, the blastopore becoming the anus, the coelom being formed by enterocoely, and a dipleurula larva; *cf.* protostome.

deuterotoky *n.* [Gk. *deuteros*, second; *tokos*, birth] reproduction of both sexes from parthenogenetic eggs; *cf.* arrhenotoky and thelytoky.

deuterotype *n.* [Gk. *deuteros*, second; *typos*, pattern] the specimen chosen to replace the original type specimen for designation of a species.

Deuterozoic *a.* [Gk. *deuteros*, second; *zoē*, life] obsolete term for the later part of the Palaeozoic era, i.e. Devonian, Carboniferous, and Permian periods.

deuterozooid *n.* [Gk. *deuteros*, second; *zōon*, animal; *eidos*, form] a zooid produced by budding from a primary zooid.

deuthyalosome *n.* [Gk. *deuteros*, second; *hyalos*, glass; *sōma*, body] the nucleus remaining in ovum after formation of 1st polar body.

deuto- *see also* deutero-.

deutoblasts *n.plu.* [Gk. *deuteros*, second; *blastos*, bud] amoeba-like bodies formed from protoblasts, and liberated to multiply in the blood.

deutobroch, deutobroque *a.* [Gk. *deuteros*, second; *brochos*, mesh] *appl.* nuclei of gonia preparing for leptotene stage; *cf.* protobroch.

deutocerebrum *n.* [Gk. *deuteros*, second; L. *cerebrum*, brain] portion of insect brain derived from fused ganglia of antennary segment of head; *alt.* deutocerebron; *cf.* deuterocerebrum.

deutomalae *n.plu.* [Gk. *deuteros*, second; *malon*, cheek] the broad plate in Chaetognatha, formed by fusion of 2nd pair of mouth appendages; 2nd pair of mouth appendages in certain Myriapoda.

deutomerite *n.* [Gk. *deuteros*, second; *meros*, part] the posterior division of certain gregarines; *cf.* primite.

deutonephros *n.* [Gk. *deuteros*, second; *nephros*, kidney] mesonephros, *q.v.*

deutonymph *n.* [Gk. *deuteros*, second; *nymphē*, chrysalis] second nymphal stage or instar, either chrysalis-like or motile, in development of Acarina; *alt.* hypopus stage, wandernymph.

deutoplasm *n.* [Gk. *deuteros*, second; *plasma*, mould] yolk or food material in cytoplasm of ovum or other cell; *alt.* deuteroplasm, metaplasm; *cf.* energid.

deutoplasmolysis *n.* [Gk. *deuteros*, second; *plasma*, mould; *lysis*, loosing] the release of surplus yolk into the perivitelline space, before 1st cleavage of ovum.

deutoscolex *n.* [Gk. *deuteros*, second; *skōlex*, worm] a secondary scolex produced by budding, in bladderworm stage of certain tapeworms.

deutosomes *n.plu.* [Gk. *deuteros*, second; *sōma*, body] granules of nucleolus cast out into cytoplasm, from which yolk is said to arise.

deutosporophyte *n.* [Gk. *deuteros*, second; *sporos*, seed; *phyton*, plant] second sporophyte phase in life cycle of Rhodophyceae; *cf.* protosporophyte.

deutosternum *n.* [Gk. *deuteros*, second; *sternon*, chest] sternite of segment bearing pedipalpi in Acarina.

deutovum *n.* [Gk. *deuteros*, second; L. *ovum*, egg] a stage in the metamorphosis of certain mites, a secondary or deutovarial membrane surrounding the embryo until the larval stage.

development *n.* [F. *développer*, to unfold] the changes undergone by an organism from its beginning to maturity.

deviation *n.* [L. *de*, away from; *via*, way] divergence from corresponding developmental stages.

Devonian *a.* [Devon, where strata were first studied] *pert.* or *appl.* Palaeozoic geological period preceding Carboniferous.

dexiotropic *a.* [Gk. *dexios*, right; *tropē*, turn] turning from left to right, as whorls, *appl.* shells, *appl.* spiral cleavage of cells, *appl.* movement of the protistan, *Volvox*; *cf.* laeotropic.

dextral *a.* [L. *dexter*, right-handed] on or *pert.* the right; *cf.* sinistral.

dextrans *n.plu.* [L. *dexter*, right-handed] polysaccharides which are made of branched chains of glucose and are storage products in bacteria and yeasts.

dextrins *n.plu.* [L. *dexter*, right-handed] a group of small soluble polysaccharides produced by hydrolysis of starch, which can be made by exposing starch to high temperature for a short time; *alt.* starch gums.

dextrorse *a.* [L. *dexter*, right; *vertere*, to turn] growing in a spiral which twines from left to right; clockwise; *alt.* eutropic, solar; *cf.* sinistrorse.

dextrose *n.* [L. *dexter*, right] glucose, *q.v.*

diabetogenic *a.* [Gk. *diabainein*, to cross over; *gennaein*, to produce] causing diabetes; *appl.* a prepituitary hormone antagonistic to insulin, affecting carbohydrate metabolism; *appl.* a hormone of sinus gland of eye-stalk in crustaceans.

diachaenium *n.* [Gk. *dis*, twice; *a*, not; *chainein*, to gape] each part of a cremocarp.

diachronous *a.* [Gk. *dia*, asunder; *chronos*, time] dating from different periods; *appl.* fossils occurring in the same geological formation, though in different areas, due, e.g., to changes in sea level.

diachyma *n.* [Gk. *dia*, throughout; *chymos*, juice] leaf parenchyma.

diacoel(e) *n.* [Gk. *dia*, through; *koilos*, hollow] third ventricle of brain.

diacranteric *a.* [Gk. *dia*, asunder; *kranterēs*, wisdom teeth] with diastema between front and back teeth, as in snakes.

diactinal *a.* [Gk. *dis*, twice; *aktis*, ray] with 2 rays pointed at ends; *cf.* biradiate.

diacytic *a.* [Gk. *dia*, through; *kytos*, hollow] *appl.* stomata of a type with one pair of subsidiary cells, with their common axis at right angles to the long axis of the guard cells, surrounding the stoma; formerly called caryophyllaceous; *cf.* anisocytic.

diadelphous *a.* [Gk. *dis*, twice; *adelphos*, brother] having stamens in 2 bundles owing to fusion of filaments; *cf.* monadelphous.

Diadematacea *n.* [Gk. *diadēma*, crown] a superorder of globular sea urchins of the subclass Euechinoidea having an Aristotle's lantern with grooved teeth.

diadematoid *a.* [Gk. *diadēma*, crown; *eidos*, shape] of Echinoidea, having 3 primary pore plates with occasionally a secondary between aboral and middle primary; as *cf.* arbacioid, one primary, with secondary on each side, and triplechinoid, 2 primaries, with one or more secondaries between.

Diadematoida *n.* [Gk. *diadēma*, crown; *eidos*, shape] an order of Diadematacea, having a flexible or rigid test.

diadinoxanthin *n.* [Gk. *dia*, through; *dinos*, rotation; *xanthos*, yellow] a xanthophyll carotinoid pigment found in algae of the classes Chrysophyceae, Bacillariophyceae, Pyrrophyceae.

diadromous *a.* [Gk. *diadromos*, wandering] having nerves or veins radiating in fan-like manner, *appl.* leaves.

diaene *n.* [Gk. *dis*, twice; an analogy of triaene, from Gk. *triaina*, trident] a form of triaene, with one of the cladi reduced or absent.

diageotropism *n.* [Gk. *dia*, through; *gē*, earth; *tropē*, turn] a growth movement in a plant organ so that it assumes a position at right angles to the direction of gravity; *alt.* geodiatropism; *a.* diageotropic.

diagnosis *n.* [Gk. *diagnōsis*, discrimination] a concise description of an organism with full distinctive characters; discrimination of a physiological or pathological condition by its distinctive signs.

diagnostic *a.* [Gk. *diagnōsis*, discrimination] distinguishing; differentiating the species or genus, etc., from others similar.

diaheliotropism *n.* [Gk. *dia*, through; *helios*, sun; *tropē*, turn] diaphototropism, *q.v.*; *alt.* diheliotropism.

diakinesis *n.* [Gk. *dia*, through; *kinēsis*, movement] the last stage of prophase of meiosis in which the nuclear membrane breaks down.

diallelic *a.* [Gk. *di*, two; *allēlōn*, one another] *appl.* polyploid with 2 different alleles at a locus.

dialycarpic *a.* [Gk. *dia*, asunder; *lyein*, to loose; *karpos*, fruit] apocarpous, *q.v.*

dialyneury *n.* [Gk. *dialyein*, to reconcile; *neuron*, nerve] in certain gastropods, condition of having pleural ganglia united to opposite visceral nerve by anastomosis with pallial nerve; *cf.* zygoneury.

Dialypetalae *n.* [Gk. *dia*, asunder; *lyein*, to loose; *petalon*, leaf] Polypetalae, *q.v.*

dialypetalous *a.* [Gk. *dia*, asunder; *lyein*, to loose; *petalon*, leaf] polypetalous, *q.v.*

dialyphyllous *a.* [Gk. *dia*, asunder; *lyein*, to loose; *phyllon*, leaf] with separate leaves.

dialysate *n.* [Gk. *dialysis*, parting] any substance which passes through a semipermeable membrane during dialysis; *alt.* diffusate; *cf.* retentate.

dialysepalous *a.* [Gk. *dia*, asunder; *lyein*, to loose; F. *sépale*, sepal] polysepalous, *q.v.*

dialysis *n.* [Gk. *dia*, asunder; *lysis*, loosing] separation of dissolved crystalloids and colloids through semipermeable membrane, crystalloids passing more readily; *alt.* permeation.

dialystely *n.* [Gk. *dia*, asunder; *lyein*, to loose; *stēlē*, post] a condition in which the steles in the stem remain more or less separate; *a.* dialystelic.

diamesogamous *a.* [Gk. *dia*, through; *mesos*, medium; *gamos*, marriage] fertilized through external agency, as by means of wind, insects, etc.

diamines *n.plu.* [Gk. *di*, two; *ammoniakon*, resinous gum] compounds containing 2 amino groups such as cadaverine and putrescine.

diancistron *n.* [Gk. *dis*, twice; *angkistron*, hook] a spicule resembling a stout sigma, but the inner margin of both hook and shaft thins out to a knife edge and is notched; *plu.* diancistra.

diandrous *a.* [Gk. *dis*, twice; *anēr*, male] having 2 free stamens, *alt.* distemonous; in a moss, having 2 antheridia surrounded by each bract.

diapause *n.* [Gk. *diapauein*, to make to cease] a spontaneous state of dormancy during development, as of insects; resting stage between anatrepsis and katatrepsis in blastokinesis; sexual rest period, *appl.* annelids; *cf.* quiescence.

diapedesis *n.* [Gk. *diapēdesis*, leaping through] emigration of white blood corpuscles through walls of capillaries into surrounding tissue; migration of cells to exterior, in certain larval sponges.

diaphototropism n. [Gk. dia, through; phôs, light; tropē, turn] a growth movement in plant organs to assume a position at right angles to rays of light; when the light is sunlight, known as diaheliotropism; a. diaphototropic.

diaphragm(a) n. [Gk. diaphragma, midriff] the wall which separates the small cell, the prothallus, from rest of megaspore in Salviniales; a septum at nodes in horsetails; a sheet of muscular tissue attached to introvert in annelids; single strongly developed septum in some polychaetes; perforated tissue that subdivides tentacle cavity in Polyzoa; a fibromuscular abdominal septum enclosing perineural sinus in certain insects; the transverse septum separating cephalothorax from abdomen in certain Arachnida; a special fan-shaped muscle spreading from anterior end of ilia to oesophagus and base of lungs in Anura; a partition partly muscular, partly tendinous, separating cavity of chest from abdominal cavity in mammals; fold of dura mater on sella turcica; a structure controlling admission of light through an aperture, as iris.

diaphysis n. [Gk. dia, through; phyein, to bring forth] shaft of limb bone, cf. epiphysis; abnormal growth of an axis or shoot.

diaplexus n. [Gk. dia, through; L. plexus, interwoven] choroid plexus of the 3rd ventricle of the brain.

diapolar a. [Gk. dia, between; polos, pole] appl. the cells between parapolar and uropolar cells in Dicyemidae.

diapophysis n. [Gk. dia, through; apo, from; phyein, to produce] lateral or transverse process of neural arch of a vertebra.

Diaporthales n. [Diaporthe, generic name] an order of Ascomycetes whose perithecia are immersed in plant tissue.

diapsid a. [Gk. dis, twice; apsis, arch] appl. skulls with supra- and infratemporal fossae distinct; cf. synapsid.

Diapsida n. [Gk. dis, twice; apsis, arch] in some classifications, a subclass of reptiles having a diapsid skull and including the Lepidosauria and Archosauria.

diarch a. [Gk. dis, twice; archē, origin] with 2 xylem and 2 phloem bundles; appl. root in which protoxylem bundles meet and form a plate of tissue across cylinder with phloem bundle on each side; appl. a bipolar type of spindle.

diarthric a. [Gk. dis, twice; arthron, joint] biarticulate, q.v.

diarthrosis n. [Gk. dis, twice; arthron, joint] an articulation allowing considerable movement, a moveable joint.

diaschistic a. [Gk. dia, through; schistos, split] appl. type of tetrads which divide once transversely and once longitudinally in meiosis; alt. pseudomitotic; cf. anaschistic.

diaspore n. [Gk. diaspora, dispersion (dia, asunder; spora, seed)] any spore, seed, fruit, or other portion of a plant when being dispersed and able to produce a new plant; alt. disseminule, propagule.

diastase n. [Gk. diastēnai, to separate] original name for a mixture of α-amylase and β-amylase, then for a time used for any enzyme, later used for α-amylase or β-amylase.

diastasis n. [Gk. diastasis, interval] rest period

preceding systole; the abnormal separation of parts that are usually joined together.

diastatic a. [Gk. dia, through; histanai, to set] pert. diastase, or having similar properties; pert. diastasis.

diastem(a) n. [Gk. diastēma, interval] a toothless space usually between 2 types of teeth; an equatorial modification of protoplasm preceding cell division.

diaster n. [Gk. dis, twice; astēr, star] the stage in mitosis where daughter chromosomes are grouped near spindle poles ready to form a new nucleus.

diastole n. [Gk. diastolē, difference] rhythmical relaxation of heart; rhythmical expansion of a contractile vacuole; cf. systole.

diastomatic a. [Gk. dia, through; stoma, mouth] through stomata or pores; giving off gases from spongy parenchyma through stomata.

diathesis n. [Gk. diathesis, disposition] a constitutional predisposition to a type of reaction, disease, or development.

diatom n. [Gk. dia, through; temnein, to cut] the common name for a member of the Bacillariophyceae, q.v.

Diatomales n. [Gk. dia, through; temnein, to cut] Bacillariales, q.v.

diatomin n. [Gk. diatemnein, to cut through] a yellow or brownish-yellow pigment found in certain algae, especially diatoms; alt. phycoxanthin.

diatoxanthin n. [Gk. dia, through; temnein, to cut; xanthos, yellow] xanthophyll pigment found in algae of the classes Chrysophyceae, Bacillariophyceae, and Phaeophyceae.

diatropism n. [Gk. dia, through; tropē, turn] the tendency of organs or organisms to place themselves at right angles to line of action of stimulus.

diauxy n. [Gk. dis, twice; auxein, to increase] the adaptation of a micro-organism to the utilization of 2 different sugars in a culture medium, having the constitutive enzymes for digestion of one sugar and inducing synthesis of the enzymes for the other.

diaxon a. [Gk. dis, twice; axōn, axis] with 2 axes, as certain sponge spicules; cf. biradiate.

diaxone n. [Gk. dis, twice; axōn, axis] a nerve cell with 2 axis-cylinder processes; a. diaxonic, alt. dineuric.

diblastula n. [Gk. dis, twice; blastos, bud] a coelenterate embryo consisting of 2 layers arranged round a central cavity.

Dibranchiata, dibranchs n., n.plu. [Gk. dis, twice; brangchia, gills] in some classifications, cephalopod molluscs with a single pair of gills and kidneys, including the Decapoda and Octopoda.

dibranchiate a. [Gk. dis, twice; brangchia, gills] with 2 gills.

dicaryo- dikaryo-, q.v.

dicellate a. [Gk. dikella, two-pronged mattock] with 2 prongs, appl. sponge spicules.

dicentral a. [Gk. dia, through; kentron, centre] appl. canal in fish vertebral centrum.

dicentric a. [Gk. dis, twice; kentron, centre] having 2 centromeres, appl. chromatids, chromosomes.

dicephalobrachial a. [Gk. dis, twice; kephalē, head; L. brachium, arm] of a chromosome, with the centromere median and both arms reduced to knob-like extensions.

dicerous *a.* [Gk. *dikerōs*, two-horned] having 2 horns; with 2 antennae.

dichasial cyme dichasium, *q.v.*

dichasium *n.* [Gk. *dichazein*, to divide in two] a cymose inflorescence in which 2 lateral branches occur about same level; *alt.* dichasial cyme, false dichotomy.

dichlamydeous *a.* [Gk. *dis*, twice; *chlamys*, cloak] having both calyx and corolla; *alt.* diplochlamydeous.

dichocarpous *a.* [Gk. *dichōs*, in two ways; *karpos*, fruit] with 2 forms of frutification, *appl.* certain fungi.

dichogamy *n.* [Gk. *dicha*, in two; *gamos*, marriage] maturing of sexual elements at different times, ensuring cross-fertilization, as in protandry protogyny; *alt.* heteracmy; *cf.* homogamy; a dichogamous.

dichophase *n.* [Gk. *dikho-*, asunder; *phasis*, appearance] the 'phase of decision' when within the cell a choice is made between synthesis leading to differentiation and synthesis leading to division.

dichophysis *n.* [Gk. *dicha*, in two; *physis*, constitution] a rigid dichotomous hypha, as in hymenium and trama of some fungi.

dichoptic *a.* [Gk. *dicha*, in two; *opsis*, sight] with eyes quite separate; *cf.* holoptic.

dichorhinic *a.* [Gk. *dicha*, differently; *rhines*, nostrils] *pert.* the nostrils separately; *appl.* different olfactory stimuli; *cf.* dirhinic.

Dichotomosiphonales *n.* [*Dichotomosiphon*, generic name] an order of Chlorophyceae having siphonaceous filaments.

dichotomy *n.* [Gk. *dicha*, in two; *temnein*, to cut] branching which results from division of growing point into 2 equal parts; repeated forking; *a.* dichotomous.

dichroic *a.* [Gk. *dis*, twice; *chrōs*, colour] exhibiting dichroism, as chlorophyll solution; *alt.* dichromic; *cf.* dichromatic.

dichroism *n.* [Gk. *dis*, twice; *chrōs*, colour] property of showing 2 colours, as one colour by transmitted and the other by reflected light.

dichromatic *a.* [Gk. *di-*, two; *chrōma*, colour] with 2 colour varieties; seeing only 2 colours; *alt.* dichromic; *cf.* dichroic.

dichromic dichroic, *q.v.*; dichromatic, *q.v.*

dichromophil *a.* [Gk. *di-*, two, *chrōma*, colour; *philein*, to love] staining with both acid and basic dyes.

dichthadiigyne *n.* [Gk. *dichthadios*, double; *gynē*, female] a gynaecoid ant with voluminous ovaries, and without eyes and wings; *a.* dichthadiiform.

diclesium *n.* [Gk. *dis*, twice; *klēsis*, a closing] a multiple fruit or anthocarp formed from an enlarged and hardened perianth; *cf.* sphalerocarp.

Diclidophoroidea *n.* [Gk. *di-*, two; *kleis*, key; *phora*, producing; *eidos*, form] an order of Polyopisthocotylea having a sucker enclosed in a pit on either side of the mouth.

diclinous *a.* [Gk. *di-*, two; *klinē*, bed] with stamens and pistils on separate flowers; with staminate and pistillate flowers on same plant; with antheridia and oogonia on separate hyphae; *alt.* unisexual; *n.* dicliny, diclinism.

Diclybothrioides *n.* [*Diclybothrium*, generic name; Gk. *eidos*, from] an order of Polyopisthocotylea having a sucker around mouth and an opisthaptor with suckers and sometimes hooks and spines.

dicoccous *a.* [Gk. *di-*, two; *kokkos*, kernel] having 2 1-seeded coherent capsules.

dicoelous *a.* [Gk. *di-*, two; *koilos*, hollow] having 2 cavities.

dicont dikont, *q.v.*

dicostalia *n.* [Gk. *di-*, two; L. *costa*, rib] the secundibrachs or 2nd brachial series in a crinoid; *alt.* distichalia.

dicot dicotyledon, *q.v.*

Dicotyledones, dicotyledons *n.*, *n.plu.* [Gk. *di-*, two; *kotylēdōn*, cup-shaped hollow] a class of angiosperms having an embryo with 2 cotyledons, parts of the flower usually in 2s or 5s or their multiples, leaves with net veins, and vascular bundles in the stem in a ring leaving a central pith; *alt.* Dicotyledon(e)ae, dicots, and in some classifications, Magnoliatae, Magnoliopsida.

dicotyledonous *pert.* dicotyledons; *pert.* embryo with 2 cotyledons.

dicoumarin, dicoumarol a substance obtained from rotting sweet clover and used as an anticoagulant.

Dicranales *n.* [*Dicranum*, generic name] an order of mosses having an upright gametophore with narrow leaves.

dicratic *a.* [Gk. *di-*, two; *kratos*, power] with 2 spores of a tetrad being of one sex, and the other 2 of the opposite sex; *cf.* monocratic.

dictyodromous *a.* [Gk. *diktyon*, net; *dramein*, to run] net-veined, when the smaller veins branch and anastomose freely.

dictyogen *n.* [Gk. *diktyon*, net; *-genēs*, producing] a net-leaved plant.

dictyokinesis *n.* [Gk. *diktyon*, net; *kinesis*, movement] the breaking up of the Golgi apparatus at mitosis and segregation of dictyosomes to daughter cells.

dictyonalia *n.plu.* [Gk. *diktyon*, net] the principal parenchyma spicules of Dictyonina and of many Lyssacina.

Dictyonina *n.* [Gk. *diktyon*, net] in some classifications, an order of sponges having hexactine spicules fused to form a rigid network; *cf.* Lyssacina.

Dictyoptera *n.* [Gk. *diktyon*, net; *pteron*, wing] an order of insects, including the cockroaches and mantids, having incomplete metamorphosis, eggs laid in oothecae and thickened fore-wings.

Dictyosiphonales *n.* [Gk. *diktyon*, net; *siphōn*, tube] an order of brown algae usually having a much-branched thallus and microscopic gametophyte.

dictyosome *n.* [Gk. *diktyon*, net; *sōma*, body] an element of the Golgi apparatus, *q.v.*; the Golgi apparatus itself in plants.

dictyospore *n.* [Gk. *diktyon*, net; *sporos*, seed] a spore, with transverse and longitudinal septa, of reticular appearance; *alt.* muriform spore.

dictyostele *n.* [Gk. *diktyon*, net; *stēlē*, post] a network formed by meristeles; a stele having large overlapping leaf gaps which dissect the vascular system into strands, each having phloem surrounding xylem.

Dictyotales *n.* [*Dictyota*, generic name] an order of brown algae having an erect thallus, isomorphic alternation of generations, and oogamous sexual reproduction.

dictyotene *n.* [Gk. *diktyon*, net; *tainia*, band] *appl.* stage of meiosis in which diplotene is prolonged, as in oocytes during yolk formation.

dictyotic *a.* [Gk. *diktyon*, net] *appl.* moment: lorication moment, *q.v.*; *appl.* stage in cell growth where chromosomes are lost to view in nuclear reticulum.

dicyclic *a.* [Gk. *di-*, two; *kyklos*, circle] having a row of perradial infrabasals, *appl.* theca of Crinoidea; with 2 whorls; biennial, *appl.* herbs.

Dicyemida(e) *n.* [*Dicyema*, generic name] a class of mesozoans having a body that is not annulated, *alt.* Rhombozoa; an order of this class having an entirely ciliated body.

dicystic *a.* [Gk. *di-*, two; *kystis*, bag] with 2 encysted stages.

didactyl(e) *a.* [Gk. *di-*, two; *daktylos*, digit] having 2 fingers, toes, or claws.

didelphic *a.* [Gk. *di-*, double; *delphys*, womb] having 2 uteri, as marsupials; amphidelphic, *q.v.*

Didesmida *n.* [Gk. *di-*, two; *desmos*, bond] a subclass of tapeworms having from 2 to 6 suckers on the scolex.

didymospore *n.* [Gk. *didymos*, twin; *sporos*, seed] a 2-celled spore.

didymous *a.* [Gk. *didymos*, twin] growing in pairs.

didynamous *a.* [Gk. *di-*, two; *dynamis*, power] with 4 stamens, 2 long, 2 short.

diecdysis *n.* [Gk. *dia*, through; *ekdysai*, to strip] in the moulting cycle of arthropods, a short period between pro- and metecdysis.

diecious dioecious, *q.v.*

diel *a.* [L. *dies*, day] during or *pert.* 24 hours; at 24-hour intervals, *appl.* life rhythms; *cf.* crepuscular, diurnal, nocturnal.

diencephalon *n.* [Gk. *dia*, between; *engkephalos*, brain] part of the fore-brain, comprising thalamencephalon, pars mamillaris hypothalami, and posterior part of 3rd ventricle; *alt.* between-brain, 'tween-brain, inter-brain.

diestrum dioestrus, *q.v.*

differentiation *n.* [L. *differre*, to differ] modification in structure and function of the parts of an organism, owing to increase in specialization.

diffluence *n.* [L. *dis*, away; *fluere*, to flow] disintegration by vacuolization.

diffraction colours colours produced not by pigments but by unevenness on the surface of the organism resulting in the diffraction of light reflected from it.

diffusate n. [L. *diffusus*, poured forth] dialysate, *q.v.*

diffuse *a.* [L. *diffundere*, to pour] widely spread; not localized; not sharply defined at margin; *appl.* placenta with villi on all parts except poles.

diffuse-porous *appl.* wood in which vessels of approximately the same diameter tend to be evenly distributed in a growth ring; *cf.* ring-porous.

diffusion *n.* [L. *diffundere*, to pour] the passage of molecules from a region where they are more concentrated to where they are less concentrated.

diffusion pressure deficit (DPD) suction pressure, *q.v.*

digametic *a.* [Gk. *dis*, twice; *gametēs*, spouse] having 2 types of gametes, one producing males, the other females; *alt.* heterogametic; *n.* digamety.

digastric *a.* [Gk. *di-*, two; *gastēr*, belly] two-bellied, *appl.* muscles fleshy at ends, tendinous in middle; *appl.* one of the suprahyoid muscles; *appl.* a branch of the facial nerve; *appl.* a lobule of cere-

bellum; *appl.* a fossa of mandible and of temporal bone; *alt.* biventral.

Digenea *n.* [Gk. *di-*, two; *genea*, descent] a subclass or order of flukes usually having suckers but no hooks, and a complex life cycle with larval stages in one or more intermediate hosts; *alt.* Malacocotylea.

digenesis *n.* [Gk. *dis*, twice; *genesis*, descent] alternation of asexual and sexual generations.

digenetic *a.* [Gk. *dis*, twice; *genesis*, descent] *pert.* digenesis; requiring an alternation of hosts, *appl.* certain parasites.

digenic *a.* [Gk. *dis*, twice; *genos*, descent] *pert.* or controlled by 2 genes.

digenoporous *a.* [Gk. *dis*, twice; *genos*, birth; *poros*, pore] with 2 genital pores, *appl.* many Turbellaria.

digeny *n.* [Gk. *dis*, double; *geneē*, descent] sexual reproduction.

digestion *n.* [L. *digestio*, digestion] the process by which nutrient materials are rendered soluble and absorbable by action of various juices containing enzymes.

digestive *a.* [L. *digestio*, digestion] *pert.* digestion, or having power of aiding in digestion.

digit *n.* [L. *digitus*, finger] terminal division of limb in any vertebrate above fishes; toe or finger; distal part of chelae and chelicerae; *alt.* digitus.

digital *a.* [L. *digitus*, finger] *pert.* finger or digit; *appl.* structures resembling a digit. *n.* Distal joint of spider's pedipalp.

digitaliform *a.* [L. *digitus*, finger; *forma*, shape] finger-shaped, *appl.* corollae which are like the finger of a glove.

digitalin a saponin glycoside obtained from the leaves of foxglove, *Digitalis purpurea*, used in treating heart disease.

digitate *a.* [L. *digitus*, finger] having parts arranged like the fingers in a hand; with fingers.

digitiform *a.* [L. *digitus*, finger; *forma*, shape] finger-shaped.

digitigrade *a.* [L. *digitus*, finger; *gradus*, step] walking with only the digits touching the ground.

digitinervate *a.* [L. *digitus*, finger; *nervus*, sinew] having veins radiating out from base like fingers of a hand, with usually 5 or 7 veins, *appl.* leaves.

digitipartite *a.* [L. *digitus*, finger; *partire*, to divide] having leaves divided up in a hand-like pattern.

digitipinnate *a.* [L. *digitus*, finger; *pinna*, feather] having digitate leaves of which the leaflets are pinnate.

digitonin a saponin glycoside in foxglove, *Digitalis purpurea*.

digitoxin a glycoside isolated from the leaves of foxglove, *Digitalis purpurea*.

digitule *n.* [L. *digitulus*, little finger] any small finger-like process; small process on insect tarsi.

digitus digit, *q.v.*

diglyphic *a.* [Gk. *dis*, twice; *glyphein*, to engrave] having 2 siphonoglyphs.

digoneutic *a.* [Gk. *dis*, twice; *goneuein*, to produce] breeding twice a year.

digonic *a.* [Gk. *dis*, twice; *gonē*, seed] producing male and female gametes in separate gones in the same individual; *cf.* amphigonic, syngonic.

digonoporous *a.* [Gk. *dis*, twice; *gonē*, seed;

poros, pore] with 2 distinct genital apertures, male and female.

digynous *a.* [Gk. *di-,* two; *gynē,* woman] having 2 carpels.

dihaploid *n.* [Gk. *dis,* twice; *haploos,* simple; *eidos,* form] an organism arising from a tetraploid but only containing half a normal tetraploid chromosome complement.

diheliotropism diaheliotropism, *q.v.*

diheterozygote *n.* [Gk. *dis,* twice; *heteros,* other; *zygōtos,* yoked together] a dihybrid, *q.v.;* a diheterozygous.

dihybrid *n.* [Gk. *dis,* twice; L. *hibrida,* mixed offspring] a cross whose parents differ in 2 distinct characters; an organism heterozygous regarding 2 pairs of alleles; *alt.* diheterozygote.

dihydrotachysterol vitamin D$_4$, irradiation product of dihydro derivative of ergosterol, which counteracts impaired parathyroid function.

dikaryo- *alt.* dicaryo-.

dikaryon *n.* [Gk. *dis,* twice; *karyon,* nucleus] a pair of nuclei which are situated close to one another and divide at the same time as in ascogenous hyphae; the hypha containing such a pair of nuclei.

dikaryophase the dikaryotic phase in the life history of a fungus.

dikaryosis *n.* [Gk. *dis,* twice; *karyon,* nucleus] heterokaryosis when there are 2 nuclei in each cell.

dikaryospore *n.* [Gk. *dis,* twice; *karyon,* nucleus; *sporos,* seed] a spore with 2 nuclei.

dikaryotic *pert.* dikaryon; diploid, *q.v.*

dikont *n.* [Gk. *dis,* twice; *kontos,* punting pole] biflagellate, *q.v.; alt.* dicont.

dilambdodont *a.* [Gk. *dis,* twice; *lambda, λ; odous,* tooth] *appl.* insectivores having molar teeth with W-shaped ridges; *cf.* zalambdodont.

dilatator dilator, *q.v.*

dilated *a.* [L. *dilatare,* to enlarge, expand] expanded, or flattened; *appl.* parts of insects, plants, etc., with a wide margin.

dilator *n.* [L. *dilatare,* to expand] name *appl.* any muscle that expands or dilates an organ; *alt.* dilatator.

dilemma *n.* [Gk. *dis,* double; *lemma,* assumption] distinction of alternative stimuli, retarding the reaction.

Dilleniales *n.* [*Dillenia,* generic name] an order of dicots (Dilleniidae) having seeds with a fleshy aril, much endosperm, and a small embryo.

Dilleniidae *n.* [*Dillenia,* generic name] a subclass of dicots having syncarpous ovaries with many ovules, usually simple leaves, and a tendency to sympetally.

dilophous *a.* [Gk. *di-,* two; *lophos,* crest] *appl.* a tetractinal spicule with 2 rays forked like a crest.

diluvial *a.* [L. *diluvium,* deluge] *pert.* a flood, especially the deluge of the Bible; produced by a flood, *appl.* soil deposits; *pert.* the geological era in which man has existed, including the present.

dimastigote *a.* [Gk. *dis,* twice; *mastix,* whip] biflagellate, *q.v.*

dimegaly *n.* [Gk. *dis,* twice; *megalos,* great] condition of having 2 sizes, *appl.* spermatozoa, ova.

dimeric *a.* [Gk. *dis,* twice; *meros,* part] having 2 parts, bilaterally symmetrical.

dimerous *a.* [Gk. *dis,* twice; *meros,* part] in 2 parts; having each whorl of 2 parts; with a 2-jointed tarsus.

dimidiate *a.* [L. *dimidius,* half] having only one-half developed; having capsule split on one side.

dimitic *a.* [Gk. *dis,* twice; *mitos,* thread] having both supporting and generative hyphae; *cf.* monomitic, trimitic.

dimixis *n.* [Gk. *dis,* twice; *mixis,* mingling] fusion of 2 kinds of nuclei in heterothallism.

dimorphic *a.* [Gk. *dis,* twice; *morphē,* shape] having or *pert.,* 2 different forms.

dimorphism *n.* [Gk. *dis,* twice; *morphē,* shape] condition of having stamens of 2 different lengths, of having 2 different kinds of leaves, flowers, etc.; state of having 2 different forms according to sex, or of one sex, 2 different kinds of zooids, or of offspring; of broods which, owing to differing conditions, differ in size or colouring; state of having reciprocally transformable unicellular and filamentous types, as in some bacteria and fungi; of polymorphic species, having 2 different forms.

dimyaric *a.* [Gk. *dis,* twice; *mys,* muscle] having both anterior and posterior adductor muscles, *appl.* some molluscs; *alt.* dimyarian.

Dinamoebidiales *n.* [*Dinamoebidium,* generic name] Rhizodiniales, *q.v.*

dinergate *n.* [Gk. *deinos,* powerful; *ergatēs,* worker] a soldier ant.

dineuric *a.* [Gk. *dis,* twice; *neuron,* nerve] having 2 axons; *alt.* diaxonic.

dineuronic *a.* [Gk. *dis,* twice; *neuron,* nerve] with double innervation; *appl.* chromatophores with concentrating and dispersing nerve fibres.

Dinocapsales *n.* [Gk. *dinos,* rotation; L. *capsa,* box] an order of Dinophyceae having palmelloid forms.

Dinocerata *n.* [Gk. *deinos,* terrible; *keras,* horn] an order of Palaeocene to Eocene placental mammals, commonly called uintatheres, which were heavily built and had horn-like protuberances.

Dinococcales *n.* [Gk. *dinos,* rotation; *kokkos,* berry] an order of Dinophyceae having coccoid forms; *alt.* Phytodiniales.

Dinoflagellata, dinoflagellates *n., n.plu.* [Gk. *dinos,* rotation; L. *flagellum,* whip] an order of Phytomastigma having 2 flagella, one pointing forwards and the other forming a girdle around the body, some naked, other having a covering of cellulose plates; many are autotrophic and found in phytoplankton and are also considered as algae in the Dinophyceae, others are heterotrophic, or both.

Dinoflagellatae, dinoflagellates *n., n.plu.* [Gk. *dinos,* rotation; L. *flagellum,* whip] Pyrrophyceae, *q.v.;* a group of Dinophyceae including motile unicellular forms, placed by zoologists in the order Dinoflagellata.

Dinokontae *n.* [Gk. *dinos,* rotation; *kontos,* punting pole] formerly a subclass of Dinophyceae, together with the Desmokontae, now raised to class status as the Dinophyceae,.*q.v.*

dinomic *a.* [Gk. *dis,* twice; *nomos,* district] *appl.* an organism restricted to 2 of the biogeographical divisions of the globe.

Dinophycales *n.* [Gk. *dinos,* rotation; *phykos,* seaweed] Peridiniales, *q.v.*

Dinophyceae *n.* [Gk. *dinos,* rotation; *phykos,* seaweed] a class of Pyrrophyta having the cell wall divided into a number of plates, 2 flagella arising laterally, and discoid chloroplasts without pyrenoids.

Dinophysidales n. [*Dinophysis*, generic name] an order of Dinophyceae having armoured forms with 2 valves.

Dinornithiformes n. [*Dinornis*, generic name of moa; L. *forma*, shape] an order of flightless birds of the subclass Neornithes from New Zealand, including the moas and kiwis; alt. Dinornithes.

dinosaur n. [Gk. *deinos*, terrible; *sauros*, lizard] a member of either of 2 orders of reptiles, the Saurischia, the lizard-hipped dinosaurs, or the Ornithischia, the bird-hipped dinosaurs.

Dinotrichales n. [*Dinothrix*, generic name] an order of Dinophyceae having filamentous forms.

dinoxanthin n. [Gk. *dinos*, rotation; *xanthos*, yellow] a xanthophyll carotinoid pigment found in algae of the class Pyrrophyceae.

dinucleotide n. [Gk. *dis*, twice; L. *nucleus*, kernel] two nucleotides linked together by a 3′,5′-phosphodiester bond.

diocoel n. [Gk. *dia*, between; *koilos*, hollow] cavity of the diencephalon particularly in the embryo.

Dioctophymatida n. [*Dioctophyma*, generic name] an order of large Aphasmidia which are parasitic in birds and mammals with fish as the intermediate host.

dioecious a. [Gk. *dis*, twice; *oikos*, house] having sexes separate; having male and female flowers on different individuals; alt. unisexual, gonochoristic, dioicious, diecious; n. dioecism, alt. gonochorism; cf. monoecious.

dioestrus n. [Gk. *dia*, between; *oistros*, gadfly] the quiescent period between heat periods in polyoestrous animals; alt. diostrum, diestrum; cf. anoestrus.

dioicous dioecious, q.v.

dionychous a. [Gk. *di-*, two; *onyx*, nail] having 2 claws, as on tarsi of certain spiders.

dioptrate a. [Gk. *dis*, twice; *ōps*, eye] having eyes or ocelli separated by a narrow line.

dioptric a. [Gk. *dia*, through; *optomai*, to see] pert. transmission and refraction of light, appl. structures in eye, as cornea, lens, aqueous and vitreous humours.

diorchic a. [Gk. *dis*, twice; *orchis*, testis] having 2 testes.

dipeptidases a group of enzymes that split dipeptides into free amino acids; EC sub-subgroup 3.4.3; r.n. dipeptide hydrolases.

dipeptide n. [Gk. *dis*, twice; *peptein*, to digest] a compound made of 2 amino acids linked by a peptide bond.

dipetalous a. [Gk. *dis*, twice; *petalon*, leaf] having 2 petals.

diphasic a. [Gk. *dis*, twice; *phainein*, to appear] appl. extended life cycle of some protozoa, including the active stage, cf. monophasic; periodically changing 2 states or appearances, as of winter and summer pelage or plumage.

diphosphopyridine nucleotide DPN, q.v.

diphycercal a. [Gk. *diphyēs*, twofold; *kerkos*, tail] with a caudal fin in which the vertebral column runs straight to tip, thereby dividing the fin symmetrically; alt. protocercal.

diphygenetic a. [Gk. *diphyes*, twofold; *genetēs*, begotten] producing embryos of 2 different types, as Dicyemidae.

diphygenic a. [Gk. *diphyēs*, twofold; *genos*, descent] with 2 types of development.

diphyletic a. [Gk. *dis*, twice; *phylon*, race] pert. or having origin in 2 lines of descent.

Diphyllidea n. [Gk. *di-*, two; *phyllidion*, little leaf] an order of segmented tapeworms with 2 suckers (bothria) and a spiny stalk to the scolex.

diphyllous a. [Gk. *dis*, twice; *phyllon*, leaf] two-leaved; alt. bifoliate.

diphyodont a. [Gk. *diphyēs*, twofold; *odous*, tooth] with deciduous and permanent sets of teeth, i.e. 2 sets of teeth.

diplanetary, diplanetic a. [Gk. *dis*, twice; *planētikos*, wandering] with 2 distinct types of zoospores; cf. monoplanetary.

diplanetism n. [Gk. *dis*, twice; *planētikos*, wandering] condition of having 2 periods of motility in one life history, as of zoospores in some fungi; cf. monoplanetism.

diplarthrous a. [Gk. *diploos*, double; *arthron*, joint] with tarsal or carpal bones of one row articulating with 2 bones in the other.

dipleurula n. [Gk. *dis*, twice; *pleuron*, side] a bilaterally symmetrical larva of echinoderms; alt. echinopaedium.

diplo- cf. haplo-.

diplobiont n. [Gk. *diploos*, double; *biōnai*, to live] an organism characterized by at least 2 kinds of individual in its life cycle, such as sexual and asexual; a. diplobiontic; cf. haplobiont.

diplobivalent n. [Gk. *diploos*, double; L. *bis*, twice; *valere*, to be strong] a bivalent comprising 2 diplochromosomes and hence 8 chromatids.

diploblastic a. [Gk. *diploos*, double; *blastos*, bud] having 2 distinct germ layers, ectoderm and endoderm; alt. bilaminar.

diplocardiac a. [Gk. *diploos*, double; *kardia*, heart] with the 2 sides of the heart quite distinct.

diplocaryon diplokaryon, q.v.

diplocaulescent a. [Gk. *diploos*, double; *kaulos*, stem] with secondary stems or branches.

diplochlamydeous a. [Gk. *diploos*, double; *chlamys*, cloak] dichlamydeous, q.v.

diplochlamydeous chimaera a periclinal chimaera in which only the innermost of the 3 tissue layers is genetically different from the others.

diplochromosome n. [Gk. *diploos*, double; *chrōma*, colour; *sōma*, body] anomalous chromosome having 4 chromatids instead of 2, attached to centromere.

diplocyte n. [Gk. *diploos*, double; *kytos*, hollow] a cell having conjugate nuclei.

diplodal a. [Gk. *diploos*, double; *hodos*, way] having both prosodus and aphodus, appl. Porifera.

diploe n. [Gk. *diploē*, double] the cancellous tissue between outer and inner lamellae of certain skull bones; tail of scorpion; mesophyll, q.v.; a. diploic.

diplogangliate a. [Gk. *diploos*, double; *gangglion*, ganglion] with ganglia in pairs.

diplogenesis n. [Gk. *diploos*, double; *genesis*, descent] supposed change in germ plasm that accompanies 'use and disuse' changes occurring in body tissues; development of 2 parts instead of usual single part.

diplohaplont n. [Gk. *diploos*, double; *haploos*, simple] an organism with alternation of diploid and haploid generations.

diploic a. [Gk. *diploos*, double] occupying channels in cancellous tissue of bones; pert. diploe.

diploid a. [Gk. *diploos*, double; *eidos*, form] having 2 sets of chromosomes, twice the haploid number; *appl.* typical zygotic number of chromosomes; *alt.* dikaryotic. *n.* A diploid organism. *Cf.* haploid, duplex.

diploid arrhenotoky the production of diploid males from unfertilized eggs which is known only in the scale insect, *Lecanium putnami.*

diploidization doubling of number of chromosomes in haploid cells or hyphae.

diplokaryon *n.* [Gk. *diploos*, double; *karyon*, nut] a nucleus with 2 diploid sets of chromosomes; *alt.* diplocaryon; *cf.* amphikaryon.

Diplomeri *n.* [Gk. *diploos*, double; *meros*, part] an order of terrestrial fossil Pennsylvanian amphibians.

Diplomonadina *n.* [Gk. *diploos*, double; *monas*, unit] an order of Zoomastigina often included in the Polymastigina.

diplomycelium *n.* [Gk. *diploos*, double; *mykēs*, fungus] diploid or dikaryotic mycelium.

diplonema *n.* [Gk. *diploos*, double; *nēma*, thread] double thread of diplotene stage in meiosis.

diplonephridia *n.plu.* [Gk. *diploos*, double; *nephros*, kidney; *idion*, dim.] nephridia derived partly from ectoderm, partly from mesoderm.

diploneural *a.* [Gk. *diploos*, double; *neuron*, nerve] supplied with 2 nerves.

diplont *n.* [Gk. *diploos*, double; *on*, being] an organism having diploid somatic nuclei; *cf.* haplont.

diploperistomous *a.* [Gk. *diploos*, double; *peri*, around; *stoma*, mouth] having a double projection or peristome.

diplophase *n.* [Gk. *diploos*, double; *phasis*, aspect] stage in life history of an organism when nuclei are diploid, *alt.* zygophase; sporophyte phase; diplotene stage in meiosis; *cf.* haplophase.

diplophyll *n.* [Gk. *diploos*, double; *phyllon*, leaf] a leaf having palisade tissue on upper and lower side with intermedial spongy parenchyma.

diplophyte *n.* [Gk. *diploos*, double; *phyton*, plant] a diploid plant or sporophyte; *cf.* haplophyte, gametophyte.

diploplacula *n.* [Gk. *diploos*, double; *plakoeis*, flat cake] a flattened blastula consisting of 2 layers of cells.

Diplopoda *n.* [Gk. *diploos*, double; *pous*, foot] a group of arthropods, commonly called millepedes, having numerous and similar body segments each actually made of 2 segments and so bearing 2 pairs of appendages; the group may be considered an order or subclass of the Myriapoda, or raised to class status.

diplopore *n.* [Gk. *diploos*, double; *poros*, passage] respiratory organ in Cystoidea.

Diploporita *n.* [Gk. *diploos*, double; *poros*, passage] an order of echinoderms of the subclass Hydrophoridea having paired thecal pores.

diploptile *a.* [Gk. *diploos*, double; *ptilon*, feather] double neossoptile, without rachis, formed by precocious development of the barbs of the teleoptile.

Diplorhina *n.* [Gk. *diploos*, double; *rhis*, nose] a subclass of Agnatha having 2 nostrils.

diplosis *n.* [Gk. *diploos*, double] doubling of the chromosome number, in syngamy.

diplosome *n.* [Gk. *diploos*, double; *sōma*, body] a double centrosome lying outside the nuclear membrane; a paired heterochromosome.

diplosomite *n.* [Gk. *diploos*, double; *sōma*, body] body segment consisting of 2 annular parts, prozonite and metazonite, in Diplopoda.

diplosphene *n.* [Gk. *diploos*, double; *sphēn*, wedge] wedge-shaped process on neural arch of certain fossil reptiles.

diplospondyly *n.* [Gk. *diploos*, double; *sphondylos*, vertebra] the condition of having 2 centra to each myotome, or with one centrum and well-developed intercentrum; *a.* diplospondylic, diplospondylous; *cf.* polyspondyly.

diplostemonous *a.* [Gk. *diploos*, double; *stēmōn*, warp] with 2 whorls of stamens in regular alternation with perianth leaves; with stamens double the number of petals.

diplostichous *a.* [Gk. *diploos*, double; *stichos*, row] arranged in 2 rows or series.

Diplostraca *n.* [Gk. *diploos*, double; *ostrakon*, shell] an order of Branchiopoda including the Conchostraca and Cladocera.

diplostromatic *a.* [Gk. *diploos*, double; *strōma*, bedding] *appl.* fungi having both entostroma and ectostroma; *cf.* haplostromatic.

diplotegia *n.* [Gk. *diploos*, double; *tegos*, roof] an inferior fruit with dry dehiscent pericarp.

diplotene *a.* [Gk. *diploos*, double; *tainia*, band] *appl.* stage in meiosis at which bivalent chromosomes split longitudinally.

diploxylic *a.* [Gk. *diploos*, double; *xylon*, wood] *appl.* leaf trace bundles with inner and outer strands of xylem, in certain extinct plants.

diplozoic *a.* [Gk. *diploos*, double; *zōon*, animal] bilaterally symmetrical.

Diplura *n.* [Gk. *diploos*, double; *oura*, tail] an order of wingless insects with a pair of cerci but no median process on the last segment, sometimes called 2-pronged bristletails.

dipnoan *a.* [Gk. *dis*, twice; *pnein*, to breathe] breathing by gills and lungs.

Dipnoi, dipnoans *n.*, *n.plu.* [Gk. *dis*, twice; *pnein*, to breathe] a group of bony fishes, sometimes placed in the Sarcopterygii, commonly called lungfish, possessing lungs and crushing teeth, the modern lungfish being considerably modified with a reduced skeleton, and fossil lungfish existing from Devonian times.

dipolar bipolar, *q.v.*

diporpa *n.* [Gk. *dis*, double; *porpē*, buckle] embryo of the trematode *Diplozoon*, which permanently unites with another.

diprotodont *a.* [Gk. *dis*, twice; *prōtos*, first; *odous*, tooth] having 2 anterior incisors large and prominent, the rest of incisors and canines being smaller or absent; *cf.* polyprotodont.

Dipsacales *n.* [*Dipsacus*, generic name] an order of dicots (Asteridae) formerly with the Gentianales making up the Rubiales, and having actinomorphic or weakly zygomorphic flowers, never with the corolla contorted.

Diptera *n.* [Gk. *dis*, twice; *pteron*, wing] an order of insects, the true flies, with complete metamorphosis, the adults having only a single pair of membranous wings, the hind-wings being modified as halteres.

dipterocecidium *n.* [Gk. *dis*, twice; *pteron*, wing; *kēkis*, gall nut; *idion*, dim.] gall caused by a dipterous insect.

dipterous *a.* [Gk. *dis*, twice; *pteron*, wing] with 2 wings or wing-like expansions; *pert.* Diptera.

directive bodies polar bodies, *q.v.*

directive mesenteries in Zoantharia, the dorsal and ventral pairs of mesenteries; *alt.* directives.

directive sphere centrosphere, *q.v.*

direct nuclear division amitosis, *q.v.*

dirhinic *a.* [Gk. *di-*, two; *rhines*, nostrils] having 2 nostrils; *pert.* both nostrils; *cf.* dichorhinic.

disaccharides *n.plu.* [Gk. *dis*, twice; L. *saccharum*, sugar] a group of carbohydrates which are the condensation products of 2 molecules of monosaccharide with the elimination of a molecule of water, e.g. sucrose, lactose, maltose; *alt.* bioses.

disarticulate *v.* [L. *dis*, asunder; *articulatus*, jointed] to separate at a joint. *a.* Separated at a joint or joints.

disassortive mating mating of organisms whose phenotypes are not alike.

disc *n.* [L. *discus*, disc] any flattened portion like a disc in shape; middle part of capitulum in Compositae; adhesive tip of tendril; base of seaweed thallus; circumoral area in many animals; circular areas at opposite poles of many animals; any modification of thalamus; area marking entrance of optic nerve into eye; cup-shaped tactile structures in skin; mass of cells of membrana granulosa which projects into cavity of egg follicle; anisotropic and isotropic parts of contractile fibrils of muscular tissue; *alt.* disk, discus.

discal *a.* [L. *discus*, disc] *pert.* any disc-like structure; *appl.* cross-vein between 3rd and 4th longitudinal veins of insect wing, *alt.* discocellular vein. *n.* A large cell at base of wing of Lepidoptera completely enclosed by wing nervures, also in some Diptera.

disc florets inner florets borne on abbreviated and reduced peduncle in many inflorescences; *cf.* ray florets.

disciflorous *a.* [L. *discus*, disc; *flos*, flower] with flowers in which receptacle is large and disc-like.

disciform *a.* [L. *discus*, disc; *forma*, shape] discoid, *q.v.*

disclimax *n.* [Gk. *dis*, double; *klimax*, ladder] a subclimax stage in plant succession replacing or modifying true climax, usually due to animal and human agency, e.g. cultivated crops; *alt.* disturbance climax, plagioclimax.

discoblastula *n.* [Gk. *diskos*, disc; *blastos*, bud] a blastula formed from a meroblastic egg with disc-like blastoderm; *a.* discoblastic.

discocarp *n.* [Gk. *diskos*, disc; *karpos*, fruit] special enlargement of thalamus below calyx; apothecium, *q.v.*

discocellular vein discal vein.

Discocephali *n.* [Gk. *diskos*, disc; *kephalē*, head] Echeneiformes, *q.v.*

discoctasters *n.plu.* [Gk. *diskos*, disc; *okto*, eight; *astēr*, star] sponge spicules with 8 rays terminating in discs, each disc corresponding in position to corners of a cube, being modified hexactines; *alt.* discoooctasters.

discodactylous *a.* [Gk. *diskos*, disc; *daktylos*, finger] with sucker at end of digit.

discohexactine *n.* [Gk. *diskos*, disc; *hex*, six; *aktis*, ray] a sponge spicule with 6 equal rays meeting at right angles.

discohexaster *n.* [G. *diskos*, disc; *hex*, six; *aster*, star] a hexactine with rays ending in discs.

discoid *a.* [Gk. *diskos*, disc; *eidos*, form] flat and circular; disc-shaped; *alt.* disciform, discous.

discoidal *a.* [Gk. *diskos*, disc; *eidos*, form] disc-like; *appl.* segmentation in which blastoderm forms a one-layered disc or cap which spreads over yolk.

Discolichenes *n.* [Gk. *diskos*, disc; *leichēn*, lichen] a group of ascolichens in which the fungus is a discomycete; *alt.* Gymnocarpeae.

Discomedusae *n.* [Gk. *diskos*, disc; *Medousa*, Medusa] a subclass of scyphozoans which are sessile only during the larval stage and have a free-swimming medusa as the main stage.

Discomycetes, Discomycetidae *n.* [Gk. *diskos*, disc; *mykēs*, fungus] a group of Euascomycetes, commonly called cup fungi, in which the ascocarp is an apothecium which is open, cup-shaped, disc-like, or subterranean.

discontinuity *n.* [O.F. *discontinuer*; from L. *dis-*, asunder; *continuare*, to continue] occurrence in 2 or more separate areas or geographical regions, *alt.* disjunction; *appl.* layer: thermocline, *q.v.*

discontinuous variation variation between individuals of a population in which differences are marked and do not grade into each other, brought about by the action of a few major genes and their mutations; *alt.* qualitative inheritance; *cf.* continuous variation.

disconula *n.* [Gk. *diskos*, disc] eight-rayed stage in larval development of certain Coelenterata.

discooctaster discooctaster, *q.v.*

discoplacenta *n.* [L. *discus*, disc; *placenta*, placenta] a placenta with villi on a circular cake-like disc.

discoplasm *n.* [Gk. *diskos*, disc; *plasma*, form] colourless framework or stroma of a red blood corpuscle.

discorhabd *n.* [Gk. *diskos*, disc; *rhabdos*, rod] a linear sponge spicule with disc-like outgrowths or whorls of spines.

discous discoid, *q.v.*

discus *n.* [L. *discus*, Gk. *diskos*, disc] disc, *q.v.*; a flat and circular structure or part.

discus proligerus granular zone in a Graafian follicle, the mass of cells of the membrana granulosa, in which the ovum is embedded; *alt.* zona granulosa, cumulus oophorus.

disjunct *a.* [L. *disiunctus*, separated] with body regions separated by deep constrictions.

disjunction *n.* [L. *disiunctus*, separated] divergence of paired chromosomes at anaphase; geographical distribution in discontinuous areas, *alt.* discontinuity.

disjunctive symbiosis a mutually helpful condition of symbiosis although there is no direct connection between the partners; *cf.* conjunctive symbiosis.

disjunctor *n.* [L. *disiunctus*, separated] weak connective structure, or an intercalary cell, and zone of separation between successive conidia; *alt.* 'bridge', connective.

disk disc, *q.v.*

disomic *a.* [Gk. *dis*, twice; *sōma*, body] *pert.* or having 2 homologous chromosomes or genes; *n.* disomy.

disoperation *n.* [L. *dis-*, asunder; *operatio*, work] coactions resulting in disadvantage to individual or to group; indirectly harmful influence of organisms upon each other.

dispermous *a.* [Gk. *dis*, twice; *sperma*, seed] having 2 seeds.

dispermy *n.* [Gk. *dis*, twice; *sperma*, seed] the en-

dispersal

trance of 2 spermatozoa into an ovum; *a.* disperm-ic.

dispersal *n.* [L. *dispergere*, to disperse] the actual scattering or distributing of organisms on earth's surface; transport of diaspores.

disphotic dysphotic, *q.v.*

dispireme *n.* [Gk. *dis*, twice; *speirēma*, skein] the stage of telophase in which each daughter nucleus has given rise to a spireme.

displacement *n.* [O.F. *desplacier*, to displace] an abnormal position of any part of a plant due to its shifting from its normal place of insertion.

displacement activity a form of behaviour when the normal behavioural response to a stimulus is prevented, and a new response is produced.

display a series of movements, postures, or sounds performed by animals, which cause specific responses in other animals, usually members of the same species, often used in courtship or territorial behaviour; *cf.* distraction display.

dispore *n.* [Gk. *dis*, twice; *sporos*, seed] one of a pair of basidial spores.

disporocystid *a.* [Gk. *dis*, twice; *sporos*, seed; *kystis*, bladder] *appl.* oocyst of Sporozoa when 2 sporocysts are present.

disporous *a.* [Gk. *dis*, twice; *sporos*, seed] with 2 spores; *alt.* bisporic.

disruptive coloration colour patterns which obscure the outline of certain animals and so act for defence.

dissected *a.* [L. *dissecare*, to cut open] having lamina cut into lobes, incisions reaching nearly to midrib; with parts displayed.

disseminule *n.* [L. *disseminare*, to scatter seed] diaspore, *q.v.*

dissepiment *n.* [L. *dissaepire*, to separate] the partition found in some compound ovaries; in corals, one of oblique calcareous partitions stretching from septum to septum and closing interseptal loculi below; trama, *q.v.*; *alt.* sepiment.

dissilient *a.* [L. *dissilire*, to burst asunder] springing open; *appl.* capsules of various plants which dehisce explosively.

dissimilation *n.* [L. *dissimilis*, different] catabolism, *q.v.*

dissoconch *n.* [Gk. *dissos*, double; *kongchē*, shell] the shell of a veliger larva.

dissogeny, dissogony *n.* [Gk. *dissos*, double; *genos*, descent] condition of having 2 sexually mature periods in the same animal—one in larva, one in adult.

dissophyte *n.* [Gk. *dissos*, twofold; *phyton*, plant] a plant with xerophytic leaves and stems, and mesophytic roots.

distad *adv.* [L. *distare*, to stand apart; *ad*, to] towards or at a position away from centre or from point of attachment; in a distal direction; *cf.* proximad.

distal *a.* [L. *distare*, to stand apart] standing far apart, distant, *appl.* bristles, etc.; *pert.* end of any structure furthest from middle line of organism or from point of attachment; *cf.* proximal.

distalia *n.plu.* [L. *distare*, to stand apart] the distal or 3rd row of carpal or of tarsal bones.

distance receptor a sense organ which reacts to stimuli emanating from distant objects, an olfactory, visual, or auditory receptor; *alt.* disticeptor, distoceptor, teleceptor, telereceptor, teloreceptor.

distemonous *a.* [Gk. *dis*, twice; *stēmōn*, spun thread] having 2 stamens; *alt.* diandrous.

disticeptor distance receptor, *q.v.*; *cf.* proximoceptor.

distichalia *n.plu.* [Gk. *distichos*, with two rows] dicostalia, *q.v.*

distichate, distichous *a.* [Gk. *distichos*, with two rows] two-ranked; *appl.* alternate leaves, so arranged that 1st is directly below 3rd; *n.* distichy.

distipharynx *n.* [L. *distans*, standing apart; Gk. *pharyngx*, gullet] a short tube formed by union of epi- and hypopharynx in some insects.

distiproboscis *n.* [L. *distans*, standing apart; Gk. *proboskis*, trunk] distal portion of insect proboscis, part of ligula.

dististyle *n.* [L. *distans*, standing apart; Gk. *stylos*, pillar] distal part or style borne on basistyle, *q.v.*, of gonostyle in mosquitoes.

distoceptor *n.* [L. *distare*, to stand apart; *recipere*, to receive] distance receptor, *q.v.*

distractile *a.* [L. *distractus*, pulled asunder] widely separate, *appl.* usually to long-stalked anthers.

distraction display a kind of behaviour found in female birds which distracts an enemy from their eggs or chicks, and often takes the form of injury feigning; *alt.* diversion behaviour.

distribution *n.* [L. *distributus*, divided] range of an organism or group in biogeographical divisions of globe.

disturbance climax disclimax, *q.v.*

disymmetrical *a.* [Gk. *dis*, twice; *syn*, with; *metron*, measure] biradial, *q.v.*

dithecal *a.* [Gk. *dis*, twice; *thēkē*, box] two-celled, as anthers.

ditokous *a.* [Gk. *dis*, twice; *tokos*, birth] producing 2 at a time, either eggs or young.

ditrematous *a.* [Gk. *dis*, twice; *trēma*, opening] with separate genital openings; with anus and genital openings separate.

ditrochous *a.* [Gk. *dis*, twice; *trochos*, runner] with a divided trochanter.

ditypism *n.* [Gk. *dis*, twice; *typos*, type] occurrence or possession of 2 types; sex differentiation, represented by + and −, of 2 apparently similar haplonts.

diuresis *n.* [Gk. *dia*, through; *ouron*, urine] increased or excessive secretion of urine.

diuretic *a.* [Gk. *dia*, through; *ouron*, urine] increasing the secretion of urine. *n.* Any agent causing diuresis.

diurnal *a.* [L. *diurnus*, pert. day] opening during the day only; active in the daytime; occurring every day.

diurnal rhythm circadian rhythm, *q.v.*

divaricate *a.* [L. *divaricatus*, stretched apart] widely divergent; bifid; forked.

divaricators *n.plu.* [L. *divaricatus*, stretched apart] muscles stretching from ventral valve to cardinal process, in brachiopods; muscles in avicularia.

divergency *n.* [L. *divergere*, to bend away] the fraction of a stem circumference, usually constant for a species, which separates 2 consecutive leaves in a spiral.

divergent *a.* [L. *divergere*, to bend away] separated from one another; having tips further apart than the bases.

diversion behaviour distraction display, *q.v.*;

behaviour likely to confuse an enemy, e.g. certain cephalopods ejecting a cloud of black ink.

diversity index of a community, the ratio between number of species and number of individuals.

diverticillate biverticillate, *q.v.*

diverticulate *a.* [L. *divertere*, to turn aside] having a diverticulum; having short offshoots approximately at right angles to axis, *appl.* certain hypae; having a projection where attached to sterigma, *appl.* certain spores.

diverticulum *n.* [L. *divertere*, to turn away] a tube or sac, blind at distal end, branching off from a canal or cavity; filament of carpogonium, giving rise to carpospore in red algae.

divided *a.* [L. *dividere*, to divide] with lamina cut by incisions reaching midrib, *appl.* leaves.

division *n.* [L. *divisio*, division] cleavage; fission; in classification of plants, a group of classes, equivalent to a phylum in animals and now often called a phylum.

division centre centriole, *q.v.*

dixenous *a.* [Gk. *di-*, two; *xenos*, host] parasitizing, or able to parasitize, 2 host species.

dizoic *a.* [Gk. *dis*, twice; *zōon*, animal] *pert.* spore containing 2 sporozoites.

dizygotic, dizygous *a.* [Gk. *dis*, twice; *zygōtos*, yoked] originating from 2 fertilized ova, *appl.* twins; *alt.* binovular, fraternal.

DNA deoxyribonucleic acid, *q.v.*

DNA-dependent RNA polymerase transcriptase, *q.v.*

DNA polymerase an enzyme catalysing the formation of DNA, using as a template one strand of DNA; EC 2.7.7.7; *r.n.* DNA nucleotidyltransferase.

DNase deoxyribonuclease, any enzyme hydrolysing DNA; EC 3.1.21.1; *r.n.* deoxyribonuclease I.

DNA virus deoxyvirus, *q.v.*

Dobie's line Z-disc, *q.v.*

Docodonta *n.* [Gk. *dokos*, shaft; *odous*, tooth] an order of Jurassic eotherians having a primitive jaw and wide molars.

docoglossate *a.* [Gk. *dokos*, shaft; *glōssa*, tongue] having an elongated radula with few marginal teeth, as limpets.

dodecagynous *a.* [Gk. *dōdeka*, twelve; *gynē*, woman] having 12 pistils.

dodecamerous *a.* [Gk. *dōdeka*, twelve; *meros*, part] having each whorl composed of 12 parts.

dodecandrous *a.* [Gk. *dōdeka*, twelve; *anēr*, man] having at least 12 stamens.

Dogiel's cells [G. S. *Dogiel*, Russian neurologist] nerve cells within spinal ganglia, with axons branching close to cell bodies.

dolabriform, dolabrate *a.* [L. *dolabra*, mattock; *forma*, shape] axe-shaped.

dolichocephalic *a.* [Gk. *dolichos*, long; *kephalē*, head] long-headed; with cephalic index of under 75; *cf.* brachycephalic.

dolichohieric *a.* [Gk. *dolichos*, long; *hieros*, sacred] with sacral index below 100; *cf.* platyhieric.

dolichoprosopic *a.* [Gk. *dolichos*, long; *prosōpon*, face] long-faced; *alt.* dolichofacial.

dolichostylous *a.* [Gk. *dolichos*, long; *stylos*, pillar] *pert.* long-styled anthers in dimorphic flowers.

dolioform *a.* [L. *dolium*, wine-cask; *forma*, shape] barrel-shaped; *alt.* doliiform.

doliolaria *n.* [L. *dolium*, wine-cask] in Hol-

othuroidea, a barrel-shaped larva which develops from an auricularia.

Doliolida(e) *n.* [L. *dolium*, wine-cask] an order of salps having a tailed larva and strong muscles forming complete rings around the body; *alt.* Cyclomyaria.

dolipore *n.* [L. *dolium*, wine-cask; Gk. *poros*, channel] the pore in the septum (cross wall) of the hyphae of Basidiomycetes, which is a barrel-shaped structure covered by a parenthosome formed from endoplasmic reticulum.

Dollo's law [L. *Dollo*, Belgian palaeontologist] the principle that evolution is not reversible.

domatium *n.* [Gk. *dōmation*, small house] a crevice or hollow in some plants, serving as lodgings for insects or mites.

dome cell the penultimate cell of a crozier, containing 2 nuclei which fuse, being the 1st stage in development of an ascus, *alt.* loop cell; in the development of certain antheridia, a cell cut off at the apex, a derivative of which forms the antheridial cap.

dominant *a.* [L. *dominans*, ruling] *appl.* plants which by their extent determine biotic conditions in a given area; *appl.* species prevalent in a particular community, or at a given period; *appl.* character possessed by one parent which in a hybrid masks the corresponding alternative character derived from the other parent, being the parental allele manifested in the F_1 heterozygote, *cf.* recessive; *appl.* stimulated part of brain when excitation is increased by stimuli usually inducing other reflexes; *appl.* parts of body controlling less active parts; *appl.* an individual which is high ranking in a hierarchy or peck order.

dominator *n.* [L. *dominator*, ruler] a broad band of the spectrum which evokes sensation of luminosity in light-adapted eye; *cf.* modulator.

dopa *n.* [Dihydroxyphenylalanine] an amino acid formed from tyrosine by action of ultraviolet rays, and oxidized by dopa-oxidase or dopase to a red precursor of melanin, as in basal layers of epidermis, etc.; it is also a precursor of adrenaline, noradrenaline, and dopamine, and in plants may be a respiratory catalyst.

dopamine a catecholamine which is an intermediate in the synthesis of adrenaline and noradrenaline, and acts as an inhibitory neurotransmitter in the corpus striatum in mammals and in other animals such as snails.

dormancy *n.* [F. *dormir*, from L. *dormire*, to sleep] resting or quiescent condition with reduced metabolic rate as in seeds; hibernation and aestivation.

dormancy callus callose deposited on sieve areas at the onset of winter.

dormin *n.* [L. *dormire*, to sleep] abscisic acid, *q.v.*

dorsad *adv.* [L. *dorsum*, back; *ad*, to] towards back or dorsal surface; *cf.* ventrad.

dorsal *a.* [L. *dorsum*, back] *pert.* or lying near back; *pert.* surface farthest from axis; upper surface of thallus or prothallus of ferns, etc.; *cf.* ventral.

dorsalis *n.* [L. *dorsum*, back] the artery which supplies the dorsal surface of any organ.

dorsiferous *a.* [L. *dorsum*, back; *ferre*, to carry] with sori on back of leaf; dorsigerous, *q.v.*

dorsifixed *a.* [L. *dorsum*, back; *fingere*, to fix] having filament attached to back of anther.

dorsigerous *a.* [L. *dorsum*, back; *gerere*, to

bear] carrying the young on the back; *alt.* dorsiferous.

dorsigrade *a.* [L. *dorsum*, back; *gradus*, step] having back of digit on the ground when walking.

dorsilateral *a.* [L. *dorsum*, back; *latus*, side] of or *pert.* the back and sides, dorsal and lateral.

dorsispinal *a.* [L. *dorsum*, back; *spina*, spine] *pert.* or referring to back and spine.

dorsiventral *a.* [L. *dorsum*, back; *venter*, belly] flattened and having upper and lower surface of distinctly different structure, *appl.* leaves; *alt.* bifacial; *cf.* dorsoventral.

dorsobronchus *n.* [L. *dorsum*, back; Gk. *bronchos*, windpipe] one of the secondary bronchi spreading dorsally from the mesobronchus in birds.

dorsocentral *a.* [L. *dorsum*, back; *centrum*, centre] *pert.* mid-dorsal surface; *pert.* aboral surface of echinoderms.

dorsolumbar *a.* [L. *dorsum*, back; *lumbus*, loin] *pert.* lumbar region of back.

dorsoumbonal *a.* [L. *dorsum*, back; *umbo*, shield-boss] lying on the back near the umbo.

dorsoventral *a.* [L. *dorsum*, back; *venter*, belly] *pert.* structures which stretch from dorsal to ventral surface; dorsiventral, *q.v.*

dorsulum *n.* [*Dim.* of L. *dorsum*, back] mesonotum, *q.v.*

dorsum *n.* [L. *dorsum*, back] the sulcular surface of Anthozoa; tergum or notum of insects and crustaceans; inner margin of insect wing; the back of higher animals; upper surface, as of tongue.

Dorylaimida *n.* [*Dorylaimus*, generic name] an order of Aphasmidia which are either predators or plant parasites and have teeth, a buccal spear, and a non-bristly cuticle.

dorylaner *n.* [Gk. *dory*, spear; *anēr*, male] an exceptionally large male ant of driver-ant group.

Dorypteriformes *n.* [*Dorypterus*, generic name; L. *forma*, shape] an order of Permian Chondrostei with a highly modified skull.

Dothidiales, Dothideales *n.* [*Dothidea*, generic name] an order of Pyrenomycetes having a mycelium almost immersed in the substratum and forming a dark coloured stroma with cavities in which asci are produced; sometimes considered a group of Loculomycetidae, and known as Dothiorales.

Dothiorales *n.* [Gk. *dothiōn*, boil, small abscess] an order of Loculomycetidae, equivalent to the Dothidiales, *q.v.*

double fertilization fusion of one of 2 gametes derived from division of the generative nucleus of the microspore with the ovum nucleus, and of the other with the primary endosperm nucleus, in angiosperms.

double helix the form of DNA suggested by Watson and Crick, made of 2 chains of nucleotides running antiparallel and arranged in 2 helices (spirals) winding round each other and joined by cross-linkages of base pairs.

double recessive a cell or organism homozygous for a recessive gene and therefore showing the recessive phenotype.

doublure *n.* [F. *doublure*, lining] the reflected margin of carapace in Trilobita and Xiphosura.

down feathers fluffy feathers which form the body covering of nestling birds, and in the adult are found between and under the contour feathers.

Doyère's cone [L. *Doyère*, French physiologist] end plate or eminence where nerve fibre branches and enters sarcolemma.

DPD diffusion pressure deficit, *q.v.*

DPN diphosphopyridine nucleotide; now called NAD, *q.v.*; formerly coenzyme I.

dragonflies the common name for the Odonata, *q.v.*

drepaniform drepanoid, *q.v.*

drepanium *n.* [Gk. *drepanē*, sickle] a helicoid cyme with secondary axes developed in a plane parallel to that of main peduncle and its 1st branch.

drepanoid *a.* [Gk. *drepanoeidēs*, sickle-shaped] sickle-shaped; *alt.* drepaniform, falcate falciform.

drift *n.* [A.S. *drifan*, to drive] *see* genetic drift, continental drift.

Drilomorpha *n.* [Gk. *drilos*, worm; *morphē*, form] an order of mud-eating, burrowing polychaetes having a small prostomium with no appendages.

drive the motivation of an animal resulting in it achieving a goal or satisfying a need.

dRNA heterogenous nuclear RNA, *q.v.*

dromaeognathous *a.* [Gk. *dramein*, to run; *gnathos*, jaw] having a palate in which palatines and pterygoids do not articulate, owing to intervention of vomer.

dromotropic *a.* [Gk. *dromos*, course; *tropē*, turn] bent in a spiral; influencing nerve conductivity.

drone *n.* [A.S. *dran*] the male bee.

drop mechanism the mechanism by which a pollen grain, trapped in a drop of liquid in gymnosperm ovules, is drawn into the micropyle.

dropper *n.* [A.S. *dreópan*, to drop] rhizomatous downward outgrowth of a bulb, which may form a new bulb.

drop roots buttress roots, *q.v.*

drosopterin one of the pteridine pigments which give the red colour to the fruit fly, *Drosophila*, eye and which are decomposed by light and may play a part in photoreception.

drupaceous *a.* [Gk. *dryppa*, olive] *pert.* drupe; bearing drupes; drupe-like.

drupe *n.* [Gk. *dryppa*, olive] a more-or-less fleshy 1-celled fruit with one or more seeds, having the pericarp differentiated into a thin epicarp, a fleshy mesocarp, and a hard stony endocarp, e.g. plum, cherry, etc.; *alt.* pyrenocarp, stone fruit.

drupel *n.* [Gk. *dryppa*, olive] one of a collection of small drupes forming an aggregate fruit in plants with an apocarpous gynaecium, e.g. raspberry, blackberry; *alt.* acinus, drupelet, druplet, drupeole.

druse *n.* [Gk. *druos*, bump] a globular compound crystal whose component crystals project from the surface.

dry *a.* [A.S. *dryge*, dry] *appl.* a fruit having a dry, hard, non-fleshy pericarp as in achenes, capsules, and schizocarps; *cf.* succulent.

drymophytes *n.plu.* [Gk. *drymos*, coppice; *phyton*, plant] small trees, bushes, and shrubs.

dryopithecines a group of fossil apes of the Miocene and Pliocene from India and Africa, including the genera *Dryopithecus* and *Proconsul*, which may be ancestral to hominids.

dsDNA double-stranded DNA.

duct *n*. [L. *ducere*, to lead] any tube which conveys fluid or other substances; a tube formed by a series of cells which have lost their walls at the points of contact; *alt.* ductus.

ductless glands glands which do not communicate with any organ directly by means of a duct such as endocrine glands.

ductule *n*. [L. *ducere*, to lead] a minute duct; fine thread-like terminal portion of a duct.

ductus *n*. [L. *ducere*, to lead] duct, *q.v.*

ductus arteriosus the connection between the pulmonary arch and dorsal aorta in mammalian foetus.

ductus deferens vas deferens, *q.v.*

ductus ejaculatorius a narrow muscular tube at end of vas deferens in various invertebrates.

Dufour's gland [L. *Dufour*, French etomologist] a gland whose duct leads into the poison sac at the base of the sting in certain Hymenoptera; *alt.* alkaline gland.

dulosis *n*. [Gk. *doulōsis*, subjugation] slavery among ants, in which those of one species are captured by another species and work for them, an extreme form of social parasitism; *alt.* helotism.

dumb-bell bone prevomer, *q.v.*

dumose *a*. [L. *dumosus*, bushy] shrub-like in appearance.

duodenal *a*. [L. *duodeni*, twelve each] *pert.* duodenum.

duodenum *n*. [L. *duodeni*, twelve each] that portion of small intestine next to pyloric end of stomach.

duplex *a*. [L. *duplex*, two-fold] double; compound, *appl.* flowers; diploid, *q.v.;* having 2 dominant genes, in polyploidy; consisting of 2 distinct structures; having 2 distinct parts.

duplex DNA Watson–Crick DNA model, *q.v.*

duplication *n*. [L. *duplex*, double] chorisis, *q.v.; a* translocated chromosome fragment attached to one of normal set.

duplicational polyploid autopolyploid, *q.v.*

duplicato- duplico-, *q.v.*

duplicature *n*. [L. *duplex*, double] a circular fold near base of protrusible portion of a polyzoan polypide.

duplicident *a*. [L. *duplex*, double; *dens*, tooth] with 2 pairs of incisors in upper jaw, one behind the other.

duplicity *n*. [L. *dupliciter*, doubly] condition of being 2-fold; *appl.* theory that cones are the photopic, or colour, receptors, and rods the scotopic, or brightness, receptors.

duplico- *alt.* duplicato-.

duplicocrenate *a*. [L. *duplex*, double; *crena*, notch] with scalloped margin, and each rounded tooth again notched, *appl.* leaf.

duplicodentate *a*. [L. *duplex*, double; *dens*, tooth] with marginal teeth on leaf bearing smaller teeth-like structures.

duplicoserrate *a*. [L. *duplex*, double; *serratus*, saw-edged] with marginal saw-like teeth and smaller teeth directed towards leaf tip.

dural *a*. [L. *dura*, hard] *pert.* dura mater.

dura mater *n*. [L. *dura*, hard; *mater*, mother] the tough membrane lining the whole cerebrospinal cavity; *alt.* pachymeninx, scleromeninx.

duramen *n*. [L. *duramen*, hardness] the hard, darker central region of a tree stem; *alt.* heartwood; *cf.* alburnum.

Duran–Reynals spreading factor hyaluronidase, *q.v.*

dura spinalis the tough membrane lining the spinal canal.

Durvilleales *n*. [*Durvillea*, generic name] an order of brown algae whose thallus has a large basal pad, with a short stipe and long flattened dissected frond.

duvet *n*. [F. *duvet*, down] downy coating, as soft matted coating by certain fungi.

dwarf male small 3- or 4-celled plant formed from androspore of the green alga *Oedogonium*; a small, usually simply formed, individual in many classes of animals, either free or carried by the female; *alt.* pigmy male.

dyad *n*. [Gk. *dyas*, two] the half of a tetrad group; a bivalent chromosome.

dymantic behaviour deimatic behaviour, *q.v.*

dynamic *a*. [Gk. *dynamis*, power] producing or manifesting activity; *appl.* specific dynamic action, the calorigenic action of food, increasing metabolism above basal rate; *cf.* static.

dynamogenesis *n*. [Gk. *dynamis*, power; *genesis*, origin] induction of motor activity by sensory stimulation.

dynamoneure *n*. [Gk. *dynamai*, to be able to do; *neuron*, nerve] a motor neurone.

dynamoplasm *n*. [Gk. *dynamis*, power; *plasm*, form] active part of the cytoplasm; *cf.* paraplasm; *a.* dynamoplastic.

dynein *n*. [Gk. *dynamis*, power] a protein found in the exoneme of cilia and flagella which appears to be involved in the production of mechanical work.

dysgenesis *n*. [Gk. *dys-*, mis-; *genesis*, descent] defective descent; infertility of hybrids in matings between themselves, though fertile with individuals of either parental stock.

dysgenic *a*. [Gk. *dysgeneia*, low birth] *pert.* tending towards, or productive of, racial degeneration; *alt.* cacogenic; *cf.* eugenic.

dysharmony *n*. [Gk. *dys-*, mis-; *harmonia*, a fitting together] allometry, *q.v.*

dysmerism *n*. [Gk. *dys-*, mis-; *meros*, part] an aggregate of unlike parts.

dysmerogenesis *n*. [Gk. *dys-*, mis-; *meros*, part; *genesis*, descent] segmentation resulting in unlike parts.

dysphotic *a*. [Gk. *dys-*, mis-; *phōs*, light] dim; *appl.* zone, waters at depths between 80 and 600 m, between euphotic and aphotic zones, *q.v.;* lower layer of photic zone; *alt.* disphotic.

dysploid aneuploid, *q.v.*

dyspnoea *n*. [Gk. *dyspnoos*, breathless] difficulty in breathing.

dysteleology *n*. [Gk. *dys-*, mis-; *teleos*, ended; *logos*, discourse] Haeckel's doctrine of purposelessness in Nature; appearance of uselessness, as of certain organs or other structures; frustration of function.

dystrophic *a*. [Gk. *dys-*, mis-; *trephein*, to nourish] wrongly or inadequately nourished; inhibiting adequate nutrition; *pert.* faulty nutrition; *appl.* lakes rich in undecomposed organic matter so nutrients are scarce, *cf.* eutrophic, oligotrophic.

Dzierzon theory [*J. Dzierzon*, Silesian apiculturist] belief that males of honey bee are always produced from unfertilized eggs.

E

ear *n.* [A.S. *éare*, ear] the auditory organ; among invertebrates, the various structures supposed to have an auditory function; the specialized tufts of hair or feathers which are close to, or similar to an external ear or pinna; an ear-shaped structure; the spike of grasses, usually of cereals.

eared *a.* [A.S. *éare*, ear] having external ears or pinnae; with tufts of feathers resembling ears; having long bristles, auricles, as in grains of corn, *alt.* auriculate.

early wood wood formed in the 1st part of an annual ring having larger and less dense cells than late wood; *alt.* spring wood.

ear sand otoconia, *q.v.*

ear stone otolith, *q.v.*

earth ball the common name for a member of the Sclerodermatales, *q.v.*

earwigs the common name for the Dermaptera, *q.v.*

Ebenales *n.* [L. *ebenus*, ebony] an order of woody dicots of the Dilleniidae or Sympetalae, used in slightly different ways by different authorities, and including many important economic plants such as ebony, persimmon, and gutta-percha.

Ebner's gland a gland in the tongue which secretes a watery fluid.

ebracteate, ebractəolate *a.* [L. *ex*, out of; *bractea*, thin plate] without bracts; without bracteoles.

ecad a plant or animal form modified by the habitat; *alt.* ecophene, ecophenotype, oecad, eocophene; *cf.* phyad.

ecalcarate *a.* [L. *ex*, out of; *calcar*, spur] having no spur or spur-like process.

ecardinal *a.* [L. *ex*, out of; *cardo*, hinge] having no hinge; *alt.* ecardinate.

Ecardines *n.* [L. *ex*, out of; *cardo*, hinge] a class of Brachiopoda having shells without a hinge and no internal skeleton.

ecarinate *a.* [L. *ex*, out of; *carina*, keel] not furnished with a keel or keel-like ridge.

ecaudate *a.* [L. *ex*, out of; *cauda*, tail] without a tail; *alt.* acaudate.

ecblastesis *n.* [Gk. *ek*, out of; *blastē*, bud] proliferation of main axis of inflorescence.

ecbolic *a.* [Gk. *ekbolē*, a throwing out] *appl.* effects of stimulation of glands on synthesis of organic substances; *cf.* hydrelatic.

eccentric excentric, *q.v.*

Eccrinales *n.* [*Eccrina*, generic name] an order of Phycomycetes that are parasites of arthropods.

eccrine *a.* [Gk. *ekkrinein*, to expel] secreting without disintegration of secretory cells; *cf.* holocrine.

eccrinology *n.* [Gk. *ekkrinein*, to expel; *logos*, discourse] the study of secretion and secretions.

eccritic *a.* [Gk. *ekkrinein*, to expel, to select] causing or *pert.* excretion; preferred, *appl.* temperature, etc. *n.* A substance or other agent which promotes excretion.

ecdemic *a.* [Gk. *ek*, out of; *dēmos*, district] not native.

ecderon *n.* [Gk. *ek*, out; *deros*, skin] the outer or epidermal layer of skin.

ecderonic *a.* [Gk. *ek*, out; *deros*, skin] ectodermal, *q.v.*

ecdysial *a.* [Gk. *ekdysai*, to strip] *pert.* ecdysis; *appl.* fluid between old and new cuticle which aids in disintegration of old cuticle, *alt.* moulting fluid; *appl.* line along which cuticle splits in moulting; *appl.* glands secreting moulting fluid, *alt.* Verson's glands, moulting glands; *appl.* excretion of calcospherites.

ecdysis *n.* [Gk. *ekdysai*, to strip] the act of moulting a cuticular layer of structures; *cf.* endysis.

ecdyson(e) *n.* [Gk. *ekdysai*, to strip] a steroid hormone produced by the prothoracic gland of insects and by Y-organs of crustaceans, which stimulates growth and moulting, and has been shown to cause the synthesis of mRNA; *alt.* moulting hormone.

ece *n.* [Gk. *oikein*, to have as one's abode] habitat, *q.v.*; *alt.* oike, oikos.

ecesis *n.* [Gk. *oikēsis*, act of dwelling] the invasion of organisms into a new habitat; *alt.* oecesis, oikesis.

echard *n.* [Gk. *echein*, to keep; *ardo*, I water] soil water not available for plant growth; *cf.* chresard, holard.

Echeneiformes *n.* [*Echeneis*, generic name; L. *forma*, shape] an order of teleosts whose dorsal fin forms a sucker and including the remora; *alt.* Discocephali.

Echinacea *n.* [Gk. *echinos*, sea urchin] a superorder of globular sea urchins of the subclass Euechinoidea with keeled teeth on the Aristotle's lantern.

echinate *a.* [Gk. *echinos*, hedgehog] furnished with spines or bristles.

echinenone *n.* [Gk. *echinos*, sea urchin] a xanthophyll carotenoid pigment found in the sea urchin gonads and in algae of the classes Cyanophyceae and Euglenophyceae.

echinidium *n.* [Gk. *echinos*, spine; *idion*, *dim.*] marginal hair, with small pointed or branched outgrowths, of pileus of fungi; *alt.* brush cell.

echinochrome *n.* [Gk. *echinos*, sea urchin; *chrōma*, colour] a red–brown respiratory pigment of echinoderms.

echinococcus *n.* [Gk. *echinos*, spine; *kokkos*, berry] a vesicular metacestode developing a number of daughter cysts, each with many heads; *see* polycercoid.

Echinodermata, echinoderms *n.*, *n.plu* [Gk. *echinos*, spine; *derma*, skin] a phylum of marine coelomate animals that are bilaterally symmetrical as larvae but show 5-rayed symmetry as adults and have a calcareous endoskeleton and a water vascular system.

echinoid *a.* [Gk. *echinos*, sea urchin; *eidos*, form] *pert.* or like sea urchins.

Echinoida *n.* [Gk. *echinos*, sea urchin; *eidos*, form] Endocyclica, *q.v.*

Echinoidea *n.* [Gk. *echinos*, sea urchin; *eidos*, form] a class of echinoderms of the subphylum Eleutherozoa, commonly called sea urchins, having a typically globular body (sometimes flattened) with skeletal plates fitting together to form a rigid test.

echinopaedium *n.* [Gk. *echinos*, sea urchin; *paidion*, young child] dipleurula, *q.v.*

echinopluteus *n.* [L. *echinus*, sea urchin; *pluteus*, shed] the pluteus larva of Echinoidea.

Echinosteliales n. [*Echinostelis*, generic name] an order of microscopic stalked slime moulds.

Echinothuroida n. [*Echinothuria*, generic name; Gk. *eidos*, form] an order of sea urchins of the superorder Diadematacea having a flexible test.

echinulate a. [Gk. *echinos*, studded over with spines] having small spines; having pointed outgrowths, *appl.* bacterial cultures; *alt.* triboloid.

Echiura n. [*Echiurus*, generic name] a phylum of marine, sedentary, unsegmented coelomate worms that were formerly classed as annelids in the class Echiuroidea, which have an extensible but not eversible proboscis with a ciliated groove for collecting food.

Echiuroidea n. [*Echiurus*, generic name; Gk. *eidos*, form] *see* Echiura.

echolocation n. [L. *echo*, echo; *locare*, to place] location of objects by means of echos, as of supersonic sounds emitted by animals, e.g. by bats.

ecize v. [Gk. *oikēn*, to settle] to invade or to be established in another habitat; *alt.* oecise.

eclipse n. [Gk. *ekleipein*, to leave incomplete] plumage assumed after spring moult, as in drake; period of multiplication of a bacterial virus during which it fails to be noticed in an infected cell.

eclosion n. [L. *e*, out; *clausus*, shut] hatching from egg or pupa case.

eco- *alt.* oeco-, oiko-.

ecobiotic a. [Gk. *oikos*, household; *biōsis*, manner of life] *appl.* adaptation to particular mode of life within a habitat.

ecoclimatic a. [Gk. *oikos*, household; *klima*, slope] *appl.* adaptation to the physical and climatic conditions in a particular region.

ecocline n. [Gk. *oikos*, household; *klinein*, to slant] a continuous variation or gradient of ecotypes in relation to variation in ecological conditions.

ecodeme n. [Gk. *oikos*, household; *dēmos*, people] a deme occupying a particular ecological habitat.

ecoid n. [Gk. *oikos*, house; *eidos*, form] the stroma of a blood corpuscle; *alt.* oecoid, oikoid.

E.coli *Escherichia coli, q.v.*

ecological a. [Gk. *oikos*, household; *logos*, discourse] *pert.* ecology.

ecological niche the status of an organism in the ecosystem, including its habitat and its effect on other organisms and on the environment; *cf.* microhabitat.

ecological pyramid a diagram showing the biomass, numbers, or energy levels of individuals of each trophic level in an ecosystem, starting with the primary producers at the base.

ecology n. [Gk. *oikos*, household; *logos*, discourse] the study of the interrelationships between organisms and their environment and each other; formerly also called bionomics, hexicology, mesology, oecology, oikology.

economic density of a population, the number of individuals per unit of the inhabited area.

ecoparasite n. [Gk. *oikos*, household; *parasitos*, parasite] a parasite that can infect a healthy and uninjured host; *alt.* oecoparasite.

ecophene n. [Gk. *oikos*, household; *phainein*, to appear] ecad, *q.v.*

ecophenotype n. [Gk. *oikos*, household; *phainein*, to appear; *typos*, pattern] ecad, *q.v.*

ecorticate a. [L. *e*, out of; *cortex*, rind] without a cortex, as certain lichens.

ecospecies n. [Gk. *oikos*, household; L. *species*, particular kind] ecotype, *q.v.*; a group of individuals associated with a particular ecological niche and behaving as a species, but capable of interbreeding with neighbouring ecospecies.

ecosphere n. [Gk. *oikos*, household; *sphaira*, globe] the planetary ecosystem, consisting of the living organisms in the world and the components of the environment with which they react.

ecostate a. [L. *e*, out; *costa*, rib] without costae; not costate.

ecosystem n. [Gk. *oikos*, household; *systēma*, composite whole] ecological system formed by the interation of coacting organisms and their environment; *alt.* biosystem.

ecotone n. [Gk. *oikos*, household; *tonos*, brace] a transitional species in intermediate area between 2 associations; the boundary line or transitional area between 2 communities.

ecotope n. [Gk. *oikos*, household; *topos*, place] a particular kind of habitat within a region.

ecotype n. [Gk. *oikos*, household; *typos*, pattern] a biotype resulting from selection in a particular habitat; *alt.* habitat type, ecospecies.

ecphoria n. [Gk. *ekphorion*, produce] the revival of a latent memory pattern or engram.

ecsoma n. [Gk. *ek*, from out of; *sōma*, body] retractile posterior part of body in certain trematodes.

ectad adv. [Gk. *ektos*, outside; L. *ad.* towards] towards the exterior; outwards, externally; *cf.* entad.

ectadenia n.plu. [Gk. *ektos*, outside; *adēn*, gland] ectodermal accessory genital glands in insects; *cf.* mesadenia.

ectal a. [Gk. *ektos*, outside] outer; external; *appl.* layer or membrane on margin of exciple; *cf.* ental.

ectamnion n. [Gk. *ektos*, outside; *amnion*, foetal membrane] ectodermal thickening in proamnion, beginning of head-fold.

ectangial a. [Gk. *ektos*, outside; *anggeion*, vessel] outside a vessel; produced outside a primary sporangium; *alt.* ectoangial; *cf.* entangial.

ectendotrophic a. [Gk. *ektos*, without; *endon*, within; *trophē*, nourishment] partly ectotrophic and partly endotrophic, *appl.* mycorrhizic fungus.

ectental a. [Gk. *ektos*, outside; *entos*, inside] *pert.* both ectoderm and endoderm; *appl.* line where ectoderm and endoderm meet at blastopore of a gastrula.

ectepicondylar a. [Gk. *ektos*, outside; *epi*, upon; *kondylos*, knuckle] *appl.* radial foramen of humerus.

ectethmoid n. [Gk. *ektos*, outside; *ēthmos*, sieve; *eidos*, form] lateral ethmoid bone; *alt.* ectoethmoid, parethmoid.

ecthoraeum n. [Gk. *ekthroskein*, to leap out] the thread of a nematocyst.

ectiris n. [Gk. *ektos*, outside; *iris*, rainbow] portion of the posterior elastic lamina of cornea anterior to the iris.

ecto- *alt.* ekto-.

ectoangial ectangial, *q.v.*

ectoascus n. [Gk. *ektos*, outside; *askos*, bag] outer membrane of an ascus in certain Ascomycetes; *cf.* endoascus.

ectobatic

ectobatic *a.* [Gk. *ektos*, outside; *bainein*, to go] efferent; exodic, centrifugal; *cf.* endobatic.

ectoblast *n.* [Gk. *ektos*, outside; *blastos*, bud] epiblast, *q.v.*

ectobronchus, ectobronchium *n.* [Gk. *ektos*, outside; *brongchos*, windpipe] lateral branch of main bronchus in birds.

Ectocarpales *n.* [*Ectocarpus*, generic name] an order of brown algae having trichothallic growth and isomorphic alternation of generations.

ectocarpous *a.* [Gk. *ektos*, outside; *karpos*, fruit] having gonads of ectodermal origin.

ectochondrostosis *n.* [Gk. *ektos*, outside; *chrondros*, cartilage; *osteon*, bone] ossification beginning in perichondrium and gradually invading cartilage.

ectochone *n.* [Gk. *ektos*, outside; *chōnē*, funnel] a funnel-shaped chamber into which lead the ostia in certain sponges.

ectochroic *a.* [Gk. *ektos*, outside; *chrōs*, complexion] having pigment on the surface of a cell or hypha; *cf.* endochroic.

ectocoelic *a.* [Gk. *ektos*, outside; *koilos*, hollow] *pert.* structures situated outside the enteron of coelenterates.

ectocommensal *n.* [Gk. *ektos*, outside; L. *cum*, with; *mensa*, table] a commensal living on the surface of another organism.

ectocondyle *n.* [Gk. *ektos*, outside; *kondylos*, knuckle] the outer condyle of a bone.

ectocranial *a.* [Gk. *ektos*, outside; *kranion*, skull] *pert.* outside of skull.

ectocrine *a.* [Gk. *ektos*, outside; *krinein*, to separate] *appl.* and *pert.* organic substances or decomposition products in the external medium which inhibit or stimulate plant life. *n.* An ectocrine compound, *alt.* environmental hormone, external diffusion hormone.

ectocuneiform *n.* [Gk. *ektos*, outside; L. *cuneus*, wedge; *forma*, shape] a bone in distal row of tarsus, the 3rd cuneiform.

ectocyst *n.* [Gk. *ektos*, outside; *kystis*, bladder] outer layer of zooecium in Polyzoa; epicyst, *q.v.*

ectoderm *n.* [Gk. *ektos*, outside; *derma*, skin] the outer of the embryonic layers in multicellular animals, and the tissues derived from it; *alt.* exodermis.

ectodermal *pert.*, or derived from ectoderm; *alt.* ecderonic.

ectoentad *adv.* [Gk. *ektos*, without; *entos*, within; L. *ad.* towards] from without inwards; *cf.* entoectad.

ectoenzyme *n.* [Gk. *ektos*, outside; *en*, in; *zymē*, leaven] exoenzyme, *q.v.*

ecto-ethmoid ectethmoid, *q.v.*

ectogenesis *n.* [Gk. *ektos*, outside; *genesis*, descent] embryonic development outside the maternal organism; development in an artificial environment; *a* ectogenetic.

ectogenic *a.* [Gk. *ektos*, outside; *gennaein*, to produce] not produced by organisms themselves; *cf.* autogenic.

ectogenous *a.* [Gk. *ektos*, outside; *genos*, birth] able to live an independent life; originating outside the organism.

ectoglia *n.* [Gk. *ektos*, outside; *glia*, glue] an outer layer in central nervous system.

ectohormone *n.* [Gk. *ektos*, outside; *hormaein*, to excite] pheromone, *q.v.*

ectolecithal *a.* [Gk. *ektos*, outside; *lekithos*, egg yolk] having yolk around the edge of the ovum.

ectoloph *n.* [Gk. *ektos*, outside; *lophos*, crest] the ridge stretching from paracone to metacone in a lophodont molar.

ectomeninx *n.* [Gk. *ektos*, outside; *meningx*, membrane] outer membrane covering embryonic brain and giving rise to dura mater.

ectomere *n.* [Gk. *ektos*, outside; *meros*, part] an epiblast cell which gives rise to ectoderm.

ectomesoderm *n.* [Gk. *ektos*, outside; *mesos*, middle; *derma*, skin] that part of the mesoderm derived from marginal cells at juncture of neural tube; *alt.* mesectoderm, mesoectoderm.

ectomesogloeal *a.* [Gk. *ektos*, outside; *mesos*, middle; *gloia*, glue] *pert.* ectoderm and mesogloea, *appl.* muscle fibre of disc of sea anemones; *alt.* meso-ectodermal.

-ectomy [Gk. *ek*, out; *temnein*, to cut] suffix signifying an excision, e.g. thyroidectomy, gonadectomy, etc.

ectoneural *a.* [Gk. *ektos*, outside; *neuron*, nerve] *appl.* system of oral ring, radial, and subepidermal nerves in echinoderms.

ectoparasite *n.* [Gk. *ektos*, outside; *parasitos*, parasite] a parasite that lives on the exterior of an organism; *alt.* ectosite, ectozoon, epiparasite, epizoon, exoparasite.

ectopatagium *n.* [Gk. *ektos*, outside; L. *patagium*, border] the part of the wing-like membrane of bats which is carried on metacarpals and phalanges; *alt.* dactylopatagium.

ectophloeodic *a.* [Gk. *ektos*, outside; *phloios*, bark] growing on plants, *appl.* lichens; *alt.* ectophloeodal, epiphloeodal.

ectophloic *a.* [Gk. *ektos*, outside; *phloios*, bark] with phloem outside xylem.

ectophyte *n.* [Gk. *ektos*, outside; *phyton*, plant] any external plant parasite of plants and animals.

ectophytic *a.* [Gk. *ektos*, outside; *phyton*, plant] *pert.* ectophytes; ectotrophic, *q.v.*

ectopic *a.* [Gk. *ek*, out of; *topos*, place] not in normal position, *appl.* organs, gestation, etc.; *cf.* entopic.

ectopic pairing pairing between bands located in different regions of a chromosome.

ectoplasm *n.* [Gk. *ektos*, outside; *plasma*, mould] the external layer of cytoplasm in a cell, next to the cell membrane and usually clear and non-granular; ectosarc, *q.v.*; *alt.* exoplasm; *cf.* endoplasm.

ectoplast *n.* [Gk. *ektos*, outside; *plastos*, formed] the plasma membrane next to the cell wall in plants.

Ectoprocta, ectoprocts *n., n.plu.* [Gk. *ektos*, outside; *prōktos*, anus] a phylum of small colonial lophophorate coelomate animals, formerly considered a class of Bryozoa, which live in horny, calcareous, or gelatinous cases.

ectopterygoid *n.* [Gk. *ektos*, outside; *pteryx*, wing; *eidos*, form] a ventral membrane bone behind palatine and extending to quadrate, *alt.* mesopterygoid; os transversum between pterygoid and maxilla in many reptiles and in some fishes; *cf.* entopterygoid.

ectoretina *n.* [Gk. *ektos*, outside; L. *rete*, net] outer pigmented layer of retina.

ectosarc *n.* [Gk. *ektos*, outside; *sarx*, flesh] the external layer of cytoplasm, especially in a protozoan; *alt.* ectoplasm.

ectosite n. [Gk. *ektos*, outside; *sitos*, food] ectoparasite, *q.v.*

ectosome n. [Gk. *ektos*, outside; *sōma*, body] pinacoderm, *q.v.*; a type of cell granule.

ectosphere n. [Gk. *ektos*, outside; *sphaira*, globe] the outer zone of attraction sphere.

ectospore n. [Gk. *ektos*, outside; *sporos*, seed] the spore formed at end of each sterigma in Basidiomycetes.

ectosporous a. [Gk. *ektos*, outside; *sporos*, seed] producing ectospores; with spores borne exteriorly.

ectostosis n. [Gk. *ektos*, outside; *osteon*, bone] formation of bone in which ossification begins under the perichondrium and either surrounds or replaces the cartilage.

ectostracum n. [Gk. *ektos*, outside; *ostrakon*, shell] outer primary layer of exocuticle of exoskeleton in Acarina.

ectostroma n. [Gk. *ektos*, outside; *strōma*, bedding] fungal tissue penetrating cortical tissue of host and bearing conidia; *alt.* epistroma; *cf.* entostroma, hypostroma.

ectotheca n. [Gk. *ektos*, outside; *thēkē*, case] outer coating of gonotheca in certain hydroids.

ectothecal *pert.* ectotheca; not enclosed by a theca.

ectotherm n. [Gk. *ektos*, outside; *thērme*, heat] a poikilothermal animal; *cf.* endotherm; *a.* ectothermal.

ectotrachea n. [Gk. *ektos*, outside; L. *trachia*, windpipe] an epithelial layer on outer side of insect tracheae.

Ectotropha n. [Gk. *ektos*, outside; *trephein*, to nourish] a superorder of wingless insects having the mouth-parts visible, and including the Thysanura.

ectotrophic a. [Gk. *ektos*, outside; *trephein*, to nourish] finding nourishment from outside; *appl.* mycorrhiza where the hyphae of the fungus are mainly superficial on the outside of the root; *alt.* ectophytic; *cf.* endotrophic.

ectotropic a. [Gk. *ektos*, outside; *trepein*, to turn] tending to curve or curving outwards.

ectoturbinal n. [Gk. *ektos*, outside; L. *turbo*, whirl] a division of the ethmoturbinal.

ectozoon n. [Gk. *ektos*, outside; *zōon*, animal] ectoparasite, *q.v.*

ecumene n. [Gk. *oikoumenē*, habitable world] biosphere, *q.v.*

edaphic a. [Gk. *edaphos*, ground] *pert.* or influenced by conditions of soil or substratum.

edaphology n. [Gk. *edaphos*, ground; *logos*, discourse] soil science, particularly the study of the influence of soil on living organisms; *cf.* pedology.

edaphon n. [Gk. *edaphos*, ground] the organisms living within the soil, i.e. soil flora and fauna; *alt.* geobios.

edeagus aedeagus, *q.v.*

Edentata, edentates n., n.plu. [L. *ex*, without; *dens*, tooth] an order of placental mammals known from the Palaeocene, including the hairy armadillo, great anteater, and 2-toed sloth, having reduced teeth and often an armoured body.

edentate a. [L. *ex*, without; *dens*, tooth] without teeth or tooth-like projections.

Edestiformes n. [Gk. *edestos*, edible; L. *forma*, shape] an order of marine Carboniferous Holocephali known only from their teeth.

edestin n. [Gk. *edestos*, edible] a plant globulin, main protein of sunflower and certain other seeds.

edge effect tendency to have greater variety and density of organisms in the boundary zone between communities or in an ecotone.

edge hair a cystidiform cell on gill margin in agarics.

edge species species living primarily or most frequently or numerously at junctions of communities.

E-disc the terminal disc between N-disc or N-band and Z-disc of a sarcomere.

Edrioasteroidea n. [Gk. *hedra*, seat; *astēr*, star; *eidos*, form] an extinct class of Palaeozoic cushion-shaped echinoderms of the subphylum Pelmatozoa.

edriophthalmic a. [Gk. *hedra*, seat; *ophthalmos*, eye] having sessile eyes, *appl.* certain Crustacea.

edwardsia larva stage in anthozoan larval development in which the larva possesses 8 mesenteries.

eelworms the common name for nematodes especially those living in the soil or as plant parasites.

effector n. [L. *efficere*, to carry out] an organ which reacts to stimulus by producing work or substance, as muscle, electric and luminous organs, chromatophores, glands; a motor end organ in muscle.

efferent a. [L. *ex*. out; *ferre*, to carry] conveying from, *appl.* vessels, lymphatics, etc.; carrying outwards, *appl.* impulses conveyed by motor nerves or these nerves themselves, *alt.* centrifugal; *appl.* ductules from rete testis opening into epididymis; *cf.* afferent.

efferent neurone motor neurone, *q.v.*

effigurate a. [L. *ex*, out; *figurare*, to shape] having a definite shape or outline; *cf.* effuse.

efflorescence n. [L. *efflorescere*, to blossom] blossoming; time of flowering; bloom.

effodient a. [L. *effodere*, to dig up] having the habit of digging.

effoliation n. [L. *ex*, out of; *folium*, leaf] shedding or removal of leaves.

effuse a. [L. *effusus*, poured out] spreading loosely, *appl.* inflorescence; spreading thinly, *appl.* bacterial cultures; *cf.* effigurate.

egesta n.plu. [L. *egestus*, discharged] the sum total of substances and fluids discharged from body, *cf.* ingesta; material passed out of the body in egestion.

egestion n. [L. *egestus*, discharged] the process of ridding the body of any waste material as by defaecation and excretion; specifically, the process of removing material that has never been taken out of the gut or digestive area, as defaecation.

egg n. [A.S. *aeg*, Icel. *egg*] ovum, *q.v.*; the reproductive body of certain animals, such as reptiles, birds, and insects, consisting of a fertilized ovum and nutritive and protective tissues, and from which a young individual emerges.

egg albumin the chief constituent of white of egg, a mixture of glucoproteins; *alt.* ovalbumin, egg albumen.

egg apparatus the 2 synergids and ovum proper, near micropyle in embryo sac of angiosperms.

egg calyx dilatation of oviduct at base of ovarioles in insects.

egg case a protective covering for eggs.

egg cell the ovum proper apart from any layer of cells derived from it or from other cells.

egg cylinder embryo after elongation of blastocyst, implanted perpendicular to uterine wall, as in rodents.

egg membrane the layer of tough tissue lining an egg shell.

egg nucleus the female pronucleus.

egg tooth a small structure on tip of upper jaw or of beak, by which the embryo breaks its shell.

eglandular *a.* [L. *ex*, out; *glandula*, small acorn] without glands.

Eichler's rule the groups of hosts with more variation are parasitized by more species than taxonomically uniform groups.

eiloid *a.* [Gk. *eilein*, to roll up; *eidos*, form] shaped like a coil.

Eimer's organs [*T. Eimer*, German zoologist] organs in the snout of moles, probably tactile organs.

ejaculate *n.* [L. *ejaculatus*, thrown out] the emitted seminal fluid.

ejaculatory *a.* [L. *ejaculare*, to throw out] throwing out, *appl.* certain ducts such as ductus ejaculatorius.

ejaculatory sac organ pumping ejaculate from vas deferens through ejaculatory duct to penis, in certain insects.

ejectisome *n.* [L. *ejectus*, thrown; Gk. *sōma*, body] trichocyst in some algae of the class Cryptophyceae.

ektexine *n.* [Gk. *ektos*, outside; *exō*, outside] outer layer of exine or extine; *cf.* endexine.

ekto- ecto-, *q.v.*

elaborate *v.* [L. *elaborare*, to work out] to change from a crude state to a state capable of assimilation; to form complex organic substances from simple materials.

Elaeagnales *n.* [*Elaeagnus*, generic name] an order of dicots (Rosidae) comprising one family, Elaeagnaceae, e.g. oleaster.

elaeo- *see also* elaio-, eleo-.

elaeoblast *n.* [Gk. *elaion*, oil; *blastos*, bud] a mass of nutrient material at posterior end of body in certain tunicates.

elaeocyte *n.* [Gk. *elaion*, oil; *kytos*, hollow] a cell containing fatty droplets, found in coelomic fluid of annelids.

elaeodochon *n.* [Gk. *elaiodochos*, oil-containing] oil gland, *q.v.*, in birds.

elaio- *see also* elaeo-, eleo-.

elaioplankton *n.* [Gk. *elaion*, oil; *plangktos*, wandering] plankton organisms rendered buoyant by oil globules.

elaioplast *n.* [Gk. *elaion*, oil; *plastos*, moulded] a plastid which forms and stores oil or fat globules; *alt.* elaiosome, oleosome, oleoplast, lipidoplast.

elaiosome *n.* [Gk. *elaion*, oil; *sōma*, body] elaioplast, *q.v.*; appendages containing oil found on seeds for dispersal by ants, such as the caruncle of castor oil.

elaiosphere *n.* [Gk. *elaion*, oil; *sphaira*, globe] an oil globule in a plant cell.

Elasipoda *n.* [Gk. *elastos*, ductile; *pous*, foot] an order of deep-sea, benthic sea cucumbers with no respiratory trees.

Elasmobranchii, elasmobranchs *n., n.plu.* [Gk. *elaunein*, to draw out; *brangchia*, gills] a subclass of Chondrichthyes, sometimes raised to class status, including the sharks, skates, and rays, having an exoskeleton of placoid scales, a spiracle, and no operculum over the gills.

elastase *n.* [Gk. *elaunein*, to draw out; *-ase*] a proteolytic enzyme secreted in pancreas and acting especially on elastin; EC 3.4.21.11; *r.n.* elastase.

elastica externa external layer of notochordal sheath.

elastica interna the epitheliomorph layer of notochordal sheath.

elastic fibres yellow fibres, *q.v.*

elastic fibrocartilage consists of cartilage cells and a matrix pervaded by a network of yellow elastic fibres which branch and anastomose in all directions; *alt.* yellow fibrocartilage.

elastin *n.* [Gk. *elaunein*, to draw out] the scleroprotein of which elastic fibres are composed.

elastoidin *n.* [Gk. *elaunein*, to draw out; *eidos*, form] substances composing elasmobranch ceratotrichia and the actinotrichia of embryonic teleosts.

elater *n.* [Gk. *elatēr*, driver] one of the filaments in the capillitium of slime moulds; one of the cells with a spiral thickening which assist in dispersing spores from capsule in liverworts; one of the spore appendages formed from epispore in horsetails; furcula or springing organ in Collembola.

elaterophore *n.* [Gk. *elatēr*, driver; *pherein*, to bear] tissue bearing elaters.

electosome *n.* [Gk. *eklektikos*, chosen; *sōma*, body] an obsolete term for a mitochondrion when it was thought to elaborate protoplasmic materials.

electric organ modifications of muscles or epithelium which discharge electric energy, mainly in certain fishes.

electroblast *n.* [Gk. *elektron*, amber; *blastos*, bud] a modified muscle fibre which gives rise to an electroplax.

electroendosmosis the bulk flow of liquid past fixed charges held in an electric field.

electroendosmotic layer a layer formed by electroendosmosis, present between 2 neurones or between neurone and muscle cell.

electrolemma *n.* [Gk. *elektron*, amber; *lemma*, skin] membrane surrounding an electroplax.

electron transfer chain the arrangement of cofactors such as NAD, FMN, cytochrome, etc., so that electrons are transferred from one to another, as in mitochondria with the production of ATP and finally the formation of water as a waste product; *alt.* electron transport chain.

electrophoresis *n.* [Gk. *elektron*, amber; *pherein*, to bear] the movement of a charged particle in response to an electric field when placed in that electric field; *alt.* cataphoresis, kataphoresis.

electrophysiology study of physiological processes in relation to electrical phenomena.

electropism electrotropism, *q.v.*

electroplax *n.* [Gk. *elektron*, amber; *plax*, plate] one of the constituent plates of an electric organ.

electrotaxis *n.* [Gk. *electron*, amber; *taxis*, arrangement] orientation of movement within an electric field; *alt.* galvanotaxis.

electrotonic *a.* [Gk. *elektron*, amber; *tonos*, tension] *pert.* a state of electric tension.

electrotonus *n.* [Gk. *elektron*, amber; *tonos*, tension] the modified condition of a nerve when subjected to a constant current of electricity.

electrotropism *n.* [Gk. *elektron*, amber; *trope*,

turn] reaction of an organism to electric stimuli; plant curvature in an electric field; *alt.* electropism, galvanotropism.

eleidin *n.* [Gk. *elaia*, olive] substance found as small granules or droplets in stratum granulosum of epidermis.

eleo- *see also* elaeo-, elaio-.

elephants *see* Proboscidea.

elephant-tooth shells a common name for the Scaphopoda, *q.v.*

elepidote alepidote, *q.v.*

eleutherodactyl *a.* [Gk. *eleutheros*, free; *daktylos*, finger] having hind-toe free.

eleutheropetalous *a.* [Gk. *eleutheros*, free; *petalon*, leaf] polupetalous, *q.v.*

eleutherophyllous *a.* [Gk. *eleutheros*, free; *phyllon*, leaf] having components of perianth whorls free.

eleutherosepalous *a.* [Gk. *eleutheros*, free; F. *sépale*, sepal] polysepalous, *q.v.*

Eleutherozoa *n.* [Gk. *eleutheros*, free; *zōon*, animal] a subphylum of free-living echinoderms without a stalk, that lie on one side or with the oral surface downwards.

eleutherozoic *a.* [Gk. *eleutheros*, free; *zōē*, life] free living.

elevator *n.* [L. *elevare*, to lift up] levator, *q.v.*

eligulate . [L. *ex*, out; *ligula*, little tongue] having no ligule, *appl.* certain club mosses.

elimination *n.* [L. *eliminare*, to turn out of doors] the expulsion of a substance, as of one which has not taken part in metabolism.

elimination bodies nucleic acid material expelled from each chromosome during meiosis, remaining in middle of spindle and disintegrating during telophase.

elittoral *n.* [L. *ex*, out; *litus*, seashore] *appl.* zone out from the coast where light ceases to penetrate to the sea bottom.

ellipsoid *a.* [Gk. *elleipsis*, a falling short; *eidos*, shape] oval. *n.* Localized thickening of coat of arterioles in spleen; Malpighian body of the spleen; fibrillar outer end of inner segment of retinal rods and cones.

elliptical *a.* [Gk. *elleipsis*, a falling short] oval-shaped; *appl.* leaves of about same breadth at equal distances from base and apex which are slightly acute.

Elopiformes *n.* [Gk. *elops*, a fish; L. *forma*, shape] the only order of the Elopomorpha when the latter is considered as a superorder.

Elopomorpha *n.* [Gk. *elops*, a fish; *morphē*, form] a primitive or aberrant group of teleosts existing from the Jurassic to the present day and including the eel.

Eltringham's organ a group of long innervated hairs on the dorsal side of head in Lepidoptera, probably responding to air currents during flight.

eluvial *a.* [L. *ex*, out; *luere*, to wash] *appl.* layer in soils which is impoverished and leached, above the illuvial layer, *alt.* A-horizon; alluvial, *q.v.*; *appl.* gravels formed by breakdown of rocks *in situ*, *cf.* alluvial.

elytra *plu.* of elytron or elytrum.

elytriform *a.* [Gk. *elytron*, sheath; L. *forma*, shape] shaped like an elytrum.

elytroid *a.* [Gk. *elytron*, sheath; *eidos*, resemblance] resembling an elytrum.

elytron elytrum, *q.v.*

elytrophore *n.* [Gk. *elytron*, covering; *pherein*, to carry] structure on prostomium of certain polychaetes, bearing an elytrum.

elytrum *n.* [Gk. *elytron*, sheath] the anterior wing of certain insects, especially beetles, which is hard and case-like, *alt.* wing sheath; one of the scales or shield-like plates found on dorsal surface of some polychaetes; *alt.* elytron; *plu.* elytra.

emarginate *a.* [L. *ex*, out; *marginare*, to delimit] having a notch at apex; having a notched margin.

emasculation *n.* [L. *ex*, out of; *mas*, male] removal of anthers from a young flower to prevent self-pollination; removal of the testis; *v.* emasculate.

Embden–Meyerhof–Parnas pathway (EMP) the usual metabolic pathway by which glycolysis proceeds from glucose to pyruvic acid; *alt.* Embden–Meyerhof pathway.

embiids the common name for the Embioptera, *q.v.*

Embioptera *n.* [*Embia*, generic name; Gk. *pteron*, wing] an order of insects with incomplete metamorphosis, commonly called embiids or foot spinners, which live in groups in silken tunnels and have wingless females and winged males.

embole *n.* [Gk. *embolē*, a throwing in] invagination; *alt.* emboly.

embolic *a.* [Gk. *embolē*, a throwing in] pushing or growing in.

embolium *n.* [Gk. *embolos*, wedge] outer or costal part of wing, or basal part of hemelytron, in certain insects.

Embolomeri, embolomeres *n.*, *n.plu.* [Gk. *embolos*, wedge; *meros*, part] a group of Devonian to Permian primitive labyrinthodonts.

embolomerous *a.* [Gk. *embolos*, wedge; *meros*, part] *appl.* type of vertebra having 2 vertebral rings in each segment, due to union of hypocentra with neural arch, and union of 2 pleurocentra below notochord.

embolus *n.* [Gk. *embolos*, wedge] a projection closing the foramen of an ovule; apical division of the palpus in certain spiders; a clot blocking a blood vessel; horn core or os cornu of ruminants.

emboly embole, *q.v.*

Embrithopoda, embrithopods *n.*, *n.plu.* [Gk. *embrithēs*, weighty; *pous*, foot] an order of extinct placental mammals known from the Oligocene fossil *Arsinoitherium* which had 2 massive nasal horns.

embryo *n.* [Gk. *embryon*, embryo] a young organism in early stages of development.

embryo cell one of 2 cells formed from 1st division of fertilized egg in certain plants, developing later into embryo, the other developing into suspensor.

embryogenesis *n.* [Gk. *embryon*, embryo; *genesis*, descent] origin of the embryo; embryogeny, *q.v.*; development from an ovum, *cf.* blastogenesis.

embryogeny *n.* [Gk. *embryon*, embryo; *gennaein*, to produce] the processes by which the embryo is formed; origin, cellular pattern, and functions of the embryo; *alt.* embryogenesis.

embryology *n.* [Gk. *embryon*, embryo; *logos*, discourse] the part of biology dealing with formation and development of the embryo.

embryonal knot inner cell mass of blastodermic vesicle.

embryonic *a.* [Gk. *embryon,* embryo] *pert.* embryo.

embryonomy *n.* [Gk. *embryon,* embryo; *nomos,* law] the laws of embryonic development; classifications of embryos; *a.* embryonomic.

embryophore *n.* [Gk. *embryon,* embryo; *pherein,* to bear] ciliated mantle enclosing embryo in many tapeworms, and formed from superficial blastomeres of embryo.

Embryophyta, embryophytes *n., n.plu.* [Gk. *embryon,* embryo; *phyton,* plant] a major division of the plant kingdom consisting of plants having an enclosed embryo, as those with an archegonium or bearing seeds, including the Bryophyta, Pteridophyta, and Spermatophyta.

embryo sac the megaspore in angiosperms, containing the female gametophyte; *alt.* gynospore.

embryotectonics *n.* [Gk. *embryon,* embryo; *tekton,* builder] the structure or cellular pattern of the embryo.

embryotega *n.* [Gk. *embyron,* embryo, *tegos,* roof] small hardened portion of testa which marks micropyle in some seeds and separates like a little lid at period of germination.

embryotroph *n.* [Gk. *embryon,* embryo; *trophē,* nourishment] cellular debris formed from uterine tissue around an implanted embryo, which provides nutrition for the embryo until it has developed its own vascular system.

embryotrophy *n.* [Gk. *embryon,* embryo; *trophē,* nourishment] nourishment of embryo, or means adapted therefor.

emergence *n.* [L. *emergere,* to come up] an outgrowth from subepidermal tissue; an epidermal appendage.

emersed *a.* [L. *emergere,* to come up] rising above surface of water, *appl.* leaves; *alt.* amphibian.

EMF erythrocyte-maturing factor, *q.v.*

eminence *n.* [L. *eminens,* eminent] ridge or projection on surface of bones, *alt.* eminentia.

emissary *a.* [L. *emittere,* to send out] coming out; *appl.* veins passing through apertures in cranial wall and establishing connection between sinuses inside and veins outside.

emmenine *n.* [Gk. *emmēnos,* monthly] a placental gonadotrophic hormone.

emmenophyte, emmophyte *n.* [Gk. *emmenein,* to abide in; *phyton,* plant] a water plant without any floating parts; *cf.* plotophyte.

EMP Embden–Meyerhof–Parnas pathway, *q.v.*

empennate pinnate, *q.v.*

empodium *n.* [Gk. *en,* in; *pous,* foot] a small variable median structure between claws of feet in many insects and spiders.

empyreumatic *a.* [Gk. *empyreuein,* to kindle] *appl.* a class of odours resembling those of burning or scorched vegetable or animal substances.

emulsin *n.* [L. *emulgere,* to milk out] a vague term for various enzymes found in certain plants, such as almonds and moulds, and in some invertebrates, which can hydrolyse glycosides and lactose.

enamel *n.* [O.F. *esmaillier,* to coat with enamel] the hard material containing over 90% calcium and magnesium salts which forms a cap over dentine or may form a complete coat to tooth or scale.

enamel cell ameloblast, *q.v.*

enamel organ a complex structure of tall columnar epithelium, lying on a mesodermal core, the

dental papilla, from which tooth enamel is developed.

enantiobiosis *n.* [Gk. *enantios,* opposite; *bios,* life] antagonistic symbiosis.

enantioblastic *a.* [Gk. *enantios,* opposite; *blastos,* bud] formed at end of seed opposite placenta.

enantiomers *n.plu.* [Gk. *enantios,* opposite; *meros,* part] optical isomers, *q.v.*

enantiomorphic *a.* [Gk. *enantios,* opposite; *morphē,* form] similar but contraposed, as mirror image, right and left hand; deviating from normal symmetry.

enantiostylous *a.* [Gk. *enantios,* opposite; *stylos,* pillar] having flowers whose styles protrude right or left of the axis, with the stamens on the other side.

enarthrosis *n.* [Gk. *en,* in; *arthron,* joint] ball-and-socket joint, *q.v.*

enation *n.* [L. *enatus,* grown from] a non-reproductive accessory part emerging from surface of telome; outgrowth from a previously smooth surface.

encapsis *n.* [Gk. *en,* in; L. *capsa,* box] the association of myofibrils in bundles, of those bundles into larger bundles, and so on.

encephalic *a.* [Gk. *engkephalos,* brain] *pert.* the brain.

encephalization *n.* [Gk. *engkephalos,* brain] brain formation by the forward-shifting and centralizing tendency of coordinating neurones.

encephalocoel(e) *n.* [Gk. *engkephalos,* brain; *koilos,* hollow] the cavity of the brain or cerebral ventricle.

encephalomere *n.* [Gk. *engkephalos,* brain; *meros,* part] a brain segment.

encephalomyelic *a.* [Gk. *engkephalos,* brain; *myelos,* marrow] *pert.* brain and spinal cord.

encephalon *n.* [Gk. *engkephalos,* brain] the brain, *q.v.*

encephalospinal *a.* [Gk. *engkephalos,* brain] cerebrospinal, *q.v.*

enchondral endochondral, *q.v.*

enchylema *n.* [Gk. *en,* in; *chylos,* juice] cell sap, *q.v.,* in Protozoa.

encretion *n.* [Gk. *en,* within; *krinein,* L. *cernere,* to put apart] endocrine secretion.

encyst *v.* [Gk. *en,* in; *kystis,* bladder] of a cell or small organism, to surround itself with an outer coat or capsule.

encystation, encystment *n.* [Gk. *en,* in; *kystis,* bladder] formation of a firm, resistant envelope or capsule.

endangium *n.* [Gk. *endon,* within; *anggeion,* vessel] innermost lining or tunica intima of blood vessels.

endarch *a.* [Gk. *endon,* within; *archē,* beginning] with central protoxylem, or several protoxylem groups surrounding pith, produced when xylem matures centrifugally, so the oldest protoxylem is closest to the centre of the axis; *cf.* exarch.

endaspidean *a.* [Gk. *endon,* within; *aspis,* shield] with scutes extending on inner surface of tarsus.

end-brain telencephalon, *q.v.*

end-brush telodendrion, *q.v.*

end-bulbs minute cylindrical or oval bodies, consisting of capsule containing a semifluid core in which axis cylinder terminates either in a bulbous extremity or in a coiled plexiform mass, being end organs in mucous and serous membranes, in skin of genitalia, and in synovial layer of certain joints; *alt.* Krause's end-bulbs, boutons.

end cell a cell incapable of further differentiation.

end-disc ring centriole, *q.v.*

endemic *a.* [Gk. *endēmos*, native] restricted to a certain region or part of a region.

endergonic *a.* [Gk. *endon*, within; *ergon*, work] absorbing energy; *cf.* exergonic.

enderon *n.* [Gk. *en*, in; *deros*, skin] the inner or endodermal layer.

enderonic *a.* [Gk. *en*, in; *deros*, skin] endodermic, *q.v.*

endexine *n.* [Gk. *endon*, within; *exō*, outside] inner membranous layer of extine; *cf.* ektexine.

endites *n.plu.* [Gk. *endon*, within] offshoots on mesial border of certain appendages of arthropods.

endo- *also* ento-, *q.v.*

endoascus *n.* [Gk. *endon*, within; *askos*, bag] inner membrane of an ascus, protruding after rupture of the ectoascus, as in certain Ascomycetes.

endobasal *a.* [Gk. *endon*, within; *basis*, base] *appl.* a body in the nucleus which acts as a centrosome, as in protozoans; *cf.* endosome.

endobasidium *n.* [Gk. *endon*, within; *basis*, base; *idion*, dim.] a basidium developing inside the fruit body, as in Gasteromycetes.

endobatic *a.* [Gk. *endon*, within; *bainein*, to go] afferent; esodic, centripetal; *cf.* ectobatic.

endobiotic *a.* [Gk. *endon*, within; *biōtikos*, pert. life] living within a substratum or within another living organism; *cf.* exobiotic.

endoblast *n.* [Gk. *endon*, within; *blastos*, bud] hypoblast, *q.v.*

endobronchus entobronchus, *q.v.*

endocardiac, endocardial *a.* [Gk. *endon*, within; *kardia*, heart] situated within the heart; *alt.* intracardiac.

endocardium *n.* [Gk *endon*, within; *kardia*, heart] the membrane which lines inner surface of heart.

endocarp *n.* [Gk. *endon*, within; *karpos*, fruit] the innermost layer of the pericarp, usually hard, fibrous, or stony, as in drupes.

endocarpic *a.* [Gk. *endon*, within; *karpos*, fruit] *pert.* endocarp; *pert.* angiocarpy.

endocarpoid *a.* [Gk. *endon*, within; *karpos*, fruit; *eidos*, form] having the disc-like ascocarps embedded in the thallus.

endocarpy *n.* [Gk. *endon*, within; *karpos*, fruit] angiocarpy, *q.v.*

endocele endocoel, *q.v.*

endochiton *n.* [Gk. *endon*, within; *chitōn*, coat] innermost layer of oogonial wall, as in Fucales; *alt.* endochite; *cf.* exochiton, mesochiton.

endochondral *a.* [Gk. *endon*, within; *chondros*, cartilage] *appl.* ossification in cartilage beginning inside and working outwards; *alt.* enchondral, intracartilaginous; *cf.* perichondral.

endochondrostosis *n.* [Gk. *endon*, within; *chondros*, cartilage; *osteon*, bone] endochondral ossification; *alt.* entochondrostosis; *cf.* perichondrostosis.

endochone *n.* [Gk. *endon*, within; *chōnē*, funnel] spacious subcortical crypt in sponge tissue, from which arise incurrent canals.

endochorion *n.* [Gk. *endon*, with; *chorion*, skin] inner lamina of chorion of insect eggs.

endochroic *a.* [Gk. *endon*, within; *chrōs*, complex-

ion] having pigment within a cell or hypha; *cf.* ectochroic.

endochrome *n.* [Gk. *endon*, within; *chrōma*, colour] any pigment within a cell, especially other than chlorophyll.

endochrome plate a band of yellowish chromatophores found in protoplasmic portion of certain diatoms.

endochromidia *n.plu.* [Gk. *endon*, within; *chrōma*, colour; *idion*, dim.] metachromatic corpuscles, formed from colloidal solution of metachromatin.

endochylous *a.* [Gk. *endon*, within; *chylos*, juice] with water cells within internal tissue.

endocoel *n.* [Gk. *endon*, within; *koilos*, hollow] coelom, *q.v.*; the cavities in proboscis, collar, and trunk of Enteropneusta; *alt.* endocele.

endocoelar *a.* [Gk. *endon*, within; *koilos*, hollow] *pert.* inner wall of coelom.

endocoelic *a.* [Gk. *endon*, within; *koilos*, hollow] in sea anemones, *appl.* radial area on disc covering space between 2 mesenteries of the same pair; *appl.* inner cycle or cycles of tentacles; *cf.* exocoelic.

endocone *n.* [Gk. *endon*, within; *kōnos*, cone] a conical structure formed in certain cephalopod shells.

endoconidium *n.* [Gk. *endon*, within; *konis*, dust; *idion*, dim.] a conidium formed within a conidiophore.

endocranium *n.* [Gk. *endon*, within; *kranion*, skull] process on inner surface of cranium of certain insects for muscle attachment; neurocranium, *q.v.*

endocrine *n.* [Gk. *endon*, within; *krinein*, to separate] a ductless gland which secretes hormones. *a.* *Appl.* or *pert.* such a gland; formerly also called cryptorhetic. *Cf.* exocrine.

endocrinology *n.* [Gk. *endon*, within; *krinein*, to separate; *logos*, discourse] study of endocrine glands and secretions, and of hormones and their effects.

endocuticle, endocuticula *n.* [Gk. *endon*, within; L. *dim.* of *cutis*, skin] the elastic inner layer of insect cuticle; inner layer of integument in spiders.

endocycle *n.* [Gk. *endon*, within; *kyklos*, circle] a layer of tissue separating internal phloem from endodermis.

endocyclic *a.* [Gk. *endon*, within; *kyklos*, circle] with the mouth remaining in axis of coil of gut, *appl.* crinoids; having an apical system with double circle of plates surrounding anus, *appl.* echinoids; *pert.* endocycle.

Endocyclica *n.* [Gk. *endon*, within; *kyklos*, circle] an order of regular Echinacea having a central mouth and anus within the apical area; *alt.* Echinoida.

endocyst *n.* [Gk. *endon*, within; *kystis*, bladder] the soft body wall in a polyzoan zooid; the membranous inner lining of a protozoan cyst; *cf.* epicyst.

endocytosis *n.* [Gk. *endon*, within; *kytos*, hollow] uptake of material by a cell by pinocytosis or phagocytosis.

endodeme *n.* [Gk. *endon*, within; *dēmos*, people] a gamodeme composed of predominantly inbreeding dioecious plants or bisexual animals.

endoderm- *alt.* entoderm-.

endoderm *n.* [Gk. *endon*, within; *derma*, skin] the

inner germ layer in a gastrula, which lines the archenteron; any tissues derived from this layer, such as the epithelium of digestive and respiratory organs, and of glands associated with digestive tract; *alt.* enderon, gastrodermis, hypoblast.

endodermal endodermic, *q.v.*

endoderm disc posterior unpaired thickening on ventral surface of blastoderm of crayfish and other Malacostraca.

endodermic *pert.* endoderm; *pert.* endodermis; *alt.* enderonic, endodermal.

endodermis *n.* [Gk. *endon*, within; *derma*, skin] innermost layer of cortex in plants; layer surrounding pericycle; *alt.* phloeoterma.

endoderm lamella a thin sheet of endoderm stretching between adjacent radial canals, and between circular canal and enteric cavity in the medusae of certain Coelenterata.

endodermoid *a.* [Gk. *endon*, within; *derma*, skin; *eidos*, form] resembling endodermis.

endoenzyme *n.* [Gk. *endon*, within; *en*, in; *zymē*, leaven] any intracellular enzyme; *cf.* exoenzyme.

endogamy *n.* [Gk. *endon*, within; *gamos*, marriage] zygote formation within the cyst by reciprocal fusion of division products of daughter nuclei; self-pollination; inbreeding, *q.v.*

endogastric *a.* [Gk. *endon*, within; *gastēr*, belly] having curvature of body with enclosing shell towards ventral side; within the stomach.

endogenote *n.* [Gk. *endos*, within; *-genēs*, producing] in bacterial sexual processes, the part of the chromosome of the recipient that is homologous to the exogenote.

endogenous *a.* [Gk. *endon*, within; *-geñes*, producing] originating within the organism; autogenic, *cf.* allogenic; developing from a deep-seated layer, *appl.* lateral roots; *appl.* metabolism concerned with tissue waste and growth; *cf.* exogenous.

endogenous multiplication *see* spore formation.

endogenous rhythm a metabolic or behavioural rhythm that originates within an organism and persists even though external conditions are kept constant, although there may be some slight change in the rhythm length, when it is said to be free running, such as a circadian rhythm changing to 23 or 25 hours, endogenous circadian rhythms being maintained at a regular 24-hour cycle by an external stimulus such as light or temperature, the zeitgeber; *alt.* internal rhythm.

endogeny *n.* [Gk. *endon*, within; *genos*, descent] development from a deep-seated layer.

endognath *n.* (Gk. *endon*, within; *gnathos*, jaw] the inner branch of oral appendages of Crustacea; *alt.* endognathite.

endognathion *n.* [Gk. *endon*, within; *gnathos*, jaw] mesial segment of human premaxilla.

endognathite endognath, *q.v.*

endogonidium *n.* [Gk. *endon*, within; *dim.* of *gonē*, seed] a gonidium formed in a gonidangium or receptacle; the colony-forming cells in such forms as the protistan, *Volvox*.

endolabium *n.* [Gk. *endon*, within; L *labium*, lip] a membranous lobe in interior of mouth on middle parts of front of labium.

endolaryngeal *a.* [Gk. *endon*, within; *laryngx*, larynx) *pert.* or in the larynx.

endolithic *a.* [Gk. *endon*, within; *lithos*, stone] burrowing or existing in stony substratum, as algal filaments.

endolymph *n.* [Gk. *endon*, within; L. *lympha*, water] the fluid in membranous labyrinth of ear.

endolymphangial *a.* [Gk. *endon*, within; L. *lympha*, water; Gk. *anggeion*, vessel] situated in a lymphatic vessel.

endolymphatic *a.* [Gk. *endon*, within; L. *lympha*, water] *pert.* lymphatics, or to ear labyrinth ducts.

endolysin *n.* [Gk. *endon*, within; *lysis*, loosing] intracellular substance of leucocytes which destroys engulfed bacteria.

endolysis *n.* [Gk. *endon*, within; *lysis*, loosing] intracellular dissolution.

endomembranes *n.plu.* [Gk. *endon*, within; L. *membrana*, membrane] the membranes inside a cell, including the endoplasmic reticulum, Golgi bodies, and various vesicles.

endomeninx *n.* [Gk. *endon*, within; *meningx*, membrane] single inner membrane covering embryonic brain, giving rise to pia mater and arachnoid.

endomere *n.* [Gk. *endon*, within; *meros*, part] a hypoblast cell which gives rise to endoderm.

endometrium *n.* [Gk. *endon*, within; *mētra*, womb] mucous membrane lining the uterus.

endomitosis *n.* [Gk. *endon*, within; *mitos*, thread] a form of mitosis occurring in endopolyploidy; multiplication of chromonemata or chromosomes without division of nucleus.

endomixis *n.* [Gk. *endon*, within; *mixis*, mixing] a stage comparable with parthenogenesis in the reproductive rhythm of some Protozoa; a type of nuclear reorganization.

Endomycetales *n.* [Gk. *endon*, within; *mykēs*, fungus] a group of lower (hemi) Ascomycetes including the yeasts, and in which plasmogamy is immediately followed by karyogamy; *alt.* Protoascales, Saccharomycetales.

endomysium *n.* [Gk. *endon*, within; *mys*, muscle] the connective tissue binding muscle fibres; *cf.* perimysium, epimysium.

endoneurium *n.* [Gk. *endon*, within; *neuron*, nerve] the delicate connective tissue holding together and supporting nerve fibres within funiculus.

endonuclease *n.* [Gk. *endon*, within; L. *nucleus*, kernel] an enzyme that hydrolyses DNA by breaking bonds in the middle of the chain.

endoparasite *n.* [Gk. *endon*, within; *parasitos*, eating at another's table] a parasite that lives inside another organism; *alt.* endosite, entoparasite.

endopeptidases *n. plu.* [Gk. *endon*, within; *pepsis*, digestion] proteinases which split a protein molecule into smaller units by attacking the interior peptide bonds; EC sub-subgroups 3.4.21–3.4.99.

endoperidium *n.* [Gk. *endon*, within; *pēridion*, little pouch] inner layer of peridium.

endoperineurial *a.* [Gk. *endon*, within; *peri*, around; *neuron*, nerve] *pert.* both endoneurium and perineurium.

endophragm *n.* [Gk. *endon*, within; *phragma*, fence] a septum formed by cephalic and thoracic apodemes in Crustacea; *a.* endophragmal.

endophyllous *a.* [Gk. *endon*, within; *phyllon*, leaf] sheathed by a leaf; living within a leaf, *appl.* parasites.

endophyte *n.* [Gk. *endon*, within; *phyton*, plant] a plant growing within another, either as a parasite or symbiont, or otherwise; *alt.* entophyte; *a.* endophytic.

endoplasm *n.* [Gk. *endon*, within; *plasma*, mould] the inner part of the cytoplasm of a cell, which is often granular; endosarc, *q.v.*; *alt.* entoplasm; *cf.* ectoplasm.

endoplasmic reticulum (epr, er, ER) a meshwork of double membranes in the cytoplasm which are continuous with the cell membrane and nuclear membrane, and may be lined with ribosomes when it is called rough (ergastoplasm) and is the site of protein synthesis, or may not have ribosomes when it is called smooth; kinoplasm, *q.v.*

endoplast *n.* [Gk. *endon*, within; *plastos*, moulded] cell nucleus; macronucleus of certain Protista.

endoplastule *n.* [Gk. *endon*, within; *plastos*, moulded] the micronucleus of certain Protista.

endopleura *n.* [Gk. *endon*, within; *pleura*, side] the inner seed coat; *alt.* tegmen.

endopleurite *n.* [Gk. *endon*, within; *pleura*, side] the epimeral portion of an apodeme; infolding between pleurites.

endoplica *n.* [Gk. *endon*, within; L. *plicare*, to fold] implex, *q.v.*

endopodite *n.* [Gk. *endon*, within; *pous*, foot] the inner or mesial branch of a biramous crustacean limb, or the only part of biramous limb remaining.

endopolyploidy *n.* [Gk. *endon*, within; *polys*, many; *aploos*, onefold; *eidos*, form] polyploidy resulting from repeated doubling of chromosome number without normal mitosis.

Endoprocta, endoprocts *n., n.plu.* [Gk. *endon*, within; *prōktos*, anus] a phylum of solitary or colonial pseudocoelomate animals, formerly considered a class of Bryozoa, having a U-shaped gut with the anus opening to the exterior within a circle of ciliated tentacles; *alt.* Entoprocta.

Endopterygota *n.* [Gk. *endon*, within; *pterygion*, little wing] a division of insects having complete metamorphosis with wings developing internally, and larvae different from adults; *alt.* Holometabola, Oligoneoptera, Oligoneuroptera; *cf.* Exopterygota.

endoral *a.* [Gk. *endon*, within; L. *os*, mouth] *pert.* structures situated in the vestibule of certain protozoans.

end organ a structure at the end of a nerve, such as a receptor or motor end plate.

endorhachis *n.* [Gk. *endon*, within; *rhachis*, backbone] a layer of connective tissue lining canal of vertebral column and cavity of skull; *alt.* endorachis.

endorhizal *a.* [Gk. *endon*, within; *rhiza*, root] with the radicle enclosed, as in seed of monocotyledons.

endorphins *n.plu.* [Gk. *endon*, within; *morphine*] a group of morphine-like chemicals in the brain, especially the mid-brain and pituitary gland, which mimic the effects of morphine.

endosarc *n.* [Gk. *endon*, within; *sarx*, flesh] the internal layer of cytoplasm, especially in a protozoan; *alt.* endoplasm.

endosclerite *n.* [Gk. *endon*, within; *sklēros*, hard] any sclerite of the endoskeleton of Arthropoda.

endoscopic *a.* [Gk. *endon*, within; *skopein*, to look] with apex directed inwards towards base of archegonium, *appl.* embryo; *cf.* exoscopic.

endosiphuncle *n.* [Gk. *endon*, within; L. *siphunculus*, little tube] the tube leading from protoconch to siphuncle in certain Cephalopoda; *alt.* prosiphon.

endosite *n.* [Gk. *endon*, within; *sitos*, food] endoparasite, *q.v.*

endoskeleton *n.* [Gk. *endon*, within; *skeletos*, dried up] internal skeleton; *alt.* neuroskeleton; *cf.* exoskeleton.

endosmosis *n.* [Gk. *endon*, within; *ōsmos*, impulse] osmosis in an inward direction, such as into a cell; *cf.* exosmosis.

endosome *n.* [Gk. *endon*, within; *sōma*, body] a chromatinic body near the centre of a vesicular nucleus, either a karyosome or plasmosome, as in certain protozoans, *cf.* endobasal body; in certain sponges, the inner part of the body with flagellated chambers but few or no spicules.

endosperm *n.* [Gk. *endon*, within; *sperma*, seed] the nutritive tissue of most seeds; nutritive residue of female prothallus surrounding an embryo; *alt.* albumen.

endospore *n.* [Gk. *endon*, within; *sporos*, seed] inner coat of sporocyst in some Protozoa; an asexual spore; a spore borne inside a sporangium; endosporium, *q.v.*

Endosporeae *n.* [Gk. *endon*, within; *sporos*, seed] A subclass of Myxomycetes having spores enclosed in a common membrane; *alt.* Myxogastres.

endosporium inner coat of a spore wall; *alt.* endospore.

endosporous *a.* [Gk. *endon*, within; *sporos*, seed] having spores borne inside an organ such as a sporangium.

endosteal *a.* [Gk. *endon*, within; *osteon*, bone] *pert.* endosteum.

endosternite *n.* [Gk. *endon*, within; L. *sternum*, sternum] internal skeletal plate for muscle attachment; median sternal apodeme; a free skeleton situated in prosoma between alimentary canal and nerve cord in arachnids; *alt.* entosternite.

endosteum *n.* [Gk. *endon*, within; *osteon*, bone] the internal periosteum lining the cavities of bones; *alt.* medullary membrane.

endostome *n.* [Gk. *endon*, within; *stoma*, mouth] inner portions of peristome, as in certain mosses.

endostosis *n.* [Gk. *endon*, within; *osteon*, bone] ossification which begins in cartilage.

endostracum *n.* [Gk. *endon*, within; *ostrakon*, shell] the inner layer of mollusc shell.

endostroma entostroma, *q.v.*

endostyle *n.* [Gk. *endon*, within; *stylos*, pillar] a band of thickened epithelium on oesophageal wall of a tornaria; 2 ventral longitudinal folds separated by a groove in pharynx of Tunicata; a longitudinal groove lined by ciliated epithelium on ventral wall of pharynx of *Amphioxus*, the lancelet; subpharyngeal gland of Cyclostomata; precursor of thyroid gland.

endotergite *n.* [Gk. *endon*, within; L. *tergum*, back] an infolding from a tergite of insects, for muscle attachment; *alt.* phragma.

endotheca *n.* [Gk. *endon*, within; *thēkē*, box] the system of dissepiments in a coral calyx; the oval surface of Cystidea.

endothecial *a.* [Gk. *endon*, within; *thēkē*, box] *pert.* endothecium; with asci in an ascocarp.

endothecium *n.* [Gk. *endon*, within; *thēkē*, box]

in bryophytes, inner tissue of sporogonium, formed from central region of embryo, cf. amphithecium; inner lining of an anther; cf. exothecium.

endotheliochorial a. [Gk. endon, within; thēlē, nipple; chorion, skin] appl. placenta with chorionic epithelium in contact with endothelium of uterine capillaries, as in carnivores and certain other mammals; alt. vasochorial; cf. epitheliochorial, haemochorial.

endotheliocyte n. [Gk. endon, within; thēlē, nipple; kytos, hollow] a mononuclear phagocyte derived from endothelium; endothelial phagocyte or primitive wandering cell; histiocyte, q.v.; macrophage, q.v.; alt. transitional cell.

endothelium n. [Gk. endon, within; thēlē, nipple] a squamous epithelium which lines internal surfaces such as serous cavities, the heart, blood and lymphatic vessels.

endotherm n. [Gk. endon, within, thērme, heat] a homoiothermal animal; cf. ectotherm; a. endothermal.

endothermic a. [Gk. endon, within; thermē, heat] absorbing and utilizing heat energy; cf. exothermic.

endothorax n. [Gk. endon, within; thōrax, chest] the apodeme system in a crustacean thorax; cf. entothorax.

endotoxin n. [Gk. endon, within; toxikon, poison] a toxin within bacterial protoplasm; cf. exotoxin.

endotrachea n. [Gk. endon, within; L. trachia, windpipe] the innermost chitinous coat of tracheal tubes of insects.

endotrophic a. [Gk. endon, within; trophē, nourishment] finding nourishment from inside; appl. space within peritrophic membrane of insects; appl. mycorrhiza where hyphae of fungus penetrate extensively into root of host; cf. ectotrophic.

endozoic a. [Gk. endon, within; zōon, animal] living within an animal or involving passage through an animal as in the distribution of some seeds; cf. epizoic, entozoic.

endozoochore n. [Gk. endon, within; zōon, animal; chōrein, to spread] any spore, seed, or organism dispersed by being carried within an animal; cf. epizoochore.

endozoochory dispersal of an endozoochore.

end-piece the sheathless axial filament which forms the tail of a spermatozoon.

end-plate potential (EPP) the potential formed across an end plate by the release of neurotransmitter at the neuromuscular junction.

end plates motor end organs, the ramified expansions within the muscle fibre which form the ends of a motor nerve.

end-product inhibition the inhibition of the synthesis of a group of enzymes in a metabolic pathway by the end-product of the activity of the enzyme which combines with and inhibits the 1st enzyme of the pathway; alt. feedback inhibition.

end-ring ring centriole, q.v.

end sac the sac-like vestigial portion of coelom in excretory glands of certain Crustacea.

end-sheath telolemma, q.v.

endysis n. [Gk. endysis, putting on] the development of a new coat; cf. ecdysis.

energesis n. [Gk. energein, to be active] the process by which energy is liberated through catabolic action.

energid n. [Gk. energos, working; idion, dim.] any living uninucleated protoplasmic unit with or without a cell wall; active protoplasm, cf. deutoplasm.

energy flow the transfer of energy from one organism to another in an ecosystem.

enervose a. [L. ex, without; nervus, sinew] having no veins, appl. certain leaves.

engram n. [Gk. en, in; graphein, to write] a character impression in the mnemic theory of heredity; a latent record, or physiological memory pattern, of stimulation in living tissue.

engraved a. [Gk. en, in; A.S. grafan, to dig] with irregular linear grooves on the surface.

enhalid a. [Gk. en, in presence of; hals, salt] containing salt water, appl. soils; growing in saltings or on loose soil in salt water, appl. plants.

enkephalin n. [Gk. en, in; kephalē, head] either of 2 molecules found in the brain of some mammals, which have pain-killing properties and seem to be composed of 2 small peptides, but these may be the breakdown products of a larger molecule.

enneagynous a. [Gk. ennea, nine; gynē, female] having 9 pistils.

enneandrous a. [Gk. ennea, nine; anēr, male] having 9 stamens.

enolase the enzyme which converts phosphoglyceric acid to phospho enol pyruvic acid; EC 4. 2. 1. 11; r.n. enolase.

Enopla n. [Gk, enoplos, armed] a class of proboscis worms having the mouth anterior to the brain, and usually an armed proboscis.

Enoplida n. (Enoplus, generic name) an order of aquatic Aphasmidia having pocket-like amphids and bristles on the head.

enphytotic a. [Gk. en, in; phyton, plant] afflicting plants, appl. diseases restricted to a locality; cf. epiphytotic.

ensiform a. [L. ensis, sword; forma, shape] sword-shaped; alt. xiphoid, gladiate.

entad adv. [Gk. entos, within; L. ad, towards] towards the interior; inwards; internally; cf. ectad.

ental a. [Gk. entos, within] inner; internal; cf. ectal.

entangial a. [Gk. entos, within; anggeion, vessel] within a vessel; produced inside a sporangium; alt. entoangial; cf. ectangial.

entelechy n. [Gk. en, in; telos, end; echein, to hold] vital principle or influence guiding living organisms in right direction.

entepicondylar a. [Gk. entos, within; epi, upon; kondylos, knuckle] pert. lower or condylar end of humerus, appl. ulnar foramen.

enteral a. [Gk, enteron, gut] within intestine; appl. the parasympathetic portion of the autonomic nervous system.

enteramine n. [Gk. enteron, gut; ammōniakon, a gum] a hormone, identical with serotonin, in chromaffin cells of mammalian intestinal tract, inducing contraction of smooth muscle.

enteric a. [Gk. enteron, gut] pert. alimentary canal; pert. intestines.

enteroblast n. [Gk. enteron, gut; blastos, bud] the hypoblast after formaton of the mesoblast; alt. enteroderm, gastrodermis.

enterocoel n. [Gk. enteron, gut; koilos, hollow] a coelom arising as a pouch-like outgrowth of archenteric cavity, or as a series of such outgrowths.

enterocoely the process of making an enterocoel.

enterocrinin n. [Gk. enteron, gut; krinein, to separate] a hormone of small intestine, which stimulates secretion of intestinal juice.

enteroderm n. [Gk. enteron, gut; derma, skin] enteroblast, q.v.

enterogastrone n. [Gk. enteron, gut; gastēr, stomach] a duodenal hormone which inhibits secretion and motility of stomach.

Enterogona n. [Gk. enteron, gut; gonē, seed] an order of solitary or colonial sea squirts that often have the body divided into regions.

enterokinase n. [Gk. enteron, gut; kinein, to move] proteinase of intestinal juice which converts trypsinogen to trypsin; EC 3.4.21.9; r.n. enteropeptidase.

enteron n. [Gk. enteron, gut] the alimentary tract; coelenteron, q.v.; any structure which corresponds to the archenteron of a gastrula.

enteronephric a. [Gk. enteron, gut; nephros, kidney] with nephridia opening into gut, appl. oligochaetes; cf. exonephric.

enteropeptidase n. [Gk. enteron, gut; peptein, to digest] enterokinase, q.v.

Enteropneusta n. [Gk. enteron, gut; pneuma, air] a class of solitary worm-like burrowing hemichordates, having many gill slits and no lophophore.

enteroproct n. [Gk. enteron, but; prōktos, anus] the opening from endodermal gut into proctodaeum.

enterostome n. [Gk. enteron, gut; stoma, mouth] the aboral opening of the actinopharynx, leading to coelenteron; the posterior opening of stomodaeum into endodermal gut.

enterosympathetic a. [Gk. enteron, gut: syn, with; pathos, feeling] appl. that part of the nervous system supplying the intestine or gut.

enteroviruses n.plu. [Gk. enteron, gut; L virus, poison] a group of RNA viruses which live in the human intestine.

enterozoon n. [Gk. enteron, gut; zōon, animal] any animal parasite inhabiting the intestines.

enthetic a. [Gk. enthetos, put in] introduced; implanted.

entire a. (O.F. entier, untouched] unimpaired; with continuous margin, appl. leaves, bacterial colony, etc.

ento- also endo-, q.v.

entoangial entangial, q.v.

entobranchiate a. [Gk. entos, within; brangchia, gills] having internal gills.

entobronchus n. [Gk. entos, within; brongchos, windpipe] the dorsal secondary branch of bronchus in birds; alt. entobronchium, endobronchus.

entochondrite n. [Gk entos, within; chondros, cartilage] plastron or endosternum of the king crab, Limulus.

entochondrostosis endochondrostosis, q.v.

entocodon n. [Gk. entos, within; kōdōn, bell] the lens-shaped mass of cells, in development of medusoid, which sinks below level of superficial ectoderm, and ultimately develops a cavity.

entocoel n. [Gk. entos, within; koilos, hollow] the space enclosed by a pair of mesenteries in Anthozoa.

entocondyle n. [Gk, entos, within; kondylos, knuckle] condyle on mesial surface of a bone.

entoconid n. [Gk. entos, within; kōnos, cone] the postero-internal cusp of a lower molar.

entocuneiform n. [Gk. entos, within; L. cuneus, wedge; form, shape) the most internal of distal tarsal bones.

entocyemate a. [Gk. entos, within; kyēma, embryo] with embryos having amnion and allantois.

entocyte n. [Gk. entos, within; kytos, hollow] the contents of a plant cell; cf. cell wall.

entoderm- endoderm-, q.v.

Entodiniomorphida n. [Entodinium, generic name; Gk. morphē, form] an order or suborder of Spirotricha which are endozoic in herbivorous mammals and have a complex body with few cilia.

entoectad adv. [Gk. entos, within; ektos, without; L. ad, towards] from within outwards; cf. ectoentad.

entogastric a. [Gk. entos, within; gastēr, belly] pert. interior of stomach or enteron, appl. gastric budding in medusae.

entoglossal a. [Gk. entos, within; glōssa, tongue] lying in substance of tongue.

entoglossum n. [Gk. entos, within; glōssa, tongue] extension of basihyal into tongue in some fishes; alt. glossohyal.

entomochoric a. [Gk. entomon, insect; chōrein, to spread] dispersed by insects; depending on insects for spreading spores, etc.; n. entomochory.

entomogenous a. [Gk. entomon, insect; genēs, born] growing in or on insects, as certain fungi.

entomology n. [Gk. entomon, insect; logos, discourse] that part of zoology which deals with insects.

entomophagous a. [Gk entomon, insect; phagein, to eat] insect-eating; alt. insectivorous.

entomophilous a. [Gk. entomon, insect; philein, to love] pollinated by agency of insects; n. entomophily.

Entomophthorales n. [Entomophthora, generic name] an order of Phycomycetes, mostly parasitic on insects, with asexual reproduction by conidiosporangia and isogamous or anisogamous sexual reproduction.

entomophyte n. [Gk. entomon, insect; phyton, plant] any fungus growing on or in insects.

Entomostraca n. (Gk. entomon, insect; ostrakon, shell] in some classifications, a group of crustaceans, including the Branchiopoda, Branchiura, Copepoda, Cirripedia, and Ostracoda.

entomo-urochrome n. [Gk. entomon, insect; ouron, urine; chrōma, colour] greenish or yellowish pigment in urine of insects.

entoneural a. (Gk. entos, within; neuron, nerve] appl. system of aboral ring and genital nerves in echinoderms.

entoparasite endoparasite, q.v.

entophyte endophyte, q.v.

entopic a. [Gk. en, in; topos, place] in normal position; cf. ectopic.

entoplasm endoplasm, q.v.

entoplastron n. [Gk. entos, within; F. plastron, breast plate] the anterior median plate in chelonian plastra, often called episternum, probably homologous with interclavicle of other reptiles; alt. entosternum.

Entoprocta n. [Gk. entos, within; prōktos, anus] Endoprocta, q.v.

entopterygoid n. [Gk. entos, within; pteryx, wing; eidos, form] a dorsal membrane bone behind the palatine in some fishes; *cf.* ectopterygoid.

entoptic *a.* [Gk. *entos*, within; *ōps*, eye] within the eye; *appl.* visual sensations caused by eye structures or processes, not by light; *alt.* intra-ocular.

entoretina *n.* [Gk. *entos*, within; L. *rete*, net] inner or neural part of retina, the retina proper.

entosphere *n.* [Gk. *entos*, within; *sphaira*, globe] the inner portion of attraction sphere.

entosternite endosternite, *q.v.*

entosternum *n.* [Gk. *entos*, within; L. *sternum*, breast bone] entoplastron, *q.v.*; an internal process of sternum of numerous arthropods.

entostroma *n.* [Gk. *entos*, within; *strōma*, bedding] stroma producing perithecia in Ascomycetes; *alt.* endostroma, hypostroma; *cf.* ectostroma.

entothorax *n.* [Gk. *entos*, within; *thōrax*, chest] an insect apophysis or sternite; endothorax, *q.v.*

Entotropha *n.* [Gk. *entos*, within; *trophē*, nourishment] a superorder of wingless insects having mouth-parts almost entirely hidden within the head and including the Collembola, Protura, and Diplura.

entoturbinals *n.plu.* [Gk. *entos*, within; L. *turbo*, whorl] a division of ethmoturbinals.

entotympanic *n.* Gk. *entos*, within; *tympanon*, drum] a separate tympanic element in some mammals; *alt.* metatympanic.

entovarial *a.* [Gk. *entos*, within; L. *ovum*, egg] inside the ovary; *pert.* canal formed in ovaries of some fishes by insinking and closure of a groove formed by covering epithelium.

entozoa *n.plu.* [Gk. *entos*, within; *zōon*, animal] internal animal parasites; *sing.* entozoon.

entozoic *a.* [Gk. *entos*, within; *zōē*, subsistence] living within the body or substance of another animal or plant; *pert.* entozoa; *cf.* endozoic.

entrainment the process by which a free-running endogenous circadian rhythm is synchronized to an exact 24-hour cycle.

entrance cone fertilization cone, *q.v.*

entrochite *n.* [Gk. *en*, in; *trochos*, wheel] the joint of fossil stem of a stalked crinoid.

enucleate *v.* [L. *e*, out of; *nucleus*, kernel] to deprive of a nucleus, as in microdissection of cells. *a.* Lacking a nucleus.

envelope *n.* [F. *enveloppe*, covering] an outer covering of an egg; any surrounding structure, e.g. floral envelope.

environment *n.* [F. *environ*, about] the sum total of external influences acting on an organism or on part of an organism.

enzootic *a.* [Gk. *en*, in; *zōon*, animal] afflicting animals; *appl.* disease restricted to a locality.

enzyme *n.* [Gk. *en*, in; *zymē*, leaven] a proteinaceous catalyst produced by living organisms and acting on one or more specific substrates; *alt.* ferment, biocatalyst, zymoprotein, zymin; *cf.* apoenzyme, coenzyme, holoenzyme.

enzyme adaptation change in intracellular enzymes caused by specific environmental changes.

enzymoid *n.* [Gk. *en*, in; *zymē*, leaven; *eidos*, form] cytotoxin, *q.v.*

enzymology *n.* [Gk. *en*, in; *zymē*, leaven; *logos*, discourse] the study of enzymes and their functions.

Eoacanthocephala *n.* [Gk. *ēos*, dawn; *akantha*, thorn; *kephalē*, head] an order of Acanthocephala which are parasitic in aquatic animals and have radial rows of proboscis spines.

Eoanthropus *n.* [Gk. *ēos*, dawn; *anthrōpos*, man] Piltdown man, *q.v.*

eobiogenesis *n.* [Gk. *ēos*, dawn; *bios*, life; *genesis*, descent] the 1st occurrence of the formation of living matter from inorganic material.

eobiont *n.* [Gk. *ēos*, dawn; *bios*, life] a hypothetical stage in biopoiesis, being a system that shows some life-like characters but not enough to be generally accepted as living.

Eocene *n.* [Gk. *ēos*, dawn; *kainos*, recent] early epoch of the Tertiary period, between Palaeocene and Oligocene.

Eocrinoidea *n.* [Gk. *ēos*, dawn; *krinon*, lily; *eidos*, form] a small class of Cambrian to Ordovician echinoderms of the subphylum Pelmatozoa having characteristics intermediate between the cystoids and crinoids, and which may be the ancestral group to the Pelmatozoa.

Eogaea *n.* [Gk. *ēos*, dawn; *gaia*, earth] a zoogeographical division including Africa, South America, and Australasia; *cf.* Caenogaea.

Eogene *a.* [Gk. *ēos*, dawn; *gennaein*, to produce] *appl.* and *pert.* the earlier epochs of the Tertiary era: Palaeocene, Eocene, and Oligocene.

eosinophil *a.* [Gk. *ēos*, dawn; *philein*, to love] *appl.* cells, particularly leucocytes or other granulocytes, whose cytoplasm stains readily with the red acidic dye eosin; since they stain with an acid dye they are also called acidophils or oxyphils.

eosinophile eosinophil leucocyte.

Eosuchia *n.* [Gk. *ēos*, dawn; *souchos*, a kind of crocodile] an extinct order of lepidosaur reptiles of Permian to Triassic age having a complete lower temporal arch.

Eotheria, eotherians *n.*, *n.plu.* [Gk. *ēos*, dawn; *thērion*, small animal] a subclass of primitive Mesozoic mammals having molar teeth usually with 3 main cusps in a line, and having a primitive brain case.

Eozoic *a.* [Gk. *ēos*, dawn; *zōē*, life] Pre-Cambrian, *q.v.*

epacme *n.* [Gk. *epi*, upon; *akmē*, prime] the stage in phylogeny of a group just previous to its highest point of development, *cf.* phyloneanic, phylonepionic; stage in development of individual before adulthood; *cf.* paracme, acme.

epactal *a.* [Gk. *epaktos*, adventitious] supernumerary; intercalary. *n.* A sutural or Wormian bone.

epalpate *a.* [L. *ex*, without; *palpus*, palp] not furnished with palpi; *alt.* expalpate.

epanthous *a.* [Gk. *epi*, upon; *anthos*, flower] living on flowers, *appl.* certain fungi.

epapillate *s.* [L. *ex*, without; *papilla*, nipple] not having papillae.

epapophysis *n.* [Gk. *epi*, upon; *apophysis*, offshoot] a median process arising from centre of vertebral neural arch.

eparterial *a.* [Gk. *epi*, upon; L. *arteria*, artery] situated above an artery, *appl.* branch of right bronchus.

epaulettes *n.plu.* [F. *épaule*, shoulder] branched or knobbed processes projecting from outer side of oral arms of many Scyphozoa; crescentic ridges of cilia in echinopluteus; tegulae of Hymenoptera.

epaxial *a.* [Gk. *epi*, upon; L. *axis*, axle] above the axis; dorsal; usually *appl.* axis formed by vertebral column.

epedaphic *a.* [Gk. *epi*, upon; *edaphos*, ground] *pert.*, or depending upon, climatic conditions.

epencephalon n. [Gk. epi, upon; engkephalos, brain] the cerebellum, q.v.

ependyma n. [Gk. ependyma, outer garment] the layer of cells lining cavities of brain and spinal cord; alt. ependyme; a. ependymal.

ephapse n. [Gk. ephaptein, to reach] region of contiguity between 2 axons lying side by side.

ephaptic a. [Gk. ephaptein, to reach] pert. an ephapse; appl. delay: the interval between stimulation of one (pre-ephaptic) axon and response of an apposed other (post-ephaptic) axon.

epharmonic a. [Gk. epi, towards; harmos, fitting] pert. epharmosis; adaptive; adapted to environment; appl. convergence: morphological resemblance of different unrelated species inhabiting the same environment.

epharmosis n. [Gk. epi, towards; harmos, fitting] the process of adaptation of organisms to new environmental conditions; attainment of the state of adaptation or epharmony.

ephebic a. [Gk. ephēbos, adult] adult, pert. stage in development of individual between childhood and old age, when individual is strongest; cf. phylephebic.

Ephedrales n. [Ephedra, generic name] an order of Gnetopsida including the single living genus Ephedra, which are densely branched shrubs with jointed stems and tiny scale-like leaves.

ephedrine n. [Ephedra, generic name] a sympathomimetic alkaloid obtained from various plants of the genus Ephedra, which causes muscle constriction and is used as a nasal decongestant and in various other medical treatments.

ephemeral n. [Gk. ephēmeros, lasting for a day] a short-lived plant or animal species. a. Short-lived; taking place once only, appl. plant movements as expanding of buds; completing life cycle within a brief period.

Ephemeroptera n. [Gk. ephēmeros, lasting for a day; pteron, wing] an order of insects with incomplete metamorphosis, commonly called mayflies, having membranous unequal wings, and aquatic nymphs with 3 long caudal appendages.

ephippial a. [Gk. ephippion, saddle cloth] pert. ephippium; appl. winter eggs, as of rotifers and daphnids.

ephippium n. [Gk. ephippion, saddle cloth] the pituitary fossa, or fossa hypophyseos of sphenoid; a thickened and indurated part of shell separating from the rest at ecdysis; a saddle-shaped modification of cuticle derived, later detached, from carapace and enclosing winter eggs in daphnids.

ephyra, ephyrula n. [Gk. Ephyra, a sea nymph] the small free-swimming jellyfish stage of certain Scyphozoa, produced by strobilation of scyphistoma.

epibasal n. [Gk. epi, upon; basis, base] upper segment of a zygote or embryo, ultimately giving rise to the shoot; alt. epibasal.

epibasidium n. [Gk. epi, upon; basis, base; idion, dim.] the part of a heterobasidium which bears sterigmata and is separated by a septum from the hypobasidium; basidium, q.v.

epibenthos n. [Gk. epi, upon; benthos, depths] fauna and flora of sea bottom between low-water mark and 200-metre line; cf. hypobenthos.

epibiotic a. [Gk. epibiōnai, to survive] surviving, appl. endemic species that are relics of a former flora or fauna; growing on the exterior of living organisms; living on a surface, as of sea bottom, cf. hypobiotic.

epiblast n. [Gk. epi, upon; blastos, bud] the outer layer of a gastrula, alt. ectoblast; in the embryo of some grasses, a small structure opposite the scutellum, thought to be a rudimentary cotyledon.

epiblem(a) n. [Gk. epiblēma, cover] the outermost layer of tissue in roots, which may be the piliferous layer, or the exodermis in an older root where the piliferous layer has worn away; alt. rhizodermis.

epibole n. [Gk. epibolē, putting on] growth of one part over another in embryonic stages; alt. epiboly.

epibolic a. [Gk. epibolē, putting on] growing so as to cover over, appl. type of gastrulation.

epiboly epibole, q.v.

epibranchial a. [Gk. epi, upon; brangchia, gills] pert. 2nd upper element in branchial arch; efferent branchial, appl. vessels.

epicalyx n. [Gk. epi, upon; kalyx, cup] a calyx-like structure formed outside but close to the actual calyx, formed from stipules fused in pairs, or by aggregation of bracts or bracteoles, or by small sepal-like structures; alt. calycle.

epicanthus n. [Gk. epi, upon; kanthos, corner of eye] a prolongation of upper eyelid over inner angle of eye, the Mongolian fold.

epicardia n. [Gk. epi, upon; kardia, stomach, heart] part of oesophagus running from diaphragm to stomach, alt. antrum cardiacum, plu. epicardiae; plu. of epicardium.

epicardium n. [Gk, epi, upon; kardia, heart, stomach] the visceral part of pericardium; tubular prolongation of branchial sac in many ascidians, which takes part in budding; plu. epicardia.

epicarp n. [Gk. epi, upon; karpos, fruit] outer layer of the pericarp; alt. exocarp.

epicentral a. [Gk. epi, upon; kentron, centre] attached to or arising from vertebral centra, appl. intermuscular bones.

epicerebral a. [Gk. epi, upon; L. cerebrum, brain] situated above the brain.

epichil(e), epichilium n. [Gk. epi, upon; cheilos, lip] in orchid flowers, the outer part of the lip where there are 2 distinct parts; cf. hypochile.

epichondrosis n. [Gk. epi, upon; chondros, cartilage] formation of cartilage on periosteum, as in production of antlers.

epichordal a. [Gk. epi, upon; chordē, cord] upon the notochord; appl. vertebrae in which ventral cartilaginous portions are almost completely suppressed; appl. upper lobe of caudal fin in fishes.

epichroic a. [Gk. epi, upon; chrōs, colour] discolouring, as after injury.

epiclinal a. [Gk. epi, upon; klinē, bed] situated on the receptacle or torus of a flower.

epicoel n. [Gk. epi, upon; koilos, hollow] cavity of mid-brain in lower vertebrates; cerebellar cavity; a perivisceral cavity formed by invagination; alt. epicoele, epicoelia.

epicondyle n. [Gk. epi, upon; kondylos, knuckle] a medial and a lateral protuberance at distal end of humerus and femur; a. epicondylar, epicondylic.

epicone n. [Gk. epi, upon; kōnos, cone] the part anterior to girdle in Dinoflagellata; alt. epivalve; cf. hypocone.

epicoracoid a. [Gk. epi, upon; korax, crow; eidos, form] pert. an element, usually cartilaginous, at sternal end of coracoid in amphibians, reptiles, and monotremes.

epicormic a. [Gk. *epi*, upon; *kormos*, trunk] growing from a dormant bud.

epicortex n. [Gk. *epi*, upon; L. *cortex*, bark] an outer layer, as of filaments, covering the cortex of certain fungi.

epicotyl n. [Gk. *epi*, upon; *kotylē*, cup] the axis of a plumule.

epicotyledonary a. [Gk. *epi*, upon; *kotylēdōn*, cup] above the cotyledons.

epicoxite n. [Gk. *epi*, upon; L. *coxa*, hip] a small process at posterior end of toothed part of coxa of 2nd to 5th pairs of appendages in Eurypterida.

epicranial a. [Gk. *epi*, upon; *kranion*, skull] pert. cranium, appl. aponeurosis, muscles, bones, suture; pert. epicranium.

epicranium n. [Gk. *epi*, upon; *kranion*, skull] the region between and behind eyes in insect head; scalp; the structures covering the cranium.

epicranius the scalp muscle, consisting of occipitalis and frontalis, connected by galea aponeurotica; alt. occipitofrontalis.

epicrine a. [Gk. *epi*, upon; *krinein*, to separate] appl. glands in which secretion is discharged without disintegration of cells.

epicritic a. [Gk. *epi*, upon; *krinein*, to judge] appl. stimuli and nerve systems concerned with delicate touch and other special sensations in skin.

epictesis n. [Gk. *epiktēsis*, further gain] capacity of a living cell to concentrate salt solutions diffusing into the cell.

epicuticle, epicuticula n. [Gk. *epi*, upon; L. dim. of *cutis*, skin] lamella or membrane external to exocuticle of insects.

epicutis n. [Gk. *epi*, upon; L. *cutis*, skin] outer layer of cutis of agarics; cf. subcutis.

epicyemate a. [Gk. *epi*, upon; *kyēma*, embryo] with embryo lying on the yolk sac.

epicyst n. [Gk. *epi*, upon; *kystis*, bladder] the external resistant cyst of an encysted protozoan; alt. ectocyst; cf. endocyst.

epicyte n. [Gk. *epi*, upon; *kytos*, hollow] the external layer of ectoplasm in certain protozoans.

epideictic displays [Gk. *epideiktos*, displayed as a sample] conventional animal displays or contests which serve as a indication of population density and the need to restore or shift the population balance.

epidemes n.plu. [Gk. *epi*, upon; *demas*, body] in certain insects, small sclerites closely related with articulation of wings.

epiderma n. [Gk. *epi*, upon; *derma*, skin] outer layer of cortex or derma in fungi.

epidermatoid a. [Gk. *epi*, upon; *derma*, skin; *eidos*, form] resembling epidermis; appl. fungal cortex made up of a single layer of cells; alt. epidermoid.

epidermis n. [Gk. *epi*, upon; *derma*, skin] the outermost protective layer of stems, roots, and leaves; scarf-skin or external layer of skin, a nonvascular stratified epithelium of ectodermic origin; single layer of ectoderm in invertebrates.

epidermophyte dermatophyte, q.v.

epididymis n. [Gk. *epi*, upon; *didymos*, testicle] a mass at back of testicle composed chiefly of vasa efferentia; the coiled anterior end of Wolffian duct.

epidural a. [Gk. *epi*, upon; L. *dura*, hard] pert.

dura mater; appl. space between dura mater and wall of vertebral canal.

epifauna n. [Gk. *epi*, upon; L. *faunus*, god of woods] animals living on the surface of the ocean bottom; any encrusting fauna.

epigaen, epigaeous epigeal, q.v.

epigamic a. [Gk. *epi*, upon; *gamos*, marriage] tending to attract opposite sex, e.g. colour displayed in courtship; alt. epigamous.

epigamous a. [Gk. *epi*, upon; *gamos*, marriage] designating that stage in polychaetes in which immature forms become heteronereid, while sexual elements are ripening; epigamic, q.v.

epigaster n. [Gk. *epi*, upon; *gastēr*, belly] that part of embryonic intestine which later develops into colon.

epigastric a. [Gk. *epi*, upon; *gastēr*, belly] pert. anterior wall of abdomen; middle region of upper zone of artificial divisions of abdomen.

epigastrium n. [Gk. *epi*, upon; *gastēr*, stomach] the epigastric region, cf. hypogastrium; sternal portions of meso- and metathorax of insects; anterior ventral portion of opisthosoma of arachnids.

epigastroid epipubis, q.v.

epigeal a. [Gk. *epi*, upon; *gē*, earth] living near the ground, appl. insects; borne above ground, appl. cotyledons when they form 1st foliage leaves; alt. epigaen, epigean, epigeic, epigeous; cf. hypogeal.

epigenesis n. [Gk. *epi*, upon; *genesis*, descent] theory of generations, now generally believed, that embryo is formed by successive changes in structure; alt. induced development; cf. preformation theory.

epigenetics n. [Gk. *epi*, upon; *genesis*, descent] study of the mechanisms causing phenotypic effects to be produced by the genes of a genotype.

epigenotype n. [Gk. *epi*, upon; *genos*, descent; *typos*, image] the chain of processes linking genotype and phenotype.

epigenous a. [Gk. *epi*, upon; *genos*, descent] developing or growing on a surface; cf. hypogenous.

epigeous epigeal, q.v.

epiglottis n. [Gk. *epi*, upon; *glōtta*, tongue] a moveable flap of fibrocartilage in front of the glottis which bends over the respiratory opening when food is being swallowed; epistome in Ectoprocta; epipharynx in insects.

epignathous a. [Gk. *epi*, upon; *gnathos*, jaw] having upper jaw longer than the lower.

epigone epigonium, q.v.

epigonial a. [Gk. *epi*, upon; *gonē*, seed] appl. sterile posterior portion of genital ridge.

epigonium n. [Gk. *epi*, upon; *gonē*, seed] the cover over the young sporogonium of a liverwort; any calyptra; alt. epigone.

epigynal a. [Gk. *epi*, upon; *gynē*, woman] pert. epigynum.

epigyne, epigynium epigynum, q.v.

epigynous a. [Gk. *epi*, upon; *gynē*, woman] having the various whorls adnate to ovary, thus apparently inserted in or above the ovary; having antheridia above oogonium; n. epigyny; alt. syngynous; cf. hypogynous, perigynous.

epigynum n. [Gk. *epi*, upon; *gynē*, woman] external female genitalia in Arachnida; alt. epigyne, epigynium, vulva.

epigyny n. [Gk. *epi*, upon; *gynē*, woman] the condition of being epigynous.

epihyal a. [Gk. *epi*, upon; *hyoeidēs*, Y-shaped] *pert.* upper portion of ventral part of hyoid arch. n. Upper element of ventral portion, a cartilage or bone in centre of stylohyoid ligament.

epihymenium n. [Gk. *epi*, upon; *hymēn*, membrane] a thin tissue of interwoven hyphae covering the hymenium, as of Basidiomycetes.

epilabrum n. [Gk. *epi*, upon; L. *labrum*, lip] a process at side of labrum in Myriapoda.

epilemmal a. [Gk. *epi*, upon; *lemma*, skin] *appl.* sensory nerve endings on surface of sarcolemma.

epilimnion n. [Gk. *epi*, upon; *limnē*, lake] upper water layer, above thermocline, in lakes; *cf.* hypolimnion.

epilithic a. [Gk. *epi*, upon; *lithos*, stone] attached on rocks, *appl.* algae, lichens.

epimandibular a. [Gk. *epi*, upon; L. *mandibulum*, jaw] *pert.* a bone in lower jaw of vertebrates.

epimatium n. [Gk. *epimation*, an outer garment] in some conifers, the ovuliferous scale when folded round the ovule to form an extra envelope.

epimeletic a. [Gk. *epimelēs*, careful] *appl.* animal behaviour relating to the care of others; *cf.* etepimeletic.

epimembranal a. [Gk. *epi*, upon; *membrana*, skin] situated or formed on the surface of a membrane, *appl.* pigmentation.

epimeral a. [Gk. *epi*, upon; *mēros*, thigh] *pert.* epimeron.

epimerases a group of enzymes causing epimerization, included in EC subgroup 5.1.

epimere n. [Gk. *epi*, upon; *meros*, part] the dorsal muscle plate of mesothelial wall.

epimerite n. [Gk. *epi*, upon; *meros*, part] prolongation of protomerite of certain gregarines for attachment to host.

epimerization n. [Gk. *epi*, upon; *meros*, part] the transformation of an organic compound, especially a sugar, into its epimer or stereoisomer.

epimeron n. [Gk. *epi*, upon; *mēros*, thigh] a portion of pleuron in insects which may be posterior or nearly as far forward as episternum; posterior pleurite of subcoxa; portion of arthropod segment between tergum and limb insertions.

epimers n.plu. [Gk. *epi*, upon; *meros*, part] molecules, especially monosaccharides, which differ only in the arrangement of atoms about a single carbon atom.

epimorpha n.plu. [Gk. *epi*, upon; *morphē*, form] larvae having all segments developed and the same form in all stages of growth, as in arthropods with incomplete metamorphosis; *cf.* anamorpha.

Epimorpha n. [Gk. *epi*, upon; *morphē*, form] a subclass of centipedes in which the eggs are laid in groups and tended by the female, and whose young are borne with the same number of legs as the adult.

epimorphic maintaining the same form in successive stages of growth.

epimorphosis n. [Gk. *epi*, upon; *morphōsis*, shaping] that type of regeneration in which proliferation of new material precedes development of new part; *cf.* morphallaxis.

epimysium n. [Gk. *epi*, upon; *mys*, muscle] the sheath of areolar tissue which invests the entire muscle; *cf.* perimysium, endomysium.

epinasty n. [Gk. *epi*, upon; *nastos*, close-pressed] the more rapid growth of upper surface of a dorsoventral organ, e.g. a leaf, thus causing unrolling or downward curvature.

epinekton n. [Gk. *epi*, on; *nēktos*, swimming] nekton which are incapable of actively swimming themselves but are attached to actively swimming organisms; *alt.* epinecton.

epinephrin(e) n. [Gk. *epi*, upon; *nephros*, kidney] adrenaline, *q.v.*

epinephros n. [Gk. *epi*, upon; *nephros*, kidney] the suprarenal or adrenal body.

epineural a. [Gk. *epi*, upon; *neuron*, nerve] arising from vertebral neural arch; *pert.* canal external to radial nerve in certain echinoderms; *appl.* sinus between embryo and yolk, beginning of body cavity in insects.

epineurium n. [Gk. *epi*, upon; *neuron*, nerve] the external fibrous connective tissue sheath of a nerve cord.

epinotum n. [Gk. *epi*, upon; *nōton*, back] propodeon, *q.v.*

epiopticon n. [Gk. *epi*, upon; *opsis*, sight] the middle zone of optic lobes of insects.

epiostracum n. [Gk. *epi*, upon; *ostrakon*, shell] thin cuticle or epicuticle covering exocuticle or ectostracum in Acarina.

epiotic a. [Gk. *epi*, upon; *ous*, the ear] *pert.* upper element of bony capsule of ear; *appl.* centre of ossification of mastoid process.

epiparasite n. [Gk. *epi*, upon; *parasitos*, eating at another's table] ectoparasite, *q.v.*

epipelagic a. [Gk. *epi*, upon; *pelagos*, sea] *pert.* deep-sea water between surface and bathypelagic zone; or, inhabiting oceanic water at depths not exceeding *ca.* 200 m i.e. above mesopelagic zone.

epiperidium exoperidium, *q.v.*

epiperipheral a. [Gk. *epi*, over; *periphereia*, circumference] located on or beyond the outer surface of the body, *appl.* source of stimuli.

epipetalous a. [Gk. *epi*, upon; *petalon*, leaf] having stamens inserted on petals.

epipetreous a. [Gk. *epi*, upon; *petraios*, pert. rock] growing on rocks.

epipharyngeal a. [Gk. *epi*, upon; *pharyngx*, throat] *pert.* upper or dorsal aspect of pharynx; *appl.* bone: fused pharyngobranchial bones.

epipharynx n. [Gk. *epi*, upon; *pharyngx*, throat] a projection on roof of mouth cavity of certain insects, *alt.* palate; membranous lining of labrum and clypeus drawn out with labrum to form a piercing organ, as in Diptera; a chitinous plate on lower surface of rostrum in certain Arachnida.

epiphloeodal, epiphloeodic a. [Gk. *epi*, upon; *phloios*, bark] *pert* epiphloem; ectophloeodic, *q.v.*

epiphloem n. [Gk. *epi*, upon; *phloios*, bark] outer bark, periderm, *q.v.*

epiphragm n. [Gk. *epiphragma*, covering] a layer of hardened mucous matter, or a calcareous plate, closing the opening of certain gastropod shells; membrane which closes the capsule in certain mosses, *alt.* tympanum; a closing membrane in sporophores of certain fungi.

epiphyll n. [Gk. *epi*, upon; *phyllon*, leaf] a plant which grows on leaves, e.g. various lichens.

epiphyllous a. [Gk. *epi*, upon; *phyllon*, leaf] growing on leaves; united to perianth, *appl.* stamens.

epiphysial a. [Gk. *epi*, upon; *phyein*, to grow] *pert.* or similar to the epiphysis; *alt.* epiphyseal.

epiphysis n. [Gk. *epi*, upon; *phyein*, to grow] any

part or process of a bone which is formed from a separate centre of ossification and later fuses with the bone, *cf.* diaphysis; pineal body; pineal and parapineal organs; stout bar firmly fused to alveolus of each jaw and articulating with rotulae in sea urchins; certain processes on tibia of insects; caruncle near hilum of seed.

epiphyte *n.* [Gk. *epi*, upon; *phyton*, plant] a plant which lives on the surface of other plants but does not derive water or nourishment from them, *alt.* aerial plant, aerophyte, air plant; a plant ectoparasite on animal or human body; any organism living on the surface of a plant; *a.* epiphytic.

epiphytotic *a.* [Gk. *epi*, upon; *phyton*, plant] *pert.* disease epidemic in plants; *cf.* enphytotic.

epiplankton *n.* [Gk. *epi*, upon; *plangktos*, wandering] that portion of plankton from surface to about 200 m.

epiplasm *n.* [Gk. *epi*, upon; *plasma*, mould] cytoplasm of a brood mother cell remaining unused in brood formation; cytoplasm of sporangium or ascus remaining after spore formation, *cf.* sporoplasm, ascoplasm.

epiplastron *n.* [Gk. *epi*, upon; F. *plastron*, breast plate] one of anterior pair of bony plates in plastron of Chelonia.

epiplectotrichoderm *n.* [Gk. *epi*, upon; *plektos*, plaited; *thrix*, hair; *derma*, skin] an epitrichoderm, *q.v.*, of interwoven hyphae.

epipleura *n.* [Gk. *epi*, upon; *pleura*, rib] epithecal part of cingulum in diatoms; one of rib-like structures in teleosts which are not preformed in cartilage; uncinate process of rib in birds; the turned down outer margin of elytra of certain beetles.

epiploic *a.* [Gk. *epiploon*, caul of entrails] *pert.* omentum.

epiploic foramen opening between bursa omentalis and large sac of peritoneum; *alt.* foramen of Winslow.

epiploon *n.* [Gk. *epiploon*, caul of entrails] greater omentum, *q.v.*; insect adipose tissue, *alt.* fatbody.

epipodial *a.* [Gk. *epi*, upon; *pous*, foot] *pert.* epipodium.

epipodite *n.* [Gk. *epi*, upon; *pous*, foot] a process arising from basal joint of crustacean limb and usually extending into gill chamber.

epipodium *n.* [Gk. *epi*, upon; *pous*, foot] the leaf blade or lamina; embryonic leaf lamina; ridge, fold, or lobe along edge of foot of Gastropoda; raised ring on an ambulacral plate in Echinoidea.

epiprecoracoid *n.* [Gk. *epi*, upon; L. *prae*, before; Gk. *korax*, crow; *eidos*, form] a small cartilage at ventral end of precoracoid in pectoral girdle in some Chelonia.

epiproct *n.* [Gk. *epi*, upon; *prōktos*, anus] a supra-anal plate representing tergum of 10th or 11th segment in some insects.

epipteric *a.* [Gk. *epi*, upon; *pteron*, wing] winged at tip; *appl.* certain seeds, *alt.* epipterous; *pert.* or shaped like, or placed above wing; *appl.* a small skull bone between parietal and sphenoidal ala. *n.* Epipteric bone.

epipterygoid *n.* [Gk. *epi*, upon; *pteryx*, wing; *eidos*, shape] a small bone extending nearly vertically downwards from pro-otic to pterygoid in some reptiles; *alt.* columella cranii.

epipubic *a.* [Gk. *epi*, upon; L. *pubes*, adult] *pert.* or borne upon pubis; *appl.* certain cartilages or bones principally in marsupials; *appl.* anterior median process of ischiopubic plate.

epipubis *n.* [Gk. *epi*, upon; L. *pubes*, adult] unpaired cartilage or bone borne anteriorly on pubis; *alt.* epigastroid; *cf.* pelvisternum.

epirhizous *a.* [Gk. *epi*, upon; *rhiza*, root] growing upon a root.

epirrhysa *n.plu.* [Gk. *epirrheein*, to flow into] inhalant canals in sponges; *cf.* aporrhysa.

episclera *n.* [Gk. *epi*, upon; *sklēros*, hard] connective tissue between sclera and conjunctiva.

episemantic *a.* [Gk. *epi*, on; *sēmantikos*, significant] *appl.* the resulting product of the action of a semantide.

episematic *a.* [Gk. *epi*, upon; *sēma*, sign] aiding in recognition, *appl.* coloration, markings; *cf.* sematic.

episeme *n.* [Gk. *epi*, upon; *sēma*, sign] a marking or colour aiding in recognition.

episepalous *a.* [Gk. *epi*, upon; F. *sépale*, sepal] adnate to sepals.

episkeletal *a.* [Gk. *epi*, upon; *skeletos*, hard] outside the endoskeleton.

episome *n.* [Gk. *epi*, in addition; *sōma*, body] a genetic unit which may multiply either as an addition to the chromosome or else independently, found in bacteria and some other organisms.

episperm *n.* [Gk. *epi*, upon; *sperma*, seed] the outer coat of seed; testa of spermoderm.

episporangium *n.* [Gk. *epi*, upon; *sporos*, seed; *anggeion*, vessel] *see* indusium.

epispore *n.* [Gk. *epi*, upon; *sporos*, seed] the outer layer of a spore wall, *alt.* episporium; perispore, *q.v.*; perinium, *q.v.*

epistasis, epistasy *n.* [Gk. *epi*, upon; *stasis*, standing] dominance of a gene over another, non-allelomorphic gene; greater degree of modification manifested by 1 or 2 related types in phylogenesis; *a.* epistatic; *cf.* hypostasis.

epistellar *a.* [Gk. *epi*, upon; L. *stella*, star] above the stellate ganglion; *appl.* neurosecretory body regulating muscular tonicity, as in Cephalopoda.

epistereom *n.* [Gk. *epi*, upon; *stereōma*, solid body] in some Cystoidea, a thin surface layer of calcite plates covering the rhombopores.

episternalia *n.plu.* [Gk. *epi*, upon; *sternon*, breast bone] two small elements preformed in cartilage frequently intervening in development between clavicles and sternum, and ultimately fusing with sternum.

episternite *n.* [Gk. *epi*, upon; *sternon*, breast bone] one of portions of an ovipositor formed from side portions of a somite.

episternum *n.* [Gk. *epi*, upon; L. *sternum*, breast bone] the interclavicle, *q.v.*; an anterior cartilaginous element of sternum; a lateral division of an arthropod somite, above sternum and in front of epimeron; anterior pleurite of subcoxa.

epistoma epistome, *q.v.*

epistome *n.* [Gk. *epi*, upon; *stoma*, mouth] a small lobe overhanging mouth in Polyzoa and containing a part of body cavity; the region between antenna and mouth in Crustacea; anterior median plate on reflected margin of carapace of certain trilobites; subcheliceral plate in certain ticks; that portion of insect head immediately behind labrum; portion of rostrum of certain Diptera; *alt.* epistoma.

epistroma *n.* [Gk. *epi*, upon; *strōma*, bedding] ectostroma, *q.v.*; *cf.* hypostroma.

epistrophe *n*. [Gk. *epistrophē*, moving about] the position assumed by chloroplasts along outer and inner cell walls when exposed to diffuse light.

epistropheus *n*. [Gk. *epistrophē*, turning] the 2nd cervical or axis vertebra.

epithalamus *n*. [Gk. *epi*, upon; *thalamos*, chamber] part of thalamencephalon, comprising trigonum habenulae, pineal body, and posterior commissure.

epithalline *a*. [Gk. *epi*, upon; *thallos*, branch] growing upon the thallus.

epithallus *n*. [Gk. *epi*, upon; *thallos*, branch] cortical layer of hyphae covering gonidia of lichens.

epitheca *n*. [Gk. *epi*, upon; *thēkē*, box] an external layer surrounding lower part of theca in many corals; theca covering epicone in Dinoflagellata; older half of frustule in diatoms.

epithecium *n*. [Gk. *epi*, upon; *thēkē*, box] the layer of hyphae on the surface of the fruit body of many fungi and lichens, in fungi being equivalent to the hymenium and in lichens to the layer over the hymenium.

epithelia *plu*. of epithelium.

epithelial *a*. [Gk. *epi*, upon; *thēlē*, nipple] *pert*. epithelium; epitheliomorph, *q.v.*

epithelial body the parathyroid, *q.v.*

epitheliochorial *a*. [Gk. *epi*, upon; *thēlē*, nipple; *chorion*, skin] *appl*. placenta with apposed chorionic and uterine epithelia, and villi pitting the uterine wall, as in marsupials and ungulates; *cf*. endotheliochorial, haemochorial.

epitheliofibrillae *n.plu*. [Gk. *epi*, upon; *thēlē*, nipple; L. *fibrilla*, small fibre] parallel or reticular fibrillae of columnar epithelium analogous to myofibrillae.

epitheliomorph *a*. [Gk. *epi*, upon; *thēlē*, nipple; *morphē*, form] resembling epithelium, *alt*. epithelioid, epithelial; *appl*. layer of cells, or elastica interna, which secretes notochordal sheath.

epithelium *n*. [Gk. *epi*, upon; *thēlē*, nipple] any cellular tissue covering a free surface or lining a tube or cavity; *plu*. epithelia.

epithem *n*. [Gk. *epi*, upon; *tithenai*, to put] a plant tissue of specialized cells and intercellular spaces forming a hydathode; the secretory layer in nectaries; an excrescence on the beak of birds; *alt*. epithema, epitheme.

epitokous *a*. [Gk. *epi*, upon; *tokos*, birth] designating the heteronereid stage of certain polychaetes.

epitope *n*. [Gk. *epi*, upon; *topos*, place] the region of an antigen at which an antibody combines with it; *cf*. paratope.

epitrematic *a*. [Gk. *epi*, upon; *trēma*, pore] *appl*. upper lateral bar of branchial basket of lamprey.

epitrichium *n*. [Gk. *epi*, upon; *thrix*, hair] an outer layer of foetal epidermis of many mammals, usually shed before birth; *alt*. periderm; *a*. epitrichial.

epitrichoderm *n*. [Gk. *epi*, upon; *thrix*, hair; *derma*, skin] a trichoderm, *q.v.*, when the coating of a pileus is 2-layered; *cf*. epiplectotrichoderm.

epitrochlea *n*. [Gk. *epi*, upon; L. *trochlea*, Gk. *trochlia*, pulley] inner condyle at distal end of humerus.

epitympanic *a*. [Gk. *epi*, upon; *tympanon*, kettle drum] situated above tympanum.

epityphlon *n*. [Gk. *epi*, upon; *typhlon*, caecum] the vermiform appendix.

epivalve *n*. [Gk. *epi*, upon; L. *valva*, fold] valve of epitheca in diatoms; epicone, *q.v.*

epixylous *a*. [Gk. *epi*, upon; *xylon*, wood] growing upon wood.

epizoic *a*. [Gk. *epi*, upon; *zōon*, animal] living on or attached to the body of an animal, *cf*. endozoic; having seeds or fruits dispersed by being attached to the surface of an animal.

epizoochore *n*. [Gk. *epi*, upon; *zōon*, animal; *chōrein*, to spread] any spore, seed, or organism dispersed by being carried upon the body of an animal; *cf*. endozoochore.

epizoochory dispersal of an epizoochore.

epizoon *n*. [Gk. *epi*, upon; *zōon*, animal] an animal living on exterior of another; ectoparasite, *q.v.*

epizootic *a*. [Gk. *epi*, upon; *zōon*, animal] common among animals; *pert*. epizoon. *n*. Disease affecting a large number of animals simultaneously, corresponding to epidemic in man.

epizygal *n*. [Gk. *epi*, upon; *zygon*, yoke] the upper ossicle in a syzygial pair of brachials or columnars in crinoids.

eplicate *a*. [L. *e*, out of; *plicatus*, folded] not folded; not plaited.

epoch *n*. [Gk. *epochē*, pause] in geological time, a subdivision of a period.

eponychium *n*. [Gk. *epi*, upon; *onyx*, nail] the thin cuticular fold which overlaps lunula of nail; dorsal portion of a neonychium.

eponym *n*. [Gk. *epi*, by; *onyma*, name] name of a person used in designation of an entity, as of a species, organ, law, disease, etc.

epoophoron *n*. [Gk. *epi*, upon; *ōon*, egg; *pherein*, to bear] a rudimentary organ (homologous with epididymis), remains of Wolffian body of embryo, lying in mesosalpinx between ovary and uterine tube; *alt*. parovarium, Rosenmüller's organ.

EPP end-plate potential, *q.v.*

epr endoplasmic reticulum, *q.v.*

EPSP excitatory postsynaptic potential, *q.v.*

epulosis *n*. [Gk. *epi*, over; *oulē*, scar] formation of a scar.

equal *a*. [L. *aequalis*, equal] having the portions of the lamina equally developed on the 2 sides of midrib, *appl*. leaves.

equation division second division in meiosis.

equatorial furrow division round equator of segmenting egg.

equatorial plate group of chromosomes lying at equator of spindle during mitosis or meiosis; locus of new cell wall after cell division; *alt*. metaphase plate, nuclear plate.

equiangular *a*. [L. *aequus*, equal; *angulus*, corner] *appl*. spicules having equal angles.

equiaxial *a*. [L. *aequus*, equal; *axis*, axle] with axes of the same length.

equibiradiate *a*. [L. *aequus*, equal; *bis*, twice; *radius*, ray] with 2 equal rays.

equicellular *a*. [L. *aequus*, equal; *cellula*, cell] composed of equal cells.

equifacial *a*. [L. *aequus*, equal; *facies*, face] having equivalent surfaces or sides, as vertical leaves.

equilateral *a*. [L. *aequus*, equal; *latus*, side] having the sides equal; *appl*. shells symmetrical about a transverse line drawn through umbo.

equilenin n. [L. *equus*, horse] an oestrogenic hormone present in urine of the pregnant mare.

equilin(e) n. [L. *equus*, horse] an oestrogenic hormone, more physiologically active than equilenin, occurring in urine of the pregnant mare.

equimolecular a. [L. *aequus*, equal; F. *molécule*, small particle] *see* isotonic.

equipotent a. [L. *aequus*, equal; *potens* powerful] totipotent, q.v.; able to perform the function of another cell, part, or organ.

equiradiate a. [L. *aequus*, equal; *radius*, ray] *appl.* spicules having equal radii.

Equisetales n. [*Equisetum*, generic name] an order of herbaceous homosporous Sphenopsida, having small leaves and peltate sporangiophores in cones.

Equisetatae n. [*Equisetum*, generic name] Sphenopsida, q.v.

Equisetineae n. [*Equisetum*, generic name] a class of the Tracheophyta included in, or equivalent to, the Sphenopsida, q.v.

equitant a. [L. *equitare*, to ride] overlapping saddlewise, as leaves in leaf bud.

equivalve a. [L. *aequus*, equal; *valva*, valve] having 2 halves of a shell alike in form and size.

ER, er endoplasmic reticulum, q.v.

era n. [L. L. *aera*, fixed date] a main division of geological time, such as Palaeozoic, Mesozoic, Caenozoic, and divided into periods.

erect a. [L. *erigere*, to raise up] directed towards summit of ovary, *appl.* ovule; not decumbent.

erectile a. [L. *erigere*, to raise up] capable of being erected; *appl.* tissue capable of being made rigid by distension of blood vessels within it.

erection n. [L. *erigere*, to raise up] the state of a part which has become swollen and distended through accumulation of blood in erectile tissue.

erector n. [L. *erigere*, to raise up] a muscle which raises up an organ or part.

ereidesm n. [Gk. *ereidein*, to support; *desma*, bond] an epithelial intracellular fibre.

Eremian a. [Gk. *erēmia*, desert] *appl.* or *pert.* part of the Palaearctic region including deserts of North Africa and Asia.

eremic a. [Gk. *erēmos*, desert] *pert.*, or living in, deserts.

eremobic a. [Gk. *erēmos*, solitary; *bios*, life] growing or living in isolation; having a solitary existence.

eremochaetous a. [Gk. *erēmos*, lonely; *chaite*, hair] having no regularly arranged system of bristles, *appl.* flies.

eremophyte n. [Gk. *erēmos*, desert; *phyton*, plant] a desert plant.

erepsin n. [L. *eripere*, to set free] a mixture of proteolytic enzymes of intestinal juice.

ergaloid a. [Gk. *ergon*, work; *eidos*, form] having the adults sexually capable though wingless.

ergastic a. [Gk. *ergastikos*, fit for working] *appl.* materials: metaplasm, q.v.

ergastoplasm n. [Gk. *ergazesthai*, to work; *plasma*, mould] archiplasm, q.v.; a former name for rough endoplasmic reticulum, and sometimes used for all endoplasmic reticulum; *alt.* ergoplasm, kinoplasm.

ergastoplasmic a. [Gk. *ergazesthai*, to work; *plasma*, mould] *appl.* fibrillae of gland cells which may induce production of secretory granules.

ergatandromorph n. [Gk. *ergatēs*, worker; *anēr*, male; *morphē*, form] an ant or other social insect in which worker and male characters are blended.

ergatandrous a. [Gk. *ergatēs*, worker; *anēr*, man] having worker-like males.

ergataner n. [Gk. *ergatēs*, worker; *anēr*, male] a male ant resembling a worker; *alt.* ergatoid or ergatomorphic male.

ergate(s) n. [Gk. *ergatēs*, worker] a worker ant.

ergatogyne n. [Gk. *ergatēs*, worker; *gynē*, female] a female ant resembling a worker; *alt.* an ergatoid or ergatomorphic female.

ergatogynous a. [Gk. *ergatēs*, worker; *gynē*, woman] having worker-like females.

ergatoid, ergatomorphic a. [Gk. *ergatēs*, worker; *eidos*, form; *morphē*, shape] resembling a worker, *appl.* ants.

ergines ergones, q.v.

ergocalciferol calciferol, q.v., especially synthetic calciferol.

ergones n.plu. [Gk. *ergon*, work] organic substances of which small amounts suffice for activation or regulation of a physiological process, as enzymes, hormones, and vitamins; *alt.* ergines.

ergonomics n. [Gk. *ergon*, work; *nomos*, law] the anatomical, physiological, and psychological study of man in his working environment.

ergonomy n. [Gk. *ergon*, work; *nomos*, law] the differentiation of functions; physiological differentiation associated with morphological specialization.

ergoplasm ergastoplasm, q.v.

ergosomes n.plu. [Gk. *ergon*, work; *sōma*, body] polysomes, q.v.

ergosterol a sterol occurring in ergot, yeast, and moulds, in certain algae, and in animal tissues, which is converted to vitamin D_2 (ergocalciferol) on irradiation with ultraviolet light.

ergot n. [O.F. *argot*, spur] rye smut disease, caused by various fungi of the genus *Claviceps*; the sclerotium of *Claviceps* which replaces the ovary of infected grasses and which yields several alkaloids, e.g. ergotoxine and ergometrine, which stimulate uterine muscle; a small bare patch on limbs of Equidae (horses, etc.) representing the last remnant of naked palm of hand and sole of foot.

Ericales n. [*Erica*, generic name] an order of dicots of the Dilleniidae or Sympetalae, used in different ways by different authorities, but always including the heathers, heaths, and wintergreen; *alt.* Bicornes.

erichthoidina n. [Gk. *erechthein*, to break; *eidos*, form] larval stage of Stomatopoda comparable with zoaea.

erichthus n. [Gk. *erechthein*, to break] larval stage of Stomatopoda, comparable with pseudozoaea.

erineum n. [Gk. *erineos*, woollen] an outgrowth of abnormal hairs produced on leaves by certain gall mites.

Eriocaulales n. [*Eriocaulon*, generic name] an order of monocots (Liliidae) comprising the mainly tropical and subtropical family Eriocaulaceae, having flowers crowded on inflorescences and somewhat similar to composites.

eriocomous a. [Gk. *erion*, wool; *komē*, hair] having woolly hair; fleece-haired.

eriophyllous a. [Gk. *erion*, wood; *phyllon*, leaf] having leaves with a cottony appearance.

erose a. [L. *erodere*, to wear away] having margin irregularly notched, *appl.* leaf, bacterial colony.

erosion n. [L. *erodere*, to wear away] decay which usually starts at apex of many gastropod shells; wearing away of soil due to the action of wind, water, or gravity.

erostrate a. [L. *ex*, without; *rostrum*, beak] having no beak, *appl.* anthers.

errantia n. [L. *errare*, to wander] mobile organisms, the term sometimes being used as a taxonomic group; *cf.* sedentaria.

ersaeome n. [Gk. *ersē*, young] the free monogastric generation of Siphonophora.

eruca n. [L. *eruca*, caterpillar] a caterpillar; an insect larva having the shape of a caterpillar.

eruciform a. [L. *eruca*, caterpillar; *forma*, shape] having the shape of, or resembling a caterpillar, *appl.* certain insect larvae, *appl.* spores of certain lichens.

erumpent a. [L. *erumpere*, to break out] breaking through suddenly, *appl.* fungal hyphae.

Erysiphales n. [*Erysiphe*, generic name] an order of parasitic Plectomycetes, commonly called powdery mildews, having asci enclosed in a cleistothecium and arising at a single level; *alt.* Perisporiales.

erythrin n. [Gk. *erythros*, red] a red pigment found in lichens and derived from lecanoric acid, a phenolcarboxylic acid.

erythrism n. [Gk. *erythros*, red] abnormal presence, or excessive amount, of red colouring matter, as in petals, feathers, hair, eggs; *alt.* erythrochroism; *cf.* rufinism.

erythroaphins n.plu. [Gk. *erythros*, red; *Aphis*, generic name of aphid] red pigments formed by postmortem enzymic transformation of the yellow pigments in aphids.

erythroblasts n.plu. [Gk. *erythros*, red; *blastos*, bud] nucleated cells, derived from mesoderm, which later contain haemoglobin and develop into red blood corpuscles; *alt.* gigantoblasts.

erythrochroism erythrism, *q.v.*

erythrocruorin n. [Gk. *erythros*, red; L. *cruor*, blood] an iron-containing haemochrome which acts as a respiratory pigment in many annelids and molluscs; *cf.* oxyerythrocruorin.

erythrocyte n. [Gk. *erythros*, red; *kytos*, hollow] a red blood corpuscle; *alt.* akaryocyte, rhodocyte.

erythrocyte-maturing factor (EMF) vitamin B_{12}, particularly cyanocobalamine.

erythrocytolysis n. [Gk. *erythros*, red; *kytos*, cell; *lysis*, loosing] haemolysis, *q.v.*

erythrogenic a. [Gk. *erythros*, red; *gennaein*, to produce] producing the sensation of redness.

erythrolabe n. [Gk. *erythros*, red; *labē*, grasping] the red-sensitive pigment of the normal human eye.

erythromycin n. [Gk. *erythros*, red; *mykēs*, fungus] an antibiotic synthesized by the actinomycete *Streptomyces erythieus.*

erythron n. [Gk. *erythros*, red; *on*, being] the red cells in bone marrow and circulating blood, collectively.

erythrophages n.plu. [Gk. *erythros*, red; *phagein*, to eat] cells which destroy red blood corpuscles, as reticulo-endothelial cells, macrophages, monocytes.

erythrophilous a. [Gk. *erythros*, red; *philein*, to love] having special affinity for red stains, *appl.* structures in a cell or to a type of cells.

erythrophore n. [Gk. *erythros*, red; *pherein*, to bear] a reddish-purple, pigment-bearing cell.

erythrophyll n. [Gk. *erythros*, red; *phyllon*, leaf] a red anthocyanin as found in some leaves and red algae.

erythropoiesis n. [Gk. *erythros*, red; *poiesis*, making] the production of red blood corpuscles.

erythropoietin n. [Gk. *erythros*, red; *poiesis*, making] a glycoprotein released from the kidney which stimulates bone marrow cells to undergo erythropoiesis.

erythropsin n. [Gk. *erythros*, red; *opsis*, sight] red colouring matter in insect eyes; rhodopsin, *q.v.*

erythropterin(e) n. [Gk. *erythros*, red; *pteron*, wing] a red pterine pigment of wings of pierid Lepidoptera and some other insects.

erythrose n. [Gk. *erythros*, red] a tetrose sugar which, when phosphorylated, forms an intermediate in photosynthesis.

erythrotin n. [Gk. *erythros*, red] cobalamine, *q.v.*

erythrozyme n. [Gk. *erythros*, red; *zymē*, leaven] an enzyme capable of decomposing ruberythric acid, and acting upon glucosides.

escape n. [O.F. *escaper*] a plant originally cultivated now found wild. a. *Appl.* behaviour in which an animal moves away from an unpleasant stimulus; *appl.* conditioning in which escape is the response to an adverse stimulus.

Escherichia coli a bacterium on which much genetic work has been done and which is a colon bacillus; abbreviated to *E.coli.*

escutcheon n. [O.F. *escuchon*, shield] area on rump of many quadrupeds which is either variously coloured or has the hair specially arranged; mesoscutellum of certain insects; ligamental area of certain bivalves.

escutellate exscutellate, *q.v.*

eseptate a. [L. *ex*, without; *septum*, enclose] without septa.

esodic a. [Gk. *eisodos*, a coming in] afferent; centripetal; *cf.* exodic.

esophag- oesophag-, *q.v.*

esoteric a. [Gk. *esōterikos*, arising within] arising within the organism.

espathate a. [L. *ex*, without; *spatha*, broad blade] having no spathe.

esquamate a. [L. *ex*, without; *squama*, scale] having no scale.

essential amino acids amino acids which are not produced in the animal body and are, therefore, necessarily obtained from the environment.

essential oils mixtures of various volatile oils derived from benzene and terpenes found in plants and producing characteristic odours, and having various functions such as attracting insects or warding off fungal attacks.

esterases hydrolysing enzymes which attack an ester, splitting it into 2 groups, one of which is an acid; EC subgroup 3.1.

esterification ester formation.

esters n.plu. [L. *aether*, ether] compounds formed by reaction of an acid with an alcohol and including the fats and oils which are esters of fatty acids and glycerol.

esth- aesth-, *q.v.*

estipulate a. [L. *ex*, without; *stipula*, stem] exstipulate, *q.v.*

estival aestival, *q.v.*

estivation aestivation, *q.v.*

estr- oestr-, *q.v.*

estriate *a.* [L. *e*, out of; *striatus*, grooved] not marked by narrow parallel grooves or lines; not streaked.

estuarine *a.* [L. *aestuarium*, estuary] *pert.* or found in an estuary.

etaerio *n.* [Gk. *etairia*, association] an aggregate fruit, composed of achenes, berries, drupels, follicles, or samaras; *alt.* eterio; *cf.* syncarp.

etepimeletic *a.* [L. *ex*, out; Gk. *epimelēs*, careful] *appl.* behaviour shown by a young animal to elicit care (epimeletic behaviour) from adults.

etheogenesis *n.* [Gk. *ētheos*, youth; *genesis*, descent] parthenogenesis producing males; development of a male gamete without fertilization.

ethereal *a.* [Gk. *aithēr*, ether] *appl.* a class of odours including those of ethers and fruits; *appl.* fragrant oils in many seed plants.

Ethiopian *a.* [Gk. *aithiops*, burned-face] *appl.* or *pert.* a zoogeographical region including Africa south of the Sahara and southern Arabia, and divisible into African and Malagasy subregions.

ethmohyostylic *a.* [Gk. *ēthmos*, sieve; ϒ; *stylos*, pillar] with mandibular suspension from ethmoid region and hyoid bar.

ethmoid *a.* [Gk. *ēthmos*, sieve; *eidos*, shape] *pert.* bones which form a considerable part of walls of nasal cavity.

ethmoidal *a.* [Gk. *ēthmos*, sieve; *eidos*, shape] *pert.* ethmoid bones or region.

ethmoidal notch a quadrilateral space separating the 2 orbital parts of the frontal bone; *alt.* incisura ethmoidalis.

ethmolysian *a.* [Gk. *ēthmos*, sieve; *lyein*, to loosen] *pert.* an apical system in which the madreporite extends backwards until it separates the 2 posterolateral genitals.

ethmopalatine *n.* [Gk. *ēthmos*, sieve; L. *palatus*, palate] *pert.* ethmoid and palatine bones, or their region.

ethmophract *a.* [Gk. *ēthmos*, sieve; *phraxai*, to fence in] *pert.* a simple, compact, apical system with pores occurring only in right anterior corner.

ethmoturbinals *n.plu.* [Gk. *ēthmos*, sieve; L. *turbo*, whorl] cartilages or bones in nasal cavity which are folded so as to increase olfactory area.

ethmovomerine *a.* [Gk. *ēthmos*, sieve; L. *vomer*, ploughshare] *pert.* ethmoid and vomer regions; *appl.* the cartilage which forms nasal septum in early embryo.

ethnography *n.* [Gk. *ethnos*, nation; *graphein*, to write] the description of the races of mankind.

ethnology *n.* [Gk. *ethnos*, nation; *logos*, discourse] science dealing with the different races of mankind, their distribution, relationship, and activities.

ethological isolation the prevention of interbreeding between species as a result of behavioural differences.

ethology *n.* [Gk. *ēthos*, custom; *logos*, discourse] study of behaviour, especially of animals in their natural habitats.

ethomerous *a.* [Gk. *ēthos*, custom; *meros*, part] having the normal number of parts or segments; with normal number of chromosomes.

ethylene a gas C_2H_4 which is produced by plants in minute amounts and has various developmental effects as a hormone, including regulation of fruit ripening.

etiolation *n.* [F. *étioler*, to blanch] the appearance of plants in the dark, having no chlorophyll, chloroplasts not developing, internodes being greatly elongated so the plant is tall and spindly, and having small, rudimentary leaves; *alt.* aetiolation.

etiolin *n.* [F. *étioler*, to blanch] protochlorophyll, *q.v.*; *alt.* aetiolin.

etiology aetiology, *q.v.*

etioplast *n.* [F. *étioler*, to blanch; Gk. *plastos*, formed] a chloroplast formed in the absence of light, found in etiolated leaves.

euapogamy *n.* [Gk. *eu*, well; *apo*, away; *gamos*, marriage] diploid apogamy, haploid apogamy being meiotic euapogamy.

Euascomycetes, Euascomycetae *n.* [Gk. *eu*, well; *askos*, bag; *mykēs*, fungus] a group of fungi, the higher Ascomycetes, having asci borne on ascogenous hyphae in a fruit body called an ascocarp; *alt.* Euascomycetidae.

euaster *n.* [Gk. *eu*, good; *astēr*, star] an aster in which the rays meet at a common centre.

eubacteria *n.* [Gk. *eu*, well; *baktērion*, small rod] true bacteria, as opposed to the Rickettsiales and Spirochaetales, sometimes considered as an order, the Eubacteriales.

Eubacteriales *n.* [Gk. *eu*, well; *baktērion*, small rod] an order of rod-shaped or spherical filamentous bacteria; the eubacteria, considered as an order of Schizomycetes.

Eubasidii *n.* [Gk. *eu*, well; *basis*, base; *idion*, *dim.*] a subclass of Basidiomycetes having basidia arising from the diploid mycelium.

Eubrya *n.* [Gk. *eu*, well; *bryon*, moss] a subclass of mosses having 'leaves' with a midrib more than one cell thick, and a complex operculum.

Eubryales *n.* [Gk. *eu*, well; *bryon*, moss] an order of mosses having perennial gametophores and pendulous capsules with well-developed peristome teeth.

Eucarida *n.* [Gk. *eu*, well; L. *caris*, a kind of crab] a superorder or order of malacostracans having a carapace covering all thoracic segments.

eucarpic *a.* [Gk. *eu*, well; *karpos*, fruit] having the fruit body formed by only a part of the thallus; *appl.* Phycomycetes having rhizoids of haustoria; *cf.* holocarpic.

eucaryo- eukaryo-, *q.v.*

eucentric *a.* [Gk. *eu*, well; *kentron*, centre] pericentric, *q.v.*

eucephalous *a.* [Gk. *eu*, good; *kephalē*, head] with well-developed head, *appl.* certain insect larvae.

Eucestoda *n.* [Gk. *eu*, well; L. *cestus*, girdle] the tapeworms, considered as a subclass of the Cestoda in classifications where the Cestodaria are also given subclass status within the Cestoda.

euchroic *a.* [Gk. *eu*, well; *chrōs*, colour] having normal pigmentation.

euchromatic *a.* [Gk. *eu*, well; *chrōma*, colour] *pert.* euchromatin; *appl.* chromosome regions which never become heteropycnotic; *cf.* heterochromatic.

euchromatin *n.* [Gk. *eu*, well; *chrōma*, colour] the chromatin which makes up and shows the staining behaviour of the bulk of the chromosomes (the euchromosomes) being uncoiled in interphase and condensing during cell division, and containing the active genes; *cf.* heterochromatin.

euchromosome *n.* [Gk. *eu*, well; *chrōma*, colour; *sōma*, body] a chromosome made mainly of euchromatin, an autosome, *q.v.*; *cf.* heterochromosome.

Euciliata *n.* [Gk. *eu*, well; L. *cilium*, eyelid] a subclass of ciliates in some classifications, including those forms with sexual reproduction by conjugation; *cf.* Protociliata.

eucoen *n.* [Gk. *eu*, well; *koinos*, common] those members of a biocoenosis which are unable to live in a different environment; *cf.* tychocoen.

eucone *a.* [Gk. *eu*, good; *kōnos*, cone] having crystalline cones fully developed in single elements of compound eye.

Eucopepoda *n.* [Gk. *eu*, well; *kopē*, oar; *pous*, foot] an order of copepods including the free-swimming forms.

eudipleural *a.* [Gk. *eu*, good; *dis*, double; *pleuron*, side] symmetrical about a median plane, i.e. bilaterally symmetrical.

eudoxome *n.* [Gk. *eudoxos*, glorious] monogastric free-swimming stage of a siphonophore without nectocalyx.

Euechinoidea *n.* [Gk. *eu*, well; *echinos*, sea urchin; *eidos*, form] a subclass of variably shaped sea urchins having 2 columns of plates on both ambulacra and interambulacra.

eugamic *a.* [Gk. *eu*, well, *gamos*, marriage] *appl.* mature period rather than youthful or senescent.

eugenic *a.* [Gk. *eugenēs*, well-born] *pert.* or tending towards racial improvement; *alt.* aristogenic; *cf.* dysgenic.

eugenics *n.* [Gk. *eu*, well; *genos*, birth] the science dealing with the factors which tend to improve or impair stock.

Euglenales *n.* [*Euglena*, generic name] an order of algae of the Euglenophyta.

euglenoid *a.* [Gk. *eu*, well; *glēnē*, eyeball, puppet; *eidos*, form] *pert.* or like the protistan, *Euglena*; *appl.* movement: metaboly, *q.v.*

Euglenoidina *n.* [*Euglena*, generic name; Gk. *eidos*, form] an order of Phytomastigina, including the *Euglenas* and their allies, having an elongated, spindle-shaped body, a reserve food of paramylum, and usually an eye-spot.

Euglenomorphales *n.* [*Euglena*, generic name; Gk. *morphē*, form] an order of Euglenophyceae that are symbiotic in the gut of tadpoles.

Euglenophyceae *n.* [*Euglena*, generic name; Gk. *phykos*, seaweed] a class of algae, in many classifications placed in the division Euglenophyta, *q.v.*

Euglenophyta *n.* [*Euglena*, generic name; Gk. *phyton*, plant] a division of unicellular flagellate algae which have no rigid cell wall, have chlorophyll and carotinoid pigments, although pigments may be absent, and store food as paramylum or fat; they are related to Protozoa, and are classed as Flagellata (Phytomastigina) by zoologists, and in some classifications the group is reduced to the class Euglenophyceae.

eugonic *a.* [Gk. *eu*, well; *gonos*, produce] prolific; growing profusely, *appl.* bacterial colonies.

euhaline *a.* [Gk. *eu*, well; *halinos*, saline] living only in saline inland waters; *cf.* euryhaline.

euhyponeuston *n.* [Gk. *eu*, well; *hypo*, under; *neustos*, floating] organisms living in the top 5 cm of water for the whole or part of their lives.

eukaryo- *alt.* eucaryo-; *cf.* prokaryo-.

eukaryon *n.* [Gk. *eu*, well; *karyon*, nucleus] the nucleus of a eukaryotic organism.

eukaryotes eukaryotic organisms, including all living organisms except bacteria and blue–green algae, the term sometimes being used as a taxonomic group.

eukaryotic *a.* [Gk. *eu*, well; *karyon*, nucleus] *appl.* cells or organisms whose DNA is organized into chromosomes with a protein coat and surrounded by a nuclear membrane.

eulamellibranch *a.* [Gk. *eu*, well; L. *lamella*, small plate; *branchiae*, gills] *appl.* gills of bivalve molluscs whose filaments are attached to adjacent ones by tissue bridges.

Eulamellibranchiata *n.* [Gk. *eu*, well; L. *lamella*, small plate; *branchiae*, gills] in some classifications an order of bivalve molluscs having eulamellibranch gills and including the Heterodonta, Schizodonta, Adapedonta, and Anomalodesmata.

Eumalacostraca *n.* [Gk. *eu*, well; *malakos*, soft; *ostrakon*, shell] a series of malacostracans having an abdomen of 6 segments, all with appendages.

eumelanin *n.* [Gk. *eu*, well; *melas*, black] black melanin; *cf.* phaeomelanin.

eumerism *n.* [Gk. *eu*, well; *meros*, part] an aggregation of like parts.

eumeristem *n.* [Gk. *eu*, well; *meristos*, divided] meristem composed of isodiametric thin-walled cells with dense cytoplasm and large nuclei.

eumerogenesis *n.* [Gk. *eu*, well; *meros*, part; *genesis*, descent] segmentation in which the units are similar at least for a certain time.

Eumetazoa *n.* [Gk. *eu*, well; *meta*, after; *zōon*, animal] the multicellular animals, excluding the sponges.

eumitosis *n.* [Gk. *eu*, well; *mitos*, thread] typical mitosis.

eumitotic *a.* [Gk. *eu*, well; *mitos*, thread] anaschistic, *q.v.*; *pert.* eumitosis.

Eumycetes *n.* [Gk. *eu*, well; *mykēs*, fungus] a group of fungi including the Ascomycetes, Basidiomycetes, and Fungi Imperfecti.

Eumycophyta *n.* [Gk. *eu*, well; *mykēs*, fungus; *phyton*, plant] the true fungi, including the Phycomycetes, Ascomycetes, Basidiomycetes, and Fungi Imperfecti, and not including the Myxomycophyta; *alt.* Eumycotina.

Eunicemorpha *n.* [*Eunice*, generic name; Gk. *morphē*, form] an order of polychaetes having all segments fairly similar, often with reduced parapodia, and a distinct prostomium bearing tentacles.

Euphausiacea *n.* [*Euphausia*, generic name] an order of malacostracans having a carapace enclosing the thorax, many of which form krill, the food of some whales.

Euphorbiales *n.* [*Euphorbia*, generic name] an order of dicots (Rosidae) which usually have unisexual flowers with no corolla; *alt.* Tricoccae.

euphotic *a.* [Gk. *eu*, well; *phōs*, light] well-illuminated, *appl.* zone: surface waters to depth of *ca.* 80–100 m; upper layer of photic zone; *cf.* dysphotic.

euphotometric *a.* [Gk. *eu*, well; *phōs*, light; *metron*, measure] *appl.* leaves oriented to receive maximum diffuse light; *cf.* panphotometric.

euplankton *n.* [Gk. *eu*, well; *plangktos*, wandering] the plankton in open water; *cf.* tychoplankton.

euplastic *a.* [Gk. *eu*, well; *plastos*, moulded] readily organized, easily forming a tissue.

euplectenchyma *n.* [Gk. *eu*, well; *plektos*, plaited; *engchyma*, infusion] fungal tissue composed of intertwined hyphae arranged in groups approximately at right angles to each other in 3 dimensions.

euploid *a.* [Gk. *eu*, well; *haploos*, onefold; *eidos*, form] polyploid when total chromosome number is an exact multiple of the haploid numbers; *cf.* aneuploid; *n.* euploidy.

eupotamic *a.* [Gk. *eu*, well; *potamos*, river] thriving both in streams and in their backwaters, *appl.* potamoplankton.

eupyrene *a.* [Gk. *eu*, well; *pyrēn*, fruit stone] *appl.* sperms of normal type; *cf.* apyrene, oligopyrene.

Euramerica a supercontinent made up of Europe and America, which is thought to have existed in past geological eras, and which separated into its components by continental drift.

Eurotiales *n.* [*Eurotium*, generic name] Plectomycetes, *q.v.*

euryadaptive *a.* [Gk. *eurys*, wide; L. *adaptare*, to fit to] capable of adapting themselves to life in many species of host, *appl.* parasites.

Euryalae *n.* [Gk. *eurys*, wide; L. *ala*, wing] an order of brittle stars having flexible branched or simple arms.

Euryapsida *n.* [Gk. *eurys*, wide; *apsis*, arch] Synaptosauria, *q.v.*

eurybaric *a.* [Gk. *eurys*, wide; *baros*, weight] *appl.* animals adaptable to great differences in altitude or pressure; *cf.* stenobaric.

eurybathic *a.* [Gk. *eurys*, wide; *bathys*, deep] having a large vertical range of distribution; *cf.* stenobathic.

eurybenthic *a.* [Gk. *eurys*, wide; *benthos*, depth of the sea] *pert.* or living within a wide range of depth of the sea bottom; *cf.* stenobenthic.

eurybiont *n.* [Gk. *eurys*, wide; *bion*, living] an organism capable of withstanding wide variations of environment.

eurychoric *a.* [Gk. *eurys*, wide; *chōrein*, to spread] widely distributed; *cf.* stenochoric.

eurycyst *n.* [Gk. *eurys*, wide; *kystis*, bladder] large cell of middle vein in mosses; *alt.* deuter cell, pointer cell; *cf.* stenocyst.

euryhaline *a.* [Gk. *eurys*, wide; *halinos*, saline] *appl.* marine organisms adaptable to a wide range of salinity; *alt.* homoiosmotic; *cf.* stenohaline, euhaline.

euryhygric *a.* [Gk. *eurys*, wide; *hygros*, wet] *appl.* organisms adaptable to a wide range of atmospheric humidity; *cf.* stenohygric.

euryoecious *a.* [Gk. *eurys*, wide; *oikos*, abode] having a wide range of habitat selection; *cf.* stenoecious.

eurypalynous *a.* [Gk. *eurys*, wide; *palynein*, to scatter] having a wide variety of types of pollen.

euryphagous *a.* [Gk. *eurys*, wide; *phagein*, to eat] subsisting on a large variety of foods; *cf.* stenophagous, omnivorous.

euryphotic *a.* [Gk. *eurys*, wide; *phōs*, light] adaptable to a wide range of illumination.

Eurypterida, eurypterids *n.*, *n.plu.* [*Eurypterus*, generic name] an order of giant fossil aquatic arthropods of the class Merostomata, or sometimes placed in the Arachnida, having a short, non-segmented prosoma, and a long segmented opisthosoma, and resembling scorpions.

eurypylous *a.* [Gk. *eurys*, broad; *pylē*, gate] wide at the opening; *appl.* canal system of sponges in which the chambers open directly into excurrent canals by wide apopyles, and receive water from incurrent canals through prosopyles.

eurysome *a.* [Gk. *eurys*, broad; *sōma*, body] short and stout; *cf.* leptosome.

eurythermic *a.* [Gk. *eurys*, wide; *thermē*, heat] *appl.* organisms adaptable to a wide range of temperature; *alt.* eurythermal, eurythermous; *cf.* stenothermic.

eurytopic *a.* [Gk. *eurys*, wide; *topos*, place] having a wide range of geographical distribution; *cf.* stenotopic.

Euselachii *n.* [Gk. *eu*, well; *selachos*, shark] in some classifications, a subclass or order of Chondrichthyes including the sharks and rays, having many teeth developed in succession.

Eusporangiatae *n.* [Gk. *eu*, well; *sporos*, seed; *anggeion*, vessel] a subclass of Pteropsida (Filicopsida) having eusporangiate sporangia with walls several cells thick and formed from a group of cells, and an erect axis; *alt.* Eusporangidae; *cf.* Leptosporangiatae.

eusporangiate *a.* [Gk. *eu*, well; *sporos*, seed; *anggeion*, vessel] having sporangia derived from the inner cell produced by periclinal division of a superficial initial, each sporangium being formed from a group of epidermal cells, *appl.* certain ferns; *cf.* leptosporangiate.

Eustachian *a.* [B. *Eustachio*, Italian physician] *appl.* tube or canal connecting tympanic cavity with pharynx, *alt.* salpinx, otosalpinx, tuba auditiva or acustica; *appl.* valve guarding orifice of inferior vena cava in atrium of heart.

eustele *n.* [Gk. *eu*, well; *stēlē*, pillar] the arrangement of vascular tissue into collateral or bicollateral bundles with conjunctive tissue between, as in gymnosperms and dicotyledons.

eusternum n. [Gk. *eu*, well; *sternon*, breast plate] antesternite, *q.v.*; any other sternal sclerite.

eustomatous *a.* [Gk. *eu*, well; *stoma*, mouth] having a distinct mouth-like opening.

eustroma *n.* [Gk. *eu*, well; *strōma*, bedding] stroma formed of fungus cells only.

Eutardigrada *n.* [Gk. *eu*, well; L. *tardigradus*, slow paced] an order of tardigrades that have no lateral cirri behind the head.

eutelegenesis *n.* [Gk. *eu*, well; *telein*, to accomplish; *genesis*, descent] improved breeding by artificial insemination.

euthenics *n.* [Gk. *euthēnein*, to thrive] the science of betterment of human race on the side of intellect and morals; the study of environmental agencies contributing to racial improvement.

Eutheria, eutherials, eutherians *n*, *n.plu.* [Gk. *eu*, well; *thērion*, small animal] an infraclass of therian mammals, the higher mammals, which are viviparous with an allantoic placenta, and have a long period of gestation, after which the young are borne as immature adults; *alt.* Placentalia, placentals, Monodelphia.

euthycomous *a.* [Gk. *euthys*, straight; *komē*, hair] straight-haired.

Euthyneura *n.* [Gk. *euthys*, straight; *neuron*, nerve] A: Opisthobranchiata, *q.v.*; B: Pulmonata, *q.v.*

euthyneurous *a.* [Gk. *euthys*, straight; *neuron*, nerve] having visceral loop of nervous system untwisted; *cf.* streptoneurous.

eutocin *n.* [Gk. *eu*, well; *tokos*, birth] a compound in human amniotic fluid, which causes contraction of uterine muscle.

eutrophic *a.* [Gk. *eu*, well; *trophē*, nourishment]

providing or *pert.* adequate nutrition, *appl.* minerals in lakes; *cf.* dystrophic, oligotrophic.

eutropic *a.* [Gk. *eu*, well; *tropikos*, turning] turning sunward; dextrorse, *q.v.*

eutropous *a.* [Gk. *eu*, well; *tropos*, direction] *appl.* insects adapted to visiting special kinds of flowers; *appl.* flowers whose nectar is available to only a restricted group of insects; *cf.* allotropous.

evaginate *v.* [L. *evaginare*, to unsheath] to evert from a sheathing structure; to protrude by eversion.

evagination *n.* [L. *e*, out; *vagina*, sheath] the process of unsheathing, or product of this process; the process of turning inside out.

evanescent *a.* [L. *evanescere*, to vanish] disappearing early; *appl.* flowers which fade quickly.

evapotranspiration loss of water from the soil by evaporation from the surface and by transpiration from the plants growing thereon; the volume of water lost in this way.

evection *n.* [L. *e*, out; *vehere*, to convey] displacement of parent cell at septum of a filament, causing dichotomous appearance, as in certain algae.

evelate *a.* [L. *e*, out of; *velatus*, veiled] without a veil or velum, *appl.* fungi.

even-toed ungulates Artiodactyla, *q.v.*

evergreen *appl.* vascular plants which do not shed all their leaves at the same time and so appear green all the year round.

eviscerate *v.* [L. *ex*, out; *viscera*, entrails] to disembowel; to eject the viscera, as do holothurians on capture.

evocation *n.* [L. *evocare*, to call forth] the biochemical process whereby induced differentiation is called forth; induction as such.

evocator *n.* [L. *evocator*, caller forth] the chemical stimulus furnished by an organizer, *q.v.*; *alt.* morphogenic hormone.

evolute *a.* [L. *evolvere*, to unroll] turned back; unfolded.

evolutility *n.* [L. *evolvere*, to unroll] capability to evolve or change in structure; capacity to change in growth and form as a result of nutritional or other environmental factors.

evolution *n.* [L. *evolvere*, to unroll] the gradual development of organisms from pre-existing organisms since the beginning of life.

evolvate *a.* [L. *e*, out of; *volva*, wrapper] without a volva.

exalate *a.* [L. *ex*, without; *ala*, wing] apterous, *q.v.*

exalbuminous *a.* [L. *ex*, without; *albumen*, white of egg] without albumen; *appl.* seeds without endosperm or perisperm; *alt.* exendospermous.

exannulate *a.* [L. *ex*, without; *annulus*, ring] having a sporangium not furnished with an annulus, *appl.* certain ferns.

exarate *a.* [L. *exaratus*, ploughed up] *appl.* a pupa with free wings and legs; *cf.* obtect.

exarch *n.* [L. *ex*, without; Gk. *archē*, beginning] with protoxylem strands to the outside of the metaxylem, produced when xylem matures centripetally so the oldest protoxylem is farthest from the centre of the axis; *cf.* endarch.

exarillate *a.* [L. *ex*, without; F. *arille*, hoop] without an aril.

exasperate *a.* [L. *exasperare*, to roughen] furnished with hard, stiff points.

excentric *a.* [L. *ex*, out of; *centrum*, centre] one-sided; having the 2 portions of lamina or pileus unequally developed; *alt.* eccentric.

exchange diffusion transport of molecules or ions across a membrane during which they are exchanged between one side and the other.

exciple, excipulum *n.* [L. *excipula*, receptacles] the marginal wall, or outer covering, of apothecium in certain lichens.

excitability capacity of a living cell, or tissue, to respond to an environmental change or stimulus.

excitation *n.* [L. *excitare*, to rouse] act of producing or increasing stimulation; immediate response of protoplasm to a stimulus.

excitation–contraction coupling the process by which the contractile fibres of a muscle are activated by the excitation of a neurone.

excitation time chronaxie, *q.v.*

excitatory cells motor cells in sympathetic nervous system.

excitatory postsynaptic potential (EPSP) a local depolarization at the postsynaptic membrane caused by release of neurotransmitter at the synapse.

excitonutrient *a.* [L. *excitare*, to rouse; *nutriens*, feeding] causing or increasing nutrient activities.

exconjugant *n.* [L. *ex*, out; *conjugare*, to yoke] an organism which is leading an independent life after conjugation with another.

excorticate decorticate, *q.v.*

excreta *n.plu.* [L. *excretum*, separated] waste material eliminated from body or any tissue thereof; deleterious substances formed within a plant.

excretion *n.* [L. *ex*, out; *cernere*, to sift] the elimination of waste material from the body of a plant or animal; specifically, the elimination of waste materials produced by metabolism within cells; *cf.* egestion.

excretophores *n.plu.* [L. *excretus*, sifted out; Gk. *pherein*, to bear] cells of coelomic epithelium in which waste substances from blood accumulate, for discharge into coelomic fluid, as, e.g. chloragogen cells of earthworm.

excurrent *a.* [L. *ex*, out; *currere*, to run] *pert.* ducts, channels, or canals in which there is an outgoing flow; with undivided main stem; having midrib projecting beyond apex.

excurvate, excurved *a.* [L. *ex*, out; *curvare*, to curve] curved outwards from centre.

excystation *n.* [L. *ex*, out of; Gk. *kystis*, bladder] emergence from encysted condition.

exendospermous *a.* [Gk. *ex*, without; *endon*, within; *sperma*, seed] without endosperm; *alt.* exalbuminous.

exergonic *a.* [Gk. *ex*, out; *ergon*, work] releasing energy; *cf.* endergonic.

exflagellation *n.* [L. *ex*, out of; *flagellum*, whip] process of microgamete formation by microgametocyte in Haemosporidia.

exfoliation *n.* [L. *ex*, out; *folium*, leaf] the shedding of leaves or scales from a bud; shedding of flakes, as of bark.

exhalant *a.* [L. *ex*, out; *halare*, to breathe] capable of carrying from the interior outwards.

exindusiate *a.* [L. *ex*, out; *indusium*, cover] having the sporangia uncovered or naked.

exine extine, *q.v.*

exinguinal *a.* [L. *ex*, out; *inguen*, groin] occurring outside the groin; *pert.* 2nd joint of arachnid leg.

exites *n.plu.* [Gk. *exō*, outside] offshoots on outer lateral border of axis of certain arthropod limbs.

Exoascales *n.* [*Exoascus*, generic name] Taphrinales, *q.v.*

Exobasidiales *n.* [Gk. *exō*, without; *basis*, base; *idion*, *dim.*] an order of Hymenomycetes that lack a basidiocarp.

exobiotic *a.* [Gk. *exō*, outside; *biōtikos*, *pert.* life] living on the exterior of a substratum or on the outside of an organism; *cf.* endobiotic.

exocardiac *a.* [Gk. *exō*, without; *kardia*, heart] situated outside the heart.

exocarp *n.* [Gk. *exō*, without; *karpos*, fruit] epicarp, *q.v.*

exoccipital *a.* [L. *ex*, without; *occiput*, back of head] *pert.* a skull bone on each side of the foramen magnum; *alt.* pleuroccipital.

exochiton *n.* [Gk. *exō*, without; *chitōn*, coat] outermost layer of oogonial wall, as in Fucales; *alt.* exochite; *cf.* endochiton, mesochiton.

exochorion *n.* [Gk. *exō*, without; *chorion*, skin] outer layer of membrane secreted by follicular cells surrounding the egg in ovary of insects.

exocoel(e) *n.* [Gk. *exō*, without; *koilos*, hollow] the space between mesenteries of adjacent couples in certain Zoantharia; exocoelom, *q.v.*

exocoelar *a.* [Gk. *exō*, without; *koilos*, hollow] *pert.* parietal wall of coelom.

exocoelic *a.* [Gk. *exō*, without; *koilos*, hollow] in Zoantharia, *pert.* space between adjacent couples of mesenteries; *appl.* radial areas on disc; *appl.* outermost cycle of tentacles; *cf.* endocoelic.

exocoelom *n.* [Gk. *exō*, without; *koilōma*, hollow] extraembryonic body cavity of embryo; *alt.* exocoel.

exocone *a.* [Gk. *exō*, outside; *kōnos*, cone] *appl.* insect compound eye with cones of cuticular origin.

exocrine *a.* [Gk. *exō*, outwards; *krinein*, to separate] *appl.* glands whose secretion is drained by ducts; *cf.* endocrine.

exocuticle, exocuticula *n.* [Gk. *exō*, without; L. *dim.* of *cutis*, skin] middle layer of insect cuticle, between endocuticle and epicuticle; outer layer of integument in spiders.

Exocyclica *n.* [Gk. *exō*, without; *kyklos*, circle] a group of irregular sea urchins including the orders Clypeasteroida and Spatangoida.

exocytosis *n.* [Gk. *exō*, outside; *kytos*, hollow] the removal of material from a cell by the reverse of endocytosis.

exoderm *n.* [Gk. *exō*, without; *derma*, skin] the dermal layer of sponges.

exodermis *n.* [Gk. *exō*, without; L. *dermis*, skin] a specialized layer below the piliferous layer; ectoderm, *q.v.*

exodic *a.* [Gk. *exodos*, a going out] efferent; centrifugal; *cf.* esodic.

exoenzyme *n.* [Gk. *exō*, outside; *en*; in; *zymē*, leaven] an extracellular enzyme; *alt.* ectoenzyme; *cf.* endoenzyme.

exogamete *n.* [Gk. *exō*, without; *gametēs*, mate] a reproductive cell which fuses with one derived from another source.

exogamy *n.* [Gk. *exō*, without; *gamos*, marriage] fusion of gametes that are not closely related, i.e. outbreeding.

exogastric *a.* [Gk. *exō*, outwards; *gastēr*, stomach] having the shell coiled towards dorsal surface of body.

exogastrula *n.* [Gk. *exō*, without; *gastēr*, stomach] an hourglass-shaped sea urchin larva induced experimentally.

exogenote *n.* [Gk. *exō*, outside; *-genēs*, producing] in bacterial sexual processes, the chromosome fragment which passes from the donor to the recipient to form part of the merozygote.

exogenous *a.* [Gk. *exō*, outside; *-genēs*, producing] originating outside the organism, *alt.* xenogenous; developed from superficial tissue, the superficial meristem; growing from parts which were previously ossified; *appl.* metabolism concerned with effector activities and temperature; *alt.* allogenic; *cf.* endogenous.

exogenous rhythm a metabolic or behavioural rhythm which is synchronized by some external factor and which ceases to occur when this factor is absent.

exognath *n.* [Gk. *exō*, outside; *gnathos*, jaw] the outer branch of oral appendages of Crustacea; *alt.* exognathite.

exognathion *n.* [Gk. *exō*, without; *gnathos*, jaw] the maxillary portion of upper jaw; the maxilla with exception of endognathion and mesognathion.

exognathite exognath, *q.v.*

exogynous *a.* [Gk. *exō*, outside; *gynē*, woman] *appl.* flower with style longer than the corolla and projecting above it.

exo-intine *n.* [Gk. *exō*, without; L. *intus*, within] middle layer of a spore-covering, between extine and intine; *alt.* medine.

exolete *a.* [L. *exolescere*, to grow out of use] disused; emptied, *appl.* capsules, perithecia, etc.

exomixis *n.* [Gk. *exō*, outside; *mixis*, mingling] union of gametes derived from different sources; *alt.* xenomixis.

exonephric *a.* [Gk. *exō*, without; *nephros*, kidney] with nephridia opening to exterior, *appl.* oligochaetes; *cf.* enteronephric.

exonuclease *n.* [Gk. *exō*, outside; L. *nucleus*, kernel] an enzyme that hydrolyses DNA by breaking bonds at the end of the chain.

exoparasite ectoparasite, *q.v.*

exopeptidases *n. plu.* [Gk. *exō*, outside; *pepsis*, digestion] peptidases which attack terminal peptide linkages, thus splitting off amino acids; EC sub-subgroups 3.4.11–3.4.17.

exoperidium *n.* [Gk. *exō*, without; *pēridion*, a small wallet] the outer layer of spore case (peridium) in certain fungi; *alt.* epiperidium.

exophylaxis *n.* [Gk. *exō*, without; *phylax*, guard] protection afforded against pathogenic organisms by skin secretions.

exophytic *a.* [Gk. *exō*, outside of; *phyton*, plant] on, or *pert.* exterior of plants, *appl.* oviposition.

exoplasm *n.* [Gk. *exō*, without; *plasma*, mould] ectoplasm, *q.v.*

exopodite *n.* [Gk. *exō*, without; *pous*, foot] the outer branch of a typical biramous crustacean limb.

Exopterygota *n.* [Gk. *exō*, outside *pterygion*, little wing] a division of insects having simple or slight metamorphosis, with wings developing externally and young stages called nymphs, which resemble adults; *alt.* Hemimetabola, Heterometabola; *cf.* Endopterygota.

exoscopic *a.* [Gk. *exō*, without; *skopein*, to

look] with apex emerging through archegonium, *appl.* embryo; *cf.* endoscopic.

exoskeleton *n.* [Gk. *exō*, without; *skeletos*, hard] a hard supporting structure secreted by and external to ectoderm or skin; *alt.* dermatoskeleton, dermoskeleton; *cf.* endoskeleton.

exosmosis *n.* [Gk. *exō*, outwards; *ōsmos*, impulse] osmosis in an outward direction, such as out of a cell; *cf.* endosmosis.

Exosporeae *n.* [Gk. *exō*, outside; *sporos*, seed] a subclass of Myxomycetes having spores produced on the surface of the fruit body.

exospore, exosporium *n.* [Gk. *exō*, outside; *sporos*, seed] conidium, *q.v.*; the outer layer of a spore, *alt.* extine.

exosporous *a.* [Gk. *exō*, outwards; *sporos*, seed] with spores borne or discharged exteriorly.

exostome *n.* [Gk. *exō*, outwards; *stoma*, mouth] outer portion of peristome in mosses; opening or foramen in outer wall of ovule.

exostosis *n.* [Gk. *exō*, outwards; *osteon*, bone] formation of knots on surface of wood; formation of knob-like outgrowths of bone at a damaged portion, or of dental tissue in a similar way.

exoteric *a.* [Gk. *exōteros*, beyond] produced or developed outside the organism.

exotheca *n.* [Gk. *exō*, outside; *thēkē*, box] the tissue outside the theca of a coral; *a.* exothecal.

exothecate *a.* [Gk. *exō*, without; *thēkē*, box] having an exotheca.

exothecium *n.* [Gk. *exō*, without; *thēkē*, case] the outer specialized dehiscing cell layer of anther; *cf.* endothecium.

exothermic *a.* [Gk. *exō*, outwards; *thermē*, heat] releasing heat energy; *cf.* endothermic.

exotic *a.* [Gk. *exōtikos*, foreign] introduced or non-endemic. *n.* A foreign plant or animal not acclimatized; an exotic organism.

exotospore *n.* [Gk. *exō*, outward; *sporos*, seed] sporozoite, *q.v.*

exotoxin *n.* [Gk. *exō*, outwards; *toxikon*, poison] a soluble toxin excreted by bacteria; *cf.* endotoxin.

exotropism *n.* [Gk. *exō*, outwards; *tropē*, turn] curvature away from axis, exhibited by a laterally geotropic organ.

expalpate epalpate, *q.v.*

expantin presumable neurohumour inducing expansion of chromatophores in crustaceans; *cf.* contractin.

expectancy the principle of animal behaviour which assumes that reinforcement must relate to the expectations of an animal.

experimental extinction blotting out an acquired behavioural response.

expiration *n.* [L. *exspirare*, to breathe out] the act of emitting air or water from the respiratory organs; emission of carbon dioxide by plants and animals; *cf.* inspiration.

expiratory *a.* [L. *exspirare*, to breathe out] *pert.* or used in expiration, *appl.* muscles.

explanate *a.* [L. *ex*, out; *planare*, to make plain] having a flat extension.

explantation *n.* [L. *ex*, out of; *plantare*, to plant] tissue culture away from organism of its origin.

exploration a pattern of animal behaviour when investigating its surroundings, which often involves latent learning and must be intrinsically rewarding, leading to an increased rate of change

of stimulation falling on the animal's sense organs.

explosive *appl.* flowers in which pollen is suddenly discharged on decompression of stamens by alighting insect, as of broom and gorse; *appl.* fruits with sudden dehiscence, seeds being discharged to some distance; *appl.* evolution, rapid formation of numerous types; *appl.* speciation, rapid formation of species from a single species in one locality.

expressivity the degree to which a gene produces a phenotypic effect.

exsculptate *a.* [L. *ex*, out; *sculpere*, to carve] having the surface marked with more-or-less regularly arranged raised lines with grooves between.

exscutellate *a.* [L. *ex*, without; *scutellum*, small shield] having no scutellum, *appl.* insects; *alt.* escutellate.

exserted *a.* [L. *exserere*, to stretch out] protruding beyond some including organ or part; *appl.* stamens which project beyond corolla.

exsertile *a.* [L. *exserere*, to stretch out] capable of extrusion.

exsiccata *n.plu.* [L. *exsiccare*, to dry up] dried specimens, as in an herbarium.

exstipulate *a.* [L. *ex*, without; *stipula*, stem] without stipules; *alt.* astipulate, estipulate, instipulate.

exstrophy *n.* [Gk. *exō*, outwards; *strophē*, turning] eversion, as normal or anomalous projection of luteal tissue to exterior of ovary.

exsuccate, exsuccous *a.* [L. *ex*, out; *succus*, juice] sapless; without juice; without latex.

exsufflation *n.* [L. *ex*, out; *sufflare*, to blow] forced expiration from lungs.

extend *v.* [L. *ex*, out; *tendere*, to stretch] to straighten out, *appl.* movement of limb; *cf.* flex.

extensor *n.* [L. *ex*, out; *tendere*, to stretch] any muscle which extends or straightens a limb or part; *cf.* flexor.

exterior *a.* [L. *externus*, on outside] situated on side away from axis or definitive plane.

external *a.* [L. *externus*, outside] outside or near the outside; away from the mesial plane.

external respiration considered in terms of the gaseous exchange between the organism and environment, and the processes taking gases to and from cells; *cf.* internal respiration.

externum *n.* [L. *externus*, outward] outer region or cortex, of a mitochondrion or of Golgi apparatus, or of acroblast.

exteroceptor *n.* [L. *exter*, outside; *capere*, to take] a receptor which receives stimuli from outside the body; a contact receptor, or a distance receptor.

extinction point the minimum percentage of illumination below which a plant species is unable to survive under natural conditions; *cf.* compensation point.

extine *n.* [L. *exter*, outside] the outer coat of spore or pollen grain; *alt.* exine, exosporium; *cf.* intine.

extra-axillary *a.* [L. *extra*, beyond; *axilla*, armpit] arising above axil of leaf, said of branches which develop from upper bud when there are more than one in connection with axil.

extrabranchial *a.* [L. *extra*, beyond: Gk. *branghia*, gills] arising outside the branchial arches.

extracapsular *a.* [L. *extra*, outside; *capsula*, small box] arising or situated outside a capsule; *appl.*

ligaments, etc., in connection with a joint; *appl.* protoplasm lying outside the central capsule in some protozoans; *appl.* dendrites.

extracellular *a.* [L. *extra*, outside; *cellula*, little cell] occurring outside the cell; diffused out of the cell.

extrachorion *n.* [L. *extra*, outside; Gk. *chorion*, skin] outermost layer, external to exochorion, of egg shell in certain insects.

extracolumella *n.* [L. *extra*, beyond; *columella*, small column] distal element of auditory skeletal structure; *alt.* hyostapes.

extracortical *a.* [L. *extra*, outside; *cortex*, bark] not within the cortex, *appl.* part of brain.

extraembryonic *a.* [L. *extra*, outside; Gk. *embryon*, foetus] situated outside the embryo proper, as portion of blastoderm, yolk sac, allantois, amnion, chorion, *appl.* membranes, *alt.* foetal membranes, especially in mammals.

extraenteric *a.* [L. *extra*, outside; Gk. *enteron*, gut] outside the alimentary tract.

extrafloral *a.* [L. *extra*, outside; *flos*, flower] situated outside the flower, *appl.* nectaries.

extrafoveal *a.* [L. *extra*, beyond; *fovea*, depression] *pert.* macula lutea surrounding fovea centralis, *appl.* rod vision; *cf.* foveal.

extrahepatic *a.* [L. *extra*, outside; Gk. *hēpar*, liver] *appl.* cystic duct and common bile duct.

extramatrical *a.* [L. *extra*, outside; *mater*, mother] located or growing on the surface of a matrix.

extranuclear *a.* [L. *extra*, outside; *nucleus*, kernel] *pert.* structures or forces acting outside the nucleus; situated outside the nucleus.

extraocular *a.* [L. *extra*, outside; *oculus*, eye] exterior to the eye, *appl.* antennae of insects.

extraperitoneal subperitoneal, *q.v.*

extrapulmonary *a.* [L. *extra*, beyond; *pulmones*, lungs] external to the lungs, *appl.* bronchial system.

extraspicular *a.* [L. *extra*, outside; *spicula*, small spike] with spicules having one end embedded in spongin and the other end free.

extrastapedial *a.* [L. *extra*, beyond; *stapes*, stirrup] extending beyond the stapediocolumellar junction.

extrastelar *a.* [L. *extra*, outside; Gk. *stēlē*, column] *pert.* ground tissue outside vascular tissue.

extravaginal *a.* [L. *extra*, outside; *vagina*, sheath] forcing a way through the sheath, as shoots of many plants.

extravasate *v.* [L. *extra*, outside; *vas*, vessel] to force its way from the proper channel into the surrounding tissue, said of blood, etc.

extraventricular *a.* [L. *extra*, beyond; *ventriculus*, belly] situated or arising beyond the ventricle.

extraxylary *a.* [L. *extra*, outside; Gk. *xylon*, wood] on the outside of the xylem, *appl.* fibres.

extremity *n.* [L. *extremitas*, limit] the limb, or distal portion of a limb; distal end of any limblike structure.

extrinsic *a.* [L. *extrinsecus*, on outside] acting from the outside; not wholly within the part, *appl.* muscles; *appl.* cycles in population of a species, due to environmental fluctuation; *appl.* brightness due to objective light intensity; *cf.* intrinsic.

extrorse *a.* [L. *extrorsus*, outwardly] turned away from axis, *appl.* dehiscence of anthers; *see* postical.

exudates substances given out by exudation.

exudation *n.* [L. *exudare*, to sweat] any discharge of material from a cell, organ, or organism, through a membrane, incision, pore, or gland, e.g. gums, resins, moisture, etc.

exumbral *a.* [L. *ex*, out; *umbra*, shade] *pert.* rounded upper surface of a jellyfish.

exumbrella *n.* [L. *ex*, out; *umbra*, shade] upper, convex surface of medusa.

exuviae *n.plu.* [L. *exuere*, to strip off] cast-off skins, shells, etc., of animals.

exuvial *a.* [L. *exuere*, to strip off] ecdysial; *appl.* insect glands whose secretion facilitates ecdysis.

eye *n.* [A.S. *éage*, eye] the organ of sight or vision; a pigment spot in various animals and in lower plants; the bud of a tuber.

eye-spots certain pigment spots in many lower plants and animals, and also in some vertebrates, which have a visual function; *alt.* ocelli.

eye teeth upper canine teeth.

F

F_1 denotes 1st filial generation, being hybrids arising from a 1st cross, successive generations arising from this one being denoted by F_2, F_3, etc. P_1 denotes parents of F_1 generation, P_2 the grandparents, etc.

Fabales *n.* [L. *faba*, bean] an order of dicots (Rosidae), also known as legumes, whose fruit is a pod and whose roots contain nitrogen-fixing bacteria of the genus *Rhizobium*, including the families Mimosaceae, Caesalpinaceae, and Papilionaceae (Fabaceae); also called Leguminosae, although this term is usually used when the group is considered a family or super-family.

fabella *n.* [*Dim.* of L. *faba*, bean] a small fibrocartilage ossified in tendon of the lateral head of the gastrocnemius.

fabiform *a.* [L. *faba*, bean; *forma*, shape] bean-shaped.

Fabrician [*J.C. Fabricius*, Danish entomologist] *appl.* a classification of the Arthropoda based on the anatomy of the mouth-parts.

facet *n.* [F. *facette*, small face] a smooth, flat, or rounded surface for articulation; ocellus, *q.v.*; corneal portion of insect eye; ommatidium, *q.v.*

facial *a.* [L. *facies*, face] *pert.* face, *appl.* artery bones, veins, etc., *appl.* 7th cranial nerve.

faciation *n.* [L. *facies*, face] formation or character of facies; a grouping of dominant species within an association; geographical differences in abundance or proportion of dominant species in a community, *cf.* lociation.

facies *n.* [L. *facies*, face] the face; a surface, in anatomy; the general aspect of a plant; aspect, as superior and inferior; a particular modification of a biotope; a grouping of dominant plants in the course of a successional series; one of different types of deposit in a geological series or system; the palaeontological and lithological character of a deposit.

facilitated diffusion transport of molecules or ions along a concentration gradient by a carrier system, but not requiring expenditure of energy.

facilitation n. [L. *facilitas*, easiness] diminution of resistance to a stimulus subsequent to previous stimulation, as of nerve cells and effector cells, *cf.* recruitment; *see* social facilitation.

faciolingual a. [L. *facies*, face; *lingua*, tongue] *pert.* or affecting face and tongue.

factor n. [L. *facere*, to make] any agent (biotic, climatic, nutritional, etc.) contributing to a result; Mendelian or genetic factor: gene, *q.v.*; a determinant.

factorial a. [L. *facere*, to make] *pert.* genetic factors or genes.

factor interaction the principle that one factor of the environment present in large amounts may affect the action of the minimum factor in the environment and so affects Liebig's law of the minimum.

facultative a. [L. *facultas*, faculty] having the power of living under different conditions; conditional; *appl.* organisms which may be normally self-dependent, but which are adaptable to a parasitic or semiparastitic mode of life; *appl.* aerobes, anaerobes; *appl.* parthenogenesis, symbionts, gametes, etc.; *cf.* obligate.

facultative parasite a saprophyte which can also survive as a parasite; *alt.* hemisaprophyte.

facultative saprophyte a parasite which can also exist as a saprophyte; *alt.* hemiparasite.

FAD flavin adenine dinucleotide, a derivative of riboflavin, important as a coenzyme in oxidation/reduction reactions as in respiratory oxidations, because it can be both oxidized and reduced.

faeces n.plu. [L. *faeces*, dregs] excrement from alimentary canal; *alt.* feces.

Fagales n. [*Fagus*, generic name] an order of dicots of the Archichlamydeae or Hamamelididae including many deciduous forest trees such as beech, sweet chestnut, birch, hazel, and hornbeam.

F-agent F-episome, *q.v.*

falcate a. [L. *falx*, sickle] sickle-shaped; hooked; *alt.* drepanoid.

falces n.plu. [L. *falces*, sickles] chelicerae of arachnids; *plu.* of falx, *q.v.*

falcial a. [L. *falx*, sickle] *pert.* falx, especially falx cerebri.

falciform a. [L. *falx*, sickle; *forma*, shape] sickle-shaped or scythe-shaped, *alt.* drepanoid; *appl.* ligament, a dorsoventral fold of peritoneum, attached to under surface of diaphragm and anterior and upper surface of liver, *alt.* mesohepar; *appl.* process, processus falciformis, a fold of choroid penetrating retina near optic disc and ending at back of lens, functioning in accommodation in teleosts; *appl.* body, a sporozoite; *appl.* young, sporocysts enclosing several spores in certain Sporozoa.

falciphore n. [L. *falx*, sickle; Gk. *pherein*, to bear] a hook-shaped conidiophore or sporangiophore.

Falconiformes n. [*Falco*, generic name of falcon; L. *forma*, shape] an order of birds of the subclass Neornithes, including the vultures, eagles, hawks, falcons.

falcula n. [L. *falcula*, little hook] a curved scythe-like claw; the falx cerebelli, *q.v.*

falcular falculate, *q.v.*; *pert.* falcula; *pert.* falx.

falculate curved, and sharp at the point, sickle-shaped; *alt.* falcular.

Fallopian tube [G. *Fallopio*, Italian anatomist] upper portion of oviduct in mammals, representing anterior portion of Müllerian duct in other vertebrates; *alt.* salpinx, uterine tube; sometimes incorrectly called oviduct.

false axis pseudaxis, *q.v.*

false dichotomy dichasium, *q.v.*

false dissepiment a partition in an ovary which has not been formed by adhesion of the edges of 2 neighbouring carpels, but is an ingrowth from the carpel wall, or some other form of cellular tissue; *alt.* spurious dissepiment.

false foot pseudopodium, *q.v.*

false fruits fruits formed from the receptacle or other parts of the flower, in addition to the ovary, or from complete inflorescences; *alt.* pseudocarps.

false germination the swelling up of a dead seed owing to uptake of water, which gives the appearance of germination.

false ribs those ribs whose cartilaginous ventral ends do not join the sternum directly; *alt.* asternal ribs, costae spuriae.

false scorpions pseudoscorpions, *q.v.*

false spiders a common name for the Solpugida, *q.v.*

false tissue pseudoparenchyma, *q.v.*

false vocal cords ventricular folds of larynx, 2 folds of mucous membrane, each covering a ligament, anterior of true vocal cords.

falx n. [L. *falx*, sickle] a sickle-shaped structure; falx cerebri: a sickle-shaped fold of the dura mater, *alt.* falcula; inguinal aponeurosis of transverse and internal oblique muscles of abdomen; *plu.* falces, *q.v.*; *a.* falcate, falcular, falciform.

familial *pert.* family; transmitted in families.

family n. [L. *familia*, household] a taxonomic group consisting of related genera, families being grouped into orders; formerly families of flowering plants were called natural orders.

famulus n. [L. *famulus*, attendant] a tarsal sensory seta in certain mites.

fan n. [A.S. *fann*, fan] a bird's tail feathers; flabellum, *q.v.*; rhipidium, *q.v.*; vannus, *q.v.*

fang n. [A.S. *fang*, grip] a long-pointed tooth, especially the poison tooth of snakes; the root of a tooth; distal joint or unguis of chelicerae in Arachnida; a canine or carnassial tooth in Carnivora.

faradization n. [*M. Faraday*, English physicist] method of stimulation inducing partial or complete tetanus.

farctate a. [L. *farctus*, stuffed] filled, not hollow; stuffed full.

farina n. [L. *farina*, flour] a substance having the consistency of flour or meal; pollen, *q.v.*; the fine mealy-like powder found on some insects.

farinaceous a. [L. *farina*, flour] containing flour; starchy; mealy, *q.v.*; farinose, *q.v.*

Farinosae n. [L. *farina*, flour] an order of monocotyledons which are usually herbaceous and have cyclic flowers with a tri- or bimerous perianth, and mealy endosperm; sometimes considered a superorder, when also called Commelinanae.

farinose a. [L. *farina*, flour] mealy, *q.v.*; producing or covered with fine powder or dust; covered with short white hairs that can be detached like dust; *alt.* farinaceous.

-farious a suffix meaning arranged in a certain number of rows, as bifarious, trifarious, etc.

far red light light of wavelength about 730 nm which reverses the effect of red light on phytochrome.

fascia *n.* [L. *fascia*, band] an ensheathing band of connective tissue; a transverse band of a different colour, as in some plants; a band-like structure.

fascia dentata one of the specialized lower borders of the hippocampus in mammals.

fascial *a.* [L. *fascia*, bundle] *pert.* a fascia, ensheathing and binding.

fasciated *a.* [L. *fascia*, bundle] banded; arranged in fascicles, *appl.* stipes; *appl.* stems or branches malformed and flattened.

fasciation *n.* [L. *fascia*, bundle] the formation of fascicles; coalescent development of branches of a shoot system, as in cauliflower.

fascicle *n.* [L. *fasciculus*, small bundle] a small bundle or tuft, as of fibres, leaves, etc.; *alt.* fasciculus.

fascicular *a.* [L. *fasciculus*, small bundle] *pert.* a fascicle; arranged in bundles or tufts; *appl.* cambium or tissue within vascular bundle, *alt.* intrafascicular.

fasciculus *n.* [L. *fasciculus*, small bundle] a bundle of muscle fibres, especially those surrounded by a perimysium; a group, bundle, or tract of nerve fibres as of medulla spinalis; a fascicle, *q.v.*

fasciola *n.* [L. *fasciola*, small bandage] a narrow colour band; a delicate lamina continuous with supracallosal gyrus and with fascia dentata of hippocampus.

fasciole *n.* [L. *fasciola*, small bandage] ciliated band on certain echinoids for sweeping water over surrounding parts.

fastigiate *a.* [L. *fastigare*, to slope up] with branches close to stem and erect, *cf.* patent; in pyramidal or conical form.

fastigium *a.* [L. *fastigium*, gable] angular top of roof of 4th ventricle, formed by contact of anterior and posterior medullary vela of cerebellum.

fat *see* fats.

fat-body one of the vascularized tissue structures filled with fat globules and associated with gonads in Amphibia; one of the subcutaneous organs along ventral sides and enlarged during breeding season in Lacertilia; tissue of indeterminate form distributed throughout body of insects and functioning as nutritive reserve; *alt.* adipose body, corpus adiposum, epiploon; fats stored as reserve food in plants especially seeds.

fat-cell lipocyte, *q.v.*

fate map a map of an embryo in an early stage of development in which the anticipated end-result of developing regions is marked.

fatigue *n.* [L. *fatigare*, to weary] effect produced by unduly prolonged stimulation on cells, tissues, or other structures so that they are less responsive.

fatigue substance a substance produced by a plant which acts as an inhibitor to its own growth.

fatigue toxin kenotoxin, *q.v.*

fatiscent *a.* [L. *fatiscere*, to crack] gaping open; cracked.

fats *n.plu.* [A.S. *faet*, fat] glycerides having a greater proportion of saturated acids and solid at 20°C, hydrolysed by lipase to fatty acids and glycerol and forming a food store in animals and plants, and including human fat, butter, etc., *cf.*

oils; loosely any substance which can be extracted from tissue with a fat solvent; adipose tissue, or any animal tissue having its cells filled with a greasy or oily reserve material.

fat-soluble *appl.* vitamins A, D, E, and K.

fatty acid oxidation a process occurring in the mitochondria of higher animals and in plants that store fat, consisting of the oxidation of fats resulting in the production of ATP; *alt.* beta oxidation.

fatty acids a group of saturated and unsaturated monobasic aliphatic carboxylic acids which form esters with glycerol and other alcohols to make fats, oils, waxes, and other lipids.

fauces *n.plu.* [L. *fauces*, throat] upper or anterior part of throat between palate and pharynx; mouth of a spirally coiled shell; throat of a corolla; *a.* faucial.

fauna *n.* [L. *faunus*, god of woods] all the animals peculiar to a country, area, or period.

faunal region an area characterized by a special group or groups of animals.

faunula *n.* [*Dim.* of *fauna*] animal population of a small unit area or microscopic niche, as of intestine, bark, etc.

favella *n.* [L. *favus*, honeycomb] a conceptacle of certain red algae.

faveolate *a.* [L. *faveolus, dim.* of *favus*, honeycomb] honeycombed; *alt.* alveolate, favose.

faveolus *n.* [L. *faveolus*, small honeycomb] a small depression or pit; *alt.* alveola.

favoid *a.* [L. *favus*, honeycomb; Gk. *eidos*, form] resembling a honeycomb.

favose *a.* [L. *favus*, honeycomb] faveolate, *q.v.*

F-duction sexduction, *q.v.*

fear the pattern behaviour produced by sudden strong unpleasant stimuli, including the responses to adrenalin activity and leading to flight; the drive underlying this behaviour.

feather *n.* [M.E. *fether*, feather] epidermal structures forming body cover of birds.

feather epithelium epithelium of cells, each having a process with numerous lateral filaments, on inner surface of nictitating membrane of many reptiles and birds, for cleaning the eye surface.

feather follicle an impushing of epidermis surrounding the base of a feather.

feather-veined *appl.* leaf in which veins run out from midrib in regular series at an acute angle; *alt.* pinnately veined.

feces faeces, *q.v.*

Fechner's Law [*G.T. Fechner*, German psychophysicist] the tendency of intensity of sensation to vary as the logarithm of the stimulus; *alt.* Weber–Fechner Law.

fecundate *v.* [L. *fecundare*, to make fruitful] to impregnate; to fertilize; to pollinate; *n.* fecundation.

fecundity *n.* [L. *fecunditas*, fruitfulness] power of a species to multiply rapidly; capacity to form reproductive elements; the number of eggs produced by an individual or a species.

feedback inhibition end-product inhibition, *q.v.*

feedback mechanism an internal regulating mechanism in which the presence of a certain substance at a certain level inhibits or promotes its further formation.

fellic *a.* [L. *fel*, bile] *pert.*, or derived from, bile.

female *n.* [L. *femina*, woman] an individual whose sex organs contain only female gametes; symbol ♀.

female pronucleus the nucleus left in the ovum after maturation.

femoral *a.* [L. *femur*, thigh] *pert.* femur; *pert.* thigh, *appl.* artery, vein, nerve, etc., *alt.* crural. *n.* Paired femoral shield of plastron in Chelonia.

femur *n.* [L. *femur*, thigh] the thigh bone, proximal bone of hindlimb in vertebrates; 3rd joint in insect, spider, and myriapod leg counting from proximal end.

fen *n.* [O.E. *fenn*, marsh] a plant community on alkaline, neutral, or slightly acid wet peat; *cf.* bog.

fenchone *n.* [Ger. *Fenchel*, fennel] a ketone, the essential oil in oil of fennel, also found in lavender and thuja oil.

fenestra *n.* [L. *fenestra*, window] an opening in a bone, or between 2 bones, or in a plant membrane; a pit on head of cockroach; fontanelle, *q.v.*, of termites; a transparent spot on wings of insects.

fenestra cochleae fenestra rotunda, *q.v.*

fenestra ovalis the upper of 2 openings in wall of bony labyrinth, between tympanic cavity and vestibule of inner ear; *alt.* fenestra vestibuli, oval window; *cf.* fenestra rotunda.

fenestra pseudorotunda opening covered by entotympanic membrane in birds, the fenestra rotunda in mammals having a different origin.

fenestra rotunda the lower of 2 openings in wall of bony labyrinth, closed by secondary tympanic membrane; *alt.* fenestra cochleae, fenestra tympani, round window; *cf.* fenestra ovalis.

fenestrate *a.* [L. *fenestra*, window] having small perforations or transparent spots, *appl.* insect wings; having numerous perforations, *appl.* leaves, dissepiments.

fenestrated membrane a close network of yellow elastic fibres resembling a membrane with perforations, as in inner tunic of arteries, *alt.* Henle's membrane; basal membrane of compound eye, penetrated by ommatidial nerve fibres.

fenestra tympani fenestra rotunda, *q.v.*

fenestra vestibuli fenestra ovalis, *q.v.*

fenestrule *n.* [*Dim.* of L. *fenestra*, window] small opening between branches of a polyzoan colony.

F-episome the episome whose presence or absence determines the sex of a bacterium, because when present the bacterium acts as a male and synthesizes the F-pilus, when absent acts as a female; *alt.* F-factor, fertility factor, F-agent, sex factor.

feral *a.* [L. *fera*, wild animal] wild, or escaped from cultivation or domestication and reverted to wild state.

ferment *n.* [L. *fermentum*, ferment] an enzyme, *q.v.*; an enzyme causing fermentation; *alt.* biocatalyst.

fermentation *n.* [L. *fermentum*, ferment] decomposition occurring in organic compounds, usually carbohydrates, brought about by enzymes either directly or in micro-organisms, resulting in the production of other compounds such as alcohol or lactic acid, often with heat and gases; often used for anaerobic respiration, especially of yeast, with the production of alcohol; *alt.* zymosis.

fern the common name for a member of the class Pteropsida (Filicopsida) in the traditional classification or of the class Filicineae in some modern classifications; formerly the term Filicales was used to include all ferns.

ferrallitic soils deep red soils, acid in reaction, found on freely drained sites in humid tropical regions.

ferredoxin *n.* [L. *ferrum*, iron; Gk. *oxys*, sharp] a protein with a simple iron prosthetic group, which in its reduced form is a very powerful biological reducing agent.

ferrichrome *n.* [L. *ferrum*, iron; Gk. *chrōma*, colour] an iron-containing nitrogenous pigment, precursor of cytochrome, found in smut fungi.

ferritin *n.* [L. *ferrum*, iron] a ferroprotein made of a protein (apoferritin) and ferric hydroxide-phosphate, found in spleen, liver (where much iron is stored as ferritin), and bone marrow, also in the intestinal mucosa wall where it is thought to aid absorption of iron.

ferrocytes *n.plu.* [L. *ferrum*, iron; Gk. *kytos*, hollow] cells formed from lymphocytes, containing iron compounds and concerned with tunicin production in ascidians.

ferroprotein *n.* [L. *ferrum*, iron; Gk. *prōteion*, first] an enzyme with an iron porphyrin prosthetic group combined with a specific protein.

ferruginous *a.* [L. *ferruginus*, rusty] having the appearance of iron rust.

Fer(r)ungulata *n.* [L. *fera*, wild animal; *ungula*, hoof] in some classifications, a group of animals including the Carnivora and Ungulata.

fertile *a.* [L. *fertilis*, fertile] producing viable gametes or spores; capable of producing living offspring; of eggs or seeds, capable of developing; *appl.* a soil containing the essential ingredients for plant growth.

fertilis- fertiliz-, *q.v.*

fertility factor F-episome, *q.v.*

fertility vitamin vitamin E, *q.v.*

fertiliz- *alt.* fertilis-.

fertilization *n.* [L. *fertilis*, fertile] the union of male and female gametes to form a zygote.

fertilization cone protuberance on ovum at point of contact and entry of spermatozoon before fertilization; *alt.* attraction cone, entrance cone.

fertilization membrane a membrane formed by an ovum in response to its penetration by a sperm, which grows rapidly from the point of penetration and covers the ovum, excluding other sperm.

fertilization tube process of an antheridium, penetrating oogonial wall, for passage of male gamete in certain fungi.

fertilizin *n.* [L. *fertilis*, fertile] a substance secreted by the ovum which may attract sperm, assist activation of sperm, cause sperm agglutination, and induce cleavage; *alt.* gynogamone II; *cf.* heteroagglutinin, isoagglutinin.

festoon *n.* [F. *feston*, garland] the margin, with rectangular divisions, of integument in ticks; rim of gum round neck of tooth.

fet-, fetus foet-, foetus, *q.q.v.*

F-factor F-episome, *q.v.*

F₁ generation *see* F₁.

F₁ hybrid in horticulture, the 1st cross between 2 pure breeding lines, brought about by controlled cross-pollination.

Fibonacci series [*L. Fibonacci*, Italian mathematician] the unending sequence 1, 1, 2, 3, 5, 8, 13, 21, 34 . . . where each term is defined as the sum of its 2 predecessors, occurring in numerators and denominators in phyllotaxy.

fibre n. [L. *fibra*, band] a strand of nerve, muscle, or connective tissue; a delicate root; an elongated tapering sclerenchyma cell for mechanical strength; *alt.* fiber.

fibre sclereid(e) a cell which is intermediate between a sclerenchyma fibre and a sclereid.

fibre tracheid(e) a cell which is intermediate between a sclerenchyma fibre and a tracheid.

fibril n. [L. *fibrilla*, small fibre] a small thread-like structure or fibre; a component part of a fibre; a root hair; a slender filiform outgrowth on some lichens; fibrilla, *q.v.*

fibrilla n. [L. *fibrilla*, small fibre] thread-like branch of root; a root hair; minute elastic fibre secreted with spongin cells; minute muscle-like thread found in various ciliates; *alt.* fibril; *plu.* fibrillae.

fibrillar *pert.*, or like, fibrils or fibrillae.

fibrillar theory the theory held in the latter part of the 19th century that cytoplasm was made up of delicate fibrils, either as separate strands (filar theory) or forming a meshwork (reticular theory), situated within a homogeneous ground substance.

fibrillate a. [L. *fibrilla*, small fibre] having fibrillae or hair-like structures.

fibrilloblast odontoblast, *q.v.*

fibrillose a. [L. *fibrilla*, small fibre] furnished with fibrils, *appl.* mycelia of certain fungi.

fibrin n. [L. *fibra*, band] an insoluble protein produced from fibrinogen in the blood by the action of thrombin, and forming a meshwork of fibres in which corpuscles are caught, causing coagulation.

fibrinogen n. [L. *fibra*, band; Gk. -*genēs*, producing] a soluble protein of blood, which, by activity of thrombin, yields fibrin and produces coagulation.

fibrinolysin n. [L. *fibrin*, band; Gk. *lysis*, loosing] an enzyme in blood plasma which dissolves fibrin; EC 3.4.21.7; *r.n.* plasmin.

fibrinolysis n. [L. *fibra*, band; Gk. *lysis*, loosing] the dissolving of blood clots as a result of fibrin degradation.

fibroblast n. [L. *fibra*, band; Gk. *blastos*, bud] a flattened, irregular shaped connective tissue cell which secretes white fibres and possibly also yellow fibres; *alt.* desmocyte, fibrocyte.

fibrocartilage n. [L. *fibra*, band; *cartilago*, gristle] a kind of cartilage whose matrix is mainly composed of fibres similar to connective tissue fibres, found at articulations, cavity margins, and osseous grooves.

fibrocyte n. [L. *fibra*, band; Gk. *kytos*, hollow] fibroblast, *q.v.*; an inactive cell produced from a fibroblast after it has finished secreting fibres.

fibrohyaline n. [L. *fibra*, band; Gk. *hyalos*, glass] chondroid, *q.v.*

fibroin n. [L. *fibra*, band] the protein of silk fibres, produced from fibroinogen.

fibroinogen n. [L. *fibra*, band; Gk. *genos*, birth] a protein secreted by the silk glands of certain insects, which is denatured to fibroin.

fibroplastin paraglobulin, *q.v.*

fibrous a. [L. *fibra*, band] composed of fibres, *appl.* tissue, roots, mycelium, etc., *appl.* proteins, as elastin, fibrin, fibroin, keratin, myosin, etc.

fibrous astrocytes astrocytes found mainly in white matter, having thick processes which branch, some having pedicles which abut on blood vessels; *alt.* spider cells; *cf.* protoplasmic astrocytes.

fibrovascular a. [L. *fibra*, fibre; *vasculum*, small vessel] *appl.* bundle of vascular tissue surrounded by sclerenchyma fibres.

fibula n. [L. *fibula*, buckle] in tetrapods, the bone posterior to the tibia in the shank of the hindlimb; outer and smaller shin bone in man; in some insects, a structure holding fore- and hind-wings together; a. fibulate.

fibulare n. [L. *fibula*, buckle] the outer bone of proximal row of tarsus.

fibularis peroneus, *q.v.*

ficin n. [*Ficus*, generic name of fig] an endopeptidase enzyme found in fig trees; EC 3.4.22.3; *r.n.* ficin.

Ficoidales n. [*Ficus*, generic name; Gk. *eidos*, form] an order of dicots (Polypetalae) usually having regular flowers and a syncarpous ovary.

fidelity n. [L. *fidelitas*, faithfulness] the degree of limitation of a species to a particular habitat.

field a dynamic system in which all the parts are interrelated and in equilibrium, so that a change in any part affects the whole.

field capacity the state of the water content of a soil when it has drained free of rain.

filament n. [L. *filum*, thread] a thread-like structure; the stalk of anther; a hypha; the stalk of a down feather; a cryptoptile, *q.v.*; slender apical end of egg tube of insect ovary; a chain of cells; in Cyanophyceae, a trichome with its enclosing sheath; a. filamentous.

Filaridea n. [*Filaria*, generic name] a class of parasitic nematodes having a small rudimentary buccal capsule.

filar theory a modification of the fibrillar theory, *q.v.*, where the fibrils do not form networks.

filator n. [L. *filum*, thread] a structure forming part of the spinning organ of silkworms and which regulates size of the silk fibre; the spinnerets of other caterpillars.

file meristem rib meristem, *q.v.*

filial generation F_1, etc., *q.v.*

filial regression tendency of offspring of outstanding parentage to revert to average for species; *alt.* Galton's law.

filibranch a. [L. *filum*, thread; *branchiae*, gills] *appl.* gills of bivalve molluscs whose filaments are attached to adjacent ones by cilia.

Filibranchiata n. [L. *filum*, thread; *branchiae*, gills] in some classifications, an order of bivalve molluscs having filibranch gills.

Filicales n. [L. *filix*, fern] an order of leptosporangiate Pteropsida (Filicopsida) including most living ferns and distinguished from the Marsileales and Salviniales by being homosporous; the term was once used more broadly to include all ferns.

filicauline a. [L. *filum*, thread; *caulis*, stalk] with a thread-like stem.

Filices n. [L. *filix*, fern] Leptosporangiatae, *q.v.*

filiciform a [L. *filix*, fern; *forma*, shape] shaped like the frond of a fern; fern-like; *alt.* filicoid.

Filicineae n. [L. *filix*, fern] a class of the Tracheophyta, commonly called ferns, and included in the Pteropsida, *q.v.*

filicinean a. [L. *filix*, fern] *appl.* or *pert.* ferns.

filicoid filiciform, *q.v.*

Filicophyta *n.* [L. *filix*, fern; Gk. *phyton*, plant] in some classifications, a division of the Tracheophyta equivalent to the Filicopsida (ferns).

Filicopsida *n.* [L. *filix*, fern; Gk. *opsis*, appearance] *see* Pteropsida.

filiform *a.* [L. *filum*, thread; *forma*, shape] thread-like.

filiform papillae papillae on the tongue ending in numerous minute slender processes.

filigerous *a.* [L. *filum*, thread; *gerere*, to carry] with thread-like outgrowths or flagella.

filipendulous *a.* [L. *filum*, thread; *pendere*, to hang] thread-like with tuberous swellings in middle or at end, *appl.* certain roots.

Filippi's glands Lyonnet's glands, *q.v.*

fillet *n.* [L. *filum*, thread] band of white matter in mid-brain and medulla oblongata; *alt.* lemniscus.

filoplasmodium *n.* [L. *filum*, thread; Gk. *plasma*, form; *eidos*, form] net plasmodium, *q.v.*

filoplume *n.* [L. *filum*, thread; *pluma*, feather] a delicate hair-like feather with long axis and a few free barbs at apex.

filopodia *n.plu.* [L. *filum*, thread; Gk *pous*, foot] protozoan thread-like pseudopodia; *cf.* reticulopodia.

filose *a.* [L. *filum*, thread] slender or thread-like, *appl.* pseudopodia of Protozoa: filopodia.

filterable of micro-organisms, filter-passers *q.v.*; *alt.* filtrable.

filterable virus a former term for a virus which would pass through a fine filter when this was the test to distinguish viruses from bacteria and other micro-organisms, and equivalent to an ultravirus.

filter feeder an animal which feeds on small organisms in water, or occasionally in air, by straining them out of the surrounding medium.

filter-passers organisms capable of passing through a filter which arrests most bacteria, including viruses and mycoplasmas; *alt.* filterable (filtrable) organisms, microhenads.

filtrable filterable, *q.v.*

filtration *n.* [F. *filtrer*, to strain] *appl.* iridial angle of cornea; straining, as of lymph through capillary walls; the separation of solid particles from a liquid medium by straining through a suitable material.

filum terminale the terminal thread, a slender grey filament, of the spinal cord; the terminal filament of the insect ovariole.

fimbria *n.* [L. *fimbria*, fringe] any fringe-like structure; a posterior prolongation of fornix to hippocampus; one of delicate processes fringing the mouth of tube or duct, as of oviduct, or of siphon of molluscs; one of the numerous filaments, smaller than flagella, fringing certain bacteria, *alt.* pilus; *plu.* fimbriae; *alt.* lacinia.

fimbriate(d) *a.* [L. *fimbriatus*, fringed] fringed at margin, as petals, tubes, ducts, antennae.

fimicolous *a.* [L. *fimus*, dung; *colere*, to dwell] inhabiting or growing on dung.

fin *n.* [A.S. *finn*, fin] a fold of skin with fin-rays in most fishes, used for locomotion or balancing; a similar structure in other aquatic animals.

finials *n.plu.* [L. *finis*, end] the ossicles of the distal rami of crinoids, which do not branch again.

fin-rays stiff rods of tissue supporting the fins.

made of bone, cartilage, or a collagen-like substance.

fire climax a plant community maintained as climax vegetation by natural or man-made fires which destroy the plants that would otherwise become dominant.

firmisternal *a.* [L. *firmus*, strong; *sternum*, breast bone] of frogs, having the epicoracoids of the pectoral girdle fused midventrally; *n.* firmisterny.

first division the earlier of the 2 meiotic divisions; *alt.* heterotypic division, reduction division.

fishes Pisces, *q.v.*; sometimes used for Pisces and Agnatha.

fish lice the common name for the Branchiura, *q.v.*

Fissidentales *n.* [*Fissidens*, generic name] an order of mosses (Musci) having 'leaves' arranged in 2 rows, being developed from a 2-sided apical cell.

fissile *a.* [L. *fissilis*, cleft] tending to split; cleavable.

fissilingual *a.* [L. *fissus*, cleft; *lingua*, tongue] with bifid tongue.

fission *n.* [L. *fissus*, cleft] cleavage of cells; division of a unicellular organism into 2 or more parts.

fission fungi bacteria, *q.v.*, especially when they are known as Schizomycetes.

fissiparous *a.* [L. *fissus*, cleft; *parere*, to beget] reproducing by fission.

fissiped *n.* [L. *fissus*, cleft; *pes*, foot] with cleft feet, i.e. with digits of feet separated.

fissirostral *a.* [L. *fissus*, cleft; *rostrum*, beak] with deeply-cleft beak.

fissure *n.* [L. *fissura*, cleft, deep groove, or furrow dividing an organ into lobes, or subdividing and separating certain areas of the lobes; sulcus, *q.v.*

fistula *a.* [L. *fistula*, pipe] pathological or artificial pipe-like opening; water-conducting vessel, *alt.* trachea.

fistular like a fistula; pipe-like; hollow and cylindrical, tube-like, *appl.* stems of Umbelliferae, *appl.* leaves surrounding stem as in some monocotyledons.

fitness a measure of the ability of a population to survive natural selection.

fix *v.* [L. *fixus*, fixed] to kill, and preserve; to establish; to retain, *appl.* nitrogen conversion into organic compounds by bacterial action, *appl.* carbon dioxide in photosynthesis.

fixation attachment of an organism to a substratum; a stereotyped behavioural response shown by an animal regardless of whether it is accompanied by positive or negative reinforcement, and often shown in an insoluble problem situation.

fixation muscles muscles which prevent disturbance of body equilibrium generally, and fix limbs in case of limb movement.

fixation of nitrogen nitrogen fixation, *q.v.*

fixed-action pattern a stereotyped and fixed response found in animal behaviour where learning has not occurred.

fixed cheeks in trilobites the part of the cranidium between the glabella and the facial sutures; *alt.* fixigenes.

fixed light position the final position of a fully developed leaf in relation to the direction of light falling on it.

fixigenes *n.plu.* [L. *fixus*, fixed; *gena*, cheek] fixed cheeks, *q.v.*

flabellate *a* [L. *flabellare*, to fan] fan-shaped, *appl.* pectinate antennae with long processes; *alt.* flabelliform.

flabelliform *a* [L. *flabellum*, fan; *forma*, shape] fan-shaped; *alt.* flabellate, rhipidate.

Flabelligerimorpha *n.* [*Flabelligera*, generic name; Gk. *morphē*, form] an order of polychaetes with a body of short similar segments covered with papillae.

flabellinerved *a.* [L. *flabellum*, fan; *nervus*, sinew] *appl.* leaves with many radiating nerves.

flabellum *n.* [L. *flabellum*, fan] any fan-shaped organ or structure; distal exite of branchiopodan limb; epipodite of certain crustacean limbs; terminal lobe of glossa in certain insects; diverging white fibres in corpus striatum.

flaccid *a.* [L. *flaccidus*, flabby] limp; *appl.* leaves that do not have enough water and are about to wilt or are wilting; *n.* flaccidity.

flagella *plu.* of flagellum.

Flagellata, flagellates *n., n.plu.* [L. *flagellum*, whip] a class of Protozoa, which are motile in the adult stage, swimming by means of flagella, and reproducing by longitudinal binary fission; *alt.* Mastigophora.

flagellate *a.* [L. *flagellum*, whip] furnished with flagella; like a flagellum; a member of the Flagellata.

flagellated chambers in sponges, cavities formed by outgrowths of the wall and lined by choanocytes.

flagelliform *a.* [L. *flagellum*, whip; *forma*, shape] lash-like; like a flagellum.

flagellispore, flagellula *n.* [L. *flagellula*, dim. of *flagellum*, whip] flagellate zoospore.

flagellum *n.* [L. *flagellum*, whip] the lash-like process of cytoplasm of many Protista and of cells, as in choanocytes and certain male gametes, being similar to cilium but longer and usually borne singly or in small groups; external structure on basal joint of chelicera of Pseudoscorpiones; telson in Pedipalpi and Palpigradi; distal part of antenna in some arthropods, as in Diptera; structure in which sperm are compacted together in genitalia of male gastropods; a long slender runner or creeping stem; *plu.* flagella.

flame cells the terminal cells of branches of excretory system (protonephridium) in platyhelminths, annelids, rotifers, and some other groups, with cavity continuous with lumen of duct, and containing a flagellum or cilium or bunch of cilia, the motion of which give a flickering appearance similar to that of a flame, *alt.* flame bulbs; solenocyte, *q.v.*

flash behaviour or colours the sudden appearance of a flash of colour during escape behaviour of prey from a predator, which may startle the predator, or when the flash disappears, may make the predator think the prey has gone.

flask fungi the common name for the Pyrenomycetes, *q.v.*

flatworms the common name for the Platyhelminthes, *q.v.*

flavedo *n.* [L. *flavus*, yellow] epicarp in hesperidium; *cf.* albedo.

flavescent *a.* [L. *flavescere*, to turn yellow] growing or turning yellow; having yellow flecks among the normal green.

flavin *see* flavins.

flavin adenine dinucleotide FAD, *q.v.*

flavin adenine mononucleotide FMN, *q.v.*

flavin mononucleotide FMN, *q.v.*

flavins *n.plu.* [L. *flavus*, yellow] yellowish pigments which are nucleotides whose nitrogen base is usually isoalloxazine, and have a greenish yellow fluorescence, occurring free in many tissues of higher animals and in plants, usually as riboflavin; *alt.* lyochromes.

flavones *n.plu.* [L. *flavus*, yellow] a group of pale yellow flavonoid plant pigments.

flavonoids any of various compounds containing a $C_6-C_3-C_6$ skeleton, the C_6 parts being benzene rings and the C_3 part varying in different compounds, and including many water-soluble plant pigments; *cf.* bioflavonoids.

flavonols *n.plu.* [L. *flavus*, yellow] a group of pale yellow flavonoid plant pigments.

flavoproteins *n.plu.* [L. *flavus*, yellow; Gk. *prōteion*, first] proteins with a firmly attached flavin prosthetic group, important in electron transfers such as respiratory oxidations, as they can be alternately reduced and oxidized, and which are yellow when oxidized but colourless when reduced and are also called the yellow, or flavin, enzymes.

flavoxanthin *n.* [L. *flavus*, yellow; Gk. *xanthos*, yellow] a yellow colouring matter in petals as of the buttercup family, Ranunculaceae.

fleas the common name for the Siphonaptera, *q.v.*

flex *v.* [L. *flectere*, to bend] to bend, *appl.* movement of limbs; *cf.* extend.

flexor n. [L. *flexus*, bent] a muscle which bends a limb, or part, by its contraction; *cf.* extensor.

flexor plate a median plate supporting praetarsus of insects for attachment of tendon of claw flexor.

flexuose, flexuous *a.* [L. *flexus*, bent] curving in a zig-zag manner, *appl.* stem or other axis.

flexure *n.* [L. *flexus*, bent] a curve or bend, *appl.* curve in embryonic brain, curve of intestine.

flies *see* true flies, mayflies, dragonflies, stoneflies, scorpionflies, caddis flies, alder flies.

flight intention the actions of one bird as it gets ready to fly, which influences the rest of the group; *cf.* intention movement.

flimmer *n.* [Ger. *Flimmern*, tinsel] minute hairs borne on certain flagella giving them the appearance of tinsel. *a. Appl.* flagella having such hairs, *alt.* tinsel.

float *n.* [A.S. *fleotan*, to float] the pneumatophore of siphonophores; 1 of 4 tracheal sacs in aquatic larva of Culicidae (mosquitoes); a pneumatocyst of bladderwrack; a large spongy mass serving as a float in some pteridophytes.

floating ribs ribs not uniting at their ventral end with the sternum.

floating tissue tissue containing air found in the seeds and fruits of plants dispersed by water currents.

floccose *a.* [L. *floccus*, a lock of wool] covered with wool-like tufts.

floccular *a.* [L. *floccus*, lock of wool] *pert.* a flocculus.

flocculation *a.* [L. *floccus*, lock of wool] the clumping of fine particles in the disperse phase of a colloidal system, such as the clumping of clay particles which can be brought about by lime.

flocculence *n.* [L. *floccus*, lock of wool] adhesion in small flakes, as of a precipitate.

flocculent *a.* [L. *floccus*, lock of wool] covered with a soft waxy substance giving appearance of wool; covered with small woolly tufts.

flocculus *n.* [L.L. *dim.* of L. *floccus*, lock of wool] a small accessory lobe on each lateral lobe of the cerebellum; a posterior hairy tuft in some Hymenoptera.

floccus *n.* [L. *floccus*, lock of wool] the tuft of hair terminating a tail; downy plumage of young birds; mass of hyphal filaments in algae and fungi; any tuft-like structure.

flor *n.* [L. *flos*, flower] a covering of yeasts and bacteria and other micro-organisms which forms on the surface of some wines during fermentation.

flora *n.* [L. *flos*, flower] the plants peculiar to a country, area, specified environment, or period; a book giving descriptions of these plants; the micro-organisms found in a particular organ.

floral *pert.* the flora of a country or area; *pert.* flowers, *appl.* leaf: a petal, sepal, or bract, *appl.* nectary, etc.

floral axis receptacle, *q.v.*

floral diagram a conventional diagram of a plan or cross-section of a flower, indicating the position of the parts as if they lay in the same plane.

floral envelope the perianth, calyx, and corolla considered together.

floral formula an expression summarizing the number and position of parts of each whorl of a flower.

floral mechanism the arrangement of flower parts to ensure either crossing or selfing.

floral stimulus florigen, *q.v.*

florescence *n.* [L. *florescere*, to begin to flower] bursting into bloom; *alt.* anthesis.

floret *n.* [L. *flos*, flower] one of the small individual flowers of a crowded inflorescence such as capitulum; flower with lemma and palea, of grasses; *alt.* floscule.

floricome *n.* [L. *flos*, flower; *coma*, hair] a form of branched hexaster spicule.

Florideae *n.* [L. *floridus*, flowery] Florideophycidae, *q.v.*

floridean starch a type of starch found in red algae which gives a brown reaction with iodine instead of blue.

Florideophycidae *n.* [L. *floridus*, flowery; Gk. *phykos*, seaweed] a subclass of red algae having a pseudoparenchymatous structure and each cell with a number of parietal chloroplasts; *alt.* Florideae.

floridoside *n.* [*Florideae*, group of red algae] a galactoside which is a food storage product in red algae.

florigen *n.* [L. *flos*, flower; *gignere*, to produce] a hypothetical, or as yet not isolated, phytohormone which causes a bud to develop into a flower, postulated as the stimulus for flowering that is perceived in the leaves and must be transmitted to the tip, and which may be a mixture of other hormones; *alt.* floral stimulus, flowering hormone; *a.* florigenic.

floristic *pert.* or *appl.* flora.

floristics *n.* [L. *flos*, flower] the study of the composition of an area of vegetation in terms of the species of plant in it.

florula *n.* [*Dim.* of *flora*] plant population of a small unit area, as of compost heap, etc.

floscelle *n.* [L. *flosculus*, little flower] flower-like structure round the mouth, composed of 5 bourrelets and 5 phyllodes, in some echinoids.

floscule, flosculus *n.* [L. *flosculus*, little flower] a small flower; a floret, *q.v.*

floss *n.* [O.F. *flosche*, down] a downy or silky substance; the loose pieces of silk in a cocoon.

flower *n.* [L. *flos*, flower] the blossom of a plant, comprising generally sepals, petals, (or perianth), stamens, and/or pistil, being basically a leafy shoot adapted for reproductive purposes.

flowering glume lemma, *q.v.*

flowering hormone florigen, *q.v.*

flowering plant angiosperm, *q.v.*

flukes the common name for the Trematoda, *q.v.*

fluorescyanine a mixture of pterins with a yellow or blue fluorescence, found in the eyes, eggs, and luminous organs of certain insects.

flush a patch of ground where water lies but does not run in a channel; a period of growth, especially in a woody plant.

flushing the washing of dissolved substances upwards in the soil so that they are deposited near the surface; *cf.* leaching.

Fluviales *n.* [L. *fluvius*, stream] Najadales, *q.v.*; Helobiae, *q.v.*

fluviatile *a.* [L. *fluviatilis*, *pert.* river] growing in or near streams; inhabiting and developing in streams, *appl.* certain insect larvae; caused by rivers, *appl.* deposits.

fluviomarine *a.* [L. *fluvius*, stream; *mare*, sea] *pert.* or inhabiting rivers and sea.

fluvioterrestrial *a.* [L. *fluvius*, stream, *terra*, land] found in streams and in the land beside them.

flux *n.* [L. *fluere*, to flow] term *appl.* species that are not yet stable.

F-mediated transduction sexduction, *q.v.*

FMN flavin mononucleotide, a substance important as a coenzyme in oxidation/reduction reactions as in respiratory oxidations, as it can be both oxidized and reduced; sometimes called flavin adenine mononucleotide, or riboflavin phosphate.

foam theory alveolar theory, *q.v.*

foet- *alt.* fet-.

foetal *a.* [L. *foetus*, offspring] embryonic; *pert.* a foetus.

foetalization the occurrence of foetal features of the development of an ancestor in the adult stage of development of a descendant.

foetal membranes the membranes which protect or nourish the foetus, such as the chorion, amnion, allantois, and yolk sac; *alt.* extraembryonic membranes.

foetation development of the foetus within the uterus.

foetid glands small sac-like glands which secrete an ill-smelling fluid, in Orthoptera.

foetoprotein *n.* [L. *foetus*, offspring; Gk. *prōteion*, first] a serum globulin found in the foetus but not in the adult of a species.

foetus *n.* [L. *foetus*, offspring] mammalian embryo after the appearance of recognizable main features of the developed animal; any vertebrate embryo in egg or uterus; *alt.* fetus.

folacin folic acid, *q.v.*

foliaceous *a.* [L. *folium*, leaf] having the form or

texture of a foliage leaf; thin and leaf-like; bearing leaves.

Folian process [*C. Folli* or *Folius*, Italian anatomist] anterior process of malleus; *alt.* processus gracilis, Ravian process.

foliar *a.* [L. *folium*, leaf] *pert.* or consisting of leaves; bearing leaves, *appl.* spurs, *cf.* brachyplast.

foliate *a.* [L. *folium*, leaf] *appl.* papillae on tongue arranged in folds and having many taste buds, few in man, conspicuous in rabbits.

foliation *n.* [L. *folium*, leaf] the production of leaves, leafing; vernation, *q.v.*

folic acid [L. *folium*, leaf] a vitamin, pteroylglutamic acid (or pteroic acid), found in yeast, liver, and green vegetables, whose deficiency causes megaloblastic anaemia and which is important in nucleic acid metabolism and required by some insects; it is sometimes considered to be a member of the vitamin B complex, vitamin B$_c$, and has also been called factor R or S, or rhizopterine, folacin, pteroyl glutamic acid, PGA, vitamin M, antianaemia vitamin.

folicaulicole, folicaulicolous *a.* [L. *folium*, leaf; *caulis*, stalk; *colere*, to inhabit] growing on leaves and stems, *appl.* certain fungi and lichens.

folicole, foliicolous *a.* [L. *folium*, leaf; *colere*, to dwell] growing on leaves, *appl.* certain fungi and lichens.

foliobranchiate *a.* [L. *folium*, leaf; *branchiae*, gills] possessing leaf-like gills.

foliolae *n.plu.* [L. *folium*, dim., leaf] leaf-like appendages of telum.

foliolate *a.* [L. *folium*, dim., leaf] *pert.*, having, or like, leaflets.

foliole *n.* [L. *folium*, dim., leaf] small leaf-like organ or appendage; a leaflet, as of a compound leaf.

foliolose *a.* [L. *folium*, dim., leaf] covered with leaflets; made up of leaflet-like lobes.

foliose *a.* [L. *folium*, leaf] with many leaves; leafy; having leaf-like lobes, *appl.* lichens and some liverworts, *cf.* frondose.

folium *n.* [L. *folium*, leaf] a flattened structure in the cerebellum, expanding laterally into superior semilunar lobules; one of the folds on sides of tongue.

follicle *n.* [L. *folliculus*, small sac] a dry dehiscent fruit formed of one carpel, dehiscing along one side; a cavity or sheath; a Graafian follicle, *q.v.*; an ovarian follicle, *q.v.*; a hair follicle, *q.v.*

follicle cells in cephalopods, peritoneal cells which surround the ova and pass on food from a special blood supply.

follicles of Langerhans [*P. Langerhans*, German anatomist] groups of cells in submucosa at junction of fore-gut and mid-gut of larval cyclostomes, secreting an insulin-like substance and being homologous to islets of Langerhans.

follicle-stimulating hormone (FSH) a glycoprotein gonadotropic hormone, secreted by the anterior part of the pituitary gland, which stimulates the growth of Graafian follicles, oestrogen secretion, and spermatogenesis; *alt.* prolan A, gametogenetic hormone, gametokinetic hormone, thylakentrin.

follicular *a.* [L. *folliculus*, a small sac] *pert.*, like, or consisting of follicles; *appl.* an ovarian hormone, oestrone, *q.v.*

folliculate containing, consisting of, or enclosed in, follicles.

folliculin *n.* [L. *folliculus*, small sac] oestrone, *q.v.*

folliculose *a.* [L. *folliculus*, small sac] having follicles.

Fontana's spaces [*F. Fontana*, Italian anatomist] spaces in trabecular tissue of angle of iris, communicating with the anterior chamber of the eye and with the sinus venosus sclerae.

fontanel(le) *n.* [F. *fontanelle*, little fountain] a gap or space between bones in the cranium, closed only by membrane; depression on head of termites; *alt.* fenestra, fonticulus.

fonticulus *n.* [L. *fonticulus*, dim. of *fons*, fountain] fontanelle, *q.v.*; depression at anterior end of sternum, the jugular notch.

food bodies excrescences formed on the seeds of certain plants and used by ants as food, so aiding dispersal; similar structures on the leaves of some plants, *alt.* Beltian or Belt's bodies.

food chain sequence of organisms in which each is food of a later member of the sequence.

food cycle food web, *q.v.*

food pollen pollen present in flowers to provide food for visiting insects, instead of or as well as nectar, which may be sterile and produced in special anthers.

food vacuole a small vacuole containing fluid and food particles, in endoplasm of many Protista.

food web interconnected food chains in an ecosystem; *alt.* food cycle.

foot *n.* [A.S. *fot*, foot] an embryonic structure in vascular cryptogams through which nourishment is obtained from prothallus; basal portion of sporophyte in bryophytes; an organ of locomotion, differing widely in different animals, from tube-foot of echinoderms, muscular foot of gastropods and other molluscs, tarsus of insects, to foot of vertebrates.

foot-jaws poison claws or 1st pair of legs in centipedes; *alt.* maxillipedes.

foot-plates terminal enlargements of processes of protoplasmic astrocytes in contact with minute blood vessels; *alt.* perivascular feet.

foot spinners a common name for the Embioptera, *q.v.*

foot-stalk pedical, *q.v.*; petiole, *q.v.*

foramen *n.* [L. *foramen*, opening] any small perforation; micropyle, *q.v.*; aperture through a shell, bone, or membranous structure; *plu.* foramina.

foramen caecum a depression behind convergence of rows of vallate papillae at back of tongue.

foramen (occipitale) magnum the opening in occipital region of skull through which passes the spinal cord; *alt.* occipital foramen.

foramen of Monro [*A. Monro (primus)*, Scottish anatomist] interventricular foramen, passage between 3rd and lateral ventricles; *alt.* porta or foramen interventriculare, interventricular foramen, prosocoel.

foramen of Winslow [*J. B. Winslow*, Danish anatomist] epiploic foramen, *q.v.*

foramen ovale opening between atria of foetal heart; aperture in great wing of sphenoid, passage for mandibular nerve.

foramen Pannizzae in crocodiles, an opening between the 2 sides of the aortic trunk.

foramen rotundum aperture in great wing of sphenoid, passage for maxillary nerve.

foramina *plu.* of foramen.

foraminate *a.* [L. *foramen*, opening] pitted, having foramina or perforations.

Foraminifera, foraminiferans *n., n.plu.* [L. *foramen*, opening; *ferre*, to carry] an order of Sarcodina having a shell of various materials and/ or reticulate pseudopodia and in pelagic species a vacuolated outer layer of protoplasm; sometimes subdivided into 2 orders the Testacea and Thalammophora.

foraminiferous *a.* [L. *foramen*, opening; *ferre*, to carry] having foramina; containing shells of Foraminifera.

forb *n.* [Gk. *phorbē*, pasture] a herbaceous plant especially a pasture herb other than a grass.

forceps *n.* [L. *forceps*, tongs] the clasper-shaped anal cercus of some insects; large fighting or seizing claws of crabs and lobsters, *alt.* cheliped; distal jaws of pedicellariae; fibres of corpus callosum, curving into frontal and occipital lobes.

forcipate *a.* [L. *forceps*, tongs] resembling forceps, or forked like forceps.

Forcipulata *n.* [*Dim.* of L. *forceps*, tongs] an order of starfish with no marginal plates and with stalked and forcipulate pedicellariae

forcipulate *a.* [*Dim.* of L. *forceps*, tongs] shaped like a small forceps, *appl.* pedicellariae of some Asteroidea.

fore-brain prosencephalon, *q.v.*

fore-gut stomodaeum, *q.v.*

fore-kidney pronephros, *q.v.*

fore-milk colostrum, *q.v.*

fore-runner tip or point a form of leaf apex which begins to photosynthesize before the basal part is mature.

forespore early stage in endospore formation, in bacteria.

forfex *n.* [L. *forfex*, shears] a pair of anal organs which open and shut transversely, occurring in certain insects.

forficate *a.* [L. *forfex*, shears] deeply notched.

forficiform *a.* [L. *forfex*, shears; *forma*, form] scissor-shaped, *appl.* type of forcipulate pedicellariae in which the jaws do not cross.

forficulate *a.* [*Dim.* of L. *forfex*, shears] scissor-shaped.

form, forma *n.* [L. *forma*, shape] taxonomic unit consisting of individuals that differ from those of a larger unit by a single character, therefore being the smallest category in classification; one of the kinds of a polymorphic species; a taxonomic group whose status is not clear but may be species or subspecies.

formation *n.* [L. *forma*, shape] structure arising from an accumulation of deposits; the vegetation proper to a definite type of habitat over a large area, as of tundra, coniferous forest, prairie and steppe, tropical rain forest, etc.

formative *a.* [L. *forma*, shape] plastic, *appl.* tissue which is living and can develop, as region of growth in stem or root; *appl.* cells of blastocyst which give rise to embryo, *cf.* trophoblast.

form genus a genus whose species may not be related by a common ancestor.

formic acid [L. *formica*, ant] an organic acid found in ants, some other insects, and some plants.

formicarian *a.* [L. *formica*, ant] *pert.* ants; *appl.*

plants which attract ants by means of sweet secretions.

formicarium, formicary *n.* [L. *formica*, ant] ants' nest, particularly an artificial arrangement for purposes of study.

form species the members of a form genus, *q.v.*

fornical *pert.* fornix.

fornicate(d) *a.* [L. *fornicatus*, vaulted] concave within, convex without, arched.

fornices *plu.* of fornix.

fornix *n.* [L. *fornix*, vault] an arched recess, as between eyelid and eyeball, or between vagina and cervix uteri; an arched sheet of white longitudinal fibres beneath corpus callosum; scutum of Cheilostomata; one of arched scales in the orifice of some flowers; *plu.* fornices; *a.* fornical.

fossa *n.* [L. *fossa*, ditch] a pit or trench-like depression; *alt.* fosse.

fosse *n.* [L. *fossa*, ditch] fossa, *q.v.*; a circular groove formed by upper part of parapet in sea anemones.

fossette *n.* [F. *fossette*, small pit, from L. *fossa*, ditch] a small pit or depression; a socket containing base of antennule in arthropods; groove for resilium in bivalve shells; depression on grinding surface of a tooth.

fossil *n.* [L. *fossilis*, dug up] the remains or traces of animal and plant life of the past, found embedded in rock either as petrified hard parts or as moulds, casts, films, tracks, etc.

fossil hominid various fossil primates thought to be involved in the evolution of man, including the dryopithecines, australopithecines, habilines, pithecanthropines, and sapients.

fossiliferous *a.* [L. *fossilis*, dug up; *ferre*, to carry] containing fossils.

fossorial *n.* [L. *fossor*, digger] adapted for digging, *appl.* animals, claws, feet.

fossula *n.* [*Dim.* of L. *fossa*, ditch] a small fossa; small pit with reduced septa on one side of a corallite cup in Rugosa.

fossulate *a.* [*Dim.* of L. *fossa*, ditch] with slight hollows or grooves.

fossulet *n.* [*Dim.* of L. *fossa*, ditch] a long narrow depression.

fourchette *n.* [F. *fourchette*, fork] furcula of birds; frog of equine hoof; junction of labia minora.

fourth ventricle the cavity of the hind-brain in vertebrates.

fovea *n.* [L. *fovea*, depression] a small pit, fossa, or depression; a small hollow at leaf base in Isoetales, containing a sporangium; pollinium base in orchids.

fovea centralis central and thinnest part of macula lutea, without rods and with long and slender cones.

fovea dentis facet on atlas, for articulation with dens of axis.

foveal *a.* [L. *fovea*, depression] *pert.* fovea; *pert.* fovea centralis, *appl.* cone vision; *cf.* extrafoveal.

foveate *a.* [L. *fovea*, depression] pitted.

foveola *n.* [L. *foveola*, small depression] a small pit; a shallow cavity in bone; a small depression just above fovea in leaf of Isoetales; *alt.* foveole.

foveolae opticae two pigmented areas in depression of neural plate of amphibian embryo, the primordia of eyes.

foveolate *a.* [L. *foveola*, small depression] having regular small depressions.

foveole foveola, *q.v.*

fovilla *n.* [L. *foveo*, I nourish] the contents of a pollen grain.

F-pilus a filamentous appendage synthesized by strains of bacteria containing the F-episome along which nucleic acid may travel under certain conditions; *alt.* sex pilus.

fraen- *see* fren-.

fragmentation *n.* [L. *frangere*, to break] division into small portions; nuclear division by simple splitting, *alt.* amitosis; the separation of a small part of a chromosome (fragment) from the rest.

frankincense *n.* [L.L. *francus*, of superior quality; L. *incensum*, something kindled] a balsam obtained from plants of the genus *Boswellia*.

fraternal dizygotic, *appl.* twins.

free *a.* [A.S. *freo*, acting at pleasure] motile; unattached; distinct; separate; of pupa, exarate, *q.v.*

free central placentation placentation in which the ovules are situated on a column arising from the base in the middle of the ovary and are not connected to the wall by septa; *cf.* axile placentation.

free cheeks in trilobites, the lateral parts of the cephalon lying outside the facial sutures.

freemartin a sterile female or intersex twin-born with a male, due to masculization of female in the uterus.

free nuclear division division of the nucleus, not followed by formation of new cell walls, as in the endosperm of angiosperm seeds.

free running *appl.* an endogenous rhythm unaffected by any external influence.

fren- *alt.* fraen-.

frenate *a.* [L. *frenare*, to bridle] having a frenum or frenulum.

frenulum *n.* [L. *frenulum*, *dim.* of *frenum*, bridle] a fold of membrane, as of tongue, clitoris, etc.; a process on hind-wing of Lepidoptera for attachment to fore-wing; a thickening of subumbrella of certain Scyphozoa; *alt.* frenum.

frenum *n.* [L. *frenum*, bridle] frenulum, *q.v.*, a fold of integument at junction of mantle and body of Cirripedia, ovigerous in Pedunculata.

frequency the relative number, usually expressed as a percentage, of species in a certain area or community.

friable easily powdered.

frigofuge *n.* [L. *frigus*, cold; *fugere*, to flee] an organism which does not tolerate cold.

frill in fungi, armilla, *q.v.*

frilled organ a structure for attachment at the anterior end of some Cestodaria.

frog an amphibian of the order Anura; a triangular mass of horny substance in the middle of the sole of the foot of a horse and related animals, *alt.* fourchette.

frond *n.* [L. *frons*, leafy branch] a leaf, especially of fern or palm; thallus of certain seaweeds or liverworts; leaf-like thalloid shoot, as of lichen; a leaf-like structure.

frondescence *n.* [L. *frondescere*, to put forth leaves] development of leaves; leafing.

frondose *a.* [L. *frondosus*, leafy] with many fronds; thalloid, *appl.* liverworts, *cf.* foliose.

frons *n.* [L. *frons*, forehead] forehead or comparable structure.

frontal *a.* [L. *frons*, forehead] in region of forehead; *appl.* artery, vein, lobe, convolution; *appl.* head organ of nemertines; a prostomial ridge of polychaetes; palps of certain nereids; specialized feeding surface in certain ciliates; ganglion, gland, and pore in insects; *appl.* plane at right angles to median longitudinal or sagittal plane. *n.* A frontal scale in reptiles; frontal bone.

frontalis *n.* [L. *frons*, forehead] frontal part of the scalp muscle or epicranius.

frontocerebellar fibres fibres passing from frontal region to cerebellum.

frontoclypeus *n.* [L. *frons*, forehead; *clypeus*, shield] frons and clypeus fused, in insects.

fronto-ethmoidal *pert.* frontal and ethmoidal bones, *appl.* suture.

frontonasal *a.* [L. *frons*, forehead; *nasus*, nose] *pert.* forehead or frontal region and nose, *appl.* ducts and process.

frontoparietal *a.* [L. *frons*, forehead; *paries*, wall] *pert.* frontal and parietal bones, *appl.* suture: the coronal suture; *cf.* parietofrontal.

frontosphenoidal *a.* [L. *frons*, forehead; Gk. *sphēn*, wedge; *eidos*, form] *pert.* frontal and sphenoid bones, *appl.* a process of zygomatic bone articulating with frontal.

fructicole *a.* [L. *fructus*, fruit; *colere*, to dwell] inhabiting fruits, *appl.* parasitic fungi.

fructification *n.* [L. *fructus*, fruit; *facere*, to make] fruit, *q.v.*; fruit formation; fruit body, *q.v.*; any spore-producing structure.

fructosan *n.* [L. *fructus*, fruit] a polysaccharide made of condensed fructose units, as inulin.

fructose *n.* [L. *fructus*, fruit] a hexose sugar found in many plants especially in fruits, and in honey, one of the products of hydrolysis of sucrose and sole product of hydrolysis of inulin, possessing a ketone group and so called a ketose sugar; *alt.* fruit sugar, laevulose, levulose.

frugivorous *a.* [L. *frux*, fruit; *vorare*, to devour] fruit-eating, *appl.* certain animals.

fruit *n.* [F. *fruit*, from L. *fructus*, fruit] the developed ovary of the flower containing ripe seeds, whether fleshy or dry, often used to include other associated parts such as a fleshy receptacle, then called a false fruit, *alt.* fructification, seed vessel (especially dry fruit); capsule of bryophytes.

fruit body the spore-bearing structure, as a sporangiocarp, basidiocarp, conidiocarp; *alt.* fructification.

fruit spot sorus, as of ferns; a fruit disease.

fruit sugar fructose, *q.v.*; a mixture of fructose and glucose or occasionally some other sugars, found in fruits.

fruit wall the outer part of the fruit, either the pericarp derived from the ovary wall, or a structure derived from the ovary wall and other parts associated with the ovary, often the receptacle.

frustose *a.* [L. *frustum*, piece] cleft into polygonal pieces; covered with markings resembling cracks.

frustration the situation where an animal cannot make an appropriate response and resulting in displacement activities, etc.

frustule *n.* [L. *frustulum*, small fragment] the siliceous 2-valved wall of a diatom, or the wall with its contained protoplasm.

frutescent *a.* [L. *frutex*, shrub] becoming shrub-like; fruticose, *q.v.*

frutex *n.* [L. *frutex*, shrub] shrub; *plu.* frutices.

frutices *plu.* of frutex.

fruticose *a.* [L. *fruticosus*, bushy] like a shrub, *appl.* thallus of certain lichens; *alt.* frutescent.

fruticulose a. [Dim. of L. fruticosus, bushy] like a small shrub.

FSH follicle-stimulating hormone, q.v.

Fucales n. [Fucus, generic name] an order of algae included in the Phaeophyceae (Cyclosporae) and having oogamous sexual reproduction, no alternation of generations in the life cycle, and a thalloid plant body which is the sporophyte, and including many brown seaweeds of the seashore, such as the wracks.

fucinic acid an acid found in the cell walls of some large species of brown algae.

fucivorous a. [L. fucus, seaweed; vorare, to devour] appl. seaweed-eating animals.

fucoid a. [L. fucus, seaweed; Gk. eidos, form] pert. or resembling seaweed. n. A fossil seaweed.

fucosan n. [L. fucus, seaweed] a polysaccharide in cells of brown algae which may be a food reserve or a waste metabolic product.

fucosan vesicles small colourless vesicles in cells of brown algae, containing fucosan.

fucosterol n. [L. fucus, seaweed; Gk. stereos, solid; alcohol] a sterol found in algae of the class Phaeophyceae, and also identified in some Bacillariophyceae and Chrysophyceae.

fucoxanthin n. [L. fucus, seaweed; Gk. xanthos, yellow] the main xanthophyll carotinoid pigment in algae of the class Phaeophyceae, also found in Bacillariophyceae and Chrysophyceae.

fugacious a. [L. fugax, fleeting] withering or falling off very rapidly; caducous, q.v.

fulcral a. [L. fulcrum, support] pert. or acting as a fulcrum; appl. triangular plates aiding in movement of stylets in Hymenoptera.

fulcrate a. [L. fulcrum, support] having a fulcrum.

fulcrum a. [L. fulcrum, support] a supporting organ such as a tendril or stipule; sporophore in lichens; in certain fungi, an outgrowth from the zygospore wall; plate supporting rami of incus in mastax of rotifers; the lower surface of a ligula; a chitinous structure in base of insect rostrum; hinge-line of brachiopods; spinelike scale on anterior fin-rays of many ganoids; the pivot of a lever, appl. points of certain bones.

fuligerous a. [L. fuligo, soot] sooty in colour.

fulturae n.plu. [L. fultura, prop] a pair of sclerites supporting the hypopharynx in myriapods.

fulvous a. [L. fulvus, tawny] deep yellow, tawny.

fumaginous a. [L. fumus, smoke] smoky in colour.

fumarase see fumaric acid; EC 4.2.1.2; r.n. fumarate hydratase.

fumaric acid [L. fumaria, fumitory] an acid of the Kreb's cycle which is hydrated to malic acid by the enzyme fumarase.

Funariales n. [Funaria, generic name] an order of mosses having annual or biennial gametophytes, hanging capsules, and the peristome double or absent.

function n. [L. functio, performance] the action proper to any organ or part.

functional a. [L. functio, performance] acting normally; acting or working part of an organ as distinct from remainder.

fundament primordium, q.v.

fundamental tissue ground tissue, q.v.

fundamentum hypocotyl, q.v.

fundatrigenia n. [L. fundare, to found; Gk. genos,

descent] one of the offspring produced by a fundatrix.

fundatrix n. [L. fundare, to found] stem-mother, a female founding a new colony by oviposition, appl. aphids.

fundic a. [L. fundus, bottom] pert. a fundus, appl. cells of stomach.

fundic gland gastric gland, q.v.

fundiform a. [L. funda, sling; forma, shape] looped, appl. a ligament of penis.

fundus n. [L. fundus, bottom] the base of an organ, as of stomach, urinary bladder, etc.; boundary between underground and above-ground portions of plant axis.

fungal a. [L. fungus, mushroom] of, or pert. fungi.

fungal cellulose a carbohydrate similar to cellulose found in the cell walls of some fungi.

Fungi n. [L. fungus, mushroom] one of the major divisions of the plant kingdom in which the plant body is made of hyphae or is occasionally unicellular, and that have no chlorophyll and live as parasites, saprophytes, or symbionts; alt. Mycophyta, Mycota, fungi.

fungicolous a. [L. fungus, mushroom; colere, to inhabit] living in or on fungi.

fungiform a. [L. fungus, mushroom; forma, shape] shaped somewhat like the fruit body of an agaric fungus, appl. certain rounded papillae scattered irregularly on the tongue and having a few taste buds in the epithelium of their walls.

Fungi Imperfecti a group of fungi which have a septate mycelium and asexual spores, but have no sexual reproduction and so cannot be assigned to any other groups; alt. adelomycetes, Deuteromycetes, Imperfect Fungi.

fungine n. [L. fungus, mushroom] chitinous substance forming cell wall of certain fungi; alt. mycin.

fungistasis n. [L. fungus, mushroom; Gk. statikos, causing to stand] inhibition of fungal growth; a. fungistatic; alt. mycostasis.

fungivore n. [L. fungus, mushroom; vorare, to devour] an organism feeding on fungi; a. fungivorous; alt. mycetophage.

fungoid, fungous a. [L. fungus, mushroom] with character or consistency of fungus; alt. mycoid, mycetoid.

funicle n. [L. funiculus, small cord] an ovule- or seed-stalk; a slender strand attaching peridiolum to peridium in bird's nest fungi; a small cord or band, as of nerve fibres; a large double strand of cells passing from aboral end of coelom to aboral wall of zooecium of molluscs; alt. funiculus, q.v.

funicular a. [L. funiculus, small cord] consisting of a small cord or band; pert. a funiculus or funicle.

funiculose n. [L. funiculus, small cord] rope-like, appl. intertwined hyphae giving the appearance of ropes.

funiculus n. [L. funiculus, small cord] funicle, q.v.; one of the ventral, lateral, and dorsal columns of white matter of the spinal cord; in Ectoprocta, a strand of mesoderm which attaches the caecum and stomach to the body wall; alt. funicle, q.v.

funiform a. [L. funis, rope; forma, shape] like a cord or rope.

funnel n. [L. fundere, to pour] internal opening of

funnelform

vasa deferentia in Oligochaeta; siphon of Cephalopoda; atriocoelomic canal in Cephalochordata.

funnelform *a.* [L. *fundere*, to pour; *forma*, shape] funnel-shaped, widening gradually from a narrow base.

furanose a monosaccharide in the form of a 5-membered ring with 4 carbon and 1 oxygen atom; *cf.* pyranose.

furca *n.* [L. *furca*, fork] the apophysis or entothorax of insect metathorax; forked intercoxal plate, as in Copepoda; any forked structure.

furcal forked, *appl.* a branching nerve of lumbar plexus.

furcasternum, furcasternite *n.* [L. *furca*, fork; *sternum*, breast bone] forked poststernite or sternellum in many insects; *alt.* intersternite.

furcate *a.* [L. *furca*, fork] branching like prongs of a fork; *alt.* furcular.

furciferous *a.* [L. *furca*, fork; *ferre*, to carry] bearing a forked appendage, as some insects.

furcula *n.* [L. *furcula*, *dim.* of *furca*, fork] a forked process of structure; the united clavicles of birds, *alt.* fourchette, merrythought; a transverse ridge in embryonic pharynx, giving rise to epiglottis; partially fused abdominal appendages forming springing organ in Collembola.

furcular furcate, *q.v.*

furfuraceous *a.* [L.L. *furfuraceus*, bran-like] scurfy, covered with scurf-like or bran-like particles.

furred *a.* [O.F. *forre*, sheath] having short decumbent hairs thickly covering the surface.

fuscin *n.* [L. *fuscus*, dusky] a brown pigment in retinal epithelium.

fuscous *a.* [L. *fuscus*, dusky] of a dark, almost black colour.

fuseau *n.* [F. *fuseau* from L. *fusus*, spindle] a spindle-shaped structure; a spindle-shaped, thick-walled spore divided by septa, in certain fungi; a fusiform macroconidium.

fusellar *a.* [L. *fusus*, spindle; *dim.*] *appl.* layer formed by half rings dovetailing to constitute a tube, as in graptolites and pterobranchs.

fusi *n.plu.* [L. *fusus*, spindle] in spiders, organs composed of 2 retractile processes which issue from mammulae and form threads.

fusiform *a.* [L. *fusus*, spindle; *forma*, shape] spindle-shaped, tapering gradually at both ends, *appl.* innermost layer of cerebral cortex, *appl.* a gyrus of temporal lobe; *alt.* atractoid.

fusiform initial in vascular cambium, a fusiform cell that gives rise to secondary xylem and phloem; *cf.* ray initial.

fusion-nucleus central nucleus of embryo sac formed by fusion of odd nuclei from each end; *alt.* secondary nucleus.

fusocellular *a.* [L. *fusus*, spindle; *cellula*, small room] having, or *pert.*, spindle-shaped cells.

fusoid *a.* [L. *fusus*, spindle, Gk. *eidos*, form] somewhat fusiform.

fusulae *n.plu.* [*Dim.* of L. *fusus*, spindle] minute tubes of spinneret; *alt.* spools.

G

GA gibberellic acid or gibberellin.

GABA gamma, γ-aminobutyric acid, a γ-amino acid found in nervous tissue, particularly the brain, and thought to act as an inhibitory neurotransmitter.

Gadiformes *n.* [*Gadus*, generic name; L. *forma*, shape] an order or suborder of teleosts having soft jointed fins and an air bladder without a duct and including cod, hake, and whiting; *alt.* Anacanthini.

galactans *n.plu.* [Gk. *gala*, milk] the condensation products of galactose, found in gums and pectins.

galactin *n.* [Gk. *gala*, milk] the prepituitary lactogenic hormone, prolactin; a polysaccharide occurring in certain plants, e.g. in lupin.

galactoblast *n.* [Gk. *gala*, milk; *blastos*, bud] a fat-containing globule or colostrum corpuscle in mammary acini.

galactobolic *a.* [Gk. *gala*, milk; *bolē*, throw] bringing about the ejection of milk by causing the mammary myoepithelium to contract, *appl.* some neurohypophysial peptides.

galactolipid(e)s *n.plu.* [Gk. *gala*, milk; *lipos*, fat] glycolipids in which the sugar is galactose; *alt.* cerebrosides.

galactophorous *a.* [Gk. *gala*, milk; *pherein*, to carry] lactiferous, *appl.* ducts of mammary glands.

galactosamine *n.* [Gk. *gala*, milk; *ammoniakon*, gum] chondrosamine, *q.v.*

galactose *n.* [Gk. *gala*, milk] a hexose sugar occurring in lactose and released with glucose when lactose is hydrolysed, also occurs in various carbohydrates such as galactans and in certain glycolipids and glycoproteins.

galactosides *n.plu.* [Gk. *gala*, milk] glycosides which on hydrolysis yield the sugar galactose and a non-sugar residue, and including the galactolipids where the non-sugar residues are fatty acids and sphingosine.

galactosis *n.* [Gk. *gala*, milk] milk secretion; *alt.* lactosis.

galactotropic *a.* [Gk. *gala*, milk; *tropē*, turn] stimulating milk secretion, *appl.* hormone: prolactin, *q.v.*

galacturonic acid an oxidation derivative of galactose.

galbulus *n.* [L. *galbulus*, cypress nut] a closed globular female cone with peltate scales which are fleshy or thickened as in cypress.

galea *n.* [L. *galea*, helmet] a helmet-shaped petal, or other similarly-shaped structure; epicranial aponeurosis, the galea aponeurotica, of the scalp muscle or occipitofrontalis; galea capitis, thin sheath covering head of spermatozoon; outer division of stipes or endopodite of 1st maxilla of insects, itself divided into basigalea and distigalea; a prominence of moveable digit of chelicerae in certain arachnids such as Pseudoscorpiones.

galeate, galeiform *a.* [L. *galeatus*, helmed] helmet-shaped; hooded.

Galen, veins of [*Galen*, Greek physician] internal cerebral veins and great cerebral vein formed by their union.

galericulate *a.* [*Dim.* of *galerum*, hide cap] bearing or covered by a small cap.

galeriform *a.* [L. *galerum*, hide cap; *forma*, form] shaped like a cap.

gall *n.* [A.S. *gealla*, gall] an excrescence on plants caused by fungi, mites, and insects, especially by

Cynipidae (gall wasps) and Cecidomyidae (gall midges), *alt.* cecidium, gall nut; bile, *q.v.*

gall bladder pear-shaped or spherical sac which stores bile; *alt.* bile cyst, cholecyst, vesica fellea.

gall flower in fig trees, an infertile female flower in which the fig wasp lays its eggs.

gallic acid an acid obtained from hydrolysis of tannin, found in tea, galls, and other plants.

gallicolous *a.* [A.S. *gealla*, gall, L. *colere*, to dwell] living in plant galls.

Galliformes *n.* [Gk. *gallus*, cock; L. *forma*, shape] an order of birds of the subclass Neornithes, including the grouse, pheasant, and domestic fowl.

gallinaceous *a.* [Gk. *gallus*, cock] resembling the domestic fowl.

gall nut *see* gall.

gallotannin a tannin found in many types of galls, especially on oak, which is a glucoside of glucose and digallic acid.

galloxanthin *n.* [L. *gallus*, cock, Gk. *xanthos*, yellow] carotenoid pigment associated with retinal cones in domestic fowl.

Galton's law [*Sir F. Galton*, English scientist] law of filial regression, *q.v.*; the idea that an individual's characteristics are derived in equal parts from each parent.

galvanotaxis, galvanotropism *n.* [*L. Galvani*, Italian physiologist] response or reaction to electrical stimulus; *alt.* electrotaxis, electrotropism.

gametal *a.* [Gk. *gametēs*, spouse] *pert.* a gamete; reproductive.

gametangiogamy *n.* [Gk. *gametēs*, spouse; *anggeion*, vessel; *gamos*, marriage] the union of gametangia.

gametangium *n.* [Gk. *gametēs*, spouse; *anggeion*, vessel] a structure producing gametes.

gametes *n.plu.* [Gk. *gametēs*, spouse] haploid cells which fuse to form a zygote in sexual reproduction, in many organisms the male gamete is called spermatozoon, spermatozoid, or antherozoid and the female gamete is called an ovum; *alt.* germ cells, sexual cells, sex cells.

gametic *a.* [Gk. *gametēs*, spouse] *pert.* gamete, *appl.* a mutation occurring before maturation of gamete, *appl.* linkage; *cf.* zygotic.

gametic number the haploid, *n*, number of chromosomes present in a gamete.

gametic phase haplophase, *q.v.*

gametide *n.plu.* [Gk. *gametēs*, spouse] primary sporoblasts destined to become gametes.

gametoblast *n.* [Gk. *gametēs*, spouse; *blastos*, bud] plasson, *q.v.*

gametocyst *n.* [Gk. *gametēs*, spouse; *kystis*, bladder] cyst surrounding 2 associated free forms in which sexual reproduction occurs in some Protozoa such as gregarines; *alt.* gamocyst.

gametocyte *n.* [Gk. *gametēs*, spouse; *kytos*, hollow] the mother cell of a gamete, *alt.* gametogonium; a phase after the trophozoite which produces gametes in certain Protozoa.

gametogamy *n.* [Gk. *gametēs*, spouse; *gamos*, marriage] the union of gametes; *alt.* syngamy.

gametogenesis *n.* [Gk. *gametēs*, spouse; *genesis*, origin] gamete formation; *alt.* gametogeny, gonogenesis.

gametogenetic *a.* [Gk. *gametēs*, spouse; *genesis*, descent] gametokinetic, *q.v.*

gametogenic *a.* [Gk. *gametēs*, spouse; *genos*, descent] arising from spontaneous changes in chromosomes of gametes, *appl.* variation.

gametogeny gametogenesis, *q.v.*

gametogonium *n.* [Gk. *gametēs*, spouse; *gonos*, offspring] *see* gametocyte.

gametoid *n.* [Gk. *gametēs*, spouse; *eidos*, form] a structure behaving like a gamete, as apocytes uniting to form a zygotoid.

gametokinetic *a.* [Gk. *gametēs*, spouse; *kinein*, to move] stimulating gamete formation, *appl.* hormones, as follicle-stimulating hormone; *alt.* gametogenetic.

gametonucleus *n.* [Gk. *gametēs*, spouse; L. *nucleus*, kernel] the nucleus of a gamete; any nucleus acting as a gamete.

gametophore *n.* [Gk. *gametēs*, spouse; *pherein*, to bear] a special part of a gametophyte on which gametangia are borne, especially the upright leafy shoot of a moss; a hyphal outgrowth which fuses with a similar neighbouring outgrowth to form a zygospore.

gametophyll *n.* [Gk. *gametēs*, spouse; *phyllon*, leaf] a modified leaf bearing sexual organs; a micro- or megasporophyll.

gametophyte *n.* [Gk. *gametēs*, spouse; *phyton*, plant] the gamete-forming haploid phase in alternation of plant generations; *alt.* haplophyte, oophyte (in lower plants); *cf.* sporophyte.

gametophytic budding the producing of gemmae on a thallus in bryophytes or prothallus in pteridophytes.

gametospore *n.* [Gk. *gametēs*, spouse; *sporos*, seed] a sporidium or spore that unites with another by means of a bridging structure.

gametothallus *n.* [Gk. *gametēs*, spouse; *thallos*, young shoot] a thallus which produces gametes; *cf.* sporothallus.

gametropic *a.* [Gk. *gamos*, marriage; *tropē*, turn] *appl.* movements of plant organs before or after fertilization.

gamic *a.* [Gk. *gamos*, marriage] fertilized.

gamma, γ-aminobutyric acid GABA, *q.v.*

gamma (γ) globulins a group of globulins in blood containing antibodies and concerned with immunity.

gammation *n.* [Gk. *gammation*, *dim.* of *gamma*] an angular bar beside the branchial arches of *Palaeospondylus*, an extinct primitive Devonian vertebrate.

gamobium *n.* [Gk. *gamos*, marriage; *bios*, life] the sexual generation in alternation of generations, i.e. the gametophyte; *cf.* agamobium.

gamocyst *n.* [Gk. *gamos*, marriage; *kystis*, bladder] oocyst, *q.v.*; sporocyst, *q.v.*; gametocyst, *q.v.*

gamodeme *n.* [Gk. *gamos*, marriage; *dēmos*, people] a deme forming a relatively isolated intrabreeding community.

gamodesmic *a.* [Gk. *gamos*, marriage; *desma*, bond] having the vascular bundles fused together instead of separated by connective tissue.

gamogastrous *a.* [Gk. *gamos*, marriage; *gastēr*, belly] *appl.* a gynaecium having ovaries fused but with styles and stigmas free.

gamogenesis *n.* [Gk. *gamos*, marriage; *genesis*, descent] sexual reproduction.

gamogenetic, gamogenic *a.* [Gk. *gamos*, marriage; *genesis*, descent] sexual, produced from union of gametes.

gamogony *n.* [Gk. *gamos*, marriage; *gonē*, descent] formation of gametes or gametocytes from a gamont; sporogony, *q.v.*

gamones *n.plu.* [Gk. *gamos*, marriage] secretions

of gametes, which act on gametes of the opposite sex, being androgamones and gynogamones.

gamont n. [Gk. *gamos*, marriage; *on*, being] in some Protozoa, a generation or individual which produces gametes which then unite in pairs to form the zygote or sporont; sporont, *q.v.*

Gamopetalae n. [Gk. *gamos*, marriage; *petalon*, leaf] Sympetalae, *q.v.*, used especially in the Bentham and Hooker system.

gamopetalous a. [Gk. *gamos*, marriage; *petalon*, leaf] having the petals joined into a tube at least at the base; *alt.* sympetalous; *cf.* monopetalous, polypetalous; n. gamopetaly.

gamophase n. [Gk. *gamos*, marriage; *phasis*, aspect] haplophase, *q.v.*; *cf.* zygophase.

gamophyllous a. [Gk. *gamos*, marriage; *phyllon*, leaf] with united perianth leaves; *alt.* monophyllous, symphyllous.

gamosepalous a. [Gk. *gamos*, marriage; F. *sépale*, sepal] having sepals joined into a tube at least at the base; *alt.* monosepalous; *cf.* polysepalous.

gamostele n. [Gk. *gamos*, marriage; *stēlē*, pillar] stele formed from fusion of several steles.

gamostely n. [Gk. *gamos*, marriage; *stēlē*, pillar] the arrangement of polystelic stems when the separate steles are fused together surrounded by pericycle and endodermis; a. gamostelic.

gamotropism n. [Gk. *gamos*, union; *tropē*, turn] tendency to mutual attraction, exhibited by movements of gametes.

ganglia *plu.* of ganglion.

gangliar a. [Gk. *gangglion*, little tumour] *pert.* a ganglion or ganglia.

gangliate a. [Gk. *gangglion*, little tumour] having ganglia; *alt.* ganglionated.

gangliform a. [Gk. *gangglion*, little tumour; L. *forma*, shape] in the form of a ganglion.

ganglioblast n. [Gk. *gangglion*, little tumour; *blastos*, bud] mother cell of gangliocyte.

gangliocyte n. [Gk. *gangglion*, little tumour; *kytos*, hollow] a ganglion cell outside the central nervous system.

ganglioid a. [Gk. *gangglion*, little tumour; *eidos*, form] like a ganglion.

ganglion n. [Gk. *gangglion*, little tumour] a mass of nerve cell bodies giving rise to nerve fibres; *plu.* ganglia.

ganglionated gangliate, *q.v.*

ganglioneural a. [Gk. *gangglion*, little tumour; *neuron*, nerve] *appl.* a system of nerves, consisting of a series of ganglia connected by nerve strands.

ganglioneuron n. [Gk. *gangglion*, little tumour; *neuron*, nerve] a nerve cell of a ganglion.

ganglionic a. [Gk. *gangglion*, little tumour] *pert.*, consisting of, or in neighbourhood of a ganglion, *appl.* layer of retina, arteries, arterial system of brain.

ganglioplexus n. [Gk. *gangglion*, little tumour; L. *plexus*, braided] a diffuse ganglion.

gangliosides *n.plu.* [Gk. *gangglion*, little tumour] a group of complex lipids, found particularly in nerve cell membranes, containing sphingosine, fatty acids, carbohydrates, N-acetylglucosamine or N-acetylgalactosamine and N-acetylneuramic acid; *cf.* cerebrosides, glycolipids.

ganoblast n. [Gk. *ganos*, sheen; *blastos*, bud] ameloblast, *q.v.*; a cell secreting ganoine which was formerly thought to be homologous with enamel.

ganoid n. [Gk. *ganos*, sheen; *eidos*, form] a primitive actinopterygian fish possessing ganoid scales. a. *Appl.* scales, *q.v.*

Ganoidei, ganoids n., n.plu. [Gk. *ganos*, sheen; *eidos*, form] a group of fish possessing ganoid scales.

ganoid scales rhomboidal scales, found in primitive actinopterygians, with many outer layers of ganoine enamel, below which is dentine sometimes called cosmine, then lamellar bone also in many layers, growth in thickness being by layers above and below; *cf.* cosmoid scale.

ganoin(e) n. [Gk. *ganos*, sheen] an enamel-like substance forming the outer layers of ganoid scales and formerly thought to be homologous to enamel; the enamel of cosmoid scales is sometimes called ganoine.

gape n. [A.S. *geapan*, to open wide] the distance between the jaws of birds, fishes, etc.

garland cells a chain of nephrocytes, in Diptera.

garland stage stage of garland-like arrangement of chromatin at poles of nucleus in prophase of meiosis.

Garryales n. [*Garrya*, generic name] an order of dicots (Archichlamydeae) which are woody plants with opposite evergreen leaves, and flowers in catkin-like panicles.

Gärtner's canal [*H. T. Gärtner*, Danish anatomist] longitudinal duct of epoophoron, representing mesonephric duct, alongside the uterus and in lateral wall of vagina.

gaseous exchange the exchange of gases between an organism and its surroundings, including uptake of oxygen and release of carbon dioxide in external respiration in animals and plants, and the uptake of carbon dioxide and release of oxygen in photosynthesis in plants.

gas gland gandular portion of air bladder of certain fishes which secretes gas into the bladder.

Gaskell's bridge [*W. H. Gaskell*, English physiologist] His' bundle, *q.v.*

gasoplankton n. [Gk. *chaos*, air; *plangktos*, wandering] plankton organisms rendered buoyant by gas-filled vesicles or sacs, as by pneumatophores.

Gasserian ganglion [*A. P. Gasser*, German anatomist] the semilunar ganglion on sensory root of 5th cranial nerve.

gaster n. [Gk. *gastēr*, stomach] an abdomen, especially a swollen one; the swollen portion of a hymenopteran's abdomen which lies behind the petiole.

gastero- *also* gastro-, *q.v.*

Gasteromycetes, Gasteromycetales n. [Gk. *gastēr*, stomach; *mykēs*, fungus] a group of Homobasidiomycetes having the hymenium enclosed until after the spores are ripe; *alt.* Gastromycetes; a. gasteromycetous.

Gasteropoda Gastropoda, *q.v.*

gasterospore n. [Gk. *gastēr*, stomach; *sporos*, seed] a thick-walled globular spore formed within a fruit body.

Gasterosteiformes n. [*Gasterosteus*, generic name; L. *forma*, shape] an order of teleosts including the sticklebacks, sometimes considered as part of the Solenichthyes.

gastraea n. [Gk. *gastēr*, stomach] a hypothetical gastrula-like animal; the ancestral metazoan, according to Haeckel.

gastraeum n. [Gk. *gastēr*, stomach] ventral side of body.

gastral *a.* [Gk. *gastēr*, stomach] *pert.* stomach, as gastral cavity, cortex, layer, etc.

gastralia *n.plu.* Gk. *gastēr*, stomach] microscleres in the gastral membranes of Hexactinellida; abdominal ribs, as in some reptiles, *alt.* parasternalia; in Chelonia, dermal ossifications contributing to the plastron.

gastral layer or membrane in sponges, a layer made of choanocytes lining the internal cavity.

gas transport the transport of gases in the body between the respiratory surface and the tissue cells, often involving a respiratory pigment.

gastric *a.* [Gk. *gastēr*, stomach] *pert.* or in region of stomach, *appl.* arteries, nerves, veins.

gastric filaments in some jellyfish, filaments of endoderm lined with nematocysts which kill any living prey entering the stomach.

gastric gland a simple or compound tubular gland at the fundic end of stomach in the wall, which secretes gastric juice and mucus; *alt.* fundic gland, principal gland.

gastric intrinsic factor a mucoprotein produced by the cells lining the stomach which is necessary for the absorption of vitamin B_{12} into the blood.

gastric juice the secretion of the gastric glands in stomach containing some salts, pepsinogen and rennin secreted by the peptic cells, and hydrochloric acid secreted by the oxyntic cells, sometimes used for secretion of peptic cells only, or used to include mucus.

gastric mill masticatory stomach, *q.v.*

gastric ossicles the teeth of the gastric mill.

gastric pit a pit in the internal stomach wall leading to a gastric gland.

gastric secretin a hormone similar in function to secretin, *q.v.*, but produced by the stomach wall.

gastric shield in bivalve molluscs, a hard structure in the stomach against which the crystalline style rubs and is worn away releasing amylase.

gastrin *n.* [Gk. *gastēr*, stomach] a hormone secreted by the pyloric mucosa which stimulates the secretion of gastric juices; *see also* histamine.

gastro- *alt.* gastero-, *q.v.*

gastrocentrous *a.* [Gk. *gastēr*, stomach; *kentron*, centre] *appl.* vertebrae with centra formed by pairs of interventralia, while the basiventralia are reduced.

gastrocnemius *n.* [Gk. *gastēr*, stomach; *knēmē*, tibia] large muscle of calf of leg.

gastrocoel *n.* [Gk. *gastēr*, stomach; *koilos*, hollow] the archenteron of a gastrula.

gastrocolic *a.* [Gk. *gastēr*, stomach; *kolon*, colon] *pert.* stomach and colon, *appl.* ligament, the greater omentum.

gastrocutaneous *a.* [Gk. *gastēr*, stomach; L. *cutis*, skin] *appl.* pores leading from intestine to surface in Hemichorda.

gastrocystis *n.* [Gk. *gastēr*, stomach; *kystis*, bladder] blastocyst, *q.v.*

gastrodermis *n.* [Gk. *gastēr*, stomach; *derma*, skin] enteroblast, *q.v.*; endoderm, *q.v.*

gastroduodenal *a.* [Gk. *gastēr*, stomach; L. *duodeni*, twelve each] *pert.* stomach and duodenum, *appl.* an artery.

gastroepiploic *a.* [Gk. *gastēr*, stomach; *epiploon*, omentum] *pert.* stomach and great omentum, *appl.* arteries, veins.

gastrohepatic *a.* [Gk. *gastēr*, stomach; *hēpar*, liver] *pert.* stomach and liver, *appl.* portion of lesser omentum, a mesentery connecting liver and stomach in reptiles; *alt.* hepatogastric.

gastrointestinal *a.* [Gk. *gastēr*, stomach; L. *intestinum*, gut] *pert.* stomach and intestines.

gastrolienal *a.* [Gk. *gastēr*, stomach; L. *lien*, spleen] *pert.* stomach and spleen; *alt.* gastrosplenic.

gastrolith *n.* [Gk. *gastēr*, stomach; *lithos*, stone] a mass of calcareous matter found on each side of gizzard of crustaceans before a moult.

Gastromycetes Gasteromycetes, *q.v.*

gastroparietal *a.* [Gk. *gastēr*, stomach; L. *paries*, wall] *pert.* stomach and body wall.

gastrophrenic *a.* [Gk. *gastēr*, stomach; *phrēn*, midriff] *pert.* stomach and diaphragm, *appl.* ligament.

Gastropoda, gastropods *n., n.plu.* [Gk. *gastēr*, stomach; *pous*, foot] class of asymmetrical molluscs having a distinct head bearing tentacles, a flattened foot, a twisted visceral hump, and a single shell often spirally coiled; *alt.* Gasteropoda.

gastropores *n.plu.* [Gk. *gastēr*, stomach; *poros*, channel] in Hydrocorallina the larger pores on the surface of the colony, through which protrude polyps with manubrium and 4 knobbed tentacles.

gastropulmonary *a.* [Gk. *gastēr*, stomach; L. *pulmo*, lung] *pert.* stomach and lungs.

gastrosplenic *a.* [Gk. *gastēr*, stomach; *splēn*, spleen] gastrolienal, *q.v.*

gastrostege, gastrostegite n. [Gk. *gastēr*, stomach; *stegē*, roof] a ventral scale of snakes.

Gastrotricha *n.* [Gk. *gastēr*, stomach; *thrix*, hair] a phylum of marine and freshwater microscopic pseudocoelomate animals which have an elongated · body and move by ventral cilia; in some classifications, considered a class of Aschelminthes.

gastrotroch *n.* [Gk. *gastēr*, stomach; *trochos*, wheel] a band of cilia, posterior to metatroch, of trochophore in certain Polychaeta.

gastrovascular *a.* [Gk. *gastēr*, stomach; L. *vasculum*, small vessel] serving both digestive and circulatory purposes, as canals of some Coelenterata; *appl.* cavity: coelenteron, *q.v.*

gastrozooid *n.* [Gk. *gastēr*, stomach; *zōon*, animal; *eidos*, form] in colonial Hydrozoa, a hydroid modified for feeding and digestion, having a normal structure with a mouth surrounded by tentacles, *cf.* dactylozooid; in tunicates of the order Doliolida, lateral buds which gather food for the community, *alt.* trophozooid; *alt.* nutrient person.

gastrula *n.* [Gk. *gastēr*, stomach] the cup- or basin-shaped structure formed by invagination of a blastula; *alt.* metembryo.

gastrulation *n.* [Gk. *gastēr*, stomach] formation of gastrula from blastula by invagination.

gas vacuoles vacuoles formed in algae of the class Cyanophyceae which appear as black bodies and may be filled with gas or a viscous substance; *alt.* pseudovacuoles.

Gause's principle [*G. F. Gause*, German geneticist] an ecological principle stating that usually only one species may occupy a specific niche in a habitat.

Gaviiformes *n.* [*Gavia*, generic name of loon; L. *forma*, shape] an order of birds of the subclass Neornithes, including the loons.

GDP guanosine diphosphate, *q.v.*

geito- *alt.* gito-.

geitonogamy *n.* [Gk. *geitōn*, neighbour; *gamos*, marriage] fertilization of a flower by another from the same plant.

gelatigenous *a.* [L. *gelare*, to congeal; Gk. *-genēs*, producing] gelatine-producing.

gelatin(e) *n.* [L. *gelare*, to congeal] a jelly-like substance obtained from animal tissue.

Gelatinosa *n.* [L. *gelare*, to congeal] a group of sponges having an outer layer and a middle gelatinous layer which contains amoebocytes; may be considered a subphylum or class or raised to the rank of phylum.

gelatinous *a.* [L. *gelare*, to congeal] jelly-like in consistency.

gelatinous fibre a fibre having little or no lignification and with a gelatinous appearance.

Gelidiales *n.* [*Gelidium*, generic name] an order of red algae having no auxiliary cells.

gemellus *n.* [L. *gemellus*, twin] either of 2 muscles, superior and inferior, from ischium to greater trochanter and to trochanteric fossa, respectively.

geminate *a.* [L. *gemini*, twins] growing in pairs, *alt.* binate; paired; *appl.* species or subspecies: corresponding forms in corresponding but separate regions, as reindeer and caribou.

gemini *n.plu.* [L. *gemini*, twins] bivalent chromosomes, pairs of paternal and maternal chromosomes at parasyndesis.

geminiflorous *a.* [L. *gemini*, twins; *flos*, flower] *appl.* a plant whose flowers are arranged in pairs.

geminous in pairs; paired.

gemma *n.* [L. *gemma*, bud] a bud or outgrowth of a plant or animal which develops into a new organism; a leaf bud rather than a flower bud; a chlamydospore, *q.v.*; a hypothetical unit, *q.v.*

gemmaceous *a.* [L. *gemma*, bud] *pert.* gemmae or buds.

gemma cup a cup-shaped or crescent-shaped structure surrounding the gemmae of some liverworts; *alt.* cyathus.

gemmate *a.* [L. *gemmare*, to bud] having or reproducing by buds or gemmae; *alt.* gemmiferous, gemmiparous.

gemmation *n.* [L. *gemma*, bud] budding; bud or gemmae formation by means of which new independent individuals are developed, *alt.* pullulation; arrangement of buds.

gemmiferous *a.* [L. *gemma*, bud; *ferre*, to bear] gemmate, *q.v.*

gemmiform *a.* [L. *gemma*, bud; *forma*, shape] shaped like a bud, *appl.* pedicellariae of some echinoderms.

gemmiparous *a.* [L. *gemma*, bud; *parere*, to produce] gemmate, *q.v.*

gemmulation *n.* [L. *gemmula*, little bud] gemmule formation.

gemmule *n.* [L. *gemmula*, little bud] hypothetical unit, *q.v.*; a bryophyte gemma; one of the internal buds of Porifera arising asexually and coming into activity on death of parent organism; one of the minute protoplasmic processes on branch of a dendrite, contact point in synapse.

gena *n.* [L. *gena*, cheek] the cheek or side part of head; anterolateral part of prosoma of trilobites, and of insect head.

genal *pert.* the cheek; *appl.* facial suture and to caeca of stomach of trilobites; *appl.* angle of cheek.

gene *n.* [Gk. *genos*, descent] a unit hereditary factor consisting of a short length of chromosome, made of DNA, and having a particular effect on the phenotype, considered to be the length of DNA concerned with making one protein or one polypeptide chain; formerly called factor, genetic factor, germinal unit, Mendelian factor.

genealogy geneology, *q.v.*

gene centres geographical regions in which certain species of cultivated plant are represented in the greatest possible diversity of forms.

genecology *n.* [Gk. *genos*, descent; *oikos*, household; *logos*, discourse] ecology in relation to genetics.

gene complex the balanced system of genes constituting the internal medium within which each gene can manifest its effect; a group of genes which combine to determine the development of a particular character.

gene conversion the phenomenon where a heterozygote, A_1A_2, undergoing meiosis does not produce $2A_1$ and $2A_2$, but $3A_1$ and $1A_2$ or $1A_1$ and $3A_2$.

gene dosage number of genes of a given sort in a cell nucleus.

gene exchange the intermixing of genes in a breeding population resulting in the production of fertile offspring, used in determining population units.

gene flow the spreading of genes resulting from outcrossing and from subsequent crossing within a group; genorheithrum, *q.v.*

gene frequency the proportion between one particular type of allele and the total of all alleles at that locus in a breeding population.

gene locus a region of a chromosome containing a single gene.

gene mutation point mutation, *q.v.*

geneogenous *a.* [Gk. *genea*, birth; *gennaein*, to produce] congenital.

geneology *n.* [Gk. *genos*, descent; *logos*, discourse] the study of development of individual and race, embryology and palaeontology combined; *alt.* genealogy.

gene pool aggregate of all the genes in an interbreeding population.

genera *plu.* of genus.

generalized *a.* [L. *generalis*, of one kind] combining characteristics of 2 or more groups, as in many fossils; not specialized.

generation *n.* [L. *generatio*, reproduction] production; formation; the individuals of a species equally remote from a common ancestor.

generation time in a population of cells, the time between one mitosis and that of the daughter cells of that mitosis.

generative *a.* [L. *generare*, to beget] concerned in reproduction; *appl.* smaller of 2 cells into which a pollen grain primarily divides.

generative apogamy the condition where the sporophyte plant is developed from the ovum or another haploid cell of the gametophyte, with no fertilization; *alt.* haploid apogamy, meiotic apogamy, meiotic euapogamy, reduced apogamy.

generator cell a cell including a dikaryon, which gives rise to aecidiospore mother cells or to probasidia.

generic a. [L. genus, race] common to all species of a genus; pert. a genus.

generitype n. [L. genus, race; typus, image] the typical species of a genus; alt. genotype.

genesiology n. [Gk. genesis, descent; logos, discourse] science dealing with reproduction.

genesis n. [Gk. genesis, descent] formation, production, or development of a cell, organ, individual, or species.

Gené's organ [C. G. Gené, Italian zoologist] subscutal or cephalic gland secreting a viscid substance used in transferring eggs to dorsal surface in ticks.

gene string a group of genes arranged like a string of beads along a chromosome.

genetic a. [Gk. genesis, descent] pert. genesis; pert. genetics.

genetic code the sequence of purine and pyrimidine bases in nucleic acids which form codons specifying amino acids, 3 bases being thought to code for each amino acid and expressed by a process of transcription and translation.

genetic drift changes in gene frequency in small isolated breeding populations owing to random fluctuations rather than natural selection; alt. Sewall Wright effect.

genetic engineering the experimental manipulation of DNA (or RNA) of different species producing recombinant DNA, which includes some genes from both species.

genetic equilibrium the condition where gene frequences in populations remain constant from generation to generation.

genetic factor gene, q.v.

genetic map chromosome map, q.v.

genetic polymorphism the long-term occurrence of 2 or more genotypes in a population in frequencies which cannot be accounted for by recurrent mutation.

genetics n. [Gk. genesis, descent] that part of biology dealing with heredity and variation.

genetic spiral in spiral phyllotaxis, imaginary spiral line following points of insertion of successive leaves.

genetic variation an inheritable variation brought about by a change in the genes.

genetype genotype, q.v.

genial a. [Gk. geneion, chin] pert. the chin; appl. chin plates of reptiles; appl. tubercles on inside of mandible, for insertion of genioglossal and geniohyoid muscles.

genic a. [Gk. genos, descent] pert. genes.

genic balance harmonious interaction of genes in such a way as to ensure normal development of the organism.

genicular a. [L. geniculum, little knee] pert. region of the knee, appl. arteries, etc.; pert. geniculum.

geniculate n. [L. geniculum, little knee] bent like a knee joint, appl. antenna; pert. geniculum, appl. a ganglion of the facial nerve; appl. bodies, lateral and medial corpora geniculata, constituting the metathalamus; having upper part of filament forming an angle more or less obtuse with lower.

geniculation n. [L. geniculum, little knee] a knee-like joint or flexure.

geniculum a. [L. geniculum, little knee] sharp bend in a nerve; part of the facial nerve in temporal bone where it turns abruptly towards stylomastoid foramen.

genioglossal, geniohyoglossal a. [Gk. geneion, chin; glõssa, tongue] connecting chin and tongue, appl. muscle which moves tongue in vertebrates.

geniohyoid a. [Gk. geneion, chin; hyoeidēs, Y-shaped] pert. chin and hyoid, appl. muscles running from hyoid to lower jaw.

genital a. [L. gignere, to beget] pert. the region of reproductive organs, appl. corpuscles, glands, ridge, tubercle, veins, etc.

genital atrium in platyhelminths and some molluscs, small cavity communicating with the outside by means of a genital pore and with the male or female reproductive duct or both opening into it.

genital bursae in brittle stars, sacs into which the gonads open, and which open on each side of the base of each arm, also concerned with respiration and sometimes with the brooding of larvae.

genital canals in crinoids, the canals in the arms along which the genital cords lie.

genital coelom in cephalopods, a coelom at the apex of the visceral hump.

genital cord cord formed by posterior ends of Müllerian and Wolffian ducts in the mammalian embryo; in crinoids, a cord passing down the arm in the genital canal and reaching the pinnule where it enlarges into a gonad.

genital disc in the fruit fly, Drosophila, imaginal disc from which external genitalia and reproductive duct system are derived.

genital duct gonoduct, q.v.

genitalia n.plu. [L. gignere, to beget] the organs of reproduction, the gonads and their accessory organs, often especially the external organs; alt. genitals.

genital operculum in some arachnids, a soft rounded median lobe divided by a cleft on the sternum of the 1st preabdominal segment, with the opening of the genital duct at its base.

genital plates the plates in echinoids bearing the opening of the gonoduct.

genital pleurae in Enteropneusta, a pair of lateral ridges or folds in the region of the gills in which the gonads lie.

genital rhachis in many echinoderms, a ring of genital cells connected with the genital stolon and on which the gonads are borne.

genital ridge germinal ridge, q.v.

genitals genitalia, q.v.

genital sinus fused male and female genital ducts in some trematodes; in cartilaginous fish, a paired sinus opening into the posterior cardinal sinuses.

genital stolon in many echinoderms, a collection of genital cells in the axil organ connected with the genital rhachis.

genital tubercle the embryonic structure which gives rise to penis or to clitoris.

genito-anal a. [L. gignere, to beget; anus, vent] in the region of genitalia and anus.

genitocrural a. [L. gignere, to beget; crus, leg] in the region of genitalia and thigh, appl. a nerve originating from 1st and 2nd lumbar nerves; alt. genitofemoral.

genito-enteric a. [L. gignere, to beget; Gk. enteron, gut] pert. genitalia and intestine.

genitofemoral a [L. gignere, to beget; femur, thigh bone] genitocrural, q.v.

genito-urinary urinogenital, q.v.

genitoventral a. [L. gignere, to beget; venter,

belly] *appl.* plate formed by fused epigynial and ventral sclerites, in certain Acarina.

Gennari's band [*F. Gennari*, Italian anatomist] a layer of white fibres in middle cell lamina of cerebral cortex, especially of occipital lobe; *alt.* line of Gennari, Gennari's line or fibres.

genoblast *n.* [Gk. *genos*, offspring; *blastos*, bud] a mature germ cell exclusively male or female.

genocentric *a.* [Gk. *genos*, offspring; *kentron*, centre] having one or more reproductive structures developing at its centre, *appl.* thallus of Chytridiales.

genocline *n.* [Gk. *genos*, race; *klinein*, to slant] a gradual reduction in the frequency of various genotypes within a population in a particular (spatial) direction.

genocopy *n.* [Gk. *genos*, race; L.L. *copia*, transcript] production of the same phenotype by mimetic non-allelic genes; *cf.* phenocopy.

genodeme *n.* [Gk. *genos*, race; *dēmos*, people] a deme differing from others genotypically but not necessarily phenotypically.

geno-ecodeme *n.* [Gk. *genos*, race; *oikos*, household; *dēmos*, people] an ecodeme differing from others genotypically.

genoholotype *n.* [Gk. *genos*, race; *holos*, whole; *typos*, image] a species defined as typical of its genus.

genom(e) *n.* [Gk. *genos*, offspring] all the genes carried by a single gamete, i.e. by the single representative of all the chromosome pairs; sometimes used for the total chromosome content of any nucleus.

genomere *n.* [Gk. *genos*, offspring; *meros*, part] a small hypothetical particle, many of which make up a gene.

genonema, genoneme *n.* [Gk. *genos*, descent; *nēma*, thread] central axial thread of a chromosome on which genes are located; *alt.* axoneme, chromonema.

genonomy *n.* [Gk. *genos*, descent; *nomos*, law] the study of laws of relationships with reference to classification of organisms; *alt.* biosystematics, taxonomy.

genophenes *n.plu.*[Gk. *genos*, offspring; *phainein*, to appear] phenotypes of the same genotype.

genophore *n.* [Gk. *genos*, descent; *pherein*, to bear] the equivalent to a chromosome in viruses and prokaryotes, containing nucleic acid but no protein.

genoplast *n.* [Gk. *genos*, race; *plastos*, formed] *see* genotype.

genorheithrum *n.* [Gk. *genos*, descent; *rheithron*, stream] the passage or descent of genes in the phylogenesis; *alt.* gene flow.

genosome *n.* [Gk. *genos*, descent; *sōma*, body] the part of the chromosome bearing the locus of a gene.

genospecies *n.* [Gk. *genos*, race; L. *species*, particular kind] a species consisting of individuals having the same genotype.

genosyntype *n.* [Gk. *genos*, race; *syn*, with; *typos*, image] a series of species together defined as typical of their genus.

genotype *n.* [Gk. *genos*, race; *typos*, image] genetic constitution of an individual, *cf.* phenotype; group of individuals with the same genetic constitution, *alt.* biotype, genoblast; type species of a genus, *alt.* generitype; *alt.* genetype.

genotypic *a.* [Gk. *genos*, race; *typos*, image] *pert.* genotype, *appl.* inherited characters determined by genes.

genovariation point mutation, *q.v.*

gens *n.* [Gk. *genos*, race] a taxonomic group used in different ways by different writers and never precisely defined; *plu.* gentes.

gentes *plu.* of gens, *q.v.*

Gentianales *n.* [*Gentiana*, generic name] an order of dicots of the Asteridae or Gamopetalae, having actinomorphic pentamerous or tetramerous flowers, often contorted in the bud; the group is also called Contortae, and with the Dipsacales makes up the Rubiales.

genu *n.* [L. *genu*, knee] knee; segment between femur and tibia in some Acarina; a knee-like bend in an organ or part; anterior end of corpus callosum.

genus *n.* [L. *genus*, race] a taxonomic group consisting of closely related species, genera being grouped into families; *plu.* genera; *a.* generic.

genys *n.* [Gk. *genys*, jaw] lower jaw.

geobionts *n.plu.* [Gk. *gē*, earth; *biōnai*, to live] organisms permanently inhabiting the soil and affecting its structure; *cf.* geocoles, geoxenes.

geobios *n.* [Gk. *gē*, earth; *bios*, life] terrestrial life; edaphon, *q.v.*

geobiotic *a.* [Gk. *gē*, earth; *bios*, life] terrestrial.

geoblast *n.* [Gk. *gē*, earth; *blastos*, bud] a germinating plumule of which the cotyledons remain underground.

geobotany phytogeography, *q.v.*

geocarpic *a.* [Gk. *gē*, earth; *karpos*, fruit] having the fruits maturing underground due to the young fruits being pushed underground by curvature of stalk after fertilization ; *n.* geocarpy.

geochronology *n.* [Gk. *gē*, earth; *chronos*, time; *logos*, discourse] the science dealing with the measurement of time in relation to earth's evolution.

geocline *n.* [Gk. *gē*, earth; *klinein*, to slant] a gradual and continuous change in a character over a considerable area as a result of its adjustment to changing geographical conditions.

geocoles *n.plu.* [Gk. *gē*, earth; L. *colere*, to dwell] organisms which spend part of their lives in the soil and affect in by aeration, drainage, etc.; *cf.* geobionts, geoxenes.

geocryptophyte *n.* [Gk. *gē*, earth; *kryptos*, hidden; *phyton*, plant] a plant with dormant parts underground; *alt.* geophyte.

geodiatropism diageotropism, *q.v.*

geogen *n.* [Gk. *gē*, earth; *genos*, birth] a geographical or geochemical aspect of an area which affects organisms living or growing in it.

geographical race an interbreeding fertile population with morphological characters identical or only varying over a narrow range, and isolated from other populations of the same species by geographical barriers.

geology *n.* [Gk. *gē*, earth; *logos*, discourse] the science dealing with structure, activities, and history of the earth.

geomalism *n.* [Gk. *gē*, earth; *omalos*, level] response to the influence of gravitation.

geonasty *n.* [Gk. *gē*, earth; *nastos*, pressed] a curvature, usually a growth curvature, in response to gravity; *a.* geonastic.

geonemy *n.* [Gk. *gē*, earth; *nemein*, to inhabit] biogeography, *q.v.*

Geophilomorpha n. [*Geophilus*, generic name; Gk. *morphē*, form] an order of long burrowing worm-like centipedes of the subclass Epimorpha, having spiracles on all except the 1st and last segments.

geophilous a. [Gk. *gē*, earth; *philein*, to love] living in or on the earth; having leaves borne at soil level on a short, stout stem.

geophyte n. [Gk. *gē*, earth; *phyton*, plant] a land plant; a plant with dormant parts (tubers, bulbs, rhizomes) underground, *alt*. geocryptophyte, *cf*. cryptophyte.

geoplagiotropism n. [Gk. *gē*, earth; *plagios*, oblique; *tropē*, turn] growing at right angles to the surface of the ground, in response to a stimulus of gravity; *alt*. plagiogeotropism.

geosere n. [Gk. *gē*, earth; L. *serere*, to put in a row] a sere originating on a clay substratum.

geotaxis n. [Gk. *gē*, earth; *taxis*, arrangement] locomotor response to gravity; *a*. geotactic.

geotonus n. [Gk. *gē*, earth; *tonos*, tension] normal position in relation to gravity.

geotropism n. [Gk. *gē*, earth; *tropē*, turn] a tropism, *q.v.*, in relation to the direction of gravity, either towards the ground (positive geotropism) or away from the ground (negative geotropism); *a*. geotropic; *alt*. helcotropism.

geoxenes n.plu. [Gk. *gē*, earth; *xenos*, strange] organisms which occur only occasionally in the soil and do not affect it very much; *cf*. geocoles, geobionts.

Gephyrea n. [Gk. *gephyra*, bridge] a group of marine worms including the Echiuroidea, Sipunculoidea, and Priapuloidea.

gephyrocercal a. [Gk. *gephyra*, bridge; *kerkos*, tail] *appl*. secondary diphycercal caudal fin brought about by reduction of extreme tip of heterocercal or homocercal fin.

Geraniales n. [*Geranium*, generic name] an order of dicots of the Rosidae, Polypetalae, or Archichlamydeae, used in different ways by different authorities, often producing explosive sling fruits; *alt*. Gruinales.

geratology n. [Gk. *gēras*, old age; *logos*, discourse] study of the factors of decadence and old age of populations; *cf*. gerontology.

germ n. [L. *germen*, bud] a unicellular microorganism; the embryo of a seed; a bud; a developing egg; *a*. see also germinal.

germarium n. [L. *germen*, bud] ovary, *q.v.*; distal portion of an ovariole.

germ band primitive streak, *q.v.*, of early embryo; in insects, the cells which give rise to the embryo.

germ cell a reproductive cell, *cf*. somatic cell; gamete, or cell giving rise to a gamete, often set apart early in embryonic life; a primitive male or female element; *alt*. germinal cell.

germ centre area of lymph corpuscle division and lymph cell production within nodule of lymph gland tissue; *alt*. Fleming's germ centre, germinal centre.

germ disc a small green cellular plate of the germ tube of liverworts; *cf*. germinal disc.

germen n. [L. *germen*, bud] a mass of undifferentiated cells, the primary form of germ cells; ovary of angiosperms; capsule of mosses.

germ gland gonad, *q.v.*

germiduct n. [L. *germen*, bud; *ducere*, to lead] oviduct of trematodes.

germigen n. [L. *germen*, bud; *generare*, to beget] ovary of trematodes.

germinal a. [L. *germen*, bud] *pert*. a seed, a germ cell, or reproduction; *cf*. somatic; *see also* germ.

germinal aperture germ pore, *q.v.*

germinal bands two sets of rows of cells in early development of annulates.

germinal cell germ cell, *q.v.*

germinal centre germ centre, *q.v.*

germinal crescent a region of blastoderm forming a crescent of primordial germ cells partially surrounding anterior end of primitive streak.

germinal disc blastodisc, *q.v.*; *cf*. germ disc.

germinal epithelium the layer of columnar epithelial cells covering the stroma of an ovary; the layer of epithelial cells lining the testis which give rise to the spermatogonia and Sertoli cells.

germinal layers primary layers of cells in a developing embryo being epiblast (ectoderm), hypoblast (endoderm), and later, mesoblast (mesoderm); histogens, *q.v.*; *alt*. germ layers.

germinal lid operculum of a pollen grain.

germinal pore germ pore, *q.v.*

germinal ridge mesodermal ridge in vertebrate embryo, into which migrate primordial germ cells, and giving rise ultimately to interstitial cells of testis or to follicle cells of ovary; *alt*. genital ridge.

germinal selection the selective process undergone by the competing genes in the germ plasm, resulting in only some of those genes being passed on to the progeny.

germinal spot the nucleolus of the germinal vesicle or of an ovum.

germinal streak primitive streak, *q.v.*

germinal unit gene, *q.v.*

germinal vesicle the nucleus of an oocyte before formation of polar bodies.

germination n. [L. *germen*, bud] beginning or process of initial development in a spore or seed from uptake of water to beginning of photosynthesis; budding; sprouting.

germinative layer Malpighian layer, *q.v.*

germiparity n. [L. *germen*, bud; *parere*, to beget] reproduction by germ formation.

germ layer an early differentiated layer of cells; germinal layer, *q.v.*

germ line a line of cells from which gametes are derived and which are continuous from generation to generation; *alt*. germ track.

germ nucleus an egg or sperm nucleus; *alt*. pronucleus.

germogen n. [L. *germen*, bud; Gk. *genos*, offspring] the central cell of gastrula-like phase, or infusorigen, in development of Rhombozoa; the residual nucleus, or unused portion, after formation of rhombogen by division of primary germogen or primitive central cell.

germ plasm a kind of protoplasm which, according to Weismann, was transmitted unchanged from generation to generation in the germ cells; *alt*. gonoplasm, idioplasm.

germ pore the thin region in a pollen grain wall through which the pollen tube emerges; a similar area in a spore wall for the exit of a germ tube; *alt*. germinal pore, germinal aperture.

germ sporangium a sporangium borne on the germ tube of a zygospore.

germ stock stolon of tunicates.

germ theory biogenesis, *q.v.*

germ track

germ track lineage of zygote in developing organism; continuity of germ cells through various cell generations, *alt.* germ line.

germ tube short filamentous tube put forth by a germinating spore.

germ vitellarium an organ, of some platyhelminths, producing both ova and vitelline material; *alt.* germ yolk gland.

germ yolk gland in some Rhabdocoelida, an embryonic structure consisting of fertile portion of egg and a sterile portion which functions as a yolk gland feeding the fertile portion; germ vitellarium, *q.v.*

gerontal *a.* [Gk. *gerōn*, old man] senile; *alt.* gerontic.

gerontic *a.* [Gk. *gerōn*, old man] *pert.* old age, *alt.* gerontal, *appl.* stage in phylogeny or ontogeny.

gerontogaeous *a.* [Gk. *gerōn*, old man; *gē*, earth] *pert.* or originating in the Old World.

gerontology *n.* [Gk. *gerōn*, old man; *logos*, discourse] the study of senescence and senility; *cf.* geratology.

gerontomorphosis *n.* [Gk. *gerōn*, old man; *morphōsis*, form] evolution involving specialization and less capacity for further evolutionary change, resulting in racial senescence.

gestalt *n.* [Ger. *Gestalt*, form] organized or unified response to an arrangement of stimuli; coordinated movements or configuration of motor reactions; a mental process considered as an organized pattern, involving explanation of parts in terms of the whole; a pattern considered in relation to background or environment, *appl.* morphology irrespective of taxonomic relationships.

gestation *n.* [L. *gestare*, to bear] the intra-uterine period in development of an embryo.

GH growth hormone, *q.v.*

ghosts empty cell membranes obtained by haemolysis of red blood cells and used in investigations of cell membranes.

giant cells large nerve cells in annelids; myeloplaxes, *q.v.*; osteoclasts, *q.v.*; megakaryocytes, *q.v.*; Langhans' cells, *q.v.*; Betz cells, *q.v.*; *alt.* gigantocytes.

giant chromosomes polytene or large chromosomes, as in salivary gland cells of larval Diptera; lampbrush chromosomes, *q.v.*, in oocytes of vertebrates.

giant fibres greatly enlarged and modified nerve fibres running longitudinally through ventral nerve cord of many invertebrates and some lower vertebrates; *alt.* neurochords.

Gianuzzi, crescents of demilunes, *q.v.*

gibberellic acid (GA) one of the gibberellins, *q.v.*, which has been isolated and whose structure is known.

gibberellins *n.plu.* [*Gibberella*, a fungal genus] a group of plant hormones, first discovered in the fungus *Gibberella fujikuroi* but which occur in many flowering plants, that accelerate shoot growth making a dwarf mutant plant grow like a normal plant, and having a range of other developmental effects, including induction of enzyme synthesis in some seeds.

gibbose, gibbous *a.* [L. *gibbus*, hump] inflated; pouched; of a solid object, any part projecting as a rounded swelling, such as lateral sepals of Cruciferae which are expanded into lateral sacs or pouches; *alt.* saccate.

gigantism *n.* [Gk. *gigas*, giant] growth of an organ or a complete organism to an abnormally great size.

gigantoblast *n.* [Gk. *gigas*, giant; *blastos*, bud] an erythroblast, especially a very large one.

gigantocyte *n.* [Gk. *gigas*, giant; *kytos*, hollow] giant cell, *q.v.*; a large erythrocyte.

Giganturiformes *n.* [*Gigantura*, generic name; L. *forma*, shape] an order of deep-sea teleosts.

Gigartinales *n.* [*Gigartina*, generic name] an order of red algae whose auxiliary cells differentiate before fusion.

gilgai *n.* [Native Australian *gilgai*] a black clay soil found in Australia, which cracks widely and the soil moves down into the crack so that it inverts itself.

gill *n.* [M.E. *gille*, gill] a plate-like or filamentous outgrowth; respiratory organ of aquatic animals; radial lamella on underside of pileus of agarics. *a.* Also called branchial.

gill arch part of the visceral skeleton in the region of the gills consisting of bony or cartilaginous arches supporting the gill bars; *alt.* branchial arch, visceral arch.

gill bar in chordates, the tissue which separates the gill slits from each other and contains blood vessels, nerves, and skeletal support, the gill arch; in cephalochordates, gill rod, *q.v.*; sometimes used to mean gill arch, *q.v.*

gill basket the cartilaginous branchial skeleton of cartilaginous fish and cyclostomes; in certain dragonfly larvae, a modification of the rectum concerned with respiration.

gill book the respiratory organ of certain Xiphosura such as *Limulus*, and probably of extinct eurypterids, consisting of a large number of leaf-like structures between which the water circulates; *alt.* book gill.

gill cavity in agarics, the cavity in the fruit body in which the gills are formed.

gill cleft visceral cleft, *q.v.*; gill slit, especially the embryonic gill slit of higher vertebrates; *alt.* branchial cleft.

gill cover operculum covering the gills in bony fish.

gill fungi members of the Basidiomycetes whose hymenium is borne on gills, including most mushrooms and toadstools.

gill helix a spirally coiled gill-like organ in certain Clupeidae (herring family).

gill plume the ctenidium of the majority of Gastropoda.

gill pouch an oval pouch containing gills and communicating with exterior, as in cyclostomes; outpushing of side wall of pharynx in all chordate embryos which develops into a gill slit in fish and some amphibians, but forms an actual slit temporarily or not at all in terrestrial vertebrates.

gill rakers small spine-like structures attached in a single or double row to the inner margins of the gill arches, preventing the escape of food.

gill rays skeletal structures on outer margins of gill arches, which stiffen gills; *alt.* gill rods.

gill remnants epithelial, postbranchial, or suprapericardial bodies arising in pharynx of higher vertebrates.

gill rods gelatinous rods supporting the pharynx in Cephalochordata, *alt.* gill bars; gill rays, *q.v.*

gill slits a series of perforations leading from pharynx to exterior, persistent in lower vertebrates, embryonic in higher; *alt.* branchial clefts, gill clefts.

gill trama the structure between the hymenial layers of a gill, as in agarics.

gimped crenate, *q.v.*

gingivae *n.plu.* [L. *gingivae*, gums] the gums; *alt.* ula.

gingival *pert.* the gums; *alt.* uletic.

Ginglymodi *n.* [Gk. *gingglymos*, hinge] an order of Ganoidei including the North American gars.

ginglymoid *a.* [Gk. *gingglymos*, hinge joint; *eidos*, form] constructed like a hinge joint.

ginglymus *n.* [Gk. *gingglymos*, hinge joint] a joint in which the articulating surfaces are so constructed to allow of motion in one plane only; *alt.* hinge joint.

Ginkgoales *n.* [*Ginkgo*, generic name] the only order of gymnosperms of the class Ginkgopsida, having one living species, *Ginkgo biloba*, and many fossils, being trees with much branched stems and variously lobed leaves, and male gamete with spiral band of flagella; in some classifications this order is included in the class Coniferopsida.

Ginkgoatae *n.* [*Ginkgo*, generic name] in some classifications, a class of Coniferophytina, including the order Ginkgoales.

Ginkgopsida *n.* [*Ginkgo*, generic name; Gk. *opsis*, appearance] a class of gymnosperms having only one order, Ginkgoales, *q.v.*; in some classifications the Ginkgoales are included in the Coniferopsida and this class does not exist.

Giraldès' organ [*J. A. C. C. Giraldès*, Portuguese surgeon] the paradidymis, *q.v.*

girdle *n.* [A.S. *gyrdan*, to gird] in appendicular skeleton, the supporting structure at shoulder and hip, each consisting typically of 1 dorsal and 2 ventral elements; spicule-bearing portion of mantle not covered by shell plates in Polyplacophora; transverse groove in Dinoflagellata, containing transverse flagellum and separating epicone and hypocone; the cingulum of diatoms.

girdle bundles leaf trace bundles which girdle the stem and converge at the leaf insertion, as in Cycadales.

girdle scar a series of scale scars on axis of twig where bud scales have fallen.

girdling the condition of a leaf making girdle bundles.

gito- geito-, *q.v.*

gizzard n. [O.F. *gezier*, gizzard] muscular, grinding chamber of alimentary canal of various animals; proventriculus, *q.v.*, of insects.

glabella *n.* [L. *glaber*, bald] the space on forehead between superciliary ridges; the elevated median region of cephalic shield of Trilobita, *alt.* mesophryon.

glabrate *a.* [L. *glaber*, smooth] becoming hairless, *alt.* glabrescent; with a nearly smooth surface.

glabrescent *see* glabrate.

glabrous *a.* [L. *glaber*, smooth] with a smooth, even surface; without hairs.

glacial *a.* [L. *glacies*, ice] *pert.* or *appl.* the Pleistocene epoch of the Quaternary period, characterized by periodic glaciation.

gladiate *a.* [L. *gladius*, sword] ensiform, *q.v.*

gladiolus *n.* [L. *gladiolus*, small sword] corpus sterni, *q.v.*

gladius *n.* [L. *gladius*, sword] the pen or chitinous

shell in Chondrophora; pro-ostracum of a phragmocone or a sepion.

glairine *n.* [F. *glaire*, white of egg] glairy film found or thermal springs and formed by pectic zoogloea.

gland *n.* [L. *glands*, acorn] single cell or mass of cells specialized for elaboration of secretions either for use in the body or for excretion, *alt.* glans; a small vesicle containing oil, resin, or other liquid on any part of a plant; *alt.* glandula.

gland cell an isolated secreting cell; a cell of glandular epithelium.

glandiform *a.* [L. *glands*, acorn; *forma*, shape] acorn-shaped.

glandilemma *n.* [L. *glans*, acorn; Gk. *lemma*, skin] the outer covering of a gland.

glandula *n.* [L. *glandula*, small acorn] gland, *q.v.*; one of the bundles of hyphae ending in basidia with a viscous secretion appearing as spots on the surface of the stipe of certain fungi; a glutinous gland subserving cohesion of pollinia; arachnoid granulation on outer surface of dura mater; *alt.* glandule.

glandulae Pacchionii Pacchionian bodies, *q.v.*

glandular *a.* [L. *glandula*, small acorn] with or *pert.* glands; with secreting function.

glandular epithelium the epithelial tissue of glands, composed of polyhedral, columnar, or cubical cells whose protoplasm contains or elaborates the material to be secreted; *alt.* glandular tissue.

glandular tissue parenchymatous tissue of single or massed cells filled with granular cytoplasm and adapted for secretion of aromatic or other substances in plants; glandular epithelium, *q.v.*

glandular vesiculosa vesicula seminalis, *q.v.*

glandule glandula, *q.v.*

glandulose-serrate *a.* [L. *glandula*, small acorn; *serratus*, sawn] having the serrations tipped with glands; *alt.* glandular-serrate.

glans *n.* [L. *glans*, acorn] nut, *q.v.*; gland, *q.v.*; glans penis, *q.v.*; glans clitoridis, *q.v.*

glans clitoridis the small rounded mass of erectile tissue at the distal end of the clitoris.

glans penis the bulbous tip of the penis in mammals; *alt.* balanus.

glareal *a.* [L. *glarea*, gravel] *pert.*, or growing on, dry gravelly ground.

Glaserian fissure (*J. H. Glaser*, Swiss anatomist] a fissue in the temporal bone of mammals which holds the Folian process of the malleus; *alt.* petrotympanic fissure.

glaucescent *a.* [L. *glaucus*, sea-green] somewhat glaucous.

glaucous *a.* [L. *glaucus*, sea-green] bluish green; covered with a pale green bloom.

gleba *n.* [L. *gleba*, clod] in Gasteromycetes, the spore-bearing tissue enclosed in the peridium of the sporophore; sometimes used for a similar structure in Tuberales; *alt.* glebe.

gleba chamber or cavity peridiolum, *q.v.*

glebe gleba, *q.v.*

glebula *n.* [L. *glebula*, small clod] a small prominence on a lichen thallus; *a.* glebulose.

glei gley, *q.v.*

glenohumeral *n.* [Gk. *glēnē*, socket; L. *humerus*, humerus] *pert.* glenoid cavity and humerus, *appl.* ligaments.

glenoid *a.* [Gk. *glēnē*, socket; *eidos*, form] like a socket, *appl.* fossa, *q.v.*, *appl.* various ligaments.

glenoidal labrum a fibrocartilaginous rim attached round the margin of glenoid cavity and of acetabulum.

glenoid fossa a cavity into which the head of the humerus fits in tetrapods, *alt.* glenoid cavity; a cavity on the squamosal bone for articulation of the lower jaw in mammals, *alt.* mandibular fossa.

gley *n.* [Russian *glei*, clay] a soil formed under conditions of poor drainage and water-logged all or part of the time; *alt.* glei.

glia, gliacyte, glial cell *n.* [Gk. *glia*, glue] neuroglia cell, *q.v.*

gliadin *n.* [Gk. *glia*, glue] a prolamine found in wheat and rye seeds; a substance interacting with glutenin to form gluten in cereals; formerly, any prolamine.

gliding growth sliding growth, *q.v.*

Glires *n.* [L. *glis*, dormouse] an infraclass or cohort of eutherians, including the living orders Rodentia and Lagomorpha.

gliosomes *n.plu.* [Gk. *glia*, glue; *sōma*, body] granules in cytoplasm of neuroglia, possibly having some relationship to mitochondria.

Glisson's capsule [*F. Glisson*, English physician] a fibrous capsule within liver, enclosing hepatic artery, portal vein, lymphatic vessels, and bile duct; *alt.* hepatobiliary capsule.

globate *a.* [L. *globus*, globe] globe-shaped, globular.

globiferous *a.* [L. *globus*, globe; *ferre*, to bear] *appl.* echinoid pedicellariae with rounded ends.

globigerina ooze sea-bottom mud which is largely composed of shells of Foraminifera, especially the calcareous shells of *Globigerina*.

globin *n.* [L. *globus*, globe] the basic protein constituent of haemoglobin.

globoid *n.* [L. *globus*, globe; Gk. *eidos*, form] a spherical body in aleurone grains, a double phosphate of calcium and magnesium combined with globulins.

globose *a.* [L. *globus*, globe] spherical or globular.

globule *n.* [L. *globulus*, small globe] any minute spherical structure; the antheridium of Charales; *alt.* globulus.

globulins *n.plu.* [L. *globus*, globe] simple proteins including ovoglobulin, lactoglobulin, antibodies, fibrinogen, myosin, legumin, edestin, and many proteins in seeds, which are water-insoluble but soluble in dilute salt solution from which they can be salted out.

globulose *a.* [L. *globulus*, small globe] spherical; consisting of, or containing globules.

globulus *n.* [L. *globulus*, small globe] globule, *q.v.*; spherical or club-shaped sensory organ at bifurcation of antenna in Pauropoda.

globus *n.* [L. *globus*, globe] a globe-shaped structure.

globus major and minor head and tail of epididymis.

globus pallidus pallidium, *q.v.*

glochid glochidium, *q.v.*

glochidiate *a.* [Gk. *glōchis*, arrow point; *idion*, dim.] covered with barbed hairs; *alt.* inuncate.

glochidium *n.* [Gk. *glōchis*, arrow point; *idion*, dim.] hairs bearing barbed processes seen on massulae of certain rhizocarps, or in areolae at base of spines in cacti of the family Opuntiae; the larva of freshwater mussels such as *Unio* and *Anodon*.

gloea *n.* [Gk. *gloia*, glue] an adhesive secretion of some Protozoa.

gloeocystidium *n.* [Gk. *gloios*, sticky; *kystis*, bag; *idion*, dim.] a cystidium containing a slimy or oily substance.

Gloger's law or rule [*C. W. L. Gloger*, German zoologist] melanin pigmentation in warm-blooded animals varies with climate, becoming darker in warmer climates and lighter in colder ones.

glomera *plu.* of glomus.

glomera carotica carotid bodies, *q.v.*

Glomerida *n.* [*Glomeris*, generic name] an order of Pentazonia having a short broad body which can be rolled into a ball; *alt.* Oniscomorpha.

Glomeridesmida *n.* [*Glomeridesmus*, generic name] an order of small Pentazonia which cannot roll into a ball; *alt.* Limacomorpha.

glomerular *a.* [L. *glomus*, ball] *pert.* or like a glomerulus.

glomerulate *a.* [L. *glomus*, ball] arranged in clusters; bearing glomerules.

glomerule *n.* [L. *glomus*, ball] a condensed cyme of almost sessile flowers; a compact cluster as of spores; *alt.* glomerulus.

glomeruliferous *a.* [L. *glomus*, ball; *ferre*, to carry] having the flowers arranged in glomerules.

glomerulus *n.* [L. *glomus*, ball] network of capillary blood vessels; blood vessel knot in Bowman's capsule; oval body terminating olfactory fibres in rhinencephalon; a mass of interlacing intracapsular dendrites, in sympathetic ganglia; excretory organ of Enteropneusta; a small mass of spores; glomerule, *q.v.*

glomus *n.* [L. *glomus*, ball] a number of glomeruli run together; coccygeal and carotid bodies, consisting largely of chromaffin cells; *plu.* glomera.

glossa *n.* [Gk. *glōssa*, tongue] a tongue-like projection in middle of labium of insects; the tongue of vertebrates; a tongue-like structure.

glossal *pert.* the tongue.

glossarium *n.* [Gk. *glōssa*, tongue] the slender-pointed glossa of certain Diptera.

glossate *a.* [Gk. *glōssa*, tongue] having a tongue or tongue-like structure.

glosso-epiglottic *a.* [Gk. *glōssa*, tongue; *epi*, upon; *glōtta*, tongue] *pert.* tongue and epiglottis, *appl.* folds of mucous membrane.

glossohyal *n.* [Gk. *glōssa*, tongue; *hyoeides*, Υ-shaped] entoglossum, *q.v.*

glossokinaesthetic area a brain area in Broca's convolution immediately connected with speech.

glossopalatine *a.* [Gk. *glōssa*, tongue; L. *palatus*, palate] connecting tongue and soft palate, *appl.* arch, nerve, etc.; *appl.* muscle: glossopalatinus, *q.v.*

glossopalatine nerve a branch of the facial nerve which supplies the tongue and palate; *alt.* nervus intermedius, nerve of Wrisberg.

glossopalatinus *n.* [Gk. *glōssa*, tongue; L. *palatus*, palate] a thin muscle which arises on each side of the soft palate and is inserted into the tongue; *alt.* glossopalatine muscle, palatoglossal muscle.

glossophagine *a.* [Gk. *glōssa*, tongue; *phagein*, to eat] securing food by means of the tongue.

glossopharyngeal *a.* [Gk. *glōssa*, tongue;

glycans

pharynx, gullet] *pert.* tongue and pharynx, *appl.* 9th cranial nerve.

glossophorous *a.* [Gk. *glōssa*, tongue; *pherein*, to bear] having a tongue or a radula.

glossopodium *n.* [Gk. *glōssa*, tongue; *pous*, foot] the sheathing leaf base of *Isoetes*, the quill-wort.

glossotheca *n.* [Gk. *glōssa*, tongue; *thēkē*, box] the proboscis-covering part of pupal integument of insects.

glottis *n.* [Gk. *glōtta*, tongue] the opening from the trachea into the pharynx.

glu glutamic acid, *q.v.*

glucagon *n.* [Gk. *glykes*, sweet; *agōn*, assembly] a polypeptide pancreatic hormone formed in α cells of islets of Langerhans, which stimulates glycogenolysis in the liver, causing increase in blood sugar; *alt.* hyperglycaemic–glycogenolytic factor, HGF.

glucan glucosan, *q.v.*

glucase *n.* [Gk. *glykes*, sweet] maltase, *q.v.*

glucocerebroside *n.* [Gk. *glykys*, sweet; L. *cerebrum*, brain] cerebroside containing glucose rather than galactose.

glucocorticoids *n.plu.* [Gk. *glykys*, sweet; L. *cortex*, bark; ster*oid*] group of steroid hormones including cortisone and hydrocortisone, secreted by the cortex of the adrenal gland, which have various effects, including formation of carbohydrate from fat and protein.

glucogenic *a.* [Gk. *glykys*, sweet; *genēs*, produced] *appl.* amino acids that yield keto acids which can be converted to glucose and so enter the normal carbohydrate metabolism of the body, e.g. glycine, alanine, aspartic acid, glutamic acid, arginine, ornithine; increasing glucose formation.

glucokinin *n.* [Gk. *glykys*, sweet; *kinein*, to move] a plant substance capable of reducing blood sugar; *alt.* vegetable insulin.

gluconeogenesis *n.* [Gk. *glykys*, sweet; *neos*, new; *genesis*, origin] the synthesis of glucose from precursors other than carbohydrates, such as pyruvic acid, Krebs' cycle intermediates, amino acids, etc., usually by reversal of glycolysis, and occurring mainly in the liver and kidneys of vertebrates.

glucoproteins glycoproteins, *q.v.*

glucosamine *n.* [Gk. *glykys*, sweet; *ammoniakon*, gum] an amino sugar based on glucose occurring in both animals and plants and forming part of the molecule of chitin and heparin and the mucopeptide forming the inner wall of blue–green algae.

glucosan *n.* [Gk. *glykys*, sweet] a polysaccharide made of condensed glucose units, such as cellulose, starch, glycogen, etc.; *alt.* glucan.

glucose *n.* [Gk. *glykys*, sweet] a hexose sugar occurring in all living cells, especially plant juices and in the blood and tissue fluids of animals, being the chief end-product of carbohydrate digestion, and possessing potentially active aldehyde and so called an aldose sugar; *alt.* dextrose, grape sugar, starch sugar, and blood sugar when found in blood.

glucosidases *n.plu.* [Gk. *glykys*, sweet] enzymes, plentiful in plants, which catalyse the conversion of glucosides to glucose and aglutones; EC 3.2.1.21 and EC 3.2.1.22.

glucosides *n.plu.* [Gk. *glykys*, sweet] a group of glycosides found in plants that on hydrolysis yield a sugar which is usually glucose.

glucuronic acid an organic acid which occurs in small amounts in the urine, and forms ethers or esters with aromatic acids or phenols; *alt.* glycuronic acid.

gluma *n.* [L. *gluma*, husk] glume, *q.v.*

glumaceous *a.* [L. *gluma*, husk] dry and scaly like glumes; formed of glumes.

glume *n.* [L. *gluma*, husk] a chaffy or membranous bract; a bract at the base of a grass inflorescence or spikelet, *alt.* barren glume, empty glume, gluma, sterile glume; flowering glume: lemma, *q.v.*

glumella palea, *q.v.*

glumellule lodicule, *q.v.*

glumiferous *a.* [L. *gluma*, husk; *ferre*, to bear] bearing or producing glumes.

Glumiflorae *n.* [L. *gluma*, husk; *flos*, flower] an order of herbaceous monocots having flowers with reduced perianth; the term formerly included the grasses and sedges, but the latter are now separated into a different order (Cyperales) and the Glumiflorae includes only grasses, and is also called Poales.

glumiflorous *a.* [L. *gluma*, husk; *flos*, flower] having flowers with glumes or bracts at their bases.

glu-NH₂ glutamine, *q.v.*

glutaeal *a.* [Gk. *gloutos*, buttock] *pert.* or in region of buttocks, *appl.* arteries, muscles, nerves, tuberosity, veins; *alt.* gluteal.

glutaeus *n.* [Gk. *gloutos*, buttock] a muscle of the buttock which moves the hindlimb; *alt.* gluteus.

glutamic acid an amino acid, α-aminoglutaric acid, non-essential in rats, which is part of the vitamin folic acid and is important in nitrogen metabolism; abbreviated to glu.

glutaminase an enzyme commonly occurring in plants and higher animals, which catalyses the hydrolysis of glutaminase to glutamic acid and ammonia; EC 3.5.1.35; *r.n.* D-glutaminase.

glutamine *n.* [Gk. *glykys*, sweet; *ammoniakon*, gum] an amino acid, the γ-monoamide of glutamic acid; abbreviated to glu-NH₂.

glutathione *n.* [L. *gluten*, glue; Gk. *theion*, sulphur] a tripeptide, made of cysteine, glutamic acid, and glycine, found in various tissues and taking part in biological oxidations because it is capable of being alternately reduced and oxidized, and also acts as a coenzyme to glyoxalase.

gluteal glutaeal, *q.v.*

glutelins *n.plu.* [L. *gluten*, glue] simple plant proteins, which are insoluble in water and salt solutions but soluble in dilute acid and alkali and include glutenin and oryzenin.

gluten *n.* [L. *gluten*, glue] a reserve protein found in cereals, a product of gliadin and glutenin, being a sticky substance which forms the binding agent in dough and flour pastes; a sticky coating on the cap of some agaric fungi.

glutenin *n.* [L. *gluten*, glue] a protein of the glutelin group, found in cereals, which interacts with gliadin to form gluten.

gluteus glutaeus, *q.v.*

glutinant *n.* [L. *gluten*, glue] a type of nematocyst in coelenterates.

glutinous *a.* [L. *gluten*, glue] having a sticky or slimy surface.

gly glycine, *q.v.*

glycans *n.plu.* [Gk. *glykys*, sweet] polysaccharides, *q.v.*

glycerides

glycerides *n.plu.* [Gk. *glykys*, sweet] esters of glycerol, including fats and oils in which 3 molecules of fatty acids combine with one molecule of glycerol.

glycerin(e) *n.* [Gk. *glykys*, sweet] the sweet syrupy liquid obtained from fats and oils by hydrolysis; *see also* glycerol.

glycerol *n.* [Gk. *glykys*, sweet] a trihydric alcohol which combines with fatty acids to produce esters which are fats and oils; *alt.* glycerine—but the term glycerol is usually used when considering the production of esters of fatty acids, and glycerine when considering the syrupy sweet liquid.

glycine *n.* [Gk. *glykys*, sweet] an amino acid, aminoacetic acid, derived from acetic acid, constituent of various proteins, particularly of collagen, elastin, and fibroin, which plays a part in the formation of creatine and other compounds, combines with cholic acid to form glycocholic acid, and with benzoic acid to form hippuric acid; abbreviated to gly; *alt.* glycocoll.

glycinin *n.* [*Glycine*, generic name of soya bean] a globulin obtained from soya bean seeds.

glycocoll glycine, *q.v.*

glycogen *n.* [Gk. *glykys*, sweet; *-genēs*, producing] a branch-chained polysaccharide, sometimes called animal starch, that is made up of glucose units and is hydrolysed via dextrins and maltose to glucose, and which acts as food storage substance in liver and muscles of vertebrates and is also found in invertebrates and some plants; *cf.* amylum.

glycogenase *n.* [Gk. *glykys*, sweet; *-genēs*, producing] a general name for 2 enzymes which catalyse synthesis of storage glycogen in liver, being glycogen (starch) synthetase, EC 2.4.1.11, and glucose-1-phosphate uridylyltransferase, EC 2.7.7.9.

glycogen body a mass of glycogen-storing cells in the sinus rhomboidalis.

glycogenesis *n.* [Gk. *glykys*, sweet; *genesis*, origin] the transformation of glucose into glycogen, as in liver and muscle.

glycogenolysis *n.* [Gk. *glykys*, sweet; *-genēs*, producing; *lysis*, loosing] the disintegration of glycogen and production of glucose, stimulated by adrenaline and glucagon, and inhibited by insulin.

glycolipid(e)s *n.plu.* [Gk. *glykys*, sweet; *lipos*, fat] complex lipids which on hydrolysis yield a carbohydrate and lipid and are found in brain tissue; *cf.* sphingolipids, glycosphingolipids.

glycollic acid an organic acid giving sourness to some unripe fruits.

glycolysis *n.* [Gk. *glykys*, sweet; *lyein*, to loosen] the anaerobic breakdown of glucose via pyruvic acid, with the production of either acetyl CoA which enters the Krebs' cycle of aerobic respiration, or lactic acid or ethyl alcohol (ethanol), and which can also include the breakdown of a more complex carbohydrate such as starch or glycogen to glucose first; *cf.* EMP pathway.

glyconeogenesis *n.* [Gk. *glykys*,, sweet; *neos*, new; *genesis*, origin] the production of glycogen from non-carbohydrate compounds.

glycophyte *n.* [Gk. *glykys*, sweet; *phyton*, plant] a plant unable to thrive on substratum containing more than 0·5% sodium chloride in solution; *cf.* halophyte.

glycoproteins *n.plu.* [Gk. *glykys*, sweet; *prōteion*, first] proteins having carbohydrates as prosthetic groups, as occurring in egg white, blood serum, and thyroid secretions and including mucins and mucoids; usually used synonymously with mucoproteins, but strictly glycoproteins contain small amounts of carbohydrate (less than 4%) and mucoproteins contain larger amounts; *alt.* glucoproteins.

glycosaminoglycans mucopolysaccharides, *q.v.*

glycosecretory *a.* [Gk. *glykys*, sweet; L. *secretus*, set apart] connected with the secretion of glycogen.

glycosides *n.plu.* [Gk. *glykys*, sweet] compounds which on hydrolysis give a sugar and non-sugar (aglutone) residue, e.g. glucosides give glucose, galactosides give galactose, widely distributed in plants and including anthocyanins and anthoxanthins.

glycosidic *a.* [Gk. *glykys*, sweet] *appl.* bonds linking monosaccharide units to form di- and polysaccharides.

glycosome *n.* [Gk. *glykys*, sweet; *sōma*, body] an organelle found in certain parasitic Protozoa, including *Trypanosoma*, and containing the enzymes concerned with glycolysis.

glycosphingolipid(e)s *n.plu.* [Gk. *glykys*, sweet; *sphingein*, to draw tight, *lipos*, fat] complex lipids which on hydrolysis yield a sugar, a fatty acid, and sphingosine, and including the cerebrosides, gangliosides, and ceramide oligosaccharides.

glycotropic, glycotrophic *a.* [Gk. *glykys*, sweet; *tropē*, turn; *trophē*, nourishment] *appl.* hormone secreted by prepituitary which inhibitis peripheral action of insulin.

glycuronic acid glucuronic acid, *q.v.*

glyoxalate reaction [Gk. *glykys*, sweet; *oxys*, sharp] the conversion of methyl glycol to lactic acid, utilizing glutathione using various enzymes including glyoxalase I, EC 4.4.1.5, and glyoxalase II, EC 3.1.2.6.

glyoxylic acid a compound important in glycine and hydroxyproline metabolism, an end-product of purine nucleotide catabolism, and used in the glyoxylic acid cycle; *alt.* glyoxylate, formylformic acid.

glyoxylic acid or glyoxylate cycle a modification of the Krebs' cycle in higher plants (especially in germinating seeds or micro-organisms, to utilize fats or 2-carbon compounds), used to oxidize glyoxylate to give ATP or to synthesize carbohydrates; *alt.* Kornberg cycle.

glyoxysomes *n.plu.* [Gk. *glukys*, sweet; *oxys*, sharp; *sōma*, body] peroxisomes, *q.v.*

GMP guanosine monophosphate, *q.v.*

gnathal, gnathic *a.* [Gk. *gnathos*, jaw] *pert.* the jaws.

Gnathiidea *n.* [Gk. *gnathos*, jaw] an order of malacostracans, sometimes considered as a suborder of isopods, having no carapace and males with very large mandibles.

gnathion *n.* [Gk. *gnathos*, jaw] lowest point of the median line of the lower jaw.

gnathism *n.* [Gk. *gnathos*, jaw] formation of jaw with reference to the degree of projection.

gnathites *n.plu.* [Gk. *gnathos*, jaw] the buccal appendages of arthropods.

gnathobase *n.* [Gk. *gnathos*, jaw; *basis*, base] an inwardly turned masticatory process on protopodite of appendages near mouth of Crustacea and trilobites; basal segment of appendages with spines directed towards mouth of arachnids.

Gnathobdellae *n.* [Gk. *gnathos*, jaw; *bdella*, leech] an order of leeches that are blood-sucking parasites on birds and mammals and possess jaws but no eversible proboscis; *alt.* Gnathobdellida.

gnathochilarium *n.* [Gk. *gnathos*, jaw; *cheilos*, lip] first maxillae and sternal plate in Pauropoda, united in Diplopoda.

gnathopod *n.* [Gk. *gnathos*, jaw; *pous*, foot] any arthropod limb in oral region modified to assist with food.

gnathopodite *n.* [Gk. *gnathos*, jaw; *pous*, foot] maxilliped, *q.v.*

gnathos *n.* [Gk. *gnathos*, jaw] a median sclerite hinged to uncus, on ventral side of 9th tergum in Lepidoptera.

gnathosoma *n.* [Gk. *gnathos*, jaw; *sōma*, body] the mouth region, including oral appendages, of some arachnids.

gnathostegites *n.plu.* [Gk. *gnathos*, jaw; *stegē*, roof] pair of covering plates for mouth-parts of some crustaceans.

Gnathostomata, gnathostomes *n., n.plu.* [Gk. *gnathos*, jaw; *stoma*, mouth] a superorder of irregular sea urchins of the subclass Euechinoidea having the mouth central in the oral surface but the anus not central; a group of vertebrates with a higher degree of cephalization than the cyclostomes and having jaws, sometimes given superclass status to compare with the Agnatha.

gnathostomatous *a.* [Gk. *gnathos*, jaw; *stoma*, mouth] with jaws at the mouth.

gnathotheca *n.* [Gk. *gnathos*, jaw; *thēkē*, case] the horny outer covering of a bird's lower jaw.

gnathothorax *n.* [Gk. *gnathos*, jaw; *thōrax*, chest] the part of the cephalothorax posterior to protocephalon, in Malacostraca.

gnesiogamy *n.* [Gk. *gnēsios*, lawful; *gamos*, marriage] fertilization by an individual of the same species; *alt.* intraspecific zygosis.

Gnetales *n.* [*Gnetum*, generic name] an order of Gnetopsida, *q.v.*, sometimes containing the genera *Gnetum*, *Welwitschia*, and *Ephedra*, but in some classifications containing only *Gnetum*, with *Welwitschia* and *Ephedra* placed in the Welwitschiales and Ephedrales respectively.

Gnetatae *n.* [*Gnetum*, generic name] in some classifications, a group of Cycadophytina equivalent to the Gnetopsida.

Gneticae *n.* [*Gnetum*, generic name] Gnetopsida, *q.v.*

Gnetopsida *n.* [*Gnetum*, generic name; Gk. *opsis*, appearance] a class of gymnosperms with the 3 living genera *Gnetum*, *Welwitschia*, and *Ephedra*, which are very different and are usually placed in one order, Gnetales, although sometimes in different orders; *alt.* Chlamydospermae, Gneticae; *cf.* Gnetatae.

gnotobiosis *n.* [Gk. *gnōtos*, known; *bios*, life] the rearing of laboratory animals in a germ-free state or containing only known micro-organisms.

gnotobiota *n.* [Gk. *gnōtos*, known; *biōtikos*, pert. life] the known micro-organisms of an animal in gnotobiosis.

gnotobiotics *n.* [Gk. *gnōtos*, known; *biōtikos*, pert. life] the study of organisms or of a species when other organisms or species are absent, or when another present organism is known; germ-free culture; study of gnotobiotes or germ-free animals.

Gobiesociformes *n.* [*Gobiesox*, generic name; L. *forma*, shape] an order of small marine teleosts in which the pelvic fins form a sucker, and including the clingfish.

goblet cells mucus-secreting cells as of columnar epithelia; *alt.* chalice cells.

golden-brown algae the common name for the Xanthophyceae, *q.v.*

Golgi apparatus or complex [*C. Golgi*, Italian histologist] a cell organelle consisting of a system of smooth-surfaced, double-membraned vesicles whose function is uncertain but may secrete giant molecules or play a role in both transport and metabolism; each element is called a **Golgi body** or dictyosome and was formerly called a batonette or pseudochromosome, while the whole Golgi apparatus was originally called apparato reticolare, canalicular system, or internal reticular apparatus, but the reticulum is probably an artefact.

Golgi–Mazzoni corpuscles [*C. Golgi* and *V. Mazzoni*, Italian histologists] cylindrical end organs or small Pacinian corpuscles at junction of tendon and muscles; *alt.* organs of Golgi.

golgiokinesis *n.* [*C. Golgi*; Gk. *kinēsis*, movement] division of the Golgi apparatus during mitosis.

Golgi, organs of Golgi–Mazzoni corpuscles, *q.v.*

golgiosomes *n.plu.* [*C. Golgi*; Gk. *sōma*, body] Golgi bodies or material produced by division of the Golgi apparatus during mitosis.

gomphosis *n.* [Gk. *gomphos*, bolt] articulation by insertion of a conical process into a socket, as of roots of teeth into their sockets.

gona *see also* gono-.

gonad *n.* [Gk. *gonē*, seed] a sexual gland, either ovary, or testes, or ovotestis; *alt.* germ gland, sex gland.

gonadal *pert.* gonads; *alt.* gonadic.

gonadectomy *n.* [Gk. *gonē*, seed; *ek*, out; *tomē*, cutting] excision of gonad, castration in the male, spaying in female.

gonadial, gonadic *pert.* gonads.

gonadin *n.* [Gk. *gonē*, seed] active principle of sex glands controlling secondary sexual characteristics.

gonadotrophic, gonadotropic *a.* [Gk. *gonē*, seed; *trophē*, nourishment; *tropē*, turn] affecting the gonad, *appl.* hormones: gonadotrophins, *q.v.*, juvenile hormone, *q.v.*

gonadotrophins *n.plu.* [Gk. *gonē*, seed; *trophē*, nourishment] two prepituitary hormones: 1. follicle-stimulating hormone, *q.v.*, 2. luteinizing hormone, *q.v.*; chorionic gonadotrophin secreted by chorionic cells of placenta and excreted in pregnancy urine, resembling but not identical with luteinizing hormone; serum gonadotrophins: follicle-stimulating hormone in blood of pregnant mare, luteinizing hormone in that of women; *alt.* gonadotropins.

gonadotropins *n.plu.* [Gk. *gonē*, seed; *tropē*, turn] gonadotrophins, *q.v.*

gonaduct gonoduct, *q.v.*

gonal *n.* [Gk. *gonē*, seed] giving rise to a gonad, *appl.* middle portion of genital ridge which alone forms functional gonad; gonidial, *q.v.*

gonangium *n.* [Gk. *gonē*, seed; *anggeion*, vessel]

any enveloping structure in which reproductive elements are produced; gonotheca, *q.v.*

gonapod gonapodium, *q.v.*

gonapophyses *n.plu.* [Gk. *gonē*, generation; *apo*, from; *phyein*, to grow] chitinous outgrowths of valves subserving copulation in insects; the component parts of a sting; any genital appendages; *sing.* gonapophysis; *alt.* rhabdites.

Gondwanaland *n.* [*Gondwanaland*, an Indian kingdom] an interconnected southern land mass of South America, Africa, India, and Australia, first postulated to account for similar fossils occurring in these continents, now thought to have existed before continental drift moved the continents to their present positions; *cf.* Laurasia.

gone *n.* [Gk. *gonē*, generation] one of the 4 haploid nuclei or cells formed after meiosis; one of the 4 daughter cells of an auxocyte; the generative portion of a gonad; an organism possessing a gone. *v.* To produce a gone.

gongylidia *n.plu.* [Gk. *gongylos*, round; *idion*, *dim.*] hyphal swellings or modifications in fungi cultivated by certain ants.

gongylus *n.* [Gk. *gongylos*, round] a globular reproductive body, as of certain algae and lichens.

gonia *n.plu* [Gk. *gonē*, seed] primitive sex cells, spermatogonia or oogonia.

gonial(e) *n.* [Gk. *gōnia*, angle] in some vertebrates a bone of lower jaw beside articular; *alt.* autarticular(e), dermarticular(e).

gonic *a.* [Gk. *gonē*, generation] *pert.* gones; *pert.* semen.

gonid gonidium, *q.v.*

gonidangium *n.* [Gk. *dim.* of *gonē*, seed; *anggeion*, vessel] a structure producing or containing gonidia.

gonidia *n.plu.* [Gk. *dim.* of *gonē*, seed] minute reproductive bodies of many bacteria; asexual non-motile reproductive cells such as gemmae produced on gametophytes; algal constituent of lichens; obsolete term for conidia; spore in some Cyanophyceae; *sing.* gonidium; *alt.* brood cells, chromidia.

gonidial *pert.* gonidia; *alt.* gonal, gonimic.

gonidiferous *a.* [Gk. *dim.* of *gonē*, seed; L. *ferre*, to carry] bearing or producing gonidia; *alt.* gonidiogenous.

gonidimium *n.* [L.L. *dim.* of Gk. *gonē*, seed] a gonidial structure smaller than a gonidium and larger than a gonimium.

gonidiogenous *a.* [Gk. *dim.* of *gonē*, seed; *-genēs*, producing] gonidiferous, *q.v.*

gonidioid *a.* [Gk. *dim.* of *gonē*, seed; *eidos*, form] like a gonidium, *appl.* certain algae.

gonidiophore *n.* [Gk. *dim.* of *gonē*, seed; *pherein*, to bear] an aerial cypha supporting a gonidangium.

gonidiophyll *n.* [Gk. *dim.* of *gonē*, seed; *phyllon*, leaf] a gametophyte leaf bearing gonidia.

gonidiospore an obsolete term for conidiospore.

gonidium *sing.* of gonidia; *alt.* gonid.

gonimic gonidial, *q.v.*

gonimium *n.* [Gk. *gonimos*, productive] one of the bluish-green gonidia of a blue–green algal cell of certain lichens.

gonimoblasts *n.plu.* [Gk. *gonimos*, productive; *blastos*, bud] filamentous outgrowths of a fertilized carpogonium of red algae.

gonimocarp *n.* [Gk. *gonimos*, productive; *karpos*, fruit] cystocarp, *q.v.*

gonimolobe *n.* [Gk. *gonimos*, productive; *lobos*, lobe] a group of carposporangia borne on a gonimoblast.

goniocyst *n.* [Gk. *gonē*, seed; *kystis*, bladder] in lichens, a cluster of gonidia; a sporangium in some plants.

gonion *n.* [Gk. *gōnia*, angle] the angle point on the lower jaw.

Goniotrichales *n.* [Gk. *gōnia*, angle; *thrix*, hair] an order of red algae that are tiny simple or branched filaments.

gono- see also gona-.

gonoblast *n.* [Gk. *gonos*, offspring; *blastos*, bud] a reproductive cell, especially in animals.

gonoblastid(ium) *n.* [Gk. *gonos*, offspring; *blastos*, bud; *idion*, *dim.*] blastostyle, *q.v.*

gonocalyx *n.* [Gk. *gonos*, offspring; *kalyx*, cup] the bell of a medusiform gonophore.

gonocheme *n.* [Gk. *gonos*, offspring; *ochēma*, support] a medusoid bearing gametes in Hydrozoa.

gonochorism *n.* [Gk. *gonos*, offspring; *chōrismos*, separation] the history or development of sex differentiation; sex determination; dioecism, *q.v.*

gonochoristic *a.* [Gk. *gonos*, offspring; *chōristos*, separated] dioecious, *q.v.*; unisexual, *q.v.*

gonocoel *n.* [Gk. *gonē*, seed; *koilos*, hollow] the cavity containing the gonads; *alt.* perigonadial cavity.

gonocoxa *n.* [Gk. *gonē*, seed; L. *coxa*, hip] base or coxite of a gonopod in insects.

gonocytes *n.plu.* [Gk. *gonē*, seed; *kytos*, hollow] in sponges, the mother cells of ova and spermatozoa or the gametes themselves.

gonodendron *n.* [Gk. *gonos*, offspring; *dendron*, tree] a branching blastostyle in some siphonophores such as *Physalia*.

gonoduct *n.* [Gk. *gonos*, birth; L. *ductus*, led] a duct leading from gonad to exterior; *alt.* genital duct, gonaduct.

gonoecium *n.* [Gk. *gonos*, begetting; *oikia*, house] a reproductive individual of a polyzoan colony.

gonogenesis *n.* [Gk. *gonē*, seed; *genesis*, descent] gametogenesis, *q.v.*

gonomere *n.* [Gk. *gonē*, seed; *meros*, part] a pronucleus persisting during early cleavage stages.

gonomery *n.* [Gk. *gonos*, descent; *meros*, part] theory that paternal and maternal chromosomes remain in separate groups throughout life; separate grouping of paternal and maternal chromosomes during cleavage stages of some organisms.

gononephrotome *n.* [Gk. *gonē*, seed; *nephros*, kidney; *temnein*, to cut] embryonic segment containing primordia of the urinogenital system.

gononucleus *n.* [Gk. *gonos*, begetting; L. *nucleus*, kernel] micronucleus, *q.v.*

gonophore *n.* [Gk. *gonos*, offspring; *pherein*, to bear] an elongation of thalamus between corolla and stamens, *cf.* gynophore; a reproductive zooid in a hydrozoan colony.

gonoplasm *n.* [Gk. *gonē*, seed; *plasma*, mould] the generative part of protoplasm; germ plasm, *q.v.*

gonopod(ium) *n.* [Gk. *gonos*, begetting; *pous*, foot] the modified anal fin serving as copulatory organ in male poeciliid fishes; clasper of male myriapods and insects; *alt.* gonapod.

gonorpore *n.* [Gk. *gonē*, seed; *poros*, channel] reproductive aperture.

Gonorhynchiformes *n.* [*Gonorhynchus*, generic name; L. *forma*, shape] an order of toothless teléosts including the milkfish and some deep-sea forms.

gonosome *n.* [Gk. *gonos*, begetting; *sōma*, body] the reproductive zooids of a coelenterate colony collectively.

gonosphaerium *n.* [Gk. *gonē*, seed; *sphaira*, globe] oosphere, *q.v.*

gonosphere *n.* [Gk. *gonos*, offspring; *sphaira*, globe] oosphere, *q.v.*; zoospore of Chytridiales.

gonospore *n.* [Gk. *gonos*, offspring; *sporos*, seed] a spore produced by meiosis.

gonostyle *n.* [Gk. *gonos*, begetting; *stylos*, pillar] the blastostyle, *q.v.*; sexual palpon or siphon of Siphonophora; gonostylus, *q.v.*

gonostylus *n.* [Gk. *gonos*, begetting; *stylos*, pillar] bristle-like process on gonocoxa of insects; clasper of Diptera; part of the sting in some Hymenoptera; *alt.* gonostyle.

gonotheca *n.* [Gk. *gonos*, birth; *thēkē*, cup] a transparent protective expansion of the perisarc round a blastostyle or gonophore; *alt.* gonangium, perigonium, teleophore.

gonotokont auxocyte, *q.v.*

gonotome *n.* [Gk. *gonos*, birth; *temnein*, to cut] an embryonic segment containing the primordium of the gonad.

gonotrema, gonotreme *n.* [Gk. *gonos*, offspring; *trēma*, hole] genital aperture, as in Arachnida.

gonotype *n.* [Gk. *gonos*, offspring; *typos*, pattern] immediate offspring of a type specimen.

gonozooid *n.* [Gk. *gonos*, birth; *zōon*, animal; *eidos*, form] a gonophore or reproductive individual of a hydrozoan colony; a zooid containing a gonad.

gonys *n.* [Gk. *genys*, lower jaw] lower part or keel of bird's bill; *a.* gonydial.

Gordioidea *n.* [*Gordius*, generic name; Gk. *eidos*, form] an order of Nematomorpha having the pseudocoel filled with mesenchyme tissue.

Gorgonacea *n.* [*Gorgonia*, generic name] an order of Octocorallia, commonly called horny corals, having an axial skeleton of calcareous spicules and/or horny material with short polyps borne on the sides.

gorgonin *n.* [Gk. *Gorgō*, from *gorgos*, terrible] a scleroprotein in axial skeleton of horny corals.

gossypine *a.* [*Gossypium*, generic name of cotton] like cotton.

Götte's larva larva with 4 ciliated lobes, of Polycladida.

G0 phase the phase of the cell cycle in which the cycle is arrested and mitosis ceases to occur.

G1 phase the phase of the cell cycle following mitosis when the diploid amount of DNA is present.

G2 phase the phase of the cell cycle following the duplication of DNA, before mitosis occurs.

Graafian follicle [*R. de Graaf*, Dutch anatomist] in the mammalian ovary, spherical vesicle containing a developing ovum and a liquid, the liquor folliculi, surrounded by numerous follicle cells; *alt.* vesicular ovarian follicle.

Graber's organ a complex larval organ in horse-flies (Tabanidae) of unknown function, but possibly sensory.

gracilis *n.* [L. *gracilis*, slender] a superficial muscle on medial side of the thigh; a fasciculus of

medulla oblongata; nucleus of grey matter ventral to clava.

grade *n.* [L. *gradus*, step] a taxonomic group between kingdom and phylum, representing a grade of organization in which a group of organisms share a number of characteristics in common but may not owe them to a common ancestor.

graduated *a.* [L. *gradus*, step] tapering; becoming longer or shorter by steps.

graduate sorus in ferns, a sorus in which sporangia develop from the tip towards the base.

graft *n.* [O.F. *graffe*, graft] a part of an organism inserted into, and uniting with a larger part of another or the same organism. *v.* To insert scion into stock, or animal tissue from donor into recipient or host. *Alt.* implant, transplant.

graft hybrid an individual formed from graft and stock, and showing characteristics of both; *alt.* graft chimaera.

grain *n.* [L. *granum*, grain] caryopsis of cereals; a granular prominence on the back of a sepal; the pattern on the surface of wood due to the nature and arrangement of the cells.

graminaceous *a.* [L. *gramen*, grass] *pert.* grasses; grass-coloured, *appl.* insects; *alt.* gramineous.

Gramineae *n.* [L. *gramen*, grass] a family of monocotyledons, consisting of the grasses.

gramineous graminaceous, *q.v.*

graminicolous *a.* [L. *gramen*, grass; *colere*, to dwell] living on grasses.

graminifolious *a.* [L. *gramen*, grass; *folium*, leaf] with grass-like leaves.

graminivorous *a.* [L. *gramen*, grass; *vorare*, to eat] grass-eating.

graminology agrostology, *q.v.*

grammate *a.* [Gk. *grammē*, line] striped; marked with lines or slender ridges.

Gram stain [*H. C. J. Gram*, Danish physician] a crystal violet stain used in bacteria which correlates well with morphological and physiological features, bacteria which can be destained with an organic solvent such as acetone being called Gram-negative, and those retaining it being Gram-positive, and thought to be related to lipid content of bacterial cell wall.

grana *n.plu.* [L. *granum*, grain] in chloroplasts, groups of disc-shaped structures (thylakoids) the membranes of which bear chlorophyll and/or other photosynthetic pigments, and which are stacked like piles of coins; *sing.* granum.

grandifoliate, grandifolious *a.* [L. *grandis*, great; *folium*, leaf] large-leaved, particularly when the leaves are the dominant organ, as in water-lilies.

grand period of growth in plants, the total period of enlargement of a cell, organ, or the whole plant, growth being slow at first, then increasing to a maximum, then slowing to zero at maturity.

Grandry's corpuscle [— *Grandry*, Belgian anatomist] an end organ of touch, being a specialized type of Merkel's corpuscle, found in skin of beak and tongue in anseriform birds.

granellae *n.plu.* [L. *dim.* of *granum*, grain] oval, refractile granules consisting chiefly of barium sulphate, found in the tubes of certain Sarcodina.

granellarium *n.* [L. *dim.* of *granum*, grain] the system of granellae-containing tubes of Sarcodina.

granose *a.* [L. *granum*, grain] in appearance like a

granular, granulate, granulose

chain of grains, like some insect antennae; *alt.* moniliform.

granular, granulate, granulose *a.* [L. *granum*, grain] consisting of grains or granules; appearing as if made up of, or covered with, granules; granular: *appl.* soil crumbs which are rounded and rather small.

granulation *n.* [L. *granum*, grain] a grain-like formation or eminence; *appl.* arachnoid elevations or Pacchionian glands on outer surface of dura mater.

granule *n.* [L. *granulum*, small grain] a small particle of matter; a small grain.

granule cells ovoid or spheroid cells formed of soft protoplasm containing basiphil granules.

granule glands the prostate glands of flatworms; skin glands of amphibians.

granules of Nissl Nissl granules, *q.v.*

granulocytes *n.plu.* [L. *granulum*, small grain; Gk. *kytos*, cell] granular white blood corpuscles, *alt.* polymorphs; myeloid cells formed in bone marrow.

granulopoiesis *n.* [L. *granulum*, small grain; Gk. *poiesis*, making] the formation of granulocytes.

granum *sing.* of grana.

grape sugar glucose, *q.v.*

Graphidales *n.* [*Graphis*, generic name] an order of crustose ascolichens in which the apothecia elongate, resembling runic inscriptions.

graphiohexaster *n.* [Gk. *graphis*, style; *hex*, six; *astēr*, star] a hexaster spicule with long outwardly-directed filamentous processes from 4 rays.

Graptolita, graptolites *n., n.plu.* [Gk. *graptos*, painted; *lithos*, stone] a group of fossil Palaeozoic animals known mainly from their external skeletons, whose taxonomic position has been in dispute and which have been classed as Coelenterata and Ectoprocta, but are now usually regarded as a class of hemichordates.

Graptoloidea *n.* [Gk. *graptos*, painted; *lithos*, stone; *eidos*, form] an order of graptolites that may be arranged in from 1 to 4 rows around stipes hanging from a common disc and connected to each other by a thread.

grasshoppers the common name for many members of the Orthoptera, *q.v.*

grasslands regions in which the climax vegetation is grass due to climate and/or the activity of herbivores, and including savannas, prairies, pampas, steppes, and veld.

grater *n.* [O.F. *grater*, to scrape] a denticle of certain polychaetes such as *Eunice*.

graveolent *a.* [L. *graveolens*, strong smelling] having a strong or offensive odour.

gravid *a.* [L. *gravidus*, loaded] *appl.* female with eggs, or pregnant uterus.

graviperception *n.* [L. *gravis*, heavy; *percipere*, to feel] perception of gravity.

graviportal *a.* [L. *gravis*, heavy; *portare*, to carry] with the legs adapted to supporting a great weight, as in elephants.

gravitational *a.* [L. *gravis*, heavy] *appl.* water in excess of soil requirements, which sinks under action of gravity and drains away.

gray grey, *q.v.*

greater omentum a fold of peritoneum attached to the colon and stomach and hanging over the small intestine; *alt.* epiploon, caul, gastrocolic omentum.

green algae the common name for the Chlorophyceae, *q.v.*

green cells chlorophyllose cells, *q.v.*; cells of the green alga *Chlorella* when living symbiotically inside other animals.

green glands the excretory antennary glands of certain Crustacea.

gregaloid *a.* [L. *grex*, flock; Gk. *eidos*, form] *appl.* colony of Protozoa of indefinite shape, usually with gelatinous base, formed by incomplete division of individuals or partial union of adults.

gregaria phase in locusts, the phase during which the locusts are gregarious and highly active; *cf.* solitaria phase.

Gregarinid(e)a, gregarines *n., n.plu.* [L. *grex*, flock] an order of Telosporidia which are parasites of invertebrates with only the adults extracellular, and with merogamous male and female gametes.

gregariniform *a.* [L. *grex*, flock; *forma*, shape] *pert.* gregarine protozoans; *appl.* spores moving with the gliding motion characteristic of gregarines, *alt.* gregarinulae.

gregarinulae *n.plu.* [*Dim.* of L. *grex*, flock] gregariniform spores, *q.v.*

gregarious *a.* [L. *grex*, flock] tending to herd together; colonial; growing in clusters.

gressorial *a.* [L. *gressus*, a stepping] adapted for walking, *appl.* certain insects and birds.

grey *alt.* gray.

grey crescent crescentic marginal zone of cytoplasm of a fertilized egg before cleavage, inductor of gastrula, as in amphibians.

grey matter tissue abundantly supplied with nerve cells, of greyish colour, internal to white matter in spinal cord, external in cerebrum.

grey nerve fibres semitransparent, grey or yellowish-grey, gelatinous non-medullated nerve fibres, comprising most of the fibres of the sympathetic system and some of the cerebrospinal; *alt.* non-medullated fibres.

Grimmiales *n.* [*Grimmia*, generic name] an order of mosses having blackish-green upright gametophores.

grinding teeth molars, *q.v.*

griseofulvin an antibiotic produced by some species of *Penicillium*, especially *P. griseofulvum*, which is toxic to some fungi by inhibiting metaphase mitosis.

grit cell a stone cell, particularly in fruit.

grooming *n.* [M.E. *grom*, groom] the cleaning of fur or feathers by an animal, performed either to itself (self grooming) or to another member of the same species (social grooming) functioning not only to keep the animal clean, but as a displacement activity (self grooming) or to improve the cohesiveness of the group (social grooming).

groove *n.* [A.S. *grafan*, to grave] any channel, furrow, or depression, as carotid, costal, optic, primitive vertebral groove.

gross efficiency a measure of the efficiency of an animal in converting its consumed food to protoplasm; *cf.* net efficiency.

gross primary production the amount of new protoplasm produced in an organism per unit time.

ground meristem a primary meristem derived from the apical meristem and giving rise to the ground tissue.

ground tissue in plants, tissues other than the epidermis or periderm and vascular tissues; *alt.* conjunctive parenchyma, fundamental tissue.

growing point a part of plant body at which cell division is localized, generally terminal and composed of meristematic cells; a hyphal tip from which spores are abstricted basipetally.

growing zone the part of a plant organ which is undergoing elongation.

growth increase in size by cell division and/or cell enlargement.

growth curvature the curved shape imparted to a plant organ by the difference in the rates of growth of its sides.

growth factor G riboflavin, *q.v.*

growth hormone (GH) in animals, any of various growth-promoting hormones, especially one of the anterior lobe of the pituitary gland, somatotrophin, *q.v.*; in plants, any of the growth-promoting phytohormones, *q.v.*, *alt.* growth substance.

growth ring annual ring, *q.v.*; layer of shell laid down in various animals such as bivalve molluscs; a layer of a scale laid down in fish.

growth substance growth hormone, especially in plants.

grub *n.* [M.E. *grobbe*, grub] a legless larval insect of certain Diptera, Coleoptera, and Hymenoptera.

Gruiformes *n.* [*Grus*, generic name of crane; L. *forma*, shape] an order of birds of the subclass Neornithes, including the rails and cranes.

Gruinales *n.* [L. *gruinalis*, shaped like a crane's bill] Geraniales, *q.v.*

grumose, grumous *a.* [L. *grumus*, hillock] clotted; knotted; collected into granule masses.

grumulus *n.* [*Dim.* of L. *grumus*, hillock] polar organ, *q.v.*

Grylloblattodea n. [*Grylloblatta*, generic name; Gk. *eidos*, form] an order of insects secondarily without wings, with incomplete metamorphosis, small or absent eyes, and fairly long, filiform antennae; *alt.* Notoptera.

gryochrome *a.* [Gk. *gry*, morsel; *chrōma*, colour] with Nissl granules irregularly scattered, *appl.* neurones, as in spinal ganglia.

GTP guanosine triphosphate, *q.v.*

guanase *n.* [Peruvian *huanu*, dung] the hydrolysing enzyme which deaminates guanine with the formation of xanthine and ammonia; EC 3.5.4.3; *r.n.* guanine deaminase.

guanidine *n.* [Peruvian *huanu*, dung] a base produced by oxidation of guanine, whose metabolism is regulated by parathyroids.

guanine *n.* [Peruvian *huanu*, dung] a purine base present in all living tissues, which is part of the genetic code of DNA and RNA and pairs with cytosine, forms part of the molecule of guanosine, and also occurs in chromatophores, *see* guanophore.

guanophore *n.* [Peruvian *huana*,, dung; Gk. *pherein*, to bear] a chromatophore containing guanine, either as pale granules usually giving a yellow colour, or as iridescent crystals as in the skin of reptiles and fishes where it is known as an iridocyte; *cf.* leucophore, ochrophore, xanthophore.

guanosine a nucleoside with guanine as its base.

guanosine diphosphate (GDP) a nucleotide cofactor analogous to ADP.

guanosine monophosphate (GMP) guanylic acid, *q.v.*, a nucleotide cofactor analogous to AMP.

guanosine triphosphate (GTP) a nucleotide cofactor analogous to ATP.

guanylic acid a nucleotide formed from guanosine and phosphoric acid, originally found in the pancrease, liver, and certain fungi; *alt.* guanosine monophosphate.

guard *n.* [O.F. *guarder*, to guard] sheath of a phragmocone; rostrum of a belemnite.

guard cells two cells surrounding stomata of aerial epidermis of plant tissue whose change in turgidity open and close the pore.

guard polyp nematocalyx, *q.v.*

gubernaculum *n.* [L. *gubernaculum*, rudder] a cord stretching from epididymis to scrotal wall and supporting testis, *alt.* mesocardial ligament; tissue between gum and dental sac of permanent teeth; strands of blastostylar ectoderm between gonophore and gonotheca in Hydrozoa; a posterior flagellum functioning as a rudder in Flagellata; in male nematodes, a sclerotized portion of the dorsal wall of the spicule pouch which helps to direct the spicules towards the anus; *a.* gubernacular.

Guérin's glands [*A. F. M. Guérin*, French surgeon] Skene's glands, *q.v.*

guest an animal living or breeding in the nest of another, especially an insect.

gula *n.* [L. *gula*, gullet] the upper part of throat; median ventral sclerite of insect head; *a.* gular.

gulamentum *n.* [L. *gula*, gullet; *mentum*, chin] plate formed by fusion of gula and submentum in insects.

gular *a.* [L. *gula*, gullet] *pert.* throat or gula; *appl.* median and lateral plates between rami of mandible in Crossopterygii and Polypterini; *appl.* pouch of skin below beak in Pelicaniformes. *n.* An anterior unpaired horny shield on plastron of Chelonia.

gullet *n.* [O.F. *goulet*, from L. *gula*, gullet] oesophagus, *q.v.*; cytopharynx, *q.v.*, in Protozoa; any cavity by which food many be taken into the body.

gullying the formation of deep gullies by erosion by rain water.

gum *see* gums.

gum benzoin a balsam obtained from plants of the genus *Styrax*.

gummiferous *a.* [L. *gummi*, gum; *ferre*, to carry] gum-producing or exuding.

gummosis *n.* [L. *gummi*, gum] condition of plant tissue when cell walls become gummy, as caused by certain bacteria.

gums *n.plu.* [L. *gummi*, gum] various celloidal materials resulting from breakdown of plant cells and exuding from wounds, mainly heterosaccharides which on hydrolysis yield pentoses, hexoses, and uronic acid; *cf.* mucilages. [A.S. *goma*, jaws] dense fibrous tissues investing jaws; *alt.* gingivae, ulae.

Gurwitsch rays [*A. G. Gurvich*, Russian biologist] mitogenetic rays, *q.v.*

gustatory *a.* [L. *gustare*, to taste] *pert.* sense of taste, *appl.* cells, hairs, pores, calyculus, nerves, stimuli, etc.

gut *n*. [A.S. *gut*, channel] alimentary canal, or part of it.

gutta *n*. [L. *gutta*, drop] a small spot of colour on surface of an animal body, such as an insect wing; an oil drop in a fungal hypha or spore; a small vacuole; *plu*. guttae; *a*. guttulate. [Mal. *gatah*, gum] latex of various trees in Malaya, including gutta percha and balata.

guttate *a*. [L. *gutta*, drop] having drop-like markings; containing small drops of matter.

guttation *n*. [L. *gutta*, drop] formation of drops of water on plants from moisture in air; exudation of aqueous solutions, as through hydathodes, or by sporangiophores, or by nectaries.

gutter pointed *appl*. leaf with a grooved tip forming a spout.

Guttiferae *n*. [L. *gutta*, drop; *ferre*, to bear] Theales, *q.v.*

Guttiferales *n*. [L. *gutta*, drop; *ferre*, to bear] an order of dicots (Polypetalae) having regular flowers.

guttiferous *a*. [L. *gutta*, drop; *ferre*, to bear] having or yielding drops; exuding a resin or gum.

guttiform *a*. [L. *gutta*, drop; *forma*, shape] droplike; in the form of a drop; *alt*. stilliform.

guttula *n*. [L. *guttula*, small drop] droplet; a small drop-like spot.

guttulate *a*. [L. *guttula*, small drop] *pert*. or containing guttae; in the form of a small drop, as markings; containing oily droplets, *appl*. spores.

guttulose covered with, or containing, droplets.

gymnanthous *a*. [Gk. *gymnos*, uncovered; *anthos*, flower] achlamydeous, *q.v.*

gymnetrous *a*. [Gk. *gymnos*, naked; *ētron*, abdomen] without an anal fin.

gymnoarian *a*. [Gk. *gymnos*, naked; *ōarion*, small egg] *appl*. gonads when naked, or not enclosed in coelomic sacs; *cf*. cystoarian.

Gymnoblastea *n*. [Gk. *gymnos*, naked; *blastos*, bud] Athecata, *q.v.*

gymnoblastic *a*. [Gk. *gymnos*, naked; *blastos*, bud] without hydrothecae and gonothecae as in certain coelenterates; *cf*. calyptoblastic.

Gymnocarpeae *n*. [Gk. *gymnos*, uncovered; *karpos*, fruit] a group of ascolichens having apothecia; *alt*. Discolichenes.

gymnocarpic, gymnocarpous *a*. [Gk. *gymnos*, uncovered; *karpos*, fruit] having the fruit naked, not covered by some kind of masking structure; having apothecia uncovered, as in certain lichens; having hymenium exposed during maturation of spores; *n*. gymnocarpy; *cf*. angiocarpic.

gymnocidium *n*. [Gk. *gymnos*, uncovered; *oikos*, house; *idion*, *dim*.] a basal swelling of certain moss capsules.

gymnocyte *n*. [Gk. *gymnos*, uncovered; *kytos*, hollow] a cell without a defining cell wall; *cf*. lepocyte.

gymnocytode *n*. [Gk. *gymnos*, naked; *kytos*, hollow; *eidos*, form] cytode without cell wall.

Gymnodiniales *n*. [*Gymnodinium*, generic name] an order of Dinophyceae having naked cells.

gymnogenous *a*. [Gk. *gymnos*, naked; *genos*, offspring] naked when born, *appl*. birds; gymnospermous, *q.v.*

gymnogynous *a*. [Gk. *gymnos*, naked; *gynē*, female] with exposed ovary.

Gymnolaemata *n*. [Gk. *gymnos*, naked; *laimos*, throat] a class of Ectoprocta having no epistome and with a circular lophophore.

Gymnophiona *n*. [Gk. *gymnos*, naked; *ophioneos*, snake-like] Apoda (amphibians) *q.v.*

gymnoplasm, gymnoplast *n*. [Gk. *gymnos*, naked; *plastos*, formed] protoplasm without definite formation or cell wall.

gymnopterous *a*. [Gk. *gymnos*, naked; *pteron*, wing] having bare wings; without scales, *appl*. insects.

gymnorhinal *a*. [Gk. *gymnos*, naked; *rhines*, nostrils] with nostril region not covered by feathers, as in some birds.

Gymnosomata *n*. [Gk. *gymnos*, naked; *sōma*, body] an order of fast-swimming planktonic pteropod opisthobranchs with no shell.

gymnosomatous *a*. [Gk. *gymnos*, naked; *sōma*, body] having no shell or mantle, as certain molluscs.

Gymnospermae, gymnosperms *n*., *n.plu*. [Gk. *gymnos*, uncovered, naked; *sperma*, seed] an important division of the plant kingdom, in some classifications considered a subdivision of the Spermatophyta, in others a subdivision of the Pteropsida, being woody plants with alternation of generations, having the gametophyte retained on the sporophyte and seeds produced on the surface of the sporophylls and not enclosed in an ovary; *alt*. Pinophyta.

gymnospermous *a*. [Gk. *gymnos*, uncovered, naked; *sperma*, seed] *pert*. gymnosperm, having seeds not enclosed in an ovary; *alt*. gymnogenous.

gymnospore *n*. [Gk. *gymnos*, naked; *sporos*, seed] a naked spore not enclosed in a protective envelope.

Gymnostomatida, Gymnostomata *n*. [Gk. *gymnos*, naked; *stoma*, mouth] an order or suborder of Holotricha being mainly fresh water and very generalized ciliates.

gymnostomatous, gymnostomous *a*. [Gk. *gymnos*, naked; *stoma*, mouth] naked-mouthed; having no peristome, *appl*. mosses.

gynaecaner *n*. [Gk. *gynē*, woman; *anēr*, man] a male ant resembling a female; *alt*. gynaecomorphic male.

gynaeceum gynaecium, *q.v.*

gynaecium *n*. [Gk. *gynaikēïe*, woman's part of a house] the female organs of the flower, consisting of one or more carpels forming one or several ovaries with their stigmas and styles, *alt*. pistil; the group of archegonia of mosses; *alt*. gynaeceum, gynecium, gynoecium.

gynaecogen *n*. [Gk. *gynaikos*, of women; *gennaein*, to produce] any female sex hormone; a feminizing agent.

gynaecoid *n*. [Gk. *gynē*, woman; *eidos*, form] an egg-laying worker ant.

gynaecomorphic male gynaecaner, *q.v.*

gynaecophore *n*. [Gk. *gynē*, woman; *pherein*, to carry] canal or groove of certain trematodes, formed by inrolling of sides, in which the female is carried; *alt*. gynaecophoric or gynaecophoral groove.

gynander *n*. [Gk. *gynē*, female; *anēr*, man] gynandromorph, *q.v.*

Gynandrae *n*. [Gk. *gynē*, women; *anēr*, man] Orchidales, *q.v.*

gynandrism *n*. [Gk. *gynē*, woman; *anēr*, man] hermaphroditism, *q.v.*

gynandromorph *n*. [Gk. *gynē*, woman; *anēr*,

man; *morphē,* form] an individual exhibiting a spatial mosaic of male and female characters; *alt.* gynander, sex mosaic; *cf.* intersex.

gynandromorphism [Gk. *gynē,* woman; *anēr,* man; *morphē,* form] condition of being a gynandromorph or manifesting a mosaic of male and female sexual characters, as having one side characteristically male, the other female.

gynandrophore *n.* [Gk. *gynē,* woman; *anēr,* man; *pherein,* to carry] an axial prolongation bearing a sporophyll; a gonophore bearing both stamens and gynaecium.

gynandrosporous *a.* [Gk. *gynē,* woman; *anēr,* man; *sporos,* seed] with androspores adjoining the oogonium, as in some algae.

gynandrous *a.* [Gk. *gynē,* woman; *anēr,* man] having stamens fused with pistils, as in some orchids.

gynantherous *a.* [Gk. *gynē,* woman; *anthēros,* flowering] having stamens converted into pistils.

gynase *n.* [Gk. *gynē,* woman] a female-determining factor in the form of an enzyme or hormone.

gynatrium *n.* [Gk. *gynē,* woman; L. *atrium,* entrance hall] female genital pouch or vestibulum, of certain insects.

gyne *n.* [Gk. *gynē,* woman] a female ant, especially a queen ant.

gynecium gynaecium, *q.v.*

gynetype *n.* [Gk. *gynē,* woman; *typos,* pattern] type specimen of the female of a species.

gynic *a.* [Gk. *gynē,* woman] female; *cf.* andric.

gynobase *n.* [Gk. *gynē,* woman; L. *basis,* base] a gynaecium-bearing receptacle in certain plants; condition in which style appears to arise from the base of the ovary; *alt.* gynobasis.

gynobasic style a style arising from the base of the ovary.

gynobasis gynobase, *q.v.*

gynodioecious *a.* [Gk. *gynē,* woman; *dis,* twice; *oikos,* house] having female and hermaphrodite flowers on different plants; *cf.* gynomonoecious.

gynoecium *n.* [Gk. *gynē,* woman; *oikos,* house] gynaecium, *q.v.*

gynogamone *n.* [Gk. *gynē,* woman; *gamos,* marriage] a gamone of ova, fertilizin, *q.v.*

gynogenesis *n.* [Gk. *gynē,* woman; *genesis,* descent] development from eggs penetrated by the spermatozoon but without fertilization.

gynogonidia *n.plu.* [Gk. *gynē,* woman; *gonidion,* small seed] female sexual elements formed after repeated division of parthenogonidia in Flagellata.

gynomerogony *n.* [Gk. *gynē,* female; *meros,* part; *gonē,* generation] the development of an egg fragment, obtained before fusion with male nucleus, and containing maternal chromosomes only.

gynomonoecious *a.* [Gk. *gynē,* woman; *monos,* alone; *oikos,* house] having female and hermaphrodite flowers on the same plant; *cf.* gynodioecious.

gynophore *n.* [Gk. *gynē,* woman; *pherein,* to carry] a stalk supporting the ovary; elongation of thalamus between stamens and pistil; *alt.* female gonophore.

gynosporangium *n.* [Gk. *gynē,* woman; *sporos,* seed; *anggeion,* vessel] female sporangium; megasporangium, *q.v.*

gynospore *n.* [Gk. *gynē,* female; *sporos,* seed] female spore; megaspore, *q.v.;* embryo sac, *q.v.*

gynostegium *n.* [Gk. *gynē,* woman; *stegē,* roof] a protective covering for a gynaecium,

especially as formed by union of stamens and style.

gynostemium *n.* [Gk. *gynē,* woman; *stēmōn,* warp] the column composed of united pistil and stamens in orchids.

gypsophil(ous) *a.* [Gk. *gypsos,* chalk, gypsum; *philein,* to love] thriving in soils containing chalk or gypsum; *alt.* calcicolous, calciphil.

gypsophyte *n.* [Gk. *gypsos,* chalk, gypsum; *phyton,* plant] a gypsophil plant; *alt.* calcicole, calcipete, calciphile, calciphyte.

gyral, gyrate *a.* [L. *gyrus,* circle] *pert.* a gyrus; *pert.* circular or spiral movement.

gyration *n.* [L. *gyrare,* to revolve] rotation, as of cells; a whorl of a spiral shell.

gyre *n.* [Gk. *gyros,* L. *gyrus,* circle] circular movement; spiral coiling, as of chromatids.

gyrencephalic, gyrencephalous *a.* [Gk. *gyros,* circle; *engkephalos,* brain] having cerebral convolutions; *cf.* lissencephalic.

gyri *plu.* of gyrus.

gyrochrome *a.* [Gk. *gyros,* circle; *chrōma,* colour] with Nissl granules arranged in a circle, *appl.* certain neurones.

Gyrocotyloidea *n.* [*Gyrocotyle,* generic name; Gk. *eidos,* form] an order of Cestodaria having a terminal sucker at the anterior end of the body.

gyrodactyloid *n.* [Gk. *gyros,* circle; *daktylos,* finger; *eidos,* form] ciliated larva of certain ectoparasitic flukes which swims to find new host.

Gyrodactyloidea *n.* [*Gyrodactylus,* generic name; Gk. *eidos,* form] an order of Monopisthocotylea that are viviparous and have one testis anterior to ovary.

gyroma *n.* [Gk. *gyros,* circle] a discoid or knob-like apothecium of certain lichens; annulus, *q.v.,* of ferns.

gyrose *a.* [L. *gyrare,* to revolve] with undulating lines; sinuous; curving.

gyrus *n.* [L. *gyrus,* circle] a cerebral convolution; a ridge winding between 2 grooves; *plu.* gyri.

H

HA hyaluronmic acid, *q.v.*

habenula *n.* [L. *habena,* strap] a name *appl.* certain band-like structures. *a.* Habenular, *appl.* a commissure of epithalamus, *appl.* nucleus or ganglion.

habilines *n.plu.* [L. *habilis,* handy] fossil hominids of the species *Homo habilis,* representing the early human phase of hominid evolution.

habit the external appearance or way of growth of a plant, e.g. climbing, erect, bushy, etc.; the normal or regular behaviour of an animal.

habitat *n.* [L. *habitare,* to inhabit] the locality or external environment in which a plant or animal lives; *alt.* ece, oike, oikos.

habitat form the way of growth of a plant which results from conditions in a particular habitat.

habitat space the habitable part of space or area available for establishing a population.

habitat type ecotype, *q.v.*

habit formation trial-and-error learning, *q.v.*

habituation *n.* [L. *habituare,* to bring into a habit] the adjustment, effected in a cell or in an organism, by which subsequent contacts of the same stimulus produce diminishing effects; a form of learning in which reflex behaviour is extinguished when the animal finds that it has no adaptive value, *cf.* association.

habitus *n*. [L. *habitus*, appearance] the general appearance or conformation characteristic of a plant or an animal; *alt.* constitutional type.

hadal *a*. [Gk. *hadēs*, unseen] *appl.* or *pert.* abyssal deeps below 6000 m; *alt.* ultra-abyssal.

hadrocentric *a*. [Gk. *hadros*, thick; *kentron*, centre] with phleom surrounding xylem.

hadromase *n*. [Gk. *hadros*, thick] a vague former term for various enzymes found in certain fungi which enables them to digest certain components of wood.

hadrom(e) *n*. [Gk. *hadros*, thick] conducting tissue of xylem; *alt.* hadromestome.

Hadromerina *n*. [Gk. *hadros*, thick; *meros*, part] an order of Monaxonida without spongin and having knobbed megascleres.

Haeckel's law [*E. H. Haeckel*, German zoologist] recapituiation theory, *q.v.*

haem *n*. [Gk. *haeima*, blood] a porphyrin containing iron, found as a prosthetic group in haemoproteins, e.g. haemoglobin, and which can be oxidized to various compounds including haemin and haematin; *alt.* heme.

haema- *see also* haemat-, haemo-; *alt.* hema-.

haemachrome *n*. [Gk. *haima*, blood; *chrōma*, colour] colouring matter found in blood.

haemacyte *n*. [Gk. *haima*, blood; *kytos*, hollow] a blood corpuscle.

haemad *adv*. [Gk. *haima*, blood; L. *ad*, to] situated on same side of vertebral column as heart; *alt.* haemal.

haemal *a*. [Gk. *haema*, blood] *pert.* blood or blood vessels, *alt.* haematal, haemic; haemad, *q.v.*

haemal arches in many vertebrates, lateral processes from caudal vertebrae which fuse ventrally and enclose an artery and a vein.

haemal canal canal formed by series of haemal arches.

haemal ridge haemapophysis, *q.v.*

haemamoeba *n*. [Gk. *haima*, blood; *amoibē*, change] protozoon with an amoeboid trophozoitic stage parasitic in a red blood corpuscle; *plu.* haemamoebae.

haemangioblast *n*. [Gk. *haima*, blood; *anggeion*, vessel; *blastos*, bud] a blood island, *q.v.*

haemapoietic *a*. [Gk. *haima*, blood; *poiein*, to form] haemopoietic, *q.v.*

haemapophysis *n*. [Gk. *haima*, blood; *apo*, from; *phyein*, to grow] one of plate-like or spine-like processes growing from the lateroventral surfaces of a vertebral centrum; *plu.* haemapophyses; *alt.* haemal ridge.

haemat- *see also* haema-, haemo-.

haematal *a*. [Gk. *haima*, blood] *pert.* blood or blood vessels; *alt.* haemal.

haematid *n*. [Gk. *haima*, blood] red blood corpuscle, *q.v.*

haematin *n*. [Gk. *haima*, blood] a blood pigment formed by the decomposition of haemoglobin, and which can be reduced to haem; *alt.* protohaem.

haematobium *n*. [Gk. *haima*, blood; *bios*, life] an organism living in blood; *a.* haematobic.

haematoblast *n*. [Gk. *haima*, blood; *blastos*, bud] a cell that will develop into a red blood cor-

puscle or into either a red or white corpuscle; thrombocyte, *q.v.*

haematochrome *n*. [Gk. *haima*, blood; *chrōma*, colour] a carotenoid red pigment of certain algae.

haematocryal *a*. [Gk. *haima*, blood; *kryos*, cold] poikilothermal, *q.v.*

haematocyanin *n*. [Gk. *haima*, blood; *kyanos*, dark blue] haemocyanin, *q.v.*

haematocytozoon *n*. [Gk. *haima*, blood; *kytos*, hollow; *zōon*, animal] an intracorpuscular blood parasite.

haematodocha *n*. [Gk. *haima*, blood; *dochē*, receptacle] a fibro-elastic bag at base of palpal organ in male spiders.

haematogen *n*. [Gk. *haima*, blood; *genos*, birth] a nucleoprotein containing iron.

haematogenesis *n*. [Gk. *haima*, blood; *genesis*, descent] haemopoiesis, *q.v.*

haematogenous *a*. [Gk. *haima*, blood; *genos*, birth] formed in blood; derived from blood.

haematoidin *n*. [Gk. *haima*, blood; *eidos*, form] an iron-free derivative of haemoglobin, forming crystals in blood clots, and identical with bilirubin.

haematology *n*. [Gk. *haima*, blood; *logos*, discourse] the study of blood and blood formation.

haematolysis *n*. [Gk. *haima*, blood; *lysis*, loosing] haemolysis, *q.v.*

haematophagous *a*. [Gk. *haima*, blood; *phagein*, to eat] feeding on blood, or obtaining nourishment from blood.

haematophyte *n*. [Gk. *haima*, blood; *phyton*, plant] any plant microorganism in blood.

haematopoiesis haemopoiesis, *q.v.*

haematoporphyrin *n*. [Gk. *haima*, blood; *porphyra*, purple] an iron-free pigment formed by decomposition of haematin.

haematosis *n*. [Gk. *haematoein*, to change to blood] haemopoiesis, *q.v.*

haematothermal *a*. [Gk. *haima*, blood; *thermos*, warm] homoiothermal, *q.v.*

haematozoon *n*. [Gk. *haima*, blood; *zōon*, animal] any animal parasitic in blood.

haemerythrin *n*. [Gk. *haima*, blood; *erythros*, red] a red haemochrome respiratory pigment found in sipunculids, certain molluscs, and crustaceans, containing iron but no porphyrin complexes; *alt.* haemoerythrin.

haemic *a*. [Gk. *haima*, blood] *pert.* blood; *alt.* haemal.

haemin *n*. [Gk. *haima*, blood] a chloride of haematin; an oxidized form of the haem group, bearing positive charge.

haemo- *see also* haema-, haemato-.

haemobilirubin *n*. [Gk. *haima*, blood; L. *bilis*, bile; *ruber*, red] a breakdown product of haemoglobin which is converted to bilirubin and biliverdin in the liver.

haemoblast *n*. [Gk. *haima*, blood; *blastos*, bud] a cell which gives rise to an erythroblast; *alt.* haematoblast.

haemochorial *a*. [Gk. *haima*, blood; *chorion*, skin] *appl.* placenta with branched chorionic villi penetrating blood sinuses after breaking down uterine tissues, as in insectivores, rodents, and primates; *cf.* endotheliochorial, epitheliochorial.

haemochromes *n.plu.* [Gk. *haima*, blood; *chrōma*, colour] chromoproteins in which the prosthetic groups are either iron, often in haem, or thiopeptide, and including the blood pigments haemoglobin, haemocyanin, haemerythrin, chlorocruorin, erythrocruorin.

haemochromogen n. [Gk. *haima*, blood; *chrōma*, colour; *genos*, birth] a compound of haem and a protein, denatured protein, or nitrogenous base, found in plant and animal tissue.

haemoclastic a. [Gk. *haima*, blood; *klastos*, broken] breaking down blood cells, *appl.* tissues; *cf.* haemoplastic.

haemocoel(e) n. [Gk. *haima*, blood; *koilos*, hollow] an expanded portion of the blood system which replaces the true coelom.

haemoconia n. [Gk. *haima*, blood; *konis*, dust] minute particles of red blood corpuscles, absorbed by reticulo-endothelial phagocytes; *alt.* blood dust, haemokonia.

haemocyanin n. [Gk. *haima*, blood; *kyanos*, dark blue] a blue haemochrome respiratory pigment found in many molluscs and arthropods and containing copper instead of iron; *alt.* haematocyanin.

haemocyte n. [Gk. *haima*, blood; *kytos*, hollow] a blood cell in insects and other invertebrates.

haemocytoblast n. [Gk. *haima*, blood; *kytos*, hollow; *blastos*, bud] primitive stem cell from which all blood cells are derived; a lymphoid haemoblast; *alt.* lymphoidocyte.

haemocytolysis n. [Gk. *haima*, blood; *kytos*, hollow; *lyein*, to dissolve] breaking up of red blood corpuscles by solution; *alt.* haemolysis.

haemocytotrypsis n. [Gk. *haima*, blood; *kytos*, hollow; *tribein*, to rub] breaking up of blood corpuscles by pressure.

haemodynamics n. [Gk. *haima*, blood; *dynamikos*, powerful] study of the principles of blood flow.

haemoerythrin n. [Gk. *haima*, blood; *erythros*, red] haemerythrin, *q.v.*

Haemoflagellata, haemoflagellates n., n.plu. [Gk. *haima*, blood; L. *flagellum*, whip] flagellates that are blood parasites, such as trypanosomes.

haemofuscin n. [Gk. *haima*, blood; L. *fuscus*, tawny] a yellow blood pigment deposited under various pathological conditions.

haemogenesis haematogenesis, *q.v.*

haemoglobin n. [Gk. *haima*, blood; L. *globus*, sphere] a red haemochrome pigment consisting of 2 pairs of polypeptide chains, 2α and 2β, each of the 4 bearing one haem group, and coiled to form an almost spherical molecule in which the haems are embedded, and which acts as a respiratory pigment in animals, the iron always being in the ferrous (reduced) state; abbreviated to Hb or Hgb; *alt.* cruorin; *cf.* leghaemoglobin, methaemoglobin, myoglobin, oxyhaemoglobin.

haemohistioblast n. [Gk. *haima*, blood; *histion*, tissue; *blastos*, bud] a free macrophage in blood, especially of veins.

haemoid a. [Gk. *haima*, blood; *eidos*, form] resembling blood.

haemokonia haemoconia, *q.v.*

haemolymph n. [Gk. *haima*, blood; L. *lympha*, water] a fluid found in coelom of some invertebrates, regarded as equivalent to blood and lymph of higher forms; *appl.* nodes: modified lymph nodes containing blood.

haemolysin n. [Gk. *haima*, blood; *lyein*, to dis-

solve] a substance developed in or added to blood serum, capable of destroying red blood corpuscles.

haemolysis n. [Gk. *haima*, blood; *lysis*, loosing] the lysis of a suspension of red blood corpuscles; *alt.* erythrocytolysis, haematolysis, haemocytolysis, laking.

haemolytic a. [Gk. *haima*, blood; *lyein*, to dissolve] *pert.*, or causing, haemolysis.

haemoparasite n. [Gk. *haima*, blood; *parasitos*, parasite] an organism parasitic in the blood of its host.

haemopathic a. [Gk. *haima*, blood; *pathein*, to suffer] affecting the circulatory system, *appl.* enzymes, as in snake venom; *cf.* neurotoxic.

haemophilia n. [Gk. *haima*, blood; *philos*, loving] absence of ready coagulation of shed blood, a sex-linked hereditary characteristic.

haemoplasmodium n. [Gk. *haima*, blood; *plasma*, mould] a unicellular parasite of blood.

haemoplastic a. [Gk. *haima*, blood; *plastos*, formed] haemopoietic, *q.v.*; *cf.* haemoclastic.

haemopoiesis n. [Gk. *haima*, blood; *poiēsis*, making] the formation of blood; *alt.* haemapoiesis, haematopoiesis, haematosis, haemtogenesis, haemogenesis.

haemopoietic a. [Gk. *haima*, blood; *poiētikos*, productive] blood-forming, *pert.* haemopoiesis; *alt.* haemapoietic, haemoplastic.

haemoproteins n.plu. [Gk. *haima*, blood, *prōteion*, first] proteins having an iron-porphyrin prosthetic group.

haemopsonin n. [Gk. *haima*, blood; *opsōnein*, to cater] an opsonin for erythrocytes.

haemorrhoidal a. [Gk. *haima*, blood; *rheein*, to flow] rectal, *appl.* blood vessels, nerve.

haemosensitins n.plu. [Gk. *haima*, blood; L. *sensus*, a sense] substances that become adsorbed onto the red blood cells, rendering them sensitive to agglutination by antibodies.

haemosiderin n. [Gk. *haima*, blood; *sidēros*, iron] a combination of protein and ferric hydroxide, a yellow pigment of blood, excess being stored in bone marrow, liver, and spleen.

Haemosporidia n. [Gk. *haima*, blood; *sporos*, seed, *idion*, dim.] a suborder of protozoans of the order Coccidiomorpha, which are true blood parasites.

haemostatic a. [Gk. *haima*, blood; *statikos*, causing to stand] *appl.* membrane crossing joint between trochanter and femur in autotomy of limb of some arthropods; *appl.* an agent that stops bleeding.

haemotoxin n. [Gk. *haima*, blood; *toxikon*, poison] a toxin which produces haemolysis.

haemotropic a. [Gk. *haima*, blood; *tropē*, turn] affecting or acting upon blood.

haemozoin n. [Gk. *haima*, blood; *zōon*, animal] granules of a black pigment, the residue from digestion of haemoglobin by malarial parasites; *cf.* melanin.

Haerangiomycetes n. [L. *haerere*, to cling; Gk. *anggeion*, vessel; *mykēs*, fungus] a class of Ascomycetes in which the walls of the asci are absent or dissolve as soon as they are formed.

haerangium n. [L. *haerere*, to cling; Gk. *anggeion*, vessel] the apparatus for collecting and dispersing spores in Haerangiomycetes, an adhesive droplet containing spores being held by the tenaculum, *q.v.*

hagfish the common name for the Myxinoidea, q.v.

hair n. [A.S. haer; hair] any epidermal filamentous outgrowth consisting of one or more cells, varied in shape; in mammals, a thread-like structure composed of cornified epidermal cells and growing by cell division in a hair follicle at its base; seta, q.v.; trichome, q.v.

hair cells sensory cells in organ of Corti.

hair follicle tubular sheath formed by invagination of epidermis and surrounding base of hair.

hair pits cryptostomata, q.v.

hair plates groups of tactile hairs near articular surfaces on many parts of the insect body, contributing to maintaining normal position of the head.

Haldane's rule [J. B. S. Haldane, Scottish geneticist] a rule stating that when offspring of one sex produced from a cross are inviable or infertile, it is the heterogametic sex.

Halecostomi, halecostomes n., n.plu. [Halec, generic name; Gk. stoma, mouth] a group of Holostei having a preoperculum which does not buttress the palate bones.

half-inferior having ovary but partially adherent to calyx.

half-sibs individuals having only one parent in common.

half-spindle unipolar spindle, as in meiosis of some insects.

half-terete rounded on one side, flat on the other.

halibios halobios, q.v.

Halichondrina n. [Halichondria, generic name] an order of Monaxonida having very little spongin and 2 or more kinds of megascleres.

haliplankton haloplankton, q.v.

hallachrome n. [Halla, an annelid; Gk. chrōma, colour] a red pigment or respiratory catalyst in skin of Halla, derived from tyrosine, formed by oxidation of dopa, and oxidized to melanin.

Haller's organ [G. Haller, German zoologist] a tarsal chemoreceptor in ticks.

hallux n. [L. hallux, great toe] first digit of hindlimb.

halm haulm, q.v.

halobenthos n. [Gk. hals, sea; benthos, depths of the sea] marine benthos.

halobiontic a. [Gk. hals, sea; bion, living] confined entirely to sea water.

halobios n. [Gk. hals, sea; bios, life] sum total of organisms living in the sea; animals living in the sea or any salt water; a. halobiotic; alt. halibios.

halodrymium n. [Gk. hals, sea; drymos, coppice] a mangrove association.

halolimnic a. [Gk. hals, sea; limnē, lake] pert. marine organisms modified to live in fresh water; alt. thalassoid.

halophilic, halophilous a. [Gk. hals, salt; philein, to love] salt-loving; thriving in presence of salt; n. halophile.

halophobe n. [Gk. hals, salt; phobos, manifest fear] a plant intolerant of salinity; a. halophobous, halophobic.

halophyte n. [Gk. hals, salt; phyton, plant] a shore plant; plant capable of thriving on salt-impregnated soils; a. halophytic; cf. glycophyte.

haloplankton n. [Gk. hals, sea; plangktos, wandering] the organisms drifting in the sea; alt. haliplankton; cf. limnoplankton.

Haloragales n. [Haloragis, generic name] an order of dicots (Rosidae) including many water plants.

halosere n. [Gk. hals, salt; L. serere, to put in a row] a plant succession originating in a saline area, as in salt marshes.

Halosphaerales n. [Halosphaera, generic name] an order of Chlorophyta (Prasinophyceae) having large uninucleate cells.

haloxene a. [Gk. hals, salt; xenos, guest] tolerating salt water.

halteres n.plu. [Gk. haltēr, weight] a pair of small capitate bodies representing rudimentary posterior wings in Diptera; alt. balancers, poisers.

Hamamelidales n. [Hamamelis, generic name] an order of dicots (Hamamelidiidae) including the families Hamamelidaceae, e.g. witch hazel, and Platanaceae, e.g. plane trees.

Hamamelidiidae n. [Hamamelis, generic name] a subclass of dicots including apetalous catkin-bearing, wind-pollinated plants formerly considered primitive; alt. Amentiflorae.

hamate a. [L. hamatus, hooked] hooked or hook-shaped at the tip; alt. hamose, hamulose, uncate, uncinate.

hamatum n. [L. hamatus, hooked] the unciform bone in the carpus, probably corresponding to 4th or 5th distalia of a typical pentadactyl limb.

hamiform a. [L. hamus, hook; forma, shape] hook-shaped; alt. unciform.

hamirostrate a. [L. hamus, hook; rostrum, beak] having a hooked beak.

hammer malleus, q.v.

hamose hamate, q.v.

hamstrings tendons of insertion of the posterior femoral muscles, i.e. of semitendinosus, semimembranosus, and biceps.

hamula n. [L. hamulus, little hook] retinaculum of insects; fused ventral appendages acting with caudal furcula in springtails; hamulus, q.v.

hamular hooked; hook-like.

hamulate a. [L. hamulus, little hook] having small hook-like processes.

hamulose a. [L. hamulus, little hook] hamate, q.v.

hamulus n. [L. hamulus, little hook] a hooklet, or hook-like process as of lacrimal, hamate, and pterygoid bones, and of osseous spiral lamina at apex of cochlea; minute hook-like process on distal barbules which aid in interlocking of feather barbs; retinaculum of Hymenoptera; alt. hamula.

hamus n. [L. hamus, hook] hooked part of uncus in male Lepidoptera.

hapanthous a. [Gk. hapax, once; anthos, flower] reproducing only once at end of plant's life; alt. hapaxanthous.

hapaxanthic, hapaxanthous a. [Gk. hapax, once; anthos, flower] with only a single flowering period, cf. pollakanthic; hapanthous, q.v.

haplo- cf. diplo-.

haplobiont n. [Gk. haploos, simple; biōnai, to live] an organism characterized by one kind of individual; cf. diplobiont.

Haplobothrioidea n. [Gk. haploos, simple; bothros, trench; eidos, form] an order of tapeworms having the scolex with 4 protrusible proboscides hooked at the base, in some classifications placed in the Pseudophyllidea.

haplocaulescent a. [Gk. haploos, simple; L.

caulis, stem] with a simple axis, i.e. capable of producing seed on the main axis.

haplocheilic *a.* [Gk. *haploos*, simple; *cheilos*, edge] *appl.* type of stomata in gymnosperms in which the subsidiary cells are not related to the guard cells ontogenetically.

haplochlamydeous *a.* [Gk. *haploos*, simple; *chlamys*, cloak] having only one whorl of perianth segments; *alt.* monochlamydeous.

haplochlamydeous chimaera one in which the epidermis forms one component, the other component being formed from all the other tissues.

haplodioecious *n.* [Gk. *haploos*, simple; *dis*, twice; *oikos*, house] heterothallic, *q.v.*

haplodiploid *a.* [Gk. *haploos*, simple; *diploos*, double; *eidos*, form] *appl.* sex differentiation in which the male is haploid, the female diploid; *n.* haplodiploidy.

haplodiplont *n.* [Gk. *haploos*, simple; *diploos*, double; *on*, being] an organism exhibiting the haplodiploid condition; a plant having haploid and diploid vegetative phases.

Haplodoci *n.* [Gk. *haploos*, simple; *dokos*, beam] Batrachoidiformes, *q.v.*

haplodont *a.* [Gk. *haploos*, simple; *odous*, tooth] having molars with simple crowns.

haploid *a.* [Gk. *haploos*, simple; *eidos*, form] having the number of chromosomes characteristic of the gametes for the organism; *appl.* the typical gametic number of chromosomes after meiosis; *alt.* hemikaryotic. *n.* A haploid organism. *Cf.* diploid.

haploid apogamy/apogamety generative apogamy, *q.v.*

haploidization in certain fungi, a phenomenon occurring in the parasexual cycle during which a diploid cell becomes haploid by loss of one chromosome after another by non-disjunction.

Haplolepiformes *n.* [Gk. *haploos*, simple; *lepis*, scale; L. *forma*, shape] an order of U. Carboniferous Chondrostei having unbranched stout fin-rays.

haplometrosis monometrosis, *q.v.*; *a.* haplometrotic.

haplomitosis *n.* [Gk. *haploos*, simple; *mitos*, thread] type of cell division where nuclear granules form chromospires which withdraw in 2 groups or divide transversely in the middle.

haplomonoecious *a.* [Gk. *haploos*, simple; *monos*, single; *oikos*, house] homothallic, *q.v.*

haplomycelium *n.* [Gk. *haploos*, simple; *mykēs*, fungus] haploid mycelium.

haploneme *n.* [Gk. *haploos*, simple; *nēma*, thread] having threads of uniform diameter, *appl.* nematocysts.

haplont *n.* [Gk. *haploos*, simple; *on*, being] an organism having haploid somatic nuclei; *alt.* monoplont; *cf.* diplont.

haploperistomic, haploperistomous *a.* [Gk. *haploos*, simple; *peri*, around; *stoma*, mouth] having a single peristome; having a peristome with a single row of teeth; *appl.* mosses.

haplopetalous *a.* [Gk. *haploos*, simple; *petalon*, leaf] with a single row of petals.

haplophase *n.* [Gk. *haploos*, simple; *phasis*, aspect] stage in life history of an organism when nuclei are haploid; *alt.* gametophyte phase, gametic phase, gamophase; *cf.* diplophase.

haplophyte *n.* [Gk. *haploos*, simple; *phyton*, plant] a haploid plant or gametophyte; *cf.* diplophyte, sporophyte.

haploptile *n.* [Gk. *haploos*, simple; *ptilon*, feather] single neossoptile, without rachis, formed by precocious development of the barbs of the teleoptile.

Haplosclerina *n.* [Gk. *haploos*, simple; *sklēros*, hard] an order of Monaxonida usually having spongin fibres and pointed megascleres.

haplosis *n.* [Gk. *haploos*, simple] halving of the chromosome number during meiosis; reduction and disjunction.

haplostele *n.* [Gk. *haploos*, simple; *stēlē*, pillar] a simple stele having a cylindrical core of xylem surrounded by phloem.

haplostemonous *a.* [Gk. *haploos*, simple; *stēmōn*, warp] having one whorl of stamens.

haplostromatic *a* [Gk. *haploos*, simple; *strōma*, bedding] fungi having little or no entostroma, perithecia being formed in ectostroma; *cf.* diplostromatic.

haplotype *n.* [Gk. *haploos*, simple; *typos*, pattern] the only species in a genus originally, and thereby becoming a genotype.

haploxylic *a.* [Gk. *haploos*, simple; *xylon*, wood] possessing only one vascular bundle.

haplozygous *a.* [Gk. *haploos*, simple; *zygon*, yoke] hemizygous, *q.v.*

hapten *n.* [Gk. *haptein*, to touch] non-protein part of certain antigens, which does not cause production of antibodies, but which can react with the appropriate antibody to form a precipitate.

hapteron *n.* [Gk. *haptein*, to fasten] holdfast, special disc-like outgrowth from the stem-like portion of certain algae which serves as an organ of attachment; a similar structure for attachment in various other lower plants; *plu.* haptera.

haptic *a.* [Gk. *haptein*, to touch] *pert.* touch, *appl.* stimuli and reactions.

haptogen *a.* [Gk. *haptein*, to fasten; *-genēs*, producing] *appl.* a limiting membrane of solidified protein which prevents miscibility.

haptoglobin *n.* [Gk. *haptein*, to touch; L. *globus*, globe] a protein in blood serum which combines with haemoglobin and so has the function of ridding the serum of free haemoglobin.

haptomonad *n.* [Gk. *haptein*, to fasten; *monas*, unit] an attached form of certain parasitic Flagellata; *cf.* nectomonad.

haptonasty *n.* [Gk. *haptein*, to touch; *nastos*, pressed close] a nastic movement to the stimulus of touch.

haptonema *n.* [Gk. *haptein*, to touch; *nēma*, thread] a distinctive kind of flagellum found in algae of the class Haptophyceae, having no locomotory function but being very long, and consisting of 3 concentric membranes surrounding a ring of fibres and a central space.

haptophores *n.plu.* [Gk. *haptein*, to fasten; *pherein*, to carry] the combining qualities of the molecule of a toxin, lysin, opsonin, precipitin, or agglutinin; *a.* haptophorous; *cf.* toxophores.

Haptophyceae *n.* [Gk. *haptein*, to touch or fasten; *phykos*, seaweed] a class of Chrysophyta often having a haptonema as well as true flagella, these being acronematic not pantonematic.

haptospore *n.* [Gk. *haptein*, to fasten; *sporos*, seed] an adhesive spore; *alt.* plasmaspore.

haptotropism *n.* [Gk. *haptein*, to touch; *tropē*, turn] response by curvature to contact stimulus.

as in tentacles, tendrils, stems; *a.* haptotropic; *alt.* thigmotropism.

haptotype *n.* [Gk. *haptein*, to touch; *typos*, pattern] an icotype collected with the holotype but possibly taken from another plant.

Harderian gland [*J. J. Harder*, Swiss anatomist] an accessory lacrimal gland of 3rd eyelid or nictitating membrane.

hard pan a hard layer developed in the B-horizon of the soil, consisting of deposited salts, which restricts drainage and root growth.

hardwood wood produced by woody dicotyledons.

Hardy–Weinberg Law in a large random mating population in the absence of mutation, migration, and selection, gene frequencies remain constant from one generation to another.

hares *see* Lagomorpha.

harlequin lobe a testicular lobe with cells differing from those of other lobes, in certain Hemiptera.

harmonic suture an articulation formed by apposition of edges or surfaces, as between palatine bones.

harmosis *n.* [Gk. *harmosis*, fitting] arrangement and adaptation in response to a stimulus.

harmozone *n.* [Gk. *harmozō*, I arrange] one of the hormones which influence growth and nutrition.

Harpacticoida *n.* [*Harpacticus*, generic name; Gk. *eidos*, form] an order of mainly benthic copepods with some parasitic members, having very short 1st antennae and biramous 2nd antennae.

harpagones *n.plu.* [L. *harpago*, hook] claspers or valves of certain male insects; a pair of sclerites between harpes and claspers in mosquitoes; harpes, *q.v.*, in Lepidoptera; *alt.* claspettes.

Harpellales *n.* [*Harpella*, generic name] an order of Trichomycetes reproducing by trichospores.

harpes *n.plu.* [Gk. *harpē*, sickle] chitinous processes between the claspers of mosquitoes; claspers or valves of Lepidoptera, *alt.* harpagones.

Hartig net [*H. J. A. R. Hartig*, German botanist] network of hyphae between cortical cells of roots in ectotrophic mycorrhiza.

harvest men the common name for the Opiliones, *q.v.*

hashish *n.* [*Ashishins*, a band of murderers in Persia in 13th century who used cannabis] the hallucinogenic drug cannabis, *q.v.*, especially that obtained as a resin from the female flowers; *alt.* charas.

Hassall's concentric corpuscles [*A. H. Hassall*, English physician] epithelial cell nests in medulla of thymus.

hastate *a.* [L. *hasta*, spear] spear-shaped, more or less triangular with the 2 basal lobes divergent.

hastate-auricled auriculate, *q.v.*

Hatch–Slack pathway a route for the fixation of carbon in photosynthesis, followed by C_4 plants, where the first products are C_4 dicarboxylic acids, first oxaloacetate, then malate and aspartate, which are transported to special chloroplasts of the bundle sheath cells and release carbon dioxide which is then fixed as in C_3 plants; *alt.* C_4 dicarboxylic acid pathway.

Hatschek's nephridium [*B. Hatschek*, Austrian zoologist] a nephridium between notochord and preoral pit in Cephalochordata.

Hatschek's pit a mucin-secreting gland in roof of oral cavity in Cephalochordata; *alt.* preoral pit.

haulm *n.* [A.S. *haelm*] the stem of such plants as peas; the stem of a grass; *alt.* halm.

haustellate *a.* [L. *haurire*, to drain] having a haustellum.

haustellum *n.* [L. *haurire*, to drain] a proboscis adapted for sucking; *a.* haustellate.

haustoria *plu.* of haustorium.

haustorial *a.* [L. *haurire*, to drink] *pert.* or resembling a haustorium.

haustorium *n.* [L. *haurire*, to drink] an outgrowth of stem, root, or hyphae of certain parasitic plants, which serves to draw food from the host plant, *alt.* sucker; an outgrowth of embryo sac which extends to nutritive tissue in certain non-parasite plants; foot in cryptogams; *plu.* haustoria.

haustra *n.plu.* [L. *haustrum*, drawer] recesses of sacculations of the colon, between plicae semilunares; *sing.* haustrum.

Haversian canals [*C. Havers*, English anatomist] small canals in bone, in which lie blood capillaries, nerve, and lymph space.

Haversian fringes villi on the surface of synovial membranes; *alt.* synovial villi.

Haversian lamellae series of concentric bony lamellae surrounding the central Haversian canal in compact bone.

Haversian spaces irregular spaces formed by the absorption of the original cartilage, in the development of bone.

Haversian system a Haversian canal, the surrounding concentric lamellae, and bone cells with canaliculi, found in compact bone; *alt.* osteone.

Hb haemoglobin, *q.v.*

H-band *see* H-line.

HbO₂ oxyhaemoglobin, *q.v.*

HCl cells oxyntic cells, *q.v.*

H-disc *see* H-line.

head-cap acrosome, *q.v.*

head-case the outer hard covering of insect head.

head cell one of the cells on manubrium of antheridium of *Chara* (Charales).

head inductor anterior part of roof of archenteron, inducing the development of head organs, and including archencephalic and deuterencephalic inductors.

head-kidney the pronephric portion of kidney, in vertebrates usually represented only in embryo; a nephridium usually developed in cephalic segment of invertebrates.

heart *n.* [A.S. *heorte*, heart] a hollow muscular organ with varying number of chambers which by rhythmic contraction keeps up circulation of blood; core or central portion of a tree or fruit.

heart-accelerating factor a substance, believed to be a protein or peptide, associated with the neurosecretory vesicles in the corpora cardiaca of some insects and accelerating the insect heart.

heartwood the darker, harder, central wood of trees, containing no living cells; *alt.* duramen; *cf.* sapwood.

heat *n.* [A.S. *haetu*, heat] a kind of energy manifested in various ways; the sensation of warmth produced by stimulation of special organs; the period of sexual urge.

heath *n.* [M.E. *heth*, heath] a lowland plant community usually found on sandy soils with some peat, and dominated by heathers.

heat spot a special area on the skin at which nerve endings sensitive to heat are found.

hebetate *a*. [L. *hebes*, dull] blunt-ended.

hebetic *a*. [Gk. *hēbētikos*, juvenile] *pert*. adolescence.

hecistotherm hekistotherm, *q.v.*

hectocotylus *n*. [Gk. *hekaton*, hundred; *kotylos*, cup] one of the arms of a male cephalopod, specialized to effect transference of sperms; *alt*. hectocotylized arm, heterocotylus.

hederiform *a*. [L. *hedera,* ivy; *forma*, shape] shaped like an ivy leaf, *appl*. nerve endings, as pain receptors in the skin.

hedonic *a*. [Gk. *hēdonē*, pleasure] *appl*. skin glands of certain reptiles, which secrete a musk-like substance and are specially active at mating season.

hedrioblast *n*. [Gk. *hedra*, seat; *blastos*, bud] medusa, *q.v.*

heel *n*. [A.S. *hela*, heel] hinder or posterior tarsal portion of foot; talon or talonid of a tooth; a spinule at base of tibia in Hymenoptera.

Heidelberg man a type of primitive man known from fossils near Heidelberg in Germany, which is somewhat intermediate between *Homo erectus* and Neanderthal man, and now usually considered a subspecies and classified as *Homo erectus heidelbergensis*.

Heidenhain, demilunes of [*R. P. Heidenhain*, German physiologist] *see* demilunes.

Heister's valve [*L. Heister*, German anatomist] spiral valve in neck of gall bladder.

hekistotherm *n*. [Gk. *hēkistos*, least; *thermē*, heat] a plant that thrives with the minimum of heat, as alpine plants; *alt*. hecistotherm.

HeLa cells an aneuploid strain of human epithelial-like cells which originated from a specimen of tissue from a patient named *Henrietta Lacks*, and has been carried in tissue culture since 1952.

helad *n*. [Gk. *helos*, marsh] helophyte, *q.v.*

helcotropism *n*. [Gk. *helkein*, to draw down; *trepein*, to turn] geotropism, *q.v.*

heleoplankton *n*. [Gk. *helos*, marsh; *plangktos*, wandering] the plankton of marshy ponds or lakes; *alt*. heloplankton.

helical *a*. [Gk. *helix*, spiral] spiral, *appl*. arrangement of myofibrils in certain smooth muscles, *appl*. cell wall thickening in xylem.

helices *plu*. of helix.

helicine *a*. [Gk. *helix*, spiral] spiral; convoluted; *appl*. certain convoluted and dilated arteries in penis; *pert*. outer rim of pinna.

helicocone *n*. [Gk. *helix*, spiral; *kōnos*, a cone] in gastropods, a cone coiled in a helical spiral.

helicoid *a*. [Gk. *helix*, spiral; *eidos*, like] spiral; shaped like a snail's shell; *pert*. type of sympodial branching in which sympodium consists of fork branches of same side.

helicoid cyme a cymose inflorescence produced by suppression of successive axes on same side, thus causing the sympodium to be spirally twisted so that the blooms are on only one side of the axis; *alt*. bostryx.

helicoid dichotomy a type of branching in which there is repeated forking but with the branches on one side uniformly more vigorous than on the other.

helicorubin *n*. [L. *helix*, spiral; *ruber*, red] a red pigment of gut of pulmonate gastropods, also in liver of certain crustaceans.

helicospore *n*. [Gk. *helix*, spiral; *sporos*, seed] a convolute or spiral spore.

helicotrema *n*. [Gk. *helix*, spiral; *trēma*, hole] a small opening near summit of cochlea by which the scalae vestibuli and tympani communicate.

heliophil, heliophilic, heliphilous *a*. [Gk. *hēlios*, sun; *philein*, to love] adapted for relatively high intensity of light; *cf*. skiophil.

heliophobic, heliphobous skiophil, *q.v.*

heliophyll *n*. [Gk. *hēlios*, sun; *phyllon*, leaf] a plant having isolateral leaves; *cf*. skiophyll.

heliophyte *n*. [Gk. *hēlios*, sun; *phyton*, plant] a plant requiring full sunlight to thrive; *cf*. skiaphyte.

heliosis *n*. [Gk. *hēlios*, sun] production of discoloured spots or markings on leaves through concentration of sun on them; solarization, *q.v.*

helioskiophyte *n*. [Gk. *hēlios*, sun; *skia*, shade; *phyton*, plant] a plant which can live in both sun and shade, but grows better in the sun.

heliotaxis *n*. [Gk. *hēlios*, sun; *taxis*, arrangement] a taxis in relation to a stimulus of sunlight; *alt*. phototaxis.

heliotropism *n*. [Gk. *hēlios*, sun; *trepein*, to turn] a tropism in response to the stimulus of sunlight; *alt*. phototropism.

helioxerophil *n*. [Gk. *hēlios*, sun; *xēros*, dry; *philein*, to love] a plant which thrives in full sunlight and under arid conditions.

Heliozoa *n*. [Gk. *hēlios*, sun; *zōon*, animal] an order of mostly freshwater Sarcodina including the sun animalicules and their allies, having a radially symmetrical body, slender stiff pseudopodia, and often a skeleton of spicules.

helix *n*. [Gk. *helix*, spiral] a spiral; the coiled spiral arrangement of certain structures in invertebrates; the outer rim of external ear; *plu*. helices.

helmet *n*. [A.S. *helm*, *helan*, to cover] the process of bill of hornbills; the bony plates covering head of certain extinct fishes; the galea of flowers and of insects.

helminthoid *a*. [Gk. *helmins*, worm; *eidos*, shape] shaped like a worm; *alt*. vermiform.

helminthology *n*. [Gk. *helmins*, worm; *logos*, discourse] the study of the natural history of worms; the study of parasitic flatworms and roundworms.

Helminthomorpha *n*. [Gk. *helmins*, worm; *morphē*, form] a superorder of long-bodied Chilognatha which can coil into a spiral and have no telopods.

Helobiae *n*. [Gk. *helos*, marsh] an order of monocots comprising water or marsh plants, *alt*. Fluviales; Alismatidae, *q.v.*

helobious *a*. [Gk. *helos*, marsh; *bios*, life] living in marshes.

helophyte *n*. [Gk. *helos*, marsh; *phyton*, plant] a marsh plant; a cryptophyte growing in soil saturated with water; *alt*. limnophyte, limnocryptophyte, helad.

heloplankton heleoplankton, *q.v.*

Helotiales *n*. [*Helotium*, generic name] an order of Discomycetes in which the ascus dehisces by a pore.

helotism *n*. [Gk. *eilōtēs*, serf, from *Helos*, Laconian town] symbiosis in which one organism enslaves another and forces it to labour on its behalf, e.g. in some lichens where the relationship is at the expense of the alga, and in some species of ant; *cf.* consortism, dulosis.

help cell synergid, *q.v.*

helper virus a virus which infects a cell already infected with a defective virus and supplies something the latter lacks, so that it can multiply.

Helvellales *n.* [*Helvella*, generic name] an order of Discomycetes having asci parallel to each other and freely exposed in the ascocarp; sometimes considered a group of Pezizales.

hema- haema-, *q.v.*

heme haem, *q.v.*

hemelytron *n.* [Gk. *hēmi*, half; *elytron*, sheath] proximally hardened fore-wing of certain insects; elytron of certain polychaetes; *alt.* hemelytrum, hemielytron.

hemera *n.* [Gk. *hēmera*, day] the time during which fossiliferous strata constituting a zone of sedimentary rocks were deposited.

hemeranthic, hemeranthous *a.* [Gk. *hēmera*, day; *anthos*, flower] flowering by day.

hemerecology *n.* [Gk. *hēmeros*, tame; *oikos*, household; *logos*, discourse] ecology of cultivated land.

hemerophilous *a.* [Gk. *hēmeros*, tame; *philein*, to love] easy to cultivate.

hemerophyte *n.* [Gk. *hēmeros*, tame; *phyton*, plant] a cultivated plant.

hemiangiocarpic *a.* [Gk. *hēmi*, half; *anggeion*, vessel; *karpos*, fruit] having an enclosed unripe hymenium exposed by rupture of the covering veil.

Hemiascomycetes, Hemiascomycetae *n.* [Gk. *hēmi*, half; *askos*, bag; *mykēs*, fungus] a group of fungi, the lower Ascomycetes, having asci usually occurring singly with no definite ascocarps; *alt.* Protoascomycetes, Protoascomycetidae, Hemiascomycetidae.

hemiascospore *n.* [Gk. *hēmi*, half; *askos*, bag; *sporos*, seed] ascospore of a hemiascus.

hemiascus *n.* [Gk. *hēmi*, half; *askos*, bag] an atypical multinucleate ascus containing hemiascospores.

hemi-autophyte *n.* [Gk. *hēmi*, half; *autos*, self; *phyton*, plant] a semiparasitic plant which produces its own chlorophyll.

Hemibasidiomycetes, Hemibasidiomycetae *n.* [Gk. *hēmi*, half; *basis*, base; *idion*, dim.; *mykēs*, fungus] Heterobasidiomycetes, *q.v.*

hemibasidium *n.* [Gk. *hēmi*, half; *basis*, base; *idion*, dim.] the promycelium of the Ustilaginales; a septate basidium as found in lower Basidiomycetes.

hemibathybial *a.* [Gk. *hēmi*, half; *bathys*, deep; *bios*, life] *pert.* plankton between littoral and bathyal zones.

hemibranch *n.* [Gk. *hēmi*, half; *brangchia*, gills] gill with gill filaments on one side only, i.e. a half gill; *alt.* demibranch.

hemicarp *n.* [Gk. *hēmi*, half; *karpos*, fruit] mericarp, *q.v.*

hemicellulases *n.plu.* [Gk. *hēmi*, half; L. *cellula*, small cell] a collection of enzymes which hydrolyse the hemicelluloses.

hemicelluloses *n.plu.* [Gk. *hēmi*, half; L. *cellula*, small cell] heterogeneous polysaccharides made of a mixture of sugars including arabinose, xylose, mannose, and galactose, forming part of the cell wall of plants and the food store of some seeds.

hemicephalous *a.* [Gk. *hēmi*, half; *kephalē*, head] *appl.* insect larvae with reduced head.

hemichlamydeous *a.* [Gk. *hēmi*, half; *chlamys*, cloak] having ovuliferous scale inverted and bearing nucellus.

Hemichorda(ta), hemichordates *n.*, *n.plu.* [Gk. *hēmi*, half; *chordē*, string] a phylum of marine, worm-like coelomates; in some classifications considered a subphylum of the Chordata or a group of protochordates, which have pharyngeal gill slits and the body and coelom divided into 3 regions, and are very different from other chordates.

hemichordate *a.* [Gk. *hēmi*, half; *chordē*, string] possessing a rudimentary notochord.

Hemicidaroida *n.* [Gk. *hēmi*, half; *Cidaris*, generic name; Gk. *eidos*, form] an order of Echinacea having perforated tubercles on the spines, a large suranal plate and the anus lying excentrically.

hemicryptophyte *n.* [Gk. *hēmi*, half; *kryptos*, hidden; *phyton*, plant] a plant with dormant buds in the soil surface, the aerial shoots surviving for a season only.

hemicyclic *a.* [Gk. *hēmi*, half; *kyklos*, round] with some floral whorls cyclic, others spiral; lacking summer stages, in life cycle of rust fungi.

hemidiscs *n.plu.* [Gk. *hēmi*, half; *diskos*, round plate] sponge microscleres of the amphidisc type but with a disc or spokes at only one end.

hemielytron hemelytron, *q.v.*

hemiepiphyte *n.* [Gk. *hēmi*, half; *epi*, upon; *phyton*, plant] a plant whose seeds germinate on another plant, but later send roots to the ground; a plant that begins life rooted but later becomes an epiphyte; *cf.* protoepiphyte.

hemigamy *n.* [Gk. *hēmi*, half; *gamos*, marriage] activation of ovum by male nucleus without nuclear fusion; *alt.* semigamy.

hemignathous *a.* [Gk. *hēmi*, half; *gnathos*, jaw] having one jaw shorter than the other, as some fishes and birds.

hemikaryon *n.* [Gk. *hēmi*, half; *karyon*, kernel] a nucleus with gametic or haploid number of chromosomes; pronucleus, *q.v.*

hemikaryotic *a.* [Gk. *hēmi*, half; *karyon*, kernel] *pert.* hemikaryon; haploid, *q.v.*

Hemimetabola *n.* [Gk. *hēmi*, half; *metabolē*, change] Exopterygota, *q.v.*; the orders of Exopterygota having nymphs adapted for aquatic life, Ephemeroptera, Odonata, Plecoptera.

hemimetabolic *a.* [Gk. *hēmi*, half; *metabolē*, change] having an incomplete or partial metamorphosis, as certain insects; *n.* hemimetabolism, *alt.* homometabolism; *cf.* holometabolic.

Hemimyaria *n.* [Gk. *hēmi*, half; *mys*, muscle] Salpidae, *q.v.*

hemiparasite *n.* [Gk. *hēmi*, half; *parasitos*, parasite] a form which is partly parasitic but can survive in the absence of its host; a parasitic plant which develops from seeds which germinate free in the soil; a parasite which can exist as a saprophyte, *alt.* facultative saprophyte.

hemipenis *n.* [Gk. *hēmi*, half; L. *penis*, penis] one of the paired copulatory organs in certain lizards and snakes; *plu.* hemipenes.

hemipneustic *a.* [Gk. *hēmi*, half; *pnein*, to breathe] with one or more pairs of spiracles closed; *cf.* holopneustic.

Hemiptera *n.* [Gk. *hēmi*, half; *pteron*, wing] an order of insects, commonly called bugs, having gradual incomplete metamorphosis, piercing and sucking mouth-parts, and 2 pairs of wings, the anterior harder than the posterior uniformly in the Homoptera and with membranous tips in the Heteroptera; in some classifications, the Homoptera and Heteroptera are given order status; *alt.* Rhynchota.

hemipterygoid n. [Gk. *hēmi*, half; *pteryx*, wing; *eidos*, form] in neognath birds, part of pterygoid which fuses with palatine.

hemisaprophyte n. [Gk. *hēmi*, half; *sapros*, decayed; *phyton*, plant] a plant living partly by photosynthesis, partly by obtaining food from humus; a saprophyte which can also survive as a parasite, *alt.* facultative parasite.

hemisome n. [Gk. *hēmi*, half; *sōma*, body] the symmetrical half of an animal about a median vertical plane.

Hemisphaeriales n. [Gk. *hēmi*, half; *sphaira*, globe] Microthyriales, *q.v.*

hemisphere n. [Gk. *hēmi*, half; *sphaira*, globe] one of the cerebral or cerebellar hemispheres.

hemispore n. [Gk. *hēmi*, half; *sporos*, seed] protoconidium, *q.v.*

hemisystole n. [Gk. *hēmi*, half; *systellein*, to contract] contraction of one ventricle of the heart.

hemitropous a. [Gk. *hēmi*, half; *tropē*, turn] turned half round, having an ovule with hilum on one side and micropyle, etc., opposite in a plane parallel to placenta; *appl.* flowers restricted to medium-length tongued insects for pollination; *appl.* insects with medium-length tongues visiting such flowers.

hemixis n. [Gk. *hēmi*, half; *mixis*, mingling] fragmentation and reorganization of macronucleus without involving micronucleus, in the ciliate *Paramecium*.

hemizonid n. [Gk. *hēmi*, half; *zōnē*, girdle] a ventrolateral commissure of the nervous system in some nematodes.

hemizygous a. [Gk. *hēmi*, half; *zygon*, yoke] *appl.* gene present in a single dose, either as a gene in a haploid organism, or a sex-linked gene in the heterogametic sex, or a gene in a segment of chromosome where its partner segment has been deleted; *alt.* haplozygous.

hemo- haemo-, *q.v.*

Henle's layer [*F. G. J. Henle*, German anatomist] outermost stratum of nucleated cubical cells in inner root sheath of a hair follicle.

Henle's loop loop of uriniferous tubule where it lies in the medulla of the kidney.

Henle's membrane fenestrated membrane of arteries.

Henle's sheath perineurium, or its prolongation surrounding branches of a nerve.

Hensen's cells [*V. Hensen*, German histologist] columnar supporting cells on basilar membrane, external to outer phalangeal cells in organ of Corti.

Hensen's line a singly refracting light disc or line dividing the darker portion of a sarcomere (the A-, or anisotropic band) into 2 parts, being a region of myosin filaments, but only very thin actin filaments; *alt.* H-band, H-disc, H-line, H-zone, M-disc, mesophragma, M-line, Q-line or disc.

Hensen's node the primitive node, *q.v.*

Hensen's stripe a band of interlacing fibrils on under surface of tectorial membrane of Corti's organ.

hepar n. [Gk. *hēpar*, liver] liver, or an organ having a similar function.

heparin n. [Gk. *hēpar*, liver] a polysaccharide containing glucosamine, glucuronic acid, and sulphuric acid, found in mast cells and other tissues, and also extracted from liver and lung, which stops blood clotting by preventing the conversion of prothrombin to thrombin, and neutralizes thrombin; *alt.* antiprothrombin, antithrombin.

heparinocytes n.plu. [Gk. *hēpar*, liver; *kytos*, hollow] mast cells containing granules in which heparin is stored.

hepatic a. [Gk. *hēpar*, liver] *pert.*, like, or associated with the liver; liver-coloured, purplish-red; *pert.* liverworts. n. A liverwort.

Hepaticae n. [Gk. *hēpar*, liver] a class of bryophytes, commonly called liverworts, characterized by leafy or thalloid gametophytes with unicellular rhizoids, no protonema, and elaters among spores; *alt.* Hepaticopsida; *cf.* Musci.

Hepaticidae n. [Gk. *hēpar*, liver] a subclass of liverworts having many chloroplasts in each cell; *cf.* Anthocerotidae.

hepaticology n. [Gk. *hēpar*, liver; *logos*, discourse] the study of Hepaticae or liverworts; *cf.* bryology.

Hepaticopsida n. [Gk. *hēpar*, liver; *opsis*, appearance] Hepaticae, *q.v.*

hepatic portal system in vertebrates, the part of the vascular system carrying blood to the liver and consisting of the hepatic portal vein and the hepatic artery; *alt.* hepatoportal system.

hepatic portal vein in vertebrates, the vein carrying blood from the gut to the liver.

hepatobiliary capsule [Gk. *hēpar*, liver; L. *bilis*, bile] Glisson's capsule, *q.v.*

hepatocolic a. [Gk. *hēpar*, liver; *colon*, large intestine] *pert.* liver and colon.

hepatocystic a. [Gk. *hēpar*, liver; *kystis*, bladder] *pert.* liver and gall bladder.

hepatoduodenal a. [Gk. *hēpar*, liver; L. *duodeni*, twelve each] *pert.* liver and duodenum.

hepatoenteric a. [Gk. *hēpar*, liver; *enteron*, gut] of or *pert.* liver and intestine.

hepatogastric a. [Gk. *hēpar*, liver; *gastēr*, stomach] *pert.* liver and stomach; *alt.* gastrohepatic.

hepatopancreas n. [Gk. *hēpar*, liver; *pan*, all; *kreas*, flesh] digestive gland in many invertebrates, supposed to perform a function similar to that of liver and of pancreas in vertebrates; *alt.* liver–pancreas.

hepatoportal system [Gk. *hēpar*, liver; L. *porta*, gate] hepatic portal system, *q.v.*

hepatorenal a. [Gk. *hēpar*, liver; L. *renes*, kidneys] *pert.* liver and kidney.

hepatoumbilical a. [Gk. *hēpar*, liver; L. *umbilicus*, navel] joining liver and umbilicus.

heptagynous a. [Gk. *hepta*, seven; *gynē*, female] with 7 pistils.

heptamerous a. [Gk. *hepta*, seven; *meros*, part] having whorls of flowers in 7; having parts in 7.

heptandrous a. [Gk. *hepta*, seven; *anēr*, man] having 7 stamens.

heptaploid a. [Gk. *hepta*, seven; *haploos*, simple; *eidos*, form] having 7 times the haploid number of chromosomes.

heptarch a. [Gk. *hepta*, seven; *archē*, beginning] having 7 initial groups of xylem.

heptastichous a. [Gk. *hepta*, seven; *stichos*, row] arranged in 7 rows, *appl.* leaves.

heptoses *n.plu.* [Gk. *hepta*, seven] monosaccharides having the formula $C_7H_{14}O_7$.

herb *n.* [L. *herba*, green crop] a seed plant with a green, non-woody stem.

herbaceous *a.* [L. *herbaceus*, grassy] *pert.* or being a herb, or similarly formed; soft, green, with little woody tissue, and having the texture of leaves, *appl.* plant organs.

herbarium *n.* [L. *herba*, herbage] a classified collection of dried or preserved plants, or of their parts; the place where they are kept.

herbivore *n.* [L. *herba*, green crop; *vorare*, to devour] a plant-eating organism; *alt.* primary consumer.

herbivorous *a.* [L. *herba*, green crop; *vorare*, to devour] eating plants, *alt.* phytophagous; originally used to mean eating herbs.

Herbst's corpuscle [*E. F. Herbst*, German anatomist] a simple type of Pacinian corpuscle, in birds.

hercogamy *n.* [Gk. *herkos*, barrier; *gamos*, union] the condition in which self-fertilization is impossible; *alt.* herkogamy.

hereditary *a.* [L. *hereditas*, heirship] transmissible from parent to offspring, as characteristics, physical or mental.

heredity *n.* [L. *hereditas*, heirship] the organic relation between successive generations; germinal constitution.

heritability *n.* [L.L. *hereditabilis*, that may be inherited] capacity for being transmitted from one generation to another; hereditary or genotypic variance expressed as percentage of total variance in the feature examined.

herkogamy hercogamy, *q.v.*

hermaphrodite *n.* [Gk. *hermaphroditos*, combining both sexes] an organism with both male and female reproductive organs; *alt.* androgyne, bisexual; *a.* hermaphroditic, hermaphrodite, *alt.* monoclinous in plants.

hermaphroditism *n.* [Gk. *hermaphroditos*, combining both sexes] the condition of having both male and female reproductive organs in one individual, in plants, also called androgynism; *alt.* gynandrism.

hermatypic *a.* [Gk. *herma*, reef; *typos*, pattern] *appl.* corals containing zooxanthellae in their endodermal tissues.

hermestism *n.* [Gk. *Hermēs*] angiocarpy, *q.v.*

heroin *n.* [Formerly trademark] an addictive alkaloid obtained from morphine by acetylation, and which acts as a narcotic.

herpes viruses a group of viruses containing double-stranded DNA, including those causing herpes and chicken pox.

herpetology *n.* [Gk. *herpeton*, reptile; *logos*, discourse] that part of zoology dealing with the structure, habits, and classification of reptiles.

hesperidin *n.* [Gk. *Hesperides*, sisters guarding the golden apples given by Gaia] a flavone derivative, the active constituent of citrin, and which affects the permeability and fragility of blood capillaries.

hesperidium *n.* [Gk. *Hesperides*] a superior, many-celled, few-seeded, indehiscent fruit, having epicarp and mesocarp joined together, and endocarp projecting into interior as membranous partitions which divide the pulp into chambers, e.g. orange.

Hesperornithiformes *n.* [*Hesperornis*, generic name; L. *forma*, shape] an order of Central and South American Cretaceous birds of the subclass Neornithes which were flightless and had teeth.

hesthogenous *a.* [Gk. *hesthos*, dress; *genēs*, born] covered with down at hatching; *alt.* dasypaedic.

het a partly heterozygous phage.

heteracanth(ous) *a.* [Gk. *heteros*, other; *akantha*, spine] having the spines in dorsal fin asymmetrical or turning alternately to one side then the other.

heteracmy *n.* [Gk. *heteros*, other; *akmē*, point] dichogamy, *q.v.*

heteractinal *a.* [Gk. *heteros*, other; *aktis*, ray] *pert.* nail-like spicules having disc of 6 to 8 rays to one plane, and a stout ray at right angles to these.

heterandrous *a.* [Gk. *heteros*, other; *anēr*, man] with stamens of different length or shape.

heterauxesis *n.* [Gk. *heteros*, other; *auxēsis*, growth] irregular or asymmetrical growth of organs; relative growth rate of parts of an organism; heterogonic or allometric growth; *cf.* bradyauxesis, isauxesis, tachyauxesis.

heterauxin heteroauxin, *q.v.*

heteraxial *a.* [Gk. *heteros*, other; L. *axis*, axis] with 3 unequal axes.

heterecious heteroecious, *q.v.*

heteroagglutinin *n.* [Gk. *heteros*, other; L. *agglutinare*, to glue to] fertilizin or agglutinin of eggs which reacts on sperm of different species; *cf.* isoagglutinin.

heteroallelic *a.* [Gk. *heteros*, other; *allēlōn*, one another] *appl.* allelic mutant genes which have mutations at different sites so that intragenic recombination can yield a functional cistron; *cf.* homoallelic.

hetero-antibody *n.* [Gk. *heteros*, other; *anti*, against; A.S. *bodig*, body] an antibody that reacts with an antigen from a member of a species other than that in which it was found.

hetero-antigen *n.* [Gk. *heteros*, other; *anti*, against; *genos*, birth] a substance that is antigenic in a species other than that from which it was obtained.

heteroauxin *n.* [Gk. *heteros*, other; *auxein*, to grow] a substance, chemically identical to β-indoleacetic acid, extracted from fungi and urine, which acts as a growth-promoting hormone; *alt.* heterauxin.

Heterobasidiomycetes, Heterobasidiomycetae *n.* [Gk. *heteros*, other; *basis*, base; *idion*, *dim.*; *mykēs*, fungus] a subclass of Basidiomycetes having septate or Y-shaped basidia; *alt.* Heterobasidiomycetidae, Hemibasidiomycetes, Phragmobasidiomycetes, Phragmobasidiomycetidae.

heterobasidium *n.* [Gk. *heteros*, other; *basis*, base; *idion*, *dim.*] a septate basidium composed of a hypobasidium and epibasidium; *cf.* homobasidium.

heteroblastic *a.* [Gk. *heteros*, other; *blastos*, bud] with indirect development; arising from dissimilar cells; *cf.* homoblastic.

heteroblastic change transition of juvenile to adult with quite abrupt changes in structure.

heterobrachial *a.* [Gk. *heteros*, other; L. *brachium*, arm] pericentric, *q.v.*

Heterocapsales *n.* [Gk. *heteros*, other; L. *capsa*, box] an order of Xanthophyceae with a palmelloid body which can return directly to a motile condition; *alt.* Heterogloeales.

heterocarpous *a.* [Gk. *heteros*, other; *karpos*, fruit] bearing more than one distinct type of fruit.

heterocaryo- heterokaryo-, *q.v.*

heterocellular *a.* [Gk. *heteros*, other; L. *cellula*, small cell] composed of cells of more than one type; *cf.* homocellular.

heterocephalous *a.* [Gk. *heteros*, other; *kephalē*, head] having pistillate flowers on separate heads from staminate; *alt.* heteroclinous.

heterocercal *a.* [Gk. *heteros*, other; *kerkos*, tail] having vertebral column terminating in upper lobe of caudal fin, which is usually larger than lower lobe; *n.* heterocercy.

heterochlamydous *a.* [Gk. *heteros*, other; *chlamys*, cloak] having a calyx differing from corolla in colour, texture, etc.; *cf.* homochlamydeous.

Heterochloridales *n.* [*Heterochloris*, generic name] an order of Xanthophyceae having a motile plant body; *alt.* Chloramoebales.

heterochore *n.* [Gk. *heteros*, other; *chorein*, to spread] a species found in more than one plant community.

heterochromatic *a.* [Gk. *heteros*, other; *chrōma*, colour] *pert.* heterochromatin, *appl.* chromosomal regions liable to become heteropycnotic; *cf.* euchromatic.

heterochromatin *n.* [Gk. *heteros*, other; *chrōma*, colour] chromatin which shows maximum staining in interphase, seems to have little genetic activity, and is the main constituent of heterochromosomes; *cf.* euchromatin.

heterochromatism *n.* [Gk. *heteros*, other; *chrōma*, colour] change of colour, as seasonal colour change in an inflorescence.

heterochromaty differential staining.

heterochromia *n.* [Gk. *heteros*, other; *chrōma*, colour] difference in colour of parts normally of one colour, as of irises of a pair of eyes.

heterochromosome *n.* [Gk. *heteros*, other; *chrōma*, colour; *sōma*, body] a chromosome made mainly of heterochromatin, an allosome, *q.v.*; *alt.* heterosome; *cf.* euchromosome.

heterochromous *a.* [Gk. *heteros*, other; *chrōma*, colour] differently coloured, *appl.* disc and marginal florets; *cf.* homochromous.

heterochronism *n.* [Gk. *heteros*, other; *chronos*, ' time] departure from typical sequence in time of formation of organs; *alt.* heterochrony.

heterochronous *pert.*, or exhibiting, heterochronism.

heterochrony heterochronism, *q.v.*

heterochrosis *n.* [Gk. *heteros*, other; *chrōs*, colouring] abnormal coloration.

heteroclinous *a.* [Gk. *heteros*, other; *klinē*, bed] heterocephalous, *q.v.*

Heterococcales *n.* [*Heterococcus*, generic name] an order of Xanthophyceae including unattached and planktonic forms with a rigid cell wall; *alt.* Mischococcales.

Heterocoela *n.* [Gk. *heteros*, other; *koilos*, hollow] an order of Calcarea having the main cavity without choanocytes which are confined to flagellated chambers in the body wall.

heterocoelous *a.* [Gk. *heteros*, other; *koilos*, hollow] *pert.* vertebrae with saddle-shaped articulatory centra; *alt.* concavoconvex.

heterocont heterokont, *q.v.*

Heterocotylea *n.* [Gk. *heteros*, other; *kotylē*, cup] Monogenea, *q.v.*

heterocotyledonous *a.* [Gk. *heteros*, other; *kotylēdōn*, cup] having cotyledons unequally developed; *alt.* heterocotylous.

heterocotylus *n.* [Gk. *heteros*, other; *kotylē*, cup] hectocotylus, *q.v.*

Heterocyemida(e) *n.* [Gk. *heteros*, other; *kyēma*, embryo] an order of mesozoans of the class Dicyemida, having a non-ciliated body in the adult stage.

heterocysts *n.plu.* [Gk. *heteros*, other; *kystis*, bladder] clear cells occurring at intervals on filaments of certain blue–green algae, whose function is uncertain, but may be associated with the formation of hormogonia, or with nitrogen fixation.

heterodactylous *a.* [Gk. *heteros*, other; *daktylous*, digit] with the 1st and 2nd toes turned backwards.

heterodermic graft a heterograft of skin.

heterodont *a.* [Gk. *heteros*, other; *odous*, tooth] having the teeth differentiated for various purposes; *alt.* anisodont; *cf.* homodont.

Heterodonta *n.* [Gk. *heteros*, other; *odous*, tooth] an order of mostly burrowing lamellibranchs with eulamellibranch gills and heterodont hinge teeth of the shell; *cf.* Eulamellibranchiata.

Heterodontiformes *n.* [Gk. *heteros*, other; *odous*, tooth; L. *forma*, shape] an order of Selachii, including the hornsharks, existing from Devonian times to the present day, having the notochord persisting in early forms but partially replaced in modern ones.

heterodromous *a.* [Gk. *heteros*, other; *dramein*, to run] having genetic spiral of stem leaves turning in different direction to that of branch leaves.

heteroduplex *n.* [Gk. *heteros*, other; L. *duplex*, twofold] a double-stranded nucleic acid with each strand from a different source.

heterodynamic *a.* [Gk. *heteros*, other; *dynamis*, power] unequal in potential performance, *appl.* dominant and recessive genes or characters.

heterodynamic hybrid one resembling one parent more than the other.

heteroecious *a.* [Gk. *heteros*, other; *oikos*, house] passing different stages of life history in different hosts; *alt.* heterecious, heteroicious, heteroxenous, metoecious, metoxenous, pleioxenous; *n.* heteroecism; *cf.* monoxenous.

heterofacial *a.* [Gk. *heteros*, other; L. *facies*, face] showing regional differentiation.

heterogameon *n.* [Gk. *heteros*, other; *gamos*, marriage] a species consisting of races which, when selfed, produce a morphologically stable population, but when crossed may produce several types of viable and fertile progeny.

heterogametangic *a.* [Gk. *heteros*, other; *gametēs*, spouse; *anggeion*, vessel] having more than one kind of gametangium.

heterogamete *n.* [Gk. *heteros*, other; *gametēs*, spouse] anisogamete, *q.v.*

heterogametic *a.* [Gk. *heteros*, other; *gametēs*, spouse] elaborating 2 kinds of gametes in equal numbers; having unequal pair of sex chromosomes, XY or WZ; *appl.* sex that is heterozygous; digametic, *q.v.*; *cf.* homogametic.

heterogamous *a.* [Gk. *heteros*, other; *gamos*,

heterogamy

marriage] having unlike gametes; having 2 or more types of flower, such as male, female, hermaphrodite, neuter; having indirect pollination methods.

heterogamy n. [Gk. *heteros*, other; *gamos*, offspring] alternation of generations, q.v.; alternation of 2 sexual generations, one being true sexual, the other parthenogenetic, alt. heterogony; anisogamy, q.v.

heterogangliate a. [Gk. *heteros*, other; *gangglion*, ganglion] with widely separated and asymmetrically placed nerve ganglia.

heterogeneity n. [Gk. *heteros*, other; *genos*, kind] heterogenous state; heterogenetic or genotypic dissimilarity.

heterogeneous a. [Gk. *heteros*, other; *genos*, a kind] consisting of dissimilar parts; cf. homogeneous.

Heterogeneratae n. [Gk. *heteros*, other; L. *generare*, to beget] a group of brown algae having heteromorphic alternation of generations with the sporophyte larger than the gametophyte.

heterogenesis n. [Gk. *heteros*, other; *genesis*, descent] abiogenesis, q.v.; alternation of generations, q.v.; alt. xenogenesis; cf. metagenesis.

heterogenetic a. [Gk. *heteros*, other; *genesis*, descent] descended from different ancestral stock; pert. heterogenesis; appl. induction or stimulation by a complex of stimuli of different origin.

heterogenous a. [Gk. *heteros*, other; *genēs*, produced] having a different origin; not originating in the body; pert. heterogeny.

heterogenous nuclear RNA (HnRNA) DNA-like RNA, being RNA which never leaves the nucleus; alt. dRNA, messenger-like RNA, pre-messenger RNA, nascent RNA.

heterogeny n. [Gk. *heteros*, other; *genos*, generation] having several distinct generations succeeding one another in a regular series.

Heterogloeales n. [Gk. *heteros*, other; *gloia*, glue] Heterocapsales, q.v.

heterogonic a. [Gk. *heteros*, other; *gonos*, produce] allometric, q.v.

heterogonous a. [Gk. *heteros*, other; *gonos*, birth] pert. heterogenesis, or heterogony.

heterogony n. [Gk. *heteros*, other; *gonos*, birth] condition of having 2 or 3 kinds of flowers differing in length of stamen; alternation of generations, q.v.; allometry, q.v.; see heterogamy.

heterograft n. [Gk. *heteros*, other; O.F. *graffe*, graft] a graft originating from a donor of a different species from the recipient; alt. heterologous graft, heteroplastic graft, xenograft.

heterogynous a. [Gk. *heteros*, other; *gynē*, woman] with 2 types of females.

heteroicous heteroecious, q.v.

heteroimmune a. [Gk. *heteros*, other; L. *immunis*, free] displaying immunity to an antigen from a different animal species.

heterokaryo- alt. heterocaryo-.

heterokaryon n. [Gk. *heteros*, other; *karyon*, nucleus] an individual having heterokaryotic cells; a cell formed by fusion of hyphal cells, the haploid nuclei remaining separate.

heterokaryosis n. [Gk. *heteros*, other; *karyon*, nucleus] presence of genetically dissimilar nuclei within individual cells; alt. heterokaryotic condition.

heterokaryote, heterokaryotic a. [Gk. *heteros*, other; *karyon*, nucleus] having genetically dissimilar nuclei, in a multinucleate cell, or in different cells of a hypha.

heterokinesis n. [Gk. *heteros*, other; *kinein*, to move] qualitative or differential division of chromosomes.

Heterokontae n. [Gk. *heteros*, other; *kontos*, punting pole] formerly a subclass of algae of the Chlorophyceae, having flagella of unequal length, now classified as Xanthophyceae.

heterokont(an) a. [Gk. *heteros*, other; *kontos*, punting pole] having flagella or cilia of unequal length; alt. heterocont(an); cf. isokont.

heterolecithal a. [Gk. *heteros*, other; *lekithos*, yolk] having unequally distributed deutoplasm.

heterologous a. [Gk. *heteros*, other; *logos*, relation] of different origin; derived from a different species; differing morphologically, appl. alternating generations; appl. various substances, e.g. agglutinins, affecting other than species of origin; appl. antigen and antibody that do not correspond; appl. graft: heterograft, q.v.; cf. homologous.

heterology n. [Gk. *heteros*, other; *logos*, relation] non-correspondence of parts owing to different origin or different elements.

heterolysis n. [Gk. *heteros*, other; *lysis*, loosing] cell or tissue disintegration by action of exogenous agents or enzymes; cf. autolysis.

heterolytic a. [Gk. *heteros*, other; *lyein*, to dissolve] causing or pert. heterolysis; cf. autolytic.

heteromallous a. [Gk. *heteros*, other; *mallos*, lock of wool] spreading in different directions.

heteromastigate, heteromastigote a. [Gk. *heteros*, other; *mastix*, whip] having 2 different types of flagella.

heteromeric a. [Gk. *heteros*, other; *meros*, part] pert. another part; appl. neuron: with axon extending to the other side of the spinal cord.

heteromerous a. [Gk. *heteros*, other; *meros*, part] having, or consisting of, an unequal number of parts, appl. whorls, tarsi, etc., cf. isomerous; having a stratified thallus, as in lichens with algae in a certain layer and not distributed equally through mycelium; composed of units, as of cells, of different types; cf. homoiomerous.

Heterometabola n. [Gk. *heteros*, other; *metabolē*, change] Exopterygota, q.v.; Exopterygota excluding the Hemimetabola.

heterometabolic a. [Gk. *heteros*, other; *metabolē*, change] having incomplete metamorphosis.

Heteromi n. [Gk. *heteros*, other; *ōmos*, shoulder] Notacanthiformes, q.v.

heteromixis n. [Gk. *heteros*, other; *mixis*, mingling] the union of genetically different nuclei, as in heterothallism.

heteromorphic a. [Gk. *heteros*, other; *morphē*, shape] having different forms at different times; appl. chromosomes of different size and shape, or chromosome pairs different in size; appl. alternation of diploid and haploid phases in morphologically dissimilar generations, alt. antithetic; heteromorphous, q.v.; n. heteromorphism; cf. homomorphic, isomorphic.

heteromorphosis n. [Gk. *heteros*, other; *morphōsis*, shaping] production of a part in an abnormal position; regeneration when the new part is different from that removed, cf. homomorphosis; alt. xenomorphosis.

heteromorphous a. [Gk. *heteros*, other; *morphē*,

shape] *pert.* an irregular structure, or departure from the normal; *alt.* heteromorphic.

heteromyaric *a.* [Gk. *heteros*, other; *mys*, muscle] with adductor muscles unequal in size; *alt.* heteromyarian.

Heteronematales *n.* [Gk. *heteros*, other; *nēma*, thread] an order of colourless Euglenophyceae.

Heteronemertini, heteronemertines *n., n.plu.* [Gk. *heteros*, other; *nēma*, thread] an order of Anopla having the body wall muscles arranged in 3 layers the innermost of which is longitudinal.

heteronereis *n.* [Gk. *heteros*, other; *Nēreis*, Nereid] a free-swimming dimorphic sexual stage of *Nereis* and other marine polychaetes.

heteronomous *a.* [Gk. *heteros*, other; *nómos*, law] subject to different laws of growth; specialized on different lines. [Gk. *heteros*, other; *nómos*, department] *appl.* segmentation into dissimilar segments.

heteropelmous *a.* [Gk. *heteros*, other; *pelma*, sole of foot] having flexor tendons of toes bifid.

heteropetalous *a.* [Gk. *heteros*, other; *petalon*, leaf] with dissimilar petals.

heterophagous *a.* [Gk. *heteros*, other; *phagein*, to eat] having young in altrices condition.

heterophil *a.* [Gk. *heteros*, other; *philos*, loving] *appl.* non-specific antigens and antibodies present in an organism, affording natural immunity; *appl.* granular leucocytes which show interspecific differences in their reaction to stains. *n.* Polymorphonuclear leucocyte, *q.v.*

heterophyadic *a.* [Gk. *heteros*, other; *phyas*, shoot] producing separate shoots, one vegetative, one reproductive.

heterophyllous *a.* [Gk. *heteros*, other; *phyllon*, leaf] bearing foliage leaves of different shape on different parts of the same plant; having lamellae of different size or shape, as some agarics.

heterophylly heterophyllous condition; *alt.* anisophylly.

heterophyte *n.* [Gk. *heteros*, other; *phyton*, plant] a saprophytic or parasitic plant; *cf.* autophyte.

heterophytic *a.* [Gk. *heteros*, other; *phyton*, plant] with 2 kinds of spores, borne by different sporophytes; *cf.* homophytic.

heteroplanogametes *n.plu.* [Gk. *heteros*, other; *planos*, wandering; *gametēs*, spouse] motile gametes that are unlike one another.

heteroplasia *n.* [Gk. *heteros*, other; *plassein*, to mould] the development of one tissue from another of a different kind.

heteroplasm *n.* [Gk. *heteros*, other; *plasma*, mould] tissue formed in abnormal places.

heteroplasma *n.* [Gk. *heteros*, other; *plasma*, mould] plasma from a different species used as a medium for tissue culture; *cf.* autoplasma, homoplasma.

heteroplastic *a.* [Gk. *heteros*, other; *plastos*, formed] *appl.* grafts of unrelated material; *appl.* grafts between individuals of different species or genera, *alt.* heterograft; *appl.* grafts from one region of the body to another; *cf.* homoplastic, xenoplastic.

heteroploid *a.* [Gk. *heteros*, other; *haploos*, one-fold] having an extra chromosome through nondisjunction of a pair in meiosis; not having a multiple of the basic haploid number of chromosomes; *alt.* aneuploid. *n.* An organism having heteroploid nuclei.

heteropolymeric *a.* [Gk. *heteros*, other; *polys*, many; *meros*, part] readily dissociated into dissimilar parts, *appl.* macromolecules.

heteroproteose *n.* [Gk. *heteros*, other; *prōteion*, first] one of primary products formed by action of gastric juices or other hydrolysing agents on proteins; *alt.* propeptone.

Heteroptera *n.* [Gk. *heteros*, other; *pteron*, wing] a group of insects including the water boatman, capsids, and bed bugs, sometimes considered as an order, or included as a suborder in the Hemiptera, *q.v.*

heteropycnosis, heteropyknosis *n.* [Gk. *heteros*, other; *pyknos*, dense] the phenomenon of certain chromosomes or regions of chromosomes having a coiling cycle out of phase with the rest of the chromosome set; *a.* heteropycnotic, heteropyknotic.

heterorhizal *a.* [Gk. *heteros*, other; *rhiza*, root] with roots coming from no determinate point.

heterosaccharides *n.plu.* [Gk. *heteros*, other; L. *saccharum*, sugar] polysaccharides composed of dissimilar structural units, e.g. gums, pectins, chitin, heparin.

heterosexual *a.* [Gk. *heteros*, other; L. *sexus*, sex] of, or *pert.* the opposite sex, *appl.* hormones, etc.

Heterosiphonales *n.* [Gk. *heteros*, other; *siphōn*, tube] an order of Xanthophyceae, comprising siphoneous forms; *alt.* Vaucheriales.

heterosis *n.* [Gk. *heteros*, other] cross-fertilization; hybrid vigour, result of heterozygosis.

heterosomal *a.* [Gk. *heteros*, other; *sōma*, body] occurring in, or *pert.* different bodies; *appl.* rearrangements in 2 or more chromosomes of a set.

Heterosomata *n.* [Gk. *heteros*, other; *sōma*, body] Pleuronectiformes, *q.v.*

heterosome *n.* [Gk. *heteros*, other; *sōma*, body] heterochromosome, *q.v.*

heterosporangic *a.* [Gk. *heteros*, other; *sporos*, seed; *anggeion*, vessel] bearing 2 kinds of spores in separate sporangia; *n.* heterosporangy.

heterospory *n.* [Gk. *heteros*, other; *sporos*, seed] the production of 2 kinds of spores, microspores and megaspores; *a.* heterosporic, heterosporous.

heterostemonous *a.* [Gk. *heteros*, other; *stēmōn*, stamen] with unlike stamens.

Heterostraci *n.* [Gk. *heteros*, other; *ostrakon*, shell] a subclass of fossil Palaeozoic ostracoderms having an exoskeleton of dermal bone and including Ordovician and Devonian forms; sometimes considered an order equivalent to the Pteraspida.

heterostrophy *n.* [Gk. *heteros*, other; *strophē*, turning] the condition of being coiled in a direction opposite to normal.

heterostyly *n.* [Gk. *heteros*, other; *stylos*, pillar] the condition of having unlike or unequal styles; *a.* heterostylic, heterostyled, heterostylous; *cf.* homostyly.

heterosynapsis *n.* [Gk. *heteros*, other; *synapsis*, union] pairing of 2 dissimilar chromosomes; *cf.* homosynapsis.

Heterotardigrada *n.* [Gk. *heteros*, other; L. *tardigradus*, slow paced] an order of tardigrades, having a pair of lateral cirri behind the head.

heterotaxis *n.* [Gk. *heteros*, other; *taxis*, arrangement] abnormal or unusual arrangement of organs or parts.

heterothallic *a.* [Gk. *heteros*, other; *thallos*, young shoot] having physiologically different strains of thalli, cells, or mycelia, so that sexual reproduction can only occur between different strains, *appl.* algae and fungi; *alt.* haplodioecious; *n.* heterothallism; *cf.* homothallic.

heterothermal *a.* [Gk. *heteros*, other; *thermē*, heat] poikilothermal, *q.v.*

heterotic *a.* [Gk. *heteros*, other] *pert.* heterosis, *appl.* vigour.

heterotomy *n.* [Gk. *heteros*, other; *temnein*, to cut] condition of having parts of perianth whorls unequal or dissimilar; irregular dichotomy in Crinoidea.

heterotopic *a.* [Gk. *heteros*, other; *topos*, place] in a different or unusual place, *appl.* transplantation; *cf.* orthotopic.

heterotopy *n.* [Gk. *heteros*, other; *topos*, place] displacement; abnormal habitat.

Heterotrichales *n.* [Gk. *heteros*, other; *thrix*, hair] an order of Xanthophyceae having cells arranged in filaments; *alt.* Tribonematales.

Heterotrichida *n.* [Gk. *heteros*, other; *thrix*, hair] an order or suborder of Spirotricha having body with normal cilia and special cilia around the mouth.

heterotrichous *a.* [Gk. *heteros*, other; *thrix*, hair] having 2 types of cilia; having thallus consisting of prostrate and erect filaments as certain algae.

heterotroph *n.* [Gk. *heteros*, other; *trophē*, nourishment] an organism using organic compounds as its source of carbon, as photoheterotrophs and chemoheterotrophs, and including holozoic animals, parasites, and saprophytes; often used synonymously with chemoheterotroph; *alt.* consumer, organotroph; *cf.* autotroph.

heterotrophic *a.* [Gk. *heteros*, other; *trophē*, nourishment] obtaining food as a heterotroph; *alt.* zootrophic, allotrophic, organotrophic.

heterotropic *a.* [Gk. *heteros*, other; *tropikos*, *pert.* turning] turning or turned in a different direction; not continued in line of axis; *alt.* heterotropous.

heterotropic chromosome sex chromosome, *q.v.*

heterotropous *a.* [Gk. *heteros*, other; *trepein*, to turn] heterotropic, *q.v.*; *pert.* ovule with hilum and micropyle at opposite ends in a plane parallel to placenta.

heterotype *n.* [Gk. *heteros*, other; *typos*, pattern] first meiotic division; *cf.* homeotype.

heterotypic *a.* [Gk. *heteros*, other; *typos*, pattern] *pert.* mitotic division in which daughter chromosomes remain united and form rings; *appl.* 1st or reduction division in meiosis; *cf.* homeotypic.

heterotypical *a.* [Gk. *heteros*, other; *typos*, pattern] *appl.* a genus comprising species that are not truly related.

heteroxenous *a.* [Gk. *heteros*, other; *xenos*, host] heteroecious, *q.v.*; *n.* heteroxeny.

heteroxylous *a.* [Gk. *heteros*, other; *xylon*, wood] *appl.* wood containing vessels and fibres as well as tracheids.

heterozygosis *n.* [Gk. *heteros*, other; *zygon*, yoke] descent from 2 different species, varieties, or races; conditions of having a pair of dissimilar alleles.

heterozygote *n.* [Gk. *heteros*, other; *zygon*, yoke] an organism or cell having 2 different alleles at corresponding loci on homologous chromosomes; *alt.* hybrid; *a.* heterozygous.

heterozygous *a.* [Gk. *heteros*, other] *pert.* heterozygote; *alt.* hybrid.

hexacanth *a.* [Gk. *hex*, six; *akantha*, thorn] having 6 hooks, *appl.* embryo or oncosphere of certain tapeworms.

hexactinal *a.* [Gk. *hex*, six; *aktis*, ray] with 6 rays; *alt.* sexradiate.

hexactine *n.* [Gk. *hex*, six; *aktis*, ray] a spicule with 6 equal and similar rays meeting at right angles.

Hexactinellida *n.* [*Hexactinella*, generic name] a class of typically radially symmetrical sponges of the subphylum Nuda, commonly called glass sponges, having a skeleton of 6-rayed large silica spicules, megascleres, and more variable microscleres.

Hexactinia *n.* [Gk. *hex*, six; *aktis*, ray] Zoantharia, *q.v.*

hexactinian *a.* [Gk. *hex*, six; *aktis*, ray] with tentacles or mesenteries in multiples of 6.

hexacyclic *a.* [Gk. *hex*, six; *kyklos*, circle] having floral whorls consisting of 6 parts.

hexaenes *n.plu.* [Gk. *hex*, six; *triaene*] in sponges, megascleres similar to triaenes but with 6 branches.

hexagynous *a.* [Gk. *hex*, six; *gynē*, woman] having 6 pistils or styles; having 6 carpels to a gynaecium.

hexamerous *a.* [Gk. *hex*, six; *meros*, part] occurring in 6s, or arranged in 6s.

Hexanchiformes *n.* [*Hexanchus*, generic name; L. *forma*, shape] an order of Selachii existing from Jurassic times to the present day, including the modern cowsharks and frilled sharks.

hexandrous *a.* [Gk. *hex*, six; *anēr*, man] having 6 stamens; *alt.* hexastemonous.

hexapetaloid *a.* [Gk. *hex*, six; *petalon*, leaf; *eidos*, form] with petaloid perianth of 6 parts.

hexapetalous *a.* [Gk. *hex*, six; *petalon*, leaf] having 6 petals.

hexaphyllous *a.* [Gk. *hex*, six; *phyllon*, leaf] having 6 leaves or leaflets.

hexaploid *a.* [Gk. *hex*, six; *haploos*, simple; *eidos*, form] with 6 sets of chromosomes. *n.* An organism having 6 times the haploid chromosome number.

hexapod *a.* [Gk. *hex*, six; *pous*, foot] having or *pert.* 6. legs. *n.* An insect.

Hexapoda *n.* [Gk. *hex*, six; *pous*, foot] Insecta, *q.v.*

hexapterous *a.* [Gk. *hex*, six; *pteron*, wing] having 6 wing-like processes or expansions.

Hexaradiata *n.* [Gk. *hex*, six; L. *radius*, ray] Zoantharia, *q.v.*

hexarch *a.* [Gk. *hex*, six; *archē*, beginning] having 6 alternating xylem and phloem groups; having 6 vascular bundles.

hexasepalous *a.* [Gk. *hex*, six; F. *sépale*, sepal] having 6 sepals.

hexaspermous *a.* [Gk. *hex*, six; *sperma*, seed] having 6 seeds.

hexasporous *a.* [Gk. *hex*, six; *sporos*, seed] having 6 spores.

hexastemonous *a.* [Gk. *hex*, six; *stēmōn*, thread] hexandrous, *q.v.*

hexaster *n.* [Gk. *hex*, six; *astēr*, star] a variety of hexactine in which the rays branch and produce star-shaped figures.

Hexasterophora n. [Gk. *hex*, six; *astēr*, star; *phora*, producing] an order of Hexactinellida in which the microscleres are all hexactines and are miniature versions of the megascleres.

hexastichous a. [Gk. *hex*, six; *stichos*, row] having the parts arranged in 6 rows.

hexicology n. [Gk. *hexis*, habit; *logos*, discourse] ecology, *q.v.*

hexokinase n. [Gk. *hex*, six; *kinein*, to move] an enzyme which causes the phosphorylation of glucose; EC 2.7.1.1; *r.n.* hexokinase.

hexosamine an amino sugar in which the sugar is a hexose.

hexosan a polysaccharide made of condensed hexose units, including starch, inulin, glycogen, cellulose.

hexose monophosphate shunt pentose phosphate pathway, *q.v.*

hexoses *n.plu.* [Gk. *hex*, six] monosaccharides having the formula $C_6H_{12}O_6$ and including glucose, fructose, galactose, and mannose.

hexuronic acid a tetrahydroxy-aldehyde acid obtained by the oxidation of hexose sugars; a historical name for vitamin C.

Hfr strain a strain of *Escherichia coli* (*E. coli*) having high frequencies of recombination (hence the abbreviation), the F-factor being integrated into the bacterial chromosome.

Hgb haemoglobin, *q.v.*

HGF hyperglycaemic-glycogenolytic factor, *q.v.*

HGH human growth hormone, *q.v.*

hiatus n. [L. *hiare*, to gape] any large gap or opening.

hibernaculum n. [L. *hibernaculum*, winter quarters] a winter bud; specially modified winter bud in freshwater Polyzoa.

hibernal a. [L. *hibernus*, wintry] of the winter; *alt.* hiemal.

hibernating glands lymph glands of richly vascularized fatty tissue occurring in some rodents and insectivores.

hibernation n. [L. *hibernus*, wintry] the condition of passing the winter in a resting state; *v.* hibernate; *cf.* aestivation.

hidrosis n. [Gk. *hidros*, sweat] excretion of sweat, perspiration.

hiemal a. [L. *hiems*, winter] *pert.* winter, *appl.* aspect of a community; *alt.* hibernal.

hierarchy n. [Gk. *hierarchēs*, rank] a social system in which the members of the group are organized in ranks so that, in any encounter, one is dominant and aggressive and the other submissive, *alt.* social hierarchy, dominance hierarchy, *cf.* peck order; a natural classification system in which organisms are grouped according to the number of characteristics they have in common, and ranked one above another.

Highmore's antrum [*N. Highmore*, English surgeon] the maxillary sinus, which communicates with the middle meatus of the nose.

Highmore's body mediastinum testis, *q.v.*

hilar a. [L. *hilum*, trifle] of or *pert.* a hilum.

hile hilum, *q.v.*, in plants.

hiliferous a. [L. *hilum*, trifle; *ferre*, to carry] having a hilum.

Hill reaction [*R. Hill*, British chemist] the reaction showing that illuminated chloroplasts isolated from green leaves could cause the reduction of certain chemical agents such as ferricyanide to ferrocyanide, and the production of oxygen.

hilum n. [L. *hilum*, trifle] scar on ovule where it was attached to placenta; the same scar on a seed; nucleus of starch grain; small notch, opening, or depression, usually where vessels, nerves, etc., enter, of kidney, lung, spleen, etc.; *alt.* hilus, umbilicus.

hilus hilum, *q.v.*

hind-brain rhombencephalon, *q.v.*

hind-gut diverticulum of yolk sac extending into tail-fold in human embryo; posterior portion of alimentary tract; proctodaeum, *q.v.*

hind-kidney metanephros, *q.v.*

hinge cells large cells which, by changes in turgor, control rolling and unrolling of a leaf.

hinge joint ginglymus, *q.v.*

hinge ligament the tough elastic substance joining the 2 valves of a bivalve shell.

hinge line the line of articulation of the 2 valves in a bivalve shell.

hinge tooth one of the projections found on the hinge line in bivalves.

hinoid a. [Gk. *his*, nerve; *eidos*, form] with parallel veins at right angles to midrib, *appl.* leaf type.

hip n. [A.S. *hēope*] cynarrhodium, *q.v.* [A.S. *hype*] coxa, *q.v.*

hip girdle pelvic girdle, *q.v.*

hip joint the ball-and-socket joint between femur and hip girdle.

hippocampal a. [Gk. *hippos*, horse; *kampē*, bend] *pert.* the hippocampus.

hippocampus n. [Gk. *hippos*, horse; *kampē*, bend] part of rhinencephalon forming an eminence extending throughout length of floor of inferior cornu of lateral ventricle; *alt.* hippocampus major.

hippocampus minor calcar avis, *q.v.*

hippocrepian, hippocrepiform a. [Gk. *hippos*, horse; *krēpis*, shoe] shaped like a horseshoe.

hippuric acid [Gk. *hippos*, horses; *ouron*, urine] benzoyl glycine, a normal constituent of the urine of herbivorous animals and first discovered in horse urine.

hippuricase an enzyme, first found in the kidneys of certain animals and in fungi, which hydrolyses an N-acyl amino acid to an amino acid plus a fatty acid, e.g. hippuric acid into glycine and benzoic acid; *alt.* histozyme; EC 3.5.1.14; *r.n.* aminoacylase.

hirsute a. [L. *hirsutus*, shaggy] covered with hair-like feathers, *appl.* birds; having stiff, hairy bristles or covering.

hirsutidin n. [L. *hirsutus*, shaggy] a blue anthocyanin pigment.

hirudin n. [L. *hirudo*, leech] a substance, obtained in solution from buccal glands of leech, which prevents clotting of blood by inhibiting action of thrombin on fibrinogen, and so acts as an anticoagulant.

Hirudinea n. [L. *hirudo*, leech] a group of carnivorous or ectoparasitic annelids, commonly called leeches, that are considered either as a class, or as a subclass of the Clitellata, and have 33 segments, circumoral and posterior suckers, and usually no chaetae.

his histidine, *q.v.*

His' bundle [*W. His, Jr.*, German anatomist] band of muscle fibres, with nerve fibres, connecting auricles and ventricles of heart; *alt.* atrioventricular or auriculoventricular bundle, Gaskell's bridge.

hispid *a.* [L. *hispidus*, rough] having stiff hairs, spines, or bristles.

histamine *n.* [Gk. *histos*, tissue; *ammōniakon*, resinous gum] a base formed in the body from histidine and also found in ergot and some plant and animal tissues, which stimulates the autonomic nervous system, capillary dilatation, and gastric hydrochloric acid and pepsin secretion.

histaminocytes *n.plu.* [Gk. *histos*, tissue; *ammōniakon*, resinous gum; *kytos*, hollow] mast cells which secrete histamine, sometimes used for any mast cell.

histidine *n.* [Gk. *histion*, tissue] an amino acid essential for many animals but probably not for man, and the precursor of histamine; abbreviated to his.

histioblast *n.* [Gk. *histion*, tissue; *blastos*, bud] an immature histiocyte; one of the cells derived from division of an archaeocyte, and capable of developing into body cells of sponges.

histiocyte *n.* [Gk. *histion*, tissue; *kytos*, hollow] a primitive blood cell giving rise to a monocyte; a monocyte of reticular origin, or a clasmatocyte derived from endothelium, a reticulo-endothelial cell; fixed macrophage in loose connective tissue; *alt.* adventitial cell, rhagiocrine cell, polyblast, leucocytoid, clasmatocyte, endotheliocyte; *a.* histiocytic.

histiogenic histogenic, *q.v.*

histioid *a.* [Gk. *histion*, web; *eidos*, form] like a web, *alt.* arachnoid; tissue-like; *alt.* histoid.

histiomonocyte *n.* [Gk. *histion*, tissue; *monos*, alone; *kytos*, hollow] an endothelial cell of certain capillaries and associated with the histiocytic metabolic system.

histiotypic *a.* [Gk. *histion*, tissue; *typos*, pattern] *appl.* uncontrolled or unorganized growth of cells, in tissue culture; *cf.* organotypic.

histoblast *n.* [Gk. *histos*, tissue; *blastos*, bud] a unit of tissue; imaginal disc, *q.v.*

histochemistry *n.* [Gk. *histos*, tissue; *chēmeia*, transmutation] the chemistry of living tissues.

histocompatibility antigen a glycoprotein antigen which is found in cell membranes especially of lymphocytes and macrophages, and initiates immunity so that a homograft is rejected; *alt.* transplantation antigen.

histocyte *n.* [Gk. *histos*, tissue; *kytos*, hollow] tissue cell as distinguished from germ cell.

histogen cell initial, *q.v.*

histogenesis *n.* [Gk. *histos*, tissue; *genesis*, descent] formation and development of tissue; *alt.* histogeny.

histogenic *a.* [Gk. *histos*, tissue; *-genēs*, producing] tissue-producing, *appl.* the separate meristematic layers in a stratified growing point; *alt.* histiogenic.

histogenous produced in or from tissue, *appl.* cavities, conidia, etc.

histogens *n.plu.* [Gk. *histos*, tissue; *gennaein*, to produce] a term formerly used for tissue-producing zones or layers in a shoot or root tip that form definite tissues, and which are now thought not to exist in a stem, but are still sometimes used for the root meristem; *see* plerome, periblem, dermatogen, calyptrogen; *alt.* germinal layers; *cf.* tunica-corpus theory.

histogeny *n.* [Gk. *histos*, tissue; *gennaein*, to produce] histogenesis, *q.v.*

histohaematins *n.plu.* [Gk. *histos*, tissue; *haima*, blood] a name for cytochromes, *q.v.*, coined by McMunn in 1855 when he discovered them in animal tissue by their absorption spectra, and which he also called myohaematin when found in muscle tissue.

histoid *a.* [Gk. *histos*, tissue; *eidos*, form] histioid, *q.v.*

histology *n.* [Gk. *histos*, tissue; *logos*, discourse] the science which treats of the detailed structure of animal or plant tissues, microscopic morphology; *alt.* histomorphology.

histolysis *n.* [Gk. *histos*, tissue; *lyein*, to dissolve] the dissolution of organic tissues; process by which most of pupal internal organs dissolve into creamy fluid, except certain cells round which new imaginal tissues are formed.

histometabasis *n.* [Gk. *histos*, tissue; *metabasis*, alteration] fossilization with retention of the detailed structure of animal or plant tissues.

histomorphology histology, *q.v.*

histones *n.plu.* [Gk. *histos*, tissue] simple basic proteins which yield large amounts of arginine and lysine on hydrolysis, found in chromosomes bound to nucleic acids, where they may be important in switching off gene action.

histophyly *n.* [Gk. *histos*, tissue; *phylē*, tribe] phylogenetic history of a group of cells.

histoteleosis *n.* [Gk. *histos*, tissue; *teleios*, full-grown] the completion of functional differentiation of tissue cells.

histotrophic *a.* [Gk. *histos*, tissue; *trephein*, to nourish] *pert.* or connected with tissue formation or repair.

histozoic *a.* [Gk. *histos*, tissue; *zōon*, animal] living within tissue, *appl.* trophozoitic stage of certain Sporozoa.

histozyme *n.* [Gk. *histos*, tissue; *zyme*, leaven] hippuricase, *q.v.*

HL-A system the group of histocompatibility antigens found in man which promote the greatest immunological response.

H-line Hensen's line, *q.v.*, it can also signify heller line, heller meaning lighter; *alt.* H-band, H-disc, H-zone.

HnRNA heterogenous nuclear RNA, *q.v.*

hoary *a.* [M.E. *hor*, grey, old] greyish-white due to fine pubescence.

hock *n.* [A.S. *hoh*, heel] the tarsal joint, or its region; *alt.* hough.

holandric *a.* [Gk. *holos*, whole; *anēr*, male] *pert.* holandry; transmitted from male to male through the Y-chromosome, *appl.* sex-linked characters.

holandry *n.* [Gk. *holos*, whole; *anēr*, male] the condition of having full number of testes, as 2 pairs in Oligochaeta; *cf.* meroandry.

Holarctic *a.* [Gk. *holos*, whole; *Arktos*, Great Bear] *appl.* or *pert.* a zoogeographical region including northern parts of the Old and New Worlds or palaearctic and nearctic subregions.

holard *n.* [Gk. *holos*, whole; *ardo*, I water] total water content of soil; *cf.* chresard, echard.

holaspidean *a.* [Gk. *holos*, whole; *aspis*, shield] with single series of large scales on posterior aspect of tarsometatarsus.

Holasteroida *n.* [Gk. *holos*, whole; *astēr*, star; *eidos*, form] an order of elongated or bottle-shaped sea urchins of the superorder Atelostomata that live in deep water.

holcodont *a.* [Gk. *holkos*, furrow; *odous*, tooth] having the teeth in a long continuous groove.

holdfast a sucker or disc-like extension of a thallus, primarily for attachment, as appressorium, hapteron, hyphopodium.

Holectypoida *n.* [*Holectypus*, generic name; Gk. *eidos*, form] an order of extinct sea urchins of the superorder Gnathostomata.

holendobiosis *n.* [Gk. *holos*, whole; *endon*, within; *biōsis*, a living] spore production in its host's tissues by a parasitic organism that cannot live outside its host; *a.* holendobiotic.

Holobasidiae *n.* [Gk. *holos*, whole; *basis*, base; *idion*, dim.] a group of fungi having a holobasidium.

Holobasidiomycetes, Holobasidiomycetidae *n.* [Gk. *holos*, whole; *basis*, base; *idion*, dim.; *mykēs*, fungus] Homobasidiomycetes, *q.v.*

holobasidium *n.* [Gk. *holos*, whole; *basis*, base; *idion*, dim.] a basidium not divided by septa; *alt.* autobasidium.

holobenthic *a.* [Gk. *holos*, whole; *benthos*, depths] living on sea bottom or in depths of sea throughout life.

holoblastic *a.* [Gk. *holos*, whole; *blastos*, bud] *pert.* eggs with total cleavage.

holobranch *n.* [Gk. *holos*, whole; *brangchia*, gills] a gill in which gill filaments are borne on both sides.

holocarpic *a.* [Gk. *holos*, whole; *karpos*, fruit] having the fruit body formed by the entire thallus, *appl.* certain algae; *appl.* fungi without rhizoids or haustoria, living in host cell, as certain Phycomycetes; *cf.* eucarpic.

holocene *a.* [Gk. *holos*, whole; *kainos*, recent] recent geological epoch following Pleistocene; *alt.* postglacial age, Recent.

Holocephali *n.* [Gk. *holos*, whole; *kephalē*, head] a subclass of Chondrichthyes, existing from Devonian times to present day, sometimes raised to class status, including the chimaeras and their fossil relatives, having holostylic jaws with crushing teeth, a whip-like tail, and an operculum covering the gills; *alt.* Bradyodonti.

holocephalous *a.* [Gk. *holos*, whole; *kephalē*, head] *appl.* rib with a single head.

holochlamydate *a.* [Gk. *holos*, whole; *chlamys*, cloak] having no notch on mantle margin.

holochroal *a.* [Gk. *holos*, whole; *chros*, skin] having eyes with globular or biconvex lenses closely crowded together, so that cornea is continuous over whole eye.

holocoen *n.* [Gk. *holox*, whole; *koinos*, common] a complete ecosystem with all its living and non-living components.

holocrine *a.* [Gk. *holos*, whole; *krinein*, to separate] *appl.* glands whose secretion is formed by complete breakdown of the secretory cells, as sebaceous glands; *cf.* apocrine, merocrine, eccrine.

holocyclic *a.* [Gk. *holos*, whole; *kyklos*, circle] *pert.* or completing alternation of sexual and parthenogenetic generations.

holodikaryotic *a.* [Gk. *holos*, whole; *dis*, double; *karyon*, nucleus] having a pair of nuclei and lacking a haploid phase.

holodont *a.* [Gk. *holos*, whole; *odous*, tooth] having no longitudinal median groove on the lamellae on the ventral side, *appl.* peristome teeth.

holoenzyme *n.* [Gk. *holos*, whole; *en*, in; *zymē*, leaven] an active enzyme consisting of an apoenzyme and coenzyme neither of which is active by itself.

hologametes *n.plu.* [Gk. *holos*, whole; *gametēs*, spouse] fully developed protistans taking part in syngamy; *cf.* merogametes.

hologamodeme *n.* [Gk. *holos*, whole; *gamos*, marriage; *dēmos*, people] a group of organisms which interbreed to produce fertile offspring, i.e. a biological species.

hologamy *n.* [Gk. *holos*, whole; *gamos*, marriage] condition of having gametes similar to somatic cells; fusion between mature individuals of a species as of the heliozoan *Actinophrys*; *alt.* macrogamy.

hologastrula *n.* [Gk. *holos*, whole; *gastēr*, stomach] gastrula formed from holoblastic egg.

holognathous *a.* [Gk. *holos*, whole; *gnathos*, jaw] having the jaw in a single piece.

hologonidium soredium, *q.v.*

hologynic *a.* [Gk. *holos*, whole; *gynē*, woman] transmitted direct from female to female, *appl.* sex-linked characters.

Holomastigina *n.* [Gk. *holos*, whole; *mastix*, whip] an order of Zoomastigina having many flagella, and whose body surface can be amoeboid.

holomastigote *a.* [Gk. *holos*, whole; *mastix*, whip] having one type of flagellum scattered evenly over the body.

Holometabola *n.* [Gk. *holos*, whole; *metabolē*, change] Endopterygota, *q.v.*

holometabolic *a.* [Gk. *holos*, whole; *metabolē*, change] having complete metamorphosis, as certain insects; *n.* holometabolism; *cf.* hemimetabolic.

holomorphosis *n.* [Gk. *holos*, whole; *morphōsis*, a shaping] regeneration in which the entire part is replaced.

holonephridia *n.* [Gk. *holos*, whole; *nephros*, kidney; *idion*, dim.] meganephridia, *q.v.*

holonephros *n.* [Gk. *holos*, whole; *nephros*, kidney] the hypothetical continuous excretory organ.

holoparasite *n.* [Gk. *holos*, whole; *parasitos*, parasite] a parasite which cannot exist independently of a host or on a dead host; *alt.* obligate parasite.

Holopeltida *n.* [Gk. *holos*, whole; *peltē*, shield] a group of arachnids, sometimes considered to be of order status or included with the whip scorpions, having a prosoma covered by a single thick chitinous plate.

holophyte *n.* [Gk. *holos*, whole; *phyton*, plant] any green, phototrophic, independent plant.

holophytic *a.* [Gk. *holos*, whole; *phyton*, plant] autotrophic, *q.v.*; sometimes used only for phototrophic *q.v.*; *cf.* holozoic.

holoplankton *n.* [Gk. *holos*, whole; *plangktos*, wandering] organisms which complete their life cycle in the plankton; *a.* holoplanktonic; *cf.* meroplankton.

holopneustic *a.* [Gk. *holos*, whole; *pnein*, to breathe] with all spiracles open for respiration; *cf.* hemipneustic.

holoptic *a.* [Gk. *holos*, whole; *ōps*, eye] having eyes of 2 sides meeting in a coadapted line of union; *cf.* dichoptic.

holorhinal *a.* [Gk. *holos*, whole; *rhines*, nostrils] having nares with posterior margin rounded; *cf.* schizorhinal.

holosaprophyte *n.* [Gk. *holos*, whole; *sapros*, rotten; *phyton*, plant] a saprophyte which cannot

holoschisis

survive on a living host, or without a host; *alt.* obligate saprophyte.

holoschisis *n.* [Gk. *holos*, whole; *schizein*, to cut] amitosis, *q.v.*

holosericeous *a.* [Gk. *holos*, whole; L.L. *sericeus*, silken] completely covered with silky hair-like structures; having a silky lustre or sheen.

Holostei *n.* [Gk. *holos*, whole; *osteon*, bone] a group of actinopterygians, intermediate between the palaeoniscids and teleosts, including many extinct forms and also the modern bow fin and bony pike.

holosteous *a.* [Gk. *holos*, whole; *osteon*, bone] having a bony skeleton.

holostomatous *a.* [Gk. *holos*, whole; *stoma*, mouth] with margin of aperture entire.

holostyly *n.* [Gk. *holos*, whole; *stylos*, pillar] the type of jaw suspension found in holocephalan fish in which the palatoquadrate is fused with the cranium without involvement of the hyoid arch; *a.* holostylic; *cf.* autostyly, hyostyly.

holosystolic *a.* [Gk. *holos*, whole; *systolē*, contraction] *pert.* complete systole.

Holothuroidea *n.* [*Holothuria*, generic name; Gk. *eidos*, form] a class of sausage-shaped echinoderms of the subphylum Eleutherozoa, commonly called sea cucumbers, having minute skeletal plates embedded in the fleshy body wall.

Holotricha *n.* [Gk. *holos*, whole; *thrix*, hair] an order of Ciliata having no obvious zone of composite cilia around the mouth, the general cilia all over the body forming the main swimming organelles; in some classifications considered a subclass of Ciliphora.

holotrichous *a.* [Gk. *holos*, whole; *thrix*, hair] having a uniform covering of cilia over the body.

holotype *n.* [Gk. *holos*, whole; *typos*, pattern] the single specimen chosen for designation of a new species; *alt.* type specimen; *cf.* hypoparatype.

holozoic *a.* [Gk. *holos*, whole; *zōon*, animal] obtaining food in the manner of animals by ingesting food material and then digesting it; *alt.* zootrophic; *n.* holozoon; *cf.* holophytic.

holozygote *n.* [Gk. *holos*, whole; *zygōtos*, yoked] a zygote containing the entire genomes of both uniting cells or gametes; *cf.* merozygote.

homacanth *a.* [Gk. *homos*, same; *akantha*, spine] having spines of dorsal fin symmetrical.

homaxial, homaxonic *a.* [Gk. *homos*, same; *axōn*, axis] built up around equal axes.

homeo- *alt.* homoeo-, homoio-, *q.v.*

homeochronous homochronous, *q.v.*

homeokinesis *n.* [Gk. *homoios*, alike; *kinein*, to move] mitosis with equal division of chromosomes to daughter nuclei.

homeostasis *n.* [Gk. *homoios*, alike; *stasis*, standing] the maintenance of the constancy of the internal environment of the body; the maintenance of equilibrium between organism and environment; the balance of nature; *alt.* homoeostasis.

homeostat *n.* [Gk. *homoios*, alike; *statos*, standing] any cytoplasmic or non-generic carrier of a heritable character.

homeosynapsis homosynapsis, *q.v.*

homeotely *n.* [Gk. *homoios*, alike; *telos*, end] evolution from homologous parts, but with less close resemblance.

homeotype *n.* [Gk. *homoios*, alike; *typos*, pattern] second meiotic division; *alt.* homotype; *cf.* heterotype.

homeotypic *a.* [Gk. *homoios*, alike; *typos*, character] *appl.* 2nd division in meiosis, similar to typical mitosis; *alt.* homotypic; *cf.* heterotypic.

homeozoic *a.* [Gk. *homoios*, alike; *zōon*, animal; or *zōē*, life] *pert.* a region or series of regions with similar fauna, or fauna and flora; *alt.* homoeozoic.

home range territory, *q.v.*

hominid *n.* [L. *homo*, man; Gk. *eidos*, form] a man-like ape; a member of the family Hominidae, including modern man.

Homo *n.* [L. *homo*, man] a genus of primates, including the living modern man *Homo sapiens*, and the fossil species *Homo habilis*, which may be an austalopithecine, *Homo erectus*, known as pithecanthropines, *q.v.*, Neanderthal man, *q.v.*, and modern man, the sapiens, *q.v.*

homoallelic *a.* [Gk. *homos*, same; *allēlon*, one another] *appl.* allelic mutant genes which have mutations at the same site, so that intragenic recombination cannot yield a functional cistron; *cf.* heteroallelic.

Homobasidiomycetes, Homobasidiomycetae *n.* [Gk. *homos*, same; *basis*, base; *idion*, *dim.*; *mykēs*, fungus] a subclass of Basidiomycetes including the mushrooms, toadstools, and puffballs, having non-septate, usually club-shaped basidia; *alt.* Homobasidiomycetidae, Holobasidiomycetes, Autobasidiomycetes.

homobasidium *n.* [Gk. *homos*, same; *basis*, base; *idion*, *dim.*] a simple non-septate basidium; *cf.* heterobasidium.

homobium *n.* [Gk. *homos*, same; *bios*, life] the interdependence and mutual life of fungus and alga in lichens.

homoblastic *a.* [Gk. *homos*, same; *blastos*, bud] having direct embryonic development; arising from similar cells; *cf.* heteroblastic.

homobrachial *a.* [Gk. *homos*, same; L. *brachium*, arm] paracentric, *q.v.*

homocarpous *a.* [Gk. *homos*, same; *karpos*, fruit] bearing one kind of fruit.

homocaryo- homokaryo-, *q.v.*

homocellular *a.* [Gk. *homos*, same; L. *cellula*, small room] composed of cells of one type only; *cf.* heterocellular.

homocercal *a.* [Gk. *homos*, same; *kerkos*, tail] having a tail with equal or nearly equal lobes, and axis ending near middle of base.

homochlamydeous *a.* [Gk. *homos*, same; *chlamys*, cloak] having outer and inner perianth whorls alike, not distinguishable as calyx and corolla; *alt.* homoiochlamydeous; *cf.* heterochlamydeous.

homochromous *a.* [Gk. *homos*, same; *chrōma*, colour] of one colour, *appl.* capitular florets; *cf.* heterochromous.

homochronous *a.* [Gk. *homos*, same; *chronos*, time] occurring at the same age or period, in successive generations; *alt.* homeochronous.

Homocoela *n.* [Gk. *homos*, same; *koilos*, hollow] an order of Calcarea having the main cavity lined with choanocytes.

homodermic *a.* [Gk. *homos*, same; *derma*, skin] sprung from same embryonic layer.

homodont *a.* [Gk. *homos*, same; *odous*, tooth] having the teeth all alike, not differentiated; *alt.* isodont; *cf.* heterodont.

homodromous *a.* [Gk. *homos*, same; *dramein*, to run] having genetic spiral alike in direction in stem and branches; moving or acting in the same direction.

homodynamic *a.* [Gk. *homos*, same; *dynamis*, power] developing without resting stages; *appl.* insects not requiring a diapause for further development; *pert.* homodynamy; acting upon the production of the same phenotypic effects at the same time, *appl.* genes.

homodynamy *n.* [Gk. *homos*, same; *dynamis*, power] metameric homology.

homoeandrous *a.* [Gk. *homoios*, alike; *anēr*, male] having uniform stamens.

homoecious *a.* [Gk. *homos*, same; *oikos*, abode] occupying the same host or shelter during the life cycle.

homoeo- *alt.* homeo-, homoio-, *q.v.*

homoeologous *a.* [Gk. *homoios*, like; *logos*, relation] *appl.* chromosomes having in part the same sequence of genes; partly homologous.

homoeologue a homoeologous chromosome.

homoeomerous homoiomerous, *q.v.*

homoeomorphic *a.* [Gk. *homoios*, like; *morphē*, form] resembling in shape or structure; exhibiting convergence.

homoeosis *n.* [Gk. *homoiōsis*, likeness] assumption by one part of likeness to another part, as modification of antenna into foot, or of petal into stamen; *alt.* metamorphy, metamorphosis.

homoeostasis homeostasis, *q.v.*

homoeotype *n.* [Gk. *homoios*, alike; *typos*, pattern] a specimen authoritatively stated to be identical with the holotype, lectotype, paratypes, or syntypes of its species.

homoeozoic *a.* [Gk. *homoios*, alike; *zōē*, life] homeozoic, *q.v.*

homogametic *a.* [Gk. *homos*, same; *gametēs*, spouse] having homogametes or gametes of one type, *appl.* sex possessing 2 X-chromosomes; *cf.* heterogametic, digametic.

homogamous *a.* [Gk. *homos*, same; *gamos*, marriage] characterized by homogamy.

homogamy *a.* [Gk. *homos*, same; *gamos*, marriage] inbreeding due to some type of isolation; condition of having flowers all alike, having stamens and pistils mature at same time; *alt.* synacme; *cf.* dichogamy.

homogangliate *a.* [Gk. *homos*, same; *ganglion*, knot] having ganglia of nerve loops symmetrically arranged.

homogen *n.* [Gk. *homos*, same; *genos*, race] one of a group having a common origin; one of a series of identically derived parts.

homogeneity homogeny, *q.v.*

homogeneous *a.* [Gk. *homos*, same; *genos*, kind] having the same kind of constituent units throughout; of the same nature; *alt.* homogeneal; *cf.* heterogeneous.

homogenesis *n.* [Gk. *homos*, same; *genesis*, descent] the type of reproduction in which like begets like.

homogenetic *a.* [Gk. *homos*, same; *genesis*, descent] having the same origin; *pert.* homogenesis; *appl.* pairing of homologous chromosomes.

homogenous *a.* [Gk. *homos*, same; *genos*, race] more or less alike owing to descent from common stock; *appl.* graft from another animal of same species.

homogeny *n.* [Gk. *homos*, same; *genos*, race] correspondence between parts due to common descent; the same genotypical structure; *alt.* homogeneity.

homoglandular *a.* [Gk. *homos*, same; L. *glandula*, small acorn] of or *pert.* the same gland.

homogony *n.* [Gk. *homos*, same; *gonos*, offspring] condition of having one type of flower with equally long stamens and pistil; *alt.* homostyly.

homograft *n.* [Gk. *homos*, same; O.F. *graffe*, graft] a graft of tissue taken from a donor of the same species as the recipient but of different genotype; *alt.* allograft.

homoio- *also* homeo-, homoeo-, *q.v.*

homoiochlamydeous *a.* [Gk. *homoios*, like; *chlamys*, cloak] homochlamydeous, *q.v.*

homoiomerous *a.* [Gk. *homoios*, like; *meros*, part] having algae distributed equally through fungoid mycelium in a lichen; composed of units, as of cells, of the same type, *cf.* heteromerous; *alt.* homoeomerous.

homoioplastic *a.* [Gk. *homoios*, like; *plastos*, formed] homoplastic, *q.v.*

homoiosmotic *a.* [Gk. *homoios*, like; *ōsmos*, impulse] *appl.* organisms with constant internal osmotic pressure; euryhaline, *q.v.*; *cf.* poikilosmotic.

homoiotherm a homoiothermal animal.

homoiothermal, homoiothermic *a.* [Gk. *homoios*, like; *thermos*, hot] having a more-or-less constant body temperature, usually above that of the surrounding medium; *alt.* homoeothermal, homothermal, homothermic, homothermous, haematothermal, endothermal, idiothermous, warm-blooded; *cf.* poikilothermal.

homoiotransplantation transplantation of tissue or organ from one organism to another, possibly unrelated; *cf.* autotransplantation.

homokaryo- *alt.* homocaryo-.

homokaryon *n.* [Gk. *homos*, same; *karyon*, nucleus] an individual having homokaryotic cells; a hypha or mycelium having more than one haploid nucleus of identical genetic constitution.

homokaryosis *n.* [Gk. *homos*, same; *karyon*, nucleus] the homokaryotic condition.

homokaryote, homokaryotic *a.* [Gk. *homos*, same; *karyon*, nucleus] having genetically identical nuclei in a multinucleate cell or in different cells of a hypha.

homolateral *a.* [Gk. *homos*, same; L. *latus*, side] on, or *pert.*, the same side.

homolecithal *a.* [Gk. *homos*, same; *lekithos*, yolk] having little deutoplasm, which is equally distributed.

homolog homologue, *q.v.*

homologous *a.* [Gk. *homologos*, agreeing] resembling in structure and origin; *appl.* alternating generations which are fundamentally alike in origin, *cf.* antithetic; *appl.* various substances, e.g. agglutinins affecting organisms of same species only; *appl.* chromosomes with the same sequence of genes; *appl.* genes determining the same character, e.g. eye colour; *cf.* heterologous, antithetic.

homologue *n.* [Gk. *homologos*, agreeing] a structure formed by homology; a homologous agent or structure; *alt.* homolog.

homology *n.* [Gk. *homologia*, agreement] the phenomenon of having the same phylogenetic origin but not necessarily the same final structure or function.

homomallous

homomallous *a.* [Gk. *homos*, same; *mallos*, lock of wool] curving uniformly to one side, *appl.* leaves.

homomixls *n.* [Gk. *homos*, same; *mixis*, mingling] the union of nuclei from the same thallus as in homothallism.

homomorphic *a.* [Gk. *homos*, same; *morphē*, form] of similar size or structure, *appl.* chromosome pairs; *pert.* or exhibiting homomorphism; *cf.* heteromorphic.

homomorphism *n.* [Gk. *homos*, same; *morphē*, shape] the condition of having perfect flowers of only one type; similarity of larva and adult, *alt.* hemimetabolism.

homomorphosis *n.* [Gk. *homos*, same; *morphōsis*, shaping] condition of having a newly regenerated part like the part removed; *cf.* heteromorphosis.

homonculus *see* homunculus.

homonomic *a.* [Gk. *homos*, same; *nômos*, law] having the same behaviour, *appl.* affinity, as of tissues combining, e.g. vascular anastomoses, or complementary affinity, as in adrenal medulla and cortex; homonomous, *q.v.*

homonomous *a.* [Gk. *homos*, same; *nomôs*, department] *appl.* segmentation into similar segments. [Gk. *nômos*, law] following same stages or process, as of development or growth; *alt.* homonomic.

homonomy *n.* [Gk. *homos*, same; *nômos*, law] the homology existing between parts arranged on transverse axes.

homonym *n.* [Gk. *homos*, same; *onyma*, name] a name already used in classification and therefore unsuitable according to law of priority.

homopetalous *a.* [Gk. *homos*, same; *petalon*, leaf] having all the petals alike.

homophyadic *a.* [Gk. *homos*, same; *phyas*, shoot] producing only one kind of shoot.

homophylic *a.* [Gk. *homos*, same; *phylē*, race] resembling one another owing to a common ancestry.

homophyllous *a.* [Gk. *homos*, same; *phyllon*, leaf] bearing leaves all of one kind.

homophytic *a.* [Gk. *homos*, same; *phyton*, plant] with 2 kinds of spores, or one bisexual type, borne by a single sporophyte; *cf.* heterophytic.

homoplasma *n.* [Gk. *homos*, same; *plasma*, mould] plasma from another animal of same species used as a medium for tissue culture; *cf.* autoplasma, heteroplasma.

homoplasmic *a.* [Gk. *homos*, same; *plasma*, mould] having the same general form, *alt.* isotelic; *pert.* homoplasma.

homoplast *n.* [Gk. *homos*, same; *plastos*, moulded] an organ or organism formed of similar cells, such as a coenobium; *alt.* catallact, homoplastid; *cf.* alloplast.

homoplastic *a.* [Gk. *homos*, same; *plastos*, moulded] *pert.* homoplasty; *appl.* graft made into individual of same species, *cf.* autoplastic; similar in shape or structure but not in origin, *cf.* homologous; *alt.* homoioplastic, isotelic; *cf.* heteroplastic.

homoplastid homoplast, *q.v.*

homoplasty, homoplasy *n.* [Gk. *homos*, same; *plastos*, moulded] resemblance in form of structure between different organs or organisms due to evolution along similar lines; *alt.* convergence, isotely.

homopolar *a.* [Gk. *homos*, same; *polos*, pole] having both ends of an axis alike.

Homoptera *n.* [Gk. *homos*, same; *pteron*, wing] a group of insects, including the plant bugs, aphids, cicadas, and scale insects, sometimes considered an order or included as a suborder in the Hemiptera.

homopterous *a.* [Gk. *homos*, same; *pteron*, wing] having the wings alike.

homorhizal *a.* [Gk. *homos*, same; *rhiza*, root] not having an antiapical root, as Pteridophyta; *cf.* allorhizal.

homosequential *a.* [Gk. *homos*, same; L.L. *sequentia*, sequence] *appl.* species of Diptera with polytene chromosomes that have exactly the same banding patterns.

homosomal *a.* [Gk. *homos*, same; *sōma*, body] occurring in, or *pert.*, the same body; *appl.* rearrangements restricted to a single chromosome.

homosporangic *a.* [Gk. *homos*, same; *sporos*, seed] having spores of one kind or of 2 kinds in one sporangium.

homosporic, homosporous *a.* [Gk. *homos*, same; *sporos*, seed] producing only one kind of spore; *alt.* isosporous; *n.* homospory.

homostyly *n.* [Gk. *homos*, same; *stylos*, pillar] the condition of having styles of the same length, and the same length as the anthers; *alt.* homogony; *a.* homostyled, homostylic, homostylous; *cf.* heterostyly.

homosynapsis *n.* [Gk. *homos*, same; *synapsis*, union] pairing of 2 homologous chromosomes; *alt.* homeosynapsis; *cf.* heterosynapsis.

homotaxis, homotaxy *n.* [Gk. *homos*, same; *taxis*, arrangement] similar assemblage or succession of species or types in different regions or strata, not necessarily contemporaneous; *a.* homotaxial.

homothallic *a.* [Gk. *homos*, same; *thallos*, young shoot] having the capacity to undergo sexual reproduction with a strain like itself, or one of its own branches, *appl.* thallus, cell, or mycelium of algae and fungi; *alt.* haplomonoecious; *cf.* heterothallic.

homothallism the homothallic condition; *alt.* homomixis.

homothallium medulla of lichens.

homotherm *n.* [Gk. *homos*, same; *thermē*, heat] any homoiothermal animal.

homothermic, homothermal, homothermous homoiothermal, *q.v.*

homotropous *a.* [Gk. *homos*, same; *tropē*, turn] turned in the same direction; having micropyle and chalaza at opposite ends, *appl.* ovules.

homotype *n.* [Gk. *homos*, same; *typos*, pattern] a structure corresponding to that on the opposite side of the body axis; an enantiomorphic structure; homeotype, *q.v.*

homotypic *a.* [Gk. *homos*, same; *typos*, pattern] homeotypic, *q.v.*; *pert.* or exhibiting homotypy.

homotypy *n.* [Gk. *homos*, same; *typos*, pattern] equality of structures along main axis of body; serial homology; reversed symmetry; enantiomorphic condition, *q.v.*

homoxylous *a.* [Gk. *homos*, same; *xylon*, wood] *appl.* wood without vessels and consisting of tracheids.

homozygosis *n.* [Gk. *homos*, same; *zygon*, yoke] the condition of being a homozygote.

homozygote *n.* [Gk. *homos*, same; *zygōtos*,

yoked] an organism or cell having 2 identical alleles at corresponding loci on homologous chromosomes; *a.* homozygous.

homozygous *a.* [Gk. *homos*, same; *zygon*, yoke] exhibiting or *pert.* homozygosis; *pert.* homozygote; *alt.* isogenic.

homunculus *n.* [L. *homunculus*, little man] the small miniature of human foetus supposed to be in spermatozoon, according to believers in preformation theory; homonculus: a human dwarf normally proportioned.

honey *n.* [M.E. *hony*, honey] a sweet secretion produced from the nectar of flowers by honey bees by enzyme action, and used as food for the larvae.

honeycomb a structure of waxen hexagonal cells formed by the honey bee for the rearing of larvae and the storage of honey; any structure of similar design.

honeycomb stomach or bag reticulum, 2nd stomach of ruminants.

honey dew a sugary exudation found on leaves of many plants; a viscous fluid secreted by mycelium of ergot; a sweet secretion produced by certain insects, e.g. by aphids.

honey guides, nectar guides, *q.v.*

honeyspoon in bees the expanded end of the united glossae.

honey-stomach in certain hymenopterans (Aculeata), an expansion of the oesophagus in the anterior portion of the abdomen, serving to store ingested liquid which is regurgitated as required.

hooded *a.* [A.S. *hōd*, hood] bearing a hood-like petal, *alt.* cucullate; rolled up like a cone of paper, as certain leaves; having head conspicuously and differently coloured from rest of body; having crests on head; having wing-shaped expansions on neck, as in cobra; in certain cephalopods (Tetrabranchiata) having a thickened anterior part of the head-foot which acts as an operculum when the animal retracts into its living chamber.

Hookeriales *n.* [*Hookeria*, generic name] an order of mosses having branched prostrate gametophores, often with 'leaves' with 2 nerves, and a capsule with a double peristome.

hook glands paired longitudinal glands uniting anteriorly to form head gland in Pentastomida.

Hoplocarida *n.* [Gk. *hoplon*, weapon; L. *caris*, a kind of crab] a superorder, or in some classifications an order, of malacostracans containing a single order, the Stomatopoda, *q.v.*

Hoplonemertini, hoplonemertines *n., n.plu.* [Gk. *hoplon*, weapon; *Nēmertēs*, a nereid] an order of Enopla having an armed proboscis.

hordeaceous *a.* [L. *hordeum*, barley] *pert.* or resembling barley.

hordein *n.* [L. *hordeum*, barley] the prolamine of barley grains.

horiodimorphism *n.* [Gk. *hōrios*, in season; *dis*, twice; *morphē*, shape] seasonal dimorphism; *alt.* horodimorphism.

horizon *n.* [Gk. *horizōn*, bounding] soil layer of a more-or-less well-defined character; a layer of deposit characterized by definite fossil species and formed at a definite time.

horizontal *a.* [Gk. *horizōn*, bounding] growing in a plane at right angles to a primary axis.

horme *n.* [Gk. *hormē*, impetus] purposive behaviour, *q.v.*; urge or drive in living cells or organisms; élan vital, *alt.* libido; *cf.* vital force.

hormesis *n.* [Gk. *hormaein*, to excite] stimulation by a non-poisonous dose of a toxic substance or agent.

hormic *a.* [Gk. *hormē*, impetus] instinctive; purposive.

hormocyst *n.* [Gk. *hormos*, chain; *kystis*, bladder] a modified thick-walled hormogonium, in some blue–green algae; in certain lichens, a soredium-like structure.

hormocystangium *n.* [Gk. *hormos*, chain; *kystis*, bladder; *anggeion*, vessel] an apothecium-like structure in which hormocysts originate, in certain lichens.

Hormogonales *n.* [Gk. *hormos*, chain; *gonē*, generation] an order of blue–green algae having cells united in trichomes and reproducing by hormogonia; *alt.* Oscillatoriales, Hormogoneales.

hormogone, hormogonium *n.* [Gk. *hormos*, chain; *gonē*, generation] that portion of an algal filament between 2 heterocysts, which, breaking away, acts as a reproductive body.

hormones *n.plu.* [Gk. *hormaein*, to excite] substances produced in cells in one part of an organism and transported to other parts where they have an effect; animal hormones (internal secretions) are produced by ductless glands of the endocrine system and pass into and travel in the blood; plant hormones include auxins, gibberellins and cytokinins, etc., *see* phytohormones; exciting agents, *cf.* chalones.

hormonic *a.* [Gk. *hormaein*, to excite] *pert.* hormones; *appl.* excitatory internal secretions, *cf.* chalonic.

hormonopoiesis *n.* [Gk. *hormaein*, to excite; *poiēsis*, making] the production of hormones.

hormoproteins *n.plu.* [Gk. *hormaein*, to excite; *prōteion*, first] proteins or protein derivatives which are hormones.

hormospore *n.* [Gk. *hormos*, chain; *sporos*, seed] a spore dividing into microgonidia, as of some lichens; multicellular spore-like hormogonium which develops at the tip of the trichomes in certain blue–green algae and germinates into a new filament.

hormotropic *a.* [Gk. *hormaein*, to excite; *tropikos*, turning] influencing endocrine glands, *appl.* secretions of anterior lobe of pituitary, as corticotropins, gonadotropins, thyrotropin.

horn *n.* [A.S. *horn*, horn] the process on head of many animals; any projection resembling a horn; anterior part of each uterus when posterior parts are united to form median corpus uteri; a tuft of feathers in an owl; a spine in fishes; a tentacle in snails; an awn; any pointed projection or process in plants; cornu, *q.v.*; keratin, *q.v.*

horn core the os cornu, fusing with frontal bone, over which fits hollow horn of ruminants.

hornworts the common name for the Anthocerotales, *q.v.*

horny corals the common name for the Gorgonacea, *q.v.*

horodimorphism horiodimorphism, *q.v.*

horological *a.* [Gk. *hōra*, right time; *logos*, discourse] *appl.* flowers opening and closing at certain times of day or night.

horotelic *a.* [Gk. *hōra*, right time; *telos*, fulfilment] evolving at the standard rate; *n.* horotely; *cf.* bradytelic, tachytelic.

horseshoe crabs a common name for the Xiphosura, *q.v.*

horsetails the common name for the Sphenopsida, *q.v.*

Hortega cells [*P. de R. Hortega*, Spanish histologist] phagocytic neuroglial cells or microglia.

host *n.* [L. *hospes*, host] any organism in which another organism spends part or the whole of its existence, and from which it derives nourishment or gets protection; an organism which receives grafted or transplanted tissue.

hough hock, *q.v.*

hourglass cell a cell in the testa of some legumes that has the lumen shaped like an hourglass due to an uneven distribution of the secondary wall; *alt.* pillar cell.

house *n.* [A.S. *hūs*, house] the external gelatinous-like covering secreted by certain tunicates; the case of some pterobranchs; the loose fitting shell of some protozoans.

Houston's valves [*J. Houston*, Irish surgeon] semilunar transverse folds of mucous membrane in the rectum; *alt.* plicae transversales recti.

Howship's lacuna [*J. Howship*, English anatomist] in bone, a lacuna containing an osteoclast.

H-substance a compound liberated by hurtful stimulation of skin tissues and acting like histamine.

hull *n.* [A.S. *helan*, to cover] outer covering or husk of cereal seeds; *appl.* cells covering a cleistothecium.

human growth hormone (HGH) the growth hormone, somatotrophin, extracted from humans.

humeral *a.* [L. *humerus*, shoulder] *pert.* shoulder region; *pert.* the anterior basal angle of insect wing; *appl.* a cross-vein; one of horny plates on plastron of chelonians.

humerus *n.* [L. *humerus*, shoulder] the bone of the proximal part of vertebrate forelimb, the upper arm.

humicole, humicolous *a.* [L. *humus*, soil; *colere*, to dwell] soil-inhabiting; growing in or on soil or humus.

humidity *n.* [L. *humere*, to be moist] the amount of water vapour in the air.

humification *n.* [L. *humus*, soil] production of humus from the decay of organic matter as a result of fungal and bacterial action.

humistratous *a.* [L. *humus*, soil; *sternere*, to spread] spreading over surface of ground.

humor *n.* [L. *humor*, moisture] any body fluid or juice; the fluid of the eye.

humoral *a.* [L. *humor*, moisture] *appl.* theory of immunity ascribing to body fluids the power to resist infection.

humulon(e) *n.* [*Humulus lupulus*, hop] a bitter antibiotic obtained from lupulin.

humus *n.* [L. *humus*, earth] a dark material formed by decomposition of vegetable or animal matter and constituting organic part of soils.

humus plant saprophyte, *q.v.*; a non-green flowering plant which obtains its food from humus with the aid of mycorrhizae.

husk *n.* [M.E. *huske*, husk] the outer coating of various seeds.

Huxley's layer [*T. H. Huxley*, English zoologist] the middle layer of polyhedral cells in inner root sheath of hair.

hyaline *a.* [Gk. *hyalos*, glass] clear; transparent; free from inclusions.

hyaline cartilage cartilage not containing obvious fibres and having a clear glassy appearance.

hyalodermis *n.* [Gk. *hyalos*, glass; *derma*, skin] tissue of large, empty and absorptive cells in sphagnum moss.

hyalogen *n.* [Gk. *hyalos*, glass; *-genēs*, producing] any of substances found in animal tissues which are insoluble and related to mucoids.

hyaloid *a.* [Gk. *hyalos*, glass; *eidos*, form] glassy; transparent.

hyaloid artery from central artery of retina through hyaloid canal to back of lens, in foetal eye.

hyaloid canal through vitreous body of eye, from optic nerve to back of lens.

hyaloid fossa anterior concavity of vitreous body, receptacle of crystalline lens.

hyaloid membrane delicate membrane enveloping vitreous body of eye.

hyalomere *n.* [Gk. *hyalos*, glass; *meros*, part] the clear homogeneous part of a blood platelet; *cf.* chromomere.

hyalomucoid *n.* [Gk. *hyalos*, glass; L. *mucus*, mucus; Gk. *eidos*, like] one of the non-phosphorized glucoproteins in vitreous humour.

hyaloplasm(a) *n.* [Gk. *hyalos*, glass; *plasma*, mould] the ground substance of cytoplasm in which organelles are found, *alt.* cytohyaloplasma; outer cytoplasm in plant cells.

hyalopterous *a.* [Gk. *hyalos*, glass; *pteron*, wing] having transparent wings.

hyalosome *n.* [Gk. *hyalos*, glass; *sōma*, body] a nucleolar body in a cell nucleus, only slightly stainable by nuclear or plasma stains.

hyalospore *n.* [Gk. *hyalos*, glass; *sporos*, seed] transparent unicellular spore, especially in Fungi Imperfecti.

hyalosporous *a.* [Gk. *hyalos*, glass; *sporos*, seed] having colourless spores or conidia.

hyaluronic acid (HA) any of a group of viscous, high molecular weight polymers of N-acetylglucosamine and glucuronic acid, found in many tissues, which act as lubricating agents in synovial fluid and form the cementing substance between animal cells.

hyaluronidase an enzyme which increases tissue permeability by diminishing viscosity of hyaluronic acid; *alt.* Duran–Reynals spreading factor; EC 3.2.1.36; *r.n.* hyaluronoglucuronidase.

hybrid *n.* [L. *hibrida*, cross] any cross-bred animal or plant; heterozygote, *q.v. a.* Cross-bred; heterozygous, *q.v. v.* To hybridize.

hybrid cline nothocline, *q.v.*

hybrid incapacitation hybrid sterility and inviability, inclusively.

hybridism *n.* [L. *hibrida*, cross] the state or quality of being a hybrid.

hybridization *n.* [L. *hibrida*, cross] act or process of hybridizing; state of being hybridized; cross-fertilization.

hybrid sterility sterility in an individual resulting from its hybrid nature.

hybrid swarms populations consisting of descendants of species hybrids, as at borders between geographical areas populated by these species.

hybrid vigour increased vigour and growth found in some hybrids and thought to be due to increased heterozygosity; *alt.* heterosis.

hybrid zone a geographic area in which 2 populations, once separated by a geographical barrier, hybridize after the barrier has broken down and in the absence of reproductive isolation.

hydathode n. [Gk. *hydatos*, of water; *hodos*, way] an epidermal structure specialized for secretion, or for exudation, of water; *alt*. water stoma, water pore.

hydatid n. [Gk. *hydatis*, watery vesicle] any vesicle or sac filled with clear watery fluid; sac containing encysted stages of larval tapeworms; vestige of Müllerian duct constituting appendix of testis: hydatid of Morgagni; stalked appendix of epididymis.

hydatiform a. [Gk. *hydatis*, watery vesicle; L. *forma*, shape] resembling a hydatid.

hydatigenous a. [Gk. *hydatis*, watery vesicle; *-genēs*, producing] producing or forming hydatids.

hydranth n. [Gk. *hydōr*, water; *anthos*, flower] a nutritive zooid in a hydrozoan colony.

hydrarch n. [Gk. *hydōr*, water; *archē*, beginning] *appl*. seres progessing from hydric towards mesic conditions; *cf*. mesarch, xerarch.

hydratase n. [Gk. *hydōr*, water; *-ase*] an enzyme catalysing the hydration of a compound, i.e. the acceptance by it of a molecule of water without splitting it, or the reverse reaction; EC sub-subgroup 4.2.1; *r.n*. hydro-lyase.

hydra-tuba a small unsegmented polyp stage in some Scyphozoa, which buds like a hydra; *cf*. scyphistoma.

hydrelatic a. [Gk. *hydōr*, water; *elaunein*, to set in motion] *appl*. effects of stimulation of glands on active transport of water and inorganic solutes; *cf*. ecbolic.

hydric a. [Gk. *hydōr*, water] characterized by an abundant supply of moisture, *appl*. plants, environment.

Hydrida n. [Gk. *hydōr*, water] Hydroida, *q.v*.

hydroanemophilous a. [Gk. *hydōr*, water; *anemos*, wind; *philos*, loving] *pert*. or having spores which are discharged after moistening of spore-producing structure, and become airborne.

hydrobilirubin a substance produced by bacteria in the intestines from hydrobilirubinogen and passed out of the body with the faeces as a brown pigment; *alt*. stercobilin.

hydrobilirubinogen a substance produced by bacteria from the bile pigments in the intestines, some converted to hydrobilirubin and the rest passed out of the body with the faeces; *alt*. stercobilinogen.

hydrobiology n. [Gk. *hydōr*, water; *bios*, life; *logos*, discourse] the study of the life of aquatic plants and animals.

hydrobiont n. [Gk. *hydōr*, water; *bion*, living] an organism living mainly in water.

hydrocarbon n. [Gk. *hydōr*, water; L. *carbo*, coal] any chemical compound of hydrogen and carbon only.

hydrocarpic a. [Gk. *hydōr*, water; *karpos*, fruit] *appl*. aquatic plants having flowers which are fertilized out of the water but submerged for development of fruit.

hydrocaulis n. [Gk. *hydōr*, water; L. *caulis*, stalk] the branching vertical portion of coenosarc in a hydroid colony; *alt*. hydrocaulus; *cf*. hydrorhiza.

Hydrocharitales n. [*Hydrocharis*, generic name]

an order of monocots (Alismatidae) comprising the single family Hydrocharitaceae, including the frogbit, water soldier, and Canadian pondweed.

hydrochoric a. [Gk. *hydōr*, water; *chōrein*, to spread] dispersed by water; dependent on water for dissemination; *n*. hydrochory.

hydrocircus n. [Gk. *hydōr*, water; *kirkos*, circle] the hydrocoelic ring surrounding mouth in echinoderms.

hydrocladia n.plu. [Gk. *hydōr*, water; *kladion*, twig] the secondary branches of a hydrocaulis.

hydrocoel n. [Gk. *hydōr*, water; *koilos*, hollow] the water vascular system in echinoderms; in embryonic echinoderms the structure which will develop into the water vascular system.

hydrocoles n.plu. [Gk. *hydōr*, water; L. *colere*, to dwell] animals living in water or a wet environment.

Hydrocorallina n. [Gk. *hydōr*, water; *korallion*, coral] an order of hydrozoans, now usually subdivided into the orders Milleporina and Stylasterina, having an external calcareous skeleton and forming fixed colonies.

hydrocortisone n. [Gk. *hydōr*, water; L. *cortex*, bark] a glucocorticoid hormone very similar to cortisone, *q.v*.; *alt*. cortisol.

hydrocryptophyte hydrophyte, *q.v*.

hydrocyst n. [Gk. *hydōr*, water; *kystis*, bladder] dactylozooid, *q.v*.

hydroecium n. [Gk. *hydōr*, water; *oikos*, house] a closed tube at upper end of some siphonophores; *alt*. infundibulum.

hydrofuge n. [Gk. *hydōr*, water; L. *fugere*, to flee] water-repelling, *appl*. insect hairs; *alt*. hydrophobe.

hydrogenase a trivial name used for 3 different oxidoreductase enzymes acting on hydrogen as a donor, being hydrogen dehydrogenase, EC 1.12.1.2, cytochrome c, EC 1.12.2.1, and ferredoxin dehydrogenase, EC 1.12.7.1.

hydroid n., a. [Gk. *hydōr*, water; *eidos*, form] elongated empty cell in central cylinder of mosses; tracheid, *q.v*.; the polyp of coelenterates.

Hydroida n. [Gk. *hydōr*, water; *eidos*, form] an order of hydrozoans now usually subdivided into the orders Athecata, Thecata, Limnomedusae, and Chondrophora; *alt*. Hydrida.

hydrolase n. [Gk. *hydōr*, water; *lyein*, to dissolve] an enzyme which catalyses a hydrolysis; EC group 3.

hydrolysis n. [Gk. *hydōr*, water; *lyein*, to dissolve] the addition of the hydrogen and hydroxyl ions of water to a molecule, with its consequent splitting into 2 or more simpler molecules; *cf*. condensation.

hydrolytic a. [Gk. *hydōr*, water; *lyein*, to dissolve] *pert*., or causing, hydrolysis, *appl*. enzymes.

hydrom(e) n. [Gk. *hydōr*, water; *mestos*, full] any tissue that conducts water.

Hydromedusae n. [Gk. *hydōr*, water; *Medousa*, Medusa, one of the Gorgons] Hydrozoa, *q.v*.

hydromegatherm n. [Gk. *hydōr*, water; *mega*, great; *thermē*, heat] a plant which must have much moisture and heat to develop fully.

hydromorph n. [Gk. *hydōr*, water; *morphē*, form] a plant having the form and structure of a hydrophyte.

hydromorphic a. [Gk. *hydōr*, water; *morphē*,

form] structually adapted to an aquatic environment, as organs of water plants; *appl.* or *pert.* soils permanently containing a surplus of water; *cf.* hygromorphic.

hydronasty *n.* [Gk. *hydōr*, water; *nastos*, closepressed] plant movement induced by changes in atmospheric humidity.

hydrophilous *a.* [Gk. *hydōr*, water; *philein*, to love] pollinated through agency of water; hydrophil, adsorbing water.

hydrophily *n.* [Gk. *hydōr*, water; *philein*, to love] water pollination.

hydrophobe *a.* [Gk. *hydōr*, water; *phobos*, fear] avoiding or repelling water, *appl.* hairs of certain accquatic insects, *alt.* hydrofuge; not readily dissolving in water.

hydrophoric *a.* [Gk. *hydrophoros*, carrying water] *appl.* canal: stone canal, *q.v.*

Hydrophoridea *n.* [Gk. *hydrophoros*, carrying water; *eidos*, form] a subclass of cystoid echinoderms having numerous irregular plates.

hydrophyllium *n.* [Gk. *hydōr*, water; *phyllon*, leaf] one of leaf-like transparent bodies arising above and partly covering the sporosacs in a siphonophore; *alt.* phyllozooid.

hydrophyte *n.* [Gk. *hydōr*, water; *phyton*, plant] an aquatic plant living on or in the water; *alt.* hydrocryptophyte; *cf.* hygrophyte.

hydrophyton *n.* [Gk. *hydōr*, water; *phyton*, plant] a complete hydroid colony, root-like organ, stem, and branches.

hydroplanula *n.* [Gk. *hydōr*, water; L. *planus*, flat] stages between planula and actinula in larval history of coelenterates.

hydropolyp *n.* [Gk. *hydōr*, water; F. *polype*, polyp] a polyp of a hydroid colony; hydrula, *q.v.*

hydroponics *n.* [Gk. *hydōr*, water; *ponos*, exertion] the cultivation of plants without soil in nutrient-rich water, which is usually irrigated over some inert medium such as sand; *cf.* water culture.

hydropore *n.* [Gk. *hydōr*, water; *poros*, opening] the opening into right hydrocoel in some echinoderm larvae and in some extinct groups.

hydropote *n.* [Gk. *hydropotēs*, water drinker] a cell or cell group, in some submerged leaves, easily permeable by water and salts.

Hydropterides *n.* [Gk. *hydōr*, water; *pteris*, fern] in some classifications, a group of Pteropsida (Filicopsida) comprising those groups which are aquatic, i.e. the Marsileales and Salviniales; *alt.* Hydropterideae.

hydropyle *n.* [Gk. *hydōr*, water; *pylē*, gate] specialized area in cuticular membrane of embryo, for passage of water, as in some Orthoptera such as grasshoppers.

hydrorhabd *n.* [Gk. *hydōr*, water; *rhabdos*, rod] a rhabdosome, in graptolites.

hydrorhiza *n.* [Gk. *hydōr*, water; *rhiza*, root] the creeping root-like portion of coenosarc of a hydroid colony; *alt.* rhizocaul; *cf.* hydrocaulis.

hydrosere *n.* [Gk. *hydōr*, water; L. *sérere*, to put in a row] a plant succession originating in a wet environment.

hydrosinus *n.* [Gk. *hydōr*, water; L. *sinus*, fold] a dorsal extension of the mouth cavity in cyclostomes.

hydrosoma, hydrosome *n.* [Gk. *hydōr*, water; *sōma*, body] the conspicuously hydra-like stage in a coelenterate life history.

hydrosphere *n.* [Gk. *hydōr*, water; *sphaira*, globe] the water on the earth's surface.

hydrospire *n.* [Gk. *hydōr*, water; L. *spira*, coil] long pouches running at the side of the ambulacral grooves of blastoids and acting as respiratory structures.

hydrospore *n.* [Gk. *hydōr*, water; *sporos*, seed] a zoospore when moving in water.

hydrostatic *a.* [Gk. *hydōr*, water; *statikos*, causing to stand] *appl.* organs of flotation, as air sacs in aquatic larvae of insects.

hydrostome *n.* [Gk. *hydōr*, water; *stoma*, mouth] the mouth of a hydroid polyp.

hydrotaxis *n.* [Gk. *hydōr*, water; *taxis*, arrangement] a taxis in response to the stimulus of water; *a.* hydrotactic.

hydrotheca *n.* [Gk. *hydōr*, water; *thēkē*, box] cup-like structure into which the polyp may withdraw in many coelenterates; type of theca in some graptolites.

hydrotropic *a.* [Gk. *hydōr*, water; *tropē*, turn] *appl.* curvature of a plant organ towards a greater degree of moisture; *appl.* substances which make insoluble substances watersoluble.

hydrotropism *n.* [Gk. *hydōr*, water; *trepein*, to turn] response to stimulus of water, especially by growth curvature.

hydroxycobalamin(e) vitamin B_{12b}, *see* cobalamine.

Hydrozoa, hydrozoans *n., n.plu.* [Gk. *hydōr*, water; *zōon*, animal] a class of cnidarian coelenterates that are typically colonial and have both polyp and medusa stages in the life history; *alt.* Hydromedusae, and sometimes called zoophytes because of their resemblance to plants; *sing.* hydrozoon, hydrozoan.

hydrula *n.* [Gk. *hydōr*, water] hypothetical simple polyp; *alt.* hydropolyp.

Hyeniales *n.* [*Hyenia*, generic name] an order of fossil Sphenopsida characterized by forked leaves and sporangiophores resembling leaves; *alt.* Protoarticulatae.

hyetal *a.* [Gk. *hyetos*, rain] *pert.* rain; *pert.* precipitation.

hygric *a.* [Gk. *hygros*, wet] humid; tolerating, or adapted to humid conditions; *cf.* xeric.

hygrochasy *n.* [Gk. *hygros*, wet; *chasis*, separation] dehiscence of fruits when induced by moisture; *cf.* xerochasy; *a.* hygrochasic.

hygrokinesis *n.* [Gk. *hygros*, wet; *kinēsis*, movement] a kinesis in response to change in humidity.

hygrometric *a.* [Gk. *hygros*, wet; *metron*, measure] *appl.* plant movements dependent on changes in the amount of moisture.

hygromorphic *a.* [Gk. *hygros*, wet; *morphē*, form] structurally adapted to a moist habitat; *cf.* hydromorphic.

hygropetric *a.* [Gk. *hygros*, wet; *petros*, stone] *appl.* fauna of submerged rocks.

hygrophanous *a.* [Gk. *hygros*, wet; *phanos*, visible] as if impregnated with water.

hygrophilic, hygrophilous *a.* [Gk. *hygros*, wet; *philein*, to love] inhabiting moist or marshy places.

hygrophyte *n.* [Gk. *hygros*, wet; *phyton*, plant] a plant which thrives in plentiful moisture; *cf.* hydrophyte.

hygroplasm *n.* [Gk. *hygros*, wet; *plasma*,

mould] the more liquid part of the cytoplasm; *cf.* stereoplasm.

hygroreceptor *n.* [Gk. *hygros*, wet; L. *recipere*, to receive] a specialized cell or structure sensitive to humidity.

hygroscopic *a.* [Gk. *hygros*, wet; *skopein*, to regard] sensitive to moisture; retaining or losing water.

hygrotaxis *n.* [Gk. *hygros* wet; *taxis*, arrangement] a taxis in response to moisture or humidity.

hygrotropism *n.* [Gk. *hygros*, wet; *tropē*, turn] a tropism in response to moisture or humidity.

hylea *n.* [Gk. *hylē*, wood] the primeval forest, especially tropical.

hylion, hylium *n.* [Gk. *hylē*, wood; *on*, being] forest climax.

hylogamy *n.* [Gk. *hylē*, material; *gamos*, marriage] the fusion of gametes, *alt.* syngamy; *cf.* somatogamy.

hylophagous *a.* [Gk. *hylē*, wood; *phagein*, to eat] eating wood, *appl.* certain insects.

hylophyte *n.* [Gk. *hylē*, wood; *phyton*, plant] a plant growing in wood.

hylotomous *a.* [Gk. *hylē*, wood; *temnein*, to cut] wood-cutting, *appl.* certain insects.

hymen *n.* [Gk. *hymēn*, membrane] thin fold of mucous membrane at orifice of vagina.

hymenial *a.* [Gk. *hymēn*, skin] *pert.* hymenium.

hymeniferous *a.* [Gk. *hymēn*, skin; L. *ferre*, to carry] having a hymenium.

hymeniform *a.* [Gk. *hymēn*, skin; L. *forma*, form] formed like a palisade of club-shaped cells, *appl.* cuticle of fleshy fungi; *alt.* hymenoid.

hymenium *n.* [Gk. *hymēn*, skin] the layer of spore-producing structures in ascomycete and basidiomycete fungi, often interspersed with barren cells or paraphyses.

Hymenogastrales *n.* [Gk. *hymēn*, skin; *gastēr*, stomach] an order of Gasteromycetes having the fruit body usually borne underground and enclosed until mature with a waxy or fleshy gleba.

hymenoid *a.* [Gk. *hymēn*, membrane; *eidos*, form] membranoid; membranous; resembling a hymenium; hymeniform, *q.v.*

Hymenolichenes, hymenolichens *n., n.plu.* [Gk. *hymēn*, skin; *leichēn*, lichen] Basidiolichenes, *q.v.*

Hymenomycetes *n.* [Gk. *hymēn*, skin; *mykēs*, fungus] a group of Homobasidiomycetes having the hymenium exposed at maturity and the basidiospores usually violently discharged.

hymenophore *n.* [Gk. *hymēn*, skin; *pherein*, to carry] the portion of the sporophore of a fungus which bears a hymenium.

hymenopodium *n.* [Gk. *hymēn*, membrane; *pous*, foot] the tissue between trama and subhymenium as in cup fungi and agarics; subhymenium, *q.v.*; hypothecium, *q.v.*

Hymenoptera *n.* [Gk. *hymēn*, membrane; *pteron*, wing] an order of insects with complete metamorphosis, including the ants, bees, and wasps, having membranous wings and with the 1st abdominal segment fused to the thorax and forming a peduncle.

Hymenostomatida, Hymenostomata *n.* [Gk. *hymēn*, skin; *stoma*, mouth] an order or suborder of Holotricha having a mouth and a complex gullet.

hyobranchial *a.* [Gk. ϒ; *brangchia*, gills] *pert.* hyoid and branchial arches.

hyoepiglottic *a.* [Gk. ϒ; *epi*, upon; *glōtta*, tongue] connecting hyoid and epiglottis.

hyoglossal *a.* [Gk. ϒ; *glōssa*, tongue] *pert.* tongue and hyoid, *appl.* membrane.

hyoglossus *n.* [Gk. ϒ; *glōssa*, tongue] an extrinsic muscle of the tongue, arising from greater cornu of hyoid bone.

hyoid *a.* [Gk. *hyoeidēs*, ϒ-shaped] *pert.* or designating a bone or series of bones lying at base of tongue and developed from hyoid arch of embryo; *appl.* a sclerite enclosing pharynx in some insects; *alt.* hyoidean. *n.* The hyoid bone.

hyoid arch second visceral arch.

hyoidean *a.* [Gk. *hyoeidēs*, ϒ-shaped] *pert.* or associated with the hyoid arch or bone; *appl.* a branch of 1st efferent branchial vessel, or of lingual artery; *appl.* nerve, the posterior post-trematic nerve.

hyoideus *n.* [Gk. *hyoeidēs*, ϒ-shaped] a nerve which supplies mucosa of mouth and muscles of hyoid region.

hyomandibular *a.* [Gk. ϒ; L. *mandibulum*, jaw] *pert.* hyoid and mandible; *pert.* dorsal segment of hyoid arch in fishes.

hyomental *a* [Gk. ϒ; L. *mentum*, chin] *pert.* hyoid and chin.

hyoplastron *n.* [Gk. ϒ; F. *plastron*, breast plate] the 2nd lateral plate in plastron of Chelonia; *alt.* hyosternum.

hyostapes *n.* [Gk. ϒ; L.L. *stapes*, stirrup] lower portion of columellar primordium which gives rise to part of columella in some reptiles; extracolumella, *q.v.*

hyosternum *n.* [Gk. ϒ; *sternon*, breast] hyoplastron, *q.v.*

hyostyly *n.* [Gk. ϒ; *stylos*, pillar] the condition of having jaw articulated to skull by hyomandibular or corresponding part; *a.* hyostylic; *cf.* autostyly, holostyly.

hyosymplecticum *n.* [Gk. ϒ; *symplektos*, plaited together] the cartilaginous primordium from which hyomandibular and symplecticum are derived.

hyothyroid *a.* [Gk. ϒ; *thyreos*, shield; *eidos*, like] *pert.* hyoid bone and thyroid cartilage of larynx, *appl.* ligaments, membrane.

hypallelomorph *n.* [Gk. *hypo*, under; *allēlōn*, of one another; *morphē*, form] allelomorphs which under certain conditions are themselves compound.

hypandrium *n.* [Gk. *hypo*, under; *anēr*, male] subgenital plate or 9th abdominal sternite of certain insects.

hypanthium *n.* [Gk. *hypo*, under; *anthos*, flower] any enlargement of the torus.

hypanthodium *n.* [Gk. *hypo*, under; *anthodēs*, like flowers] an inflorescence with concave capitulum on whose walls the flowers are arranged.

hypantrum *n.* [Gk. *hypo*, under; *antron*, cave] notch on vertebrae of certain reptiles for articulation with hyposphene.

hypapophysis *n.* [Gk. *hypo*, under; *apo*, upon; *phyein*, to grow] a ventral process on a vertebral centrum.

hyparchic *a.* [Gk. *hypo*, under; *archon*, ruler] *appl.* genes in mosaic organisms which are inhibited from expressing their effect by the presence of neighbouring genes of a different genotype; *cf.* autarchic.

hyparterial *a.* [Gk. *hypo*, under; L. *arteria*, artery] situated below an artery, *appl.* branches of bronchi below pulmonary artery.

hypascidium *n.* [Gk. *hypo*, under; *askidion*, little bag] a funnel-like growth of a leaf, the inner surface of the funnel representing the lower surface of the leaf.

hypaxial *a.* [Gk. *hypo*, under; L. *axis*, axis] ventral or below vertebral column, *appl.* muscles.

hyperapophysis *n.* [Gk. *hyper*, above; *apo*, from; *phyein*, to grow] a posterolateral process of dorsal side of vertebra.

hyperchimaera *n.* [Gk. *hyper*, above; L. *chimaera*, monster] a chimaera with a mosaic arrangement of its genetically different components.

hyperchromasy *n.* [Gk. *hyper*, above; *chrōma*, colour] a relatively superabundant supply of chromatin to cytoplasm in a cell.

hyperchromatosis *n.* [Gk. *hyper*, above; *chrōma*, colour] excess of nuclear substance in a cell previous to division.

hypercoracoid *a.* [Gk. *hyper*, above; *korax*, crow; *eidos*, form] *pert.* or designating upper bone at base of pectoral fin in fishes; *alt.* paraglenal.

hypercyesis *n.* [Gk. *hyper*, above; *kyēsis*, conception] additional fertilization in a mammal already pregnant; *alt.* superfoetation.

hyperdactyly *n.* [Gk. *hyper*, above; *dactylos*, finger] polydactyly, *q.v.*

hyperdiploids *n.plu.* [Gk. *hyper*, above; *diploos*, double; *eidos*, form] cells or organisms which as a result of translocation have more than 2 identical chromosome segments in the genome.

hyperfeminization condition of a feminized male with female characteristics exaggerated, as in small size and weight.

hypergamesis *n.* [Gk. *hyper*, above; *gamos*, marriage] process of absorption by female of excess spermatozoa.

hyperglyc(a)emic-glycogenolytic **factor (HGF)** glucagon, *q.v.*

hyperhaploids *n.plu.* [Gk. *hyper*, above; *haploos*, simple; *eidos*, form] haploid cells or organisms containing supernumerary chromosomes.

hyperhydric *a.* [Gk. *hyper*, exceeding; *hydōr*, water] absorbing or containing an excess of water, *appl.* cells; watery.

hyperhypophysism *n.* [Gk. *hyper*, above; *hypo*, under; *physis*, growth] hyperpituitarism, *q.v.*

hypermasculinization condition of a masculinized female with male characteristics exaggerated, as in large proportions, appearance of male secondary sexual characters.

Hypermastigina *n.* [Gk. *hyper*, above; *mastix*, whip] an order of Zoomastigina being small with numerous flagella, living as symbionts in the gut of insects, and assisting in the digestion of wood in gut of termites; the group is often included in the Polymastigina or with the Polymastigina in the Metamonadina.

hypermetamorphosis *n.* [Gk. *hyper*, above; *meta*, after; *morphōsis*, shaping] a protracted and thorough-going metamorphosis; metamorphosis involving 2 or more distinct types of larval instar, in certain insects.

hypermorph *n.* [Gk. *hyper*, above; *morphē*, form] a mutant allele whose effect is greater than the normal wild-type gene.

hypermorphosis *n.* [Gk. *hyper*, above; *morphōsis*, shaping] the development of additional characters, in comparison with the adult ancestral stage.

Hyperoartia, hyperoartians *n., n.plu.* [Gk. *hyperōia*, palate; *artios*, complete] an order of cyclostomes comprising the lampreys as distinct from hagfish; *alt.* Petromyzontia; *cf.* Hyperotreta.

Hyperotreta, hyperotretans *n., n.plu.* [Gk. *hyperōia*, palate; *trētos*, perforated] an order of cyclostomes comprising the hagfish as distinct from lampreys; *alt.* Myxinoidea, Myxinoidei; *cf.* Hyperoartia.

hyperparasite *n.* [Gk. *hyper*, above; *para*, beside; *sitos*, food] an organism which is parasitic on or in another parasite; *alt.* superparasite.

hyperphalangy *n.* [Gk. *hyper*, above; *phalangx*, line of battle] condition of having digits with more than normal number of phalanges.

hyperpharyngeal *a.* [Gk. *hyper*, above; *pharyngx*, gullet] dorsal to the pharynx, *appl.* gill or bar in salps.

hyperpituitarism *n.* [Gk. *hyper*, above; L. *pituita*, phlegm] overaction of pituitary gland, resulting in gigantism or giantism; *alt.* hyperhypophysism.

hyperplasia *n.* [Gk. *hyper*, above; *plassein*, to mould] overgrowth; excessive or hyperplastic development due to increase in number of cells; *cf.* hypertrophy.

hyperploid *a.* [Gk. *hyper*, above; *haploos*, onefold; *eidos*, form] aneuploid with extra chromosomes; *cf.* hypoploid.

hyperpneustic *a.* [Gk. *hyper*, above; *pneuma*, breath] possessing more than the usual number of spiracles, as in some Thysanura.

hyperpnoea *n.* [Gk. *hyper*, above; *pnoē*, breath] rapid breathing due to insufficient supply of oxygen.

hyperpolarization an increased potential difference across a membrane.

hyperpolyploid *n.* [Gk. *hyper*, above; *polys*, many; *aploos*, onefold; *eidos*, form] a polyploid containing one or more extra chromosomes in its sets.

hypersensitivity *n.* [Gk. *hyper*, above; L. *sentire*, to feel] a condition of being unduly sensitive to a stimulus.

hyperstomatous *a.* [Gk. *hyper*, above; *stoma*, mouth] with stomata on the upper surface, *appl.* leaves.

hypertely, hypertelia *n.* [Gk. *hyper*, above; *telos*, end] excessive imitation in colour or pattern, being of problematical utility; overdevelopment, as canines of babirusa, an East Indian pig, the male of which has 4 large tusks.

hypertensin *n.* [Gk. *hyper*, above; L. *tensio*, stretching] angiotonin, *q.v.*

hypertensinogen *n.* [Gk. *hyper*, above; L. *tensio*, stretching; Gk. *genos*, birth] a blood protein which forms angiotonin by the action of renin.

hypertonia *n.* [Gk. *hyper*, above; *tonos*, tone] excessive tonicity.

hypertonic *a.* [Gk. *hyper*, exceeding; *tonos*, intensity] having a higher osmotic pressure than another fluid; *cf.* hypotonic, isotonic.

hypertrophy *n.* [Gk. *hyper*, above; *trophē*, nourishment] excessive growth due to increase in size of cells; *cf.* hyperplasia.

hypha *n.* [Gk. *hyphē*, web] the thread-like element or filament of vegetative mycelium of a fungus; filamentous cell in medulla of an algal thallus.

hyphal of, or *pert.*, hyphae or a hypha.

hyphal bodies in certain fungi, short thick hyphae which produce fruiting hyphae.

hyphasma *n.* [Gk. *hyphasma*, thing woven] a barren mycelium; a cord of mycelium.

hyphidium *n.* [Gk. *hyphē*, web; *idion*, dim.] spermatium in certain lichens; a sterile paraphysis-like hypha in Basidiomycetes.

Hyphochytriales *n.* [*Hypochytrium*, generic name] an order of Phycomycetes which are mainly aquatic parasites on algae or other fungi, and whose motile stages have one anterior tinsel flagellum; *alt.* Anisochytridiales.

Hyphochytridiomycetes *n.* [*Hyphochytrium*, generic name; Gk. *mykēs*, fungus] in some classifications, a class of fungi containing the order Hyphochytriales, *q.v.*

hyphodrome *a.* [Gk. *hyphē*, web; *dromos*, course] running throughout the tissues; *appl.* thick leaves where veins are not visible from surface.

Hyphomicrobiales *n.* [*Hyphomicrobium*, generic name] an order of bacteria having rigid cell walls, often stalked, and reproducing by budding.

Hyphomycetes *n.* [Gk. *hyphē*, web; *mykēs*, fungus] Moniliales, *q.v.*

hyphopodium *n.* [Gk. *hyphē*, web; *pous*, foot] a hyphal branch with enlarged terminal cell or haustorium for attaching the hypha, as in some Ascomycetes; *alt.* holdfast.

hyphostroma *n.* [Gk. *hyphē*, web; *strōma*, bedding] mycelium, *q.v.*

hypnobasidium *n.* [Gk. *hypnos*, sleep; *basis*, base; *idion*, dim.] sclerobasidium, *q.v.*

Hypnobryales *n.* [*Hypnum*, generic name; Gk. *bryon*, moss] an order of mosses having typically branched gametophores and 'leaves' with 2 nerves.

hypnocyst *n.* [Gk. *hypnos*, sleep; *kystis*, bladder] cyst in which contained organism simply rests; dormant cyst.

hypnody *n.* [Gk. *hypnōdia*, sleepiness] the long resting period of certain larval forms.

hypnogenic *a.* [Gk. *hypnos*, sleep; *-genēs*, producing] sleep-inducing, *appl.* influences which tend to produce hypnosis.

hypnosperm *n.* [Gk. *hypnos*, sleep; *sperma*, seed] hypnospore, *q.v.*

hypnosporangium *n.* [Gk. *hypnos*, sleep; *sporos*, seed; *anggeion*, vessel] a sporangium containing resting spores.

hypnospore *n.* [Gk. *hypnos*, sleep; *sporos*, seed] a resting spore; a zygote that remains in a quiescent condition during winter; *alt.* hypnosperm.

hypnote *n.* [Gk. *hypnos*, sleep] an organism in a dormant condition.

hypnozygote *n.* [Gk. *hypnos*, sleep; *zygōtos*, yoked] a zygote that becomes encysted, thereby constituting a hypnospore, e.g. oöspore, zygospore.

hypoachene *n.* [Gk. *hypo*, under; *a*, not; *chainein*, to gape] achene developed from an inferior ovary.

hypoarion *n.* [Gk. *hypo*, under; *ōarion*, little egg] a small lobe below the optic lobes of most teleosts; *plu.* hypoaria.

hypobasal *n.* [Gk. *hypo*, under; *basis*, base] the lower segment of a developing ovule, which ultimately gives rise to the root; *cf.* epibasal.

hypobasidium *n.* [Gk. *hypo*, under; *basis*, base; *idion*, dim.] basal cell or part of a heterobasidium, in which nuclei unite, and which gives rise to an epibasidium.

hypobenthos *n.* [Gk. *hypo*, under; *benthos*, depths of the sea] the fauna of the sea bottom below 1000 m; *cf.* epibenthos.

hypobiotic *a.* [Gk. *hypo*, under; *biōnai*, to live] living under objects or projections, as on sea bottom; *cf.* epibiotic.

hypoblast *n.* [Gk. *hypo*, under; *blastos*, bud] the inner germ layer of a gastrula, capable of becoming endoderm and to some extent mesoderm; endoderm, *q.v.*; *alt.* endoblast, hypodermis; *a.* hypoblastic.

hypobranchial *a.* [Gk. *hypo*, under; *brangchia*, gills] *pert.* lower or 4th segment of branchial arch; *appl.* groove: endostyle of Tunicata and Cephalochordata such as *Amphioxus*; *appl.* space under gills in decapod crustaceans.

hypocalcaemic *a.* [Gk. *hypo*, under; L. *calx*, lime; Gk. *haima*, blood] reducing the blood calcium level.

hypocarp *n.* [Gk. *hypo*, under; *karpos*, fruit] a fleshy modified peduncle of certain fruits, as cashew-apple.

hypocarpogenous *a.* [Gk. *hypo*, under; *karpos*, fruit; *-genēs*, producing] having both flowers and fruits borne underground.

hypocentrum *n.* [Gk. *hypo*, under; *kentron*, centre] a transverse cartilage that arises below nerve cord and forms part of vertebral centrum.

hypocercal *a.* [Gk. *hypo*, under; *kerkos*, tail] having notochord terminating in lower lobe of caudal fin.

hypocerebral *a.* [Gk. *hypo*, under; L. *cerebrum*, brain] *appl.* ganglion of stomatogastric system, linked to frontal and ventricular ganglia, also to corpora cardiaca.

hypochile, hypochilium *n.* [Gk. *hypo*, under; *cheilos*, lip] in orchid flowers, the inner or basal part of the lip when there are 2 distinct parts to this organ; *cf.* epichile.

hypochondrium *n.* [Gk. *hypo*, under; *chondros*, cartilage] abdominal region lateral to epigastric and above lumbar.

hypochord *n.* [Gk. *hypo*, under; *chordē*, cord] subnotochord, *q.v.*

hypochordal *a.* [Gk. *hypo*, under; *chordē*, cord] below the notochord, *appl.* lower lobe of caudal fin, *appl.* bar of mesodermal tissue developing into ventral arch of atlas and amalgamating with fibrocartilages in other cervical vertebrae.

hypocleidium *n.* [Gk. *hypo*, under; *kleidion*, little key] interclavicle, *q.v.*

hypocone *n.* [Gk. *hypo*, under; *kōnos*, cone] posterointernal cusp of upper molar; the part posterior to girdle in Dinoflagellata, *cf.* epicone; *alt.* hypovalve.

hypoconid *n.* [Gk. *hypo*, under; *kōnos*, cone] posterobuccal cusp of lower molar.

hypoconule *n.* [Gk. *hypo*, under; *kōnos*, cone] fifth or distal cusp of upper molar.

hypoconulid *n.* [Gk. *hypo*, under; *kōnos*, cone] posteromesial cusp of lower molar.

hypocoracoid *a.* [Gk. *hypo*, under; *korax*, crow; *eidos*, form] *pert.* lower bone at base of pectoral fin in fishes.

hypocotyl *n.* [Gk. *hypo*, under; *kotylē*, cup] that portion of stem below cotyledons in an embryo; *alt.* fundamentum.

hypocotyledonary *a.* [Gk. *hypo*, under; *kotylē-dōn*, hollow] below the cotyledons.

hypocrateriform *a.* [Gk. *hypo*, under; *kratēr*, bowl; L. *forma*, shape] saucer-shaped; having a gamopetalous corolla with long narrow tube, and limbs at right angles to tube; *alt.* hypocraterimorphous.

Hypocreales *n.* [*Hypocrea*, generic name] an order of Euascomycetes in which the ascocarp is a perithecium, and perithecia are surrounded by pale periderm different from the rest of the mycelium.

hypodactylum *n.* [Gk. *hypo*, under; *daktylos*, digit] the under surface of a bird's toes.

hypoderm(a) *n.* [Gk. *hypo*, under; *derma*, skin] hypodermis, *q.v.*

hypodermal *pert.* hypodermis.

hypodermalia *n.plu.* [Gk. *hypo*, under; *derma*, skin] sponge spicules situated just below the derma or skin.

hypodermic *a.* [Gk. *hypo*, under; L. *dermis*, skin] *pert.* parts just under the skin.

hypodermis *n.* [Gk. *hypo*, under; L. *dermis*, skin] tissue just under the epidermis in plants; the cellular layer lying beneath and secreting the cuticle of annelids, arthropods, and other invertebrates; hypoblast, *q.v.*; *alt.* hypoderm, hypodermis.

hypodicrotic *a.* [Gk. *hypo*, under; *di*, two; *krotein*, to beat] having 2 arterial beats for the one cardiac.

hypodiploids *n.plu.* [Gk. *hypo*, under; *diploos*, double; *eidos*, form] individuals or cells having less than the diploid number of chromosomes.

hypogaean, hypogaeous hypogeal, *q.v.*

hypogastric *a.* [Gk. *hypo*, under; *gastēr*, stomach] *pert.* lower median region of abdomen, *appl.* artery, vein, plexus, etc.

hypogastrium *n.* [Gk. *hypo*, under; *gastēr*, stomach] lower median region of abdomen; *cf.* epigastrium.

hypogastroid hypoischium, *q.v.*

hypogeal, hypogean *a.* [Gk. *hypo*, under; *gē*, earth] living or growing underground, *appl.* stems, insects, etc.; borne underground, *appl.* germination, when cotyledons remain below ground; *alt.* hypogaean, hypogaeous; *cf.* epigeal.

hypogenesis *n.* [Gk. *hypo*, under; *genesis*, origin] development without occurrence of alternation of generations.

hypogenous *a.* [Gk. *hypo*, under; *-genēs*, producing] growing on lower surface of anything; *cf.* epigenous.

hypogeous *a.* [Gk. *hypo*, under; *gē*, earth] growing or maturing under the soil surface.

hypoglossal *n.* [Gk. *hypo*, under; *glōssa*, tongue] the 12th paired cranial nerve, distributed to base of tongue, in Amniota; a spinal nerve in Anamniota.

hypoglottis *n.* [Gk. *hypo*, under; *glōtta*, tongue] the under part of tongue; a division of labium of beetles.

hypoglyc(a)emic *a.* [Gk. *hypo*, under; *glykys*, sweet; *haima*, blood] causing or *pert.* decrease in blood sugar; *appl.* hormone: insulin, *q.v.*

hypognathous *a.* [Gk. *hypo*, under; *gnathos*, jaw] having the lower jaw slightly longer than the upper; with mouth-parts ventral, *appl.* head of insects; *cf.* opisthognathous, orthognathous, prognathous.

hypogynium *n.* [Gk. *hypo*, under; *gynē*, female] structure supporting ovary in such plants as sedges.

hypogynous *a.* [Gk. *hypo*, under; *gynē*, female] inserted below the gynaecium, and not adherent; immediately below oogonium, *appl.* antheridium, as in some Peronosporales, *cf.* epigynous, perigynous; *n.* hypogyny.

hypohaploids *n.plu.* [Gk. *hypo*, under; *haploos*, simple; *eidos*, form] individuals or cells with one or several chromosomes missing from their haploid complement.

hypohyal *n.* [Gk. *hypo*, under; *hyoeidēs*, Y-shaped] the hyoid element lying between ceratohyal and basihyal.

hypohypophysism *n.* [Gk. *hypo*, under; *hypo*, under; *physis*, growth] hypopituitarism, *q.v.*

hypoischium *n.* [Gk. *hypo*, under; *ischion*, hip] a small bony rod passing backwards from ischiadic symphysis and supporting ventral cloacal wall; *alt.* hypogastroid, os cloacae.

hypolemmal *a.* [Gk. *hypo*, under; *lemma*, peel] beneath the sarcolemma, *appl.* arborization of an axis cylinder in a motor plate.

hypolimnion *n.* [Gk. *hypo*, under; *limnē*, lake] the water between the thermocline and bottom of lakes; *cf.* epilimnion.

hypolithic *a.* [Gk. *hypo*, under; *lithos*, stone] found or-living beneath stones.

hypomeral, hypomeric *a.* [Gk. *hypo*, under; *meros*, part] *appl.* slender bones among lower trunk muscles in some fishes.

hypomere *n.* [Gk. *hypo*, under; *meros*, part] lower or lateral plate zone of coelomic pouches.

hypomeron *n.* [Gk. *hypo*, under; *meros*, part] the lateral inflexed side of a coleopterous prothorax.

hypomorph *n.* [Gk. *hypo*, under; *morphē*, form] a mutant allele which behaves in a similar way to the normal gene but with a weaker effect; *alt.* hypomorphic gene, leaky gene.

hyponasty *n.* [Gk. *hypo*, under; *nastos*, closepressed] the state of growth in a flattened structure in which the under surface grows more vigorously than the upper.

hyponeural *a.* [Gk. *hypo*, under; *neuron*, nerve] *appl.* system of radial and transverse motor nerves in echinoderms.

hyponeuston *n.* [Gk. *hypo*, under; *neustos*, floating] organisms swimming or floating immediately under the water surface.

hyponome *n.* [Gk. *hypo*, under; *hyponomos*, water pipe] the funnel of Cephalopoda.

hyponychium *n.* [Gk. *hypo*, under; *onyx*, nail] layer of epidermis on which nail rests; *a.* hyponychial.

hyponym *n.* [Gk. *hypo*, under; *onyma*, name] a generic name not founded on a type species; a provisional name for a specimen.

hypoparatype *n.* [Gk. *hypo*, under; *para*, beside; *typos*, pattern] a specimen originally indicating a new species, but not chosen as a type specimen; *cf.* holotype, paratype.

hypopetalous *a.* [Gk. *hypo*, under; *petalon*, leaf] having corolla inserted below, and not adherent to, gynaecium.

hypophamine *see* oxytocin, vasopressin.

hypophare *n.* [Gk. *hypo*, under; *pharos*, cloth]

lower part of sponge, in which there are no chambers; cf. spongophare.

hypopharyngeal a. [Gk. hypo, under; pharyngx, pharynx] pert. or situated below or on lower surface of pharynx; appl. bone formed by 5th branchial arch.

hypopharynx n. [Gk. hypo, under; pharyngx, pharynx] in many insects, a tongue-like lobe which arises from the mouth-cavity floor, alt. lingua; in many dipterans, an outgrowth from base of labium which bears the salivary groove or duct; in spiders, a chitinous plate on surface of labium.

hypophloeodal a. [Gk. hypo, under; phloios, bark] living or growing under bark.

hypophragm n. [Gk. hypo, under; phragma, protection] operculum or epiphragm closing the opening of shell in some gastropods.

hypophyllium n. [Gk. hypo, under; phyllon, leaf] a scale-like leaf below a cladophyll; base of stipulate leaf, forming abscission layer.

hypophyllous a. [Gk. hypo, under; phyllon, leaf] located or growing under a leaf.

hypophysectomy n. [Gk. hypo, under; physis, growth; ek, out; temnein, to cut] excision or removal of the pituitary gland.

hypophyseos see hypophysis.

hypophysial a. [Gk. hypo, under; physis, growth] pert. the hypophysis.

hypophysin n. [Gk. hypo, under; physis, growth] pituitrin, q.v.

hypophysis n. [Gk. hypo, under; physis, growth] the pituitary body, q.v., alt. hypophyseos; the olfactory pit in some Cephalochordata; the last cell of the suspensor; the cell from which the root cap and part of the root tip arises in dicotyledons.

hypopituitarism n. [Gk. hypo, under; L. pituita, phlegm] deficiency of pituitary gland, resulting in a type of infantilism; alt. hypohypophysism; cf. apituitarism.

hypoplasia n. [Gk. hypo, under; plasis, formation] developmental deficiency; deficient growth; a. hypoplastic.

hypoplastron n. [Gk. hypo, under; F. plastron, breast plate] the 3rd lateral bony plate in plastron of Chelonia.

hypopleuron n. [Gk. hypo, under; pleuron, side] region below metapleuron in insects.

hypoploid a. [Gk. hypo, under; haploos, onefold; eidos, form] aneuploid with few chromosomes; lacking one chromosome of the complement; cf. hyperploid.

hypopneustic a. [Gk. hypo, under; pnein, to breathe] having a reduced number of spiracles, appl. modified tracheal system in certain insects.

hypopodium n. [Gk. hypo, under; podion, little foot] basal portion of a leaf, including stalk; style of carpel.

hypopolyploids n.plu. [Gk. hypo, under; polys, many; aploos, onefold; eidos, form] polyploids lacking one or more chromosomes.

hypoproct n. [Gk. hypo, under; prōktos, anus] medial prolongation of terminal abdominal segment beneath the anus, in Diplopoda and some Insecta.

hypopteron n. [Gk. hypo, under; pteron, feather] axillary feather in birds.

hypoptilum n. [Gk. hypo, under; ptilon, down] the aftershaft of a feather.

hypopus n. [Gk. hypo, under; pous, foot] deutonymph, q.v.

hypopyge, hypopygium n. [Gk. hypo, under; pygē, rump] clasping organ of male Diptera.

hyporachis hyporhachis, q.v.

hyporadiolus n. [Gk. hypo, under; L. radiolus, small rod] a barbule of aftershaft of a feather.

hyporadius n. [Gk. hypo, under; L. radius, rod] a barb of aftershaft of a feather.

hyporhachis n. [Gk. hypo, under; rhachis, spine] the stem of aftershaft of a feather; alt. hyporachis.

hyposkeletal a. [Gk. hypo, under; skeletos, dried] lying beneath or internally to endoskeleton.

hyposomite n. [Gk. hypo, under; sōma, body] ventral part of a body segment, as in certain cephalochordates such as Amphioxus.

hyposperm n. [Gk. hypo, under; sperma, seed] the lower region of an ovule or seed, below the level at which the integument or testa is free from the nucellus.

hyposphene n. [Gk. hypo, under; sphēn, wedge] a wedge-shaped process on neural arch of vertebra of certain reptiles, which fits into hypantrum.

hypostasis n. [Gk. hypo, under; stasis, standing] recessiveness of non-allelomorphic characters, cf. epistasis; sediment or deposit, as of blood; a. hypostatic.

hypostereom n. [Gk. hypo, under; stereōma, basis] the 3rd or inner layer of thecal plates of Cystoidea; the inner layer of integument of Crinoidea.

hypostoma n. [Gk. hypo, under; stoma, mouth] the fold bounding posterior margin of oral aperture in crustaceans; labrum or median preoral plate in trilobites; oral projection or manubrium of a hydrozoan; anteroventral region of insect head; ventral mouth-part of ticks; alt. hypostome.

hypostomatic a. [Gk. hypo, under; stoma, mouth] situated beneath stomata of plant epidermis, appl. chamber or cavity; alt. substomatal.

hypostomatous a. [Gk. hypo, under; stoma, mouth] having stomata on under surface; having mouth placed on lower or ventral side.

hypostome hypostoma, q.v.

Hypostomides n. [Hypostoma, generic name] Pegasiformes, q.v.

hypostracum n. [Gk. hypo, under; ostrakon, shell] inner primary layer or endocuticle of exoskeleton in Acarina.

hypostroma n. [Gk. hypo, under; strōma, bedding] basal part of a fungal stroma; enterostroma, q.v.; cf. epistroma.

hyposyndesis n. [Gk. hypo, under; syndēsai, to bind together] the formulation in a hybrid in meiosis of a smaller number of bivalents than in the parental forms.

hypotarsus n. [Gk. hypo, under; L.L. tarsus, ankle] the calcaneum of a bird.

hypothalamus n. [Gk. hypo, under; thalamos, chamber] region below thalamus, and structures forming greater part of floor of 3rd ventricle; alt. subthalamus.

hypothallus n. [Gk. hypo, under; thallos, young shoot] a thin layer under sporangia in Myxomycetes; sclerotium, q.v.; undifferentiated hyphal growth, or marginal outgrowth in lichens.

hypotheca n. [Gk. *hypo*, under; *thēkē*, box] theca covering hypocone in Dinoflagellata; younger or inner half of frustule in diatoms, *alt.* hypovalve.

hypothecium n. [Gk. *hypo*, under; *thēkē*, box] the layer of dense hyphal threads below the hymenium in lichens and fungi; *alt.* subhymenium, hymenopodium.

hypothenar a. [Gk. *hypo*, under; *thenar*, palm of hand] *pert.* the prominent part of palm of hand above base of little finger.

hypothermia n. [Gk. *hypo*, under; *thermos*, hot] a condition in a homoiothermic animal in which its body temperature is lower than normal and cannot be rapidly restored to normal without exogenous heat.

hypothetical units an obsolete term for ultimate component parts of protoplasm, being ultracellular, loosely defined units ranking between the molecule and the cell; also called variously Altmann's granules, bioblasts, biogens, biophores, gemmae, gemmules, genes, gens, idioblasts, idiosomes, ids, inotagmata, micellae, microzymas, pangens, physiological units, plastidules, plasomes, primordia, sphaeroplasts, spheroplasts, somacules.

hypotonic a. [Gk. *hypo*, under; *tonos*, tension] having a lower osmotic pressure than that of another fluid; *cf.* hypertonic, isotonic.

Hypotremata n. [Gk. *hypo*, under; *trēma*, hole] a group of Chondrichthyes including the rays; *cf.* Pleurotremata.

hypotrematic a. [Gk. *hypo*, under; *trēma*, pore] *appl.* the lower lateral bar of branchial basket of Cyclostomata.

Hypotrichida n. [Gk. *hypo*, under; *thrix*, hair] an order or suborder of Spirotricha having a flattened body, few simple cilia, but having compound cilia called cirri; *alt.* Hypotricha.

hypotrichous a. [Gk. *hypo*, under; *thrix*, hair] having cilia mainly restricted to under surface; with deficient hair.

hypotrochanteric a. [Gk. *hypo*, under; *trochanter*, runner] beneath the trochanter.

hypotrophy n. [Gk. *hypo*, under; *trophē*, nourishment] the condition where wood or cortex grows more thickly on the underside of a horizontal plant organ; the condition where stipules or buds form on the underside.

hypotympanic a. [Gk. *hypo*, under; L. *tympanum*, drum] situated below the tympanum; *pert.* quadrate.

hypotype n. [Gk. *hypo*, under; *typos*, pattern] any specimen described or figured in order to amplify or correct the identification of a species; a species related to a genotype found in a different region or geological formation; *alt.* apotype, plesiotype, supplementary type.

hypovalve n. [Gk. *hypo*, under; L. *valva*, fold] the antapical part of envelope in certain Dinoflagellata; hypocone, *q.v.*; hypotheca, *q.v.*; in diatoms.

hypoxanthine n. [Gk. *hypo*, under; *xanthos*, yellow] a purine derivative formed by enzymatic deamination and hydrolysis of the purines of nucleic acids, and first found in glandular and muscle tissue and in some seeds.

hypozygal n. [Gk. *hypo*, under; *zygon*, yoke] lower ossicle of a syzgial pair bearing no pinnule.

hypselodont hypsodont, *q.v.*

hypsiloid ypsiloid, *q.v.*

hypsodont a. [Gk. *hypsos*, height; *odous*, tooth] *pert.* or designating teeth with high crowns and short roots; *alt.* hypselodont.

hypsophyll n. [Gk. *hypsi*, high; *phyllon*, leaf] any leaf beneath the sporophylls or perianth, and inserted at a high level near the flower, such as a floral bract or bracteole; *cf.* cataphyll.

hypural a. [Gk. *hypo*, under; *oura*, tail] *pert.* a bony structure formed by fused haemal spines of last few vertebrae, which supports the caudal fin in certain fishes.

Hyracoidea, hyracoids n., n.plu. [*Hyrax*, generic name; Gk. *eidos*, form] an order of placental mammals known from Oligocene times to the present day, including the hyrax, and having a rodent-like body and skull, but digits over a pad and bearing nails like elephants.

hysteranthous a. [Gk. *hysteros*, coming after; *anthos*, flower] leafing after appearance of flowers.

hysteresis n. [Gk. *hysterēsis*, late arrival] lag in one of 2 associated processes or phenomena; lag in adjustment of external form to internal stresses, as in chromosome during spiralization.

Hysteriales n. [*Hysterium*, generic name] an order of Ascomycetes having elongated ascomata which open by a longitudinal slit.

hysterochroic a. [Gk. *hysteros*, later; *chrōs*, colour] gradually discolouring from base to tip, *appl.* ageing fruit bodies.

hysterogenic a. [Gk. *hysteros*, later; *genos*, birth] of later development or growth.

hysterophyte n. [Gk. *hysteros*, inferior; *phyton*, plant] saprophyte, *q.v.*; any parasitic fungus.

hysterosoma n. [Gk. *hysteros*, after; *sōma*, body] part of body posterior to proterosoma and comprising metapodosoma and opisthosoma in Acarina.

hysterotely n. [Gk. *hysteros*, after; *telos*, completion] the retention or manifestation of larval characteristics in pupa or imago, or of pupal characters in imago; *alt.* metathetely; *cf.* prothetely.

hysterothecium n. [Gk. *hysteros*, after; *thēkē*, case] an apothecium with slits opening in moist conditions, closing in drought, as in certain fungi and lichens.

hyther n. [Gk. *hydōr*, water; *thermē*, heat] combined effect of moisture and temperature on an organism.

H-zone *see* H-line.

I

IAA indole-acetic acid, *q.v.*; *alt.* indolyl acetic acid.

IAN indole-acetonitrile, *q.v.*; *alt.* indolyl acetonitrile.

IBA indole-butyric acid, *q.v.*; *alt.* indolyl butyric acid.

I-band n. [*I*, isotropic] the isotropic or singly refracting band of a sarcomere which appears light and is made only of actin filaments; *alt.* I-disc, J-disc.

Ice Age Pleistocene, *q.v.*

I-cells interstitial cells, as in coelenterates.

ichnite, ichnolite n. [Gk. *ichnos*, track; *lithos*, stone] a fossil footprint.

ichthulin *n.* [Gk. *ichthys*, fish] a phosphoprotein found in carp roes.

ichthyic *a.* [Gk. *ichthys*, fish] *pert.*, characteristic of, or resembling fishes; *alt.* ichthyoid.

ichthyodont *n.* [Gk. *ichthys*, fish; *odous*, tooth] a fossil tooth of fish.

ichthyodorulite *n.* [Gk. *ichthys*, fish; *dory*, spear; *lithos*, stone] a fossil dermal or fin spine of fish.

ichthyofauna *n.* [Gk. *ichthys*, fish; L. *fauna*, god of woods] fish fauna.

ichthyoid *a.* [Gk. *ichthys*, fish; *eidos*, form] ichthyic, *q.v.*

ichthyolite *n.* [Gk. *ichthys*, fish; *lithos*, stone] a fossil fish or part of one.

ichthyology *n.* [Gk. *ichthys*, fish; *logos*, discourse] the study of fishes.

Ichthyopsida *n.* [Gk. *ichthys*, fish; *opsis*, appearance] collectively, agnathous vertebrates, fishes, and amphibians; *cf.* Sauropsida.

Ichthyopterygia *n.* [Gk. *ichthys*, fish; *pteryx*, wing or fin] a subclass of aquatic reptiles containing only one order, the Ichthyosauria, *q.v.*

ichthyopterygium *n.* [Gk. *ichthys*, fish; *pteryx*, wing or fin] the vertebrate limb in the form of a fin, especially as one of the paired fins of fish, but also in the course of evolutionary or an individual's development; *cf.* cheiropterygium.

Ichthyornithiformes *n.* [*Ichthyornis*, generic name; L. *forma*, shape] an order of gull- or tern-like birds of the subclass Neornithes existing in the Cretaceous in North America.

Ichthyosauria, ichthyosaurs *n.*, *n.plu.* [Gk. *ichthys*, fish; *sauros*, lizard] the only order of the Ichthyopterygia, being Mesozoic aquatic reptiles with a spindle-shaped body with fins and fin-like limbs.

Ichthyostegalia, ichthyostegids *n.*, *n.plu.* [Gk. *ichthys*, fish; *stegos*, roof] a group of primitive extinct Amphibia existing from Devonian to Carboniferous times and having many fish-like characteristics, sometimes considered to be an order of labyrinthodonts.

iconotype *n.* [Gk. *eikōn*, image; *typos*, pattern] representation, drawing or photograph, of a type.

icosandrous *a.* [Gk. *eikosi*, twenty; *anēr*, man] having 20 or more stamens.

icotype *n.* [Gk. *eikōs*, to be like; *typos*, pattern] a representative specimen used for identification of a species.

ICSH interstitial cell stimulating hormone, *q.v.*

id *n.* [Gk. *idios*, distinct] hypothetical unit, *q.v.*; chromomere, *q.v.*; the instincts, collectively.

idant *n.* [Gk. *idios*, distinct] a unit resulting from an aggregation of ids; the chromosome.

ideal angle in phyllotaxis, the angle between successive leaf insertions on a stem when no leaf would be exactly above any lower leaf, 137° 30′ 28″.

identical *a.* [L. *idem*, the same] *appl.* progeny having the same genes, as monozygotic twins; *appl.* points on retina corresponding to those of the other eye.

ideoglandular *a.* [Gk. *idein*, to see; L. *glandula*, small acorn] *pert.* glandular activity induced by a mental image.

ideomotor *a.* [Gk. *idein*, to see; L. *movere*, to move] *pert.* unwilled movement in response to a mental image.

ideotype *n.* [Gk. *idein*, to see; *typos*, pattern] specimen other than a topotype, named by the author who has described the species to which it belongs.

ideovascular *a.* [Gk. *idein*, to see; L. *vasculum*, small vessel] *pert.* circulatory changes induced by a mental image.

idioandrosporous *a.* [Gk. *idios*, distinct; *anēr*, male; *sporos*, seed] with androspores formed in filaments that do not bear oogonia.

idiobiology *n.* [Gk. *idios*, personal; *bios*, life; *logos*, discourse] biology of an individual organism; *alt.* autobiology.

idioblast *n.* [Gk. *idios*, distinct; *blastos*, bud] a hypothetical unit, *q.v.*; plant cell containing oil, gum, calcium carbonate, or other product and which differs from the surrounding parenchyma.

idiocalyptrosome *n.* [Gk. *idios*, distinct; *kalyptra*, covering; *sōma*, body] outer zone derived from idiosphaerosome in sperm cells.

idiochromatin *n.* [Gk. *idios*, distinct; *chrōma*, colour] temporarily dormant generative chromatin which controls reproduction; *cf.* tropochromatin.

idiochromidia *n.plu.* [Gk. *idios*, distinct; *chrōma*, colour] chromidia of generative chromatin which can be reformed into a nucleus or replace the nucleus; *alt.* sporetia; *cf.* trophochromidia.

idiochromosome *n.* [Gk. *idios*, distinct; *chrōma*, colour; *sōma*, body] sex chromosome, *q.v.*

idiocryptosome *n.* [Gk. *idios*, distinct; *kryptos*, hidden; *sōma*, body] inner zone derived from idiosphaerosome in sperm cells.

idiocuticular *a.* [Gk. *idios*, personal; L. *cuticula*, cuticle] characteristic of a cuticle; produced in a cuticle, as microtrichia of epicuticle in insects.

idiogamy *n.* [Gk. *idios*, personal; *gamos*, marriage] self-fertilization.

idiogram *n.* [Gk. *idios*, distinct; *gramma*, drawing] a diagrammatic representation of a characteristic chromosomal constitution.

idiomuscular *a.* [Gk. *idios*, peculiar; L. *musculus*, muscle] *appl.* contraction of a fatigued or degenerated muscle artifically stimulated.

idiophthartosome *n.* [Gk. *idios*, distinct; *phthartos*, transitory; *sōma*, body] the idiozome remnant.

idioplasm *n.* [Gk. *idios*, distinct; *plasma*, mould] chromatin, *q.v.*; the generative or germinal part of a cell, *alt.* germ plasm, *cf.* trophoplasm.

idiosoma *n.* [Gk. *idios*, distinct; *sōma*, body] the body, prosoma and opisthosoma, of Acarina.

idiosome *n.* [Gk. *idios*, distinct; *sōma*, body] a hypothetical unit, *q.v.*; sphere or region of cytoplasm differing in viscosity from remainder of cell and surrounding the centriole or centrosome, *alt.* cytocentrum; idiozome, *q.v.*; *alt.* archiplasm.

idiosphaerosome *n.* [Gk. *idios*, distinct; *sphaira*, globe; *sōma*, body] acrosome, central granule of idiosphaerotheca.

idiosphaerotheca *n.* [Gk. *idios*, distinct; *sphaira*, globe; *thēkē*, case] acroblast, vesicle containing acrosome in sperm cells.

idiothalamous *a.* [Gk. *idios*, distinct; *thalamos*, room] *appl.* lichens in which various parts are differently coloured from thallus.

idiothermous *a.* [Gk. *idios*, personal; *thermos*, hot] homoiothermal, *q.v.*

idiotrophic *a.* [Gk. *idios*, personal; *trophē*, nourishment] capable of selecting food.

idiotype n. [Gk. *idios*, personal; *typos*, pattern] individual genotype.

idiovariation mutation, *q.v.*

idiozome n. [Gk. *idios*, distinct; *zōma*, girdle] in spermatogenesis a separated portion of archoplasm which ultimately becomes head-cap of spermatozoon, *alt.* centrotheca; a cell body of auxocytes containing the centrioles; *alt.* idiosome.

I-disc I-band, *q.v.*

idorgan n. [Gk. *idios*, distinct; *organon*, instrument] a purely morphological multicellular unit which does not possess the features of a soma.

IEP isoelectric point, *q.v.*

ileac, ileal a. [Gk. *eilō*, to roll up] *pert.* ileum, *appl.* arteries, lymph glands, etc.

ileocaecal a. [Gk. *eilō*, to roll; *caecus*, blind] *pert.* ileum and caecum, *appl.* fossae, folds, valve.

ileocolic a. [Gk. *eilo*, to roll; *kolon*, colon] *pert.* ileum and colon, *appl.* artery, lymph glands, etc.

ileu isoleucine, *q.v.*

ileum n. [Gk. *eilō*, to roll] lower part of small intestine; anterior part of hind-gut in insects.

iliac a. [L. *ilia*, flanks] *pert.* ilium; *appl.* artery, fossa, furrow, tuberosity, vein, etc.; *appl.* muscle: iliacus, from upper part of iliac fossa to side of tendon of psoas major; *appl.* processes of ischio-pubic plate forming base for pelvic fins.

iliocaudal a. [L. *ilia*, flanks; *cauda*, tail] connecting ilium and tail, *appl.* muscle.

iliococcygeal a. [L. *ilia*, flanks; Gk. *kokkyx*, cuckoo] *pert.* ilium and coccyx, *appl.* a muscle.

iliocostal a. [L. *ilia*, flanks; *costa*, rib] in region of ilia and ribs, *appl.* several muscles.

iliofemoral a. [L. *ilia*, flanks; *femur*, thigh] *pert.* ilium and femur, *appl.* a ligament, *alt.* Y-ligament.

iliohypogastric a. [L. *ilia*, flanks; Gk. *hypo*, under; *gastēr*, stomach] *pert.* ilium and lower anterior part of abdomen, *appl.* a nerve.

ilio-inguinal a. [L. *ilia*, flanks; *inguen*, groin] in the region of ilium and groins, *appl.* a nerve.

ilio-ischiadic a. [L. *ilia*, flanks; Gk. *ischion*, hip] *appl.* fenestra between ilium and ischium when these are fused at both ends.

iliolumbar a. [L. *ilia*, flanks; *lumbus*, loin] in region of ilium and loins, *appl.* artery, ligament, vein.

iliopectineal a. [L. *ilia*, flanks; *pecten*, crest] *appl.* an eminence marking point of union of ilium and pubis; *appl.* fascia.

iliopsoas n. [L. *ilia*, flanks; Gk. *psoa*, loins] iliacus (iliac muscle) and psoas major considered as one muscle.

iliotibial a. [L. *ilia*, flanks; *tibia*, shin] *appl.* tract or band of muscle at lower end of thigh.

iliotrochanteric a. [L. *ilia*, flanks; Gk. *trochanter*, runner] uniting ilium and trochanter of femur, *appl.* a ligament.

ilium n. [L. *ilium*, flank] dorsal bone of pelvic girdle.

illicium n. [L. *illix*, decoy] anterior dorsal spine with modified tip for luring prey of Lophiidae (angler fish).

illuvial a. [L. *in*, into; *luere*, to wash] *appl.* layer of deposition and accumulation below the alluvial layer in soils; *alt.* B-horizon.

imaginal a. [L. *imago*, image] *pert.* an imago; *appl.* larval discs, patches of cells from which new organs develop, *alt.* histoblasts.

imago n. [L. *imago*, image] the last or adult stage in insect metamorphosis, the perfect insect.

imbibition n. [L. *in*, into; *bibere*, to drink] the non-active (i.e. non-energy requiring) absorption of water especially by certain substances such as cellulose and starch, as in uptake of water by seeds before germination.

imbricate(d) a. [L. *imbricare*, to tile] having parts overlapping each other like roof tiles, *appl.* scales, plates, bud scales, bracts.

imbrication lines parallel growth lines of dentine; *alt.* contour lines of Owen.

imitative a. [L. *imitari*, to imitate] *appl.* form, structure, habit, colouring, etc., assumed for protection or aggression; *appl.* behaviour of an animal where it repeats the activity of another animal.

immaculate a. [L. *in*, not; *macula*, spot] without spots or marks of different colour.

immarginate a. [L. *in*, not; *margo*, edge] without a distinct margin.

immobilization n. [L. *immobilis*, immobile] assimilation of inorganic compounds into protoplasm; *cf.* mineralization.

immune body heat-stable antibody or lysin; amboceptor, *q.v.*

immune response immunity, *q.v.*

immune serum globulin immunoglobulin, *q.v.*

immunity n. [L. *immunis*, free] an organism's resistance, natural or acquired, to the onset of pathological conditions from infection, natural or artifical, by micro-organisms or their products; *alt.* immune response.

immunize v. [L. *immunis*, free] to render invulnerable to a toxin, usually by injecting the toxin in small quantities at short intervals, without appearance of severe symptoms.

immunogenic a. [L. *immunis*, free; *genos*, birth] causing formation of antibodies.

immunoglobulin n. [L. *immunis*, free; *globus*, globe] any globulin capable of acting as an antibody; *alt.* immune serum globulin.

immunoprotein n. [L. *immunis*, free; *prōteion*, first] antibody, *q.v.*

immunotoxin n. [L. *immunis*, free; Gk. *toxikon*, poison] any antitoxin in blood which confers immunity against a disease.

IMP inosine monophosphate, inosinic acid, *q.v.*

impar a. [L. *impar*, unequal] not paired; not existing in pairs; *alt.* azygous.

imparidigitate a. [L. *impar*, unequal; *digitus*, finger] having an odd number of digits.

imparipinnate a. [L. *impar*, unequal; *pinna*, wing] pinnate with an odd terminal leaflet; *alt.* unequally pinnate; *cf.* paripinnate.

impedicellate a. [L. *in*, not; *pediculus*, small foot] without short or slender stalks; not having pedicels.

Impennae n. [L. *in*, not; *penna*, feather or wing] the order or superorder of birds including the penguins, having wings modified for swimming, short legs, and webbed feet.

imperfect a. [L. *imperfectus*, unfinished] incomplete; *appl.* fungi lacking the sexual spore stage; *appl.* flowers lacking functional stamens or pistils.

Imperfect Fungi Fungi Imperfecti, *q.v.*

imperforate a. [L. *in*, not; *per*, through; *foratus*, bored] not pierced; *appl.* foraminiferous shells without fine pores in addition to principal opening; *appl.* certain spiral shells, with occlusion of umbilicus.

impervious *a.* [L. *in*, not; *pervius*, passable] not permeable; *appl.* nostrils with septum between nasal cavities.

implacental *a.* [L. *in*, not; *placenta*, cake] aplacental, *q.v.*

implant *n.* [L. *in*, into; *plantare*, to plant] an organ or part transplanted to an abnormal position; graft, *q.v.*

implantation *n.* [L. *in*, into; *planta*, plant] the act of inserting or grafting; embedding of fertilized ovum in lining of uterus, *alt.* nidation.

implantation cone cone of origin, *q.v.*

implex *n.* [L. *implexus*, plaited] infolding of integument for muscle attachment in insects; *alt.* endoplica.

importation *n.* [L. *importare*, to carry into] ingestion by sinking of food into protoplasm of captor, as in certain Protozoa.

impregnation *n.* [L. *impraegnare*, to fertilize] transference of spermatozoa from male to body of female; *alt.* insemination.

impressio *n.* [L. *impressio*, impression] impression or concavity in one organ or structure where in contact with another, as of surface of liver in contact with stomach, etc.

imprinting *n.* [L. *imprimere*, to imprint] a kind of learning which usually occurs shortly after birth in which an animal recognizes that an object has a certain function, such as young birds acquiring a following response to a large moving object which is normally the parent bird.

impuberal *a.* [L. *impubes*, under age] prepubertal, sexually immature.

impulse *n.* [L. *impulsus*, driven] the message conducted along nerve fibres in the form of a wave of electric discharge.

inantherate *a.* [L. *in*, not; Gk. *antheros*, flowering] ananantherous, *q.v.*

inappendiculate *a.* [L. *in*, not; *appendicula*, small appendage] without appendages.

Inarticulata *n.* [L. *in*, not; *articulatus*, jointed] a class of brachiopods in which the valves of the shell are held together by muscles only.

inarticulate *a.* [L. *in*, not; *articulatus*, jointed] not segmented; not jointed.

inaxon *n.* [Gk. *is*, fibre; *axon*, axis] a neurone with the axon branching at a distance from the cell body; *cf.* dendraxon.

inborn behaviour species-specific behaviour, *q.v.*

inbreeding breeding through a succession of parents belonging to the same stock, or very nearly related; *alt.* endogamy.

inbreeding depression loss of vigour following inbreeding.

Inca bones distinct portions of interparietal, found in skulls of former Peruvians; *alt.* os interparietale.

incaliculate *a.* [L. *in*, not; *caliculus*, small flower cup] lacking a calicle.

incasement theory preformation theory, *q.v.*

inception anlage, *q.v.*

incisal *a.* [L. *incidere*, to cut into] cutting, as edge of a tooth.

incised *a.* [L. *incisus*, cut into] with deeply notched margin.

incisiform *a.* [L. *incisus*, cut into; *forma*, shape] incisor-shaped.

incisive *a.* [L. *incisus*, cut into] *pert.* or in region of incisors, *appl.* bones, foramina, fossa.

incisor *a.* [L. *incisus*, cut into] adapted for cutting, *appl.* mammalian premaxillary teeth, *appl.* process on mandible of malacostracans. *n.* A crest or ridge on palatine process of maxilla.

incisura *n.* [L. *incidere*, to cut into] notch, depression, or indentation, as in bone, stomach, liver, etc.

included *a.* [L. *includere*, to shut in] having stamens and pistils not protruding beyond corolla, not exserted; *appl.* secondary phloem included in the secondary xylem of certain dicotyledons, *alt.* interxylary phloem.

inclusion bodies intracellular particles, as pigment granules, mitochondria, Golgi bodies, microsomes, viruses, etc.

incompatibility *n.* [L. *in*, not; *compatibilis*, compatible] genetically determined inability to mate successfully; inability of scion to join with stock.

Incompletae *n.* [L. *in*, not; *completus*, complete] Monochlamydeae, *q.v.*

incomplete metamorphosis insect metamorphosis in which young are hatched in general adult form (but without wings, which develop outside body, and without mature sexual organs) and develop without quiescent stage; *cf.* complete metamorphosis.

incongruent *a.* [L. *incongruens*, not suiting] not suitable or fitting; *appl.* surface of joints which do not fit properly.

incoordination *n.* [L. *in*, not; *cum*, together; *ordo*, order] want of coordination; irregularity of movement due to loss of muscle control.

incrassate *a.* [L. *incrassare*, to thicken] thickened; becoming thicker.

incretion *n.* [L. *in*, into; *cretus*, separated] autacoid, *q.v.*

incrustation *n.* [L. *in*, into; *crusta*, shell] encasement in mineral substance as in fossilization or in walls of certain algae.

incubation *n.* [L. *incubare*, to lie on] the hatching of eggs by means of heat, natural or artificial; period between infection and appearance of symptoms induced by parasitic organisms.

incubatorium *n.* [L. *incubare*, to lie on] temporary pouch surrounding mammary area, in which egg of the echidna, spiny anteater, is hatched.

incubous *a.* [L. *incubare*, to lie on] *appl.* leaves so arranged that the base of each is covered by upper portion of next lower; *cf.* succubous.

incudal *a.* [L. *incus*, anvil] *pert.* the incus, *appl.* fold, fossa.

incudate *a.* [L. *incus*, anvil] *appl.* type of rotifer mastax with large and hooked rami and reduced mallei.

incudes *plu.* of incus.

incumbent *a.* [L. *incumbere*, to lie upon] lying upon; bent downwards to lie along a base; *appl.* cotyledons so folded that flat sides are next radicle, *alt.* notorhizal; *appl.* hairs or spines applied lengthwise to their base; *appl.* insect wings resting on abdomen.

incurrent *a.* [L. *in*, into; *currere*, to run] leading into; afferent; *appl.* ectoderm-lined canals which admit water, in sponges; *appl.* inhalant siphons of molluscs; *appl.* ostia in insects heart which admit blood.

incurvate *a.* [L. *incurvus*, bent] curved inwards or bent back; *alt.* incurved, inflected.

incurvation *n*. [L. *incurvare*, to curve] the doubling back on itself of a structure or organ, as of a spirochaete about to divide.

incus *n*. [L. *incus*, anvil] part of a rotifer mastax; the anvil-shaped ear ossicle of mammals; *plu*. incudes.

incyathescence *n*. [L. *in*, not; *cyathus*, cup; in*florescence*] a group of reduced cyathia in some Euphorbiaceae.

indeciduate *a*. [L. *in*, not; *decidere*, to fall down] non-caducous; with maternal part of placenta not coming away at birth; *alt*. non-deciduate, adeciduate.

indeciduous *a*. [L. *in*, not; *decidere*, to fall down] persistent; not falling off at maturity; everlasting; evergreen.

indefinite *a*. [L. *in*, not; *definitus*, limited] not limited; not determinate; of no fixed number; racemose, *q.v.*

indehiscent *a*. [L. *in*, not; *dehiscens*, gaping] *appl*. fruits which do not open to release seeds, but whole fruit is shed from the plant; not opening to release spores.

independent assortment segregation of one pair of alleles occurring independently of another pair, due to them being on different chromosomes.

indeterminate *a*. [L. *in*, not; *determinare*, to limit] indefinite; undefined; not classified.

indeterminate growth *appl*. apical meristem which forms an unrestricted number of lateral organs such as stems, branches, or shoots indefinitely, not limited or stopped by the development of terminal bud, as in vegetative apical meristem; indefinite prolongation and subdivision of an axis.

indeterminate inflorescence growth of a floral axis by indefinite branching because unlimited by development of a terminal bud; *alt*. racemose inflorescence.

index the forefinger or digit next to the thumb; a number or formula expressing ratio of one quantity to another; *appl*. fossil characterizing a geological horizon.

Indian hemp the cannabis plant, *q.v.*

indicators species characteristic of climatic, soil, and other conditions in a particular region or habitat; dominant species in a biotope; dyes which dissociate to produce different colours, over a range of hydrogen ion concentrations of a solution.

indicator yellow a product of the retinal transient orange, *q.v.*, pale yellow in alkaline, deep yellow in acid, solutions of rhodopsin.

indigenous *a*. [L. *indigena*, native] belonging to the locality; not imported; native.

indirect nuclear division mitosis, *q.v.*

individual *a*. [L. *in*, not; *dividuus*, divisible] *pert*. a single example or unit, as individual variations of colour. *n*. A person or zooid of distinctive function in a hydrozoan colony.

individual distance the area around a bird in a flock, which it defends while feeding, etc.

individualism symbiosis in which the 2 parties together from what appears to be a single organism.

individuation development of interdependent functional units, as in colony formation; organization of morphogenetic processes; regional or tissue differentiation; process of developing into an individual.

indole a compound present in various plants and certain essential oils, produced from tryptophane by bacteria in large intestine.

indole-acetic acid (IAA) natural auxin, *q.v.*; heteroauxin, *q.v.*; *alt*. indolyl acetic acid.

indole-acetonitrile (IAN) a natural auxin that has been isolated; *alt*. indolyl acetonitrile.

indole-butyric acid (IBA) a synthetic auxin; *alt*. indolyl butyric acid.

induced development epigenesis, *q.v.*

induced movement movement dictated and influenced by external stimulus, as plant curvature.

inducer a compound which causes the synthesis of an enzyme; *cf*. repressor.

inducible of enzymes, adaptive, *q.v.*

induction *n*. [L. *inducere*, to lead in] act or process of causing to occur; process whereby a cell or tissue influences neighbouring cells or tissues; lowering by one reflex of the threshold of another, spinal induction.

inductive stimulus an external stimulus which influences growth or behaviour of an organism.

inductor organizer, *q.v.*

indumentum *n*. [L. *indumentum*, covering] the plumage of birds; a hairy covering in plants or animals.

induplicate *a*. [L. *in*, in; *duplex*, double] in vernation, having bud leaves bent or rolled without overlapping; in aestivation, having bud sepals or petals folded inwards at points of contact.

induplicative *a*. [L. *in*, in; *duplex*, double] *appl*. vernation or aestivation with induplicate foliage or floral leaves respectively.

indurated, indurescent *a*. [L. *indurescere*, to harden] becoming firmer or harder.

indusia *plu*. of indusium.

indusial *a*. [L. *induere*, to put on] containing larval insect cases, as certain limestones; *pert*. the indusium.

indusiate *a*. [L. *induere*, to put on] having an enveloping case, *appl*. insect larvae; having an indusium.

indusiform *a*. [L. *induere*, to put on; *forma*, shape] resembling an indusium.

indusium *n*. [L. *induere*, to put on] an outgrowth of plant epidermis covering and protecting a sorus, as in ferns, *alt*. episporangium; outgrowth hanging from top of stipe in certain fungi, *alt*. veil; cup-like fringe of hairs surrounding a stigma; an insect larva case; supracallosal gyrus of the rhinencephalon, indusium griseum; *plu*. indusia.

industrial melanism the appearance of dark or melanistic forms of a species in industrial regions.

induviae *n.plu*. [L. *induviae*, garments] scale leaves; leaves which remain attached to stem after withering.

induviate *a*. [L. *induviae*, garments] covered with induviae.

inequilateral *a*. [L. *in*, not; *aequus*, equal; *latus*, side] having 2 sides unequal; having unequal portions on either side of a line drawn from umbo to gape of a bivalve shell.

inequilobate *a*. [L. *in*, not; *aequus*, equal; *lobus*, lobe] with lobes of unequal size.

inequivalve *a*. [L. *in*, not; *aequus*, equal; *valvae*, folding doors] having the valves of shell unequal, *appl*. molluscs.

inerm(ous) *a*. [L. *inermis*, unarmed] without means of defence and offence; without spines.

inert *a.* [L. *iners*, inactive] physiologically inactive; *appl.* heterochromatic region of chromosome with paucity of active genes.

infauna *n.* [L. *in*, into; *faunus*, god of woods] bottom-dwelling and burrowing animals.

infection *n.* [L. *inficere*, to taint] invasion, or condition caused by endoparasites; *cf.* infestation.

inferior *a.* [L. *inferior*, lower] *appl.* lower placed of 2, farther down axis; growing or arising below another organ; *appl.* ovary having perianth inserted round the top; *appl.* vena cava: postcava, *q.v.*; *cf.* superior.

inferoanterior *a.* [L. *inferus*, beneath; *anterior*, in front] below and in front.

inferobranchiate *a.* [L. *inferus*, beneath; Gk. *brangchia*, gills] with gills under margin of mantle, as in certain molluscs.

inferolateral *a.* [L. *inferus*, beneath; *latus*, side] below and at or towards the side.

inferomedian *a.* [L. *inferus*, beneath; *medius*, middle] below and about the middle.

inferoposterior *a.* [L. *inferus*, beneath; *posterior*, behind] below and behind.

inferradial *n.* [L. *inferus*, beneath; *radius*, radius] lower part of transversely bisected radials of certain fossil crinoids.

infestation *n.* [L. *infestare*, to be hostile] invasion by exterior organisms, as by ectoparasites; *cf.* infection.

inflected, inflexed *a.* [L. *inflectere*, to bend in] curved or abruptly bent inwards or towards the axis; *cf.* incurvate.

inflorescence *n.* [L. *inflorescere*, to begin to blossom] a flowering or putting forth blossoms; method in which flowers are arranged on an axis; a flowering branch; in bryophytes, the area bearing antheridia and archegonia.

inflorescent fruit anthocarp, *q.v.*

influents *n.plu.* [L. *influere*, to flow into] the animals present in a plant community, or those primarily dependent and acting upon the dominant plant species.

informosome *n.* [L. *informare*, to shape; Gk. *sōma*, body] a hypothetical mRNA molecule surrounded by a protein coat to protect it against nuclease digestion, and passing from nucleus to cytoplasm.

infra- in classification, a group just below the status of a subgroup of the taxon following it, as in infraclass, the group below subclass.

infra-axillary *a.* [L. *infra*, below; *axilla*, armpit] branching off below the axil.

infrabasal *n.* [L. *infra*, below; *basis*, base] one of a series of plates, perradial in position, below the basals in crinoids; *plu.* infrabasalia.

infrabranchial *a.* [L. *infra*, below; *branchiae*, gills] below the gills, *appl.* part of pallial chamber.

infrabuccal *a.* [L. *infra*, below; *bucca*, cheek] below the cheeks; beneath the buccal mass, in molluscs.

infracentral *a.* [L. *infra*, below; *centrum*, centre] below a vertebral centrum.

infraciliature *n.* [L. *infra*, below; *cilia*, eyelashes] the structures or organellae just below the cilia, consisting of kinetia, in Ciliata.

infraclavicle *n.* [L. *infra*, below; *clavicula*, little key] membrane bone occurring in pectoral girdle of some fishes.

infraclavicular *a.* [L. *infra*, below; *clavicula*, small key] beneath the clavicle, *appl.* branches of brachial plexus, *appl.* fossa or triangle between deltoid and pectoralis major.

infracortical *a.* [L. *infra*, below; *cortex*, bark] beneath the cortex.

infracostal *a.* [L. *infra*, below; *costa*, rib] beneath the ribs, *appl.* muscles.

infradentary *a.* [L. *infra*, below; *dens*, tooth] beneath the dentary bone.

infraepimeron *n.* [L. *infra*, below; Gk. *epi*, upon; *meros*, thigh] katepimeron, *q.v.*

infraepisternum *n.* [L. *infra*, below; Gk. *epi*, upon; L. *sternum*, breast bone] katepisternum, *q.v.*

infra-ergatoid *n.* [L. *infra*, below; Gk. *ergatēs*, worker; *eidos*, form] phthisergate, *q.v.*

infraglenoid *a.* [L. *infra*, below; Gk. *glēnē*, socket; *eidos*, like] below glenoid cavity, *appl.* a tuberosity.

infrahyoid *a.* [L. *infra*, below; Gk. *hyoeidēs*, Υ-shaped] beneath the hyoid, *appl.* muscles.

infralabial *a.* [L. *infra*, below; *labium*, lip] beneath the lower lip.

inframammary *a.* [L. *infra*, below; *mamma*, breast] between mammary and hypochondriac regions.

inframarginal *a.* [L. *infra*, below; *margo*, margin] under the margin, or marginal structure, *appl.* a cerebral convolution; *appl.* certain plates on carapace of Chelonia below marginals; *appl.* lower of 2 series of plates around margin of stelleroid arms and discs.

inframaxillary *a.* [L. *infra*, below; *maxilla*, jaw] beneath maxilla, *appl.* nerves.

infranasal *n.* [L. *infra*, below; *nasus*, nose] an additional nasal element in some Theromorpha.

infraneuston *n.* [L. *infra*, below; Gk. *neustos*, floating] the animals living on the underside of the surface film of water, such as some mosquito larvae.

infraorbital *a.* [L. *infra*, below; *orbis*, eye socket] beneath the orbit, *appl.* artery, canal, foramen, groove, nerve, glands, etc.

infrapatellar *a.* [L. *infra*, below; *patella*, knee cap] *appl.* pad of fat beneath patella; *appl.* bursa between tibia and ligamentum patellae.

infraprotein *n.* [L. *infra*, below; *prōteion*, first] metaprotein, *q.v.*

infrarostral *a.* [L. *infra*, below; *rostrum*, snout] beneath a rostrum; *appl.* paired cartilages, derived from Meckel's cartilage, of lower part of suctorial mouth of tadpoles.

infrascapular *a.* [L. *infra*, below; *scapula*, shoulder blade] beneath the scapula, *appl.* artery.

infraspecific *a.* [L. *infra*, below; *species*, particular kind] *pert.* a subdivision of a species, as subspecies and varieties.

infraspinatous, infraspinous *a.* [L. *infra*, below; *spina*, spine] beneath the spine; beneath scapular spine; *appl.* muscle, fossa.

infrastapedial *a.* [L. *infra*, below; *stapes*, stirrup] beneath stapes of ear, *appl.* part of columella.

infrasternal *a.* [L. *infra*, below; *sternum*, breast bone] below the breast bone; *appl.* notch superficially at lower end of sternum.

infratemporal *a.* [L. *infra*, below; *tempora*, temples] beneath the temporal bone, *appl.* a crest and fossa.

infratrochlear a. [L. *infra*, below; *trochlea*, pulley] beneath the trochlea, *appl.* nerve given off from nasociliary nerve.

infructescence n. [L. *in*, into; *fructus*, fruit] anthocarp, q.v.

infundibula *plu.* of infundibulum; passages surrounded by air cells in the lung.

infundibular a. [L. *infundibulum*, funnel] funnel-shaped, *appl.* an abdominal muscle, *appl.* corolla, *alt.* infundibuliform; choanoid, q.v.; *pert.* infundibulum.

infundibulin n. [L. *infundibulum*, funnel] pituitrin, q.v.

infundibulum n. [L. *infundibulum*, funnel] any funnel-shaped organ or structure; *appl.* part of ethmoid bone, of right ventricle, etc.; outpushing of floor of brain which develops with hypophysis into pituitary body; conus arteriosus; a cephalopod siphon; part of bird's oviduct; flattened stomach-like cavity of ctenophores; septal funnel in Scyphozoa; hydroecium, q.v.

infundin pituitrin, q.v.

infuscate a. [L. *in*, into; *fuscus*, dark] tinged to appear dark, as insect wings.

infusoria, infusorians n., n.plu. [L. *infusus*, poured into] a term used for a group of animals, formerly including those microscopic animals such as protozoans and rotifers found in hay infusions, but today used only for ciliate protozoans.

infusoriform a. [L. *infusus*, poured into; *forma*, shape] resembling an infusorian, *appl.* embryonic forms of Coelenterata, *appl.* male form of Dicyemidae.

infusorigen n. [L. *infusus*, poured into; *genos*, offspring] a gastrula-like phase in development of certain Mesozoa.

ingest v. [L. *ingestus*, taken in] to convey food material into the alimentary canal or food cavity.

ingesta n.plu. [L. *ingestus*, taken in] the sum total of substances taken in by the body; *cf.* egesta.

ingestion n. [L. *ingestus*, taken in] the swallowing or taking in of food material to gut or food cavity.

ingluvies n. [L. *ingluvies*, crop] the crop of a bird or insect; a dilatation of oesophagus; rumen, q.v.

inguinal a. [L. *inguen*, groin] in region of groin; *pert.* groin.

inguinal ring abdominal ring, q.v.

inguino-abdominal in region of abdomen and groin.

inguino-crural in region of groin and leg.

inhalent a. [L. *in*, into; *halere*, to breathe] adapted for inspiring or drawing in, as terminal pores of incurrent canals in sponges, or siphons in molluscs.

inhibin n. [L. *inhibere*, to restrain] a testicular hormone depressing gonadotrophic activity of pre-hypophysis.

inhibition n. [L. *inhibere*, to restrain] prohibition, or checking, of an action or process.

inhibitor n. [L. *inhibere*, to restrain] any agent which checks or prevents an action or process; a substance which prevents the normal action of enzymes without actually destroying them, such as salts of certain heavy metals, *cf.* antienzyme.

inhibitory a. [L. *inhibere*, to restrain] *appl.* nerves which control movement or secretion by decreasing these activities.

inhibitory postsynaptic potential (IPSP) the inhibition of potentials at synapses due to a negative charge building up in the postsynaptic nerve cell.

Iniomi n. [Gk. *inion*, back of the head; *ōmos*, shoulder] Scopeliformes, q.v.

inion n. [Gk. *inion*, back of head] the external protuberance of occipital bone.

initial n. [L. *initium*, beginning] a cell which initiates differentiation of tissues, as in apical meristem, vascular cambium, etc.; *alt.* histogen cell, primordial cell.

injury feigning a kind of distraction display in which a female bird behaves as though it is injured, to draw an enemy away from its eggs or chicks.

ink sac in certain cephalopods such as *Sepia*, a pear-shaped body in wall of mantle cavity which contains the ink gland, secreting a black substance, ink or sepia, ejection of which is a means of defence.

innate a. [L. *innatus*, inborn] inherited; *appl.* behaviour occurring instinctively, without learning, *alt.* species-specific behaviour; basifixed, q.v.

innate releasing mechanism (IRM) an internal and instinctive mechanism in an animal which releases its response to sign stimuli.

inner glume in grasses, palea, q.v.

innervation n. [L. *in*, into; *nervus*, sinew] the nerve supply to an organ or part.

innidiation n. [L. *in*, into; *nidus*, nest] colonization or development of cells or organisms in a part of the body to which they have been transferred by metastasis; q.v.

innominate a. [L. *in*, not; *nomen*, name] nameless, *appl.* various arteries and veins.

innominate artery an artery which gives rise to arteries to the head and forelimb; *alt.* brachiocephalic artery.

innominate bone the hip bone or lateral half of pelvic girdle; *alt.* os coxae, os innominatum.

innominate vein a vein which joins up veins from the head and forelimb; *alt.* brachiocephalic vein.

innovation n. [L. *innovare*, to renew] a growth or shoot of mosses which develops into a new plant by dying off of portion of parent plant behind it; basal vegetative shoot of grasses; a shoot carrying on the growth of a plant.

inocomma n. [Gk. *is*, fibre; *komma*, clause] sarcomere, q.v.; *alt.* inokomma.

inocular a. [L. *in*, into; *oculus*, eye] *appl.* antennae inserted close to eye.

inoculation n. [L. *inoculare*, to engraft] grafting by insertion of a single bud; the placing of a microorganism on a nutrient medium to obtain a culture of it; the introduction of material containing micro-organisms into a tissue, to produce antibodies.

inoculum n. [L. *inoculare*, to engraft] the cells, bacteria, spores, etc., introduced into a medium for cultures.

inocyte n. [Gk. *is*, fibre; *kytos*, hollow] elongated cell of fibrous tissue.

inogen n. [Gk. *is*, fibre; *gennaein*, to produce] a hypothetical nitrogen-containing substance said to be continuously reproduced and decomposed in the muscles and which acts as an oxygen reserve.

inokomma inocomma, q.v.

inoperculate a. [L. in, un-; operculum, lid] without a lid or operculum.

inophragma n. [Gk. is, fibre; phragma, fence] the transverse membrane through adjacent myofibrillae, being collectively the mesophragma and telophragma, the M-and Z-lines bisecting A- and I-discs.

inordinate a. [L. in, not; ordinatus, arranged] not showing any regular arrangement, appl. spores in an ascus.

inosculate v. [L. in, in; osculari, to kiss] to intercommunicate or unite, as vessels, ducts, etc.; to anastomose.

inosinic acid (IMP) a nucleotide, inosine monophosphate, from which AMP and GMP are formed.

inosine a nucleoside made up of hypoxanthine and ribose.

inosine monophosphate (IMP) inosinic acid, q.v.

inositol any of various cyclic hexahydroxyalcohols of formula $C_6H_6(OH)_6$ especially: (a) a component of the vitamin B complex and a lipotropic agent which occurs widely in plants in phytic acid, in micro-organisms, and in higher animals combined in phosphatides as in heart and brain; alt. i-inositol, mesoinositol, myoinositol; (b) a sweet dextrorotatory alcohol dextroinositol; (c) a sweet laevorotatory alcohol, laevoinositol; (d) scyllitol, q.v.; alt. muscle sugar, anti-alopecia factor.

inotagmata n.plu. [Gk. is, fibre; tagma, arrangement] hypothetical units, q.v.; sing. inotagma.

inquiline n. [L. inquilinus, tenant] animal living in home of another and getting share of its food; partner in commensalism; an insect developing in gall produced by an insect of another species, being detrimental to the latter.

inscriptions, tendinous three fibrous bands crossing the rectus abdominis muscles.

insculpt(ate) a. [L. in, in; sculpere, to carve] embedded in rocks, appl. lichens.

Insecta, insects n., n.plu. [L. insectum, cut into] a class of arthropods having the body divided into head, thorax, and abdomen, the thorax bearing 3 pairs of walking legs and usually 1 or 2 pairs of wings, and with the life history usually showing metamorphosis; alt. Hexapoda.

Insectivora, insectivores n., n.plu. [L. insectum, cut into; vorare, to devour] an order of primitive placental mammals known from Cretaceous times to the present day and including the hedgehogs, moles, and shrews.

insectivorous a. [L. insectum, cut into; vorare, to devour] insect-eating, appl. certain animals and carnivorous plants; alt. entomophagous.

insectorubins n.plu. [L. insectum, cut into; ruber, red] red or red–brown eye pigments of insects, derived from an oxidation product of tryptophane.

insectoverdins n.plu. [L. insectum, cut into; viridis, green] green pigments, mixtures of carotenoids and biliverdin, of insects.

insemination n. [L. in, in; seminatio, sowing] the introduction of semen or spermatozoa into female genital tract, alt. impregnation; transfer of a fertilized ovum from one female to another; cf. semination.

inserted a. [L. in, in; serere, to join] united by natural growth.

insertion n. [L. insertus, joined] point of attachment of organs, as of muscles, leaves; point on which force of a muscle is applied.

insertional appl. translocation in which the portion between 2 breaks of a chromosome is transferred to a break in another chromosome; cf. shift.

insessorial a. [L. insidere, to sit upon] adapted for perching.

insight a form of learning where an animal perceives a solution to a problem by an immediate and clear understanding of how the goal can be achieved without using trial-and-error methods.

insistent a. [L. insistere, to stand upon] appl. hind-toe, of certain birds, whose tip only reaches the ground.

insolation n. [L. in, into; sol, sun] exposure to sun's rays.

inspiration n. [L. inspirare, to inhale] the act of drawing air or water into the respiratory organs, cf. expiration; absorption of oxygen by plants and animals.

instaminate a. [L. in, not; stamen, thread] not bearing stamens.

instar n. [L. instar, form] insect at a particular stage between moults.

instinct n. [L. instinctus, impulse] a group of reflex actions which occur as a whole in response to an appropriate stimulus, sometimes equivalent to species-specific behaviour.

instipulate a. [L. in, not; stipula, stalk] exstipulate, q.v.

insula n. [L. insula, island] a triangular eminence lying deeply in lateral fissure of temporal lobe, alt. island of Reil; islet of Langerhans, q.v.; blood island, q.v.

insulin n. [L. insula, island] a proteinaceous hormone produced by the β-cells of islets of Langerhans which decreases the amount of glucose in the blood, its action being antagonistic to glucagon, adrenal glucocorticoids and adrenaline, and which was the 1st protein whose structure was determined and consists of 2 chains of amino acids linked by disulphide bridges, the arrangement differing in different species; alt. hypoglycemic hormone.

intectate a. [L. in, not; tectum, roof] without a tectum.

integrifolious a. [L. integer, whole; folium, leaf] with entire leaves.

integripallial, integripalliate a. [L. integer, whole; pallium, mantle] having an unbroken pallial line, appl. shells of molluscs with small or no siphons; cf. sinupalliate.

integument(um) n. [L. integumentum, covering] a covering, investing, or coating structure or layer; coat of ovule, alt. tegument; a. integumentary.

intemperate phage virulent phage, q.v.

intention movement a behavioural response which regularly occurs before another response in a series of activities; cf. flight intention.

interacinar a. [L. inter, between; acinus, grape] among alveoli of a racemose gland; alt. interacinous.

interallantoic a. [L. inter, between; Gk. allas, sausage] appl. septum formed by fusion of adjoining allantoic lobes.

interalveolar a. [L. inter, among; alveolus, small cavity] among alveoli, appl. cell islets.

interamb *n.* [L. *inter*, between; *ambulare*, to walk] interambulacral area.

interambulacral *a.* [L. *inter*, between; *ambulare*, to walk] *appl.* area of echinoderm test between 2 ambulacral areas. *n.* A plate of that area.

interambulacrum *n.* [L. *inter*, between; *ambulare*, to walk] the area between 2 ambulacral areas.

interarticular *a.* [L. *inter*, between; *articulus*, joint] between articulating parts of bones, *appl.* certain ligaments and fibrocartilages.

interatrial *a.* [L. *inter*, between; *atrium*, hall] *appl.* groove and septum separating the 2 atria of the heart.

interauricular *a.* [L. *inter*, between; *auricula*, little ear] between auricles of heart.

interaxillary *a.* [L. *inter*, between; *axilla*, armpit] placed between the axils.

interbands in polytene chromosomes, the regions between the conspicuous bands, containing little DNA.

interbrachial *a.* [L. *inter*, between; *brachium*, arm] between arms, rays, or brachial plates.

interbrain diencephalon, *q.v.*

interbranchial *a.* [L. *inter*, between; *branchiae*, gills] *appl.* septum between successive gill slits.

interbreed *v.* [L. *inter*, between; A.S. *brod*, brood] to cross different varieties, species, or genera of plants or animals.

intercalare *n.* [L. *intercalaris*, inserted] in many fishes and fossil amphibians, an additional element in the vertebra.

intercalarium *n.* [L. *intercalaris*, inserted] the 3rd Weberian ossicle.

intercalary *a.* [L. *intercalaris*, inserted] inserted between others; *appl.* meristematic layers between masses of permanent tissue; *appl.* growth elsewhere than at growing point; *appl.* veins between main veins of insect wings; *appl.* plates in Dinoflagellata; *appl.* bands in diatoms; *appl.* cartilage between neural arches, *alt.* interneural or interdorsal plate; *appl.* discs: transverse wavy bands formed by boundaries of sarcomeres in heart muscles; *alt.* intercalated.

intercalated intercalary, *q.v.*

intercapitular *a.* [L. *inter*, between; *capitulum*, little head] between capitula, *appl.* veins of fingers and toes.

intercarotid *a.* [L. *inter*, between; Gk. *karos*, deep sleep] between carotid arteries.

intercarpal *a.* [L. *inter*, between; *carpus*, wrist] among or between carpal bones, *appl.* joints.

intercarpellary *a.* [L. *inter*, between; Gk. *karpos*, fruit] between the carpels.

intercartilaginous *a.* [L. *inter*, between; *cartilago*, gristle] between cartilages.

intercavernous *a.* [L. *inter*, between; *caverna*, cavern] *appl.* sinuses connecting cavernous sinuses, part of ophthalmic veins.

intercellular *a.* [L. *inter*, between; *cellula*, little cell] among or between cells, as spaces in meristem, biliary passages among liver cells, plexus of dendrites between sympathetic ganglion cells, substances such as middle lamella, etc.

intercentral *a.* [L. *inter*, between; *centrum*, centre] uniting, or between, 2 centra.

intercentrum *n.* [L. *inter*, between; *centrum*, centre] a 2nd central ring in embolomerous vertebra.

interchange mutual or reciprocal translocation in chromosomes.

interchondral *a.* [L. *inter*, between; Gk. *chondros*, cartilage] *appl.* articulations and ligaments between costal cartilages.

interchromomeres *n.plu.* [L. *inter*, between; Gk. *chrōma*, colour; *meros*, part] in chromonemata, the lighter staining portions between chromomeres.

interchromosomal *a.* [L. *inter*, between; Gk. *chrōma*, colour; *sōma*, body] between chromosomes; *appl.* fibrils playing part in the beginning of cell wall formation in plants.

intercingular *a.* [L. *inter*, between; *cingulum*, girdle] *appl.* area of longitudinal groove between parts of a spiral girdle, in certain Dinoflagellata.

interclavicle *n.* [L. *inter*, between; *clavicula*, small key] a median ventral bone between clavicles; *alt.* episternum, hypocleidium.

interclavicular *a.* [L. *inter*, between; *clavicula*, small key] between the clavicles, *appl.* a ligament.

interclinoid *a.* [L. *inter*, between; Gk. *klinē*, bed; *eidos*, form] joining clinoid processes, *appl.* fibrous process or ligament.

intercolumnar *a.* [L. *inter*, between; *columna*, column] between columnar structures, as certain abdominal muscle fibres.

intercondyloid *a.* [L. *inter*, between; Gk. *kondylos*, knuckle; *eidos*, form] between condyles, *appl.* an eminence of tibia, and fossae of femur and tibia.

intercostal *a.* [L. *inter*, between; *costa*, rib] between the ribs, as arteries, glands, membranes, nerves, veins, muscles; between ribs of leaf, mericarp, etc.

intercostobrachial *a.* [L. *inter*, between; *costa*, rib; *brachium*, arm] *appl.* lateral branch of 2nd intercostal nerve which supplies upper arm; *alt.* intercostohumeral.

intercoxal *a.* [L. *inter*, between; *coxa*, hip] between the coxae or proximal limb joints of arthropods, *appl.* plate, etc.

intercrescence *n.* [L. *inter*, between; *crescere*, to grow] a growing into each other, as of tissues.

intercross the crossing of heterozygotes.

intercrural *a.* [L. *inter*, between; *crus*, leg] *appl.* intercolumnar tendinous fibres arching across external oblique muscles.

intercuneiform *a.* [L. *inter*, between; *cuneus*, wedge; *forma*, shape] connecting the 3 cuneiform bones of the ankle, *appl.* articulations and ligaments.

interdeferential *a.* [L. *inter*, between; *deferre*, to carry down] between the vasa deferentia.

interdigital *a.* [L. *inter*, between; *digitus*, finger] between digits, *appl.* glands.

interdorsal plate intercalary cartilage, *q.v.*

interfascicular *a.* [L. *inter*, between; *fasciculus*, small bundle] situated between the fascicles or vascular bundles, *appl.* cambium.

interfemoral *a.* [L. *inter*, between; *femur*, thigh bone] between the thighs.

interference *n.* [L. *inter*, between; *ferire*, to strike] the effect by which the occurrence of one cross-over reduces the probability of another occurring in its vicinity; in virology, prevention of superinfection with or multiplication of one virus as a result of the presence of another.

interference colours those produced by optical

interference between reflections from a series of superimposed laminae or ribs.

interferon *n.* [L. *inter*, between; *ferire*, to strike] a small protein which inhibits viral action and is produced by host cells in response to infection.

interfertile *a.* [L. *inter*, between; *fertilis*, fertile] able to interbreed.

interfilamentar *a.* [L. *inter*, between; F. *filament*, from L. *filum*, thread] *appl.* junctions or horizontal bars connecting molluscan gill filaments.

interfilar *a.* [L. *inter*, between; *filum*, thread] *appl.* ground substance of protoplasm, as opposed to reticulum; *alt.* paramitome.

interfoliaceous, interfoliar *a.* [L. *inter*, between; *folium*, leaf] situated or arising between 2 opposite leaves.

interfrontal *n.* [L. *inter*, between; *frons*, forehead] an unpaired median bone between frontals and nasals in labyrinthodont amphibians such as *Eryops*.

interganglionic *a.* [L. *inter*, between; Gk. *ganglion*, little tumour] connecting 2 ganglia, as nerve cords or strands.

intergemmal *a.* [L. *inter*, between; *gemma*, bud] between taste buds, *appl.* nerve fibres.

intergeneric *a.* [L. *inter*, between; *genus*, kind] between genera, *appl.* hybridization.

intergenic change a mutation involving more than one gene.

intergenital *a.* [L. *inter*, between; *genitalis*, generative] between the genitals, *appl.* certain echinoderm plates.

interglacial *a.* [L. *inter*, between; *glacies*, ice] *appl.* or *pert.* ages between glacial ages, particularly of the Pleistocene epoch.

interglobular *a.* [L. *inter*, between; *globulus*, small globe] *appl.* a series of spaces towards outer surface of dentine, due to imperfect calcification.

intergular *n.* [L. *inter*, between; *gula*, gullet] a paired or unpaired plate in front of gulars in Chelonia.

interhyal *n.* [L. *inter*, between; Gk. *hyoeidēs*, Υ-shaped] a small bone between hyomandibular and rest of hyoid of some vertebrates.

interkinesis *n.* [L. *inter*, between; Gk. *kinēsis*, movement] interphase, *q.v.*; the short interphase between the 1st and 2nd meiotic divisions and in which no DNA replication occurs.

interlamellar *a.* [L. *inter*, between; *lamella*, thin plate] *appl.* vertical bars of tissue joining gill lamellae of molluscs; *appl.* compartments of lung book in scorpions and spiders; *appl.* spaces between lamellae or gills of agarics.

interlaminar *a.* [L. *inter*, between; *lamina*, thin plate] uniting laminae; between laminae.

interlobar *a.* [L. *inter*, between; L.L. *lobus*, lobe] between lobes; *appl.* sulci and fissures dividing cerebral hemispheres into lobes.

interlobular *a.* [L. *inter*, between; *lobulus*, small lobe] occurring between lobules, *appl.* kidney arteries, vessels of liver, etc.

interlocular *a.* [L. *inter*, between; *loculus*, compartment] between loculi.

interloculus *n.* [L. *inter*, between; *loculus*, compartment] space between 2 loculi.

intermalar *a.* [L. *inter*, between; *mala*, cheek bone] situated between the cheek bones.

intermandibular *a.* [L. *inter*, between; *mandibulum*, jaw] between rami of mandibles.

intermaxilla *n.* [L. *inter*, between; *maxilla*, jaw] bone between maxillae; premaxilla, *q.v.*

intermaxillary *a.* [L. *inter*, between; *maxilla*, jaw] between maxillae, *pert.* premaxillae; *appl.* gland in nasal septum of certain amphibians and reptiles.

intermediary *a.* [L. *inter*, between; *medius*, middle] acting as a medium; *appl.* nerve cells receiving impulses from afferent cells and transmitting them to efferent cells.

intermediate *a.* [L. *inter*, between; *medius*, middle] occurring between 2 points or parts, *appl.* a nerve mass, certain areas of brain, ribs, etc.

intermediate disc Z-disc, *q.v.*

intermediate host a host in which a parasite lives for part of its life cycle, but in which it does not become sexually mature; *alt.* secondary host; *cf.* primary host.

intermedin *n.* [L. *inter*, between; *medius*, middle] a hormone produced by the pars intermedia of the pituitary gland, causing darkening of the animal by expansion of chromatophores; *alt.* B-substance, melanocyte-stimulating hormone, chromatophorotropic hormone.

intermedium *n.* [L. *inter*, between; *medius*, middle] a small bone of carpus and tarsus; *alt.* lunar bone.

intermesenteric *a.* [L. *inter*, between; Gk. *mesos*, middle; *enteron*, gut] occurring between mesenteries, *appl.* spaces in sea anemones.

intermetatarsal *a.* [L. *inter*, between; Gk. *meta*, after; *tarsos*, flat of the foot] between metatarsal bones, *appl.* articulations.

intermitosis *n.* [L. *inter*, between; Gk. *mitos*, thread] *see* interphase.

intermitotic *n.* [L. *inter*, between; Gk. *mitos*, thread] a cell with individual life between mitoses causing its origin and division into daughter cells; *cf.* postmitotic.

intermuscular *a.* [L. *inter*, between; *musculus*, muscle] between or among muscle fibres.

intermyotomic *a.* [L. *inter*, between; Gk. *mys*, muscle; *tomē*, cutting] *appl.* vertebra formed of caudals of one somite and cranials of next posterior, *cf.* intrasegmental; between myotomes, *appl.* septa.

internal *a.* [L. *internus*, within] located on inner side; nearer middle axis; located or produced within.

internal fertility control the raising or lowering of the birth rate by some animals in response to favourable or unfavourable external conditions.

internal phloem primary phloem found internal to primary xylem; *alt.* intraxylary phloem.

internal respiration respiration considered in terms of the biochemical processes occurring in all living cells, resulting in energy release; *cf.* external respiration.

internal rhythm endogenous rhythm, *q.v.*

internal secretion hormone, *q.v.*; *alt.* endocrine secretion.

internal spiral coil within a single chromatid as between prophase and anaphase.

internarial *a.* [L. *inter*, between; *nares*, nostrils] between the nostrils, *appl.* septum.

internasal *a.* [L. *inter*, between; *nasus*, nose] between nasal cavities, *appl.* plate, septum, gland.

International Unit (IU) the unit used to measure vitamin content of foods.

internema

internema n. [L. *inter*, between; Gk. *nēma*, thread] the segment of chromosome between synaptomeres.

interneural a. [L. *inter*, between; Gk. *neuron*, nerve] between neural processes, arches, or spines; *appl.* sharp bones attached to dorsal fin-rays; *appl.* intercalary cartilages, *q.v.*

interneuron(e) n. [L. *inter*, between; Gk. *neuron*, nerve] internuncial neurone, *q.v.*

internodal a. [L. *inter*, between; *nodus*, knot] *pert.* part between 2 nodes.

internode n. [L. *inter*, between; *nodus*, knot] the part between 2 nodes or joints; *appl.* plant stem, *alt.* merithallus; *appl.* medullated nerve fibre; *appl.* stolon of an ectoproct; non-genetic segment of a chromosome.

internodia n.plu. [L. *inter*, between; *nodus*, knot] phalanges, *q.v.*

internum n. [L. *internus*, inward] inner region or medulla of a mitochondrion, or of Golgi apparatus, or of acroblast.

internuncial a. [L. *inter*, between; *nuntius*, messenger] intercommunicating, as paths of transmission or nerve fibres, *appl.* neurone interposed between afferent and efferent neurones; *alt.* association neurone, connector neurone, interneurone, relay cell.

interoceptor n. [L. *internus*, inside; *capere*, to take] a receptor which receives stimili from within the body; end organ for visceral sensibility.

interocular a. [L. *inter*, between; *oculus*, eye] placed between the eyes.

interoperculum, interopercle n. [L. *inter*, between; *operculum*, lid] a membrane bone of operculum of Teleostei and Dipnoi, attached to mandible.

interoptic a. [L. *inter*, between; Gk. *optikos*, *pert.* sight] between optic lobes.

interorbital a. [L. *inter*, between; *orbis*, eye socket] between the orbits, *appl.* septum of tropibasic skull, *appl.* sinus.

interosculant a. [L. *inter*, between; *osculari*, to kiss] possessing characters common to 2 or more groups or species.

interosseous a. [L. *inter*, between; *os*, bone] occurring between bones, *appl.* arteries, ligaments, membranes, muscles, nerves.

interparietal a. [L. *inter*, between; *paries*, wall] in many vertebrates a bone arising between parietals and supraoccipital.

interpeduncular a. [L. *inter*, between; *pedunculus*, little foot] *appl.* fossa between cerebral peduncles, and to ganglion.

interpetaloid a. [L. *inter*, between; Gk. *petalon*, leaf; *eidos*, form] between petaloid areas of an echinoderm test.

interpetiolar a. [L. *inter*, between; *petiolus*, little foot] situated petioles or bases of opposite leaves.

interphalangeal a. [L. *inter*, between; Gk. *phalangx*, line of battle] *appl.* articulations between successive phalanges.

interphase n. [L. *inter*, between; Gk. *phasis*, aspect] the period between 2 mitoses, *alt.* intermitosis; the period sometimes occurring between 1st and 2nd meiotic division; *cf.* interkinesis.

interplacental a. [L. *inter*, between; *placenta*, flat cake] between placentae.

interpleural a. [L. *inter*, between; Gk. *pleuron*, side] between pleurae.

interpleurite n. [L. *inter*, between; Gk. *pleuron*, side] a small sclerite between sclerites of the pleura; *alt.* intersegmental pleural sclerite.

interpositional growth of cells, by interposition between neighbouring cells without loss of contact; *alt.* intrusive growth; *cf.* sliding growth.

interpubic a. [L. *inter*, between; *pubes*, mature] *appl.* the fibrocartilaginous lamina between pubic bones.

interracial a. [L. *inter*, between; *radix*, root] between races or breeds, *appl.* hybridization, differences, etc.

interradial a. [L. *inter*, between; *radius*, radius] *pert.* an interradius.

interradium n. [L. *inter*, between; *radius*, radius] the area between 2 radii of any radially symmetrical animal.

interradius n. [L. *inter*, between; *radius*, radius] the radius of a radiate animal half way between 2 perradii.

interramal a. [L. *inter*, between; *ramus*, branch] between branches or rami.

interramicorn n. [L. *inter*, between; *ramus*, branch; *cornu*, horn] a piece of a bird's bill beyond mandibular rami forming the gonys.

interrenal a. [L. *inter*, between; *renes*, kidneys] between the kidneys, *appl.* veins.

interrenal body a gland, situated between kidneys of elasmobranchs, representing the adrenal cortex of higher vertebrates.

interrupted a. [L. *inter*, between; *rumpere*, to break] with continuity broken; irregular; asymmetrical.

interruptedly pinnate pinnate with pairs of small leaflets occurring between larger ones.

interscapular a. [L. *inter*, between; *scapula*, shoulder blade] between the shoulder blades; *appl.* feathers; *appl.* brown fatty tissue, so-called hibernating gland, as in some rodents.

interscutal a. [L. *inter*, between; *scutum*, shield] between scuta or scutes.

intersegmental a. [L. *inter*, between; *segmentum*, piece] between segments; between spinal segments, *appl.* axons, septa.

intersegmentalia n.plu. [L. *inter*, between; *segmentum*, piece] sclerites between adjacent body segments in insects, as intertergites, interpleurites, intersternites.

interseminal a. [L. *inter*, between; *semen*, seed] between seeds or ovules, *appl.* scales in certain gymnosperms.

interseptal a. [L. *inter*, between; *septum*, fence] *pert.* spaces between septa or partitions.

intersex n. [L. *inter*, between; *sexus*, sex] an organism with characteristics intermediate between typical male and typical female of its species; an organism first developing as a male or female, then as an individual of the opposite sex; a sex mosaic in time; *cf.* gynandromorph, freemartin.

intersomitic a. [L. *inter*, between; Gk. *sōma*, body] between somites or body segments.

interspecific a. [L. *inter*, between; *species*, kind] between distinct species, *appl.* crosses, as mule, hinny, cattalo, tigron, *appl.* selection, *appl.* competition.

intersphincteric a. [L. *inter*, between; Gk. *sphingktēr*, tight band] between sphincters, *appl.* groove of anal canal.

interspicular a. [L. inter, between; spiculum, sharp point] occurring between spicules.

interspinal a. [L. inter, between; spina, spine] occurring between spinal processes or between spines, appl. bones, muscles, ligaments; alt. interspinous. n. Axonost, q.v.

interspinous see interspinal.

interstapedial a. [L. inter, between; stapes, stirrup] appl. a part of columella of ear.

intersterility n. [L. inter, between; sterilis, unfruitful] incapacity for interbreeding.

intersternal a. [L. inter, between; sternum, breast bone] between the sterna; appl. ligaments connecting manubrium and body of sternum.

intersternite n. [L. inter, between; sternum, breast bone] a sternal sclerite between thoracic segments of insects, alt. intersegmental sternite; furcasternite, q.v.

interstitial a. [L. inter, between; sistere, to set] occurring in interstices or spaces; appl. growth; appl. lamellae between Haversian systems; appl. flora and fauna living between sand grains, alt. psammon; or soil particles, appl. soil water.

interstitial cells connective tissue cells producing testosterone lying between the tubules in vertebrate testis; small cells lying in the interstices between other cells in coelenterates, and giving rise to various cell types, alt. I-cells.

interstitial cell stimulating hormone (ICSH) luteinizing hormone, q.v.

interstitium interstitial tissue; intertubular tissue.

intertemporal n. [L. inter, between; tempora, temples] a paired membrane bone, part of sphenoid complex, fusing with alisphenoids; dermosphenotic, q.v.

intertentacular a. [L. inter, between; tentaculum, feeler] between tentacles; appl. a ciliated tube opening at base of tentacles and connecting coelom and exterior, found in Ectoprocta.

intergal a. [L. inter, between; tergum, back] between tergites or dorsal sclerites.

intertergite n. [L. inter, between; tergum, back] a small sclerite between dorsal sclerites; alt. intersegmental tergal sclerite.

intertidal a. [L. inter, between; A.S. tid, time] appl. shore organisms living between high- and low-water marks.

intertrabecula n. [L. inter, between; trabecula, little beam] a separate plate between the trabeculae anteriorly, in some birds.

intertragic a. [L. inter, between; Gk. tragos, goat] appl. notch between tragus and antitragus.

intertrochanteric a. [L. inter, between; Gk. trochanter, runner] between trochanters, appl. crest, line.

intertrochlear a. [L. inter, between; trochlea, pulley] appl. an ulnar ridge fitting into a groove of the humerus.

intertubercular a. [L. inter, between; tuberculum, small hump] transtubercular, q.v.; appl. sulcus between tubercles of humerus.

intertubular a. [L. inter, between; tubulus, small tube] between tubules; between kidney tubules, appl. capillaries; between seminiferous tubules.

intervaginal a. [L. inter, between; vagina, sheath] between sheaths, appl. space.

intervarietal a. [L. inter, between; varius, diverse] appl. crosses between 2 distinct varieties of a species.

interventricular a. [L. inter, between; ventricula, small cavity] between ventricles, appl. foramen: foramen of Monro, q.v.

intervertebral a. [L. inter, between; vertebra, vertebra] occurring between vertebrae, appl. discs, fibrocartilages, foramina, veins.

intervillous a. [L. inter, between; villi, hairs] occurring between villi; appl. spaces in trophoblastic network filled with maternal blood.

interxylary a. [L. inter, between; Gk. xylon, wood] between xylem strands, appl. phloem, alt. included phloem, appl. cork.

interzonal a. [L. inter, between; zona, belt] between 2 zones; appl. spindle fibres uniting groups of daughter chromosomes in anaphase of mitosis.

interzooecial a. [L. inter, between; zōon, animal; oikos, house] occurring among zooecia.

intestinal a. [L. intestina, entrails] pert. intestines, appl. glands, villi, etc.

intestinal juice succus entericus, q.v.

intestine n. [L. intestina, entrails] part of alimentary canal from pylorus to anus, or part corresponding to this.

intextine n. [L. intus, within; exter, without] an inner membrane of an extine.

intima n. [L. intimus, innermost] the innermost lining membrane of a part or organ; alt. tunica intima.

intine n. [L. intus, within] the inner covering membrane of a pollen grain, or of a spore; cf. extine.

intrabiontic a. [L. intra, within; Gk. bios, life; on, being] appl. a process of selection occurring in a living unit.

intrabulbar intragemmal, q.v.

intracapsular a. [L. intra, within; capsula, small chest] contained within a capsule, appl. protoplasm of Radiolaria, appl. dendrites, etc.

intracardiac a. [L. intra, within; Gk. kardia, heart] endocardiac, q.v.

intracartilaginous a. [L. intra, within; cartilago, gristle] inside the cartilage, appl. ossification; alt. endochondral.

intracellular a. [L. intra, within; cellula, small room] within the cell.

intrachange a chromosomal aberration, in the form of loss, duplication, or rearrangement of material, occurring within one chromosome.

intraclonal a. [L. intra, within; Gk. klōn, twig] within a clone, appl. differentiation.

intracortical a. [L. intra, within; cortex, rind] within the cortex; uniting parts of brain cortex.

intradermal a. [L. intra, within; Gk. derma, skin] within the structure of the skin.

intra-epithelial a. [L. intra, within; Gk. epi, upon; thēlē, nipple] occurring in epithelium, appl. glands, usually mucous.

intrafascicular a. [L. intra, within; fasciculus, little bundle] within a vascular bundle; alt. fascicular.

intrafoliaceous a. [L. intra, within; folium, leaf] appl. stipules encircling stem and forming a sheath; alt. ocreate.

intrafusal a. [L. intra, within; fusus, spindle] appl. fasciculi and fibres connected respectively with neurotendinous and neuromuscular spindles.

intragemmal a. [L. intra, within; gemma, bud] within a taste bud, appl. nerve fibres, spaces; alt. intrabulbar.

intrageneric *a.* [L. *intra*, within; *genus*, race] among members of the same genus.

intragenic *a.* [L. *intra*, within; Gk. *genos*, descent] occurring within the same gene.

intraglobular *a.* [L. *intra*, within; *globulus*, globule] occurring within a globule or corpuscle.

intrajugular *a.* [L. *intra*, within; *jugulum*, throat] *appl.* a process in middle of a jugular notch of occipital bone.

intralamellar *a.* [L. *intra*, within; *lamella*, thin plate] within a lamella; *appl.* trama of gill-bearing fungi.

intralobular *a.* [L. *intra*, within; *lobulus*, small lobe] occurring within lobules; *appl.* veins draining liver lobules.

intramatrical *a.* [L. *intra*, within; *matrix*, from *mater*, mother] within a matrix; within a substrate.

intramembranous *a.* [L. *intra*, within; *membrana*, film] within a membrane, *appl.* bone development.

intramolecular *a.* [L. *intra*, within; F. *molécule*, small particle] occurring or existing within the molecule.

intramolecular respiration the formation of carbon dioxide and organic acid by normally aerobic organisms or tissues if they are deprived of oxygen.

intranarial *a.* [L. *intra*, within; *nares*, nostrils] inside the nostrils, *appl.* larynx in cetaceans.

intranuclear *a.* [L. *intra*, within; *nucleus*, kernel] within the nucleus, *appl.* spindles, fibres, etc.

intra-ocular *a.* [L. *intra*, within; *oculus*, eye] entoptic, *q.v.*

intraparietal *a.* [L. *intra*, within; *paries*, wall] enclosed within an organ; within parietal lobe, as sulcus, etc.

intrapetalous *a.* [L. *intra*, within; Gk. *petalon*, leaf] situated in a petaloid area, in echinoderms.

intrapetiolar *a.* [L. *intra*, within; *petiolus*, little foot] within the petiole base expansion.

intrapleural *a.* [L. *intra*, within; Gk. *pleuron*, side] within the thoracic cavity; *appl.* pressure in space between parietal and visceral pleura.

intrasegmental *a.* [L. *intra*, within; *segmentum*, part] *appl.* vertebra formed of cranial and caudal elements of same original myotome; *cf.* intermyotomic.

intraselection *n.* [L. *intra*, within; *selectio*, choice] selection within an organ, of cells fittest to survive.

intrasexual *a.* [L. *intra*, within; *sexus*, sex] *appl.* selection of competing individuals of the same sex.

intraspecific *a.* [L. *intra*, within; *species*, particular kind; *facere*, to make] within a species, *appl.* selection of individuals, *appl.* competition; *appl.* zygosis: gnesiogamy, *q.v.*

intraspicular *a.* [L. *intra*, within; *spicula*, small spike] having spicules completely embedded in spongin.

intrastelar *a.* [L. *intra*, within; Gk. *stēlē*, pillar] within the stele of a stem or root, *appl.* ground tissue, bundles, etc.

intratarsal *a.* [L. *intra*, within; *tarsus*, ankle] within the tarsus; *appl.* joint of reptilian limb between rows of tarsal bones.

intrathecal *a.* [L. *intra*, within; Gk. *thēkē*, case] within the meninges of the spinal cord.

intrathyroid *a.* [L. *intra*, within; Gk *thyreos*, shield; *eidos*, form] within the thyroid; *appl.* a cartilage joining laminae of thyroid cartilage during infancy.

intrauterine *a.* [L. *intra*, within; *uterus*, womb] within the uterus.

intravaginal *a.* [L. *intra*, within; *vagina*, sheath] within vagina; contained within a sheath, as grass leaves or branches.

intravascular *a.* [L. *intra*, within; *vasculum*, small vessel] within blood vessels.

intraventricular *a.* [L. *intra*, within; *ventriculus*, small cavity] within a ventricle; *appl.* caudate nucleus of corpus striatum, seen within ventricle of brain.

intravesical *a.* [L. *intra*, within; *vesica*, bladder] within the bladder.

intravitelline *a.* [L. *intra*, within; *vitellus*, egg yolk] within the yolk of an egg.

intraxylary *a.* [L. *intra*, within; Gk. *xylon*, wood] within xylem, *appl.* phloem: internal phloem, *q.v.*

intrazonal *a.* [L. *intra*, within; *zona*, belt] within a zone; *appl.* soils characteristic of locally limited soil-forming conditions, differing from prevalent or normal soils of the region or zone.

intrinsic *a.* [L. *intrinsecus*, inwards] inward; inherent; *appl.* inner muscles, as of tongue, of syrinx, etc.; *appl.* cycles, in population of a species, owing to coaction within or between species; *appl.* rate of natural increase in a stabilized population having a balanced age distribution; *appl.* brightness sensation due to differential retinal response to different wavelengths; *cf.* extrinsic.

intrinsic factor gastric intrinsic factor, *q.v.*

introduced *a.* [L. *introducere*, to introduce] not native but thought to have been brought into a country by man.

introgression *n.* [L. *intro*, within; *gressus*, walked] introgressive hybridization, *q.v.*

introgressive hybridization hybridization between 2 species in which the genes of one species gradually diffuse into the gene pool of another across an incomplete isolation barrrier between them; *alt.* introgression.

introitus *n.* [L. *introitus*, entry] an opening or orifice.

intromittent *a.* [L. *intro*, within; *mittere*, to send] adapted for inserting, *appl.* male copulatory organs.

introrse *a.* [L. *introsus*, inwards] turned inwards or towards axis; of anthers, opening towards the centre of the flower.

introvert *n.* [L. *intro*, within; *vertere*, to turn] that which is capable of being drawn inwards, as anterior region of body in certain zooids, of certain annulates, mouth extremity of certain molluscs. *v.* To turn, bend, or draw inwards.

intrusive growth interpositional growth, *q.v.*

intumescence *n.* [L. *intumescere*, to swell up] the process of swelling up, *cf.* detumescence; a swollen or tumid condition.

intussusception *n.* [L. *intus*, within; *suscipere*, to receive] growth in surface extent or volume by intercalation of materials among those already present; *cf.* accretion, apposition.

inulase *n.* [L. *inula*, elecampane] inulinase, *q.v.*

inulin *n.* [L. *inula*, elecampane] a storage polysaccharide made up of fructose units into which it can be hydrolysed by dilute acids or the enzyme inulinase, and found in roots, rhizomes, and tubers of many Compositae; *alt.* dahlia starch.

inulinase *n.* [L. *inula*, elecampane] the enzyme which hydrolyses inulin to fructose; *alt.* inulase; EC 3.2.1.7; *r.n.* inulinase.

inuncate *a.* [L. *inuncatus*, hooked together] glochidiate, *q.v.*

invaginate *v.* [L. *in*, into; *vagina*, sheath] to involute or draw into a sheath; *appl.* insinking of wall of cavity or vessel. *a.* Introverted; enclosed in a sheath; concave.

invagination *n.* [L. *in*, into; *vagina*, sheath] involution; introversion; gastrula formation by infolding of blastula wall; ingestion by temporarily transformed periplast-like ectoplasm in certain Protozoa.

inversion *n.* [L. *invertere*, to turn upside down] reversal in order of genes, or reversal of a chromosome segment, within the chromosome as a whole; a turning inward, or inside out, or upside down of a part; *alt.* resupination.

invertase *n.* [L. *invertere*, to turn into] the enzyme which hydrolyses terminal non-reducing β-D-fructofuranoside residues in β-D-fructofuranosides, e.g. sucrose to glucose and fructose; *alt.* invertin, saccharase; sometimes wrongly called sucrase; EC 3.2.1.26; *r.n.* β-D-fructofuranosidase.

Invertebrata, invertebrates *n., n.plu.* [L. *in*, not; *vertebra*, joint] a general term for all animal groups except the vertebrates, i.e. all animals without a backbone.

invertin invertase, *q.v.*

invert sugar sucrose, *q.v.*

investing bone membrane bone, *q.v.*

investment *n.* [L. *in*, in; *vestire*, to clothe] outer covering of a part, organ, animal, or plant.

in vitro [L. literally: in glass] *appl.* biological processes occurring experimentally in isolation from an organism.

in vivo [L. literally: in something alive] *appl.* biological processes occurring within the living organism.

involucel *n.* [*Dim.* of L. *involucrum*, covering] the small bracts at base of secondary umbel or capitulum as in teasels; a partial involucre; *alt.* involucellum, involucret.

involucellate *a.* [*Dim.* of L. *involucrum*, covering] bearing involucels.

involucellum involucel, *q.v.*

involucral *a.* [L. *involucrum*, covering] *pert.* or like an involucre.

involucrate *a.* [L. *involucrum*, covering] bearing involucres.

involucre *n.* [L. *involucrum*, covering] bracts forming whorl at base of a condensed inflorescence, as of capitulum and umbel; a group of leaves surrounding antheridial and archegonial groups in bryophytes, *cf.* perichaetium; *alt.* involucrum.

involucret involucel, *q.v.*

involucrum *n.* [L. *involucrum*, covering] in some Hydrozoa, protective cup into which nematocysts can be spirally retracted; metanotum of Orthoptera; periosteal layer formed around dead portion of bone, in certain diseased conditions; involucre, *q.v.*

involuntary *a.* [L. *in*, not; *voluntas*, wish] not under control of will.

involuntary muscle non-striated muscle, composed of fusiform cells, and not under control of will, such as those of the alimentary canal; *alt.*

unstriped, plain, smooth muscle; *cf.* voluntary muscle.

involute *a.* [L. *involutus*, rolled up] of leaves, having the edges rolled inwards at each side, *cf.* revolute; of shells, closely coiled.

involution *n.* [L. *involutus*, rolled up] reduction to normal of enlarged, modified, or deformed conditions; decrease in size, or structural and functional changes, as in old age; *appl.* forms that have become deformed in structure, but not to such an extent as to be incapable of recovery; a rolling inwards, as of leaves; movement of cells to interior in a certain type of gastrulation; resting, *appl.* spores, stage, etc.

iodinophilous iodophilic, *q.v.*

iodogorgoic acid an amino acid (3,5-diiodotyrosine) which contains iodine, and from which thyroxine is derived.

iodophilic *a.* [Gk. *ioeidēs*, violet-like; *philos*, loving] staining darkly in iodine solution; *appl.* certain cytoplasmic inclusions and vacuoles; *appl.* bacteria that stain blue with iodine; *alt.* iodinophilous.

iodopsin *n.* [Gk. *ioeidēs*, violet; *opsis*, sight] a photosensitive protein, vitamin A compound of retinal cones; *alt.* visual violet.

iodothyrin *n.* [Gk. *ioeidēs*, violet; *thyreos*, shield] an iodine compound in the colloid material of thyroid gland.

iodothyroglobulin *n.* [Gk. *ioeidēs*, violet; *thyreos*, shield; L. *globus*, globe] compound of iodothyrin and nucleoprotein, extractable hormone of the thyroid gland.

ion exchange the adsorption of some ions from solution onto a solid, such as clay particles in the soil, in exchange for others that are lost into solution.

ipsilateral *a.* [L. *ipse*, same; *latus*, side] *pert.* or situated on the same side; *cf.* contralateral.

IPSP inhibitory postsynaptic potential, *q.v.*

iridal, iridial *a.* [Gk., L. *iris*, rainbow] *pert.* the iris.

iridial angle filtration angle of eye; an angular recess between cornea and anterior surface of iris.

iridocytes *n.plu.* [Gk. *iris*, rainbow; *kytos*, hollow] guanine granules, bodies or plates of which the reflecting tissue of skin of fishes and reptiles is composed; iridescent cells in integument of some cephalopods as *Sepia*; *alt.* iridophores, guanophores, leucophores, ochrophores.

iridomotor *a.* [L. *iris*, rainbow; *movere*, to move] connected with movements of iris.

iridophores iridocytes, *q.v.*

iris *n.* [L. *iris*, rainbow] a thin circular contractile and vascular disc of eye between cornea and lens, and surrounding the pupil; a marking immediately encircling the pupil of an ocellus, as on wing of some Lepidoptera; *a.* iridial; *plu.* irises, irides.

iris cells pigment cells surrounding cone and retinula of an ommatidium.

iris frill collarette, *q.v.*

IRM innate releasing mechanism, *q.v.*

irradiation *n.* [L. *in*, into; *radius*, ray] treatment with ray, as ultraviolet rays, X-rays, etc.; the spreading of an effect of a stimulus; spreading of an excitatory process; apparent enlargement of objects, due to difference in illumination.

irreciprocal *a.* [L. *in*, not; *reciprocus*, going backwards and forwards] not reversible; one-way, *appl.* conduction, as in an axon or in a reflex arc.

irregular *n.* [L. *in,* not; *regula,* rule] zygomorphic, *q.v.*

irritability *n.* [L. *irritare,* to provoke] power of receiving external impressions, and reacting to them, inherent to living matter.

irritant *n.* [L. *irritare,* to provoke] an external stimulus which provokes a response.

irrorate *a.* [L. *irrorare,* to bedew] covered as if by minute droplets; dotted with minute colour markings, as wings of certain butterflies. *v.* Ability of white tissue in a variegated plant to turn green under certain conditions of illumination.

isadelphous *a.* [Gk. *isos,* equal; *adelphos,* brother] with equal number of stamens in 2 phalanges.

isandrous *a.* [Gk. *isos,* equal; *aner,* male] having similar stamens, their number equalling that of the sections of the corolla.

isantherous *a.* [Gk. *isos,* equal; *antheros,* flowering] having equal anthers.

isanthous *a.* [Gk. *isos,* equal; *anthos,* flower] having uniform or regular flowers.

isauxesis *n.* [Gk. *isos,* equal; *auxesis,* growth] isometry, *q.v.*; *cf.* bradyauxesis, tachyauxesis.

ischiadic *a.* [Gk. *ischion,* hip] *pert.* or in region of hip, *appl.* artery, vein, process of ischiopubic plate; sciatic, *appl.* nerve.

ischiatic sciatic, *q.v.*

ischiocapsular *a.* [Gk. *ischion,* hip; L. *capsula,* little chest] *appl.* a ligament joining capsular ligament and hip.

ischiocavernosus *a.* [Gk. *ischion,* hip; L. *cavus,* hollow] *appl.* muscle between hip and corpora cavernosa: erector of penis, or of clitoris.

ischioflexorius *n.* [Gk. *ischion,* hip; L. *flexus,* bent] posterior thigh muscle in salamander, corresponding to semimembranosus.

ischiopodite *n.* [Gk. *ischion,* hip; *pous,* foot] proximal joint of walking legs of certain Crustacea, or of maxillipedes; *alt.* ischium.

ischiopubic *a.* [Gk. *ischion,* hip; L. *pubes,* adult] *appl.* a gap or fenestra between ischium and pubis; *appl.* a median cartilaginous plate with median and lateral processes, in Dipnoi.

ischiopubis *n.* [Gk. *ischion,* hip; L. *pubes,* adult] the ischium of pterodactyls, pubis being excluded from acetabulum; a fused ischium and pubis; *alt.* puboischium.

ischiorectal *a.* [Gk. *ischion,* hip; L. *rectus,* straight] *pert.* ischium and rectum, *appl.* fossa and muscles.

ischium *n.* [Gk. *ischion,* hip] the ventral and posterior bone of each half of pelvic girdle of vertebrates except fishes; ischiopodite, *q.v.*

Ischnacanthiformes *n.* [Gk. *ischion,* hip; *akantha,* thorn; L. *forma,* shape] an order of acanthodians existing in Silurian to Carboniferous times, having 2 dorsal fins and no spines between pelvic and pectoral fins.

isidia *n.plu.* [Gk. *isis,* plant; *idion,* dim.] coral-like soredia on surface of some lichens; *sing.* isidium.

isidiferous *a.* [Gk. *isis,* plant; L. *ferre,* to bear] bearing isidia; *alt.* isidophorous.

isidioid *a.* [Gk. *isis,* plant; *idion,* dim.; *eidos,* like] like an isidium.

isidium *sing.* of isidia.

isidophorous *a.* [Gk. *isis,* plant; *idion,* dim.; *pherein,* to bear] isidiferous, *q.v.*

island of Reil [*J. C. Reil,* German anatomist] *see* insula.

islets of Langerhans [*P. Langerhans,* German anatomist] spherical or oval groups of cells scattered throughout the pancreas acting as an endocrine organ, being concerned with metabolism of sugar and including α (A) cells which secrete glucagon and β (B) cells which secrete insulin; *alt.* insula.

isoagglutination *n.* [Gk. *isos,* equal; L. *agglutinare,* to glue to] agglutination of the erythrocytes of an animal by transfusion from another member of the same species or the same blood group; agglutination of sperm due to a substance produced by an ovum of the same species.

isoagglutinin *n.* [Gk. *isos,* equal; L. *agglutinare,* to glue to] fertilizin or agglutinin of eggs which reacts on sperm of same species; *cf.* heteroagglutinin.

isoagglutinogen *n.* [Gk. *isos,* equal; L. *agglutinare,* to glue to; Gk. *genos,* birth] substance producing agglutination of erythrocytes within the same blood group; *alt.* isohaemagglutinogen.

isoalleles *n.plu.* [Gk. *isos,* equal; *allelon,* one another] alleles so similar in their effects that special techniques are needed to distinguish them.

isoantigen *n.* [Gk. *isos,* equal; *anti,* against; *genos,* birth] an antigen which stimulates antibody production in different individuals of a species; *alt.* alloantigen.

isobilateral *a.* [Gk. *isos,* equal; L. *bis,* twice; *latus,* side] *appl.* a form of bilateral symmetry where a structure is divisible in 2 planes at right angles; *appl.* leaf: *see* isolateral.

isoblabe *n.* [Gk. *isos,* equal; *blabe,* damage] a line connecting points, on a map, indicating the same degree of damage, infestation, or infection by a harmful agent or pathogenic species.

isobrachial *a.* [Gk. *isos,* equal; L. *brachium,* arm] *appl.* chromosome in which the 2 arms are of equal length, the centromere occupying the median position.

Isobryales *n.* [Gk. *isos,* equal; *bryon,* moss] an order of mosses having a prostrate gametophore and capsule with a double peristome.

isobryonic, isobryous *a.* [Gk. *isos,* equal; *bryein,* to proliferate] developing equally, as lobes of a dicotyledonous embryo.

isocarpous *a.* [Gk. *isos,* equal; *karpos,* fruit] having carpels and perianth divisions equal in number.

isocercal *a.* [Gk. *isos,* equal; *kerkos,* tail] with vertebral column ending in median line of caudal fin.

isochela *n.* [Gk. *isos,* equal; *chele,* claw] a chela with 2 parts equally developed; a 2-pronged or anchor-shaped spicule in certain sponges.

isochromatic isochromous, *q.v.*

isochromosome *n.* [Gk. *isos,* equal; *chroma,* colour; *soma,* body] chromosome with identical arms united in a median centromere; metacentric derived from telocentric chromosome.

isochromous *a.* [Gk. *isos,* equal; *chroma,* colour] equally tinted; uniformly coloured; *alt.* isochromatic, isochrous.

isochronic *a.* [Gk. *isos,* equal; *chronos,* time] having an equal duration; occurring at the same rate; having an equal chronaxy; *alt.* isochronal, isochronical, isochronous.

isochrous isochromous, *q.v.*

Isochrysidales n. [Gk. *isos*, equal; *chrysos*, gold] an order of Haptophyceae with a flagellate stage and no haptonema.

isocitric acid an acid of the Krebs' cycle which is oxidized to oxalosuccinic acid, using NAD as a hydrogen acceptor, by the enzyme isocitric dehydrogenase.

isocitric dehydrogenase see isocitric acid; EC 1.1.^{s}1.41 (NAD$^+$); r.n. isocitrate dehydrogenase (NAD$^+$).

isocont isokont, q.v.

isocortex n. [Gk. *isos*, equal; L. *cortex*, bark] the part of cerebral cortex made up of 6 layers of nerve cells; cf. allocortex.

isocytic a. [Gk. *isos*, equal; *kytos*, hollow] with all cells equal.

isodactylous a. [Gk. *isos*, equal; *daktylos*, finger] having all digits of equal size.

isodemic a. [Gk. *isos*, equal; *dēmos*, people] with, or *pert.*, populations composed of an equal number of individuals; *appl.* lines on a map which pass through points representing equal population density.

isodiametric a. [Gk. *isos*, equal; *dia*, through; *metron*, measure] having equal diameters, *appl.* cells or other structures; *appl.* rounded or polyhedral cells.

isodont a. [Gk. *isos*, equal; *odous*, tooth] homodont, q.v.; cf. anisodont.

isodynamic a. [Gk. *isos*, equal; *dynamis*, power] of equal strength; providing the same amount of energy, *appl.* foods.

isoelectric point (IEP) the point at which an amphoteric molecule carries no net charge, being a definite value for each protein.

isoenzymes n.plu. [Gk. *isos*, equal; *en*, in; *zymē*, leaven] different forms in which a protein may exist with the same enzyme specificity but differing in properties such as optimum pH or isoelectric point; *alt.* isozymes.

Isoetales n. [*Isoetes*, generic name] an order of Lycopsida having linear leaves and a 'corm' with complex secondary thickening, and including the genus *Isoetes*, commonly called quillworts.

isoflavones n.plu. [Gk. *isos*, equal; L. *flavus*, yellow] a group of colourless flavonoids.

isoflors n.plu. [Gk. *isos*, equal; L. *flos*, flower] lines enclosing areas with the same number of species within a section or genus.

isogametangiogamy n. [Gk. *isos*, equal; *gametēs*, spouse; *anggeion*, vessel; *gamos*, marriage] the union of similar gametangia.

isogamete n. [Gk. *isos*, equal; *gametēs*, spouse] one of a pair of undifferentiated gametes.

isogamy n. [Gk. *isos*, equal; *gamos*, marriage] union of similar gametes, or of similar unicells; *alt.* isomerogamy, zygogamy; cf. anisogamy, oogamy; a. isogamous.

isogeneic isogenic, q.v.

Isogeneratae n. [Gk. *isos*, equal; L. *generare*, to beget] a group of brown algae having a life cycle with isomorphic alternation of generations.

isogenes n.plu. [Gk. *isos*, equal; *genos*, descent] lines on a map which connect points where same gene frequency is found.

isogenetic a. [Gk. *isos*, equal; *genesis*, descent] arising from the same or a similar origin, *alt.* isogenous; of the same genotype.

isogenic a. [Gk. *isos*, equal; *genos*, race] having the same genetic constitution, *alt.* homozygous;

with all or certain specified genes identical, *appl.* separate individuals; *alt.* isogeneic, syngeneic, syngenic.

isogenomatic, isogenomic a. [Gk. *isos*, equal; *genos*, race] containing similar sets of chromosomes, *appl.* nuclei.

isogenous a. [Gk. *isos*, equal; *genēs*, produced] of the same origin; *alt.* isogenetic.

isognathous a. [Gk. *isos*, equal; *gnathos*, jaw] having both jaws alike.

isogonal a. [Gk. *isos*, equal; *gōnia*, angle] forming equal angles, *appl.* branching; *alt.* isogonic.

isogonic a. [Gk. *isos*, equal; *gonos*, offspring] producing similar individuals from differing stocks. [Gk. *gōnia*, angle] isogonal, q.v.

isograft n. [Gk. *isos*, equal; O.F. *graffe*, graft] a graft of tissue taken from another animal of the same genotype as the recipient; *alt.* syngraft.

isogynous a. [Gk. *isos*, equal; *gynē*, woman] having similar gynaecia.

isohaemagglutinogen n. [Gk. *isos*, equal; *haima*, blood; L. *agglutinare*, to glue to; Gk. *genos*, birth] isoagglutinogen, q.v.

Isokontae n. [Gk. *isos*, equal; *kontos*, punting pole] formerly a subclass of algae of the Chlorophyceae, having flagella of equal length.

isokont(an) a. [Gk. *isos*, equal; *kontos*, punting pole] having flagella or cilia of the same length; *alt.* isocont(an); cf. heterokont, isomastigote.

isolate n. [It. *isola*, from L. *insula*, island] a breeding group restricted by isolation; the 1st pure or single-spore isolation of a fungus.

isolateral a. [Gk. *isos*, equal; L. *latus*, side] having equal sides; *appl.* leaves with palisade tissue on both sides, *alt.* isobilateral.

isolation n. [L. *insula*, island] separation from others; prevention of mating between breeding groups owing to spatial, topographical, ecological, phenological, physiological, genetic, behavioural, or other barriers.

isolecithal a. [Gk. *isos*, equal; *lekithos*, yolk] *appl.* ova with yolk granules distributed nearly equally throughout egg substances.

isoleucine n. [Gk. *isos*, equal; *leukos*, white] an amino acid, essential to man and other mammals; abbreviated to ileu.

isolichenin n. [Gk. *isos*, equal; *leichēn*, lichen] a polysaccharide, found in hyphal cell walls of some lichens.

isomar isophane, q.v.

isomastigote a. [Gk. *isos*, equal; *mastix*, whip] having flagella of equal length; *alt.* isokont.

isomer n. [Gk. *isos*, equal; *meros*, part] one of the chemical compounds having the same kind and number of atoms, but differing in properties and in arrangement of the atoms.

isomerases enzymes which convert one isomer to another; EC group 5.

isomere n. [Gk. *isos*, equal; *meros*, part] a homologous structure or part.

isomerism existence of compounds as isomers.

isomerogamy isogamy, q.v.

isomerous a. [Gk. *isos*, equal; *meros*, part] having equal numbers of different parts; *appl.* flowers with the same number of parts in each whorl; cf. heteromerous.

isomery the condition of being isomerous.

isometric a. [Gk. *isos*, equal; *metron*, measure] of equal measure or growth rate; *appl.* contraction of muscle under tension without change in length, *cf.* isotonic.

isometry *n.* [Gk. *isos*, equal; *metron*, measure] growth of a part at the same rate as the standard or the whole; *alt.* isauxesis.

isomorphic *a.* [Gk. *isos*, equal; *morphē*, shape] superficially alike, *alt.* isomorphous; *appl.* alternation of diploid and haploid phases in morphologically similar generations; *cf.* heteromorphic.

isomorphism *n.* [Gk. *isos*, equal; *morphē*, shape] apparent similarity of individuals of different race or species.

isomorphous *see* isomorphic.

isomyarian, isomyaric *a.* [Gk. *isos*, equal; *mys*, muscle] with adductor muscles equal in size.

isonym *n.* [Gk. *isos*, equal; *onyma*, name] a new name, of species, etc., based upon the oldest name or basinym.

iso-osmotic *see* isotonic.

isopedin *n.* [Gk. *isopedos*, level] inner layer of laminated bony material in cosmoid and ganoid fish scales.

isopetalous *a.* [Gk. *isos*, equal; *petalon*, leaf] having similar petals.

isophagous *a.* [Gk. *isos*, equal; *phagein*, to eat] feeding on one or allied species, *appl.* fungi.

isophane *n.* [Gk. *isos*, equal; *phainein*, to show] a line connecting all places within a region at which a biological phenomenon, e.g. flowering of a plant, occurs at the same time; *alt.* isomar, phenocontour.

isophene *n.* [Gk. *isos*, equal; *phainein*, to show] a contour line delimiting area corresponding to a given frequency of a variant form; *alt.* phenocontour.

isophenous *a.* [Gk. *isos*, equal; *phainein*, to show] being of the same phenotype.

isophyllous *a.* [Gk. *isos*, equal; *phyllon*, leaf] having uniform foliage leaves, on the same plant.

isophytoid *a.* [Gk. *isos*, equal; *phyton*, plant; *eidos*, form] an 'individual' of a compound plant not differentiated from the rest.

isoplankt *n.* [Gk. *isos*, equal; *plangktos*, wandering] line representing, on a map, distribution of equal amounts of plankton, or of particular plankton species.

isoploid *a.* [Gk. *isos*, equal; *aploos*, onefold] with an even number of chromosome sets in somatic cells. *n.* An isoploid individual.

Isopoda, isopods *n., n.plu.* [Gk. *isos*, equal; *pous*, foot] an order of freshwater, marine, or terrestrial malacostracans, including the woodlice, having a dorsoventrally flattened body, a reduced abdomen, and no carapace.

isopodous *a.* [Gk. *isos*, equal; *pous*, foot] having the legs alike and equal.

isopogonous *a.* [Gk. *isos*, equal; *pōgōn*, beard] of feathers, having the 2 webs equal and similar.

isopolyploid *a., n.* [Gk. *isos*, equal; *polys*, many; *aploos*, onefold; *eidos*, form] polyploid with an even number of chromosome sets, as tetraploid, hexaploid, octoploid, etc.

isoprenoids chemical compounds found as fossils possibly related to chlorophyll but which may be synthesized in non-living systems, and that are important in cholesterol structure.

Isoptera *n.* [Gk. *isos*, equal; *pteron*, wing] an order of social insects, commonly called termites, which have slight incomplete metamorphosis and live in large communities, including reproductive winged forms and sterile, wingless soldiers and workers.

isopycnosis, isopyknosis *n.* [Gk. *isos*, equal; *pyknos*, dense] the condition of those chromosomes or regions of chromosomes that do not show heteropycnosis, i.e. the majority of chromosomes.

isopygous *a.* [Gk. *isos*, equal; *pygē*, rump] with pygidium and cephalon of equal size, *appl.* trilobites.

isorhiza *n.* [Gk. *isos*, equal; *rhiza,* root] a type of nematocyst in coelenterates.

Isospondyli *n.* [Gk. *isos*, equal; *sphondylos*, vertebra] Clupeomorpha, *q.v.*

isospore *n.* [Gk. *isos*, equal; *sporos*, seed] a spore produced by a homosporous organism; a spore without sexual dimorphism; an agamete produced by schizogamy; *cf.* anisospore.

isosporous *a.* [Gk. *isos*, equal; *sporos*, seed] homosporous, *q.v.*

isostemonous *a.* [Gk. *isos*, equal; L. *stēmōn*, warp] having stamens equal in number to that of sepals or of petals.

isotelic *a.* [Gk. *isos*, equal; *telos*, end] exhibiting, or tending to produce, the same effect; homoplastic, *q.v.*; *appl.* food factors that can replace each other; *pert.* isotely.

isotels *n.plu.* [Gk. *isos*, equal; *telos*, end] substances having the same physiological, e.g. nutritional, effect.

isotely homoplasty, *q.v.*

isotomy *n.* [Gk. *isos*, equal; *temnein*, to cut] bifurcation repeated in a regular manner, as in crinoid brachia.

isotonic *a.* [Gk. *isos*, equal; *tonos*, strain] of equal tension; having equal osmotic pressure, *appl.* solution, *alt.* iso-osmotic, equimolecular, *cf.* hypotonic, hypertonic; *appl.* contraction of muscle with change in length, *cf.* isometric contraction.

isotonicity *n.* [Gk. *isos*, equal; *tonos*, tone] normal tension under pressure or stimulus.

isotope *n.* [Gk. *isos*, equal; *topos*, place] chemical elements having the same atomic number and identical chemical properties, but differing in atomic weight; *a.* isotopic.

isotropic, isotropous *a.* [Gk. *isos*, equal; *tropikos*, turning] singly refracting in polarized light, *appl.* the light stripes of voluntary muscle fibres, *alt.* monorefringent, *cf.* anisotropic; *appl.* chitin, *cf.* actinochitin; symmetrical around longitudinal axis; not influenced in any one direction more than another, *appl.* growth rate; without predetermined axes, as eggs; *n.* isotropy.

isotype *n.* [Gk. *isos*, equal; *typos*, pattern] a specimen collected from the same plant as the holotype and at the same time; type of plant or animal common to 2 or more areas or regions.

isoxanthopterin(e) *n.* [Gk. *isos*, equal; *xanthos*, yellow; *pteron*, wing] a colourless pteridine in wings of cabbage butterflies and in eyes and bodies of other insects; *alt.* leucopterin B.

isozoic *a.* [Gk. *isos*, equal; *zōon*, animal] inhabited by similar forms of animal life.

isozooid *n.* [Gk. *isos*, equal; *zōon*, animal; *eidos*, like] a zooid similar to parent stock.

isozygoty *n.* [Gk. *isos*, equal; *zygon*, yoke] possession of a genotype homozygous in regard to all loci; *a.* isozygotic.

isozyme *n.* [Gk. *isos*, equal; *zymē*, leaven] isoenzyme, *q.v.*

isthmiate *a.* [Gk. *isthmos*, neck] connected by an isthmus-like part.

isthmus *n.* [Gk. *isthmos*, neck] a narrow structure connecting 2 larger parts, as those of aorta, acoustic meatus, limbic lobe, prostate, thyroid, etc., or between semicells; junction between perikaryon and axon base; of Pander: column of white yolk forming neck of latebra.

iter *n.* [L. *iter*, way] a passage or canal, as those of middle ear, brain, etc.; *alt.* aqueduct.

iteration *n.* [L. *iteratio*, repetition] repetition, as of similar trends in successive branches of a taxonomic group.

IU International Unit, *q.v.*

ivory *n.* [L. *ebur*, ivory, through F. *ivoire*] dentine of teeth, usually that of elephant's tusks and similar structures, formed from odontoblasts.

ixocomous *a.* [Gk. *ixos*, mistletoe; *komē*, hair] *pert.* or formed by viscous or slimy hyphae, as surface of certain fungi.

ixoderm *n.* [Gk. *ixos*, mistletoe; *derma*, skin] a layer of hyphae that have become viscous, covering the pileus of certain fungi; *alt.* ixotrichoderm.

ixotrichoderm *n.* [Gk. *ixos*, mistletoe; *thrix*, hair; *derma*, skin] ixoderm, *q.v.*

J

jacket cell a cell in the antheridium which gives rise to the wall, not to the androcytes; *cf.* androgonium.

Jacob's membrane [*A. Jacob*, Irish opthalmologist] layer of rods and cones of retina; *alt.* bacillary layer.

Jacobson's cartilage [*L. L. Jacobson*, Danish anatomist] vomeronasal cartilage supporting Jacobson's organ.

Jacobson's nerve tympanic branch of the glossopharyngeal nerve.

Jacobson's organ a diverticulum of olfactory organ in many vertebrates, often developing into an epithelium-lined sac opening into mouth; *cf.* vomeronasal organ.

jactitation *n.* [L. *jactare*, to toss] process of scattering seeds by censer mechanism, *q.v.*

jaculator *n.* [L. *jaculator*, shooter] a placental process, usually hooked, of certain fruits.

jaculatory *a.* [L. *jaculatorius*, throwing] darting out; capable of being emitted.

jaculatory duct portion of vas deferens which is capable of being protruded, in many animals.

jaculiferous *a.* [L. *jaculum*, a dart; *ferre*, to carry] bearing dart-like spines.

jarovization *n.* [Russ. *yarovizatsya*, from *yarovoi*, vernal] vernalization, *q.v.*

Java man a fossil hominid found in Java and originally named *Pithecanthropus erectus*, now called *Homo erectus erectus*, and coming from M. Pleistocene times; *cf.* Peking man.

jaw *n.* [Akin to *chaw*, chew] a structure, of vertebrates, supported by bone or cartilage, naked or sheathed in horn, or bearing teeth or horny plates, forming part of mouth, and helping to open or shut it; a similarly placed structure in invertebrates.

jaw-foot maxillipede, *q.v.*

J-disc I-band, *q.v.*

jecoral *a.* [L. *jecur*, liver] of or *pert.* the liver.

jecorin *n.* [L. *jecur*, liver] a lecithin-like substance in liver and other organs of the body.

jejunum *n.* [L. *jejunus*, empty] part of small intestine between duodenum and ileum in mammals.

jellyfish the common name for the Scyphozoa, *q.v.*

jelly fungi the common name for Basidiomycetes of the orders Tremellales, Auriculariales, and Dacromycetales, which are sometimes included in the single order Tremellales, and have a gelatinous basidiocarp.

jelly of Wharton [*T. Wharton*, English anatomist] the gelatinous connective tissue surrounding the vessels of umbilical cord.

Johnston's organ [*C. Johnston*, British entomologist] a statical or chordotonal organ in 2nd segment of insect antenna.

joint *n.* [O.F. *joindre*, from L. *jungere*, to join] place of union or separation of 2 parts, as between bones; articulation, *q.v.*; node, *q.v.*; portion between 2 nodes or joints.

jordanon *n.* [*K. Jordan*, zoologist] a true breeding genetic unit below the species level, with little variability, such as a race, subspecies, or variety; *alt.* Jordan's species, microspecies; *cf.* linneon.

Jordan's organ [*K. Jordan*, zoologist] chaetosema, *q.v.*

Jordan's rule fish in cold waters tend to have more vertebrae than those in warmer waters.

Jordan's species jordanon, *q.v.*

juba *n.* [L. *juba*, mane] a mane; a loose panicle.

jubate *a.* [L. *jubatus*, maned] with mane-like growth.

jugal *n.* [L. *jugum*, yoke] a membrane bone of the zygoma of the skull between the maxilla and squamosal; *alt.* malar, zygomatic bone. *a. Pert.* a jugum.

jugate *a.* [L. *jugum*, yoke] having pairs of leaflets; furnished with a jugum.

Juglandales *n.* [*Juglans*, generic name] an order of dicots of the Archichlamydeae or Hamamelididae, having a simple perianth and pinnate leaves and comprising the family Juglandaceae which includes the walnut.

jugofrenate *a.* [L. *jugum*, yoke; *frenum*, bridle] having the jugal lobe of the fore-wing able to engage in the frenulum of hind-wing, *appl.* certain Lepidoptera.

jugular *a.* [L. *jugulum*, collar bone] *pert.* neck or throat, *appl.* veins, foramen, fossa, etc.; *appl.* nerve: the hyoidean or posterior post-trematic nerve; *appl.* ventral fish fins beneath and in front of pectoral fins.

jugulum *n.* [L. *jugulum*, collar bone] the fore-neck region of a bird's breast; in insects, the jugum of wing.

jugum *n.* [L. *jugum*, yoke] a pair of opposite leaflets or leaves; ridge on mericarp of umbelliferous plants; small lobe on posterior border of forewing of certain moths; ridge or depression connecting 2 structures; structure connecting the 2 halves of a brachidium; union of lesser sphenoidal wings in 1st year after birth.

Julianiales *n.* [*Juliania*, generic name] an order of woody dicots of the Archichlamydeae, having alternate usually pinnate leaves and dioecious flowers, the male flowers in a dense panicle and the female in 4s at the end of a downwardly projecting spike.

Juliformia *n.* [*Julus* (*Iulus*), generic name; L. *forma*, shape] an order of Helminthomorpha having at least 40 rings to the body and no silk glands, and including the common millipede, *Iulus*.

Juncales *n.* [*Juncus*, generic name] an order of monocots (Liliidae) comprising the family Juncaceae, the rushes and woodrushes.

junctional complex in many epithelia, the region at which neighbouring cells attach.

Jungermanniales *n.* [*Jungermannia*, generic name] an order of liverworts having a leafy or thalloid gametophyte, and the capsule of the sporophyte dehiscing by 4 valves.

Jurassic *a.* [*Jura* mountains] *pert.* or *appl.* Mesozoic period between Triassic and Cretaceous.

juvenal *a.* [L. *juvenalis*, youthful] youthful; *appl.* plumage replacing nestling down of 1st plumage.

juvenile hormone one of several hormones produced by the corpus allatum of insects, which maintains larval characteristics in the young insect and is inhibited at the final moult, ensures deposition of yolk in the developing egg, induces behaviour appropriate to the type of growth and reproduction that is to follow, and has some general effects on metabolism; *alt.* allatum hormone, neotenin, gonadotropic hormone, status quo hormone, vitellogenic hormone, nymphal hormone.

juvenile leucocyte a metamyelocyte in circulation before maturation.

juxta *n.* [L. *juxta*, close to] a ring-walled structure supporting sheath of aedeagus.

juxta-articular *a.* [L. *juxta*, close to; *articulus*, joint] near a joint or articulation.

juxtaglomerular *a.* [L. *juxta*, close to; *glomerare*, to form into a ball] *appl.* cells surrounding arteriole of glomerulus of kidney.

juxtamedullary *a.* [L. *juxta*, close to; *medulla*, marrow] near medulla; *appl.* inner portion of zona reticularis of adrenal glands.

juxtanuclear *a.* [L. *juxta*, close to; *nucleus*, kernel] *appl.* bodies: basophil deposits in cytoplasm of vitamin D deficient parathyroid cells.

K

kaino- *caeno-*, *q.v.*

kairomone *n.* [Gk. *kairos*, fitness; hor*mone*] a chemical messenger emitted by one species which has an effect on a member of another species, sometimes to the detriment of the transmitter, such as a chemical which attracts a male to a female also attracting a predator.

kako- caco-, *q.v.*

kalidium *n.* [Dim. of Gk. *kalia*, hut] a form of sporocarp, or cystocarp.

kalymma *n.* [Gk. *kalymma*, covering] vacuolated part of outer layer of certain radiolarians; *alt.* calymma.

kalymmocytes *n.plu.* [Gk. *kalymma*, covering; *kytos*, cell] in ascidians, certain follicle cells which migrate into the egg after maturation; *alt.* calymmocytes.

kanamycins antibiotics produced by *Streptomyces kanamyceticus*, which interfere with protein synthesis in bacteria.

kappa *n.* [Gk. κ, kappa] a self-reproducing particle containing DNA found in the cytoplasm of certain strains of the ciliate *Paramecium*, which confers the ability to kill other *Paramecia* on the strains possessing it.

karya *plu.* of karyon.

karyapsis *n.* [Gk. *karyon*, nut, nucleus; *apsis*, juncture] karyogamy, *q.v.*

karyaster *n.* [Gk. *karyon*, nut, nucleus; *astēr*, star] a star-shaped group of chromosomes; *alt.* aster.

karyenchyma *n.* [Gk. *karyon*, nucleus; *engchyma*, infusion] nuclear sap, *q.v.*

karyo- also caryo-, *q.v.*

karyochylema *n.* [Gk. *karyon*, nucleus; *chylos*, juice] nuclear sap, *q.v.*

karyoclasis *n.* [Gk. *karyon*, nucleus; *klasis*, breaking] breaking down of a cell nucleus.

karyogamy *n.* [Gk. *karyon*, nucleus; *gamos*, marriage] the fusion of the 2 nuclei of gametes, with interchange of nuclear material, after cytoplasmic fusions; *alt.* karyapsis; *cf.* plasmogamy.

karyokinesis *n.* [Gk. *karyon*, nucleus; *kinēsis*, movement] mitosis, nuclear division; *cf.* cytokinesis.

karyolemma *n.* [Gk. *karyon*, nucleus; *lemma*, skin] nuclear membrane, *q.v.*

karyology *n.* [Gk. *karyon*, nucleus; *logos*, discourse] nuclear cytology, especially concerned with the chromosomes.

karyolymph *n.* [Gk. *karyon*, nucleus; L. *lympha*, water] nuclear sap, *q.v.*

karyolysis *n.* [Gk. *karyon*, nucleus; *lyein*, to loosen] supposed dissolution of the nucleus in mitosis; liquefaction of nuclear membrane; *alt.* nucleolysis; *a.* karyolytic.

karyomere, **karyomerite** *n.* [Gk. *karyon*, nucleus; *meros*, part] in mitosis, a small vesicle into which a chromosome is converted in one type of nuclear construction; *alt.* chromosomal vesicle.

karyomicrosome *n.* [Gk. *karyon*, nucleus; *mikros*, small; *sōma*, body] a nuclear granule; *cf.* cytomicrosome.

karyomite *n.* [Gk. *karyon*, nucleus; *mitos*, thread] chromosome, *q.v.*

karyomitome *n.* [Gk. *karyon*, nucleus; *mitōma*, network] the nuclear threadwork.

karyomitosis *n.* [Gk. *karyon*, nucleus; *mitos*, thread] mitosis, *q.v.*

karyomixis *n.* [Gk. *karyon*, nucleus; *mixis*, mixing] mingling or union of nuclear material of gametes.

karyon *n.* [Gk. *karyon*, nucleus] the cell nucleus; *plu.* karya.

karyophans *n.plu.* [Gk. *karyon*, nucleus; *phainein*, to appear] microsomes or nucleus-like granules surrounded by an ovoid matrix, which form the spironeme and axoneme in stalk of certain Protozoa.

karyophore *n.* [Gk. *karyon*, nucleus; *pherein*, to bear] system of ectoplasmic fibrils or membranes for mooring the nucleus, in certain ciliates.

karyoplasm *n.* [Gk. *karyon*, nucleus; *plasma*, mould] nucleoplasm, *q.v.*

karyorhexis *n.* [Gk. *karyon*, nucleus; *rhēxis*, breaking] fragmentation of the cell nucleus; *alt.* karyoschisis.

karyoschisis karyorhexis, q.v.

karyosome n. [Gk. karyon, nucleus; sōma, body] a nucleolus of the 'net-knot' type; a chromosome, q.v.; a special aggregation of chromatin in resting nucleus; the cell nucleus itself; cf. plasmosome.

karyosphere n. [Gk. karyon, nucleus; sphaira, globe] the large nucleolus from which arise all or most of the chromosomes of Protista; a condensed mass of DNA seen in the mature primary oocyte of the fruit fly Drosophila, from which chromosomes emerge and enter metaphase I of meiosis.

karyota n.plu. [Gk. karyon, nucleus] nucleated cells.

karyotheca n. [Gk. karyon, nucleus; thēkē, covering] the nuclear membrane, q.v.

karyotin n. [Gk. karyon, nucleus] chromatin, q.v.; nuclear substance.

karyotype n. [Gk. karyon, nucleus; typos, pattern] group of individuals with the same chromosome number and similar linear arrangement of genes in homologous chromosomes; chromosome complement of such a group.

kat- also cat-, q.v.

katabolism catabolism, q.v.

katabolite catabolite, q.v.

katagenesis n. [Gk. kata, down; genesis, descent] retrogressive evolution.

katakinetic a. [Gk. kata, down; kinein, to move] appl. process leading to discharge of energy; cf. anakinetic.

katakinetomeres n.plu. [Gk. kata, down; kinein, to move; meros, part] unreactive, stable atoms, molecules, or protoplasm; cf. anakinetomeres.

kataphase n. [Gk. kata, down; phasis, appearance] the stages of mitosis from formation of chromosomes to division of cell; cf. anaphase.

kataphoresis electrophoresis, q.v.

kataphoric a. [Gk. kata, down; pherein, to carry] appl. passive action, the result of lethargy.

kataplexy n. [Gk. kata, down; plessein, to strike] condition of an animal feigning death; maintenance of a postural reflex induced by restraint or shock; alt. cataplexis; cf. catalepsis.

katastate catastate, q.v.

katatrepsis n. [Gk. kata, down; trepein, to turn] stage of decreasing movement in blastokinesis; cf. anatrepsis.

katatropic a. [Gk. kata, down; tropikos, turning] turning downwards.

katepimeron n. [Gk. kata, down; epi, upon; mēros, thigh] lower sclerite of epimeron; alt. infraepimeron.

katepisternum n. [Gk. kata, down; epi, upon; L. sternum, breast bone] lower part of insect episternum; alt. infraepisternum.

katharobic a. [Gk. katharos, pure; bios, life] living in clean waters, appl. Protista; cf. saprobic.

kathodic a. [Gk. kathodos, descent] not arising in conformity with genetic spiral, appl. leaves.

kation cation, q.v.

KB cells a culture of cells derived from carcinomal epidermal tissue of the nasopharynx.

Keber's organ [G. A. F. Keber, German zoologist] pericardial glands in lamellibranchs.

keel n. [A.S. ceol, ship] the carina on breast bone of flying birds; boat-shaped structure formed by 2 anterior petals of Leguminosae; ridge on blade or on other parts of grasses.

Keith and Flack's node sinu-atrial node, q.v.

kelp n. [M.E. culp, kelp] the common name for a member of the Laminariales, sometimes used for any large seaweed.

kenanthy n. [Gk. kenos, empty; anthos, flower] non-development of stamens and pistils of a flower; alt. cenanthy.

kenenchyma n. [Gk. kenos, empty; engchyma, infusion] a tissue devoid of its living contents, as cork.

kenosis n. [Gk. kenos, empty] process of voiding, or condition of having voided; exhaustion, inanition.

kenotoxin n. [Gk. kenos, empty; toxikon, poison] a hypothetical toxin said to form in the body after muscular effort; alt. fatigue toxin.

kentragon n. [Gk. kentron, sharp point; gonia, angle] a larval stage following the cypris larva of parasitic cirripedes, which penetrates into the body of the host.

Kenyapithecus n. [Kenya; Gk. pithēkos, ape] a genus of fossil apes from the Miocene, found in Kenya and having hominid teeth, possibly ancestral to, or being, early hominids.

keph-, kephal- ceph-, cephal-, q.v.

ker- also cer-, q.v.

keraphyllous a. [Gk. keras, horn; phyllon, leaf] appl. layer of a hoof between horny and sensitive parts.

kerasin n. [Gk. keras, born] an important glycolipid (cerebroside) obtained from the brain which on hydrolysis yields a fatty acid (lignoceric acid), galactose, and sphingosine.

keratin n. [Gk. keras, horn] a scleroprotein forming the basis of epidermal structures such as horns, nails, hairs, and normally containing a large amount of cystine.

keratinization n. [Gk. keras, horn] state of becoming horny; appl. cells of epidermis developing in a horny material, alt. cornification.

keratinolytic a. [Gk. keras, horn; lyein, to dissolve] hydrolysing keratin, appl. enzymes.

keratinophilic a. [Gk. keras, horn; philos, loving] growing on a horny or keratinized substrate, appl. certain fungi.

keratinous a. [Gk. keras, horn] horny; pert., containing, or formed by, keratin.

keratogenous a. [Gk. keras, horn; -genēs, producing] horn-producing.

keratohyalin n. [Gk. keras, horn; hyalos, glass] substance contained in stratum lucidum of skin.

keratoid a. [Gk. keras, horn; eidos, form] horny; resembling horn; alt. keroid, ceratoid.

Keratosa n. [Gk. keras, horn] a subclass of Demospongea including the bath sponges, without spicules and with the skeleton composed entirely of spongin fibres.

keratose a. [Gk. keras, horn] having horny fibres in skeleton, as certain sponges.

Kerckring's valves valvulae conniventes, q.v.

kernel n. [A.S. cyrnel, small grain] the inner part of a seed containing the embryo.

keroid keratoid, q.v.

ketogenesis n. [Ketone; Gk. genesis, descent] the production of ketone bodies.

ketogenic a. [Ketone; Gk. gennaein, to produce] appl. amino acids which are catabolized to form the ketone bodies β-hydroxybutyric acid and acetoacetic acid during their oxidation to carbon dioxide and water, e.g. leucine, phenylalanine, and tyrosine.

ketogenic hormone a prepituitary hormone which influences fat metabolism.

α-ketoglutaric acid an acid of the Krebs' cycle which undergoes oxidative decarboxylation to yield succinic acid, and is an important intermediate in that it can be aminated to make an amino acid, glutamic acid.

ketone bodies compounds such as acetoacetate, β-hydroxybutyrate, and acetone, produced in the liver, resulting from the condensation of acetoacetyl coenzyme A, formed by fatty acid oxidation, with acetyl coenzyme A.

ketose a sugar containing a ketone ($-C=O$) group; cf. aldose.

ketosis the production of more ketone bodies than the organism can use.

key n. [M.E. key, key] a method of identifying objects or organisms in the form of a dichotomic table of alternative questions.

key fruit winged achenes (samaras) hanging in clusters as in ash.

key gene oligogene, q.v.

kidney n. [A. S. cwith, womb; neere, kidney] one of a pair of organs of nitrogenous excretion and osmoregulation in vertebrates; alt. nephros.

kinaesth- alt. kinesth-.

kinaesthesis n. [Gk. kinein, to move; aisthēsis, perception] perception of movement due to stimulation of muscles, tendons, and joints; alt. proprioception.

kinaesthetic a. [Gk. kinein, to move; aisthēsis, perception] pert. sense of movement or muscular effort, appl. sense, area.

kinase n. [Gk. kinein, to move] a substance which transforms zymogens to enzymes; a phosphotransferase without regeneration of the donor, EC sub-subgroups 2.7.1–2.7.4.

kinesiodic a. [Gk. kinēsis, movements; hodos, way] pert. motor nerve paths.

kinesis n. [Gk. kinēsis, movement] random movement; an orientation movement in which an organism moves at random until it reaches a better environment, the movement depending on the intensity and not the direction of the stimulus; variation in linear or angular velocity; cf. nasty, taxis, tropism.

kinesodic a. [Gk. kinēsis, movement; hodos, way] conveying motor impulses.

kinesth- kinaesth-, q.v.

kinetia plu. of kinetium.

kinetic a. [Gk. kinein, to move] active, appl. function of movement, cf. static; appl. energy employed in producing or changing motion; appl. division centre in cell division.

kinetin n. [Gk. kinētēs, mover] a cytokinin, a derivative of adenine.

kinetium n. [Gk. kinein, to move] a row of kinetosomes with a kinetodesma; plu. kinetia; alt. kinety.

kinetoblast n. [Gk. kinein, to move; blastos, bud] outer ciliated investment of aquatic larvae with special locomotor properties.

kinetochore n. [Gk. kinein, to move; chōros, place] centromere, q.v.

kinetodesma n. [Gk. kinein, to move; desma, bond] a fibril alongside a row of kinetosomes in Ciliata.

kinetogenesis n. [Gk. kinein, to move; genesis, descent] the evolution theory that animal structures have been produced by animal movements.

kinetomeres n.plu. [Gk. kinein, to move; meros, part] molecules or atoms, reactive or stable, ana- and katakinetomeres; centromeric chromomeres.

kinetonema n. [Gk. kinein, to move; nēma, thread] part of the chromonema associated with spindle-attachment region or centromere.

kinetonucleus n. [Gk. kinein, to move; L. nucleus, kernel] a kinetoplast or parabasal body in haemoflagellates, in close connection with the flagellum and undulating membrane, formerly thought to be a secondary nucleus and concerned with motor activities; cf. trophonucleus.

kinetoplasm n. [Gk. kinein, to move; plasma, something formed] an iron-containing nucleoprotein forming a source of energy to Nissl granules.

kinetoplast n. [Gk. kinein, to move; plastos, formed] composite body formed by union of parabasal body with blepharoplast in some Flagellata.

kinetosome n. [Gk. kinein, to move; sōma, body] one of a group of granules occupying the polar plate region in moss sporogenesis; a self-duplicating granule at the base of a cilium in Ciliata.

kinetospore n. [Gk. kinein, to move; sporos, seed] a zoospore in its physiological aspect.

kinety kinetium, q.v.

king in social Hymenoptera or Isoptera, a male reproductive individual; used in compound names of animals thought king-like in appearance or habits.

king crabs a common name for the Xiphosura, q.v.

kingdom the largest taxonomic group, such as the plant kingdom and animal kingdom.

kinins n.plu. [Gk. kinein, to move] cytokinins, q.v.; bradykinins, q.v.

kinomere n. [Gk. kinein, to move; meros, part] centromere, q.v.

kinoplasm n. [Gk. kinein, to move; plasma, mould] archiplasm, q.v.

kinoplasmosomes n.plu. [Gk. kinein, to move; plasma, form; sōma, body] phragmoplast fibres seen at periphery of cell plate.

Kinorhyncha n. [Gk. kinein, to move; rhyngchos, snout] a phylum of marine microscopic pseudocoelomate animals having a body of jointed spiny segments and spiny head; in some classifications, considered a class of the Aschelminthes.

kin selection the evolution of characteristics which favour the survival of close relatives of an individual.

klado- clado-, q.v.

klasma plates n.plu. [Gk. klasma, fragment; platys, flat] small parts of compound ambulacral plates separated by growth pressure, in some echinoids.

kleisto- cleisto-, q.v.

kleptobiosis n. [Gk. klepēnai, to steal; biōsis, manner of life] thievery by ants.

kleptoparasite n. [Gk. klepēnai, to steal; parasitos, parasite] an animal that steals scraps of food falling from the mouth of another animal.

kleronomous a. [Gk. klēronomos, heir] inherited, appl. paths in nervous system.

klin- also clin-, q.v.

klinokinesis n. [Gk. klinein, to slope; kinēsis, movement] a kinesis in which an organism continues to move in a straight line until it meets an unfavourable environment, when it turns, resulting in it remaining in a favourable environment, the frequency of turning depending on the intensity of environmental stimuli (formerly called an avoiding reaction or phobotaxis in Protozoa); change in rate of change of direction or angular velocity due to intensity of stimulation; *cf.* orthokinesis.

klinotaxis *n.* [Gk. *klinein*, to slope; *taxis*, arrangement] a taxis in which an organism orients itself in relation to a stimulus by moving its head or whole body from side to side symmetrically in moving towards the stimulus, and so compares the intensity of the stimulus on either side; formerly called phobotaxis in Protozoa; *cf.* telotaxis, tropotaxis.

klon clone, *q.v.*

knee *n.* [A.S. *cneow*, knee] genu, *q.v.*; joint between femur and tibia; root process of certain swamp-inhabiting trees; joint in stem of certain grasses.

knephoplankton *n.* [Gk. *knephas*, twilight; *plangktos*, wandering] plankton living at depths between 30 and 500 metres; *cf.* phaoplankton, skotoplankton.

knot *n.* [A.S. *cnotta*, knot] in wood, base of branch surrounded by new layers of wood and hardened by pressure; in nuclear meshwork, small particles of chromatin where meshes cross.

koino- coeno-, *q.v.*

Kölliker's canal [*R. A. von Kölliker*, Swiss zoologist] a canal leading from otocyst towards exterior, as in certain Cephalopoda.

Kölliker's pit a ciliated preoral pit somewhat to the left side, chemoreceptor in Cephalochordata.

kolyone *n.* [Gk. *kōlyon*, hindrance] substance elaborated in, and conveyed from, a tissue or organ, which lessens or inhibits function of other tissues; *alt.* colyone.

kolytic *a.* [Gk. *kōlytikos*, hindering] inhibiting; inhibitory.

komma *n.* [Gk. *komma*, clause] sarcomere, *q.v.*

koniocortex *n.* [Gk. *konis*, dust; L. *cortex*, bark] granular part of cortex. characteristic of sensory areas of brain.

Korff's fibres *see* predentine.

Kornberg cycle glyoxylate cycle, *q.v.*

Kornberg enzyme an enzyme isolated in 1958 from *E. coli* by a team led by A. Kornberg, which is a DNA polymerase acting in repair synthesis.

Kovalevsky's canal [*P. Kovalevskii*, Russian embryologist] the neurenteric canal.

krasnozems deep friable red loamy soils found in the subtropics and developed from base-rich parent materials.

Krause's end-bulbs *see* end-bulbs.

Krauses' glands [*K. F. T. Krause*, German anatomist] accessory lacrimal glands with ducts opening into fornix of conjunctiva.

Krause's membrane [*W. J. F. Krause*, German anatomist] Z-disc, *q.v.*

Krebs' cycle [*H. Krebs*, German biochemist] an important, mainly respiratory metabolic cycle associated with mitochondria in animals and plants, involving the interconversions of various acids, with decarboxylations and oxidations, and which results in the reduction of NAD and FAD (the $NADH_2$ and $FADH_2$ being used to produce ATP in oxidative phosphorylation) and produces acids which act as intermediates in other syntheses; *alt.* citric acid cycle, tricarboxylic acid cycle.

Krebs'–Henseleit cycle urea cycle, *q.v.*

kremastoplankton *n.* [Gk. *kremastos*, hung up; *plangktos*, wandering] plankton consisting mainly of organisms floating by means of hairs, bristles, and similar appendages.

Kulchitsky's cells argentaffin connective tissue found between the crypts of Lieberkühn in the intestinal musosa.

Kupffer cells [*K. W. von Kupffer*, German anatomist] stellate phagocytic cells found in liver sinuses which ingest red blood cells; *alt.* star cells, stellate cells.

Kurloff body an inclusion of unknown origin sometimes found in the mononuclear leucocytes of some animals.

kyano- cyano-, *q.v.*

kybernetics *n.* [Gk. *kybernētikos*, skilled in governing] cybernetics *appl.* living structures; regulation of homeostasis.

kynurenine a metabolic product derived from tryptophan in certain animals, a genetically controlled precursor of some ommatochromes and other pigments in insects.

kyogenic *a.* [Gk. *kyēsis*, pregnancy; *genos*, descent] *appl.* prepituitary hormone stimulating secretion of progestin by corpora lutea.

kyto- cyto-, *q.v.*

L

labella *n.* [L. *labellum*, small lip] in some Diptera, the pair of lobes that are the expanded labium; paraglossa, *q.v.*, of some insects; *plu.* of labellum.

labellate *a.* [L. *labellum*, small lip] furnished with labella or small lips.

labelled *appl.* an element made identifiable by a radioactive or heavy isotope.

labelloid *a.* [L. *labellum*, small lip; Gk. *eidos*, form] like a labellum.

labellum *n.* [L. *labellum*, small lip] the lower petal, morphologically posterior, of an orchid, *alt.* mesopetalum; 2 fused lateral staminodes, as in flower of the family Zingiberaceae; small lobe beneath labrum, or labial palp, in insects, *alt.* proboscis lobe.

labia *n.plu.* [L. *labium*, lip] lips; lip-like structures; *plu.* of labium, *q.v.*

labia cerebri margins of cerebral hemispheres overlapping corpus callosum.

labial *a.* [L. *labium*, lip] *pert.* or resembling a lip, or labium.

labial palp lobe-like structure near mouth of molluscs; jointed appendage on labium of insects, *alt.* labipalp.

labia majora outer lips of vulva.

labia minora inner lips of vulva; *alt.* nymphae.

Labiatae *n.* [L. *labium*, lip] a family of dicotyledons, commonly called labiates, having typically square stems, opposite decussate leaves, and a corolla divided into 2 lips, sometimes with the upper lip reduced.

labiate *a.* [L. *labium*, lip] lip-like; possessing lips or thickened margins; having limb of calyx or corolla so divided that one portion overlaps the other; labiatiflorous, *q.v. n.* A member of the Labiatae.

labiatiflorous *a.* [L. *labium*, lip; *flos*, flower]

labidophorous

having the corolla divided into 2 lip-like portions; *alt.* labiate.
labidophorous *a.* [Gk. *labis*, forceps; *pherein*, to carry] possessing pincer-like organs.
labiella *n.* [L. *labium*, lip] a mouth-part of Myriapoda.
labile *a.* [L. *labilis*, apt to slip] readily undergoing change; unstable; *appl.* genes that are constantly mutating.
labiodental *a.* [L. *labium*, lip; *dens*, tooth] *pert.* lip and teeth, *appl.* an embryonic lamina, *appl.* labial surface of tooth.
labiosternite *n.* [L. *labium*, lip; *sternum*, breast bone] a median area between palpigers of insect head.
labiostipes *n.* [L. *labium*, lip; *stipes*, stalk] a portion of basal part of insect labium.
labipalp(us) *n.* [L. *labium*, lip; *palpare*, to feel] labial palp of insects.
labium *n.* [L. *labium*, lip] a lip or lip-shaped structure; in insects the fused 2nd maxillae forming the lower lip; a small plate attached to the sternum of spiders; inner margin of mouth of gastropod shell; *plu.* labia.
Laboulbeniales [*Laboulbenia*, generic name] an order of very small Pyrenomycetes which are ectoparasites on insects and have an ascogonium with a trichogyne and fertilization by spermatia.
labral *a.* [L. *labrum*, lip] *pert.* a labrum.
labrocyte *n.* [Gk. *labros*, greedy; *kytos*, hollow] mast cell, *q.v.*
labrum *n.* [L. *labrum*, lip] anterior lip of certain arthropods; hypostoma of trilobites; outer margin of mouth of gastropod shell; ambon, *q.v.*
labyrinth *n.* [L. *labyrinthus*, labyrinth] the complex internal ear, bony or membranous; lateral mass of air cells of ethmoidal bone; portions of kidney cortex with uriniferous tubules; a modified dilatation near root of carotid artery, regulating blood flow, as in some amphibians; any of various other convoluted structures.
labyrinthine *a.* [L. *labyrinthus*, labyrinth] *pert.* labyrinth of internal ear, *appl.* sense of equilibrium.
labyrinthodont *a.* [Gk. *labyrinthos*, labyrinth; *odous*, tooth] having teeth with great complexity of dentine arrangement.
Labyrinthodontia, labyrinthodonts *n.*, *n.plu.* [Gk. *labyrinthos*, labyrinth; *odous*, tooth] a primitive and extinct group of swat amphibians that were the dominant amphibians of late Palaeozoic and early Mesozoic times.
Labyrinthulae, Labyrinthulales *n.* [Gk. *labyrinthos*, labyrinth] a group of slime moulds which form a net plasmodium, and may be more closely related to the algae.
lac *n.* [Persian *lak*, lacquer] a resinous secretion of lac glands of certain insects of the family Coccidae, the composition depending on the food plant, some types used to make shellac.
laccate *a.* [It. *lacca*, varnish] appearing as if varnished.
lacerated *a.* [L. *lacerare*, to tear] having margin or apex deeply cut into irregular lobes.
lacertiform *a.* [L. *lacerta*, lizard; *forma*, shape] having the shape of a lizard.
lacertus *n.* [L. *lacertus*, arm muscle] lacertus fibrosus, aponeurosis of tendon of biceps muscle of the arm; *alt.* bicipital fascia.
lacewings *see* Neuroptera.

lachri- lacri-, *q.v.*
lachry- lacri-, *q.v.*
lacinia *n.* [L. *lacinia*, flap] segment of an incised leaf or petal; slender projection from margin of a thallus; extension of posterior margin of proglottis over anterior part of following proglottis; inner division of endopodite or stipes of maxilla of insects; fimbria, *q.v.*
laciniate *a.* [L. *lacinia*, flap] irregularly incised, as petals; fringed; *appl.* a ligament of the ankle, the internal annular ligament.
laciniform *a.* [L. *lacinia*, flap; *forma*, shape] shaped like lacinia; fringe-like.
laciniolate *a.* [L. *lacinia*, dim., flap] minutely incised or fringed.
lacinula *n.* [L. *lacinia*, dim., flap] small lacinia; inflexed sharp point of petal.
lacinulate *a.* [L. *lacina*, dim., flap] having lacinulae.
lacri- *alt.* lachry-, lacry-, lachri-.
lacrimal *a.* [L. *lacrima*, tear] secreting or *pert.* tears; *pert.* or situated near tear gland, *appl.* artery, bone, duct, nerve, papillae; *appl.* sac: dacryocyst, *q.v.*
lacrimal bone a bone in skull near tear gland; *alt.* unguis.
lacrimiform *a.* [L. *lacrima*, tear; *forma*, shape] tear-shaped, *appl.* spores; *alt.* dacryoid, lacrimaeform, lacrioid.
lacrimonasal *a.* [L. *lacrima*, tear; *nasus*, nose] *pert.* lacrimal and nasal bones or duct.
lacrimose *a.* [L. *lacrimosus*, tearful] bearing tear-shaped appendages, as gills of certain fungi.
lacrioid *a.* [L. *lacrima*, tear; Gk. *eidos*, form] lacrimiform, *q.v.*
lacry- lacri-, *q.v.*
lactalbumin *n.* [L. *lac*, milk; *albumen*, egg white] an albumin found in milk.
lactase *n.* [L. *lac*, milk] the enzyme which hydrolyses terminal non-reducing β-D-galactose residues in β-D-galactosides, e.g. lactose to glucose and galactose; EC 3.2.1.23; *r.n.* β-D-galactosidase.
lactation *n.* [L. *lac*, milk] secretion of milk in mammary glands, *alt.* galactosis; period during which milk is secreted.
lactation vitamins vitamin L_1 present in liver, and L_2 in yeast, promoting milk section.
lacteals *n.plu.* [L. *lac*, milk] chyliferous or lymphatic vessels of small intestine; ducts which carry latex.
lacteous *a.* [L. *lac*, milk] milky in appearance or texture.
lactescent *a.* [L. *lactescere*, to turn to milk] producing milk; yielding latex.
lactic *a.* [L. *lac*, milk] *pert.* milk.
lactic acid an organic acid which accumulates in muscle at the end of anaerobic respiration if oxygen is not supplied to the cells fast enough, and is also produced by fermentation from sugars by certain bacteria.
lactifer laticifer, *q.v.*
lactiferous *a.* [L. *lac*, milk; *ferre*, to carry] forming or carrying milk, *alt.* galactophorous; carrying latex, *appl.* vessels: milk tubes.
lactific *a.* [L. *lac*, milk; *facere*, to make] milk-producing.
lactobiose lactose, *q.v.*
lactochrome *n.* [L. *lac*, milk; Gk. *chrōma*, colour] riboflavin, *q.v.*

lactoflavin n. [L. lac, milk; flavus, yellow] riboflavin, q.v.

lactogenesis n. [L. lac, milk; Gk. genesis, descent] the initiation of milk secretion.

lactogenic a. [L. lac, milk; Gk. -genēs, producing] pert. or stimulating, secretion of milk; appl. hormone: prolactin, q.v.; appl. interval between parturition and ovulation, or between parturition and menstruation.

lactoglobulin n. [L. lac, milk; globulus, dim. of globus, globe] a protein fraction obtained from milk which is soluble in ammonium sulphate but insoluble in pure water.

lactoproteid n. [L. lac, milk; Gk. prōtos, first; eidos, form] any milk proteid.

lactoprotein n. [L. lac, milk; Gk. prōteion, first] any of the proteins in milk.

lactose n. [L. lac, milk] a reducing disaccharide occurring in the milk of mammals, which can be hydrolysed by dilute acids or lactase to yield gluctose and galactose; alt. milk sugar, lactobiose.

lactosis n. [L. lac, milk] galactosis, q.v.

lacuna n. [L. lacuna, cavity] a space between cells; sinus, q.v.; urethral follicle; cavity in cartilage or bone; small cavity or depression on surface in lichens; leaf gap, q.v.; plu. lacunae.

lacunar having, resembling, or pert. lacunae.

lacunate a. [L. lacuna, cavity] pert. lacunae, alt. lacunar; appl. collenchyma with cell walls thickened where bordering intercellular spaces.

lacunose a. [L. lacuna, cavity] having many cavities; pitted.

lacunosorugose a. [L. lacuna, cavity; rugosus, wrinkled] having deep furrows or pits, as some seeds and fruits.

lacunula n. [L. dim. of lacuna, cavity] a minute cavity or lacuna; a minute air space, as in grey hair.

lacus lacrimalis the triangular space between eyelids which contain lacrimal caruncle and receives tears from orifices of the lacrimal ducts.

lacustrine a. [L. lacus, lake] pert., or living in or beside, lakes.

laeotropic a. [Gk. laios, left; tropē, turning] inclined, turned, or coiled to the left; alt. laeotropous, laiotropic, leiotropic, leotropic, sinistral; cf. dexiotropic.

laetrile a term first used to describe an acid derived from amygdalin, and now also used for amygdalin itself, especially when derived from apricot stones and used in cancer therapy where its value is a subject of controversy; alt. vitamin B$_{17}$.

laevigate levigate, q.v.

laevulose n. [L. laevus, left] fructose, q.v.; alt. levulose.

lagena n. [L. lagena, flask] apical portion of the cochlear duct or scala media; alt. lagoena.

Lagenidiales n. [Lagenidium, generic name] an order of Phycomycetes having a simple mycelium with cellulose cell walls, and biflagellate zoospores with one whiplash and one tinsel flagellum; formerly called Ancylistales.

lageniform a. [L. lagena, flask; forma, shape] shaped like a flask.

lagenostome n. [L. lagena, flask; Gk. stoma, mouth] a pollen-catching device in some Palaeozoic pteridosperms.

lagging slower movement of certain chromosomes towards the poles at anaphase so that they are not included in the daughter nuclei.

lagoena lagena, q.v.

Lagomorpha, lagomorphs n., n.plu. [Gk. lagos, hare; morphē, form] an order of placental mammals including rabbits, hares, and pikas, being herbivorous and having skulls similar to rodents with a 2nd pair of upper incisors behind the 1st pair, and hindlimbs modified for leaping; formerly classified with rodents but they are a clearly distinct group with fossils different from the Eocene.

lagopodous a. [Gk. lagos, hare; pous, foot] possessing hairy or feathery feet.

lag phase the 1st phase of bacterial multiplication, in which there is no appreciable reproduction of the bacteria.

laiotropic laeotropic, q.v.

laking haemolysis, q.v.

Lamarckian a. [J.-B. de Lamarck, French biologist] of or pert. theories put forward by Lamarck.

Lamarckism the evolution theory of Lamarck, embodying the principle that acquired characteristics are inherited.

lambda n. [Gk. Λ, lambda] the junction of lambdoid and sagittal sutures.

lambda particles cytoplasmic inclusions seen in the ciliate Paramecium.

lambdoid a. [Gk. Λ, lambda; eidos, form] Λ-shaped, appl. the cranial suture joining occipital and parietal bones.

lamella n. [L. lamella, small plate] any thin plate- or scale-like structure; the gill of an agaric; a plate of cells.

lamellar, lamellate a. [L. lamella, small plate] composed of, or possessing thin plates.

lamellasome n. [L. lamella, small plate; Gk. soma, body] a membranous inclusion in the cytoplasm of blue–green algae, consisting of a series of invaginated lamellae enclosed by a common membrane.

lamellated corpuscles Pacinian corpuscles, q.v.

Lamellibranchia(ta), lamellibranchs n., n.plu. [L. lamella, small plate; branchiae, gills] Bivalvia, q.v.; a subclass of bivalves, in some classifications, in which the labial palps are small and the ctenidia are the main feeding organs.

lamellibranch(iate) a. [L. lamella, small plate; branchiae, gills] having plate-like gills on each side; with bilaterally compressed symmetrical body, like a bivalve.

lamellicorn a. [L. lamella, small plate; cornu, horn] having antennal joints expanded into flattened plates; alt. lamelliform.

lamelliferous a. [L. lamella, small plate; ferre, to carry] having small plates or scales.

lamelliform a. [L. lamella, small plate; forma, shape] plate-like, alt. lamelloid; lamellicorn, q.v.

lamellirostral a. [L. lamella, small plate; rostrum, beak] having inner edges of bill bearing lamella-like ridges; alt. odontorhynchous.

lamelloid lamelliform, q.v.

lamellose a. [L. lamella, small plate] containing lamellae; having a lamellar structure.

Lamiales n. [Lamium, generic name] an order of dicots of the Asteridae or Gamopetalae, having opposite leaves, a 2-lipped corolla, and 4 downwardly directed ovules in the ovary, each of which forms a nutlet.

lamina n. [L. lamina, plate] a thin layer or scale; the blade of a leaf or petal; the flattened part of a thallus; one of the thin plate-like expansions of sensitive tissue which fit into grooves on inside of horses hoof.

lamina basalis Bruch's membrane, q.v.

lamina choriocapillaris capillary plexus constituting inner layer of choroid.

lamina cribrosa cribriform plate, q.v.; membraneous portion of sclera at site of attachment of optic nerve and with perforations for axons of ganglion cells of retina.

lamina fusca inner layer of sclera, adjoining lamina suprachoroidea.

lamina papyracea plate or os planum of ethmoidal bone, forming part of medial wall of orbit.

lamina perpendicularis median process of mesethmoid or ethmoid forming proximal or bony part of nasal septum.

lamina propria a layer of loose connective tissue in the mucosa of gut wall, which accommodates glands and contains blood and lymph vessels.

laminar a. [L. lamina, plate] consisting of plates or thin layers; alt. laminiform, laminous.

Laminariales n. [Laminaria, generic name] an order of brown algae commonly called kelps having a large sporophyte differentiated into holdfast, stipe, and lamina with meristematic cells between the stipe and lamina.

laminarian a. [Laminaria, a genus of brown algae) appl. zone between low water to about 30 m.

laminarin n. [Laminaria, genus of brown algae] any of various carbohydrates which are the main food reserves in brown algae and are stored in solution, consisting mainly of glucose units but some containing mannitol.

lamina suprachoroidea delicate tissue of membrane between choroid and sclera.

laminated a. [L. lamina, plate] composed of thin plates, appl. plant cuticle.

lamina terminalis thin layer of grey matter forming anterior boundary of 3rd ventricle of brain.

lamination n. [L. lamina, plate] the formation of thin plates or layers; arrangement in layers, as nerve cells of cerebral cortex.

lamina vasculosa outer layer of choroid beneath suprachoroid membrane.

lamina vitrea Bruch's membrane, q.v.

laminiform a. [L. lamina, plate; forma, shape] like a thin layer or layers; like a leaf blade; laminar, q.v.

laminiplanter a. [L. lamina, plate; planta, sole of foot] having scales of metatarsus meeting behind in a smooth ridge; alt. caligate.

laminous laminar, q.v.

Lamniformes n. [Lamna, generic name; L. forma, shape] an order of Selachii existing from Jurassic times to the present day and including typical sharks.

lampbrush chromosomes long chromosomes in nuclei of oocytes of many vertebrates and some other places, which are covered with fine hair-like loops; alt. giant chromosomes.

lampreys the common name for the Petromyzontia, q.v.

Lampridiformes n. [Lampris, generic name; L. forma, shape] an order or suborder of teleosts having a protractile mouth and including the

ribbon fish, dealfish, and many deep-sea fishes; alt. Allotriognathi.

lamp shells the common name for the Brachiopoda, q.v.

lanate a. [L. lana, wool] woolly; covered with short hair-like processes giving woolly appearance to surface.

lancelets the common name for the Cephalochordata, q.v.

lance-linear a. [L. lanceae, lance; linea, line] between lanceolate and linear in form.

lance-oblong a. [L. lancea, lance; oblongius, oblong] oblong with tapering ends.

lanceolate a. [L. lanceola, little lance] slightly broad, or tapering, at base and tapering to point; lance-shaped.

lance-oval a. [L. lancea, lance; ovalis, oval] having a shape intermediate between lanceolate and oval; alt. lance-ovate.

lancet n. [F. lancette, from L. lancea, lance] one of the paired parts, ventral to stylet, of sting in Hymenoptera.

lancet plates plates supporting water vascular vessels of Blastoidea.

Landolt's fibre [E. Landolt, French ophthalmologist] free end of outer processes of cone-bipolar cells in inner nuclear layer of retina.

Langerhans' cell [P. Langerhans, German anatomist] melanoblast, q.v.

Langerhans, follicles of see follicles.

Langerhans, islets of see islets.

Langhans' cells [T. Langhans, German histologist] giant cells of cytotrophoblast.

languet, languette n. [F. languette, small tongue] a process of branchial sac of ascidians; cells from neural crest, between gill clefts, later converted into cartilage to form branchial arches.

laniary a. [L. laniare, to tear to pieces] modified for tearing, appl. canine tooth.

laniferous, lanigerous a. [L. lana, wool; ferre, gerare, to bear] wool-bearing; fleecy.

lantern n. [L. lanterna, lantern] Aristotle's lantern, q.v.; a photophore, as of lantern fishes (Scopeliformes).

lanthanin n. [Gk. lanthanein, to be unnoticed] see linin.

lanuginose, lanuginous a. [L. lanugo, down] covered with down.

lanugo n. [L. lanugo, down] the downy covering on a foetus which begins to be shed before birth.

lapidicolous a. [L. lapis, stone; colere, to dwell] appl. animals that live under stones.

lapillus n. [L. lapillus, pebble] a small otolith in utriculus of teleosts.

lappaceous a. [L. lappa, bur] like a bur; prickly.

lappet n. [A.S. laeppa, loose hanging part] any of various hanging, lobe-like structures; one of the paired lobes extending downwards from the distal end of stomadaeum of jellyfish; lobe of a sea anemone gullet; ciliated lobe of actinotroch larva of Phoronida; wattle of a bird.

large intestine in some vertebrates, the caecum, colon, and appendix; sometimes used for the colon only.

larmier n. [F. larme, tear] dacryocyst, q.v.

larva n. [L. larva, ghost] an embryo which becomes self-sustaining and independent before it has assumed the characteristic features of its parents; a. larval.

Larvacea n. [L. larva, ghost] a class of small free-swimming urochordates which retain larval characteristics in adult stage; *alt.* Appendicularia.

larviform *a.* [L. *larva*, ghost; *forma*, shape] shaped like a larva.

larviparous *a.* [L. *larva*, ghost; *parere*, to produce] producing offspring at the larval stage.

larvivorous *a.* [L. *larva*, ghost; *vorare*, to devour] larva-eating.

larvule *n.* [L. *larvula*, small larva] young larva.

laryngeal *a.* [Gk. *laryngx*, larynx] *pert.* or near the larynx, *appl.* artery, vein, nerve, etc.

laryngeal prominence in primates, a subcutaneous projection of the thyroid cartilage in front of the throat, causing a ridge on the ventral surface of the neck, and more pronounced in males; *alt.* Adam's apple, pomum Adami.

larynges *plu.* of larynx.

laryngopharynx *n.* [Gk. *laryngx*, larynx; *pharyngx*, gullet] part of pharynx between soft palate and oesophagus.

laryngotracheal *a.* [Gk. *laryngx*, larynx; L. *trachia*, trachea] *pert.* larynx and trachea, *appl.* embryonic groove and tube, *appl.* chamber into which lungs open in amphibians.

larynx *n.* [Gk. *laryngx*, larynx] the organ of voice in most vertebrates except birds, situated at the beginning of the trachea; *plu.* larynges; *alt.* voice box.

lash flagellum one in which the main filament ends in a thinner end-piece.

lasso *n.* [Sp. *lazo*, noose] a contractile filamentous noose used in trapping nematodes by certain soil fungi; a contractile filament tethering a nematocyst.

lasso cells colloblasts, *q.v.*

lata-type a mutant with one or more suprenumerary chromosomes as compared with its parent (from *Oenothera lata*).

latebra *n.* [L. *latebra*, hiding place] the bulb or flask-shaped mass of white yolk in eggs.

latebricole *a.* [L. *latebra*, hiding place; *colere*, to inhabit] inhabiting holes.

latent *a.* [L. *latens*, hidden] lying dormant but capable of development under certain circumstances, *appl.* buds, resting stages, *appl.* characteristics which will become apparent under certain conditions.

latent bodies the resting stage of certain Haemoflagellata.

latent learning learning that occurs without reward.

latent period or time the time interval between completion of presentation of a stimulus and the beginning of a reaction; *alt.* reaction time.

laterad *adv.* [L. *latus*, side; *ad.* towards] towards the side; away from the axis; *cf.* mediad.

lateral *a.* [L. *latus*, side] *pert.* or situated at a side, or at a side of an axis.

lateral-chain theory side-chain theory, *q.v.*

lateralia *n.plu.* [L. *latus*, side] the lateral plates of Cirripedia.

lateralis nervus lateralis, *q.v.*

lateralis organ in fishes, any of the cutaneous sensory cells of the lateral line system.

lateral line longitudinal line at each side of body of fishes, marking position of sensory cells, neuromasts; longitudinal excretory vessel in nematodes.

lateral meristems any dividing tissue in the plant other than the apical meristems, including the cambium and cork cambium.

lateral mesenteries the mesenteries of Zoantharia, excluding directive or doral and ventral pairs.

laterigrade *a.* [L. *latus*, side; *gradus*, step] walking sideways, as a crab.

laterinerved *a.* [L. *latus*, side; *nervus*, sinew] with lateral veins.

laterite *n.* [L. *later*, brick] *appl.* tropical red soils containing alumina and iron oxides and little silica owing to leaching under hot moist conditions.

laterobronchi *n.plu.* [L. *latus*, side; Gk. *brongchos*, windpipe] secondary bronchi arising from the mesobronchus in birds.

laterocranium *n.* [L. *latus*, side; L.L. *cranium*, skull] area of insect head comprising genae and postgenae.

laterosensory *a.* [L. *latus*, side; *sensus*, sense] *appl.* system of lateral sense organs in fishes, or lateral-line system.

laterosphenoid *n.* [L. *latus*, side; Gk. *sphēn*, wedge; *eidos*, form] bone on the mid-line of the reptilian skull, behind the orbit and above the palate.

laterosternites *n.plu.* [L. *latus*, side; *sternum*, breast bone] sclerites at side of eusternum, as in Dermaptera and Isoptera.

laterostratum *n.* [L. *latus*, side; *stratum*, layer] a hyphal layer diverging from mediostratum into subhymenium of agarics.

laterotergites *n.plu.* [L. *latus*, side; *tergum*, back] small sclerites adjoining tergum of abdominal segments in some crustaceans and insects.

late wood wood formed in the later part of an annual ring, having denser and smaller cells than early wood; *alt.* summer wood.

latex *n.* [L. *latex*, a liquid] a milky, or clear, sometimes coloured, juice or emulsion of diverse composition found in some plants, as in spurges, rubber trees, certain agarics, etc.; *plu.* latices.

latices *plu.* of latex.

laticifer *n.* [L. *latex*, a liquid; *ferre*, to carry] any latex-containing cell, series of cells, or duct; *alt.* lactifer.

laticiferous *a.* [L. *latex*, a liquid; *ferre*, to carry] conveying latex, *appl.* cells, tissue, vessels.

latifoliate *a.* [L. *latus*, wide; *folium*, leaf] with broad leaves; *cf.* angustifoliate.

latiplantar *a.* [L. *latus*, broad; *planta*, sole of foot] having hinder tarsal surface rounded.

latirostral *a.* [L. *latus*, broad; *rostrum*, beak] broad-beaked; *alt.* latirostrate; *cf.* angustirostral.

latiseptate *a.* [L. *latus*, broad; *septum*, septum] having a broad septum in the silicula; *cf.* angustiseptate.

latitudinal furrow one running round a segmenting egg above and parallel to the equatorial.

latosol *n.* [L. *later*, brick; Russ. *zolit'*, to leach] leached yellow or red tropical soils.

Latreille's segment propodeon, *q.v.*

Laurasia *n.* [St. *Lawrence*, N. America; *Asia*] an interconnected northern land mass of North America, Europe, and N. Asia, first postulated to account for similar fossils occurring in these continents, now thought to have existed before continental drift moved these continents to their present positions; *cf.* Gondwanaland.

Laurer–Stieda canal a canal leading from junction of oviduct and vitelline duct to opening on dorsal surface in trematodes; *alt.* Laurer's canal.

laurinoxylon *n.* [L. *laurus*, laurel; Gk. *xylon*, wood] any fossil wood; *alt.* lithoxyle.

law of acceleration the generalization that organs of greater importance develop more quickly.

law of independent assortment Mendel's 2nd law which states that each pair of factors (genes) segregate independently of all other pairs, and which is only true for genes on different chromosomes.

law of segregation Mendel's 1st law which states that factors (genes) present in pairs in the somatic cell become segregated during meiosis, so that each gamete contains one of a pair of factors.

lax *a.* [L. *laxus*, loose] loose, as *appl.* panicle.

layer of Langhans [*T. Langhans*, German histologist] cytotrophoblast, *q.v.*

lea ley, *q.v.*

leaching *n.* [O.E. *leccan*, causing to leak] washing ions out of the soil downwards during drainage; *cf.* flushing.

leader *n.* [A.S. *leadan*, to lead] highest shoot or part of trunk of a tree.

leaf *n.* [A.S. *lēaf*, leaf] an expanded outgrowth of a stem, usually green, and usually containing a bud in its axil.

leaf area ratio the photosynthetic surface per unit of dry weight.

leaf buttress lateral prominence on shoot axis, due to underlying leaf primordium, representing leaf base.

leaf cushions prominent persistent leaf bases, furnishing diagnostic characters in certain extinct plants.

leaf divergence the fixed proportion of the circumference of the stem by which every leaf is separated from the next in phyllotaxis.

leaf fibres economic fibres derived mainly from leaves of monocotyledons.

leaf gap mesh of stelar network, corresponding to site of leaf attachment in ferns; gap in vascular cylinder of stem, a parenchymatous region associated with leaf traces; *alt.* lacuna.

leaf insects the common name for many members of Phasmida, *q.v.*

leaflet a small leaf; individual unit of a compound leaf.

leaf mosaic an arrangement of leaves on plant or shoot which results in minimum overlap and maximum exposure to sunlight.

leaf scar the trace (usually covered by a suberized layer) left on a stem after a leaf has fallen.

leaf sheath extension of leaf base sheathing the stem, as in grasses.

leaf stalk petiole, *q.v.*

leaf trace vascular bundles extending from stem bundles to leaf base; *cf.* girdle bundles.

leaky gene hypomorph, *q.v.*

learning *n.* [M.E. *lernen*, to learn] a process in which animal behaviour becomes modified as a result of experience, as in conditioning, habituation, imprinting.

leberidocytes *n.plu.* [Gk. *leberis*, exuvia; *kytos*, hollow] cells containing glycogen, and developing from and regressing to leucocytes, found in blood of Arachnida at moulting.

Lecanicephaloidea *n.* [*Lecanicephalum*, generic

name; Gk. *eidos*, form] an order of segmented tapeworms having a scolex with 4 strong suckers (acetabula), in some classifications included in the Tetraphyllidea.

Lecanorales *n.* [*Lecanora*, generic name] an order of ascolichens having green algae extending to the edge of the apothecium.

lecanoric acid a common lichen acid.

lechriodont *a.* [Gk. *lechrios*, crosswise; *odous*, tooth] with vomerine and pterygoid teeth in a row nearly transverse.

Lecideales *n.* [*Lecidea*, generic name] an order of ascolichens having green algae, but with the margin of the apothecia free of them.

lecithalbumin *n.* [Gk. *lekithos*, egg yolk; L. *albumen*, white of egg] a substance, consisting of albumin and lecithin, of various body organs.

lecithelles *n.plu.* [Gk. *lekithos*, egg yolk] yolk granules in hypoblastic or other lecithoblasts.

lecithinases *n.plu.* [Gk. *lekithos*, egg yolk] enzymes that hydrolyse lecithins by attacking ester linkages, found in venoms and in various plant and animal extracts, being phospholipase A_2, EC 3.1.1.4, lysophospholipase, EC 3.1.1.5, phospholipase C, EC 3.1.4.3, and phospholipase D, EC 3.1.4.4.

lecithins *n.plu.* [Gk. *lekithos*, egg yolk] phospholipids containing glycerol, aliphatic acids, phosphoric acid, and choline, which are widely distributed and especially common in embryonic and nervous tissue, egg yolk, and bone marrow; *alt.* phosphatidylcholines.

lecithoblast *n.* [Gk. *lekithos*, egg yolk; *blastos*, bud] in developing eggs, the yolk-containing blastomeres.

lecithocoel *n.* [Gk. *lekithos*, egg yolk; *koilos*, hollow] segmentation cavity of holoblastic eggs.

lecithoproteins *n.plu.* [Gk. *lekithos*, egg yolk; *prōteion*, first] compound proteins in which lecithin is the prosthetic group.

lecithotrophic *a.* [Gk. *lekithos*, egg yolk; *trophē*, nourishment] feeding on stored yolk as in some echinoid larvae.

lecithovitellin *n.* [Gk. *lekithos*, L. *vitellus*, egg yolk] a lipoprotein, composed of lecithin and vitellin, in egg yolk.

lectin *n.* [L. *legere*, to select] a protein found in plants which acts like an antibody in animals causing agglutination of red blood cells, and in plants may be concerned with the recognition of host roots by symbiotic bacteria.

lectoallotype *n.* [Gk. *lektos*, chosen; *allos*, other; *typos*, pattern] a specimen of the opposite sex to that of the lectotype and subsequently chosen from the original materials.

lectotype *n.* [Gk. *lektos*, chosen; *typos*, pattern] a specimen chosen from syntypes to designate type of species.

leeches the common name for the Hirudinea, *q.v.*

leghaemoglobin *n.* [L. *legumen*, pulse; Gk. *haima*, blood; L. *globus*, sphere] an iron-containing red pigment, resembling haemoglobin, in root nodules of Leguminosae.

legio, legion *n.* [L. *legio*, legion] a taxonomic group used in different ways by different writers and never precisely defined.

legume *n.* [L. *legumen*, pulse] the dehiscent 1-celled fruit of the Leguminosae made of a single carpel and splitting into 2 valves, as pod of pea or bean; lomentum, *q.v.*; a member of the family Leguminosae.

legumin *n.* [L. *legumen*, pulse] a globulin in seeds of Leguminosae; *alt.* vegetable casein.

Leguminosae *n.* [L. *legumen*, pulse] a large family of dicotyledons, commonly called legumes or leguminous plants, including trees, shrubs, herbs, and climbers, typically having irregular flowers and the fruit a legume or lomentum; *see also* Fabales.

leguminous *a.* [L. *legumen*, pulse] *pert.* Leguminosae; *pert.* legumes; *pert.* or consisting of peas, beans, or other legumes.

leimocolous *a.* [Gk. *leimōn*, meadow; L. *colere*, to dwell] inhabiting damp meadows.

leiosporous *a.* [Gk. *leios*, smooth; *sporos*, seed] with smooth spores.

leiothric, leiotrichous *a.* [Gk. *leios*, smooth; *thrix*, hair] having straight hair.

leiotropic laeotropic, *q.v.*

leipsanenchyma *n.* [Gk. *leipsanon*, remnant; *engchyma*, infusion] part of primordial tissue of a carpophore, located between stipe and pileus; *alt.* lipsanenchyma.

Leitneriales [*Leitneria*, generic name] an order of woody dicots (Archichlamydeae) having alternate entire leaves and spikes of dioecious flowers, the female flowers developing into drupes.

lek *n.* [Swedish *lek*, game] the display ground on which lekking takes place.

lekking *n.* [Swedish *lek*, game] sexual display, often highly ritualized, by some species of birds preceding mating.

lemma *n.* [Gk. *lemma*, husk] the lower of 2 membranous bracts enclosing the flower in grasses; *alt.* flowering glume, lower pale or palea, outer pale or palea, valve; *cf.* palea.

lemniscus *n.* [Gk. *lēmniskos*, ribbon] one of paired club-shaped organs at base of acanthocephalan proboscis; fillet, *q.v.*

lenitic *a.* [L. *lenis*, smooth] lentic, *q.v.*

lens *n.* [L. *lens*, lentil] a transparent part of eye, which focuses rays of light on retina, *alt.* crystalline lens; modified portion of cornea in front of each element of a compound eye; modified cells of luminescent organ in certain fishes.

lens pit depression formed by the invagination of the lens placode.

lens placode local thickening of the ectoderm opposite the optic vesicle, which invaginates to form the lens pit; *alt.* lens rudiment.

lens rudiment lens placode, *q.v.*

lens vesicle a vesicle formed by closure of the external lips of the lens pit, the inner cells elongating and filling the vesicle.

lentic *a.* [L. *lentus*, slow] *appl.* or *pert.* standing water; living in swamp, pond, or lake; *alt.* lenitic; *cf.* lotic.

lenticel *n.* [L. *lens*, lentil] ventilating pore in periderm in stems and roots distinguished from phellem by having intercellular spaces, *alt.* air pore; canal in cork; a lenticular gland; *alt.* lenticula.

lenticula *n.* [L. *lenticula*, dim. of *lens*, lentil] a spore case in certain fungi; lenticel, *q.v.*; a lentigo or freckle.

lenticular shaped like a double-convex lens, *appl.* glands, lymphoid structures between pyloric glands, *alt.* lentiform, *q.v.*; *pert.* lenticels, *appl.* transpiration, *cf.* stomatal. *n.* Tip of incus articu-

lating with stapes, often ossified as a separate unit.

lenticulate *a.* [L. *lens*, lentil] meeting in a sharp point; depressed, circular, and frequently ribbed.

lentiform *a.* [L. *lens*, lentil; *forma*, shape] lentil-shaped, *appl.* nucleus, the extraventricular portion of corpus striatum; *alt.* lenticular.

lentigerous *a.* [L. *lens*, lentil; *gerere*, to bear] furnished with a lens.

lentiginose, lentiginous *a.* [L. *lentigo*, freckle] freckled; speckled; bearing numerous small dots.

lentocapillary point point, just above wilting coefficient, at which flow of water towards root hairs is impeded on account of surface tension resistance.

leotropic laeotropic, *q.v.*

lepidic *a.* [Gk. *lepis*, scale] consisting of scales; *pert.* scales.

Lepidodendrales *n.* [Lepidodendron, generic name] an order of heterosporous fossil Lycopsida, having tree-like sporophytes with secondary thickening and small leaves which produce scale-like leaf scars.

lepidodendroid *a.* [Gk. *lepis*, scale; *dendron*, tree; *eidos*, form] *pert.* Lepidodendron or Lepidodendrales; having scale-like leaf scars.

lepidoid *a.* [Gk. *lepis*, scale; *eidos*, from] resembling a scale or scales.

lepidomorium *n.* [Gk. *lepis*, scale; *morion*, constituent part] small scale, or unit of composite scale, with bony base and conical or conoid crown of dentine, containing pulp cavity and sometimes covered with enamel; *plu.* lepidomoria; *a.* lepidomorial.

Lepidophyta *n.* [Gk. *lepis*, scale; *phyton*, plant] Lycopsida, *q.v.*

lepidophyte *n.* [Gk. *lepis*, scale; *phyton*, plant] a fossil fern.

Lepidopleurida *n.* [*Lepidopleurus*, generic name] an order of primitive chitons of the subclass Polyplacophora.

Lepidoptera *n.* [Gk. *lepis*, scale; *pteron*, wing] an order of insects including butterflies and moths, having complete metamorphosis, 2 pairs of membranous wings covered with scales, a sucking proboscis, and caterpillars as larvae; *a.* lepidopterous.

Lepidosauria, lepidosaurs *n.*, *n.plu.* [Gk. *lepis*, scale; *sauros*, lizard] a subclass or superorder of reptiles with a diapsid skull having limbs and girdles unspecialized, reduced, or absent.

lepidosis *n.* [Gk. *lepis*, scale] character and arrangement of scales of animals.

lepidosteoid *a.* [Gk. *lepis*, scale; *osteon*, bone; *eidos*, form] *appl.* a ganoid scale lacking cosmine.

lepidote *a.* [Gk. *lepidōtos*, scaly] covered with minute scales.

lepidotic *a.* [Gk. *lepidōtos*, scaly] *appl.* an acid found in wings of some Lepidoptera.

lepidotrichia *n.plu.* [Gk. *lepis*, scale; *thrix*, hair] the bony actinotrichia of teleosts; *cf.* ceratotrichia.

lepocyte *n.* [Gk. *lepis*, husk; *kytos*, hollow] a cell with a defining cell wall; *cf.* gymnocyte.

Lepospondyli *n.* [Gk. *lepis*, husk; *sphondylos*, vertebra] a group of amphibians having vertebral centra formed as single structures and often spool-shaped; the Urodela and Apoda may be included in this group or in the Lissamphibia.

lepospondylous *a.* [Gk. *lepis*, husk; *sphondylos*, vertebra] having amphicoelous, or hourglass-shaped, vertebrae.

leptocaul *a.* [Gk. *leptos*, slender; L. *caulis*, stalk] having a thin or slender primary stem.

leptocentric *a.* [Gk. *leptos*, slender; *kentron*, centre] *appl.* concentric bundle with central leptome.

leptocephaloid *a.* [Gk. *leptos*, slender; *kephalē*, head; *eidos*, form] resembling or having the shape of eel larvae.

leptocephalus *n.* [Gk. *leptos*, slender; *kephalē*, head] translucent larva of certain eels, before the elver stage.

leptocercal *a.* [Gk. *leptos*, slender; *kerkos*, tail] with long slender tapering tail, *appl.* some fishes, *appl.* Protozoa; *alt.* leptocercous.

leptocystidium *n.* [Gk. *leptos*, thin; *kystis*, bladder; *idion*, *dim.*] a thin-walled cystidium, as in many agarics.

leptodactylous *a.* [Gk. *leptos*, slender; *daktylos*, finger] having slender digits.

leptodermatous *a.* [Gk. *leptos*, thin; *derma*, skin] thin-skinned, *appl.* various thecae; *alt.* leptodermic, leptodermous.

lepto-forms *n.plu.* [Gk. *leptos*, slender] rusts with a teleutostage and sometimes a pycnial stage, germinating without rest.

leptoid *n.* [Gk. *leptos*, slender; *eidos*, form] one of the living elongated thin-walled cells forming simple conducting tissue in bryophytes; a tubular cell in stem of certain pteridophytes.

Leptolepidiformes *n.* [Gk. *leptos*, slender; *lepis*, scale; L. *forma*, shape] the only order of the Leptolepidomorpha, *q.v.*

Leptolepidomorpha *n.* [Gk. *leptos*, slender; *lepis*, scale; *morphē*, shape] a superorder of teleosts which are sometimes classified as halecostomes and contain the single order Leptolepidiformes.

leptology *n.* [Gk. *leptos*, small; *logos*, discourse] the study of minute particles or structures.

leptom(e) *n.* [Gk. *leptos*, slender] sieve elements and parenchyma of phloem; bast, *q.v.*; *alt.* leptomestome.

Leptomedusae *n.* [Gk. *leptos*, thin; *Medousa*, Medusa, one of the Gorgons] Thecata, *q.v.*

leptomeninges *n.plu.* [Gk. *leptos*, thin; *meningx*, membrane] the pia mater and arachnoid membrane; *sing.* leptomeninx.

leptomestome *n.* [Gk. *leptos*, slender; *mestos*, filled] leptome, *q.v.*

Leptomitales *n.* [*Leptomitus*, generic name] an order of Phycomycetes having a mycelial thallus with a basal holdfast bearing erect hyphae, asexual reproduction by biflagellate zoospores, and oogamous sexual reproduction.

leptonema *n.* [Gk. *leptos*, slender; *nēma*, thread] fine unpaired chromosome thread at leptotene.

leptophloem *n.* [Gk. *leptos*, slender; *phloios*, smooth bark] rudimentary bast tissue.

leptophragmata *n.plu.* [Gk. *leptos*, thin; *phragma*, fence] in some cryptonephridial insects, small thin areas adjoining the haemocoel which give strong reaction to chlorides; *sing.* leptophragma.

leptophyllous *a.* [Gk. *leptos*, slender; *phyllon*, leaf] with slender leaves; having a small leaf area, under 25 mm².

leptosome *a.* [Gk. *leptos*, slender; *sōma*, body] tall and slender; *alt.* asthenic; *cf.* eurysome.

Leptosporangiatae *n.* [Gk. *leptos*, slender; *sporos*, seed; *anggeion*, vessel] subclass of Pteropsida (Filicopsida) having leptosporangiate sporangia with walls one cell thick and formed from a single cell, a flat prothallus and an erect or creeping stem; *alt.* Leptosporangidae, Filices; *cf.* Eusporangiatae.

leptosporangiate *a.* [Gk. *leptos*, slender; *sporos*, seed; *anggeion*, vessel] having sporangia derived from the outer cell produced by periclinal division of a superficial initial, each sporangium being formed from a single epidermal cell, *appl.* certain ferns; *cf.* eusporangiate.

Leptostraca *n.* [Gk. *leptos*, slender; *ostrakon*, shell] a group of marine malacostracans having a carapace as a bivalved shell and a moveable head plate; they may be considered a series containing one superorder, Phyllocarida, and one order, Nebaliacea, or as an order.

leptostroterate *a.* [Gk. *leptos*, slender; *strōtos*, covered] with ambulacral plates narrow and crowded together, as in certain Stelleroidea.

leptotene *n.* [Gk. *leptos*, slender; *tainia*, band] early stage of the prophase of meiosis where chromatin is in form of fine threads.

leptotichous *a.* [Gk. *leptos*, thin; *teichos*, wall] thin-walled, *appl.* plant tissue.

leptotrombicula *n.* [Gk. *leptos*, slender; It. *tromba*, trumpet] the larval form of a trombicula, a mite transmitting tsutsugamushi disease.

leptoxylem *n.* [Gk. *leptos*, slender; *xylon*, wood] rudimentary wood tissue.

lepto-zygotene *a.* [Gk. *leptos*, slender; *zygon*, yoke; *tainia*, band] *appl.* transition stage between leptonema and zygonema.

leptus *n.* [Gk. *leptos*, small] the 6-legged larva of mites.

Lernaeopodoida *n.* [*Lernaea*, generic name; Gk. *pous*, foot; *eidos*, form] an order of copepods that are ectoparasites on fish and attach to the host by the 2nd maxillae.

lesser omentum a fold of peritoneum which connects the stomach and liver and supports the hepatic vessels.

lestobiosis *n.* [Gk. *lēstēs*, plunderer; *biōsis*, manner of living] brigandage by ants.

lethal *a.* [L. *letalis*, deadly] causing death; of a parasite, fatal or deadly in relation to a particular host; *appl.* gene which so influences development that the individual is rendered non-viable. *n.* A lethal factor or gene.

lethality *n.* [L. *letalis*, deadly] the ratio of fatal cases to total number of cases affected by a disease or other harmful agency.

leu leucine, *q.v.*

Leucettida *n.* [*Leucetta*, generic name] an order of Calcinea having a distinct dermal membrane or cortex, and the choanocytes in flagellated chambers not lining the central cavity.

leucine *n.* [Gk. *leukos*, white] an essential amino acid, α-amino isocaproic acid, found in various tissues of animals and some plants as a constituent of proteins; abbreviated to leu.

leucism *n.* [Gk. *leukos*, white] the presence of white plumage or pelage in animals with pigmented eyes and skin.

leucite *n.* [Gk. *leukos*, white] a colourless plastid; *alt.* leucoplast.

leuco- *alt.* leuko-.

leucoagglutination n. [Gk. *leukos*, white; *agglutinare*, to cement] agglutination of white blood corpuscles.

leucoanthocyanidins n.plu. [Gk. *leukos*, white; *anthos*, flower; *kyanos*, blue] a group of colourless flavonoids.

leucoanthocyanins n.plu. [Gk. *leukos*, white; *anthos*, flower; *kyanos*, blue] precursors of anthocyanins which possess marked growth-promoting qualities.

leucoblast n. [Gk. *leukos*, white; *blastos*, bud] a leucocyte in development; *alt.* proleucocyte.

leucocarpous a. [Gk. *leukos*, white; *karpos*, fruit] with the fruit white.

leucocidin n. [Gk. *leukos*, white; L. *caedere*, to kill] a leucocyte-destroying toxin elaborated by some staphylococci.

leucocyan n. [Gk. *leukos*, white; *kyanos*, blue] a pigment found in certain algae.

leucocyte n. [Gk. *leukos*, white; *kytos*, hollow] a colourless often amoeboid cell of blood; *alt.* white blood cell, plasmocyte, amoebocyte.

leucocytogenesis n. [Gk. *leukos*, white; *kytos*, hollow; *genesis*, descent] leucopoiesis, *q.v.*

leucocytoid histiocyte, *q.v.*

leucocytolysis n. [Gk. *leukos*, white; *kytos*, hollow; *lysis*, loosing] the breakdown or dissolution of white blood corpuscles.

leucon grade a grade of sponge structure in which the flagellated chambers are small and round, and the canals are extensively branched.

leucophore n. [Gk. *leukos*, white; *pherein*, to bear] a white chromatophore; guanophore, *q.v.*

leucoplast(id)s n.plu. [Gk. *leukos*, white; *plastos*, formed; *idion*, *dim.*] colourless plastids from which amylo-, chloro-, and chromoplasts arise; *alt.* leucites, anaplasts; colourless granules of plant cytoplasm, *cf.* chromoplasts.

leucopoiesis n. [Gk. *leukos*, white; *poiēsis*, making] the formation of white blood corpuscles; *alt.* leucocytogenesis.

leucopsin n. [Gk. *leukos*, white; *opsis*, sight] visual white, *q.v.*

leucopterin(e) n. [Gk. *leukos*, white; *pteron*, wing] a pterin that constitutes the white pigment of cabbage butterflies and other Lepidoptera and wasps, and can be reduced to xanthopterine; *cf.* isoxanthopterine.

leucosin n. [Gk. *leukos*, white] a substance thought to be a polysaccharide which occurs as whitish lumps in the cells of algae of the Chrysophyta, and acts as a food reserve; *alt.* chrysolaminarin.

Leucosoleniidae n. [*Leucosolenia*, generic name] an order of Calcaronea having no true dermal membrane or cortex, and the central cavity always lined with choanocytes.

leucotaxin(e) n. [Gk. *leukos*, white; *taxis*, arranging] a crystalline nitrogenous substance formed in response to tissue injury and recovered from inflamed areas, which decreases capillary resistance and promote diapedesis of leucocytes.

leucoviruses n.plu. [Gk. *leukos*, white; L. *virus*, poison] a group of oncornaviruses that bud from cytoplasmic membranes.

leukin n. [Gk. *leukos*, white] a basic polypeptide extracted from leucocytes and active against Gram-positive bacteria.

leuko- leuco-, *q.v.*

levator n. [L. *levare*, to raise] a name given to muscles serving to raise an organ or part; *alt.* elevator; *cf.* depressor.

levigate v. [L. *levigare*, to make smooth] to smoothen. a. Made smooth, polished; *alt.* laevigate.

levulose fructose, *q.v.*

ley n. [L. *lucus*, grove] a temporary pasture or hayfield which is ploughed up and sown with another crop after a few years; *alt.* lea.

Leydig's cells [F. *von Leydig*, German anatomist] cells in testicular interstitial tissue.

Leydig's duct Wolffian duct, *q.v.*

Leydig's organs minute organs on antennae of arthropods, supposed to be organs of smell.

L-form [*Lister* Institute] a filterable form of certain bacteria which resemble mycoplasmas and are thought to be specialized reproductive bodies produced when the environment is unfavourable.

LH luteinizing hormone, *q.v.*

liana n. [F. *liane*, from L. *ligare*, to bind] any woody climbing plant of tropical or semitropical forests; *alt.* liane.

Lias n. [M.E. *lyas*, a kind of limestone] marine and estuarine deposits of Jurassic period, containing remains of cycads, insects, ammonites, saurians, and other fossils.

liber n. [L. *liber*, inner bark] inner bark; *alt.* bast.

libido n. [L. *libido*, desire] excitation within body associated with instinct; sexual energy; psychic energy; horme, *q.v.*; élan vital.

libriform a. [L. *liber*, inner bark; *forma*, shape] resembling bast; *appl.* woody fibres with thick walls and simple pits.

lice the common name for various insects of the orders Psocoptera, Mallophaga, Anoplura, and for crustaceans of the orders Branchiura and Isopoda.

Liceales n. [*Licea*, generic name] an order of slime moulds growing typically on dead wood or bark.

lichen acids a number of acids produced by lichens and deposited extracellularly on the surface of the hyphae rather than inside cells, many of which are found exclusively in lichens.

lichenase the enzyme which breaks down lichenin to glucose, and so digests lichens, as found in the gut of reindeer and caribou and some gastropods; EC 3.2.1.6; *r.n.* endo-1,3(4)-β-D-glucanase.

Lichenes, lichens n., n.plu. [Gk. *leichēn*, lichen] a large group of plants which consist of a symbiotic association between an alga and a fungus, and are classified as the entire plant.

lichenicole, lichenicolous a. [Gk. *leichēn*, lichen; L. *colere*, to dwell] living or growing on lichens.

lichenin(e) n. [Gk. *leichēn*, lichen] a polysaccharide containing 1,4-β glucoside linkages, found in the hyphal cell walls of lichens and also in some seeds, especially oat seeds, which can be hydrolysed by lichenase to glucose; *alt.* lichen starch, moss starch.

lichenism n. [Gk. *leichēn*, lichen] symbiotic relationship between fungi and algae.

lichenization production of a lichen by alga and fungus; spreading or coating of lichens over a substrate; effect of lichens on their substrates.

lichenoid a. [Gk. *leichēn*, lichen; *eidos*, form] resembling a lichen.

lichenology n. [Gk. *leichēn*, lichen; *logos*, discourse] the study of lichens.

lichen starch lichenin, *q.v.*

lid cell cell at apex of an archegonium or of antheridium.

Lieberkühn's crypts crypts of Lieberkühn, *q.v.*

Liebig's law [*J. von Liebig*, German chemist] the food element least plentiful in proportion to the requirements of plants limits their growth; law of the minimum, *q.v.*

lien *n.* [L. *lien*, spleen] spleen, *q.v.*

lienal *a.* [L. *lien*, spleen] *pert.* spleen, *appl.* artery, vein, nerve plexus; *alt.* splenic.

lienculus *n.* [*Dim.* of L. *lien*, spleen] an accessory spleen.

lienogastric *a.* [L. *lien*, spleen; Gk. *gastēr*, belly] *pert.* spleen and stomach; *appl.* artery supplying spleen and parts of stomach and pancreas; *appl.* vein of hepatic portal system.

lienorenal phrenicolienal, *q.v.*

life cycle the various phases through which an individual species passes to maturity.

life form the body shape of an organism or species at maturity, e.g. trees are usually the dominant life form of cool wet areas; a way of classifying Cormophyta according to the length of shoots, position of resting buds, and way they are protected in unfavourable conditions, e.g. phanerophytes, cryptophytes, etc.

life zone a biome, *q.v.*; a subdivision of a biome, as temperature, distribution, community, etc., zones.

ligament *n.* [L. *ligamentum*, bandage] a strong fibrous band of tissue connecting 2 or more moveable bones or cartilages; band of conchiolin forming hinge of bivalve shells.

ligamenta flava yellow elastic ligaments connecting laminae of adjoining vertebrae.

ligand *n.* [L. *ligare*, to bind] a molecule that binds to a particular site on another molecule, as an enzyme does to a substrate.

ligases *n.plu.* [L. *ligare*, to bind] synthetases, *q.v.*

light compass reaction a menotaxis where the stimulus is light.

light line in sections through certain legume seeds, a continuous line through the epidermis thought to be an impermeable region, formed as a result of the matching of regions of a high degree of refraction in the walls of adjacent epidermal cells.

light mitochondria lysosomes, *q.v.*

light reaction in photosynthesis, the reactions occurring in the grana of the chloroplasts, directly concerned with light absorption and resulting in the formation of ATP and reducing power; *cf.* dark reaction.

ligneous *a.* [L. *lignum*, wood] woody or resembling wood in structure; *alt.* xyloid.

lignescent *a.* [L. *lignescere*, to become woody] developing the characters of woody tissue.

lignicole, lignicolous *a.* [L. *lignum*, wood; *colere*, to inhabit] growing or living on or in wood.

lignification *n.* [L. *lignum*, wood; *facere*, to form] wood formation; thickening of plant cell walls by deposition of lignin.

lignin *n.* [L. *lignum*, wood] a colloidal polymer with no clear chemical structure and varying from species to species but based on coniferyl alcohol, used as secondary wall material in xylem vessels, tracheids, and sclerenchyma fibres; *alt.* xylogen.

ligniperdous *a.* [L. *lignum*, wood; *perdere*, to ruin] causing destruction of wood, *appl.* certain fungi, insects, etc.

lignivorous *a.* [L. *lignum*, wood; *vorare*, to devour] eating wood, *appl.* various insects, etc.

lignocellulose *n.* [L. *lignum*, wood; *cellula*, little cell] essential constituent of woody tissue, lignin and cellulose combined.

lignose *n.* [L. *lignum*, wood] a constituent of lignin.

ligula *n.* [L. *ligula*, little tongue] a band or taenia of white matter in dorsal wall of 4th ventricle; median structure between labial palps of insects; lobe of parapodium in certain annelids; ligule, *q.v.*; lingula, *q.v.*

ligular tongue-shaped; *pert.* ligulae or ligules; *appl.* pit on leaf base above leaf scar, as in extinct Lycopodiales.

ligulate *a.* [L. *ligula*, little tongue] having or *pert.* ligules; strap-shaped, as ray florets of Compositae; *appl.* capitulum of strap-shaped florets; *alt.* lingulate.

ligule *n.* [L. *ligula*, little tongue] a membranous outgrowth at junction of blade and leaf sheath or petiole; small scale on upper surface of leaf base in Lepidodendrales, Selaginellales, and Isoetales; a tongue-shaped corolla, as of certain florets; *alt.* ligula.

liguliflorous *a.* [L. *ligula*, little tongue; *flos*, flower] having ligulate flowers only.

Ligustrales *n.* [*Ligustrum*, generic name] Oleales, *q.v.*

Liliales *n.* [*Lilium*, generic name] an order of monocots (Liliidae) having actinomorphic flowers and syncarpous ovaries; *alt.* Liliiflorae.

Liliatae *n.* [*Lilium*, generic name] in some classifications, Monocotyledones, *q.v.*; *alt.* Liliopsida.

Liliidae *n.* [*Lilium*, generic name] a subclass of monocots having cyclic flowers and 3 more-or-less united carpels.

Liliiflorae *n.* [*Lilium*, generic name; L. *flos*, flower] an order of monocots used in different ways by different authorities but now usually used as an alternative term for Liliales, *q.v.*

Liliopsida Liliatae, *q.v.*

limacel(le) *n.* [F., from L. *limax*, slug] concealed vestigial shell of slugs.

limaciform *a.* [L. *limax*, slug; *forma*, shape] like a slug; slug-shaped.

limacine *a.* [L. *limax*, slug] *pert.* slugs.

Limacomorpha *n.* [L. *limax*, slug; Gk. *morphē*, form] Glomeridesmida, *q.v.*

limb *n.* [A.S. *lim*, limb] branch; arm; leg; wing; expanded part of calyx or corolla, the base of which is tubular.

limbate *a.* [L. *limbus*, border] with a border; bordered and having a differently coloured edge.

limbic *a.* [L. *limbus*, border] bordering, *appl.* a cerebral lobe, including hippocampal and cingulate gyri.

limbous *a.* [L. *limbus*, border] overlapping, *appl.* sutures.

limbus *n.* [L. *limbus*, border] any border if distinctly marked off by colour or structure; transition zone between cornea and sclera.

lime *n.* [L. *limus*, mud] calcium oxide; calcium hydroxide; any calcium salt.

limen *n.* [L. *limen*, threshold] threshold, minimum stimulus, or quantitative difference in stimulation, that is perceptible; boundary, as between vestibule of nostril and nasal cavity: limen nasi; *cf.* subliminal.

limicolous *a.* [L. *limus*, mud; *colere*, to dwell] living in mud.

liminal *a.* [L. *limen*, threshold] *pert.* a threshold, *appl.* stimulus, *appl.* sensation.

limited *appl.* chromosomes in germinal, not in somatic, nuclei.

limiting factor the factor present in an environment in so short supply that it limits growth or some other life process.

limiting membranes of retina, *see* membrana limitans.

limitrophic *a.* [Gk. *limos*, hunger; *trophē*, nourishment] *pert.* or controlling nutrition.

limivorous *a.* [L. *limus*, mud; *vorare*, to devour] mud-eating, *appl.* certain aquatic animals.

limnetic *a.* [Gk. *limnē*, marshy lake] living in, or *pert.*, marshes or lakes; zone of deep water between surface and compensation depth.

limnium *n.* [Gk. *limnē*, lake] a lake community.

limnobiology *n.* [Gk. *limnē*, lake; *bios*, life; *logos*, discourse] the study of life in standing fresh waters.

limnobios *n.* [Gk. *limnē*, lake; *bios*, life] life in fresh water; freshwater plants and animals collectively.

limnobiotic *a.* [Gk. *limnē*, lake; *bios*, life] limnophilous, *q.v.*

limnocryptophyte *n.* [Gk. *limnē*, marsh; *kryptos*, hidden; *phyton*, plant] helophyte, *q.v.*

limnology *n.* [Gk. *limnē*, marshy lake; *logos*, discourse] science dealing with biological and other phenomena *pert.* inland waters; the study of standing waters.

Limnomedusae *n.* [Gk. *limnē*, lake; *Medousa*, Medusa, one of the Gorgons] an order of mostly freshwater hydrozoans having medusae with hollow tentacles and polyps often reduced, formerly included in the Thecata, Athecata and Trachymedusae.

limnophilous *a.* [Gk. *limnē*, marsh; *philein*, to love] living in freshwater marshes; *alt.* limnobiotic.

limnophyte *n.* [Gk. *limnē*, marshy lake; *phyton*, plant] a pond plant; a helophyte, *q.v.*

limnoplankton *n.* [Gk. *limnē*, marshy lake; *plangktos*, wandering] the floating animal and plant life in freshwater lakes, ponds, and marshes; *cf.* haloplankton.

limosphere *n.* [Gk. *limēn*, receptacle; *sphaira*, globe] a spherical body containing a vacuole, situated near blepharoplast in spermiogenesis of some mosses.

linea *n.* [L. *linea*, line] a line-like structure or mark.

linea alba tendinous medial line separating recti abdominis, from xiphoid process to symphysis pubis.

linea aspera longitudinal crest on dorsal side of femur.

linear *a.* [L. *linea*, line] *pert.* or in a line; tape- or thread-like; asthenic, *appl.* constitutional type.

linear-ensate between linear and ensiform in shape.

linear-lanceolate between linear and lanceolate in shape.

linear-oblong between linear and oblong in shape.

linea spendens a longitudinal fibrous line along middle of ventral surface of spinal pia mater.

linellae *n.plu.* [L. *linella*, fine thread] a system of filaments in certain Sarcodina holding together the xenophya.

lineolate *a.* [L. *linea*, line] marked by fine lines or striae.

line transect the recording of types of plants or plant communities along a measured line.

lingua *n.* [L. *lingua*, tongue] the floor of mouth in mites; hypopharynx, *q.v.*, of insects; a tongue, or tongue-like structure.

lingual *pert.* tongue, *appl,* artery, gyrus, nerve, vein, etc.; *appl.* radula of molluscs; *pert.* lingua.

Linguatulida *n.* [*Linguatula*, generic name] Pentastomida, *q.v.*

linguiform *a.* [L. *lingua*, tongue; *forma*, shape] tongue-shaped; *alt.* lingulate.

lingula *n.* [L. *lingula*, little tongue] a small tongue-like process of bone or other tissue, as of cerebellum or sphenoid; a genus of brachiopods; ligula, *q.v.*

lingulate tongue-shaped, but being shorter and broader than ligulate; ligulate, *q.v.*; *alt.* linguiform.

linin *n.* [L. *linum*, flax] the fibrillar reticulum in cell nuclei which does not take up nuclear stains, *alt.* lanthanin, oxychromatin; a protein in flax seed; a bitter purgative substance obtained from purging flax.

lininoplast *n.* [L. *linum*, flax; Gk. *plastos*, moulded] plasmosome, *q.v.*

linkage *n.* [A.S. *hlince*, link] tendency of certain genes to remain associated through several generations because they are situated on the same chromosome.

linkage group group of genes lying on the same chromosome.

linkage map chromosome map, *q.v.*

Linnaean *a.* [C. *Linné*, or *Linnaeus*, Swedish naturalist] *pert.* or designating the system of classification established by Linnaeus.

Linnaean species linneon, *q.v.*

linneon *n.* [C. *Linné*, Swedish naturalist] a taxonomic species distinguished on morphological grounds, especially one of the large species described by Linnaeus or other early naturalists; macrospecies, *q.v.*; *alt.* Linnaean species; *cf.* jordanon.

linoleic acid [Gk. *linon*, flax] a fatty acid essential for growth of mammals and necessary in the diet.

linolenic acid [Gk. *linon*, flax] a fatty acid necessary for growth mammals, γ-linolenic acid can be synthesized from linoleic acid.

lipase *n.* [Gk. *lipos*, fat] an enzyme which hydrolyses a triacylglycerol to a diacylglycerol plus a fatty acid anion, *alt.* steapsin, EC 3.1.1.3, *r.n.* triacylglycerol lipase; sometimes wrongly used for lipolytic enzymes.

lip cell a sporangium cell at the point of dehiscence.

lipid(e)s *n.plu.* [Gk. *lipos*, fat] compounds that are esters of higher aliphatic alcohols, insoluble in water but soluble in organic solvents, and including the simple lipids (oils, fats, and waxes), chromolipids, complex lipids (lipins), and lipoids; the part of any biological material which can be extracted with low polarity solvents.

lipidoplast

lipidoplast n. [Gk. *lipos*, fat; *plastos*, formed] elaioplast, *q.v.*

lipin(e)s *n.plu.* [Gk. *lipos*, fat] complex lipids, including the phospholipids and glycolipids.

lipochondria *n.plu.* [Gk. *lipos*, fat; *chondros*, grain] lipoid granules in the Golgi zone; Golgi presubstance.

lipochrin *a.* [Gk. *lipos*, fat; *ōchros*, sallow] *appl.* yellow lipoid droplet, fading by light, in unpigmented base of retinal cell.

lipochroic *a.* [Gk. *lipos*, fat; *chrōs*, colour] with pigment in oil droplets.

lipochromes *n.plu.* [Gk. *lipos*, fat; *chrōma*, colour] a more-or-less indefinite group of plant and animal pigments, as carotins, luteins, chromophanes, zoonerythrin, etc.

lipoclastic *a.* [Gk. *lipos*, fat; *klastos*, broken] lipolytic, *q.v.*

lipocyte *n.* [Gk. *lipos*, fat; *kytos*, hollow] a cell containing lipids; *alt.* fat-cell.

lipofuscin *n.* [Gk. *lipos*, fat; L. *fuscus*, dusky] a yellowish-brown pigment in cytoplasm of some nerve cells.

lipogastry *n.* [Gk. *leipesthai*, to be lacking; *gastēr*, stomach] temporary obliteration of gastral cavity, as in some sponges.

lipogenesis *n.* [Gk. *lipos*, fat; *genesis*, origin] the synthesis of fatty acids in cells.

lipogenous *a.* [Gk. *lipos*, fat; *gennaein*, to produce] fat-producing.

lipohumor *n.* [Gk. *lipos*, fat; L. *humor*, moisture] a fat-soluble substance produced by nerves and acting on chromatophores.

lipoic acid [Gk. *lipos*, fat] a substance which includes a fatty acid and disulphide, 1,2-dithiolane-3-valeric acid, formerly 6,8-dithiooctanoic acid, required for the growth of various micro-organisms and concerned with carbohydrate metabolism.

lipoid *a.* [Gk. *lipos*, fat; *eidos*, form] resembling a fatty substance.

lipoids *n.plu.* [Gk. *lipos*, fat; *eidos*, form] substances which are not true lipids but resemble fats in certain physical properties and are extracted in fat solvents, including sterols and steroids; the term is also used for true lipids.

lipolysis *n.* [Gk. *lipos*, fat; *lysis*, loosing] breakdown of fats by enzymes, such as during digestion; *alt.* adipolysis.

lipolytic *a.* [Gk. *lipos*, fat; *lyein*, to dissolve] fatreducing; capable of digesting or dissolving fat, *appl.* enzymes, *alt.* lipoclastic; *appl.* hormone (LPH): lipotrophin, *q.v.*

lipomerism *n.* [Gk. *leipesthai*, to be lacking; *meros*, part] suppression of segmentation, or coalescence of segments, as in crustaceans.

lipopalingenesis *n.* [Gk. *leipesthai*, to be lacking; *palin*, anew; *genesis*, descent] the omission of some stage or stages in phylogeny.

lipopexia *n.* [Gk. *lipos*, fat; *pexis*, fixing] deposition and storage of fats in tissues.

lipophanerosis *n.* [Gk. *lipos*, fat; *phanerōsis*, manifestation] the appearance of lipids in metabolism, as during spermatogenesis in mammals.

lipophore *n.* [Gk. *lipos*, fat; *-phoros*, -bearing] a wandering cell originating in neural crest and containing a lipochrome; xanthophore, *q.v.*

lipopolysaccharide *n.* [Gk. *lipos*, fat; *polys*, many; L. *saccharum*, sugar] a molecule consisting of a lipid joined to a polysaccharide, which is one

of the main constituents of the cell wall of Gram-negative bacteria (the other being lipoprotein).

lipoproteins *n.plu.* [Gk. *lipos*, fat; *prōteion*, first] proteins combined with lipids.

liporhodine *n.* [Gk. *lipos*, fat; *rhodon*, rose] a red lipochrome, as in certain fungi.

liposome *n.* [Gk. *lipos*, fat; *sōma*, body] a fatty droplet in cytoplasm, especially of an egg.

lipostomy *n.* [Gk. *leipesthai*, to be lacking; *stoma*, mouth] temporary obliteration of mouth or osculum.

Lipostraca *n.* [Gk. *lipos*, fat; *ostrakon*, shell] an extinct order of Branchiopoda having no carapace or eyes.

lipotrop(h)in *n.* [Gk. *lipos*, fat; *trophē*, nourishment] in mammals, either of 2 peptides produced by the pars distalis of the pituitary gland, which stimulate lipolysis; *alt.* lipolytic hormone (LPH), lipotropic hormone.

lipotropic *a.* [Gk. *lipos*, fat; *trope*, turn] concerned with fat movement; influencing fat metabolism; accelerating removal of fat, *appl.* hormone: lipotrophin, *q.v.*

lipovitellin *n.* [Gk. *lipos*, fat; L. *vitellus*, yolk] a lipoprotein which is part of amphibian egg yolk.

lipoxanthins *n.plu.* [Gk. *lipos*, fat; *xanthos*, yellow] yellow lipochromes.

lipoxenous *a.* [Gk. *lipein*, to abandon; *xenos*, host] leaving the host before completion of development.

lipoxidase *n.* [Gk. *lipos*, fat; *oxys*, sharp] an enzyme which catalyses the addition of a molecule of oxygen to the double bonds of certain unsaturated fatty acids; EC 1.13.11.12; *r.n.* lipoxygenase.

lipsanenchyma leipsanenchyma, *q.v.*

liquor folliculi the liquid surrounding the ovum in the Graafian follicle.

lirella *n.* [L. *lira*, furrow] a linear apothecium of lichens.

Lissamphibia *n.* [Gk. *lissos*, smooth; *amphi*, both; *bios*, life] a subclass of amphibians, not recognized in all classifications, which includes all the living amphibians, i.e. the orders Anura, Urodela, Apoda, having scales reduced or absent and the skin respiratory.

lissencephalous *a.* [Gk. *lissos*, smooth; *engkephalos*, brain] having few or no convolutions of the brain; *alt.* lissencephalic; *cf.* gyrencephalic.

Lissoflagellata, lissoflagellates *n., n.plu.* [Gk. *lissos*, smooth; L. *flagellum*, whip] a group of flagellates, sometimes considered as an order, which do not have a protoplasmic collar around the base of the flagellum; *cf.* Choanoflagellata.

Lithistida *n.* [Gk. *lithizein*, to resemble a stone] in earlier classifications, an order of fossil sponges with a massive net-like skeleton made of fused silica spicules.

lithite *n.* [Gk. *lithos*, stone] a calcareous secretion found in connection with ear, or with otocysts, lithocysts, and tentaculocysts, sensory organs of many invertebrates; *alt.* statolith.

Lithobiomorpha *n.* [*Lithobius*, generic name; Gk. *morphē*, form] an order of centipedes of the subclass Anamorpha having paired spiracles.

lithocarp *n.* [Gk. *lithos*, stone; *karpos*, fruit] carpolith, *q.v.*

lithocysts *n.plu.* [Gk. *lithos*, stone; *kystis*, bladder] minute sacs or grooves, containing lithites, found in various invertebrates; enlarged cells of plant epidermis, in which cystoliths are formed; *alt.* statocysts.

lithodesma *n.* [Gk. *lithos*, stone; *desma*, bond] a small plate, shelly in nature, found in certain bivalves; *alt.* ossiculum.

lithodomous *a.* [Gk. *lithos*, stone; *domos*, house] living in rock holes or clefts.

lithogenous *a.* [Gk. *lithos*, stone; *-genēs*, producing] rock-forming, or rock-building, as certain corals.

lithophagous *a.* [Gk. *lithos*, stone; *phagein*, to eat] stone-eating, as birds; rock-burrowing, as some molluscs and sea urchins, *alt.* saxicavous.

lithophilous *a.* [Gk. *lithos*, stone; *philein*, to love] growing or living among stones or rocks; *alt.* saxicoline, saxatile.

lithophyll *n.* [Gk. *lithos*, stone; *phyllon*, leaf] a fossil leaf, or leaf impression.

lithophyte *n.* [Gk. *lithos*, stone; *phyton*, plant] plant growing on rocky ground.

lithosere *n.* [Gk. *lithos*, stone; L. *serere*, to put in a row] a plant succession originating on rock surfaces.

lithosols *n.plu.* [Gk. *lithos*, stone; L. *solum*, soil] soils which develop at high altitudes on resistant parent materials which withstand weathering and result in a humus-rich, shallow, stony soil; *cf.* regosols.

lithosphere *n.* [Gk. *lithos*, stone; *sphaira*, globe] the crust of the earth.

lithotomous *a.* [Gk. *lithos*, stone; *temnein*, to cut] stone-boring, as certain molluscs.

lithotroph *n.* [Gk. *lithos*, stone; *trophē*, nourishment] autotroph, *q.v.*

lithoxyle *n.* [Gk. *lithos*, stone; *xylon*, wood] laurinoxylon, *q.v.*

Litopterna, litopterns *n.*, *n.plu.* [Gk. *litos*, smooth; *pterna*, heel] an extinct order of South American placental mammals of the Palaeocene to Pleistocene that in some ways were similar to perissodactyls.

litoral *a.* [L. *litus*, seashore] growing or living at or near the seashore; *appl.* zone between high-and low-water marks; *appl.* zone of shallow water and bottom above compensation depth in lakes, *cf.* profundal; *appl.* cells, fixed macrophages, lining sinuses of reticular tissues and the wall of lymph channels; *alt.* littoral.

litter *n.* [M.E. *litere*, litter] partly decomposed plant residues on the surface of the soil, mainly in woodlands; animals produced at a single multiple birth.

littoral litoral, *q.v.*

Littré's glands [A. *Littré*, French surgeon] urethral mucous glands.

lituate *a.* [L. *lituus*, augur's staff] forked, with prongs curving outwards.

liver *n.* [A.S. *lifer*, liver] in vertebrates a gland attached to the gut, which secretes bile and is important in the storage and metabolism of foodstuffs; digestive gland of some invertebrates.

liver factor vitamin B_{12}, cobalamine, *q.v.*

liver-pancreas an organ in molluscs and crustaceans, combining functions of liver and pancreas; *alt.* hepatopancreas.

liverwort the common name for a member of the Hepaticae, *q.v.*

living fossil a living organism whose related organisms are known only as fossils and which

was itself thought to be extinct, such as the coelacanth and ginkgo.

lizards *see* Squamata.

loam *n.* [M.E. *lome*, loam, clay] a rich friable soil consisting of a fairly equal mixture of sand and silt and a smaller proportion of clay.

lobar *a.* [L.L. *lobus*, lobe] of or *pert.* a lobe.

Lobata *n.* [L.L. *lobus*, lobe] an order of Ctenophora of the class Tentaculata having 2 large oral lobes.

lobate *a.* [L.L. *lobus*, lobe] divided into lobes; *alt.* lobose.

lobe *n.* [L.L. *lobus*, from Gk. *lobos*, lobe] any rounded projection of an organ, *alt.* lobus; a flap-like structure on toes of certain birds.

lobed *a.* [Gk. *lobos*, lobe] having margin cut up into rounded divisions by incisions which reach less than halfway to midrib.

lobe-finned fishes a common name for the Crossopterygii, *q.v.*

lobopodia *n.plu.* [Gk. *lobos*, lobe; *pous*, foot] blunt pseudopodia of certain Protozoa.

Lobosa *n.* [Gk. *lobos*, lobe] an order of Protozoa with thick pseudopodia.

lobose *a.* [Gk. *lobos*, lobe] lobate, *q.v.*

lobular *a.* [Gk. *lobos*, lobe] like or *pert.* small lobes.

lobulate *a.* [Gk. *lobos*, lobe] divided into small lobes.

lobule, lobulus *n.* [Dim. of L.L. *lobus*, lobe] a small lobe or subdivision of a lobe; a polyhedral division of the liver in vertebrates; an adventitious outgrowth from a lichen thallus, often originating along the margins of lobes.

lobus *n.* [L.L. *lobus*, lobe] lobe, *q.v.*; portion of an organ, as of glands and brain, delimited by fissures or septa.

lobus osphradicus olfactory spindle, *q.v.*

local sign characteristic quality of a tactile or other sensation associated with point of stimulation.

localization *n.* [L. *localis*, local] determination of a position; restriction to a limited area; restriction of pairing and chiasma formation at pachytene to one part of the chromosome.

localization of function reference to different parts of brain as communicating centres of various senses.

localization of sensation identification on surface of body of exact spot affected.

locellate loculate, *q.v.*

locellus *n.* [L. *locellus*, from *locus*, place] a small compartment of an ovary.

loci *plu.* of locus; *cf.* compound loci.

lociation *n.* [L. *locus*, place] local differences in abundance or proportion of dominant species; *alt.* local faciation.

lock-and-key theory the theory that an enzyme forms an interlocking complex with its substrate like a lock and key; the theory that insect genitalia interlock so exactly from a structural point of view that even minor variations make copulation impossible.

locomotor rods hooked or knobbed rods for crawling, on ventral surface of certain Nematoda.

locular *a.* [L. *loculus*, little place] loculate, *q.v.*

loculate *pert.* loculi; containing, or composed of loculi; *alt.* locellate, locular.

locule loculus, *q.v.*

loculi *plu.* of loculus.

loculicidal *a.* [L. *loculus*, compartment; *caedere*, to cut] dehiscent dorsally down middle of carpels.

Loculoascomycetidae *n.* [L. *loculus*, compartment; Gk. *askos*, bag; *mykēs*, fungus] Loculomycetes, *q.v.*

Loculomycetes, Loculomycetidae *n.* [L. *loculus*, compartment; Gk. *mykēs*, fungus] in some classifications, a group of Ascomycetes in which the asci develop in cavities; *alt.* Loculoascomycetidae.

loculose *a.* [L. *loculus*, compartment] having several loculi; partitioned into small cavities.

loculus *n.* [L. *loculus*, compartment] a small chamber or cavity; cavity in stroma, containing asci; cavity of an ovary or of an anther; cavity of a septate sporangium or spore; cavity between septa in certain Coelenterata; chamber of foraminiferal shell; *alt.* locule; *plu.* loculi.

locus *n.* [L. *locus*, place] position of gene in the chromosome; location of a stimulus; hilum of starch grain; *plu.* loci.

locus coeruleus a pair of minute bodies lying just below the floor of the mid-brain, whose function is unknown, but which may be important in the control of a range of behaviours.

locusta *n.* [L. *locusta*, locust] spikelet of grasses; a locust.

locusts the common name for many members of the Orthoptera, *q.v.*

lodicule *n.* [L. *lodicula*, coverlet] a scale at base of ovary in grasses, supposed to represent part of a perianth; *alt.* glumellule, paleola, periphyllum, squamula.

lodix *n.* [L. *lodix*, blanket] a ventral sclerite of 7th abdominal segment, covering genital plate, in Lepidoptera.

loess *n.* [Ger. *lösen*, to loosen] a clay soil formed from wind-blown particles which is very fertile when mixed with humus; *alt.* löss.

logarithmic phase the 2nd stage in bacterial multiplication and the most rapid, with the increase tending to occur in geometric progression.

logotype *n.* [Gk. *logos*, word; *typos*, pattern] a genotype by subsequent designation, not originally described as such.

loma *n.* [Gk. *lōma*, hem] a thin membranous flap forming a fringe round an opening; fringe of toe in birds.

lomasomes *n.plu.* [Gk. *lōma*, hem; *sōma*, body] invaginations of cell membranes found in certain algae, higher plants, and fungi.

lomastome *a.* [Gk. *lōma*, hem; *stoma*, mouth] having margin of lip recurved or reflected.

loment lomentum, *q.v.*

lomentaceous *a.* [L. *lomentum*, bean meal] *pert.*, resembling, or having lomenta.

lomentum *n.* [L. *lomentum*, bean meal] a legume or pod constricted between seeds; *alt.* loment.

long-day *appl.* plants in which the flowering period is hastened by a relatively long photoperiod, ordinarily more than 12 hours, and a correspondingly short dark period, *cf.* short-day, day-neutral.

longicorn *a.* [L. *longus*, long; *cornu*, horn] having long antennae, *appl.* certain beetles.

longipennate *a.* [L. *longus*, long; *penna*, wing] having long wings, or long feathers.

longirostral, longirostrate *a.* [L. *longus*, long; *rostrum*, beak] with a long beak or rostrum.

longisection *n.* [L. *longus*, long; *sectio*, cut] section along or parallel to a longitudinal axis; *alt.* longitudinal section; *cf.* transection.

loop cell dome cell, *q.v.*

loph *n.* [Gk. *lophos*, crest] crest which may connect cones in teeth and so form a ridge.

Lophiiformes *n.* [*Lophius*, generic name; L. *forma*, shape] Pediculati, *q.v.*

lophiostomate *a.* [Gk. *lophion*, small crest; *stoma*, mouth] with crested conceptacle-opening.

lophium *n.* [Gk. *lophion*, small crest] a ridge community.

lophobranchiate *a.* [Gk. *lophos*, crest; *brangchia*, gills] with tufted gills.

lophocaltrops *n.* [Gk. *lophos*, crest; A.S. *coltraeppe*, kind of thistle] a sponge spicule with rays crested or branched; *alt.* lophotriaene.

lophocercal *a.* [Gk. *lophos*, crest; *kerkos*, tail] having a ray-less caudal fin like a ridge round end of vertebral column.

lophodont *a.* [Gk. *lophos*, crest; *odous*, tooth] having transverse ridges on the cheek-teeth grinding surface.

lophophorate bearing a lophophore.

lophophore *n.* [Gk. *lophos*, crest; *pherein*, to carry] a horseshoe-shaped, tentacle-supporting organ in Ectoprocta, Phoronida, and Brachiopoda.

lophoselenodont *a.* [Gk. *lophos*, crest; *selēnē*, moon; *odous*, tooth] having cheek teeth ridged with crescentic cuspid ridges on grinding surface.

lophosteon *n.* [Gk. *lophos*, crest; *osteon*, bone] the keel-ridge of a sternum.

lophotriaene *n.* [Gk. *lophos*, crest; *triaina*, trident] lophocaltrops, *q.v.*

lophotrichous *a.* [Gk. *lophos*, tuft; *thrix*, hair] having long whip-like flagella; with a tuft of flagella at one pole; *alt.* lophotrichate, lophotrichic.

loral *a.* [L. *lorum*, thong] *pert.* or situated at the lore.

lorate *a.* [L. *lorum*, thong] strap-shaped.

lore *n.* [L. *lorum*, thong] space between bill and eyes in birds; anterior part of gena, in some insects.

loreal *a.* [L. *loreus*, *pert.* thongs] *appl.* scale between nasal and preocular scales, as in Ophidia.

Lorenzini's ampullae ampullary receptors of rostrum in elasmobranchs, at present considered to be electroreceptors, but formerly thought to be sensitive to temperature or pressure.

lorica *n.* [L. *lorica*, corselet] a protective external case as in rotifers, protozoans, and diatoms; *alt.* cuirass.

Loricata *n.* [L. *lorica*, corselet] any of various groups of animals having a lorica.

loricate *a.* [L. *lorica*, corselet] covered with protective shell or scales.

lorication moment the occasion of deposition of silica or calcium carbonate for an entire skeleton or shell at one time, as of siliceous skeleton of radiolarians; *alt.* dictyotic moment.

lorulum *n.* [L. *dim.* of *lorum*, thong] the small strap-shaped and branched thallus of certain lichens.

lorum *n.* [L. *lorum*, thong] the piece of under jaw on which submentum lies in certain insects; dorsal plate protecting pedicle in spiders.

löss loess, *q.v.*

lotic *a.* [L. *lotum*, flowed over] *appl.* or *pert.* running water; living in brook or river; *cf.* lentic.

Lycoperdales

Louis, angle of [*A. Louis*, French surgeon] angulus Ludovici, *see* angulus.

loxodont *a.* [Gk. *loxos*, oblique; *odous*, tooth] having molar teeth with shallow grooves between the ridges.

LPH lipolytic hormone: lipotrophin, *q.v.*

LS antigen a complex antigen comprising an L (labile) component destroyed at 70°C and an S (stable) component that resists temperatures up to 90°C.

LSD lysergic acid diethylamide, *q.v.*

LTH luteotrophic hormone, *q.v.*

Luciae *n.* [L. *lux*, light] Pyrosomida, *q.v.*

luciferase *n.* [L. *lux*, light; *ferre*, to carry] an enzyme found in all luminescent organisms which appears to activate the substrate luciferin to cause luminescence, but the enzyme itself may be the light-emitting molecule; *alt.* photogenin.

luciferin *n.* [L. *lux*, light; *ferre*, to carry] intracellular or extracellular substance oxidized by luciferase, causing luminescence; *alt.* photphelein.

lucifugal, lucifugous *a.* [L. *lucifugus*, avoiding the light] shunning light, *appl.* fruit body of certain fungi; *alt.* photophobic; *cf.* lucipetal.

lucipetal *a.* [L. *lux*, light; *petere*, to seek] requiring light; *alt.* photophilous; *cf.* lucifugal.

Luganoiiformes *n.* [*Luganois*, generic name; L. *forma*, shape] an order of probably predaceous Chondrostei which have some characteristics of the Holostei.

lumbar *a.* [L. *lumbus*, loin] *pert.* or near the region of the loins, *appl.* artery, vein, vertebrae, plexus, gland, etc.

lumbarization *n.* [L. *lumbus*, loin] fusion of lumbar and sacral vertebrae.

lumbocostal *a.* [L. *lumbus*, loin; *costa*, rib] *pert.* loins and ribs, *appl.* arch, ligament.

lumbosacral *a.* [L. *lumbus*, loin; *sacrum*, sacred] *pert.* loins and sacrum, *appl.* nerve and trunk, plexus.

lumbrical *a.* [L. *lumbricus*, earthworm] lumbriciform, *q.v.*, *appl.* 4 small muscles in palm of hand and in sole of foot: lumbricales, *sing.* lumbricalis.

lumbriciform *a.* [L. *lumbricus*, earthworm; *forma*, shape] like a worm in appearance; *alt.* lumbricoid.

lumbricoid lumbriciform, *q.v.*

lumen *n.* [L. *lumen*, light] the cavity of a tubular part or organ; central cavity of a plant cell.

luminal *a.* [L. *lumen*, light] within or *pert.* a lumen.

luminate having a lumen.

luminescent organs specialized organs for the production of light, found in various plants and animals.

lumirhodopsin *n.* [L. *lumen*, light; Gk. *rhodon*, rose; *opsis*, sight] transient orange–red product of the bleaching of rhodopsin by light, which is converted into metarhodopsin.

lunar, lunate *a.* [L. *luna*, moon] somewhat crescent-shaped; *alt.* semilunar.

lunar bone a carpal bone, the middle of the 3 proximal carpals; *alt.* os lunare, lunatum, semilunar, intermedium.

lunatum lunar bone, *q.v.*

lunette *n.* [F. *lunettes*, spectacles] transparent lower eyelid of snakes.

lung *n.* [A.S. *lunge*, lung] the paired or single respiratory organ of air-breathing vertebrates.

lung book the respiratory organ of some arachnids, formed like a purse with numerous compartments; *alt.* book lung.

lungfish the common name for the Dipnoi, *q.v.*

lunula *n.* [L. *lunula*, small moon] *see* lunule.

lunular with crescent-shaped marking.

lunulate shaped like a small crescent.

lunule *n.* [L. *lunula*, small moon] a crescent-shaped structure or marking, *alt.* lunula; small crescentic sclerite, the frontal lunule, above antennal bases in certain Diptera; white opaque portion of nail near root.

lunulet *n.* [L. *lunula*, small moon] a small bundle.

lupulin *n.* [L. *lupus*, hop] the resinous glandular scales of hops; a yellow resinous powder on the flower scales of hops, containing humulone and lupulone.

lupuline, lupulinous *a.* [L. *lupus*, hop] resembling a group of hop flowers.

lupulon(e) *n.* [L. *lupus*, hop] a bitter antibiotic obtained from lupulin, which is effective against fungi and various bacteria.

luteal *a.* [L. *luteus*, orange–yellow] *pert.* or like cells of corpus luteum; *appl.* lutein and paralutein cells; *appl.* hormones: progesterone, relaxin.

lutein *n.* [L. *luteus*, orange–yellow] a yellow lipochrome, being a xanthophyll carotinoid, found in the leaves and petals of various plants, in algae especially of the classes Chlorophyceae and Phaeophyceae, in animals in egg yolk and in the corpus luteum, and which is isomeric with zeaxanthin.

lutein cells yellow cells found in the corpus luteum of the ovary and formed from either the follicular cells or the lutein cells of the theca interna.

luteination, luteinization *n.* [L. *luteus*, orange–yellow] the formation of corpus luteum.

luteinizing hormone (LH) a gonadotrophic pituitary hormone which stimulates theca-lutein cell formation and interstitial cells of testis; *alt.* ICSH, prolan B, metakentrin.

luteofuscous *a.* [L. *luteus*, orange–yellow; *fuscus*, dusky] darkish yellow.

luteol *n.* [L. *luteus*, orange–yellow] a carotenoid in leaves.

luteosterone progesterone, *q.v.*; progestin, *q.v.*

luteotrop(h)ic *a.* [L. *luteus*, orange–yellow; Gk. *trophē*, nourishment] *appl.* hormone: prolactin, *q.v.*, *alt.* LTH.

luteotrop(h)in prolactin, *q.v.*

lutin *n.* [L. *luteus*, orange–yellow] progesterone, *q.v.*

Luys, nucleus of corpus subthalamicum of hypothalamus.

lyases *n.plu.* [Gk. *lyein*, to loose] a group of enzymes with split C–C, C–O, C–N, and other bonds by oxidation or hydrolysis, having 2 substrates concerned with one reaction direction, but only one with the opposite direction; EC group 4.

lychnidiate *a.* [Gk. *lychnidion*, small lamp] luminous.

lycopene, lycopin *n.* [L.L. *lycopersicum*, tomato, from Gk. *lykopersikon*] the red carotinoid pigment of tomatoes and other fruits, and also found in certain classes of algae.

Lycoperdales *n.* [*Lycoperdon*, generic name] an order of Gasteromycetes, commonly called puffballs, having basidia enclosed in sterile tissues of a fruit body which sometimes ruptures when the spores are ripe.

Lycophyta *n.* [*Lycopodium*, generic name; Gk. *phyton*, plant] in some classifications, a division of the Tracheophyta equivalent to the Lycopsida, *q.v.*

Lycopodiales *n.* [*Lycopodium*, generic name] an order of homosporous Lycopsida, commonly called club mosses, having sporophylls usually differing from leaves.

Lycopodiatae *n.* [*Lycopodium*, generic name] Lycopsida, *q.v.*

Lycopodineae *n.* [*Lycopodium*, generic name] a class of the Tracheophyta included in or equivalent to the Lycopsida, *q.v.*

Lycopodiopsida *n.* [*Lycopodium*, generic name; Gk. *opsis*, appearance] Lycopsida, *q.v.*

Lycopsida *n.* [*Lycopodium*, generic name; Gk. *opsis*, appearance] a class of pteridophytes (in some classifications a subdivision of the Tracheophyta), characterized by a sporophyte with roots, stems, and small leaves, microphylls, arranged spirally on the stem, with the sporangia solitary and borne on or associated with a sporophyll; they are commonly known as club mosses, although this term is sometimes restricted to the Lycopodiales and/or Selaginellales; *alt.* Lycopodiatae, Lycopodiopsida, Lepidophyta.

Lyginopteridales *n.* [*Lyginopteris*, generic name] an order of pteridosperms including many important Carboniferous members; *alt.* Cycadofilicales, Pteridospermales.

Lyginopteridatae *n.* [*Lyginopteris*, generic name] in some classifications, Pteridospermae, *q.v.*

lygophil *a.* [Gk. *lygē*, shadow; *philos*, loving] prefering shade or darkness.

lymph *n.* [L. *lympha*, water] an alkaline colourless fluid, similar to blood plasma, contained in lymphatic vessels or forming tissue fluid.

lymphatic *a.* [L. *lympha*, water] *pert.* or conveying lymph.

lymph gland lymph node, *q.v.*

lymph heart contractile expansion of a lymph vessel where it opens into a vein, in many vertebrates.

lymph nodes structures made of lymphoid tissue lying in the lymphatic vessels, whose functions are to make lymphocytes and to filter foreign bodies such as bacteria from lymph and prevent them entering blood system; *alt.* lymph gland.

lymphoblast *n.* [L. *lympha*, water; Gk. *blastos*, bud] a primitive lymphocyte, normally found in small numbers in the germinal centres in lymphatic tissue.

lymphocyte *n.* [L. *lympha*, water; Gk. *kytos*, hollow] any of various small mononuclear leucocytes of blood and lymph which are responsible for the synthesis of antibodies; *alt.* achroacyte.

lymphogenic *a.* [L. *lympha*, water; Gk. *-genēs*, producing] produced in lymph glands.

lymphogenous *a.* [L. *lympha*, water; Gk. *-genēs*, producing] lymph-forming.

lymphoid *a.* [L. *lympha*, water; Gk. *eidos*, form] *appl.* tissue in which lymphocytes are produced, consisting of aggregations of lymphocytes either as discrete organs such as tonsils and lymph nodes, or as diffuse aggregations in connective tissue; *alt.* adenoid, lymphatic.

lymphoidocyte *n.* [L. *lympha*, water; Gk. *eidos*, form; *kytos*, hollow] haemocytoblast, *q.v.*

lymphomonocyte *n.* [L. *lympha*, water; Gk. *monos*, single; *kytos*, hollow] a large mononuclear leucocyte.

lymphomyelocyte *n.* [L. *lympha*, water; Gk. *myelos*, marrow; *kytos*, hollow] myeloblast, *q.v.*

lyochromes *n.plu.* [Gk. *lyein*, to loose; *chrōma*, colour] flavins, *q.v.*

lyocytosis *n.* [Gk. *lyein*, to loose; *kytos*, hollow] histolysis by extracellular digestion, as in insect metamorphosis.

lyoenzyme *n.* [Gk. *lyein*, to loose; *en*, in; *zymē*, leaven] any of various intracellular enzymes dissolved directly in the cytoplasm and therefore relatively easy to extract.

Lyomeri *n.* [Gk. *lyein*, to dissolve; *meris*, part] Saccopharyngiformes, *q.v.*

Lyon(n)et's glands [*P. Lyon(n)net*, Dutch entomologist] paired accessory silk glands in lepidopterous larvae; *alt.* Filippi's glands.

lyophil *a.* [Gk. *lyein*, to loose; *philos*, loving] *appl.* solutions which, after evaporation to dryness, go readily into solution again on addition of fluid; *cf.* lyophobe.

lyophobe *a.* [Gk. *lyein*, to loose; *phobos*, fear] *appl.* solutions which, after evaporation to dryness, remain as a solid; *cf.* lyophil.

lyosphere *n.* [Gk. *lyein*, to loose; *sphaira*, globe] a thin film of water surrounding a colloidal particle.

lyotropic *a.* [Gk. *lyein*, to loose; *tropē*, turn] *appl.* solutions which are dependent on changes in the solvent itself.

lyra *n.* [Gk. *lyra*, lyre] triangular lamina or psalterium joining lateral part of fornix, marked with fibres as a lyre; a lyrate pattern as on some bones; a series of chitinous rods forming part of the stridulating organ in certain spiders.

lyrate *a.* [Gk. *lyra*, lyre] lyre-shaped, *appl.* certain leaves.

lyriform *a.* [L. *lyra*, lyre; *forma*, shape] lyre-shaped, *appl.* a tarsal sensory organ, the lyra, in some arachnids.

lys lysine, *q.v.*

lysactinic *a.* [Gk. *lysis*, loosing; *aktis*, ray] having podia limited to lower half of body instead of continued to apical plates in Asteroidea; *cf.* desmactinic.

Lysenkoism the doctrine promoted by Trofin Lysenko in the USSR, being a form of Lamarckism, i.e. based on the inheritance of acquired characteristics.

lysergic acid diethlylamide (LSD) a powerful hallucinogenic drug, whose effects sometimes mimic the symptoms of some psychotic conditions, especially schizophrenia.

lysigenic, lysigenous *a.* [Gk. *lysis*, loosing; *-genēs*, producing] *appl.* formation of tissue cavities caused by degeneration and breaking down of cell walls in centre of mass; *alt.* lysogenous, rhexigenous.

lysin *n.* [Gk. *lysis*, loosing] any substance capable of causing dissolution or lysis of cells or bacteria; *alt.* amboceptor.

lysine *n.* [Gk. *lysis*, loosing] an essential amino acid, diaminocaproic acid, responsible for the base-neutralizing powers of proteins due to its free -NH_2 group; abbreviated to lys.

lysis *n.* [Gk. *lysis*, loosing] breaking down or dissolution of compounds or cells.

lysogenesis *n.* [Gk. *lysis*, loosing; *genesis*, descent] the action of lysins.

lysogenic *a.* [Gk. *lysis*, loosing; *gennaein*, to produce] producing lysis; *appl.* bacteria carrying temperate phage; *appl.* clone formed by the associated multiplication of bacterium and phage; *appl.* virus which is a temperate phage.

lysogenous lysigenous, *q.v.*

lysogeny *n.* [Gk. *lysis*, loosing; *gennaein*, to produce] the situation where the genetic material of a virus is integrated into that of the host.

Lysorophia *n.* [*Lysorophus*, generic name] an order of lepospondyl amphibians, possibly degenerate microsaurs, having reduced limbs.

lysosomes, lysozomes *n.plu.* [Gk. *lysis*, loosing; *sōma*, body] nearly spherical organelles, smaller than mitochondria, which are thought to isolate autolysing enzymes of the cell, and so prevent self-digestion; *alt.* light mitochondria, mitochrondria B.

lysozyme *n.* [Gk. *lysis*, loosing; *zymē*, leaven] an enzyme found in mammalian tissue secretions, such as tears, white of egg, and some micro-organisms, and having mucolytic and bacteriolytic properties; EC 3.2.1.17; *r.n.* lysozyme.

Lyssacina *n.* [Gk. *lyssa*, madness; *akis*, needle] in some classifications, an order of sponges whose spicules are not fused, or are only partially fused; *cf.* Dictyonina.

lytic *a.* [Gk. *lyein*, to break down] *pert.* lysis; *pert.* a lysin.

lytic virus a virus which multiplies inside a host cell and then causes its lysis.

lytta *n.* [Gk. *lytta*, madness] a vermiform structure of muscle, fatty and connective tissue, or cartilage, under the tongue of carnivores such as dog, *alt.* worm; cantharis, a blister beetle.

M

macerate *v.* [L. *macerare*, to soften] to wear away or to isolate parts of a tissue or organ; to soften and wear away by digestion or other means. *n.* A softened tissue in which cells have separated.

maceration *n.* [L. *macerare*, to soften] the production of a macerate.

Machaeridea *n.* [L. *machaera*, dagger] an extinct class of M. Cambrian echinoderms of the subphylum Pelmatozoa of doubtful affinities, which have formerly been classified as fossil Cirripedia.

machopolyp *n.* [Gk. *machē*, fight; *polys*, many; *pous*, foot] a nematophore of certain Hydrozoa, provided with cnidoblasts or adhesive globules.

macrander *n.* [Gk. *makros*, large; *anēr*, male] a large male plant.

macrandrous *a.* [Gk. *makros*, large; *anēr*, male] having large male plants or elements.

macraner *n.* [Gk. *makros*, large; *anēr*, male] male ant of unusually large size.

macrergate *n.* [Gk. *makros*, large; *ergatēs*, worker] worker ant of unusually large size.

macro- *see also* mega-; *cf.* micro-.

macrobiota *n.* [Gk. *makros*, large; *bios*, life] all soil organisms of larger size than those of the mesobiota including plant roots, larger insects, earthworms, etc.; *cf.* mesobiota, microbiota.

macrobiotic *a.* [Gk. *makros*, long; *biōtikos*, lively] long-lived; life-prolonging; *n.* macrobios.

macroblast *n.* [Gk. *makros*, large; *blastos*, bud] a large cell or corpuscle; a young normoblast.

macrocarpous *a.* [Gk. *makros*, large; *karpos*, fruit] producing large fruit.

macrocentrosome *n.* [Gk. *makros*, large; *kentron*, centre; *sōma*, body] centrosphere, *q.v.*

macrocephalous *a.* [Gk. *makros*, large; *kephalē*, head] having the cotyledons thickened; bigheaded.

macrochaeta *n.* [Gk. *makros*, large; *chaitē*, hair] a large bristle, as on body of certain insects.

macrochromosomes *n.plu.* [Gk. *makros*, large; *chrōma*, colour; *sōma*, body] relatively large chromosomes in a nucleus, and usually on periphery of equatorial plate during metaphase; *cf.* microchromosomes.

macrocnemic *a.* [Gk. *makros*, large; *knēmē*, tibia] *appl.* Zoantharia having the 6th protocneme or primary pair of mesenteries perfect.

macroconidium *n.* [Gk. *makros*, large; *konis*, dust; *idion*, *dim.*] a large asexual spore or conidium.

macroconjugant *n.* [Gk. *makros*, large; L. *conjugare*, to unite] the larger individual of a conjugating pair.

macrocyclic *a.* [Gk. *makros*, large; *kyklos*, circle] having a complete or a long cycle; with both gametophyte and sporophyte stages; *cf.* microcyclic.

macrocyst *n.* [Gk. *makros*, large; *kystis*, bladder] a large reproductive cell of certain fungi; a large cyst or case, as for spores.

macrocystidium *n.* [Gk. *makros*, large; *kystis*, bladder; *idion*, *dim.*] a long cystidium-like structure in some Gasteromycetes.

macrocytase *n.* [Gk. *makros*, large; *kytos*, hollow] the enzyme of macrophages or endothelial cells.

macrodactylous *a.* [Gk. *makros*, long; *daktylos*, finger] with long digits.

Macrodasyoidea *n.* [*Macrodasys*, generic name; Gk. *eidos*, form] an order of marine Gastrotricha that have a long worm-like body and move by cilia or looping.

macrodont *a.* [Gk. *makros*, large; *odous*, tooth] with large teeth.

macro-elements macronutrients, *q.v.*

macroevolution *n.* [Gk. *makros*, large; L. *evolvere*, to unroll] evolutionary processes extending through geological eras; large-scale evolution of new genera and species owing to mutations resulting in marked changes in chromosome pattern and reaction system; *cf.* microevolution.

macrofauna *n.* [Gk. *makros*, large; L. *faunus*, god of woods] animals whose length is measured in centimetres, rather than microscopic units.

macrofibrils *n.plu.* [Gk. *makros*, large; L. *fibrilla*, small fibre] fibrils of cellulose made of bundles of microfibrils, arranged scattered in the primary cell wall and more ordered in the secondary wall.

macrogamete *n.* [Gk. *makros*, large; *gametēs*, spouse] the larger of 2 gametes in a heterogamous organism, usually considered equivalent to the ovum or female gamete; *alt.* megagamete.

macrogametocyte *n.* [Gk. *makros*, large; *gametēs*, spouse; *kytos*, hollow] the mother cell of a macrogamete, especially in Protista; *alt.* mega-gametocyte.

macrogamy *n.* [Gk. *makros*, large; *gamos*, marriage] hologamy, *q.v.*

macroglia *n.plu.* [Gk. *makros*, large; *glia*, glue] a general term for certain neuroglia cells which may apply to astrocytes and oligodendroglia, or also to ependyma.

macroglossate *a.* [Gk. *makros*, large; *glōssa*, tongue] furnished with a large tongue.

macrognathic *a.* [Gk. *makros*, large; *gnathos*, jaw] having specially developed jaws.

macrogonidium *n.* [Gk. *makros*, large; *gonē*, generation; *idion*, dim.] megalogonidium, *q.v.*

macrogyne *n.* [Gk. *makros*, large; *gynē*, woman] female ant of unusually large size.

macrolecithal megalolecithal, *q.v.*

macroleucocyte *n.* [Gk. *makros*, large; *leukos*, white; *kytos*, hollow] a chromophil leucocyte, developed from a proleucocyte.

macrolymphocyte *n.* [Gk. *makros*, large; L. *lympha*, water; Gk. *kytos*, hollow] any large lymphocyte.

macromere *n.* [Gk. *makros*, large; *meros*, part] in cleavage of telolecithal eggs, a larger cell of lower hemisphere; *alt.* megamere.

macromerozoite *n.* [Gk. *makros*, large; *meros*, part; *zōon*, animal] one of many divisions produced by macroschizont stage of Sporozoa.

macromesentery *n.* [Gk. *makros*, large; *mesos*, middle; *enteron*, gut] one of the larger complete mesenteries of Anthozoa.

macromitosome *n.* [Gk. *makros*, large; *mitos*, thread; *sōma*, body] the paranucleus, as in Lepidoptera.

macromolecule *n.* [Gk. *makros*, large; F. *molécule*, small particle] a very large molecule such as protein, nucleic acid, etc.

macromutation *n.* [Gk. *makros*, large; L. *mutare*, to change] simultaneous mutation of a number of different characters.

macromyelon *n.* [Gk. *makros*, long; *myelos*, marrow] the medulla oblongata, *q.v.*

macronotal *a.* [Gk. *makros*, large; *nōton*, back] with large thorax, as a queen ant.

macront *n.* [Gk. *makros*, large; *on*, being] the larger of 2 sets of cells formed after schizogony in Neosporidia, the macront giving rise to macrogametes.

macronucieocyte *n.* [Gk. *makros*, large; L. *nucleus*, kernel; Gk. *kytos*, hollow] a leucocyte having a relatively large nucleus; chromophil leucocyte of insects.

macronucleus *n.* [Gk. *makros*, large; L. *nucleus*, kernel] meganucleus, *q.v.*

macronutrients elements required and occurring in relatively large quantities as natural constituents of living organisms or tissues; *alt.* major elements, macro-elements; *cf.* trace elements.

macrophage *n.* [Gk. *makros*, large; *phagein*, to eat] a large phagocytic cell, fixed or wandering; a large mononuclear leucocyte; a histiocyte, clasmatocyte, pericyte, endotheliocyte, etc.

macrophagous *a.* [Gk. *makros*, large; *phagein*, to eat] feeding on relatively large masses of food; *cf.* microphagous.

macrophanerophytes *n.plu.* [Gk. *makros*, large; *phaneros*, manifest; *phyton*, plant] trees.

macrophyll *n.* [Gk. *makros*, large; *phyllon*, leaf] megaphyll, *q.v.*

macrophyllous *a.* [Gk. *makros*, large; *phyllon*, leaf] megaphyllous, *q.v.*

macroplankton *n.* [Gk. *makros*, large, *plankton*, wandering] the larger organisms drifting with the surrounding water, as jellyfish, sargassum, etc.; *alt.* megaplankton.

macropodous *a.* [Gk. *makros*, long; *pous*, foot] having a long stalk, as a leaf or leaflet; having hypocotyl large in proportion to rest of embryo; long-footed.

macropterous *a.* [Gk. *makros*, long; *pteron*, wing or fin] with unusually large fins or wings; fully winged, *cf.* brachypterous.

macropyrenic *a.* [Gk. *makros*, large; *pyrēn*, fruit stone] with nuclei markedly larger than average for the species or other group. *n.* A macropyrenic individual.

Macroscelidea *n.* [*Macroscelides*, generic name] an order of placental mammals including the elephant shrews.

macroschizogony *n.* [Gk. *makros*, large; *schizein*, to cleave; *gonē*, generation] method of multiplication of macroschizonts; schizogony giving rise to large merozoites.

macroschizont *n.* [Gk. *makros*, large; *schizein*, to cleave; *on*, being] stage in life cycle of certain Haemosporidia developed from sporozoite, and giving rise to macromerozoites.

macrosclere megasclere, *q.v.*

macrosclereids *n.plu.* [Gk. *makros*, large; *sklēros*, hard; *eidos*, form] relatively large columnar sclereids, as in coat of certain seeds of Leguminosae; *alt.* Malpighian cells.

macroscopic *a.* [Gk. *makros*, large; *skopein*, to view] visible by the naked eye; *alt.* megascopic.

macrosepalous *a.* [Gk. *makros*, large; F. *sépale*, sepal] with specially large sepals.

macroseptum *n.* [Gk. *makros*, large; L. *septum*, partition] a primary or perfect septum of Anthozoa.

macrosiphon *n.* [Gk. *makros*, large; *siphōn*, tube] large internal siphon of certain cephalopods.

macrosmatic *a.* [Gk. *makros*, large; *osmē*, smell] with well-developed sense of smell.

macrosomatous *a.* [Gk. *makros*, large; *sōma*, body] possessing abnormally large body.

macrosome *n.* [Gk. *makros*, large; *sōma*, body] a large alveolar sphere or granule in protoplasm; *alt.* megasome.

macrospecies *n.* [Gk. *makros*, large; L. *species*, particular kind] a large polymorphic species usually with several to many subdivisions; *cf.* Linnaean species, microspecies.

macrospheric *a.* [Gk. *makros*, large; *sphaira*, globe] megalospheric, *q.v.*

macrosplanchnic *a.* [Gk. *makros*, large; *splanchnon*, entrail] large-bodied and short-legged.

macrosporangiophore *n.* [Gk. *makros*, large; *sporos*, seed; *anggeion*, vessel; *pherein*, to bear] a structure bearing a macrosporangium.

macrosporangium *n.* [Gk. *makros*, large; *sporos*, seed; *anggeion*, vessel] megasporangium, *q.v.*

macrospore *n.* [Gk. *makros*, large; *sporos*, seed] a large anisospore or gamete of Sarcodina; megaspore, *q.v.*

macrosporogenesis *n.* [Gk. *makros*, large; *sporos*, seed; *genesis*, descent] production of macrospores.

macrosporophore n. [Gk. makros, large; sporos, seed; pherein, to bear] a leafy megasporophyll.

macrosporophyll n. [Gk. makros, large; sporos, seed; phyllon, leaf] megasporophyll, q.v.

macrosporozoite n. [Gk. makros, large; sporos, seed; zōon, animal] a large endogamous sporozoite of Sporozoa.

macrostomatous a. [Gk. makros, large; stoma, mouth] with very large mouth.

macrostylospore n. [Gk. makros, large; stylos, pillar; sporos, seed] a large spore-like stalked body.

macrostylous a. [Gk. makros, long; stylos, pillar] with long styles.

macrosymbiote n. [Gk. makros, large; symbiōtēs, companion] the larger of 2 symbiotic organisms; alt. macrosymbiont.

macrotherm(ophyte) n. [Gk. makros, large; thermē, heat] megatherm, q.v.

macrotous a. [Gk. makros, large; ous, ear] with large ears.

macrotrichia n.plu. [Gk. makros, large; thrix, hair] the larger setae on body or wings of insects.

macrotype n. [Gk. makros, large; typos, a type] a modified arrangement of mesenteries containing more macromesenteries than normal microtype, in Anthozoa.

macrozoogonidium n. [Gk. makros, large; zōon, generation; idion, dim.] a large zoogonidium.

macrozoospore n. [Gk. makros, large; zōon, animal; sporos, seed] large motile spore.

macruric a. [Gk. makros, long; oura, tail] longtailed; alt. macrural, macrurous.

macula n. [L. macula, spot] a spot or patch of colour; a small pit or depression; a tubercle; neuroepithelial area of membranous labyrinth, as in sacculus, utriculus, ampullae, and cochlear duct; plu. maculae.

macula adherans desmosome, q.v.

macula cribrosa area on wall of vestibule of ear, perforated for passage of auditory nerve filaments.

maculae plu. of macula.

macula germinitiva the germinal spot, nucleolus of an ovum.

macula lutea an oval yellowish area in centre of posterior part of retina at point of most perfect vision; alt. yellow spot.

macular a. [L. macula, spot] pert. a macula; pert. macula lutea.

maculate, maculiferous, maculose a. [L. macula, spot] spotted.

maculation n. [L. maculare, to spot] the arrangement of spots on a plant or an animal.

madescent a. [L. madescere, to become wet] becoming moist; slightly moist.

madid a. [L. madidus, moist] moist; wet.

Madreporaria n. [L. mater, mother; Gk. pōros, friable stone] Scleractinia, q.v.

madrepore n. [F. madrépore, from L. mater, mother; Gk. pōros, friable stone] a branching, stony, reef-building coral of the order Madreporaria.

madreporic a. [F. madrépore, madrepore] pert. a madrepore or madreporite.

madreporic canal stone canal, q.v.

madreporite n. [F. madrépore, madrepore] in echinoderms, a perforated plate at the end of the stone canal of the water vascular system, that may be internal or inconspicuous, as in holothurians, or a conspicuous structure near the centre of the aboral surface as in starfish; alt. sieve plate.

Magendie's foramen [F. Magendie, French physiologist] in brain, median aperture in roof of 4th ventricle, connecting the latter with subarachnoid cavities; alt. metapore.

maggot n. [M.E. magot, grub] the worm-like insect larva, without appendages or distinct head, as that of some Diptera such as the blowfly.

magnetotropism n. [Gk. lithos magnētis, magnet; tropē, turning] a tropism in response to lines of magnetic force.

Magnoliales n. [Magnolia, generic name] a primitive order of dicots (Magnoliidae).

Magnoliatae n. [Magnolia, generic name] in some classifications, Dicotyledones, q.v.; alt. Magnoliopsida.

Magnoliidae n. [Magnolia, generic name] a subclass of dicots having great diversity and many primitive characteristics, such as an apocarpous gynaecium and spiral arrangement of floral parts which are in large and indefinite numbers; alt. Polycarpicae.

Magnoliophytina n. [Magnolia, generic name; Gk. phyton, plant] in some classifications, Angiospermae, q.v.

Magnoliopsida Magnoliatae, q.v.

magnum capitatum, q.v.

maintenance behaviour behaviour which allows an animal to carry out its day-to-day activities such as search for food, mating, reproduction, or avoidance of extreme environments.

maiosis meiosis, q.v.

major element macronutrient, q.v.

major gene a gene having obvious pronounced phenotypic effects as distinguished from modifying genes.

mala n. [L. mala, cheek] part of maxilla of some insects; part of mandible of certain myriapods; part of exterior of lower jaw of birds; cheek; malar bone.

Malacocotylea n. [Gk. malakos, soft; kotylē, cup] Digenea, q.v.

malacoid a. [Gk. malakos, soft; eidos, form] soft in texture.

malacology n. [Gk. malakos, soft; logos, discourse] the study of molluscs.

malacophilous a. [Gk. malakos, soft; philein, to love] pollinated by agency of gastropods; n. malacophily.

malacophyllous a. [Gk. malakos, soft; phyllon, leaf] with soft or fleshy leaves.

malacopterous a. [Gk. malakos, soft; pteron, wing or fin] soft-finned.

malacospermous a. [Gk. malakos, soft; sperma, seed] having the seeds covered by a soft coat.

Malacostraca, malacostracans n., n.plu. [Gk. malakos, soft; ostrakon, shell] a subclass of crustaceans including crabs, lobsters, shrimps, and woodlice having carapace developed to a variable extent.

malacostracous a. [Gk. malakos, soft; ostrakon, shell] soft-shelled.

Malagasy appl. or pert. the zoogeographical subregion including Madagascar and adjacent islands.

malar a. [L. mala, cheek bone] pert. mala; pert, or in region of cheek; alt. zygomatic. n. The jugal bone.

malaxation

malaxation n. [Gk. malassein, to soften] compression of mandibles, or chewing, as by wasps.

Malayan appl. and pert. the zoogeographical subregion including Malaya, Indonesia west of Wallace's line, and the Philippines.

male n. [L. mas, male] an individual whose sex organs contain only male gametes; symbol ♂.

malella n. [L.L. dim. of L. mala, jaw] distal toothed process of outer stipes of deutomala in certain Myriapoda.

male pronucleus nucleus of spermatozoon; alt. sperm nucleus.

malic acid [L. malum, apple] an acid of the Krebs' cycle and various other metabolic cycles, which can be oxidized to oxaloacetic acid by malic dehydrogenases using NAD as a hydrogen acceptor EC 1.1.1.37, 38, 39, or NADP, EC 1.1.1.40, and also gives the taste to some plants such as apples.

malic enzyme collectively, the malic dehydrogenases, see malic acid.

malleate a. [L. malleus, hammer] hammer-shaped; appl. a type of trophi of rotifer gizzard.

malleoincudal a. [L. malleus, hammer; incus, anvil] pert. malleus and incus of ear.

malleolar n. [L. dim. of malleus, hammer] the vestigal fibula of ruminants. a. Pert. or in region of malleolus, appl. arteries, folds, sulcus.

malleolus n. [L. dim. of malleus, hammer] medial and lateral malleolus, lower extremity prolongations of tibia and fibula respectively; one of the club- or racket-shaped appendages on basal segments of hindlegs of Solpugida.

malleoramate a. [L. malleus, hammer; ramus, branch] appl. type of trophi with looped manubrium and toothed incus in rotifer gizzard.

malleus n. [L. malleus, hammer] a part of rotifer mastax or gizzard; auditory ossicle attached to tympanum of mammals, alt. hammer, plectrum; one of the Weberian ossicles of fishes; any of various other hammer-shaped structures.

mallochorion n. [Gk. mallos, wool; chorion, skin] the primitive mammalian chorion.

Mallophaga n. [Gk. mallos, wood; phagein, to eat] an order of insects, commonly called bird lice or biting lice, which are ectoparastic usually on birds, and have biting mouth-parts, secondarily no wings, and slight metamorphosis, sometimes included with the Anoplura in the Phthiraptera.

malloplacenta n. [Gk. mallos, wool; L. placenta, flat cake] non-deciduate placenta with villi evenly distributed, as in cetaceans and some ungulates.

malonic acid a specific metabolic reversible inhibitor of succinic dehydrogenase which has been used in investigations of the Krebs' cycle, and also occurs as an end-product of metabolism in some plants, such as beetroot.

Malpighian a. [M. Malpighi, Italian anatomist] discovered by or named after Malpighi.

Malpighian body or corpuscle in spleen, a nodular mass of lymphoid tissue ensheathing the smaller arteries, alt. ellipsoid, splenic nodule; in kidney, a glomerulus of convoluted capillary blood vessels enclosed in a dilatation of uriniferous tubule, the Bowman's capsule, alt. renal corpuscle.

Malpighian cell macroscereid, q.v.

Malpighian layer in skin of vertebrates, the innermost layer of epidermal cells, which undergo mitosis and sometimes contain melanin, alt. stratum germinativum, germinative layer; sometimes considered to be made up of 2 layers, the stratum germinativum and an overlying stratum spinosum, whose cells contain filaments of keratin; alt. rete Malpighii, rete mucosum.

Malpighian pyramids conical structures in medulla of kidney; alt. medullary pyramids.

Malpighian tubules thread-like excretory tubes leading into posterior part of gut of insects, arachnids, and myriapods.

maltase n. [A.S. mealt, malt] the enzyme which hydrolyses non-reducing α-D-glucose residues with the release of glucose, e.g. hydrolyses maltose to glucose; EC 3.2.1.20; r.n. α-D-glucosidase.

maltobiose maltose, q.v.

maltodextrins n.plu. [A.S. mealt, malt; L. dexter, right-handed] polysaccharides formed during incomplete hydrolysis of starch.

maltose n. [A.S. mealt, malt] a reducing sugar, a disaccharide produced by hydrolysis of starch by amylase, which does not occur widely in the free state but is produced by germinating barley, and is hydrolysed to glucose by the enzyme maltase and by dilute acids; alt. malt sugar, maltobiose.

malt sugar maltose, q.v.

Malvales n. [Malva, generic name] an order of typically woody dicots of the Archichlamydeae, Polypetalae, or Dilleniidae, having a valvate calyx but convolute corolla in bud; alt. Columniferae.

malvidin n. [L. malva, mallow] a mauvish anthocyanin pigment.

mamelon n. [F. mamelon, from L. mamilla, nipple] small pimple-like structure in centre of tubercle of echinoid interambulacral plate; papilla forming nucellus in cycads.

mamilla n. [L. mamilla, nipple] a nipple; a nipple-shaped structure; alt. mammilla.

mamillary bodies corpora mamillaria, q.v.

mamillary process or tubercle superior tubercle connected with transverse process of lower thoracic vertebrae; alt. metapophysis.

mamillate a. [L. mamilla, nipple] studded with small protuberances.

mamillation formation or presence of nipple-like protuberances on a surface.

mamillothalamic tract bundles of Vicq d-Azyr, q.v.

mamma n. [L. mamma, breast] milk-secreting organ of female mammals; alt. mammary gland.

Mammalia, mammals n., n.plu. [L. mamma, breast] a class of vertebrates having the body covered with hair, possessing mammary glands, and a 4-chambered heart, and with the female that suckles her young.

mammal-like reptiles the common name for the Synapsida, q.v.

mammalogy n. [L. mamma, breast; Gk. logos, discourse] the study of mammals.

mammary a. [L. mamma, breast] pert. the breast, appl. arteries, veins, tubules, etc.

mammary gland mamma, q.v.

mammiferous a. [L. mamma, breast; ferre, to bear] developing mammae; milk-secreting; alt. mammalian.

mammiform a. [L. mamma, breast; forma, shape] breast-shaped; appl. pileus of certain fungi.

mammilla mamilla, q.v.

mammogenic a. [L. mamma, breast; Gk. gen-

naein, to produce] *appl*. pituitary hormone complex which promotes growth of the lobe-alveolar and duct systems of the mammary gland.

mammose *a*. [L. *mammōsus*, full-breasted] shaped like a breast; having breast-shaped protuberances.

manchette *n*. [F. *manchette*, cuff] membrane enveloping the cytoplasm surrounding the axial filament of a spermatid; in agarics, the frill on the stipe left attached on expansion of pileus, and which at first forms a covering of the hymenium, *alt*. armilla, superior annulus.

mandible *n*. [L. *mandibulum*, jaw] the lower jaw of vertebrates, either a single bone or composed of several; a paired mouth appendage of arthropods; *alt*. mandibulum, submaxilla.

mandibular *a*. [L. *mandibulum*, jaw] *pert*. the lower jaw, *appl*. arch, canal, foramen, nerve, notch, *alt*. submaxillary; *appl*. fossa: glenoid fossa, *q.v.*

Mandibulata *n*. [L. *mandibulum*, jaw] a subphylum of arthropods in which the body is of variable forms, but the head appendages consist of antennae, mandibles, and maxillae.

mandibulate *a*. [L. *mandibulum*, jaw] having a lower jaw; having functional jaws; having mandibles.

mandibuliform *a*. [L. *mandibulum*, jaw; *forma*, shape] resembling, or used as a mandible, *appl*. certain insect maxillae.

mandibulohyoid *a*. [L. *mandibulum*, jaw; Gk. *hyoeidēs*, Y-shaped] in region of mandible and hyoid.

mandibulomaxillary *a*. [L. *mandibulum*, jaw; *maxilla*, jaw] *pert*. maxillae and mandibles of arthropods.

mandibulum mandible, *q.v.*

manducation *n*. [L. *manducare*, to chew] mastication, *q.v.*

manicate *a*. [L. *manicatus*, sleeved] covered with entangled hairs or matted scales.

manna *n*. [Gk. *manna*, manna] hardened exudation of bark of certain trees such as the European flowering ash, *Fraxinus ornus*; a similar substance in other plants such as tamarisk, where its production is caused by scale insects; honey dew secreted by certain scale insects. *a*. *Appl*. any of various lichens used as food by man and animals.

mannan *n*. [Gk. *manna*, manna] a polysaccharide made of condensed mannose units, as found in the cell walls of some plants; *alt*. mannosan.

mannitol *n*. [Gk. *manna*, manna; alcho*l*] a polyhydroxy alcohol derived from mannose or fructose, found in many plants, including some algae.

mannoglycerate *n*. [Gk. *manna*, manna; *glykys*, sweet] a food storage product in red algae.

mannosan *n*. [Gk. *manna*, manna] mannan, *q.v.*

mannose *n*. [Gk. *manna*, manna] a hexose sugar found in certain glycosides and the polysaccharide mannan.

manocyst *n*. [L. *manare*, to proceed from; Gk. *kystis*, pouch] a receptive oogonial papilla reaching the antheridium, as in *Phytophthora* (Peronosporales).

manoxylic *a*. [Gk. *manos*, slack; *xylon*, wood] having soft loose wood containing much parenchyma, as Cycadales; *cf*. pycnoxylic.

mantids the common name for many members of the Dictyoptera, *q.v.*

mantle *n*. [L. *mantellum*, cloak] outer soft fold of integument next to shell of molluscs, cirripedes, and brachiopods, *alt*. pallium; sheath of spongoblast cells; body wall of ascidians; scapulars and wing coverts of birds; ocrea, *q.v.*

mantle cavity a space between the mantle and body proper.

mantle cell a cell of tapetum or investing tissue of a sporangium.

mantle fibres the spindle fibres of a fully formed spindle.

mantle layer a layer of embryonic medulla spinalis representing the future grey columns.

mantle lobes dorsal and ventral flaps of mantle in bivalves.

manual *n*. [L. *manus*, hand] a wing quill borne on manus of birds; *alt*. remex primarius, primary feather.

manubrial *a*. [L. *manubrium*, handle] *pert*. a manubrium; handle-shaped.

manubrium *n*. [L. *manubrium*, handle] a cell projecting inwards from shield of an antheridial globule in certain algae; conical elevation at distal end of hydrozoan polyp, *alt*. hypostome, oral cone; clapper-like portion hanging down from under surface of medusae; handle of malleus of mastax; basal part of furcula in Collembola; presternum or anterior part of sternum; handle-like part of malleus of ear.

manus *n*. [L. *manus*, hand] hand, or part of forelimb corresponding to it, as found in amphibians, reptiles, and mammals.

manyplies omasum, *q.v.*, so called from its folded structure.

Marattiales *n*. [*Marattia*, generic name] an order of eusporangiate Pteropsida (Filicopsida) that are mostly large ferns with no secondary thickening and with sporangia in sori on the dorsal surface of the lamina.

marcescent *a*. [L. *marcescere*, to wither] withering but not falling off; *appl*. a calyx or corolla persisting after fertilization.

Marchantiales *n*. [*Marchantia*, generic name] an order of liverworts having a thalloid gametophyte with photosynthetic tissue confined to the dorsal region, sex organs sometimes raised in the air, and the capsule not dehiscing by 4 valves.

marcid *a*. [L. *marcidus*, withered] withered; shrivelled.

marginal *a*. [L. *margo*, edge] *pert*., at, or near the margin, edge, or border; *appl*. veil, a secondary growth of edge of pileus, in agarics and boletes; *appl*. a form of nervation; *appl*. a convolution of frontal lobe; *appl*. a type of placenta; *appl*. plates round margin of chelonian carapace.

marginalia *n.plu*. [L. *margo*, edge] a body part in various organisms that is marginal in relation to another part; prostalia, *q.v.*; *sing*. marginal.

marginal meristem in a leaf, a meristem located along the margin of the leaf primordium and forming the blade.

marginate *a*. [L. *margo*, edge] having a distinct margin in structure or colouring.

marginella *n*. [*Dim*. of L. *margo*, edge] ring formed by part of cutis proliferating beyond margin of lamellae, in certain fungi with an exposed hymenium.

marginicidal *a*. [L. *margo*, edge; *caedere*, to cut] dehiscing by line of union of carpels.

marginiform *a*. [L. *margo*, edge; *forma*, shape]

like a margin or border in appearance or structure.

marginirostral a. [L. margo, edge; rostrum, beak] forming the edges of a bird's bill.

marijuana, marihuana n. [Sp. Mary Jane] the drug, cannabis, q.v.; sometimes used for the whole cannabis plant.

marita n. [L. maritus, conjugal] sexually mature stage in trematode life history.

marital a. [L. maritus, conjugal] pert. marita; producing fertilized eggs, appl. trematodes.

marker an identifying factor; a gene of known location and effect which makes possible the determination of the distribution of other, less conspicuously effective, genes.

marmorate a. [L. marmor, marble] of marbled appearance.

marrow n. [A.S. mearg, pith] connective tissue filling up cylindrical cavities in bodies of long bones, and spaces of cancellous tissue, differing in composition in different bones, alt. medulla ossium; pith of certain plants; vegetable marrow.

marrow brain myelencephalon, q.v.

marsh n. [M.E. mersh, meadowland] an area or group of plants on wet but not peaty soil.

Marsileales n. [Marsilea, generic name] an order of heterosporous leptosporangiate Pteropsida (Filicopsida) which are shallow water plants with creeping rhizomes and erect leaves.

Marsipobranchia n. [Gk. marsypos, pouch; brangchia, gills] Cyclostomata (Agnatha), q.v.

marsupial a. [L. marsupium, pouch] pert. a marsupium; pouch-bearing, as a kangaroo; appl. bones of pelvic girdle in certain mammals.

Marsupialia, marsupials n., n.plu. [L. marsupium, pouch] the only order of the Metatheria, q.v.

marsupium n. [L. marsupium, pouch] any pouch-like structure in which the young of an animal complete their development, such as abdominal pouch of marsupials; gill cavities of bivalves; recess formed by diverging spines and a supporting membrane in stelleroids; structure protecting the acrocyst in Sertularia (Calyptoblastea); a nursing sac surrounding certain archegonia.

Martinotti cells [G. Martinotti, Italian physician] pyramidal nerve cells of cerebral cortex, with axons directed to the peripheral plexiform or molecular layer.

mask n. [F. masque, mask] a hinged prehensile structure, corresponding to adult labium, peculiar to dragonfly nymph.

masked a. [F. masque, mask] personate, appl. corolla; concealed, appl fat of cell which is not evident microscopically.

masked virus a virus which gives no evidence of its presence because of auto-interference.

massa intermedia grey matter connecting thalami across 3rd ventricle; alt. middle commissure.

masseter n. [Gk. masētēr, one that chews] muscle which raises lower jaw and assists in chewing.

masseteric a. [Gk. masētēr, one that chews] pert. or near masseter muscle of cheek, appl. artery, vein, nerve.

mass flow a theory of the way in which materials are translocated through the phloem, which states that the cause of movement is a difference in the hydrostatic pressure at each end of a sieve tube, resulting in flow of contents along the tube.

massive a. [L. massa, mass] bulky; heavy; compacted; appl. nuclei deficient in nuclear sap.

mass meristem a meristem in which the derivative tissue increases in volume by division of cells in various planes.

mass provisioning providing food at one session to last developing progeny throughout larval life, as in certain insects.

massula n. [L. massula, small mass] a mass of microspores in a sporangium of certain pteridophytes; a massed group of pollen grains in orchids.

mast n. [O.E. mete, food] the fruit of beech and some similar related trees.

Mastacembeliformes n. [Mastacembelus, generic name; L. forma, shape] an order or suborder of small teleosts including the spiny eels; alt. Opisthomi.

mastax n. [Gk. mastax, jaws] the gizzard or pharyngeal mill of rotifers.

mast cells [Ger. mästen, to feed] spheroid, ovoid, or amoeboid cells of very granular cytoplasm and large nucleus, found in tissue where fat is laid down, and in connective tissue, especially areolar connective where they secrete the ground substance, heparin (from heparinocytes), and histamine (from histaminocytes); alt. Mastzellen of Ehrlich, basicytes, basiphils, clasmatoblasts, labrocytes, mastocytes, mucinoblasts.

mastication n. [L. masticare, to chew] process of chewing food with teeth until reduced to small pieces or to a pulp; alt. manduction.

masticatory stomach in decapod crustaceans, a structure made of the hard lining of the gizzard which grinds and strains food; alt. gastric mill, stomodaeal apparatus.

mastidion n. [Gk. mastos, breast; idion, dim.] nipple-like protuberance on paturon, in some spiders.

mastigium n. [Gk. mastigion, little whip] defensive posterior lash of certain larvae.

mastigobranchia n. [Gk. mastix, whip; brangchia, gills] epipodite of adult Decapoda, a bilobed membranous lamina extending upwards between gills.

mastigonemes n.plu. [Gk. mastix, whip; nēma, thread] filaments extending laterally from the flagellar sheath.

Mastigophora n. [Gk. mastix, whip; pherein, to bear] Flagellata, q.v.

mastigosome n. [Gk. mastix, whip; sōma, body] blepharoplast, q.v.

mastocytes n.plu. [Ger. mästen, to feed; Gk. kytos, hollow] mast cells, q.v.

mastoid a. [Gk. mastos, breast; eidos, form] nipple-shaped, appl. a process of temporal bone, cells, foramen, fossa, notch.

mastoideosquamous a. [Gk. mastos, breast; eidos, like; L. squama, scale] pert. mastoid and squamous parts of temporal bone.

mastoidohumeralis a. [Gk. mastos, breast; eidos, like; L. humerus, humerus] a muscle of certain quadrupeds, connecting mastoid and humerus.

masto-occipital a. [Gk. mastos, breast; L. occiput, occiput] pert. occipital bone and mastoid process of temporal.

mastoparietal a. [Gk. mastos, breast; L. paries, wall] pert. parietal bone and mastoid process of temporal.

mastotympanic a. [Gk. *mastos*, breast; *tympanon*, drum] *appl.* part of tympanic cavity's boundary in certain reptiles.

maternal inheritance differences in phenotype in cells or organisms of the same genotype due to effects of substances synthesized in the mother and found in the egg cytoplasm.

mating types groups, the individuals of which do not conjugate with individuals of other groups, as of ciliates.

matriclinous a. [L. *mater*, mother; Gk. *klinein*, to bend] with hereditary characteristics more maternal than paternal; *alt.* matroclinic, matroclinal, matroclinous; *cf.* patriclinal.

matrilinear inheritance the inheritance of cytoplasmic particles only through the female line.

matrix n. [L. *mater*, mother] medium in which a structure is embedded; ground substance of connective tissue; part beneath body and root of nail; uterus, *q.v.*; body upon which lichen or fungus grows; envelope of chromatid; substance in which a fossil is embedded.

matroclinal, matroclinic, matroclinous matriclinous, *q.v.*

matromorphic a. [L. *mater*, mother; Gk. *morphē*, form] resembling the female parent in morphological characters.

mattula n. [L. *matta*, mat] fibrous network covering petiole bases of palms.

maturation n. [L. *maturus*, ripe] ripening; *appl.* divisions by which gametes are produced from primary gametocytes, during which meiosis occurs with reduction of chromosome number from diploid to haploid.

Mauthner's cells [L. *Mauthner*, Austrian physician] a layer between medullary sheath and neurilemma of nerve fibre.

maxilla n. [L. *maxilla*, jaw] the upper jaw; part of upper jaw behind premaxilla; an appendage of most arthropods, posterior to mandible, modified in various ways in adaptation to function and requirements.

maxillary a. [L. *maxilla*, jaw] *pert.* or in region of maxilla or upper jaw, *appl.* artery, nerve, process, sinus, tuberosity, vein, etc.

maxillary glands paired excretory organs opening at base of maxilla in Crustacea.

maxilliferous a. [L. *maxilla*, jaw; *ferre*, to carry] bearing maxillae.

maxilliform a. [L. *maxilla*, jaw; *forma*, shape] like a maxilla.

maxilliped(e) a. [L. *maxilla*, jaw; *pes*, foot] an appendage, in 1, 2, or 3 pairs, posterior to maxillae in arthropods; *alt.* gnathopodite, jaw-foot, footjaw.

maxillodental a. [L. *maxilla*, jaw; *dens*, tooth] *pert.* jaws and teeth.

maxillojugal a. [L. *maxilla*, jaw; *jugum*, yoke] *pert.* jaw and jugal bone.

maxillolabial a. [L. *maxilla*, jaw; *labium*, lip] *pert.* maxilla and labium, *appl.* dart in ticks.

maxillomandibular a. [L. *maxilla*, jaw; *mandibulum*, jaw] *appl.* arch forming jaws of primitive fishes; *pert.* maxilla and mandible.

maxillopalatal, maxillopalatine a. [L. *maxilla*, jaw; *palatus*, palate] *pert.* jaw and palatal bones, *appl.* a maxillary process of birds; n. maxillopalatine.

maxillopharyngeal a. [L. *maxilla*, jaw; Gk. *pharyngx*, gullet] *pert.* lower jaw and pharynx.

maxillopremaxillary a. [L. *maxilla*, jaw; *pre*, before] *pert.* whole of upper jaw; *appl.* jaw when maxilla and premaxilla are fused.

maxilloturbinal a. [L. *maxilla*, jaw; *turbo*, whorl] *pert.* maxilla and turbinals. n. A bone arising from lateral wall of nasal cavity, which supports sensory epithelium.

maxillula n. [L. *dim.* of *maxilla*, jaw] a 1st maxilla in Crustacea when there are more pairs than one; an appendage between mandible and 1st maxilla in primitive insects.

maxim n. [L. *maximus*, greatest] an ant of the large worker type or of the soldier caste; *cf.* minim.

mayflies the common name for the Ephemeroptera, *q.v.*

mazaedium, mazedium n. [Gk. *maza*, cake; *idion*, dim.] a coat formed by ends of paraphyses and their secretions, covering hymenium of certain Ascomycetes; a fruit body of certain lichens.

mazic a. [Gk. *maza*, cake] placental, *q.v.*

M-chromosome a microchromosome; a mediocentric chromosome.

M-disc median or intermediate disc, Hensen's line, *q.v.*

mealy a. [L. *molere*, to grind] covered with a powder resembling flour or coarse ground cereal (meal); *alt.* farinaceous, farinose.

meatus n. [L. *meatus*, passage] a passage or channel, as acoustic, nasal, etc.

mechanical tissue supporting tissue, *q.v.*

mechanism n. [Gk. *mēchanē*, machine] the view that all vital phenomena are due to physical and chemical laws.

mechanocyte n. [Gk. *mēchanē*, contrivance; *kytos*, hollow] a cell derived from bone, cartilage, connective tissue, tendon, or muscle; a supporting cell.

mechanoreceptor n. [Gk. *mēchanē*, contrivance; L. *recipere*, to receive] a specialized structure sensitive to mechanical stimuli such as contact, pressure, or gravity.

Meckel's also called Meckelian.

Meckel's cartilage or rod [J. F. Meckel, junior, German anatomist] the lower jaw of lower vertebrates, and in higher vertebrates, the axis around which membrane bones of jaw are arranged and formed.

Meckel's ganglion [J. F. Meckel, senior, German anatomist] the sphenopalatine ganglion, *q.v.*

meconidium n. [Gk. *mēkōn*, poppy; *idion*, dim.] sessile or pedicellate extracapsular medusa usually lying on top of gonangium of certain hydroids.

meconium n. [Gk. *mēkōn*, poppy] waste products of a pupa or other embryonic form; contents of intestine of a newborn mammal.

Mecoptera n. [Gk. *mekos*, length; *pteron*, wing] an order of slender carnivorous insects with complete metamorphosis, commonly called scorpion flies, having biting mouth-parts, long slender legs, and membranous wings lying along the body in repose.

media n. [L. *medius*, middle] a middle structure, such as a layer of tissue, a central nervure, *plu.* mediae; *plu.* of medium; medial, *q.v.*

mediad adv. [L. *medius*, middle; *ad*, to] towards but not quite in the middle line or axis; *alt.* mesad, admedial; *cf.* laterad.

medial a. [L. *medius*, middle] situated in the middle. n. The middle vein of wing of insects; *alt.* media, mesal.

median

median *a.* [L. *medius*, middle] lying or running in axial plane; intermediate; middle. *n.* The middle variate when variates are arranged in order of magnitude.

median nerve nerve arising from union of medial and lateral cord of brachial plexus, with branches in forearm.

median segment propodeon, *q.v.*

mediastinal *a.* [L. *mediastinus*, medial] *pert.* or in region of mediastinum, *appl.* cavity, arteries, glands, pleura.

mediastinum *n.* [L. *mediastinus*, medial] space between right and left pleura in and near median sagittal thoracic plane; incomplete vertical septum of testis; *alt.* Highmore's body, corpus highmoreanum.

mediator *n.* [L. *medius*, middle] a nerve cell maintaining relation between receptor and effector; a chemical such as an enzyme or hormone which mediates in a metabolic process.

medidural *a.* [L. *medius*, middle; *durus*, hard] *pert.* medial part of dura mater.

medifurca *n.* [L. *medius*, middle; *furca*, fork] mesofurca, *q.v.*

medine *n.* [L. *medius*, middle] exo-intine, *q.v.*

mediocentric *a.* [L. *medius*, middle; *centrum*, centre] *appl.* chromosomes having a medial or mediad centromere and therefore having 2 arms, *alt.* M-chromosomes, V-chromosomes.

Medio-Columbian Sonoran, *q.v.*

mediocubital *n.* [L. *medius*, middle; *cubitalis*, of elbow] a cross-vein between posterior media and cubitus of insect wing.

mediodorsal *a.* [L. *medius*, middle; *dorsum*, back] in the dorsal middle line.

mediolecithal *a.* [L. *medius*, middle; Gk. *lekithos*, egg yolk] medium-yolked.

mediopalatine *a.* [L. *medius*, middle; *palatus*, palate] between palatal bones, *appl.* a cranial bone of some birds.

mediopectoral *a.* [L. *medius*, middle; *pectus*, breast] *appl.* middle part of sternum.

mediostapedial *n.* [L. *medius*, middle; *stapes*, stirrup] *pert.* that portion of columella of the ear external to stapes.

mediostratum *n.* [L. *medius*, middle; *stratum*, layer] inner tissue of trama in agarics.

mediotarsal *a.* [L. *medius*, middle; *tarsus*, ankle] between tarsal bones.

medioventral *a.* [L. *medius*, middle; *venter*, belly] in the middle ventral line.

mediproboscis *n.* [L. *medius*, middle; Gk. *proboskis*, trunk] middle portion of insect proboscis, part of ligula.

medithorax *n.* [L. *medius*, middle; Gk. *thōrax*, chest] middle part of the thorax; the mesothorax of insects.

medium *n.* [L. *medium*, middle] any of the structures through which a force acts, as refracting media of eyeball; substance in which cultures are reared or tissues propagated.

medulla *n.* [L. *medulla*, marrow, pith] central part of an organ or tissue; marrow of bones, *alt.* medulla ossium; pith or central region of stem; medullary layer, *q.v.*; loose hyphae in a tangled fungal structure such as rhizomorphs or fruit bodies.

medulla oblongata posterior portion of brain continuous with medulla spinalis or spinal cord; *alt.* macromyelon, oblongata, bulb.

medullary *a.* [L. *medulla*, pith] *pert.* or in region of medulla, *appl.* axis, artery, lamina, membrane, bone, spaces, canal, vascular bundles, etc.

medullary canal hollow cylindrical portion of a long bone containing marrow; neurocoel, *q.v.*; neural tube, *q.v.*

medullary folds neural folds, *q.v.*

medullary groove a groove on surface of medullary plate, bounded by folds which grow and coalesce, converting groove into a canal, the neurocoel; *alt.* neural groove.

medullary keel a downward growth towards archenteron, the rudiment of central nervous system in development of certain primitive vertebrates.

medullary layer a thick subcortical layer of the thallus of some lichens; *alt.* medulla.

medullary membrane endosteum, *q.v.*

medullary phloem internal phloem in a bicollateral bundle, as in Cucurbitaceae, marrow family.

medullary plate plate-like formation of ectoderm cells bordering blastopore of early embryo; earliest rudiment of nervous system, *alt.* neural plate.

medullary pyramids Malpighian pyramids, *q.v.*

medullary rays tissue extending between pith and pericycle, *alt.* interfascicular region; groups of uriniferous tubules in medulla of kidney.

medullary sheath a ring of protoxylem around the edge of the pith of certain stems, *alt.* perimedullary region; a white layer of myelin surrounding axis cylinder of medullated nerve fibre, *alt.* myelin sheath.

medullary velum valve of Vieussens, *q.v.*

medulla spinalis spinal cord, *q.v.*

medullated *a.* [L. *medulla*, pith] having pith; having a medullary sheath.

medullated nerve fibres fibres of brain and spinal cord, consisting of axis cylinder surrounded by medullary sheath, in turn covered by delicate neurilemma.

medulliblasts *n.plu.* [L. *medulla*, marrow; Gk. *blastos*, bud] cells of embryonic nervous tissue which give rise to neuroblasts and spongioblasts.

medullispinal *a.* [L. *medulla*, pith; *spina*, spine] of the spinal cord.

medusa *n.* [Gk. *Medousa*, Medusa, one of the Gorgons] in coelenterates, a free-swimming sexual jellyfish-like organism in hydrozoans, or a jellyfish in scyphozoans; *alt.* medusoid.

medusiform *a.* [Gk. *Medousa*, Medusa; L. *forma*, shape] like a medusa; *alt.* medusoid.

medusoid *n.* [Gk. *Medousa*, Medusa; *eidos*, like] medusa, *q.v.*, *alt.* hedrioblast; *a.* medusiform, *q.v.*

medusome *n.* [Gk. *Medousa*, Medusa; *sōma*, body] medusoid stage in life history of some colonial hydrozoans such as *Obelia*.

mega- *alt.* macro-, *q.v.*; *cf.* micro-.

megacephalic *a.* [Gk. *megas*, large; *kephalē*, head] with abnormally large head; having a cranial capacity of over 1450 cm³; *cf.* mesocephalic, microcephalic.

megachromosomes *n.plu.* [Gk. *megas*, large; *chrōma*, colour; *sōma*, body] large chromosomes forming an outer set in certain sessile Ciliata; *cf.* microchromosomes.

megacins *n.plu.* [*megatherium*; L. *caedere*, to kill] a group of bacteriocins found in certain strains of *Bacillus megatherium*.

Megadrili *n*. [Gk. *megas*, large; *drilos*, worm] a group of large terrestrial oligochaetes having the nephridium with a capillary network; *a*. megadriline.

megagamete *n*. [Gk. *megas*, large; *gametēs*, spouse] a rounded cell regarded as an ovum or its equivalent, developed from a megagametocyte after a process akin to maturation; macrogamete, *q.v.*

megagametocyte *n*. [Gk. *megas*, large; *gametēs*, spouse; *kytos*, hollow] a cell developed from a merozoite, and itself giving rise to a megagamete; macrogametocyte, *q.v.*

megagametogenesis *n*. [Gk. *megas*, great; *gametēs*, spouse; *genesis*, descent] development of megagametes or ova.

megagametophyte *n*. [Gk. *megas*, large; *gametēs*, spouse; *phyton*, plant] the female gametophyte developed from a megaspore; *cf.* microgametophyte.

megaheterochromatic *a*. [Gk. *megas*, great; *heteros*, other; *chrōma*, colour] containing large amounts of heterochromatin.

megakaryoblast *n*. [Gk. *megas*, large; *karyon*, nut; *blastos*, bud] precursor of a megakaryocyte found in bone marrow.

megakaryocyte *n*. [Gk. *megas*, large; *karyon*, nut; *kytos*, hollow] an amoeboid giant cell of bone marrow with one large annular lobulated nucleus, containing a number of nucleoli, thought to give rise to blood platelets; *alt.* myeloplax.

megalaesthetes *n.plu.* [Gk. *megalon*, great; *aisthētēs*, perceiver] sensory organs, sometimes in form of eyes, in Amphineura.

megalecithal megalolecithal, *q.v.*

megaloblast *n*. [Gk. *megalos*, greatly; *blastos*, bud] a primitive large erythroblast.

megalogonidium *n*. [Gk. *megalos*, greatly; *gonos*, offspring; *idion*, *dim.*] a large gonidium; *alt.* macrogonidium.

megalolecithal *a*. [Gk. *megalos*, greatly; *lekithos*, yolk] containing much yolk as telolecithal eggs; *alt.* macrolecithal, megalecithal.

megalopa megalops, *q.v.*

megalopic *a*. [Gk. *megalos*, greatly; *ōps*, eye] belonging to the megalops stage.

megalopore *n*. [Gk. *megalon*, great; *poros*, channel] pore in dorsal plates of amphineurans, for placing a megalaesthete in direct communication with exterior.

megalops *n*. [Gk. *megalos*, greatly; *ōps*, eye] a larval stage of certain Crustacea such as crabs, having large stalked eyes and a crab-like cephalothorax; *alt.* megalopa.

Megaloptera *n*. [Gk. *megalos*, greatly; *pteron*, wing] an order of insects, commonly called alder flies, sometimes included in the Neuroptera with the Planipennia.

megalospheric *a*. [Gk. *megalos*, greatly; *sphaira*, globe] of polythalamous foraminifer shells, having a megalosphere or large initial chamber; *alt.* megaspheric, macrospheric.

megamere *n*. [Gk. *megas*, large; *meros*, part] macromere, *q.v.*

megameric *a*. [Gk. *megas*, large; *meros*, part] with relatively large parts; *appl.* chromosomes with large heterochromatic regions; *pert.* megameres.

meganephridia *n.plu.* [Gk. *megas*, large; *nephros*, kidney; *idion*, *dim.*] large nephridia, occurring as one pair per segment; *alt.* holonephridia.

meganucleus *n*. [Gk. *megas*, large; L. *nucleus*, kernel] the larger nucleus, thought to be of vegetative nature, in ciliates; *alt.* macronucleus, trophonucleus, trophic nucleus, vegetative nucleus; *cf.* micronucleus.

megaphanerophyte *n*. [Gk. *megas*, large; *phaneros*, manifest; *phyton*, plant] tree exceeding 30 m in height.

megaphyll *n*. [Gk. *megas*, large; *phyllon*, leaf] a large leaf, especially as produced by pteridophytes; *alt.* macrophyll.

megaphyllous *a*. [Gk. *megas*, large; *phyllon*, leaf] having relatively large leaves or leaflets; *alt.* macrophyllous.

megaplankton *n*. [Gk. *megas*, large; *plangktos*, wandering] macroplankton, *q.v.*

megasclere *n*. [Gk. *megas*, large; *sklēros*, hard] skeletal spicule of general supporting framework of sponges; *alt.* macrosclere; *cf.* microsclere.

megascopic *a*. [Gk. *mega*, large; *skopein*, to view] macroscopic, *q.v.*

megasome macrosome, *q.v.*

megasorus *n*. [Gk. *megas*, large; *sōros*, heap] a sorus containing megasporangia; *cf.* microsorus.

megaspheric megalospheric, *q.v.*

megasporangium *n*. [Gk. *megas*, large; *sporos*, seed; *anggeion*, vessel] a megaspore-producing sporangium; the nucellus of an ovule; sometimes used incorrectly for the whole ovule; *alt.* gynosporangium, macrosporangium.

megaspore *n*. [Gk. *megas*, great; *sporos*, seed] a larger-sized spore of dimorphic forms in reproduction by spore formation; the larger spore of heterosporous plants which gives rise to the female gametophyte and in seed plants is also called embryo sac, gynospore, macrospore; a fossil spore exceeding 0·2 mm in diameter, *cf.* miospore.

megasporocyte *n*. [Gk. *megas*, large; *sporos*, seed; *kytos*, hollow] the embryo sac mother cell, diploid cell in ovary that undergoes meiosis, producing 4 haploid megaspores.

megasporophyll *n*. [Gk. *megas*, great; *sporos*, seed; *phyllon*, leaf] a leaf or structure derived from a leaf on which megasporangia develop, in flowering plants also known as a carpel; *alt.* carpophyll, macrosporophyll.

megatherm *n*. [Gk. *megas*, great; *thermē*, heat] a tropical plant; a plant requiring moist heat; *alt.* macrotherm, macrothermophyte.

megazooid *n*. [Gk. *megas*, great; *zōon*, animal; *eidos*, form] the larger zooid resulting from binary or other fission.

megazoospore *n*. [Gk. *megas*, great; *zōon*, animal; *sporos*, seed] a large zoospore, as in reproduction of certain Radiolaria; a zoogonidium of certain algae.

megistotherm *n*. [Gk. *megistos*, greatest; *thermē*, heat] a plant that thrives at a more-or-less uniformly high temperature.

Mehlis' glands glands surrounding the ootype of trematodes which play part in egg shell formation.

Meibomian glands [*H. Meibom*, German anatomist] modified sebaceous glands of the eyelids, the ducts opening on the free margins; *alt.* tarsal glands.

meiocyte *n*. [Gk. *meion*, smaller; *kytos*, hollow] a reproductive cell prior to meiosis; auxocyte, *q.v.*

meiofauna *n*. [Gk. *meion*, smaller; L. *faunus*, god

of the woods] small invertebrates found on the sea bottoms; mesofauna, *q.v.*

meiogenic *a.* [Gk. *meion*, smaller; *gennaein*, to produce] promoting nuclear division.

meiogyrous *a.* [Gk. *meion*, less; *gyros*, circle] slightly coiled inwards.

meiolecithal *a.* [Gk. *meion*, less; *lekithos*, yolk] having little yolk, as homolecithal and isolecithal eggs.

meiomery *n.* [Gk. *meion*, smaller; *meros*, part] condition of having fewer than the normal number of parts.

meiophase *n.* [Gk. *meion*, less; *phaein*, to appear] the stage during which meiosis occurs so that the chromosome number is reduced from diploid to haploid.

meiophylly *n.* [Gk. *meion*, smaller; *phyllon*, leaf] suppression of one or more leaves in a whorl.

meiosis *n.* [Gk. *meion*, smaller] a type of nuclear and cell division in which the chromosome number is reduced from diploid to haploid; *alt.* maiosis, miosis; *plu.* meioses; *cf.* reduction division, mitosis.

meiosporangium *n.* [Gk. *meion*, less; *sporos*, seed; *anggeion*, vessel] a thick-walled diploid sporangium, producing haploid zoospores; *cf.* mitosporangium.

meiospore *n.* [Gk. *meion*, less; *sporos*, seed] a uninucleate haploid zoospore produced in a meiosporangium; *cf.* mitospore.

meiostemonous *a.* [Gk. *meion*, smaller; *stēmōn*, spun thread] having fewer stamens than petals or sepals; *alt.* miostemonous.

meiotaxy *n.* [Gk. *meion*, smaller; *taxis*, arrangement] suppression of whorl or set of organs.

meiotherm *n.* [Gk. *meion*, less; *thermē*, heat] a plant that thrives in a cool temperate environment.

meiotic *a.* [Gk. *meion*, smaller] *pert.* meiosis; *alt.* maiotic, miotic.

meiotic apogamy or euapogamy generative apogamy, *q.v.*

meiotrophic *a.* [Gk. *meion*, less; *trophē*, nourishment] requiring fewer amino acids or other nutrients than the parent strain as in certain bacterial mutants.

Meissner's corpuscles [G. *Meissner*, German histologist] tactile corpuscles, associated with sense of pain, in skin of digits, lips, nipple, and certain other areas; *alt.* Wagner's corpuscles.

Meissner's plexus a gangliated plexus of nerve fibres in submucous coat of small intestine; *alt.* Remak's plexus.

Melanconiales *n.* [*Melanconium*, generic name] an order of parasitic Fungi Imperfecti having conidiophores borne in a subepidermal or subcortical acervulus.

melanins *n.plu.* [Gk. *melas*, black] a range of pigments, usually black or dark brown, produced from tyrosine by the enzyme tyrosinase, and giving colour to animal skin, etc., and also some plants; *cf.* eumelanin, phaeomelanin, dopa, haemozoin.

melaniridosome *n.* [Gk. *melas*, black; *iris*, rainbow; *sōma*, body] a pigment body consisting of a melanophore and associated iridocytes in corium of fishes.

melanism *n.* [Gk. *melas*, black] excessive development of black pigment; *cf.* industrial melanism.

melanoblast *n.* [Gk. *melas*, black; *blastos*, bud] a cell in which melanin is formed in the Malpighian layer of epidermis; *alt.* Langerhans' cell.

melanocyte *n.* [Gk. *melas*, black; *kytos*, hollow] a black pigmented lymphocyte.

melanocyte-stimulating hormone (MSH) intermedin, *q.v.*

melanogen *n.* [Gk. *melas*, black; *gennaein*, to produce] a colourless compound formed by reduction of the red oxidation product of tyrosine, and oxidized to melanin.

melanogenesis *n.* [Gk. *melas*, black; *genesis*, origin] the formation of melanin.

melanoids *n.plu.* [Gk. *melas*, black; *eidos*, form] dark-brown or black pigments related to melanin.

melanophore *n.* [Gk. *melas*, black; *pherein*, to bear] a pigment cell containing melanin, especially black melanin.

melanosoma, melanosome *n.* [Gk. *melas*, black; *sōma*, body] dark pigment mass associated with ocellus, as in certain Dinoflagellata.

melanospermous *a.* [Gk. *melas*, black; *sperma*, seed] *appl.* seaweeds with dark-coloured spores.

melanosporous *a.* [Gk. *melas*, black; *sporos*, seed] having black or dark coloured spores.

melanotic *a.* [Gk. *melas*, black] having black pigment unusually developed.

melatonin *n.* [Gk. *melas*, black; *tonos*, tension] a substance secreted by the pineal gland, which causes melanin to be more narrowly dispersed in melanophores in the skin.

Meliolales *n.* [*Meliola*, generic name] an order of Ascomycetes having a dark mycelium growing superficially on stems and leaves of vascular plants.

meliphagous *a.* [Gk. *meli*, honey; *phagein*, to eat] feeding on honey; *alt.* mellivorous.

melliferous *a.* [L. *mel*, honey; *ferre*, to carry] honey-producing.

mellisugent *a.* [L. *mel*, honey; *sugere*, to suck] honey-sucking.

mellivorous *a.* [L. *mel*, honey; *vorare*, to devour] meliphagous, *q.v.*

member *n.* [L. *membrum*, member] a limb or organ of the body; a well-defined part or organ of a plant.

membrana *n.* [L. *membrana*, membrane] membrane, *q.v.*

membranaceous *a.* [L. *membrana*, membrane] of the consistency, or having the structure, of a membrane.

membranal *a.* [L. *membrana*, membrane] *pert.*, or within membranes.

membrana limitans interna beneath stratum opticum of retina, and externa at bases of rods and cones, formed, respectively, by contiguous bases and processes of sustentacular fibres; *alt.* limiting membrane.

membrana propria basement membrane, *q.v.*

membrane *n.* [L. *membrana*, membrane] a thin film, skin, or layer of tissue covering a part of an animal or plant; a thin covering of cells or of unicellular organisms; *alt.* membrana.

membrane bone a bone developing directly in membrane without passing through a cartilage stage; *alt.* investing bone, allostosis.

membranella *n.* [L. *membrana*, membrane; *dim.*] an undulating membrane formed by fusion of rows of cilia, in some Protozoa; ciliated band, in tornaria.

membraniferous *a.* [L. *membrana*, membrane; *ferre*, to carry] enveloped in or bearing a membrane.

membranoid *a.* [Gk. *membrana*, membrane; *eidos*, form] resembling a membrane.

membranous *a.* [L. *membrana*, membrane] resembling or consisting of membrane; thin, dry, pliable, and semitransparent.

membranous cranium a mesenchymal investment enclosing brain.

membranous labyrinth internal ear, separated from bony cavities by perilymph, and itself containing endolymph.

membranous vertebral column continuous sheath of mesoderm enveloping notochord and neural tube.

membranula *n.* [L. *dim.* of *membrana*, membrane] a concrescence of cilia forming a paddle-like organ in certain Ciliata.

membranule *n.* [L. *dim.* of *membrana*, membrane] a small opaque space close to body of insect, in anal area of wing of some dragonflies.

MeNA α-naphthylacetic acid methyl ester, a volatile artificial auxin used as a gas.

menacme *n.* [Gk. *mēn*, month; *akmē*, prime] the interval between first and final menstruation, i.e. life between menarche and menopause.

menadione a vitamin K_2 analogue (2-methyl-1,4-naphthoquinone) sometimes known as vitamin K_3, *see* vitamin K.

menarche *n.* [Gk. *mēn*, month; *archē*, beginning] first menstruation; age at 1st menstruation.

Mendelian *a.* [G. *Mendel*, Czechoslovakian monk] *pert.* character which behaves according to results of Mendel's laws, manifesting allelomorphic inheritance.

Mendelian population a group of interbreeding individuals which share a common gene pool, the species being the most extensive.

Mendelism a law or rule governing inheritance of characters in plants and animals, discovered by Gregor Mendel. This principle deals with inheritance of 'unit characters' presence or absence of one or other of a pair of contrasting characters, dominant and recessive. It also shows that offspring of organisms with a pair of contrasting characters will exhibit these in a definte ratio, and it is extended to deal with groups of characters.

Mendel's laws laws of inheritance put forward by Gregor Mendel; the 1st law is called the law of segregation, *q.v.*; the 2nd law is called the law of independent assortment, *q.v.*

meningeal *a.* [Gk. *meningx*, membrane] *pert.* or in region of meninges, *appl.* arteries, veins, nerves, etc.

meninges *n.plu.* [Gk. *meningx*, membrane] the 3 membranes enclosing brain and spinal cord, from without inwards: dura mater, arachnoid, and pia mater; *sing.* meninx.

meningocyte *n.* [Gk. *meningx*, membrane; *kytos*, hollow] a phagocytic cell of the subarachnoid space.

meningosis *n.* [Gk. *meningx*, membrane] attachment by means of membranes.

meningospinal *a.* [Gk. *meningx*, membrane; L. *spina*, spine] *pert.* spinal cord membranes.

meningovascular *a.* [Gk. *meningx*, membrane; L. *vasculum*, small vessel] *pert.* meningeal blood vessels.

meninx *sing.* of meninges.

meninx primaria membrane representing dura mater, as in Anura.

meninx primitiva a single membrane surrounding the central nervous system, as in Cyclostomata (chordates) and Elasmobranchii.

meninx secundaria a pigmented membrane representing pia mater and arachnoid, as in Anura.

meniscus *n.* [Gk. *mēniskos*, a crescent] interarticular fibrocartilage found in joints exposed to violent concussion; semilunar cartilage; intervertebral disc; a tactile disc, being terminal expansion of axis cylinder and tactile corpuscles; *plu.* menisci.

menognathous *a.* [Gk. *menein*, to remain; *gnathos*, jaw] with persistent biting jaws, *appl.* insects.

menopause *n.* [Gk. *mēn*, month; *pausi*, ending] climacterical cessation of menstruation; *cf.* climacteric.

menorhynchous *a.* [Gk. *menein*, to remain; *rhynchos*, snout] with persistent suctorial mouthparts, *appl.* insects.

menotaxis *n.* [Gk. *menein*, to remain; *taxis*, arrangement] compensatory movements to maintain a given direction of body axis in relation to sensory stimuli, especially light, but not necessarily moving towards or away from it, *cf.* light compass reaction; maintenance of visual axis during locomotion.

mensa *n.* [L. *mensa*, table] grinding surface of a tooth.

menses *n.plu.* [L. *menses*, months] the fluid discharged during menstruation; *alt.* catamenia.

menstrual *a.* [L. *menstrualis*, monthly] monthly; catamenial: of or *pert.* menses; lasting for a month, as flower.

menstruation *n.* [L. *mensis*, month; *struere*, to flow] periodic discharge from uterus of various vertebrates, chiefly higher mammals.

mental *a.* [L. *mentum*, chin] *pert.* or in region of chin, *appl.* foramen, nerve, spines, tubercle, muscle, *appl.* scale of plate of fish and of reptile; *pert.* mentum of insects. [L. *mens*, mind] *pert.* the mind.

mentigerous *a.* [L. *mentum*, chin; *gerere*, to carry] supporting or bearing the mentum.

mentomeckelian *a.* [L. *mentum*, chin; J. F. *Meckel*, junior, German anatomist] *appl.* a cartilage bone, present in a few lower vertebrates, at either side of mandibular symphysis, produced by ossification of the tip of Meckel's cartilage.

mentum *n.* [L. *mentum*, chin] the chin; medial part of gnathochilarium in Diplopoda; region of labium between prementum and submentum in insects; projection between head and foot of some gastropods.

mere *n.* [Gk. *meros*, part] a part; blastomere, *q.v.*

mericarp *n.* [Gk. *meris*, part; *karpos*, fruit] a 1-seeded unit which breaks off from a syncarpous ovary at maturity; *alt.* hemicarp; *cf.* cremocarp, schizocarp.

mericlinal *a.* [Gk. *meris*, part; *klinein*, to bend] partly periclinal, *appl.* chimaera with inner tissue of one species only partly surrounded by outer tissue of the other.

meridional *a.* [L. *meridies*, noon, south] running from pole to pole along a meridian.

meridional canal in ctenophores, a canal into which adradial canals open.

meridional furrow a longitudinal furrow extending from pole to pole of a segmenting egg.

merisis *n.* [Gk. *merizein*, to divide] increase in size owing to cell division; *cf.* auxesis.

merism metamerism, *q.v.*

merismatic *a.* [Gk. *merismos*, partition] dividing or separating into cells or segments; meristematic, *q.v.*

merismatoid, merismoid *a.* [Gk. *merismos*, partition; *eidos*, like] with branched pileus.

merispore *n.* [Gk. *meris*, part; *sporos*, seed] a segment or spore of a multicellular spore body.

meristele *n.* [Gk. *meris*, part; *stēlē*, pillar] a separate part of a monostelic stem passing outwards from stele to leaves; the branch of a stele supplying a leaf.

meristem *n.* [Gk. *meristos*, divided] plant tissue capable of undergoing mitosis and so giving rise to new cells and tissues, as at growing tips (apical meristem), and cork cambium and cambium (lateral meristems); *appl.* spore formed and abstricted at tip or growing point of a hypha or sterigma.

meristematic *pert.* meristem; *alt.* merismatic.

meristematic ring tube of meristematic tissue between cortex and pith, subtending the apical meristem and giving rise to vascular tissues; *alt.* residual meristem.

meristic *a.* [Gk. *meristos*, divided] segmented; divided off into parts; differing in number of parts.

meristic variation changes in number of parts or segments, and in geometrical relations of the parts; *cf.* substantive variation.

meristoderm *n.* [Gk. *meristos*, divided; *derma*, skin] outer layer of thallus when thickened by meristematic tissue.

meristogenetic, meristogenous *a.* [Gk. *meristos*, divided; *genesis*, descent] developing from meristem; developing from a single hyphal cell or a group of contiguous cells.

merithallus *n.* [Gk. *meris*, part; *thallos*, young shoot] a stem unit, an internode.

Merkel's corpuscle [*F. S. Merkel*, German anatomist] a tactile receptor, in skin and in submucosa of mouth; *cf.* Grandry's corpuscle.

mermaid's purse horny, floating or fixed, egg envelope of elasmobranchs.

Mermis a genus of nematodes parasitic on ants and giving them abnormal characteristics.

mermithaner *n.* [Gk. *mermis*, cord; *anēr*, male] male ant parasitized by *Mermis*.

mermithergate *n.* [Gk. *mermis*, cord; *ergatēs*, worker] an enlarged worker ant parasitized by *Mermis*.

mermithogyne *n.* [Gk. *mermis*, cord; *gynē*, female] female ant parasitized by *Mermis*.

Mermithoidea *n.* [*Mermis*, generic name; Gk. *eidos*, like] Trichosyringida, *q.v.*

meroandry *n.* [Gk. *meros*, part; *anēr*, male] the condition of having a reduced number of testes, as a single pair in certain Oligochaeta; *cf.* holandry.

meroblast *n.* [Gk. *meros*, part; *blastos*, bud] intermediate stage between schizont and merozoite in some Sporozoa; a meroblastic ovum.

meroblastic *a.* [Gk. *meros*, part; *blastos*, bud] *appl.* ova which undergo only partial segmentation or cleavage in development; developing from part of the oosphere only.

merocerite *n.* [Gk. *meros*, thigh; *keras*, horn] the 4th segment of crustacean antennae.

merocrine *a.* [Gk. *meros*, part; *krinein*, to separate] *appl.* glands whose secretion accumulates below the free surface of the cell through which it is released, with no loss of cytoplasm so the cell can function repeatedly, e.g. goblet cells and sweat glands; *cf.* apocrine, holocrine.

merocytes *n.plu.* [Gk. *meros*, part; *kytos*, hollow] nuclei formed by repeated division of supernumerary sperm nuclei, as in egg of selachians, reptiles, and birds; schizonts, *q.v.*

merogametes *n.plu.* [Gk. *meros*, part; *gametēs*, spouse] protistan individuals smaller than normal cells and specialized for syngamy; *alt.* microgametes; *cf.* hologametes.

merogamy microgamy, *q.v.*

merogastrula *n.* [Gk. *meros*, part; *dim.* of *gastēr*, stomach] the gastrula formed from a meroblastic ovum.

merogenesis *n.* [Gk. *meros*, part; *genesis*, descent] formation of parts; segmentation.

merogeny merogony, *q.v.*

merognathite *n.* [Gk. *meros*, thigh; *gnathos*, jaw] fourth segment of crustacean mouth-part.

merogony *n.* [Gk. *meros*, part; *gonē*, generation] development of normal young of small size, from part of an egg, in which there was no female pronucleus; schizogony, *q.v.*; *alt.* merogeny.

merohyponeuston *n.* [Gk. *meros*, part; *hypo*, under; *neustos*, floating] organisms which live in the top 5 cm of water in the larval stage only.

meroistic *a.* [Gk. *meros*, part; *ōon*, egg] ovariole containing nutritive or nurse cells; *cf.* acrotrophic, polytrophic, panoistic.

merokinesis *n.* [Gk. *meros*, part; *kinēsis*, movement] formation and division of a thread-like chromosome in the karyomeres.

merome *n.* [Gk. *meros*, part] merosome, *q.v.*

meromixis *n.* [Gk. *meros*, part; *mixis*, a mixing] a type of genetic exchange in bacteria where transfer of genetic material is in one direction only.

meromorphosis *n.* [Gk. *meros*, part; *morphōsis*, shaping] regeneration of a part with the new part less than that lost.

meromyarian *a.* [Gk. *meros*, part; *mys*, muscle] with only a few longitudinal rows of muscle cells as in some nematodes.

meromyosin *n.* [Gk. *meros*, part; *mys*, muscle] a constituent unit in the myosin molecule.

meron *n.* [Gk. *mēros*, upper thigh] posterior portion of coxa of insects; sclerite between middle and hind coxae, or immediately above hind coxa, in Diptera; *alt.* meseusternum.

meronephridia micronephridia, *q.v.*

meront *n.* [Gk. *meros*, part; *on*, being] any unit produced by cleavage or schizogony; a uninucleate schizont stage in Neosporidia, succeeding the planont stage.

meroplankton *n.* [Gk. *meros*, part; *plangktos*, wandering] plankton living only part-time near the surface; temporary plankton, consisting mainly of eggs and larvae; seasonal plankton; *cf.* holoplankton.

meropodite *n.* [Gk. *mēros*, upper thigh; *pous*, foot] fourth segment of thoracic appendage in crustaceans; femur in spiders; *alt.* meros, merus.

meros meropodite, *q.v.*

merosomatous *a.* [Gk. *meros*, part; *sōma*, body]

appl. ascidiozooids divided into 2 regions, thorax and abdomen.

merosome *n.* [Gk. *meros*, part; *sōma*, body] a body segment or somite; *alt.* merome, metamere.

merospermy *n.* [Gk. *meros*, part; *sperma*, seed] the condition where the nucleus of a sperm does not fuse with that of the egg, and development of the new individual is by parthenogenesis (gynogenesis).

merosporangium *n.* [Gk. *meros*, part; *sporos*, seed; *anggeion*, vessel] outgrowth from the apex of a sporangiophore, producing a row of spores, as in certain Mucorales.

merosthenic *a.* [Gk. *mēros*, upper thigh; *sthenos*, strength] with unusually developed hindlimbs.

merostomata *n.* [Gk. *meros*, part; *stoma*, mouth] a class of aquatic chelicerates which breathe by gills and have chelicerae and walking legs on the prosoma, and an opisthosoma with some segments lacking appendages.

merotomy *n.* [Gk. *meros*, part; *temnein*, to cut] segmentation or division into parts.

merotype *n.* [Gk. *meros*, part; *typos*, pattern] part of the same perennial plant or vegetatively propagated animal from which a holotype was taken.

Merozoa *n.* [Gk. *meros*, part; *zōon*, animal] a group of tapeworms which undergo strobilization and so produce chains of proglottides; *cf.* Monozoa.

merozoite *n.* [Gk. *meros*, part; *zōon*, animal] schizozoite, *q.v.*

merozoon *n.* [Gk. *meros*, part; *zōon*, animal] a fragment of a unicellular animal containing part of the macronucleus, obtained by artificial division.

merozygote *n.* [Gk. *meros*, part; *zygōtos*, yoked] a zygote containing only part of the genome of one of the 2 uniting cells or gametes; *cf.* holozygote.

merrythought furcula of birds, formed by coalesced clavicles.

merus meropodite, *q.v.*

Méry's glands [*J. Méry*, French anatomist] bulbo-urethral glands, *q.v.*

mesad mediad, *q.v.*; mesiad, *q.v.*

mesadenia *n.plu.* [Gk. *mesos*, middle; *adēn*, gland] mesodermal accessory genital glands in insects; *cf.* ectadenia.

mesal medial, *q.v.*; mesial, *q.v.*

mesameoboid *a.* [Gk. *mesos*, middle; *amoibē*, change; *eidos*, form] *appl.* nucleated cells of blood islands from which blood corpuscles are derived.

mesanepimeron *n.* [Gk. *mesos*, middle; *ana*, up; *mēros*, upper thigh] sclerite above mesepimeron and below wing base, in Diptera.

mesanepisternum mesepisternum, *q.v.*

mesarch *a.* [Gk. *mesos*, middle; *archē*, beginning] *appl.* xylem having metaxylem developing in all directions from the protoxylem, characteristic of ferns; having the protoxylem surrounded by metaxylem; beginning in a mesic environment; *appl.* seres, *cf.* hydrarch, xerarch.

mesaticephalic *a.* [Gk. *mesatos*, mid; *kephalē*, head] having a cephalic index of from 75 to 80.

mesaxon *n.* [Gk. *mesos*, middle; *axōn*, axis] an ultramicroscopic membranous structure joining axon and Schwann cell.

Mesaxonia *n.* [Gk. *mesos*, middle; *axōn*, axis] a superorder of ungulates containing one order, the Perissodactyla, *q.v.*

mesaxonic *a.* [Gk. *mesos*, middle; *axōn*, axis] with the line dividing the foot passing up the middle digit, as in Perissodactyla; *cf.* paraxonic.

mescaline *n.* [Sp. *mescal*, liquor] a hallucinogenic alkaloid obtained from the dried tops of the mescal cactus, *Lophophora williamsii*, and sometimes used in psychiatry.

mesectoderm *n.* [Gk. *mesos*, middle; *ektos*, outside; *derma*, skin] epiblast of germinal disc before separation of mesoderm from ectoderm; parenchyma formed of descendants of ectodermal cells which migrated inwards; ectomesoderm, *q.v.*

mesembryo *n.* [Gk. *mesos*, middle; *embryon*, embryo] blastula, *q.v.*

mesencephalon *n.* [Gk. *mesos*, middle; *en*, in; *kephalē*, head] the 2nd primary brain vesicle; mid-brain, comprising corpora quadrigemina (bigemina), cerebral peduncles, and aqueduct of Sylvius.

mesenchyma, mesenchyme *n.* [Gk. *mesos*, middle; *engchein*, to pour in] a mass of tissue, intermediate between ectoderm and endoderm of a gastrula.

mesendoderm *n.* [Gk. *mesos*, middle; *endon*, within; *derma*, skin] cells lying posteriorly to lip of blastopore, partly invaginated with endoderm in gastrulation, in development of some molluscs; an embryonic cell layer which will give rise to both endoderm and mesoderm; *alt.* mesentoderm.

mesenterial *a.* [Gk. *mesos*, middle; *enteron*, gut] *pert.* a mesentery, *appl.* filaments of Actinozoa.

mesenteric *a.* [Gk. *mesos*, middle; *enteron*, gut] *pert.* a mesentery, *appl.* arteries, glands, nerves, veins, etc.; *pert.* mesenteron.

mesenteriole *n.* [L. *dim.* of *mesenterium*, mesentery] a fold of peritoneum derived from mesentery and retaining vermiform process or appendix in position; *alt.* mesoappendix.

mesenterium mesentery, *q.v.*

mesenteron *n.* [Gk. *mesos*, middle; *enteron*, gut] the main digestive cavity of Actinozoa as distinct from spaces between mesenteries, *cf.* metenteron; portion of alimentary canal lined by endoderm and derived from archenteron, *alt.* mid-gut.

mesentery *n.* [L. *mesenterium*, mesentery] a peritoneal fold serving to hold viscera in position; a muscular partition extending inwards from body wall in coelenterates; *alt.* mesenterium.

mesentoderm mesendoderm, *q.v.*

mesepimeron *n.* [Gk. *mesos*, middle; *epi*, upon; *mēros*, upper thigh] the epimeron of insect mesothorax; meskatepimeron in Diptera.

mesepisternum *n.* [Gk. *mesos*, middle; *epi*, upon; *sternon*, chest] mesoepisternum, sclerite below anterior spiracle in Diptera; *alt.* mesanepisternum, meso-episternum.

mesepithelium mesothelium, *q.v.*

mesethmoid *a.* [Gk. *mesos*, middle; *ēthmos*, sieve; *eidos*, form] between the 2 ectethmoid bones; *appl.* ethmoid plate of cranium when it ossifies; median cranial bone of vertebrates.

meseusternum *n.* [Gk. *mesos*, middle; *eu*, well; *sternon*, breast plate] meron, *q.v.*

mesiad *adv.* [Gk. *mesos*, middle; L. *ad*, to] towards or near the middle plane; *alt.* mesad.

mesial, mesian *a.* [Gk. *mesos*, middle] in the middle vertical or longitudinal plane; *alt.* mesal.

mesic *a.* [Gk. *mesos*, middle] conditioned by temperate moist climate, neither xeric nor hydric; *alt.* mesophil, mesophilic.

meskatepimeron *n.* [Gk. *mesos*, middle; *kata*, down; *epi*, upon; *mēros*, upper thigh] sclerite posterior to mesosternal area of Diptera; *alt.* mesepimeron.

meskatepisternum *n.* [Gk. *mesos*, middle; *kata*, down; *epi*, upon; *sternon*, chest] sclerite between root of wing and underside of mesothorax of Diptera; *alt.* sternopleurite, mesosternal area.

mesoappendix mesenteriole, *q.v.*

mesoarion mesovarium, *q.v.*

mesobenthos *n.* [Gk. *mesos*, middle; *benthos*, depths] animal and plant life of sea bottom when depth is between 200 and 1000 m.

mesobiota *n.* [Gk. *mesos*, middle; *bios*, life] soil organisms ranging in size from those just visible with a hand lens to those several centimetres long, i.e. the middle size range; *cf.* macrobiota, microbiota.

mesoblast(ema) *n.* [Gk. *mesos*, middle; *blastos*, bud] mesoderm, *q.v.*

mesoblastic *a.* [Gk. *mesos*, middle; *blastos*, bud] *pert.* or developing from mesoderm of an embryo.

mesobranchial *a.* [Gk. *mesos*, middle; *brangchia*, gills] *pert.* middle gill region, as in Crustacea.

mesobronchus *n.* [Gk. *mesos*, middle; *brongchos*, windpipe] in birds, the main trunk of a bronchus giving rise to secondary bronchi; *alt.* mesobronchium.

mesocaecum *n.* [Gk. *mesos*, middle; L. *caecus*, blind] the mesentery connected with the caecum.

mesocardium *n.* [Gk. *mesos*, middle; *kardia*, heart] an embryonic mesentery binding heart to pericardial walls; part of pericardium enclosing veins (venous m.) or aorta (arterial m.); mesocardial ligament: *see* gubernaculum.

mesocarp *n.* [Gk. *mesos*, middle; *karpos*, fruit] the middle layer of the pericarp; *cf.* sarcocarp.

mesocentrous *a.* [Gk. *mesos*, middle; *kentron*, centre] ossifying from a median centre.

mesocephalic *a.* [Gk. *mesos*, middle; *kephalē*, head] having a cranial capacity of between 1350 and 1450 cm^3; *cf.* megacephalic, microcephalic.

mesocercaria *n.* [Gk. *mesos*, middle; *kerkos*, tail] trematode larval stage between cercaria and metacercaria.

mesocerebrum [Gk. *mesos*, middle; L. *cerebrum*, brain] deuterocerebrum, *q.v.*

mesochilium *n.* [Gk. *mesos*, middle; *cheilos*, lip] the middle portion of labellum of orchids.

mesochimaera *n.* [Gk. *mesos*, middle; L. *chimaera*, monster] a periclinal chimaera in which only the middle of the 3 tissue layers is genetically different from the others.

mesochiton, mesochite *n.* [Gk. *mesos*, middle; *chitōn*, coat] middle layer of oogonial wall, between endochiton and exochiton, as in Fucales.

mesocoel *n.* [Gk. *mesos*, middle; *koilos*, hollow] middle portion of coelomic cavity; the 2nd of 3 main parts of coelom of molluscs; cavity of mesencephalon, *alt.* aqueduct of Sylvius, iter, midventricle, cerebral aqueduct.

mesocole *n.* [Gk. *mesos*, middle; L. *colere*, to dwell] an animal living in conditions in which there is neither an excess nor a deficiency of water.

mesocolic *a.* [Gk. *mesos*, middle; *kolon*, large intestine] *pert.* mesocolon, *appl.* lymph glands.

mesocolon *n.* [Gk. *mesos*, middle; *kolon*, large intestine] a mesentery or fold of peritoneum attaching colon to dorsal wall of abdomen.

mesocoracoid *a.* [Gk. *mesos*, middle; *korax*, crow; *eidos*, form] situated between hyper- and hypocoracoid, *appl.* middle part of coracoid arch of certain fishes.

mesocotyl *n.* [Gk. *mesos*, middle; *kotylē*, cup] internode between scutellum and coleoptile in grass seeds.

mesocycle *n.* [Gk. *mesos*, middle; *kyklos*, circle] a layer of tissue between xylem and phloem of a monostelic stem; part of conjunctive tissue of stele.

mesodaeum *n.* [Gk. *mesos*, middle; *hodaios*, pert. way] endodermal part of embryonic digestive tract, between stomodaeum and proctodaeum.

mesoderm *n.* [Gk. *mesos*, middle; *derma*, skin] embryonic layer lying between ectoderm and endoderm; *alt.* mesoblast, mesoblastema.

mesodermal, mesodermic *a.* [Gk. *mesos*, middle; *derma*, skin] *pert.*, derived, or developing from mesoderm.

mesodesm *n.* [Gk. *mesos*, middle; *desma*, bond] part of mesocycle.

mesodont *a.* [Gk. *mesos*, middle; *odous*, tooth] *appl.* stag beetles having a medium development of mandible projections.

meso-ectodermal *pert.* ectomesoderm, *alt.* ectomesodermal; ectomesogloeal, *q.v.*

meso-episternum mesepisternum, *q.v.*

mesofauna *n.* [Gk. *mesos*, middle; L. *faunus*, god of woods] animals of intermediate size, from 200 μm to 1 cm; *alt.* meiofauna.

mesofurca *n.* [Gk. *mesos*, middle; L. *furca*, fork] a forked apodeme of mesothorax in insects; *alt.* medifurca.

mesogamy *n.* [Gk. *mesos*, middle; *gamos*, marriage] entrance of the pollen tube into an ~vule through the funicle or the integument.

mesogaster *n.* [Gk. *mesos*, middle; *gastēr*, stomach] the mesentery or fold of peritoneum supporting the stomach.

mesogastric *a.* [Gk. *mesos*, middle; *gastēr*, stomach] *pert.* a mesogaster or mesogastrium, or to middle gastric region.

mesogastrium *n.* [Gk. *mesos*, middle; *gastēr*, stomach] mesentery connecting stomach with dorsal abdominal wall in embryo; middle abdominal region.

Mesogastropoda *n.* [Gk. *mesos*, middle; *gastēr*, stomach; *pous*, foot] an order of usually carnivorous prosobranchs with monopectinate ctenidia.

mesogenous *a.* [Gk. *mesos*, middle; *gennaein*, to produce] produced at or from the middle.

mesoglia *n.* [Gk. *mesos*, middle; *gloia*, glue] mesoglia of Hortega: microglia, *q.v.*; mesoglia of Robertson: oligodendroglia, *q.v.*

mesogloea *n.* [Gk. *mesos*, middle; *gloia*, glue] an intermediate non-cellular gelatinous layer in sponges and coelenterates.

mesognathion *n.* [Gk. *mesos*, middle; *gnathos*, jaw] the lateral segment of premaxilla, bearing lateral incisor.

mesohepar *n.* [Gk. *mesos*, middle; *hēpar*, liver]

mesentery supporting liver; *alt.* falciform ligament.

mesohydrophytic *a.* [Gk. *mesos*, middle; *hydōr*, water; *phyton*, plant] growing in temperate regions but requiring much moisture.

mesolamella *n.* [Gk. *mesos*, middle; L. *lamella*, thin plate] a thin mesogloeal layer between ocellus and gastrodermis in jellyfish.

mesolecithal *a.* [Gk. *mesos*, middle; *lekithos*, yolk] having a moderate yolk content; *cf.* centrolecithal, oligolecithal, telolecithal.

mesology *n.* [Gk. *mesos*, middle; *logos*, discourse] ecology, *q.v.*

mesome *n.* [Gk. *mesos*, middle] the axis regarded as a morphological unit of plants.

mesomere *n.* [Gk. *mesos*, middle; *meros*, part] middle zone of coelomic pouches in embryo; mesoblastic somite or protovertebra; medial branch of phallic lobe in insects.

mesometrium *n.* [Gk. *mesos*, middle; *mētra*, uterus] the mesentery of uterus and connecting tubes.

mesomitosis *n.* [Gk. *mesos*, middle; *mitos*, thread] mitosis within nuclear membrane, without cooperation of cytoplasmic elements; *cf.* metamitosis.

mesomorph *n.* [Gk. *mesos*, middle; *morphē*, form] a mesomorphic animal; a mesomorphic plant, usually a mesophyte.

mesomorphic *a.* [Gk. *mesos*, middle; *morphē*, form] having form, structure, or size average, normal, or intermediate between extremes as in mesophytes or mesomorphs; *alt.* mesoplastic.

mesomyodian *a.* [Gk. *mesos*, middle; *mys*, muscle; *eidos*, form] *appl.* birds with muscles of syrinx attached to middle of bronchial semirings.

meson *n.* [Gk. *meson*, the middle] the central plane, or region of it.

mesonephric *a.* [Gk. *mesos*, middle; *nephros*, kidney] *pert.* mesonephros, *appl.* tubules, *appl.* duct: Wolffian duct, *q.v.*

mesonephridium *n.* [Gk. *mesos*, middle; *nephros*, kidney; *idion*, dim.] a nephridium or excretory organ of certain invertebrates, derived from mesoblast.

mesonephros *n.* [Gk. *mesos*, middle; *nephros*, kidney] one of the middle of the 3 pairs of renal organs of vertebrate embryos which persist as the adult kidney of Anamniota; *alt.* Wolffian body, deutonephros; *cf.* metanephros.

mesonotum *n.* [Gk. *mesos*, middle; *nōton*, back] dorsal part of insect mesothorax; *alt.* dorsulum.

mesoparapteron *n.* [Gk. *mesos*, middle; *para*, beside; *pteron*, wing] a small sclerite of mesothorax of some insects.

mesopelagic *a.* [Gk. *mesos*, middle; *pelagos*, sea] *pert.* or inhabiting, the ocean at depths between 200 and 1000 m, i.e. between the epipelagic and bathypelagic zones.

mesoperidium *n.* [Gk. *mesos*, middle; *pēridion*, small wallet] a middle layer, between endoperidium and exoperidium, of the coat investing the sporophore of certain fungi.

mesopetalum *n.* [Gk. *mesos*, middle; *petalon*, leaf] labellum of an orchid.

mesophanerophyte *n.* [Gk. *mesos*, middle; *phaneros*, manifest; *phyton*, plant] tree from 8 to 30 m in height.

mesophil(ic) *a.* [Gk. *mesos*, middle; *philein*, to love] thriving at moderate temperatures, at be-

tween 20°C and 40°C, when *appl.* bacteria; mesic, *q.v.*; *n.* mesophile.

mesophloem *n.* [Gk. *mesos*, middle; *phloios*, smooth bark] middle or green bark; *alt.* mesophloeum.

mesophragma *n.* [Gk. *mesos*, middle; *phragma*, fence] a chitinous piece descending into interior of insect body with postscutellum for base; M- or Hensen's line, *q.v.*

mesophryon *n.* [Gk. *mesos*, middle; *ophrys*, eyebrow] glabella of trilobites.

mesophyll *n.* [Gk. *mesos*, middle; *phyllon*, leaf] the internal parenchyma of a leaf, usually photosynthetic; *alt.* diploe.

mesophyllous *a.* [Gk. *mesos*, middle; *phyllon*, leaf] having leaves of moderate size, between microphyllous and macrophyllous.

mesophyte *n.* [Gk. *mesos*, middle; *phyton*, plant] a plant thriving in temperate climate with normal amount of moisture.

mesoplankton *n.* [Gk. *mesos*, middle; *plangktos*, wandering] drifting animal and plant life from 200 m downwards; drifting organisms of medium size.

mesoplast *n.* [Gk. *mesos*, middle; *plastos*, moulded] a cell nucleus.

mesoplastic *a.* [Gk. *mesos*, middle; *plastos*, moulded] mesomorphic, *appl.* constitutional type.

mesoplastron *n.* [Gk. *mesos*, middle; F. *plastron*, breast plate] plate between hyo- and hypoplastron of certain turtles.

mesopleurite, mesopleuron *n.* [Gk. *mesos*, middle; *pleura*, side] lateral sclerite of mesothorax as in Diptera.

mesopodial *a.* [Gk. *mesos*, middle; *pous*, foot] having a supporting structure, such as a stipe, in a central position; *pert.* a mesopodium.

mesopodium *n.* [Gk. *mesos*, middle; *pous*, foot] leaf stalk or petiole region of leaf; middle part of molluscan foot; the metacarpus or metatarsus.

mesopore *n.* [Gk. *mesos*, middle; *poros*, passage] opening between zooecia, containing a minute zooecium with transverse partitions, in fossil Polyzoa.

mesopostscutellum *n.* [Gk. *mesos*, middle; L. *post*, after; *scutellum*, small shield] postscutellum of mesothorax in insects.

mesopraescutum *n.* [Gk. *mesos*, middle; L. *prae*, before; *scutum*, shield] praescutum of mesothorax in insects; *alt.* mesoprescutum.

mesopsammon *n.* [Gk. *mesos*, between; *psammos*, sand; *on*, being] psammon, *q.v.*

mesopterygium *n.* [Gk. *mesos*, middle; *pterygion*, little wing or fin] the middle of 3 basal pectoral fin cartilages in recent elasmobranchs; *cf.* metapterygium, propterygium.

mesopterygoid *n.* [Gk. *mesos*, middle; *pteryx*, wing; *eidos*, form] the middle of 3 pterygoid bone elements of teleosts; the ectopterygoid, *q.v.*

mesoptile *n.* [Gk. *mesos*, middle; *ptilon*, feather] prepenna following prototptile and succeeded by metaptile or by teleoptile.

mesorchium *n.* [Gk. *mesos*, middle; *orchis*, testicle] mesentery supporting testis.

mesorectum *n.* [Gk. *mesos*, middle; L. *rectus*, straight] mesentery supporting rectum.

mesorhinal *a.* [Gk. *mesos*, middle; *rhines*, nostrils] between nostrils.

mesorhinium *n.* [Gk. *mesos*, middle; *rhis*, nose] the internarial surface region of a bird's bill.

mesosalpinx n. [Gk. *mesos*, middle; *salpingx*, trumpet] the portion of broad ligament enclosing uterine tube.

mesosaprobic a. [Gk. *mesos*, middle; *sapros*, rotten; *bios*, life] *appl.* aquatic habitats midway between polysaprobic and katharobic, having a decreased quantity of oxygen and substantial organic decomposition.

Mesosauria, mesosaurs n., *n.plu.* [Gk. *mesos*, middle; *sauros*, lizard] an order of lower Permian anapsid reptiles which were fish-eating and slender with long jaws.

mesoscapula n. [Gk. *mesos*, middle; L. *scapula*, shoulder blade] the spine on the scapula when considered a distinct unit.

mesoscutellum n. [Gk. *mesos*, middle; L. *scutellum*, small shield] scutellum of insect mesothorax.

mesoscutum n. [Gk. *mesos*, middle; L. *scutum*, shield] scutum of insect mesothorax.

mesosoma n. [Gk. *mesos*, middle; *sōma*, body] middle part of body of certain invertebrates, especially when their primitive segmentation is obscured; anterior segments of opisthosoma in arachnids; *alt.* mesosome; *cf.* praeabdomen.

mesosome n. [Gk. *mesos*, middle; *sōma*, body] an invagination of the cell membrane near the nuclear area of some prokaryotic cells, to which DNA molecules may be attached, or which may be concerned with cell wall formation or respiration; phallosome, *q.v.*; mesosoma, *q.v.*

mesosperm n. [Gk. *mesos*, middle; *sperma*, seed] integument investing nucellus of ovule.

mesospore n. [Gk. *mesos*, middle; *sporos*, seed] a unicellular teleutospore in certain rust fungi; amphispore, *q.v.*; mesosporium, *q.v.*

mesosporium n. [Gk. *mesos*, middle; *sporos*, seed] the intermediate of 3 spore coats; *alt.* mesospore.

mesostate n. [Gk. *mesos*, middle; *stasis*, standing] intermediate stage in metabolism.

mesostereom n. [Gk. *mesos*, middle; *stereos*, solid] the middle layer of thecal plates of Cystoidea.

mesosternebra n. [Gk. *mesos*, middle; *sternon*, breast bone] a part of developing mesosternum.

mesosternellum n. [Gk. *mesos*, middle; *dim.* of L. *sternum*, breast bone] a small rod-like plate articulating posteriorly with the insect mesosternum.

mesosternum n. [Gk. *mesos*, middle; L. *sternum*, breast bone] middle part of sternum of vertebrates, *alt.* corpus sterni; sternum of mesothorax of insects; mesosternal area, episternum of mesothorax, or meskatepisternum of Diptera; *alt.* mesostethium.

mesostethium n. [Gk. *mesos*, middle; *stēthos*, chest] mesosternum, *q.v.*

mesostylous a. [Gk. *mesos*, middle; *stylos*, pillar] having styles of intermediate length, *appl.* heterostylus flowers.

mesotarsus n. [Gk. *mesos*, middle; L.L. *tarsus*, ankle joint] a middle-limb tarsus; a. mesotarsal.

mesotergum n. [Gk. *mesos*, middle; L. *tergum*, back] median arched portion or axis of trilobite body.

mesothecium n. [Gk. *mesos*, middle; *thēkē*, box] the middle investing layer of an anther sac; lichen thecium.

mesotheic a. [Gk. *mesos*, middle; *theinai*, to settle] neither highly susceptible nor entirely resistant to parasites or infection.

mesothelium n. [Gk. *mesos*, middle; *thēlē*, nipple] mesoderm bounding primitive coelom and giving rise to muscular and connective tissue; epithelium of mesoblastic origin; *alt.* coelarium, mesepithelium.

mesotherm n. [Gk. *mesos*, middle; *thermē*, heat] plant thriving in moderate heat, as in a warm temperate climate.

mesothoracic a. [Gk. *mesos*, middle; *thōrax*, chest] *pert.* or in region of mesothorax, *appl.* spiracle of insects.

mesothorax n. [Gk. *mesos*, middle; *thorax*, chest] the middle segment of thoracic region of insects; *alt.* medithorax; *cf.* prothorax, metathorax.

mesotic a. [Gk. *mesos*, middle; *ous*, ear] *appl.* paired chondrocranial cartilages in birds, between parachordal and acrochordal; *alt.* basiotic.

mesotriaene n. [Gk. *mesos*, middle; *triaina*, trident] aberrant type of triaene spicule.

mesotrochal a. [Gk. *mesos*, middle; *trochos*, wheel] *appl.* an annulate larva with circlet of cilia round middle of body.

mesotrophic a. [Gk. *mesos*, middle; *trophē*, nourishment] mixotrophic, *q.v.*; providing a moderate amount of nutrition, *appl.* environment.

mesotropic a. [Gk. *mesos*, middle; *tropikos*, turning] turning or directed toward the middle or toward the median plane.

mesotympanic n. [Gk. *mesos*, middle; *tympanon*, drum] a bone in suspensory apparatus of lower jaw in fishes, between the quadrate and hyomandibular; *alt.* symplectic.

mesovarium n. [Gk. *mesos*, middle; L. *ovarium*, ovary] mesentery of ovary; suspensory mesentery in fishes; *alt.* mesoarion.

mesoventral a. [Gk. *mesos*, middle; L. *venter*, belly] in middle ventral region.

Mesozoa, mesozoans n., *n.plu.* [Gk. *mesos*, middle; *zōon*, animal] a subkingdom, phylum, or subphylum of small multicellular animals which are parasitic in marine organisms and have the body composed of 2 layers of cells; they possibly should be included in the Platyhelminthes.

Mesozoic a. [Gk. *mesos*, middle; *zōē*, life] *appl.* or *pert.* secondary geological era, the age of reptiles.

messenger-like RNA (ml RNA) heterogenous nuclear RNA, *q.v.*

messenger RNA (mRNA) a type of RNA complementary to DNA from which it is made by transcription, and which functions in translation by lining up amino acids in the correct order on the ribosomes to form polypeptides; *alt.* template RNA.

mestom(e) n. [Gk. *mestos*, filled] a vascular bundle, including hadrome and leptome.

mestom(e) sheath a sheath with thickened walls around a vascular bundle; the inner of 2 sheaths in some grasses, such as festucoids; any endodermal sheath.

met methionine, *q.v.*

metabasis n. [Gk. *metabasis*, alteration] transition; change, as of symptoms; transfer of energy.

metabiosis n. [Gk. *meta*, after; *biōsis*, a living] condition in which one organism lives only after another has prepared its environment and has died; changed condition of living resulting from an external cause, as bacterial mutations due to radiation.

Metabola *n.* [Gk. *metabolē*, change] Pterygota, *q.v.*

metabolic *a.* [Gk. *metabolē*, change] *pert.* metabolism.

metabolin metabolite, *q.v.*

metabolism *n.* [Gk. *metabolē*, change] the chemical change, constructive and destructive, occurring in living organisms; *alt.* metastasis, transformation.

metabolite *n.* [Gk. *metabolē*, change] any product of metabolism; *alt.* metabolin.

metaboly *n.* [Gk. *metabolē*, change] change of shape resulting in movement as in euglenoids and other flagellates; *alt.* euglenoid movement.

metabranchial *a.* [Gk. *meta*, after; *brangchia*, gills] *pert.* or in region of posterior gill region.

metacarpal *a.* [Gk. *meta*, after; *karpos*, wrist] *pert.* metacarpus, *appl.* bones, articulations, etc. *n.* A primary wing quill in the metacarpal region, *alt.* metacarpale.

metacarpophalangeal *a.* [Gk. *meta*, after; *karpos*, wrist; *phalanges*, ranks] *appl.* articulations between metacarpal bones and phalanges.

metacarpus *n.* [Gk. *meta*, after; *karpos*, wrist] the skeletal part of hand between wrist and fingers, consisting typically of 5 cylindrical bones.

metacele metacoel, *q.v.*

metacentric *a.* [Gk. *meta*, among; *kentron*, centre] having the centromere at or near the middle, *appl.* chromosomes, *cf.* acrocentric, telocentric. *n.* A metacentric or V-shaped chromosome, *alt.* isochromosome.

metacercaria *n.* [Gk. *meta*, after; *kerkos*, tail] encysted stage after cercaria; *alt.* adolescaria.

metacerebrum tritocerebrum, *q.v.*

metacestode *n.* [Gk. *meta*, after; *kestos*, girdle; *eidos*, form] a larval tapeworm, found in an intermediate host; *cf.* plerocestoid.

Metachlamydeae *n.* [Gk. *meta*, among; *chlamys*, cloak] Sympetalae, *q.v.*

metachroic *a.* [Gk. *meta*, change of; *chrōs*, colour] changing colour, as older tissue in fungi.

metachromasis, metachromasy *n.* [Gk. *meta*, change of; *chrōma*, colour] condition of certain tissues and cell components which, treated with basic aniline stains, show other than the fundamental colour constituent.

metachromatic *a.* [Gk. *meta*, change of; *chrōma*, colour] *appl.* substances characterized by metachromasis; *appl.* various bodies that show metachromasis and are made of metachromatin, such as Woronin bodies.

metachromatin *n.* [Gk. *meta*, change of; *chrōma*, colour] a substance densely staining with basic dyes, that is probably a nucleic acid and may be found in the cytoplasm, when it is also known as volutin, or in plant vacuoles.

metachromatinic grains grains of metachromatin, *q.v.*

metachrome a metachromatic grain or granule.

metachromy *n.* [Gk. *meta*, change of; *chrōma*, colour] change in colour, as of flowers.

metachronal, metachronic *a.* [Gk. *metachronos*, done afterwards] one acting after the other, *appl.* rhythm of movement of cilia.

metachrosis *n.* [Gk. *meta*, change of; *chrōsis*, colouring] ability to change skin colour by expansion or contraction of pigment cells.

metacneme *n.* [Gk. *meta*, after; *knēmē*, tibia] a secondary mesentery of Zoantharia.

metacoel(e) *n.* [Gk. *meta*, after; *koilos*, hollow] the posterior part of coelom as of molluscs; anterior extension of 4th ventricle of brain; *alt.* metacele.

metacone *n.* [Gk. *meta*, after; *kōnos*, cone] postero-external cusp of upper molar tooth.

metaconid *n.* [Gk. *meta*, after; *kōnos*, cone] postero-internal cusp of lower molar tooth.

metaconule *n.* [Gk. *meta*, after; *kōnos*, cone] posterior secondary cusp of upper molar tooth.

metacoracoid *n.* [Gk. *meta*, after; *korax*, crow; *eidos*, form] posterior part of coracoid.

metacromion *n.* [Gk. *meta*, after; *akros*, summit; *ōmos*, shoulder] posterior branch process of acromion process of scapular spine.

metacyclic *a.* [Gk. *meta*, after; *kyklos*, circle] *appl.* final infective forms, of certain parasitic Protozoa, which pass on to next host.

metadiscoidal *a.* [Gk. *meta*, after; *diskos*, disc; *eidos*, form] *appl.* placenta in which villi are at first scattered and later restricted to a disc, as in primates.

metadromous *a.* [Gk. *meta*, after; *dromos*, running] with primary veins of segment arising from upper side of midrib.

meta-epimeron metepimeron, *q.v.*

meta-episternum metepisternum, *q.v.*

metaesthetism *n.* [Gk. *meta*, after; *aisthētos*, perceptible by senses] doctrine that 'consciousness is a product of evolution of matter and force'.

metafemale *n.* [Gk. *meta*, change of; L. *femina*, woman] a female *Drosophila* with a normal diploid set of autosomes but having 3 X-chromosomes; *alt.* superfemale.

metagastric *a.* [Gk. *meta*, after; *gastēr*, stomach] *pert.* posterior gastric region.

metagastrula *n.* [Gk. *meta*, after; *gastēr*, stomach] a modified form of gastrula.

metagenesis *n.* [Gk. *meta*, after; *genesis*, descent] alternation of generations, especially of sexual and asexual generations, and especially in animals; *cf.* heterogenesis.

metagnathous *a.* [Gk. *meta*, change of; *gnathos*, jaw] having mouth-parts for biting in the larval stage and for sucking in the adult, as certain insects; having the points of the beak crossed, as crossbills.

metagon *n.* [Gk. *meta*, after; *gonos*, offspring] in the ciliate *Paramecium aurelia*, a type of RNA which behaves like mRNA, but when the *Paramecium* is digested by another ciliate, *Didinium*, it behaves like an RNA virus.

metagyny *n.* [Gk. *meta*, afterwards; *gynē*, female] protandry, *q.v.*

metakentrin *n.* [Gk. *meta*, after; *kentron*, sharp point] luteinizing hormone, *q.v.*

metakinesis *n.* [Gk. *meta*, after; *kinēsis*, movement] metaphase, *q.v.*; prometaphase, *q.v.*; hypothetical quality of organisms which has the potentiality of evolving into consciousness.

metaleptic *a.* [Gk. *metalēpsis*, participation] associated in a process or action; operating together; *alt.* synergic.

metallic *a.* [Gk. *metallon*, mine] irridescent, *appl.* colours due to interference by fine striae or thin lamellae, as in insects.

metalloenzyme an enzyme made of a protein combined with metal ions.

metaloph *n.* [Gk. *meta*, after; *lophos*, crest] the

posterior crest of a molar tooth, uniting meta-cone, metaconule, and hypocone.

metamale n. [Gk. meta, change of; L. mas, male] a male Drosophila with one X-chromosome and 3 sets of autosomes; alt. supermale.

metamere n. [Gk. meta, after; meros, part] mero-some, q.v.

metameric a. [Gk. meta, after; meros, part] pert. metamerism or segmentation.

metamerism a. [Gk. meta, after; meros, part] the condition of a body divided up into segments more-or-less alike; alt. segmentation, zonal sym-metry, merism, metameric segmentation.

metamerized a. [Gk. meta, after; meros, part] segmented.

metamitosis n. [Gk. meta, after; mitos, thread] mitosis in which cytoplasm and nucleus are both involved, so the nuclear membrane disappears and the chromosomes lie in the cytoplasm; cf. mesomitosis.

Metamonadina n. [Gk. meta, after; monas, unit] an order of Zoomastigina including the Poly-mastigina and Hypermastigina.

metamorphosis n. [Gk. meta, change of; mor-phōsis, form] change of form and structure under-gone by an animal from embryo to adult stage, as in insects and amphibians; transformation of one structure into another, as of stamens into petals, alt. homoeosis; interference with normal sym-metry in flowers; internal chemical change.

metamorphy n. [Gk. meta, change of; morphē, form] homoeosis, q.v.

metamps n.plu. [Gk. meta, change of; morphē, form] different forms of same species, as in cer-tain sponges.

metamyelocyte n. [Gk. meta, beyond; myelos, marrow; kytos, hollow] a myelocyte with horse-shoe-shaped nucleus before transformation into a leucocyte.

metanauplius n. [Gk. meta, after; L. nauplius, kind of shellfish] larval stage of Crustacea, suc-ceeding nauplius stage.

metandry n. [Gk. meta, after; anēr, male] mero-andry with retention of posterior pair of testes only, cf. proandry; protogyny, q.v., cf. protandry.

Metanemertini, Metanemertea n. [Gk. meta, among; Nēmertēs, a nereid] an order of Nemer-tini, including those in which the lateral nerves and brain lie inside the somatic muscles.

metanephric a. [Gk. meta, after; nephros, kidney] pert. or in region of hind-kidney.

metanephridium n. [Gk. meta, after; nephros, kidney; idion, dim.] a nephridial tubule with opening into the coelom.

metanephromixium n. [Gk. meta, after; nephros, kidney; mixis, mingling] a metanephridium when opening into the coelomoduct; cf. mixonephri-dium.

metanephros n. [Gk. meta, after; nephros, kid-ney] the organ arising behind mesonephros and replacing it as functional kidney of fully de-veloped Amniota; alt. hind-kidney.

metanotum n. [Gk. meta, after; nōton, back] not-um or tergum of insect metathorax; a small scler-ite between postnotum and 1st abdominal tergite; formerly postnotum.

metanucleus n. [Gk. meta, after; L. nucleus, ker-nel] egg nucleolus after extrusion from germinal vesicle.

metapeptone n. [Gk. meta, after; peptos, diges-ted] a product of action of gastric juice on albu-mins.

metaphase n. [Gk. meta, after; phainein, to ap-pear] stage in mitosis or meiosis when the chromo-somes have reached an equilibrium position on the equatorial plate with all the centromeres lying either along the spindle equator (mitosis and meiosis 2nd metaphase) or equidistant from it (meiosis 1st metaphase); alt. aster phase; cf. meta-kinesis.

metaphase plate equatorial plate, q.v.

metaphery n. [Gk. metapherein, to transfer] dis-placement of organs.

metaphloem n. [Gk. meta, after; phloios, inner bark] part of the primary phloem developing after the protophloem and before the secondary phloem if this is present.

metaphragma n. [Gk. meta, after; phragma, fence] an internal metathoracic septum in in-sects.

metaphysis n. [Gk. meta, besides; physis, growth] paraphysis, q.v., of fungi; vascular part of diaphysis adjoining epiphyseal cartilage.

metaphyte n. [Gk. meta, after; phyton, plant] a multicellular plant; cf. protophyte.

metaplasia n. [Gk. meta, change of; plasis, mould-ing] conversion of tissue from one form to another, as in ossification.

metaplasis n. [Gk. meta, after; plasis, mould-ing] the mature period in life of an individual.

metaplasm n. [Gk. meta, after; plasma, mould] non-living materials produced by proto-plasmic activity, such as starch, yolk, fat, cell walls; alt. ergastic materials, deutoplasm; cf. organ-oids.

metaplastic a. [Gk. meta, after; plastos, moulded] pert. metaplasia; pert. metaplasm.

metaplastic or metaplasmic bodies grains of protoplasm which are stages or products of meta-bolism and not true protoplasm.

metapleural a. [Gk. meta, after; pleura, side] posteriorly and laterally situated; pert. meta-pleure; pert. metapleuron.

metapleure n. [Gk. meta, after; pleura, side] an abdominal or ventrolateral fold of integument of certain primitive Chordata.

metapleuron n. [Gk. meta, after; pleura, side] the pleuron of insect metathorax.

metapneustic a. [Gk. meta, after; pneuma, breath] appl. insect larva with only the terminal pair of spiracles; cf. peripneustic.

metapodeon n. [Gk. meta, after; podeōn, neck] that part of insect abdomen behind petiole or podeon; alt. metapodeum, metapodium.

metapodial a. [Gk. meta, after; pous, foot] pert. a metapodeon or to a metapodium.

metapodium n. [Gk. meta, after; pous, foot] pos-terior portion of molluscan foot; portion of foot between tarsus and digits, cf. acropodium; in 4-footed animals, metacarpus and metatarsus; meta-podeon, q.v.

metapodosoma n. [Gk. meta, after; pous, foot; sōma, body] body region bearing 3rd and 4th pair of legs in Acarina.

metapolar cells second circlet of cells of polar cap or calotte of rhombogen of Rhombozoa.

metapophysis n. [Gk. meta, after; apo, from; phyein, to grow] a prolongation of a vertebral articular process especially in lumbar region, developed in certain vertebrates; *alt*. mamillary process.

metapore *n*. [Gk. *meta*, after; *poros*, channel] Magendie's foramen, *q.v.*

metapostscutellum *n*. [Gk. *meta*, after; L. *post*, after; *scutellum*, small shield] postscutellum of insect metathorax.

metapraescutum *n*. [Gk. *meta*, after; L. *prae*, before; *scutum*, shield] prescutum of insect metathorax; *alt*. metaprescutum.

metaprotein *n*. [Gk. *meta*, after; *prōteion*, first] a 1st fraction product of the hydrolysis of protein, consisting of large-sized molecules, which are further hydrolysed to proteoses; *alt*. infraprotein.

metapterygium *n*. [Gk. *meta*, after; *pterygion*, little wing or fin] the posterior basal fin cartilage, pectoral or pelvic, of recent elasmobranchs; *cf*. propterygium, mesopterygium.

metapterygoid *n*. [Gk. *meta*, after; *pteryx*, wing; *eidos*, form] posterior of 3 pterygoid elements in certain lower vertebrates.

metaptile *n*. [Gk. *meta*, after; *ptilon*, feather] a plumose penna or feather; *cf*. mesoptile, teleoptile.

metarachis *n*. [Gk. *meta*, after; *rhachis*, spine] face of Pennatulacea which coincides with sulcus of terminal zooid—so-called dorsal surface.

metarchon *n*. [Gk. *meta*, change of; *archōn*, ruler] an external stimulus artificially introduced into the environment of an organism for the purpose of modifying its behaviour and eliciting an inappropriate response or inhibiting an appropriate one.

metarhodopsin *n*. [Gk. *meta*, after; *rhodon*, rose; *opsis*, sight] transient orange product of lumirhodopsin, dissociating into *trans*-vitamin A, aldehyde, and scotopsin.

metarteriole *n*. [Gk. *meta*, besides; L.L. *arteriola*, small artery] branch of an arteriole between arteriole and arterial capillaries.

metarubricyte *n*. [Gk. *meta*, after; L. *ruber*, red; Gk. *kytos*, hollow] normoblast, *q.v.*

metascolex *n*. [Gk. *meta*, after; *skōlex*, a worm] a massive organ formed by enlargement of the neck area directly behind the scolex in some cestodes.

metascutellum *n*. [Gk. *meta*, after; L. *scutellum*, small shield] scutellum of insect metathorax.

metascutum *n*. [Gk. *meta*, after; L. *scutum*, shield] scutum of insect metathorax.

metaseptum *n*. [Gk. *meta*, after; L. *septum*, partition] a secondary or subsequently formed septum; a protoplasmic partition.

metasicula *n*. [Gk. *meta*, after; L. *sicula*, small dagger] part of the sicula from which the 1st theca buds laterally, in graptolites.

metasitism *n*. [Gk. *meta*, after; *sitos*, food] a cannibalistic mode of life.

metasoma *n*. [Gk. *meta*, after; *sōma*, body] the posterior region of body; posterior region of opisthosoma of arachnids and other chelicerates and some crustaceans; *see* postabdomen; abdomen, as of woodlice; *alt*. metasome.

metasomatic *a*. [Gk. *meta*, after; *sōma*, body] *pert*. or situated in metasoma.

metasome metasoma, *q.v.*

metasperm *n*. [Gk. *meta*, after; *sperma*, seed] angiosperm, *q.v.*

metasporangium *n*. [Gk. *meta*, after; *sporos*,

seed; *anggeion*, vessel] a sporangium containing resting spores.

metastasis *n*. [Gk. *metastasis*, removal] metabolism, *q.v.*; transference of function from one organ to another; transport of bacteria or malignant cells by the circulatory system.

metastatic life history that of certain Trematoda in which the young form, after entering intermediate host, metamorphoses into adult, after which intermediate host is swallowed by final host.

metasternum *n*. [Gk. *meta*, after; L. *sternum*, breast bone] the sternum of insect metathorax; sternum of 4th segment of podosoma in Acarina; posterior sternal part, or xiphisternum, of Anura; xiphoid or ensiform process, posterior part of sternum of higher vertebrates.

metasthenic *a*. [Gk. *meta*, after; *sthenos*, strength] with well-developed posterior part of body.

metastigmate *a*. [Gk. *meta*, after; *stigma*, mark] having posterior tracheal openings or stigmata, as in mites.

metastoma *n*. [Gk. *meta*, after; *stoma*, mouth] the 2-lobed lip of Crustacea; the hypopharynx of Myriapoda; median plate behind mouth in Palaeostraca; *alt*. metastome.

metastomial *a*. [Gk. *meta*, after; *stoma*, mouth] behind the mouth region; *appl*. segment posterior to peristomium or buccal segment in annelids; *pert*. metastoma.

metastructure *n*. [Gk. *meta*, after; L. *struere*, to build] ultramicroscopic organization; a suggested ultrastructure of cytoplasm.

metasyndesis *n*. [Gk. *meta*, after; *syndesis*, bond] telosyndesis, *q.v.*

metatarsal *a*. [Gk. *meta*, after; *tarsos*, foot] in region of metatarsus, *appl*. arteries, veins, etc.; *pert*. metatarsal bones.

metatarsophalangeal *a*. [Gk. *meta*, after; *tarsos*, foot; *phalangx*, troop] *appl*. articulations between metatarsus and phalanges of foot.

metatarsus *n*. [Gk. *meta*, after; *tarsos*, foot] part of foot between tarsus and toes; 1st joint of tarsus in insects; 1st dactylopodite or basitarsus in spiders.

metathalamus *n*. [Gk. *meta*, after; *thalamos*, chamber] the geniculate bodies of the thalamencephalon.

Metatheria, metatherians *n*., *n.plu* [Gk. *meta*, beyond; *thērion*, small animal] an infraclass of therian mammals known from U. Cretaceous times, and including one order, Marsupialia, the marsupials or pouched mammals, having the young borne in a very immature state and migrating to a pouch (marsupium) where they are suckled.

metathetely *n*. [Gk. *metatheein*, to run behind; *telos*, completion] hysterotely, *q.v.*

metathorax n. [Gk. *meta*, after; *thōrax*, chest] posterior segment of insect thorax; *cf*. mesothorax, prothorax.

metatracheal *a*. [Gk. *meta*, between; L.L. *trachia*, windpipe] *appl*. wood, with xylem parenchyma located independently of the vessels, and scattered through the annual ring.

metatroch *n*. [Gk. *meta*, after; *trochos*, wheel] in a trochophore, a circular band of cilia behind the mouth.

metatrophic *a*. [Gk. *meta*, change of; *trophē*,

nourishment] requiring organic sources of both carbon and nitrogen.

metatympanic entotympanic, *q.v.*

metatype *n.* [Gk. *meta*, after; *typos*, image] a topotype of the same species as the holotype or lectotype.

metaxenia *n.* [Gk. *meta*, after; *xenia*, hospitality] physiological effect of pollen upon maternal tissue.

metaxylem *n.* [Gk. *meta*, after; *xylon*, wood] part of the primary xylem developing after the protoxylem and before the secondary xylem if that is present, and which is distinguished by wider vessels and tracheids.

Metazoa, metazoans *n., n.plu.* [Gk. *meta*, after; *zōon*, animal] a subkingdom of animals having the body made of cells organized into tissues and coordinated by a nervous system, and including all animals except the Protozoa and Parazoa, and possibly also the Mesozoa.

metazoaea *n.* [Gk. *meta*, after; *zōē*, life] a larval stage of Crustacea between zoaea and megalops stages.

metazonite *n.* [Gk. *meta*, after; *zōnē*, girdle] the posterior ring of a diplosomite; *cf.* prozonite.

metazoon *n.* [Gk. *meta*, after; *zōon*, animal] a member of the Metazoa, *q.v.*

metecdysis *n.* [Gk. *meta*, after; *ekdysai*, to strip] in arthropods, the period following a moult when the new cuticle hardens.

metembryo *n.* [Gk. *meta*, towards; *embryon*, embryo] gastrula, *q.v.*

metencephalon *n.* [Gk. *meta*, after; *en*, in; *kephalē*, head] part of hind-brain, consisting of cerebellum, pons, and intermediate part of 4th ventricle; the whole hind-brain.

metenteron *n.* [Gk. *meta*, after; *enteron*, gut] intermesenteric part of gastric cavity in Actinozoa; *cf.* mesenteron.

metepencephalon rhombencephalon, *q.v.*

metepimeron *n.* [Gk. *meta*, after; *epi*, upon; *mēros*, upper thigh] epimeron of insect metathorax; *alt.* meta-epimeron.

metepisternum *n.* [Gk. *meta*, after; *epi*, upon; *sternon*, breast bone] episternum of insect metathorax; *alt.* meta-episternum.

metestrum metoestrus, *q.v.*

methaemoglobin *n.* [Gk. *meta*, after; *haima*, blood; L. *globus*, globe] oxidized haemoglobin, usually present in the blood only in a small amount and useless as an oxygen carrier, produced by the action of oxidizing agents such as nitrite and chlorate poisons, and containing iron in the ferric (oxidized) state.

methionine *n.* [methyl; Gk. *theion*, sulphur] an essential amino acid which contains sulphur, and provides the sulphur and methyl groups necessary for various metabolic processes; abbreviated to met.

metochy *n.* [Gk. *metochē*, sharing] relationship between a neutral guest insect and its host.

metoecious *a.* [Gk. *meta*, after; *oikos*, house] heteroecious, *q.v.*

metoestrus *n.* [Gk. *meta*, after; *oistros*, gadfly] the luteal phase, period when activity subsides after oestrus; *alt.* metoestrum, metestrum, postoestrus, postestrum.

metope *n.* [Gk. *metōpon*, forehead] the middle frontal portion of a crustacean.

metopic *a.* [Gk. *metōpon*, forehead] *pert.* forehead; *appl.* frontal suture.

metopion *n.* [Gk. *metōpion*, forehead] point on forehead where mid-sagittal plane intersects line connecting frontal eminences.

metosteon *n.* [Gk. *meta*, after; *osteon*, bone] a posterior sternal ossification in birds.

metotic *a.* [Gk. *meta*, after; *ous*, ear] *appl.* 2 transitory somites of early vertebrate embryo; *appl.* postotic somite; *appl.* cartilage fusing later with otic capsule.

metovum *n.* [Gk. *meta*, with; L. *ovum*, egg] an egg cell surrounded by nutritive material.

metoxenous *a.* [Gk. *meta*, after; *xenos*, guest] heteroecious, *q.v.*

metra *n.* [Gk. *mētra*, womb] uterus, *q.v.*

metrameristem *n.* [Gk. *mētra*, womb; *meristos*, divided] promeristem, *q.v.*

metraterm *n.* [Gk. *mētra*, womb; *terma*, end] terminal portion of uterus in trematodes.

metrocyte *n.* [Gk. *mētēr*, mother; *kytos*, hollow] a cell that has given rise to other cells by division; *alt.* mother cell, precursory cell.

metrogonidium *n.* [Gk. *mētēr*, mother; *dim.* of *gonē*, seed] a gonidium which produces new gonidia by division in lichens.

metula *n.* [*Dim.* of L. *meta*, end post] a spore-bearing branch having flask-shaped outgrowths, as in certain fungi; *cf.* phialide.

M-factor a certain antigen in erythrocytes of higher animals.

micella, micelle *n.* [L. *dim.* of *mica*, morsel] hypothetical unit, *q.v.*; an orderly aggregate of chain-like molecules of colloidal size; compressions of the cell membrane into a series of pillars, as seen under the electron microscope; parts of cellulose microfibrils in which the cellulose molecules are arranged parallel to each other.

Michurinism [*Ivan V. Michurin*, Russian horticulturalist] genetic theories put forward by Michurin, a Russian horticulturalist, which were later extended by Lysenko to form Lysenkoism.

micraerophiles micro-aerophiles, *q.v.*

micraesthetes *n.plu.* [Gk. *mikros*, small; *aisthētēs*, perceiver] the smaller sensory organs of Amphineura.

micrander *n.* [Gk. *mikros*, small; *anēr*, male] a dwarf male, as of certain green algae.

micraner *n.* [Gk. *mikros*, small; *anēr*, male] a dwarf male ant.

micrergate *n.* [Gk. *mikros*, small; *ergatēs*, worker] a dwarf worker ant.

micro- *cf.* macro-, mega-.

micro-aerophiles organisms requiring less oxygen than is present in the air; *alt.* micraerophiles.

Microascales *n.* [Gk. *mikros*, small; *askos*, bag] an order of Ascomycetes including serious plant parasites.

microbe *n.* [Gk. *mikros*, small; *bios*, life] a microorganism, especially a bacterium; *alt.* microbion.

microbiology *n.* [Gk. *mikros*, small; *bios*, life; *logos*, discourse] biology of microscopic organisms.

microbion microbe, *q.v.*

microbiophagy *n.* [Gk. *mikros*, small; *bios*, life; *phagein*, to consume] destruction or lysis of micro-organisms by a phage.

microbiota *n.* [Gk. *mikros*, small; *biōnai*, to

live] organisms of microscopic size, especially in the soil; *cf.* macrobiota, mesobiota.

microbivorous *a.* [Gk. *mikros*, small; *bios*, life; L. *vorare*, to devour] eating microbes.

microblast *n.* [Gk. *mikros*, small; *blastos*, bud] an erythroblast smaller than normal.

microbodies *n.plu.* [Gk. *mikros*, small; A.S. *bodig*, body] peroxisomes, *q.v.*

microcaltrops *n.* [Gk. *mikros*, small; A.S. *coltraeppe*, kind of thistle] a primitive tetraxon, or euaster with 4 persistent rays.

microcentrosome centriole, *q.v.*

microcentrum *n.* [Gk. *mikros*, small; *kentron*, centre] centrosome, *q.v.*

microcephalic *a.* [Gk. *mikros*, small; *kephalē*, head] with abnormally small head; having a cranial capacity of under 1350 cm³; *cf.* megacephalic, mesocephalic.

microchaeta *n.* [Gk. *mikros*, small; *chaetē*, hair] a small bristle, as on body of certain insects.

microchromosomes *n.plu.* [Gk. *mikros*, small; *chrōma*, colour; *sōma*, body] chromosomes considerably smaller than the other chromosomes of the same type of nucleus, and usually centrally placed in the equatorial plate during metaphase; *alt.* M-chromosomes; *cf.* macrochromosomes.

microclimate *n.* [Gk. *mikros*, small; *klima*, slope] the climate in a very small habitat or area.

microconidium *n.* [Gk. *mikros*, small; *konis*, dust; *idion*, *dim.*] a comparatively small conidium; aleurospore, *q.v.*

microconjugant *n.* [Gk. *mikros*, small; L. *conjugare*, to unite] a motile ciliated free-swimming conjugant or gamete which attaches itself to a macroconjugant and fertilizes it.

microcyclic *a.* [Gk. *mikros*, small; *kyklos*, circle] having a simple or short cycle; with haplophase or gametophyte stage only; *cf.* macrocyclic.

Microcyprini *n.* [Gk. *mikros*, small; L. *cyprinus*, carp] Cyprinodontiformes, *q.v.*

microcyst *n.* [Gk. *mikros*, small; *kystis*, bladder] a resting-spore stage of slime moulds.

microcytase *n.* [Gk. *mikros*, small; *kytos*, hollow] the enzyme of microphages or small leucocytes.

microcytes *n.plu.* [Gk. *mikros*, small; *kytos*, hollow] blood corpuscles about half the size of erythrocytes, numerous in diseased conditions; *alt.* schistocytes.

microdont *a.* [Gk. *mikros*, small; *odous*, tooth] with comparatively small teeth.

Microdrili *n.* [Gk. *mikros*, small; *drilos*, worm] a group of slender, elongated aquatic oligochaetes having the nephridium without a capillary network; *a.* microdriline.

micro-elements trace elements, *q.v.*

micro-endemic *a.* [Gk. *mikros*, small; *endēmos*, dwelling in] restricted to a very small area.

microevolution *n.* [Gk. *mikros*, small; L. *evolvere*, to unroll] evolutionary processes that can be noticed within a relatively brief period, as during a human lifetime; evolution due to gene mutation and recombination; *cf.* macroevolution.

microfauna *n.* [Gk. *mikros*, small; L. *faunus*, god of woods] animals less than 200 μm long, (the limit of comfortable visibility with the naked eye) such as protozoa.

microfibrils *n.plu.* [Gk. *mikros*, small; L. *fibrilla*,

small fibre] fibrils composed of chains of cellulose molecules, visible only with the electron microscope, and aggregated to form macrofibrils.

microfilaments *n.plu.* [Gk. *mikros*, small; L. *filum*, thread] protein filaments found in certain eukaryotic cells, in bundles or scattered.

microfilaria *n.* [Gk. *mikros*, small; L. *filum*, thread] the embryo of certain parasitic hematodes such as *Filaria*.

microflora *n.* [Gk. *mikros*, small; L. *flos*, flower] the microscopic flora of a region; the micro-organisms present in an organism or organ; the dwarf flora of high mountains.

microgamete *n.* [Gk. *mikros*, small; *gametēs*, spouse] the smaller of 2 conjugant gametes, regarded as male; merogamete, *q.v.*

microgametoblast *n.* [Gk. *mikros*, small; *gametēs*, spouse; *blastos*, bud] intermediate stage between microgametocyte and microgamete in certain Sporozoa.

microgametocyte *n.* [Gk. *mikros*, small; *gametēs*, spouse; *kytos*, hollow] cell developed from merozoite in certain protistans, giving rise to microgametes.

microgametogenesis *n.* [Gk. *mikros*, small; *gametēs*, spouse; *genesis*, descent] development of microgametes or spermatozoa.

microgametophyte *n.* [Gk. *mikros*, small; *gametēs*, spouse; *phyton*, plant] the male gametophyte developed from a microspore; *cf.* megagametophyte.

microgamy *n.* [Gk. *mikros*, small; *gamos*, marriage] in certain protistans, syngamy between individuals much smaller than vegetative cells; *alt.* merogamy.

microglia *n.plu.* [Gk. *mikros*, small; *glia*, glue] neuroglia found in greater numbers in grey than white matter, having small elongated cell bodies, and which are phagocytic and can be spherical or amoeboid; *alt.* mesoglia of Hortega.

microgonidium *n.* [Gk. *mikros*, small; *gonos*, offspring; *idion*, *dim.*] a comparatively small gonidium; a male gamont or gametocyte.

microgyne *n.* [Gk. *mikros*, small; *gynē*, female] dwarf female ant.

microhabitat *n.* [Gk. *mikros*, small; L. *habitare*, to inhabit] the immediate special environment of an organism, a small place in the general habitat; *alt.* biotope; *cf.* niche.

microhenad *n.* [Gk. *mikros*, small; *henas*, unit] filter-passer, *q.v.*

microlecithal *a.* [Gk. *mikros*, small; *lekithos*, yolk] containing little yolk.

microleucoblast *n.* [Gk. *mikros*, small; *leukos*, white; *blastos*, bud] myeloblast, *q.v.*

microleucocyte *n.* [Gk. *mikros*, small; *leukos*, white; *kytos*, hollow] a small amoebocyte.

micromere *n.* [Gk. *mikros*, small; *meros*, part] a cell of upper or animal hemisphere in meroblastic and other eggs.

micromerozoite *n.* [Gk. *mikros*, small; *meros*, part; *zōon*, animal] cell derived from microschizont and developing into gametocyte in Haemosporidia.

micromesentery *n.* [Gk. *mikros*, small; *mesos*, middle; *enteron*, gut] a secondary incomplete mesentery in Zoantharia.

micrometre *n.* [Gk. *mikros*, small; *metron*, measure] a unit of microscopic measurement, being

10^{-6} m, one-thousandth mm; formerly also called micron; symbol μm or formerly μ.

micromutation *n*. [Gk. *mikros*, small; L. *mutare*, to change] point mutation, *q.v.*

micromyelocyte *n*. [Gk. *mikros*, small; *myelos*, marrow; *kytos*, hollow] a small heterophil myelocyte.

micron *n*. [Gk. *mikros*, small] a former term for micrometre, one-thousandth mm; symbol μ.

microne any particle not less than 0·2 μm in diameter, i.e. visible with ordinary microscopy.

micronemic, micronemeous *a*. [Gk. *mikros*, small; *nēma*, thread] *pert.* or having small hyphae.

micronephridia *n.plu.* [Gk. *mikros*, small; *nephros*, kidney; *idion*, *dim.*] small nephridia; *alt.* meronephridia.

micront *n*. [Gk. *mikros*, small; *on*, being] a small cell formed by schizogony, itself giving rise to microgametes.

micronucleocyte *n*. [Gk. *mikros*, small; L. *nucleus*, kernel; Gk. *kytos*, hollow] an amoebocyte with a relatively small nucleus.

micronucleus *n*. [Gk. *mikros*, small; L. *nucleus*, kernel] the smaller, reproductive nucleus of many Protozoa in close proximity to meganucleus; *alt.* gononucleus, paranucleus; *cf.* meganucleus.

micronutrients trace elements, *q.v.*

micro-organism *n*. [Gk. *mikros*, small; *organon*, instrument] a microscopic organism such as a bacterium, protistan, or virus; *alt.* microbe, microbion.

microparasite *n*. [Gk. *mikros*, small; *parasitos*, parasite] a parasite of microscopic size.

microphages *n.plu.* [Gk. *mikros*, small; *phagein*, to eat] small phagocytes of blood, chiefly polymorphonuclear heterophil leucocytes; *alt.* microphagocytes.

microphagic, microphagous *a*. [Gk. *mikros*, small; *phagein*, to eat] feeding on minute organisms or particles; feeding on small prey; *cf.* macrophagous.

microphagocyte *n*. [Gk. *mikros*, small; *phagein*, to eat; *kytos*, hollow] microphage, *q.v.*

microphanerophyte *n*. [Gk. *mikros*, small; *phaneros*, manifest; *phyton*, plant] tree or shrub from 2 to 8 m in height.

microphil(ic) *a*. [Gk. *mikros*, small; *philein*, to love] tolerating only a narrow range of temperature, *appl.* certain bacteria; *n.* microphile.

microphyll *n*. [Gk. *mikros*, small; *phyllon*, leaf] a small leaf, especially as produced by pteridophytes.

microphyllous *a*. [Gk. *mikros*, small; *phyllon*, leaf] with small leaves.

microphyte *n*. [Gk. *mikros*, small; *phyton*, plant] any microscopic plant.

microphytology *n*. [Gk. *mikros*, small; *phyton*, plant; *logos*, discourse] science of microphytes; bacteriology.

microplankton *n*. [Gk. *mikros*, small; *plangktos*, wandering] small organism drifting with the surrounding water, somewhat larger than those of nanoplankton, *q.v.*; *alt.* seston.

Micropodiformes *n*. [Gk. *mikros*, small; *pous*, foot; L. *forma*, shape] Apodiformes, *q.v.*

micropodous *a*. [Gk. *mikros*, small; *pous*, foot] with rudimentary or small foot or feet.

micropore *n*. [Gk. *mikros*, small; *poros*, channel] a small pole in shell of Amphineura containing a sense organ (micraesthete).

micropterous *a*. [Gk. *mikros*, small; *pteron*, wing] having small hind-wings invisible until tegmina are expanded, as in some insects; with small or rudimentary fins; *n.* micropterism.

micropyle *n*. [Gk. *mikros*, small; *pylē*, gate] aperture for admission of pollen tube at ovule apex, *alt.* foramen; the corresponding aperture in testa of seed between hilum and point of radicle; small opening in cyst wall of macrogamete, for entry of microgamete; pore of oocyst; aperture in egg membrane for admission of spermatozoon; pore in spongin coat of sponges for escape of gemmules.

micropyle apparatus raised processes or porches, sometimes of elaborate structure, developed round micropyle of certain insect eggs.

micropyrenic *a*. [Gk. *mikros*, small; *pyrēn*, fruit stone] with nuclei markedly smaller than average for the species or other groups. *n.* A micropyrenic individual.

microrhabdus *n*. [Gk. *mikros*, small; *rhabdos*, rod] minute monaxon or rod-like spicule.

Microsauria, microsaurs *n.*, *n.plu.* [Gk. *mikros*, small; *sauros*, lizard] an extinct order of lepospondyl amphibians having an elongated or reptilian shape and developed limbs each with 4 or fewer fingers.

microschizogony *n*. [Gk. *mikros*, small; *schizein*, to cleave; *gonos*, birth] schizogony resulting in small merozoites.

microschizont *n*. [Gk. *mikros*, small; *schizein*, to cut; *onta*, beings] a male schizont of certain Protozoa.

microsclere *n*. [Gk. *mikros*, small; *sklēros*, hard] one of small spicules found lying scattered in tissues of sponges; *cf.* megasclere.

microsclerotium *n*. [Gk. *mikros*, small; *sklēros*, hard] a microscopic sclerotium.

microseptum *n*. [Gk. *mikros*, small; L. *septum*, partition] an incomplete mesentery of Zoantharia.

microsere *n*. [Gk. *mikros*, small; L. *serere*, to put in a row] a successional series of plant communities in a microhabitat.

microsmatic *a*. [Gk. *mikros*, small; *osmē*, smell] with feebly-developed sense of smell.

microsomes *n.plu.* [Gk. *mikros*, small; *sōma*, body] the smallest size particles spun down in an ultracentrifuge from cell homogenates and including broken parts of other fractions, especially endoplasmic reticulum with ribosomes concerned with protein synthesis; formerly used by early cytologists, for any small granules in the cytoplasm, *alt.* cytosomes; peroxisomes, *q.v.*; ribosomes, *q.v.*; *cf.* plasmosome.

microsomia *n*. [Gk. *mikros*, small; *sōma*, body] nanism, *q.v.*

microsorus *n*. [Gk. *mikros*, small; *sōros*, heap] a sorus containing microsporangia; *cf.* megasorus.

microspecies *n*. [Gk. *mikros*, small; L. *species*, particular kind] jordanon, *q.v.*; *cf.* macrospecies.

Microspermae *n*. [Gk. *mikros*, small; *sperma*, seed] a group of monocots, now usually considered equivalent to the Orchidales, *q.v.*

microsphere *n*. [Gk. *mikros*, small; *sphaira*, globe] the initial chamber of Foraminifera when very small; centrosphere, *q.v.*; structures of organic material about the size of bacteria, formed by heating polypeptides, which can absorb various organic molecules from aqueous solution, and may have once played a part in making cells.

microspheric *a.* [Gk. *mikros*, small; *sphaira*, globe] *appl.* Foraminifera when initial chamber of shell is small.

microsplanchnic *a.* [Gk. *mikros*, small; *splangchnon*, entrail] small-bodied and long-legged.

microsporangium *n.* [Gk. *mikros*, small; *sporos*, seed; *anggeion*, vessel] a sporangium containing a number of microspores; pollen sac in spermatophytes; *alt.* microsporophore.

microspore *n.* [Gk. *mikros*, small; *sporos*, seed] the spore developed in a microsporangium in heterosporous plants, giving rise to male gametophyte; the cell from which a pollen grain develops; the pollen grain itself; *alt.* androspore; the smaller anisospore of Sarcodina.

microsporocyte *n.* [Gk. *mikros*, small; *sporos*, seed; *kytos*, hollow] the microspore mother cell or pollen mother cell producing microspores or pollen grains.

microsporophore *n.* [Gk. *mikros*, small; *sporos*, seed; *pherein*, to bear] microsporangium, *q.v.*

microsporophyll *n.* [Gk. *mikros*, small; *sporos*, seed; *phyllon*, leaf] a microsporangium-bearing leaf; stamen, *q.v.*; *alt.* androphyll.

microsporozoite *n.* [Gk. *mikros*, small; *sporos*, seed; *zōon*, animal] a smaller endogenous sporozoite of Sporozoa.

microstome *n.* [Gk. *mikros*, small; *stoma*, mouth] a small opening or orifice.

microstrobilus *n.* [Gk. *mikros*, small; *strobilos*, cone] a male cone made of microsporophylls as in gymnosperms.

microstylospore *n.* [Gk. *mikros*, small; *stylos*, pillar; *sporos*, seed] a comparatively small stylospore.

microstylous *a.* [Gk. *mikros*, small; *stylos*, pillar] having short styles, *appl.* heterostylous flowers.

microsymbiote, microsymbiont *n.* [Gk. *mikros*, small; *symbiōtēs*, companion] the smaller of 2 symbiotic organisms.

Microtabiotes Microtatobiotes, *q.v.*

Microtatobiotes *n.plu* [Gk. *mikros*, small; *tatos*, stretched; *bios*, life] a taxonomic group, sometimes considered a superphylum, consisting of submicroscopic obligate parasites and including the rickettsias and viruses; *alt.* Microtabiotes.

microtaxonomy *n.* [Gk. *mikros*, small; *taxis*, arrangement; *nomos*, law] classification and its principles as applied to subspecies, varieties, or races.

microteliospore *n.* [Gk. *mikros*, small; *telos*, end; *sporos*, seed] a spore produced in a microtelium.

microtelium *n.* [Gk. *mikros*, small; *telos*, end] sorus of microcyclic rust fungi.

microtherm *n.* [Gk. *mikros*, small; *thermē*, heat] a plant of the cold temperate zone.

microthorax *n.* [Gk. *mikros*, small; *thōrax*, chest] in insects, term applied to the cervix or neck in the belief that it represents a reduced prothorax.

Microthyriales *n.* [*Microthyrium*, generic name] an order of Ascomycetes having peltate fruit bodies; *alt.* Hemisphaeriales.

microtomy *n.* [Gk. *mikros*, small; *tomē*, a cutting] the cutting of thin sections of objects, as of tissues, or cells, in preparing specimens for microscopic or ultramicroscopic examination.

microtrichium *n.* [Gk. *mikros*, small; *thrix*, hair] one of the small hairs without basal articulations in insect wings; *alt.* aculeus; *plu.* microtrichia.

microtubules *n.plu.* [Gk. *mikros*, small; L. *tubulus*, small tube] tubular structures in the cytoplasm seen with the electron microscope in many plant and animal cells.

microtype *n.* [Gk. *mikros*, small; *typos*, a type] normal mesentery arrangement of Anthozoa; *cf.* macrotype.

microvilli *n.plu.* [Gk. *mikros*, small; L. *villus*, shaggy hair] villi-like foldings on individual cell membranes, observed with the electron microscope, which on intestinal epithelium form brush borders.

microzoid *n.* [Gk. *mikros*, small; *zōon*, animal; *idion*, *dim.*] male gamete, as in algae.

microzooid *n.* [Gk. *mikros*, small; *zōon*, animal; *eidos*, form] a free-swimming motile ciliated bud of *Vorticella* and other Protozoa.

microzoon *n.* [Gk. *mikros*, small; *zōon*, animal] a microscopic animal.

microzoospore *n.* [Gk. *mikros*, small; *zōon*, animal; *sporos*, seed] small planogamete; small anisospore of Radiolaria.

microzyma *n.* [Gk. *mikros*, small; *zymē*, leaven] hypothetical unit, *q.v.*

microzyme *n.* [Gk. *mikros*, small; *zymē*, leaven] a micro-organism of fermenting or decomposing liquids.

micton *n.* [Gk. *miktos*, mixed; *on*, being] a species resulting from interspecific hybridization and of which the individuals are interfertile.

micturition *n.* [L. *mingere*, to void water] urination, the act of voiding contents of urinary bladder.

mid-body a cell plate or group of granules in equatorial region of spindle in anaphase of mitosis.

mid-brain middle zone of primitive or embryonic brain; mesencephalon of adults.

middle lamella the layer derived from the cell plate, made of pectic substances, and covered on both sides by cellulose in formation of the wall of a plant cell; a thin mesogloeal layer between ectoderm and endoderm, as in hydra.

mid-gut mesenteron, *q.v.*

midrib the large central vein of a leaf.

midriff *n.* [A.S. *mid*, middle; *hrif*, belly] the diaphragm or muscular partition between thoracic and abdominal cavities.

mid-ventricle *see* mesocoel.

Miescher's tubes [*J. F. Miescher*, Swiss pathologist] Rainey's tubes, *q.v.*

migration *n.* [L. *migrare*, to transfer] change of habitat, according to season, climate, food supply, etc., of birds, reindeer, bats, certain fishes, insects, etc.; movements of plants into a new area.

migratory cell an amoeboid leucocyte; *alt.* wandering cell, planocyte.

miliary *a.* [L. *millium*, millet] of granular appearance; consisting of small and numerous grain-like parts.

milk ridges two ventral ectodermal bands in mammalian embryo, converging from bases of forelimbs towards inguinal region, from which mammae are developed.

milk sugar lactose, *q.v.*

milk teeth first dentition of mammals, shed after or before birth; *alt.* deciduous teeth.

milk tubes laticiferous vessels.

Milleporina *n.* [*Millepora,* generic name] an order of hydrozoans often included in the Athecata having colonial forms with 2 kinds of polyps living in pits in the surface of a massive calcareous skeleton (corallium) and a brief medusoid stage.

millimicron *n.* [L. *mille,* one thousand; Gk. *mikros,* small] a former term for nanometre, being one-thousandth of a micron (micrometre); symbol mμ.

millipedes the common name for the Diplopoda, *q.v.*

milt *n.* [A.S. *milte,* spleen] an old or colloquial term for the spleen; testis or sperm of fishes.

mimesis *n.* [Gk. *mimēsis,* imitation] mimicry, *q.v.*; social facilitation, *q.v.*

mimetic *a.* [Gk. *mimētikos,* imitative] *pert.* or exhibiting mimicry.

mimic *v.* [Gk. *mimikos,* imitating] to assume, usually for protection, the habits, colour, or structure of another organism.

mimicry *n.* [Gk. *mimikos,* imitating] the resemblance of one animal (the mimic) to another, so that a 3rd animal is deceived into confusing them; the resemblance of an animal or plant to an inanimate object; *alt.* mimesis; *a.* mimetic.

mineralization *n.* [L. *minera,* ore, mine] the production of inorganic ions in the soil by the oxidation of organic compounds; *cf.* immobilization.

mineralocorticoids a group of corticosteroid hormones secreted by the cortex of the adrenal bodies including aldosterone, and which control the salt and water balance of the body.

mingin *n.* [L. *mingere,* to make water] a trypsin inhibitor found in human urine.

minim *n.* [L. *minimus,* least] an ant of the smallest worker caste; *cf.* maxim.

minimum, law of the that factor for which an organism or species has the narrowest range of tolerance or adaptability limits its existence; extension of Liebig's law, *q.v.*

minimus *n.* [L. *minimus,* least] fifth digit of hand or foot.

minor elements trace elements, *q.v.*

Miocene *n.* [Gk. *meion,* less; *kainos,* recent] a Tertiary geological epoch, between Oligocene and Pliocene.

miolecithal *a.* [Gk. *meiōn,* less; *lekithos,* egg yolk] poorly yolked.

miosis meiosis, *q.v.*; myosis, *q.v.*

miospore *n.* [Gk. *meion,* less; *sporos,* seed] a fossil spore less than 0·2 mm in diameter; *cf.* megaspore.

miostemonous meiostemonous, *q.v.*

miotic meiotic, *q.v.*; myotic, *q.v.*

miracidium *n.* [Gk. *dim.* of *meirakion,* stripling] the ciliated embryo or youngest stage in life history of a trematode.

miscegenation *n.* [L. *miscere,* to mix; *genus,* race] interbreeding between races or varieties.

Mischococcales *n.* [*Mischococcus,* generic name] Heterococcales, *q.v.*

misogamy *n.* [Gk. *misein,* to hate; *gamos,* mar-

riage] antagonism to mating; reproductive isolation.

mispairing the condition of having a base in the DNA chain which does not pair with the base of the other chain of the double helix.

missense mutant a mutant in which a codon codes for a different amino acid, forming a nonactive enzyme.

Mississippian Lower Carboniferous in North America.

miter mitra, *q.v.*

mites a common name for many members of the Acari, *q.v.*

mitochondria *n.plu.* [Gk. *mitos,* thread; *chondros,* grain] double membraned organelles in the cytoplasm of all eukaryotes, having the inner membrane invaginated and being the site of the Krebs' cycle and oxidative phosphorylation of respiration; *alt.* Altmann's granules, bioblasts, chondrioconts, chondriomites, chondriosomes, chondriospheres, chondrioplasts, plastosomes, plastochondria, spheroplasts; *sing.* mitochondrion.

mitochondria B lysosomes, *q.v.*

mitochondrial sheath an envelope containing mitochondrial granules sheathing spiral thread of spermatozoan body or connecting-piece.

mitochondrion *sing.* of mitochondria.

mitogen *n.* [Gk. *mitos,* thread; *genos,* birth] any substance that causes mitosis.

mitogenetic *a.* [Gk. *mitos,* thread; *genesis,* descent] inducing cell division; *appl.* influence, probably hormones, inducing mitosis in apical meristem and emanating from the same or another apical meristem; *appl.* radiation, Gurwitsch or M-rays, from living matter and supposed to induce mitosis.

mitome *n.* [Gk. *mitos,* thread] reticulum of cell protoplasm, consisting of the cytomitome and karyomitome; *alt.* spongioplasm.

mitoschisis *n.* [Gk. *mitos,* thread; *schizein,* to cleave] mitosis, *q.v.*

mitosis *n.* [Gk. *mitos,* thread] the typical process of nuclear division with chromosome formation and spindle formation, which results in each daughter nucleus containing the same number of chromosomes as each other and as the parent nucleus, and which is followed by cell division; *alt.* cytodiaeresis, karyokinesis, karyomitosis, indirect nuclear division, mitoschisis; *plu.* mitoses; *cf.* amitosis, meiosis.

mitosome *n.* [Gk. *mitos,* thread; *sōma,* body] a body arising from spindle fibres of secondary spermatocytes, eventually said to form connecting-piece and tail envelope of spermatozoon; the spindle remnant.

mitosporangium *n.* [Gk. *mitos,* thread; *sporos,* seed; *anggeion,* vessel] a thin-walled diploid sporangium, producing zoospores by mitoses; *cf.* meiosporangium.

mitospore *n.* [Gk. *mitos,* thread; *sporos,* seed] a uninucleate diploid zoospore produced in a mitosporangium; *cf.* meiospore.

mitotic *a.* [Gk. *mitos,* thread] *pert.* or produced by mitosis, *appl.* division, figure, etc.

mitotic index the number of cells simultaneously in the process of division, out of a total of 1000 cells.

mitotin *n.* [Gk. *mitos,* thread] substance supposed to act with an enzyme mitotase in generating mitogenetic radiation.

mitra *n.* [L. *mitra*, head band] a helmet-shaped part of calyx or corolla; the mitriform pileus of certain fungi; *alt.* miter.

mitral mitriform, *q.v.*

mitral cells pyramidal cells with thick basal dendrites, found in molecular layer of olfactory bulb.

mitral valve valve of the left auriculoventricular orifice of the heart; *alt.* bicuspid valve.

mitraria *n.* [L. *mitra*, head band] lobed type of trochophore found in the life cycle of the annelid *Owenia*.

mitriform *a.* [L. *mitra*, head band; *forma*, shape] mitre-shaped; *alt.* mitral.

mixipterygium *n.* [Gk. *mixis*, mixing; *pterygion*, little wing or fin] clasper of male elasmobranchs, medial lobe of pelvic fin; *alt.* mixopterygium, myxopterygium.

mixis *n.* [Gk. *mixis*, mingling] the fusion of gametes; karyogamy and karyomixis; fertilization.

mixochimaera *n.* [Gk. *mixis*, mixing; *chimaira*, monster] a heterokaryotic hypha.

mixochromosome *n.* [Gk. *mixis*, mixing; *chrōma*, colour; *sōma*, body] the new chromosome formed by fusion of a pair, in syndesis; *alt.* zygosome.

mixonephrium *n.* [Gk. *mixis*, mixing; *nephros*, kidney; *dim.*] a type of metanephromixium in which the nephridium and coelomoduct form a single organ.

mixoploidy *n.* [Gk. *mixis*, mixing; *haploos*, onefold; *eidos*, form] condition of having cells or tissues with different chromosome numbers in the same individual, as in a chimaera or mosaic.

mixopterygium mixipterygium, *q.v.*

mixote *n.* [Gk. *mixis*, mingling] the product of fusion of reproductive cells whether of gametes or of gametoids; zygote, *q.v.*; zygotoid, *q.v.*

mixotrophic *a.* [Gk. *mixis*, mixing; *trephein*, to nourish] combining holophytic with saprophytic nutrition; obtaining part of nourishment from an outside source; partly parasitic; *alt.* mesotrophic.

M-line Hensen's line, *q.v.*

MLO mycoplasma-like organism, *q.v.*

ml RNA messenger-like RNA, *q.v.*

mnemic *a.* [Gk. *mnēmē*, memory] *appl.* theory which attributes hereditary phenomena to latent memory of past generations; *cf.* engram.

mnemotaxis *n.* [Gk. *mnēmē*, memory; *taxis*, arrangement] locomotion directed by memory stimulus, as returning to a feeding place and homing; *cf.* pharotaxis.

modality *n.* [L. *modus*, manner] manner or quality, as type of stimulus or of sensation.

moderator *n.* [L. *moderator*, regulator] band of muscle checking excessive distention of right ventricle, as in heart of some mammals.

modification *n.* [L. *modus*, measure; *facere*, to make] a phenotypic change due to environment or use.

modifier *n.* [L. *modus*, measure; *facere*, to make] a factor which modifies the effect of another factor; a gene which modifies function of a gene at a different locus.

modiolus *n.* [L. *modiolus*, small measure] the conical central axis of cochlea of ear; the convergency of muscle fibres close to the angle of the mouth.

modulating codon a codon which codes for rare tRNAs.

modulation *n.* [L. *modulatus*, measured] de-

differentiation and redifferentiation of cells during definitive tissue development; alteration in cells, produced by environmental stimuli, without impairment of their essential character.

modulator *n.* [L. *modulatus*, measured] a band of the spectrum, localized in the red–yellow, green, and blue regions, which evokes colour sensation; a physiological unit of colour reception; *cf.* dominator; the component in processes essential for maintaining a steady state which controls specific reactions, as a catalyst, gene, brain, etc.; the agency which selects the appropriate way of transmission between receptor and effector.

molar *a.* [L. *molere*, to grind] adapted for grinding, as *appl.* teeth, *appl.* buccal glands, *appl.* process on mandible of malacostracans. [L. *moles*, mass] in, or *pert.*, a mass; containing one gram molecule or mole per litre, *appl.* solutions. *n.* A grinding tooth.

molecular biology study of biological phenomena at the molecular level.

molecular hypothesis the supposition that muscle and nerve are composed of molecules or particles, like the molecules of a magnet, with positive and negative surfaces.

molecular layer external layer of cortex of cerebrum and cerebellum; *alt.* plexiform layer.

Mollicutes *n.* [L. *molliculus*, tender] a class of Protista including the Mycoplasmatales.

Moll's glands modified sudoriferous glands between follicles of eyelashes; *alt.* ciliary glands.

Mollusca, molluscs *n., n.plu.* [L. *molluscus*, soft] a phylum of usually unsegmented coelomate animals (but with a reduced coelom) the body consisting of a head, foot, and visceral mass covered by a mantle, all of which show great variety of form.

molluscoid *a.* [L. *molluscus*, soft; Gk. *eidos*, like] *pert.* or resembling a mollusc.

Molpadonia, Molpadida *n.* [*Molpadia*, generic name] an order of burrowing sea cucumbers having no podia except as anal papillae.

molt moult, *q.v.*

moltinism *n.* [L. *mutare*, to change] a polymorphism in which different strains undergo a different number of larval moults; *alt.* moultinism.

monacanthid *a.* [Gk. *monos*, single; *akantha*, thorn] with one row of ambulacral spines, as certain starfishes.

monactine *n.* [Gk. *monos*, single; *aktis*, ray] a single-rayed spicule; *a.* monactinal.

monactinellid *a.* [Gk. *monos*, single; *aktis*, ray] containing uniaxial spicules only, as certain sponges.

monad *n.* [Gk. *monas*, unit] a primitive organism or organic unit; flagellula form of a protozoan; single cell, instead of tetrad, produced by a spore mother cell owing to meiotic anomaly.

monadelphous *a.* [Gk. *monos*, single; *adelphos*, brother] having stamens united into one bundle by union of filaments; *alt.* monodelphous; *cf.* diadelphous.

monadiform *a.* [Gk. *monas*, unit; L. *forma*, shape] like a flagellate protozoan.

monamniotic *a.* [Gk. *monos*, single; *amnion*, foetal membrane] having one amnion; *appl.* uniovular twins.

monandrous *a.* [Gk. *monos*, alone; *anēr*, male] having only one stamen; having one antheridium; having only one male mate.

monanthous *a.* [Gk. *monos*, single; *anthos*, flower] having or bearing only one flower.

monarch *a.* [Gk. *monos*, single; *archē*, beginning] with only one protoxylem strand; with only one vascular bundle.

monaster *n.* [Gk. *monos*, single; *astēr*, star] the single aster of monocentric mitosis.

monaxial *a.* [Gk. *monos*, single; *axōn*, axis] having one line of axis, *alt.* uniaxial; having inflorescence developed on primary axis.

monaxon *n.* [Gk. *monos*, single; *axōn*, axis] a type of spicule built upon a single axis; a monaxonic nerve cell.

monaxonic *a.* [Gk. *monos*, single; *axōn*, axis] elongate; *appl.* types of Protozoa with one long body axis; with one axon, *appl.* nerve cell.

Monaxonida *n.* [Gk. *monos*, single; *axōn*, axis] a subclass or order of very common sponges of the class Demospongea having megascleres as monaxons and sometimes having spongin fibres.

monecious monoecious, *q.v.*

monembryonic *a.* [Gk. *monos*, single; *embryon*, foetus] producing one embryo at a time.

monergic *a.* [Gk. *monos*, single; *energos*, active] having one energid; consisting of one nucleated cell.

monestrous monoestrous, *q.v.*

Monhysterida *n.* [*Monhystera*, generic name] an order of Aphasmidia which are free living and often predatory and have teeth, no buccal spear, and a bristly cuticle.

Moniliales *n.* [*Monilia*, generic name] an order of Fungi Imperfecti having conidia borne on conidiophores or mycelia, and including some economically important parasites and saprophytes and also predacious members which capture eelworms and other small animals; *alt.* Hyphomycetes.

monilicorn *a.* [L. *monile*, necklace; *cornu*, horn] having antennae with appearance of a chain of beads.

moniliform *a.* [L. *monile*, necklace; *forma*, shape] arranged like a chain of beads; *appl.* spores, hyphae, antennae, *alt.* monilioid, toruloid; constricted at regular intervals, *appl.* nucleus of certain Ciliata; with contractions and expansions alternately, as branches of certain roots; granose, *q.v.*

moniliospore *n.* [L. *monile*, necklace; Gk. *sporos*, seed] any spore of a moniliform series.

monimostylic *a.* [Gk. *monimos*, fixed; *stylos*, pillar] having quadrate united to squamosal, and sometimes to other bones, as in certain reptiles; *cf.* streptostylic; *n.* monimostyly.

monkeys *see* Primates.

monoallelic *a.* [Gk. *monos*, single; *allēlōn*, one another] *appl.* polyploid in which all the alleles at a locus are the same.

mono-aminoxidase an enzyme which breaks down noradrenaline.

monoblast *n.* [Gk. *monos*, single; *blastos*, bud] a cell, as in spleen, that develops into a monocyte.

monoblastic *a.* [Gk. *monos*, single; *blastos*, bud] with a single undifferentiated germinal layer.

Monoblepharidales *n.* [*Monoblepharis*, generic name] an order of Phycomycetes having a well-developed mycelium that produces sporangia and sex organs, and oogamous sexual reproduction.

monocardian *a.* [Gk. *monos*, single; *kardia*, heart] having one auricle and ventricle.

monocarp *n.* [Gk. *monos*, single; *karpos*, fruit] a monocarpic plant.

monocarpellary *a.* [Gk. *monos*, single; *karpos*, fruit] containing or consisting of a single carpel.

monocarpic *a.* [Gk. *monos*, single; *karpos*, fruit] dying after bearing fruit once; *alt.* monotocous.

monocarpous *a.* [Gk. *monos*, single; *karpos*, fuit] having only one carpel or ovary, *appl.* gynaecium.

monocaryon monokaryon, *q.v.*

monocellular unicellular, *q.v.*

monocentric *a.* [Gk. *monos*, single; *kentron*, centre] having, derived from, or *pert.* a single centre; with a single centromere; *cf.* polycentric.

monocephalous *a.* [Gk. *monos*, single; *kephalē*, head] with one capitulum only.

monocercous *a.* [Gk. *monos*, single; *kerkos*, tail] with one flagellum, as certain Protista; *alt.* uniflagellate.

monocerous *a.* [Gk. *monos*, single; *keras*, horn] having one horn only.

monochasium *n.* [Gk. *monos*, single; *chasis*, division] a cymose inflorescence with main axis producing one branch each; *alt.* pseudaxis.

Monochlamydeae *n.* [Gk. *monos*, single; *chlamys*, cloak] a subdivision of dicots in which the perianth is absent or only partially developed; *alt.* Apetalae, Incompletae.

monochlamydeous *a.* [Gk. *monos*, single; *chlamys*, cloak] haplochlamydeous, *q.v.*; having a calyx but no corolla, *alt.* apetalous.

monochorionic *a.* [Gk. *monos*, single; *chorion*, skin] having a single chorion, *appl.* uniovular twins.

monochromatic *a.* [Gk. *monos*, single; *chrōma*, colour] having only one colour, *alt.* unicoloured; colour blind, seeing brightness but no colours.

monochronic *a.* [Gk. *monos*, single; *chronos*, time] occurring or originating only once.

monociliated *a.* [Gk. *monos*, single; L. *cilium*, eyelid] uniflagellate, *q.v.*

monoclinous *a.* [Gk. *monos*, single; *klinē*, couch] hermaphrodite, having stamens and pistil in each flower; having antheridium and oogonium originating from the same hypha.

monocondylar *a.* [Gk. *monos*, single; *kondylos*, knuckle] having a single occipital condyle as skull of reptiles and birds; *alt.* monocondylic, monocondylous.

monocont monokont, *q.v.*

monocot monocotyledon, *q.v.*

Monocotyledones, monocotyledons *n., n.plu.* [Gk. *monos*, single; *kotylēdōn*, cup-shaped hollow] a class of angiosperms having an embryo with only one cotyledon, parts of the flower usually in 3s, leaves with parallel veins, and scattered vascular bundles in the stem; *alt.* Monocotyledon(e)ae, monocots, and in some classifications, Liliatae, Liliopsida.

monocotyledonous *a.* [Gk. *monos*, single; *kotylēdōn*, cup-shaped hollow] *pert.* monocotyledons; *pert.* embryo with only one cotyledon, *alt.* unicotyledonous.

monocratic *a.* [Gk. *monos*, single; *kratos*, powcr] with the 4 spores of a tetrad being of the same sex; *cf.* dicratic.

monocrepid *a.* [Gk. *monos*, single; *krēpis*, foundation] *appl.* a desma formed by secondary silica deposits on a monaxial spicule.

monocule n. [Gk. *monos*, single; L. *oculus*, eye] a 1-eyed animal as certain insects and crustaceans.

monoculture n. [Gk. *monos*, single; L. *cultus*, tilling] a large area covered by a single plant species, usually a crop grown year after year, but can be applied to natural communities.

monocyclic a. [Gk. *monos*, single; *kyklos*, circle] having one cycle; with a single whorl; annual, *appl.* herbs.

monocystic a. [Gk. *monos*, single; *kystis*, bag] with one stage of encystation.

monocytes n.plu [Gk. *monos*, single; *kytos*, hollow] a group of large leucocytes which are phagocytic and have an oval or horseshoe-shaped nucleus.

monodactylous a. [Gk. *monos*, single; *daktylos*, finger] with one digit, or one claw, only; *alt.* unidactyl.

Monodelphia n. [Gk. *monos*, single; *delphys*, womb] Eutheria, *q.v.*

monodelphic a. [Gk. *monos*, single; *delphys*, womb] having uteri more or less united, as in placental mammals; having a single uterus, as *appl.* certain nematodes.

monodelphous monadelphous, *q.v.*

monodesmic a. [Gk. *monos*, single; *desmos*, bond] *appl.* scales formed of fused lepidomoria with continuous covering layer of dentine, as some placoid scales, *cf.* adesmic, polydesmic; having a single vascular bundle.

monodont a. [Gk. *monos*, single; *odous*, tooth] having one persistent tooth, as male narwhal with one long tusk.

monoecious a. [Gk. *monos*, single; *oikos*, house] having unisexual male and female flowers on the same plant; with male and female sex organs on one gametophyte; having microsporangia and megasporangia on the same sporophyte; in animals, hemaphrodite, *q.v.*; *alt.* ambisexual, monecious, autoicious; *cf.* dioecious; n. monoecism.

monoestrous a. [Gk. *monos*, single; *oistros*, gadfly] having one oestrous period in a sexual season; *alt.* monestrous; *cf.* polyoestrous.

monofactorial unifactorial, *q.v.*

monogamous a. [Gk. *monos*, single; *gamos*, marriage] consorting with one mate only.

monoganglionic a. [Gk. *monos*, single; *ganglion*, little tumour] having a single ganglion.

monogastric a. [Gk. *monos*, single; *gastēr*, stomach] with only one gastric cavity; with one venter, *appl.* muscles.

Monogenea n. [Gk. *monos*, single; *genea*, descent] a class of flatworms that are usually ectoparasites of fish, or sometimes amphibians and chelonians, and have a simple life cycle with only a single host; in former classifications considered an order or subclass of the Trematoda; *alt.* Heterocotylea.

monogenesis n. [Gk. *monos*, single; *genesis*, descent] asexual reproduction; theory of development of all organisms from single cells; origin of a new form at one place or period; the condition of parasites having a direct life cycle, i.e. having only one host; *alt.* unigenesis; *cf.* polygenesis.

monogenetic a. [Gk. *monos*, single; *genesis*, descent] *pert.* monogenesis.

monogenic a. [Gk. *monos*, single; *genos*, sex] producing offspring consisting of one sex, either arrhenogenic or thelygenic; controlled by a single gene, *alt.* unifactorial.

monogenomic a. [Gk. *monos*, single; *genos*, offspring] having a single set of chromosomes.

monogenous a. [Gk. *monos*, alone; *genos*, offspring] asexual, as *appl.* reproduction.

monogeny n. [Gk. *monos*, single; *genos*, sex] production of offspring consisting of one sex, including arrhenogeny and thelygeny.

monogoneutic a. [Gk. *monos*, single; *goneuein*, to produce] breeding once a year.

Monogononta n. [Gk. *monos*, single; *gonos*, offspring] an order of marine and freshwater rotifers having a reduced male form.

monogonoporous a. [Gk. *monos*, single; *gonos*, offspring; *poros*, channel] having one genital pore common to both male and female organs, as in certain Turbellaria.

monogony n. [Gk. *monos*, alone; *gonos*, offspring] asexual reproduction, including schizogony and gemmation.

monogynaecial a. [Gk. *monos*, single; *gynē*, female; *oikos*, house] developing from one pistil; *alt.* monogynoecial.

monogynous a. [Gk. *monos*, single; *gynē*, female] having one pistil; having one carpel to gynaecium; consorting with only one female.

monohybrid n. [Gk. *monos*, alone; L. *hybrida*, mongrel] a hybrid offspring of parents differing in one character. a. Heterozygous for a single pair of factors.

monokaryon n. [Gk. *monos*, single; *karyon*, nut] a nucleus with a single centriole; cells of a hypha containing one nucleus; *alt.* monocaryon.

monokont a. [Gk. *monos*, single; *kontos*, punting pole] uniflagellate, *q.v.*; *alt.* monocont.

monolayer n. [Gk. *monos*, single; A.S. *lecgan*, to lie] a single homogeneous layer of units, as of molecules, cells, etc.; *alt.* monomolecular layer.

monolepsis n. [Gk. *monos*, single; *lēpsis*, receiving] transmission of characteristics from only one parent to progeny.

monolocular unilocular, *q.v.*

monolophous a. [Gk. *monos*, alone; *lophos*, crest] *appl.* spicules with one ray forked or branched like a crest.

monomastigote a. [Gk. *monos*, single; *mastix*, whip] having one flagellum, as certain Protista.

monomeniscous a. [Gk. *monos*, single; *mēniskos*, small moon] having an eye with only one lens.

monomer n. [Gk. *monos*, single; *meros*, part] a chemical compound made up of single unit molecules, many monomers being joined together to make a polymer.

monomeric a. [Gk. *monos*, single; *meros*, part] *pert.* one segment; derived from one part; bearing a dominant gene at only 1 of 2 loci.

monomerosomatous a. [Gk. *monos*, alone; *meros*, part; *sōma*, body] having body segments all fused together, as in certain arthropods.

monomerous a. [Gk. *monos*, single; *meros*, part] consisting of one part only, *appl.* flower whorls.

monometrosis n. [Gk. *monos*, alone; *mētēr*, mother] colony foundation by one female, as by queen in some social Hymenoptera; *alt.* haplometrosis; *cf.* pleometrosis; a. monometrotic.

monomial a. [Gk. *monos*, single; L. *nomen*, name] *appl.* a name or designation consisting of one term only; *alt.* uninomial, uninomial, unitary; *cf.* binomial.

monomitic a. [Gk. *monos*, single; *mitos*, thread]

having only generative hyphae; *cf.* dimitic, trimitic.

monomorphic *a.* [Gk. *monos*, single; *morphē*, form] developing with no or very slight change of form from stage to stage, as certain protozoans and insects, *cf.* polymorphic; producing spores of one kind only; all members of the same species looking alike.

monomyaric *a.* [Gk. *monos*, alone; *mys*, muscle] with posterior adductor only, anterior adductor being aborted, *appl.* certain bivalves; *alt.* monomyarian.

mononeme *n.* [Gk. *monos*, single; *nēma*, thread] *appl.* suggested chromatid structure consisting of a single strand of DNA; *alt.* unineme; *cf.* bineme, polyneme.

mononeuronic *a.* [Gk. *monos*, single; *neuron*, nerve] with one nerve; *appl.* chromatophores with single type of innervation.

monont *n.* [Gk. *monos*, alone; *on*, being] a single individual reproducing without conjugation; *cf.* sporont, zygote.

mononuclear *a.* [Gk. *monos*, single; L. *nucleus*, kernel] with one nucleus only, *alt.* uninuclear. *n.* A mononuclear leucocyte.

mononucleotide *n.* [Gk. *monos*, single; L. *nucleus*, kernel] a nucleotide or its derivatives, occurring free in a cell.

mononychous *a.* [Gk. *monos*, single; *onyx*, claw] having a single or uncleft claw.

mononym *n.* [Gk. *monos*, single; *onyma*, name] a designation consisting of one term only; name of a monotypic genus.

monopectinate *a.* [Gk. *monos*, single; L. *pecten*, comb] having one margin furnished with teeth like a comb; *alt.* pectinibranch.

monopetalous *a.* [Gk. *monos*, single; *petalon*, leaf] having one petal only, *alt.* unipetalous; having petals united all round; *cf.* gamopetalous, polypetalous; *n.* monopetaly.

monophagous *a.* [Gk. *monos*, single; *phagein*, to eat] subsisting on one kind of food; *appl.* Sporozoa living permanently in a single cell; *appl.* caterpillars feeding on plants of one genus only; *cf.* stenophagous; *appl.* insects restricted to one species or variety of food plant, *cf.* oligophagous.

monophasic *a.* [Gk. *monos*, alone; *phainein*, to appear] *appl.* condensed life cycle of some trypanosomes, lacking the active stage; *cf.* diphasic.

monophyletic *a.* [Gk. *monos*, single; *phylē*, tribe] derived from a single common parent form; *cf.* oligophyletic; polyphyletic.

monophyllous *a.* [Gk. *monos*, single; *phyllon*, leaf] having one leaf only, *alt.* unifoliate; gamophyllous, *q.v.*; having a 1-piece leaf on calyx.

monophyodont *a.* [Gk. *monos*, alone; *phyein*, to produce; *odous*, tooth] having only one set of teeth, the milk dentition being absorbed in foetal life or absent altogether.

Monopisthocotylea *n.* [Gk. *monos*, single; *opisthe*, behind; *kotylē*, cup] a subclass of Monogenea having an oral sucker and 2 muscular suckers or adhesive organs and a single opisthaptor.

monoplacid *a.* [Gk. *monos*, single; *plax*, flat plate] with one plate only, of any kind.

Monoplacophora *n.* [Gk. *monos*, single; *plax*, flat plate; *pherein*, to bear] a mainly extinct class of molluscs with a shell like a limpet, the living forms being known only from the deep-sea bed, and placed in the single order Tryblidioidea; the group is sometimes considered an order of the Amphineura.

monoplacula *n.* [Gk. *monos*, alone; *plax*, flat plate] a single-layered placula.

monplanetary, monoplanetic *a.* [Gk. *monos*, alone; *planētēs*, wanderer] with one stage of motility in life history, *appl.* formation of zoospores in certain fungi; *cf.* diplanetary.

monoplanetism *n.* [Gk. *monos*, single; *planētēs*, wanderer] condition of having one period of motility in one life history, as of zoospores in some fungi; *cf.* diplanetism.

monoplastic *a.* [Gk. *monos*, alone; *plastos*, formed] persisting in one form; with one chloroplast, *appl.* cell.

monoploid *a.* [Gk. *monos*, single; *haploos*, simple; *eidos*, form] having one set of chromosomes, true haploid; in a polyploid series, having the basic haploid chromosome number. *n.* A monoploid organism.

monoplont haplont, *q.v.*

monopodal *a.* [Gk. *monos*, single; *pous*, foot] having one supporting structure; with one pseudopodium.

monopodial *a.* [Gk. *monos*, alone; *pous*, foot] branching from one primary axis acropetally.

monopodium *n.* [Gk. *monos*, single; *pous*, foot] a single main or primary axis from which all main lateral branches develop; *cf.* sympodium.

monopolar unipolar, *q.v.*

monopyrenous *a.* [Gk. *monos*, single; *pyrēn*, kernel] single-stoned, as a fruit.

monorchic *a.* [Gk. *monos*, single; *orchis*, testis] having one testis.

monorefringent *a.* [Gk. *monos*, single; L. *refringere*, to break off] singly refracting; *alt.* isotropic.

Monorhina *n.* [Gk. *monos*, single; *rhines*, nostrils] a subclass of Agnatha having one nostril.

monorhinal *a.* [Gk. *monos*, single; *rhines*, nostrils] having only one nostril, as Cyclostomata (chordates); *pert.* one nostril.

monosaccharides *n.plu.* [Gk. *monos*, single; L. *saccharum*, sugar] carbohydrates with the general formula $C_nH_{2n}O_n$, all being reducing sugars, including the trioses, tetroses, pentoses, hexoses, etc.; *alt.* monoses.

monosepalous *a.* [Gk. *monos*, single; F. *sépale*, sepal] having a single sepal, *alt.* unisepalous; gamosepalous, *q.v.*

monoses *n.plu.* [Gk. *monos*, single] monosaccharides, *q.v.*

monosiphonic *a.* [Gk. *monos*, alone; *siphōn*, tube] having tubes of a hydrocaulis distinct from one another, as in certain Hydromedusae; having a single central tube in filament, as in certain algae; *alt.* monosiphonous.

monosome *n.* [Gk. *monos*, alone; *sōma*, body] an unpaired X-chromosome; a chromosome without its homologous partner; a single ribosome which is not incorporated in a polyribosome.

monosomic *a.* [Gk. *monos*, alone; *sōma*, body] diploid with one chromosome missing.

monospermous *a.* [Gk. *monos*, single; *sperma*, seed] one-seeded; *pert.* monospermy; *alt.* monospermic.

monospermy *n.* [Gk. *monos*, single; *sperma*,

seed] normal fertilization by entrance of one sperm only into an ovum.

monospondylic *a.* [Gk. *monos*, alone; *sphondylos*, vertebra] *appl.* vertebrae without intercentra.

monosporangium *n.* [Gk. *monos*, alone; *sporos*, seed; *anggeion*, vessel] a sporangium producing simple spores.

monospore *n.* [Gk. *monos*, alone; *sporos*, seed] a simple or undivided spore.

monosporic *a.* [Gk. *monos*, single; *sporos*, seed] *pert.* or originating from a single spore; *alt.* monosporial, monosporous.

monosporous *a.* [Gk. *monos*, single; *sporos*, seed] having only one spore or a simple spore; monosporic, *q.v.*

monostachyous *a.* [Gk. *monos*, single; *stachys*, corn ear] with only one spike.

monostele *n.* [Gk. *monos*, alone; *stēlē*, column] protostele, *q.v.*; *a.* monostelic.

monosterigmatic *a.* [Gk. *monos*, alone; *stērigma*, support] having a single sterigma, *appl.* fungi.

monostichous *a.* [Gk. *monos*, single; *stichos*, row] arranged in one row; along one side of an axis.

monostigmatous *a.* [Gk. *monos*, single; *stigma*, mark] with one stigma only.

monostromatic *a.* [Gk. *monos*, single; *strōma*, bedding] of algae, having a single-layered thallus.

monostylous *a.* [Gk. *monos*, single; *stylos*, pillar] having one style only.

monosy *n.* [Gk. *monos*, alone] separation of parts normally fused.

monosymmetrical *a.* [Gk. *monos*, alone; *symmetria*, due proportion] zygomorphic, *q.v.*

monotaxic *a.* [Gk. *monos*, single; *taxis*, arrangement] belonging to the same taxonomic group.

monothalamous *a.* [Gk. *monos*, single; *thalamos*, chamber] unilocular, *q.v.*; single-chambered; *appl.* fruits formed from single flowers; having one gynaecium; *appl.* shells of Foraminifera and other Protozoa with a single chamber; *alt.* monothalamic.

monothecal *a.* [Gk. *monos*, single; *thēkē*, box] having one loculus; single-chambered.

monothelious *a.* [Gk. *monos*, alone; *thēlys*, female] *appl.* a female consorting with more than one male.

monothetic *a.* [Gk. *monos*, single; *thetos*, placed] *appl.* a classification based on only one or a few characteristics, such as of plants based on the number of stamens; *cf.* polythetic.

monotocous *a.* [Gk. *monos*, single; *tokos*, offspring] uniparous, having one offspring at a birth; monocarpic, *q.v.*

Monotremata, monotremes *n., n.plu.* [Gk. *monos*, single; *trēmatōdēs*, having holes] an order of living Prototheria, *q.v.*

monotrichous *a.* [Gk. *monos*, single; *thrix*, hair] having only one flagellum at one pole; *alt.* monotrichic, monotrichate.

monotrochal *a.* [Gk. *monos*, single; *trochos*, wheel] having a prototroch only, as trochosphere of certain Polychaeta.

monotrochous *a.* [Gk. *monos*, alone; *trochos*, wheel] having a single-piece trochanter, as in most stinging Hymenoptera.

monotrophic *a.* [Gk. *monos*, only; *trophē*, nourishment] subsisting on one kind of food.

monotropic *a.* [Gk. *monos*, single; *tropē*, turn] turning in one direction; visiting only one kind of flower, *appl.* insects.

monotype *n.* [Gk. *monos*, only; *typos*, type] single type which constitutes species or genus; a unique holotype.

monotypic *a.* [Gk. *monos*, only; *typos*, type] *pert.* monotype; having only one species, *appl.* genus; having no subspecies, *appl.* species; *cf.* polytypic.

monovalent univalent, *q.v.*

monovoltine univoltine, *q.v.*

monovular uniovular, *q.v.*

monoxenous *a.* [Gk. *monos*, only; *xenos*, host] inhabiting one host only, *appl.* parasites; *cf.* heteroecious.

monoxylic *a.* [Gk. *monos*, only; *xylon*, wood] having wood formed as a continuous ring, *appl.* stems.

Monozoa *n.* [Gk. *monos*, single; *zōon*, animal] a group of tapeworms which do not undergo strobilization and so do not produce proglottides; *cf.* Merozoa.

monozoic *a.* [Gk. *monos*, single; *zōon*, animal] producing one sporozoite only, *appl.* archispores forming only one sporozoite on liberation from cyst.

monozygotic, monozygous *a.* [Gk. *monos*, alone; *zygōtos*, yoked] developing from one fertilized ovum, as identical twins; *alt.* uniovular.

Monro, foramen of [*A. Monro*, Scottish anatomist] foramen of Monro, *q.v.*

mons pubis, mons Veneris prominence due to subcutaneous fatty tissue in front of symphysis pubis.

Monstrilloida *n.* [*Monstrilla*, generic name; Gk. *eidos*, form] an order of copepods with marine members whose larvae are parasitic in polychaetes.

montane *a.* [L. *montanus*, *pert.* mountains] *pert.* mountains; monticolous, *q.v.*

Montgomery's glands [*W. F. Montgomery*, Irish physician] areolar glands of nipple, prominent during lactation.

monticolous *a.* [L. *mons*, mountain; *colere*, to inhabit] inhabiting mountainous regions; *alt.* montane.

monticule monticulus, *q.v.*; a group of small modified zooecia forming a protuberance on surface of a polyzoan colony.

monticulus *n.* [L. *dim.*, *mons*, mountain] largest part of superior vermis of cerebellum; *alt.* monticule.

moor *n.* [M.E. *mor*, moor] a plant community found on uplands on peat, often dominated by heather.

mor *n.* [Dan. *mor*, humus] acid humus of cold wet soils which inhibits action of soil organisms and may form peat; *cf.* mull.

mores *n.plu.* [L. *mos*, wont] groups of organisms preferring the same habitat, having the same reproductive season, and agreeing in their general reactions to the physical environment; *alt.* mune.

Morgagni, columns of [*G. B. Morgagni*, Italian anatomist] rectal columns, *q.v.*

Morgagni, hydatid of *see* hydatid.

morgan *n.* [*T. H. Morgan*, geneticist] a unit of map distance on a chromosome in which the mean number of recombinations is unity.

Morganucodonta, morganucodonts *n., n.plu.* [*Morganucodon*, generic name] a primitive order of Triassic and Jurassic eotherians having primitive 3-cusped molars, and a somewhat reptilian jaw joint.

moriform *a.* [L. *morum*, mulberry; *forma*, form] formed in a cluster resembling aggregate fruit; shaped like a mulberry.

Mormyriformes *n.* [*Mormyrus*, generic name; L. *forma*, shape] an order of teleosts including the elephant fish or mormyrs.

morph *n.* [Gk. *morphē*, form] one of the members of a polymorphic population.

morphactins *n.plu.* [*Morpho*logically *active* substances] a group of substances derived from fluorine-9-carboxylic acids, which affect plant growth and development.

morphallaxis *n.* [Gk. *morphē*, form; *allaxis*, changing] transformation of one part into another, in regeneration of parts, *cf.* epimorphosis; gradual growth or development into a particular form.

morphine *n.* [L. *Morpheus*, god of sleep] the main alkaloid of opium, which is used as a hypnotic to relieve pain.

morphism polymorphism, *q.v.*

morphogenesis *n.* [Gk. *morphē*, form; *genesis*, descent] the development of shape; origin and development of organs or parts of organisms; *alt.* morphogeny.

morphogenetic *a.* [Gk. *morphē*, form; *genesis*, descent] *pert.* morphogenesis; *appl.* internal secretions which influence growth and nutrition of organs or organisms; *appl.* movement of parts of a developing embryo.

morphogenic hormone evocator, *q.v.*

morphogens *n.plu.* [Gk. *morphē*, form; *gennaein*, to produce] substances interacting in presence of an evocator, and determining the pattern of embryonic development.

morphogeny morphogenesis, *q.v.*

morphologic index ratio expressing relation of trunk to limbs.

morphology *n.* [Gk. *morphē*, form; *logos*, discourse] the science of form and structure of plants and animals, as distinct from consideration of functions.

morphon *n.* [Gk. *morphē*, form; *on*, being] a definitely formed individual.

morphoplankton *n.* [Gk. *morphē*, form; *plangktos*, wandering] plankton organisms rendered buoyant by small size, or body shape, oily globules, mucilage, gas-containing structures, etc.

morphoplasm *n.* [Gk. *morphē*, form; *plasma*, formation] protoplasm, as distinguished from cell sap, and including kinoplasm and trophoplasm.

morphoplasy *n.* [Gk. *morphē*, form; *plassein*, to mould] formative potentiality of a growing organism.

morphosis *n.* [Gk. *morphōsis*, form] the manner of development of part of organism; the formation of tissues; *a.* morphotic.

morphospecies *n.* [Gk. *morphē*, form; L. *species*, particular kind] a group of individuals which are considered to belong to the same species on morphological grounds alone.

morphotype *n.* [Gk. *morphē*, form; *typos*, pattern] type specimen of one of the forms of a polymorphic species.

Morren's glands [C. F. A. Morren, Belgian zoologist] calciferous glands of earthworms.

morula *n.* [L. *morum*, mulberry] a solid cellular globular mass, the 1st result of ovum segmentation; stage in development preceding gastrula; a globular aggregation of developing male gametes, *alt.* sperm morula; a coelomocyte containing refractive globules, *alt.* morula-shaped cell; *alt.* mulberry body.

morulation *n.* [L. *morum*, mulberry] morula formation by segmentation.

morulit *n.* [L. *morum*, mulberry] nucleolus, *q.v.*

mosaic *n.* [It. *mosaica*, mosaic] hybrid having unblended parental allelomorphic characters; chimaera, *q.v.*; a virus disease of plants; *appl.* theory that each ommatidium in compound eye of arthropods receives a portion of an image, the several portions being integrated as the total image by the brain.

moschate *a.* [Gk. *moschos*, musk] having or resembling the odour of musk; musky.

moss the common name for a member of the Musci, *q.v.*; also used for members of other groups, such as reindeer moss (lichen), club moss (pteridophyte), Spanish moss (angiosperm).

moss animals the common name for the Bryozoa, *q.v.*

moss fibres nerve fibres branching around cells of internal layer of cerebellar cortex.

moss starch lichenin, *q.v.*

mossy cells protoplasmic astrocytes.

mother cell metrocyte, *q.v.*

moths the common name for many members of the Lepidoptera having antennae tapering to a point and not clubbed.

motile *a.* [L. *movere*, to move] capable of spontaneous movement.

motivation *n.* [L. *movere*, to move] the internal factors controlling behaviour, including drive.

motoneuron(e) *n.* [L. *movere*, to move; Gk. *neuron*, nerve] motor neurone, *q.v.*

motor *a.* [L. *movere*, to move] *pert.* or connected with movement, *appl.* nerves, etc.

motor areas areas of the brain where muscular activity is correlated and controlled; *alt.* motorium.

motor cell bulliform cell, *q.v.*

motor end organ or end plate the structure in which a motor nerve fibre terminates with branching processes in a striated muscle fibre.

motorium *n.* [L. *movere*, to move] motor areas, *q.v.*; *cf.* sensorium.

motor neurones nerve cells concerned with carrying impulses away from the CNS to an effector organ such as a muscle or gland; *alt.* efferent neurones, motoneurones; *cf.* sensory neurones, myoneures.

motor oculi oculomotor nerve, the 3rd cranial nerve.

motor unit a motor neurone and associated muscle fibres.

moult *v.* [L. *mutare*, to change] to cast or shed periodically the outer covering, whether of feathers, hair, skin, or horns. *n.* The process of shedding, *alt.* ecdysis. *Alt.* molt.

moulting glands ecdysial glands, *q.v.*

moulting hormone ecdysone, *q.v.*

moultinism moltinism, *q.v.*

mouth-part a head or mouth appendage of arthropods.

M-phase the phase of the cell cycle in which mitosis occurs.

M-rays mitogenetic rays, *q.v.*

mRNA messenger RNA, *q.v.*

MSH melanocyte-stimulating hormone, *q.v.*

multangular

mucedinous *a.* [L.L. *mucedo*, mould, from L. *mucus*, mucus] having loosely spaced white filaments, like a mould fungus; *alt.* mucedineous.

mucid *a.* [L. *mucidus*, mouldy] mouldy; slimy.

mucific *a.* [L. *mucus*, mucus; *facere*, to make] mucus-secreting; *alt.* muciparous.

muciform *a.* [L. *mucus*, mucus; *forma*, shape] resembling mucus.

mucigen *n.* [L. *mucus*, mucus; Gk. *-genēs*, producing] the substance occurring in mucous gland cells as granules or globules, and which produces mucin; *alt.* mucinogen.

mucilages *n.plu.* [L. *mucus*, mucus] heterosaccharides widely occurring in plants which are hard when dry, but capable of absorbing water, swelling and becoming slimy, and which on acid hydrolysis produce mixtures of hexoses, pentoses, and uronic acids; *cf.* gums.

mucilaginous *a.* [L. *mucus*, mucus] *pert.*, containing, resembling, or composed of mucilage, *appl.* certain glands of joints, *appl.* cells, ducts, canals, slits; *alt.* muculent.

mucin *n.* [L. *mucus*, mucus] any of a group of widely distributed glycoproteins including mucus, secreted by various cells or glands.

mucinoblast *n.* [L. *mucus*, mucus; Gk. *blastos*, bud] mast cell, *q.v.*

mucinogen *n.* [L. *mucus*, mucus; Gk. *-genēs*, producing] mucigen, *q.v.*

muciparous *a.* [L. *mucus*, mucus; *parere*, to beget] mucific, *q.v.*

mucivorous *a.* [L. *mucus*, mucus; *vorare*, to devour] feeding on plant juices, *appl.* insects.

mucocellulose *n.* [L. *mucus*, mucus; *cellula*, small cell] cellulose mixed with mucous substance, as in some seeds and fruits.

mucocutaneous, mucodermal *a.* [L. *mucus*, mucus; *cutis*, skin; Gk. *derma*, skin] *pert.* skin and mucous membrane.

mucoid *a.* [L. *mucus*, mucus; Gk. *eidos*, like] *pert.* or caused by mucus or mucilage, *appl.* degeneration, tissue. *n.* A mucoprotein of cartilage, bone, tendon, etc.

mucolytic *a.* [L. *mucus*, mucus; *lysis*, loosing] breaking down or dissolving mucus.

mucopeptide *n.* [L. *mucus*, mucus; Gk. *peptein*, to digest] a molecule similar to but simpler than murein, *q.v.*; sometimes used as an alternative name to murein.

mucopolysaccharides *n.plu.* [L. *mucus*, mucus; Gk. *polys*, many; L. *saccharum*, sugar] a group of polysaccharides containing sugar derivatives such as amino sugars and uronic acids, and including heparin; *alt.* glycosaminoglycans.

mucoproteins *n.plu.* [L. *mucus*, mucus; Gk. *prōteion*, first] *see* glycoproteins.

Mucorales *n.* [*Mucor*, generic name] an order of Phycomycetes, often called black moulds, including saprophytic and weakly parasitic species, which have a mostly unbranched mycelium, asexual reproduction by aplanospores produced in sporangia and sexual reproduction by aplanogametes resulting in a zygospore.

mucosa *n.* [L. *mucus*, mucus] the innermost layer of gut wall consisting of 3 layers, namely the inner epithelium with digestive glands, the lamina propria, and the muscularis mucosa; the epithelium and lamina propria are together called the mucous membrane (mucous coat) or the whole mucosa can be called the mucous membrane.

mucosanguineous *a.* [L. *mucus*, mucus; *sanguis*, blood] containing mucus and blood, *appl.* faeces.

mucoserous *a.* [L. *mucus*, mucus; *serum*, whey] secreting mucus and body fluid.

mucous *n.* [L. *mucus*, mucus] secreting, containing, or *pert.* mucus, *appl.* glands.

mucous alveoli alveoli which secrete a thick, viscid, mucus-containing saliva; *cf.* serous alveoli.

mucous membrane *see* mucosa; *alt.* mucous coat.

mucro *n.* [L. *mucro*, sharp point] a stiff or sharp point abruptly terminating an organ; a small awn; pointed keel or sterile 3rd carpel, as in pine; projection below orifice in Polyzoa; distal part of furcula in Collembola; posterior tip of cuttlebone.

mucronate, mucroniferous *a.* [L. *mucro*, sharp point; *ferre*, to bear] abruptly terminated by a sharp spine.

mucronulate *a.* [L. *mucro*, sharp point] tipped with small mucro.

mucronule *n.* [L. *mucro*, sharp point] a small mucro.

muculent *a.* [L. *mucus*, mucus] like mucus; containing mucus; mucilaginous, *q.v.*

mucus *n.* [L. *mucus*, mucus] a slimy glairy substance rich in mucins, secreted by goblet cells of a mucous membrane or by mucous cells of a gland; a similar slimy secretion produced by the external body surface of many animals.

Mugiliformes *n.* [*Mugil*, generic name; L. *forma*, shape] an order of teleosts including the grey mullets, sand smelts, and barracudas.

mulberry body morula, *q.v.*

mull *n.* [Dan. *muld*, mould] humus of well-aerated moist soils, formed by action of soil organisms on plant debris, and favouring plant growth; *cf.* mor.

Müllerian bodies [F. *Müller*, German naturalist] structures containing albuminous and oily substances in trichilium, eaten by tropical ants.

Müllerian ducts [J. *Müller*, German anatomist] ducts arising on the lateral sides of the mesonephric or Wolffian ducts, and giving rise to the oviducts; *alt.* paramesonephric ducts.

Müllerian eminence [J. *Müller*, German anatomist] a colliculus or elevation of ventral part of cloaca at entrance of Müllerian ducts and between openings of Wolffian ducts.

Müllerian mimicry [F. *Müller*, German naturalist] the resemblance of 2 animals to their mutual advantage, especially of insects; *cf.* Batesian mimicry.

Müller's fibres [H. *Müller*, German anatomist] neuroglial fibres forming framework supporting nervous layers of retina; *alt.* sustentacular or radial fibres of Müller.

Müller's larva [J. *Müller*, German zoologist] larva of Polycladida having 8 ciliated processes around mouth; *alt.* cephalotrocha.

Müller's law [F. *Müller*, German zoologist] a modified restatement of von Baer's view of the recapitulation theory, *q.v.*

Müller's muscle [H. *Müller*, German anatomist] the circular muscle fibres of the ciliary muscles of the vertebrate eye.

multangular *a.* [L. *multus*, many; *angulus*, angle] *appl.* 2 carpal bones, greater and lesser multangulum, respectively trapezium and trapezoid.

multaxial multiaxial, *q.v.*

multiarticulate *a.* [L. *multus*, many; *articulus*, joint] with many articulations; many-jointed; *alt.* polyarthric.

multiaxial *a.* [L. *multus*, many; *axis*, axis] having or *pert.* several axes; allowing movement in many planes, *appl.* articulations; *alt.* multaxial.

multicamerate *a.* [L. *multus*, many; *camera*, chamber] with many chambers; *alt.* multilocular.

multicapsular *a.* [L. *multus*, many; *capsula*, little chest] with many capsules.

multicarinate *a.* [L. *multus*, many; *carina*, keel] having many carinae or ridges.

multicarpellary *a.* [L. *multus*, many; Gk. *karpos*, fruit] polycarpellary, *q.v.*

multicauline *a.* [L. *multus*, many; *caulis*, stalk] with many stems.

multicellular *a.* [L. *multus*, many; *cella*, cell] many-celled; consisting of more than one cell.

multicentral *a.* [L. *multus*, many; *centrum*, centre] with more than one centre of growth or development.

multiciliate *n.* [L. *multus*, many; *cilium*, eyelid] with some or many cilia.

multicipital *a.* [L. *multus*, many; *caput*, head] with many heads or branches arising from one point.

multicostate *a.* [L. *multus*, many; *costa*, rib] with many ribs or veins; with many ridges.

multicuspid(ate) *a.* [L. *multus*, many; *cuspis*, spearhead] with several cusps or tubercles, *appl.* molar teeth.

multidentate *a.* [L. *multus*, many; *dens*, tooth] with many teeth, or indentations.

multidigitate *a.* [L. *multus*, many; *digitus*, finger] many-fingered.

multifactorial *a.* [L. *multus*, many; *facere*, to make] *pert.* or controlled by a number of genes; *alt.* polygenic.

multifarious *a.* [L. *multifarius*, manifold] polystichous, *q.v.*

multifascicular *a.* [L. *multus*, many; *fasciculus*, small bundle] containing, or *pert.* many fasciculi.

multifid *a.* [L. *multus*, many; *findere*, to cleave] having many clefts or divisions.

multifidus the musculotendinous fasciculi lateral to spinous processes from sacrum to axis vertebra.

multiflagellate *a.* [L. *multus*, many; *flagellum*, whip] furnished with several or many flagella; *alt.* polymastigote, polykont.

multiflorous *a.* [L. *multus*, many; *flos*, flower] bearing many flowers.

multifoliate *a.* [L. *multus*, many; *folium*, leaf] with many leaves.

multifoliolate *a.* [L. *multus*, many; *foliolum*, small leaf] with many leaflets.

multiform *a.* [L. *multus*, many; *forma*, form] occurring in, or containing, different forms; *appl.* layer: inner cell lamina of cerebral cortex, *alt.* polymorphous.

multiganglionate *a.* [L. *multus*, many; Gk. *ganglion*, small tumour] with several or many ganglia.

multigyrate *a.* [L. *multus*, many; *gyrus*, circle] with many gyri; tortuous.

multijugate *a.* [L. *multus*, many; *jugum*, yoke] having many pairs of leaflets.

multilacunar *a.* [L. *multus*, many; *lacuna*, cavity] with many lacunae; having a number of leaf gaps, *appl.* nodes.

multilaminate *a.* [L. *multus*, many; *lamina*, plate] composed of several or many laminae.

multilobar, multilobate *a.* [L. *multus*, many; *lobus*, lobe] composed of many lobes.

multilobulate *a.* [L. *multus*, many; *lobulus*, small lobe] having many lobules.

multilocular, multiloculate *a.* [L. *multus*, many; *loculus*, compartment] having many cells or chambers; *appl.* spore; sporodesm, *q.v.*; containing a number of oil droplets, as cells in brown fat; plurilocular, *q.v.*; *cf.* unilocular, multicamerate.

multimer *n.* [L. *multus*, many; Gk. *meros*, part] a protein molecule made of several polypeptide chains.

multinervate *a.* [L. *multus*, many; *nervus*, sinew] with many nerves or nervures.

multinodal, multinodate *a.* [L. *multus*, many; *nodus*, knot] with many nodes.

multinomial *a.* [L. *multus*, many; *nomen*, name] *appl.* a name or designation composed of several names or terms; *cf.* binomial, trinomial.

multinucleate *a.* [L. *multus*, many; *nucleus*, kernel] with several or many nuclei; *alt.* polykaric, polynuclear, polynucleate.

multinucleolate *a.* [L. *multus*, many; *nucleolus*, small kernel] with more than one nucleolus.

multiovulate *a.* [L. *multus*, many; *ovum*, egg] with several or many ovules.

multiparous *a.* [L. *multus*, many; *parere*, to beget] bearing several, or more than one, offspring at a birth; developing several or many lateral axes, *alt.* pleiochasial.

multipennate *a.* [L. *multus*, many; *penna*, feather] *appl.* muscle containing a number of extensions of its tendon of insertion.

multiperforate *a.* [L. *multus*, many; *perforare*, to bore through] having more than one perforation, *appl.* perforation plate.

multipinnate *a.* [L. *multus*, many; *pinnatus*, feathered] divided into many lateral processes or leaflets; many times pinnate.

multiple alleles a series of more than 2 alternative forms of a gene at a single locus; *alt.* multiple allelomorphs.

multiple corolla a corolla with 2 or more whorls of petals.

multiple diploid allopolyploid, *q.v.*

multiple factors genes having a joint or cumulative effect.

multiple fission repeated division; division into a large number of parts or spores.

multiple fruit anthocarp, *q.v.*

multiplicate *a.* [L. *multiplicare*, to make manifold] consisting of many; having many folds or plicae.

multipolar *a.* [L. *multus*, many; *polus*, axis end] *appl.* nerve cells with more than 2 axis-cylinder processes; involving more than 2 poles, *appl.* mitosis, normal in certain Sporozoa, but usually pathological; pluripolar, *q.v.*

multiporous *a.* [L. *multus*, many; Gk. *poros*, passage] having many pores.

multipotent *a.* [L. *multus*, many; *potens*, able] capable of giving rise to several kinds of structures; *appl.* primordia, as in meristem.

multiradiate *a.* [L. *multus*, many; *radius*, ray] many-rayed; *appl.* spicule: polyaxon, *q.v.*

multiradicate *a.* [L. *multus*, many; *radix*, root] polyrhizal, *q.v.*

multiramose *a.* [L. *multus*, many; *ramus*, branch] much branched.

multiseptate *a.* [L. *multus*, many; *septum*, partition] having numerous partitions.

multiserial, multiseriate *a.* [L. *multus*, many; *series*, row] arranged in many rows; *appl.* xylem rays more than one cell wide; *appl.* ascospores in rows in ascus.

multispiral *a.* [L. *multus*, many; *spira*, coil] with many coils or whorls.

multisporous polysporous, *q.v.*

multistaminate *a.* [L. *multus*, many; *stamen*, thread] having several or many stamens.

multisulcate *a.* [L. *multus*, many; *sulcus*, furrow] much furrowed.

multitentaculate *a.* [L. *multus*, many; *tentaculum*, feeler] having many tentacles.

Multituberculata, multituberculates *n., n.plu.* [L. *multus*, many; *tuberculum*, small hump] the only order of the Allotheria, *q.v.*, in some classifications considered to be an infraclass of monotremes.

multituberculate *a.* [L. *multus*, many; *tuberculum*, small hump] having several or many small prominences.

multituberculy *n.* [L. *multus*, many; *tuberculum*, small hump] the theory that molar teeth are derived from forms with a number of tubercles.

multivalent *n.* [L. *multus*, many; *valere*, to be strong] the association between more than 2 chromosomes during meiosis in polyploids.

multivalve *n.* [L. *multus*, many; *valvae*, folding doors] a shell composed of more valves or pieces than 2.

multivincular *a.* [L. *multus*, many; *vinculum*, fetter] *appl.* hinge of bivalve shell with several ligaments.

multivoltine *a.* [L. *multus*, many; It. *volta*, turn] having more than one brood in a year, *appl.* silkworms, certain birds.

multocular *a.* [L. *multus*, many; *oculus*, eye] many-eyed.

multungulate *a.* [L. *multus*, many; *ungula*, hoof] having the hoof in more than 2 parts.

mune *n.* [L. *munus*, function] a group of organisms with a characteristic behaviour response; mores, *q.v.*

Munsell colour notation a way of expressing the colour of a soil by matching it against a Munsell colour chart.

mu particles particles borne by some *Paramecium aurelia*, a ciliate, which kill or injure susceptible partners with which they conjugate.

mural *a.* [L. *muralis*, of walls] constituting or *pert.* a wall, as cells or membranes.

muralium *n.* [L. *muralis*, *pert.* a wall] a structure formed by layers one cell thick, as of liver cells.

muramic acid a material found in bacterial cell wall consisting of N-acetylglucosamine linked to lactic acid, and which forms part of the molecule of mucopeptide.

murein *n.* [L. *murus*, wall] a peptidoglycan found in the cell walls of bacteria and blue–green algae, consisting of a few amino acids bridging chains composed of N-acetylglucosamine and N-acetylmuramic acid and linking to other peptide bridges and so forming a 3-dimensional network; *cf.* mucopeptide.

muricate *a.* [L. *muricatus*, having sharp points] formed with sharp points; covered with short sharp outgrowths; studded with oxalate crystals, *appl.* cystidia.

muriform *a.* [L. *murus*, wall; *forma*, shape] like a brick wall; *appl.* a parenchyma so arranged, occurring in medullary ray of dicotyledons and in cork; *appl.* arrangement of germinating spores; *appl.* spores: dictyospores, *q.v.* [F. *mûriforme*, mulberry-shaped] shaped like a morula, *appl.* coelomocytes.

muscarine a ptomaine base, found in the fly agaric toadstool, *Amanita muscaria*, and other plants.

Musci *n.* [L. *muscus*, moss] a class of bryophytes, commonly called mosses, characterized by leafy gametophytes, called gametophores, with multicellular rhizoids, spores producing a filamentous protonema, and no elaters; *alt.* Bryopsida; *cf.* Hepaticae.

muscicoline *a.* [L. *muscus*, moss; *colere*, to inhabit] living or growing among or on mosses; *alt.* muscicolous.

muscle *n.* [L. *musculus*, muscle] a mass of contractile fibres with motorial function; an organ of the body made of contractile fibres or cells.

muscle banners folds or plaits of mesogloea on sulcus of anthozoan mesenteries, supporting retractor muscles.

muscle column sarcostyle, *q.v.*

muscle segment myomere, *q.v.*

muscle spindle a sensory receptor in muscle consisting of a spindle-shaped connective tissue sheath containing small modified fibres, with nerve fibres entering each spindle and forming spirals or arborizations around individual muscle fibres; *alt.* spindle organ.

muscle sugar inositol, *q.v.*

muscle tail a muscular projection at the base of a musculo-epithelial cell of coelenterates.

muscoid *a.* [L. *muscus*, moss; Gk. *eidos*, form] moss-like; mossy; *alt.* muscous.

muscology *n.* [L. *muscus*, moss; Gk. *logos*, discourse] the study of Musci, mosses; *cf.* bryology.

muscous muscoid, *q.v.*

muscular *a.* [L. *musculus*, muscle] *pert.* or consisting of muscle, *appl.* sense, excitability, fibres, tissue, process, triangle, stomach, etc.

muscularis externa a layer of gut wall between the submucosa and serosa consisting of a sheet of longitudinal and a sheet of circular muscle fibres, with Auerbach's plexus lying between them.

muscularis mucosa the outermost layer of the mucosa of the gut wall, made of smooth muscle fibres.

musculature *n.* [L. *musculus*, muscle] the system or arrangement of muscles as a whole.

musculocutaneous *a.* [L. *musculus*, muscle; *cutis*, skin] *pert.* muscles and skin; *appl.* limb veins and nerves supplying muscles and skin.

musculoepithelial myoepithelial, *q.v.*

musculophrenic *a.* [L. *musculus*, muscle; Gk. *phrēn*, midriff] supplying diaphragm and body wall muscles; *appl.* artery: a branch of the internal mammary artery.

musculospiral *a.* [L. *musculus*, muscle; *spira*, coil] *appl.* radial nerve which passes spirally down humerus; *appl.* spiral arrangement of muscle fibres.

musculotendinous *a.* [L. *musculus*, muscle;

tendo, tendon] *pert.* muscle and tendon, or to their fibrils; *alt.* myotendinal.

mushroom bodies corpora pedunculata, *q.v.*

mushroom gland the large mushroom-shaped seminal vesicles of certain insects, as Dictyoptera.

mutafacient *a.* [L. *mutare*, to change; *facere*, to make] inducing or aiding the creation of a mutation, as intracellular agents, mainly.

mutagen *n.* [L. *mutare*, to change; Gk. *gennaein*, to generate] a mutagenic agent.

mutagenic *a.* [L. *mutare*, to change; Gk. *gennaein*, to generate] capable of inducing a mutation, as radiation, chemicals, or other extracellular agents.

mutant *n.* [L. *mutare*, to change] an individual with transmissible characteristics different from those of the parent form. *a.* Exhibiting mutation.

mutate *v.* [L. *mutare*, to change] to undergo or exhibit mutation.

mutation *n.* [L. *mutare*, to change] a change in the amount or structure of the genetic material (usually DNA) in an organism, resulting in a change in the characteristics of the organism; *alt.* idiovariation.

mutator *appl.* genes which increase the general mutation rate.

Mutica *n.* [L. *muticus*, docked, curtailed] an infraclass or cohort of eutherians including the order Cetacea.

muticate, muticous *a.* [L. *muticus*, curtailed] without a point, as of an awn or mucro; without defensive structures.

mutein *n.* [L. *mutare*, to change] a protein produced from a mutated gene.

mutilation *n.* [L. *mutilare*, to maim] loss of an essential part of a structure; amputation.

mutilous *a.* [L. *mutilus*, maimed] without defensive structures, as clawless, harmless, toothless, blunt.

muton *n.* [L. *mutare*, to change] the smallest unit within a gene, which when altered may cause a mutation, equivalent to a single purine or pyrimidine base.

mutualism *n.* [L. *mutuus*, reciprocal] a form of symbiosis in which both parties derive advantage without sustaining injury.

myarian, myaric *a.* [Gk. *mys*, muscle] *appl.* classification according to musculature; *pert.* musculature.

Mycelia Sterila an order of Fungi Imperfecti having no spores but with the mycelium having a characteristic structure or mode of growth.

mycelioid *a.* [Gk. *mykes*, fungus; *eidos*, form] like mycelium.

mycelium *n.* [Gk. *mykes*, fungus] network of filaments, hyphae, forming typical vegetative structure of fungi; *alt.* mycele, hyphostroma.

myceloconidium stylospore, *q.v.*

myceto- *see also* myco-.

mycetocyte *n.* [Gk. *mykes*, fungus; *kytos*, hollow] one of follicle cells at posterior oocyte pole through which the egg of aphids and other bugs is infected by symbionts.

mycetogenetic *a.* [Gk. *mykes*, fungus; *genesis*, descent] produced by a fungus; *alt.* mycetogenic.

mycetoid *a.* [Gk. *mykes*, fungus; *eidos*, form] mycoid, *q.v.*

mycetology mycology, *q.v.*

mycetoma, mycetome *n.* [Gk. *mykes*, fungus] the mycetocytes collectively.

mycetophage *n.* [Gk. *mykes*, fungus; *phagein*, to eat] fungivore *q.v.*; *a.* mycetophagous.

Mycetozoa *n.* [Gk. *mykes*, fungus; *zoon*, animal] an order of Sarcodina commonly called slime moulds or slime fungi and classed by botanists as Myxomycetes, which form encrusting masses on rotten wood, etc., produce amoeba-like colonies but also have spores produced in sporangia as in fungi.

mycin fungine, *q.v.*

mycina *n.* [Gk. *mykes*, fungus] a spherical stalked apothecium of certain lichens.

myco- *see also* myceto-.

mycobiont *n.* [Gk. *mykes*, fungus; *bion*, living] the fungal component of a lichen; *cf.* phycobiont.

mycobiota *n.* [Gk. *mykes*, fungus; *bionai*, to live] the fungi of an area or region.

mycocecidium *n.* [Gk. *mykes*, fungus; *kekis*, gall nut; *idion*, *dim.*] any gall caused by fungi; *alt.* mycodomatium.

mycoclera *n.* [Gk. *mykes*, fungus; *kleros*, portion] the mycelial covering of ectotrophic mycorrhiza.

mycocriny *n.* [Gk. *mykes*, fungus; *krinein*, to separate] chemical decomposition of plant debris by fungi.

mycoderm *n.* [Gk. *mykes*, fungus; *derma*, skin] a superficial bacterial or yeast film during alcoholic fermentation.

mycodomatium *n.* [Gk. *mykes*, fungus; *domation*, little house] root nodule, *q.v.*; mycocecidium, *q.v.*

mycoecotype *n.* [Gk. *mykes*, fungus; *oikos*, household; *typos*, pattern] the habitat type of mycorrhizal and parasitic fungi.

mycoflora *n.* [Gk. *mykes*, fungus; L. *flos*, flower] all the fungi growing in a specified area.

mycogenetics *n.* [Gk. *mykes*, fungus; *genesis*, descent] genetics of fungi.

mycoid *a.* [Gk. *mykes*, fungus; *eidos*, form] like a fungus; *alt.* fungoid, fungous, mycetoid.

mycology *n.* [Gk. *mykes*, fungus; *logos*, discourse] that part of botany which deals with fungi; *alt.* mycetology.

mycolysis *n.* [Gk. *mykes*, fungus; *lysis*, loosing] the lysis of fungi, as by bacteria.

mycophthorous *a.* [Gk. *mykes*, fungus; *phthoros*, destruction] fungus-destroying; *appl.* or *pert.* fungi parasitizing other fungi.

Mycophyta *n.* [Gk. *mykes*, fungus; *phyton*, plant] Fungi, *q.v.*

mycoplasm *n.* [Gk. *mykes*, fungus; *plasma*, form] a resting phase of a rust fungus in cereal seeds, giving rise to mycelium in the developing host plant; bacterial content of root nodules.

mycoplasma-like organism (MLO) any of a group of organisms which resemble mycoplasmas and are found in plants.

mycoplasmas *n.plu.* [Gk. *mykes*, fungus; *plasma*, form] a group of bacteria including the genus *Mycoplasma*, which are highly pleomorphic, have complex life cycles and are in some respects intermediate between bacteria and viruses, and were formerly also called pleuropneumonia-like organisms; they are mostly parasitic in animals, while similar organisms in plants are called either mycoplasmas or mycoplasma-like organisms (MLOs).

Mycoplasmatales mycoplasmas, *q.v.*, when considered as an order of bacteria.

mycopremna *n*. [Gk. *mykēs*, fungus; *premnon*, stem] a rhizome containing symbiotic fungi, as in some orchids.

mycorrhiz- mycorrhiz-, *q.v.*

mycorrhiza *n*. [Gk. *mykēs*, fungus; *rhiza*, root] association of fungal mycelium with roots of a higher plant; *alt*. mycorrhiza.

mycorrhizal *pert*. mycorrhiza.

mycorrhizic *a*. [Gk. *mykēs*, fungus; *rhiza*, root] exhibiting the features of a mycorrhiza; partially symbiotic; *alt*. mycorhizic.

mycorrhizoma *n*. [Gk. *mykēs*, fungus; *rhizōma*, root] the association of fungi with rhizomes; *plu*. mycorrhizomata.

mycorrhizome *n*. [Gk. *mykēs*, fungus; *rhizōma*, root] in orchids, protocorm, *q.v.*

mycose *n*. [Gk. *mykēs*, fungus] trehalose, *q.v.*

mycosin *n*. [Gk. *mykēs*, fungus] a nitrogenous substance found in fungal cell walls and resembling chitin.

mycostasis *n*. [Gk. *mykēs*, fungus; *stasis*, standing] fungistasis, *q.v.*

mycosterols *n.plu*. [Gk. *mykēs*, fungus; *stereos*, solid; alcohol] sterols that were originally discovered in cryptogams, especially fungi, but have now been found more widely distributed, such as ergosterol, fucosterol, etc.; *cf*. phytosterols, zoosterols.

Mycota *n*. [Gk. *mykēs*, fungus] Fungi, *q.v.*

mycothallus *n*. [Gk. *mykēs*, fungus; *thallos*, young shoot] the assimilative body of fungi; association of fungi with thallus of liverworts.

mycotic *a*. [Gk. *mykēs*, fungus] caused by fungi.

mycotoxins *n.plu*. [Gk. *mykēs*, fungus; *toxikon*, poison] toxins produced by fungi.

mycotrophic *a*. [Gk. *mykēs*, fungus; *trophē*, nourishment] *appl*. plants living symbiotically with fungi.

mycteric *a*. [Gk. *myktēr*, nose] *pert*. nasal cavities.

Myctophiformes *n*. [*Myctophum*, generic name; L. *forma*, shape] an order of marine teleosts including the lizard fish, lantern fish, and deep-sea spider fish.

mydriasis *n*. [Gk. *mydriasis*] dilatation of pupil of the eye; *cf*. myosis.

myelencephalon *n*. [Gk. *myelos*, marrow; *engkephalos*, brain] the posterior part of hindbrain, comprising medulla oblongata and lower part of 4th ventricle; *alt*. after-brain, marrow brain.

myelic *a*. [Gk. *myelos*, marrow] *pert*. spinal medulla.

myelin *n*. [Gk. *myelos*, marrow] a highly refracting fatty material forming medullary sheath of nerve fibres.

myelination *n*. [Gk. *myelos*, marrow] acquisition of a medullary sheath; *alt*. myelinization.

myelin sheath medullary sheath, *q.v.*

myeloblast *n*. [Gk. *myelos*, marrow; *blastos*, bud] an undifferentiated non-granular lymphoid cell of bone marrow; *alt*. lymphomyelocyte, microleucoblast.

myelobranchium restibranchium, *q.v.*

myelocoel *n*. [Gk. *myelos*, marrow; *koilos*, hollow] the spinal cord canal.

myelocyte *n*. [Gk. *myelos*, marrow; *kytos*, hollow] an amoeboid cell of bone marrow.

myeloic *a*. [Gk. *myelos*, marrow] *appl*. and *pert*. cells which give rise to neutrophil or polymorphonuclear leucocytes.

myeloid *a*. [Gk. *myelos*, marrow; *eidos*, form] like marrow in appearance or structure, *appl*. cells, as megakaryocytes, monocytes, and parenchyma cells; resembling myelin, *appl*. granules at base of retinal pigment cells.

myelomere *n*. [Gk. *myelos*, marrow; *meros*, part] a segment of the spinal cord.

myelon *n*. [Gk. *myelos*, marrow] spinal cord of Vertebrata.

myeloplast *n*. [Gk. *myelos*, marrow; *plastos*, formed] a leucocyte of bone marrow.

myeloplax *n*. [Gk. *myelos*, marrow; *plax*, something flat] a giant cell of marrow and blood-forming organs; megalokaryocyte and osteoclast.

myelopoiesis *n*. [Gk. *myelos*, marrow; *poiēsis*, making] the formation and development of cells of bone marrow, as of granulocytes.

myelospongium *n*. [Gk. *myelos*, marrow; *spongia*, sponge] interconnected spongioblasts which give rise to neuroglia.

myenteric *a*. [Gk. *mys*, muscle; *enteron*, gut] *appl*. nerve plexus: Auerbach's plexus, *q.v.*; *appl*. reflex.

myenteron *n*. [Gk. *mys*, muscle; *enteron*, gut] the muscular coat of intestine.

myiasis *n*. [Gk. *myia*, fly] the invasion of living tissues by larvae of Diptera.

mylohyoid *a*. [Gk. *myle*, mill; *hyoeidēs*, Υ-shaped] in the region of hyoid bone and posterior part of mandible, *appl*. artery, groove, muscle, nerve.

myoalbumin *n*. [Gk. *mys*, muscle; L. *albumen*, white of egg] an albumin found in muscle.

myoblast *n*. [Gk. *mys*, muscle; *blastos*, bud] a cell which develops into muscle fibre.

myocardium *n*. [Gk. *mys*, muscle; *kardia*, heart] the muscular walls of the heart.

myochemistry *n*. [Gk. *mys*, muscle; *chēmeia*, transmutation] the chemistry of muscle.

myochrome *n*. [Gk. *mys*, muscle; *chrōma*, colour] any muscle pigment.

myocoel *n*. [Gk. *mys*, muscle; *koilos*, hollow] part of the coelom enclosed in a myotome.

myocomma *n*. [Gk. *mys*, muscle; *komma*, clause] a ligamentous connection between successive myomeres; *alt*. myoseptum.

myocyte *n*. [Gk. *mys*, muscle; *kytos*, hollow] contractile inner layer of ectoplasm of gregarines; a contractile or muscle cell.

Myodocopa *n*. [Gk. *myōdēs*, mouse-like] an order of ostracods having 2 pairs of trunk legs and biramous 2nd antenna with a large basal segment.

myodome *n*. [Gk. *mys*, muscle; *domos*, chamber] a chamber containing the eye muscles in some teleosts.

myodynamic *a*. [Gk. *mys*, muscle; *dynamis*, power] *pert*. muscular force or contraction.

myoelastic *a*. [Gk. *mys*, muscle; *elaunein*, to draw out] *appl*. tissue composed of unstriped muscle fibres and elastic connective tissue fibres.

myoepicardial *a*. [Gk. *mys*, muscle; *epi*, upon; *kardia*, heart] *appl*. a mantle consisting of the mesocardium walls, destined to form the muscular and epicardial walls of the heart.

myoepithelial *a.* [Gk. *mys*, muscle; *epi*, upon; *thēlē*, nipple] *pert.* muscle and epithelium; *appl.* epithelium cells with contractile outgrowths, as in coelenterates; *appl.* contractile cells of epithelial origin in salivary and sweat glands; *alt.* musculoepithelial.

myofibrillae, myofibrils *n.plu.* [Gk. *mys*, muscle; L. *fibrilla*, small fibre] contractile fibrils of muscular tissue; *alt.* sarcostyles.

myofilaments *n.plu.* [Gk. *mys*, muscle; L. *filum*, thread] thin thread-like components of a myofibrila.

myogen *n.* [Gk. *mys*, muscle; *gennaein*, to produce] any of a mixture of albumins found in muscle.

myogenesis *n.* [Gk. *mys*, muscle; *genesis*, origin] the origin and development of muscle fibres.

myogenic *a.* [Gk. *mys*, muscle; *gennaein*, to produce] having origin in muscle cells; *appl.* contractions arising in muscle cells spontaneously and independent of nervous stimulation, as heart beat; *cf.* neurogenic, propriogenic.

myoglobin *n.* [Gk. *mys*, muscle; L. *globus*, globe] a type of respiratory haem protein, found in muscle and concerned with oxygen transport and storage; formerly called myohaematin and now also called myohaemoglobin.

myoglobulin *n.* [Gk. *mys.* muscle; L. *globulus*, small globe] a globulin of muscle.

myohaematin *n.* [Gk. *mys*, muscle; *haima*, blood] *see* histohaematins; *see* myoglobin.

myohaemoglobin *n.* [Gk. *mys*, muscle; *haima*, blood; L. *globus*, sphere] myoglobin, *q.v.*

myoid *a.* [Gk. *mys*, muscle; *eidos*, form] resembling or composed of muscular fibres; *appl.* striated cells or sarcolytes of thymus. *n.* Contractile proximal part of filament of rods and cones of retina.

myolemma *n.* [Gk. *mys*, muscle; *lemma*, skin] sarcolemma, *q.v.*

myology *n.* [Gk. *mys*, muscle; *logos*, discourse] the branch of anatomy dealing with muscles.

myomere *n.* [Gk. *mys*, muscle; *meros*, part] a muscle segment divided off by connective tissue insertions or myocommata; *alt.* myotome, muscle segment.

myometrial *a.* [Gk. *mys*, muscle; *mētra*, uterus] *pert.* myometrium; *appl.* glandular tissue of uterus, supposed to produce a hormone affecting growth of mammary glands.

myometrium *n.* [Gk. *mys*, muscle; *mētra*, uterus] the muscular uterine wall.

myone *n.* [Gk. *myōn*, muscular part] unit of muscle; individual muscle fibre.

myonema, myoneme *n.* [Gk. *mys*, muscle; *nēma*, thread] a minute contractile fibril of Protista.

myoneural *a.* [Gk. *mys*, muscle; *neuron*, nerve] neuromuscular, *q.v.*

myoneure *n.* [Gk. *mys*, muscle; *neuron*, nerve] motor neurone, *q.v.*, especially when concerned with regulation of movement.

myonicity *n.* [Gk. *mys*, muscle] the contracting power of muscular tissue.

myophan *a.* [Gk. *mys*, muscle; *phainein*, to appear] muscle-like; *appl.* striations in Protista.

myophore *n.* [Gk. *mys*, muscle; *pherein*, to bear] a structure adapted for muscle attachment.

myophrisks *n.plu.* [Gk. *mys*, muscle; *phrix*, ripple] in certain radiolarians, specialized myonemes used in regulation of diameter of body.

myoplasm *n.* [Gk. *mys*, muscle; *plasma*, mould]

the contractile portion of a muscle cell or fibre; *cf.* sarcoplasm.

myopolar *a.* [Gk. *mys*, muscle; *polos*, axle end] *pert.* muscular polarity.

myoproteid *n.* [Gk. *mys*, muscle; *prōteion*, first] a globulin-like substance of fish muscle.

myoseptum *n.* [Gk. *mys*, muscle; L. *septum*, partition] myocomma, *q.v.*

myosin *n.* [Gk. *mys*, muscle] a protein of muscle which combines with actin to form actomyosin, and which with actin gives the dark colour to the A-band.

myosis *n.* [Gk. *myein*, to close] contraction of pupil of the eye; *alt.* miosis; *cf.* mydriasis.

myotasis *n.* [Gk. *mys*, muscle; *tasis*, tension] muscular tension or tonicity; *alt.* myotonia.

myotatic *a.* [Gk. *mys*, muscle; *tasis*, tension] causing or *pert.* myotasis, *appl.* stretch reflex.

myotendinal musculotendinous, *q.v.*

myotic *a.* [Gk. *myein*, to close] causing or *pert.* myosis or pupillary contraction; *alt.* miotic.

myotome *n.* [Gk. *mys*, muscle; *tomē*, cutting] one of a series of hollow cubes formed in early vertebrate embryo; a muscular metamere of primitive vertebrates and segmented invertebrates; myomere, *q.v.*

myotonia *n.* [Gk. *mys*, muscle; *tonos*, tension] myotasis, *q.v.*

myotube *n.* [Gk. *mys*, muscle; L. *tubus*, tube] an elongated myoblast in which longitudinal filaments surround a cytoplasmic axis, a stage in development of myofibrillae.

Myriangiales *n.* [*Myriangium*, generic name] an order of Loculomycetidae which are mainly tropical and parasitic on plants and insects.

Myriapoda, myriapods *n., n.plu.* [Gk. *myrios*, numberless; *pous*, foot] a class of arthropods having a clear head with one pair of antennae, then numerous similar segments bearing legs and including the Chilopoda, Diploda, and sometimes Pauropoda and Symphyla; *alt.* Myriopoda.

Myricales *n.* [*Myrica*, generic name] an order of dicots of the Archichlamydeae or Hamamelididae comprising the single family Myricaceae, e.g. bog myrtle.

myriophylloid *a.* [Gk. *myrios*, numberless; *phyllon*, leaf; *eidos*, form] having a much-divided thallus, *appl.* certain algae.

Myriopoda Myriapoda, *q.v.*

myriospored, myriosporous *a.* [Gk. *myrios*, numberless; *sporos*, seed] having very numerous spores, extremely polysporous.

myrmecochore *n.* [Gk. *myrmēx*, ant; *chorē*, farm] an oily seed modified to attract and be spread by ants.

myrmecochory dispersal of a myrmecochore.

myrmecodomatium *n.* [Gk. *myrmēx*, ant; *dōmation*, small house] an ant-inhabited cavity in plant tissue.

myrmecole *n.* [Gk. *myrmēx*, ant; L. *colere*, to inhabit] an organism occupying ants' nests.

myrmecology *n.* [Gk. *myrmēx*, ant; *logos*, discourse] the study of ants.

myrmecophagous *a.* [Gk. *myrmēx*, ant; *phagein*, to eat] ant-eating.

myrmecophil *n.* [Gk. *myrmēx*, ant; *philos*, loving] a guest insect in a nest of ants.

myrmecophilous *a.* [Gk. *myrmēx*, ant; *philos*, loving] pollinated by agency of ants; *appl.* fungi serving as food for ants; living with, or preying on, or mimicking ants, *appl.* spiders.

myrmecophobic *a.* [Gk. *myrmēx*, ant; *phobeisthai*, to flee] repelling ants; *appl.* certain plants equipped with glands, hairs, etc., that check ants.

myrmecophyte *n.* [Gk. *myrmēx*, ant; *phyton*, plant] a myrmecophilous plant, or one that benefits from ant inhabitants and has special adaptations for housing them.

myrosin *n.* [Gk. *myron*, unguent] a glucoside occurring in the plants of the family Cruciferae and other Capparales, which is acted upon by the enzyme myrosinase to release substances giving the pungent taste to capers, mustard, etc.; *alt.* sinigrin.

myrosinase the enzyme which acts upon myrosin; *alt.* sinigrinase.

myrrh *n.* [Gk. *murr(h)a*, a transparent gum resin] a resin obtained from plants of the genus *Commiophora*.

Myrtales *n.* [*Myrtus*, generic name] an order of dicots of the Rosidae or Polypetalae usually having united styles and bicollateral vascular bundles.

Myrtiflorae *n.* [*Myrtus*, generic name; L. *flos*, flower] an order of dicots of the Archichlamydeae, being more-or-less equivalent to the Myrtales.

myrtiform *a.* [L. *myrtus*, myrtle; *forma*, shape] *appl.* incisive fossa.

Mysidacea *n.* [*Mysis*, generic name] an order or suborder of Peracarida that are mainly marine and pelagic and have most of the thorax covered by a carapace.

mystacial *a.* [Gk. *mystax*, moustache] *appl.* a pad of thickened skin on side of snout, and to tactile hairs or vibrissae.

Mystacocarida *n.* [Gk. *mystax*, upper lip; *L. caris*, a kind of crab] a subclass of small crustaceans, similar to the copepods, living in marine sands and having no carapace, no clear divisions of the body, and simple appendages, containing only one order, the Derocheilocarida.

mystax *n.* [Gk. *mystax*, moustache] a group of hairs above mouth of certain insects; mystacial hairs.

Mysticeti *n.* [*Mysticetus*, generic name] an order of placental mammals, the baleen or whalebone whales.

myxamoeba *n.* [Gk. *myxa*, slime; *amoibē*, change] in slime moulds, a non-flagellated naked amoeboid cell produced from a germinating spore, which may develop a flagellum and swim as a zoospore.

Myxinoidea, Myxinoidei, myxinoids *n.*, *n.plu.* [*Myxine*, generic name; Gk. *eidos*, form] Hyperotreta, *q.v.*

Myxobacteriales, myxobacteria *n.* [Gk. *myxa*, slime; *baktērion*, small rod] an order of colonial bacteria that do not have a rigid wall, and which creep about over a layer of slime they secrete.

myxocyte *n.* [Gk. *myxa*, slime; *kytos*, hollow] cell of mucous tissue.

myxoflagellate *n.* [Gk. *myxa*, slime; L. *flagellum*, whip] a flagellula or zoospore following myxamoeba stage in development of Myxomycetes or Mycetozoa.

Myxogastres *n.* [Gk. *myxa*, slime; *gastēr*, stomach] Endosporeae, *q.v.*; *alt.* Myxogastromycetidae.

Myxomycetae *n.* [Gk. *myxa*, slime; *mykēs*, fungus] a class of Myxomycophyta equivalent to the Myxomycetales, *q.v.*

Myxomycetales *n.* [Gk. *myxa*, slime; *mykēs*, fungus] an order of Myxomycophyta in which the vegetative phase is a plasmodium; *cf.* Myxomycetae.

Myxomycetes *n.* [Gk. *myxa*, slime; *mykēs*, fungus] Myxomycophyta, *q.v.*, usually when reduced to the status of a group of fungi; in some classifications, a class of Myxomycophyta equivalent to the Myxomycetales; *cf.* Mycetozoa.

Myxomycophyta *n.* [Gk. *myxa*, slime; *mykēs*, fungus; *phyton*, plant] a group of very simple organisms, commonly called slime moulds and sometimes classified as fungi, which have some plant and some animal characteristics, with a body called a plasmodium (a naked creeping mass of protoplasm containing many nuclei), or a pseudoplasmodium (an aggregation of many cells), and developing sporangia which liberate spores that germinate to form myxamoebae; *alt.* Myxomycotina, Myxomycetes, Myxothallophyta, Mycetozoa.

Myxomycotina *n.* [Gk. *myxa*, slime; *mykēs*, fungus] Myxomycophyta, *q.v.*

Myxophyceae *n.* [Gk. *myxa*, slime; *phykos*, seaweed] Cyanophyceae, *q.v.*

Myxophyta *n.* [Gk. *myxa*, slime; *phyton*, plant] Cyanophyta, *q.v.*

myxopodium *n.* [Gk. *myxa*, slime; *pous*, foot] a slimy pseudopodium.

myxopterygium mixipterygium, *q.v.*

Myxospongida, Myxospongiae *n.* [Gk. *myxa*, slime; L. *spongia*, sponge] an order of simple sponges without spicules of the subclass Tetractinellida.

myxosporangium *n.* [Gk. *myxa*, slime; *sporos*, seed; *anggeion*, vessel] a sporangium producing spores embedded in a slimy substance; fruit body of Myxomycetes.

myxospore *n.* [Gk. *myxa*, slime; *sporos*, seed] a spore separated by a slimy disintegration of the hypha; plasmaspore, *q.v.*; a spore of slime moulds; *alt.* slime spore.

Myxothallophyta *n.* [Gk. *myxa*, slime; *thallos*, young shoot; *phyton*, plant] Myxomycophyta, *q.v.*

myxoviruses *n.plu.* [Gk. *myxa*, slime; L. *virus*, poison] a group of viruses containing RNA including those causing influenza, mumps, measles, and rabies.

myxoxanthin *n.* [Gk. *myxa*, slime; *xanthos*, yellow] a xanthophyll carotinoid pigment found in blue–green algae.

myxoxanthophyll *n.* [Gk. *myxa*, slime; *xanthos*, yellow; *phyllon*, leaf] a xanthophyll carotinoid pigment found in blue–green algae.

myzesis *n.* [Gk. *myzein*, to suck] suction; sucking.

Myzostomaria *n.* [*Myzostomum*, generic name] a class of annelids that are ectoparasitic on echinoderms and almost circular in shape; *alt.* Myzostomida.

N

NAA α-naphthaleneacetic acid, *q.v.*

nacre n. [L. *nacara*, a kind of drum] mother-of-pearl, the iridescent inner layer of mollusc shells, made mainly of calcium carbonate.

nacré *a*. [F. *nacré*, having a pearly lustre] having a pearly lustre, *appl.* the thick wall of sieve elements; *alt.* nacreous.

nacreous *a*. [Ar. *nakir*, hollowed] yielding or resembling nacre; nacré, *q.v.*

nacrine *n*. [Ar. *nakir*, hollowed] mother-of-pearl colour; *pert.* nacre.

NAD nicotinamide adenine dinucleotide, a cofactor (coenzyme) involved in various metabolic pathways as a hydrogen acceptor forming $NADH_2$, accepting hydrogen from a number of substrates and dehydrogenases; formerly called DPN and coenzyme I.

NADP nicotinamide adenine dinucleotide phosphate, a cofactor (coenzyme) similar to NAD, *q.v.*; formerly called TPN and coenzyme II.

naiad *n*. [Gk. *Naias*, water nymph] the nymph stage of hemimetabolic insects.

Naiadales Najadales, *q.v.*

nail *n*. [A.S. *naegel*, nail] terminal horny plate of finger or toe, or of beak; *alt.* unguis.

nail bone terminal bone of finger or toe; *alt.* ungual phalanx.

Najadales [*Najas*, generic name] an order of monocots (Alismatidae) which are water or marsh plants and whose perianth is single or lacking; *cf.* Fluviales, Naiadales.

naked *a*. [A.S. *nacod*, naked] without a covering; *appl.* spores, seeds, etc.; *appl.* non-nuclear genes, as virus.

nan-, nana-, nani-, nano- *alt.* nann-, nanna-, nanni-, nanno-.

nanander *n*. [Gk. *nanos*, dwarf; *anēr*, male] a dwarf male, *appl.* plants; *a*. nanandrous.

nanism *n*. [Gk. *nanos*, dwarf] dwarfishness; *alt.* microsomia.

nanno- nano-, *q.v.*

nano- *alt.* nanno-.

nanocyte *n*. [Gk. *nanos*, dwarf; *kytos*, hollow] in some algae, a modified endospore, being a small naked protoplast produced by division of cell contents with no enlargement of vegetative cell.

nanoid *a*. [Gk. *nanos*, dwarf; *eidos*, dwarfish].

nanometre *n*. [Gk. *nanos*, dwarf; *metron*, measure] a unit of microscopic measurement and of wavelength of some electromagnetic radiation, being 10^{-9} m, a thousandth of a micrometre, 10 Ångstrom units; symbol nm; formerly called millimicron.

nanophanerophyte *n*. [Gk. *nanos*, dwarf; *phaneros*, manifest; *phyton*, plant] shrub under 2 m in height.

nanophyllous *a*. [Gk. *nanos*, dwarf; *phyllon*, leaf] with small leaves, of an area between that of leptophyllous and microphyllous leaves.

nanoplankton *n*. [Gk. *nanos*, dwarf; *plangktos*, wandering] microscopic floating plant and animal organisms; *cf.* microplankton.

nanous *a*. [L. *nanus*, dwarf] dwarfed; dwarfish.

α-naphthaleneacetic acid (NAA) a synthetic auxin.

naphthaquinone a derivative of quinone from which vitamin K is derived.

napiform *a*. [L. *napus*, turnip; *forma*, shape] turnip-shaped, *appl.* roots.

Narcomedusae *n*. [Gk. *narkē*, numbness; *Medousa*, Medusa, one of the Gorgons] an order

of hydrozoans having medusae with solid tentacles, lobed umbrella margin, and no radial canals; sometimes included with the Trachymedusae in the order Trachylina.

narcosis *n*. [Gk. *narkē*, numbness] state of unconsciousness or stupor produced by a drug.

narcotic *n*. [Gk. *narkē*, numbness] a drug which produces numbness, sleep or unconsciousness. *a*. *Pert.* or producing narcosis.

narcotine *n*. [Gk. *narkē*, numbness] an alkaloid found in opium and having antispasmodic properties.

nares *n.plu*. [L. *nares*, nostrils] nostrils, *q.v.*; *sing.* naris.

nares, anterior openings of olfactory organ to exterior; *alt.* nostrils.

nares, posterior openings of olfactory organ into pharynx or throat; *alt.* choanae.

narial *a*. [L. *nares*, nostrils] *pert.* the nostrils; *appl.* septum, the partition between nostrils; *alt.* narine.

naricorn *n*. [L. *nares*, nostrils; *cornu*, horn] terminal horny part of nostril of certain birds such as albatross; a nasal scale.

nariform *a*. [L. *nares*, nostrils; *forma*, shape] shaped like nostrils.

narine narial, *q.v.*

naris *sing.* of nares.

nasal *a*. [L. *nasus*, nose] *pert.* the nose. *n*. Nasal scale, plate, or bone.

nasalis *n*. [L. *nasus*, nose] muscle drawing alae of the nose towards septum: compressor naris.

nascent RNA heterogenous nuclear RNA, *q.v.*

nasion *n*. [L. *nasus*, nose] middle point of nasofrontal suture.

Nasmyth's membrane [*A. Nasmyth*, Scottish dentist] a transparent membrane over enamel of crown of a mammalian tooth; *alt.* cuticula dentis, primary enamel cuticle.

nasoantral *a*. [L. *nasus*, nose; *antrum*, cavity] *pert.* nose and maxillary cavity.

nasobuccal *a*. [L. *nasus*, nose; *bucca*, cheek] *pert.* nose and cheek; *pert.* nose and mouth cavity.

nasociliary *a*. [L. *nasus*, nose; *cilia*, eyelashes] *appl.* branch of ophthalmic nerve, with internal and external nasal branches, and giving off the long ciliary and other nerves.

nasofrontal *a*. [L. *nasus*, nose; *frons*, forehead] *appl.* part of superior ophthalmic vein which communicates with the angular vein.

nasolabial *a*. [L. *nasus*, nose; *labium*, lip] *pert.* nose and lip, *appl.* muscle, *appl.* groove and glands in Plethodontidae, a family of salamanders.

nasolacrimal *a*. [L. *nasus*, nose; *lacrima*, tear] *appl.* canal from lacrimal sac to inferior meatus of nose through which tear duct passes.

nasomaxillary *a*. [L. *nasus*, nose; *maxilla*, jaw] *pert.* nose and upper jaw.

naso-optic *a*. [L. *nasus*, nose; Gk. *optikos*, relating to sight] *appl.* an embryonic groove between nasal and maxillary processes.

nasopalatine *a*. [L. *nasus*, nose; *palatus*, palate] *pert.* nose and palate; *appl.* groove of vomer, recess in nasal septum, nerve, canal communicating with vomeronasal organ; *alt.* nasopalatal.

nasopharyngeal *a*. [L. *nasus*, nose; Gk. *pharyngx*, gullet] *pert.* nose and pharynx, or nasopharynx.

nasopharynx *n*. [L. *nasus*, nose; Gk. *pharyngx*,

gullet] that part of pharynx continuous with posterior nares; *alt.* rhinopharynx.

nasoturbinal *a.* [L. *nasus*, nose; *turbo*, whorl] *appl.* outgrowths from lateral wall of nasal cavity increasing area of sensory surface.

nastic movement nasty, *q.v.*

nasty *n.* [Gk. *nastos*, pressed close] a plant movement caused by a diffuse non-directional stimulus, usually a growth movement, but may be a change in turgidity, as in the sensitive plant that droops on contact; *alt.* nastic movement; *cf.* kinesis.

nasus *n.* [L. *nasus*, nose] nose; clypeus of insect head.

nasute, nasutus *a.* [L. *nasutus*, large-nosed] *appl.* a soldier termite with rostrum.

natal *a.* [L. *natalis*, *pert.* birth] *pert.* birth. [L. *nates*, buttocks] *pert.* the buttocks.

natality *n.* [L. *natalis*, *pert.* birth] birth rate.

natant *a.* [L. *natare*, to swim] floating on water surface.

natatorial *a.* [L. *natare*, to swim] formed or adapted for swimming.

natatory *a.* [L. *natator*, swimmer] swimming habitually; *pert.* swimming.

nates *n.plu.* [L. *natis*, rump] buttocks, *alt.* clunes; superior colliculi of corpora quadrigemina; umbones of bivalves.

native *a.* [L. *natus*, born] *appl.* animals and plants which originated in district or area in which they live.

natriferic *a.* [L. *natrium*, sodium; *ferre*, to carry] transporting sodium.

natural classification a classification that groups organisms or objects together on the basis of the sum total of all their characteristics, and tries to indicate evolutionary relationships; *cf.* artificial classification.

natural group a group of organisms having a common ancestor.

natural order an old term for family of flowering plants.

natural selection processes occurring in nature which result in survival of fittest and elimination of individuals less well adapted to their environment.

nature *n.* [L. *natura*, nature] sum total of inheritance, genotype, as distinct from nurture or environment.

naupliiform *a.* [L. *nauplius*, shellfish; *forma*, shape] superficially resembling a nauplius, *appl.* larvae of certain Hymenoptera; *alt.* naupliform, cyclopoid.

nauplius *n.* [L. *nauplius*, shellfish] the earliest larval stage of certain crustaceans.

nautiliform *a.* [L. *nautilus*, nautilus; *forma*, shape] shaped like a nautilus shell; *alt.* nautiloid.

Nautiloidea, nautiloids *n., n.plu.* [L. *nautilus*, nautilus; Gk. *eidos*, form] a mainly extinct subclass of tetrabranch cephalopods with only one living genus, *Nautilus*, having an external many-chambered shell and a membranous protoconch.

navel *n.* [A.S. *nafela*, navel] place of attachment of umbilical cord to body of embryo; *alt.* umbilicus.

navicular *a.* [L. *navis*, *dim.*, ship] boat-shaped; *alt.* scaphoid, cymbiform.

naviculare *a.* [L. *navis*, *dim.*, ship] the scaphoid radiale of mammalian carpus; tarsal bone between talus and cuneiform bones.

N-discs discs or bands on either side of Z-disc; *alt.* accessory discs.

neala vannus, *q.v.*, of insect wing.

neallotype *n.* [Gk. *neos*, new; *allos*, other; *typos*, pattern] a type specimen of the opposite sex to that of the specimen previously chosen for designation of a new species.

nealogy *n.* [Gk. *nealēs*, youthful; *logos*, discourse] the study of young animals.

Neanderthal man a subspecies of *Homo sapiens*, *H. sapiens neanderthalensis*, living in the Old World during the Pleistocene.

neanic *a.* [Gk. *neanikos*, youthful] *appl.* adolescent phase of life history of individual, *cf.* phyloneanic; brephic, *q.v.*

Nearctic *a.* [Gk. *neos*, new; *Arktos*, Great Bear] *appl.* or *pert.* a zoogeographical region, or subregion of the holarctic region, comprising Greenland and North America, and including northern Mexico.

Nebaliacea *n.* [*Nebalia*, generic name] an order of malacostracans of the series Leptostraca, *q.v.*

nebenkern *n.* [Ger. *neben*, near; *Kern*, nucleus] a structure in the cytoplasm thought to be made of concentric layers of endoplasmic reticulum arranged like the layers of an onion bulb; a spherical mass of mitochondria, *alt.* paranucleus.

nebenkörper *n.* [Ger. *neben*, near; *Körper*, body] a body surrounded by oil drops at hinder pole of *Pyrodinium*.

neck canal the hollow within the elongated part of an archegonium.

neck canal cell one of the cells occupying the neck canal in an archegonium.

neck cell one of the cells making up the wall of the neck of an archegonium.

necrobiosis *n.* [Gk. *nekros*, dead; *biōsis*, manner of life] the activity of cells after death of an organism leading to their degeneration; continuance of certain vital functions after disorganization of a cell.

necrocytosis *n.* [Gk. *nekros*, dead; *kytos*, hollow] death of cells.

necrogenic *a.* [Gk. *nekros*, dead; *genos*, birth] promoting necrosis in the host, *appl.* certain parasitic fungi.

necrogenous *a.* [Gk. *nekros*, dead; *genos*, offspring] living or developing in dead bodies.

necrohormone *n.* [Gk. *nekros*, dead; *hormaein*, to excite] substance in tissue extracts or dead cells which may either kill living cells or induce mitosis.

necron *n.* [Gk. *nekros*, dead] plant material that has died but has not yet decomposed.

necrophagous *a.* [Gk. *nekros*, dead; *phagein*, to eat] feeding on dead bodies; *alt.* necrophilous.

necrophilous *a.* [Gk. *nekros*, dead; *philein*, to love] necrophagous, *q.v.*

necrophoric *a.* [Gk. *nekros*, dead; *pherein*, to carry] containing dead cells, *appl.* water-storing layers in lichens; carrying away dead bodies, *appl.* certain beetles; *alt.* necrophoral, necrophorous.

necrosis *n.* [Gk. *nekrōsis*, deadness] the death of cells or of tissues; *a.* necrotic.

necrotoxin *n.* [Gk. *nekros*, dead; *toxikon*, poison] a toxin leading to necrosis of tissue.

nectar *n.* [Gk. *nektar*, nectar] sweet substance secreted by special glands, nectaries, in flowers and in certain leaves to attract insects for pollination; substance containing spores and attracting insects, produced by certain fungi, as on pycnidia.

nectar guides series of markings on petals of flowers, aiding insects in finding nectar, and at same time facilitating cross-fertilization; *alt.* honey guides.

nectariferous *a.* [L. *nectar*, nectar; *ferre*, to carry] producing nectar, or having nectar-secreting structures.

nectarivorous *a.* [L. *nectar*, nectar; *vorare*, to devour] nectar-eating.

nectary *n.* [Gk. *nektar*, nectar] a group of modified subepidermal cells of no definite position in a flower, less commonly in leaves, secreting nectar, *alt.* nectar gland; honey tube of aphids.

nectocalyx *n.* [Gk. *nektos*, swimming; *kalyx*, cup] a modified medusiform person adapted for swimming purposes found as part of a siphonophore colony; *alt.* nectozooid, nectophore, bell, swimming bell; *plu.* nectocalyces.

nectochaeta *n.* [Gk. *nektos*, swimming; *chaitē*, hair] a free-swimming larva in some polychaetes, bearing rings of cilia and 3 pairs of parapodia.

nectocyst *n.* [Gk. *nektos*, swimming; *kystis*, bladder] the cavity of a nectocalyx; *alt.* nectosac.

nectomonad *n.* [Gk. *nektos*, swimming; *monas*, unit] a free form of certain parasitic flagellates; *cf.* haptomonad.

necton nekton, *q.v.*

Nectonematoidea *n.* [*Nectonema*, generic name; Gk. *eidos*, form] an order of Nematomorpha whose pseudocoel is not filled up with tissue.

nectophore *n.* [Gk. *nektos*, swimming; *pherein*, to carry] nectocalyx, *q.v.*; that portion of common coenosarc on which nectocalyces are borne.

nectopod *n.* [Gk. *nektos*, swimming; *pous*, foot] an appendage modified for swimming.

nectosac nectocyst, *q.v.*

nectosome *n.* [Gk. *nektos*, swimming; *sōma*, body] upper or swimming part of a siphonophore.

nectozooid *n.* [Gk. *nektos*, swimming; *zōon*, animal; *eidos*, form] nectocalyx, *q.v.*

Nectridea *n.* [Gk. *nēktris*, female swimmer] an order of Carboniferous to Permian lepospondyl amphibians with developed limbs, a primitive roof to the skull, and vertebrae with extra articulations.

Needham's sac spermatophore sac, formed by dilatation of male genital duct, in certain cephalopods.

neencephalon neoencephalon, *q.v.*

negative reinforcement a stimulus or series of stimuli which are unpleasant to an animal and so diminish the response to the stimulus or cause avoidance reactions; *alt.* punishment.

negative tropism tendency to move away from the source of a stimulus.

nekton *n.* [Gk. *nektos*, swimming] the organisms swimming actively in water; *alt.* necton; *cf.* seston.

nema *n.* [Gk. *nēma*, thread] a thread-like tubular projection at apex of graptolite sicula; a filament; a nematode; *plu.* nemata.

Nemalionales *n.* [*Nemalion*, generic name] an order of red algae which usually have meiotic division of the zygote nucleus, and the carposporophyte developing from the carpogonium.

nemata *plu.* of nema.

nemathecium *n.* [Gk. *nēma*, thread; *thēkē*,

box] a protuberance on thallus of certain algae, bearing reproductive organs.

Nemathelminthes *n.* [Gk. *nēma*, thread; *helmins*, worm] a former term for the Nematoda, Nematomorpha, and Acanthocephala when they were thought to belong to the same phylum.

nemathybomes *n.plu.* [Gk. *nēma*, thread; *hybos*, humped] mesogloeal parts containing developing nematocysts, as in certain actinozoans such as *Edwardsia*.

nematoblast *n.* [Gk. *nēma*, thread; *blastos*, bud] the cell from which a nematocyst develops; *alt.* cnidoblast.

nematocalyx *n.* [Gk. *nēma*, thread; *kalyx*, cup] in certain colonial hydrozoans, a small polyp which has no mouth but engulfs organisms by pseudopodia; *alt.* guard polyp, nematophore.

nematocerous *a.* [Gk. *nēma*, thread; *keras*, horn] possessing thread-like antennae.

nematocyst *n.* [Gk. *nēma*, thread; *kystis*, bladder] a stinging cell of cnidarian coelenterates consisting of a sac full of fluid in which lies a long coiled thread which is shot out on stimulation; the term is sometimes used only for the contents of the cell; *alt.* cnida, cnidoblast, nettling cell, thread capsule or cell, urticator.

Nematoda, nematodes *n.*, *n.plu.* [Gk. *nēma*, thread; *eidos*. form] a phylum of pseudocoelomate unsegmented worms, commonly called roundworms, that have a thick cuticle and include free-living species, ones living in the soil or as plant parasites (eelworms), and animal parasites; in some classifications, considered a class of Aschelminthes.

nematodology nematology, *q.v.*

nematogen(e) *n.* [Gk. *nēma*, thread; *genos*, offspring] *appl.* phase of Dicyemidae when their vermiform embryos escape from parent by perforating body wall; *cf.* rhombogene.

nematogone *n.* [Gk. *nēma*, thread; *gonē*, seed] a thin-walled propagative cell in gemma of certain mosses.

nematoid *a.* [Gk. *nēma*, thread; *eidos*, form] thread-like; filamentous; *alt.* nemeous.

nematology *n.* [Gk. *nēma*, thread; *logos*, discourse] the study of Nematoda; *alt.* nematodology.

Nematomorpha *n.* [Gk. *nēma*, thread; *morphē*, form] a phylum of pseudocoelomate worm-like animals that are free living in soil or fresh water as adults and parasitic in arthropods when young; in some classifications, considered a class of Aschelminthes.

nematophore *n.* [Gk. *nēma*, thread; *pherein*, to carry] nematocalyx, *q.v.*; *a.* nematophorous.

nematosphere *n.* [Gk. *nēma*, thread; *sphaira*, globe] the capitate end of a tentacle in certain sea anemones.

nematotheca *n.* [Gk. *nēma*, thread; *thēkē*, case] a theca specialized for lodging nematocysts in graptolites.

nematozooid *n.* [Gk. *nēma*, thread; *zōon*, animal; *eidos*, form] a defensive zooid in Hydrozoa.

nemeous nematoid, *q.v.*

Nemertea *n.* [Gk. *Nēmertēs*, a nereid] Nemertini, *q.v.*

Nemertini, nemertines *n.*, *n.plu.* [Gk. *Nēmertēs*, a nereid] a phylum of acoelomate mostly marine animals, commonly called proboscis worms, with a long, sometimes thread-like, body, and an eversible proboscis lying in a cylindrical cavity (the rhynchocoel) dorsal to the gut; *alt.* Nemertea, Rhynchocoela.

nemic *a.* [Gk. *nēma*, thread] *pert.* a nema; *pert.* Nematoda.

nemorose, nemoricole *a.* [L. *nemorosus*, sylvan] inhabiting open woodland places.

neobiogenesis *n.* [Gk. *neos*, new; *bios*, life; *genesis*, origin] theory of renewability of origination of living organisms; repetition of biopoiesis.

neoblast *n.* [Gk. *neos*, new; *blastos*, bud] one of the undifferentiated cells forming primordium of regeneration tissue in response to a wound stimulus.

neocarpy *n.* [Gk. *neos*, young; *karpos*, fruit] production of fruit by an otherwise immature plant.

neocerebellum *n.* [Gk. *neos*, new; L. *dim.* of *cerebrum*, brain] cerebellar region which receives pontine fibres predominantly; *cf.* palaeocerebellum.

neocortex neopallium, *q.v.*

neocyte *n.* [Gk. *neos*, young; *kytos*, hollow] an immature leucoblast.

Neo-Darwinism modern evolutionary theory based on Darwinism and incorporating the findings from Mendelian genetics.

neoencephalon *n.* [Gk. *neos*, young; *engkephalos*, brain] the telencephalon or latest evolved anterior portion of brain; *alt.* neencephalon.

Neogaea *n.* [Gk. *neos*, new; *gaia*, earth] zoogeographical area comprising the neotropical region.

neogamous *a.* [Gk. *neos*, young; *gamos*, marriage] *appl.* forms of Protozoa exhibiting precocious association of gametocytes.

Neogastropoda *n.* [Gk. *neos*, new; *gastēr*, stomach; *pous*, foot] an order of carnivorous siphonate prosobranchs; *alt.* Stenoglossa.

Neogene *a.* [Gk. *neos*, young; *genos*, age] *pert.* or *appl.* the later Tertiary period, Miocene and Pliocene epochs.

neogenesis *n.* [Gk. *neos*, new; *genesis*, origin] new tissue formation; regeneration.

Neognathae *n.* [Gk. *neos*, new; *gnathos*, jaw] an order or superorder of birds including the majority of modern birds, having typical bird features, and being capable of flight; *alt.* Carinatae.

Neo-Lamarckism modification of Lamarck's doctrine of evolution, that inherited acquired characters formed inception of specific differences.

Neolaurentian *a.* [Gk. *neos*, young; *St Lawrence* River] *pert.* or *appl.* early Proterozoic era.

Neolithic *a.* [Gk. *neos*, young; *lithos*, stone] *appl.* or *pert.* the New or polished Stone Age, having cultivator economy and polished stone tools and weapons.

Neomeniomorpha *n.* [*Neomenia*, generic name; Gk. *morphē*, form] an order of chitons of the subclass Aplacophora which live on corals and have a median groove containing a vestigial foot.

neomorph *n.* [Gk. *neos*, new; *morphē*, form] a structural variation from type; an allele which induces new reactions in developmental processes.

neomorphosis *n.* [Gk. *neos*, new; *morphōsis*, change] regeneration in case where new part is unlike anything in body.

neomycin *n.* [Gk. *neos*, new; strepto*mycin*] an antibiotic synthesized by the actinomycete *Streptomyces fradiae*.

neonatal *a.* [Gk. *neos*, new; L. *natus*, born] newborn; recently hatched or born.

neontology *n.* [Gk. *neos*, young; *on*, being; *logos*, discourse] the science of present organic life; *cf.* palaeontology.

neonychium *n.* [Gk. *neos*, young; *onyx*, nail] a soft pad enclosing each claw of embryo of unguiculate vertebrates and of some other mammals, to prevent tearing of foetal membranes; horny claw pad in birds before hatching.

neopallium *n.* [Gk. *neos*, young; L. *pallium*, cloak] in mammalian brain, the cerebral cortex, excluding hippocampus and pyriform lobe; *alt.* neocortex; *cf.* archipallium.

neoplasm *n.* [Gk. *neos*, new; *plasma*, formation] new or added tissue, generally pathological.

Neoptera *n.* [Gk. *neos*, young; *pteron*, wing] a division of insects with incomplete metamorphosis, including the Polyneoptera and Paraneoptera.

Neopterygii *n.* [Gk. *neos*, new; *pterygion*, little wing or fin] a subclass of Osteichthyes, in some classifications, including the higher bony fish, i.e. the Holostei and Teleostei; *cf.* Palaeopterygii.

neoptile *n.* [Gk. *neos*, young; *ptilon*, feather] neossoptile, *q.v.*

Neornithes *n.* [Gk. *neos*, new; *ornis*, bird] a subclass of birds that are more advanced than the Archaeornithes, having fused metacarpals, a short tail, teeth present only in extinct forms, claws only rarely present on wing digits, and having a keel on the sternum.

Neosporidia *n.* [Gk. *neos*, new; *sporos*, seed; *idion*, *dim.*] a subclass of Sporozoa having composite spores and an adult with more than one nucleus.

neossoptile *n.* [Gk. *neossos*, nestling; *ptilon*, feather] feather of nestlings; down feather; down; *alt.* neoptile, nessoptile; *cf.* teleoptile.

neoteinia neoteny, *q.v.*

neoteinic *a.* [Gk. *neos*, young; *teinein*, to stretch] *appl.* substitution royalties of termites which remain undeveloped in certain respects.

neotenia neoteny, *q.v.*

neotenin *n.* [Gk. *neos*, young; *teinein*, to extend, stretch] juvenile hormone, *q.v.*

neoteny *n.* [Gk. *neos*, young; *teinein*, to extend, stretch] retention of larval characters beyond normal period, or occurrence of adult characteristics in larva; attainment of sexual maturity by larval form as in axolotl; *alt.* neotenia, neoteinia; *cf.* paedogenesis.

neothalamus *n.* [Gk. *neos*, new; *thalamos*, chamber] the part of the thalamus with nuclei connected with association areas of the cerebral cortex.

Neotremata *n.* [Gk. *neos*, new; *trēma*, hole] an order of inarticulate brachiopods having a notch in the ventral valve only for the stalk to pass.

Neotropical *a.* [Gk. *neos*, new; *tropikos*, tropic] *appl.* or *pert.* a zoogeographical region consisting of Southern Mexico, Central and South America, and the West Indies; *appl.* phytogeographical realm, including tropical and subtropical regions of America; *alt.* Austro-Columbian.

neotype *n.* [Gk. *neos*, new; *typos*, pattern] a new type; a new type specimen from the original type locality.

neovirus *n.* [Gk. *neos*, new; L. *virus*, poison] a virus having a double strand of RNA, which is rare; a virus directly formed by a mutant viroid, *cf.* palaeovirus.

neoxanthin *n.* [Gk. *neos*, new; *xanthos*, yellow] a xanthophyll carotenoid pigment found in algae, especially of the classes Chlorophyceae and Euglenophyceae.

Neozoic *a.* [Gk. *neos*, young; *zōē*, life] *pert.* period from end of Mesozoic to present day.

nephric *a.* [Gk. *nephros*, kidney] *pert.* kidney; *alt.* renal.

nephridial *a.* [Gk. *nephros*, kidney; *idion*, dim.] nephric, usually *appl.* the small excretory tubules in kidney; *pert.* excretory organ or nephridium of invertebrates.

nephridioblast *n.* [Gk. *nephros*, kidney; *idion*, *dim.*; *blastos*, bud] an ectodermal cell which gives rise to a nephridium.

nephridiopore *n.* [Gk. *nephros*, kidney; *idion*, *dim.*; *poros*, passage] the external opening of a nephridium; *alt.* nephropore.

nephridiostome *n.* [Gk. *nephros*, kidney; *idion*, *dim.*; *stoma*, mouth] ciliated coelomic opening of a nephridium.

nephridium *n.* [Gk. *nephros*, kidney; *idion*, *dim.*] an excretory organ, usually that of invertebrates; embryonic kidney tubule of vertebrates, *alt.* segmental organ.

nephroblast *n.* [Gk. *nephros*, kidney; *blastos*, bud] one of the embryonic cells which give rise ultimately to nephridia.

nephrocoel(e) [Gk. *nephros*, kidney; *koilos*, hollow] the cavity of a nephrotome.

nephrocoelostoma *n.* [Gk. *nephros*, kidney; *koilos*, hollow; *stoma*, mouth] opening of the nephrocoel into the coelom.

nephrocytes *n.plu.* [Gk. *nephros*, kidney; *kytos*, hollow] cells in sponges and insects which store waste and then migrate to surface of body to discharge; brown cells for storage and removal of waste products, as in ascidians.

nephrodinic *a.* [Gk. *nephros*, kidney; *ōdis*, labour] having one duct serving for both excretory and genital purposes.

nephrogenic *a.* [Gk. *nephros*, kidney; *gennaein*, to produce] *pert.* development of kidney; *appl.* cord or column of fused mesodermal cells giving rise to tubules of mesonephros.

nephrogenous *a.* [Gk. *nephros*, kidney; *gennaein*, to produce] produced by the kidney.

nephrogonoduct *n.* [Gk. *nephros*, kidney; *gonos*, seed; L. *ducere*, to lead] excretory and genital duct in one.

nephroi *plu.* of nephros.

nephroid *a.* [Gk. *nephros*, kidney; *eidos*, form] reniform, *q.v.*

nephrolytic *a.* [Gk. *nephros*, kidney; *lyein*, to dissolve] *pert.* or designating enzymatic action destructive to kidneys.

nephromere *n.* [Gk. *nephros*, kidney; *meros*, part] nephrotome, *q.v.*

nephromixium *n.* [Gk. *nephros*, kidney; *mixis*, mixing] a compound excretory organ comprising flame cells and coelomic funnel, and acting as both an excretory organ and genital duct.

nephron *n.* [Gk. *nephros*, kidney] structural and functional unit of a kidney, including the Malpighian corpuscle, convoluted tubules, and Henle's loop.

nephropore *n.* [Gk. *nephros*, kidney; *poros*, passage] nephridiopore, *q.v.*

nephros *n.* [Gk. *nephros*, kidney] a kidney, or usually the functional portion of a kidney; *plu.* nephroi.

nephrostoma, nephrostome *n.* [Gk. *nephros*, kidney; *stoma*, mouth] the opening of a nephridial tubule into body cavity.

nephrotome *n.* [Gk. *nephros*, kidney; *temnein*, to cut] that part of a somite developing into an embryonic excretory organ; *alt.* nephromere.

nepionic *a.* [Gk. *nēpios*, infant] postembryonic; infantile; during infancy; *appl.* phase in development of individual; *cf.* phylonepionic.

nepionotype *n.* [Gk. *nēpios*, infant; *typos*, pattern] type or type specimen of a larva of a species.

Neritacea *n.* [*Nerita*, generic name] an order of freshwater and terrestrial prosobranchs having a single bipectinate ctenidium.

neritic *a.* [Gk. *nērites*, a mussel] *pert.* or living only in coastal waters, as distinct from oceanic.

neritopelagic *a.* [Gk. *Nēreis*, Nereid; *pelagos*, sea] *pert.*, or inhabiting, the sea above continental shelf.

nervate *a.* [L. *nervus*, sinew] having nerves or veins.

nervation, nervature *n.* [L. *nervus*, sinew] the disposition of nervures; *alt.* nervuration, neuration, venation.

nerve *n.* [L. *nervus*, sinew] one of numerous fibrous stimuli-transmitting cords connecting CNS with all other parts of body; vein of insect wing; a vein of leaf.

nerve canal a canal for passage of nerve to pulp of a tooth.

nerve cell neurone, *q.v.*

nerve centre collection of nerve cells associated with a particular function.

nerve eminence neuromast, *q.v.*

nerve ending the terminal distal portion of a nerve, modified in various ways.

nerve fibre an axon or dendrite with or without a sheath or neurilemma.

nerve net a reticulum of nerve cells and their processes connecting sensory cells and muscular elements, in coelenterates.

nerve pentagon five-sided nerve ring around mouth of echinoderms.

nervicole, nervicolous *a.* [L. *nervus*, sinew; *colere*, to dwell] inhabiting or growing on leaf veins.

nerviduct *n.* [L. *nervus*, sinew; *ducere*, to lead] passage for nerves in cartilage or bone.

nervimotion *n.* [L. *nervus*, sinew; *movere*, to move] motion due to direct stimulus from nerves.

nervi nervorum branching nerve fibres with end-bulbs in epineurium.

nervous *a.* [L. *nervus*, sinew] *pert.* nerves; *appl.* tissue composed of nerve fibres.

nervous system brain, spinal cord, nerves, and all their branches taken collectively.

nervule *n.* [L. *dim.* of *nervus*, sinew] branch or terminal portion of nervure of insect wing.

nervuration *n.* [L. *nervus*, sinew] nervation, *q.v.*

nervure *n.* [L. *nervus*, sinew] one of rib-like structures which support membranous wings of insects, branches of tracheal system; a leaf vein.

nervus lateralis *n.* [L. *nervus*, sinew; *lateralis*, *pert.* side] a branch of vagus nerve in fishes, connecting sensory lateral line with brain; *alt.* lateralis.

nervus terminalis n. [L. nervus, sinew; terminalis, bounding] a cranial nerve associated with vomeronasal organ and ending in nasal mucosa; alt. preoptic nerve, terminal nerve.

nesidioblast n. [Gk. nēsidion, islet; blastos, bud] a cell which gives rise to an islet cell of pancreas.

nessoptile neossoptile, q.v.

nest epiphyte an epiphyte which builds up a store of humus round itself for its growth.

nest provisioning returning regularly to nests to bring food to developing offspring, as in some solitary wasps.

net efficiency a measure of the efficiency of an organism converting its assimilated food to protoplasm; cf. gross efficiency.

net knots aggregations of chromatin at the intersections of the net-like formation making up the nuclear reticulum; alt. karyosomes.

net plasmodium the kind of plasmodium found in slime moulds of the group Labyrinthulales where the cells are connected by cytoplasmic strands forming a net; alt. filoplasmodium.

net production the amount of food available for the primary consumers, being the gross primary production minus the amount of protoplasm used in respiration by primary consumers.

netrum n. [A.S. net, meshwork] the initial spindle of a dividing cell arising within the centrosome when the centriole divides.

netted reticulate, q.v.

netted-veined with veins in form of a network.

nettling cells nematocysts, q.v.

Neumann's sheath dentinal sheath surrounding dental canaliculi.

neurad adv. [Gk. neuron, nerve; L. ad, to] dorsally, being on the same side of the backbone as the spinal cord.

neural a. [Gk. neuron, nerve] pert. or closely connected with nerves or nervous system or tissues; alt. neuric.

neural arc the afferent and efferent nervous connection between receptor and effector.

neural arch arch formed on dorsal surface of vertebral centrum, by neural plates and neural spine, for passage of spinal cord.

neural axis spinal cord, q.v.

neural canal canal formed by neural arches.

neural folds two ectodermal ridges of the medullary plate, forming the incipient neural tube; alt. medullary folds.

neural gland a body on ventral side of nerve ganglion in ascidians, presumable homologue of hypophysis in Craniata.

neural groove medullary groove, q.v.

neural lobe pars nervosa of neurohypophysis.

neural plates lateral members of a neural arch; median row, usually of 8 bony plates, in carapace of turtle; neural plate, see medullary plate.

neural shields horny shields above neural plates of turtles.

neural spine spinous process of vertebra.

neural stalk infundibulum of neurohypophysis.

neural tube the tube formed in vertebrate embryos from the medullary or neural plate by development, convergence, and fusion of the 2 neural folds; alt. medullary canal.

neuraminic acid a component of phospholipid membranes, at which viruses may attach, being a 9-carbon, 3-deoxy-5-amino sugar acid, an aldol condensation of mannosamine and pyruvic acid.

neurapophysis n. [Gk. neuron, nerve; apo, from; phyein, to grow] one of the 2 plates growing from the centrum of a vertebra and meeting over the spinal cord forming the neural spine.

neuration n. [Gk. neura, sinew] nervation, q.v.

neuraxis n. [Gk. neuron, nerve; L. axis, axle] the cerebrospinal axis; neuraxon, q.v.

neuraxon n. [Gk. neuron, nerve; axōn, axle] axis cylinder, q.v.; axon, q.v.; alt. neuraxis.

neure neuron, q.v.

neurectoderm n. [Gk. neuron, nerve; ektos, outside; derma, skin] the ectodermal cells forming the earliest rudiment of the nervous system as distinct from skin ectoderm.

neurenteric a. [Gk. neuron, nerve; enteron, gut] pert. neurocoel and enteric cavity; appl. canal, temporarily connecting posterior end of neural tube with posterior end of archenteron, alt. Kovalevsky's canal.

neuric a. [Gk. neuron, nerve] neural, q.v.

neuricity n. [Gk. neuron, nerve] property peculiar to nerves.

neurilemma n. [Gk. neuron, nerve; lemma, skin] a thin sheath investing the medullary sheath of a nerve fibre; alt. sheath of Schwann, primitive sheath; occasionally used for Henle's sheath, q.v.; alt. neurolemma.

neurilemmal pert. neurilemma; appl. cell: Schwann cell, q.v.

neurility n. [Gk. neuron, nerve] the stimuli-transmitting capacity of nerves.

neurine n. [Gk. neuron, nerve] a ptomaine that has a fishy smell and is obtained mainly from brain, bile, and egg yolk, and from choline in putrefying meat.

neurite n. [Gk. neuron, nerve] the axis cylinder process; axon, q.v.

neurobiotaxis n. [Gk. neuron, nerve; bios, life; taxis, arrangement] tendency of nerve fibres or ganglion cell groups to migrate, or growth of dendrites, towards source of most frequent stimulus.

neuroblasts n.plu. [Gk. neuron, nerve; blastos, bud] special epithelial cells from which nerve cells are formed.

neurocele neurocoel, q.v.

neurocentral a. [Gk. neuron, nerve; L. centrum, centre] appl. 2 vertebral synchondroses persisting during first few years of human life.

neurochord n. [Gk. neuron, nerve; chordē, string] a giant fibre, q.v.; primitive tubular nerve cord.

neurocirrus n. [Gk. neuron, nerve; L. cirrus, curl] the cirrus of neuropodium of a polychaete annelid.

neurocoel n. [Gk. neuron, nerve; koilos, hollow] the cavity of central nervous system; alt. neurocele, medullary canal.

neurocranium n. [Gk. neuron, nerve; kranion, skull] the cartilaginous or bony case containing the brain and capsules of special sense organs; alt. endocranium; cf. viscerocranium.

neurocrine a. [Gk. neuron, nerve; krinein, to separate] pert. secretory function of nervous tissue or cells, alt. neurosecretory. n. Neurohumor, q.v.

neurocyte n. [Gk. neuron, nerve; kytos, hollow] neurone, q.v.

neurocyton n. [Gk. neuron, nerve; kytos, hollow] the cell body of a nerve cell; *alt.* cyton, centron.

neurodendron *n.* [Gk. *neuron*, nerve; *dendron*, tree] dendrite, *q.v.*

neurodokon *n.* [Gk. *neuron*, nerve; *dokos*, shaft] the portion of a nerve within a foramen or channel in cartilage or bone.

neuro-ectoderm *see* neuro-epithelium.

neuroendocrine systems the nervous system and endocrine system considered together for bringing about communication.

neuro-epithelium *n.* [Gk. *neuron*, nerve; *epi*, upon; *thēlē*, nipple] superficial layer of cells where specialized for a sense organ; part of the ectoderm giving rise to the nervous system, *alt.* neuro-ectoderm.

neurofibrils *n.plu.* [Gk. *neuron*, nerve; L. *fibrilla*, fine fibre] very fine fibres running longitudinally in axons and dendrons and forming a complex meshwork in the cell body; *alt.* neurofibrillae.

neurogenesis *n.* [Gk. *neuron*, nerve; *genesis*, descent] nerve production.

neurogenic *a.* [Gk. *neuron*, nerve; *gennaein*, to produce] depending on nervous stimuli, such as certain muscular contractions and the activity of certain glands, *cf.* myogenic; giving rise to nervous tissue or system.

neuroglandular *a.* [Gk. *neuron*, nerve; L. *glandula*, small acorn] having both nervous and glandular functions; *pert.* relation between nervous system and glands.

neuroglia *n.plu.* [Gk. *neuron*, nerve; *glia*, glue] cells other than neurones found in the central nervous system, including astrocytes, oligodendroglia, microglia, and ependyma; *alt.* glia, gliacytes, glial cells.

neurogram *n.* [Gk. *neuron*, nerve; *gramma*, inscription] modification of nerve cells due to previous activity, providing a record for future recall; *alt.* neural engram.

neurohaemal *a.* [Gk. *neuron*, nerve; *haima*, blood] composed of a group of nerve endings bearing a close spatial relationship to the blood vascular system and discharging neurosecretory material into the blood.

neurohormone *n.* [Gk. *neuron*, nerve; *hormaein*, to excite] any hormone produced by neurosecretory nerve cells, usually in the brain; sometimes also called neurohumor, and has been confused with this term and used to mean neurotransmitter.

neurohumor *n.* [Gk. *neuron*, nerve; L. *humor*, moisture] neurohormone, *q.v.*; neurotransmitter, *q.v.*; *alt.* neurocrine.

neurohypophysis *n.* [Gk. *neuron*, nerve; *hypo*, under; *phyein*, to grow] the part of the pituitary gland developed from the posterior part of the fore-brain and comprising the pars nervosa (neural lobe) and infundibulum (neural stalk); *cf.* adenohypophysis.

neuroid *a.* [Gk. *neuron*, nerve; *eidos* form] like a nerve; *appl.* intercellular conduction by non-nervous tissue; *appl.* intracellular transmission of stimuli, as in Protozoa.

neurokeratin *n.* [Gk. *neuron*, nerve; *keras*, horn] a protein of nervous tissue.

neurolemma neurilemma, *q.v.*

neurology *n.* [Gk. *neuron*, nerve; *logos*, discourse] the study of the morphology, physiology, and pathology of the nervous system.

neurolymph *n.* [Gk. *neuron*, nerve; L. *lympha*, water] cerebrospinal fluid, *q.v.*

neurolysis *n.* [Gk. *neuron*, nerve; *lysis*, loosing] destruction of nerve tissue.

neuromasts *n.plu.* [Gk. *neuron*, nerve; *mastos*, knoll] groups of sensory cells in lateral-line system of some fishes, and also found in other lower vertebrates; *alt.* nerve eminences, sense hillocks.

neuromere *n.* [Gk. *neuron*, nerve; *meros*, part] a spinal segment corresponding in length to extent of attachment of pair of spinal nerves, a division of convenience, not structural; a segment between transitory shallow constrictions of embryonic medulla oblongata; segmental ganglion of annelids and arthropods; *alt.* neurotome.

neuromerism *n.* [Gk. *neuron*, nerve; *merismos*, partition] appearance of segmentation in developing nervous system; *alt.* neuromery.

neuromery neuromerism, *q.v.*

neuromuscular *a.* [Gk. *neuron*, nerve; L. *musculus*, muscle] *pert.* or involving both nerve and muscles; *appl.* junction of nerve end plate and muscle as a functional unit; *alt.* neuromyal, myoneural.

neuromyal *a.* [Gk. *neuron*, nerve; *mys*, muscle] neuromuscular, *q.v.*

neuron(e) *n.* [Gk. *neuron*, nerve] the nerve cell with its outgrowths, being the structural unit of the nervous system; *alt.* neure, neurocyte.

neuroneme *n.* [Gk. *neuron*, nerve; *nēma*, thread] a nerve fibril running parallel to a myoneme, as in some Ciliata.

neuronephroblast *n.* [Gk. *neuron*, nerve; *nephros*, kidney; *blastos*, bud] one of cells derived from one of megameres, as in segmenting egg of the leech *Clepsine*, which later give rise to part of germinal bands from which nerve cord and nephridia develop.

neurophags *n.plu.* [Gk. *neuron*, nerve; *phagein*, to eat] phagocytic cells that encroach upon and destroy nerve cells in old age.

neurophan *a.* [Gk. *neuron*, nerve; *phainein*, to appear] nervous, sensory, *appl.* supposed nervous fibrils of Ciliata.

neuropil(e) *n.* [Gk. *neuron*, nerve; *pilos*, felt] in ganglia, as of earthworm, a network of processes of association, motor, and sensory neurones, *alt.* neuropileus; punctate or plexiform intercellular substance of grey matter, forming layer of glial expansions and dendrites constituting the synaptic field, *alt.* neuropilema.

neuroplasm *n.* [Gk. *neuron*, nerve; *plasma*, form] the undifferentiated portion of interfibrillar substance of cytoplasm of a neurone.

neuropodium *n.* [Gk. *neuron*, nerve; *pous*, foot] ventral lobe of polychaetan parapodium; terminal fibril of non-medullated nerve fibre.

neuropore *n.* [Gk. *neuron*, nerve; *poros*, passage] opening of neural tube or neurocoel to exterior.

Neuroptera *n.* [Gk. *neuron*, nerve; *pteron*, wing] an order of insects, with complete metamorphosis, including alder flies, lacewings, and ant lions, having long antennae, biting mouth-parts and 2 pairs of wings which are membranous and held roof-like over the abdomen in repose; alder flies are sometimes separated into the order Megaloptera, and the lacewings and ant lions in the order Planipennia.

Neuropteroidea *n.* [Gk. *neuron*, nerve; *pteron*,

wing; *eidos*, form] in some classifications, a super-order of insects including the Neuroptera and various other groups.

neurose *a.* [Gk. *neuron*, nerve] having numerous veins, *appl.* leaves; with many nervures, *appl.* wings of insects.

neurosecretory *a.* [G. *neuron*, nerve; L. *secernere*, to separate] *appl.* or *pert.* gland-like nerve cells which secrete hormones; *alt.* neurocrine.

neurosensory *a.* [Gk. *neuron*, nerve; L. *sensus*, sense] *appl.* epithelial sensory cells with basal neurite, in coelenterates.

neuroskeleton *n.* [Gk. *neuron*, nerve; *skeletos*, dried up] endoskeleton, *q.v.*

neurosomes *n.plu.* [Gk. *neuron*, nerve; *sōma*, body] mitochondria of nerve cells.

neurosynapse *n.* [Gk. *neuron*, nerve; *synapsis*, union] synapse, *q.v.*

neurotendinous *a.* [Gk. *neuron*, nerve; L. *tendere*, to stretch] *pert.* nerves and tendons; *appl.* nerve endings in tendons.

neurotome neuromere, *q.v.*

neurotoxic *a.* [Gk. *neuron*, nerve; *toxikon*, arrow poison] affecting nervous tissue, *appl.* enzymes, as in snake venom; *cf.* haemopathic.

neurotransmitter *n.* [Gk. *neuron*, nerve; L. *transmittere*, to send across] a chemical such as acetylcholine or noradrenaline that is secreted at a nerve ending and allows the transmission of a nervous impulse across a synapse; *alt.* neuro-humor, transmitter.

neurotrophic *a.* [Gk. *neuron*, nerve; *trephein*, to nourish] nourishing the nervous system.

neurotropic *a.* [Gk. *neuron*, nerve; *trepein*, to turn] *pert.* neurotropism; acting upon nervous tissue, *appl.* viruses, bacteria, toxins, stains.

neurotropism *n.* [Gk. *neuron*, nerve; *trepein*, to turn] the attraction exerted by nervous tissue upon developing nerve tissue.

neurotubules delicate structures within axons, observed with the aid of an electron microscope.

neurovascular *a.* [Gk. *neuron*, nerve; L. *vasculum*, small vessel] *pert.* nerves and vessels, *appl.* hilum, *q.v.*

neurula *n.* [Gk. *neuron*, nerve] the stage in development of Chordata which coincides with formation of the neural tube.

neurulation the formation of the neural tube.

neuston *n.* [Gk. *neustos*, floating] organisms floating or swimming in surface water, or inhabiting surface film; *cf.* seston.

neuter *a.* [L. *neuter*, of neither sex] sexless, neither male nor female; having neither functional stamens nor pistils; *alt.* neutral. *n.* A non-fertile female of social insects; a castrated animal.

neutral *a.* [L. *neuter*, neither] neuter, *q.v.*; neither acid nor alkaline, pH 7; achromatic, as white, grey, and black; day-neutral, *q.v.*

neutrocyte *n.* [L. *neutro*, to neither side; Gk. *kytos*, hollow] a neutrophil leucocyte.

neutrophil *a.* [L. *neuter*, neither; Gk. *philein*, to love] *appl.* white blood corpuscles whose granules stain only with neutral stains; *alt.* neutrophilic; *cf.* amphophil. *n.* A polymorphonuclear leucocyte.

newts *see* Urodela.

nexus *n.* [L. *nexus*, tying] region of fusion of the plasma membranes of 2 excitable cells.

N-factor an antigen in erythrocytes of higher animals.

niacin nicotinic acid, *q.v.*

niacinamide nicotinamide, *q.v.*

niche *n.* [F. *niche*, from It. *nicchia*, recess in wall] ecological niche, *q.v.*

nicotinamide an amide found in living organisms and interconvertible with nicotinic acid, usually occurring as a constituent of coenzymes such as NAD and NADP, and also being used similarly to nicotinic acid; *alt.* niacinamide, vitamin B₇, pellagra-preventive factor, P-P factor, vitamin P-P.

nicotinamide adenine dinucleotide NAD, *q.v.*

nicotinamide adenine dinucleotide phosphate NADP, *q.v.*

nicotine an alkaloid obtained from *Nicotiana tabacum.*

nicotinic acid a member of the vitamin B complex, vitamin B₇, often occurring as a complex of nicotinamide in various parts of plants and animals, such as blood, liver, legumes, and yeast, made by oxidation of nicotine and various other compounds, and used to treat pellagra in people and black-tongue in dogs; *alt.* niacin, pellagra-preventive factor, P-P factor, vitamin P-P, vitamin B₇.

nictitant *a.* [L. *nictare*, to wink] *appl.* an ocellus with central lunate spot.

nictitating membrane third eyelid, a membrane which assists in keeping eye clean, in reptiles, birds, mammals.

nidamental *a.* [L. *nidamentum*, material for a nest] *appl.* glands which secrete material for an egg covering.

nidation *n.* [L. *nidus*, nest] the renewal of uterus lining between menstrual periods; embedding of fertilized ovum in uterine mucous membrane, *alt.* implantation.

nidicolous *a.* [L. *nidus*, nest; *colere*, to dwell] living in the nest for a time after hatching.

nidification *n.* [L. *nidus*, nest; *facere*, to make] the excavation or construction of nests and the behaviour connected with it.

nidifugous *a.* [L. *nidus*, nest; *fugere*, to flee] leaving the nest soon after hatching.

nidulant *a.* [L. *dim.* of *nidus*, nest] partially surrounded or lying free in a hollow or cup-like structure.

Nidulariales *n.* [L. *nidulus*, small nest] an order of Gasteromycetes, commonly called bird's nest fungi, having fruit bodies resembling minute bird's nests full of eggs.

nidulus *n.* [L. *nidulus*, small nest] a group of nerve-cell bodies in central nervous system; the nucleus from which a nerve originates.

nidus *n.* [L. *nidus*, nest] a nest; a nest-like hollow; a nucleus; a cavity for development of spores; nest of cells replacing epithelial cells of mid-gut in Orthoptera; focus or primary site of an infection.

nidus hirundinalis a fossa of cerebellum; *alt.* nidus avis.

nigrescent *a.* [L. *nigrescere*, to turn black] nearly black; blackish; turning black.

nigropunctate *n.* [L. *niger*, black; *punctum*, point] black-spotted.

Nilssoniales *n.* [*Nilssonia*, generic name] in some classifications, an extinct Mesozoic order of Cycadatae.

nimbospore *n.* [L. *nimbus*, cloud; Gk. *sporos*, seed] a spore having a gelatinous coat, as of certain fungi lacking an ascocarp.

nipple *n.* [*Dim.* of A.S. *nib*, for *neb*, nose] teat of mammary gland bearing opening from milk glands; *alt.* mammary papilla, mamilla.

Nippotaeniidea *n.* [*Nippotaenia*, generic name] an order of segmental tapeworms having an apical sucker on the scolex as the only adhesive organ.

Nissl granules or bodies [*F. Nissl*, German neurologist] granules of ribonucleoprotein found in the cell bodies of nerve cells, which extend into dendrons but are absent from the axon hillock, and are thought to be concerned with protein synthesis; *alt.* chromophil or tigroid bodies.

nisus *n.* [L. *nisus*, effort] strong tendency; effort; muscular contraction for expulsion of eggs, young, or excreta.

nisus formativus *n.* [L. *nisus*, effort; *formare*, to form] the tendency to reproduce.

nitid, nitidous *a.* [L. *nitidus*, shining] glossy.

nitrate bacteria bacteria in the soil that convert nitrites to nitrates.

nitrate-reducing bacteria denitrifying bacteria, *q.v.*

nitrate reductase a general name for enzymes that catalyse the conversion of nitrate to nitrite; EC 1.6.6.1, 1.6.6.2, 1.6.6.3, 1.9.6.1, and 1.7.99.4.

nitration the conversion of nitrites to nitrates by soil bacteria, followed by the uptake of the nitrates by higher plants and their incorporation into amino acids, etc.; a general term for the process in which a substance is nitrated, i.e. has an NO_2 group added to it.

nitrification *n.* [Gk. *nitron*, soda; L. *facere*, to make] oxidation of ammonia to nitrites and of nitrites to nitrates, as by action of bacteria.

nitrifiers nitrifying bacteria, *q.v.*

nitrifying bacteria those causing nitrification, including nitrate bacteria and nitrite bacteria; *alt.* nitrifiers.

nitrite bacteria bacteria in the soil which convert ammonium compounds to nitrites.

nitrite reductase an enzyme that catalyses the conversion of nitrite to ammonium hydroxide; EC 1.6.6.4; *r.n.* nitrite reductase.

nitrocabalamin(e) vitamin B_{12c}, *see* cobalamine.

nitrogenase a multienzyme complex found in nitrogen-fixing organisms which reduces nitrogen to ammonium ions.

nitrogen cycle the incorporation of inorganic nitrogen in the form of ions or gas into organic compounds, the decomposition of these and subsequent production and recycling of ions, brought about by bacteria and fungi in the soil, and by plants and animals.

nitrogen fixation the conversion of atmospheric nitrogen gas to inorganic or organic compounds by micro-organisms or lighting.

nitrogenous *a.* [Gk. *nitron*, soda; *genos*, descent] *pert.* or containing nitrogen.

nitrogenous equilibrium equilibrium of body maintained by equality of income and output of nitrogen.

nitrophilous *a.* [Gk. *nitron*, soda; *philein*, to love] thriving in nitrogenous soils; *n.* nitrophile.

nitrophyte *n.* [Gk. *nitron*, soda; *phyton*, plant] a nitrophilous plant; *alt.* nitrophile.

nitrozation the conversion of ammonia to nitrites by bacteria in the soil.

nociceptive *a.* [L. *nocere*, to hurt; *capere*, to take] *appl.* stimuli which tend to injure tissue or induce pain; *appl.* reflexes to which protect from injury.

nociceptor *n.* [L. *nocere*, to hurt; *capere*, to take] a receptor sensitive to injurious stimuli.

noctilucent *a.* [L. *nox*, night; *lucere*, to shine] phosphorescent; luminescent.

nocturnal *a.* [L. *nox*, night] seeking food and moving about at night only; occurring at night.

nodal *a.* [L. *nodus*, knob] *pert.* a node or nodes.

nodal membrane myelin-free surface of an axon at nodes of Ranvier.

node *n.* [L. *nodus*, knob] the knob or joint of a stem at which leaves arise; aggregation of specialized cardiac cells, as atrioventricular and sinuatrial nodes; a lymph gland; one of the constrictions of medullary sheath, *alt.* node of Ranvier; transverse septum in stolon of ectoprocts; *alt.* nodus.

nodose *a.* [L. *nodus*, knob] having intermediate and terminal joints thicker than remainder; having knots or swellings.

nodular *a.* [L. *nodulus*, dim. of *nodus*, knob] *pert.*, or like a nodule or knot.

nodulation formation of root nodules.

nodule *n.* [L. *nodulus*, dim. of *nodus*, knob] a small knob-like structure, as root nodule, lymphatic nodule; anterior part of inferior vermis of cerebellum; *alt.* nodulus.

noduliferous *a.* [L. *nodulus*, dim. of *nodus*, knob; *ferre*, to carry] bearing nodules.

nodulus nodule, *q.v.*

nodus *n.* [L. *nodus*, knob] node, *q.v.*; indentation near middle of anterior or costal margin of wing in Odonata; joint or swelling in insect wing.

neomatic *a.* [Gk. *noēma*, thought] *pert.* mental processes.

nomenclature *n.* [L. *nomen*, name; *calare*, to call] the making and giving distinguishing names to all groups of plants and animals; *see also* binomial nomenclature.

nomen nudum a name not valid since when it was originally published the organism to which it referred was not adequately described, defined, or sketched; *alt.* bare name.

nominifer *a.* [L. *nomen*, name; *ferre*, to bear] the constituent element of a taxon to which the name of the taxon is permanently attached, e.g. the type species of a genus.

nomogenesis *n.* [Gk. *nomos*, law; *genesis*, descent] view that development and evolution are governed by laws of development and not by environment.

nomology *n.* [Gk. *nomos*, law; *logos*, discourse] the study and formulation of principles or laws discovered in any science.

non-autogenous anautogenous, *q.v.*

non-conjunction failure of chromosome pairing.

non-cyclic photophosphorylation *appl.* a type of photophosphorylation in which both ATP and $NADPH_2$ are generated using light energy; *cf.* cyclic photophosphorylation.

non-deciduate indeciduate, *q.v.*

non-disjunction failure of a normal pair of chromosomes to separate at meiosis.

non-essential *appl.* amino acids which can be synthesized in the body and are not required in the diet.

non-inducible *appl.* enzyme: constitutive, *q.v.*

non-medullated, non-myelinated fibres nerve fibres without myelin medullary sheath, which are therefore grey; *alt.* amyelinic, amyelinate fibres, Remak's fibres, grey fibres.

nonoses *n.plu.* [L. *nona*, ninth] a group of monosaccharides having the general formula $C_9H_{18}O_9$.

non-porous wood secondary xylem with no vessels.

nonsense codon a DNA triplet which does not appear to code for an amino acid.

non-storied *appl.* cambium in which initials are not arranged in horizontal series on tangential surfaces; *alt.* non-stratified; *cf.* storied.

non-stratified non-storied, *q.v.*

non-striated *appl.* muscle: involuntary muscle, *q.v.*

non-viable incapable of surviving or developing.

noradrenalin(e) a substance which is closely related to adrenaline, secreted with it by the suprarenal medulla and acting as a neurotransmitter in the sympathetic nervous system and secreted at the synapses of adrenergic nerves, also found in effectors supplied by the sympathetic nervous system, such as smooth muscle, and broken down by the enzyme mono-aminoxidase; *alt.* norepinephrine, arterenol; *cf.* acetylcholine.

norepinephrin(e) noradrenaline, *q.v.*

norma *n.* [L. *norma*, rule] view of the skull as a whole from certain points: basal, vertical, frontal, occipital, and lateral.

normoblasts *n.plu.* [L. *norma*, rule; Gk. *blastos*, bud] immature nucleated red blood corpuscles, derived from erythroblasts; *alt.* metarubricytes, sauroid cells.

normocyte *n.* [L. *norma*, rule; Gk. *kytos*, hollow] the fully developed red blood corpuscle.

nosogenic *a.* [Gk. *nosos*, disease; *gennaein*, to produce] pathogenic, *q.v.*

nosology *n.* [Gk. *nosos*, disease; *logos*, discourse] pathology, *q.v.*

Nostocales *n.* [*Nostoc*, generic name] an order of blue–green algae having a filamentous plant body and producing hormogonia.

nostrils *n.plu.* [A.S. *nosthyrl*, nostril] the external openings of the nose; *alt.* nares (external).

Notacanthiformes *n.* [*Notacanthus*, generic name; L. *forma*, shape] an order of deep-sea teleosts; *alt.* Heteromis.

notal *a.* [Gk. *nōton*, back] dorsal; *pert.* the back; *pert.* notum; *cf.* tergal.

Notaspidea *n.* [Gk. *nōton*, back; *aspis*, shield] an order of flattened slug-like opisthobranchs having a reduced or internal shell; *alt.* Pleurobranchomorpha.

notate *a.* [L. *notatus*, marked] marked with lines or spots.

nothocline *n.* [Gk. *nothos*, illegimate; *klinein*, to slant] the serial arrangement of characters or forms produced by crossing species; *alt.* hybrid cline.

Nothosauria, nothosaurs *n., n.plu.* [Gk. *nothos*, spurious; *sauros*, lizard] an order of Triassic sauropterygians, that are tending towards an aquatic form, a slender body, and shortened distal parts to the limbs.

Notidani *n.* [*Notidanus*, generic name] in some classifications, an order of elasmobranchs having 6 or 7 pairs of gill slits and embolomerous vertebrae.

notocephalon *n.* [Gk. *nōton*, back; *kephalē*, head] dorsal shield of leg-bearing segments in certain Acarina; *alt.* podosomatal plates; *cf.* notogaster.

notochord *n.* [Gk. *nōton*, back; *chordē*, cord] a rod made of large vacuolated cells in a firm sheath, found in chordates and lying lengthways between neural tube and gut, and in vertebrates replaced by skull and vertebral column.

notochordal *a.* [Gk. *nōton*, back; *chordē*, cord] *pert.* or enveloping notochord, *appl.* sheath, tissue, etc.

notocirrus *n.* [Gk. *nōton*, back; L. *cirrus*, curl] cirrus of notopodium of Polychaeta.

Notodelphyoida *n.* [*Notodelphys*, generic name; Gk. *eidos*, form] an order of copepods living as commensals in tunicates.

Notogaea *n.* [Gk. *nōtos*, south; *gaia*, earth] zoogeographical area comprising Australian, New Zealand, and Pacific Ocean Islands regions, and formerly, Neotropical region.

notogaster *n.* [Gk. *nōton*, back; *gastēr*, belly] posterior dorsal shield in certain Acarina; *alt.* opisthosomatal plate; *cf.* notocephalon.

notogenesis *n.* [Gk. *nōton*, back; *genesis*, origin] development of the notochord, and the associated stage of mesoderm differentiation.

notonectal *a.* [Gk. *nōton*, back; *nektos*, swimming] swimming back downwards.

notopleural suture in Diptera, a lateral suture separating the mesonotum from the pleuron, running from the humeral callus to the wing base.

notopleuron *n.* [Gk. *nōton*, back; *pleuron*, side] in Diptera, a triangular plate at the end of the transverse suture, behind the humerus.

notopodium *n.* [Gk. *nōton*, back; *pous*, foot] dorsal lobe of polychaetan parapodium.

Notoptera *n.* [Gk. *nōton*, back; *pteron*, wing] Grylloblattodea, *q.v.*

notorhizal *a.* [Gk. *nōton*, back; *rhiza*, root] incumbent, *q.v.*; campylotropous, *q.v.*

Notostraca *n.* [Gk. *nōton*, back; *ostrakon*, shell] an order of Branchiopoda having carapace forming a dorsal shield, sessile eyes, and vestigial antennae.

nototribe *a.* [Gk. *nōton*, back; *tribein*, to rub] *appl.* flowers whose anthers and stigma touch back of insect as it enters calyx, a device for securing cross-fertilization.

Notoungulata, notoungulates *n., n.plu* [Gk. *nōton*, back; L. *ungula*, hoof] an extinct order of South American placental mammals of the Palaeocene to Pleistocene, that in some ways were similar to perissodactyls.

notum *n.* [Gk. *nōton*, back] the dorsal portion of insect segment; *alt.* tergum.

novobiocin *n.* [L. *novus*, new; antio*biotic*; strepto*mycin*] an antibiotic synthesized by the actinomycete *Streptomyces niveus*.

nucellus *n.* [L. *dim.* of *nux*, nut] parenchymatous tissue between megaspore (embryo sac) of ovule and its inner integument, and extending from chalaza at base to micropyle at apex; the tissue in the ovule in which meiosis occurs, equivalent to megasporangium.

nuchal *a.* [L.L. *nucha*, nape of neck] *pert.* nape of the neck; *appl.* 2 sense organs, regarded as olfactory, on prostomium of Chaetopoda; *appl.* thin cartilage between head and anterior dorsal part of mantle in decapod Cephalopoda; *appl.* anterior plate of chelonian carapace; *appl.* flexure of medulla oblongata. *n.* An unpaired posterior dorsal skull bone in Chondrostei.

nuciferous *a.* [L. *nux*, nut; *ferre*, to carry] nut-bearing.

nucivorous *a.* [L. *nux*, nut; *vorare*, to devour] nut-eating.

nuclear *a.* [L. *nucleus*, kernel] *pert.* a nucleus.

nuclear disc a star-like structure formed by chromosomes in equator of spindle during mitosis.

nuclear layer internal layer of cerebellar cortex; inner n.l. of retina, between inner and outer plexiform layers, and outer n.l., between outer plexiform layer and limiting membrane of layer of rods and cones.

nuclear membrane delicate membrane bounding a nucleus; *alt.* karyolemma, karyotheca.

nuclear plate equatorial plate, *q.v.*

nuclear reticulum net-like pattern of precipitated chromatin in the nucleus of a cell that has been fixed and stained; *alt.* skein.

nuclear sap the colloidal, non-staining fluid in the nucleus; *alt.* achromatin, karyenchyma, karyochylema, karyolymph, nucleoplasm, nucleohyaloplasm, nucleochylema, nucleochyme.

nuclear spindle a spindle-shaped structure formed of fine fibrils in cytoplasm surrounding nucleus, a stage in mitosis and meiosis.

nuclease *n.* [L. *nucleus*, kernel] any enzyme which hydrolyses nucleic acids to nucleotides.

nucleate *a.* [L. *nucleus*, kernel] having a nucleus. *v.* To form into a nucleus. *n.* An ester or salt of nucleic acid.

nucleation *a.* [L. *nucleus*, kernel] nucleus formation.

nuclei *plu.* of nucleus.

nucleic acids [L. *nucleus*, kernel] polymers of nucleotides; *see* deoxyribonucleic acid (DNA) and ribonucleic acid (RNA).

nucleiform *a.* [L. *nucleus*, kernel; *forma*, shape] shaped like a nucleus; *alt.* pyrenoid.

nuclein *n.* [L. *nucleus*, kernel] any of several substances obtained from nuclei, especially from enzymatic digestion of nucleoproteins and particularly DNA or a mixture of nucleic acids and proteins.

nucleochylema *n.* [L. *nucleus*, kernel; Gk. *chylos*, juice] nuclear sap, *q.v.*

nucleochyme nuclear sap, *q.v.*

nucleohistone *n.* [L. *nucleus*, kernel; Gk. *histos*, tissue] a nucleoprotein whose protein is derived from a histone.

nucleohyaloplasm *n.* [L. *nucleus*, kernel; Gk. *hyalos*, glass; *plasma*, mould] nuclear sap, *q.v.*

nucleoid *a.* [L. *nucleus*, kernel; Gk. *eidos*, form] resembling a nucleus. *n.* A nucleus-like body occurring in certain blood corpuscles; in prokaryotes, a region containing nucleic acid equivalent to the nucleus of eukaryotes.

nucleolar *a.* [L. *dim.* of *nucleus*, kernel] *pert.* a nucleolus.

nucleolar organizer a region of a chromosome with which the nucleolus is associated.

nucleolinus *n.* [*Dim.* of L. *nucleus*, kernel] a small deeply staining granule within the nucleolus.

Nucleolitoida *n.* [*Nucleolites*, generic name; Gk. *eidos*, form] an order of sea urchins of the superorder Atelostomata.

nucleolo-centrosome a nuclear body which may act as a centrosome during mitosis or meiosis.

nucleolonema [*Dim.* of L. *nucleus*, kernel; Gk. *nēma*, thread] a structure seen in the nucleolus under the electron microscope, consisting of a network made of coarse strands.

nucleolus *n.* [L. *nucleolus, dim.* of *nucleus*, little kernel] a spherical body found in the nucleus associated with the nucleolar organizer, and consisting mainly of molecules of ribosomal RNA and some protein; *alt.* chromatospherite, morulit; *cf.* plasmosome, karyosome.

nucleolysis karyolysis, *q.v.*

nucleome *n.* [L. *nucleus*, kernel] collectively, the nuclear material in a cell.

nucleomicrosomes *n.plu.* [L. *nucleus*, kernel; Gk. *mikros*, small; *sōma*, body] nuclear chromatin granules.

nucleoplasm *n.* [L. *nucleus*, kernel; Gk. *plasma*, mould] the protoplasm making up the nucleus of the cell, *alt.* karyoplasm, *cf.* cytoplasm; nuclear sap, *q.v.*

nucleoplasmic ratio the ratio of the volume of the nucleus to the volume of cytoplasm.

nucleoprotein *n.* [L. *nucleus*, kernel; Gk. *prōteion*, first] any of various compounds consisting of a nucleic acid linked to protein.

nucleosidase an enzyme which splits certain nucleosides into a nitrogenous base and a pentose; EC 3.2.2.1; *r.n.* nucleosidase.

nucleosides *n.* [L. *nucleus*, kernel] a compound made of a purine or pyrimidine base linked to ribose or deoxyribose.

nucleosome *n.* [L. *nucleus*, kernel; Gk. *sōma*, body] a chromomere of a chromosome when considered to be made of DNA and protein.

nucleotidase an enzyme which converts nucleotides to nucleosides and phosphoric acid, as 5'-nucleotidase, EC 3.1.3.5, and 3'-nucleotidase, EC 3.1.3.6.

nucleotides *n.plu.* [L. *nucleus*, kernel] compounds made of nucleoside linked to phosphoric acid, which when polymerized form nucleic acids.

nucleus *n.* [L. *nucleus*, kernel] complex spheroidal mass essential to life of most cells; mass of grey matter in central nervous system, *alt.* nidulus; centre of origin of hilum of starch grain; centre around which are formed the growth rings of cycloid and ctenoid fish scales; centre of perithecium in certain fungi; protoconch, *q.v.*; *plu.* nuclei.

nucleus ambiguus cells in medulla oblongata from which originate the motor fibres of glossopharyngeal and vagus, and of cerebral part of spinal accessory nerves.

nucleus pulposus the soft core of an intervertebral disc, remnant of notochord.

nucleus suprachiasmaticus the part of the brain which may contain a biological clock.

nuculanium *n.* [L. *nucula*, small nut] a berry formed from a superior ovary; *cf.* bacca, uva.

nucule *n.* [L. *nucula*, small nut] nutlet, *q.v.*; oogonium in Charales.

Nuda *n.* [L. *nudus*, naked] a group of sponges having no true epidermis, with the external surface composed of a net derived from amoebocytes, and no middle gelatinous layer, considered as a subphylum or class or raised to the rank of phylum; a class of Ctenophora without tentacles.

nudation *n.* [L. *nudus*, naked] the occurrence of areas bare of plants, as a result of natural or artificial factors.

Nudibranchia *n.* [L. *nudus*, naked; *branchiae*, gills] Acoela, *q.v.*

nudibranchiate *a.* [L. *nudus*, naked; *branchiae*,

gills] having gills not covered by a protective shell or membrane in a branchial chamber.

nudicaudate a. [L. *nudus*, naked; *cauda*, tail] having a tail not covered by hair or fur.

nudicaulous a. [L. *nudus*, naked; *caulis*, stem] *appl.* or having stems without leaves.

nudiflorous a. [L. *nudus*, naked; *flos*, flower] having flowers without glands or hairs.

nudum n. [L. *nudus*, naked] small bared area, as sensitive portion of antenna of butterflies.

Nuhn, glands of [*A. Nuhn*, German anatomist] Blandin's glands, *q.v.*

nullipennate a. [L. *nullus*, none; *penna*, feather] without flight feathers.

nulliplex a. [L. *nullus*, none; *plexus*, interwoven] having recessive but no dominant genes for a given character, in polyploidy.

nullisomic a. [L. *nullus*, none; Gk. *sōma*, body] *appl.* an organism or cell which has both members of a pair of chromosomes missing.

number sense the ability of some animals to recognize a certain abstract number when presented to them in different objects.

numerical *appl.* hybrid of parents that have different chromosome numbers.

numerical taxonomy taxometrics, *q.v.*

nummulation n. [L. *nummus*, coin] the tendency of red blood corpuscles to adhere together like piles of coins (rouleaux).

nummulitic a. [L. *nummus*, coin] like, *pert.*, or containing the fossil foraminifers, nummulites.

nunatak n. [Eskimo, *nunatak*, nunatak] a refugium on a mountain or plateau.

nuptial flight flight taken by queen bee when fertilization takes place.

nurse cells single cells or layers of cells attached to or surrounding an egg cell, for elaboration of food material; *alt.* trophocytes.

nurse generation an asexual budding generation of some Tunicata, in which phorozooids act as foster parents to later formed buds, the gonozooids.

nurture n. [O.F. *noriture*, nursing] the sum total of environmental influences; *cf.* nature.

nut n. [A.S. *hnutu*, nut] dry, indehiscent 1- or 2-seeded, 1-celled fruit with a hard, woody pericarp, such as acorn; *alt.* glans.

nutant a. [L. *nutare*, to nod] bent downwards; drooping.

nutation n. [L. *nutare*, to nod] rotational curvature of growing tip of a plant; slow rotating movement by pseudopodia.

nutlet n. [*Dim.* of *nut*] a small nut or nut-like fruit; achene of a schizocarp; the stone of a drupel; *alt.* nucule.

nutramin vitamin, *q.v.*

nutricism n. [L. *nutrix*, nurse] symbiotic relationship with all the benefits to one partner.

nutrient a. [L. *nutrire*, to nourish] nourishing; *appl.* artery to marrow of bone, and foramen of entry. n. Food substance.

nutrient person gastrozooid, *q.v.*

nutrilites n.plu. [L. *nutrire*, to nourish] accessory organic food substances such as bios or vitamins.

nutrition n. [L. *nutrire*, to nourish] the ingestion, digestion, and assimilation of food materials by animals and heterotrophic plants; the production of organic compounds from inorganic ones in autotrophic plants.

nutritive a. [L. *nutrire*, to nourish] concerned in function of nutrition, *appl.* yolk, polyp, zooid, plasma, etc.

nyctanthous a. [Gk. *nyktos*, by night; *anthos*, flower] flowering at night.

nyctinasty n. [Gk. *nyktios*, nightly; *nastos*, pressed close] sleep movements of plants, involving a change in position of leaves, petals, etc., in response to stimuli of change in light or temperature; *alt.* nyctitropism; *a.* nyctinastic.

nyctipelagic a. [Gk. *nyktios*, nightly; *pelagos*, sea] rising to surface of sea only at night.

nyctitropism n. [Gk. *nyktios*, nightly; *trepein*, to turn] nyctinasty, *q.v.*; *a.* nyctitropic.

nyctoperiod n. [Gk. *nyktos*, by night; *periodos*, circuit] daily period of exposure to darkness.

nymph n. [Gk. *nymphē*, chrysalis] a juvenile form without wings or with incomplete wings in insects with incomplete metamorphosis; juvenile form in mites; formerly: a pupa.

nymphae n.plu. [Gk. *nymphē*, bride] the labia minora, *q.v.*; shell edges to which the hinge ligaments are attached, in bivalves; a pair of sclerites beneath epigynal plate in mites.

Nymphaeales n. [*Nymphaea*, generic name] an order of dicots (Magnoliidae) which are herbaceous aquatics rooted in the mud in shallow water.

nymphal a. [Gk. *nymphē*, chrysalis] *pert.* a nymph; *appl.* hormone: juvenile hormone, *q.v.*

nymphiparous pupiparous, *q.v.*

nymphochrysalis n. [Gk. *nymphē*, pupa; *chrysallis*, from *chrysos*, gold] pupa-like resting stage between larval and nymphal form in certain mites.

Nymphonomorpha n. [*Nymphon*, generic name; Gk. *morphē*, form] an order of Pycnogonida having a clear neck separating the cephalon and the trunk.

nymphosis n. [Gk. *nymphē*, chrysalis] the process of changing into a nymph or a pupa.

nystatin n. [*New York State*, where it first originated] an antibiotic which interferes with synthesis of cell wall in eukaryotic plants.

O

oar feathers the wing feathers used in flight.

oarium ovary, *q.v.*

obcompressed a. [L. *ob*, towards; *comprimere*, to compress] flattened in a vertical direction.

obconic a. [L. *ob*, against; *conus*, cone] shaped like a cone, but attached at its apex.

obcordate, obcordiform a. [L. *ob*, against; *cor*, heart] inversely heart-shaped; *appl.* leaves which have stalk attached to apex of heart.

obcurrent a. [L. *ob*, against; *currere*, to run] converging, and attaching at point of contact.

obdeltoid a. [L. *ob*, against; Gk. Δ, delta; *eidos*, form] more-or-less triangular with apex at point of attachment, i.e. inversely deltoid.

obdiplostemonous a. [L. *ob*, against; Gk. *diploos*, double; *stēmōn*, warp] with stamens in 2 whorls, the inner opposite the sepals and the outer opposite the petals.

obelion n. [Gk. *obelos*, a spit] the point between parietal foramina, on sagittal suture.

obex n. [L. *obex*, obstacle] a triangular layer of

grey matter, also a membranous ependymal layer, in roof of 4th ventricle; a limiting factor, *appl.* plant distribution; *plu.* obices.

obices *plu.* of obex.

obimbricate *a.* [L. *ob*, reversely; *imbrex*, tile] with regularly overlapping scales, with the overlapped ends downwards.

oblanceolate *a.* [L. *ob*, reversely; *lancea*, spear] inversely lanceolate.

obligate *a.* [L. *obligatus*, bound] obligatory; limited to one mode of life or action; not optional; *appl.* aerobes, anaerobes; *appl.* sexual reproduction; *appl.* parthenogenesis; *appl.* saprophytes, *alt.* holosaprophytes; *appl.* parasites which cannot exist independently of a host, *alt.* holoparasites; *appl.* symbionts; *cf.* facultative.

oblique *a.* [L. *obliquus*, slanting] placed obliquely, *appl.* septum forming ventral wall of thoracic air sac in birds, *appl.* vein of left atrium, etc.; asymmetrical, *appl.* leaves; *appl.* cleavage: alternating or spiral.

obliquus *n.* [L. *obliquus*, slanting] an oblique muscle, as of ear, eye, head, abdomen.

obliterate *a.* [L. *obliteratus*, erased] indistinct or profuse, *appl.* markings on insects; suppressed.

obliterative coloration type in which the parts of an animal exposed to the most intense light are shaded more darkly, thus ensuring that it blends with its background more effectively.

oblongata medulla oblongata, *q.v.*

obovate *a.* [L. *ob*, against; *ovum*, egg] inversely egg-shaped; *appl.* leaf with narrow end attached to stalk; *appl.* spores.

obovoid *a.* [L. *ob*, against; *ovum*, egg; Gk. *eidos*, shape] inversely ovoid; roughly egg-shaped, with narrow end downwards.

obpiriform obpyriform, *q.v.*

obpyriform *a.* [L. *ob*, against; *pyrum*, pear; *forma*, form] inversely pear-shaped; *alt.* obpiriform.

obsolescence *n.* [L. *obsolescere*, to wear out] the gradual reduction and consequent disappearance of a species; gradual cessation of a physiological process; a blurred portion of a marking on animal.

obsolete *a.* [L. *obsolescere*, to wear out] wearing out or disappearing; *appl.* any character that is becoming less and less distinct in each succeeding generation; *appl.* calyx united with ovary or reduced to a rim.

obsubulate *a.* [L. *ob*, against; *subula*, awl] reversely awl-shaped or subulate; narrow and tapering from tip to base.

obtect *a.* [L. *obtectus*, covered over] *appl.* pupa with wings and legs held to body; *alt.* agglutinate; *cf.* exarate.

obturator *n.* [L. *obturare*, to close] any of various structures which close off a cavity. *a. Pert.* any structure in the neighbourhood of obturator foramen.

obturator foramen an oval foramen between ischium and os pubis.

obtuse *a.* [L. *obtusus*, blunt] with blunt or rounded end, *appl.* leaves, *appl.* left margin of heart, etc.

obtusilingual *a.* [L. *obtusus*, blunt; *lingua*, tongue] short- and blunt-tongued.

obumbrate *a.* [L. *obumbrare*, to over-shadow] with some structure overhanging the parts so as partially to conceal them.

obverse *a.* [L. *obvertere*, to turn round] with base narrower than apex.

obvolute *a.* [L. *obvolvere*, to wrap round] overlapping; *appl.* vernation when half of one leaf is wrapped round half of another similar leaf.

obvolvent *a.* [L. *obvolvere*, to wrap round] bent downwards and inwards, *appl.* wings or elytra of some insects, etc.

Occam's razor [*William of Occam*, English philosopher] the principle that, where several hypotheses are possible, the simplest is chosen; *alt.* Ockham's razor.

occasional species one which is only found from time to time in a community, but is not a regular member of it.

occipital *a.* [L. *occiput*, back of head] *pert.* back part of head; *appl.* bones: occipitalia, *q.v.*

occipital foramen posterior opening of head in insects; foramen magnum of skull in vertebrates.

occipitalia *n.plu.* [L. *occiput*, back of head] the group of parts of cartilaginous brain case forming back part of head; *alt.* occipital bones.

occipito-atlantal *a.* [L. *occiput*, back of head; Gk. *Atlas*, a Titan] *appl.* membrane closing gap between skull and neural arch of atlas in amphibians; *appl.* dorsal (posterior) and ventral (anterior) membranes between margin of foramen magnum and atlas in mammals; *alt.* atlantooccipital.

occipito-axial *a.* [L. *occiput*, back of head; *axis*, axis] *appl.* ligament or membrana tectoria connecting occipital bone with axis or epistropheus.

occipitofrontal *a.* [L. *occiput*, back of head; *frons*, forehead] *appl.* longitudinal arc of skull; *appl.* fasciculus of long association-fibres between frontal and occipital lobes of cerebral hemispheres; *appl.* muscle, the epicranius, *q.v.*, *alt.* occipitofrontalis.

occiput *n.* [L. *occiput*, back of head] occipital region of skull; dorsolateral region of insect head.

occlusal *a.* [L. *occludere*, to shut in] contacting the opposing surface; *appl.* surface of teeth which touch those of the other jaw when jaws are closed.

occlusion *n.* [L. *occludere*, to shut in] overlapping of activation of motor neurones by simultaneous stimulation of several afferent nerves; closure of a structure, duct, or opening.

occlusor *n.* [L. *occludere*, to shut in] a closing muscle. *a. Appl.* muscles of a operculum or movable lid.

oceanic inhabiting the open sea.

oceanodromous *a.* [Gk. *ōkeanos*, L. *oceanus*, ocean; Gk. *dromos*, running] migrating only in the ocean, *appl.* fishes; *cf.* potamodromous.

ocellar *a.* [L. *ocellus*, little eye] of, or *pert.*, ocelli.

ocellate *a.* [L. *ocellus*, little eye] like an eye or eyes, *appl.* markings.

ocellated having ocelli; having eye-like spots or markings; *alt.* ocelliferous.

ocellation *n.* [L. *ocellus*, little eye] condition of having ocelli, or of having ocellate markings; ocellate marking.

ocelli *plu.* of ocellus.

ocelliferous ocellated, *q.v.*

ocellus *n.* [L. *ocellus*, little eye] a simple eye or eye-spot found in many lower animals; a dorsal eye in insects; an eye-like marking as in many insects, fishes, on feathers, etc.; a large cell of leaf epidermis, specialized for reception of light; swelling on sporangiophore of some fungi for perception of light; *alt.* facet, speculum; *plu.* ocelli.

ochraceous, ochreous *a.* [Gk. *ōchros*, yellow] ochre-coloured.

ochrea ocrea, *q.v.*

ochroleucous *a.* [Gk. *ōchros*, pale yellow; *leukos*, white] yellowish-white, buff.

Ochromonadales *n.* [Gk. *ōchros*, pale yellow; *monas*, unit] an order of Chrysophyta having unicellular and colonial motile forms.

ochrophore *n.* [Gk. *ōchros*, pale yellow; *pherein*, to bear] a yellow pigment-bearing cell, *alt.* xanthophore, guanophore; iridocyte, *q.v.*

ochrosporous *a.* [Gk. *ōchros*, pale yellow; *sporos*, seed] having ochre-coloured spores.

Ockham's razor Occam's razor, *q.v.*

ocrea *n.* [L. *ocrea*, greave] a tubular sheath-like expansion at base of petiole; a sheath; partial covering of a stipe, formed by fragments of the disintegrated universal veil; *alt.* ochrea, mantle.

ocreaceous *a.* [L. *ocrea*, greave] ocrea-like, *appl.* various structures in plants and animals.

ocreate *a.* [L. *ocrea*, greave] having an ocrea; booted, sheathed; intrafoliaceous, *q.v.*

octactine *n.* [Gk. *okta*, eight; *aktis*, ray] a sponge spicule with 8 rays, a modification of a hexactine.

octad *n.* [Gk. *okta*, eight] a group of 8 cells, originating by division of a single cell.

octagynous *a.* [Gk. *okta*, eight; *gynē*, woman] having 8 pistils or styles; having 8 carpels to the gynaecium; *alt.* octogynous.

octamerous *a.* [Gk. *okta*, eight; *meros*, part] *appl.* organs or parts of organs when arranged in 8s; *appl.* parts of whorls of certain plants; *appl.* parts of certain Alcyonaria.

octandrous *a.* [Gk. *okta*, eight; *anēr*, man] having 8 stamens.

octant *n.* [L. *octo*, eight] one or all of the 8 cells formed by division of the fertilized ovum in plants and animals. *a. Appl.* division of an embryo into an octant.

octarch *a.* [Gk. *okta*, eight; *archē*, element] with 8 alternating xylem and phloem groups; with 8 vascular bundles.

Octocorallia *n.* [Gk. *oktō*, eight; *korallion*, coral] a subclass of anthozoans which are colonial with 8 tentacles and 8 complete septa and a ventral siphonoglyph; *alt.* Alcyonaria, Octoradiata.

octogynous octagynous, *q.v.*

octopetalous *a.* [Gk. *oktō*, eight; *petalon*, leaf] having 8 petals.

octophore *n.* [Gk. *oktō*, eight; *-phoros*, -bearing] a modified ascus with 8 spores arranged radially, as in fungi of the Haerangiomycetes.

octoploid *a.* [Gk. *oktō*, eight; *aploos*, onefold; *eidos*, form] having 8 haploid chromosome sets in somatic cells. *n.* An octoploid organism.

octopod *a.* [Gk. *oktō*, eight; *pous*, foot] having 8 tentacles, feet, or arms.

Octopoda *n.* [Gk. *oktō*, eight; *pous*, foot] an order of cephalopods of the subclass Coleoidea, including the octopuses, having 8 equal arms and no internal shell.

Octoradiata *n.* [L. *octo*, eight; *radius*, ray] Octocorallia, *q.v.*

octoradiate *a.* [L. *octo*, eight; *radius*, ray] having 8 rays or arms.

octosepalous *a.* [L. *octo*, eight; F. *sépale*, sepal] having 8 sepals.

octospore *n.* [Gk. *oktō*, eight; *sporos*, seed] one of 8 spores, as formed at end of carpogonial filaments, or in an octophore.

octosporous *a.* [Gk. *oktō*, eight; *sporos*, seed] having 8 spores.

octostichous *a.* [Gk. *oktō*, eight; *stichos*, row] arranged in 8 rows; having leaves in 8s, in phyllotaxis.

octozoic *a.* [Gk. *oktō*, eight; *zōon*, animal] *appl.* a spore, or sporocyst, containing 8 sporozoites.

ocular *a.* [L. *oculus*, eye] *pert.*, or perceived by, the eye.

ocular lobe projecting thoracic lobe in some beetles.

ocular plates plates at end of ambulacral areas in sea urchins.

oculate *a.* [L. *oculus*, eye] having eyes, or eye-like spots.

oculiferous, oculigerous *a.* [L. *oculus*, eye; *ferre, gerere*, to carry] bearing eyes.

oculofrontal *a.* [L. *oculus*, eye; *frons*, forehead] *pert.* region of forehead and eye.

oculomotor *a.* [L. *oculus*, eye; *movere*, to move] causing movements of eyeball, *appl.* 3rd cranial nerve: motor oculi.

oculonasal *a.* [L. *oculus*, eye; *nasus*, nose] *pert.* eye and nose.

oculus *n.* [L. *oculus*, eye] the eye; a leaf bud in a tuber.

Oddi's sphincter [R. *Oddi*, Italian anatomist] muscle fibres surrounding duodenal end of common bile duct.

odd-pinnate pinnate with one terminal leaflet.

odd-toed ungulates Perissodactyla, *q.v.*

Odonata *n.* [Gk. *odous*, tooth] an order of predaceous insects, commonly called dragonflies, having incomplete metamorphosis, biting mouthparts, 2 pairs of equal elongated wings, and aquatic nymphs with a prehensile labium.

odontoblast *n.* [Gk. *odous*, tooth; *blastos*, bud] one of columnar cells on outside of dental pulp that form dentine; one of the cells giving rise to teeth of a radula; *alt.* fibrilloblast, odontoplast.

odontobothrion *n.* [Gk. *odous*, tooth; *bothros*, pit] tooth socket; *alt.* alveolus dentis, phatne.

Odontoceti *n.* [Gk. *odous*, tooth; *ketos*, whale] an order of placental mammals, the toothed whales.

odontoclast *n.* [Gk. *odous*, tooth; *klan*, to break] one of the large multinucleate cells that absorb roots of milk teeth and destroy dentine.

odontogeny *n.* [Gk. *odous*, tooth; *gennaein*, to produce] the origin and development of teeth; *alt.* odontosis.

Odontognathae *n.* [Gk. *odous*, tooth; *gnathos*, jaw] an order or superorder of extinct primitive Cretaceous birds adapted to life on or near water, whose jaws have teeth, wings are poorly developed, and legs are strong and used for swimming.

odontoid *a.* [Gk. *odous*, tooth; *eidos*, form] tooth-like; *pert.* the odontoid process.

odontoid process a tooth-like peg on axis around which atlas rotates, the centrum of atlas, which has first become free and finally fused with axis; *alt.* dens.

odontology *n.* [Gk. *odous*, tooth; *logos*, discourse] dental anatomy, histology, physiology, and pathology.

odontophore *n.* [Gk. *odous*, tooth; *pherein*, to

carry] the tooth-bearing organ in molluscs, including the radula, radula sac, cartilage, and muscles.

odontoplast *n*. [Gk. *odous*, tooth; *plastos*, moulded] odontoblast, *q.v.*

odontorhynchous *a*. [Gk. *odous*, tooth; *rhyngchos*, snout] lamellirostral, *q.v.*

odontosis *n*. [Gk. *odous*, tooth] dentition; odontogeny, *q.v.*

odontostomatous *a*. [Gk. *odous*, tooth; *stoma*, mouth] having tooth-bearing jaws.

odorimetry *n*. [L. *odor*, smell; Gk. *metron*, measure] measurement of the strength of the sense of smell, using substances of known ability to stimulate olfaction.

odoriphore *n*. [L. *odor*, smell; Gk. *pherein*, to carry] a group of atoms responsible for the odour of a compound; *alt*. osmophore.

oecad *n*. [Gk. *oikade*, to one's home] ecad, *q.v.*

oecesis ecesis, *q.v.*

oecise ecize, *q.v.*

oecium *n*. [Gk. *oikion*, abode] the calcareous or chitinoid covering of a polyzooid.

oeco- *see also* eco-, oiko-.

oecoid ecoid, *q.v.*

oecology *n*. [Gk. *oikos*, household; *logos*, discourse] ecology, *q.v.*

oecoparasite ecoparasite, *q.v.*

oecophene *n*. [Gk. *oikos*, household; *phainein*, to appear] ecad, *q.v.*

oecotrophobiosis *n*. [Gk. *oikos*, household; *trophē*, food; *biōsis*, a living] trophallaxis, *q.v.*

oedematin *n*. [Gk. *oidema*, swelling] the microsomes of ground substance of nucleus.

Oedogoniales *n*. [*Oedogonium*, generic name] an order of green algae having cylindrical uninucleate cells united end to end in simple or branched filaments and a unique cell division with annular splitting of the lateral cell wall.

oenocyte *n*. [Gk. *oinos*, wine; *kytos*, hollow] one of large cells from clusters which surround trachea and fat-body of insects and undergo changes in relation to moulting cycle.

oenocytoid *n*. [Gk. *oinos*, wine; *kytos*, hollow; *eidos*, form] one of rounded acidophil leucocytes in haemolymph of insects.

oesophageal *a*. [Gk. *oisophagos*, gullet] *pert*. or near oesophagus, as ganglia; *alt*. esophageal.

oesophageal bulbs two swellings on the oesophagus in nematodes, the posterior of which, the pharynx, exhibits rhythmical pumping movements.

oesophageal glands or pouches in earthworms, diverticula attached to the oesophagus, the cells of which secrete calcium carbonate.

oesophagus *n*. [Gk. *oisophagos*, gullet] that part of alimentary canal between pharynx and stomach, or part equivalent thereto; *alt*. esophagus, gullet.

oestr- *alt*. estr-.

oestradiol *n*. [Gk. *oistros*, gadfly; *diolou*, together] the principle oestrogen found especially in the follicular fluid of the ovary, and also in the urine of pregnant mares; *alt*. dehydrotheelin.

oestrin oestrone, *q.v.*

oestriol an oestrogen usually obtained from the urine of pregnant women; *alt*. theelol.

oestrogen *n*. [Gk. *oistros*, gadfly; *gennaein*, to produce] any of several steroid sex hormones such as oestradiol, oestriol, and oestrone, produced mainly in the ovaries, and which cause the

production of female secondary sexual characteristics and promote oestrus; any of several substances found in plants that have similar effects.

oestrogenic *a*. [Gk. *oistros*, gadfly; *gennaein*, to produce] inducing oestrus; *appl*. hormones; *pert*. oestrogen.

oestrone *n*. [Gk. *oistros*, gadfly] an oestrogen which is the excretion product of oestradiol and is obtained from the urine of pregnant females such as women and mares; *alt*. follicular hormone, folliculin, oestron, theelin.

oestrous *a*. [Gk. *oistros*, gadfly] *pert*. oestrus; *alt*. oestrual.

oestrus *n*. [Gk. *oistros*, gadfly] the sexual heat of female animals; rut, *q.v.*; *alt*. oestrum, oestruation.

official *a*. [L. *officium*, office] officinal, *q.v.*

officinal *a*. [L. *officinalis*, of a storeroom] used medicinally; *alt*. official.

offset a short prostrate branch which takes root at apex and develops new individuals.

offshoot lateral shoot from main stem.

oidia *plu*. of oidium.

oidiophore *n*. [Gk. *ōon*, egg; *idion*, *dim*.; *pherein*, to bear] a hypha or hyphal structure bearing oidia.

oidiospore oidium, *q.v.*

oidium *n*. [Gk. *ōon*, egg; *idion*, *dim*.] a spore formed by transverse segmentation of a hypha; *alt*. oidiospore, arthrospore; *plu*. oidia.

oike *n*. [Gk. *oikein*, to have as one's abode] habitat, *q.v.*

oikesis *n*. [Gk. *oikēsis*, act of dwelling] ecesis, *q.v.*

oiko- *see also* oeco-, eco-.

oikoid ecoid, *q.v.*

oikology ecology, *q.v.*

oikoplast *n*. [Gk. *oikos*, house; *plastos*, moulded] one of large glandular ectoderm cells which form gelatinous layer of appendicularians.

oikosite *n*. [Gk. *oikos*, house; *sitos*, food] a stationary or attached commensal or parasite.

oil gland a gland which secretes oil; in birds, a cutaneous oil-secreting gland used in preening feathers, *alt*. uropygial gland, preen gland, elaeodochon.

oils *n.plu*. [L. *oleum*, oil] glycerides and esters of fatty acids including palm oils, coconut oils, etc., which are liquid at $20°C$, and can be hydrolysed by lipolytic enzymes to fatty acid and glycerol.

Olacales *n*. [*Olax*, generic name] an order of dicots (Polypetalae) which are trees and shrubs and have hermaphrodite or unisexual flowers.

Old Red Sandstone a northern facies of Devonian deposits, with characteristic fossils.

oleaginous *a*. [L. *oleaginus*, *pert*. olive] oily; *pert*., containing, or producing oil.

Oleales *n*. [*Olea*, generic name] an order of dicots (Asteridae) comprising the single family Oleaceae, including lilac, olive, and ash; *alt*. Ligustrales.

olecranon *n*. [Gk. *olekranon*, point of elbow] a large process at upper end of ulna.

oleic acid a widely distributed fatty acid.

oleiferous *a*. [L. *oleum*, oil; *ferre*, to carry] producing oil.

olein *n*. [L. *oleum*, oil] a fat, liquid at ordinary temperatures, found in animal and plant tissues.

oleocyst *n*. [L. *oleum*, oil; Gk. *kystis*, bladder] a diverticulum of the nectocalyx containing oil.

oleoplast *n*. [L. *oleum*, oil; Gk. *plastos*, formed] elaioplast, *q.v.*

oleosome elaioplast, *q.v.*

olfaction *n*. [L. *olfacere*, to smell] the sense of smell; the process of smelling.

olfactory *a*. [L. *olfacere*, to smell] *pert*. sense of smell, *appl*. stimuli, structures, reaction, etc.; *appl*. 1st cranial nerve, to olfactory organs.

olfactory lobe projecting from anterior lower margin of cerebral hemispheres, concerned with sense of smell.

olfactory pit an olfactory organ of nature of a small pit or hollow; embryonic ectodermal pit or sac which gives rise to nasal cavity, olfactory epithelium, and vomeronasal organ.

olfactory spindle sensory cell structure associated with olfactory nerve in antennule of decapod crustaceans; *alt*. lobus osphradicus.

oligacanthous *a*. [Gk. *oligos*, few; *akantha*, spine] bearing few spines.

oligandrous *a*. [Gk. *oligos*, few; *anēr*, man] having few stamens; *alt*. oligostemonous.

oligarch *a*. [Gk. *oligos*, few; *archē*, element] having few vascular elements or bundles.

oligocarpous *a*. [Gk. *oligos*, few; *karpos*, fruit] having few carpels.

Oligocene *n*. [Gk. *oligos*, few; *kainos*, recent] a Tertiary geological epoch between Eocene and Miocene.

Oligochaeta *n*. [Gk. *oligos*, few; *chaitē*, hair] a group of annelids which includes the earthworms, considered either as a class, or subclass of the Clitellata, and having few chaeta, no parapodia, and a small prostomium.

oligodendrocytes *n.plu*. [Gk. *oligos*, few; *dendron*, tree; *kytos*, hollow] oligodendroglia, *q.v.*

oligodendroglia *n.plu*. [Gk. *oligos*, few; *dendron*, tree; *glia*, glue] a type of small neuroglia cells found in both grey and white matter, having few very fine processes which show little branching; *alt*. oligoglia, oligodendrocytes, mesoglia of Robertson.

oligodynamic *a*. [Gk. *oligos*, few; *dynamis*, power] caused by small or minute forces; functioning in minute quantities.

oligogene *n*. [Gk. *oligos*, few; *genos*, descent] a gene having a pronounced phenotypic effect; *alt*. key gene; *cf*. polygene.

oligogenic *a*. [Gk. *oligos*, few; *genos*, descent] controlled by a few genes responsible for major heritable changes, *appl*. characters.

oligoglia *n*. [Gk. *oligos*, few; *glia*, glue] oligodendroglia, *q.v.*

oligolecithal *a*. [Gk. *oligos*, few; *lekithos*, egg yolk] containing not much yolk; *cf*. mesolecithal, telolecithal.

oligolectic *a*. [Gk. *oligos*, few; *lektos*, chosen] selecting only a few; *appl*. insects visiting only a few different food plants or flowers.

oligolobate *a*. [Gk. *oligos*, few; L.L. *lobus*, lobe] divided into only a small number of lobes, as in leucocytes.

oligomer *n*. [Gk. *oligos*, few; *meros*, part] a molecule composed of only a few monomer units.

oligomerous *a*. [Gk. *oligos*, few; *meros*, part] having one or more whorls with fewer members than the rest.

Oligoneoptera, Oligoneuroptera *n*. [Gk.

oligos, few; *neos*, new; *neuron*, nerve; *pteron*, wing] Endopterygota, *q.v.*

oligonephrous *a*. [Gk. *oligos*, few; *nephros*, kidney] having few Malpighian tubules, *appl*. certain insects; *alt*. oligonephric.

oligonucleotide *n*. [Gk. *oligos*, few; L. *nucleus*, kernel] a few nucleotides joined together.

oligophagous *a*. [Gk. *oligos*, few; *phagein*, to eat] restricted to a single order, family or genus of food plants, *appl*. insects; *cf*. monophagous.

oligophyletic *a*. [Gk. *oligos*, few; *phylē*, tribe] derived from a few ancestral forms; *cf*. monophyletic, polyphyletic.

oligopneustic *a*. [Gk. *oligos*, few; *pneustikos*, disposed to breathe] *appl*. insect respiratory system in which only a few spiracles are functional.

oligopod *a*. [Gk. *oligos*, few; *pous*, foot] having few legs or feet; having thoracic legs fully developed; *cf*. polypod, protopod.

oligopyrene *a*. [Gk. *oligos*, few; *pyrēn*, fruit stone] *appl*. certain spermatozoa with reduced number of chromosomes; *cf*. apyrene, eupyrene.

oligorhizous *a*. [Gk. *oligos*, few; *rhiza*, root] having few roots, *appl*. certain marsh plants.

oligosaccharide *n*. [Gk. *oligos*, few; *sakchar*, sugar] a molecule made of only a few polymerized monosaccharide units.

oligosaprobic *a*. [Gk. *oligos*, few; *sapros*, rotten; *bios*, life] *appl*. aquatic environment having high dissolved oxygen content and little organic decomposition.

oligospermous *a*. [Gk. *oligos*, few; *sperma*, seed] bearing few seeds.

oligosporous *a*. [Gk. *oligos*, few; *sporos*, seed] producing or having few spores.

oligostemonous *a*. [Gk. *oligos*, few; *stēmōn*, thread] oligandrous, *q.v.*

oligotaxy *n*. [Gk. *oligos*, few; *taxis*, arrangement] diminution in number of whorls.

oligothermic *a*. [Gk. *oligos*, little; *thermē*, heat] tolerating relatively low temperatures.

oligotokous *a*. [Gk. *oligos*, few; *tokos*, offspring] bearing few young.

Oligotrichida, Oligotricha *n*. [Gk. *oligos*, few; *thrix*, hair] an order or suborder of Spirotricha having cilia restricted to certain areas of the body.

oligotrophic *a*. [Gk. *oligos*, little; *trophē*, nourishment] providing or *pert*. inadequate nutrition; *appl*. lakes deficient in organic matter, so with a low supply of nutrient minerals; *cf*. eutrophic, dystrophic.

oligotrophophyte *n*. [Gk. *oligos*, few; *trophē*, nourishment; *phyton*, plant] a plant which will grow on poor soil.

oligotropic *a*. [Gk. *oligos*, few; *tropikos*, turning] visiting only a few allied species of plant during their lifetime, *appl*. insects.

oligoxenous *a*. [Gk. *oligos*, few; *xenos*, host] *appl*. parasites adapted to life in only a few species of host; *n*. oligoxeny.

olistherochromatin *n*. [Gk. *olisthēros*, sliding; *chrōma*, colour] the substance of which the olistherozones are composed.

olistherozones *n.plu*. [Gk. *olisthēros*, sliding; *zōnē*, girdle] regions of incomplete splitting of chromatids, possibly due to nucleic acid deficiency; zones of differential reactivity.

oliva *n*. [L. *oliva*, olive] a prominence on each side

of anterior end of medulla just below pons; *alt.* olive, olivary body.

olivary *a.* [L. *oliva*, olive] *pert.* the oliva; *pert.* certain nuclei of grey matter.

olive oliva, *q.v.*

omasum *n.* [L. *omasum*, paunch] in ruminants, the 3rd chamber of the stomach, which acts as a filter; *alt.* manyplies, psalterium.

ombrophil(e) *a.* [Gk. *ombros*, rain; *philein*, to love] thriving in rain, adapted to living in a rainy place, *appl.* plants, leaves.

ombrophobe *n.* [Gk. *ombros*, rain; *phobos*, fear] a plant which does not thrive under conditions of heavy rainfall.

ombrophyte *n.* [Gk. *ombros*, rain; *phyton*, plant] a plant adapted to rainy conditions.

omental *a.* [L. *omentum*, caul] *pert.* omentum or omenta.

omentum *n.* [L. *omentum*, caul] a fold of peritoneum either free or acting as connecting link between viscera; *alt.* caul.

ommachromes ommatochromes, *q.v.*

ommateum *n.* [Gk. *ommation*, little eye] a compound eye.

ommatidium *n.* [Gk. *ommation*, little eye; *idion*, *dim.*] one of component elements of a compound eye; *alt.* facet, stemma.

ommatin(e)s *n.plu.* [Gk. *omma*, eye] a group of ommatochrome pigments having small molecules and being dialysable; *cf.* ommines.

ommatochromes *n.plu.* [Gk. *omma*, eye; *chrōma*, colour] a group of pigments in Arthropoda, derivatives of tryptophane, first recognized as eye pigments, but now also found in the body, being yellow, red, or brown, and including the ommines and ommatines; *alt.* ommachromes, ommochromes.

ommatoids *n.plu.* [Gk. *omma*, eye; *eidos*, form] two or 4 light-coloured spots of disputed function on last opisthosomal segment of some arachnids.

ommatophore *n.* [Gk. *omma*, eye; *pherein*, to bear] a movable process bearing an eye; *alt.* ophthalmophore.

ommin(e)s *n.plu.* [Gk. *omma*, eye] a group of ommatochrome pigments having large molecules and being non-dialysable; *cf.* ommatines.

ommochromes ommatochromes, *q.v.*

omnicolous *a.* [L. *omnis*, all; *colere*, to dwell] capable of growing on different substrata, *appl.* lichens.

omnivorous *a.* [L. *omnis*, all; *vorare*, to devour] eating both plants and animals; *n.* omnivore.

omohyoid *a.* [Gk. *ōmos*, shoulder; *hyoeidēs*, Υ-shaped] *pert.* shoulder and hyoid, *appl.* a muscle.

omoideum *n.* [Gk. *ōmos*, shoulder; *eidos*, shape] pterygoid bone of bird's skull.

omosternum *n.* [Gk. *ōmos*, shoulder; L. *sternum*, breast bone] anterior element of amphibian sternum; cartilage or bone between sternum and clavicle of certain mammals.

omphalic *a.* [Gk. *omphalos*, navel] umbilical, *q.v.*

omphalodisc *n.* [Gk. *omphalos*, navel; *diskos*, disc] an apothecium with a small central protuberance, as in certain lichens.

omphalodium omphaloidium, *q.v.*

omphalogenesis *n.* [Gk. *omphalos*, navel; *genesis*, descent] development of the umbilical vesicle and cord.

omphaloid *a.* [Gk. *omphalos*, navel; *eidos*, form] like a navel; umbilicate, *q.v.*

omphaloidium *n.* [Gk. *omphalos*, navel; *idion*, *dim.*] the scar at hilum of a seed, or hilum itself; *alt.* omphalodium.

omphalomesenteric *a.* [Gk. *omphalos*, navel; *mesenteron*, mid-gut] *pert.* umbilicus and mesentery, *appl.* arteries, veins, ducts.

oncho- *see* onco-, *q.v.*

onchosphere, oncosphere *n.* [Gk *ongkos*, hook; *sphaira*, globe] the hexacanth embryo and its surrounding egg shell; the hexacanth embryo only.

oncogene *n.* [Gk. *onkos*, bulk, mass; *genos*, descent] a gene found in some viruses, and which is thought to cause cancer.

oncogene theory the theory that genes for oncornaviruses are part of the genome of animals and may have a role in normal development of the embryo as well as causing cancer.

oncogenic *a.* [Gk. *onkos*, bulk, mass; *genos*, descent] causing tumours.

oncolytic *a.* [Gk. *onkos*, bulk, mass; *lyein*, to break down] capable of killing cancer cells.

oncornavirus acronym for *oncogenic RNA virus.*

onisciform *a.* [*Oniscus*, generic name of woodlouse; L. *forma*, shape] like a woodlouse in shape, *appl.* some lepidopterous larvae.

Oniscomorpha *n.* [*Oniscus*, generic name of woodlouse; Gk. *morphē*, form] Glomerida, *q.v.*

onomatography *n.* [Gk *onoma*, name; *graphein*, to write] the practice of the correct writing of plant or animal names, analogous to bibliography.

ontocycle *n.* [Gk. *on*, being; *kyklos*, circle] evolution which in its later stages tends to produce forms exactly like those in the early stages.

ontogenesis, ontogeny *n.* [Gk. *on*, being; *genesis*, descent] the history of development and growth of an individual; *cf.* phylogeny; *a.* ontogenetic.

onychium *n.* [Gk. *onyx*, nail] the layer below the nail; pulvillus, *q.v.*; a special false articulation to bear claws at end of tarsus, in some spiders.

onychogenic *a.* [Gk. *onyx*, nail; *-genēs*, producing] capable of producing a nail or nail-like substance; *appl.* material in nail matrix, and cells forming fibrous substance and cuticula of hairs.

Onychophora, onychophorans *n.*, *n.plu.* [Gk. *onyx*, nail; *phora*, producing] a group of animals including *Peripatus*, which are intermediate between annelids and arthropods, having features of both, and may be considered as a separate phylum, or as a class or subphylum of arthropods; *alt.* Prototracheata.

Onygenales *n.* [*Onygena*, generic name] an order of Ascomycetes in which asci are formed in a mazaedium.

onymy *n.* [Gk. *onyma*, name] nomenclature; applying onyms or technical names.

ooangium archegonium, *q.v.*

ooapogamy *n.* [Gk. *ōon*, egg; *apo*, away; *gamos*, marriage] parthenapogamy, *q.v.*

ooblast *n.* [Gk. *ōon*, egg; *blastos*, bud] in certain red algae, a tubular outgrowth from the carpogonium through which the fertilized nucleus or its derivatives pass into the auxiliary cell.

ooblastema *n.* [Gk. *ōon*, egg; *blastēma*, bud] oosperm, *q.v.*

oocarp n. [Gk. ōon, egg; karpos, fruit] oospore, q.v.

oocyst n. [Gk. ōon, egg; kystis, bladder] cyst formed around 2 conjugating gametes in Sporozoa; alt. pseudonavicella, gamocyst, ovocyst.

oocyte n. [Gk. ōon, egg; kytos, hollow] an egg before formation of 1st polar body; in Protozoa, a stage in 'female' conjugant before it prepares for fertilization; alt. ovocyte.

oocytin n. [Gk. ōon, egg; kystos, hollow] substance extracted from spermatozoa which has a fertilizing and agglutinating effect on ova of same species.

ooecium n. [Gk. ōon, egg; oikos, house] ovicell, q.v.

oogamete n. [Gk. ōon, egg; gametēs, spouse] a female gamete, especially a large non-motile one with food material for the zygote.

oogamy n. [Gk. ōon, egg; gamos, marriage] the union of unlike gametes, usually a large non-motile female gamete and a small motile male gamete; cf. anisogamy, isogamy; a. oogamous.

oogenesis n. [Gk. ōon, egg; genesis, descent] formation, development, and maturation of the female gamete or ovum; alt. ovogenesis.

oogloea n. [Gk. ōon, egg; gloia, glue] egg cement.

oogone oogonium, q.v.

oogonial a. [Gk. ōon, egg; gonos, begetting] pert. the oogonium.

oogonium n. [Gk. ōon, egg; gonos, begetting] the female reproductive organ in algae and fungi; formerly a mother egg cell (oocyte); alt. oogone, oosporangium.

ooid a. [Gk. ōon, egg; eidos, form] egg-shaped; oval.

ookinesis n. [Gk. ōon, egg; kinēsis, movement] the mitotic stages of nucleus in maturation and fertilization of eggs.

ookinete n. [Gk. ōon, egg; kinein, to move] the motile worm-shaped stage of the zygote in certain Protozoa; alt. vermicule.

oolemma n. [Gk. ōon, egg; lemma, husk] zona pellucida, q.v.

oology n. [Gk. ōon, egg; logos, discourse] the study of eggs, particularly those of birds.

Oomycetales n. [Gk. ōon, egg; mykēs, fungus] the Oomycetes when considered as an order of Phycomycetes.

Oomycetes n. [Gk. ōon, egg; mykēs, fungus] a group of fungi, sometimes placed in the Phycomycetes, or sometimes raised to class status, having each zoospore with 2 flagella, one whiplash and one tinsel, and oogamous sexual reproduction.

oophore n. [Gk. ōon, egg; pherein, to bear] ovary, q.v.; oophyte, q.v.

oophoridium n. [Gk. ōon, egg; pherein, to bear; idion, dim.] the megasporangium in certain plants.

oophoron ovary, q.v.

oophyte n. [Gk. ōon, egg; phyton, plant] gametophyte in lower plants; alt. oophore.

ooplasm n. [Gk. ōon, egg; plasma, mould] cytoplasm of an egg; alt. ovoplasm.

ooplast n. [Gk. ōon, egg; plastos, formed] an unfertilized ovum; alt. oosphere.

oopod n. [Gk. ōon, egg; pous, foot] a component part of sting or ovipositor.

ooporphyrin n. [Gk. ōon, egg; porphyra, purple] a protoporphyrin, the pigment of egg shell in birds.

oosperm n. [Gk. ōon, egg; sperma, seed] a fertilized egg or zygote; alt. oospore, ooblastema, oocarp.

oosphere n. [Gk. ōon, egg; sphaira, globe] an egg before fertilization, alt. ooplast; a female gamete, especially as produced in an oogonium or archegonium; alt. gonosphaerium, gonosphere.

oosporangium n. [Gk. ōon, egg; sporos, seed; anggeion, vessel] oogonium, q.v.

oospore n. [Gk. ōon, egg; sporos, seed] the zygote or fertilized egg cell, alt. oosperm, ooblastema; encysted zygote in certain Protozoa; alt. oocarp.

oostegite n. [Gk. ōon, egg; stegē, roof] a plate-like structure on basal portion of thoracic limb in certain Crustacea, which helps to form a receptacle for the egg and acts as a brood pouch.

oostegopod n. [Gk. ōon, egg; stegē, roof; pous, foot] a thoracic foot bearing an oostegite.

ootheca n. [Gk. ōon, egg; thēkē, case] sporangium, q.v.; an egg case as in certain insects, alt. ovicapsule.

ootid n. [Gk. ōon, egg; idion, dim.] on analogy of spermatid, one of 4 parts into which egg divides at maturation.

ootocoid a. [Gk. ōon, egg; tokos, delivery; eidos, form] giving birth to young at a very early stage, and then carrying them in a marsupium.

ootocous a. [Gk. ōon, egg; tokos, delivery] egg-laying.

ootype n. [Gk. ōon, egg; typos, mould] part of oviduct receiving ducts from shell- and yolk-glands, in flatworms.

ooze n. [A.S. wáse, mud] a deposit containing skeletal parts of minute organisms and covering large areas of ocean bottom; soft mud.

oozoite n. [Gk. ōon, egg; zōon, animal] asexual parent, in tunicates.

oozooid n. [Gk. ōon, egg; zōon, animal; eidos, form] any individual developed from an egg; cf. blastozooid.

OP osmotic pressure, q.v.

Opalina n. [Opalina, generic name] an order of Zoomastigina having many flagella, never being amoeboid, and with no extranuclear division centre; sometimes considered a separate class, the Opalinata.

open appl. aestivation where perianth leaves do not meet at the edges, as in Cruciferae; appl. plant community which does not completely colonize the ground but leaves bare areas.

operant conditioning trial-and-error conditioning, q.v.

operational taxonomic unit (OTU) any group such as genus, species, etc., evaluated on taxometric methods.

operator gene a gene in an operon that controls the functioning of adjacent structural genes.

opercle operculum, q.v.

opercula plu. of operculum.

opercular a. [L. operculum, lid] posterior bone of fish operculum. a. Pert. operculum; appl. dehiscing antheridial cell, as in ferns; appl. fold of skin covering gills in tadpoles.

operculate a. [L. operculum, lid] having a lid, as sporangia of certain fungi, the capsule of mosses, etc., alt. calyptrate; having a covering for gills, as most fishes; alt. operculiferous.

operculation formation or existence of an operculum.

operculiferous *a.* [L. *operculum*, lid; *ferre*, to carry] operculate, *q.v.*

operculiform *a.* [L. *operculum*, lid; *forma*, shape] lid-like.

operculigenous *a.* [L. *operculum*, lid; Gk. *gennaein*, to produce] producing or forming a lid.

operculum *n.* [L. *operculum*, lid] a lid or covering flap, as at apex of an ascus, or of capsules of mosses; sepaline and petaline bud cover, shed at flowering as in eucalyptus; a convolution covering island of Reil; gill cover of fishes; flap covering of nostrils and ears in some birds; lid-like structure closing mouth of shell in some gastropods, *alt.* epiphragm; movable plates in shell of barnacle; 1st pair of abdominal appendages in some arachnids; small plate covering opening of a lung book in spiders; egg cap, opened by emerging insect; chitinous lid of orifice in Polyzoa; *alt.* opercle; *plu.* opercula.

operon *n.* [L. *opus* work] a section of DNA consisting of an operator gene, sometimes a repressor gene, and several structural genes, which act as a unit.

opesia *n.* [Gk. *opē*, hole] membranous aperture below orifice in Polyzoa.

Ophidia *n.* [Gk. *ophis*, serpent] a division of Squamata, consisting of the snakes; *alt.* Serpentes.

Ophiocephaliformes *n.* [*Ophiocephalus*, generic name; L. *forma*, shape] an order of teleosts having an accessory air-breathing suprabranchial organ.

ophiocephalous *a.* [Gk. *ophis*, serpent; *kephalē*, head] snake-headed, *appl.* small pedicellariae of some echinoderms having broad jaws with toothed edges.

Ophioglossales *n.* [*Ophioglossum*, generic name] an order of eusporangiate Pteropsida (Filicopsida) which are small plants with fertile leaves of 2 differentiated parts, a flattened sterile part and spike-like fertile part.

ophiopluteus *n.* [Gk. *ophis*, serpent; L. *pluteus*, shed] the pluteus larva of Ophiuroidea.

Ophiurae *n.* [Gk. *ophis*, serpent; *oura*, tail] an order of brittle stars having fairly stiff simple arms.

ophiuroid *a.* [Gk. *ophis*, serpent; *oura*, tail; *eidos*, form] resembling or *pert.* a brittle star; *appl.* cells: multiradiate or spiculate sclereids, astrosclereids.

Ophiuroidea *n.* [Gk. *ophis*, serpent; *oura*, tail; *eidos*, form] a class of echinoderms of the subphylum Eleutherozoa, commonly called brittle stars, having a star-shaped body with the arms clearly marked off from the central disc.

ophryon *n.* [Gk. *ophrys*, brow] point of junction of median line of face with a line across narrowest part of forehead.

ophthalmic *a.* [Gk. *ophthalmos*, eye] *pert.* eye; *appl.* a division of trigeminal nerve; *appl.* an artery arising from internal carotid; *appl.* inferior and superior veins of orbit.

ophthalmogyric *a.* [Gk. *ophthalmos*, eye; L. *gyrus*, circle] bringing about movement of the eye.

ophthalmophore ommatophore, *q.v.*

ophthalmopod *n.* [Gk. *ophthalmos*, eye; *pous*, foot] eye-stalk, as of decapod crustaceans.

Opiliones *n.* [*Opilia*, generic name] an order of arachnids, commonly known as harvest men, having very long legs and with the prosoma and opisthosoma forming a single structure; *alt.* Phalangida.

opisthaptor *n.* [Gk. *opisthe*, behind; *haptein*, to fasten] posterior sucker or disc in trematodes; *alt.* opisthohaptor.

opisthe *n.* [Gk. *opisthe*, behind] the posterior daughter individual produced when a protozoan divides transversely; *cf.* proter.

opisthial *a.* [Gk. *opisthe*, behind] posterior, *appl.* pore or stomatal margin.

opisthion *n.* [Gk. *opisthion*, behind] median point of posterior margin of foramen magnum.

Opisthobranchia(ta), **opisthobranchs** *n.*, *n.plu.* [Gk. *opisthe*, behind; *brangchia*, gills] a subclass, or in some classifications an order of gastropods having a reduced or lost shell and reduced torsion resulting in apparent bilateral symmetry; *alt.* Euthyneura A.

opisthocoelus *a.* [Gk. *opisthe*, behind; *koilos*, hollow] having the centrum concave behind, *appl.* vertebrae.

opisthocont opisthokont, *q.v.*

opisthodetic *a.* [Gk. *opisthe*, behind; *detos*, bound] lying, posterior to beak or umbo, *appl.* ligaments in some bivalve shells; *cf.* amphidetic, paravincular.

opisthogenesis *n.* [Gk. *opisthe*, behind; *genesis*, origin] development of segments or markings proceeding forward from the posterior end of the body.

opisthoglossal *a.* [Gk. *opisthe*, behind; *glōssa*, tongue] having tongue fixed in front, free behind.

opisthognathous *a.* [Gk. *opisthe*, behind; *gnathos*, jaw] having retreating jaws; with mouthparts directed backward, *appl.* head of insects; *cf.* hypognathous, prognathous, orthognathous.

Opisthogoneata *n.* [Gk. *opisthe*, behind; *gonē*, generation] in some classifications, a group of arthropods with opisthogoneate genital apertures, including insects and centipedes, or sometimes only centipedes.

opisthogoneate *a.* [Gk. *opisthe*, behind; *gonē*, generation] having the genital aperture at hind end of body, as in Opisthogoneata; *cf.* progoneate.

opisthohaptor opisthaptor, *q.v.*

opisthokont *a.* [Gk. *opisthe*, behind; *kontos*, punting pole] with flagellum or flagella at posterior end; *alt.* opisthocont.

opisthomere *n.* [Gk. *opisthe*, behind; *meros*, part] a terminal plate on the abdomen of a female earwig.

Opisthomi *n.* [Gk. *opisthe*, behind; *ōmos*, shoulder] Mastacembeliformes, *q.v.*

opisthonephros *n.* [Gk. *opisthe*, behind; *nephros*, kidney] a renal organ of embryo, consisting of meso- and metanephric series of tubules.

Opisthopora *n.* [Gk. *opisthe*, behind; *poros*, channel] an order of oligochaetes whose male gonopores are some way behind the last pair of testes.

opisthosoma *n.* [Gk. *opisthe*, behind; *sōma*, body] posterior body region in trilobites and arachnids.

opisthotic *a.* [Gk. *opisthe*, behind; *ous*, ear] *pert.* inferior posterior bony element of otic capsule.

opisthure *n.* [Gk. *opisthe*, behind; *oura*, tail] the projecting tip of vertebral column.

opium *n.* [Gk. *opion*, poppy juice] an addictive drug obtained from the opium poppy, *Papaver somniferum*, consisting of the dried milky juice from the slit poppy capsules, and acting as a stimulant, narcotic, and hallucinogen, formerly widely used to ease pain, but now largely replaced by its derivative alkaloids.

opophilous *a.* [Gk. *opos*, sap; *philein*, to love] feeding on sap.

opponens *a.* [L. *opponere*] *appl.* muscles which cause digits to approach one another.

opposite *a.* [L. *opponere*, to oppose] *appl.* leaves or other organs which are opposite one another at same level on stem; *appl.* member of a whorl in a flower opposite rather than alternate to, a member of the whorl next to it; *appl.* pits in horizontal pairs or rows.

opsiblastic *a.* [Gk. *opsi*, late; *blastē*, growth] with delayed cleavage, *appl.* eggs having a dormant period before hatching; *cf.* tachyblastic.

opsigenes *n.plu.* [Gk. *opsi*, late; *-genēs*, born] structures formed or becoming functional long after birth.

opsin *n.* [Gk. *opsis*, sight] a protein which combines with retinal to form rhodopsin.

opsis-forms *n.plu.* [Gk. *opsis*, sight] rusts whose developmental cycle includes spermogonia, aecidia, and teleutospores but not uredospores.

opsonic *a.* [Gk. *opsōnein*, to cater] *pert.* or affected by opsonin.

opsonin *n.* [Gk. *opsōnein*, to cater] a constituent of blood which helps phagocytes to destroy invading bacteria; *alt.* bacteriotropin, tropine.

optic *a.* [Gk. *opsis*, sight] *pert.* vision.

optical activity the property of a substance of rotating the plane of polarized light passing through a solution of it.

optical isomers isomers which are mirror images of each other and differ in optical activity, one turning the plane of polarized light to the left (the L-isomer) and the other to the right (the D-isomer), *alt.* enantiomers.

optic axis line between central points of anterior and posterior curvature or poles of eyeball.

optic bulb peripheral expansion of the embryonic optic vesicle, later invaginated to form the optic cup which gives rise to the retina.

optic disc blind spot, *q.v.*

optic lobes part of brain intimately connected with optic tracts; corpora bigemina, *q.v.*

optic nerves second pair of cranial nerves.

opticocilary *a.* [Gk. *opsis*, sight; L. *cilia*, eyelashes] *pert.* optic and ciliary nerves.

opticon *n.* [Gk. *opsis*, sight] inner zone of optic lobes of insects.

opticopupillary *a.* [Gk. *opsis*, sight; L. *pupilla*, pupil of eye] *pert.* optic nerve and pupil.

optic rod *see* rhabdome.

optimum *n.* [L. *optimum*, best] the most suitable degree of environmental factor for full development of organism concerned; point at which best response can be obtained.

optoblast *n.* [Gk. *opsis*, sight; *blastos*, bud] nerve cell of ganglionic layer of retina.

optocoel(e) *n.* [Gk. *opsis*, sight; *koilos*, hollow] the cavity in optic lobes of brain.

optogram *n.* [Gk. *opsis*, sight; *graphein*, to write] the image impressed on retina by action of light on visual purple.

optokinetic *a.* [Gk. *opsis*, sight; *kinein*, to move] *pert.* movement of the eyes.

optomotor *a.* [Gk. *opsis*, sight; L. *movere*, to move] *appl.* reflex of turning head or body in response to a stimulus of moving stripes.

Opuntiales *n.* [*Opuntia*, generic name] an order of dicots (Archichlamydeae) comprising the family Cactaceae, the cacti; *alt.* Cactales, Cactiflorae.

ora *n.* [L. *ora*, boundary] a margin; ora serrata: wavy border of retina, where nervous elements cease. *n.plu.* [L. *os*, mouth] mouths.

orad *adv.* [L. *os*, mouth; *ad*, to] towards the mouth or mouth region.

oral *a.* [L. *os*, mouth] *pert.* or belonging to mouth; on side on which mouth lies, *cf.* aboral.

oral cone in hydrozoan polyps, a conical projection surrounded by tentacles with the mouth at the apex; *alt.* manubrium, hypostome.

oral disc in actinozoan polyps, a circular flattened area surrounded by tentacles with the mouth at the centre.

oral siphon in urochordates, a narrowing of the body near the mouth.

oral valve in crinoids, 1 of 5 low triangular flaps separating the ambulacral grooves.

orbicular *a.* [L. *orbis*, orb] round or shield-shaped with petiole attached to centre, *appl.* leaves; surrounding, *appl.* eye muscles; annular, *appl.* ligament of head of radius.

orbicularis *a.* [L. *orbis*, orb] *appl.* a muscle whose fibres surround an opening.

orbiculate *a.* [L. *orbiculatus*, rounded] nearly circular in outline, *appl.* leaves, pileus.

orbit *n.* [L. *orbis*, eye socket] bony cavity in which eye is situated; skin around eye of bird; hollow in arthropod cephalothorax where eye-stalk arises; conspicuous zone, or rim, or head-capsule, around compound eye of insects.

orbital *pert.* the orbit.

orbitomalar *a.* [L. *orbis*, eye socket; *mala*, cheek] *pert.* orbit and malar bone.

orbitonasal *a.* [L. *orbis*, eye socket; *nasus*, nose] *pert.* orbit and nasal portions of adjoining bones.

orbitosphenoid *a.* [L. *orbis*, eye socket; Gk. *sphēn*, wedge; *eidos*, form] *pert.* paired cranial elements lying between presphenoid and frontal; *appl.* bone with foramen for optic nerve.

Orchidales *n.* [*Orchis*, generic name] an order of monocots (Liliidae) comprising the family Orchidaceae, the orchids; *alt.* Gynandrae, Microspermae.

orchitic *a.* [Gk. *orchis*, testis] testicular, *q.v.*

orculaeform *a.* [L.L. *orcula*, *dim.* of L. *orca*, cask; *forma*, shape] cask-shaped, *appl.* spores of certain lichens.

order *n.* [L. *ordo*, order] a taxonomic group of related organisms ranking between family and class.

ordinate *a.* [L. *ordinatus*, arranged] having markings arranged in rows.

ordinatopunctate *a.* [L. *ordinatus*, arranged; *punctum*, prick] indicating serial presence of dots, etc.

Ordovician *a.* [L. *Ordovices*, tribe of North Wales] *pert.* or *appl.* period of Palaeozoic era between Cambrian and Silurian.

organ *n.* [Gk. *organon*, implement] any part or

structure of an organism adapted for a special function or functions.

organelles, organellae *n.plu.* [Gk. *organon*, instrument] the various inclusions in a cell which have a special function; *alt.* organoid, cell organ.

organic *a.* [Gk. *organon*, instrument] *pert.*, derived from, or showing the peculiarities of a living organism; *pert.* carbon compounds.

organicism *n.* [Gk. *organon*, instrument] the cooperation or competition of cells, tissues, and organs and their reciprocal modifying action; the integration of an organism as a unit; the interblending of events within the organism.

organific *a.* [L. *organum*, instrument; *facere*, to make] producing an organism; making an organized structure.

organism *n.* [Gk. *organon*, instrument] any animal or plant; anything capable of carrying on life processes.

organismic *a.* [Gk. *organon*, instrument] *appl.* or *pert.* factors or processes involved in integrating and maintaining individuality of an organism.

organization centre organizer, *q.v.*

organized *a.* [Gk. *organon*, instrument] exhibiting characteristics of, or behaving like an organism; *appl.* growth resembling normal growth, in tissue culture, as distinct from unorganized growth of cells migrating from cut tissue.

organizer *n.* [Gk. *organos*, fashioning] a part of an embryo which provides a stimulus for the direction of morphological development and differentiation of other parts; *alt.* organization centre, inductor; *cf.* evocator.

organ of Corti Corti's organ, *q.v.*

organ of Tömösvary postantennal organ, *q.v.*

organ of Valenciennes paired lamellated organ in female nautiloids.

organogen *n.* [Gk. *organon*, instrument; *gennaein*, to produce] any of the elements C, H, O, N, also S, P, Cl.

organogenesis *n.* [Gk. *organon*, instrument; *genesis*, descent] formation and development of organs; *alt.* organogeny.

organogenic *a.* [Gk. *organon*, instrument; *genēs*, produced] due to the activity of an organ; *pert.* organogenesis.

organogeny organogenesis, *q.v.*

organography *n.* [Gk. *organon*, instrument; *graphein*, to write] the description of organs in a living organism.

organoid *n.* [Gk. *organon*, instrument; *eidos*, form] an organelle in the cytoplasm, *alt.* cell organ, *cf.* metaplasm *a.* Having a definite organized structure, *appl.* certain plant galls.

organoleptic *a.* [Gk. *organon*, instrument; *lambanein*, to take hold of] capable of receiving, or of making, an impression.

organology *n.* [Gk. *organon*, instrument; *logos*, discourse] the study of organs of plants and animals.

organonomy *n.* [Gk. *organon*, instrument; *nomos*, law] the laws that deal with life or living organisms.

organon spirale Corti's organ, *q.v.*

organonymy *n.* [Gk. *organon*, instrument; *onyma*, name] the nomenclature of organs.

organophyly *n.* [Gk. *organon*, instrument; *phylē*, tribe] the phylogeny of organs.

organoplastic *a.* [Gk. *organon*, organ; *plassein*, to

form] capable of forming, or producing, an organ; *pert.* formation of organs.

organotroph *n.* [Gk. *organon*, implement; *trophē*, nourishment] heterotroph, *q.v.*

organotrophic *a.* [Gk. *organon*, instrument; *trophē*, nourishment] *pert.* formation and nourishment of organs; heterotrophic, *q.v.*

organotypic *a.* [Gk. *organon*, instrument; *typos*, pattern] *appl.* organized growth under somatic control, as in tissue culture; *cf.* histiotypic.

organule *n.* [L. *organum*, instrument] a cell or element of an organism, or of an organ.

orgasm *n.* [Gk. *orgasmos*, swelling] immoderate excitement, especially sexual; turgescence of an organ.

Oriental *a.* [L. *orientalis*, eastern] *appl.* or *pert.* a zoogeographical region including India and Indo-China south to Wallace's line.

orientation *n.* [L. *oriens*, rising of sun] arrangement of chromosomes with centromeres lying axially in relation to spindle; alteration in position shown by organs or organisms under stimulus, including such acts as balancing and migration.

orienting curvature *see* tropism.

orifice *n.* [L. *os*, mouth; *facere*, to make] mouth or aperture; opening of a tube, duct, etc.; *alt.* orificium.

original *a.* [L. *origo*, origin] *pert.* beginning; *appl.* wild species from which cultivated species have been derived.

ornis *n.* [Gk. *ornis*, bird] avifauna, *q.v.*

ornithic *a.* [Gk. *ornis*, bird] *pert.* birds.

ornithichnite *n.* [Gk. *ornis*, bird; *ichnos*, track] the fossil track or footprint of a bird.

ornithine *n.* [Gk. *ornis*, bird] an amino acid, diaminovaleric acid, concerned with urea formation in the urea cycle, and excreted with one of its derivatives, ornithuric acid, in birds.

ornithine cycle urea cycle, *q.v.*

Ornithischia *n.* [Gk. *ornis*, bird; *ischion*, hip] an order of Mesozoic archosaurs, commonly called bird-hipped dinosaurs, having a pelvis resembling that of a bird and all being herbivorous and most quadrupedal.

ornithocopros *n.* [Gk. *ornis*, bird; *kopros*, dung] the dung of birds.

Ornithodelphia *n.* [Gk. *ornis*, bird; *delphys*, womb] in some classifications, an infraclass of mammals including the order Monotremata.

Ornithogaea *n.* [Gk. *ornis*, bird; *gaia*, earth] the zoogeographical region which includes New Zealand and Polynesia.

ornithology *n.* [Gk. *ornis*, bird; *logos*, discourse] the study of birds.

ornithophilous *a.* [Gk. *ornis*, bird; *philein*, to love] bird-loving, *appl.* flowers pollinated through agency of birds; *n.* ornithophily.

ornithuric acid an excretory substance, dibenzoylornithine, a constituent of bird excrement.

oroanal *a.* [L. *os*, mouth; *anus*, anus] serving as mouth and anus; connecting mouth and anus.

orobranchial *a.* [L. *os*, mouth; *branchiae*, gills] *pert.* mouth and gills, *appl.* epithelium.

oronasal *a.* [L. *os*, mouth; *nasus*, nose] *pert.* or designating groove connecting mouth and nose.

oropharynx *n.* [L. *os*, mouth; Gk. *pharynx*, gullet] the cavity of the mouth and pharynx; the space between the glossopalatine and pharyngopalatine arches or anterior and posterior pillars of the fauces.

orphan virus a virus which is not known to cause disease.

orrhoid a. [Gk. orrhos, serum; eidos, form] serous, q.v.

ortet n. [L. ortus, origin] the original single ancestor of a clone; cf. ramet.

orthal a. [Gk. orthos, straight] straight up and down, appl. jaw movement; cf. palinal, proral.

orthaxial a. [Gk. orthos, straight; L. axis, axle] with a straight axis, or vertebral axis, appl. caudal fin.

orthoblastic a. [Gk. orthos, straight; blastos, bud] with a straight germ band; cf ankyloblastic.

orthochromatic a. [Gk. orthos, straight; chrōma, colour] appl. large oval erythrocytes with nuclear strands passing out to nuclear membrane; of the same colour as that of the stain; staining positively.

orthochromatin n. [Gk. orthos, straight; chrōma, colour] normal stable chromatin; cf. parachromatin.

orthocladous a. [Gk. orthos, straight; klados, sprout] straight-branched.

orthordentine n. [Gk. orthos, straight; L. dens, tooth] dentine pierced by numerous more-or-less parallel dentinal tubules; inner layer of circumpulpar dentine and outer layer of pallial dentine; alt. tubular dentine; cf. osteodentine.

orthodromic a. [Gk. orthos, right; dromos, running] moving in the normal direction, appl. conduction of impulse; cf. antidromic.

orthoenteric a. [Gk. orthos, straight; enteron, intestine] having alimentary canal along internal ventral body surface, appl. certain Tunicata.

orthogamy autogamy, q.v.

orthogenesis n. [Gk. orthos, straight; genesis, descent] evolution along some apparently predestined line and independent of natural selection or other external forces; alt. determinate evolution.

orthognathous a. [Gk. orthos, straight; gnathos, jaw] having straight jaws; having axis of head at right angles to body, as in some insects; cf. prognathous, opisthognathous, hypognathous.

orthograde a. [Gk. orthos, straight; L. gradus, step] walking with the body in a vertical position.

orthokinesis n. [Gk. orthos, straight; kinēsis, movement] a kinesis in which an organism changes its speed of movement when it meets an unfavourable environment, the speed depending on the intensity of the environmental stimulus, resulting in the dispersal or aggregation of organisms; variation in linear velocity due to intensity of stimulation; cf. klinokinesis.

Orthonectida n. [Gk. orthos, straight; nektos, swimming] a class of mesozoans having an annulated body.

orthophyte n. [Gk. orthos, straight; phyton, plant] the plant in the interval between megaspore and megaspore production; sporophyte and gametophyte.

orthoplasis n. [Gk. orthos, right; plassein, to form] determinate formation due to organic selection in orthogenesis.

orthoploid a. [Gk. orthos, straight; haploos, onefold; eidos, form] with even chromosome number; polyploid with complete and balanced genomes.

Orthoptera n. [Gk. orthos, straight; pteron, wing] an order of insects including the grasshoppers, locusts, and crickets, having straight folded posterior wings, incomplete metamorphosis, and usually enlarged hindlegs and stridulation organs; a. orthopterous.

Orthopteroidea n. [Gk. orthos, straight; pteron, wing; eidos, form] in some classifications, a superorder of insects including the Orthoptera and various other groups.

orthoradial a. [Gk. orthos, straight; L. radius, ray] appl. cleavage where divisions are symmetrically disposed around egg axis.

orthoselection n. [Gk. orthos, straight; L. selectio, choice] natural selection acting continuously in the same direction over long periods.

orthosomatic a. [Gk. orthos, straight; sōmatikos, of the body] having a straight body, appl. certain larval insects.

orthospermous a. [Gk. orthos, straight; sperma, seed] with straight seeds.

orthospiral a. [Gk. orthos, straight; speira, coil] appl. coiling of parallel chromatids, interlocked at each twist; alt. plectonemic, relational coiling; cf. anorthospiral, paranemic.

orthostichous a. [Gk. orthos, straight; stichos, row] arranged in a vertical row, appl. leaves; appl. fin skeleton when peripheral somactids are parallel.

orthostichy n. [Gk. orthos, straight; stichos, row] vertical line on which a row of leaves or scales is found; arrangement of leaves or scales in this row.

orthotopic a. [Gk. orthos, true; topos, place] in the proper place, appl. transplantation; cf. heterotopic.

orthotopy n. [Gk. orthos, true; topos, place] natural placement; existence in a normal habitat.

orthotriaene n. [Gk. orthos, straight; triaina, trident] a triaene with cladi directed outwards at right angles to shaft.

orthotropal orthotropous, q.v.

orthotropism n. [Gk. orthos, straight; tropē, turn] growth in a vertical line, as main stems or tap roots, cf. plagiotropism; condition of tending to be oriented in the line of action of a stimulus; alt. parallelotropism; a. orthotropic.

orthotropous a. [Gk. orthos, straight; tropē, turn] appl. ovules having chalaza, hilum, and micropyle in a straight line so that ovule is not inverted; alt. atropous, atropal, orthotropal.

orthotype n. [Gk. orthos, straight; typos, pattern] genotype originally designated.

oryctics, oryctology n. [Gk. oryktos, dug out] the study of fossils; alt. palaeontology.

oryzenin n. [Oryza, generic name of rice] a glutelin protein of rice.

os n., ora plu. [L. os, mouth] a mouth; mouths.

os n., ossa plu. [L. os, bone] a bone; bones.

oscheal a. [Gk. oschē, scrotum] scrotal; pert. scrotum.

Oscillatoriales n. [Oscillatoria, generic name] Hormogonales, q.v.

oscilloxanthin n. [Oscillatoria, generic name; Gk. xanthos, yellow] a xanthophyll carotinoid pigment found in blue–green algae.

oscitate v. [L. oscitare, to yawn] to yawn; to gape.

oscula plu. of osculum.

osculant a. [L. osculans, kissing] closely adherent;

intermediate in character between 2 groups, genera, or species; *alt.* osculate.

oscular *a.* [L. *osculum,* small mouth] *pert.* an osculum.

osculate *v.* [L. *osculari,* to kiss] to have characters intermediate between 2 groups. *a.* Osculant, *q.v.*

oscule osculum, *q.v.;* pore of a spore of Uredinales.

osculiferous *a.* [L. *osculum,* small mouth; *ferre,* to bear] having oscula.

osculum *n.* [L. *osculum,* small mouth] an excurrent opening in a sponge; *plu.* oscula; *alt.* oscule.

osmatic *a.* [Gk. *osmē,* smell] having a sense of smell.

osmeterium *n.* [Gk. *osmē,* smell; *tērein,* to keep] a forked protrusible organ borne on 1st thoracic segment of larva of some butterflies, emitting a smell.

osmics *n.* [Gk. *osmē,* smell] the study of olfactory organs and the sense of smell, and of odoriferous organs and substances.

osmiophil(ic) *a.* [*Osmium,* from Gk. *osmē,* smell; *philein,* to love] staining readily with osmic acid, as olein in tissues, and as externum of Golgi bodies; *appl.* globules: plastoglobuli, *q.v.*

osmomorphosis *n.* [Gk. *ōsmos,* impulse; *morphōsis,* a shaping] change in shape or in structure due to differences in osmotic pressure, such as due to changes in salinity.

osmophore *n.* [Gk. *osmē,* smell; *phora,* producing] odoriphore, *q.v.*

osmoreceptors *n.plu.* [Gk. *ōsmos,* impulse; L. *recipere,* to receive] receptor cells stimulated by changes in osmotic pressure; cells reacting to osmotic changes in blood, and, via parasympathetic fibres innervating the posterior lobe of pituitary gland, controlling secretion of the antidiuretic hormone.

osmoregulation *n.* [Gk. *ōsmos,* impulse; L. *regulatus,* regulated] in animals, regulation of the osmotic pressure in the body by controlling the amount of water and/or salts in the body.

osmosis *n.* [Gk. *ōsmos,* impulse] diffusion of a solvent, usually water, through a semipermeable membrane from a dilute to a concentrated solution, or from the pure solvent to a solution.

osmosium *n.* [Gk. *ōsmos,* impulse] the part of nematode intestine connecting with demanian vessels.

osmotaxis *n.* [Gk. *ōsmos,* impulse; *taxis,* arrangement] a taxis in response to changes in osmotic pressure; *cf.* tonotaxis.

osmotic *a.* [Gk. *ōsmos,* impulse] *pert.* osmosis.

osmotic pressure (OP) the pressure concerned with osmosis exerted by a solution due to movement of its molecules and therefore depending on the concentration of the solution and its temperature measured either as the pressure developed in a solution, or as the pressure that must be applied to such a solution to just prevent osmosis.

osmotroph *n.* [Gk. *ōsmos,* impulse; *trophē,* nourishment] a heterotrophic organism which absorbs organic substances in solution, such as bacteria and fungi; *cf.* phagotroph.

Osmundales *n.* [*Osmunda,* generic name] the only order of the Osmundidae, *q.v.,* sometimes considered an order of Leptosporangiatae.

Osmundidae *n.* [*Osmunda,* generic name] a subclass of Pteropsida (Filicopsida) that are inter-

mediate between Eusporangiatae and Leptosporangiatae.

osmyl *n.* [Gk. *ōsme,* smell; *hylē,* matter] any odorous substance.

osphradium *n.* [Gk. *osphradion,* strong scent] a chemical sense organ associated with visceral ganglia in many molluscs.

osphresiology *n.* [Gk. *osphrēsis,* sense of smell; *logos,* discourse] the study of the sense of smell.

osphresis *n.* [Gk. *osphrēsis,* sense of smell] the sense of smell.

ossa bones, *plu.* of os.

ossa suturara or triquetra sutural bones, *q.v.*

ossein *n.* [L. *osseus,* bony] bone collagen, the most abundant organic constituent of bone.

osseoalbuminoid *a.* [L. *osseus,* bony; *albumen,* white of egg; Gk. *eidos,* form] an albuminous protein extracted from bone.

osseous *a.* [L. *osseus,* bony] composed of or resembling bone.

osseous labyrinth vestibule, semicircular canals, and cochlea, in petrous part of temporal bone and containing the membranous labyrinth.

ossicle *n.* [*Dim.* of L. *os,* bone] any small bone; one of those in ear, or in sclerotic; one of those in gastric mill of Crustacea; a plate of skeleton of echinoderms; *alt.* ossiculum.

ossicone *n.* [L. *os,* bone; *conus,* cone] the os cornu or horn core of ruminants.

ossicular *a.* [*Dim.* of L. *os,* bone] *pert.* ossicles.

ossiculate having ossicles.

ossiculum *n.* [*Dim.* of L. *os,* bone] ossicle, *q.v.;* lithodesma, *q.v.;* a partly calcified byssus; pyrene, *q.v.*

ossification *n.* [L. *os,* bone; *facere,* to make] the formation of bone, *alt.* osteogenesis; replacement of cartilage by bone, *see* autostoses, allostoses.

ossify *v.* [L. *os,* bone; *fieri,* to become] to change to bone.

Ostariophysi *n.* [Gk. *ostarion,* small bone; *physa,* bladder] a large group of freshwater teleosts existing from the Palaeocene to the present day and including the carp and minnow, having primitive features such as a connection between the swim bladder and inner ear called the Weberian ossicle; *alt.* Cypriniformes.

Osteichthyes *n.* [Gk. *osteon,* bone; *ichthys,* fish] a class of fish whose endoskeleton is of bone, so commonly called bony fishes, and usually having an air bladder or lungs, and gills covered by an operculum.

osteoblast *n.* [Gk. *osteon,* bone; *blastos,* bud] a bone-forming cell which secretes the matrix; *alt.* bone cell, osteoplastic cell.

osteochondral *a.* [Gk. *osteon,* bone; *chondros,* cartilage] *pert.* bone and cartilage.

osteochondrous consisting of bone and cartilage.

osteoclasis *n.* [Gk. *osteon,* bone; *klan,* to break] the destruction of bone by osteoclasts.

osteoclast *n.* [Gk. *osteon,* bone; *klan,* to break] a cell which destroys bone or any preceding matrix either cartilaginous or calcified, during bone formation; *alt.* giant cell.

osteocomma *n.* [Gk. *osteon,* bone; *komma,* piece] osteomere, *q.v.*

osteocranium *n.* [Gk. *osteon,* bone; *kranion,* skull] bony skull as distinguished from cartilaginous or chondrocranium.

osteocyte *n.* [Gk. *osteon*, bone; *kytos*, hollow] a bone cell, developed from osteoblast.

osteodentine *n.* [Gk. *osteon*, bone; L. *dens*, tooth] a variety of dentine which closely approaches bone in structure; *cf.* orthodentine.

osteodermis *n.* [Gk. *osteon*, bone; *derma*, skin] a dermis which is more-or-less ossified; a bony dermal plate.

osteogen *n.* [Gk. *osteon*, bone; *gennaein*, to produce] the tissue which alters and forms bone.

osteogenesis *n.* [Gk. *osteon*, bone; *genesis*, descent] bone formation; *alt.* ossification.

osteogenetic, osteogenic *a.* [Gk. *osteon*, bone; *genesis*, descent] *pert.* or causing formation of bone.

Osteoglossiformes *n.* [Gk. *osteon*, bone; *glōssa*, tongue; L. *forma*, shape] an order of freshwater teleosts including the mooneyes, knife fish, and bony tongues.

Osteoglossomorpha *n.* [Gk. *osteon*, bone; *glōssa*, tongue; *morphē*, form] a group of primitive freshwater teleosts existing from Cretaceous times to the present day, sometimes considered as a superorder, containing the orders Osteoglossiformes and Mormyriformes.

osteoid *a.* [Gk. *osteon*, bone; *eidos*, form] bonelike.

osteology *n.* [Gk. *osteon*, bone; *logos*, discourse] the part of zoology dealing with structure, nature, and development of bones.

osteolysis *n.* [Gk. *osteon*, bone; *lysis*, loosing] breakdown and dissolution of bone.

osteomere *n.* [Gk. *osteon*, bone; *meros*, part] a segment of the vertebral skeleton; *alt.* osteocomma.

osteone Haversian system, *q.v.*

osteoplastic *a.* [Gk. *osteon*, bone; *plastos*, moulded] producing bone, *appl.* cells: osteoblasts, *q.v.*

osteosclereid(e) *n.* [Gk. *osteon*, bone; *sklēros*, hard; *eidos*, shape] a sclereid with both ends knobbed; *alt.* pillar cell.

osteoscute *n.* [Gk. *osteon*, bone; L. *scutum*, shield] a bony external scale or plate, as in labyrinthodonts and armadillos.

Osteostraci *n.* [Gk. *osteon*, bone; *ostrakon*, shell] a group of fossil Palaeozoic ostracoderms having an exoskeleton of dermal bone and including Silurian and Devonian forms; sometimes considered an order equivalent to the Cephalaspida.

ostia *plu.* of ostium.

ostial of or *pert.* ostia or an ostium.

ostiate *a.* [L. *ostium*, door] furnished with ostia.

ostiolar *a.* [L. *ostiolum*, little door] *pert.* an ostiole.

ostiolate *a.* [L. *ostiolum*, little door] provided with ostioles.

ostiole *n.* [L. *ostiolum*, little door] a small opening, as of conceptacle, perithecium, stoma, anther sac, etc.; inhalant aperture of sponge.

ostium *n.* [L. *ostium*, door] any mouth-like opening; opening of Fallopian tube; opening between atria of foetal heart; opening in arthropod heart by which bloods enters from pericardium; opening from flagellate canal into paragastric cavity in sponges; *plu.* ostia.

Ostracoda, ostracods *n., n.plu.* [Gk. *ostrakon*, shell; *-ōdēs*, like] a subclass of small crustaceans having a bivalved carapace enclosing head and body, and reduced trunk and abdominal limbs.

Ostracodermi, ostracoderms *n., n.plu.* [Gk.

ostrakon, shell; *derma*, skin] a group of Palaeozoic Agnatha, including the Osteostraci and Heterostraci, which having an exoskeleton of dermal bone.

Ostropales *n.* [*Ostropa*, generic name] an order of Ascomycetes whose ascocarp is open and often cup-like.

otic *a.* [Gk. *ous*, ear] *pert.* ear; *pert.* region of auditory capsule; *appl.* ganglion on mandibular nerve; *appl.* vesicle: otocyst, *q.v.*

otidium *n.* [Gk. *ous*, ear; *idion*, dim.] the otocyst of a mollusc.

otoconia, otoconites *n.plu.* [Gk. *ous*, ear; *konia*, sand] minute crystals of calcium carbonate found in membranous labyrinth of inner ear; *alt.* ear sand.

otocrypt *n.* [Gk. *ous*, ear; *kryptos*, hidden] an open invagination of integument of foot in certain molluscs.

otocyst *n.* [Gk. *ous*, ear; *kystis*, bladder] a sac containing fluid and otoliths which move under gravity and stimulate sensory cells, giving the animal its position with respect of gravity and allowing it to balance, formerly thought to be auditory, *cf.* statocyst; the utriculus of vertebrate ear; formerly, embryonic auditory vesicle; *alt.* otic vesicle.

otolith *n.* [Gk. *ous*, ear; *lithos*, stone] calcareous particle or plate-like structure found in otocyst of many animals; *alt.* ear stone, statolith.

oto-occipital *n.* [Gk. *ous*, ear; L. *occiput*, back of head] bone formed by fusion of opisthotic with exoccipital.

otoporpae *n.plu.* [Gk. *ous*, ear; *porpē*, brooch] stripes of cnidoblasts on exumbrella of Hydrozoa.

otosalpinx *n.* [Gk. *ous*, ear; *salpingx*, trumpet] Eustachian tube, *q.v.*

otostapes *n.* [Gk. *ous*, ear; L.L. *stapes*, stirrup] otic portion of columellar primordium which in adult may give rise to stapes and part of columella.

otosteon *n.* [Gk. *ous*, ear; *osteon*, bone] an auditory ossicle.

OTU operational taxonomic unit, *q.v.*

outbreeding the breeding together of individuals that are not closely related; cross-fertilization.

ova *plu.* of ovum.

oval *a.* [L. *ovum*, egg] egg-shaped; *pert.* an egg.

ovalbumin *n.* [L. *ovum*, egg; *albumen*, white of egg] egg albumin, *q.v.*

oval window fenestra ovalis, *q.v.*

ovarian *a.* [L. *ovarium*, ovary] *pert.* an ovary.

ovarian follicle in many animals, a group of cells investing a developing oocyte, and probably concerned with its nutrition; *cf.* Graafian follicle.

ovariole *n.* [L. *ovarium*, ovary] egg tube of insect ovary.

ovariotestis *n.* [L. *ovarium*, ovary; *testis*, testicle] generative organ when both male and female elements are formed, as in case of sex reversal; *cf.* ovotestis.

ovarium *n.* [L. *ovarium*, ovary] ovary, *q.v.*

ovary *n.* [L. *ovarium*, ovary] the enlarged basal portion of the gynaecium containing the ovules, *alt.* ovulary; the essential female reproductive gland containing ova; *alt.* ovarium, oarium, oophoron, oophore, germarium.

ovate *a.* [L. *ovum*, egg] egg-shaped; egg-shaped

and attached by the broader end, *appl.* leaves; *cf.* ovoid.

ovate-acuminate having an ovate lamina with very sharp point, *appl.* leaves.

ovate-ellipsoidal ovate, approaching ellipsoid, *appl.* leaves.

ovate-lanceolate having a form of lamina intermediate between ovate and lanceolate, *appl.* leaves.

ovate-oblong having an oblong lamina with one end narrower, *appl.* leaves.

ovejector *n.* [L. *ovum*, egg; *ejectum*, thrown out] the muscular terminal part of female genital tract considered as a functional unit, in nematodes.

ovenchyma *n.* [L. *ovum*, egg; Gk. *engchyma*, infusion] a connective tissue with ovoid cells.

overdominance the condition where a heterozygote has a more extreme phenotype than either of the homozygotes from which it was formed; *alt.* superdominance.

overflow *appl.* behaviour in which an inappropriate response occurs to a certain stimulus in order to satisfy certain drives, such as a dog displaying maternal care to a bone.

overlapping the situation where one branch of a dichotomy grows more than another.

ovicapsule *n.* [L. *ovum*, egg; *capsula*, small box] an egg case or ootheca.

ovicell *n.* [L. *ovum*, egg; *cella*, cell] a dilatation of a zooecium, serving as a brood pouch; *alt.* ooecium.

oviducal *a.* [L. *ovum*, egg; *ducere*, to lead] *pert.* oviduct.

oviduct *n.* [L. *ovum*, egg; *ducere*, to lead] the tube which carries eggs from ovary to exterior; *see* Müllerian duct; *cf.* Fallopian tube.

oviferous *a.* [L. *ovum*, egg; *ferre*, to carry] serving to carry eggs; *alt.* ovigerous.

oviform *a.* [L. *ovum*, egg; *forma*, shape] egg-shaped; oval.

oviger *n.* [L. *ovum*, egg; *gerere*, to bear] egg-carrying leg of Pycnogonida.

ovigerous oviferous, *q.v.*

oviparous *a.* [L. *ovum*, egg; *parere*, to bring forth] egg-laying; *cf.* viviparous, ovoviviparous; *n.* oviparity.

oviposit *v.* [L. *ovum*, egg; *ponere*, to place] to lay eggs.

ovipositor *n.* [L. *ovum*, egg; *ponere*, to place] a specialized structure in insects for placing eggs in a suitable place; a tubular extension of genital orifice in fishes; *alt.* oviscapte.

ovisac *n.* [L. *ovum*, egg; *saccus*, bag] an egg capsule or receptacle; a brood pouch, *q.v.*

oviscapte *n.* [L. *ovum*, egg; F. *capter*, from L. *captare*, to conduct] ovipositor, *q.v.*

ovism *n.* [L. *ovum*, egg] theory held by ovists that the egg contained the germ with germs of all future generations within it; *cf.* spermism.

ovocentre *n.* [L. *ovum*, egg; *centrum*, centre] the egg centrosome during fertilization.

ovocyst, ovocyte, ovogenesis oocyst, oocyte, oogenesis, *q.v.*

ovoid *a.* [L. *ovum*, egg; Gk. *eidos*, form] somewhat egg-shaped; egg-shaped in 3 dimensions; *cf.* ovate.

ovomucoid *n.* [L. *ovum*, egg; *mucus*, mucus; Gk. *eidos*, form] a mucoid in white of egg.

ovoplasm ooplasm, *q.v.*

ovotestis *n.* [L. *ovum*, egg; *testis*, testicle] the

hermaphrodite reproductive gland of certain gastropods; *cf.* ovariotestis.

ovovitellin *see* vitellin.

ovoviviparous *a.* [L. *ovum*, egg; *vivus*, living; *parere*, to bring forth] *pert.* forms which produce an egg with persistent membranes, which hatches in maternal body; *cf.* oviparous, viviparous.

ovular *a.* [*Dim.* of L. *ovum*, egg] like or *pert.* an ovule.

ovulary *n.* [L. *ovum*, egg] ovary in plants.

ovulate *a.* [L. *ovum*, egg] containing an egg or ovule *v.* To emit egg or eggs from ovary or ovarian follicles.

ovulation *n.* [L. *ovum*, egg; *latum*, borne away] the emission of the egg or eggs from the ovary.

ovulatory *a.* [L. *ovum*, egg; *latum*, borne away] *pert.* ovulation.

ovule *n.* [L. *ovum*, egg] a structure in seed plants which contains the megasporangium (nucellus), megaspore (embryo sac), a food store, and a coat, and develops into a seed after fertilization; used loosely for a small egg or egg-like structure; *alt.* seed bud.

ovuliferous *a.* [L. *ovum*, egg; *ferre*, to carry] ovule-producing; containing ovules; *appl.* scales, each bearing one or more ovules, developed on bract scales, as in Coniferae.

ovulophore a gynaecium bearing ovules.

ovum *n.* [L. *ovum*, egg] a female gamete; *alt.* egg; *plu.* ova.

Oweniimorpha *n.* [*Owenia*, generic name; Gk. *morphē*, form] an order of polychaetes whose body is composed of only a few long segments, and whose prostomium usually ends in a frilled organ for collecting food.

oxalates *n.plu.* [Gk. *oxys*, sharp] salts of oxalic acid, occurring as metabolic by-products in various plant tissues and in urine; also found in mantle of certain bivalves.

oxaloacetic acid an acid of the Krebs' cycle, which accepts acetyl groups from acetyl coenzyme A forming citric acid.

oxalosuccinic acid an acid of the Krebs' cycle which is decarboxylated to α-ketoglutaric acid in the presence of the enzyme oxalosuccinic decarboxylase.

oxea *n.* [Gk. *oxys*, sharp] a sponge spicule, rod-shaped and sharp at both ends.

oxeote *a.* [Gk. *oxys*, sharp] in form of a simple rod; like an oxea, *appl.* sponge spicules.

oxidases *n. plu.* [Gk. *oxys*, sharp] enzymes which catalyse oxido-reductions using molecular oxygen as an acceptor; *cf.* reductases.

oxidation *n.* [Gk. *oxys*, sharp] any process involving the addition of oxygen, loss of hydrogen, or loss of electrons from a compound; *cf.* reduction.

oxidation–reduction enzymes oxido-reductases, *q.v.*

oxidative phosphorylation making ATP using energy released by aerobic respiration; *cf.* photophosphorylation.

oxido-reductases enzymes which catalyse the oxidation of one compound with the reduction of another, including the dehydrogenases, oxidases, peroxidases, and catalases; *alt.* oxidation–reduction enzymes.

oxyaster *n.* [Gk. *oxys*, sharp; *astēr*, star] stellate sponge spicule with sharp-pointed rays.

oxybiotic *a.* [Gk. *oxys*, sharp; *biōtos*, means of life] living in presence of oxygen; *alt.* aerobic.

oxychlorocruorin *n.* [Gk. *oxys*, sharp; *chlōros*, green; L. *cruor*, blood] chlorocruorin combined with oxygen as in the aerated blood of certain polychaetes.

oxychromatin *n.* [Gk. *oxys*, sharp; *chrōma*, colour] *see* linin.

oxydactyl(ous) *a.* [Gk. *oxys*, sharp; *daktylos*, finger] having slender tapering digits.

oxydiact *a.* [Gk. *oxys*, sharp; *di-*, two; *aktis*, ray] having 3 rays with 2 fully developed, *appl.* sponge spicules.

oxyerythrocruorin *n.* [Gk. *oxys*, sharp; *erythros*, red; L. *cruor*, blood] erythrocruorin combined with oxygen as in the aerated blood of many annelids and molluscs.

oxygen debt a deficit in oxygen which builds up when an organism or tissue is working with inadequate oxygen supply, such as muscle tissue, which then consumes oxygen above the normal rate for some time, until the debt is repaid.

oxygen dissociation curve a graph of percentage saturation of haemoglobin with oxygen against concentration of oxygen, which gives information about dissociation of oxyhaemoglobin under different environmental conditions or in different animals; any graph showing dissociation of oxygen from a substance.

oxygenotaxis oxytaxis, *q.v.*; *a.* oxygenotactic.

oxygenotropism oxytropism, *q.v.*; *a.* oxygenotropic.

oxygen quotient (Q_{O_2}) the volume of oxygen in microlitres gas at normal temperature and pressure taken in per hour per milligram dry weight.

oxygnathous *a.* [Gk. *oxys*, sharp; *gnathos*, jaw] having more-or-less sharp jaws.

oxyhaemocyanin *n.* [Gk. *oxys*, sharp; *haima*, blood; *kyanos*, blue] haemocyanin combined with oxygen as in the aerated blood of many molluscs and arthropods.

oxyhaemoerythrin *n.* [Gk. *oxys*, sharp; *haima*, blood; *erythros*, red] haemoerythrin combined with oxygen as in the aerated blood of some annelids.

oxyhaemoglobin (HbO₂) *n.* [Gk. *oxys*, sharp; *haima*, blood; L. *globus*, globe] haemoglobin combined with oxygen, formed where the concentration of oxygen is high, and dissociating into its component parts where the concentration of oxygen is low.

oxyhexactine *n.* [Gk. *oxys*, sharp; *hex*, six; *aktis*, ray] a hexactine with rays ending in sharp points.

oxyhexaster *n.* [Gk. *oxys*, sharp; *hex*, six; *astēr*, star] a hexaster with rays ending in sharp points.

oxyluciferin *n.* [Gk. *oxys*, sharp; L. *lux*, light; *ferre*, to carry] the substance formed by action of luciferase on luciferin, emitting light in photogenic organs.

oxyneurine betaine, *q.v.*

oxyntic *a.* [Gk. *oxynein*, to sharpen] acid-secreting.

oxyntic cells cells in the gastric gland of stomach which secrete hydrochloric acid; *alt.* parietal cells, HCl cells, acid gland; *cf.* peptic cells.

oxyphil *a.* [Gk. *oxys*, sharp; *philein*, to love] having strong affinity for acidic stains; *alt.* oxyphilic. *n.* Oxyphil cell or tissue element; *alt.* acidophil.

oxyphilous *a.* [Gk. *oxys*, sharp; *philein*, to love] tolerating only acid soils and substrates; *alt.* basifuge.

oxyphobe *a.* [Gk. *oxys*, sharp; *phobos*, flight] unable to tolerate soil acidity.

oxyphyte *n.* [Gk. *oxys*, sharp; *phyton*, plant] a plant thriving on acid soils; *alt* calcifuge, oxytophyte.

oxysome *n.* [Gk. *oxys*, sharp; *sōma*, body] the functional unit of electron transfer and oxidative phosophorylation.

oxytaxis *n.* [Gk. *oxys*, sharp; *taxis*, arrangement] a taxis in response to the stimulus of oxygen; *alt.* oxygenotaxis; *cf.* oxytropism; *a.* oxytactic.

oxytocic *a.* [Gk. *oxys*, sharp; *tokos*, birth] accelerating parturition, *appl.* pituitary hormone: oxytocin, *q.v.*

oxytocin *n.* [Gk. *oxys*, sharp; *tokos*, birth] polypeptide hormone secreted by posterior lobe of pituitary gland which, in mammals, induces contraction of smooth muscle, particularly uterine muscle; *alt.* α-hypophamine, pitocin.

oxytophyte oxyphyte, *q.v.*

oxytropism *n.* [Gk. *oxys*, sharp; *tropē*, turn] tendency of organisms or organs to be attracted by oxygen; *alt.* oxygenotropism; *cf.* oxytaxis.

oxytylote *n.* [Gk. *oxys*, sharp; *tylos*, knob] a slender straight sponge spicule, sharp at one end, knobbed at the other.

ozonium *n.* [Gk. *ozos*, twig] barren mycelium; a dense mycelium, as at base of a stipe.

P

P₁ denoting 1st parental generation, P₂ the grandparents, etc., in Mendel's laws; *cf.* F₁.

PABA para-amino-benzoic acid, *q.v.*

Pacchionian bodies [*A. Pacchioni*, Italian anatomist] eminences of subarachnoid tissue covered by arachnoid membrane and pressing into dura mater; *alt.* arachnoidal granulations, glandulae Pacchionii, villi.

pacemaker a part or region determining rate of activity in other parts of the body; the sinu-auricular node, which initiates the normal heart beat.

pachycarpous *a.* [Gk. *pachys*, thick; *karpos*, fruit] with a thick pericarp.

pachycaul(ous) *a.* [Gk. *pachys*, thick; *kaulos*, stem] with thick or massive primary stem and root.

pachycladous *a.* [Gk. *pachys*, thick; *klados*, young shoot] with thick shoots.

Pachycormiformes *n.* [Gk. *pachys*, thick; *kormos*, trunk; L. *forma*, shape] an order of Jurassic to Cretaceous holosteans having a long snout like a swordfish.

pachydermatous *a.* [Gk. *pachys*, thick; *derma*, skin] with thick skin or covering.

pachymeninx *n.* [Gk. *pachys*, thick; *meningx*, membrane] the dura mater, *q.v.*

pachynema *n.* [Gk. *pachys*, thick; *nēma*, thread] chromosome thread at the pachytene stage.

pachynesis *n.* [Gk. *pachynesis*, thickening] thickening, as of mitochondria.

pachynosis *n.* [Gk. *pachynesis*, thickening] growth in thickness, as of plants.

pachyphyllous *a.* [Gk. *pachys*, thick; *phyllon*, leaf] thick-leaved.

pachytene *a.* [Gk. *pachys*, thick; *tainia*, band] *appl.* prophase stage in meiosis during which homologous chromosomes are associated as bivalents.

pacifarins *n.plu.* [L. *pacifacare*, to make peace] a group of ecological ectocrines, distinct from vitamins and antibiotics, which limit infectious diseases.

Pacinian bodies or corpuscles [F. *Pacini*, Italian anatomist] distal nerve endings, consisting of lamellated connective tissue capsule with core of nucleated protoplasmic cells containing ramifications of a medullated nerve fibres, thought to be sensitive to pressure; *alt.* corpusculum lamellosum, lamellated corpuscles, Vater's corpuscles.

paedogamy *n.* [Gk. *pais*, child; *gamos*, marriage] type of autogamy in Protozoa where gametes are formed after multiple division of nucleus; *alt.* pedogamy.

paedogenesis *n.* [Gk. *pais*, child; *genesis*, descent] reproduction in young or larval stages, as axolotl; spore production in immature fungi; *alt.* pedogenesis; *cf.* neoteny.

paedomesoblast *n.* [Gk. *pais*, child; *mesos*, middle; *blastos*, bud] portions of primitive mesoblast destined to form transitory larval structures.

paedomorphosis *n.* [Gk. *pais*, child; *morphōsis*, form] evolutionary change in which primitive or embryonic structures appear in adult animals; *a.* paedomorphic.

Paenungulata *n.* [L. *paene*, nearly; *ungula*, hoof] a superorder of ungulates containing the orders Hyracoidea, Proboscoidea, Pantodonta, Dinocerata, Pyrotheria, Embrithopoda, and Sirenia.

paired bodies small bodies lying close to sympathetic chain in Elasmobranchii, representing the adrenal medulla.

paired fins pectoral and pelvic fins of fishes.

pairing process of attraction between homologous chromosomes during zygotene.

Palade granules ribosomes, *q.v.*

Palaeacanthocephala *n.* [Gk. *palaios*, ancient; *akantha*, thorn; *kephalē*, head] an order of Acanthocephala which are parasitic mainly in aquatic animals and have alternating radial rows of proboscis spines.

Palaearctic *a.* [Gk. *palaios*, ancient; *Arktos*, Great Bear] *appl.* or *pert.* zoogeographical region, or subregion of the holarctic region, including Europe, North Africa, Western Asia, Siberia, northern China, and Japan.

palaeo- *alt.* paleo-.

palaeobiology *n.* [Gk. *palaios*, ancient; *bios*, life; *logos*, discourse] biology of extinct plants and animals.

palaeobotany *n.* [Gk. *palaios*, ancient; *botanē*, pasture] botany of fossil plants and plant impressions; *alt.* palaeophytology.

Palaeocene *a.* [Gk. *palaios*, ancient; *kainos*, recent] *appl.* and *pert.* the earliest epoch of the Tertiary period, before Eocene.

palaeocerebellum *n.* [Gk. *palaios*, ancient; L. *dim.* of *cerebrum*, brain] phylogenetically older region of cerebellum, receiving spinal and vestibular afferent fibres; *cf.* neocerebellum.

palaeocranium *n.* [Gk. *palaios*, ancient; *kranion*, skull] type of skull or stage in development extending no further back than vagus nerve.

palaeodendrology *n.* [Gk. *palaios*, ancient; *dendron*, tree; *logos*, discourse] botany of fossil trees and tree impressions.

palaeo-ecology *n.* [Gk. *palaios*, ancient; *oikos*, household; *logos*, discourse] the study of the relationship between extinct organisms and their lifetime environment.

palaeo-encephalon *n.* [Gk. *palaios*, ancient; *engkephalos*, brain] the segmental or primitive vertebrate brain.

Palaeogaea *n.* [Gk. *palaios*, ancient; *gaia*, the earth] the area comprising the Palaearctic, Ethiopian, Indian, and Australian zoogeographical regions.

Palaeogene *a.* [Gk. *palaios*, ancient; *genos*, an age] *pert.* the early Tertiary period, Eocene and Oligocene.

palaeogenetic *a.* [Gk. *palaios*, ancient; *genesis*, descent] *appl.* atavistic features fully developed, which are usually characteristically embryonic; *alt.* recapitulatory.

palaeogenetics *n.* [Gk. *palaios*, ancient; *genesis*, descent] genetics as applied to palaeontology; genetic interpretation of fossil structures or species.

Palaeognathae *n.* [Gk. *palaios*, ancient; *gnathos*, jaw] an order or superorder of birds including the large running birds such as the ostrich, rhea, and emu, most being large, heavy, and unable to fly, with small wings and no keel to sternum; *alt.* Ratitae.

Palaeolaurentian *a.* [Gk. *palaios*, ancient; River St *Lawrence*] *pert.* or *appl.* Archaeozoic era.

Palaeolithic *a.* [Gk. *palaios*, ancient; *lithos*, stone] *appl.* or *pert.* the Older or chipped Stone Age, characterized by a hunter–gatherer economy and rough chipped stone tools.

Palaeonemertini, palaeonemertines *n.*, *n.plu.* [Gk. *palaios*, ancient; *Nēmertēs*, a nereid] an order of Anopla having the body wall muscles in 2 or 3 layers, where 3 layers, the innermost is circular.

Palaeonisciformes *n.* [*Palaeoniscus*, generic name; L. *forma*, shape] a subgroup of palaeoniscids including only fossil members from the Devonian to Cretaceous times.

Palaeonisciodei, palaeoniscids *n.*, *n.plu.* [*Palaeoniscus*, generic name; Gk. *eidos*, form] a group of actinopterygians, existing from the Devonian to the present day but most extinct, including the bichirs of the Nile, which are carnivorous with large sharp teeth and usually a very heterocercal tail.

palaeontology *n.* [Gk. *palaios*, ancient; *on*, being; *logos*, discourse] the science of past organic life, based on fossils and fossil impressions; *alt.* oryctics, oryctology; *cf.* neology.

palaeophytology palaeobotany, *q.v.*

Palaeoptera *n.* [Gk. *palaios*, ancient; *pteron*, wing] a group of winged insects in which metamorphosis is incomplete, the nymphs aquatic, and the wings do not fold over the body in repose.

Palaeopterygii *n.* [Gk. *palaios*, ancient; *pterygion*, little wing or fin] a subclass of Osteichthyes in some classifications, including the lower bony fish, the Cladistia and Chondrostei; cf. Neopterygii.

palaeospecies n. [Gk. palaios, ancient; L. species, particular kind] a group of extinct organisms which are assumed to have been capable of interbreeding and so are placed in the same species.

Palaeospondylus n. [Gk. palaios, ancient; sphondylos, vertebra] a genus of primitive Devonian vertebrates known only from a single fossil from Scotland and thought to be a placoderm or cyclostome, in some classifications placed in the order Palaeospondyliformes.

Palaeostraca n. [Gk. palaios, ancient; ostrakon, shell] a group of arthropods including the king crabs, eurypterids, and trilobites.

Palaeotropical a. [Gk. palaios, ancient; tropikos, pert. tropics] appl. or pert. floristic region including African, Indo-Malaysian, and Polynesian subregions.

palaeovirus n. [Gk. palaios, ancient; L. virus, poison] a virus evolved from a more-or-less remote viroid ancestor; cf. neovirus.

Palaeozoic a. [Gk. palaios, ancient; zōon, animal] appl. era comprising the Proterozoic and Deuterozoic faunal epochs, preceding the Mesozoic era; Cambrian to Permian periods; alt. Primary era.

palaeozoology n. [Gk. palaios, ancient; zōon, animal; logos, discourse] zoology of fossil animals and animal impressions.

palama n. [Gk. palamē, the palm] foot-webbing of aquatic birds.

palatal a. [L. palatum, palate] pert. palate, appl. bone, sinus, etc.; alt. palatine.

palate n. [L. palatum, palate] roof of mouth in vertebrates; insect epipharynx; projection of lower lip of personate corolla.

palatine a. [L. palatum, palate] pert. or in region of palate, appl. artery, bone, foramen, nerves, suture, tonsil; alt. palatal.

palatoglossal a. [L. palatum, palate; Gk. glōssa, tongue] pert. palate and tongue; appl. a muscle: glossopalatinus, q.v.

palatonasal a. [L. palatum, palate; nasus, nose] pert. palate and nose.

palatopharyngeal a. [L. palatum, palate; Gk. pharyngx, pharynx] pharyngopalatine, q.v.

palatopterygoid a. [L. palatum, palate; Gk. pteryx, wing; eidos, form] pert. palate and pterygoid.

palatoquadrate a. [L. palatum, palate; quadratus, squared] connecting palatine and quadrate, appl. dorsal cartilage of mandibular arch.

pale n. [L. palea, chaff] palea, q.v.; upper or inner pale: palea, q.v.; lower or outer pale: lemma, q.v.

palea n. [L. palea, chaff] the upper of 2 membranous bracts enclosing the flower in grasses; alt. glumella, inner glume, pale, palet, upper or inner pale or palea, squamella, valvule; lower palea: lemma, q.v.; a small bract on floret of Compositae; ramentum of ferns.

paleaceous a. [L. palea, chaff] chaffy; appl. a capitulum furnished with small scaly bracts or paleae.

paleo- palaeo-, q.v.

paleola n. [L. palea, chaff] lodicule, q.v.

palet palea, of grasses.

palette n. [F. palette from L. pala, spade] the modified cupule-bearing tarsus of anterior leg, in male beetles.

pali n.plu. [L. palus, stake] a series of small pillars projecting upwards from the theca base towards stomadaeum of madrepore corals; sing. palus, q.v.

paliform a. [L. palus, stake; forma, shape] like an upright stake.

palinal a. [Gk. palin, reversely] from behind forwards, appl. jaw movement, as in elephants; cf. proral, orthal.

palingenesis n. [Gk. palin, anew; genesis, descent] abrupt metamorphosis; rebirth of ancestral characters; recapitulation, q.v.

palingenetic a. [Gk. palin, anew; genesis, descent] of remote or ancient origin; pert. palingenesis.

palisade n. [F. palissade, from L. palus, stake] arrangement of apposed elongated cellular structures; appl. fungi, the Basidiomycetes; appl. cells, of ends of cortical hyphae in lichens; appl. tissue, the layer or layers of photosynthetic cells beneath the epidermis of many foliage leaves; appl. nerve fibrils in inner surface of electric layer in ray-fish; appl. tissue derived from neurilemma at neuromuscular junction in end plates.

pallaesthesia n. [Gk. pallein, to quiver; aisthēsis, sensation] vibratory sensation, alt. palmaesthesia; bone sensibility.

pallet n. [L. pala, spade] a shelly plate on a bivalve siphon.

pallial a. [L. pallium, mantle] pert. molluscan pallium or mantle, appl. line, groove, sinus, muscles, ganglion.

palliate a. [L. pallium, mantle] having a mantle or similar structure.

pallidum n. [L. pallidus, pale] the median portion of the lentiform nucleus; alt. globus pallidus.

palliopedal a. [L. pallium, mantle; pes, foot] pert. molluscan mantle and foot.

pallioperitoneal a. [L. pallium, mantle; Gk. periteinein, to stretch around] appl. complex in some molluscs including heart, renal organs, gonads, and ctenidia.

pallium n. [L. pallium, mantle] a mollusc or brachiopod mantle; portion of cerebral wall.

palmaesthesia n. [Gk. palmos, quivering; aisthēsis, sensation] see pallaesthesia.

palmar a. [L. palma, palm of hand] pert. palm of hand, appl. aponeurosis, nerve, muscle, reflex; alt. thenal.

palmaria n.plu. [L. palmaris, pert. palm] the 3rd brachials of Crinoidea.

palmate a. [L. palma, palm] appl. leaves divided into lobes arising from a common centre; appl. hand-like tuber, as in certain orchids; appl. folds of cervix uteri; having anterior toes webbed, as in most aquatic birds.

palmatifid a. [L. palma, palm; findere, to cleave] appl. leaves divided into lobes to about the middle, at acute angles to each other.

palmatilobate a. [L. palma, palm; lobus, lobe] palmate with rounded lobes and divisions halfway to base.

palmatipartite a. [L. palma, palm; partitus, divided] palmate with divisions more than halfway to base.

palmatisect a. [L. palma, palm; sectus, cut] palmate with divisions nearly to base.

palmella n. [Gk. palmos, quivering] a sedentary stage of certain algae, the cells dividing within a jelly-like mass and producing motile gametes; a. palmelloid.

palmigrade

palmigrade plantigrade, *q.v.*

palmiped *a.* [L. *palma*, palm; *pes*, foot] web-footed. *n.* A web-footed bird.

palmitic acid a common fatty acid.

palmitin *n.* [Gk. *palma*, palm tree] a fat occurring in adipose tissue, milk, and palm oil.

palmula *n.* [L. *palma*, palm] terminal lobe or process between paired claws of insect feet.

palp palpus, *q.v.*

palpacle *n.* [L. *palpare*, to touch softly] the tentacle of a dactylozooid (palpon) of Siphonophora.

palpal *a.* [L. *palpare*, to stroke] *pert.* a palpus.

palpate *a.* [L. *palpare*, to stroke] provided with palpus or palpi, *v.* To examine by touch.

palpebra *n.* [L. *palpebra*, eyelid] an eyelid; *plu.* palpebrae.

palpebral *pert.* eyelids, *appl.* arteries, ligament, nerves, etc.; *appl.* a lobe on which the eye of trilobites rests.

palpi *n.plu.* [L. *palpare*, to stroke] labial feelers of insects; sensory appendages on prostomium of polychaetes or on head of some arthropods; pedipalps, *q.v.*; *alt.* palps; *sing.* palpus; *a.* palpal.

palpifer, palpiger *n.* [L. *palpare*, to stroke; *ferre, gerere*, to carry] in insects a lobe of maxilla or prementum bearing a palp.

palpiform *a.* [L. *palpare*, to stroke; *forma*, shape] resembling a palp.

Palpigradi, Palpigrada *n.* [L. *palpare*, to stroke; *gradus*, step] an order of very small arachnids having a long jointed flagellum on the last segment of the opisthosoma.

palpimacula *n.* [L. *palpare*, to stroke; *macula*, spot] sensory area on labial palps of certain insects.

palpocil *n.* [L. *palpare*, to touch; *cilium*, eyelash] a stiff sensory filament attached to sense cells of some coelenterates.

palpon *n.* [L. *palpare*, to stroke] dactylozooid, *q.v.*

palpulus *n.* [L. *palpare*, to stroke] a small palp of feeler.

palpus *sing.* of palpi, *q.v.*

paludal *a.* [L. *palus*, marsh] marshy; *pert.*, or growing in, marshes or swamps; *alt.* paludine, paludinous, paludose, palustral, palustrine, uliginose, uliginous.

paludicole *a.* [L. *palus*, marsh; *colere*, to inhabit] living in marshes; *alt.* paludal, palustral.

palule *n.* [L. *palus*, stake] an unattached calcareous process of corals; a small palus.

palus *n.* [L. *palus*, stake] a stake-like structure; *plu.* pali, *q.v.*

palustral paludal, *q.v.*; paludicole, *q.v.*

palynology *n.* [Gk. *palynein*, to scatter (*palē*, pollen); *logos*, discourse] the study of pollen and of its distribution, *alt.* pollen analysis; the study of spores.

pampiniform *a.* [L. *pampinus*, tendril; *forma*, shape] tendril-like; *appl.* a convoluted vein plexus of spermatic cord; *appl.* body: the parovarium, *q.v.*

pamprodactylous *a.* [Gk. *pan*, all; *pro*, in front; *daktylos*, digit] with all toes pointing forward.

Pancarida *n.* [Gk. *pan*, all; L. *caris*, a kind of crab] a superorder of malacostracans containing the order Thermosbaenacea, *q.v.*

pancolpate *a.* [Gk. *pan*, all; *kolpos*, fold] of pollen, having many furrows.

pancreas *n.* [Gk. *pan*, all; *kreas*, flesh] a compound racemose gland, with exocrine and endocrine functions, of most vertebrates.

pancreatic *a.* [Gk. *pan*, all; *kreas*, flesh] *pert.* pancreas, *appl.* artery, duct, vein, enzymes, hormones, juice.

pancreatic juice a digestive juice secreted by the pancreas into the gut and containing several enzymes and enzyme precursors including trypsinogen, chymotrypsinogen, amylase, and lipolytic enzymes; *alt.* pancreatin.

pancreaticoduodenal *pert.* pancreas and duodenum, *appl.* arteries, veins.

pancreatin pancreatic juice, *q.v.*

pancreatrophic *a.* [Gk. *pan*, all; *kreas*, flesh; *trophē*, nourishment] *appl.* prepituitary hormone causing increase in secretion of insulin.

pancreozymin *n.* [Gk. *pan*, all; *kreas*, flesh; *zymē*, leaven] duodenal hormone which stimulates production of pancreatic enzymes.

Pandales *n.* [*Panda*, generic name] an order of dicots (Archichlamydeae) having cyclic dioecious flowers with a superior ovary made of 3 carpels which develops into a drupe.

Pandanales *n.* [*Pandanus*, generic name] an order of monocots (Arecidae) used in different ways by different authorities, but always including the family Pandanaceae, e.g. the screw pines.

pandemic *a.* [Gk. *pandēmos*, common] epidemic everywhere; very widely distributed, *alt.* cosmopolitan.

Pander, nucleus of [*K. G. Pander*, Russian zoologist] the white yolk underlying the blastodisc in a bird's egg, the column forming neck of latebra being termed Pander's isthmus.

panduriform *a.* [Gk. *pandoura*, lute; L. *forma*, shape] fiddle-shaped, *appl.* leaves.

Paneth cells [*J. Paneth*, Austrian physician] enzyme-producing cells at base of crypts of Lieberkühn.

Pangaea *n.* [Gk. *pan*, all; *gaia*, earth] the supercontinent made of our present continents fitted together, which is thought to have been made and later split up in past geological ages due to continental drift.

pangamic *a.* [Gk. *pan*, all; *gamos*, marriage] *appl.* indiscriminate or random mating; *n.* pangamy.

pangen *n.* [Gk. *pan*, all; *genos*, offspring] hypothetical unit, *q.v.*

pangenesis *n.* [Gk. *pan*, all; *genesis*, descent] the gemmule theory, that hereditary characteristics are carried by germs from individual body cells.

panicle *n.* [L. *panicula*, tuft] a branched racemose inflorescence; often applied more widely to any branched inflorescence.

paniculate *a.* [L. *panicula*, tuft] having flowers arranged in panicles.

panmeristic *a.* [Gk. *pan*, all; *meros*, part] *appl.* an ultimate protoplasmic structure of independent units.

panmictic *a.* [Gk. *pan*, all; *miktos*, mixed] characterized by, or resulting from, random matings; *pert.* panmixia.

panmixia *n.* [Gk. *pan*, all; *mixis*, mixing] indiscriminate interbreeding consequent on suspension of influence of natural selection.

panniculus *n.* [L. *dim.* of *pannus*, cloth] a layer of tissue, as superficial fascia.

panniculus carnosus a thin layer of muscle fibres inserted in dermis and moving or twitching the skin.

pannose *a.* [L. *pannosus*, from *pannus*, cloth] like cloth.

panoistic *a.* [Gk. *pan*, all; *ōon*, egg] *appl.* ovariole in which nutritive cells are absent, egg yolk being formed by epithelium of follicle; *cf.* meroistic.

panphotometric *a.* [Gk. *pan*, all; *phōs*, light; *metron*, measure] *appl.* leaves oriented to avoid maximum direct sunlight; *cf.* euphotometric.

panphytotic *n.* [Gk. *pan*, all, *phyton*, plant] a pandemic affecting plants.

panspermia *n.* [Gk. *pan*, all; *sperma*, seed] a theory popular in the 19th century, that life did not evolve from inorganic matter on earth, but reached as fully developed in the form of a bacterial spore which had escaped from a distant planet.

pansporoblast *n.* [Gk. *pan*, all; *sporos*, seed; *blastos*, bud] a cell complex, of Neosporidia, producing sporoblasts and spores.

panthalassic *a.* [Gk. *pan*, all; *thalassa*, sea] living both in coastal and offshore waters, neritic and oceanic.

panting centre a hypothalamic region whose stimulation causes the rate of respiration to quicken.

Pantodonta, pantodonts *n.*, *n.plu.* [Gk. *panto-*, all-; *odous*, tooth] a varied order of Palaeocene to Oligocene North American and Asian herbivorous placental mammals.

pantonematic *a.* [Gk. *panto-*, all-; *nēma*, thread] *appl.* flagella that have longitudinal rows of fine hairs along their axis.

Pantopoda *n.* [Gk. *panto-*, all-; *pous*, foot] Pycnogonida, *q.v.*

pantostomatic *a.* [Gk. *panto-*, all-; *stoma*, mouth] capable of ingesting food at any part of the surface, as amoeboid organisms.

pantothenic acid [Gk. *pantothen*, from everywhere] an acid of the vitamin B complex, sometimes called vitamin B_3 or B_5, that is found in all living tissues, often combined as in coenzyme A, and is necessary for growth in various animals, and is the rat anti-grey hair and chick antidermatitis factor.

Pantotheria, pantotheres *n.*, *n.plu.* [Gk. *panto-*, all-; *thērion*, small animal] an order of Jurassic trituberculate mammals, thought to be the ancestors of the living therians, having molar teeth showing the basic pattern found in living forms.

pantropic *a.* [Gk. *pan*, all; *tropikos*, turning] turning to any direction; invading many different tissues, *appl.* viruses; *alt.* polytropic.

pantropical distributed throughout the tropics, *appl.* species.

papain *n.* [Sp. *papaya*, papaw] an endopeptidase found in the fruit juice and leaves of pawpaws (papayas) *Carica papaya*, and used commercially as a meat tenderizer; EC 3.4.22.2; *r.n.* papain.

Papaverales *n.* [*Papaver*, generic name] an order of dicots (Magnoliidae) formerly included with the Capparales in the Rhoeadales, having a dimerous cyclic calyx and corolla and an ovary with parietal placentation, and including the families Papaveraceae (poppies) and Fumariaceae (fumitory).

papilionaceous *a.* [L. *papilio*, butterfly] resembling a butterfly; *appl.* a corolla of 5 petals, one enlarged posterior standard or vexillum, 2 united anterior forming a keel or carina, and 2 lateral, the wings or alae, as in the family Papilionaceae; *see* Fabales.

papilla *n.* [L. *papilla*, nipple] a glandular hair with one secreting cell above the epidermis level; an accessory adhesive organ with retractile tip, of some trematodes; a conical dermal structure on birds, the beginning of a feather; one of various small projections of corium of tongue, and eminences on skin; a conical structure, as nipple, apex of renal pyramid, lacrimal papilla, etc.

papillary *a.* [L. *papilla*, nipple] *pert.* or with papillae; *appl.* a dermal layer; *appl.* a process of caudate lobe of liver; *appl.* muscles between walls of ventricles of heart and chordae tendineae.

papillate *a.* [L. *papilla*, nipple] covered by papillae, *alt.* papillose; like a papilla; *appl.* petals with external cells projecting slightly above surface.

papilliform *a.* [L. *papilla*, nipple; *forma*, shape] like a papilla in shape.

papillose *see* papillate.

papovaviruses *n.plu.* [*pa*pilloma virus, *po*lyoma virus, *va*cuolating *virus*] a group of viruses containing double-stranded DNA, including those causing warts in man.

pappiferous *a.* [L. *pappus*, down; *ferre*, to carry] pappus-bearing.

pappose, pappous *a.* [L. *pappus*, down] having limb of calyx developed as a tuft of hairs or bristles (pappus); downy, or covered with feathery processes.

pappus *n.* [L. *pappus*, down] a circle or tuft of bristles, hairs, or feathery processes in place of a calyx.

Papuan *appl.* subregion of Australian zoogeographical region: New Guinea and islands westward to Wallace's line.

papulae *n.plu.* [L. *papula*, pimple] hollow contractile skin processes of some echinoderms such as Asteroidea, with respiratory function, *alt.* dermal gills, skin gills; pustules on skin.

papyraceous, papyritious *a.* [L. *papyrus*, papyrus rush] chartaceous. *q.v.*

para-amino-benzoic acid (PABA) a substance which has a wide distribution in living organisms and is ingested as part of the folic acid molecule, does not seem to be required by humans but affects hair pigmentation in rats.

para-aortic *a.* [Gk. *para*, beside; *aortē*, great artery] *appl.* chromaffin bodies or paraganglia alongside the abdominal aorta.

parabasal *a.* [Gk. *para*, beside; *basis*, base] *appl.* a striated apparatus surrounding the calyx of certain protozoans; *appl.* body or granule in flagellates: kinetonucleus, *q.v.*

parabasalia *n.plu.* [Gk. *para*, beside; *basis*, base] the basalia of crinoids when a circlet of perradial infrabasalia occurs beneath them; *sing.* parabasal.

parabiosis *n.* [Gk. *para*, beside; *biōsis*, manner of life] the condition of being conjoined either from birth, as Siamese twins, or experimentally as laboratory animals; phylacobiosis, *q.v.*; temporary inhibition of activity.

parabiotic *a.* [Gk. *para*, beside; *bios*, life] conjoined to greater or lesser extent; phylacobiotic, in ants; living amicably in compound nest, as ants of different species or genera.

parablast *n.* [Gk. *para*, beside; *blastos*, bud] the yolk of meroblastic eggs; large nuclei of cells laden with yolk granules, in development of higher mammals; *a.* parablastic.

parabranchia *n.* [Gk. *para*, beside; *brangchia*, gills] a much-plumed mollusc osphradium.

parabronchi *n.plu.* [Gk. *para*, beside; *brongchos*, windpipe] the tertiary lung tubes of birds, their terminations being embedded in lung mesenchyme.

Paracanthopterygii *n.* [Gk. *para*, beside; *akantha*, thorn; *pterygion*, little wing or fin] an advanced group of teleosts existing from the Eocene to the present day and including the cod.

paracardial *a.* [Gk. *para*, beside; *kardia*, stomach] near, or surrounding cardia or neck of stomach, *appl.* lymph glands.

paracardo *n.* [Gk. *para*, beside; L. *cardo*, hinge] part of the cardo of insect maxilla.

paracasein *see* casein.

paracele paracoel, *q.v.*

paracentral *a.* [Gk. *para*, beside; L. *centrum*, centre] situated near the centre, *appl.* lobule. gyrus, fissure; *appl.* retinal area surrounding fovea centralis.

paracentric *a.* [Gk. *para*, beside; *kentron*, centre] on same side of centromere; *appl.* rearrangements in same chromosome arm; *appl.* inversions not including the centromere; *alt.* homobrachial; *cf.* pericentric.

parachordal *a.* [Gk. *para*, beside; *chordē*, cord] on either side of notochord; *appl.* paired horizontal cartilage plates on sides of chondrocranium.

parachromatin *n.* [Gk. *para*, beside; *chrōma*, colour] achromatic nuclear substance giving rise to spindle fibres; hypothetical chromatin said to be sensitive to changes in cellular environment and so capable of controlling normal development; *cf.* orthochromatin.

parachrosis *n.* [Gk. *para*, proceeding from; *chrōs*, colour; *parachroos*, changing colour] process or condition of changing colour; discoloration; fading.

parachute *n.* [F. *parer*, from L. *parare*, to prepare; F. *chute*, fall] a special structure of seeds as aril, caruncle, pappus, wing, which assists dispersal; a patagium, *q.v.*

paracme *n.* [Gk. *parakmē*, decadence] the decline of a species or race after reaching highest point of development, *alt.* phylogenetic period; the declining or senescent period in the life history of an individual after the adult stage; *cf.* epacme, acme.

paracoel(e) *n.* [Gk. *para*, beside; *koilos*, hollow] lateral ventricle or cavity of cerebral hemisphere; *alt.* paracele.

paracondyloid *a.* [Gk. *para*, beside; *kondylos*, knuckle; *eidos*, form] *appl.* process of occipital occurring beside condyles of some mammals.

paracone *n.* [Gk. *para*, beside; *kōnos*, cone] antero-external cusp of upper molar tooth.

paraconid *n.* [Gk. *para*, beside; *kōnos*, cone] antero-internal cusp of lower molar tooth.

paracorolla *n.* [Gk. *para*, beside; L. *corolla*, small crown] a corolla appendage such as a corona.

Paracrinoidea *n.* [Gk. *para*, beside; *krinon*, lily; *eidos*, form] an extinct class of Ordovician echinoderms of the subphylum Pelmatozoa with crinoid and cystoid features.

Paractinopoda *n.* [Gk. *para*, beside; *aktis*, ray; *pous*, foot] Apoda (sea cucumbers) *q.v.*

paracutis *n.* [Gk. *para*, beside; L. *cutis*, skin] a fungal cutis consisting of more-or-less isodiametric cells.

paracymbium *n.* [Gk. *para*, beside; *kymbion*, L. *cymbium*, small cup] accessory part of cymbium, between tibia and tarsus, in some spiders.

paracyst *n.* [Gk. *para*, beside; *kystis*, bladder] the antheridium of the discomycete fungus *Pyronema*.

paracyte *n.* [Gk. *para*, beside; *kytos*, hollow] a modified cell extruded from embryonic tissue into yolk, as in some insects.

paracytic *a.* [Gk. *para*, beside; *kytos*, hollow] *appl.* stomata of a type in which subsidiary cells lie alongside the stoma parallel to the long axis of the guard cells; formerly called rubiaceous; *cf.* anisocytic.

paracytoids *n.plu.* [Gk. *para*, beside; *kytos*, hollow; *eidos*, shape] coherent minute chromatin pieces cast out from nuclei of embryonic tissue cells, with cytoplasmic envelope, into the blood, as in certain insects.

parademe *n.* [Gk. *para*, beside; *demas*, body] a secondary apodeme arising from edge of a sclerite.

paraderm *n.* [Gk. *para*, beside; *derma*, skin] a derm composed of isodiametric hyphae; the delicate limiting membrane of a pronymph.

paradermal *a.* [Gk. *para*, beside; *derma*, skin] *appl.* section cut parallel to the surface of a flat organ such as a leaf; *alt.* tangential.

paradesmus *n.* [Gk. *para*, beside; *desmos*, bond] secondary connection between centrioles outside nucleus in mitosis of flagellates; *alt.* paradesmore, paradesm.

paradidymis *n.* [Gk. *para*, beside; *didymos*, testicle] a body of convoluted tubules anterior to lower part of spermatic cord, representing posterior part of embryonic mesonephros; *alt.* Giraldès' organ.

para-esophageal para-oesophageal, *q.v.*

parafacialia *n.plu.* [Gk. *para*, beside; L. *facies*, face] narrow parts of head capsule between frontal suture and eyes, as in certain Diptera.

parafibula *n.* [Gk. *para*, beside; L. *fibula*, buckle] an accessory element outside fibula at proximal end, seen in some Lacertilia and young marsupials.

paraflagellum *n.* [Gk. *para*, beside; L. *flagellum*, whip] a subsidiary flagellum.

paraflocculus *n.* [Gk. *para*, beside; L. *floccus*, lock of wool] cerebellar lobule lateral to flocculus.

parafollicular cells C-cells, *q.v.*

parafrons *n.* [Gk. *para*, beside; L. *frons*, forehead] area between eyes and frontal suture in certain insects.

parafrontals *n.plu.* [Gk. *para*, beside; L. *frons*, forehead] the continuation of genae between eyes and frontal suture in insects.

paraganglia *n.plu.* [Gk. *para*, beside; *ganglion*, swelling] scattered cell clusters along aorta and in other parts of body, which are thought to secrete adrenaline and are chromophil or phaeochrome cells.

paragaster *n.* [Gk. *para*, beside; *gastēr*, stomach] a central cavity lined with choanocytes into which gastric ostia open, in sponges.

paragastric *a.* [Gk. *para*, beside; *gastēr*, stomach] *pert.* a paragaster, *appl.* passages or cavities in branches of sponge; *appl.* paired blind canals from infundibulum to oral cone of ctenophores.

paragastrula *n.* [Gk. *para*, beside; *gastēr*, stomach] stage of amphiblastula of sponge when flagellated cells are invaginated into dome of rounded cells.

parageneon *n.* [Gk. *para*, beside; *genos*, descent] a little-changing species but containing some aberrant genotypes.

paragenesis *n.* [Gk. *para*, beside; *genesis*, descent] hybrids' fertility with parent species but not between themselves; a subsidiary mode of reproduction.

paragenetic *a.* [Gk. *para*, beside; *genesis*, descent] *appl.* chromosome change influencing the expression rather than structure of a gene.

paraglenal *a.* [Gk. *para*, beyond; *glēnē*, socket] hypercoracoid, *q.v.*

paraglobulin *n.* [Gk. *para*, beside; L. *globus*, globe] globulin of blood serum; *alt.* fibroplastin.

paraglossa *n.* [Gk. *para*, beside; *glōssa*, tongue] an appendage on either side of the labium of insects, *alt.* labella; a paired cartilage of chondrocranium.

paraglossum median cartilaginous or bony prolongation of copula supporting the tongue, as in birds.

paraglycogen *n.* [Gk. *para*, beside; *glykys*, sweet; *genēs*, produced] a reserve carbohydrate in Protozoa, resembling glycogen; *alt.* zooamylon.

paragnatha *n.plu.* [Gk. *para*, beside; *gnathos*, jaw] paired, delicate, unjointed processes of maxilla of certain arthropods; the buccal denticles of certain polychaetes.

paragnathous *a.* [Gk. *para*, beside; *gnathos*, jaw] mandibles of equal length, *appl.* birds.

para-Golgi apparatus small constituents of cell, in spaces between parts of Golgi apparatus.

paragula *n.* [Gk. *para*, beside; L. *gula*, gullet] a region beside gula on insect head.

paragynous *a.* [Gk. *para*, beside; *gynē*, female] *appl.* antheridia lateral to oogonium, as in some Peronosporales.

paraheliode *n.* [Gk. *para*, against; *hēlios*, sun] a special arrangement of spines in certain cacti; *alt.* parasol.

paraheliotropism *n.* [Gk. *para*, against; *hēlios*, sun; *tropē*, turn] tendency of plants to turn edges of leaves towards intense illumination, thus protecting surfaces; *alt.* paraphototropism.

parahormone *n.* [Gk. *para*, beside; *hormaein*, to arouse] a substance which acts like a hormone but is a product of ordinary metabolism of cells.

parahypophysis *n.* [Gk. *para*, beside; *hypo*, under; *phyein*, to grow] vestigial structure below pituitary gland.

paralactic sarcolactic, *q.v.*

paralectotype *n.* [Gk. *para*, beside; *lektos*, chosen; *typos*, pattern] specimen of a series used to designate a species, which is later designated as a paratype.

paralimnic *a.* [Gk. *para*, beside; *limnē*, lake] *pert.* or inhabiting shore of lakes; *n.* paralimnion.

paralinin *n.* [Gk. *para*, beside; *linon*, linen thread] nuclear ground-substance.

parallel descent or evolution, parallelism evolution in a similar direction in different groups.

parallel flow the flow of 2 fluids in the same direction.

parallelinervate, parallelodrome *appl.* leaves with veins or nerves parallel.

parallelotropism orthotropism, *q.v.*

paralogy *n.* [Gk. *para*, beside; *logos*, discourse] similarities in anatomy that are not related to common descent or similar function.

paralutein *n.* [Gk. *para*, beside; L. *luteus*, golden-yellow] *appl.* epithelioid luteal cells of theca interna, as distinct from epithelial follicular luteal cells; *alt.* theca lutein.

paramastigote *a.* [Gk. *para*, beside; *mastix*, whip] having one long principal flagellum and a short accessory one, as certain Mastigophora.

paramastoid *a.* [Gk. *para*, beside; *mastos*, breast: *eidos*, form] beside the mastoid; *appl.* 2. paroccipital processes of exoccipitals; *appl.* a process projecting from the jugular process.

paramecin a toxin produced by the ciliate *Paramecium* containing the kappa particle, which enables it to kill sensitive *Paramecia*.

paramere *n.* [Gk. *para*, beside; *meros*, part] half of a bilaterally symmetrical structure; one of paired lobes exterior to penis in some insects.

paramesonephric ducts [Gk. *para*, beside; *mesos*, middle; *nephros*, kidney] Müllerian ducts, *q.v.*

parametrium *n.* [Gk. *para*, beside; *mētra*, womb] fibrous tissue partly surrounding uterus.

paramitome *n.* [Gk. *para*, beside; *mitos*, thread] interfilar substance of protoplasm.

paramitosis *n.* [Gk. *para*, beside; *mitos*, thread] nuclear division, as in Protozoa, in which the chromosomes are not regularly arranged on equator of spindle and tend to cohere at one end when separating.

paramorph *n.* [Gk. *para*, beside; *morphē*, form] any variant form or variety; a form induced by environmental factors without genetically produced changes; *cf.* phenocopy.

paramutation *n.* [Gk. *para*, beside; L. *mutare*, to change] the phenomenon where one allele (the paramutagenic allele) influences the expression of another (the paramutable allele) at the same locus when they are combined in a heterozygote.

paramutualism *n.* [Gk. *para*, beside; L. *mutuus*, reciprocal] facultative symbiosis.

paramylon *n.* [Gk. *para*, beside; *amylon*, starch] a substance allied to starch, occurring in certain algae and flagellates; *alt.* paramylum, zooamylon.

paramyosin *n.* [Gk. *para*, beside; *mys*, muscle] a protein, in filaments cross-linked to form ribbons, in unstriated muscle, as of molluscs.

paranasal *a.* [Gk. *para*, beside; L. *nasus*, nose] *appl.* air sinuses in maxilla, frontal, ethmoid, sphenoid, and palatine bones; *appl.* cartilage connecting transverse laminae of embryonic nasal capsule, as in some marsupials and rodents.

paranema *n.* [Gk. *para*, beside; *nēma*, thread] paraphysis of cryptogams; *plu.* paranemata.

paranemic *a.* [Gk. *para*, beside; *nēma*, thread] having spirals not interlocked, as in sister chromatids; *alt.* anorthospiral; *cf.* plectonemic, orthospiral.

Paraneoptera, Paraneuroptera *n.* [Gk. *para*, beside; *neos*, new; *neuron*, nerve; *pteron*, wing] a group of winged insects with incomplete metamorphosis, in which the wings are folded in repose, and that have few Malpighian tubules.

paranephric *a.* [Gk. *para*, beside; *nephros*, kidney] beside the kidney; *appl.* a fatty body behind renal fascia.

paranephrocyte *see* athrocyte.

paranephros *n.* [Gk. *para*, beside; *nephros*, kidney] adrenal body, *q.v.*

paranota *n.plu.* [Gk. *para*, beside; *nōton*, back] lateral expansions of arthropod notum or tergum, believed to have developed into wings during evolution of insects; *sing.* paranotum; *a.* paranotal.

Paranthropus *n.* [Gk. *para*, beside; *anthrōpos*, man] a genus of fossil hominids including the species *P. robustus*, from southern Africa, and subsequently renamed *Australopithecus robustus.*

paranuchal *a.* [Gk. *para*, beside; L. L. *nucha*, nape of neck] *appl.* bone on each side of nuchal bone of skull in placoderms.

paranuclein *n.* [Gk. *para*, beside; L. *nucleus*, kernel] pyrenin, *q.v.*

paranucleus *n.* [Gk. *para*, beside; L. *nucleus*, kernel] micronucleus, *q.v.*; a spherical mass of mitochondria, formerly nebenkern; an aggregation of mitochondria in the spermatid, destined to form axial filament envelope.

para-oesophageal *n.* [Gk. *para*, beside; *oisophagos*, gullet] *appl.* nerves connecting tritocerebrum with suboesophageal ganglion; *alt.* para-esophageal.

parapatric *a.* [Gk. *para*, beside; *patēr*, father] *appl.* distribution of taxa which meet in a very narrow zone of overlap.

parapet *n.* [It. *parare*, to guard; *petto*, breast] a circular fold of body wall below margin of disc in sea anemones.

paraphototropism paraheliotropism, *q.v.*

paraphyll(ium) *n.* [Gk. *para*, beside; *phyllon*, leaf] one of the branching chlorophyll-containing outgrowths arising between leaves or from their bases, in mosses; *alt.* stipule.

paraphysis *n.* [Gk. *para*, beside; *physis*, growth] a slender filamentous epidermal outgrowth occurring among sporogenous organs in cryptogams, *alt.* paranema; a protective or nutritive interascal hypha; a non-sexual hypha; basidiolum, *q.v.*; one of the marginal projections of the pygidium in Coccidae (mealy bugs); a non-nervous outgrowth on top of brain of nearly all vertebrates; *plu.* paraphyses.

paraphysoid *n.* [Gk. *paraphysis*, side shoot; *eidos*, form] tissue between asci; a modified cystidium; *alt.* pseudoparaphysis.

parapineal *a.* [Gk. *para*, beside; L. *pinea*, pine cone] *appl.* parietal organ of epiphysis, eye-like in cyclostomes and some reptiles, pineal body of other vertebrates.

paraplasm *n.* [Gk. *para*, beside; *plasma*, mould] vegetative or less active part of cytoplasm, *cf.* dynamoplasm; formerly used for ectoplasm; *a.* paraplastic.

paraplectenchyma *n.* [Gk. *para*, beside; *plektos*, twisted; *engchyma*, infusion] pseudoparenchyma, *q.v.*

parapleuron *n.* [Gk. *para*, beside; *pleuron*, side] episternum of metathorax, or of mesothorax and metathorax, in insects; parapteron of insects; *alt.* parapleurum.

parapodium *n.* [Gk. *para*, beside; *pous*, foot] a paired lateral locomotory structure on body segments of polychaetes; a process on basal segment of leg, as in Symphyla; lateral extension of foot,

for propulsion, as in Pteropoda and certain Nudibranchiata.

parapolar *a.* [Gk. *para*, beside; *polos*, pivot] beside the pole, *appl.* 1st 2 trunk cells in development of Rhombozoa.

parapophysis *n.* [Gk. *para*, beside; *apo*, from; *physis*, growth] a transverse process arising from a vertebral centrum.

parapostgenal *a.* [Gk. *para*, beside; L. *post*, after; *gena*, cheek] *appl.* thickened portion of occiput in insects.

paraproct *n.* [Gk. *para*, beside; *prōktos*, anus] a plate situated on each side of anus in Diplopoda and some insects; *alt.* podical plate.

paraprostate *n.* [Gk. *para*, beside; L. *pro*, before; *stare*, to stand] anterior bulbo-urethral glands; *alt.* superior Cowper's glands of Leydolph.

parapsid *a.* [Gk. *para*, beside; (*h*)*apsis*, arch] *appl.* skull with single vacuity, bounded by parietal, postorbital, and squamosal.

Parapsida *n.* [Gk. *para*, beside; (*h*)*apsis*, arch] a subclass of extinct Mesozoic reptiles in some classifications, including the Ichthyopterygia and Sauropterygia.

parapsidal *pert.* parapsis; *appl.* furrows or sutures between dorsal portion of mesonotum and the parapsides in Hymenoptera.

parapsides *plu.* of parapsis.

parapsis *n.* [Gk. *para*, beside; (*h*)*apsis*, arch] lateral portion of mesonotum, as in ants; *plu.* parapsides.

parapteron *n.* [Gk. *para*, beside; *pteron*, wing] a small sclerite like a shoulder lappet on the side of the mesothorax of some insects, especially the tegula of Hymenoptera and Lepidoptera; penna on humerus of birds; *alt.* parapterum.

parapyles *n.plu.* [Gk. *para*, beside; *pylēs*, gate] two accessory openings in certain developing Radiolaria.

paraquadrate *n.* [Gk. *para*, beside; L. *quadratus*, squared] squamosal, *q.v.*

pararectal *a.* [Gk. *para*, beside; L. *rectus*, straight] beside rectum, *appl.* fossa, lymph glands.

parasematic *a.* [Gk. *para*, beside; *sēma*, sign] *appl.* markings, structures, or behaviour tending to mislead or deflect attack by an enemy; *cf.* sematic.

paraseme *n.* [Gk. *para*, beside; *sēma*, sign] misleading appearance or markings, as an ocellus near tail of fishes.

Parasemionotiformes *n.* [*Parasemionotus*, generic name; L. *forma*, shape] an order of Triassic Chondrostei which are very close to the Holostei in structure.

paraseptal *a.* [Gk. *para*, beside; L. *septum*, partition] *appl.* cartilage more or less enclosing vomeronasal organ.

parasexual *a.* [Gk. *para*, compared with; L. *sexus*, sex] *appl.* or *pert.* the operation of genetic recombination other than by means of the alternation of karyogamy and meiosis characteristic of sexual reproduction.

parasite *n.* [Gk. *parasitos*, from *para*, beside; *sitos*, food] an organism which at some stage of its life history makes a connection with the tissues of an organism of a different species (the host) from which it derives its food, the host usually being harmed to some extent by the association; *a.* parasitic.

parasite chain a food chain passing from large to small organisms; *cf.* predator chain, saprophyte chain.

parasitic castration castration caused by presence of a parasite, as in male crabs infested by the barnacle, *Sacculina*; sterility in various other plants or animals caused by a parasite attacking the sex organs.

parasitic male a dwarf male which is parasitic on its female and has a reduced body in all but the sex organs, as in some deep-sea fish.

parasitism *n.* [Gk. *parasitos*, parasite] a form of symbiosis in which one symbiont, or parasite, receives advantage to detriment of the other, or host.

parasitocoenosis *n.* [Gk. *parasitos*, parasite; *koinos*, joint, common] the whole complex of parasites living in any one host.

parasitoid *n.* [Gk. *parasitos*, parasite; *eidos*, form] an organism alternately parasitic and free living.

parasitology *n.* [Gk. *parasitos*, parasite; *logos*, discourse] the study of parasites, especially animal parasites.

parasol *n.* [Gk. *para*, against; L. *sol*, sun] paraheliode, *q.v.*

parasphenoid *n.* [Gk. *para*, beside; *sphēn*, wedge; *eidos*, form] membrane bone forming floor of cranium in certain vertebrates.

parasporal *a.* [Gk. *para*, beside; *sporos*, seed] *appl.* bodies: protein particles formed within cytoplasm during sporulation of some bacilli.

parasporangium *n.* [Gk. *para*, beside; *sporos*, seed; *anggeion*, vessel] a sporangium containing paraspores.

paraspore *n.* [Gk. *para*, beside; *sporos*, seed] a spore formed from a cortical cell, in certain algae.

parastamen, parastemon *n.* [Gk. *para*, beside; *stēmōn*, thread] staminode, *q.v.*

parasternalia *n.plu.* [Gk. *para*, beside; *sternon*, breast] abdominal ribs, gastralia, *q.v.*

parasternum *n.* [Gk. *para*, beside; L. *sternum*, breast bone] the sum total of abdominal ribs in certain reptiles, also in Stegocephali and *Archaeopteryx*.

parastichy *n.* [Gk. *para*, beside; *stichos*, row] a secondary spiral in phyllotaxis.

parastipes *n.* [Gk. *para*, beside; L. *stipes*, stalk] subgalea, *q.v.*

parasymbiosis *n.* [Gk. *para*, beside; *symbiōnai*, to live with] the living together of organisms without mutual harm or benefit.

parasympathetic *a.* [Gk. *para*, beside; *sympathēs*, of like feelings] enteral, *appl.* the craniosacral portion of the autonomic nervous system, having cholinergic nerve endings.

parasynapsis parasyndesis, *q.v.*

parasyndesis *n.* [Gk. *para*, beside; *syndesis*, binding together] normal side-to-side pairing of chromosomes in syndesis, rather than end-to-end union; *alt.* parasynapsis; *cf.* telosyndesis.

parately *n.* [Gk. *para*, beside; *telos*, end] evolution from material unrelated to that of type, but resulting in superficial resemblance.

paraterminal *a.* [Gk. *para*, beside; L. *terminus*, boundary] *appl.* bodies constituting part of anterior median wall of lateral ventricles, in amphibians and reptiles.

paratestis n. [Gk. *para*, beside; L. *testis*, testicle] small reddish-yellow fatty body in some male newts and salamanders which produces autacoids regulating appearance of nuptial apparel.

parathecium *n.* [Gk. *para*, beside; *thēkē*, box] peripheral layer of apothecium, as in cup fungi; peripheral hyphal layer in lichens.

parathormone parathyrin, *q.v.*

parathyreoid parathyroid, *q.v.*

parathyrin *n.* [Gk. *para*, beside; *thyreos*, shield] proteinaceous hormone secreted by parathyroids, which regulates calcium and phosphorus metabolism; *alt.* parathormone, parathyroid hormone (PTH).

parathyroid *n.* [Gk. *para*, beside; *thyreos*, shield; *eidos*, form] one of 4 small brownish-red endocrine glands near or within the thyroid, producing the hormone parathyrin; *alt.* parathyreoid, epithelial body.

parathyroid hormone (PTH) parathyrin, *q.v.*

paratoid *a.* [Gk. *parateinein*, to extend along] *appl.* a double row of poison glands extending along back of certain amphibians, as of salamanders.

paratomium *n.* [Gk. *para*, beside; *tomos*, cutting] side of a bird's beak, between tomium and culmen.

paratomy *n.* [Gk. *para*, beside; *tomē*, cutting] reproduction by fission with antecedent regeneration, in certain annelids; *cf.* architomy.

paratonic *a.* [Gk. *para*, beside; *tonos*, strain] stimulating or retarding; *appl.* movements induced by external stimuli, as tropisms and nastic movements; *cf.* autonomic.

paratope *n.* [Gk. *para*, beside; *topos*, place] the region of an antibody that combines with an antigen; *cf.* epitope.

paratracheal *a.* [Gk. *para*, beside; L.L. *trachia*, windpipe] with xylem parenchyma cells around or close to vascular tissue.

paratrophic *a.* [Gk. *para*, beside; *trephein*, to nourish] *appl.* method of nutrition of obligate parasites.

paratympanic *a.* [Gk. *para*, beside; *tympanon*, kettle drum] medial and dorsal to tympanic cavity; *appl.* a small organ, with sensory epithelium innervated from geniculate ganglion, in many birds.

paratype *n.* [Gk. *para*, beside; *typos*, pattern] specimen described at same time as the one regarded as type or holotype, of a new genus or species; aggregate of external factors affecting manifestation of a genetic character; abnormal type of a species, as of bacterial colony; *cf.* hypoparatype.

para-urethral *a.* [Gk. *para*, beside; *ourēthra*, from *ouron*, urine] *appl.* racemose glands of the urethra, Littre's glands, Skene's glands.

paravertebral *a.* [Gk. *para*, beside; L. *vertebra*, vertebra] alongside the spinal column, *appl.* sympathetic nerve trunk.

paravesical *a.* [Gk. *para*, beside; L. *vesica*, bladder] beside the bladder, *appl.* a fossa or depression of peritoneum.

paraxial *a.* [Gk. *para*, beside; L. *axis*, axle] alongside the axis, *appl.* a medial column of mesoderm.

paraxon *n.* [Gk. *para*, beside; *axōn*, axle] a lateral branch of the axis-cylinder process of a nerve cell.

Paraxonia *n.* [Gk. *para*, beside; *axōn*, axle] a

superorder of ungulates containing one order, the Artiodactyla, *q.v.*

paraxonic *a.* [Gk. *para*, beside; *axōn*, axle] *pert.* or having an axis outside the usual axis; with axis of foot between 3rd and 4th digits, as in Artiodactyla, *cf.* mesaxonic.

Parazoa [Gk. *para*, beside; *zōon*, animal] a subkingdom of multicellular animals, containing the sponges, having a loose organization of cells which never form tissues; *cf.* Mesozoa, Metazoa, Protozoa.

parazoon *n.* [Gk. *para*, beside; *zōon*, animal] a member of the Parazoa, *q.v.*

parencephalon *n.* [Gk. *para*, beside; *engkephalos*, brain] one of paired cerebral hemispheres.

parenchyma *n.* [Gk. *para*, beside; *engchyma*, infusion] plant tissue, generally soft and of thin-walled relatively undifferentiated cells, which may vary in structure and function, as pith, of mesophyll, etc.; groundwork tissue of organs; ground tissue of triploblastic acoelomates.

parenchymalia *n.plu.* [Gk. *para*, beside; *engchyma*, infusion] spicules of parenchyma of Hexactinellida.

parenchymatous *a.* [Gk. *para*, beside; *engchyma*, infusion] *pert.* or found in parenchyma.

parenchymula *n.* [Gk. *para*, beside; *engchyma*, infusion] a flagellate sponge larva with cavity filled with gelatinous connective tissue.

parental generation *see* P₁.

parenthosome *n.* [Gk. *parentithenai*, to insert; *sōma*, body] a structure formed from endoplasmic reticulum and covering the dolipore, *q.v.*

parethmoid *n.* [Gk. *para*, beside; *ēthmos*, sieve; *eidos*, form] ectethmoid, *q.v.*

parhomology *n.* [Gk. *para*, beside, *homos*, alike; *logos*, discourse] apparent similarity of structure.

parichnos *n.* [Gk. *para*, beside; *ichnos*, trace] two lateral scars at sides of vascular bundle trace in certain extinct ferns.

paries *n.* [L. *paries*, wall] the central division of a compartment of Cirripedia; wall of a hollow structure, as of tympanum, or of honeycomb; *plu.* parietes.

parietal *a.* [L. *paries*, wall] *pert.*, next to, or forming part of wall of a structure, *appl.* cytoplasm, membrane, layer, lobe, placentation, area between frons and occiput in insects, etc.

parietal bone a paired bone of roof of skull.

parietal cells oxyntic cells, *q.v.*

Parietales *n.* [L. *paries*, wall] an order of dicots of the Archichlamydeae or Polypetalae having parietal placentation, but used in different ways by different authorities.

parietal organ epiphyseal photoreceptor in lower vertebrates; *alt.* parapineal organ.

parietal region pineal region of brain.

parietal vesicle dilated distal part of pineal stalk.

parietes *n.plu.* [L. *parietes*, walls] *plu.* of paries; walls or sides of structure.

parietobasilar *a.* [L. *paries*, wall; *basis*, base] *appl.* muscles between pedal disc and lower part of body wall in sea anemones.

parietofrontal *a.* [L. *paries*, wall; *frons*, forehead] *appl.* a skull bone, in place of parietals and frontals, as in Dipnoi; *cf.* frontoparietal.

parietomastoid *a.* [L. *paries*, wall; Gk. *mastos*, breast; *eidos*, form] connecting mastoid with parietal, *appl.* a suture.

parieto-occipital *a.* [L. *paries*, wall; *occiput*, back of head] *appl.* fissure between parietal and occipital lobes of cerebrum.

parietotemporal *a.* [L. *paries*, wall; *tempora*, the temples] *pert.* parietal and temporal regions; *appl.* a branch of the middle cerebral artery.

parietovaginal *a.* [L. *paries*, wall; *vagina*, sheath] *appl.* paired muscle for retracting introvert and tentacles in Bryozoa.

paripinnate *a.* [L. *par*, equal; *pinna*, wing] pinnate without a terminal leaflet; *cf.* imparipinnate.

parity *n.* [L. *parere*, to give birth] the number of times a female has given birth, regardless of the number of offspring produced at any one birth.

parivincular *a.* [L. *par*, equal; *vinculum*, bond] *appl.* bivalve hinge ligament attached to nymphae; *cf.* opisthodetic, amphidetic.

paroccipital *a.* [Gk. *para*, beside; L. *occiput*, back of head] *appl.* ventrally directed processes of exoccipitals.

parocciput *n.* [Gk. *para*, beside; L. *occiput*, back of head] in insects, a thickening of the occiput for articulation of neck sclerites.

paroecious *a.* [Gk. *para*, beside; *oikia*, house] with antheridium and archegonium close to one another; *alt.* paroicous.

parolfactory *a.* [Gk. *para*, beside; L. *olfactorius*, olfactory] *appl.* an area and sulcus adjoining olfactory trigone of rhinencephalon.

paronychia *n.plu.* [Gk. *para*, beside; *onyx*, nail] bristles on pulvillus of insect foot; whitlow.

paroophoron *n.* [Gk. *para*, beside; *ōon*, egg; *pherein*, to bear] a few scattered rudimentary tubules, remnants of Wolffian body in female, in broad ligament between uterus and epoophoron.

parosteal *a.* [Gk. *para*, beside; *osteon*, bone] *appl.* abnormal bone formations.

parosteosis *n.* [Gk. *para*, beside; *osteon*, bone] bone formation in tracts normally fibrous.

parotic *n.* [Gk. *para*, beside; *ous*, ear] a process formed by fusion of exoccipital and opisthotic in some reptiles; a similar process in some fish.

parotid glands paired salivary glands opening into mouth cavity of mammals.

parotoid glands in some amphibians, large swellings on side of head, formed of aggregated cutaneous glands, sometimes poisonous.

parovarium *n.* [Gk. *para*, beside; L. *ovarium*, ovary] a small collection of tubules anterior to ovary, the remnant in adult of embryonic mesonephros, *alt.* pampiniform body; epoophoron, *q.v.*

pars *n.* [L. *pars*, part] a part of an organ, as pars glandularis, nervosa, intermedia, tuberalis, of pituitary gland; *plu.* partes.

pars amorpha a spherical body in the nucleolus composed of fine granular low-density material.

pars distalis a part of the adenohypophysis.

pars intercerebralis in an insect's fore-brain, the region containing neurosecretory cells.

pars intermedia a part of the adenohypophysis.

pars nervosa a part of the neurohypophysis; *alt.* neural lobe.

pars tuberalis a part of the adenohypophysis.

partes *plu.* of pars.

parthenapogamy *n.* [Gk. *parthenos*, virgin; *apo*, away; *gamos*, marriage] diploid or somatic parthenogenesis; *alt.* ooapogamy.

parthenita *n.* [Gk. *parthenos*, virgin] unisexual stage of trematodes in intermediate host.

parthenocarpy n. [Gk. parthenos, virgin; karpos, fruit] condition of producing fruit without seeds; a. parthenocarpic.

parthenocaryogamy parthenokaryogamy, q.v.

parthenogamy n. [Gk. parthenos, virgin; gamos, marriage] parthenomixis, q.v.

parthenogenesis n. [Gk. parthenos, virgin; genesis, descent] reproduction without fertilization by male gamete.

parthenogenetic appl. organisms developed by parthenogenesis; appl. reagents which can activate ovum; cf. zygogenetic.

parthenogone n. [Gk. parthenos, virgin; gonē, offspring] an organism produced by parthenogenesis.

parthenogonidia n.plu. [Gk. parthenos, virgin; gonos, offspring; idion, dim.] zooids of a protozoan colony, with function of asexual reproduction.

parthenokaryogamy n. [Gk. parthenos, virgin; karyon, nucleus; gamos, marriage] the fusion of 2 female haploid nuclei; alt. parthenocaryogamy.

parthenomixis n. [Gk. parthenos, virgin; mixis, mingling] the mingling of 2 nuclei produced within one gamete or gametangium; alt. parthenogamy.

parthenosperm n. [Gk. parthenos, virgin; sperma, seed] a sperm produced without fertilization, but resembling a zygote.

parthenospore n. [Gk. parthenos, virgin; sporos, seed] azygospore, q.v.

parthenote n. [Gk. parthenos, virgin] a parthenogenetically produced haploid organism.

partial parasite semiparasite, q.v.

partial veil inner veil of certain Basidiomycetes, growing from stipe towards edge of pileus and becoming separated to constitute the cortina or superior annulus; alt. velum partiale.

particulate inheritance inheritance in one organism of distinctive paternal and maternal characteristics.

partite a. [L. partitus, divided] divided nearly to base.

parturition n. [L. parturire, to bring forth] the act or process of birth.

parumbilical a. [Gk. para, beside; L. umbilicus, navel] beside the navel; appl. small veins from anterior abdominal wall to portal and iliac veins.

parvicellular a. [L. parvus, small; cellula, small cell] consisting of small cells.

pascual a. [L. pascuum, pasture] pert. pastures or ground for grazing, appl. flora.

pasculomorphosis n. [L. pascuum, pasture; Gk. morphōsis, shaping] changes in the structural features of plants as a result of grazing.

passage cells thin-walled endodermal or exodermal cells of root which permit passage of solutions, and are usually associated with cells with thick walls.

Passeres n. [Passer, generic name of sparrow] a large suborder of Passeriformes consisting of the typical singing birds.

Passeriformes n. [Passer, generic name; L. forma, shape] the largest order of birds, including the songbirds, crows, etc.

Passiflorales n. [Passiflora, generic name] an order of dicots (Polypetalae) not recognized by all authorities.

Pasteur effect [L. Pasteur, French biochemist] the observation that glycolysis is inhibited by oxygen; sometimes incorrectly said to be the ability of a normally anaerobic organism to oxidize sugar completely to carbon dioxide and water in the presence of oxygen, which is a later interpretation of the effect.

patagial a. [L. patagium, border] of or pert. a patagium.

patagiate a. [L. patagium, border] furnished with a patagium.

patagium n. [L. patagium, border] membranous expansion between fore- and hindlimbs of bats; extension of skin between fore- and hindlimbs of flying lemurs and flying squirrels; similar expansion in lizards and on bird's wing, alt. parachute; tegula, or dorsal process of prothorax in certain Lepidoptera; anterior pronotum in Diptera; cf. prepatagium, plagiopatagium, uropatagium.

patella n. [L. patella, small pan] the knee cap or elbow cap; segment between femur and tibia in Pycnogonida; 4th segment of spider's leg, alt. carpopodite; a genus of limpet; a rounded apothecium of lichens.

patellar pert. a patella.

patelloid, patelliform a. [L. patella, small pan; Gk. eidos, L. forma, shape] shaped like a patella; pan-shaped; like a bordered disc.

patent a. [L. patens, lying open] open; spreading widely, cf. fastigiate; expanded.

pateriform a. [L. patera, flat dish; forma, shape] saucer-shaped.

pathetic a. [Gk. pathos, feeling] appl. nerve: trochlear nerve, q.v.; appl. superior oblique muscle of eye.

pathogen n. [Gk. pathos, suffering; -genēs, producing] any disease-producing micro-organism.

pathogenic a. [Gk. pathos, suffering; -genēs, producing] disease-producing, appl. a parasite in relation to a particular host; alt. nosogenic.

pathology n. [Gk. pathos, suffering; logos, discourse] science dealing with disease and with morbid structures and functions; alt. nosology.

patina n. [L. patina, dish] circles of plates around calyx of crinoids.

patobionts n.plu. [Gk. patos, bottom, path; bion, living] true cryptozoa, spending their entire lives in the cryptosphere.

patocoles n.plu. [Gk. patos, bottom, path; L. colere, to dwell] cryptozoa that spend part of their time in the cryptosphere but also emerge to hunt and mate.

patoxenes n.plu. [Gk. patos, bottom, path; xenos, strange] animals which from time to time bury or conceal themselves in the cryptosphere.

patriclinal, patriclinous a. [Gk. patēr, father; klinein, to incline] with hereditary characteristics more paternal than maternal; alt. patroclinic, patroclinal; cf. matriclinal.

patristic a. [Gk. patēr, father] in plants, pert. similarity due to common ancestry; cf. cladistic.

patroclinic patriclinous, q.v.

patulent, patulose, patulous a. [L. patulus, standing open] spreading open; expanding.

paturon n. [Gk. patein, to trample on; oura, after part] basal joint of arachnid chelicerae, used for crushing and expressing fluids of insects; alt. tige.

paucilocular a. [L. pauci, few; loculus, compartment] containing, or composed of, few small cavities or loculi.

paucispiral *a.* [L. *pauci*, few; *spira*, coil] with few coils or whorls.

paulospore *n.* [Gk. *paula*, rest; *sporos*, seed] a resting stage in development, as a cyst; chlamydospore, *q.v.*

paunch *n.* [L. *pantex*, paunch] rumen, *q.v.*

paurometabolism *n.* [Gk. *pauros*, brief; *metabolē*, change] incomplete metamorphosis in which the nymph resembles the adult.

Pauropoda *n.* [Gk. *pauros*, brief; *pous*, foot] a class of arthropods, sometimes considered as myriapods, having 12 body segments, 9 of which have appendages.

pavement *n.* [L. *pavimentum*, from *pavire*, to ram down] a flat structure of compact units; *appl.* epithelium of flat, nucleated scales in mosaic pattern, simple squamous epithelium; *appl.* teeth, as in certain sharks; *a.* pavimental.

Pavlovian conditioning [*I. P. Pavlov*, Russian physiologist] a form of learning in which a neutral stimulus (the conditioned stimulus) is accompanied over a period of time with a stimulus (the unconditioned stimulus) that normally elicits a certain response, and eventually the neutral stimulus elicits the same response as the unconditioned stimulus; *alt.* classical conditioning.

paxilla *a.* [L. *paxillus*, peg] thick plate supporting calcareous pillars, summit of each covered by group of small spines, in certain Asteroidea; *alt.* paxillus; *a.* paxillar.

paxillate having paxillae; *alt.* paxilliferous, paxillose.

paxilliform *a.* [L. *paxillus*, peg; *forma*, shape] shaped like a paxilla.

paxillus paxilla, *q.v.*

pearl *n.* [F. *perle*, pearl] in shells of some Mollusca, an abnormal growth formed with a grain of foreign matter or a minute organism for nucleus and many thin layers of nacre surrounding it.

peat *n.* [M.E. *pete*, peat] dead marsh vegetation undergoing bacterial decay in stagnant deoxygenated water.

peck order a social hierarchy, especially in birds, ranging from the most dominant and aggressive animal down to the most submissive.

pectase pectinesterase, *q.v.*

pecten *n.* [L. *pecten*, comb] any comb-like structure; a process of inner retinal surface in reptiles, expanded into a folded quadrangular plate in birds; a ridge of superior ramus of os pubis; part of anal canal between internal sphincter and anal valves; a part of stridulating organ of certain spiders; sensory abdominal appendage of scorpions; a genus of scallops; a comb-like assemblage of sterigmata; *plu.* pectines.

pectic acids colloidal acidic pectic substances with few or no methyl ester groups.

pectic enzymes those which hydrolyse pectic substances, including pectase, pectinase, and protopectinase.

pectic substances *a.* [Gk. *pēktos*, congealed] a group of complex colloidal carbohydrates mainly derived from galacturonic acids and including the protopectins, pectins, pectinic acids, and pectic acids, which form the middle lamella of plant cell walls.

pectinal *a.* [L. *pecten*, comb] *pert.* a pecten.

pectinase *n.* [Gk. *pēktos*, congealed] enzyme catalysing the hydrolysis of certain pectic substances; EC 3.2.1.15; *r.n.* polygalacturonase.

pectinate *a.* [L. *pecten*, comb] comb-like, *appl.* leaves, arrangement of sporangia, pedicellariae of Asteroidea, a ligament of iris, gills as of certain molluscs, a septum between corpora cavernosa, fibres, muscles of crista terminalis of right atrium, etc.; *alt.* pectiniform.

pectineal *a.* [L. *pecten*, comb] comb-like; *appl.* process on pubis of birds; *appl.* a ridge-line on femur and attached muscle, the pectineus muscle.

pectinellae *n.plu.* [L. *pectinella*, small comb] transverse comb-like membranellae constituting adoral ciliary spiral of some Ciliata.

pectines *n.plu.* [L. *pecten*, comb] *plu.* of pecten.

pectinesterase *n.* [Gk. *pēktos*, congealed] a hydrolysing enzyme which catalyses the conversion of pectin into pectic acid and methyl alcohol, and so causes the formation of vegetable jelly; *alt.* pectase; EC 3.1.1.11; *r.n.* pectinesterase.

pectineus a flat muscle between pecten pubis and upper medial part of femur.

pectinibranch *a.* [L. *pecten*, comb; *branchiae*, gills] in mollusc gills, monopectinate, *q.v.*

pectinic acids colloidal acidic pectic substances obtained by partial hydrolysis of protopectins and having a methyl ester content between that of pectic acids and pectins.

pectiniform pectinate, *q.v.*

pectinirhomb *n.* [L. *pecten*, comb; Gk. *rhombos*, wheel] a type of stereom-folding in Cystoidea.

pectins *n.plu.* [Gk. *pēktos*, congealed] pectic substances obtained from protopectin, which on hydrolysis yield pectic acid and methanol and are found in certain fruits and succulent vegetables.

pectocellulose *n.* [Gk. *pēktos*, congealed; L. *cellula*, small cell] cellulose mixed with pectose, as in fleshy roots and fruits.

pectolytic *a.* [Gk. *pēktos*, congealed; *lyein*, to break down] destroying pectin, *appl.* enzymes.

pectoral *a.* [L. *pectus*, breast] *pert.* chest; in chest region; *appl.* arch, fins, etc.

pectoral girdle in vertebrates, a skeletal support in the shoulder region for attachment of fore-fins or -limbs, made of a hoop of cartilages or bones, usually the scapular, clavicle, and coracoid; *alt.* shoulder girdle.

pectoralis major and minor outer and inner chest muscles connecting ventral chest wall with shoulder and humerus; *plu.* pectorales.

pectose *n.* [Gk. *pēktos*, congealed] protopectin, *q.v.*

pectosinase protopectinase, *q.v.*

pectus *n.* [L. *pectus*, breast] the chest or breast region; fused pleuron and sternum of arthropods.

ped *n.* [L. *pes*, foot] soil aggregate, *q.v.*

pedal *a.* [L. *pes*, foot] *pert.* foot or feet, *appl.* cords, ganglia, glands, etc.; *appl.* disc: base of sea anemones; *cf.* podal.

pedalfer *n.* [*pedon*; *alumen*; *ferrum*] any of a group of soils, in humid regions, usually characterized by the presence of aluminium and iron compounds, and by the absence of carbonates.

pedate *a.* [L. *pes*, foot] with toe-like parts; pedatipartite, *q.v.*

pedatipartite *a.* [L. *pes*, foot; *partitus*, divided] *appl.* a variety of palmate leaf having 3 main divisions and the 2 outer divisions subdivided one or more times; *alt.* pedate.

pedatisect *a.* [L. *pes*, foot; *sectus*, cut] in pedate arrangement and with divisions nearly to midrib.

pedes *plu.* of pes.

pedicel *n.* [L. *pediculus*, small foot] a small, short foot-stalk of leaf, fruit, or sporangium foot-stalk or stem of stationary or fixed organism, or of organ; 2nd segment of insect antenna; modified 2nd abdominal segment of abdomen in Hymenoptera; *alt.* pedicellus.

pedicellariae *n.plu.* [L. *pediculus*, small foot] minute pincer-like structures studding the surface of certain echinoderms.

pedicellate *a.* [L. *pediculus*, small foot] supported by a pedicel or petiole; *appl.* Hymenoptera with stalked abdomen; *cf.* pseudosessile.

pedicellus *n.* [L. *pediculus*, small foot] pedicel, *q.v.*

pedicle *n.* [L. *pediculus*, small foot] a short stem; backward-projecting vertebral process; dilated end of branch of an astrocyte in contact with a blood vessel; narrow stalk uniting cephalothorax with abdomen in arachnids.

Pediculati, pediculates *n., n.plu.* [L. *pediculus*, small foot] an order of very specialized marine teleosts including anglers and batfishes having anterior portion of dorsal fin modified as a lure; *alt.* Lophiiformes.

pediferous *a.* [L. *pes*, foot; *ferre*, to carry] having feet; having a foot-stalk; *alt.* pedunculate.

Pedinoida *n.* [*Pedina*, generic name; Gk. *eidos*, form] an order of sea urchins of the superorder Diadematacea having a rigid test, sometimes included in the Echinothuroida.

pedipalp(us) *n.* [L. *pes*, foot; *palpare*, to feel] in arrachnids and other chelicerates, 2nd cephalothoracic paired appendage, variously a pincer-like claw or a simple or leg-like appendage; *alt.* palp(us).

pedocal *n.* [Gk. *pedon*, ground; *calcium*] any of a group of soils, of semiarid and arid regions, characterized by the presence of carbonate of lime.

pedogamy *n.* [Gk. *pais*, child; *gamos*, union] paedogamy, *q.v.*

pedogenesis paedogenesis, *q.v.*

pedogenic *a.* [Gk. *pedon*, soil; *gennaein*, to produce] *pert.* the formation of soil.

pedology *n.* [Gk. *pedon*, soil; *logos*, discourse] soil science; *cf.* edaphology.

pedonic *a.* [Gk. *pedon*, ground] *appl.* organisms of freshwater lake bottom.

peduncle *n.* [L.L. *pedunculus*, small foot] an inflorescence stalk; a band of white fibres joining different parts of brain; stalk of crinoids, brachiopods, and barnacles; link between thorax and abdomen in arthropods; stalk of sedentary protozoans.

pedunculate *a.* [L.L. *pedunculus*, small foot] growing on or having a peduncle; *appl.* bodies: corpora pedunculata, *q.v.*; *appl.* hydatid or appendage of epididymis; *alt.* pediferous.

Pegasiformes *n.* [*Pegasus*, generic name; L. *forma*, shape] an order of teleosts including the sea moths of the Indo-Pacific; *alt.* Hypostomides.

Pekin(g) man an extinct form of fossil hominid found near Peking and at first called *Sinanthropus pekinensis*, then known as *Pithecanthropus pekinensis*, and now classified as *Homo erectus pekinensis*; *cf.* Java man.

pelage *n.* [F. *pelage*, fur] the hairy, furry, or woolly coat of mammals.

pelagic *a.* [Gk. *pelagos*, sea] living in the sea or ocean at the middle or surface levels.

Pelagothurida *n.* [*Pelagothuria*, generic name] an order of sea cucumbers in some classifications, including the only pelagic members.

pelargonidin a red anthocyanin pigment found in the dicot, *Pelargonium*.

pelasgic *a.* [Gk. *Pelasgikos*, *pert.* Pelasgians] moving from place to place.

Pelecaniformes *n.* [*Pelecanus*, generic name of pelican; L. *forma*, shape] an order of birds of the subclass Neornithes, including the pelicans, cormorants, and boobies.

Pelecypoda *n.* [Gk. *pelekys*, hatchet; *pous*, foot] Bivalvia, *q.v.*

pellagra-preventive factor nicotinamide or nicotinic acid, *q.v.*; *alt.* P-P factor, vitamin P-P.

pellicle *n.* [L. *pellicula*, small skin] the delicate protective investment of Protozoa; any filmy protective covering; *alt.* pellicula.

pelliculate *a.* [L. *pellicula*, small skin] having a pellicle on external surface.

pellions *n.plu.* [Gk. *pella*, cup; *dim.*] ring of plates supporting suckers of echinoids; *alt.* rosettes.

pelma *n.* [Gk. *pelma*, sole] the sole of foot; *alt.* planta.

Pelmatozoa *n.* [Gk. *pelma*, sole of foot; *zōon*, animal] a subphylum of mostly extinct echinoderms that are attached to the substratum at least as young forms, either directly or by a stalk.

pelophilous *a.* [Gk. *pēlos*, clay; *philein*, to love] growing on clay; *n.* pelophile.

peloria *n.* [Gk. *pelōrios*, monstrous] condition of abnormal regularity; a modification of structure from irregularity to regularity; *alt.* pelory.

peloric *a.* [Gk. *pelōrios*, monstrous] *appl.* a flower which, normally irregular, becomes regular.

pelory peloria, *q.v.*

peloton *n.* [F. *peloton*, ball of thread] a knot or skein of hyphae, as in some mycorrhizae.

pelta *n.* [Gk. *peltē*, shield] the shield-like apothecium of certain lichens.

peltate *a.* [Gk. *peltē*, shield] shield-shaped; having the stalk inserted at or near the middle of the under surface, not at the edge.

peltinervate *a.* [Gk. *peltē*, shield; L. *nervus*, nerve] having veins radiating from near the centre, as of a peltate leaf.

Peltopleuriformes *n.* [Gk. *peltē*, shield; *pleura*, side; L. *forma*, shape] an order of marine Triassic Chondrostei with large eyes and some members probably feeding on plankton.

pelvic *a.* [L. *pelvis*, basin] *pert.* or situated at or near pelvis, *appl.* cavity, fin, limbs, plexus. etc.

pelvic girdle in vertebrates, a skeletal support in the hip region for attachment of hind-fins or -limbs, made of a hoop of cartilages or bones, usually the ilium, and ischiopubis or ischium and pubis; *alt.* hip girdle.

pelvis *n.* [L. *pelvis*, basin] the bony cavity formed by pelvic girdle along with coccyx and sacrum; expansion of ureter at its junction with kidney; basal portion of cup of crinoids.

pelvisternum *n.* [L. *pelvis*, basin; L. *sternum*, breast bone] epipubis separate from pubis.

Pelycosauria, pelycosaurs *n., n.plu.* [Gk. *pelyx*, wooden bowl; *sauros*, lizard] an order of aberrant and primitive mammal-like reptiles of the

Carboniferous to Permian age, having a primitive sprawling gait, and including the sail-back lizards; *alt.* Theromorpha.

pen *n.* [L. *penna*, feather] a leaf midrib; proostracum of certain cephalopods; primary wing feather or remex; a female swan.

pendent *a.* [L. *pendens*, hanging down] hanging down, as certain lichens, leaves, flowers, etc.; *alt.* pensile.

pendulous *a.* [L. *pendere*, to hang] bending downwards from point of origin; overhanging; *appl.* ovules, branches, flowers, etc.

penes *plu.* of penis.

penetrance *n.* [L. *penetrare*, to penetrate] the frequency, measured as a percentage, with which a gene shows any effect.

penetrant *n.* [L. *penetrare*, to penetrate] stenotele, *q.v.*

penetration path copulation path of spermatozoon in ooplasm to the female pronucleus.

penial *a.* [L. *penis*, penis] of or *pert.* penis.

penial setae paired needle-like chitinoid bodies at nematode anus; setae near aperture of vas deferens in earthworms; *alt.* copulatory spicules.

penicillate, penicilliform *a.* [L. *penicillum*, painter's brush] pencil-shaped; tipped with hairs; having a structure like a camel-hair or bottle brush.

penicillin any of various antibiotics derived from the fungus *Penicillium notatum* and related species, which inhibits the synthesis of bacterial cell wall, resulting in weakening of the wall and osmotic lysis.

pencillinase an enzyme produced by penicillin-resistant bacteria which catalyses the hydrolysis of penicillin to penicilloic acid; *alt.* cephalosporinase; EC 3.5.2.6; *r.n.* penicillinase.

penicillus *n.* [L. *penicillus*, painter's brush] a brush-shaped structure as certain type of nematocyst; tuft of hairs of tegumen; a tuft of arterioles, in spleen, *alt.* Ruysch's penicilli; a tuft of conidiophores; *plu.* pe... ..lli.

penis *n.* [L. *penis*, penis] the male copulatory organ; *alt.* phallus, thyrsus; *plu.* penes.

penna *n.* [L. *penna*, feather] a contour feather of birds, as distinguished from plume or down feathers.

pennaceous *a.* [L. *penna*, feather] like a plume or feather, *alt.* penniform; *appl.* feathers with hamuli on barbules, *cf.* plumose.

Pennales *n.* [L. *penna*, feather] an order of diatoms having bilateral symmetry; *cf.* Centrales.

pennate *a.* [L. *penna*, feather] pinnate, *q.v.*; having a wing; feathered; in the shape of a wing; *pert.* Pennales.

Pennatulacea *n.* [*Pennatula*, generic name] an order of Octocorallia commonly called sea pens, having one long axial polyp and a skeleton of separate calcareous spicules.

Pennsylvania *a.* [Pennsylvania] *appl.* and *pert.* an epoch of the Carboniferous era; *appl.* coal measures in North America.

pensile *a.* [L. *pensilis*, hanging down] pendent, hanging down, *appl.* some bird's nests.

pentacapsular *a.* [Gk. *pente*, five; L. *capsula*, capsule] with 5 capsules.

pentacarpellary *a.* [Gk. *pente*, five; *karpos*, fruit] with 5 carpels.

pentachenium *n.* [Gk. *pente*, five; *a*, not; *chainein*, to gape] a form of schizocarp with 5 carpels.

pentacoccous *a.* [Gk. *pente*, five; *kokkos*, kernel] with 5 seeds or carpels.

pentacrinoid *a.* [Gk. *pente*, five; *krinon*, lily; *eidos*, form] resembling a *Pentacrinus*, *appl.* larval stage of feather stars.

pentactinal *a.* [Gk. *pente*, five; *aktis*, ray] five-rayed; 5-branched.

pentacula *n.* [Gk. *pente*, five; L. *aculeus*, prickle] in life history of echinoderms, the stage with 5 tentacles.

pentacyclic *a.* [Gk. *pente*, five; *kylos*, circle] arranged in 5 whorls.

pentadactyl *a.* [Gk. *pente*, five; *daktylos*, finger] having all 4 limbs normally terminating in 5 digits.

pentadactyl limb a limb with 5 digits characteristic of tetrapods; *alt.* cheiropterygium.

pentadelphous *a.* [Gk. *pente*, five; *adelphos*, brother] having 5 clusters of more-or-less united filaments.

pentafid *a.* [Gk. *pente*, five; L. *findere*, to cleave] in 5 divisions or lobes.

pentagonal *a.* [Gk. *pente*; five; *gōnia*, angle] having 5 angles; quinary, *q.v.*

pentagynous *a.* [Gk. *pente*, five; *gynē*, woman] having 5 pistils or styles; having 5 carpels to the gynaecium.

pentalith *n.* [Gk. *pente*, five; *lithos*, stone] a 5-sided coccolith.

pentamerous *a.* [Gk. *pente*, five; *meros*, part] composed of 5 parts; in whorls of 5 or a multiple of 5.

pentandrous *a.* [Gk. *pente*, five; *anēr*, male] having 5 stamens; *alt.* pentastemonous.

pentangular *a.* [Gk. *pente*, five; L. *angulus*, angle] having 5 angles.

pentapetalous *a.* [Gk. *pente*, five; *petalon*, leaf] having 5 petals.

pentaphyllous *a.* [Gk. *pente*, five; *phyllon*, leaf] having 5 leaves or leaflets.

pentaploid *a.* [Gk. *pente*, five; *haploos*, simple; *eidos*, form] with 5 sets of chromosomes; having 5 times the haploid chromosome number.

pentapterous *a.* [Gk. *pente*, five; *pteron*, wing] with 5 wings, as some fruits.

pentaradiate *a.* [Gk. *pente*, five; L. *radius*, ray] *appl.* body built on a 5-rayed plan.

pentarch *a.* [Gk. *pente*, five; *archē*, beginning] with 5 alternating xylem and phloem groups; with 5 vascular bundles.

pentasepalous *a.* [Gk. *pente*, five; F. *sépale*, sepal] having 5 sepals.

pentaspermous *a.* [Gk. *pente*, five; *sperma*, seed] with 5 seeds.

pentastemonous *a.* [Gk. *pente*, five; *stēmōn*, stamen] pentandrous, *q.v.*

pentasternum *n.* [Gk. *pente*, five; *sternon*, chest] sternite of 5th segment of prosoma or 3rd segment of podosoma in Acarina.

pentastichous *a.* [Gk. *pente*, five; *stichos*, row] arranged in 5 vertical rows, *appl.* leaves; *n.* pentastichy.

Pentastomida *n.* [Gk. *pente*, five; *stoma*, mouth] a phylum of animals that are parasitic in the respiratory passages of air-breathing vertebrates and have a mouth with claws, an annulated abdomen, and a chitinous cuticle, and which have also been classed as arachnids; *alt.* Linguatulida.

Pentazonia *n.* [Gk. *pente*, five; *zōne*, girdle] a

superorder of millipedes of the subclass Chilognatha having 1 or 2 pairs of telopods.

pentosans *n.plu.* [Gk. *pente*, five] polysaccharides made of condensed pentose units, the most important being xylan and arabinan.

pentose phosphate pathway (PPP) a group of cyclic reactions which produce energy as $NADPH_2$ from the partial oxidation of glucose-6-phosphate to triose phosphate and carbon dioxide, many of the intermediates being pentose phosphates; *alt.* Warburg–Dickens pathway, hexose monophosphate shunt, pentose oxidation cycle, pentose cycle.

pentoses *n.plu.* [Gk. *pente*, five] monosaccharides having the formula $C_5H_{10}O_5$, including arabinose, xylose, ribose.

Pentoxylales *n.* [*Pentoxylon*, generic name] an order of fossil Cycadopsida, probably shrubs or very small trees, having long and short shoots, the latter bearing terminal reproductive organs and spirally arranged leaves, sometimes considered to be a subgroup of Bennettitatae.

peonidin *n.* [*Paeonia*, generic name of paeony] an anthocyanin pigment found in paeonies and other plants.

PEP phospho enol pyruvate, *q.v.*

pepo *n.* [Gk. *pepōn*, melon] an inferior 1-celled, many-seeded pulpy fruit, being a special kind of berry, as of the Cucurbitaceae, such as cucumber.

pepsin *n.* [Gk. *pepsis*, digestion] an endopeptidase enzyme which hydrolyses proteins to peptides, works at a low pH, and is formed from pepsinogen using hydrochloric acid, found in the stomach of vertebrates and secreted by some carnivorous plants; EC 3.4.23.1; *r.n.* pepsin A.

pepsinogen *n.* [Gk. *pepsis*, digestion; *gennaein*, to produce] an enzyme precursor secreted in the gastric mucosa and activated by hydrochloric acid from oxyntic cells to form pepsin.

peptic *a.* [Gk. *peptein*, to digest] relating to or promoting digestion.

peptic cells cells in the gastric glands of the stomach which secrete pepsinogen and prorennin; *alt.* chief cells, principal cells; *cf.* oxyntic cells.

peptidases *n.plu.* [Gk. *peptein*, to digest] formerly enzymes which attack the peptide bond in proteins, including endopeptidases and exopeptidases; now used only for enzymes which split off only 1 or 2 units at the end of a protein; EC 3.4.11–17.

peptide *n.* [Gk. *peptein*, to digest] a chain of amino acids linked by peptide bonds, the product of partial hydrolysis of protein, formerly used for a unit smaller than a peptone.

peptide bond a covalent bond joining 2 amino acids, with the amino group of one acid bonded to the carboxyl group of the other with the elimination of water.

peptide hydrolase a general term for enzymes hydrolysing peptide bonds, including the proteinases and peptidases; EC subgroup 3.4.

peptidoglycan *n.* [Gk. *peptein*, to digest; *glykys*, sweet] a molecule made of a sugar and a peptide, e.g. murein.

peptidyl transferase an enzyme which catalyses the formation of peptide bonds during translation, and is one of the proteins of the ribosome; EC sub-subgroup 2.3.2; *r.n.* amino-acyltransferases.

peptonephridia *n.plu.* [Gk. *pepsis*, digestion; *nephros*, kidney] the anterior nephridia which function as digestive glands, of some Oligochaeta.

peptones *n.plu.* [Gk. *peptein*, to digest] polypeptides, the product of hydrolysis of proteins by enzymes such as pepsin, formerly used to mean a fraction intermediate in size between proteoses and the smaller peptides.

Peracarida *n.* [Gk. *pēra*, pouch; L. *caris*, a kind of crab] a superorder or in some classifications an order of malacostracans having a small carapace and the 1st thoracic segment fused with the head.

peraeopods pereiopods, *q.v.*

Perciformes *n.* [L. *perca*, perch; *forma*, shape] a very large order of teleosts, being typical acanthopterygian fishes and including the perch, mackerel, blennies, etc.; *alt.* Percomorphi.

percnosome *n.* [Gk. *perknos*, dark; *sōma*, body] deeply staining granule of an androcyte, possibly a chromatoid accessory body.

Percomorphi *n.* [L. *perca*, perch; Gk. *morphē*, form] Perciformes, *q.v.*

Percopsiformes *n.* [L. *perca*, perch; Gk. *opsis*, appearance; L. *forma*, shape] a small order of teleosts comprising North American freshwater fishes; *alt.* Salmopercae.

percurrent *a.* [L. *percurrens*, running through] extending throughout length, or from base to apex.

perdominant *n.* [L. *per*, throughout; *dominans*, ruling] a species present in almost all the associations in a given formation.

pereion *n.* [Gk. *peraioun*, to convey] the thorax of Crustacea.

pereiopods *n.plu.* [Gk. *peraioun*, to convey; *pous*, foot] the locomotory thoracic limbs of Malacostraca; *alt.* trunk legs, peraeopods, pereopods.

perennation *n.* [L. *per*, through; *annus*, year] survival for a number of years.

perennial *a.* [L. *per*, through; *annus*, year] *appl.* plants living for more than 2 years, often for a number of years, and usually flowering each year.

perennibranchiate *a.* [L. *per*, through; *annus*, year; *branchiae*, gills] having gills persisting throughout life, as certain amphibians.

pereopods pereiopods, *q.v.*

perfect *a.* [L. *perfectus*, finished] complete; *appl.* flower with both stamens and pistil; *appl.* seed with radicle, cotyledons, and plumule; *appl.* fungi producing sexual spores.

perfoliate *a.* [L. *per*, through; *folium*, leaf] *appl.* a leaf with basal lobes so united as to appear as if stem ran through it; *appl.* antennae with expanded joints apparently surrounding the connecting axis, as in Lamellicornes, stag and dung beetles, etc.

perforate *a.* [L. *perforare*, to bore through] having pores, as corals, foraminiferans, some leaves; *appl.* certain areas of brain perforated by small blood vessels.

perforation plate perforate septum or area of contact between cells or elements of wood vessels.

perforator *n.* [L. *perforare*, to bore through] a barbed spear-like head and process of some spermatozoa, as of salamander.

perforatorium *n.* [L. *perforare*, to bore

through] the acrosome, *q.v.*; acrosome with galea capitis.

perianth *n.* [Gk. *peri*, around; *anthos*, flower] the outer whorl of floral leaves of a flower, when not clearly divided into calyx and corolla; collectively, the calyx and corolla; the cover or sheath surrounding the archegonia in some bryophytes.

periaxial *a.* [Gk. *peri*, around; *axōn*, axis] surrounding an axis or an axon; *appl.* space between axolemma and sheath of Schwann.

periblast *n.* [Gk. *peri*, around; *blastos*, bud] the outside layer, epiblast, or blastoderm of an insect embryo; syncytium formed by fusion of small marginal blastomeres and not forming part of mammalian embryo.

periblastesis *n.* [Gk. *peri*, around; *blastē*, growth] envelopment by surrounding tissue, as of lichen gonidia.

periblastic *a.* [Gk. *peri*, around; *blastos*, bud] *pert.* periblast; superficial, as *appl.* segmentation.

periblastula *n.* [Gk. *peri*, around; *blastos*, bud] a blastula resulting from periblastic segmentation.

periblem *n.* [Gk. *peri*, around; *blēma*, coverlet] the meristem which produces the cortex; a histogen between the dermatogen and plerome.

peribranchial *a.* [Gk. *peri*, around; *brangchia*, gills] around gills; *appl.* type of gemmation in ascidians; *appl.* atrial cavity in ascidians and lancelet; *appl.* circular spaces surrounding basal parts of papulae of Asteroidea.

peribulbar *a.* [Gk. *peri*, around; L. *bulbus*, bulb] surrounding the eyeball; perigemmal, *q.v.*

pericambium *n.* [Gk. *peri*, around; L. *cambium*, change] the layer of cells around the stele, also called pericycle, phloem sheath, or periphloem.

pericapillary *a.* [Gk. *peri*, around; L. *capillus*, hair] *appl.* cells in contact with outer surface of wall of capillaries, as fibroblasts, histiocytes, pericytes, Rouget cells.

pericardiac, pericardial *a.* [Gk. *peri*, around; *kardia*, heart] *pert.* pericardium; surrounding heart, *appl.* cavity, septum; *appl.* paired excretory glands in lamellibranchs, *alt.* Keber's organ; *appl.* cells: cords of nephrocytes in certain insects.

pericardium *n.* [Gk. *peri*, around; *kardia*, heart] the cavity containing heart; membrane enveloping heart.

pericarp *n.* [Gk. *peri*, around; *karpos*, fruit] the fruit wall which has developed from the ovary wall; sometimes used for any fruit covering.

pericaryon perikaryon, *q.v.*

pericellular *a.* [Gk. *peri*, around; L. *cellula*, small cell] surrounding a cell; *appl.* net of glial origin surrounding a neurocyton; *alt.* pericytial.

pericemental periodontal, *q.v.*

pericentral *a.* [Gk. *peri*, around; L. *centrum*, centre] around or near centre; *appl.* auxiliary cells, as in certain algae.

pericentric *a.* [Gk. *peri*, around; *kentron*, centre] *pert.* or involving the centromere of a chromosome; *appl.* breaks in arms of a chromosome on either side of the centromere; *appl.* inversions including the centromere; *alt.* heterobrachial, eucentric; *cf.* paracentric.

perichaetial *a.* [Gk. *peri*, around; *chaitē*, foliage] *pert.* perichaetium.

perichaetine *a.* [Gk. *peri*, around; *chaitē*, hair] having a ring of chaetae or setae encircling the body.

perichaetium *n.* [Gk. *peri*, around; *chaitē*, foliage] one of membranes or leaves enveloping archegonia or antheridia of bryophytes; *alt.* perigamium; *cf.* involucre; *a.* perichaetial.

perichondral *a.* [Gk. *peri*, around; *chondros*, cartilage] *appl.* ossification in cartilage beginning outside and working inwards; *cf.* endochondral.

perichondrium *n.* [Gk. *peri*, around; *chondros*, cartilage] a fibrous membrane that covers cartilages.

perichondrostosis *n.* [Gk. *peri*, around; *chondros*, cartilage; *osteon*, bone] perichondral ossification; *cf.* endochondrostosis.

perichordal *a.* [Gk. *peri*, around; *chordē*, cord] enveloping or near the notochord.

perichoroidal *a.* [Gk. *peri*, around; *chorion*, skin; *eidos*, form] surrounding the choroid, *appl.* lymph space; *alt.* perichorioidal.

perichrome *a.* [Gk. *peri*, around; *chrōma*, colour] having Nissl bodies arranged near periphery of nerve cell body, as in molecular layer of cerebellar cortex.

perichylous *a.* [Gk. *peri*, around; *chylos*, juice] with water-storage cells outside chlorenchyma.

pericladium *n.* [Gk. *peri*, around; *klados*, branch] the lowermost clasping portion of a sheathing petiole.

periclinal *a.* [Gk. *peri*, around; *klinein*, to bend] *appl.* system of cells parallel to surface of apex of a growing point; *appl.* graft hybrids or chimaeras with inner tissue of one species surrounded by epidermis of the other.

periclinium *n.* [Gk. *peri*, round; *klinē*, bed] the involucre of a composite flower; *alt.* periphoranthium.

pericranium *n.* [Gk. *peri*, around; *kranion*, skull] fibrous membrane investing skull; periosteum of skull.

pericycle *n.* [Gk. *peri*, around; *kyklos*, circle] the external layer of stele, the layer between endodermis and conducting tissues; *alt.* pericambium.

pericyclic *a.* [Gk. *peri*, around; *kyklos*, circle] *appl.* fibre situated on the outer edge of the vascular region and usually arising in the primary phloem, although sometimes outside it.

pericyte *n.* [Gk. *peri*, around; *kytos*, hollow] a macrophage in adventitia of small blood vessels, in some species also called Rouget cell; *alt.* pericapillary cell.

pericytial *a.* [Gk. *peri*, around; *kytos*, hollow vessel] pericellular, *q.v.*

peridental *a.* [Gk. *peri*, around; L. *dens*, tooth] periodontal, *q.v.*

periderm *n.* [Gk. *peri*, around; *derma*, skin] phellogen, phellem, and phelloderm collectively, *alt.* bark, outer bark, epiphloem, rhytidome; perisarc of Hydrozoa; outer cell layer of epidermis, shed later; epitrichium, *q.v.*

peridesm *n.* [Gk. *peri*, around; *desmē*, bundle] tissue surrounding a vascular bundle.

peridesmium *n.* [Gk. *peri*, around; *desmos*, bond] tissue surrounding a ligament.

peridial *a.* [Gk. *pēridion*, small wallet] *pert.* a peridium.

perididymis *n.* [Gk. *peri*, around; *didymos*, testicle] the tunica albuginea of testis.

Peridiniales *n.* [*Peridinium*, generic name] an order of Dinophyceae having armoured forms; *alt.* Dinophycales.

peridinin *n.* [*Peridinium*, generic name] a xanthophyll pigment found in algae of the class Pyrrophyceae.

peridiole, peridiolum *n.* [*Dim.* of Gk. *pēridion*, small wallet] one of the small portions, 'eggs', into which the gleba of a bird's nest fungus is broken up, lying inside a cup-like peridium; *alt.* gleba chamber or cavity.

peridium *n.* [Gk. *pēridion*, small wallet] the outer layer or membrane investing the sporophore of a fungus and completely or partially surrounding the spore-bearing organs; a wall-like layer around sporangia in slime moulds; cortex of sterile hyphae; *alt.* saccule, sacculus.

peridural *a.* [Gk. *peri*, around; L. *durus*, hard] *appl.* perimeningeal space at later stage of development.

perienteric *a.* [Gk. *peri*, around; *enteron*, gut] surrounding the enteron.

perienteron *n.* [Gk. *peri*, around; *enteron*, gut] a cavity surrounding the enteron; visceral cavity in embryo.

perifibrillar *a.* [Gk. *peri*, around; L. *fibrilla*, small fibre] surrounding a fibril; *appl.* substance: axoplasm surrounding neurofibrils.

perifoliary *a.* [Gk. *peri*, around; L. *folium*, leaf] around a leaf margin.

perigamium perichaetium, *q.v.*

periganglionic *a.* [Gk. *peri*, around; *gangglion*, little tumour] surrounding a ganglion; *appl.* glands: Swammerdam's glands, *q.v.*

perigastric *a.* [Gk. *peri*, around; *gastēr*, stomach] surrounding the viscera, *appl.* abdominal cavity.

perigastrium *n.* [Gk. *peri*, around; *gastēr*, stomach] body cavity, *q.v.*

perigastrula *n.* [Gk. *peri*, around; *gastēr*, stomach] the gastrula resulting after superficial segmentation.

perigemmal *a.* [Gk. *peri*, around; L. *gemma*, bud] surrounding a taste bud, *appl.* nerve fibres, spaces; *alt.* peribulbar.

perigenous *a.* [Gk. *peri*, around; *-genēs*, producing] *see* amphigenous.

perigonadial *a.* [Gk. *peri*, around; *gonē*, seed] surrounding the gonads; *appl.* cavity: the gonocoel, *q.v.*

perigonium *n.* [Gk. *peri*, around; *gonē*, seed] a floral envelope or perianth; involucre around antheridium of mosses; gonotheca, *q.v.*; *alt.* perigone.

perigynium *n.* [Gk. *peri*, around; *gynē*, female] membranous envelope or marsupium of archegonium in liverworts; involucre in mosses; fruit investing utricle of the sedge, *Carex.*

perigynous *a.* [Gk. *peri*, around; *gynē*, female] *appl.* flowers in which the receptacle develops into a flat or concave structure between the base of the gynaecium and the insertion of the perianth and stamens; *n.* perigyny; *cf.* epigynous, hypogynous.

perihaemal *a.* [Gk. *peri*, around; *haima*, blood] *appl.* blood vascular system of canals and spaces of Echinodermata; *appl.* dorsal outgrowths of 3rd body cavity of Enteropneusta.

perikaryon *n.* [Gk. *peri*, around; *karyon*, nucleus] cytoplasm surrounding nucleus in nerve cell body; the whole cell body of nerve cell; *alt.* pericaryon.

perilymph *n.* [Gk. *peri*, around; L. *lympha*, water] a fluid separating membranous from osseous labyrinth of ear.

perimedullary *a.* [Gk. *peri*, around; L. *medulla*, marrow] surrounding the pith in a stem, *appl.* a zone, also called medullary sheath.

perimeningeal *a.* [Gk. *peri*, around; *meningx*, membrane] *appl.* a space between endorhachis and meninx primitiva or spinal cord envelope.

perimetrium *n.* [Gk. *peri*, around; *mētra*, womb] the peritoneal covering of the uterus.

perimysium *n.* [Gk. *peri*, around; *mys*, muscle] connective tissue binding numbers of fibres into bundles and muscles, and continuing into tendons; alternatively, *appl.* only to fasciculi envelopes; *cf.* epimysium, endomysium.

perinaeal *a.* [Gk. *perinaion*, part between anus and scrotum] *pert.* perinaeum, *appl.* artery, body, nerve, gland; *alt.* perineal.

perinaeum *n.* [Gk. *perinaion*, part between anus and scrotum] a surface of body limited by scrotum or vulva in front, anus behind, and laterally by medial side of thigh; *alt.* perineum.

perine perinium, *q.v.*

perineal perinaeal, *q.v.*

perinemata *n.plu.* [Gk. *peri*, around; *nēma*, thread] mitochondria in certain cells of connective tissue; *sing.* perinema.

perinephrium *n.* [Gk. *peri*, around; *nephros*, kidney] the enveloping adipose and connective tissue of kidney.

perineum perinaeum, *q.v.*

perineural *a.* [Gk. *peri*, around; *neuron*, nerve] surrounding a nerve or nerve cord; *appl.* a ventral sinus in some insects.

perineurium *n.* [Gk. *peri*, around; *neuron*, nerve] the tubular sheath of a small bundle of nerve fibres.

perineuronal *a.* [Gk. *peri*, around; *neuron*, nerve] surrounding a nerve cell or nerve cells.

perinium *n.* [Gk. *peri*, around; *is*, fibre] outer microspore coating of certain pteridophytes; *alt.* perine, epispore.

periocular *a.* [Gk. *peri*, around; L. *oculus*, eye] surrounding the eyeball within the orbital cavity.

period *n.* [Gk. *periodos*, circuit] in geological time, a subdivision of an era, e.g. the Jurassic is a period of the Mesozoic era.

periodicity *n.* [Gk. *periodos*, circuit] the fulfilment of functions at regular periods or intervals; rhythm, *q.v.*

periodontal *a.* [Gk. *peri*, around; *odous*, tooth] covering or surrounding a tooth, *appl.* membrane, etc.; *alt.* pericemental, peridental.

perioesophageal *a.* [Gk. *peri*, around; *oisophagos*, gullet] surrounding oesophagus, *appl.* a nerve ring.

periople *n.* [Gk. *peri*, around; *hoplē*, hoof] thin outer layer of the hoof of equines.

periopticon *n.* [Gk. *peri*, around; *opsis*, sight] in insects, the zone of optic lobes nearest the eye.

periorbital *a.* [Gk. *peri*, around; L. *orbis*, eye socket] surrounding the orbit of the eye.

periosteum *n.* [Gk. *peri*, around; *osteon*, bone] the fibrous membrane investing the surface of bones.

periostracum *n.* [Gk. *peri*, around; *ostrakon*, shell] the external layer of most mollusc and brachiopod shells.

periotic *n.* [Gk. *peri*, around; *ous*, ear] a cranial bone enclosing parts of membranous labyrinth of internal ear; *alt.* petrosal or petrous bone.

peripetalous a. [Gk. peri, around; petalon, leaf] surrounding petals or petaloid structure.

peripharyngeal a. [Gk. peri, around; pharyngx, gullet] encircling or surrounding pharynx, appl. cilia of ascidians and cephalochordates.

peripheral a. [Gk. peripherein, to move around] distant from centre or near circumference.

peripheral nervous system the nervous system other than the CNS, made up mostly of nerves.

peripherical a. [Gk. peripherein, to move around] appl. an embryo more-or-less completely surrounding endosperm in seed.

periphloem n. [Gk. peri, around; phloios, smooth bark] phloem sheath, q.v.; alt. pericambium.

periphloic a. [Gk. peri, around; phloios, inner bark] pert. periphloem; having phloem outside centric xylem, appl. bundles; cf. perixylic.

periphoranthium n. [Gk. peri, around; pherein, to bear; anthos, flower] periclinium, q.v.

periphorium n. [Gk. peri, around; pherein, to bear] fleshy structure supporting ovary, and to which stamens and corolla are attached.

periphyllum n. [Gk. peri, around; phyllon, leaf] lodicule, q.v.

periphysis n. [Gk. peri, around; physis, growth] in certain fungi, a filament branching from an hymenium without asci.

periphyton n. [Gk. peri, around; phyton, plant] the plants and animals adhering to parts of rooted aquatic plants.

peripileic a. [Gk. peri, around; L. pileus, cap] pert. or arising from the marginal region of a pileus.

periplasm n. [Gk. peri, around; plasma, form] the region of an oogonium outside the oosphere, in fungi; centroplasm or zone around the aster; cytoplasm surrounding yolk of centrolecithal ova.

periplasmodium n. [Gk. peri, around; plasma, form, eidos, form] protoplasmic mass, derived from tapetal cells and enclosing developing spores.

periplast n. [Gk. peri, around; plastos, moulded, formed] centrosome, q.v.; ectoplasm of flagellates; pellicle covering ectoplasm; intercellular substance or stroma of tissues.

peripneustic a. [Gk. peri, around; pneustikos, pert. breathing] having spiracles arranged all along sides of body, normal in insect larvae; cf. amphipneustic, metapneustic.

peripodial a. [Gk. peri, around; pous, foot] surrounding an appendage; appl. membrane covering wing bud of insects.

periportal a. [Gk. peri, around; L. porta, gate] pert. transverse fissure of the liver; appl. connective tissue partially separating lobules and forming part of the hepatobiliary capsule of Glisson.

periproct n. [Gk. peri, around; prōktos, anus] the surface immediately surrounding anus of echinoids.

peripyle n. [Gk. peri, around; pylē, gate] one of the apertures, additional to astropyle, of the central capsule in certain Radiolaria.

perisarc n. [Gk. peri, around; sarx, flesh] the chitinous outer covering of the colony of some hydrozoans; cf. coenosarc.

Perischoechinoidea n. [Gk. perischesis, surrounding; echinos, sea urchin; eidos, form] a subclass of mostly extinct globular sea urchins having up to 14 columns of simple plates in the interambulacra and from 2 to 20 in the ambulacra.

periscleral a. [Gk. peri, around; sklēros, hard] appl. lymph space external to sclera of eye.

perisome n. [Gk. peri, around; sōma, body] a body wall; integument of echinoderms.

perisperm n. [Gk. peri, around; sperma, seed] nutritive tissue in certain seeds derived from the nucellus rather than from the endosperm and deposited outside the embryo sac; sometimes used for any nutritive tissue in a seed including that formed from both endosperm and nucellus.

perisphere n. [Gk. peri, around; sphaira, ball] outer region of centrosphere.

perispiracular a. [Gk. peri, around; L. spiraculum, air hole] surrounding a spiracle; appl. glands with oily secretion, in certain aquatic insect larva; alt. peristigmatic.

perisporangium n. [Gk. peri, around; sporos, seed; anggeion, vessel] membrane covering a sorus; indusium of ferns.

perispore n. [Gk. peri, around; sporos, seed] spore-covering; transient outer membrane enveloping a spore; alt. perisporium; mother cell in algal spores; alt. epispore.

Perisporiales n. [Gk. peri, around; sporos, seed] Erysiphales, q.v.

perissodactyl a. [Gk. perissos, odd; daktylos, finger] with uneven number of digits; cf. artiodactyl.

Perissodactyla, perissodactyls n., n.plu. [Gk. perissos, odd; daktylos, finger] an order of placental mammals known from the Eocene, commonly called odd-toed ungulates and including the horse, tapir, and rhinoceros, which have a large caecum but a simple stomach, teeth adapted for grinding, and a large 3rd digit; cf. Artiodactyla.

peristalsis n. [Gk. peri, around; stellein, to draw in] movement of muscular tubes, as of digestive tract, by means of successive contractions in a definite, usually anteroposterior, direction; a. peristaltic.

peristasis n. [Gk. peri, around; stasis, standing] environment, including physiological action within the organism, vital to development of a particular genotype; a. peristatic.

peristethium n. [Gk. peri, around; stēthos, chest] an insect mesosternum.

peristigmatic perispiracular, q.v.

peristome n. [Gk. peri, around; stoma, mouth] the region surrounding the mouth, appl. moss capsule, ciliates such as Vorticella, Actinozoa, annulates, insects, echinoderms, etc.; alt. peristomium.

peristomium peristome, q.v.

perisystole n. [Gk. peri, around; systolē, drawing together] the interval elapsing between diastole and systole of heart.

perithecium n. [Gk. peri, around; thēkē, case] a flask-shaped asocarp with a terminal ostiole as in Pyrenomycetes; alt. pyrenocarp.

perithelium n. [Gk. peri, around; thēlē, nipple] connective tissue associated with capillaries.

peritoneal a. [Gk. periteinein, to stretch around] pert. peritoneum, appl. cavity, fossa, membrane, etc.; appl. funnel: coelostome of archinephros.

peritoneum n. [Gk. periteinein, to stretch around] a serous membrane partly applied to abdominal walls, partly reflected over contained viscera; *cf.* serosa.

peritreme, peritrema *n.* [Gk. *peri,* around; *trēma,* hole] margin of a shell opening; small plate perforated by spiracle opening in ticks and insects.

Peritricha, Peritrichida *n.* [Gk. *peri,* around; *thrix,* hair] an order of Ciliata which are usually permanently fixed by the aboral surface and have reduced cilia.

peritrichous *a.* [Gk. *peri,* around; *thrix,* hair] having adoral band of cilia arranged in a spiral as in the ciliate *Vorticella*; having several flagella attached laterally, as certain bacteria; surrounding a hair follicle, *appl.* nerve endings; *alt.* peritrichal, peritrichic.

peritrochium *n.* [Gk. *peri,* around; *trochos,* wheel] a ciliary band; a circularly ciliated larva.

peritrophic *a.* [Gk. *peri,* around; *trophē,* food] *appl.* a fold of membrane in mid-gut of insects and to space between it and gut lining; *appl.* mycorrhiza with special fungal populations on root surfaces.

perittogamy *n.* [Gk. *perittos,* extraordinary; *gamos,* marriage] random plasmogamy of undifferentiated cells in gametophytes.

periurethral *a.* [Gk. *peri,* around; *ourēthra,* from *ouron,* urine] surrounding the urethra, *appl.* glands, homologues of prostate.

perivascular *a.* [Gk. *peri,* around; L. *vasculum,* small vessel] surrounding the vascular cylinder, *appl.* fibres; surrounding the blood vessels, *appl.* lymph channels, fibres; *appl.* feet: foot-plates, *q.v.*

perivisceral *a.* [Gk. *peri,* around; L. *viscera, bowels*] surrounding the viscera, *appl.* body cavity.

perivitelline *a.* [Gk. *peri,* around; L. *vitellus,* yolk of egg] surrounding the yolk of an egg; *appl.* space between ovum and zona pellucida.

perixylic *a.* [Gk. *peri,* around; *xylon,* wood] having xylem outside centric phloem, *appl.* bundles; *cf.* periphloem.

perizonium *n.* [Gk. *peri,* around; *zōnē,* girdle] the membrane or siliceous wall enveloping the auxospore or zygote in diatoms.

Perleidiformes *n.* [*Perleidus,* generic name; L. *forma,* shape] an order of Triassic Chondrostei with ganoid scales.

permafrost the permanently frozen ground under the soil in polar regions.

permanent cartilage cartilage which remains unossified throughout life as opposed to temporary cartilage which is ossified to bone.

permanent hybrid a heterozygote which breeds true because of the elimination of certain homozygous genotypes by lethal factors in the genotype.

permanent teeth or dentition a set of teeth developed after milk or deciduous dentition; 2nd set of most, 3rd set of some, 1st set of other mammals.

permanent tissue tissue consisting of cells which have completed their period of growth, are fully differentiated, and subsequently change little until they lose their protoplasm and die.

permeability vitamin vitamin P, *q.v.*

permeants *n.plu.* [L. *permeare,* to pass through] animals which move freely from one community or habitat to another.

permease *n.* [L. *permeare,* to pass through] any of a hypothetical group of proteins associated with cell membranes that permit facilitated diffusion or active transport of materials across the membranes and into cells; a protein bound to the membrane in the bacterium *Escherichia coli* which is often considered an enzyme and is responsible for carrying lactose into the cell.

permeation *n.* [L. *permeare,* to pass through] dialysis, *q.v.*

Permian *a.* [*Perm.* E. Russia] *pert.* late period of Palaeozoic era, following the Carboniferous.

peronaeus peroneus, *q.v.*

peronate *a.* [L. *peronatus,* hide-booted] covered with woolly hairs; surrounded by volva, *appl.* stipe; powdery or mealy externally; *alt.* caligate.

peroneal *a.* [Gk. *peronē,* fibula] *pert.*, or lying near, the fibula, *appl.* artery, nerve, retinacula, tubercle.

peroneotibial *a.* [Gk. *peronē,* fibula; L. *tibia,* tibia] in region of fibula and tibia, *appl.* certain muscles.

peroneus *n.* [Gk. *peronē,* fibula] any of several muscles arising from the fibula in the lower leg, as 2 lateral muscles, longus and brevis, and an anterior muscle, tertius; *alt.* peronaeus, fibularis.

peronium *n.* [Gk. *peronē,* fibula] in Trachymedusae, one of the mantle rivets, or cartilaginous processes ascending from disc margin towards centre.

Peronosporales *n.* [*Peronospora,* generic name] an order of Phycomycetes, including many plant parasites, having asexual reproduction by biflagellate zoospores and oogamous sexual reproduction.

peropod *a.* [Gk. *pēros,* defective; *pous,* foot] with rudimentary limbs.

peroral *a.* [L. *per,* through; *os,* mouth] *appl.* a membrane formed by concrescence of rows of cilia around cytopharynx in ciliates.

peroxidases *n.plu.* [L. *per,* through; Gk. *oxys,* sharp] enzymes occurring in plants and animal tissues, especially mammalian spleen and lung, which act on hydrogen peroxide or organic peroxidases only in the presence of a substrate which accepts oxygen, EC sub-subgroup 1.11.1; specifically, an enzyme using several oxidizing donors with hydrogen peroxide, EC 1.11.1.7, *r.n.* peroxidase; *cf.* catalase.

peroxisomes, peroxysomes *n.plu.* [L. *per,* through; Gk. *oxys,* sharp; *sōma,* body] small organelles with a single membrane, sometimes associated with smooth endoplasmic reticulum, and containing a number of enzymes including those of the glyoxylate cycle, catalase, and peroxidases; *alt.* glyoxysomes, microbodies, microsomes.

perradius *n.* [L. *per,* through; *radius,* radius] one of 4 primary radii of coelenterates.

perseveration *n.* [L. *perseverare,* to persist] tendency of a set of neurones to remain in a state of excitation; persistent response after cessation of original stimulus.

persistent *a.* [L. *persistere,* to persevere] remaining attached until maturation, as a corolla or perianth; *appl.* teeth with continuous growth; *appl.* organs or parts in adult which normally disappear in the larval stage or youth, such as gills.

person *n.* [L. *persona,* person] an individual or zooid of a colony.

Personales, Personatae n. [L. *personatus*, masked] Scrophulariales, *q.v.*

personate a. [L. *personatus*, masked] *appl.* a corolla of 2 lips, closely approximated and with a projection of the lower closing the throat of the corolla; *alt.* masked.

perthophyte n. [Gk. *perthai*, to destroy; *phyton*, plant] a parasitic fungus that obtains nourishment from host tissues after having killed them by a poisonous secretion.

pertusate a. [L. *pertusus*, thrust through] pierced at apex.

perula n. [L. *perula*, little wallet] a leaf-bud scale.

pervalvar a. [L. *per*, through; *valvae*, folding doors] dividing a valve longitudinally.

pervious a. [L. *pervius*, passable] perforated; permeable; *appl.* nostrils with no septum between nasal cavities.

pes n. [L. *pes*, foot] a foot, base, or foot-like structure, as certain parts of brain, branches of facial nerve; *plu.* pedes.

pessulus n. [L. *pessulus*, bolt] an internal dorsoventral rod at lower end of trachea in syrinx of some birds.

petal n. [Gk. *petalon*, leaf] one of the inner series of perianth segments (the corolla) if differing from the outer series (the calyx), especially if brightly coloured; expanded part of ambulacral areas of certain echinoderms.

petaled a. [Gk. *petalon*, leaf] petaliferous, *q.v.*

petaliferous a. [Gk. *petalon*, leaf; L. *ferre*, to carry] bearing petals; *alt.* petaled; *cf.* apetalous.

petaliform a. [Gk. *petalon*, leaf; L. *forma*, shape] petal-shaped; petal-like, *alt.* petaloid, petaline.

Petalodontiformes n. [Gk. *petalon*, leaf; *odous*, tooth; L. *forma*, shape] an order of marine Carboniferous to Permian Holocephali known only from their teeth and of uncertain taxonomic position.

petalody n. [Gk. *petalon*, leaf; *eidos*, form] conversion of other parts of a flower into petals.

petaloid a. [Gk. *petalon*, leaf; *eidos*, form] like a petal, *appl.* perianth, *appl.* pileus, *appl.* ambulacral areas of certain echinoderms; *alt.* petaliform.

petaloideous a. [Gk. *petalon*, leaf; *eidos*, form] petaloid, *appl.* monocotyledons with coloured perianth.

petalomania n. [Gk. *petalon*, leaf; *mania*, madness] an unusual multiplication of petals.

petasma n. [Gk. *petasma*, anything spread out] a complicated membranous plate on inner side of peduncle with interlocking coupling hooks, an apparatus of certain Crustacea.

petiolar a. [L. *petiolus*, small foot] *pert.*, having, or growing on, a small stalk.

petiolate a. [L. *petiolus*, small foot] growing on, or provided with, a petiole; having thorax and abdomen connected by a petiole.

petiole n. [L. *petiolus*, small foot] the stalk of a leaf, *alt.* foot-stalk, leaf stalk; a slender stalk connecting the thorax and abdomen in certain insects such as Hymenoptera, *alt.* podeon; a small sclerite at base of palpal organ in spiders; flattened and modified barb base in feathers.

petiolule n. [*Dim.* of L. *petiolus*, small foot] the stalk of a leaflet of a compound leaf.

petites *n.plu.* [F. *petite*, small] colonies of slow-growing yeast, *Saccharomyces cerevisiae*, that are dwarf and lack certain respiratory enzymes, vegetative petites containing mutations in mitochondria, and segregational petites containing mutations in nuclear genes.

Petit's canal spatia zonularia, *q.v.*

petrification n. [L. *petra*, rock; *facere*, to make] fossilization through saturation by mineral matter in solution, subsequently turned to solid form.

petrohyoid a. [Gk. *petros*, stone; *hyoeidēs*, Y-shaped] *pert.* hyoid and petrous part of temporal.

petromastoid a. [Gk. *petros*, stone; *mastos*, breast; *eidos*, form] *pert.* mastoid process and petrous portion of temporal.

Petromyzontia n. [*Petromyzon*, generic name] an order of freshwater and anadromous cyclostomes (chordates), commonly called lampreys, with sucking and rasping mouth-parts; *alt.* Hyperoartia.

petro-occipital a. [Gk. *petros*, stone; L. *occiput*, back of head] *pert.* occipital and petrous part of temporal, *appl.* a fissure.

petrophyte n. [Gk. *petros*, stone; *phyton*, plant] a rock plant.

petrosal a. [Gk. *petros*, stone] of compact bone; *appl.* otic bones of fishes; *appl.* a sphenoidal process, to a ganglion of glossopharyngeal, to nerves and sinus in region of petrous portion of temporal bone; *appl.* bone, the periotic, *q.v.*; *alt.* petrous.

petrosphenoidal a. [Gk. *petros*, stone; *sphēn*, wedge; *eidos*, form] *pert.* sphenoid and petrous part of temporal, *appl.* fissure.

petrosquamosal, petrosquamous a. [Gk. *petros*, stone; L. *squama*, scale] *pert.* squamosal and petrous part of temporal, *appl.* sinus and suture.

petrotympanic a. [Gk. *petros*, stone; *tympanon*, drum] *pert.* tympanum and petrous portion of temporal; *appl.* a fissure: Glaserian fissure, *q.v.*

petrous a. [Gk. *petros*, stone] very hard or stony; *appl.* a pyramidal portion of temporal bone between sphenoid and occipital; *appl.* a ganglion on its lower border; *alt.* petrosal.

petunidin n. [*Petunia*, generic name] a purple anthocyanin pigment found in petunias and other plants.

Peyer's glands or patches [*J. C. Peyer*, Swiss anatomist] white roundish patches of aggregated lymphatic nodules scattered along the inner lining of the intestine walls; *alt.* agminated glands.

Pezizales n. [*Peziza*, generic name] an order of Discomycetes in which asci dehisce by a lid.

P_F, P_{FR} the pigment phytochrome in its active form causing developmental responses, produced by the action of red light on P_R; *alt.* P_{730}.

Pflüger's cords [*E. F. W. Pflüger*, German physiologist] cell columns growing from the germinal epithelium into the stroma, and which give rise to gonads.

PGA phosphoglyceric acid, *q.v.*; pteroylglutamic acid, *q.v.*

pH the negative value of the power to which 10 is raised in order to obtain the concentration of hydrogen ions in gram molecules per litre, pH of a neutral solution being 7; pH of acid solutions being smaller than 7, pH of alkaline solutions being greater than 7.

phacea n. [Gk. *phakos*, lentil] the crystalline lens of the eye.

phacella n. [Gk. *phakelos*, bundle] a delicate filament with mesogloea core, and supplied with stinging capsules, occurring in rows in enteron of certain coelenterates; *alt.* gastric filament.

Phacidiales *n.* [*Phacidium*, generic name] an order of Discomycetes in which the asci are developed in cavities and form a hymenium later.

phacocyst *n.* [Gk. *phakos*, lentil, lens; *kystis*, bladder] transparent sac enclosing lens of eye; capsule of the lens, *alt.* capsula lentis.

phacoid *a.* [Gk. *phakos*, lentil; *eidos*, form] lentil-shaped.

phaeic phaeochrous, *q.v.*

phaeism *n.* [Gk. *phaios*, dusky] duskiness; *appl.* colouring of butterflies due to incomplete melanism.

phaenantherous *a.* [Gk. *phainein*, to show; *anthēros*, flowering] with anthers exserted; with stamens exserted.

phaeno- *alt.* pheno-, *q.v.*

Phaenogamia, phaenogams *n., n.plu.* [Gk. *phainein*, to show; *gamos*, marriage] Phanerogamia, *q.v.*

phaeo- *alt.* pheo-.

phaeochrome *n.* [Gk. *phaios*, dusky; *chrōma*, colour] chromaffin, *q.v.*

phaeochromoblast *n.* [Gk. *phaios*, dusky; *chrōma*, colour; *blastos*, bud] cell which develops into a phaeochromocyte or chromaffin cell.

phaeochrous *a.* [Gk. *phaios*, dusky; *chrōs*, colour] of dusky colour; *alt.* phaeic.

phaeodium *n.* [Gk. *phaios*, dusky; *eidos*, form] in certain Radiolaria, an aggregation of food and excretory substances forming a mass around the central capsule aperture.

phaeomelanin *n.* [Gk. *phaios*, dusky; *melas*, black] a brownish melanin; *cf.* eumelanin.

phaeophore phaeoplast, *q.v.*

Phaeophyceae *n.* [Gk. *phaios*, dusky; *phykos*, seaweed] a class of algae, commonly called brown algae or brown seaweeds, which have chlorophyll somewhat masked by the brown pigment fucoxanthin so they appear brownish and are mainly marine; in some classifications they are raised to division status called Phaeophyta.

phaeophyll *n.* [Gk. *phaios*, dusky; *phyllon*, leaf] the colouring matter of brown algae, a mixture of fucoxanthin, xanthophyll, chlorophyll, and carotene.

Phaeophyta *n.* [Gk. *phaios*, dusky; *phyton*, plant] in some classifications the Phaeophyceae, *q.v.*, raised to division status.

phaeophytin *n.* [Gk. *phaios*, dusky; *phyton*, plant] either of 2 blue–black pigments derived from chlorophylls a and b by removing magnesium.

phaeoplast *n.* [Gk. *phaios*, dusky; *plastos*, formed] chromoplast of brown algae; *alt.* phaeophore.

phaeospore *n.* [Gk. *phaios*, dusky; *sporos*, seed] a spore containing phaeoplasts.

Phaeothamniales *n.* [*Phaeothamnion*, generic name] Chrysotrichales, *q.v.*

phage *n.* [Gk. *phagein*, to eat] bacteriophage, *q.v.*

phagocytable *a.* [Gk. *phagein*, to eat; *kytos*, hollow] *appl.* bacteria rendered more easily ingested by leucocytes.

phagocyte *n.* [Gk. *phagein*, to eat; *kytos*, hollow] an amoeboid white blood corpuscle which ingests foreign particles; a root cell, with lobed nucleus, capable of digesting endotrophic fungal filaments.

phagocytic *pert.* phagocytes; *pert.* or effecting phagocytosis.

phagocytosis *n.* [Gk. *phagein*, to eat; *kytos*, hollow] uptake of solid particles in a vesicle and into a cell, *cf.* pinocytosis, cytopempsis; ingestion and destruction of microparasites by phagocytes.

phagolysis *n.* [Gk. *phagein*, to eat; *lysis*, loosing] dissolution of phagocytes.

phagolysosome *n.* [Gk. *phagein*, to eat; *lysis*, loosing; *sōma*, body] a cell organelle produced by the fusion of a phagosome and a lysosome.

phagosome *n.* [Gk. *phagein*, to eat; *sōma*, body] a membrane-bound vesicle in the cytoplasm, formed by budding off of an invagination of the cell membrane and containing material taken up by phagocytosis.

phagotrophs *n.plu.* [Gk. *phagein*, to eat; *trophē*, nourishment] heterotrophic organisms which ingest solid particles of food and digest them, as many animals; *cf.* osmotroph.

phagozoite *n.* [Gk. *phagein*, to eat; *zōon*, animal] an animal which feeds on disintegrating or dead tissue.

phalange *n.* [Gk. *phalangx*, line of battle] phalanx, *q.v.*

phalangeal *a.* [Gk. *phalangx*, line of battle] *pert.* or resembling, phalanges, *appl.* bones, *appl.* processes of rods of Corti, of Deiters' cells, etc.

phalanges *n.plu.* [Gk. *phalangx*, line of battle] segments of digits of vertebrates; segments of tarsus of insects; rows of phalangeal processes forming reticular lamina of Corti's organ; *plu.* of phalanx, *q.v.*; *alt.* internodia.

Phalangida *n.* [Gk. *phalangx*, line of battle] Opiliones, *q.v.*

phalanx *n.* [Gk. *phalangx*, line of battle] a bundle of stamens united by filaments; a taxonomic group, never precisely defined, but usually used for a group resembling a subfamily; *alt.* phalange; *plu.* phalanges, *q.v.*

Phallales *n.* [*Phallus*, generic name] an order of Gasteromycetes, commonly known as stinkhorns, having a foetid odour to the gleba and a horn-like mature fruit body.

phallic *a.* [Gk. *phallos*, penis] *pert.* phallus; *appl.* gland secreting substance for spermatophores, as in certain insects.

phallomere *n.* [Gk. *phallos*, penis; *meros*, part] penis valve, in insects.

phallosome *n.* [Gk. *phallos*, penis; *sōma*, body] a structure of tissue from inner surface of basistyles and penis valves in mosquitoes; *alt.* mesosome.

phallus *n.* [Gk. *phallos*, penis] the embryonic structure which becomes penis or clitoris; the penis; external genitalia of male insect; a genus of Basidiomycetes.

phanerocodonic *a.* [Gk. *phaneros*, manifest; *kōdōn*, bell] *appl.* detached and free-swimming medusa of a hydroid colony; *cf.* adelocodonic.

Phanerogamia, Phanerogamae, phanerogams *n., n.plu.* [Gk. *phaneros*, manifest; *gamos*, marriage] Spermatophyta, *q.v.*; formerly sometimes used for plants with conspicuous flowers; *alt.* Anthophyta, Phaenogamia, Phenogamia; *cf.* Cryptogamia.

phanerogamous *a.* [Gk. *phaneros*, manifest; *gamos*, marriage] *pert.* phanerogam; *alt.* phanerogamic.

phanerophyte *n.* [Gk. *phaneros*, manifest; *phyton*, plant] tree or shrub with aerial dormant buds;

plant whose size is not appreciably less during cold or dry season.

phaneroplasmodium n. [Gk. *phaneros*, manifest; *plasma*, form; *eidos*, form] in slime moulds, a thick opaque plasmodium with veins having clearly defined endoplasm and ectoplasm.

phanerozoa n. [Gk. *phaneros*, manifest; *zōon*, animal] animals which have relatives in the cryptozoa, but do not live in the cryptosphere themselves, including winged insects, spiders, and mites.

Phanerozonia n. [Gk. *phaneros*, manifest; *zōnē*, girdle] an order of starfish having rather rigid arms edged with plates.

phaoplankton n. [Gk. *phaos*, light; *plangktos*, wandering] surface plankton, living at depths to which light penetrates; *cf.* knephoplankton, skotoplankton.

phaosome n. [Gk. *phaos*, light; *sōma*, body] an optic organelle in certain epidermal cells of annelids.

pharate a. [Gk. *pharos*, loose mantle] *appl.* instar within previous cuticle prior to ecdysis.

Pharetronida n. [L. *pharetra*, quiver] an order of Calcinea, sometimes considered a separate sub-class, having the skeleton made of a calcareous network not composed of spicules, or of 4-rayed spicules joined by calcareous cement.

pharmacodynamics n. [Gk. *pharmakon*, drug; *dynamis*, power] the science of the action of drugs; *alt.* pharmacology.

pharmacogenetics n.plu. [Gk. *pharmakon*, drug; *genesis*, descent] the study of genetically controlled responses to drugs.

pharmacology n. [Gk. *pharmakon*, drug; *logos*, discourse] pharmacodynamics, q.v.

pharmacophore n. [Gk. *pharmakon*, drug; *pherein*, to carry] the part of a molecule causing the specific physiological reaction of a drug.

pharotaxis n. [Gk. *pharos*, lighthouse; *taxis*, arrangement] movement of an animal towards a definite place, the stimulus for which is acquired by conditioning or learning; *cf.* mnemotaxis.

pharyngeal a. [Gk. *pharyngx*, gullet] *pert.* pharynx, *appl.* artery, membrane, nerve, tonsil, tubercle, veins, etc., *appl.* nephridia, in certain worms.

Pharyngobdellae n. [Gk. *pharyngx*, gullet; *bdella*, leech] an order of carnivorous leeches without teeth or jaws.

pharyngobranchial a. [Gk. *pharyngx*, gullet; *brangchia*, gills] *pert.* pharynx and gills, *appl.* certain bones of fishes.

pharyngopalatine a. [Gk. *pharyngx*, gullet; L. *palatum*, palate] *pert.* pharynx and palate, *appl.* arch and muscle; *alt.* palatopharyngeal.

pharyngotympanic a. [Gk. *pharyngx*, gullet; *tympanon*, drum] *appl.* tube connecting pharynx and tympanic cavity, the auditory or Eustachian tube.

pharynx n. [Gk. *pharyngx*, gullet] a musculo-membranous tube extending from under surface of skull to level of 6th cervical vertebra; gullet or anterior part of alimentary canal following buccal cavity.

phaseolin a globulin obtained from seeds of the bean, *Phaseolus*.

Phasmida n. [Gk. *phasma*, apparition; *dim.*] an order of insects with incomplete metamorphosis, commonly called stick or leaf insects, that may be winged or secondarily wingless and have an elongated or flattened body; *alt.* Cheleutoptera.

Phasmidia n. [Gk. *phasma*, apparition; *dim.*] a class of nematodes having phasmids, and amphids opening on to lateral lips; *alt.* Secernentea.

phasmids n.plu. [Gk. *phasma*, apparition; *dim.*] caudal papillae in some Nematoda, bearing pores connecting with glandular pouch.

phatne n. [Gk. *phatnē*, manger] odontobothrion, q.v.

phe phenylalanine, q.v.

phellem(a) n. [Gk. *phellos*, cork] cork; cork and non-suberized layers forming the external zone of periderm produced by action of phellogen; *alt.* phellum.

phelloderm n. [Gk. *phellos*, cork; *derma*, skin] the secondary parenchymatous suberous cortex of trees, formed by and on inner side of cork cambium; *alt.* secondary cortex.

phellogen n. [Gk. *phellos*, cork; *gennaein*, to generate] the cork cambium of woody stems, arising as a secondary meristem and giving rise to cork and phelloderm.

phelloid a. [Gk. *phellos*, cork; *eidos*, form] cork-like. n. Non-suberized cell layer in outer periderm.

phellum phellema, q.v.

phene n. [Gk. *phainein*, to appear] a character in the phenotype under the control of genes.

phenetic a. [Gk. *phainein*, to appear] *appl.* classification based on appearances of organisms rather than on evolution from a common ancestor.

phengophil a. [Gk. *phenggos*, light; *philos*, friend] preferring light, *appl.* animals.

phengophobe a. [Gk. *phenggos*, light; *phobos*, fear] shunning light, *appl.* animals.

pheno- *alt.* phaeno-, q.v.

phenocontour n. [Gk. *phainein*, to appear; It. *contornare*, to outline] a contour line on a chart or map showing the distribution of a certain phenotype; *alt.* isophane, isophene, isomar.

phenocopy n. [Gk. *phainein*, to appear; F. *copie*, copy, from L.L. *copia*, transcript] a modification induced by environmental factors which simulates a genetically produced change; *cf.* paramorph, genocopy.

Phenogamia, phenogams n., n.plu. [Gk. *phainein*, to appear; *gamos*, marriage] Phanerogamia, q.v.

phenogenetics n.plu. [Gk. *phainein*, to appear; *genesis*, descent] the genetics of development.

phenogram n. [Gk. *phainein*, to appear; *graphein*, to write] a tree-like diagram which shows the conclusions of numerical taxonomy.

phenogroup n. [Gk. *phainein*, to appear; F. *groupe*, group] a group of antigenic responses determined at a certain gene locus.

phenological a. [Gk. *phainein*, to appear; *logos*, discourse] *pert.* phenology; *appl.* isolation of species owing to differences in flowering or breeding season.

phenology n. [Gk. *phainein*, to appear; *logos*, discourse] recording and study of periodic biotic events, as of flowering, breeding, migration, etc., in relation to climatic and other factors; *alt.* phenomenology.

phenom(e) n. [Gk. *phainein*, to appear] all the phenotypic characteristics of an organism.

phenomenology phenology, q.v.

phenon *n.* [Gk. *phainein*, to appear] a group of organisms placed together by numerical taxonomy.

phenotype *n.* [Gk. *phainein*, to appear; *typos*, image] the characters of an organism due to the interaction of genotype and environment; a group of individuals exhibiting the same phenotypic characters; *alt.* reaction type.

phenotypic *a.* [Gk. *phainein*, to show; *typos*, image] *pert.* phenotype, *appl.* characters arising from reaction to environmental stimulus.

phenotypic plasticity the range of variability shown by the phenotype in response to environmental fluctuations.

phenylalanine an essential amino acid derived from propionic acid and important in the formation of certain hormones; abbreviated to phe.

pheo- phaeo-, *q.v.*

pheromone *n.* [Gk. *pherein*, to bear; hor*mone*] a chemical given out by one animal that acts as a signal to another of the same species, such as sexual attractants given out by certain female insects to attract males, queen-bee substance, etc., or a substance signalling danger, *see* alarm pheromone; *alt.* ectohormone, socio-hormone.

pheron *n.* [Gk. *pherein*, to bear] the colloidal bearer of the active principle of an enzyme: *cf.* agon, symplex.

phialide *n.* [Gk. *phialē*, bowl; *eidos*, form] a flask-shaped outgrowth of spore-bearing hypha, in certain fungi; *alt.* sterigma; *cf.* metula.

phialiform *a.* [L. *phiala*, shallow cup; *forma*, form] cup-shaped; saucer-shaped; *alt.* phialaeform.

phialophore *n.* [Gk. *phialē*, bowl; *pherein*, to bear] a hypha which bears a phialide.

phialopore *n.* [Gk. *phialē*, bowl; *poros*, channel] the opening in the hollow daughter colony or gonidium of the protistan *Volvox*.

phialospore *n.* [Gk. *phialē*, bowl; *sporos*, seed] a spore or conidium borne at tip of a phialide.

Philadelphia chromosome an abnormality of chromosome 22 associated with a kind of leukaemia.

philopatry *n.* [Gk. *philos*, loving; *patris*, fatherland] the tendency of an organism to stay in, or return to, its home area.

philotherm *n.* [Gk. *philos*, loving; *thermē*, heat] a plant which completes life cycle only in a warm environment; *cf.* thermophil, thermophyte, therophyte.

philtrum *n.* [Gk. *philtron*, philtre] the depression on upper lip beneath septum or nose; *cf.* prolabium.

phlebenterism *n.* [Gk. *phleps*, vein; *enteron*, intestine] condition of having branches of the intestine extending into other organs, as arms or legs.

phleboedesis *n.* [Gk. *phleps*, vein; *oidein*, to swell] condition of having circulatory system cavity so distended and insinuated as to diminish the coelom, especially so in molluscs and arthropods.

phlobaphenes *n.plu.* [Gk. *phloios*, inner bark; *baphē*, dye] phenolic compounds, derivatives of tannins, producing yellow, red, or brown colours in fern ramenta, roots, and sections of wood.

phloem *n.* [Gk. *phloios*, inner bark] the tissue conducting sugars, and some amino acids, in a vascular plant, being composed of sieve elements, parenchyma cells, and sometimes also fibres and sclereids; bast, *q.v.*; *alt.* phloeum.

phloem parenchyma thin-walled parenchyma associated with sieve tubes of phloem.

phloem sheath pericycle, together with inner layer of a bundle sheath where latter consists of 2 layers; *alt.* periphloem, pericambium.

phloeodic *a.* [Gk. *phloios*, inner bark; *eidos*, form] having the appearance of bark.

phloeoterma *n.* [Gk. *phloios*, inner bark; *terma*, boundary] endodermis, *q.v.*

phloeum phloem, *q.v.*

phloic *a.* [Gk. *phloios*, inner bark] *pert.* phloem; *appl.* procambium that gives rise to phloem.

phobotaxis *n.* [Gk. *phobos*, manifest fear; *taxis*, arrangement] avoiding reaction in some Protozoa, now known as klinokinesis, *q.v.*; trial-and-error reaction.

Phoenicopteriformes *n.* [*Phoenicopterus*, generic name of flamingo; L. *forma*, shape] an order of birds of the subclass Neornithes, including the flamingoes.

pholadophyte *n.* [Gk. *phōlas*, lurking; *phyton*, plant] a plant living in hollows, shunning bright light.

Pholidophoriformes *n.* [Gk. *pholis*, scale; *pherein*, to bear; L. *forma*, shape] an order of Triassic to Cretaceous Holostei similar to some primitive teleosts.

Pholidopleuriformes *n.* [Gk. *pholis*, scale; *pleura*, side; L. *forma*, shape] an order of Triassic Chondrostei having a vertical or oblique jaw support.

pholidosis *n.* [Gk. *pholis*, scale] scale arrangement as of scaled animals; *alt.* scutellation, squamation.

Pholidota *n.* [Gk. *pholis*, scale] an order of placental mammals known from the Pleistocene or possibly Oligocene, the only living member being the pangolin (scaly anteater), having no teeth, and the body being covered with imbricated scales.

phonation *n.* [Gk. *phōnē*, sound] production of sounds, e.g. by insects.

phonoreceptor *n.* [Gk. *phōnē*, sound; L. *receptor*, receiver] a receptor of sound waves, as ear, certain sensillae.

phoranth(ium) *n.* [Gk. *pherein*, to bear; *anthos*, flower] the receptacle of composite plants.

phoresia, phoresy *n.* [Gk. *pherein*, to bear] the carrying of one organism by another, without parasitism, as in certain insects.

Phoronida, Phoronidea *n.* [*Phoronis*, generic name] a phylum of small marine worm-like animals that live in membranous secreted tubes, have a lophophore, and are coelomate with mesocoels and metacoels.

phoront *n.* [Gk. *phora*, producing; *ontos*, being] encysted stage produced by tomite and leading to formation of trophont in life cycle of Holotricha.

phorozooid *n.* [Gk. *pherein*, to bear; *zōon*, animal; *eidos*, form] foster forms of buds in Doliolidae, never sexually mature but set free with gonozooids attached to a ventral outgrowth.

phorozoon *n.* [Gk. *pherein*, to bear; *zōon*, animal] an asexual organism or larval stage preceding the sexual.

phosphagen creatine phosphate, a high-energy compound in vertebrate muscle which is chiefly used to transfer a phosphate group to ADP to form ATP during muscle contraction, and is re-formed during recovery, and so can be considered an energy store, *alt.* phosphocreatine; arginine phosphate in muscles of certain invertebrates, having a similar function.

phosphatases a large group of enzymes in all living organisms that catalyse the hydrolysis and synthesis of organic esters of phosphoric acid, transferring phosphate groups from one com-pound to another, and which are important in the formation of bones and teeth, in lactation and muscle metabolism; bone phosphatase was for-merly called bone enzyme; EC sub-subgroup 3.1.3; *r.n.* phosphoric monoester hydrolase.

phosphatides phospholipids, *q.v.*

phosphatidylcholines lecithins, *q.v.*

phosphatidylethanolamine cephalin, *q.v.*

phosphene *n.* [Gk. *phos*, light; *phainein*, to show] a light impression on retina due to stimulus other than rays of light.

phosphocreatine *see* phosphagen.

phospho enol pyruvate or pyruvic acid (PEP) a compound occurring as an intermediate in glycolysis, formed by the action of the enzyme enolase on 2-phosphoglyceric acid, and converted to pyruvic acid by pyruvate kinase.

phosphoglucomutase a phosphotransferase which catalyses the change of glucose-1-phos-phate to glucose-6-phosphate using glucose 1,6-bisphosphate (diphosphate) as a donor; EC 2.7.5.1; *r.n.* phosphoglucomutase.

phosphoglyceric acid (PGA) either of 2 iso-mers, monophosphates of glyceric acid, which are important intermediates in photosynthesis, respir-ation, and carbohydrate metabolism.

phosphoglycerides a group of phospholipids, derivatives of glycerol phosphate, including ceph-alin and lecithin which are abundant in the brain, and the plasmalogens found in the brain and heart.

phosphoglyceromutase a phosphotransferase which catalyses the change of 3-phosphoglyceric acid to 2-phosphoglyceric acid, using 2,3-bisphos-phoglycerate as a donor; EC 2.7.5.3; *r.n.* phos-phoglyceromutase.

phosphohexose isomerase an isomerase en-zyme which catalyse the change of glucose-6-phosphate into fructose-6-phosphate; EC 5.3.1.9; *r.n.* glucosephosphate isomerase.

phospholipid(e)s *n.plu.* [Gk. *phŏsphoros*, bring-ing light; *lipos*, fat] phosphorylated lipids, e.g. lecithins and cephalins; *alt.* phosphatides, phos-pholipins.

phospholipins phospholipids, *q.v.*

phosphoproteins *n.plu.* [Gk. *phŏsphoros*, bring-ing light; *prŏteion*, first] conjugated proteins whose prosthetic group contains phosphoric acid, including caseinogen and vitellin.

phosphorescence *n.* [Gk. *phŏsphoros*, bringing light] the state of being luminous without sen-sible heat, common in marine protozoans, some copepods, and the majority of deep-sea animals; *alt.* bioluminescence.

phosphorolysis *n.* [Gk. *phŏsphoros*, bringing light; *lysis*, loosing] a process similar to hydroly-sis of carbohydrate except that the bonds are broken by phosphoric acid instead of water, and which occurs in the early stages of glycolysis.

phosphorylase an enzyme which catalyses the

transfer of glucose residues from polyglucose polymers to the phosphate to give glucose 1-phos-phate; EC 2.4.1.1; *r.n.* phosphorylase.

phosphorylation the addition of a phosphate group to a molecule, including high-energy phos-phate groups to cofactors such as ADP to make ATP.

phosphotransferases a group of enzymes which transfer phosphate, diphosphate (pyrophos-phate), and nucleotide residues; EC subgroup 2.7; *r.n.* phosphotransferases.

phosphovitin *n.* [Gk. *phŏsphoros*, bringing light; L. *vita*, life] a phosphoprotein which is part of amphibian egg yolk.

photic *a.* [Gk. *phŏs*, light] *pert.* light; *appl.* zone, the surface waters penetrated by sunlight; *appl.* euphotic and dysophotic zones; *cf.* aphotic.

photoautotrophs *n.plu.* [Gk. *phŏs*, light; *autos*, self; *trophē*, nourishment] organisms using light as an energy source and carbon dioxide as the main source of carbon including the green plants and some bacteria; *alt.* photolithotrophs.

photoceptor photoreceptor, *q.v.*

photochemical *a.* [Gk. *phŏs*, light; *chēmeia*, transmutation] *appl.* and *pert.* chemical changes produced by light.

photochromatic *a.* [Gk. *phŏs*, light; *chrōma*, colour] *appl.* interval between achromatic and chromatic thresholds.

photochromic effect change of colour brought about by exposure to light.

photodinesis *n.* [Gk. *phŏs*, light; *dinē*, eddy] pro-toplasmic streaming induced by light.

photodynamics *n.* [Gk. *phŏs*, light; *dynamis*, strength] the study of the effects of light stimula-tion on plants.

photogen *n.* [Gk. *phŏs*, light; *-genēs*, producing] a light-producing organ, or substance.

photogenesis bioluminescence, *q.v.*

photogenic *a.* [Gk. *phŏs*, light; *-genēs*, produc-ing] light-producing; luminescent.

photogenin luciferase, *q.v.*

photoheterotrophs *n.plu.* [Gk. *phŏs*, light; *hete-ros*, other; *trophē*, nourishment] organisms using light as a source of energy, but deriving much of their carbon from organic compounds, and in-cluding a group of photosynthetic bacteria called the non-sulphur purple bacteria; *alt.* photo-organotrophs.

photoinhibition *n.* [Gk. *phŏs*, light; L. *inhibere*, to restrain] inhibition by light, e.g. of germina-tion.

photokinesis *n.* [Gk. *phŏs*, light; *kinēsis*, move-ment] a kinesis in response to stimulation by cer-tain regions of the visible spectrum.

photolabile *a.* [Gk. *phŏs*, light; L. *labare*, to waver] modified by light, as retinal pigments, etc.

photolithotrophs *n.plu.* [Gk. *phŏs*, light; *lithos*, stone; *trophē*, nourishment] photoautotrophs, *q.v.*

photolysis *n.* [Gk. *phŏs*, light; *lysis*, loosing] split-ting by the action of light, as of water in photo-synthesis.

photomorphogenesis *n.* [Gk. *phŏs*, light; *morphē*, form; *genesis*, descent] the effect of light in causing normal growth and reversing etiola-tion; any change in plant growth due to light; *alt.* photomorphosis.

photomorphosis *n.* [Gk. *phŏs*, light; *morphōsis*,

phragmosome

form] the form of organisms as affected by illumination; photomorphogenesis, *q.v.*; *cf.* chromomorphosis.

photomotor reflex change in size of the pupil of eye with sudden change in light intensity.

photonasty n. [Gk. *phōs*, light; *nastos*, close pressed] response of plants to diffuse light stimuli, or to variations in illumination.

photoorganotrophs *n.plu.* [Gk. *phōs*, light; *organon*, instrument; *trophē*, nourishment] photoheterotrophs, *q.v.*

photopathy n. [Gk. *phōs*, light; *pathos*, feeling] a pronounced movement in relation to light, usually away from it, as in negative phototaxis or phototropism.

photoperiod n. [Gk. *phōs*, light; *periodos*, circuit] duration of daily exposure to light; length of day favouring optimum functioning of an organism.

photoperiodism n. [Gk. *phōs*, light; *periodos*, circuit] response of an organism to the relative duration of day and night; *alt.* photoperiodicity; *cf.* thermoperiodism.

photophase n. [Gk. *phōs*, light; *phainein*, to appear] developmental stage during which the plant, after thermophase, shows definite requirements as to duration and intensity of light and temperature.

photophelein n. [Gk. *phōs*, light; *phēlos*, deceiving] luciferin, *q.v.*; a substance in plant and animal cells which may produce luciferin.

photophilous a. [Gk. *phōs*, light; *philos*, loving] seeking, and thriving in, strong light; *alt.* lucipetal.

photophobic a. [Gk. *phōs*, light; *phobos*, fear] not tolerating light, shunning light; *alt.* lucifugal.

photophore n. [Gk. *phōs*, light; *pherein*, to bear] a light-emitting organ which directs light ventrally in some bathypelagic fish, crustaceans, and cephalopods and so camouflages the silhouette from ventral view; any light-emitting organ.

photophosphorylation the formation of ATP from ADP using energy from light during photosynthesis; *cf.* oxidative phosphorylation.

photophygous a. [Gk. *phos*, light; *phygē*, flight] avoiding strong light.

photopia n. [Gk. *phōs*, light; *ōps*, eye] adaptation of the eye to light; *cf.* scotopia; a. photopic.

photopigment n. [Gk. *phōs*, light; L. *pingere*, to paint] a light-sensitive pigment.

photoplagiotropy n. [Gk. *phōs*, light; *plàgios*, oblique; *tropē*, turn] tendency to take up a transverse position in relation to incident light rays.

photopsin n. [Gk. *phōs*, light; *opsis*, sight] the protein component of the violet retinal cone pigment iodopsin.

photoreactivation the repair of ultraviolet radiation damage to biological systems by means of light of a longer wavelength than that of the radiation that caused the damage, and initiated by a light-dependent enzyme; the use of light to reverse some inactivation; *alt.* photorecovery, photorestoration, photoreversal.

photoreceptor n. [Gk. *phōs*, light; L. *receptus*, received] terminal organ receiving light stimuli; *alt.* photoceptor.

photorespiration n. [Gk. *phōs*, light; L. *respiration*, breathing] a type of respiration in plants that occurs only in the light and is different from

normal respiration, by various pathways. e.g. using glycollate produced from a carbohydrate from photosynthesis, as a substrate; *alt.* photosynthetic respiration.

photospheres *n.plu.* [Gk. *phōs*, light; *sphaira*, globe] luminous organs of Crustacea.

photosynthesis n. [Gk. *phōs*, light; *synthesis*, putting together] in green plants the synthesis of carbohydrates in chloroplasts from carbon dioxide, as a source of carbon, and water as a hydrogen donor, using sunlight energy trapped by chlorophyll, catalysed by various enzymes in the chloroplasts, and with oxygen produced as a waste product; in certain bacteria, a similar process but sometimes using hydrogen donors other than water, and producing waste products other than oxygen; a. photosynthetic.

photosynthetic quotient the ratio between the volume of oxygen produced and the volume of carbon dioxide used in photosynthesis; *cf.* respiratory quotient.

photosynthetic zone of sea or lakes, between surface and compensation point, *q.v.*

phototaxis n. [Gk. *phōs*, light; *taxis*, arrangement] a taxis in response to stimulus of light, where the stimulus is sunlight also called heliotaxis; a. phototactic.

phototonus n. [Gk. *phōs*, light; *tonos*, tension] sensitiveness to light; condition of a plant or plant organ induced by light.

phototroph n. [Gk. *phōs*, light; *trophē*, nourishment] an organism using sunlight as a source of energy, as photoheterotrophs and photoautotrophs; *cf.* chemotroph.

phototrophic a. [Gk. *phōs*, light; *trophē*, nourishment] requiring light as a source of energy in nutrition; *alt.* holophytic.

phototropism n. [Gk. *phōs*, light; *tropē*, turn] a growth movement of plants in response to stimulus of light, where the stimulus is sunlight known as heliotropism; a. phototropic.

phototropy n. [Gk. *phōs*, light; *tropē*, turn] a reversible change in the colour of a substance while it is illuminated.

phragma n. [Gk. *phragma*, fence] a spurious dissepiment; a septum; an endotergite or dorsal apodeme of thorax and abdomen in Diplopoda and insects; *plu.* phragmata.

Phragmobasidiomycetes, Phragmobasidiomycetidae [Gk. *phragmos*, fence; *basis*, base; *idion, dim.*; *mykēs*, fungus] Heterobasidiomycetes, *q.v.*

phragmobasidium n. [Gk. *phragmos*, fence; *basis*, base; *idion, dim.*] a septate basidium forming 4 cells.

phragmocone n. [Gk. *phragmos*, fence; *kōnos*, cone] in belemnites and other molluscs, a cone divided internally by a series of septa perforated by a siphuncle.

phragmocyttarous a. [Gk. *phragmos*, fence; *kyttaros*, honeycomb cell] building, or *pert.*, combs attached to supporting surface, as of certain wasps; *cf.* stelocyttarous.

phragmoplast n. [Gk. *phragmos*, fence; *plastos*, moulded] barrel-shaped stage of spindle at telophase of mitosis, which plays a role in the formation of the cell plate and later spreads to a ring-like shape.

phragmosome n. [Gk. *phragmos*, fence; *sōma*, body] a disc of cytoplasm formed in the equatorial plane during mitosis, before the phragmoplast forms, and in which the cell plate is formed.

phragmospore *n*. [Gk. *phragmos*, fence; *sporos*, seed] a septate spore.

phratry *n*. [Gk. *phratrē*, a subdivision of a tribe] a loose term in classification, never generally adopted or precisely defined, but often used to mean a subtribe; *alt*. clan.

phreatophyte *n*. [Gk. *phreatia*, tank; *phyton*, plant] plant with very long roots reaching water table.

phrenic *a*. [Gk. *phrēn*, diaphragm, mind] *pert*. or in region of diaphragm, *appl*. artery, ganglion, nerve, plexus, vein; *pert*. mind.

phrenicocolic *a*. [Gk. *phrēn*, diaphragm; *kolon*, lower part of intestine] *appl*. a ligament or a fold of peritoneum from left colic flexure to diaphragm.

phrenicocostal *a*. [Gk. *phrēn*, diaphragm; L. *costa*, rib] *appl*. a narrow slit or sinus between costal and diaphragmatic pleurae.

phrenicolienal *a*. [Gk. *phrēn*, diaphragm; L. *lien*, spleen] *appl*. ligament forming part of peritoneum reflected over spleen and extending to diaphragm; *alt*. lienorenal.

phrenicopericardiac *a*. [Gk. *phrēn*, diaphragm; *peri*, around; *kardia*, heart] *appl*. a ligament extending from diaphragm to pericardium.

phrenosin *n*. [Gk. *phrēn*, mind] an important glycolipid (cerebroside) obtained from the brain, which on hydrolysis yields a fatty acid (phrenosinic acid), galactose, and sphingosine.

phthinoid *a*. [Gk. *phthisthai*, to wither; *eidos*, form] withered; weak; underdeveloped.

Phthiraptera *n*. [Gk. *phtheir*, louse; *pteron*, wing] an order of lice including the Anoplura and Mallophaga.

phthisaner *n*. [Gk. *phthisis*, wasting; *anēr*, male] pupal male ant parasitized by an *Orasema* larva.

phthisergate *n*. [Gk. *phthisis*, wasting; *ergatēs*, worker] pupal worker ant parasitized by an *Orasema* larva; *alt*. infra-ergatoid.

phthisogyne *n*. [Gk. *phthisis*, wasting; *gynē*, female] pupal female ant parasitized by an *Orasema* larva.

phyad *n*. [Gk. *phya*, nature] a plant or animal form due to inheritance rather than environment; *cf*. ecad.

phycobilins *n.plu*. [Gk. *phykos*, seaweed; L. *bilis*, bile] biloproteins, *q.v*.

phycobilisome *n*. [Gk. *phykos*, seaweed; L. *bilis*, bile; Gk. *sōma*, body] one of a number of small particles which are found on photosynthetic lamellae of red and blue–green algae and contain phycobilin.

phycobiont *n*. [Gk. *phykos*, seaweed; *bion*, living] the algal component of a lichen; *cf*. mycobiont.

phycochrome *n*. [Gk. *phykos*, seaweed; *chrōma*, colour] a pigment of algae.

phycochrysin *n*. [Gk. *phykos*, seaweed; *chrysos*, gold] golden-yellow pigment present in chromophores in algae of the class Chrysophyceae.

phycocoenology *n*. [Gk. *phykos*, seaweed; *koinos*, common; *logos*, discourse] the study of algal communities.

phycocyanin *n*. [Gk. *phykos*, seaweed; *kyanos*, dark blue] a biloprotein pigment which gives the blue–green colour to algae of the class

Cyanophyceae, and is also present in other algae, especially Rhodophyceae and Cryptophyceae.

phycoerythrin *n*. [Gk. *phykos*, seaweed; *erythros*, red] a biloprotein pigment which gives the red colour to algae of the class Rhodophyceae, and is also found in some other algae especially Cyanophyceae and Cryptophyceae.

phycology *n*. [Gk. *phykos*, seaweed; *logos*, discourse] algology, *q.v*.

Phycomycetes, Phycomycetae *n*. [Gk. *phykos*, seaweed; *mykēs*, fungus] a class of fungi usually having an aseptate mycelium, with many species aquatic and reproducing by zoospores, and having isogamous or heterogamous sexual reproduction; formerly called algal fungi and lower fungi.

phycophaein *n*. [Gk. *phykos*, seaweed; *phaios*, dusky] a brown pigment in brown algae, now thought to be an oxidation product of fucosan.

Phycophyta *n*. [Gk. *phykos*, seaweed; *phyton*, plant] Algae, *q.v*.

phycoxanthin *n*. [Gk. *phykos*, seaweed; *xanthos*, yellow] diatomin, *q.v*.

phyla *n.plu*. [Gk. *phylon*, race] *plu*. of phylum, *q.v*.

phylacobiosis *n*. [Gk. *phylax*, guard; *biōsis*, manner of living] mutual or unilateral protective behaviour, as of certain ants; *alt*. parabiosis; *a*. phylacobiotic.

phylactocarp *n*. [Gk. *phylaktikos*, guarding; *karpos*, fruit] a modification of hydrocladium in Hydrozoa, for protection of gonophore.

Phylactolaemata *n*. [Gk. *phylaktikos*, guarding; *laimos*, throat] a class of Ectoprocta with an epistome and a horseshoe-shaped lophophore.

phylembryo *n*. [Gk. *phylon*, race; *embryon*, embryo] stage in development of Brachiopoda, at completion of protegulum.

phylephebic *a*. [Gk. *phylon*, race; *ephēbeia*, manhood] *appl*. adult stage of maximum vigour, in race history; *alt*. phyloephebic; *cf*. ephebic.

phyletic *a*. [Gk. *phylon*, race] *pert*. a phylum or race.

phyllade *n*. [Gk. *phyllas*, foliage] a reduced scale-like leaf.

phyllary *n*. [Gk. *phyllon*, leaf] a bract of the involucre of Compositae.

phyllidium *n*. [Gk. *phyllidion*, little leaf] bothridium, *q.v*.

phyllobranchia *n*. [Gk. *phyllon*, leaf; *brangchia*, gills] a gill consisting of numbers of lamellae or thin plates.

phyllocaline *n*. [Gk. *phyllon*, leaf; *kalein*, to summon] a plant hormone, or complex of similar substances, thought to stimulate growth of mesophyll of a leaf.

Phyllocarida *n*. [Gk. *phyllon*, leaf; L. *caris*, a kind of crab] a superorder of malacostracans of the series Leptostraca, *q.v*.

phylloclade, phyllocladium *n*. [Gk. *phyllon*, leaf; *klados*, sprout] a green flattened photosynthetic stem; in cacti, a green flattened or rounded stem functioning as a leaf; an assimilative branch of a fruticose thallus in lichens; *cf*. cladode.

phyllocyst *n*. [Gk. *phyllon*, leaf; *kystis*, bladder] the rudimentary cavity of a hydrophyllium or protective medusoid.

phyllode *n*. [Gk. *phyllon*, leaf; *eidos*, form] winged petiole with flattened surfaces placed laterally to stem, functioning as leaf; in echinoderms, oral ambulacral areas modified to give the appearance of radiating petals.

Phyllodocemorpha n. [*Phyllodoce*, generic name; Gk. *morphē*, form] an order of polychaetes having all segments fairly similar with well-developed parapodia and a prostomium with sensory appendages.

phyllody n. [Gk. *phyllon*, leaf; *eidos*, form] metamorphosis of an organ into a foliage leaf; *alt.* phyllomorphosis.

phylloerythrin n. [Gk. *phyllon*, leaf; *erythros*, red] a red pigment derived from chlorophyll and occurring in bile of herbivorous mammals; *alt.* bilipurpurin, cholohaematin.

phyllogen n. [Gk. *phyllon*, leaf; *gennaein*, to produce] meristematic cells which give rise to a primordial leaf.

phyllogenetic a. [Gk. *phyllon*, leaf; *genesis*, descent] producing or developing leaves; *cf.* phylogenetic.

phylloid a. [Gk. *phyllon*, leaf; *eidos*, form] leaf-like. n. The leaf regarded as a flattened branch, or as a telome.

phyllomania n. [Gk. *phyllon*, leaf; *mania*, madness] abnormal leaf-production.

phyllome n. [Gk. *phyllon*, leaf] the leaf structures of a plant as a whole.

phyllomorphosis n. [Gk. *phyllon*, leaf; *morphōsis*, form] phyllody, *q.v.*; variation of leaves at different seasons.

phyllophagous a. [Gk. *phyllon*, leaf; *phagein*, to eat] feeding on leaves.

phyllophore n. [Gk. *phyllophoros*, leaf-bearing] terminal bud or growing point of palms.

phyllophorous a. [Gk. *phyllophoros*, leaf-bearing] bearing or producing leaves.

phyllopode n. [Gk. *phyllon*, leaf; *pous*, foot] a sheathing leaf base of Isoetales.

phyllopodium n. [Gk. *phyllon*, leaf; *pous*, foot] the axis of a leaf; the stem regarded as pseudo-axis formed of fused leaf bases; a leaf-like swimming appendage in certain crustaceans such as Branchiopoda; a. phyllopodous.

phylloptosis n. [Gk. *phyllon*, leaf; *ptōsis*, falling] the fall of the leaf.

phylloquinone n. [Gk. *phyllon*, leaf; *quinone*] *see* vitamin K; α-phylloquinone; vitamin K_1; β-phylloquinone: vitamin K_2.

phyllorhiza n. [Gk. *phyllon*, leaf; *rhiza*, root] a young leaf with a root.

phyllosiphonic a. [Gk. *phyllon*, leaf; *siphōn*, tube] with insertion of leaf trace disturbing axial stele tissue; *cf.* cladosiphonic.

phyllosoma n. [Gk. *phyllon*, leaf; *sōma*, body] a larval stage of *Palinurus* (the crawfish or spiny lobster) being a broad, thin schizopod larva.

phyllosperm n. [Gk. *phyllon*, leaf; *sperma*, seed] a plant having seeds or spores borne on the leaves.

phyllosphere n. [Gk. *phyllon*, leaf; *sphaira*, sphere] the leaf surfaces.

phyllospondylous a. [Gk. *phyllon*, leaf; *sphondylos*, vertebra] *appl.* vertebrae consisting of hypocentrum and neural arch, both contributing to hollow transverse process, as in Stegocephali.

phyllosporous a. [Gk. *phyllon*, leaf; *sporos*, seed] having sporophylls like foliage leaves.

phyllotactic a. [Gk *phyllon*, leaf; *taktikos*, fit for arrangement] *pert.* phyllotaxis, *appl.* fraction of circumference of a stem between successive leaves, representing the angle of their divergence.

phyllotaxis, phyllotaxy n. [Gk. *phyllon*, leaf; *taxis*, arrangement] the arrangement of leaves on an axis or stem.

phylloxanthin xanthophyll, *q.v.*

phyllozooid n. [Gk. *phyllon*, leaf; *zōon*, animal; *eidos*, form] a shield-shaped medusoid of protective function; hydrophyllium, *q.v.*

phylobiology n. [Gk. *phylon*, race; *bios*, life; *logos*, discourse] the study of reactions or behaviour of organisms in relation to their racial history.

phyloephebic phylephebic, *q.v.*

phylogenesis n. [Gk. *phylon*, race; *genesis*, descent] phylogeny, *q.v.*

phylogenetic a. [Gk. *phylon*, race; *genesis*, descent] *pert.* phylogeny, *appl.* reproductive cells, *cf.* autogenetic or body cells; *cf.* phyllogenetic.

phylogeny n. [Gk. *phylon*, race; *genesis*, descent] history of development of species or other group; *alt.* race history, phylogenesis; *cf.* ontogeny.

phylogerontic a. [Gk. *phylon*, race; *gerōn*, old man] *appl.* decadent or senescent stage in phylogeny, *alt.* paracme, typolysis.

phylon phylum, *q.v.*

phyloneanic a. [Gk. *phylon*, race; *neanikos*, youthful] *appl.* youthful or adolescent stage in phylogeny; *cf.* neanic.

phylonepionic a. [Gk. *phylon*, race; *nēpios*, infant] *appl.* post-embryonic stage in phylogeny; *cf.* nepionic.

phylum n. [Gk. *phylon*, race or tribe] in classification, a primary division consisting of plants or animals constructed on a similar general plan, and thought to be related; in plants, also called a division; *alt.* phylon; *plu.* phyla.

phyma n. [Gk. *phyma*, tumour] an excrescence not containing gonidia, on podetium of lichens.

Phymosomatoida n. [Gk. *phyma*, tumour; *sōma*, body; *eidos*, form] an order of Echinacea having unperforated tubercles on the spines.

phyon(e) n. [Gk. *phyein*, to make to grow] somatotrophin, *q.v.*

physa n. [Gk. *physa*, bellows] the modified rounded base of burrowing sea anemones.

Physarales n. [*Physarum*, generic name] an order of stalked or sessile slime moulds whose assimilative stage is a phaneroplasmodium.

physicist n. [Gk. *physikos*, physical] from biological standpoint, an upholder of theory that vital phenomena are explicable on a physicochemical basis.

physiogenesis n. [Gk. *physis*, nature; *genesis*, descent] the development of vital activities; ontogenesis in its physiological aspect; *alt.* physiogeny.

physiogenic a. [Gk. *physis*, nature; *-genēs*, producing] caused by functioning of an organ or part; *pert.* physiogenesis; caused by environmental factors.

physiogeny physiogenesis, *q.v.*

physiology n. [Gk. *physis*, nature; *logos*, discourse] that part of biology dealing with functions and activities of organisms.

physoclistous a. [Gk. *physa*, bladder; *kleiein*, to close] having no channel connecting swim bladder and digestive tract, as in most teleosts; *cf.* physostomous.

physodes n.plu. [Gk. *physa*, bubble] spindles of phloroglucin contained in plasmodium of certain Sarcodina.

physogastry n. [Gk. *physan*, to blow up; *gastēr*, belly] excessive fat-body and enlargement of abdomen in insects.

physostomous a. [Gk. *physa*, bladder; *stoma*, mouth] having swim bladder and digestive tract connected throughout life by pneumatic duct, as in ganoids; *cf.* physoclistous.

phytalfauna n. [Gk. *phyton*, plant; L. *faunus*, god of woods] cryptofauna, *q.v.*

phytamins n.plu. [Gk. *phyton*, plant; *ammonia*] phytohormones, originally applied to auxins.

phytic acid [Gk. *phyton*, plant] hexaphosphoinositol, a phosphate-containing substance found in seeds, the quantity reflecting the amount of phosphate present.

phytin n. [Gk. *phyton*, plant] calcium phytate, a substance stored in seeds, which on hydrolysis yields phosphate and inositol.

phytoactive a. [Gk. *phyton*, plant; L. *activus*, active] stimulating plant growth.

phytoalexins n.plu. [Gk. *phyton*, plant; *alexein*, to ward off] a group of antibiotics produced by plants which help disease resistance to bacteria or environmental accidents.

phytobiology n. [Gk. *phyton*, plant; *vios*, life; *logos*, discourse] botany, particularly the life history of plants.

phytobiotic a. [Gk. *phyton*, plant; *bios*, life] living within plants, *appl.* some protozoans.

phytochemistry n. [Gk. *phyton*, plant; *chēmeia*, transmutation] the chemistry of plants.

phytochoria n.plu. [Gk. *phyton*, plant; *chōria*, countries] phytogeographical realms and regions.

phytochory n. [Gk. *phyton*, plant; *chōrein*, to spread] dissemination of pathogens through the agency of plants.

phytochrome n. [Gk. *phyton*, plant; *chrōma*, colour] a blue protein pigment in plants, closely related to phycocyanin, which regulates many developmental phenomena such as photomorphogenesis, photoperiodism, germination of some seeds, some fruit ripening, and can detect daylength by existing into 2 interconvertible forms, P_R and P_{FR}.

phytocoenology n. [Gk. *phyton*, plant; *koinos*, common; *logos*, discourse] study of plant communities.

phytocoenosis n. [Gk. *phyton*, plant; *koinōs*, in common] the assemblage of plants living in a particular locality; *cf.* synusia.

Phytodiniales n. [Gk. *phyton*, plant; *dinos*, rotation] Dinococcales, *q.v.*

phytoedaphon n. [Gk. *phyton*, plant; *edaphos*, ground] microscopic soil flora.

phytogenesis n. [Gk. *phyton*, plant; *genesis*, descent] evolution, or development, of plants; *alt.* phytogeny.

phytogenetics n. [Gk. *phyton*, plant; *genesis*, descent] plant genetics.

phytogenous a. [Gk. *phyton*, plant; *genos*, generation] of vegetable origin; produced by plants.

phytogeny phytogenesis, *q.v.*

phytogeography n. [Gk. *phyton*, plant; *gē*, earth; *graphein*, to write] study of the geographical distribution of plants; *alt.* geobotany.

phytography n. [Gk. *phyton*, plant; *graphein*, to write] descriptive botany.

phytohaemagglutinin n. [Gk. *phyton*, plant; *haima*, blood; L. *ad*, to; *glutinare*, to glue] a mucoprotein extracted from the bean *Phaseolus vulgaris* which causes changes in mammalian leucocytes, such as causing them to undergo mitosis; *alt.* phytohemagglutinin.

phytohormones n.plu. [Gk. *phyton*, plant; *hormaein*, to excite] plant hormones, including auxins, gibberellins, cytokinins, abscissic acid, traumatins; *alt.* phytamins; *cf.* florigen.

phytoid a. [Gk. *phyton*, plant; *eidos*, form] plantlike. n. An individual in a plant colony; *cf.* zooid.

phytokinins n.plu. [Gk. *phyton*, plant; *kinein*, to move] cytokinins, *q.v.*

phytol n. [Gk. *phyton*, plant] a product of hydrolysis of chlorophyll used in the synthesis of vitamins E and K, being a long-chain alcohol forming the tail of the chlorophyll molecule.

phytolith n. [Gk. *phyton*, plant; *lithos*, stone] mineral particle, as hydrate of silica, in plant tissue, particularly of grasses.

phytology n. [Gk. *phyton*, plant; *logos*, discourse] botany.

phytoma n. [Gk. *phyton*, plant] vegetative plant substances.

Phytomastigina, Phytomastigophora n. [Gk. *phyton*, plant; *mastix*, whip] a subclass of Flagellata having many plant characteristics, including presence of chloroplasts, many being classed as algae by botanists.

phytome n. [Gk. *phyton*, plant] plants considered as an ecological unit; vegetation.

phytomer n. [Gk. *phyton*, plant; *meros*, part] a structural unit of a plant; a bud-bearing node; *alt.* phyton.

phytomitogen n. [Gk. *phyton*, plant; *mitos*, thread; *genos*, birth] a mitogen extracted from plant tissue.

Phytomonadina n. [Gk. *phyton*, plant; *monas*, unit] Volvocina, *q.v.*

phytomorphic a. [Gk. *phyton*, plant; *morphē*, form] with plant-like structure.

phytomorphosis n. [Gk. *phyton*, plant; *morphōsis*, shaping] changes in structural features of plants as a result of fungal and bacterial infections.

phyton n. [Gk. *phyton*, plant] a rudimentary plant; propagation unit, smallest detached part which can form another plant; phytomer, *q.v.*

phytonadione vitamin K₁, *see* vitamin K.

phytonomy n. [Gk. *phyton*, plant; *nomos*, law] the laws of origin and development of plants.

phytoparasite n. [Gk. *phyton*, plant; *parasitos*, parasite] any parasitic plant organism.

phytopathology n. [Gk. *phyton*, plant; *pathos*, suffering; *logos*, discourse] study of plant diseases.

phytophagous, phytophilous a. [Gk. *phyton*, plant; *phagein*, to eat; *philos*, loving] *see* herbivorous.

phytophysiology n. [Gk. *phyton*, plant; *physis*, nature; *logos*, discourse] plant physiology.

phytoplankton n. [Gk. *phyton*, plant; *plangkton*, wandering] plant plankton.

phytoplasm n. [Gk. *phyton*, plant; *plasma*, mould] plant protoplasm.

Phytosauria, phytosaurs n., n.plu. [Gk. *phyton*, plant; *sauros*, lizard] an order of Triassic archosaurs, commonly called plant lizards, because they were thought, wrongly, to be herbivorous.

phytosis n. [Gk. *phyton*, plant] production of disease by vegetable parasites, as by fungi; any disease so caused.

phytosociology n. [Gk. *phyton*, plant; L. *socius*, companion; Gk. *logos*, discourse] the branch of botany comprising ecology, biogeography, and genetics of plant associations.

phytosterols n.plu. [Gk. *phyton*, plant; *stereos*, solid; alcoho*l*] sterols, such as sitosterol, that were originally found in plants but now have sometimes also been found in animals; *cf.* mycosterols, zoosterols.

phytosuccivorous a. [Gk. *phyton*, plant; L. *succus*, juice; *vorare*, to devour] living on plant juices.

phytotomy n. [Gk. *phyton*, plant; *tomē*, cutting] the dissection of plants; plant anatomy.

phytotoxin n. [Gk. *phyton*, plant; *toxikon*, poison] any toxin originating in plants.

phytotrophic a. [Gk. *phyton*, plant; *trephein*, to nourish] autotrophic, *q.v.*

phytotype n. [Gk. *phyton*, plant; *typos*, pattern] representative type of plant.

pia mater n. [L. *pia mater*, tender mother] a delicate vascular membrane investing brain and spinal cord, the inner of 3 membranes.

Piciformes n. [*Picus*, generic name of woodpecker; L. *forma*, shape] an order of birds of the subclass Neornithes, including the woodpeckers.

picodna viruses [Sp. *pico*, small; *DNA*] a group of very small viruses, containing a DNA core and a protein coat.

picornavirus n. [Sp. *pico*, small; *RNA*, L. *virus*, poison] any of a group of very small RNA viruses including polio viruses.

pigment n. [L. *pingere*, to paint] colouring matter in plants and animals.

pigmentation n. [L. *pingere*, to paint] disposition of colouring matter in an organ or organism.

pigment cell chromatophore, *q.v.*; chromatocyte, *q.v.*

pigmy male complemental male, *q.v.*; dwarf male, *q.v.*; *alt.* pygmy male.

pikas *see* Lagomorpha.

pilea *plu.* of pileum.

pileate a. [L. *pileatus*, wearing a cap] having a pileus.

pileated crested, *appl.* birds.

pilei *plu.* of pileus.

pileocystidium n. [L. *pileus*, cap; Gk. *kystis*, bag; *idion*, *dim.*] one of the cystidium-like structures on pileus of certain Basidiomycetes; *alt.* pilocystidium, pilotrichome, dermatocystidium.

pileolated a. [L. *pileolus*, small cap] furnished with a small cap or caps.

pileolus n. [L. *pileolus*, small cap] a small pileus.

pileorhiza n. [L. *pileus*, cap; Gk. *rhiza*, root] a root covering; root cap, *q.v.*

pileum n. [L. *pileum*, cap] top of head region of bird; *plu.* pilea.

pileus n. [L. *pileus*, cap] umbrella-shaped structure of mushroom or toadstool, or of jellyfish; *plu.* pilei.

pili *plu.* of pilus.

pilidium n. [Gk. *pilidion*, small cap] the characteristic helmet-shaped larva of Nemertea; a hemispherical apothecium of certain lichens.

pilifer n. [L. *pilus*, hair; *ferre*, to carry] part of labrum of some Lepidoptera.

piliferous a. [L. *pilus*, hair; *ferre*, to carry] bearing or producing hair; *appl.* outermost layer of root or epiblema which gives rise to root hairs; *alt.* piligerous.

piliform a. [L. *pilus*, hair; *forma*, shape] resembling hair; *alt.* trichoid.

piligerous piliferous, *q.v.*

pillar cell hourglass cell, *q.v.*; osteosclereid, *q.v.*

pilocystidium pileocystidium, *q.v.*

pilomotor a. [L. *pilus*, hair; *movere*, to move] causing hairs to move; *appl.* non-myelinated fibres innervating muscles to hair follicles.

pilose a. [L. *pilosus*, hairy] hairy; downy.

pilotrichome pileocystidium, *q.v.*

Piltdown man a fossil hominid skull found at Piltdown in Sussex and named *Eoanthropus*, which was found to be a hoax consisting of a jawbone of a modern ape and a skull of *Australopithecus*.

pilus n. [L. *pilus*, hair] one of slender hair-like structures covering some plants; fimbria of bacteria; *plu.* pili.

pinacocytes n.plu. [Gk. *pinax*, tablet; *kytos*, hollow] the flattened plate-like cells of dermal epithelium of sponges.

pinacoderm n. [Gk. *pinax*, tablet; *derma*, skin] the outer layer of cells in sponges, consisting of pinacocytes; *alt.* ectosome, pinnacoderm; *cf.* choanosome.

Pinales n. [*Pinus*, generic name] an order of conifers in some classifications.

Pinatae n. [*Pinus*, generic name] in some classifications, a class of Coniferophytina, including the Cordaitales, Coniferales, and Taxales.

pincers prehensile claws, as of lobster; chelae of insects; chelicerae of arachnids.

pineal a. [L. *pineus*, of the pine] cone-shaped; *pert.* pineal gland.

pineal gland or body in vertebrates, a median outgrowth from the 1st cerebral vesicle, which may have endocrine functions and is known to produce vasotocin in mammals and melatonin, in lower vertebrates externally visible as the pineal or median eye, in higher vertebrates embedded in nervous tissue, but possibly sensitive to light passing into the skull; *alt.* conarium, epiphysis cerebri.

pineal region portion of brain giving rise to pineal and parapineal organs.

pineal sac end vesicle of epiphysis, as in *Sphenodon*, the tuatara.

pineal system the parietal organ and associated structures, as pineal sac, stalk, and nerves, parapineal organ, epiphysis.

pin-eyed having stigma at mouth of tubular corolla, with shorter stamens; *cf.* thrum-eyed.

Pinicae n. [*Pinus*, generic name] Coniferophytina, *q.v.*

pin feather a young feather, especially one just emerging through the skin and still enclosed in a sheath; *alt.* stipule.

pinna n. [L. *pinna*, feather] a leaflet of a pinnate leaf; auricula or outer ear; a bird's feather or wing; a fish fin; a branch of a pinnate thallus.

pinnacoderm pinacoderm, *q.v.*

pinnaglobulin a brown respiratory pigment containing manganese, in certain bivalves.

pinnate a. [L. *pinnatus*, feathered] divided in a feathery manner; with lateral processes; of a compound leaf, having leaflets on each side of an axis or midrib; *alt.* pennate, empennate.

pinnatifid a. [L. *pinna*, feather; *findere*, to cleave] *appl.* leaves lobed halfway to midrib.

pinnatilobate a. [L. *pinna*, feather; *lobus*, lobe] with leaves pinnately lobed.

pinnation n. [L. *pinna*, feather] pinnate condition.

pinnatipartite a. [L. *pinna*, father; *partitus*, divided] with leaves lobed three-quarters of way to midrib.

pinnatiped a. [L. *pinna*, feather; *pes*, foot] having lobed toes, as certain birds.

pinnatisect a. [L. *pinna*, feather; *sectus*, cut] with leaves lobed almost to base or midrib.

pinnatodentate a. [L. *pinna*, feather; *dens*, tooth] pinnate with toothed lobes.

pinnatopectinate a. [L. *pinna*, feather; *pecten*, comb] pinnate, with pectinate lobes.

pinniform a. [L. *pinna*, feather; *forma*, shape] feather-shaped or fin-shaped.

pinninervate a. [L. *pinna*, feather; *nervus*, sinew] with veins disposed like parts of feather.

Pinnipedia n. [L. *pinna*, feather; *pes*, foot] a suborder, or in some classifications an order, of carnivorous eutherian mammals, including the seals, sealions, and walruses.

pinnulary n. [L. *pinnula*, dim. of *pinna*, feather] any of the ossicles of a pinnule of Crinoidea.

pinnule n. [L. *pinnula*, dim. of *pinna*, feather] a secondary leaflet of a bipinnate or of a pinnately compound leaf; a reduced parapodium in certain Polychaeta; in Crinoidea, one of the side branches, 2 rows of which fringe arms.

pinocytosis n. [Gk. *piein*, to drink; *kytos*, hollow] uptake or ingestion of liquid droplets in a vesicle by a cell; *cf.* phagocytosis, cytopempsis.

Pinophyta n. [*Pinus*, generic name; Gk. *phyton*, plant] Gymnospermae, *q.v.*

pinosome n. [Gk. *piein*, to drink; *sōma*, body] a membrane-bound vesicle in the cytoplasm formed by budding off or an invagination of the cell membrane, and containing material taken up by pinocytosis.

pinulus n. [L. *pinulus*, small fir] a spicule resembling a fir tree owing to development of small spines from one ray.

pioneer community the organisms which first establish themselves on bare ground at the start of a primary succession, including plants, microorganisms, and small animals.

Piperales n. [*Piper*, generic name] an order of dicots of the Archichlamydeae or Magnoliidae, including tropical woody plants and herbs, used in different ways by different authorities, but always including the Piperaceae, peppers.

piperidine an alkaloid obtained from *Piper nigrum*, pepper.

piriform a. [L. *pirum*, pear; *forma*, shape] pear-shaped, *appl.* a muscle of gluteal region, musculus piriformis; pyriform, *q.v.*

Pisces n. [L. *piscis*, fish] a group of jawed vertebrates, the fishes, formerly given class status, but now subdivided into the classes Placodermi, Chondrichthyes, and Osteichthyes; sometimes also includes the Agnatha.

piscicolous a. [L. *piscis*, fish; *colere*, to inhabit] living within fishes, as certain parasites.

pisciform a. [L. *piscis*, fish; *forma*, shape] shaped like a fish.

piscine a. [L. *piscis*, fish] *pert.* fishes.

piscivorous a. [L. *piscis*, fish; *vorare*, to devour] fish-eating.

pisiform a. [L. *pisum*, pea; *forma*, shape] pea-shaped, *appl.* a carpel bone, os pisiforme.

pisohamate a. [L. *pisum*, pea; *hamus*, hook] *appl.* a ligament connecting pisiform and hamate bones.

pisometacarpal a. [L. *pisum*, pea; Gk. *meta*, beyond; L. *carpus*, wrist] *appl.* a ligament connecting pisiform bone with 5th metacarpal.

pistil n. [L. *pistillum*, pestle] gynaecium, *q.v.*, in a syncarpous flower; each separate carpel, forming a stigma, style, and ovary in an apocarpous gynaecium.

pistillate a. [L. *pistillum*, pestle] bearing pistils; *appl.* flower bearing pistils but no functional stamens, *cf.* staminate.

pistillidium n. [L. *pistillum*, pestle; Gk. *idion*, dim.] archegonium, *q.v.*

pistillode n. [L. *pistillum*, pestle; Gk. *eidos*, form] a rudimentary or non-functional pistil.

pistillody n. [L. *pistillum*, pestle; Gk. *eidos*, form] the conversion of any organ of a flower into carpels.

pistillum n. [L. *pistillum*, pestle] a mass of muscle in a chitinous tube in aurophore of Siphonophora.

pit n. [A.S. *pyt*, pit] a depression formed in course of cell wall thickening in plant tissue; embryonic olfactory depression.

pit chamber the cavity of a bordered pit below the overarching border.

pitcher n. [L.L. *picarium*, beaker] a modification of a leaf or part of a leaf for insect-catching purposes, as pitcher-shaped leaf of certain insectivorous plants such as *Nepenthes, Sarracenia*, etc.

pit fields areas of depressions in primary cell walls.

pith n. [A.S. *pitha*, pith] the medulla or central region of a dicotyledonous stem; *alt.* stelar parenchyma.

pithecanthropines n.plu. [Gk. *pithēkos*, ape; *anthrōpos*, man] fossil hominids formerly placed in the genus *Pithecanthropus* and now considered to be the species *Homo erectus* and including Java man and Peking man.

pit lines superficial grooves on dermal bones of primitive fishes, formed by laterosensory system.

pit membrane middle lamella of plant cell wall forming floor of pits of adjacent cells.

pitocin oxytocin, *q.v.*

pitressin vasopressin, *q.v.*

pituicyte n. [L. *pituita*, phlegm; Gk. *kytos*, hollow] a glial cell in pars nervosa of pituitary gland.

pituitary a. [L. *pituita*, phlegm] *appl.* gland: pituitary gland, *q.v.*; formerly *appl.* membrane: nasal mucous membrane.

pituitary gland in vertebrates, an endocrine gland attached to the brain by a short stalk below the hypothalamus, secreting a number of important hormones, and consisting of 2 parts, the adenohypophysis and neurohypophysis; *alt.* hypophysis.

pituitrin n. [L. *pituita*, phlegm] an extract from the posterior lobe of the pituitary gland containing oxytocin and vasopressin, which causes a decrease in renal water excretion; *alt.* antidiuretin, hypophysin, infundibulin, infundin.

pivot joint a trochoid joint, i.e. one in which movement is limted to rotation.

placenta *n.* [L. *placenta*, flat cake] ovule-bearing part of carpel, *alt.* spermaphore; a sporangium-bearing area; in eutherian mammals, a double vascular spongy structure formed by interlocking of foetal and maternal tissue in uterus, and in which maternal and foetal blood vessels are in close proximity, allowing nutritive, respiratory, and other exchange.

placental *a.* [L. *placenta*, flat cake] *pert.* a placenta or similar structure; *appl.* mammals which develop a placenta; secreted by placenta, *appl.* anterior pituitary-like hormone; *alt.* mazic.

Placentalia, placentals *n., n.plu.* [L. *placenta*, flat cake] Eutheria, eutherials, *q.v.*

placentate *a.* [L. *placenta*, flat cake] having a placenta developed; *alt.* placentiferous, placentigerous.

placentation *n.* [L. *placenta*, flat cake] the manner in which the ovules are attached to the ovary wall; the way in which an embryo is attached to a uterus; formation of placenta; structural type of placenta.

placentiferous, placentigerous placentate, *q.v.*

placentotrophin *n.* [L. *placenta*, flat cake; Gk. *trophē*, nourishment] a pituitary glycoprotein hormone which during pregnancy stimulates the placenta to produce chorionic gonadotrophin.

placids *n.plu.* [Gk. *plax*, plate] in Kinorhyncha, large plates in the neck or 2nd zonite, arranged differently in the subgroups.

placochromatic *a.* [Gk. *plax*, plate; *chrōma*, colour] with plate-like arrangement of chromatophores.

placode *n.* [Gk. *plax*, plate; *eidos*, form] a localized thickening of ectoderm forming a neural primordium; a plate-like structure.

Placodermi, placoderms *n., n.plu.* [Gk. *plax*, plate; *derma*, skin] a class of Silurian to Carboniferous jawed primitive fish with archaic jaw suspension and dermal plates on the head and thorax in primitive forms; *alt.* Aphetohyoidea.

Placodontia, placodonts *n., n.plu.* [Gk. *plax*, plate; *odus*, tooth] a Triassic order of sauropterygians that are fully aquatic with a short armoured body, some with a turtle-like carapace.

placoid *a.* [Gk. *plax*, plate; *eidos*, form] plate-like; *appl.* sensillae, possibly reacting to differences in air pressure, in insects.

placoid scale tooth-like scale present in many fish and covering the whole of the body surface in elasmobranchs, having a structure similar to teeth; *alt.* denticle.

placula *n.* [Gk. *plax*, plate] a flattened blastula with small segmentation cavity, an embryonic stage of Urochordata; a stage in the protistan *Volvox*.

plagioclimax *n.* [Gk. *plagios*, athwart; *klimax*, ladder] climax of a plagiosere, disclimax, *q.v.*

plagiogeotropism geoplagiotropism, *q.v.*

plagiopatagium *a.* [Gk. *plagios*, sideways; L. *patagium*, border] part of the patagium between fore-and hind-feet in flying lemurs.

Plagiosauria, plagiosaurs *n., n.plu.* [Gk. *plagios*, oblique, sideways; *sauros*, lizard] the most advanced order of labyrinthodonts, existing from Permian to Triassic times and having a very wide flat body in advanced forms, and a body armour of interlocking plates.

plagiosere *n.* [Gk. *plagios*, athwart; L. *serere*, to put in a row] plant succession deviating from its course owing to external intervention, as by human activity; a deflected sere.

plagiotropism *n.* [Gk. *plagios*, oblique; *tropē*, turn] growth tending to incline a structure from the vertical plane to the oblique or horizontal as in lateral roots and branches; *a.* plagiotropic; *cf.* orthotropism.

plagiotropous *a.* [Gk. *plagios*, oblique; *tropē*, turn] obliquely inclined, *appl.* the asymmetrical polar cap of Rhombozoa.

plagula *n.* [L. *plagula*, curtain] ventral plate protecting the pedicle in spiders.

plain muscle involuntary muscle, *q.v.*

plakea *n.* [Gk. *plakoeis*, flat cake] plate-like early stage in formation of a coenobium.

plakin *n.* [Gk. *plax*, plate] a thermostable substance extracted from platelets and possessing bactericidal properties.

planaea blastaea, *q.v.*

planarians *n.plu.* [*Planaria*, generic name] the common name for the Tricladida, *q.v.*

planation *n.* [L. *planus*, flat] the flattening of branched structures, e.g. in evolution of fronds of pteridophytes.

planetism *n.* [Gk. *planētēs*, wanderer] the character of having motile or swarm stages.

planidium *n.* [Gk. *planos*, wandering; *idion*, dim.] active migratory larva of certain insects.

planiform *a.* [L. *planus*, level; *forma*, shape] with nearly flat surface, *appl.* certain articulation surfaces.

Planipennia *n.* [L. *planus*, flat; *penna*, wing] an order of insects, including the lacewings and ant lions, sometimes included in the Neuroptera with the Megaloptera.

plankter *n.* [Gk. *plangktēr*, a roamer] an individual plankton organism.

planktohyponeuston *n.* [Gk. *plangktos*, wandering; *hypo*, under; *neustos*, floating] aquatic organisms which gather near the surface at night but spend their days in the main water mass.

plankton *n.* [Gk. *plangktos*, wandering] the usually small marine or freshwater plants (phytoplankton) and animals (zooplankton) drifting with the surrounding water, including animals with weak locomotory power; *cf.* seston.

Planktosphaeroidea *n.* [Gk. *plangktos*, wandering; *sphaira*, globe; *eidos*, form] a class of hemichordates that are pelagic and thought to be larvae of an unknown hemichordate, having a spherical shape, ciliary bands over the surface, and a U-shaped gut.

planktotrophic *a.* [Gk. *plangktos*, wandering; *trophē*, nourishment] feeding on plankton.

planoblast *n.* [Gk. *planos*, wandering; *blastos*, bud] the free-swimming medusa form of a hydrozoan.

planoconidium *n.* [Gk. *planos*, wandering; *konis*, dust; *idion*, dim.] zoospore of certain fungi.

planocyte *n.* [Gk. *planos*, wandering; *kytos*, hollow] a wandering or migratory cell; planospore, *q.v.*; swarm spore of certain fungi.

planogamete *n.* [Gk. *planos*, wandering; *gametēs*, spouse] a motile, usually ciliated, gamete especially one of algae and fungi; *alt.* microzoospore, zoogamete, zoozygosphere, swarm cell or spore.

planont

planont n. [Gk. *planos*, wandering; *on*, being] any motile spore, gamete, or zygote; the initial amoebula stage of Neosporidia; a swarm spore produced in thick-walled or resting sporangia of certain Phycomycetes.

planosome n. [Gk. *planos*, wandering; *sōma*, body] a supernumerary chromosome due to non-disjunction of pairs in meiosis.

planospore n. [Gk. *planos*, wandering; *sporos*, seed] a motile spore; *alt.* planocyte, zoospore; *cf.* aplanospore.

planozygote n. [Gk. *planos*, wandering; *zygōtos*, yoked] a motile zygote.

planta n. [L. *planta*, sole of foot] the sole of foot, *alt.* pelma; 1st tarsal joint of insects; apex of proleg.

Plantaginales n. [*Plantago*, generic name] an order of dicots (Sympetalae) being usually herbaceous plants or sometimes shrubs with alternate leaves and hermaphrodite or unisexual flowers with the parts arranged in 4s.

plantar a. [L. *planta*, sole of foot] *pert.* sole of foot, *appl.* arteries, ligaments, muscles, nerves, veins, etc.

plantigrade a. [L. *planta*, sole of foot; *gradus*, step] walking with whole sole of foot touching the ground; *alt.* palmigrade.

plantula n. [L. *plantula*, small sole] a pulvillus-like adhesive pad on tarsal joints of some insects.

planula n. [L. *planus*, flat] the ovoid young free-swimming ciliated larva of coelenterates.

planum n. [L. *planus*, flat] a plane or area, *appl.* certain cranial bone surfaces.

plaque n. [F. *plaque*, plate] area cleared by a phage in a bacterial growth.

plasm plasma, *q.v.*

plasma n. [Gk. *plasma*, form] the liquid part of various body fluids such as blood, lymph, milk, as opposed to suspended material such as cells or fat globules; protoplasm, *q.v.*; *alt.* plasm.

plasma cells plasmacytes, *q.v.*

plasmacytes n.plu. [Gk. *plasma*, form; *kytos*, hollow] cells derived from mesenchyme and formed from lymphocytes or other cells, in lymph nodes, spleen, and mammary gland, synthesizing ribonucleic acid and other nutritive substances; *alt.* plasma cells, plasmacytes, *q.v.*

plasmagel n. [Gk. *plasma*, form; L. *gelare*, to congeal] the more solid (gel) part of the cytoplasm, usually ectoplasm, which can be reversibly converted to plasmasol.

plasmagene n. [Gk. *plasma*, form; *genos*, descent] any hereditary determinant found in the cytoplasm rather than on chromosomes in the nucleus; *alt.* plasmid, blastogene, cytogene, cytoplasmic determiner.

plasma growth factor somatomedin, *q.v.*

plasmalemma n. [Gk. *plasma*, form; *lemma*, skin] plasma membrane, *q.v.*

plasmalemmosomes n.plu. [Gk. *plasma*, form; *lemma*, skin; *sōma*, body] cytoplasmic membranes of various sizes and shapes continuous with the plasmalemma and found in certain micro-organisms.

plasmalogens n.plu. [Gk. *plasma*, form; alkali; *genos*, birth] a group of phospholipids found in the head, especially in the brain.

plasma membrane the cell membrane consisting of phospholipid and protein; *alt.* cytolemma, plasmalemma; *cf.* ectoplast.

plasmasol n. [Gk. *plasma*, form; L. *solutio*, solution] the more fluid (sol) part of the cytoplasm, usually endoplasm, which may be reversibly converted to plasmagel.

plasmaspore n. [Gk. *plasma*, form; *sporos*, seed] an adhesive spore in a sporangium; *alt.* haptospore, myxospore.

plasmatic a. [Gk. *plasma*, form] *pert.* plasma; protoplasmic.

plasmatogamy plasmogamy, *q.v.*

plasmatoönkosis n. [Gk. *plasma* form; *ongkos*, bulk] a thickened storage organ or toruloid structure of zoosporangium, as in Peronosporales.

plasmatoparous a. [Gk. *plasma*, form; L. *parere*, to beget] developing directly into a mycelium upon germination instead of into zoospores, *appl.* spores of some fungi as grape mildew.

plasmid n. [Gk. *plasma*, form] plasmagene, *q.v.*; any structure in the cytoplasm thought capable of self-replication such as mitochondria; a cytoplasmic replicating element made of nucleic acid in a bacterial cell.

plasmin n. [Gk. *plasma*, form] fibrinolysin, *q.v.*

plasmocyte n. [Gk. *plasma*, form; *kytos*, hollow] leucocyte, *q.v.*; plasmacyte, *q.v.*

plasmodesm(a) [Gk. *plasma*, form; *desma*, bond] cytoplasmic threads penetrating cell wall and forming intercellular bridge; *alt.* plasmodesmid; *plu.* plasmodesmata.

plasmodial a. [Gk. *plasma*, form; *eidos*, form] *pert.* a plasmodium.

plasmodiocarp n. [Gk. *plasma*, form; *eidos*, form; *karpos*, fruit] a modification of a plasmodium in some slime moulds.

Plasmodiophorales, Plasmodiophorae n. [*Plasmodiophora*, generic name] an order of phycomycete fungi which are obligate plant parasites and are sometimes classed as slime moulds as they have a plasmodial vegetative stage and no mycelium.

Plasmodiophoromycetes n. [*Plasmodiophora*, generic name; Gk. *mykēs*, fungus] a class of fungi including the order Plasmodiophorales.

plasmoditrophoblast syntrophoblast, *q.v.*

plasmodium n. [Gk. *plasma*, form; *eidos*, form] a collection of amoeboid masses without nuclear fusion; a multinucleate mass of protoplasm without cell wall, of Myxomycetes; syncytium, *q.v.*; a. plasmodial.

plasmogamy n. [Gk. *plasma*, form; *gamos*, marriage] in Protozoa, fusion of several individuals into a multinucleate mass; fusion of cytoplasm without nuclear fusion; *alt.* plasmatogamy; *cf.* karyogamy.

plasmolysis n. [Gk. *plasma*, form; *lysis*, loosing] the withdrawal of water from a plant cell due to osmosis, resulting in contraction of cytoplasm away from cell walls.

plasmomites n.plu. [Gk. *plasma*, form; *mitos*, thread] minute fibrillae forming with plasmosomes the intergranular substance of a cell; *cf.* plasmosome.

plasmon n. [Gk. *plasma*, form; *on*, being] collectively, all the cytoplasmic hereditary determinants, i.e. those that are not situated on chromosomes.

plasmonema n. [Gk. *plasma*, form; *nēma*, thread] protoplasmic thread in connection with plastids; *plu.* plasmonemata.

plasmophore n. [Gk. *plasma*, form; *phora*, carrying] Z-disc, *q.v.*

plasmoptysis n. [Gk. *plasma*, form; *ptysis*, expectoration] emission of cytoplasm from tips of hyphae in host cells, in certain endotrophic mycorrhizae, *cf.* ptyosome; localized extrusion of cytoplasm through cell wall of bacteria.

plasmosome n. [Gk. *plasma*, form; *sōma*, body] the true nucleolus, *cf.* karyosome; a minute cytoplasmic granule, *cf.* plasmomite, microsome; *alt.* lininoplast.

plasmotomy n. [Gk. *plasma*, form; *tomē*, cutting] division of plasmodium by cleavage into multinucleate parts.

plasome n. [Gk. *plasma*, form; *sōma*, body] a hypothetical unit, *q.v.*

plasson n. [Gk. *plassein*, to form] the formative substances which may give rise to cellular elements; undifferentiated protoplasm; *alt.* gametoblast.

plastic a. [Gk. *plastos*, formed] formative; *appl.* substances used in forming or building up tissues or organs; *appl.* force which gives matter definite form; *appl.* tonus: producibility of different degrees of tension in the same length of skeletal muscle.

plastid n. [Gk. *plastos*, formed; *idion*, dim.] a cell body other than nucleus or centrosome, *alt.* trophoplast; a cellular organelle containing pigment, especially in plants.

plastidogen organ the axial organ of echinoderms.

plastidome n. [Gk. *plastos*, formed; *idion*, dim.; *domos*, chamber] in a cell, the plastids as a whole; cytoplasmic inclusions which give rise to plastids.

plastidule n. [Gk. *plastos*, formed; *idion*, dim.] a hypothetical unit, *q.v.*

plastin n. [Gk. *plastos*, formed] a substance found in reticulum of cells.

plastochondria mitochondria, *q.v.*

plastochron(e) n. [Gk. *plastos*, formed; *chronos*, time] time interval between successive stages in development, as between appearance of successive primordia in spiral systems of phyllotaxis.

plastocont chondriocont, *q.v.*

plastocyanin n. [Gk. *plastos*, formed; *kyanos*, blue] a blue protein in the chloroplasts, which contains copper and is an electron carrier.

plastodeme n. [Gk. *plastos*, formed; *dēmos*, people] a deme, which differs from others phenotypically but not genotypically.

plastodynamia n. [Gk. *plastos*, formed; *dynamis*, power] plastic or formative force.

plastogamy n. [Gk. *plastos*, formed; *gamos*, marriage] union of distinct unicellular individuals with fusion of cytoplasm but not of nuclei.

plastogenes n.plu. [Gk. *plastos*, formed; *gennaein*, to produce] cytoplasmic factors, controlled by or interacting with the nucleus, which determine differentiation of plastids.

plastoglobuli n.plu. [Gk. *plastos*, formed; L. *globulus*, small globe] globules made mainly of lipids found in stroma of chloroplasts which are osmiophilic, and enlarge and become full of carotinoid when the chloroplast ages or is converted to a chromoplast; *alt.* osmiophilic globules.

plastokont chondriocont, *q.v.*

plastolysis n. [Gk. *plastos*, formed; *lysis*, loosing] dissolution of mitochondria.

plastomere n. [Gk. *plastos*, formed; *meros*, part] the region of sperm containing many mitochondria, *alt.* chondriomere; the non-nuclear portion of a sperm, *alt.* cytomere.

plastoquinone any of various quinones in the chloroplasts which transport electrons during photosynthesis.

plastorhexis n. [Gk. *plastos*, formed; *rhēxis*, breaking] the breaking up of mitochondria into granules.

plastosomes n.plu. [Gk. *plastos*, formed; *sōma*, body] mitochondria, *q.v.*

plastral a. [F. *plastron*, breast plate] *pert.* a plastron.

plastron n. [F. *plastron*, breast plate] ventral bony shield of tortoises and turtles; other corresponding structure; sternum of arachnids; film of gas, or layer of gas bubbles retained by hairs, covering epicuticle of aquatic insects.

plate n. [F. *plat*, Gk. *platys*, flat] a flat, broad, plate-like structure or surface; a lamina, scale, disc, etc.

platelet a minute plate; blood platelet, *q.v.*

plate meristem a ground meristem in plant parts with a flat form such as a leaf, consisting of parallel layers of cells dividing only anticlinally in relation to the surface.

platy *appl.* soil crumbs in which the vertical axis is shorter than the horizontal.

platybasic a. [Gk. *platys*, flat; *basis*, base] *appl.* the primitive chondrocranium with wide hypophysial fenestra; *cf.* tropibasic.

Platycopa n. [Gk. *platys*, flat; *-ōdēs*, like] an order of ostracods consisting of only one marine genus having one pair of trunk legs and biramous 2nd antennae.

Platyctenia n. [Gk. *platys*, flat; *kteis*, comb] an order of Ctenophora of the class Tentaculata having the body flattened in the oral–aboral plane and combs often in the larva only.

platydactyl a. [Gk. *platys*, flat; *daktylos*, finger] with flattened-out fingers and toes as certain tailless amphibians.

Platyhelminthes n. [Gk. *platys*, flat; *helmins*, worm] a phylum of metazoan acoelomate animals, commonly called flatworms, which are dorsoventrally flattened and bilaterally symmetrical, and have the epidermis (ectoderm) and endoderm separated by a solid mass of mesoderm; *alt.* platyhelminths.

platyhieric a. [Gk. *platys*, flat; *hieros*, sacred] having sacral index above 100; *cf.* dolichohieric.

platymyarian a. [Gk. *platys*, flat; *mys*, muscle] having flat muscle cells, *appl.* some nematodes.

Platysiagiformes n. [*Platysiagus*, generic name; L. *forma*, shape] an order of marine Triassic to Jurassic Chondrostei with an elongated body.

platysma n. [Gk. *platysma*, flat piece] broad sheet of muscle between superficial fascia of neck.

platyspermic a. [Gk. *platys*, flat; *sperma*, seed] having seed which is flattened in transverse section.

play n. [M.E. *play*, play] behaviour exhibited especially by young animals in which they explore the environment and learn by trial and error during the times when conditions of life are fairly easy for them.

Plecoptera n. [Gk. *plekein*, to fold; *pteron*, wing] an order of insects commonly called stoneflies, having incomplete metamorphosis, aquatic nymphs, biting mouth-parts, and membranous wings.

Plectascales *n.* [Gk. *plektos*, twisted; *askos*, bag] an order of Plectomycetes having the asci enclosed in a cleistothecium and arising at various levels; *alt.* Aspergillales.

plectenchyma *n.* [Gk. *plektos*, twisted; *engchyma*, infusion] a tissue of interwoven cell filaments or hyphae in algae and fungi.

plectoderm *n.* [Gk. *plektos*, plaited; *derma*, skin] outer tissue of a fruit body, when composed of densely interwoven branched hyphae.

Plectognathi *n.* [Gk. *plektos*, twisted; *gnathos*, jaw] Tetraodontiformes, *q.v.*

Plectomycetes *n.* [Gk. *plektos*, twisted; *mykēs*, fungus] a group of Euascomycetes in which the ascocarp is a cleistothecium; *alt.* Eurotiales.

plectonemic *a.* [Gk. *plektos*, twisted; *nēma*, thread] having orthospirals interlocked at each twist as of sister chromatids; *cf.* paranemic.

plectonephridia, plectonephria *n.plu.* [Gk. *plektos*, twisted; *nephros*, kidney] nephridia of diffuse type formed of networks of fine excretory tubules lying on body wall and septa of certain oligochaetes.

plectostele *n.* [Gk. *plektos*, plaited; *stēlē*, pillar] a modified actinostele found in some species of *Lycopodium* (club mosses) being deeply fissured in cross-section.

plectron *n.* [Gk. *plēktron*, instrument to strike with] hammer-like form of certain bacilli during sporulation.

plectrum *n.* [L. *plectrum*, instrument to strike with] styloid process of temporal bone; *alt.* malleus.

plegetropism *n.* [Gk. *plēgē*, shock; *tropē*, turn] a movement of an organ, resulting from redistribution of particles in protoplasm, in response to change in velocity.

pleio- *alt.* pleo-, *q.v.*

pleioblastic *a.* [Gk. *pleion*, more; *blastos*, bud] having several buds; germinating at several points, as spores of certain lichens; *alt.* pleioblastous.

pleiochasium *n.* [Gk. *pleion*, more; *chasis*, division] axis of a cymose inflorescence bearing more than 2 lateral branches; *alt.* pleiochasial cyme.

pleiocotyl *n.* [Gk. *pleion*, more; *kotylē*, cup] a plant having more than 2 cotyledons.

pleiocotyledony *n.* [Gk. *pleion*, more; *kotylēdōn*, cup-shaped hollow] the condition of having more than 2 seed leaves or cotyledons.

pleiocyclic *a.* [Gk. *pleion*, more; *kyklos*, circle] living through more than one cycle of activity, as a perennial plant.

pleiomerous *a.* [Gk. *pleion*, more; *meros*, part] having more than normal number of parts as of petals or sepals in a whorl; *n.* pleiomery.

pleiomorphous, pleiomorphism pleomorphic, *q.v.*; pleomorphism, *q.v.*

pleiopetalous *a.* [Gk. *pleion*, more; *petalon*, leaf] having more than the normal number of petals; having double flowers.

pleiophyllous *a.* [Gk. *pleion*, more; *phyllon*, leaf] having more than normal number of leaves or leaflets.

pleiosporous polysporous, *q.v.*

pleiotaxy *n.* [Gk. *pleion*, more; *taxis*, arrangement] a multiplication of whorls, as in double flowers.

pleiotropy, pleiotropism *n.* [Gk. *pleion*, more; *tropē*, turn] multiple effects of a single gene, influencing more than one character; *alt.* polypheny; *a.* pleiotropic.

pleioxenous *a.* [Gk. *pleion*, more; *xenos*, host] heteroecious, *q.v.*; *n.* pleioxeny.

Pleistocene *a.* [Gk. *pleistos*, most; *kainos*, recent] *pert.* or *appl.* glacial and postglacial epoch following the Tertiary period, and merging into the Psychozoic. *n.* The great Ice Age, with 4 glacial and 3 interglacial phases.

pleiston *n.* [Gk. *pleistos*, very many] a group of very similar strains of bacteria.

pleo- *alt.* pleio-, *q.v.*

pleochroic *a.* [Gk. *pleon*, more; *chrōs*, colour] with various colours.

pleochromatic *a.* [Gk. *pleon*, more; *chrōma*, colour] exhibiting different colours under different environmental or physiological conditions.

pleogamy *n.* [Gk. *pleon*, more; *gamos*, marriage] maturation, therefore pollination, at different times, as of flowers of one plant.

pleometrosis *n.* [Gk. *pleon*, more; *mētēr*, mother] colony foundation by more than one female, as in some social Hymenoptera; *cf.* monometrosis; *a.* pleometrotic.

pleomorphic *a.* [Gk. *pleon*, more; *morphē*, form] being able to change shape; polymorphous, *q.v.*; *alt.* pleomorphous, pleiomorphous.

pleomorphism the ability to assume several different shapes; polymorphism, *q.v.*; *alt.* pleiomorphism.

pleon *n.* [Gk. *plein*, to swim] the abdominal region of Crustacea.

pleophagous *a.* [Gk. *pleon*, more; *phagein*, to eat] having several hosts; having a variety of foods.

pleophyletic *a.* [Gk. *pleon*, more; *phylon*, race] originating from several lines of descent; *alt.* polyphyletic.

pleopod *n.* [Gk. *plein*, to swim; *pous*, foot] an abdominal appendage or swimming leg of Crustacea.

Pleosporales *n.* [Gk. *pleon*, more; *sporos*, seed] an order of Ascomycetes whose asci are borne among pseudoparaphyses in a basal layer.

pleotrophic *a.* [Gk. *pleon*, more; *trophē*, nourishment] eating various kinds of food.

plerergates *n.plu.* [Gk. *plērēs*, full; *ergatēs*, worker] repletes, *q.v.*

plerocercoid *n.* [Gk. *plērēs*, full; *kerkos*, tail; *eidos*, form] a solid elongated metacestode of certain tapeworms, especially when found in fish muscle; *alt.* plerocestoid.

plerocestoid *n.* [Gk. *plērēs*, full; *kestos*, girdle; *eidos*; form] metacestode, *q.v.*; plerocercoid, *q.v.*

plerome *n.* [Gk. *plērōma*, a filling] the core or central part of an apical meristem; 1 of the 3 histogens of a stem meristem, thought to form the central region of the stem.

plerotic *a.* [Gk. *plēroun*, to fill] completely filling a space; *appl.* oospore filling oogonium; *cf.* aplerotic.

plesiobiotic *a.* [Gk. *plēsios*, near; *biōtikos*, pert. life] living in close proximity, *appl.* colonies of ants of different species; building contiguous nests, *appl.* ants and termites.

plesiometacarpal *a.* [Gk. *plēsios*, near; *meta*, after; *karpos*, wrist] *appl.* condition of retaining proximal elements of metacarpals, as in many Cervidae (deer); *cf.* telemetacarpal.

plesiomorphous *a.* [Gk. *plēsios*, near; *morphē*, form] having a similar form.

Plesiopora *n.* [Gk. *plēsios*, near; *poros*, channel] an order of small aquatic oligochaetes having the male gonopores opening on the segment immediately behind the one containing the testis.

Plesiosauria, plesiosaurs *n., n.plu.* [Gk. *plēsios*, near; *sauros*, lizard] an order of Mesozoic sauropterygians that are fully aquatic with a barrel-shaped body and paddle-like limbs.

plesiotype *n.* [Gk. *plēsios*, near; *typos*, pattern] hypotype, *q.v.*

pleura *n.* [Gk. *pleura*, side] a serous membrane lining thoracic cavity and investing lung, *plu.* pleurae; *plu.* of pleuron.

pleural *a.* [Gk. *pleura*, side] *pert.* a pleura or pleuron, as pleural ganglia; *appl.* recesses: spaces within pleural sac not occupied by lung; *appl.* costal plates of chelonian carapace.

pleuralia *n.plu.* [Gk. *pleura*, side] defensive spicules scattered over general body surface.

pleuranthous *a.* [Gk. *pleura*, side; *anthos*, flower] having inflorescences on lateral axes, not on main axis; *cf.* acranthous.

pleurapophysis *n.* [Gk. *pleura*, side; *apo*, from; *physis*, growth] a lateral vertebral process or true rib.

pleurethmoid *n.* [Gk. *pleura*, side; *ēthmos*, sieve; *eidos*, form] the compound ectethmoid and prefrontal of some fishes.

pleurite *n.* [Gk. *pleura*, side] a sclerite of the pleuron.

pleuroblastic *a.* [Gk. *pleura*, side; *blastos*, bud] producing, having, or *pert.* lateral buds or outgrowths; *appl.* haustoria of some Peronosporales.

pleurobranchiae *n.plu.* [Gk. *pleura*, side; *brangchia*, gills] gills springing from lateral walls of thorax of certain arthropods especially crustaceans; *alt.* pleurobranchs.

Pleurobranchomorpha *n.* [Gk. *pleura*, side; *brangchia*, gills; *morphē*, form] Notaspidea, *q.v.*

Pleurocapsales *n.* [Gk. *pleura*, side; L. *capsa*, box] an order of blue–green algae which have a filamentous plant body, lack hormogonia, and reproduce by endospores.

Pleurocarpi *n.* [Gk. *pleura*, side; *karpos*, fruit] a group of mosses comprising pleurocarpous forms; *cf.* Acrocarpi.

pleurocarpic, pleurocarpous *a.* [Gk. *pleura*, side; *karpos*, fruit] with lateral fructifications; *appl.* mosses bearing archegonia and therefore capsules on a small side branch and not at the tip of the stem or main branches; *cf.* acrocarpous.

pleuroccipital exoccipital, *q.v.*

pleurocentrum *n.* [Gk. *pleura*, side; L. *centrum*, centre] a lateral element of centrum of many fishes and fossil amphibians.

pleurocerebral *a.* [Gk. *pleura*, side; L. *cerebrum*, brain] *pert.* pleural and cerebral ganglia, in molluscs.

pleurocystidium *n.* [Gk. *pleura*, side; *kystis*, bag; *idion*, dim.] a cystidium in hymenium of surface of gill lamella; *alt.* pleurotrichome; *cf.* cheilocystidium.

pleurodont *a.* [Gk. *pleura*, side; *odous*, tooth] having teeth fixed by sides to lateral surface of jaw ridge, as in some lizards.

pleurogenous *a.* [Gk. *pleura*, side; *gennaein*, to

produce] originating or growing from the side or sides.

Pleurogona *n.* [Gk. *pleura*, side; *gonē*, seed] an order of solitary or colonial sea squirts in which the body is not divided into separate regions.

pleuron *n.* [Gk. *pleuron*, side] one of the external lateral pieces of body segments of arthropods; *alt.* pleurum; *plu.* pleura.

Pleuronectiformes *n.* [*Pleuronectes*, generic name; L. *forma*, shape] an order of teleosts which are asymmetrical and have both eyes on one side, including the plaice, sole, and halibut; *alt.* Heterosomata.

pleuropedal *a.* [Gk. *pleura*, side; L. *pes*, foot] *pert.* pleural and pedal ganglia of molluscs.

pleuroperitoneum *n.* [Gk. *pleura*, side; *periteinein*, to stretch around] pleura and peritoneum combined, body-lining membrane of animals without diaphragm.

pleuropneumonia-like organisms (PPLO) an early name for mycoplasmas, *q.v.*

pleuropodium *n.* [Gk. *pleura*, side; *pous*, foot] a lateral glandular process of abdomen of some insect embryos.

pleurosphenoid *n.* [Gk. *pleura*, side; *sphēn*, wedge] sphenolateral, *q.v.*

pleurospore *n.* [Gk. *pleura*, side; *sporos*, seed] spore formed on sides of a basidium.

pleurosteon *n.* [Gk. *pleura*, side; *osteon*, bone] lateral process of sternum in young birds, afterwards costal process.

pleurosternal *a.* [Gk. *pleuron*, side; *sternon*, chest] connecting or *pert.* pleuron and sternum, *appl.* thoracic muscles in insects.

Pleurotremata *n.* [Gk. *pleura*, side; *trēma*, hole] a group of Chondrichthyes including the sharks; *cf.* Hypotremata.

pleurotribe *a.* [Gk. *pleura*, side; *tribein*, to rub] *appl.* flowers whose anthers and stigma are so placed as to rub sides of insects entering—a device for securing cross-pollination.

pleurotrichrome *n.* [Gk. *pleura*, side; *trichōma*, growth of hair] pleurocystidium, *q.v.*

pleurovisceral *a.* [Gk. *pleura*, side; L. *viscera*, intestines] *pert.* pleural and visceral ganglia, of molluscs.

pleurum pleuron, *q.v.*

pleuston *n.* [Gk. *pleustikos*, ready for sailing] free-floating organisms especially those possessing a gas-filled bladder or float.

plexiform *a.* [L. *plexus*, interwoven; *forma*, shape] entangled or complicated; like a network, *appl.* layers of retina, *appl.* peripheral layer of grey matter of cerebral cortex, *alt.* molecular layer.

plexiform gland the axial organ of echinoderms.

plexus *n.* [L. *plexus*, interwoven] a network of interlacing vessels, nerves, fibres, etc.; *alt.* rete.

plexus myentericus Auerbach's plexus, *q.v.*

plica *n.* [L. *plicare*, to fold] a fold of skin, membrane, or lamella; a corrugation of brachiopod shell.

plicate *a.* [L. *plicare*, to fold] folded like a fan, as a leaf; folded or ridged.

pliciform *a.* [L. *plicare*, to fold; *forma*, shape] resembling a fold; disposed in folds.

Pliocene *n.* [Gk. *pleion*, more; *kainos*, recent] the geological epoch which followed the Miocene and preceded the Pleistocene.

plotophyte n. [Gk. *plōtos*, floating; *phyton*, plant] a plant adapted for floating.

ploughshare bone pygostyle, q.v.; vomer, q.v.

pluma n. [L. *pluma*, feather] a contour feather of birds.

plumate a. [L. *pluma*, feather] plume-like.

Plumbaginales n. [*Plumbago*, generic name] an order of dicots of the Sympetalae or Caryophyllidae, comprising the family Plumbaginaceae which are xerophytes or halophytes of steppes, semideserts, and coasts, and include sea lavender and thrift.

plume n. [L. *pluma*, feather] a feather, or feather-like structure.

plumicome n. [L. *pluma*, feather; *coma*, hair] a spicule with plume-like tufts.

plumicorn n. [L. *pluma*, feather; *cornu*, horn] horn-like tuft of feathers on bird's head.

plumigerous a. [L. *pluma*, feather; *gerere*, to carry] feathered.

plumiped n. [L. *pluma*, feather; *pes*, foot] a bird with feathered feet.

plumose a. [L. *pluma*, feather] feathery; having feathers; feather-like, *appl.* a type arrangement of skeletal fibre in sponges, *appl.* gills, as in cephalopods; *appl.* feathers without hamuli on barbules, cf. pennaceous.

plumula n. [L. *plumula*, small feather] an adult down feather, succeeding preplumula; plumule, q.v.

plumulaceous plumulate, q.v.

plumular *pert.* a plumule.

plumulate a. [L. *plumula*, small feather] downy; with a downy covering; *alt.* plumulaceous.

plumule n. [L. *plumula*, small feather] a primary bud on epicotyl of seed and which develops into the shoot; androconia, q.v.; plumula, q.v.

plumule sheath coleoptile, q.v.

pluriascal a. [L. *plus*, more; Gk. *askos*, bag] *pert.* or containing several asci.

pluriaxial a. [L. *plus*, more; *axis*, axle] having flowers developed on secondary shoots.

plurilocular a. [L. *plus*, more; *loculus*, little place] having 2 or more loculi or compartments; *alt.* multilocular, pluriloculate.

plurinuclear a. [L. *plus*, more; *nucleus*, kernel] having several nuclei.

pluriparous a. [L. *plus*, more; *parere*, to bring forth] giving or having given birth to a number of offspring.

pluripartite a. [L. *plus*, more; *partitus*, divided] with many lobes or partitions.

pluripolar a: [L. *plus*, more; *polus*, axis end] having several poles, *appl.* ganglion cells, etc.; *alt.* multipolar.

pluripotent a. [L. *plus*, more; *potens*, powerful] *pert.* embryonic tissue before determination of its ultimate developmental fate has taken place.

plurisegmental a. [L. *plus*, more; *segmentum*, a slice] *pert.*, or involving, a number of segments, *appl.* conduction, reflexes.

pluriseptate a. [L. *plus*, more; *septum*, partition] with multiple septa.

pluriserial a. [L. *plus*, more; *series*, row] arranged in 2 or more rows.

pluristratose a. [L. *plus*, more; *stratum*, layer] arranged in a number of layers; much stratified.

plurivalent a. [L. *plus*, more; *valere*, to be worth] *appl.* chromatin rods of several associated homologous chromosomes.

plurivorous a. [L. *plus*, more; *vorare*, to devour] feeding on several substrates or hosts.

pluteus n. [L. *pluteus*, shed] the free-swimming larva of echinoids (sea urchins) and ophiurans (brittle stars) which resembles an upturned easel; cf. echinopluteus, ophiopluteus; a. pluteal.

pneumathode n. [Gk. *pneuma*, breath, air; *hodos*, way] an aerial or respiratory root.

pneumatic a. [Gk. *pneuma*, air] *appl.* bones penetrated by canals connected with respiratory system, in birds, *alt.* aerostatic; *appl.* duct between swim bladder and alimentary tract, in physostomous fishes.

pneumaticity n. [Gk. *pneuma*, air] state of having air cavities, as bones of flying birds.

pneumatized a. [Gk. *pneuma*, air] furnished with air cavities.

pneumato- also pneumo-, q.v.

pneumatocyst n. [Gk. *pneuma*, air; *kystis*, bladder] the air bladder or swim bladder of fishes; air cavity used as float; air bladder of bladderwrack.

pneumatophore n. [Gk. *pneuma*, air; *pherein*, to bear] the air sac or float of Siphonophora; an air bladder of marsh or shore plants; aerating outgrowth in certain ferns; an aerating root.

pneumatopyle n. [Gk. *pneuma*, air; *pylē*, gate] a pore of a pneumatophore, opening above to exterior in certain Siphonophora.

pneumatotaxis pneumotaxis, q.v.

pneumo- also pneumato-, q.v.

pneumogastric a. [Gk. *pneuma*, air; *gastēr*, stomach] *appl.* 10th cranial or vagus nerve, supplying pharynx, larynx, heart, lungs, and viscera.

pneumostome n. [Gk. *pneuma*, breath; *stoma*, mouth] the pulmonary aperture, through which air passes to and from respiratory mantle cavity in terrestrial gastropods; in some arachnids, a small opening to a pit containing gill books.

pneumotaxis n. [Gk. *pneuma*, air; *taxis*, arrangement] reaction to stimulus by gases, especially by carbon dioxide in solution; *alt.* pneumatotaxis.

poad n. [Gk. *poa*, meadow] a meadow plant.

Poales n. [*Poa*, generic name] an order of monocots (Liliidae) comprising the large family Poaceae (Gramineae), the grasses; *alt.* Glumiflorae.

poculiform a. [L. *poculum*, cup; *forma*, shape] cup-shaped; goblet-shaped.

pod n. [M.E. *pod*, bag] a superior 1-celled, 1- or many-seeded fruit made from a single carpel and splitting into 2 valves at maturity, *alt.* legume; a husk; the cocoon in which eggs are laid in locust.

podal a. [Gk. *pous*, foot] *pert.* feet, *alt.* pedal; *pert.* parapodia, *appl.* membrane.

podeon n. [Gk. *podeōn*, neck] the slender middle part of abdomen of Hymenoptera, uniting propodeon and metapodeon; *alt.* podeum, petiole.

podetiiform a. [Gk. *pous*, foot; L. *forma*, shape] resembling a podetium.

podetium n. [Gk. *pous*, foot] a stalk-like elevation; outgrowth of thallus bearing apothecium in certain lichens.

podeum podeon, q.v.

podex n. [L. *podex*, rump] the region about the anus; pygidium, q.v.

podia *plu.* of podium.

podical a. [L. *podex*, rump] in anal region; *appl.* each of a pair of small plates: paraproct, q.v.

Podicipitiformes n. [L. *podex*, rump; *pes*, foot;

forma, shape] Colymbiformes, *q.v.*; *alt.* Podici-pediformes.

podite *n.* [Gk. *pous,* foot] a crustacean walking leg.

podium *n.* [Gk. *pous,* foot] a foot or foot-like structure such as tube foot; a stem axis.

podobranchiae *n.plu.* [Gk. *pous,* foot; *brangchia,* gills] foot-gills springing from coxopodites of thoracic appendages of certain arthropods; *alt.* podobranchs.

podocephalous *a.* [Gk. *pous,* foot; *kephalē,* head] having head of flowers on long stalk.

podoconus *n.* [Gk. *pous,* foot; *kōnos,* cone] a conical mass of endoplasm connecting the central capsule with the disc of Sarcodina.

Podocopa *n.* [Gk. *pous,* foot; *-ōdēs,* like] an order of marine and freshwater ostracods having 2 pairs of trunk legs and uniramous 2nd antennae.

podocyst *n.* [Gk. *pous,* foot; *kystis,* bladder] a pedal sinus or caudal vesicle in certain Gastropoda.

podocyte *n.* [Gk. *pous,* foot; *kytos,* hollow] a flat blood cell with a few pointed outgrowths, in insects; an epithelial cell of the Bowman's capsule of the kidney, having numerous processes which rest on the basement membrane.

podocytosis *n.* [Gk. *pous,* foot; *kytos,* hollow] cytopempsis, *q.v.*

pododerm *n.* [Gk. *pous,* foot; *derma,* skin] dermal layer of a hoof, within horny layer.

podogynium *n.* [Gk. *pous,* foot; *gynē,* female] a stalk supporting the gynaecium; *alt.* basigynium.

podomere *n.* [Gk. *pous,* foot; *meros,* part] a limb segment of arthropods.

podophthalmite *n.* [Gk. *pous,* foot; *ophthalmos,* eye] in crustaceans, eye-stalk segment farthest from head.

podosoma *n.* [Gk. *pous,* foot; *sōma,* body] the body region of Acarina which bears the 4 pairs of walking legs.

Podostemales, Podostemonales *n.* [*Podostemon,* generic name] an order of dicots (Rosidae) comprising one family, Podostemaceae, whose vegetative parts are reduced to thallus-like structures and which live in fast-flowing tropical waters.

podotheca *n.* [Gk. *pous,* foot; *thēkē,* box] a foot covering, as of birds or reptiles; pupal leg sheath.

podsol podzol, *q.v.*

podzol *n.* [Russ. *pod,* under; *zolit',* to leach] grey forest soil, the soil type of cold temperate regions, and formed on heathlands and under coniferous forest; *alt.* podsol.

poecilo- *see also* poikilo-.

Poecilosclerina *n.* [Gk. *poikilos,* various; *sklēros,* hard] a large order of sponges of the subclass Monaxonida having various kinds of microscleres and projecting megascleres joined by spongin fibres.

pogonion *n.* [Gk. *pōgōnion,* little beard] most prominent point of chin as represented on mandible.

Pogonophora *n.* [Gk. *pōgōnophoros,* wearing a beard] a group of protochordates similar to the hemichordates but now usually separated from them into a separate phylum.

poikilo- *alt.* poecilo-, *q.v.*

poikilochlorophyllous *a.* [Gk. *poikilos,* various; *chloros,* grass green; *phyllon,* leaf] completely losing and regaining their chlorophyll in response to changes in environmental conditions, *appl.* some angiosperms.

poikilocyte *n.* [Gk. *poikilos,* various; *kytos,* cell] a distorted form of erythrocyte present in certain pathological conditions; *alt.* schistocyte.

poikilogony *n.* [Gk. *poikilos,* various; *gonē,* generation] intraspecific variation in duration of embryological processes, due to environmental factors.

poikilohydrous *a.* [Gk. *poikilos,* various; *hydōr,* water] becoming dormant in the dry season after losing most of their water, *appl.* some angiosperms.

poikilokont *n.* [Gk. *poikilos,* various; *kontos,* punting pole] the flagellum and all homologous organelles of higher organisms.

poikilosmotic *a.* [Gk. *poikilos,* various; *osmos,* impulse] having internal osmotic pressure varying with that of the surrounding medium, as with salinity; *cf.* homoiosmotic.

poikilotherm a poikilothermal organism.

poikilotherm(al) *a.* [Gk. *poikilos,* various; *thermē,* heat] *appl.* animals whose temperature varies with that of the surrounding medium; *alt.* poikilothermous, cold-blooded, ectothermal, heterothermal, haematocryal; *cf.* homoiothermal.

pointer cell eurycyst, *q.v.*

point mutation a mutation occurring at a single gene locus; *alt.* gene mutation, genovariation, micromutation, transgenation.

poisers halteres, *q.v.*

poison canal duct between stylet and lancets of sting of Hymenoptera, conveying secretion of poison glands from poison sac outwards.

polar *a.* [Gk. *polos,* pole] in region of end of an axis; at, or *pert.,* a pole.

polar body one of 2 cells divided off from ovum during maturation, before germ nuclei fuse; *alt.* polocyte, polar globule, directive body.

polar capsules of spores containing coiled extrusible filaments, in Cnidosporidia; *alt.* pole capsules.

polar cartilage posterior portion of trabecula, or independent cartilage in that region.

polar corpuscle centrosome, *q.v.*

polar globule polar body, *q.v.*

polar granule centromere, *q.v.*

polarilocular *n.* [L. *polaris,* polar; *loculus,* compartment] *appl.* a cask-shaped spore with 2 cells separated by a partition having a perforation, of certain lichens; *alt.* polaribilocular.

polarity *n.* [Gk. *polos,* axis] the tendency of plants to develop from the poles, roots downwards, stems upwards; the tendency of an ovum to place itself with axis corresponding to that of mother; differential distribution on gradation along an axis; condition of having opposite poles; existence of opposite qualities.

polarization *n.* [Gk. *polos,* pole] the setting up of a potential difference across a membrane.

polar nuclei nuclei at each end of angiosperm embryo, which later form secondary nucleus.

polaron *n.* [Gk. *polos,* pole] a region in a chromosome in which gene conversion has taken place resulting in polarized genetic recombination.

polar organ a caudal cell-cluster in early embryo of insects; *alt.* grumulus.

polar plates two narrow ciliated areas produced in transverse plane, part of equilibrium apparatus of certain Coelenterata; cytoplasmic areas without centrosomes beyond spindle poles of a dividing nucleus in certain Protista.

polar rays a ray arising from the aster as distinguished from a spindle fibre; *alt.* astral rays.

polar rings two ring-shaped cytoplasmic masses near ovum poles formed after union of germ nuclei.

polar translocation or transport movement of materials through plant tissues in one direction only.

pole capsules polar capsules, *q.v.*

pole cell teloblast of annelids and molluscs.

Polemoniales *n.* [*Polemonium*, generic name] an order of dicots of the Gamopetalae or Asteridae, used in different ways by different authorities.

pole plates end plates or achromatic masses at spindle poles in protozoan mitosis.

Polian vesicles [*G.S. Poli*, Italian naturalist] interradial vesicles opening into ring vessel of ambulacral system of most Asteroidea and Holothuroidea.

Polioplasm *n.* [Gk. *polios*, grey; *plasma*, form] spongioplasm, *q.v.*; granular protoplasm.

pollakanthic *a.* [Gk. *pollakis*, many times; *anthos*, flower] having several flowering periods; *cf.* hapaxanthic.

pollen *n.* [L. *pollen*, fine flour] the powder, produced by anthers, consisting of pollen grains; *alt.* farina.

pollen analysis qualitative and quantitative determination of the occurrence of pollen in deposits, as in peat; palynology, *q.v.*

pollen basket the pollen-transporting hairs at back of tibia of worker bees; *alt.* corbiculum.

pollen brush enlarged hairy tarsal joint of bees; *alt.* sarothrum, scopa.

pollen case *see* theca.

pollen chamber pit formed at apex of nucellus below micropyle.

pollen comb comb-like structure of bristles on leg of bee below the pollen basket.

pollen flower a flower without nectar, attracting pollen-feeding insects.

pollen grain the haploid microspore of seed plants.

pollen profile the vertical distribution of pollen grains in a deposit.

pollen sac loculus of anther in which pollen is produced, being the microsporangium of seed plants.

pollen spectrum the relative numerical distribution or percentage of pollen grains of different species in a sample of deposit.

pollen tube a tubular process developed from pollen grain after attachment to stigma, and growing towards ovule carrying male nuclei to embryo sac.

pollex *n.* [L. *pollex*, thumb] the thumb, or innermost digit of the normal 5 in anterior limb.

pollinarium *n.* [L. *pollen*, fine flour] the pollinium with its caudicle and adhesive disc.

pollination *n.* [L. *pollen*, fine flour] transfer of pollen from anther to stigma prior to fertilization in angiosperms; transfer of pollen from male to female cone prior to fertilization in gymnosperms.

pollination drop mucilaginous drop exuded from micropyle and which detains pollen grains, as in gymnosperms.

polliniferous, pollinigerous *a.* [L. *pollen*, fine

flour; *ferre*, *gerere*, to carry] pollen-bearing; adapted for transferring pollen.

pollinium *n.* [L. *pollen*, fine flour] an agglutinated pollen mass in orchids and other plants.

pollinodium *n.* [L. *pollen*, fine flour; Gk. *hodos*, way] an obsolete term for antheridium, especially of algae and fungi.

pollinoid *n.* [L. *pollen*, fine flour; Gk. *eidos*, form] a male gamete; spermatium, *q.v.*

polocyte *n.* [Gk. *polos*, axis; *kytos*, hollow] polar body, *q.v.*

polospore *n.* [L. *pollen*, fine flour; Gk. *sporos*, seed] a fossil pollen grain or spore.

polster *n.* [Ger. *Polster*, pad] a low compact perennial or cushion plant.

polyadelphous *a.* [Gk. *polys*, many; *adelphos*, brother] having stamens united by filaments into more than 2 bundles.

polyamines *n.plu.* [Gk. *polys*, many; *ammoniakon*, gum] compounds containing 2 (diamines) or more amino groups.

polyandrous *a.* [Gk. *polys*, many; *anēr*, male] having 20 or more stamens, *alt.* polystemonous; mating with more than one male; *n.* polyandry.

polyanisomere *n.* [Gk. *polys*, many; *anisos*, unequal; *meros*, part] a structural unit composed of polyisomeres and anisomeres, e.g. vertebral column.

polyarch *a.* [Gk. *polys*, many; *archē*, beginning] having many protoxylem bundles; *appl.* multipolar spindle in higher plants.

polyarthric *a.* [Gk. *polys*, many; *arthron*, joint] multiarticulate, *q.v.*

polyaxon *n.* [Gk. *polys*, many; *axōn*, axle] type of spicule laid down along numerous axes; *alt.* multiradiate spicule.

polyblast *n.* [Gk. *polys*, many; *blastos*, bud] histiocyte, *q.v.*

polyblastic *a.* [Gk. *polys*, many; *blastos*, bud] having spores divided by a number of septa, *appl.* lichens; *alt.* polyblastous.

polycarp *n.* [Gk. *polys*, many; *karpos*, fruit] a gonad of some ascidians, on inner surface of mantle.

polycarpellary *a.* [Gk. *polys*, many; *karpos*, fruit] with compound gynaecium consisting of many carpels; *alt.* multicarpellary.

polycarpic, polycarpous *a.* [Gk. *polys*, many; *karpos*, fruit] with numerous, usually free carpels; producing seed season after season, *appl.* perennials.

Polycarpicae *n.* [Gk. *polys*, many; *karpos*, fruit] Magnoliidae, *q.v.*

polycaryo- polykaryo-, *q.v.*

polycentric *a.* [Gk. *polys*, many; *kentron*, centre] with several growth centres; with several centromeres, *appl.* chromosome, *cf.* monocentric. *n.* A polycentric chromosome.

polycercous, polycercoid *a.* [Gk. *polys*, many; *kerkos*, tail] *appl.* bladderworms developing several cysts, each with head, as in an echinococcus.

Polychaeta, polychaetes *n.*, *n.plu.* [Gk. *polys*, many; *chaitē*, hair] a class of annelids, commonly called bristle worms, that have parapodia with numerous chaetae, and a pronounced head bearing tentacles, palps, and often eyes.

polychasium *n.* [Gk. *polys*, many; *chasis*, division] a cymose branch system when more than 2 branches arise about the same point.

polychromasy, -ie *n.* [Gk. *polys*, many; *chrōma*, colour] multiple and differential tinting with one staining mixture.

polychromatic *a.* [Gk. *polys*, many; *chrōma*, colour] with several colours, as pigment areas; *appl.* 2 forms of erythrocytes with well-defined chromatin.

polychromatocyte *n.* [Gk. *polys*, many; *chrōma*, colour; *kytos*, hollow] a blood cell developed from a normoblast and which becomes a normocyte or mature erythrocyte; *alt.* polychromatophil erythrocyte or rubricyte.

polychromatophil *a.* [Gk. *polys*, many; *chōma*, colour; *philein*, to love] having a staining reaction, characterized by varying colours; *appl.* erythroblasts with small haemoglobin content.

polycistronic message a giant mRNA molecule which codes for 2 or more proteins by adjacent structural genes or cistrons in the same operon.

Polycladida, polyclads *n., n.plu.* [Gk. *polys*, many; *klados*, sprout] an order of large marine turbellarians having a gut with many caeca which ramify all through the body.

polyclimax *n.* [Gk. *polys*, many; *klimax*, ladder] a climatic climax formation consisting of several different climax associations or consociations, none of which shows a tendency to give way to any other.

polycotyledon *n.* [Gk. *polys*, many; *kotylēdōn*, hollow vessel] a plant with more than 2 cotyledons; *a.* polycotyledonous.

polycotyledonary *a.* [Gk. *polys*, many; *kotylēdōn*, hollow vessel] having placenta in many divisions.

polycotyledony *n.* [Gk. *polys*, many; *kotylēdōn*, hollow vessel] a great increase in number of cotyledons.

polycrotism *n.* [Gk. *polys*, many; *krotos*, beating] condition of having several secondary elevations in pulse curve.

polycyclic *a.* [Gk. *polys*, many; *kyklos*, circle] having many whorls or ring structures; with vascular system forming several concentric cylinders, *appl.* stele.

polycystid *a.* [Gk. *polys*, many; *kystis*, bladder] septate, *q.v.*; partitioned off.

polydactyly *n.* [Gk. *polys*, many; *daktylos*, finger] condition of having more than 5 fingers or toes; *alt.* polydactylism, hyperdactyly; *a.* polydactyl.

polydelphic *a.* [Gk. *polys*, many; *delphys*, womb] having more than one set of ovaries, oviducts, and uteri, *appl.* nematodes.

polyderm *n.* [Gk. *polys*, many; *derma*, skin] a protective tissue in which layers of endodermal cells and parenchyma cells alternate.

polydesmic *a.* [Gk. *polys*, many; *desmos*, bond] *appl.* cyclomorial scales made up of monodesmic scales; *cf.* synpolydesmic, deuteropolydesmic.

Polydesmida *n.* [Gk. *polys*, many; *desmos*, bond] an order of Helminthomorpha without silk glands or eyes.

polyembryony *n.* [Gk. *polys*, many; *embryon*, foetus] formation of several embryos in one ovule; instance of a zygote giving rise to more than one embryo, e.g. identical twins, offspring of armadillos, certain insects, etc.

polyenergid *a.* [Gk. *polys*, many; *energos*, active] *appl.* nuclei with more than one centriole.

polyestrous polyoestrous, *q.v.*

Polygalales, Polygalinae *n.* [*Polygala*, generic name] an order of dicots of the Rosidae or Polypetalae, the term being used in different ways by different authorities.

polygamy *n.* [Gk. *polys*, many; *gamos*, marriage] the condition of bearing male, female, and hermaphrodite flowers on the same or different plants; condition of consorting with more than one mate at a time; *a.* polygamous.

polygene *n.* [Gk. *polys*, many; *genos*, descent] a gene controlling quantitative characters; *alt.* buffering gene; *cf.* oligogene.

polygenesis *n.* [Gk. *polys*, many; *genesis*, descent] derivation from more than one source; origin of a new type at more than one place or time; *cf.* monogenesis.

polygenetic *a.* [Gk. *polys*, many; *genesis*, descent] derived from more than one source; *alt.* polyphyletic, polygenic.

polygenic *a.* [Gk. *polys*, many; *genos*, descent] controlled by a number of genes, *alt.* multifactorial; *pert.* polygenes; polygenetic, *q.v.*

polygenomatic *a.* [Gk. *polys*, many; *genos*, offspring] having nuclei containing more than one genome.

polygerm *n.* [Gk. *polys*, many; L. *germen*, bud] an isolated group of morulae.

polyglucosan *n.* [Gk. *polys*, many; *glykys*, sweet] a polymer made of a chain of glucose units.

Polygonales *n.* [*Polygonum*, generic name] an order of dicots of the Archichlamydeae or Caryophyllidae, comprising the single family Polygonaceae which has alternate leaves and stipules that unite to form a hood or ochrea, and includes the docks, rhubarb, and buckwheat.

polygoneutic *a.* [Gk. *polys*, many; *goneuein*, to beget] rearing more than one brood in a season.

polygynaecial *a.* [Gk. *polys*, many; *gynē*, woman; *oikos*, house] having multiple fruits formed by united gynaecia.

polygynous *a.* [Gk. *polys*, many; *gynē*, female] consorting with more than one female at a time; with many styles.

polyhaploid *a.* [Gk. *polys*, many; *haploos*, simple] *pert.* the gametic chromosome number of a polyploid organism. *n.* A haploid organism derived from a normally polyploid species.

polyhybrid *n.* [Gk. *polys*, many; L. *hybrida*, mongrel] a hybrid heterozygous for many genes.

polyhydric *a.* [Gk. *polys*, many; *hydōr*, water] *appl.* alcohol or acid with 3, 4, or more hydroxyl groups.

polyisomeres *n.plu.* [Gk. *polys*, many; *isos*, equal; *meros*, part] parts all homologous with each other, as leaves of plants of the same species; *cf.* anisomeres, polyanisomere.

polykaric *a.* [Gk. *polys*, many; *karyon*, nut] multinucleate, *q.v.*

polykaryo- *alt.* polycaryo-.

polykaryocyte *n.* [Gk. *polys*, many; *karyon*, nut; *kytos*, hollow] a multinucleate cell, of bone marrow.

polykaryon *n.* [Gk. *polys*, many; *karyon*, nut] a polyenergid nucleus.

polykont *a.* [Gk. *polys*, many; *kontos*, pole] multiflagellate, *q.v.*

polylecithal *a.* [Gk. *polys*, many; *lekithos*, yolk] containing relatively much yolk, as centrolecithal eggs; *cf.* megalolecithal, mesolecithal, meiolecithal.

polylepidous *a.* [Gk. *polys*, many; *lepis*, scale] having many scales.

polymastia polymastism, *q.v.*

Polymastigina *n.* [Gk. *polys*, many; *mastix*, whip] an order of non-amoeboid Zoomastigina having 2 to many flagella and an extranuclear division centre; used strictly to include flagellates with 3 or more flagella, or more widely to include these and also the Hypermastigina and Diplomonadina, or the group may be placed with the Hypermastigina in the Metamonadina.

polymastigote *a.* [Gk. *polys*, many; *mastix*, whip] having flagella arranged in a tuft; *alt.* multiflagellate.

polymastism *n.* [Gk. *polys*, many; *mastos*, breast] occurrence of more than normal number of mammae; *alt.* polymastia.

polymegaly *n.* [Gk. *polys*, many; *megalē*, large] occurrence of more than 2 sizes of sperm in one animal.

polymeniscous *a.* [Gk. *polys*, many; *mēniskos*, small moon] having many lenses, as compound eye.

polymer *n.* [Gk. *polys*, many; *meros*, part] a chemical compound made up of many monomer units put together during polymerization.

polymerase any enzyme which causes polymerization, especially of nucleic acids, all being EC group 2 enzymes.

polymeric *a.* [Gk. *polys*, many; *meros*, part] *appl.* system of independently segregating genes, additive in affecting the same phenotypic character.

polymerization *n.* [Gk. *polys*, many; *meros*, part] the formation of a polymer, a molecule which is an exact multiple of a smaller molecule and directly obtainable from it; often used biologically as a synonym for condensation.

polymerous *a.* [Gk. *polys*, many; *meros*, part] consisting of many parts or members; *n.* polymery.

polymitosis *n.* [Gk. *polys*, many; *mitos*, thread] excessive cell division.

Polymixiiformes *n.* [*Polymixia*, generic name; L. *forma*, shape] an order of deep-water marine teleosts existing from Cretaceous times to the present, and including the barbudos.

polymorph *n.* [Gk. *polys*, many; *morphē*, form] a polymorphonuclear leucocyte; *alt.* granulocyte.

polymorphic polymorphous, *q.v.*

polymorphism *n.* [Gk. *polys*, many; *morphē*, form] occurrence of different forms of individuals in same species; occurrence of different forms or different forms of organs, in same individual at different periods of life; *alt.* morphism, pleomorphism.

polymorphonuclear *a.* [Gk. *polys*, many; *morphē*, form; L. *nucleus*, kernel] *appl.* amoeboid leucocytes with multipartite nuclei connected by fine threads of chromatin; *alt.* polynuclear; *cf.* heterophil, neutrophil.

polymorphous *a.* [Gk. *polys*, many; *morphē*, form] showing a marked degree of variation in body form, during the life history, or within the species; *pert.* or containing variously shaped units; *appl.* layer, the inner cell lamina of cerebral cortex, *alt.* multiform; *alt.* polymorphic, pleomorphic, pleiomorphic; *cf.* monomorphic.

polymyarian *a.* [Gk. *polys*, many; *mys*, muscle] having more than 5 longitudinal rows of muscle cells, *appl.* some nematodes.

polyneme *a.* [Gk. *polys*, many; *nēma*, thread] *appl.* suggested chromatid structure consisting of many-stranded DNA; *cf.* mononeme, bineme.

Polyneoptera, Polyneuroptera *n.* [Gk. *polys*, many; *neos*, new; *neuron*, nerve; *pteron*, wing] a group of winged insects with incomplete metamorphosis in which the wings fold over the body in repose, and having many Malpighian tubules.

polynuclear polynucleate, *q.v.*; polymorphonuclear, *q.v.*

polynucleate *a.* [Gk. *polys*, many; L. *nucleus*, kernel] multinucleate, *q.v.*

polynucleotide *n.* [Gk. *polys*, many; L. *nucleus*, kernel] a polymer of many nucleotides.

polyoestrous *a.* [Gk. *polys*, many; *oistros*, gadfly] having a succession of oestrous periods in one sexual season; *alt.* polyestrous; *cf.* monoestrous.

Polyopisthocotylea *n.* [Gk. *polys*, many; *opisthe*, behind; *kotylē*, cup] a subclass of Monogenea having a funnel-shaped mouth with an adhesive organ on either side of it, but usually no oral sucker.

polyoses polysaccharides, *q.v.*

polyp *n.* [L. *polypus*, polyp] a zooid or individual of a colonial animal; in coelenterates, an individual having a tubular body, usually with mouth and ring of tentacles at the top, *alt.* hydroid; a small stalked outgrowth of proliferating cells from a mucous surface such as intestine.

polyparium, polypary *n.* [L. *polypus*, polyp] the common base and connecting tissue of a colony of polyps; *alt.* zoarium.

polypeptide *n.* [Gk. *polys*, many; *peptein*, to digest] a condensation product of many amino acids linked together by peptide bonds.

Polypetalae *n.* [Gk. *polys*, many; *petalon*, leaf] a group of dicots used especially in the Bentham and Hooker system, having 2 whorls to the perianth and being polypetalous; *alt.* Dialypetalae.

polypetalous *a.* [Gk. *polys*, many; *petalon*, leaf] having separate, free or distinct petals; *alt.* apopetalous, choripetalous, dialypetalous, eleutheropetalous; *cf.* gamopetalous, monopetalous.

polyphagous *a.* [Gk. *polys*, many; *phagein*, to eat] eating various kinds of food; of insects, using many different food plants; *cf.* monophagous, oligophagous, stenophagous; of Sporozoa, passing different phases of life history in different cells; nourished by a number of hosts or host cells.

polyphenism *n.* [Gk. *polys*, many; *phainein*, to appear] the occurrence in a population of several phenotypes which are not genetically controlled.

polyphenol oxidase any of several enzymes which catalyse the oxidation of tyrosine to dioxophenylalanine using molecular oxygen and dihydrophenylalanine; *alt.* tyrosinase, polyphenolase; EC 1.14.18.1; *r.n.* monophenol monooxygenase.

polypheny *n.* [Gk. *polys*, many; *phainein*, to appear] pleiotropy, *q.v.*

polyphyletic *a.* [Gk. *polys*, many; *phylon*, race] convergent as *appl.* group; combining characteristics of more than one ancestral type; having origin from several lines of descent, *alt.* pleophyletic, polygenetic; *cf.* oligophyletic, monophyletic.

polyphyllous *a.* [Gk. *polys*, many; *phyllon*, leaf] many-leaved.

polyphyodont *a.* [Gk. *polyphyes*, manifold;

odous, tooth] having many successive sets of teeth.

polypide, polypite *n.* [L. *polypus,* polyp] an individual or person of a zooid colony.

Polyplacophora *n.* [Gk. *polys,* many; *plax,* plate; *phora,* producing] a subclass, or in some classifications an order, of chitons living in the intertidal zone, which are flattened with a broad foot, and have a shell of 8 plates.

polyplanetic *a.* [Gk. *polys,* many; *planētēs,* wanderer] having several motile phases with intervening resting stages.

polyplastic *a.* [Gk. *polys,* many; *plastos,* formed] capable of assuming many forms.

polyploid *a.* [Gk. *polys,* many; *aploos,* onefold; *eidos,* form] with a reduplication of the chromosome number, as triploid, tetraploid, etc., having 3, 4, etc., times the normal haploid or gametic number; exhibiting polyploidy. *n.* An organism with more than 2 chromosome sets.

polyploidogen any substance inducing polyploidy, as colchicine, β-naphthol, etc.

polyploidy the polyploid condition.

polypneustic *a.* [Gk. *polys,* many; *pnein,* to breathe] *appl.* lateral lobes bearing multiple spiracle pores, in certain insects.

Polypodiales *n.* [*Polypodium,* generic name] in some classifications, an order of ferns including many common ferns.

Polypodiopsida *n.* [*Polypodium,* generic name; Gk. *opsis,* appearance] *see* Pteropsida.

polypod(ous) *a.* [Gk. *polys,* many; *pous,* foot] furnished with many feet or legs, *appl.* larva as of Lepidoptera; *cf.* oligopod, protopod.

polypoid *a.* [Gk. *polypous,* polyp; *eidos,* form] polyp-like.

Polyporales *n.* [Gk. *polys,* many; *poros,* channel] Poriales, *q.v.*

polyprotodont *a.* [Gk. *polys,* many; *prōtos,* first; *odous,* tooth] with 4 or 5 incisors on each side of upper jaw, and 1 or 2 fewer on lower; *cf.* diprotodont.

Polypterini *n.* [Gk. *polys,* many; *pteron,* fin] a group of palaeoniscids including the living bichir, having heavy ganoid scales, well-developed ventral lungs and a fleshy lobe on the pectoral fins; *alt.* Brachiopterygii, Polypteriformes.

polyrhizal *a.* [Gk. *polys,* many; *rhiza,* root] with many roots or rootlets; *alt.* polyrhizous, multiradicate.

polyribosomes *n.plu.* [Gk. *polys,* many; *ribonucleic* acid; Gk. *sōma,* body] polysomes, *q.v.*

polysaccharides *n.plu.* [Gk. *polys,* many; L. *saccharum,* sugar] carbohydrates of high molecular weight formed by the condensation of a large number of monosaccharide units, including the pentosans made of condensed pentose units and hexosans of condensed hexose units, and sometimes also used to include heterosaccharides; *alt.* polyoses, glycans.

polysaprobic *a.* [Gk. *polys,* many; *sapros,* rotten; *bios,* life] *appl.* aquatic habitats with heavy pollution by organic matter with little or no dissolved oxygen, the formation of sulphides, abundant bacteria, but few animals feeding on them or on the decaying matter.

polysepalous *a.* [Gk. *polys,* many; F. *sépale,* sepal] having free or distinct sepals; *alt.* dialysepalous, chorisepalous, eleutherosepalous; *cf.* gamosepalous.

polysiphonic *a.* [Gk. *polys,* many; *siphōn,* tube] consisting of several rows of cells, *appl.* thallus of red or brown algae; consisting of several tubes, *appl.* hydrocaulis of some hydrozoan colonies.

polysomaty *n.* [Gk. *polys,* many; *sōma,* body] polyploid condition in somatic cells; *alt.* somatic polyploidy.

polysome *n.* [Gk. *polys,* many; *sōma,* body] the group of ribosomes concerned with the synthesis of a protein or polypeptide chain; *alt.* ergosome, polyribosome.

polysomic *a.* [Gk. *polys,* many; *sōma,* body] having one or more chromosomes, not the entire set, in the polyploid state; *pert.* a number of homologous genes.

polysomitic *a.* [Gk. *polys,* many; *sōma,* body; *temnein,* to cut] having many body segments; formed from fusion of primitive body segments.

polysomy *n.* [Gk. *polys,* many; *sōma,* body] the polysomic condition.

polyspermous *a.* [Gk. *polys,* many; *sperma,* seed] having many seeds.

polyspermy *n.* [Gk. *polys,* many; *sperma,* seed] entry of several sperms into one ovum.

polyspondyly *n.* [Gk. *polys,* many; *sphondylos,* vertebra] condition of having vertebral parts multiple where myotome has been lost; *cf.* diplospondyly.

polysporic polysporous, *q.v.*

polysporocystid *n.* [Gk. *polys,* many; *sporos,* seed; *kystis,* bladder; *eidos,* form] *appl.* oocyst of Sporozoa when more than 4 sporocysts are present.

polysporous *a.* [Gk. *polys,* many; *sporos,* seed] many-seeded; many-spored; *alt.* multisporous, pleiosporous, multisporic; *n.* polyspory.

polystachyous *a.* [Gk. *polys,* many; *stachys,* ear of corn] with numerous spikes.

polystely *n.* [Gk. *polys,* many; *stēlē,* post] arrangement of axial vascular tissue in several steles, each containing more than one vascular bundle; *a.* polystelic.

polystemonous *a.* [Gk. *polys,* many; *stēmōn,* warp] having stamens more than double the number of petals or sepals; *alt.* polyandrous.

polystichous *a.* [Gk. *polys,* many; *stichos,* row] arranged in numerous rows or series; *alt.* multifarious.

Polystomatoidea *n.* [Gk. *polys,* many; *stoma,* mouth; *eidos,* form] an order of Polyopisthocotylea that are parasitic on reptiles and amphibians and have an opisthaptor with from 1 to 3 pairs of large suckers.

polystomatous *a.* [Gk. *polys,* many; *stoma,* mouth] having many pores, mouths, opening, or suckers; many-mouthed, as Discomedusae and sponges.

polystomium *n.* [Gk. *polys,* many; *stoma,* mouth] a suctorial mouth of Discomedusae.

polystylar *a.* [Gk. *polys,* many; *stylos,* pillar] many-styled.

polysymmetrical *a.* [Gk. *polys,* many; *symmetria,* due proportion] divisible through several planes into bilaterally symmetrical portions.

polytene *a.* [Gk. *polys,* many; *tainia,* band] *appl.* giant chromosomes in certain somatic cells of Diptera, e.g. in salivary glands, with numerous parallel chromonemata fused in homologous regions, and with deeply staining transverse bands separated by regions relatively lacking nucleic acids.

polyteny n. [Gk. *polys*, many; *tainia*, band] the replication of chromosomes during interphase to produce giant chromosomes.

polythalamous a. [Gk. *polys*, many; *thalamos*, chamber] aggregate or collective, as *appl.* fruits, galls; *appl.* shells made up of many chambers formed successively.

polythelia n. [Gk. *polys*, many; *thēlē*, nipple] the occurrence of supernumerary nipples.

polythermic a. [Gk. *polys*, much; *thermē*, heat] tolerating relatively high temperatures.

polythetic a. [Gk. *polys*, many; *thetos*, placed] *appl.* a classification based on many characteristics, not all of which are necessarily shown by every member of the group; *cf.* monothetic.

polytocous, polytokous a. [Gk. *polys*, many; *tokos*, offspring] prolific; producing several young at a birth; fruiting repeatedly, *alt.* caulocarpous.

polytomous a. [Gk. *polys*, many; *tomē*, cutting] having more than 2 secondary branches; with a number of branches originating in one place, *appl.* podetium.

polytopic a. [Gk. *polys*, many; *topos*, place] occurring or originating in several places.

Polytrichales n. [*Polytrichum*, generic name] an order, or in some classifications superorder, of mosses of the subclass Bryidae.

polytrichous a. [Gk. *polys*, many; *thrix*, hair] having the body covered with an even coat of cilia, as certain ciliates; having many hair-like outgrowths.

polytrochal a. [Gk. *polys*, many; *trochos*, wheel] having several circlets of cilia between mouth and posterior end, as in certain annulates; *alt.* polytrochous.

polytrophic a. [Gk. *polys*, many; *trophē*, nourishment] *appl.* ovariole in which nutritive cells are enclosed in oocyte follicles; nourished by more than one organism or substance; obtaining food from many sources; *cf.* meroistic, acrotrophic.

polytropic a. [Gk. *polys*, many; *tropikos*, turning] pantropic, *q.v.*

polytypic a. [Gk. *polys*, many; *typos*, type] having or *pert.* many types; *appl.* species having geographical subspecies; *appl.* genus having several species; *cf.* monotypic.

polyvoltine a. [Gk. *polys*, many; It. *volta*, time] producing several broods in one season.

Polyxenida n. [*Polyxena*, generic name] the only order of the Pselaphognatha, *q.v.*

polyxenous a. [Gk. *polys*, many; *xenos*, host] adapted to life in many different species of host, *appl.* parasites; n. polyxeny.

polyxylic a. [Gk. *polys*, many; *xylon*, wood] having many xylem strands and several concentric vascular rings.

Polyzoa, polyzoans n., n.plu. [Gk. *polys*, many; *zōon*, animal] Bryozoa, *q.v.*

polyzoarium n. [Gk. *polys*, many; *zōon*, animal] the skeletal system of a polyzoan colony; the colony itself.

polyzoic a. [Gk. *polys*, many; *zōon*, animal] *appl.* a colony of many zooids; *appl.* a spore containing many sporozoites.

polyzooid n. [Gk. *polys*, many; *zōon*, animal; *eidos*, form] an individual in a polyzoan colony.

pome n. [L. *pomum*, apple] a fruit such as apple in which the seeds are surrounded by a tough layer derived from the pericarp, which is fused to the deeply cup-shaped fleshy receptacle.

pompetta n. [It. *pompetta*, little pump] an organ forcing spermatozoa into penis, as in flies of the genus *Phlebotomus*; *alt.* sperm pump.

pomum Adami laryngeal prominence, *q.v.*

ponderal a. [L. *ponderare*, to weigh] *pert.* weight; *appl.* growth by increase in mass.

ponogen n. [Gk. *ponos*, toil; *gennaein*, to produce] waste matter produced by exertion; fatigue poison.

pons n. [L. *pons*, bridge] a structure connecting 2 parts, as pons Varolii.

pons Varolii n. [L. *pons*, bridge; *C. Varolio*, or *Varolius*, Italian anatomist] broad band of white fibres connecting cerebrum, cerebellum, and medulla oblongata, and including the pontine nuclei of grey matter.

pontal, pontic *pert.* a pons; *alt.* pontile.

ponticulus n. [L. *ponticulus*, small bridge] a vertical ridge on auricular cartilage; propons, *q.v.*

pontile pontal, *q.v.*

pontine a. [L. *pons*, bridge] *pert.* pons Varolii, *appl.* branches of basilar artery, flexure of embryonic brain, nuclei of basilar part.

pool feeder telmophage, *q.v.*

popliteal a. [L. *poples*, ham] *pert.* region behind and above knee joint, *appl.* artery, glands, vein, muscle, etc.

popliteal nerve internal or medial, the tibial nerve; external or lateral, the common peroneal nerve.

population n. [L. *populus*, people] a group of individuals of a species living in a certain area.

porate a. [Gk. *poros*, channel] set with pores, *appl.* pollen grains.

porcellanous a. [F. *porcelaine*, from It. *porcellana*, Venus shell] resembling porcelain, white and opaque, *appl.* calcareous shells, as of Foraminifera, certain Mollusca, etc.

pore n. [Gk. *poros*, channel] a minute opening or passage, as of the skin, sieve plates, stomata, etc.

pore canals minute spiral tubules passing through the cuticle, but not the epicuticle, of insects.

pore cell in sponges, cells of the outer layer that are perforated by a pore through which water enters; *alt.* porocyte, pylocyte.

pore chains in wood, extensive radial, oblique, tangential groups of vessels.

pore organ structure surrounding canal for excretion of mucilage through pores, in desmids.

pore rhombs canals grouped in half rhombs on each of 2 adjoining plates of calyx in Cystidea.

pore space all the spaces between particles of soil.

Poriales n. [Gk. *poros*, channel] an order of Hymenomycetes having the basidia lining tubes or pores; *alt.* Aphyllophorales, Polyporales.

poricidal a. [Gk. *poros*, channel; L. *caedere*, to cut] dehiscing by valves or pores.

Porifera n. [Gk. *poros*, channel; L. *ferre*, to bear] a phylum of radially symmetrical or asymmetrical animals, commonly called sponges, whose body is penetrated by cavities some of which are lined with choanocytes.

poriferous a. [Gk. *poros*, channel; L. *ferre*, to bear] furnished with numerous openings.

poriform a. [Gk. *poros*, channel; L. *forma*, shape] *see* poroid.

Porocephalida n. [Gk. *poros*, channel; *kephalē*,

head] an order of Pentastomida in which the ventral ganglia are fused with the supraoesophageal ganglion to form a single mass.

porocyte n. [Gk. poros, channel; kytos, hollow] pore cell, q.v.

porogam n. [Gk. poros, channel; gamos, marriage] a plant whose pollen tube enters ovule by micropyle; cf. chalazogam.

porogamy n. [Gk. poros, channel; gamos, marriage] entrance of a pollen tube into ovule by micropyle to secure fertilization; cf. aporogamy, chalazogamy.

poroid a. [Gk. poros, channel; eidos, shape] like a pore or pores; having pore-like depressions; alt. poriform. n. Minute depression in theca of dinoflagellates and diatoms.

porophyllous a. [Gk. poros, channel; phyllon, leaf] having, or appl. leaves with numerous transparent spots.

porose a. [Gk. poros, channel] having or containing pores.

Porphyridiales n. [Porphyra, generic name] an order of red algae which are unicellular and may be aggregated into gelatinous colonies.

porphyrins n.plu. [Gk. porphyra, purple] organic compounds occurring in plants and animals, based on a structure consisting of 4 pyrrole rings, usually with a metal such as iron or magnesium, including chlorophyll, and found in the haemochromes and cytochromes associated with protein.

porphyrophore n. [Gk. porphyra, purple; pherein, to bear] a reddish-purple pigment-bearing cell.

porphyropsin n. [Gk. porphyra, purple; opsis, sight] a retinal pigment in freshwater vertebrates; alt. visual purple.

porrect a. [L. porrectus, stretched out] extended outwards.

porta n. [L. porta, gate] a gate-like structure as transverse fissure of liver; alt. hilum.

portal a. [L. porta, gate] appl. a system of veins draining alimentary canal, spleen, and pancreas to the liver; also a system to kidney in lower vertebrates.

portio n. [L. portio, portion] a part or portion of a nerve, blood vessel, etc.

positional alleles alleles which have changed their function under the influence of the position effect.

position effect effect due to relative position of a gene or genes within the chromosome.

positive reinforcement a stimulus, or series of stimuli, which is pleasant to an animal and increases its response; alt. reward.

positive taxis or tropism tendency to move or grow towards the source of a stimulus.

postabdomen n. [L. post, after; abdomen, belly] in scorpions, metasoma or posterior narrower 5 segments of abdomen; anal tubercle in spiders.

postalar a. [L. post, after; ala, wing] situated behind the wings, appl. callus, membrane in Diptera.

postanal a. [L. post, after; anus, vent] situated behind anus.

postantennal a. [L. post, after; antenna, sailyard] situated behind antennae; appl. a sensory organ in Myriapoda and Collembola, alt. organ of Tömösvary.

postarticular n. [L. post, after; articulus, joint] posterior process of surangular, behind articulation with quadrate.

postaxial a. [L. post, after; axis, axle] on posterior side of axis; as on fibular side of leg.

postbacillary a. [L. post, after; bacillum, small staff] having nuclei behind sensory zone of retinal cells, appl. ocellus, inverted eye, as of spiders; cf. prebacillary.

postbranchial a. [L. post, after; branchiae, gills] behind gill clefts; appl. a structure arising in pharynx; appl. bodies: the ultimobranchial bodies; cf. post-trematic.

postcardinal a. [L. post, after; Gk. kardia, heart] behind region of heart, appl. a dorsal vein.

postcava, postcaval vein n. [L. post, after; cavus, hollow] the vein bringing blood from the posterior part of body to heart in all tetrapod vertebrates; alt. inferior or posterior vena cava.

postcentral a. [L. post, after; centrum, centre] behind central region; appl. a cerebral sulcus, part of intraparietal sulcus.

postcentrum n. [L. post, after; centrum, centre] the posterior part of vertebral centrum of certain vertebrates.

postcerebral a. [L. post, after; cerebrum, brain] posterior to the brain; appl. cephalic salivary glands, as in Hymenoptera.

postcingular a. [L. post, behind; cingulum, girdle] posterior to cingulum; appl. a plate of hypovalve in certain Dinoflagellata.

postclavicle n. [L. post, after; clavicula, small key] a membrane bone occurring in shoulder girdle of some higher ganoids and teleosts; alt. postcleithrum.

postcleithrum postclavicle, q.v.

postclitellian a. [L. post, after; clitellae, pack saddle] situated behind clitellum.

postclival a. [L. post, after; clivus, hill] appl. fissure behind clivus of cerebellum.

postclypeus n. [L. post, after; clypeus, shield] the posterior part of clypeus of an insect; cf. anteclypeus.

postcolon n. [L. post, after; colon, colon] part of gut between colon and rectum in certain mites.

postcornual a. [L. post, after; cornu, horn] appl. glands situated behind horns, as in chamois.

postcranial a. [L. post, after; L.L. cranium, skull] appl. area of posterior head region.

postdicrotic a. [L. post, after; Gk. dis, twice; krotein, to beat] appl. a secondary wave of a pulse, or that succeeding the dicrotic.

postembryonic a. [L. post, after; Gk. embryon, foetus] pert. the age or stages succeeding the embryonic.

posterior a. [L. posterior, latter] situated behind; dorsal, in human anatomy; behind the axis; superior, or next the axis.

posterior lobe of pituitary gland pars intermedia and pars nervosa.

posterior vena cava post cava, q.v.

posterolateral a. [L. posterus, following; latus, side] placed posteriorly and towards the side, appl. arteries.

posteromedial a. [L. posterus, following; medius, middle] placed posteriorly and medianly, appl. arteries.

postesophageal postoesophageal, q.v.

postestrum metoestrus, q.v.

postflagellate a. [L. post, after; flagellum, lash] appl. forms of trypanosome intermediate between flagellates and cyst.

postfrons n. [L. post, after; frons, forehead] portion of frons posterior to antennary base line in insects.

postfrontal a. [L. post, after; frons, forehead] appl. a bone occurring behind orbit of some vertebrates.

postfurca n. [L. post, after; furca, fork] forked sternal process or apodeme of metathorax in insects.

postganglionic a. [L. post, after; Gk. gangglion, tumour] appl. autonomic nerve fibres issuing from ganglia; cf. preganglionic.

postgena n. [L. post, after; gena, cheek] posterior portion of insect gena.

postgenital a. [L. post, after; genitalis, of generation] situated behind the genital segment.

postglacial age Holocene, q.v.

postglenoid a. [L. post, after; Gk. glēnē, socket] behind the glenoid fossa, appl. a process or tubercle.

posthepatic a. [L. post, after; Gk. hēpar, liver] appl. latter part of alimentary canal, that from liver to end.

postheterokinesis n. [L. post, after; Gk. heteros, other; kinēsis, movement] case of meiosis in which the sex chromosome passes undivided to one pole in the 2nd division; cf. preheterokinesis.

posthypophysis postpituitary, q.v.

postical, posticous a. [L. posticus, behind] an outer or posterior surface, alt. extrorse; appl. lower or back surface of a thallus, leaf, or stem, especially in liverworts, cf. antical.

postischium n. [L. post, after; Gk. ischion, hip] a lateral process on hinder side of ischium of some reptiles.

postlabrum n. [L. post, after; labrum, lip] posterior portion of insect labrum, where differentiated.

postmentum n. [L. post, after; mentum, chin] the united cardines constituting the base of labium in insects.

postminimus n. [L. post, after; minimus, smallest] a rudimentary additional digit occurring occasionally in amphibians and mammals.

postmitotic n. [L. post, after; Gk. mitos, thread] a cell with individual life originating in mitosis and ending at death; cf. intermitotic.

postneural a. [L. post, after; Gk. neuron, nerve] pygal, appl. plates of chelonian carapace.

postnodular a. [L. post, after; nodulus, small knot] appl. a cerebellar fissure between nodule and uvula.

postnotum n. [L. post, after; Gk. nōton, back] postscutellum, q.v.; metanotum, q.v.

postocular a. [L. post, backwards; oculus, eye] posterior to the eye, appl. scales.

postoesophageal a. [L. post, after; Gk. oisophagos, gullet] appl. commissure connecting ganglia of tritocerebrum; alt. postesophageal.

postoestrus metoestrus, q.v.

postoral a. [L. post, after; os, mouth] behind the mouth.

postorbital a. [L. post, after; orbis, eye socket] behind the orbit; appl. bone forming part of posterior wall of orbit; appl. luminescent organ in certain fishes.

postotic a. [L. post, after; Gk. ous, ear] behind the ear.

postparietal a. [L. post, after; paries, wall] appl. paired bones sometimes occurring between parietal and interparietal.

postpatagium n. [L. post, after; patagium, border] in birds, small fold of skin extending between upper arm and trunk.

postpermanent a. [L. post, after; permanens, remaining] appl. traces of a dentition succeeding the permanent.

postpetiole n. [L. post, after; petiolus, stalk of a fruit] in ants, 2nd segment of abdominal stalk.

postphragma n. [L. post, after; Gk. phragma, fence] a phragma developed in relation with a postnotum in insects.

postpituitary a. [L. post, after; pituita, phlegm] pert. or secreted by posterior lobe of the pituitary gland; alt. posthypophysis.

postpubic a. [L. post, after; pubes, adult] at posterior end of pubis; appl. processes of pubis parallel to ischium.

postpubis n. [L. post, after; pubes, adult] a ventral process or bone of pelvic girdle in some Sauropsida.

postpyramidal a. [L. post, after; pyramis, pyramid] behind the pyramid, appl. a cerebellar fissure.

postreduction n. [L. post, after; reductus, reduced] halving the chromosome number in the 2nd meiotic division instead of the 1st; cf. prereduction.

postretinal a. [L. post, after; retina, from rete, net] situated behind the retina; appl. nerve fibres connecting periopticon and inner ends of ommatidia.

postscutellum n. [L. post, after; scutellum, small shield] a projection under mesoscutellar lobe of insects, the base of mesophragma; sclerite behind scutellum; alt. postnotum, pseudonotum.

postsegmental a. [L. post, after; segmentum, piece] posterior to body segments or somites; cf. presegmental.

postsphenoid n. [L. post, after; Gk. sphēn, wedge; eidos, form] the posterior part of sphenoid.

poststernellum n. [L. post, after; sternum, breast bone] most posterior portion of an insect sternite.

poststernite n. [L. post, after; sternum, breast bone] posterior sternal sclerite of insects; sternellum, q.v.

postsynaptic a. [L. post, after; Gk. synapsis, union] appl. membrane of the muscle cell at a motor end plate; appl. membrane or cell after a synapse.

post-temporal a. [L. post, after; tempora, temples] behind temporal bone, appl. bone and fossa.

post-trematic a. [L. post, after; Gk. trēma, hole] behind an opening such as a gill cleft; appl. nerves running in posterior wall of 1st gill cleft in pharynx; cf. postbranchial.

postzygapophysis n. [L. post, after; Gk. zygon, yoke; apo, from; physis, growth] an articular process on posterior face of neural arch for articulation with following vertebra.

potamodromous a. [Gk. potamos, river; dromos, running] migrating only in fresh water; cf. oceanodromous, anadromous, catadromous.

potamoplankton n. [Gk. potamos, river; planktos, wandering] the plankton of rivers, streams, and their backwaters.

potential a. [L. potens, powerful] latent, as appl. characteristics.

Pottiales n. [*Pottia*, generic name] an order of mosses having erect gametophores.

pouch n. [O.F. *poche*, bag] a bag-like structure; a sac or bladder, as pharyngeal pouches, marsupial pouch, etc.

Poupart's ligament [F. *Poupart*, French anatomist] the inguinal ligament.

powder-down feathers those which do not develop beyond the early stage, and in which the tips of barbs disintegrate into powder; *alt*. pulviplumes.

powdery mildews the common name for the Erysiphales, *q.v.*

poxviruses a group of viruses containing double-stranded DNA, including those causing smallpox and myxomatosis.

P-particle a kappa particle which has been liberated into the medium and that releases paramecin which kills sensitive *Paramecia.*

P-P factor pellagra-preventive factor, *q.v.*

PPLO pleuropneumonia-like organism, *q.v.*

PPP pentose phosphate pathway, *q.v.*

P$_R$ the pigment phytochrome in its inactive form, produced by the action of far-red light on P$_{FR}$; *alt*. P$_{660}$.

prae- *also* pre-, *q.v.*

praeabdomen n. [L. *prae*, before; *abdomen*, belly] the anterior, broader part of abdomen of scorpions; *alt*. mesosoma.

prae-auricular a. [L. *prae*, before; *auricula*, small ear] *appl*. a sulcus at anterior part of auricular surface of hip bone.

praecentrum n. [L. *prae*, before; *centrum*, centre] the anterior part of the vertebral centrum of certain lower vertebrates.

praecoces n.plu. [L. *prae*, before; *coquere*, to cook] newly hatched birds able to take care of themselves; *cf*. altrices.

praecostal a. [L. *prae*, before; *costa*, rib] *appl*. short spurs on basal portion of hind wing of Lepidoptera.

praecrural a. [L. *prae*, before; *crus*, leg] on anterior side of leg or thigh.

praecuneus n. [L. *prae*, before; *cuneus*, wedge] the medial surface of parietal lobe, or quadrate lobe of cerebrum.

praeoccipital a. [L. *prae*, before; *occiput*, back of head] preoccipital, *q.v.*

praeoral a. [L. *prae*, before; *os*, mouth] preoral, *q.v.*

praeputium n. [L. *praeputium*, foreskin] foreskin, part of integument of penis which leaves surface at neck and is folded upon itself; fold of labia minora over glans clitoridis; *alt*. prepuce.

Prasinophyceae n. [*Prasinocladus*, generic name; Gk. *phykos*, seaweed] a small class of algae having scales on the body surface and scales or hairs on the surface of flagella.

Prasiolales n. [*Prasiola*, generic name] an order of green algae with a unique life history; *alt*. Schizogoniales.

pratal a. [L. *pratum*, meadow] *pert*. meadows; *appl*. flora of rich humid grasslands.

pre- *also* prae-, *q.v.*

preadaptation n. [L. *prae*, before; *ad*, to; *aptare*, to fit] a change in an organism occurring before a change in the environment for which it will be needed; adaptation of a mutant to particular conditions.

preanal a. [L. *prae*, before; *anus*, anus] anterior to anus.

preantenna n. [L. *prae*, before; *antenna*, sailyard] one of the pair of feelers on the 1st segment in Onychophora.

preaxial a. [L. *prae*, before; *axis*, axle] in front of the axis; on anterior border or surface.

prebacillary a. [L. *prae*, before; *bacillum*, small staff] having nuclei distal to sensory zone of retinal cells, *appl*. ocellus, converted or erect eye, as of spiders; *cf*. postbacillary.

prebasilare n. [L. *prae*, before; *basis*, base] transverse sclerite between mentum of gnathochilarium and 1st body sternite, in certain Diplopoda.

prebiotic a. [L. *prae*, before; *biotikos*, *pert*. life] before life appeared on earth.

prebranchial a. [L. *prae*, before; *branchiae*, gills] pretrematic, *q.v.*

Pre-Cambrian n. [L. *prae*, before; *Cambria*, Wales] the oldest geological era, before the Cambrian; *alt*. Archaean, Eozoic.

precapillary a. [L. *prae*, before; *capillus*, hair] *appl*. arterioles having an incomplete muscular layer. n. A small vessel conducting blood from arteriole to capillary.

precartilage n. [L. *prae*, before; *cartilago*, gristle] type of cartilage preceding formation of other kinds, or persisting as in fin-rays of certain fishes.

precava, precaval vein n. [L. *prae*, before; *cavus*, hollow] the vein bringing blood from the anterior part of body to heart; *alt*. superior or anterior vena cava.

precentral a. [L. *prae*, before; *centrum*, centre] anteriorly to centre; *appl*. sulcus parallel to central sulcus of cerebrum; *appl*. gyrus.

precheliceral a. [L. *prae*, before; Gk. *chēlē*, claw; *keras*, horn] anterior to chelicerae, *appl*. segment of mouth region or gnathosoma in Arachnida.

prechordal a. [L. *prae*, before; Gk. *chordē*, cord] anterior to notochord or to spinal cord; *appl*. part of base of skull.

precingular a. [L. *prae*, before; *cingulum*, girdle] anterior to cingulum; *appl*. a plate of epivalve in certain Dinoflagellata.

precipitins specific antibodies in immune serum which form precipitates with their respective antigens, e.g. bacterio-, haemato-, lact-, myco-, phyto-, zooprecipitin.

preclavia n. [L. *prae*, before; *clavis*, key] an element of pectoral girdle.

preclimacteric n. [L. *prae*, before; Gk. *klimaktēr*, step of a staircase] period before climacteric.

preclimax n. [L. *prae*, before; Gk. *klimax*, ladder] the plant community immediately preceding the climax community; *cf*. proclimax, subclimax.

preclival a. [L. *prae*, before; *clivus*, hill] *appl*. fissure in front of clivus of cerebellum.

preconnubia n. [L. *prae*, before; *connubium*, marriage] gatherings of animals before the mating season.

precoracoid n. [L. *prae*, before; Gk. *korax*, crow; *eidos*, shape] an anterior ventral bone of pectoral girdle in amphibians and reptiles.

precoxa n. [L. *prae*, before; *coxa*, hip] subcoxa, *q.v.*

precoxal *pert*. precoxa; *appl*. bridge: junction of laterosternite and episternum.

precursor n. [L. *praecursor*, forerunner] the substance which precedes the formation of a compound.

precursory cell metrocyte, *q.v.*

precystic *a.* [L. *prae*, before; Gk. *kystis*, bladder] *appl.* small forms appearing before the encystment stage in some protozoans.

predator *n.* [L. *praedator*, hunter] an animal that kills other animals for food.

predator chain a food chain which starts from a plant and passes from herbivores to carnivores; *cf.* parasite chain, saprophyte chain.

predelineation *n.* [L. *prae*, before; *de*, down; *linea*, line] formation and individualization of various physiological molecules in definite areas and substances of undeveloped egg—theory of germinal localization.

predentary *n.* [L. *prae*, before; *dens*, tooth] a bone at tip of jaw of many dinosaurs.

predentin(e) *n.* [L. *prae*, before; *dens*, tooth] immature dentine which is not yet calcified but made mainly of fibrils known as Korff's fibres.

predigital *n.* [L. *prae*, before; *digitus*, finger] a primary wing quill connected with distal phalanx of 2nd digit.

preen gland oil gland, *q.v.*

pre-epistome *n.* [L. *prae*, before; Gk. *epi*, upon; *stoma*, mouth] a plate covering basal portion of epistome of certain arachnids.

prefemur *n.* [L. *prae*, before; *femur*, thigh] second trochanter, as in walking legs of Pycnogonida.

preflagellate *a.* [L. *prae*, before; *flagellum*, lash] *appl.* forms of trypanosomes intermediate between cyst and elongate flagellates.

prefloration *n.* [L. *prae*, before; *flos*, flower] the form and arrangement of floral leaves in the flower bud; *alt.* ptyxis, aestivation.

prefoliation *n.* [L. *prae*, before; *folium*, leaf] the form and arrangement of foliage leaves in the bud; *see* ptyxis and vernation.

preformation theory theory according to which it was supposed that each ovum of an animal contained a miniature adult, and that nourishment only was required to develop it into the perfect form; *alt.* incasement theory; *cf.* epigenesis.

prefrontal *a.* [L. *prae*, before; *frons*, forehead] *appl.* a bone anterior to frontal of certain vertebrates; *appl.* paired plates or scales anterior to frontal scale in some reptiles.

pregammation *n.* [L. *prae*, before; *gammation*, *dim.* of Γ] a bar in front of the gammation in *Palaeospondylus*, an extinct primitive Devonian vertebrate.

preganglionic *a.* [L. *prae*, before; Gk. *gangglion*, tumour] *appl.* medullated fibres from spinal cord, ending in synapses around sympathetic ganglion cells; *cf.* postganglionic.

pregenital *a.* [L. *prae*, before; *genitalis*, *pert.* generation] situated anterior to genital opening; *appl.* segment behind 4th pair of walking legs in arachnids.

preglobulin *n.* [L. *prae*, before; *globulus*, small globe] a compound protein of white blood corpuscles.

pregnancy cells modified oxyphil cells of anterior lobe of hypophysis, multiplying during pregnancy.

prehallux *n.* [L. *prae*, before; *hallus*, great toe] a rudimentary additional digit on hindlimb; *alt.* calcar.

prehalteres *n.plu.* [L. *prae*, before; Gk. *haltēr*, weight] the squamae of Diptera.

prehaustorium *n.* [L. *prae*, before; *haurire*, to drink] a rudimentary root-like sucker.

prehensile *a.* [L. *prehendere*, to seize] adapted for holding.

prehepatic *a.* [L. *prae*, before; Gk. *hēpar*, liver] *appl.* part of digestive tract anterior to liver.

preheterokinesis *n.* [L. *prae*, before; Gk. *heteros*, other; *kinēsis*, movement] case of meiosis in which the sex chromosome passes undivided to one pole in the 1st division; *cf.* postheterokinesis.

prehyoid *a.* [L. *prae*, before; Gk. *hyoeidēs*, Y-shaped] mandibulohyoid; *appl.* cleft between mandible and ventral parts of hyoid arch.

prehypophysis *n.* [L. *prae*, before; Gk. *hypo*, under; *physis*, growth] prepituitary, *q.v.*

preimaginal *a.* [L. *prae*, before; *imago*, image] preceding the imaginal or adult stage.

preimaginal conditioning insects, in a response learnt in a larva which is remembered by the adult.

preinterparietal *n.* [L. *prae*, before; *inter*, between; *paries*, wall] one of 2 small upper membranous centres of formation of supraoccipital.

prelacteal *a.* [L. *prae*, before; *lac*, milk] *pert.* a dentition which may occur previous to the milk dentition.

prelocalization *n.* [L. *prae*, before; *locus*, place] the theory that certain portions of ovum are predestined to develop into certain organs or parts.

premandibular *a.* [L. *prae*, before; *mandibulum*, jaw] anterior to mandible; *appl.* somites of *Amphioxus*, the lancelet; *appl.* a bone of certain reptiles.

premaxilla *n.* [L. *prae*, before; *maxilla*, jaw] a paired bone anterior to maxilla in most vertebrates; *alt.* os incisivum, intermaxilla.

premaxillary *a.* [L. *prae*, before; *maxilla*, jaw] anterior to maxilla; *pert.* premaxilla.

premedian *a.* [L. *prae*, before; *medius*, middle] anterior to middle of body or part; *appl.* a head plate in certain primitive fishes; *appl.* vein in front of median vein of certain insect wings.

prementum *n.* [L. *prae*, before; *mentum*, chin] the united stipites bearing ligula and labial palps of insects.

premetaphase *n.* [L. *prae*, before; Gk. *meta*, after; *phasis*, appearance] prometaphase, *q.v.*

pre-messenger RNA (pre-mRNA) heterogenous nuclear RNA, *q.v.*

premolar *a.* [L. *prae*, before; *mola*, mill] *appl.* teeth developed between canines and molars; *alt.* bicuspid teeth.

premorse *a.* [L. *praemorsus*, bitten off] with irregular and abrupt termination, as if end were bitten off.

pre-mRNA pre-messenger RNA, *q.v.*

premyoblast *n.* [L. *prae*, before; Gk. *mys*, muscle; *blastos*, bud] a slightly differentiated embryonic cell which gives rise to a myoblast.

prenasal *a.* [L. *prae*, before; *nasus*, nose] *appl.* a bone developed in septum in front of mesethmoid in certain skulls; *alt.* rostral, telethmoid.

preoccipital *a.* [L. *prae*, before; *occiput*, back of head] *appl.* an indentation or notch in front of posterior end of cerebral hemispheres; *alt.* praeoccipital.

preocular *a.* [L. *prae*, before; *oculus*, eye] anterior to the eye, as antennae, scales.

preopercle preoperculum, *q.v.*

preopercular *a.* [L. *prae*, before; *operculum*, cover] anterior to gill cover; *appl.* luminescent organ in certain fishes; *appl.* bone: preoperculum, *q.v.*

preoperculum *n.* [L. *prae*, before; *operculum*, cover] anterior membrane bone of gill cover of fishes; *alt.* preopercle, preopercular bone.

preoptic nerve nervus terminalis, *q.v.*

preoral *a.* [L. *prae*, before; *os*, mouth] situated in front of mouth; *appl.* cilia, etc.; *appl.* food cavity, the anterior part of the 'buccal cavity', between labrum, prementum, and mandibles, in insects; *appl.* segment: prostomium, *q.v.*; *appl.* pit: Hatschek's pit, *q.v.*; *alt.* praeoral.

preorbital *a.* [L. *prae*, before; *orbis*, eye socket] anterior to orbit, *appl.* a membrane bone of teleosts, *appl.* glands in ruminants.

preparietal *n.* [L. *prae*, before; *paries*, wall] a bone in front of parietals, in some extinct reptiles.

prepatagium *n.* [L. *prae*, before; *patagium*, border] fold of skin extending between upper arm and forearm of birds, *alt.* alar membrane; part of the patagium between neck and fore-feet in flying lemurs; *alt.* propatagium.

prepatellar *a.* [L. *prae*, before; *patella*, small pan] *appl.* bursa between lower part of patella and the skin.

prepenna *n.* [L. *prae*, before; *penna*, feather] a nestling down feather which is succeeded by adult contour feather; protoptile and mesoptile.

prepharynx *n.* [L. *prae*, before; Gk. *pharyngx*, gullet] narrow thin-walled structure connecting oral sucker and pharynx, in trematodes.

prephragma *n.* [L. *prae*, before; Gk. *phragma*, fence] a phragma developed in relation with the notum of insects.

prepituitary *n.* [L. *prae*, before; *pituita*, phlegm] anterior lobe of the pituitary gland; *alt.* prehypophysis.

preplacental *a.* [L. *prae*, before; *placenta*, flat cake] occurring before placenta formation or development.

preplumula *n.* [L. *prae*, before; *plumula*, small feather] a nestling down feather which is succeeded by adult down feather.

prepollex *n.* [L. *prae*, before; *pollex*, thumb] a rudimentary additional digit occurring sometimes preaxially to thumb of certain amphibians and mammals.

prepotency *n.* [L. *prae*, before; *potens*, powerful] the fertilization of a flower by pollen from another flower in preference to pollen from its own stamens, when both are offered simultaneously; capacity of one parent to transmit more characteristics to offspring than the other parent.

prepotent *a.* [L. *prae*, before; *potens*, powerful] transmitting the majority of characteristics; *appl.* a flower exhibiting a preference for cross-pollination; having priority, as one reflex among other reflexes.

prepuberal *a.* [L. *prae*, before; *pubes*, mature] anterior to pubis; prepubertal, *q.v.*

prepubertal *a.* [L. *prae*, before; *pubertas*, adult state] *pert.* age or state before puberty; *alt.* prepuberal.

prepubic *a.* [L. *prae*, before; *pubes*, mature] *pert.* prepubis; *appl.* processes of pelvic arch, in certain fishes; on anterior part of pubis; *appl.* elongated processes of pubis of certain vertebrates.

prepubis *n.* [L. *prae*, before; *pubes*, mature] part of pelvic girdle of certain reptiles and birds, anterior to os pubis.

prepuce praeputium, *q.v.*

prepupa *n.* [L. *prae*, before; *pupa*, puppet] a quiescent stage preceding the pupal in some insects; *alt.* propupa.

preputial *pert.* the prepuce, *appl.* glands, sac.

prepygidial *a.* [L. *prae*, before; Gk. *pygidion*, narrow rump] anterior to pygidium, *appl.* growth zone in polychaetes.

prepyloric *a.* [L. *prae*, before; *pylōros*, gate keeper] *appl.* ossicle hinged to pyloric ossicle in gastric mill of Crustacea.

prepyramidal *a.* [L. *prae*, before; *pyramis*, pyramid] in front of pyramid, *appl.* a cerebellar fissure, *appl.* tract: the rubrospinal fasciculus.

prereduction *n.* [L. *prae*, before; *reductus*, reduced] halving the chromosome number in the 1st meiotic division; *cf.* postreduction.

prescutum *n.* [L. *prae*, before; *scutum*, shield] anterior sclerite of insect notum.

presegmental *a.* [L. *prae*, before; *segmentum*, piece] anterior to body segments or somites; *cf.* postsegmental.

presentation time minimum duration of continuous stimulation necessary for production of a response.

prespermatid *n.* [L. *prae*, before; Gk. *sperma*, seed] secondary spermatocyte.

presphenoid *n.* [L. *prae*, before; Gk. *sphēn*, wedge; *eidos*, form] in many vertebrates, a cranial bone anterior to the basisphenoid; the anterior part of the sphenoid.

prespiracular *a.* [L. *prae*, before; *spiraculum*, air hole] pretrematic, *q.v.*

pressor *a.* [L. *pressare*, to press] causing a rise of arterial pressure, *appl.* stimuli, nerve fibres.

presternal *a.* [L. *prae*, before; *sternum*, breast bone] situated in front of sternum; *pert.* anterior part of sternum; *appl.* jugular notch, on superior border of sternum.

presternum *n.* [L. *prae*, before; *sternum*, breast bone] the manubrium or anterior part of sternum; anterior sclerite of insect sternum; *alt.* prosternum.

prestomium *n.* [L. *prae*, before; Gk. *stoma*, mouth] in biting and sucking insects, the aperture between the tips of the mouth-parts serving for the intake of food.

presumptive *a.* [L. *praesumere*, to infer beforehand] *appl.* the name of the tissue or organ eventually arising by normal development from a particular cell, tissue, or region in the embryo.

presynaptic *a.* [L. *prae*, before; Gk. *synapsis*, union] *appl.* membrane at terminal arborization of an axon; *appl.* vesicles liberating neurotransmitter at the terminal arborizations of an axon.

pretarsus *n.* [L. *prae*, before; Gk. *tarsos*, sole of foot] terminal part of, or outgrowth on, leg or claw of insects and spiders.

pretrematic *a.* [L. *prae*, before; Gk. *trēma*, hole] anterior to an opening such as a gill cleft or spiracle; *appl.* nerves running in anterior wall of 1st gill cleft to pharynx; *alt.* prebranchial, prespiracular.

pretrochantin *n.* [L. *prae*, before; Gk. *trochanter*, runner] subcoxa, *q.v.*

prevenules *n.* [L. *prae*, before; *venula*, small vein] small vessels conducting blood from capillaries to venules.

prevernal *a.* [L. *prae*, before; *vernus*, spring] *pert.*, or appearing in, early spring.

prevertebral *a.* [L. *prae*, before; *vertebra*, vertebra] *pert.* or situated in region in front of vertebral column; *appl.* portion of base of skull; *appl.* ganglia of sympathetic system.

previtamin *n.* [L. *prae*, before; *vita*, life; *ammoniacum*, resinous gum] provitamin, *q.v.*

prevomer *n.* [L. *prae*, before; *vomer*, ploughshare] a bone anterior to pterygoid in some vertebrates; vomer of non-mammalian vertebrates; in Monotremata, a membrane bone in floor of nasal cavities; *alt.* dumb-bell or paradoxical bone.

prey *n.* [L. *praeda*, prey] animals killed and eaten by predators.

prezygapophysis *n.* [L. *prae*, before; Gk. *zygon*, yoke; *apo*, from; *physis*, growth] a process on anterior face of neural arch, for articulation with vertebra in front.

Priapulida *n.* [*Priapulus*, generic name] a phylum of burrowing and marine worm-like pseudocoelomate animals with a warty and superficially annulated body with circumoral spines.

prickle *n.* [A.S. *prica*, point] a pointed process arising through epidermal tissue, as of bramble; a modified trichome.

prickle cells cells of deeper layers of stratified squamous epithelium, having short, fine, marginal connecting keratin fibrils, prickle-like when broken.

primary *a.* [L. *primus*, first] first; principal; original; Palaeozoic, *q.v.*

primary cell wall under the light microscope, the 1st cell wall laid down while cell is growing, made entirely of cellulose; under the electron microscope, cell wall in which cellulose microfibrils are laid down with various orientations.

primary centre part of central nervous system directly linked by nerve fibres with a peripheral organ.

primary consumer herbivore, *q.v.*

primary epithelium blastoderm, *q.v.*

primary feather a flight feather on the distal joint of a bird's wing, attached to the bones of the hand and fingers, as compared with the secondaries borne on the forearm; *alt.* manual, remex primarius.

primary growth growth of roots and shoots from the time of their initiation at the apical meristem to when their expansion and differentiation is completed.

primary host a host in which a parasite lives for much of its life cycle and in which it becomes sexually mature; *cf.* intermediate host.

primary meristem each of the meristematic tissues derived from the apical meristem, being the ground meristem, procambium, and protoderm.

primary mycelium haploid mycelium originating from a basidiospore.

primary organizer anterior part of primitive streak, or dorsal lip of blastopore.

primary phloem collectively, protophloem and metaphloem, the phloem derived from the procambium during primary growth.

primary plant body the plant body formed from growth at the apical meristems.

primary producer autotroph, *q.v.*

primary root the tap root, developed as a continuation of the radicle of the embryo.

primary sere prisere, *q.v.*

primary succession a succession that begins on bare ground.

primary vascular tissue xylem and phloem differentiating from procambium during primary growth.

primary xylem collectively, protoxylem and metaxylem, the xylem derived from the procambium during primary growth.

Primates *n.* [L. *primas*, one of the first] an order of placental mammals known from the Palaeocene to the present day, including the tree shrews, bush babies, monkeys, apes, and man, being mostly arboreal with limbs modified for climbing, leaping, or brachiating, a shortening of the snout and elaboration of visual apparatus, often with stereoscopic vision; *alt.* Primata.

primaxil *n.* [L. *primus*, first; *axilla*, armpit] the 1st axillary arm of a crinoid.

primer DNA single-stranded DNA necessary for DNA polymerase to work *in vitro.*

primibrachs *n.plu.* [L. *primus*, first; *brachia*, arms] in crinoids, all brachials up to and including the 1st axillary.

primine *n.* [L. *primus*, first] the external integument of an ovule, *cf.* secundine; occasionally *appl.* first-formed or internal coat.

primite *n.* [L. *primus*, first] the first of any pair of individuals of a caternoid colony in pseudoconjugation of Gregarinida, in which protomerite of one (the satellite) becomes attached to deutomerite of another (the primite).

primitive *a.* [L. *primitivus*, original] of earliest origin, *appl.* groove, streak, etc.; *appl.* sheath: neurilemma, *q.v.*; not differentiated or specialized.

primitive knot primitive node, *q.v.*

primitive node area of proliferating cells in which the primitive streak begins, thickened anterior wall of primitive pit; *alt.* primitive knot, Hensen's node.

primitive pit enclosure at anterior end of the confluent primitive folds.

primitive plate floor of the primitive groove.

primitive streak two primary embryonic folds, between which lies the primitive groove; *alt.* germ band, germinal streak.

Primofilices *n.* [L. *primus*, first; *filix*, fern] a subclass of Pteropsida (Filicopsida) known only as fossils, consisting of non-fern-like early forms with leaves and branches not well differentiated, and fern-like later forms, and which may be the ancestors of modern ferns; *alt.* Primofilicidae.

primordial *a.* [L. *primordium*, beginning] primitive; original, first commenced; first formed; *appl.* ova, utricle, veil, etc.

primordial cell initial, *q.v.*

primordium *n.* [L. *primordium*, beginning] original form; a structure when first indicating assumption of form; *alt.* anlage, fundament.

Primulales *n.* [*Primula*, generic name] an order of dicots of the Dilleniidae, Sympetalae, or Gamopetalae, having central placentation in the ovary and including the woody tropical families Theophrastaceae and Myrsinaceae and the temperate family Primulaceae.

principal cells peptic cells, *q.v.*

principal gland gastric gland, *q.v.*

Principes *n.* [L. *princeps*, prince] Arecales, *q.v.*

priodont *a.* [Gk. *priōn*, saw; *odous*, tooth] saw-toothed; *appl.* stag beetles with smallest development of mandible projections.

prisere *n.* [L. *primus*, first; *serere*, to put in a row] plant succession on area previously without vegetation, from bare ground to climax community; *alt.* primary sere.

prismatic *a.* [L. *prisma*, prism] like a prism, *appl.* cells, leaves; consisting of prisms, as prismatic layer of shells; *appl.* soil crumbs in which the vertical axis is longer than the horizontal.

Pristiophoriformes *n.* [*Pristiophorus*, generic name; L. *forma*, shape] an order of Selachii existing from Cretaceous times to the present day and including the saw sharks.

pro proline, *q.v.*

pro-acrosome *n.* [Gk. *pro*, before; *akros*, tip; *sōma*, body] structure in spermatid, which develops into acrosome.

proamnion *n.* [Gk. *pro*, before; *amnion*, foetal membrane] an area of blastoderm in front of head of early embryos of higher vertebrates.

proandry *n.* [Gk. *pro*, before; *anēr*, male] meroandry with retention of anterior pair of testes only; *cf.* metandry.

proangiosperm *n.* [Gk. *pro*, before; *anggeion*, vessel; *sperma*, seed] a fossil type of angiosperm.

Proanura *n.* [L. *pro*, before; Gk. *a*, without; *oura*, tail] an order of fossil amphibia known from one form from the Triassic of Madagascar, having a frog-like skull.

proatlas *n.* [Gk. *pro*, before; *Atlas*, a Titan] a median bone intercalated between atlas and skull in certain reptiles.

probasidium *n.* [Gk. *pro*, before; *basis*, base; *idion*, *dim.*] a thick-walled resting spore, as of Uredinales, Ustilaginales, Auriculariales; the cell which gives rise to a heterobasidium; an immature basidium, before forming sterigmata or basidiospores.

Proboscidea, proboscideans *n.*, *n.plu.* [Gk. *proboskis*, trunk] an order of placental mammals known from the Eocene to the present day, including the elephants, having a massive skeleton and the incisors modified as tusks.

proboscidiform *a* [Gk. *proboskis*, trunk; L. *forma*, shape] proboscis-like.

proboscis *n.* [Gk. *proboskis*, trunk] a trunk-like process of head, as of insects, annelids, nemerteans, elephants, etc.

proboscis worms the common name for the Nemertini, *q.v.*

probud a larval bud from the stolon in Doliolidae, which moves by pseudopodia to the cadophore and there divides to produce definitive buds.

procambial strand a strand of procambial tissue; *alt.* desmogen strand.

procambium *n.* [L. *pro*, before; L.L. *cambium*, change] the meristematic tissue from which vascular bundles are developed; *alt.* provascular tissue; *a.* procambial.

procarp *n.* [Gk. *pro*, before; *karpos*, fruit] the female organ of red seaweeds, a 1- or more-celled structure, consisting of the carpogonium, trichogyne, and auxiliary cells.

procartilage *n.* [L. *pro*, before; *cartilago*, gristle] the early stage of cartilage.

procary- prokary-, *q.v.*

Procellariiformes *n.* [*Procellaria*, generic name of petrel; L. *forma*, shape] an order of birds of the subclass Neornithes, including the albatrosses, shearwaters, and petrels.

procercoid *n.* [Gk. *pro*, before; *kerkos*, tail; *eidos*, form] early larval form of certain cestodes in 1st intermediate host.

procerebrum *n.* [L. *pro*, before; *cerebrum*, brain] the fore-brain, developed in preantennary region of insects.

procerus *n.* [Gk. *pro*, before; *keras*, horn] pyramidal muscle of the nose.

processus gracilis Folian process, *q.v.*

prochorion *n.* [Gk. *pro*, before; *chorion*, skin] an enveloping structure of blastodermic vesicle preceding formation of chorion.

prochromatin *n.* [Gk. *pro*, before; *chrōma*, colour] pyrenin, *q.v.*

prochromosome *n.* [Gk. *pro*, before; *chrōma*. colour; *sōma*, body] a discrete mass of basichromatin, primordium of the future chromosome.

proclimax *n.* [Gk. *pro*, before; *klimax*, ladder] stage in a sere appearing instead of usual climatic climax and not determined by climate; *alt.* subclimax; *cf.* preclimax.

procoelous *a.* [Gk. *pro*, before; *koilos*, hollow] with concave anterior face, as vertebral centra.

procollagen *n.* [Gk. *pro*, before; *kolla*, glue; *genos*, descent] collagen precursor found in the skin and other organs of many vertebrates.

Procolophonia *n.* [*Procolophon*, generic name] one of the orders into which the cotylosaurs are sometimes divided, having a large pineal opening and often a massive skull.

procoracoid *n.* [Gk. *pro*, before; *korax*, crow; *eidos*, form] an anteriorly directed process from glenoid fossa of urodeles.

procruscula *n.plu.* [L. *pro*, for; *dim.* of *crus*, leg] a pair of blunt locomotory outgrowths on posterior half of a redia.

procrypsis *n.* [Gk. *pro*, for; *krypsis*, concealment] shape, pattern, colour, or behaviour tending to make animals less conspicuous in their normal environment; camouflage; *a.* procryptic.

proctal *a.* [Gk. *prōktos*, anus] anal, *appl.* fish fins.

proctiger *n.* [Gk. *prōktos*, anus; L. *gerere*, to bear] anal portion of terminalia in Diptera; anal lobe.

proctodaeum *n.* [Gk. *prōktos*, anus; *hodos*, way] the latter part of embryonic alimentary canal, formed by anal invagination of ectoderm; a similar ectoderm-lined part in certain invertebrates; *alt.* hind-gut.

proctodone *n.* [Gk. *prōktos*, anus; horm*one*] an insect hormone thought to be secreted by anterior part of intestine and which ends diapause.

procumbent *a.* [L. *pro*, forward; *cumbens*, lying down] trailing on the ground; lying loosely along a surface; *cf.* prostrate; *appl.* ray cells elongated in radial direction.

procuticle, procuticula *n.* [L. *pro*, before; *cuticula*, thin outer skin] the colourless cuticle of insects, composed of protein and chitin, before differentiation into endocuticle and exocuticle.

prodeltidium *n.* [Gk. *pro*, before; Δ, delta; *idion*, *dim.*] a plate which develops into a pseudodeltidium.

prodentin(e) *n.* [L. *pro*, before; *dens*, tooth] a layer of uncalcified matrix capping tooth cusps before formation of dentine.

prodrome *n.* [Gk. *prodromos*, running before] a preliminary process, indication, or symptom.

producer *n.* [L. *producere*, to produce] an autotrophic organism, usually a photosynthetic green plant in an ecosystem, which synthesizes organic matter from inorganic materials.

productivity *n.* [L. *producere*, to produce] the amount of organic matter fixed by an ecosystem per unit time.

proecdysis *n.* [Gk. *pro*, before; *ekdysai*, to strip] in arthropods, the period of preparing for moulting with the laying down of new cuticle and the detachment of the older one from it.

proembryo *n.* [Gk. *pro*, before; *embryon*, foetus] an embryonic structure preceding true embryo.

proenzyme *n.* [Gk. *pro*, before; *en*, in; *zymē*, leaven] zymogen, *q.v.*

proepimeron *n.* [Gk. *pro*, before; *epi*, upon; *mēros*, upper thigh] a sclerite posterior to propleura; posterior pronotal lobe of Diptera.

proerythroblast *n.* [Gk. *pro*, before; *erythros*, red; *blastos*, bud] rubriblast, *q.v.*

proerythrocyte *n.* [Gk. *pro*, before; *erythros*, red; *kytos*, hollow] an immature red blood corpuscle; *alt.* pronormocyte, reticulocyte.

proestrum pro-oestrus, *q.v.*

proeusternum *n.* [Gk. *pro*, before; *eu*, well; *sternon*, chest] sclerite between propleura, forming ventral part of prothorax in Diptera.

proferment *n.* [Gk. *pro*, before; L. *fermentum*, ferment] zymogen, *q.v.*

profile transect stratum transect, *q.v.*

profunda *a.* [L. *profundus*, deep] deep-seated, *appl.* a branch of brachial, femoral, or costocervical artery, to the ranine artery, terminal part of lingual artery, and to a vein of femur. *n.* A deep artery or vein.

profundal *appl.* or *pert.* zone of deep water and bottom below compensation depth in lakes; *cf.* litoral.

progamete *n.* [Gk. *pro*, before; *gametēs*, spouse] a structure giving rise to gametes by abstriction, in certain fungi.

progamic *a.* [Gk. *pro*, before; *gamos*, marriage] *appl.* brood division for gamete production.

progastrin *n.* [Gk. *pro*, before; *gastēr*, stomach] precursor of gastric secretion in mucous membrane of stomach.

progenesis *n.* [Gk. *pro*, before; *genesis*, origin] the maturation of gametes before completion of body growth; in trematodes precocious sexual reproduction, in which larval forms such as metacercariae or cercariae may lay eggs which repeat the life cycle.

progeotropism *n.* [Gk. *pro*, for; *gē*, earth; *tropē*, turn] positive geotropism.

progestational *a.* [L. *pro*, before; *gestare*, to bear] *appl.* phase of oestrous cycle during luteal and endometrial activity; *appl.* hormones controlling uterine cycle and preparing uterus for nidation.

progesterone *n.* [L. *pro*, before; *gestare*, to bear] a steroid progestational hormone produced mainly by the corpus luteum, that prepares the uterus to receive the fertilized egg, maintains the uterus during pregnancy, and is the precursor of some other hormones; sometimes also called corporin, luteal hormone, lutin, luteosterone, progestone, progestin.

progestin *n.* [L. *pro*, before; *gestare*, to bear] progestational hormone of corpus luteum containing progesterone; a brand of progesterone; *alt.* luteosterone.

progestogens *n.plu.* [L. *pro*, before; *gestare*, to bear; Gk. *genos*, birth] a group of steroid compounds that have effects like progesterone.

proglottides *n.plu.* [Gk. *pro*, for; *glōtta*, tongue] the body segments of a tapeworm, formed by strobilization from neck; *sing.* proglottis.

prognathous *a.* [Gk. *pro*, forth; *gnathos*, jaw] having prominent or projecting jaws; with mouthparts projecting downwards, *appl.* insects, *cf.* hypognathous; with projecting anthers; *alt.* prognathic.

progonal *a.* [Gk. *pro*, before; *gonos*, begetting] *appl.* sterile anterior portion of genital ridge.

Progoneata *n.* [Gk. *pro*, before; *gonē*, generation] in some classifications, a group of arthropods with progoneate genital apertures, including the Chilopoda, Pauropoda, and Symphyla.

progoneate *a.* [Gk. *pro*, before; *gonē*, generation] having the genital aperture anteriorly, as in Progoneata; *cf.* opisthogoneate.

progressive provisioning in some insects and spiders, feeding the young regularly during their development.

Progymnospermopsida *n.* [Gk. *pro*, before; *gymnos*, naked; *sperma*, seed; *opsis*, appearance] a group of fossil plants whose anatomical characteristics are similar to those of the gymnosperms, but are thought to be below full gymnosperm status; *alt.* progymnosperms, Progymnospermae.

prohaemocyte *n.* [Gk. *pro*, before; *haima*, blood; *kytos*, hollow] a cell that develops into a haemocyte; proleucocyte, *q.v.*

prohaptor *n.* [Gk. *pro*, before; *haptein*, to fasten] anterior adhesive organ in Trematoda, as sucker, suctorial grooves, or glands.

prohormone *n.* [Gk. *pro*, before; *hormaein*, to excite] a precursor of a hormone especially of a polypeptide hormone.

prohydrotropism *n.* [Gk. *pro*, for; *hydōr*, water; *tropē*, turn] positive hydrotropism.

proiospory *n.* [Gk. *prōios*, early; *sporos*, seed] premature development of spores; *alt.* prospory.

projectile *a.* [L. *pro*, forth; *jacere*, to throw] protrusible; that can be thrust forward.

projection *n.* [L. *pro*, forth; *jacere*, to throw] the referring of stimulations to end organs of sense by means of connecting projection nerve fibres; the throwing forth by a plant of pollen, spores, or seeds.

projicient *a.* [L. *projiciens*, projecting] *appl.* sense organs reacting to distant stimuli, as light, sound.

prokary- *alt.* procary-; *cf.* eukary-.

prokaryocyte *n.* [Gk. *pro*, before; *karyon*, nucleus; *kytos*, hollow] prorubricyte, *q.v.*

prokaryon *n.* [Gk. *pro*, before; *karyon*, nucleus] the nucleus of a prokaryotic organism or cell; formerly also called protokaryon.

prokaryotes *n.plu.* [Gk. *pro*, before; *karyon*, nucleus] prokaryotic organisms, the bacteria and blue–green algae; the term is sometimes used as a taxonomic group.

prokaryotic *a.* [Gk. *pro*, before; *karyon*, nucleus] *appl.* cells or organisms whose DNA is not organized into chromosomes or surrounded by a nuclear membrane, but is a simple strand in the cell.

prolabium n. [L. pro, in front of; labium, lip] middle part of upper lip; cf. philtrum.

prolactin n. [L. pro, for; lac, milk] a proteinaceous hormone secreted by the anterior lobe of the pituitary gland which stimulates the secretion of milk in mammals and of crop milk in birds such as the pigeon, assists in maintaining the corpus luteum in mammals, and has a range of effects in lower vertebrates; alt. lactogenic hormone, luteotrophic hormone, luteotrophin, galactotropic hormone.

prolamines simple proteins found in plants especially in seeds of cereals, including gliadin from wheat, zein from maize, hordein from barley.

prolan gonadotropic hormones occurring in various tissues and body fluids during pregnancy in some mammals; prolan A is FSH and prolan B is luteinizing hormone.

prolarva n. [Gk. pro, before; L. larva, ghost] a newly hatched larva during the first few days when it feeds on its supply of embryonic yolk, as in some fish.

proleg n. [L. pro, for; M.E. leg, leg] an unjointed abdominal appendage of larvae of Lepidoptera and some other arthropods; alt. propes.

proles n. [L. proles, offspring] a taxonomic group used in different ways by different writers and never precisely defined.

proleucocyte n. [Gk. pro, before; leukos, white; kytos, hollow] a small leucocyte with basophil cytoplasm and large nuclei, and developing into macronucleocyte, in insects; leucoblast, q.v.; alt. prohaemocyte.

proliferation n. [L. proles, offspring; ferre, to bear] increase by frequent and repeated reproduction; growth by cell division; alt. prolification.

proliferous a. [L. proles, offspring; ferre, to bear] multiplying quickly, appl. bud-bearing leaves; developing supernumerary parts abnormally.

prolification n. [L. proles, offspring; facere, to make] proliferation, q.v.; shoot development from a normally terminal structure.

proline an amino acid obtained by hydrolysis from many proteins such as gliadin, casein, zein; abbreviated to pro.

proloculus n. [L. pro, before; loculus, compartment] first chamber, microspheric when formed by conjugation of swarm spores, megalospheric when formed asexually by fission, in polythalamous Foraminifera.

prolymphocyte n. [Gk. pro, before; L. lympha, water; kytos, hollow] an immature lymphocyte.

promegaloblast n. [Gk. pro, before; megalos, large; blastos, bud] a cell which develops into a megaloblast; alt. rubriblast.

promeristem n. [Gk. pro, before; meristos, divided] the part of the apical meristem consisting of the actively dividing cells and their most recent derivatives; alt. protomeristem, metrameristem.

prometaphase n. [Gk. pro, before; meta, after; phasis, appearance] stage between prophase and metaphase in mitosis and meiosis; alt. metakinesis, premetaphase.

promine a postulated substance that promotes cell growth in multicellular organisms.

promitochondria n. [Gk. pro, before; mitos, thread; chondros, grain] abnormal mitochondria which are found in yeast grown anaerobically and lacking some cristae and cytochromes.

promitosis n. [Gk. pro, before; mitos, thread] a simple type of mitosis, exemplified in nuclei of prokaryon type; protomitosis, q.v.

promonocyte n. [Gk. pro, before; monos, single; kytos, hollow] a cell developed from a monoblast and developing into a monocyte.

promontory n. [L. pro, forth; mons, mountain] prominence or projection, as of cochlea and sacrum.

promorphology n. [Gk. pro, before; morphē, form; logos, discourse] morphology from the geometrical standpoint.

promotor n. [L. promovere, to move forward] a protractor muscle, cf. remotor; in an operon, a region of DNA between the operator gene and 1st structural gene, which is needed for the operon to work, alt. promoter.

promuscis n. [L. promuscis, proboscis] the proboscis of Hemiptera.

promycelium n. [Gk. pro, before; mykēs, mushroom] a short hypha produced from some fungal spores, on which a different kind of spore develops; protobasidium, q.v.

promyelocyte n. [Gk. pro, before; myelos, marrow; kytos, hollow] amoeboid marrow cell which develops into a myelocyte or granulocyte.

pronase an enzyme which is one of a series of microbial metallo-proteinases classified as EC 3.4.24.4.

pronate a. [L. pronare, to bend forward] prone; inclined.

pronation n. [L. pronare, to bend forward] act by which palm of hand is turned downwards by means of pronator muscles; cf. supination.

pronephric n. [Gk. pro, before; nephros, kidney] pert. or in region of pronephros, appl. duct, tubules.

pronephros n. [Gk. pro, before; nephros, kidney] the fore-kidney of embryonic or larval life.

pronormocyte n. [Gk. pro, before; L. norma, rule; Gk. kytos, hollow] proerythrocyte, q.v.

pronotum n. [Gk. pro, before; nōton, back] the dorsal part of prothorax of insects.

pronucleus n. [L. pro, before; nucleus, kernel] egg- or sperm-nucleus after maturation; alt. germ nucleus, hemikaryon.

pronymph n. [Gk. pro, before; nymphē, pupa] the stage in metamorphosis preceding nymph stage.

pro-oestrus n. [Gk. pro, before; oistros, gadfly] period of preparation for pregnancy; phase before oestrus or heat; alt. pro-oestrum, proestrum.

pro-ostracum n. [Gk. pro, before; ostrakon, shell] the horny pen of a decapod dibranchiate shell or belemnite; anterior phragmocone; alt. gladius.

pro-otic n. [Gk. pro, before; ous, ear] the anterior bone of otic capsule in vertebrates. a. Pert. a centre of ossification of petromastoid part of temporal bone.

propagate v. [L. propagare, to propagate] to cause multiplication, as of plants; to impel, as excitation along a nerve fibre.

propagative a. [L. propagare, to propagate] reproductive, appl. a cell, a phase, an individual of a colony.

propagule, propagulum n. [L. propagare, to propagate] diaspore, q.v.

propatagium prepatagium, q.v.

properithecium n. [Gk. pro, before; peri, around: thēkē, case] a young perithecium which contains

a single zygote giving rise ultimately to ascospores.

propes *n.* [L. *pro*, before; *pes*, foot] proleg, *q.v.*

prophage *n.* [Gk. *pro*, before; *phagein*, to eat] a
temperate bacteriophage, *q.v.*; the nucleic acid of
such a bacteriophage, which carries information
necessary for production of phage and confers
certainly hereditary properties on the host bacterium.

prophase *n.* [Gk. *pro*, before; *phasis*, appear-
ance] the 1st stage in mitosis or meiosis, during
which chromosomes become distinct.

prophialide *n.* [Gk. *pro*, before; *phialē*, bowl;
eidos, form] a hyphal structure giving rise to
phialides; *alt.* sporocladium.

prophloem protophloem, *q.v.*

prophlogistic *a.* [Gk. *pro*, very; *phlogistos*, set on
fire] stimulating vasodilatation and capillary permeability; inflammatory.

prophototropism *n.* [Gk. *pro*, for; *phōs*, light;
trope, turn] positive phototropism.

prophyl(lum) *n.* [Gk. *pro*, before; *phyllon*, leaf] a
small bract or bracteole; first foliage leaf, at base
of branch.

proplastid *n.* [Gk. *pro*, before; *plastos*, formed;
idion, *dim.*] an immature plastid, as in meristematic cells.

propleuron *n.* [Gk. *pro*, before; *pleura*, side] a
lateral plate of prothorax of insects.

propneustic *a.* [Gk. *pro*, before; *pnein*, to breathe]
with only prothoracic spiracles open for respiration.

propodeon, propodeum *n.* [Gk. *pro*, before;
podeōn, neck] an abdominal segment in front of
petiole or podeon of Hymenoptera; *alt.* median
segment, Latreille's segment, epinotum.

propodite *n.* [Gk. *pro*, before; *pous*, foot] foot
segment 6th from body, in Malacostraca; tibia in
Arachnida; *alt.* propus.

propodium *n.* [Gk. *pro*, before; *pous*, foot] the
small anterior part of a molluscan foot.

propodosoma *n.* [Gk. *pro*, before; *pous*, foot;
sōma, body] body region bearing 1st and 2nd legs
in Acarina.

propolar *a.* [Gk. *pro*, before; *polos*, axis] *appl.*
anterior cells of calotte in Dicyemida.

propolis *n.* [Gk. *pro*, for; *polis*, city] resinous
substance from buds of leaf axils of certain trees,
utilized by worker bees to fasten comb portions
and fill up crevices; *alt.* bee glue.

propons *n.* [L. *pro*, before; *pons*, bridge] alae
pontis, delicate bands of white matter crossing
anterior end of pyramid below pons Varolii; *alt.*
ponticulus.

proprioception *n.* [L. *proprius*, one's own;
capere, to take] the reception of stimuli originating within the organism; *cf.* kinaesthesis.

proprioceptor *n.* [L. *proprius*, one's own; *capere*,
to take] a receptor sensitive to internal stimuli
such as in muscle, tendon, etc.; *alt.* proprioreceptor.

propriogenic *a.* [L. *proprius*, one's own; *genus*,
kind] *appl.* effectors other than muscle, or organs which are both receptors and effectors; *cf.*
myogenic.

proprioreceptor proprioceptor, *q.v.*

propriospinal *a.* [L. *proprius*, one's own; *spina*,
spine] *pert.* wholly to the spinal cord; *appl.* fibres,
etc., found only in spinal cord.

prop roots adventitious aerial roots growing

downwards from stem, as in mangrove and
maize, and helping to hold up the stem; *alt.* buttress roots.

propterygium *n.* [Gk. *pro*, before; *pterygion*, little
wing or fin] the foremost of 3 basal cartilages
supporting pectoral fin of elasmobranchs; *cf.*
mesopterygium, metapterygium.

propulsive pseudopodium in some Neosporidia, a pseudopodium developed posteriorly
which by its elongation pushes the body forward.

propupa *n.* [L. *pro*, before; *pupa*, puppet]
prepupa, *q.v.*

propus *n.* [Gk. *pro*, before; *pous*, foot] propodite,
q.v.

propygidium *n.* [Gk. *pro*, before; *pygidion*, small
rump] the dorsal plate anterior to pygidium in
Coleoptera and some other insects.

prorachis *n.* [Gk. *pro*, before; *rhachis*, spine] the
face of Pennatulacea which is sterile and coincides
with aspect of terminal zooid away from the
sulcus.

proral *a.* [Gk. *prōra*, prow] from front backwards,
appl. jaw movement, as in rodents; *cf.* palinal,
orthal.

prorennin *n.* [Gk. *pro*, before; A.S. *rennan*, to
cause to run] the precursor of rennin, secreted by
the peptic cells and converted to rennin by the
action of hydrochloric acid.

Prorocentrales *n.* [*Prorocentrum*, generic
name] an order of Desmophyceae having one flagellum projecting forwards and the other at right
angles.

prorsad *adv.* [L. *prorsus*, forwards; *ad*, to] anteriorly; forward.

prorsal *a.* [L. *prorsus*, forward] anterior.

prorubricyte *n.* [L. *pro*, before; *ruber*, red; Gk.
kytos, hollow vessel] a basophil erythroblast; *alt.*
prokaryocyte.

proscapula *n.* [L. *pro*, before; *scapula*, shoulder
blade] clavicle, *q.v.*

proscolex *n.* [Gk. *pro*, before; *skōlēx*, worm] cysticercus, *q.v.*; the inverted scolex inside a cysticercus; *alt.* protoscolex.

prosecretin(e) *n.* [L. *pro*, before; *secretus*, separ-
ated] the precursor of secretin.

prosencephalization *n.* [Gk. *pros*, before; *engkephalos*, brain] the progressive shifting of controlling centres towards the fore-brain and the
increasing complexity of cerebral cortex in the
course of evolution.

prosencephalon *n.* [Gk. *pros*, before; *engkephalos*, brain] the fore-brain, comprising telencephalon and diencephalon; the 1st primary brain vesicle.

prosenchyma *n.* [Gk. *pros*, near; *engchyma*, in-
fusion] elongated pointed cells, with thin or thick
cell walls, as in mechanical and vascular tissues of
plants; *a.* prosenchymatous.

prosethmoid *n.* [Gk. *pros*, near; *ēthmos*, sieve;
eidos, form] an anterior cranial bone of teleosts.

prosicula *n.* [L. *pro*, before; *sicula*, small dagger]
distal part of sicula, bearing the nema in graptolites.

prosiphon *n.* [Gk. *pro*, for; *siphōn*, tube] endosiphuncle, *q.v.*

Prosobranchia(ta), prosobranchs *n.*, *n.plu.*
[Gk. *prosō*, forward; *brangchia*, gills] a subclass,
or in some classifications an order, of aquatic
gastropods which have a spiral shell closed by

an operculum, and a mantle cavity with 1 or 2 ctenidia; *alt.* Streptoneura.

prosocoel *n.* [Gk. *proso*, forward; *koilos*, hollow] a narrow cavity in epistome of molluscs, the 1st main part of coelom; median cavity between 3rd and lateral ventricles of brain, *alt.* foramen of Monro.

prosodetic *a.* [Gk. *proso*, forward; *detos*, bound] anterior to beak, *appl.* certain bivalve ligaments.

prosodus *n.* [Gk. *prosodos*, advance] a delicate canal between chamber and incurrent canal in some sponges.

prosoma *n.* [Gk. *pro*, before; *sōma*, body] the anterior part of the body, especially a cephalothorax, in various invertebrates; *alt.* proterosoma.

prosoplectenchyma *n.* [Gk. *prosō*, forward; *plektos*, twisted; *engchyma*, infusion] a false tissue of fungal hyphae where the cells are oriented parallel and the walls are indistinct, as in some lichens.

Prosopora *n.* [Gk. *prosō*, forward; *poros*, channel] an order of small aquatic oligochaetes having male gonopores on the segment containing the testes.

prosopyle *n.* [Gk. *prosō*, forward; *pylē*, gate] the aperture of communication between adjacent incurrent and flagellate canals in some sponges.

prosorus *n.* [Gk. *pro*, before; *sōros*, heap] the cell from which a sorus is derived.

prospory *n.* [Gk. *pro*, before; *sporos*, seed] proiospory, *q.v.*; seed production in plants that are not fully developed.

prostaglandin *n.* [L. *pro*, before; *stare*, to stand; *glans*, acorn] any of a group of compounds formed from C_{20} fatty acids containing a 5-membered ring, which are found in many mammalian tissues, especially human seminal fluid, whose activity includes stimulating contractions of the uterus and causing abortion and labour, regulating blood flow, and that have effects in various hormone activities.

prostalia *n.plu.* [L. *pro*, forth; *stare*, to stand] sponge spicules which project beyond the body surface; *alt.* marginalia; *sing.* prostal.

prostate *a.* [L. *pro*, before; *stare*, to stand] *appl.* a muscular and glandular organ around commencement of male urethra in pelvic cavity. *n.* The prostate gland; the spermiducal gland in annelids.

prostatic *a.* [L. *pro*, before; *stare*, to stand] *pert.* prostate gland, *appl.* duct, nerve, sinus, utricle, hormone, etc.

prostemmate *a.* [Gk. *pro*, before; *stemma*, wreath] *appl.* an ante-ocular structure or organ of some Collembola, of doubtful function; *alt.* prostemmatic.

prosternum *n.* [L. *pro*, before; *sternum*, breast bone] ventral part of prothorax of insects; presternum, *q.v.*; proeusternum of Diptera; ventral part of cheliceral segment in some arachnids.

prostheca *n.* [Gk. *prosthēkē*, appendage] movable inner lobe of mandibles in certain insect larvae.

prosthetic group [Gk. *prosthetos*, added] an accessory substance, not a protein, necessary for the protein part of the enzyme to work and which is strongly bound to the protein; *cf.* coenzyme.

prosthion *n.* [Gk. *prosthios*, foremost] the alveolar point, middle point of the upper alveolar arch.

prosthomere *n.* [Gk. *prosthen*, forward; *meros*, part] most anterior or preoral somite.

prostomiate *a.* [Gk. *pro*, before; *stoma*, mouth] having a portion of head in front of mouth.

prostomium *n.* [Gk. *pro*, before; *stoma*, mouth] in some annelids and molluscs, part of head anterior to mouth; *alt.* preoral segment; *a.* prostomial.

prostrate *a.* [L. *prostratus*, thrown down] trailing on the ground; lying closely along a surface; *cf.* procumbent.

Protacanthopterygii *n.* [Gk. *prōtos*, first; *akantha*, thorn; *pterygion*, little wing or fin] a group of generalized and fairly primitive teleosts existing from Cretaceous times to the present day, including the pike and stickleback.

protamines *n.plu.* [Gk. *prōtos*, first; *ammoniakon*, gum] small, simple proteins such as salmine, associated with nucleic acids in fish testes, which are strongly basic, soluble in water, and not coagulated by heat.

protandrism *n.* [Gk. *prōtos*, first; *anēr*, male] protandry, sometimes exclusively in zoological application.

protandry *n.* [Gk. *prōtos*, first; *anēr*, male] condition of hermaphrodite plants and animals where male gametes mature and are shed before female gametes mature; *alt.* proterandry, metagyny; *a.* protandrous, proterandrous, proterandric; *cf.* protandrism, metandry, protogyny.

protaphin *n.* [Gk. *prōtos*, first; *Aphis*, generic name of aphid] a natural yellow pigment which is water-soluble and deep magenta in alkaline solution and is readily converted to the aphins.

protaspis *n.* [Gk. *prōtos*, first; *aspis*, shield] developmental stage or larva of trilobites.

protaxis *n.* [Gk. *prōtos*, first; L. *axis*, axle] primordial filament or axis in evolution of plant stem.

protaxon *n.* [Gk. *prōtos*, first; *axōn*, axle] axon base.

Proteales *n.* [*Protea*, generic name] an order of dicots of the Rosidae or Archichlamydeae having a homochlamydeous and brightly coloured tetramerous perianth, comprising the single family Proteaceae.

protease *n.* [Gk. *proteion*, first] any proteolytic enzyme.

protective layer a layer of cells in the abscission zone which form a protective layer over a scar when a part of a plant, such as leaf or fruit, falls off.

protegulum *n.* [L. *pro*, before; *tegulum*, covering] the semicircular or semielliptical embryonic shell of brachiopods.

proteid *n.* [Gk. *prōteion*, first; *eidos*, form] an obsolete term for any nitrogen-containing compound, or particularly for a protein.

protein *n.* [Gk. *prōteion*, first] any of a large group of compounds made of linked amino acids containing carbon, hydrogen, oxygen, nitrogen, and sulphur, and occasionally some other elements, essential in cells of all plants and animals, including simple proteins which do not have a prosthetic group, conjugated proteins which contain one or more prosthetic groups, and derived proteins which are obtained by denaturation or cleavage of simple or conjugated proteins.

proteinaceous *pert.* or composed of protein.

proteinase *n.* [Gk. *prōteion*, first] any of a group

of proteolytic enzymes, especially endopeptidases, that cause the splitting of proteins into peptides; EC sub-subgroups 3.4.21–99.

proteinoid n. [Gk. *prōteion*, first; *eidos*, form] a molecule like a protein, produced when trying to mimic primaeval conditions, by heating or other treatment of amino acids.

proteinoplast n. [Gk. *prōteion*, first; *plastos*, formed] a plastid that accumulates protein.

proteism n. [L. *Proteus*, a sea god] the capacity to change shape, as of amoeba and some other Protista.

protembryo n. [Gk. *prōtos*, first; *embryon*, embryo] the fertilized ovum and its cleavage stages preceding formation of blastula.

protenchyma n. [Gk. *prōtos*, first; *engchyma*, infusion] zone of primordial tissue of a carpophore below origin of the universal veil.

protentomon n. [Gk. *prōtos*, first; *entomon*, insect] the hypothetical archetype of insects.

Proteocephaloidea n. [Gk. *prōteion*, first; *kephalē*, head; *eidos*, form] an order of segmented tapeworms having a scolex with 4 strong suckers and an apical adhesive organ.

proteoclastic a. [Gk. *prōteion*, first; *klan*, to break] proteolytic, q.v.

proteolysis n. [Gk. *prōteion*, first; *lysis*, loosing] the disintegration of proteins, as by proteolytic enzymes.

proteolytic enzymes [Gk. *prōteion*, first; *lysis*, loosing] formerly enzymes which hydrolyse proteins by breaking peptide bonds, including endopeptidases and exopeptidases, *alt.* proteases, proteoclastic enzymes; now specifically used for proteinases (endopeptidases), EC sub-subgroups 3.4.21–99.

proteose n. [Gk. *prōteion*, first] a product of the hydrolysis of a protein being a fraction intermediate in size between metaproteins and the smaller peptones.

proter n. [Gk. *proteros*, in front] the anterior daughter individual produced when a protozoan divides transversely; cf. opisthe.

proterandry n. [Gk. *proteros*, earlier; *anēr*, male] protandry, q.v.; a. proterandrous, proterandric.

proteranthous a. [Gk. *proteros*, earlier; *anthos*, flower] flowering before foliage leaves appear.

proterogenesis n. [Gk. *proteros*, forward; *genesis*, descent] foreshadowing of adult or later forms by youthful or earlier forms.

proteroglyph a. [Gk. *proteros*, in front; *glyphein*, to carve] with specialized canine teeth in anterior upper jaw region.

proterogyny n. [Gk. *proteros*, earlier; *gynē*, woman] protogyny, q.v.; a. proterogynous.

proterokont n. [Gk. *proteros*, earlier; *kontos*, punting pole] a flagellum in prokaryotes not homologous with those of eukaryotes.

proterosoma n. [Gk. *proteros*, forward; *sōma*, body] body region comprising gnathosoma and propodosma in Acarina; prosoma, q.v.

proterotype n. [Gk. *proteros*, earlier; *typos*, pattern] original or primary type, as holotype, paratype, syntype.

Proterozoic a. [Gk. *proteros*, earlier; *zōon*, animal] a geological era of the Pre-Cambrian after the Archaeozoic and just above the Cambrian, whose rocks contain a few fossils, mainly blue–

green algae and soft-bodied invertebrates such as annelids.

prothallial a. [Gk. *pro*, before; *thallos*, young shoot] pert. a prothallus; appl. cell in microspore of gymnosperms and some pteridophytes, considered as a vestige of a prothallus.

prothallium prothallus, q.v.

prothalloid a. [Gk. *pro*, before; *thallos*, young shoot; *eidos*, form] like a prothallus.

prothallus n. [Gk. *pro*, before; *thallos*, young shoot] the hyphae of lichens during the initial growth stages; a small haploid gametophyte as in pteridophytes and gymnosperms, bearing antheridia or archegonia or both, and developed from a spore; *alt.* prothallium.

protheca n. [Gk. *pro*, before; *thēkē*, box] the basal part of a coral calicle, which is formed first.

prothecium n. [Gk. *pro*, before; *thēkē*, box] a primary perithecium of certain fungi.

prothetely n. [Gk. *protheein*, to run before; *telos*, completion] the development or manifestation of pupal or of imaginal characters in insect larva; cf. hysterotely.

prothoracic a. [Gk. *pro*, before; *thōrax*, chest] pert. prothorax; appl. glands secreting ecdysone; appl. anterior lobe of pronotum.

prothorax n. [Gk. *pro*, before; *thōrax*, chest] anterior segment of thorax of insects; cf. mesothorax, metathorax.

prothrombin n. [Gk. *pro*, before; *thrombos*, clot] a protein in blood plasma, produced in the liver in the presence of vitamin K, which is converted to thrombin at a wound in the presence of calcium ions by the action of activators such as thrombokinase; *alt.* serozyme, thrombogen, thrombinogen.

prothyalosome n. [Gk. *prōtos*, first; *hyalos*, glass; *sōma*, body] the area surrounding germinal spot in germinal vesicle.

Protista n. [Gk. *prōtistos*, first of all] originally a kingdom of living organisms showing little differentiation into tissues and including bacteria, protozoans, algae, and fungi; now used for unicellular organisms, Schizomycetes and Mollicutes, unicellular and colonial algae, and some unicellular fungi, or for organisms showing both plant and animal characteristics, especially the Flagellata; also used as a collective term for the groups Protozoa and Protophyta.

protistology n. [Gk. *prōtistos*, first of all; *logos*, discourse] the science dealing with the Protista.

proto-aecidium n. [Gk. *prōtos*, first; *oikidion*, small house] a cell mass surrounded by hyphal layers, containing cells eventually producing aecidiospores and disjunctor cells; *alt.* protoaecium, aecial primordium, primordial aecidium.

protoaphins n.plu. [Gk. *prōtos*, first; *Aphis*, generic name aphid] yellow pigments found in some aphids and precursors of erythroaphins.

Protoarticulatae n. [Gk. *prōtos*, first; L. *articulus*, joint] Hyeniales, q.v.

Protoascales n. [Gk. *prōtos*, first; *askos*, bag] Endomycetales, q.v.

Protoascomycetes, Protoascomycetidae n. [Gk. *prōtos*, first; *askos*, bag; *mykes*, fungus] Hemiascomycetes, q.v.

protobasidium n. [Gk. *prōtos*, first; *basidion*, small pedestal] a basidium producing a mycelium of 4 cells from each of which a sporidium is developed by abstriction; promycelium, *q.v.*; *alt.* apobasidium.

protobiology *n.* [Gk. *prōtos*, first; *bios*, life; *logos*, discourse] the study of ultramicroscopic organisms.

protobiont *n.* [Gk. *prōtos*, first; *biōnai*, to live] a protist, protophyton or protozoon.

protobios ultramicroscopic life, such as ultraviruses.

protoblast *n.* [Gk. *prōtos*, first; *blastos*, bud] a naked cell, devoid of membrane; 1st or single-cell stage of an embryo; a blastomere which develops into a definite organ or part; internal bud stage in life history of Neosporidia.

protoblem(a) *n.* [Gk. *prōtos*, first; *blēma*, coverlet] a layer of flaky tissue covering the teleoblema and constituting the primary or primordial veil of certain fungi; *alt.* subuniversal veil.

protobranch *a.* [Gk. *prōtos*, first; *brangchia*, gills] *appl.* gills of bivalve molluscs having flat non-reflected filaments.

Protobranchia(ta), protobranchs *n., n.plu.* [Gk. *prōtos*, first; *brangchia*, gills] an order, or in some classifications also a subclass of bivalve molluscs which usually feed by enlarged labial palps rather than ctenidia.

protobroch, protobroque *a.* [Gk. *prōtos*, first; *brochos*, mesh] *appl.* nuclei of gonia in resting stage; *cf.* deutobroch.

protocephalic *a.* [Gk. *prōtos*, first; *kephalē*, head] *appl.* or *pert.* primary head region of insect embryo; *pert.* protocephalon.

protocephalon *n.* [Gk. *prōtos*, first; *kephalē*, head] head part of cephalothorax in Malacostraca; first of 6 segments composing insect head.

protocercal *a.* [Gk. *prōtos*, first; *kerkos*, tail] diphycercal, *q.v.*

protocerebron, protocerebrum *n.* [Gk. *prōtos*, first; L. *cerebrum*, brain] anterior part of ganglionic centres of crustaceans; anterior part of insect brain, formed by fused ganglia of optic segment of head.

protochlorophyll *n.* [Gk. *prōtos*, first; *chlōros*, green; *phyllon*, leaf] a yellowish pigment found in chloroplasts of plants grown in darkness, which is converted to chlorophyll by the agency of light; *alt.* etiolin, aetiolin.

Protochordata, protochordates *n., n.plu.* [Gk. *prōtos*, first; *chordē*, string] Acrania, *q.v.*; in some classifications, a phylum of animals including the Hemichordata, Urochordata, and Cephalochordata, having gills slits, a dorsal hollow CNS, a notochord, and a postanal tail.

Protociliata *n.* [Gk. *prōtos*, first; L. *cilium*, eyelid] a subclass of ciliates in some classifications, including those forms with sexual reproduction by fusion of gametes; *cf.* Euciliata.

protocnemes *n.plu.* [Gk. *prōtos*, first; *knēmē*, wheel spoke] the 6 primary pairs of mesenteries of Zoantharia.

Protococcales *n.* [*Protococcus*, generic name] Chlorococcales, *q.v.*

protocoel *n.* [Gk. *prōtos*, first; *koilos*, hollow] the front portion of the coelomic cavity.

protoconch *n.* [Gk. *prōtos*, first; *kongchē*, shell] the larval shell of molluscs, indicated by scar on adult shell.

protocone *n.* [Gk. *prōtos*, first; *kōnos*, cone] inner cusp of upper molar tooth.

protoconid *n.* [Gk. *prōtos*, first; *kōnos*, cone; *eidos*, form] external cusp of lower molar tooth.

protoconidium *n.* [Gk. *prōtos*, first; *konis*, dust; *idion*, *dim.*] a rounded or club-shaped cell at the tip of a filament, giving rise to deuteroconidia, as in dermatophytes; *alt.* hemispore.

protoconule *n.* [Gk. *prōtos*, first; *kōnos*, cone] anterior intermediate cusp of upper molar tooth.

protocorm *n.* [Gk. *prōtos*, first; *kormos*, trunk] swelling of rhizophore, preceding root formation as in certain club mosses that have mycorrhizal fungi in early stages; in orchids, structure produced usually underground from germinating seedling and heavily infected with mycorrhizal fungus, *alt.* mycorrhizome; undifferentiated cell mass of archegonium in Ginkgoales; the posterior portion of germ band, which gives rise to trunk segments in insects; *a.* protocormic.

protocranium *n.* [Gk. *prōtos*, first; *kranion*, skull] posterior part of insect epicranium.

protoderm *n.* [Gk. *prōtos*, first; *derma*, skin] the outer cell layer of apical meristem; primordial epidermis of plants; superficial dermatogen.

protoecium *n.* [Gk. *prōtos*, first; *oikos*, house] the 2 valves of a polyzoan larva, adhering to a substrate.

protoepiphyte *n.* [Gk. *prōtos*, first; *epi*, upon; *phyton*, plant] a plant which is an epiphyte all its life and does not start life rooted in the ground or come to root in the ground later; *cf.* hemiepiphyte.

protofibrils *n.plu.* [Gk. *prōtos*, first; L. *fibrilla*, small fibre] minute threads seen in ground substance between submicroscopic fibrils, in connective tissue.

protogene *n.* [Gk. *prōtos*, first; *genos*, descent] a dominant allele; *cf.* allogene.

protogenesis *n.* [Gk. *prōtos*, first; *genesis*, origin] first embryonic stage, including development of archenteron; *cf.* deuterogenesis.

protogenic *a.* [Gk. *prōtos*, first; *genos*, offspring] persistent from beginning of development.

protogyny *n.* [Gk. *prōtos*, first; *gynē*, woman] condition of hermaphrodite plants and animals in which female gametes mature and are spent before maturation of male gametes; *alt.* proterogyny, metagyny; *a.* protogynous, proterogynous; *cf.* protandry.

Protogyrodactyloidea *n.* [*Protogyrodactylus*, generic name; Gk. *eidos*, form] an order of Monopisthocotylea having a single testis and an opisthaptor with about 12 spines and 2 linked central hooks.

protohaem *n.* [Gk. *prōtos*, first; *haima*, blood] haematin, *q.v.*

protokaryon *n.* [Gk. *prōtos*, first; *karyon*, nut] prokaryon, *q.v.*

Protolepidodendrales *n.* [*Protolepidodendron*, generic name] an order of fossil Lycopsida, known from Cambrian times but mainly found in Devonian strata.

protologue *n.* [Gk. *prōtos*, first; *logos*, discourse] the printed matter accompanying the 1st description of a name.

protoloph *n.* [Gk. *prōtos*, first; *lophos*, crest] anterior transverse crest of upper molar teeth.

protomala *n.* [Gk. *prōtos*, first; L. *mala*, cheek] a mandible of myriapods.

protomeristem n. [Gk. prōtos, first; meristos, divided] promeristem, q.v.

protomerite n. [Gk. prōtos, first; meros, part] anterior part of medullary protoplasm of adult gregarines; cf. primite.

protomite n. [Gk. prō, early; tomē, cutting; mitos, thread] stage between tomont and tomite in life cycle of Holotricha.

protomitosis n. [Gk. prōtos, first; mitos, thread] primitive mitosis; cruciform division, as in slime fungi; alt. promitosis.

Protomonadina n. [Gk. prōtos, first; monas, unit] an order of Zoomastigina having 1 or 2 flagella and only part of body surface able to be amoeboid.

protomonostelic a. [Gk. prōtos, first; monos, alone; stēlē, column] appl. stem or root with protostele or central cylinder.

protomont n. [Gk. prō, early; tomē, cutting; onta, beings] transitory stage, between trophont and tomont, with condensed central nucleus, in life cycle of Holotricha.

protomorphic a. [Gk. prōtos, first; morphē, form] first-formed; primordial; primitive.

Protomycetales n. [Gk. prōtos, first; mykēs, fungus] a small group of Ascomycetes which are plant parasites with a synascus.

protonema n. [Gk. prōtos, first; nema, thread] the filamentous structure of mosses from which the moss plant buds; early filamentous stage in development of certain algae.

protonematoid a. [Gk. prōtos, first; nēma, thread; eidos, form] like a protonema.

protonephridial a. [Gk. prōtos, first; nephros, kidney] pert., resembling, or functioning like a protonephridium.

protonephridium n. [Gk. prōtos, first; nephros, kidney; idion, dim.] a nephridium having a solenocyte.

protonephromixium n. [Gk. prōtos, first; nephros, kidney; mixis, mingling] a protonephridium with opening into the coelomoduct.

protoneurone n. [Gk. prōtos, first; neuron, nerve] the primitive intermediary cell connecting receptor with effector; cellular unit of nerve net; a unipolar ganglion cell.

protonymph n. [Gk. prōtos, first; nymphē, chrysalis] first nymphal size as in some Acarida.

protopathic a. [Gk. prōtos, first; pathos, feeling] appl. stimuli and nerve systems concerned with sensation of pain and of marked variations in temperature.

protopectin n. [Gk. prōtos, first; pektos, congealed] any of a group of pectic substances, insoluble in water, found in plants and yielding pectin or pectinic acids on hydrolysis, as by protopectinase during fruit ripening, alt. pectose.

protopectinase the enzyme that catalyses the change of protopectin to soluble pectin or pectinic acids and results in the separation of plant cells; alt. pectosinase, propectinase, cytase.

protopepsia n. [Gk. prōtos, first; pepsis, digestion] solution and alteration of food material accomplished in stomach.

protoperithecium n. [Gk. prōtos, first; peri, around; thēkē, case] primary haploid perithecium, as in certain Pyrenomycetes.

protophloem n. [Gk. prōtos, first; phloios, inner bark] the 1st phloem elements of a vascular bundle, the 1st part of the primary phloem; alt. prophloem.

Protophyta n. [Gk. prōtos, first; phyton, plant] in some classifications, a grade of organization including all unicellular plants and those which are loose aggregations with little differentiation.

protophyte n. [Gk. prōtos, first; phyton, plant] a primitive unicellular plant, alt. protophyton, cf. metaphyte; the gametophyte in antithetic alternation of generations, cf. antiphyte; a member of the Protophyta.

protoplasm n. [Gk. prōtos, first; plasma, form] living cell substance, cytoplasm and nucleoplasm; alt. bioplasm, plasm, plasma, sarcode; a. protoplasmic.

protoplasmic astrocytes astrocytes found in grey matter having thick branched processes similar to pseudopodia; cf. fibrous astrocytes.

protoplasmic bead structure on anterior part of middle piece of mammalian spermatozoon.

protoplast n. [Gk. prōtos, first; plastos, formed] a living uninucleate protoplasmic unit; protoplasm of a plant cell without a cell wall.

protopod a. [Gk. prōtos, first; pous, foot] with feet or legs on anterior segments, not on abdomen, appl. insect larvae; cf. oligopod, polypod.

protopodite n. [Gk. prōtos, first; pous, foot] basal segment of arthropod limb; alt. sympodite.

protoporphyrin n. [Gk. prōtos, first; porphyra, purple] a precursor of porphyrin without its metallic ion, which may act as a pigment such as ooporphyrin.

Protopteridales n. [Gk. prōtos, first; pteris, fern] in some classifications, an order of Primofilices, q.v.

protoptile n. [Gk. prōtos, first; ptilon, feather] the primary prepenna, succeeded by mesoptile.

Protorosauria, protorosaurs n., n.plu. [Protorosaurus, generic name] a group of extinct Synaptosauris that include various primitive and aberrant types.

Pretorthoptera n. [Gk. prōtos, first; orthos, straight, pteron, wing] a fossil order of insects including Carboniferous and Permian forms that are probably related to the Orthoptera.

protoscolex proscolex, q.v.

Protoselachii n. [Gk. prōtos, first; selachos, shark] Notidani, q.v.

protosoma, protosome n. [Gk. prōtos, first; sōma, body] prosoma or anterior part of body; part of body bearing cephalic lobe dorsally and tentacles ventrally in Pogonophora.

protospore n. [Gk. prōtos, first; sporos, seed] a spore of 1st generation; a mycelium-producing spore.

protosporophyte n. [Gk. prōtos, first; sporos, seed; phyton, plant] the filament produced by the fertilized female cell, 1st sporophyte stage in life cycle of Rhodophyceae; cf. deutosporophyte.

protostele n. [Gk. prōtos, first; stēlē, column] the simplest kind of stele, being a solid rod of vascular tissue, with the phloem surrounding the xylem; alt. monostele.

protosterigma n. [Gk. prōtos, first; stērigma, prop] basal portion of a sterigma.

protosternum n. [Gk. prōtos, first; sternon, chest] sternite of cheliceral segment of prosoma in Acarina.

protostigmata n.plu. [Gk. prōtos, first; stigma,

pricked mark] two primary gill slits of embryo; *sing.* protostigma.

protostoma *n.* [Gk. *prōtos*, first; *stoma*, mouth] blastopore, *q.v.*

protostome *n.* [Gk. *prōtos*, first; *stoma*, mouth] *appl.* coelomates with a true coelom, having spiral cleavage of the egg, with the blastopore forming mouth, or mouth and anus, and having a trochophore as a larva; *cf.* deuterostome.

protostylic *a.* [Gk. *prōtos*, first; *stylos*, column] having lower jaw connected with cranium by original dorsal end of arch; *n.* protostyly.

protothallus *n.* [Gk. *prōtos*, first; *thallos*, young shoot] first-formed structure which gives rise to a prothallus.

prototheca *n.* [Gk. *prōtos*, first; *thēkē*, box] a skeletal cup-shaped plate at aboral end of coral embryo, the 1st skeletal formation.

Prototheria, prototherians *n., n.plu.* [Gk. *prōtos*, first; *thērion*, small animal] a subclass of primitive mammals that lay eggs, have mammary glands without nipples, and no external ears, and includes the duck-billed platypus and spiny anteater which are placed in the order Monotremata.

Prototracheata *n.* [Gk. *prōtos*, first; L.L. *trachia*, windpipe] Onychophora, *q.v.*

prototroch *n.* [Gk. *prōtos*, first; *trochos*, wheel] a preoral circlet of cilia of a trochosphere larva.

prototrophic *a.* [Gk. *prōtos*, first; *trophē*, nourishment] nourished from one supply or in one manner only; feeding on inorganic matter, *appl.* iron, sulphur, and nitrifying bacteria, *appl.* plants; *alt.* protrophic.

prototype *n.* [Gk. *prōtos*, first; *typos*, model] an original type species or example; an ancestral form.

Protoungulata *n.* [Gk. *prōtos*, first; L. *ungula*, hoof] a superorder of ungulates containing the orders Condylarthra, Notoungulata, Litopterna, Astrapotheria, and Tubulidentata.

protovertebrae *n.plu.* [Gk. *prōtos*, first; L. *vertebra*, vertebra] a series of primitive mesodermal segments in a vertebrate embryo; *alt.* mesomere.

protoxylem *n.* [Gk. *prōtos*, first; *xylon*, wood] the 1st xylem elements in a plant organ, being the first-formed primary xylem.

Protozoa, protozoans *n., n.plu.* [Gk. *prōtos*, first; *zōon*, animal] a subkingdom and phylum of microscopic animals whose body is equivalent to a single cell; also protozoa.

protozoaea *n.* [Gk. *prōtos*, first; *zōon*, animal] stage in life history of certain arthropods, succeeding free-swimming nauplius.

protozoology *n.* [Gk. *prōtos*, first; *zōon*, animal; *logos*, discourse] the branch of zoology dealing with Protozoa.

protozoon *n.* [Gk. *prōtos*, first; *zōon*, animal] a member of the phylum Protozoa.

protozygote *n.* [Gk. *prōtos*, first; *zygōtos*, yoked] a homozygote having dominant characters; *cf.* allozygote.

protractor *n.* [L. *pro*, forth; *tractus*, drawn out] a muscle which extends a part or draws it out or away from the body; *cf.* retractor.

protriaene *n.* [Gk. *pro*, before; *triaina*, trident] a triaene with anteriorly directed branches.

protrophic prototrophic, *q.v.*

Protura *n.* [Gk. *prōtos*, first; *oura*, tail] an order of minute wingless insects having a telson on the last segment but no compound eyes or antennae.

provascular tissue procambium, *q.v.*

proventriculus *n.* [L. *pro*, before; *ventriculus*, small stomach] in decapod crustaceans, the so-called stomach containing gastric mill; in insects, the digestive chamber anterior to stomach, *alt.* gizzard; in annelids that anterior to gizzard; in birds, the glandular stomach anterior to gizzard.

provinculum *n.* [L. *pro*, before; *vinculum*, bond] a primitive hinge of young stages of certain Lamellibranchiata.

provirus *n.* [L. *pro*, before; *virus*, poison] a virus which does not cause lysis and has become part of the host cell chromosome, and is transmitted from one cell generation to the next in the chromosome.

provisioning *n.* [L. *provisio*, foresight] providing food for young, as in mass provisioning, nest provisioning, progressive provisioning.

provitamin *n.* [L. *pro*, before; *vita*, life; *ammoniacum*, resinous gum] precursor of a vitamin; *alt.* previtamin.

provitamin A carotene, *q.v.*

proximad *adv.* [L. *proximus*, next; *ad*, towards] towards, or placed nearest the body or base of attachment; *cf.* distad.

proximal *a.* [L. *proximus*, next] nearest body or centre or base of attachment; *cf.* distal.

proximoceptor *n.* [L. *proximus*, next; *recipere*, to receive] a receptor which reacts only to nearby stimuli, as a contact receptor; *cf.* disticeptor.

prozonite *n.* [Gk. *pro*, before; *zōne*, girdle] the anterior ring of a diplosomite; *cf.* metazonite.

prozymogen *n.* [Gk. *pro*, before; *zymē*, leaven; *-genēs*, producing] precursor of zymogen.

pruinose *a.* [L. *pruina*, hoar frost] covered with whitish particles or globules; covered by bloom.

Prymnesiales *n.* [*Prymnesium*, generic name] an order of Haptophyceae having a haptonema and including the coccolithophorids.

psalterium *n.* [L. *psalterium*, psalter] omasum, *q.v.*; the lyra, a thin triangular lamina joining lateral portions of fornix.

Psammodontiformes *n.* [Gk. *psammos*, sand; *odous*, tooth; L. *forma*, shape] an order of marine Carboniferous Holocephali known only from their teeth and of uncertain taxonomic position.

Psammodrilomorpha *n.* [Gk. *psammos*, sand; *drilos*, worm; *morphē*, form] an order of minute polychaetes that have a unique pumping apparatus in the pharynx region.

psammon *n.* [Gk. *psammos*, sand; *on*, being] the organisms living between sand grains, as of freshwater and marine shores; *alt.* mesopsammon.

psammophilous *a.* [Gk. *psammos*, sand; *philos*, loving] thriving in sandy places; *alt.* arenicolous; *n.* psammophile.

psammophore *n.* [Gk. *psammos*, sand; *phora*, carrying] one of rows of hairs under mandibles and sides of head in desert ants, used for removal of sand grains.

psammophyte *n.* [Gk. *psammos*, sand; *phyton*, plant] a plant growing in sandy or gravelly ground.

psammosere *n.* [Gk. *psammos*, sand; L. *serere*, to put in a row] a plant succession originating in a sandy area, as on dunes.

Pselaphognatha *n.* [Gk. *psēlaphan*, to grope about; *gnathos*, jaw] a subclass of small milli-

pseudambulacrum

pedes with an uncalcified integument, and which have hairs and special barbed bristles called trichomes.

pseudambulacrum *n.* [Gk. *pseudēs*, false; L. *ambulare*, to walk] the lancet plate with adhering side plates and covering plates, of Blastoidea.

pseudannual *n.* [Gk. *pseudēs*, false; L. *annus*, year] a plant which completes its growth in one year but provides a bulb or other means of surviving winter.

pseudanthium *n.* [Gk. *pseudēs*, false; *anthos*, flower] an inflorescence condensed to such an extent that it looks like a simple flower.

pseudapogamy *n.* [Gk. *pseudēs*, false; *apo*, away; *gamos*, marriage] fusion of pair of vegetative nuclei, as in certain fungi and in fern prothallus.

pseudaposematic *a.* [Gk. *pseudēs*, false; *apo*, from; *sēma*, sign] imitating warning coloration or other protective features of harmful or distasteful animals, i.e. showing Batesian mimicry; *alt.* pseudoaposematic.

pseudapospory *n.* [Gk. *pseudēs*, false; *apo*, away; *sporos*, seed] spore formation without meiosis, resulting in a diploid spore which gives rise to the gametophyte.

pseudaxis *n.* [Gk. *pseudēs*, false; L. *axis*, axle] an apparent main axis which really consists of a number of lateral branches running parallel; *alt.* false axis, monochasium, sympodium.

pseudepipodite *n.* [Gk. *pseudēs*, false; *epi*, upon; *pous*, foot] in Cephalocarida, a flattened outer region of the 2nd maxilla and trunk appendages, giving the limb a triramous structure.

pseudepisematic *a.* [Gk. *pseudēs*, false; *epi*, upon; *sēma*, sign] having false coloration or markings, as in protective mimicry or for alluring or aggressive purposes; *alt.* pseudo-episematic; pseudosematic.

pseudergates *n.* [Gk. *pseudēs*, false; *ergatēs*, worker] in some species of termite, a caste consisting of large blind wingless workers.

pseudhaemal *a.* [Gk. *pseudēs*, false; *haima*, blood] *appl.* the vascular system of echinoderms.

pseudholoptic *a.* [Gk. *pseudēs*, false; *holos*, whole; *optikos*, relating to sight] intermediate between holoptic and dichoptic, conditions in eyes of Diptera.

pseudimago *n.* [Gk. *pseudēs*, false; L. *imago*, image] subimago, *q.v.*

pseudoacrorhagus *n.* [Gk. *pseudēs*, false; *akros*, summit; *rhax*, grape] a structure resembling an acrorhagus, but containing ordinary ectodermal nematocysts, in certain Actiniaria.

pseudo-aethalium *n.* [Gk. *pseudēs*, false; *aithalos*, soot] a dense aggregation of distinct sporangia, as in Myxomycetes; *cf.* aethalium.

pseudoalleles *n.plu.* [Gk. *pseudēs*, false; *allēlōn*, one another] subdivisions of a gene due to crossing-over at compound loci.

pseudoalveolar *a.* [Gk. *pseudēs*, false; L. *alveus*, hollow] *appl.* a structure of cytoplasm containing starch grains or deutoplasm spheres.

pseudoangiocarpic *a.* [Gk. *pseudēs*, false; *anggeion*, vessel; *karpos*, fruit] with an exposed hymenium temporarily enclosed by incurved edge of pileus or by a secondary pseudovelum.

pseudoaposematic pseudaposematic, *q.v.*

pseudoaquatic *a.* [Gk. *pseudēs*, false; L. *aqua*, water] thriving in wet ground.

pseudoarticulation *n.* [Gk. *pseudēs*, false; L. *articulus*, joint] incomplete subdivision of a segment, or groove having the appearance of a joint, as in limbs of arthropods.

pseudobasidium *n.* [Gk. *pseudēs*, false; *basis*, base; *idion*, dim.] a large basidium with thickened wall, constituting a resting spore.

pseudoblepharoplast *n.* [Gk. *pseudēs*, false; *blepharis*, eyelash; *plastos*, formed] temporary concentration of chromatiin near centriole in sperm formation of certain insects.

pseudobrachium *n.* [Gk. *pseudēs*, false; *brachion*, arm] appendage for locomotion on a substratum, formed from elongated pterygials of pectoral fin of pediculates.

pseudobranch *n.* [Gk. *pseudēs*, false; *brangchia*, gills] an accessory gill of some fishes, not respiratory in function; spiracular or vestigial hyoidean gill.

pseudobulb *n.* [Gk. *pseudēs*, false; L. *bulbus*, bulb] a thickened internode of orchids and some other plants, for storage of water and reserves.

pseudobulbil *n.* [Gk. *pseudēs*, false; L. *bulbus*, bulb] an outgrowth of some ferns, a substitute for sporangia.

pseudobulbous *a.* [Gk. *pseudēs*, false; L. *bulbus*, bulb] adapted to xerophytic conditions through development of pseudobulbs.

pseudocarp *n.* [Gk. *pseudēs*, false; *karpos*, fruit] false fruit, *q.v.*

pseudocartilage *n.* [Gk. *pseudēs*, false; L. *cartilago*, cartilage] a cartilage-like substance serving as support in certain invertebrates and lower vertebrates.

pseudocellus *n.* [Gk. *pseudēs*, false; L. *ocellus*, little eye] one of scattered sense organs of unknown function in certain insects.

pseudocentrous *a.* [Gk. *pseudēs*, false; L. *centrum*, centre] *appl.* vertebra composed of 2 pairs of arcualia meeting and forming a suture laterally.

pseudochromatin *n.* [Gk. *pseudēs*, false; *chrōma*, colour] pyrenin, *q.v.*

pseudochromosomes *n.plu.* [Gk. *pseudēs*, false; *chrōma*, colour; *sōma*, body] elements of the Golgi apparatus, *q.v.*

pseudocilia *n.plu.* [Gk. *pseudēs*, false; L. *cilium*, eyelid] cytoplasmic threads projecting from cell through surrounding sheath of mucilage as in algae of the Tetrasporaceae.

pseudocoel(e) *n.* [Gk. *pseudēs*, false; *koilos*, hollow] the narrow cavity between the 2 laminae of septum lucidum, so-called 5th ventricle of brain; space between mesodermal tissue of the body wall and gastrodermis, derived from blastocoel, as in Aschelminthes; haemocoel of Arthropoda.

pseudocoelomate *a.* [Gk. *pseudēs*, false; *koilōma*, hollow] a metazoan animal in which the body cavity between the gut and body wall is not a true coelom but a derivative of a persistent blastocoel.

pseudoconch *n.* [Gk. *pseudēs*, false; *kongchē*, shell] a structure developed above and behind the true concha in crocodiles; in some gastropod molluscs, a non-spiral shell.

pseudocone *a.* [Gk. *pseudēs*, false; *kōnos*, cone] *appl.* insect compound eye having cone cells filled with transparent gelatinous material.

pseudoconidium *n.* [Gk. *pseudēs*, false; *konis*, dust; *idion*, dim.] one of the spores formed on lateral projections of pseudomycelium of certain yeasts.

pseudoconjugation n. [Gk. *pseudēs*, false; L. *cum*, with; *jugum*, yoke] conjugation of Sporozoa in which 2 individuals, temporarily and without true fusion, join end to end, protomerite to deutomerite, or side to side.

pseudocopulation n. [Gk. *pseudēs*, false; L. *copula*, bond] in orchids, a process in which the resemblance of an orchid to the female of an insect species leads to an attempt by a male to copulate with it and effect pollination.

pseudocortex n. [Gk. *pseudēs*, false; L. *cortex*, bark] a cortex composed of gelatinous hyphae, as in certain lichens.

pseudocostate a. [Gk. *pseudēs*, false; L. *costa*, rib] false-veined, having a marginal vein uniting all others.

pseudoculus n. [Gk. *pseudēs*, false; L. *oculus*, eye] an oval area on each side of head of Pauropoda, possibly a receptor for mechanical vibrations.

pseudocyesis n. [Gk. *pseudēs*, false; *kyēsis*, conception] pseudopregnancy, q.v.

pseudocyphella n. [Gk. *pseudēs*, false; *kyphella*, hollow of ear] a structure in lichens like a cyphella but smaller, also thought to be used in aeration of the thallus.

pseudocyst n. [Gk. *pseudēs*, false; *kystis*, bladder] a residual protoplasmic mass which swells and ruptures, liberating spores of Sporozoa.

pseudodeltidium n. [Gk. *pseudēs*, false; Δ, delta; *idion*, dim.] a plate partly or entirely closing deltidial fissue in ventral valve of certain Testicardines.

pseudoderm n. [Gk. *pseudēs*, false; *derma*, skin] a kind of skin-like covering of certain compact sponges, formed also towards pseudogaster.

pseudodominance n. [Gk. *pseudēs*, false; L. *dominans*, ruling] expression of a recessive gene in the absence of its dominant allele.

pseudodont a. [Gk. *pseudēs*, false; *odous*, tooth] having horny pads or ridges instead of teeth, as monotremes.

pseudo-elater n. [Gk. *pseudēs*, false; *elatēr*, driver] one of the chains of cells in sporogonium of some liverworts, probably functioning as true elaters.

pseudo-episematic pseudepisematic, q.v.

pseudofaeces n. [Gk. *pseudēs*, false; L. *faeces*, dregs] excess particles of food discharged by bivalve molluscs.

pseudofoliaceous a. [Gk. *pseudēs*, false; L. *folium*, leaf] with expansions resembling leaves.

pseudogamy n. [Gk. *pseudēs*, false; *gamos*, marriage] union of hyphae from different thalli; activation of ovum by a spermatozoon or pollen grain which plays no part thereafter; pseudomixis, q.v.

pseudogaster n. [Gk. *pseudēs*, false; *gastēr*, stomach] an apparent gastral cavity of certain sponges, opening to exterior by pseudo-osculum and having true oscula opening into itself.

pseudogastrula n. [Gk. *pseudēs*, false; *gastēr*, stomach] the stage in development of certain sponges, in which the archaeocytes become completely enclosed by flagellate cells.

pseudogyne n. [Gk. *pseudēs*, false; *gynē*, female] a worker ant with female thoracic characters.

pseudohaptor n. [Gk. *pseudēs*, false; *haptein*, to fasten] a large discoidal organ in some trematodes with a ventral armature of spines arranged in radial rows.

pseudoheart the axial organ of echinoderms; one of the contractile vessels pumping blood from dorsal to ventral vessel in annelids.

pseudoidium n. [Gk. *pseudēs*, false; *ōon*, egg; *idion*, dim.] a separate hyphal cell which may germinate; plu. pseudoidia.

pseudolamina n. [Gk. *pseudēs*, false; L. *lamina*, plate] expanded apical portion of a phyllode.

pseudomanubrium n. [Gk. *pseudēs*, false; L. *manubrium*, handle] the manubrium considered as a process of subumbrella where the former contains the gastric cavity in certain Trachylina.

pseudometamerism n. [Gk. *pseudēs*, false; *meta*, after; *meros*, part] apparent serial segmentation; an approximation to metamerism, as in certain cestodes.

pseudomitotic a. [Gk. *pseudēs*, false; *mitos*, thread] diaschistic, q.v.

pseudomixis n. [Gk. *pseudēs*, false; *mixis*, mingling] fusion of vegetative cells instead of gametes, leading to embryo formation; alt. pseudogamy, somatogamy; cf. syngamy.

Pseudomonadales n. [Gk. *pseudēs*, false; *monas*, unit] an order of rod-shaped or spiral Gram-negative bacteria, which do not form spores.

pseudomonocarpous a. [Gk. *pseudēs*, false; *monos*, alone; *karpos*, fruit] with seeds retained in leaf bases until liberated, as in cycads.

pseudomonocotyledonous a. [Gk. *pseudēs*, false; *monos*, alone; *kotylēdōn*, cup-like hollow] with 2 cotyledons coalescing to appear as one.

pseudomonocyclic a. [Gk. *pseudēs*, false; *monos*, alone; *kyklos*, circle] appl. crinoids with infrabasals absent in adults but present in young or in near ancestors.

pseudomorph n. [Gk. *pseudēs*, false; *morphē*, form] a structure having an indefinite form; a fungal stroma composed of parts of plants and interwoven hyphae.

pseudomycelium n. [Gk. *pseudēs*, false; *mykēs*, fungus] an assemblage of chains or groups of adherent cells, of yeasts; alt. sprout mycelium.

pseudomycorrhiza n. [Gk. *pseudēs*, false; *mykēs*, fungus; *rhiza*, root] association of short roots of conifers with parasitic fungi in the absence of mycorrhizal fungi.

pseudonavicella n. [Gk. *pseudēs*, false; L. *navicella*, small boat] a small boat-shaped spore containing sporozoites, in Sporozoa; alt. oocyst.

pseudonotum n. [Gk. *pseudēs*, false; *nōton*, back] postscutellum, q.v.

pseudonuclein n. [Gk. *pseudēs*, false; L. *nucleus*, kernel] pyrenin, q.v.

pseudonucleoli n.plu. [Gk. *pseudēs*, false; L. *nucleus*, kernel] knots or granules in nuclear reticulum not true nucleoli.

pseudonychium n. [Gk. *pseudēs*, false; *onyx*, claw] a lobe or process between claws of insects.

pseudo-osculum n. [Gk. *pseudēs*, false; L. *osculum*, small mouth] the exterior opening of a pseudogaster; alt. pseudostoma.

pseudo-ostiolum n. [Gk. *pseudēs*, false; L. *ostiolum*, small door] a small opening formed by breaking down of cell walls or tissues, in certain fungi without perithecia; alt. pseudo-ostiole, pseudostiole.

pseudopallium *n.* [Gk. *pseudēs*, false; L. *pallium*, mantle] in some gastropods parasitic on echinoderms, a ring-like skin fold developing at the base of the proboscis and eventually extending like a sac over the whole parasite.

pseudoparaphysis *n.* [Gk. *pseudēs*, false; *para*, beside; *phyein*, to grow] basidiolum, *q.v.*; paraphysoid, *q.v.*

pseudoparasitism *n.* [Gk. *pseudēs*, false; *parasitos*, parasite] accidental entry of a free-living organism into the body of another and its survival there.

pseudoparenchyma *n.* [Gk. *pseudēs*, false; *para*, beside; *engchyma*, infusion] a tissue-like mass of interwoven hyphae which resembles parenchyma; in red algae, an elaborate plant body resembling parenchyma made from aggregated filaments; *alt.* paraplectenchyma, false tissue.

pseudopenis *n.* [Gk. *pseudēs*, false; L. *penis*, penis] the protruded evaginated portion of male deferent duct, in certain Oligochaeta; copulatory structure in Orthoptera.

pseudoperculum *n.* [Gk. *pseudēs*, false; L. *operculum*, lid] a structure resembling an operculum or closing membrane.

pseudoperianth *n.* [Gk. *pseudēs*, false; *peri*, around; *anthos*, flower] an archegonium-investing envelope of certain liverworts.

pseudoperidium *n.* [Gk. *pseudēs*, false; *pēridion*, small wallet] the aecidiospore envelope of certain fungi.

Pseudophyllidea *n.* [Gk. *pseudēs*, false; *phyllidion*, little leaf] an order of tapeworms having a scolex with either no adhesive organs or bothria, in some classifications, separated into the Didesmida, the rest of the tapeworms forming the Tetradesmida.

pseudoplasmodium *n.* [Gk. *pseudēs*, false; *plasma*, form] an aggregation of myxamoebae without fusion of their protoplasm.

pseudopod *n.* [Gk. *pseudēs*, false; *pous*, foot] a foot-like body-wall process of certain larvae; pseudopodium, *q.v.*

pseudopodiospore *n.* [Gk. *pseudēs*, false; *pous*, foot; *sporos*, seed] *see* amoebula.

pseudopodium *n.* [Gk. *pseudēs*, false; *pous*, foot; *eidos*, form] a protrusion of cytoplasm serving for locomotion and prehension in Rhizopoda and amoeboid cells of metazoans; in certain mosses, the sporogonium-supporting pedicel; slender branch of the gametophyte that bears gemmae in some mosses; *alt.* pseudopod, false foot.

pseudopore *n.* [Gk. *pseudēs*, false; *poros*, channel] a small orifice between outermost tube and intercanal system of certain sponges.

pseudopregnancy *n.* [Gk. *pseudēs*, false; L. *praegignere*, to bring forth] condition of development of accessory reproductive organs simulating true pregnancy, although fertilization has not taken place; *alt.* pseudocyesis.

pseudopupa *n.* [Gk. *pseudēs*, false; L. *pupa*, puppet] coarctate pupa, *q.v.*

pseudoramose *a.* [Gk. *pseudēs*, false; L. *ramus*, branch] having false branches.

pseudoramulus *n.* [Gk. *pseudēs*, false; L. *ramulus*, small branch] a spurious branch of certain algae.

pseudoraphe *n.* [Gk. *pseudēs*, false; *rhaphē*, seam] a smooth axial area in some diatoms.

pseudoreduction the pairing of chromosomes in early stages of meiosis so apparently halving the chromosome number.

pseudorhabdites *n.plu.* [Gk. *pseudēs*, false; *rhabdos*, rod] granular masses of formed secretion produced by gland cells of Rhabdocoela.

pseudorhiza *n.* [Gk. *pseudēs*, false; *rhiza*, root] a root-like structure connecting mycelium in the soil with the fruit body of a fungus; *alt.* storage trunk; *plu.* pseudorhizae.

pseudorumination reingestion or refection; *q.v.*

pseudosacral *a.* [Gk. *pseudēs*, false; L. *sacrum*, sacred] *appl.* sacral vertebra attached to pelvis by transverse process and not by sacral rib.

pseudoscolex *n.* [Gk. *pseudēs*, false; *skōlēx*, worm] modified anterior proglottides of certain cestodes where true scolex is absent.

Pseudoscorpiones, pseudoscorpions *n.*, *n.plu.* [Gk. *pseudēs*, false; L. *scorpio*, scorpion] an order of small arachnids, commonly called false scorpions, which resemble scorpions but whose opisthosoma is not divided into 2 regions.

pseudosematic *a.* [Gk. *pseudēs*, false; *sēma*, sign] pseudepisematic, *q.v.*

pseudoseptate *a.* [Gk. *pseudēs*, false; L. *septum*, division] apparently, but not morphologically septate.

pseudoseptum *n.* [Gk. *pseudēs*, false; L. *septum*, partition] a perforated or incomplete septum; septum with pores, as in certain fungi.

pseudosessile *a.* [Gk. *pseudēs*, false; L. *sedere*, to sit] *appl.* abdomen of petiolate insects when petiole is so short that abdomen is close to thorax; *cf.* pedicellate.

pseudosperm *n.* [Gk. *pseudēs*, false; *sperma*, seed] a small indehiscent fruit resembling a seed.

Pseudosphaeriales *n.* [Gk. *pseudēs*, false; *sphaira*, globe] an order of Loculomycetidae including many plant pathogens, in which the fruit body opens by a pore or canal and is often surrounded by bristles.

pseudospore *n.* [Gk. *pseudēs*, false; *sporos*, seed] an encysted resting myxamoeba; formerly a basidiospore.

pseudostele *n.* [Gk. *pseudēs*, false; *stēlē*, pillar] an apparently stelar structure, as midrib of leaf.

pseudostigma *n.* [Gk. *pseudēs*, false; *stigma*, mark] a cup-like pit of integument, as the socket of a sensory seta in Acarina.

pseudostiole pseudo-ostiolum, *q.v.*

pseudostipe *n.* [Gk. *pseudēs*, false; L. *stipes*, stem] a stem-like structure formed by presumptive spore-producing tissue, as in Gasteromycetes.

pseudostipula, pseudostipule *n.* [L. *pseudēs*, false; L. *stipula*, stalk] part of lamina at the base of a leaf stalk, which resembles a stipule.

pseudostoma *n.* [Gk. *pseudēs*, false; *stoma*, mouth] a temporary mouth or mouth-like opening; pseudo-osculum, *q.v.*

pseudostroma *n.* [Gk. *pseudēs*, false; *strōma*, bedding] a mass of mixed fungous and host cells.

Pseudosuchia *n.* [Gk. *pseudēs*, false; *souchos*, a kind of crocodile] an extinct order of lizard-like archosaurs.

pseudosuckers *n.plu.* [Gk. *pseudēs*, false; A.S. *sucan*, to suck] powerful organs of attachment, with gland cells, on either side of the oral sucker in some trematodes.

pseudothecium *n.* [Gk. *pseudēs*, false; *thēkē*,

case] a spherical fruit body resembling a perithecium.

pseudotrachea *n.* [Gk. *pseudēs*, false; L. *trachia*, windpipe] a trachea-like structure; one of the trachea-like food channels of labellum, as in Diptera.

pseudotroch *n.* [Gk. *pseudēs*, false; *trochos*, wheel] an arc of stiff cirri in the supra-oral region of the buccal field of the corona of some rotifers.

pseudotrophic *a.* [Gk. *pseudēs*, false; *trophē*, nourishment] *appl.* mycorrhiza when the fungus is parasitic.

pseudo-unipolar *a.* [Gk. *pseudēs*, false; L. *unus*, one; *polus*, pole] *appl.* unipolar nerve cells with a T-shaped or Y-shaped axom, formed by partial fusion of axons of originally bipolar cells.

pseudovacuole *n.* [Gk. *pseudēs*, false; L. *vacuus*, empty] gas vacuole, *q.v.*

pseudovarium *n.* [Gk. *pseudēs*, false; L. *ovarium*, ovary] ovary producing pseudova.

pseudovelum *n.* [Gk. *pseudēs*, false; L. *velum*, covering, veil] velum without muscular and nervous cells, in Scyphozoa; in agarics, a structure formed by union of contemporaneous outgrowths from pileus and stipe, protecting immature hymenium, *alt.* pseudoveil.

pseudovitellus *n.* [Gk. *pseudēs*, false; L. *vitellus*, egg yolk] a mycetome in aphids, consisting of a mass of fatty cells in the abdomen.

pseudovum *n.* [Gk. *pseudēs*, false; L. *ovum*, egg] an ovum that can develop without fertilization, i.e. a parthenogenetic ovum; the earlier condition of viviparously produced aphids.

pseudozoaea *n.* [Gk. *pseudēs*, false; *zōon*, animal] a larval stage of stomatopods, so-called from its resemblance to zoaea stage of decapods.

pseudozygospore *n.* [Gk. *pseudēs*, false; *zygon*, yoke; *sporos*, seed] azygospore, *q.v.*

Psilophyta in some classifications, a division of Tracheophyta equivalent to the Psilopsida, and including the Psilophytales and Psilotales.

Psilophytales *n.* [*Psilophyton*, generic name] the only order of the Psilophytopsida, *q.v.*

Psilophytatae Psilopsida, *q.v.*

psilophyte *n.* [Gk. *psilos*, without trees; *phyton*, plant] any plant of savanna; a Palaeozoic vascular plant; a member of the Psilophyta.

Psilophytopsida *n.* [*Psilophyton*, generic name; Gk. *opsis*, appearance] a class of pteridophytes known only as rootless fossil sporophytes of the Silurian and Devonian, having rhizomes and dichotomously branched upright stems with or without small leaf-like appendages, and sporangia borne on naked branches.

Psilopsida a group of pteridophytes which includes the Psilotopsida and Psilophytopsida; *alt.* Psilophytatae; *cf.* Psilophyta.

Psilotales *n.* [*Psilotum*, generic name] the only order of Psilotopsida, *q.v.*

Psilotopsida *n.* [*Psilotum*, generic name; Gk. *opsis*, appearance] a class of pteridophytes found mostly in the tropics and subtropics, having a very simple structure with rootless sporophyte, dichotomously branching rhizomes, and aerial branches with small appendages; thought to be closely related to the Psilophytopsida.

Psittaciformes *n.* [*Psittacus*, generic name of parrot; L. *forma*, shape] an order of birds of the subclass Neornithes, including the parrots.

psoas *n.* [Gk. *psoa*, loins] name of 2 loin muscles, major and minor, formerly magnus and parvus.

Psocoptera *n.* [*Psocus*, generic name; Gk. *pteron*, wing] an order of small insects, commonly called book lice and bark lice, having incomplete metamorphosis, a globular abdomen, and often no wings; *alt.* Corrodentia.

psorosperms *n.plu.* [Gk. *psōra*, itch; *sperma*, seed] the resistant encysted stages of Sporozoa; minute parasitic organisms generally.

psychogenetic *a.* [Gk. *psychē*, soul; *genesis*, descent] *pert.* mental development; caused by the mind.

psychogenic *a.* [Gk. *pyschē*, mind; *gennaein*, to produce] of mental origin, *appl.* physiological and somatic changes.

psychon *n.* [Gk. *psychē*, mind] synapse during passage of impulse from one nerve cell to the next.

psychophysiology *n.* [Gk. *psychē*, mind; *physis*, nature; *logos*, discourse] physiology in relation to mental processes.

psychosomatic *a.* [Gk. *psychē*, mind; *sōma*, body] *pert.* relationship between mind and body; *pert.* or having body reactions to mental stimuli.

Psychozoic *a.* [Gk. *psychē*, mind; *zōon*, animal] *pert.* or *appl.* geological era in which man predominates; *alt.* Anthropozoic.

psychrophil(ic) *a.* [Gk. *psychros*, cold; *philein*, to love] thriving at relatively low temperatures, at below 20°C, *appl.* certain bacteria; *alt.* crymophil; *n.* psychrophile.

psychrophyte *n.* [Gk. *psychros*, cold; *phyton*, plant] a plant which grows on a cold substratum.

pteralia *n.plu.* [Gk. *pteron*, wing] axillary sclerites forming articulation of wing with processes of mesonotum in insects.

Pteraspida *n.* [*Pteraspis*, generic name] an order of Ordovician and Devonian ostracoderms (Heterostraci) having a head shield made of large plates.

pterate pterote, *q.v.*

pterergate *n.* [Gk. *pteron*, wing; *ergatēs*, worker] a worker or a soldier ant with vestigial wings.

pteridine *n.* [Gk. *pteron*, wing] a nitrogen-containing organic compound important in certain natural compounds, such as folic acid, leucopterin, xanthopterin.

pteridines *n.plu.* [Gk. *pteron*, wing] pterins, *q.v.*

pteridium pterodium, *q.v.*

pteridology *n.* [Gk. *pteris*, fern; *logos*, discourse] the branch of botany dealing with ferns.

Pteridophyta, pteridophytes *n., n.plu.* [Gk. *pteris*, fern; *phyton*, plant] a major division of the plant kingdom, having clear alternation of generations with a dominant vascular sporophyte initially dependent upon the gametophyte which is very reduced, represented as fossils from the Silurian and including the classes Psilophytopsida, Psilotopsida, Lycopsida, Sphenopsida, and Pteropsida (Filicopsida).

Pteridospermae, Pteridospermales, pteridosperms *n., n.plu.* [Gk. *pteris*, fern; *sperma*, seed] an order of fossil gymnosperms of class Cycadopsida, commonly called seed ferns, having fern-like leaves with ovules and seeds borne on them, *alt.* Lyginopteridales, Cycadofilicales; also classed as a division of equal status to gymnosperms in the Spermatophyta or as a subgroup of Cycadophytina, *alt.* Lyginopteridatae.

pterins *n.plu.* [Gk. *pteron*, wing] a group of pigments, derivatives of the pteridine ring, which are widespread in insects giving colour to their wings, and are also found in vertebrates and in plants; *alt.* pteridines.

pterion *n.* [Gk. *pteron*, wing] the point of junction of parietal, frontal, and great wing of sphenoid, *appl.* ossicle, a sutural bone.

Pterobranchia, pterobranchs *n., n.plu.* [Gk. *pteron*, wing; *brangchia*, gills] a class of colonial hemichordates that live in secreted tubes and have a U-shaped gut and a lophophore with ciliated tentacles for feeding.

pterocardiac *a.* [Gk. *pteron*, wing; *kardia*, stomach] *appl.* ossicles with curved ends in gastric mill of Crustacea.

pterocarpous *a.* [Gk. *pteron*, wing; *karpos*, fruit] with winged fruit.

pterodactyls *n.plu.* [Gk. *pteron*, wing; *daktylos*, finger] the common name for the pterosaurs, *q.v.*

pterodium *n.* [Gk. *pteron*, wing] a winged fruit or samara; *alt.* pteridium.

pteroic acid an amino acid, which reacts with glutamic acid to form pteroylglutamic acid (folic acid).

pteroid *a.* [Gk. *pteron*, wing; *pteris*, fern; *eidos*, form] resembling a wing; like a fern.

pteromorphae *n.plu.* [Gk. *pteron*, wing; *morphē*, shape] outgrowths from notogaster which cover sides of podosoma and 3rd and 4th pair of legs in certain Acarina.

pteropaedes *n.plu.* [Gk. *pteron*, wing; *pais*, child] birds able to fly when newly hatched.

pteropegum *n.* [Gk. *pteron*, wing; *pēgē*, source] an insect's wing socket.

Pterophyta *n.* [Gk. *pteris*, fern; *phyton*, plant] *see* Pteropsida.

pteropleurite *n.* [Gk. *pteron*, wing; *pleura*, side] thoracic sclerite between wing insertion and mesopleurite, in Diptera.

Pteropoda, pteropods *n., n.plu.* [Gk. *pteron*, wing; *pous*, foot] a group of pelagic opisthobranchs having wing-like lobes to the foot, and commonly called sea butterflies.

pteropodial *a.* [Gk. *pteron*, wing; *pous*, foot] *appl.* wing-like lobes of mid-foot of Pteropoda.

pteropodium *n.* [Gk. *pteron*, wing; *pous*, foot] a winged foot, as of certain bats.

Pteropsida *n.* [Gk. *pteris*, fern; *opsis*, appearance] a class of pteridophytes, commonly called ferns, having a sporophyte with roots, stems, and large leaves, megaphylls, with sporangia borne on the leaves, *alt.* Filicopsida, Pterophyta, Polypodiopsida; in some classifications a subdivision of the Tracheophyta which includes Gymnospermae and Angiospermae as well as ferns which are then referred to the class Filicineae.

Pterosauria, pterosaurs *n., n.plu.* [Gk. *pteron*, wing; *sauros*, lizard] an order of Jurassic and Cretaceous archosaurs, flying reptiles commonly called pterodactyls, which have a membranous wing supported by a greatly elongated 4th finger.

pterospermous *a.* [Gk. *pteron*, wing; *sperma*, seed] with winged seeds.

pterostigma *n.* [Gk. *pteron*, wing; *stigma*, mark] an opaque cell on insect wings.

pterote *a.* [Gk. *pterōtos*, winged] winged; having wing-like outgrowths; *alt.* alate, pterate.

pterotheca *n.* [Gk. *pteron*, wing; *thēkē*, case] the wing-case of pupae.

pterothorax *n.* [Gk. *pteron*, wing; *thōrax*, chest] a fused mesothorax and metathorax, as in Odonata.

pterotic *n.* [Gk. *pteron*, wing; *ous*, ear] a cranial bone overlying horizontal semicircular canal of ear. *a. Appl.* bone between pro-otic and epiotic.

pteroylglutamic acid (PGA) a polypeptide of glutamic acid with pteroic acid as the prosthetic group, commonly called folic acid, *q.v.*

pterygia *plu.* of pterygium.

pterygial *a.* [Gk. *pteryx*, wing] *pert.* a wing or fin; *appl.* a bone supporting a fin-ray; *pert.* a pterygium.

pterygiophore *n.* [Gk. *pterygion*, little wing or fin; *pherein*, to bear] one of the cartilaginous fin-rays; an actinost, *q.v.*; *alt.* pterygophore.

pterygium *n.* [Gk. *pterygion*, little wing] a prothoracic process of weevils; a small lobe on base of under-wings in Lepidoptera; a vertebrate limb; *plu.* pterygia.

pterygobranchiate *a.* [Gk. *pteryx*, wing; *brangchia*, gills] having spreading or feathery gills, as certain Crustacea.

pterygoda *n.plu.* [Gk. *pteryx*, wing; *eidos*, form] the tegulae of an insect.

pterygoid *n.* [Gk. *pteryx*, wing; *eidos*, form] a cranial bone. *a.* Wing-like, *appl.* processes of sphenoid, canal, fissure, fossa, plexus, muscles.

pterygoideus externus and internus, muscles causing protrusion and raising of mandible.

pterygomandibular *a.* [Gk. *pteryx*, wing; L. *mandibulum*, jaw] *pert.* pterygoid and mandible; *appl.* a tendinous band or raphe of buccopharyngeal muscle.

pterygomaxillary *a.* [Gk. *pteryx*, wing; L. *maxilla*, jaw] *appl.* a fissure between maxilla and pterygoid process of sphenoid.

pterygopalatine *a.* [Gk. *pteryx*, wing; L. *palatus*, palate] *pert.* region of pterygoid and palatal cranial bones, *appl.* canal, fossa, groove; *appl.* ganglion: sphenopalatine ganglion, *q.v.*; *alt.* pterygopalatal.

pterygophore pterygiophore, *q.v.*

pterygopodial *a.* [Gk. *pteryx*, wing; *pous*, foot] *appl.* mucous glands associated with claspers, in elasmobranchs.

pterygoquadrate *a.* [Gk. *pteryx*, wing; L. *quadratus*, squared] *appl.* a cartilage constituting dorsal half of mandibular arch of certain fishes.

pterygospinous *a.* [Gk. *pteryx*, wing; L. *spina*, spine] *appl.* a ligament between lateral pterygoid plate and spinous process of sphenoid.

Pterygota *n.* [Gk. *pteryx*, wing] a subclass of insects that have wings, or may be secondarily wingless, and have metamorphosis of some sort, and an abdomen with no organs of locomotion in the adult; *alt.* Metabola; *a.* pterygotous.

pterylae *n.plu.* [Gk. *pteron*, feather; *hylē*, a wood] a bird's feather tracts, skin areas on which feathers grow; *cf.* apteria.

pterylosis *n.* [Gk. *pteron*, feather; *hylē*, a wood] arrangement of pterylae and apteria in birds; *alt.* ptilosis.

PTH parathyroid hormone: parathyrin, *q.v.*

ptilinum *n.* [Gk. *ptilon*, feather] a head vesicle or bladder-like expansion of head of a fly emerging from pupa.

ptilopaedic *a.* [Gk. *ptilon*, feather; *pais*, child] covered with down when hatched.

ptilosis pterylosis, *q.v.*

ptomaines *n.plu.* [Gk. *ptōma*, dead body] a group of amino compounds, usually poisonous, produced during the decomposition of proteins in the putrefaction of dead animal matter and including choline, neurine, putrescine, cadavarine, and muscarine.

ptyalin *n.* [Gk. *ptyalon*, saliva] salivary amylase, found in man and some herbivores, which acts on broken starch grains, and catalyses the internal hydrolysis of 1,4-α-glucosidic linkages in polysaccharides containing 3 or more 1,4-α linked glucose units; EC 3.2.1.1; *r.n.* α-amylase.

Ptycholepiformes *n.* [*Ptycholepis*, generic name; L. *forma*, shape] an order of Triassic to Jurassic Chondrostei having some characteristics of the Holostei and probably being plankton feeders.

ptyophagous *a.* [Gk. *ptyein*, to spit; *phagein*, to eat] digesting, by host cells, the cytoplasm emitted by tips of hyphae, *appl.* a type of mycorrhiza; *cf.* plasmoptysis.

ptyosome *n.* [Gk. *ptyein*, to spit; *sōma*, body] cytoplasmic mass formed by plasmoptysis, *q.v.*, in ptyophagous mycorrhiza; *alt.* sporangiolum.

ptyxis *n.* [Gk. *ptyx*, fold] the form in which young leaves are folded or rolled on themselves in the bud; *cf.* prefoliation.

puberty *n.* [L. *pubertas*, adult state] beginning of sexual maturity.

puberty gland interstitial tissue of testis.

puberulent *a.* [L. *pubes*, adult] covered with down or fine hair.

pubes *n.* [L. *pubes*, adult] the lower portion of the hypogastric region of the abdomen, *alt.* pubic region; the hair appearing on that region at puberty.

pubescence *n.* [L. *pubescere*, to become mature] downy or hairy covering on some plants and certain insects.

pubescent *a.* [L. *pubescere*, to become mature] covered with soft hair or down.

pubic *a.* [L. *pubes*, mature] in region of pubes, *appl.* arch, ligament, symphysis, tubercle, vein; *pert.* pubis.

pubis *n.* [L. *pubes*, mature] anterior part of hip bone, consisting of body and rami; anteroventral portion of pelvic girdle; *alt.* os pubis.

puboischium *n.* [L. *pubes*, adult; Gk. *ischion*, hip] fused pubis and ischium, bearing acetabulum and ilium on each side; ischiopubis, *q.v.*

pudenda *n.plu.* [L. *pudere*, to be ashamed] external genitalia, as of primates.

pudendal in region of pudendum, *appl.* artery, cleft nerve, veins; *alt.* pudic.

pudendum *n.* [L. *pudere*, to be ashamed] vulva, or external female genitalia.

pudic pudendal, *q.v.*

puffball the common name for a member of the Lycoperdales, *q.v.*

puffing ejection of a cloud of spores from a ripe ascocarp or apothecium; the formation of puffs, *q.v.*

puffs bulbous enlargements of certain parts of polytene chromosomes, correlated with nucleic acid and protein synthesis; *alt.* Balbiani rings.

pullulation *n.* [L. *pullulare*, to sprout] reproduction by vegetative budding, as in yeast cells; *alt.* gemmation.

pulmobranch(ia) *n.* [L. *pulmo*, lung; Gk. *branchia*, gills] a gill-like organ adapted to air-breathing conditions; a lung book, as of spiders.

pulmogastric *a.* [L. *pulmo*, lung; Gk. *gastēr*, stomach] *pert.* lungs and stomach.

pulmonary *a.* [L. *pulmo*, lung] *pert.* lungs, *appl.* artery, ligament, valves, veins, pleura, etc.

pulmonary cavity or sac the mantle cavity of molluscs without ctenidia.

Pulmonata, pulmonates *n.*, *n.plu.* [Gk. *pulmo*, lung] a subclass, or in some classifications an order of gastropods that have no ctenidia but have a mantle cavity modified as a lung, usually have a spiral shell and visceral mass, but these may be reduced and the body slug-like; *alt.* Euthyneura B.

pulmones *n.plu.* [L. *pulmo*, lung] lungs; *sing.* pulmo.

pulp *n.* [L. *pulpa*, flesh] soft, fleshy part of fruit; the dental papilla; soft mass of splenic tissue; mesodermal core of feather cylinder; any of various other soft spongy tissues.

pulsating vacuole contractile vacuole, *q.v.*

pulse *n.* [L. *pulsus*, driven] the beat or throb observable in arteries, due to action of heart. [O.F. *pols*, from L. *puls*, pottage] a legume; a leguminous plant; the edible seed of such a plant.

pulsellum *n.* [L. *pulsare*, to beat] a flagellum situated at posterior end of protozoan body.

pulse wave a wave of increased pressure over arterial system, started by ventricular systole.

pulverulent *a.* [L. *pulverulentus*, dusty] powdery; powdered.

pulvillar *a.* [L. *pulvillus*, small cushion] *pert.* or at a pulvillus.

pulvilliform *a.* [L. *pulvillus*, small cushion; *forma*, shape] like a small cushion.

pulvillus *n.* [L. *pulvillus*, small cushion] pad, process, or membrane on foot or between claws, sometimes serving as an adhesive organ, in insects; lobe beneath each claw; *alt.* onychium, pulvinulus, cushion; *cf.* plantula.

pulvinar *n.* [L. *pulvinar*, couch] an angular prominence on thalamus. [L. *pulvinus*] *a.* Cushion-like; *pert.* a pulvinus.

pulvinate *a.* [L. *pulvinus*, cushion] cushion-like, *appl.* a repugnatorial gland in ants; having a pulvinus.

pulvinoid *a.* [L. *pulvinus*, cushion; Gk. *eidos*, form] resembling a pulvinus, *appl.* modified petiole.

pulvinulus *n.* [L. *pulvinus*, cushion] pulvillus, *q.v.*

pulvinus *n.* [L. *pulvinus*, cushion] a cellular swelling at junction of axis and leaf stalk, which plays a part in leaf or leaflet movement.

pulviplume *n.* [L. *pulvis*, powder; *pluma*, feather] powder-down feather, *q.v.*

punctae *n.plu.* [L. *punctum*, point] small pores, holes, or dots on a surface, especially the markings on valves of diatoms; *sing.* puncta.

Punctariales *n.* [*Punctaria*, generic name] an order of brown algae having a parenchymatous sporophyte.

punctate *a.* [L. *punctum*, point] dotted; having surface covered with small holes, pores, or dots; having a dot-like appearance.

punctiform *a.* [L. *punctum*, point; *forma*, form] having a dot-like appearance; *appl.* distribution, as of cold, warm, and pain spots on skin.

punctulate a. [L. dim. of punctum, point] covered with very small dots or holes.

punctum n. [L. punctum, point] a minute dot, point, or orifice, as puncta lacrimalia, puncta vasculosa; apex of a growing point, punctum vegetationis.

puncture n. [L. punctura, prick] a small round surface depression; a perforation.

pungent a. [L. pungere, to prick] producing a prickling sensation, appl. stimuli affecting chemical sense receptors; bearing a sharp point, appl. apex of leaf or leaflet.

punishment negative reinforcement, q.v.

Punnett square the chequerboard used to determine the number and type of progeny genotypes and phenotypes produced on fusion of certain gametes.

pupa n. [L. pupa, puppet] in insects with complete metamorphosis, resting stage in life cycle where insect is enclosed in a case and tissues are broken down and reorganized, cf. chrysalis; embryo with series of transverse rings of cilia in Holothuria; a. pupal.

puparium n. [L. pupa, puppet] the casing of a coarctate pupa, formed from the last larval skin, especially in Diptera; pupal instar.

pupate v. [L. pupa, puppet] to pass into the pupal stage.

pupiform a. [L. pupa, puppet; forma, shape] pupa-shaped; pupa-like; alt. pupoid.

pupigerous a. [L. pupa, puppet; gerere, to bear] containing a pupa.

pupil n. [L. pupilla, pupil of eye] aperture of iris through which rays pass to retina; central spot of an ocellus.

pupillary a. [L. pupilla, pupil of eye] pert. pupil of eye, appl. a membrane; appl. reflex: variation in aperture due to change in illumination.

pupillate a. [L. pupilla, pupil of eye] with a differently coloured central spot, appl. eye-like marking or ocellus.

pupiparous a. [L. pupa, puppet; parere, to beget] bringing forth young already developed to the pupa stage, as certain parasitic insects; alt. nymphiparous.

pupoid pupiform, q.v.

pure line a series of generations of organisms originating from a single homozygous ancestor, or identical homozygous ancestors, and therefore themselves homozygous.

purines n.plu. [Gk. pyren, nucleus] nitrogen-containing organic bases including adenine and guanine, which pair with pyrimidines in DNA and RNA, and whose derivatives are important in metabolism.

Purkinje cells [J. E. Purkinje, Bohemian physiologist] an incomplete stratum of flask-shaped cells between the molecular and nuclear layers of cerebellar cortex.

Purkinje fibres muscle fibres in atrioventricular bundle and its terminal strands, differing from typical cardiac fibres, especially in a higher rate of conduction of the contractile impulse.

puromycin an antibiotic which becomes incorporated into a polypeptide chain and causes release of incomplete polypeptide from ribosomes.

purposive behaviour goal-related behaviour, although the animal may not be aware of the goal; alt. horme.

pustule n. [L. pustula, blister] a blister-like prominence.

pusule n. [L. pusula, blister] non-contractile vacuole containing watery fluid, filling or emptying by duct, found in many Dinoflagellata; a contractile vacuole in some lower plants; alt. pusula.

putamen n. [L. putamen, nutshell] the hard endocarp or stone of drupes, alt. pyrene; lateral part of lentiform nucleus of cerebrum; shell membrane of bird's egg.

putrefaction n. [L. putrefacere, to make rotten] decomposition of organic material especially the usually anaerobic splitting of proteins by microorganisms, resulting in the formation of incompletely oxidized products which produce bad smells, such as mercaptans and alkaloids.

putrescine n. [L. putrescere, to grow rotten] an amine (diamine) produced by decaying animal protein, derived from ornithine.

pycn- alt. pykn-.

pycnia plu. of pycnium.

pycnial pycnidial, q.v.

pycnic a. [Gk. pyknos, thick] thick-set; appl. type of body build, short, stocky, with broad face and head.

pycnid pycnidium, q.v.

pycnidia plu. of pycnidium.

pycnidial pert. pycnidia; appl. drops: fungal nectar; alt. pycnial.

pycnidiophore n. [Gk. pyknos, dense; idion, dim.; pherein, to bear] a conidiophore producing pycnidia.

pycnidiospore n. [Gk. pyknos, dense; idion, dim.; sporos, seed] the spore produced in a pycnidium; alt. pycnidial conidium, pycnospore, pycnoconidium, pycnogonidium, spermatium, stylospore.

pycnidium n. [Gk. pyknos, dense; idion, dim.] a small flask-shaped organ or spermagonium containing slender filaments which form pycnidiospores or spermatia by abstriction in rust fungi; receptacle for stylospores in fungi and lichens; alt. pycnid, pycnium, conidiocarp, clinosporangium, spermogonium; plu. pycnidia.

pycnium n. [Gk. pyknos, dense] pycnidium, q.v.; plu. pycnia.

pycnoconidangium spermogonium, q.v.

pycnoconidium, pycnogonidium pycnidiospore, q.v.

Pycnodontiformes n. [Gk. pyknos, dense; odous, tooth; L. forma, shape] a widespread order of marine holosteans existing from the Jurassic to at least Eocene times.

Pycnogonida n. [Gk. pyknos, dense; gonē, seed, dim.] a class of chelicerates having a long slender body consisting of an anterior cephalon, a trunk with long walking legs, and a short segmented abdomen; in some classifications considered an order of arachnids; alt. Pantopoda.

Pycnogonomorpha n. [Pycnogōnum, generic name; Gk. morphē, form] an order of Pycnogonida having no chelicerae or palps.

pycnoplasson n. [Gk. pyknos, dense; plassein, to mould] an unexpanded form of plasson.

pycnosis n. [Gk. pyknos, dense] cell degeneration including nuclear condensation, resulting in formation of intensely staining clump of chromosomes; thickening of thallus, as in certain Ascomycetes.

pycnospore pycnidiospore, q.v.

pycnotic a. [Gk. pyknos, dense] characterized by,

or *pert.* pycnosis, *appl.* small irregular nucleus of degenerated cells.

pycnoxylic *a.* [Gk. *pyknos*, dense; *xylon*, wood] having compact wood; *cf.* manoxylic.

pygal *a.* [Gk. *pygē*, rump] situated at or *pert.* posterior end of back; *appl.* certain plates of chelonian carapace.

pygidial *a.* [Gk. *pygidion*, narrow rump] *pert.* pygidium; *appl.* paired repugnatorial glands in certain beetles.

pygidium *n.* [Gk. *pygidion*, narrow rump] a caudal shield covering abdomen of certain arthropods; terminal uncovered abdominal segment of a beetle; compound terminal segment of a scale insect; sensory dorsal plate of 9th abdominal segment of fleas; anal segment of annelids; tail shield of trilobites; *alt.* podex.

pygmy male pigmy male, *q.v.*

pygochord *n.* [Gk. *pygē*, rump; *chordē*, cord] a ventral median ridge-like outgrowth of intestinal epithelium in certain Enteropneusta.

pygofer *n.* [Gk. *pygē*, rump; L. *ferre*, to carry] the last abdominal segment in leaf hoppers.

pygostyle *n.* [Gk. *pygē*, rump; *stylos*, column] an upturned compressed bone at end of vertebral column of birds, formed by fusion of hindmost vertebrae; *alt.* ploughshare bone.

pykn- pycn-, *q.v.*

pylangium *n.* [Gk. *pylē*, gate; *anggeion*, vessel] proximal portion of a truncus arteriosus.

pylocyte *n.* [Gk. *pylos*, gateway; *kytos*, hollow] a pore cell at inner end of small funnel-shaped depression, the porocyte of certain sponges.

pylome *n.* [Gk. *pylōma*, gate] in certain Sarcodina, an aperture for emission of pseudopodia and intake of food.

pyloric *a.* [Gk. *pylōros*, gate-keeper] *pert.* or in region of pylorus; *appl.* artery, antrum, orifice, valve, vein; *appl.* sphincter between mid-gut and hind-gut of insects, between stomach and duodenum of vertebrates; *appl.* gland secreting mucus; *appl.* posterior region of gizzard in decapod crustaceans, and to ossicle in gastric mill.

pylorus *n.* [Gk. *pylōros*, gate-keeper] lower orifice of stomach, communicating with duodenum.

pyogenetic, pyogenic *a.* [Gk. *pyon*, pus; *gennaein*, to produce] pus-forming, *appl.* bacteria.

pyramid *n.* [L. *pyramis*, pyramid] a conical structure, protuberance, eminence, as of cerebellum, medulla oblongata, temporal bone, vestibule, kidney; pyramidal cell of cerebral cortex; a piece of the dental apparatus of echinoids.

pyramidal *a.* [L. *pyramis*, pyramid] conical, like a pyramid, *appl.* leaves, a carpal bone, brain cells, tract, lobes, processes, muscles.

pyramid of biomass a diagram of the total biomass at each level of a food chain, forming a pyramid.

pyramid of energy a diagram showing the energy available per unit time at each trophic level in an ecosystem, usually expressed in kilocalories per square metre per year.

pyramid of numbers the numerical relationships of a food chain, represented by numbers of organisms at different levels.

pyranose a monosaccharide in the form of 6-membered ring of 5 carbons and one oxygen atom; *cf* furanose.

pyrene *n.* [Gk. *pyrēn*, fruit stone] a seed surrounded by a hard body, forming a fruit stone or

kernel, often with several in one fruit; *alt.* ossiculum, putamen.

pyrenin *n.* [Gk. *pyrēn*, fruit stone] the substance of a true nucleolus; *alt.* paranuclein, prochromatin, pseudochromatin, pseudonuclein.

pyrenocarp *n.* [Gk. *pyrēn*, fruit stone; *karpos*, fruit] perithecium, *q.v.*; drupe, *q.v.*

Pyrenocarpeae *n.* [Gk. *pyrēn*, fruit stone; *karpos*, fruit] a group of ascolichens having perithecia.

pyrenoid *n.* [Gk. *pyrēn*, fruit stone; *eidos*, form] in certain algae and some liverworts, a proteinaceous body in the chloroplast, and which is the centre of starch formation. *a.* Nucleiform, *q.v.*

Pyrenomycetes *n.* [Gk. *pyrēn*, fruit stone; *mykēs*, fungus] a group of Euascomycetes, commonly called flask fungi, in which the ascocarp is a perithecium and is closed except for an aperture at the top; *alt.* Pyrenomycetidae.

pyrenophore *n.* [Gk. *pyrēn*, fruit stone; *pherein*, to bear] part of cytoplasm which contains the nucleus.

pyretic *a.* [Gk. *pyretos*, fever] increasing heat production; causing rise in body temperature.

pyridoxal phosphate a coenzyme which takes part in transaminase reactions and is derived from pyridoxine.

pyridoxin(e) a member of the vitamin B_6 group found in cereals, being a phenolic alcohol derived from pyridine, which can be converted to pyridoxal and pyridoxamine within the organism; broadly vitamin B_6, *q.v.*; *alt.* adermin.

pyriform *a.* [L. *pyrum*, pear; *forma*, shape] pear-shaped, *appl.* cells, spores, a muscle, a larval sensory organ in Bryozoa, an organ of larval molluscs, vestigial left vesicula seminalis of a nautiloid, a type of silk gland in spiders, etc.; piriform, *q.v.*

pyrimidines *n.plu.* [Altered form of *purine*] nitrogen-containing organic bases including thymine, cytosine, and uracil, which pair with purines in DNA and RNA, and whose derivatives are important in metabolism.

pyrophosphate two phosphate groups linked together.

Pyrosomida, Pyrosomatidae *n.* [L. *pyrum*, pear; Gk. *sōma*, body] an order of colonial salps having no larval stage and with only feeble muscles developing in complete rings around body; *alt.* Luciae.

Pyrotheria, pyrotheres *n.*, *n.plu.* [Gk. *pyr*, fire; *theriōn*, animal] an order of Eocene to Oligocene South American placental mammals with tusk-like teeth and tending to large size like elephants.

Pyrrophyceae *n.* [Gk. *pyrrhos*, tawny-red; *phykos*, seaweed] a class of algae having the carotenoids dinoxanthin and peridinin as well as chlorophyll, biflagellate cells, most cells containing membrane-bound pusules; *alt.* Dinoflagellatae.

Pyrrophyta *n.* [Gk. *pyrrhos*, tawny-red; *phyton*, plant] in some classifications, the Pyrrophyceae, *q.v.*, raised to division status.

pyruvic acid a 3-carbon acid produced at the end of the glycolysis process, which is an important intermediate in metabolic cycles and is the starting point of the Krebs' cycle.

pyxidiate *a.* [Gk. *pyxis*, box; *idion*, dim.] opening like a box by transverse dehiscence; *pert.*, or like, a pyxidium or a pyxis.

pyxidium *n.* [Gk. *pyxis*, box; *idion*, dim.] a cap-

pyxis

sular fruit which dehisces transversely, the top coming off as a lid; *alt.* pyxis.

pyxis *n.* [Gk. *pyxis*, box] a dilatation of podetium in lichens; pyxidium, *q.v.*

Q

Q_{10} temperature coefficient, *q.v.*

Q_{CO_2} the volume of carbon dioxide in microlitres gas at normal temperature and pressure given out per hour per milligram dry weight.

Q-disc anisotropic or A-disc, *q.v.*; Hensen's line, *q.v.*

Q-enzyme an enzyme which catalyses the transfer of a segment of 1,4-α D-glucan chain to primary hydroxyl group in a similar chain, e.g. the formation of amylopectin from amylose; EC 2.4.1.18; *r.n.* 1.4-α D-glucan branching enzyme.

Q_{O_2} oxygen quotient, *q.v.*

quadrangular *a.* [L. *quadrangulus*] *appl.* lobes or lobules of cerebellar hemispheres, connected by monticulus.

quadrant *n.* [L. *quadrans*, fourth part] all the cells derived by divisions from one of the first 4 cleavage cells or blastomeres.

quadrat *n.* [L. *quadratus*, squared] a sample area enclosed within a frame, usually a square, within which a plant community, or sometimes an animal community, is analysed; the frame itself.

quadrate *n.* [L. *quadratus*, squared] the bone with which lower jaw articulates in birds, reptiles, amphibians, and fishes; ligament extending from annular ligament to neck of radius; one of lobes of liver; lobe of cerebrum, the praecuneus. *a. Appl.* plates: paired sclerites at base of sting in Hymenoptera; *appl.* foramen in central tendon of diaphragm, for posterior or inferior vena cava.

quadratojugal *n.* [L. *quadratus*, squared; *jugum*, yoke] membrane bone connecting quadrate and jugal bones; *alt.* quadratomaxillary.

quadratomandibular *a.* [L. *quadratus*, squared; *mandibulum*, jaw] *pert.* quadrate and mandibulum.

quadratomaxillary quadratojugal, *q.v.*

quadratus *n.* [L. *quadratus*, squared] name of several rectangular muscles such as quadratus femoris, labii, lumborum, plantae.

quadricarpellary *a.* [L. *quattuor*, four; Gk. *karpos*, fruit] containing 4 carpels.

quadriceps *n.* [L. *quattuor*, four; *caput*, head] muscle in front of thigh, extending lower leg and divided into 4 portions at upper end.

quadrifarious *a.* [L. *quadrifariam*, fourfold] in 4 rows, *appl.* leaves.

quadrifid *a.* [L. *quattuor*, four; *findere*, to cleave] deeply cleft into 4 parts.

quadrifoliate *a.* [L. *quattuor*, four; *folium*, leaf] four-leaved; *appl.* compound palmate leaf, with 4 leaflets arising at a common point; *alt.* quadrinate, tetraphyllous.

quadrigeminal bodies corpora quadrigemina, *q.v.*

quadrihybrid *n.* [L. *quattuor*, four; *hibrida*, mongrel] a cross whose parents differ in 4 distinct characters. *a.* Heterozygous for 4 pairs of alleles.

quadrijugate *a.* [L. *quattuor*, four; *jugum*, yoke] *appl.* pinnate leaf having 4 pairs of leaflets.

quadrilateral *n.* [L. *quattuor*, four; *latus*, side] the discal cell in Zygoptera, dragonflies.

quadrilobate *a.* [L. *quattuor*, four; *lobus*, lobe] four-lobed.

quadrilocular *a.* [L. *quattuor*, four; *loculus*, compartment] having 4 loculi or chambers, as ovary, or anthers, of certain plants; *alt.* tetrathecal.

quadrimaculate *a.* [L. *quattuor*, four; *macula*, spot] having 4 spots.

quadrimanous quadrumanous, *q.v.*

quadrinate quadrifoliate, *q.v.*

quadripennate *a.* [L. *quattuor*, four; *penna*, wing] tetrapterous, *q.v.*

quadripinnate *a.* [L. *quattuor*, four; *pinnatus*, feathered] divided pinnately 4 times.

quadriradiate *a.* [L. *quattuor*, four; *radius*, ray] four-rayed; *alt.* tetractinal.

quadriserial *a.* [L. *quattuor*, four; *series*, row] arranged in 4 rows or series; *alt.* quadriseriate, tetraseriate, tetrastichous.

quadritubercular *a.* [L. *quattuor*, four; *tuberculum*, small hump] *appl.* teeth with 4 tubercles or cusps; *alt.* quadrituberculate.

quadrivalent *n.* [L. *quattuor*, four; *valere*, to be strong] association of 4 chromosomes held together by chiasmata between diplotene and metaphase of 1st division in meiosis.

quadrivoltine *a.* [L. *quattuor*, four; It. *volta*, time] having 4 broods in a year.

quadrulus *n.* [L. *quattuor*, four] a quadripartite ciliary organelle found in the gullet of certain ciliates such as *Paramecium*.

quadrumanous *a.* [L. *quattuor*, four; *manus*, hand] having hind-feet, as well as front-feet, constructed like hands, as most Primates except man; *alt.* quadrimanous.

quadrupedal *a.* [L. *quadrupes*, four-footed] having and walking on, 4 feet; *pert.* 4-footed animals.

quadruplex *a.* [L. *quadruplex*, fourfold] having 4 dominant genes, in polyploidy.

qualitative inheritance discontinuous variation, *q.v.*

quantasomes quantosomes, *q.v.*

quantitative inheritance continuous variation, *q.v.*

quantosomes *n.plu.* [L. *quantus*, how much; Gk. *sōma*, body] in chloroplasts, globular particles on the thylakoid membranes which may represent the functional units of photosynthesis; *alt.* quantasomes.

quartet *n.* [L. *quartus*, fourth] a group of 4 nuclei or cells resulting from meiosis; 4 cells derived from a segmenting ovum during cleavage; *cf.* tetrad.

quaternary *a.* [L. *quaterni*, four each] *appl.* flower symmetry when there are 4 parts in a whorl; tetragonal, *q.v.*

Quaternary *appl.* or *pert.* period or era after Tertiary, comprising Pleistocene and Holocene epochs.

quaternate *a.* [L. *quaterni*, four each] in sets of 4; *appl.* leaves growing in 4s from one point.

queen *n.* [A.S. *cwen*, woman] the reproductive female in colonies of social Hymenoptera.

queen-bee substance a pheronome given out by a developing queen bee, which inhibits the development of other queens.

quiescence *n.* [L. *quiescere*, to become still] temporary cessation of development, or of other activity, owing to unfavourable environment; *cf.* diapause.

quiescent centre a group of cells, with a low DNA content and few mitoses, between root meristem and root cap.

quill *n.* [M.E. *quille*, feather] the calamus or barrel of a feather; the calamus and rachis; a hollow spine, as of porcupine.

quill feathers feathers of wings (remiges) and tail (rectrices) of bird.

quill knobs tubercles or exostoses on ulna of birds, for attachment of fibrous ligaments connecting with quill follicle.

quillworts *see* Isoetales.

quinary *a.* [L. *quini*, five each] *appl.* flower symmetry when there are 5 parts in a whorl; *alt.* pentagonal.

quinate *a.* [L. *quini*, five each] *appl.* 5 leaflets growing from one point; *alt.* quinquefoliolate.

quincuncial *a.* [L. *quinque*, five; *uncia*, twelfth part] *pert.* or arranged in quincunx.

quincunx *n.* [L. *quinque*, five; *uncia*, twelfth part] arrangement of 5 structures of which 4 are at corners of a square and 1 at centre; arrangements of 5 petals or leaves, of which 2 are exterior, 2 interior, and the 5th partly exterior, partly interior, *alt.* quincuncial aestivation.

quinine *n.* [Sp. *quina*, cinchona bark] an alkaloid produced from the bark of species of *Cinchona* and used medicinally as an antimalarial drug and as a febrifuge.

quinone any of various compounds derived from benzene, which function in biological oxidation-reduction systems.

quinquecostate *a.* [L. *quinque*, five; *costa*, rib] having 5 ribs on the leaf.

quinquefarious *a.* [L. *quinque*, five; *fariam*, in rows] in 5 directions, rows, or parts.

quinquefid *a.* [L. *quinque*, five; *findere*, to cleave] cleft in 5 parts.

quinquefoliate *a.* [L. *quinque*, five; *folium*, leaf] with 5 leaves.

quinquefoliolate quinate, *q.v.*

quinquelobate *a.* [L. *quinque*, five; L.L. *lobus*, lobe] with 5 lobes.

quinquelocular *a.* [L. *quinque*, five; *loculus*, compartment] having 5 cavities or loculi.

quinquenerved *a.* [L. *quinque*, five; *nervus*, sinew] having the midrib divided into 5, giving 5 main veins; *alt.* quintuplinerved.

quinquepartite *a.* [L. *quinque*, five; *partitus*, divided] divided into 5 parts.

quinquetubercular *a.* [L. *quinque*, five; *tuberculum*, small hump] *appl.* molar teeth with 5 tubercles or cusps; *alt.* quinquetuberculate.

quintuplinerved quinquenerved, *q.v.*

R

rabbits *see* Lagomorpha.

race *n.* [F. *race*, family] a group of individuals within a species forming a permanent variety; a particular breed; jordanon, *q.v.*; [O.F. *rais*, from L. *radix*, root] a rhizome as of Zingiberaceae.

race history phylogeny, *q.v.*

racemation *n.* [L. *racemus*, bunch] a cluster, as of grapes.

raceme *n.* [L. *racemus*, bunch] inflorescence having a common axis and stalked flowers in acropetal succession, as hyacinth.

racemiferous *a.* [L. *racemus*, bunch; *ferre*, to carry] bearing racemes.

racemiform *a.* [L. *racemus*, bunch; *forma*, shape] in the form of a raceme.

racemose *a.* [L. *racemus*, bunch] *appl.* inflorescence whose growing points continue to add to the inflorescence and in which there are no terminal flowers, and the branching is monopodial, as racemes, spikes, *alt.* indefinite; *appl.* glands with many branches whose shape suggests a raceme; *alt.* botryose.

racemule *n.* [L. *racemulus*, small bunch] a small raceme.

racemulose *a.* [L. *racemulus*, small bunch] in small clusters.

rachi- *alt.* rhachi-.

rachial *a.* [Gk. *rhachis*, spine] *pert.* a rachis; *alt.* rachidial.

rachides *plu.* of rachis.

rachidial *a.* [Gk. *rhachis*, spine] rachial, *q.v.*

rachidian placed at or near a rachis; *appl.* median tooth in row of teeth of radula.

rachiform *a.* [Gk. *rhachis*, spine; L. *forma*, shape] in the form of a rachis.

rachiglossate *a.* [Gk. *rhachis*, spine; *glōssa*, tongue] having a radula with pointed teeth, as whelks.

rachilla *n.* [Gk. *rhachis*, spine; *dim.*] a small or secondary rachis; axis of spikelet, as in grasses.

rachiodont *a.* [Gk. *rhachis*, spine; *odous*, tooth] *appl.* egg-eating snakes with well-developed hypophyses of anterior thoracic vertebrae, which function as teeth.

rachiostichous *a.* [Gk. *rhachis*, spine; *stichos*, row] having a succession of somactids as axis of fin skeleton, as in dipnoans.

rachis *n.* [Gk. *rhachis*, spine] the vertebral column; a stalk or an axis; the shaft of a feather; median dorsal elevation of opisthosoma in trilobites.

rachitomous *a.* [Gk. *rhachis*, spine; *tomos*, cut] temnospondylous, *q.v.*

racket cells raquet mycelium, *q.v.*

radial *a.* [L. *radius*, ray] *pert.* radius; *pert.* ray of an echinoderm; *appl.* plates supporting oral disc of crinoids; *appl.* fibres supporting retina; *appl.* cleavage: with radial symmetry of blastomeres; *appl.* leaves or flowers growing out like rays from a centre. *n.* Somactid, *q.v.*; cross-vein of wing in insects.

radial apophysis a process on palp of male Arachnida, inserted into groove of epigynum during mating.

radiale *n.* [L. *radius*, ray] a carpal bone in line with radius.

radial notch lesser sigmoid cavity of coronoid process of ulna.

radial symmetry having a plane of symmetry about each radius or diameter; in flowers also called actinomorphic or regular; *cf.* bilateral symmetry.

radiant *a.* [L. *radians*, radiating] emitting rays; radiating; *pert.* radiants; *pert.* radiation. *n.* An organism or group of organisms dispersed from an original geographical location.

radiate

radiate *a.* [L. *radius*, ray] radially symmetrical; radiating, *appl.* sternocostal ligaments; stellate, *appl.* ligament connecting head of rib with 2 vertebrae and their intervertebral disc. *v.* To diverge or spread from a point; to emit rays.

radiate-veined veined in a palmate manner.

radiatiform *a.* [L. *radius*, ray; *forma*, shape] with radiating marginal florets.

radical *a.* [L. *radix*, root] arising from root close to ground, as basal leaves and peduncles. *n.* A group of atoms that does not exist in the free state but as a unit in a compound, as OH, NH_4, C_6H_5, etc.

radicant *a.* [L. *radicari*, to take root] with roots developing from stem; rooting.

radicate *a.* [L. *radicatus*, rooted] rooted; possessing root-like structures; fixed to substrate as if rooted.

radication *n.* [L. *radix*, root] the rooting pattern of a plant.

radicel *n.* [*Dim.* of L. *radix*, root] a small root; rootlet, *q.v.*

radicicolous radicolous, *q.v.*

radiciferous *a.* [L. *radix*, root; *ferre*, to carry] bearing roots.

radiciflorous *a.* [L. *radix*, root; *flos*, flower] with flowers arising at extreme base of stem, so apparently arising from root; *alt.* rhizanthous.

radiciform, radicine *a.* [L. *radix*, root; *forma*, shape] resembling a root.

radicivorous *a.* [L. *radix*, root; *vorare*, to devour] root-eating.

radicle *n.* [L. *radix*, root] the embryonic root below hypocotyl, being the lower part of tigellum; a small root.

radicle sheath coleorhiza, *q.v.*

radicolous *a.* [L. *radix*, root; *colere*, to inhabit] inhabiting roots; *alt.* radicicolous.

radicose *a* [L. *radix*, root] with large root.

radicular *a.* [L. *radix*, root] *pert.* a radicule or radicle.

radicule *n.* [L. *radix*, root] rootlet, *q.v.*

radiculose *a.* [L. *radix*, root] having many rootlets or rhizoids.

radii *plu.* of radius.

radiobiology *n.* [L. *radius*, ray; Gk. *bios*, life; *logos*, discourse] the study of the effects of radioactivity on living cells and organisms.

radiocarbon *n.* [L. *radius*, ray; *carbo*, charcoal] a radioactive isotope of carbon ^{14}C, used in chronological and physiological research.

radiocarpal *a.* [L. *radius* ray; L.L. *carpus*, wrist] *pert.* radius and wrist.

radioecology *n.* [L. *radius*, ray; Gk. *oikos*, household; *logos*, discourse] the study of radiation as affecting the relationship between living organisms and environment, and of the ecological effects and destination of radioisotopes; *alt.* radiation ecology.

radioiodine *n.* [L. *radius*, ray; Gk. *io-eidēs*, violet-coloured] a radioactive isotope of iodine, ^{131}I, used in studying the thyroid gland.

Radiolaria *n.* [L. *radiolus*, feeble sunbeam] an order of marine planktonic Sarcodina having usually symmetrical bodies, and a skeleton of siliceous spicules.

radiole *n.* [L. *radiolus*, small shuttle] a spine of sea urchins.

radiomedial *n.* [L. *radius*, ray; *medius*, middle] a cross vein between radius and medius of insect wing.

radiomimetic *a.* [L. *radius*, ray; Gk. *mimētikos*, imitative] resembling the effects of radiation, *appl.* chemicals inducing mutations.

radiophosphorous *n.* [L. *radius*, ray; Gk. *phôsphoros*, bringing light] a radioactive isotope of phosphorus, ^{32}P, used in physiological research and therapeutics.

radioreceptor *n.* [L. *radius*, ray; *receptor*, receiver] a terminal receptor organ for receiving light, or temperature, stimuli.

radioresistant offering a relatively high resistance to the effects of radiation.

radiosensitive sensitive to the effects of radiation.

radiospermic *a.* [L. *radius*, ray; Gk. *sperma*, seed] having seed which is circular in transverse section; *appl.* plants, especially fossils, having such seeds.

radiosymmetrical *a.* [L. *radius*, ray; Gk. *syn*, with; *metron*, measure] having similar parts similarly arranged around a central axis.

radioulna *n.* [L. *radius*, ray; *ulna*, elbow] radius and ulna combined, as a single bone.

radioulnar *a.* [L. *radius*, ray; *ulna*, elbow] *pert.* radius and ulna.

radius *n.* [L. *radius*, ray] a bone of arm or forelimb between humerus and carpals, in some vertebrates fused with ulna; barbule, of feather; one of radial depressions or markings on fish scales; a plate of Aristotle's lantern; an insect wing vein; radial area of disc in sea anemones; ray floret of composite flower; in radially symmetrical animals, a primary axis of symmetry; *plu.* radii.

radix *n.* [L. *radix*, root] a root; point of origin of a structure, as of aorta.

radula *n.* [L. *radere*, to scrape] a short and broad strip of membrane with longitudinal rows of chitinous teeth in mouth of most gastropods, *cf.* odontophore; a hyphal structure with numerous short lateral sterigmata bearing spores.

radulate, raduliferous *a.* [L. *radere*, to scrape] having a radula.

raduliform *a.* [L. *radere*, to scrape; *forma*, shape] like a radula or flexible file.

raffinose *n.* [F. *raffiner*, to refine] a trisaccharide found in sugar beet, cereals, and some fungi, which yields glucose, fructose, and galactose on hydrolysis.

Rafflesiales *n.* [*Rafflesia*, generic name] an order of dicots (Magnoliidae) comprising total parasites of the familes Rafflesiaceae and Hydnoraceae.

Rainey's corpuscles [G. *Rainey*, English morphologist] spores of *Sarcocystis*, an elongated sporozoan found in voluntary muscle fibres.

Rainey's tubes elongated sacs found in voluntary muscle which are the adult stages of certain Neosporidia; *alt.* Miescher's tubes.

raised bog a lens-shaped bog developed in fen basins where the climate is not too dry.

Rajiformes *n.* [*Raja*, generic name; L. *forma*, shape] an order of Selachii existing from Jurassic to recent times, and including the rays, banjo fishes, and saw fishes.

ramal *a.* [L. *ramus*, branch] belonging to branches; originating on a branch; *alt.* rameal.

Ramapithecus *n.* [*Rama*, Hindu epic hero; Gk. *pithēkos*, ape] a genus of fossil apes from the Miocene, which have hominid features and may be ancestral to the hominids.

ramate *a.* [L. *ramus*, branch] branched.

rameal ramal, q.v.

ramellose a. [L. *ramus*, branch] having small branches.

rament, ramenta see ramentum.

ramentaceous a. [L. *ramenta*, shavings] like a ramentum; covered with ramenta.

ramentiferous a. [L. *ramenta*, shavings; *ferre*, to carry] bearing ramenta.

ramentum n. [L. *ramenta*, shavings] one of the brown scale-like structures found on fern leaves, stems, and petioles, *alt.* palea; an elongated epidermal hair; *alt.* rament; *plu.* ramenta.

rameous a. [L. *rameus*, pert. branches] branched; *pert.* a branch.

ramet n. [L. *ramus*, branch] an individual member of a clone; *cf.* ortet.

rami plu. of ramus.

rami communicantes nerve fibres connecting sympathetic ganglia and spinal nerves; *sing.* ramus communicans.

ramicorn a. [L. *ramus*, branch; *cornu*, horn] having branched antennae, as some insects. n. Lateral horny sheath of mandible in birds.

ramiferous a. [L. *ramus*, branch; *ferre*, to bear] branched.

ramification n. [L. *ramus*, branch; *facere*, to make] branching; a branch of a tree, nerve, artery, etc.

ramiflory n. [L. *ramus*, branch; *flos*, flower] the state of flowering from the branches; *cf.* cauliflory; a. ramiflorous.

ramiform a. [L. *ramus*, branch; *forma*, shape] branch-like.

ramigenous ramiparous, q.v.

ramigerous a. [L. *ramus*, branch; *gerere*, to carry] bearing branches.

ramiparous a. [L. *ramus*, branch; *parere*, to beget] producing branches; *alt.* ramigenous.

ramoconidium n. [L. *ramus*, branch; Gk. *konis*, dust; *idion*, dim.] a fungal spore produced from a portion of a conidiophore.

ramose a. [L. *ramosus*, branching] much branched.

ramule n. [L. *ramulus*, twig] a small branch; *alt.* ramulus, ramuscule.

ramuliferous a. [L. *ramulus*, twig; *ferre*, to bear] bearing small branches.

ramulose, ramulous a. [L. *ramulus*, twig] with many small branches.

ramulus n. [L. *ramulus*, twig] ramule, q.v.

ramus n. [L. *ramus*, branch] any branch-like structure; part of chewing apparatus of rotifers; barb of feathers; mandible, or its proximal part, of vertebrates; branch of spinal nerve; in springtails, an individual part of the retinaculum, q.v.; *plu.* rami.

ramuscule ramule, q.v.

Ranales n. [*Ranunculus*, generic name] an order of dicots of the Archichlamydeae or Polypetalae having spiral, spirocyclic, or cyclic flowers and a large number of stamens and carpels.

rangiferoid a. [*Rangifer*, generic name of reindeer; Gk. *eidos*, form] branching like a reindeer's antlers.

ranine a. [L. *rana*, frog] pert. under surface of tongue, *appl.* artery and vein.

ranivorous a. [L. *rana*, frog; *vorare*, to devour] feeding on frogs.

Ranunculales n. [*Ranunculus*, generic name] an order of dicots of the Magnoliidae having many primitive features and including the families Ranunculaceae and Berberidaceae.

ranunculaceous a. [*Ranunculus*, generic name] an old term for anomocytic, *appl.* stomata; *pert.* a member of the dicot flower family Ranunculaceae.

Ranvier's nodes [*L.-A. Ranvier*, French histologist] constrictions or interruptions of medullary sheath of a nerve fibre.

raphe n. [Gk. *rhaphē*, seam] a seam-like suture, as junction line of some fruits; line of fusion of funicle and anatropous ovule; a slit-like line in diatom valves; line, or ridge, of perineum, scrotum, hard palate, medulla oblongata, etc.

raphides n.plu. [Gk. *rhaphis*, needle] minute needle-shaped crystals, frequently of calcium oxalate, formed as metabolic by-products in plant cells.

raphidiferous a. [Gk. *rhaphis*, needle; L. *ferre*, to carry] containing raphides.

Raphidioptera n. [*Raphida*, generic name; Gk. *pteron*, wing] an order of insects, commonly called snake flies, that have complete metamorphosis and a long prothorax.

raptatory a. [L. *raptare*, to rob] preying.

raptorial a. [L. *raptor*, robber] adapted for snatching or robbing, *appl.* birds of prey, *appl.* 2nd thoracopod of Malacostraca.

raquet mycelium hyphae enlarged at one end of each segment, small and large ends alternating; *alt.* racquet or racket mycelium or cells.

rasorial a. [L. *radere*, to scratch] adapted for scratching or scraping the ground, as fowls.

rassenkreis n. [Ger. *Rasse*, race; *Kreis*, circle] polytypic species; *alt.* rheogameon.

rastellus n. [L. *rastellus*, rake] a group of teeth on paturon of arachnid chelicera.

rate gene a gene which influences the rate of a developmental process; *alt.* rate factor.

Rathke's pouch [*M. H. Rathke*, German anatomist] diverticulum of buccal ectoderm in vertebrates, the commencement of prepituitary gland formation; sometimes called craniobuccal or neurobuccal pouch.

Ratitae n. [L. *ratis*, raft] Palaeognathae, q.v.

ratite a. [L. *ratis*, raft] having an unkeeled sternum; *cf.* carinate.

rattle n. [M.E. *ratelen*, to clatter] the sound-producing series of horny joints at end of rattlesnake's tail; *alt.* crepitaculum.

Rauber's layer or cells [*A. Rauber*, Estonian anatomist] covering layer of cells formed by part of trophoblast on embryonic ectoderm.

Raunkiaer's classification of life forms a method of classifying plants according to the position of their winter perennating buds, including chamaephytes, geophytes, helophytes, hemicryptophytes, hydrophytes, phanerophytes, therophytes.

Ravian or Rau's process [*J. J. Rau* or *Ravius*, Dutch anatomist] Folian process, q.v.

ray n. [L. *radius*, ray] a parenchymatous band penetrating from cortex towards centre of stem especially one formed by cambium and extending into secondary xylem and phloem; the stalk of a group of flowers in an umbel; one of bony spines supporting fins; division of a radiate animal, as arm of asteroid; one of straight uriniferous tubules passing from medulla through cortex of kidney (medullary rays).

ray-finned fishes the common name for the Actinopterygii, q.v.

ray florets the outermost florets of certain inflorescences, especially of some Compositae; cf. disc floret.

ray initial in vascular cambium, a cell that gives rise to medullary rays; cf. fusiform initial.

rDNA DNA which codes for rRNA.

RDP ribulose diphosphate, q.v.

reaction time latent period, q.v.

reaction type phenotype, q.v.

reaction wood wood modified by bending of stem or branches, apparently trying to restore the original vertical position, and including compression wood in conifers and tension wood in dicotyledons.

read abomasum, q.v.

recalcitrant a. [L. recalcitrare, to kick back] non-biodegradable, appl. organic, usually man-made compounds in the soil.

recapitulation theory theory that ontogeny tends to recapitulate phylogeny, that individual life history reproduces certain stages in life history of race; alt. biogenetic law, Haeckel's law, palingenesis, von Baer's law.

recapitulatory a. [L. re, again; L.L. capitulatus, arranged under heads] atavistic, q.v.; palaeogenetic, q.v.

Recent Holocene, q.v.

receptacle n. [L. recipere, to receive] an organ used as a repository; peduncle of a racemose inflorescence; point from which floral organs of a flower arise, alt. thalamus, torus, floral axis; modified end of thallus branch containing conceptacles in algae, or soredia in lichens; pycnidium, q.v.; sporophore, q.v.; terminal disc of mosses.

receptacular a. [L. recipere, to receive] pert. a receptacle of any kind; largely composed of the receptacle, as certain fruits.

receptaculum n. [L. receptaculum, reservoir] a receptacle of any kind.

receptaculum chyli the cavity in lower part of thoracic duct receiving lymph and chyle from vessels of hindlimbs and abdomen; alt. cisterna chyli, chylocyst; cf. thoracic duct.

receptaculum ovarum an internal sac in which ova are collected in oligochaetes such as earthworms.

receptaculum seminis spermatheca, q.v.

receptive spot small mucilaginous area adjacent to aperture in an ovum at which sperm enters; point of sperm entry into ovum; antheridial wall at point of contact with oogonium and of penetration of oosphere by fertilization tube; alt. receptive papilla.

receptor n. [L. receptor, receiver] part of cell which functions as an antibody in combining with outside molecules or haptophores; specialized tissue or cell sensitive to a specific stimulus, often in a sense organ; alt. ceptor.

recess n. [L. recessus, withdrawn] a fossa, sinus, cleft, or hollow space, as omental, optic, pineal recess; alt. recessus.

recessive a. [L. recessus, withdrawn] appl. character or gene possessed by one parent which in a hybrid is masked by the corresponding alternative or dominant character derived from the other parent; an allele which is not manifest in the phenotype of a heterozygote.

recessus recess, q.v.

reciprocal cross two crosses between the same pair of genotypes or phenotypes in which the sources of gametes are reversed in one cross.

reciprocal feeding trophallaxis, q.v.

reciprocal hybrids two hybrids, one descended from male of one species and female of another, the other from a female of first and a male of second, such as the mule and hinney.

reciprocal translocation a mutual exchange of sections between 2 chromosomes.

reclinate a. [L. reclinare, to lean] curved downwards from apex to base; appl. an ovule suspended from a funicle.

reclining a. [L. reclinare, to lean] leaning over; not perpendicular.

recombinant the organism or nucleus produced by recombination.

recombinant DNA DNA produced from recombination, especially that produced artificially from 2 different species by genetic engineering.

recombination n. [L. re, again; L.L. combinare, to combine] a rearrangement of linked genes as a result of crossing-over.

recombination frequency or value the total number of recombined gametes divided by the total number of gametes.

recon the smallest mutable unit in a gene, separable by recombination.

reconstitution n. [L. re, again; constituere, to put together] reassembly of isolated differentiated cells to form a new individual, as experimentally in sponges; alt. reconstruction.

recretion n. [L. re, again; secretion] the elimination by plants of substances in the same form as that in which they were absorbed from the soil.

recrudescence n. [L. re, again; crudescere, to become violent] state of breaking out into renewed activity; fresh growth from ripe part; a relapse.

recruitment n. [O.F. recruter from L. recrescere, to grow again] activation of additional motor neurones, causing increased reflex when stimulus of same intensity is continued; cf. facilitation.

rectal a. [L. rectus, straight] pert. rectum; appl. gland: a small vascular sac of unknown significance near end of gut in certain fishes; appl. columns: longitudinal folds of mucous membrane of anal canal, alt. anal columns, columns of Morgagni.

rectigradation n. [L. rectus, straight; gradatio, flight of steps] adaptive evolutionary tendency; a structure exhibiting an adaptive trend or sequence in evolution.

rectinerved a. [L. rectus, straight; nervus, nerve] with veins or nerves straight or parallel, appl. leaves.

rectipetality n. [L. rectus, straight; petere, to seek] autotropism, q.v.

rectirostral a. [L. rectus, straight; rostrum, beak] straight-beaked.

rectiserial a. [L. rectus, straight; series, row] arranged in straight or vertical rows.

rectivenous a. [L. rectus, straight; vena, vein] with straight veins.

rectogenital a. [L. rectus, straight; genitalia, genitals] pert. rectum and genital organs.

recto-uterine a. [L. rectus, straight; uterus, womb] appl. posterior ligaments of uterus.

rectovesical a. [L. rectus, straight; vesica, bladder] pert. rectum and bladder.

rectrices n.plu. [L. regere, to rule] the stiff tail

feathers of a bird, used in steering; *sing.* retrix; *a.* retrical.

rectum *n.* [L. *rectus*, straight] the posterior terminal part of the alimentary canal, leading to the anus.

rectus *n.* [L. *rectus*, straight] a name for a rectilinear muscle, as rectus femoris, rectus abdominis, etc.

recurrent *a.* [L. *re*, back; *currere*, to run] returning or re-ascending towards origin; reappearing at intervals.

recurrent sensibility sensibility shown by motor roots of spinal cord due to sensory fibres of sensory roots.

recurved *a.* [L. *recurvus*, bent back] bent backwards; *alt.* recurvate, retrocurved.

recurvirostral *a.* [L. *recurvus*, bent back; *rostum*, beak] with beak bent upwards.

recutite *a.* [L. *recutitus*, skinned] seemingly devoid of epidermis.

red algae the common name for the Rhodophyceae, *q.v.*

red blood corpuscle a blood corpuscle of vertebrates, non-nucleated in mammals, containing haemoglobin which colours it red, and which carries oxygen, in the form of oxyhaemoglobin, from the respiratory surface to tissue cells; *alt.* erythrocyte, sometimes called haematid, akaryocyte.

red body rete mirabile, *q.v.*

Redfieldiiformes *n.* [*Redfieldia*, generic name; L. *forma*, shape] an order of Triassic Chondrostei resembling the Perleidiformes.

red gland rete mirabile, *q.v.*

redia *n.* [F. *Redi*, Italian scientist] a larval stage of certain Trematoda.

redifferentiation *n.* [L. *re*, again; *differre*, to differ] in a cell or tissue, a reversal of differentiation and then differentiation into another type.

redintegration *n.* [L. *redintegrare*, to make whole again] restoration or regeneration of an injured or lost part.

red light light of wavelength of about 660 nm which is active in promoting responses of phytochrome.

red nucleus collection of nerve cells in tegmentum of mid-brain.

redox *a.* [reduction–oxidation] *pert.* mutual reduction and oxidation.

red spot rete mirabile, *q.v.*

reduced apogamy generative apogamy, *q.v.*

reducing power a general term for the presence of a cofactor in a cell in a reduced form, such as $NADH_2$, $NADPH_2$, which can give up its hydrogen to reduce organic compounds.

reductases enzymes which catalyse reductions; a group of oxidoreductases where hydrogen transfer from the donor is not readily demonstrated; *cf.* oxidases.

reduction *n.* [L. *reductus*, reduced] halving of number of chromosomes at meiosis; structural and functional development less complex than that of ancestry, *cf.* amplification; decrease in size, as in old age; decreasing the oxygen content or increasing the proportion of hydrogen in a chemical compound or adding an electron to an atom or ion; *cf.* oxidation.

reduction division the 1st meiotic division; sometimes used for the whole of meiosis.

reductionism *n.* [L. *reductus*, reduced] the doctrine that all biological events can be explained in terms of physical chemistry.

reduplicate *a.* [L. *re*, again; *duplicare*, to repeat] *appl.* aestivation in which margins of bud sepals or petals turn outwards at points of contact.

reduviid *a.* [L. *reduvia*, hangnail] *appl.* eggs of certain insects, protected by micropyle apparatus with porches.

reed abomasum, *q.v.*

refaunation *n.* [L. *re*, again; *faunus*, god of woods] reintroduction of animals, especially of a protozoan symbiotic fauna into an insect from which the fauna has been removed.

refection *n.* [L. *refectio*, restoration] feedback; control of output to cause inverse changes in input, negative when stabilizing, as parasympathetic system, positive when causing trend to maximum or zero, as sympathetic system; reingestion of incompletely digested food by certain animals, as eating faecal pellets, or in rumination, *alt.* pseudorumination.

referred *a.* [L. *referre*, to carry back] *appl.* sensation in a part of the body remote from the part acted upon primarily.

reflected *a.* [L. *reflectere*, to turn back] turned or folded back on itself.

reflector layer layer of cells on inner surface of photogenic tissue, as in fireflies; silvery reflecting plates above the argenteum in fishes.

reflex *a.* [L. *reflectere*, to turn back] reflected; involuntary, *appl.* reaction to stimulus. *n.* Function of reflex arc or arcs, being unit reaction or reaction pattern.

reflex action simplest expression of principles according to which nervous system acts, involuntary action on activation of reflex arc.

reflex arc the unit mechanism of nervous system, consisting of the organ in which the reaction starts (receptor), nervous path (conductor), and gland cells or muscle cells (effector).

reflex chain chain behaviour, *q.v.*

reflexed *a.* [L. *reflectere*, to turn back] curved or turned backwards.

refracted *a.* [L. *re*, back; *frangere*, to break] bent backwards at an acute angle.

refractory *a.* [L. *refractarius*, obstinate] unresponsive; *appl.* period after excitation during which repetition of stimulus fails to induce a response.

refugium *n.* [L. *refugere*, to flee away] an area in which a change affecting the surrounding region did not take place such as a mountain area that was not covered by ice in the Pleistocene, and so in which the flora and fauna survived.

regeneration *n.* [L. *re*, again; *generare*, to beget] renewal of a portion of body which has been injured or lost; reconstitution of a compound after dissociation; e.g. of rhodopsin.

regma *n.* [Gk. *rhēma*, fracture] a dry dehiscent fruit whose valves open by elastic movement, as in *Geranium* species.

regosols *n.plu.* [Gk. *rhēgos*, blanket; L. *solum*, soil] soils which are developed on fairly deep unconsolidated parent material such as dune sands or volcanic ash; *cf.* lithosols.

regression *n.* [L. *regressus*, returned] a reversal in the direction of evolution, sometimes causing progeny to revert to a less extreme condition than their parents, or sometimes used to explain the extinction of groups such as graptolites.

regular *a.* [L. *regula*, rule] actinomorphic, *q.v.*

regulator genes genes that control the action of other genes that are not closely linked to them by producing repressors that can block their expression.

regulon *n.* [L. *regulare*, to regulate] a group of genes which regulate the production of a protein, but are not usually linked as they are in an operon.

Reil, island of [*J. C. Reil*, German physician] *see* insula.

reinforcement *n.* [L. *re*, again; F. *enforcier*, to strengthen] an event that alters an animal's response to a stimulus, positive reinforcement being reward and negative reinforcement being punishment.

Reissner's membrane [*E. Reissner*, German physiologist] the membrana vestibularis, stretching from lamina spiralis ossea to outer cochlear wall of ear.

rejungant *a.* [L. *re*, again; *jungere*, to join] coming together again, *appl.* related but hitherto separated taxa when in the course of time their ranges come to rejoin.

rejuvenescence *n.* [L. *re*, again; *juvenescere*, to grow young] a renewal of youth; regrowth from injured or old parts; in cells, renewed life and vigour following on conjugation and interchange and fusion of nuclear and protoplasmic material; *alt.* rejuvenation.

relational coiling or spiral plectonemic coiling around one another of 2 chromosomes or chromatids; *alt.* orthospiral.

relaxation-time the period during which excitation subsides after removal of stimulus.

relaxin *n.* [L. *relaxare*, to loosen] a luteal hormone which produces relaxation of pelvic ligaments during pregnancy.

relay cell internuncial neurone, *q.v.*

release factors proteins that read codons for the ending of a sequence, and cause release of the polypeptide chain.

releaser *n.* [L. *relaxare*, to unloose] a stimulus or group of stimuli which activates an inborn tendency or pattern of behaviour, as of species-specific behaviour; *cf.* sign stimulus.

relic coil or spiral surviving cell of chromosome at telophase and prophase.

relict *a.* [L. *relictus*, abandoned] not functional but originally adaptive, *appl.* structures; surviving in an area isolated from main distribution area, owing to intervention of environmental events, e.g. of glaciation, *appl.* species.

Remak's fibres [*R. Remak*, German anatomist] non-medullated fibres, *q.v.*

Remak's plexus Meissner's plexus, *q.v.*

remex *n.*, **remiges** *plu.* [L. *remex*, rower] the large feathers or quills of a bird's wing, comprising primaries and secondaries.

remiped *n.* [L. *remus*, oar; *pes*, foot] having feet adapted for rowing motion.

remotor *n.* [L. *removere*, to draw back] a retractor muscle; *cf.* promotor.

renal *a.* [L. *ren*, kidney] *pert.* kidneys; *alt.* nephric.

renal columns cortical tissue between medullary pyramids of kidney; *alt.* Bertini's columns.

renal corpuscle Malpighian corpuscle, *q.v.*

renal portal *appl.* a system of circulation in which some returning blood passes through kidneys; *alt.* reniportal.

renaturation *n.* [L. *re*, again; *natura*, birth] the return of a denatured compound such as a protein or nucleic acid to its original configuration.

rendzina *n.* [Polish, rich limy soil] any of a group of rich, dark greyish-brown, limy soils of humid or subhumid grasslands, having a brown upper layer and yellowish-grey lower layers.

renes *n.plu.* [L. *ren*, kidney] kidneys.

renette a glandular excretory cell in nematodes.

reniculus *n.* [*Dim.* of L. *ren*, kidney] kidney lobe, comprising papillae, pyramid, and surrounding part of cortex.

reniform *a.* [L. *ren*, kidney; *forma*, shape] shaped like a kidney; *alt.* nephroid.

renin *n.* [L. *ren*, kidney] a protein formerly thought to be a hormone, now known to be an enzyme, secreted by the kidney which converts angiotensinogen to angiotensin and so raises the blood pressure so that ultrafiltration is carried out efficiently; EC 3.4.99.19; *r.n.* renin.

reniportal renal portal, *q.v.*

Renner complex a group of chromosomes which pass from generation to generation as a unit, as in the evening primrose, *Oenothera.*

rennet-stomach abomasum *q.v.*

rennin *n.* [A.S. *rennan*, to cause to run] a proteinase which catalyses the precipitation of cascinogen (casein) in milk to insoluble casein (paracasein), found in gastric juice and secreted by glandular hairs of some insectivorous plants; *alt.* chymase; EC 3.4.23.4; *r.n.* chymosin.

renopericardial *a.* [L. *ren*, kidney; Gk. *peri*, around; *kardia*, heart] *appl.* a ciliated canal connecting kidney and pericardium in higher molluscs.

repand *a.* [L. *repandus*, bent backwards] with undulated margin, *appl.* leaf; wrinkled, *appl.* colony of bacteria.

repandodentate *a.* [L. *repandus*, bent backwards; *dens*, tooth] varying between undulated and toothed.

repandous curved convexly.

reparative *a.* [L. *reparare*, to mend] restoring, *appl.* buds developing after injury to leaf.

repeat *n.* [L. *repetere*, to fetch back] duplication or further repetition of a chromosome segment owing to unequal crossing-over.

repent *a.* [L. *repens*, crawling] creeping along the ground.

repetitious DNA DNA consisting of groups of repeated nucleotide sequences; *alt.* repetitive DNA.

replacement name scientific name adopted as a substitute for one found invalid under the rule of the International Codes of Nomenclature.

repletes *n.plu.* [L. *repletus*, filled up] workers with distensible crops for storing and regurgitating honey dew and nectar, and constituting a physiological caste of honey ants; *alt.* plerergates.

replicase *n.* [L. *replicare*, to fold back] RNA replicase, an enzyme which causes RNA to be formed from precursors, but needs RNA to do so.

replicate *a.* [L. *replicare*, to fold back] doubled over on itself.

replicatile *a.* [L. *replicare*, to fold back] *appl.* wings folded back on themselves when at rest.

replication *n.* [L. *replicatio*, a folding back] duplication of a molecule such as DNA by copying from an existing molecule which acts as a tem-

plate; duplication of organelles such as mitochondria, chloroplasts, or nuclei.

replicator a segment of nucleic acid which controls a replicon.

replicon a segment of nucleic acid which replicates sequentially.

replum *n.* [L. *replum*, bolt] a wall, not the carpellary wall, formed by ingrowths from the placenta
and dividing a fruit into sections.

repressor *n.* [L. *repressus*, restrained] a compound which prevents the synthesis of a protein;
cf. inducer.

reproduction *n.* [L. *re*, again; *producere*, to lead
forth] the process of forming new individuals of a
species by sexual or non-sexual methods.

reptant *a.* [L. *reptare*, to creep] creeping, *appl.*
polyzoan colony with zooecia lying on substrate,
appl. gastropods.

Reptilia, reptiles *n., n.plu.* [L. *repere*, to crawl] a
class of amniote vertebrates having a dry horny
skin with scales, plates, or scutes, functional lungs
throughout life, one occipital condyle and a 4-
chambered heart.

reptiloid *a.* [L. *repere*, to crawl; Gk. *eidos*, form]
with characteristics of a reptile.

repugnatorial *a.* [L. *repugnare*, to resist] defensive or offensive, *appl.* glands and other structures.

reserve cellulose cellulose found in plant storage tissue, subsequently used for nutrition following germination.

reservoir *n.* [F. from L. *reservare*, to keep back] a
non-contractile space discharging into gullet of
some Mastigophora.

residual air volume of air remaining in lungs
after strongest possible breathing out.

residual meristem meristematic ring, *q.v.*

resilifer *n.* [L. *resilire*, to leap back; *ferre*, to carry]
projection of valve carrying the resilium; *alt.*
resiliophore.

resilium *n.* [L. *resilire*, to leap back] the horny
flexible hinge of a bivalve.

resin *n.* [L. *resina*, resin] any of various substances
of high molecular weight, including resin acids,
esters, and terpenes, which are found in mixtures
in plants and often exuded from wounds where
they may protect against insects and fungi as
they harden to amorphous vitreous solids; *cf.*
balsam.

resin canals ducts in bark, wood, mesophyll,
etc., particularly of conifers, lined with glandular
epithelium excreting essential oils, e.g. terpenes,
forming oxidation products, such as resin.

resiniferous *a.* [L. *resina*, resin; *ferre*, to carry]
resin-secreting.

resinous *pert.*, of, or like resin, *appl.* a class of
odours.

resistance factor a group of episomes conferring resistance to antibiotics on bacteria; *alt.* R-
factor.

respiration *n.* [L. *respiratio*, breathing] the processes by which energy is acquired in a living
organism or cell, by the breakdown of organic
molecules, especially hexose sugars with the release of waste carbon dioxide; *see also* aerobic
and anaerobic respiration.

respiratory centres medullary areas involved in
respiration.

respiratory enzymes enzymes involved in physiological oxidation–reduction processes, e.g. oxi-

dases, dehydrogenases, hydrases, peroxidases,
catalases.

respiratory index the amount of carbon dioxide
produced per unit of dry weight per hour.

respiratory heart a name given to auricle and
ventricle of right side of heart where there is no
direct communication between right and left
sides; *cf.* systemic heart.

respiratory movements any movements connected with the supply of oxygen to respiratory
surfaces and the removal of carbon dioxide, such
as the movements of the thorax and diaphragm in
mammals.

respiratory pigments pigments which form a
loose association with oxygen and carry it from
the respiratory surface to tissue cells, such as haemoglobin and other haemochromes; pigments
concerned with oxidation–reduction processes
such as cytochrome.

respiratory quotient (RQ) the ratio between
the volume of carbon dioxide produced and the
volume of oxygen used in respiration; *cf.* photosynthetic quotient.

respiratory sac a backward extension of the
suprabranchial chamber, its lumen dependent on
the cucullaris muscle, in certain air-breathing
teleosts.

respiratory substrates substances which can be
broken down by living organisms to yield
energy.

respiratory surface the surface at which gaseous exchange occurs between the environment
and the body, such as gill lamellae, alveoli of
lungs, etc.

respiratory trees in some echinoderms, a respiratory system consisting of a series of tubules arising just inside the anus into which water can be
drawn and expelled.

response *n.* [L. *respondere*, to answer] the activity
of an organism in terms of movement, hormone
secretion, enzyme production, etc., as a result of a
stimulus; the behaviour of an organism as a result
of fluctuations in the environment.

response latency the time interval between a
stimulus and response.

restibrachium *n.* [L. *restis*, rope; *brachium*,
arm] restiform body or inferior peduncle of cerebellum; *alt.* myelobrachium.

restiform *a.* [L. *restis*, rope; *forma*, shape] having
appearance of a rope, *appl.* 2 bodies of nerve fibres
on medulla oblongata, the inferior cerebellar
peduncles.

resting cell or nucleus one that is not undergoing mitosis or meiosis.

resting potential the potential difference existing across a nerve or muscle cell membrane when it
is not being stimulated, and which is usually about
-70 millivolts in a nerve cell; *cf.* action potential.

Restionales *n.* [*Restio*, generic name] an order of
monocots (Liliidae) having wind pollinated,
usually unisexual, reduced flowers.

restitution *n.* [L. *restitutio*, restoration] the formation of a single body by union of separate
pieces of tissue; the union of separated breaks;
regeneration; *appl.* nucleus resulting from failure
of 1st meiotic division.

resupinate *a.* [L. *resupinare*, to bend back] so
twisted that parts are upside down.

resupination *n.* [L. *resupinare*, to bend back] inversion, *q.v.*

rete n. [L. *rete*, net] a net or network; plexus, *q.v.*

retecious a. [L. *rete*, net] in form of a network; *alt.* retiary.

reteform retiform, *q.v.*

rete Malpighii Malpighian layer, *q.v.*

rete mirabile a small dense network of blood vessels formed by breaking up of chiefly arterial vessels, thought to be important in secretion and storage of oxygen, as in the swim bladder of fishes, and in certain mammals; *alt.* red body, gland, or spot, vasoganglion.

rete mucosum Malpighian layer, *q.v.*

retentate n. [L. *retenare*, to hold back] any substance retained by a semipermeable membrane during dialysis; *cf.* dialysate.

retial a. [L. *rete*, net] *pert.* a rete.

retiary a. [L. *rete*, net] making or having, a netlike structure; constructing a web; net-like, *alt.* retecious, retiform.

reticle a. [L. *reticulum*, small net] reticulum, *q.v.*

reticular a. [L. *reticulum*, small net] having interstices like network; *pert.* a reticulum.

reticular activating system a system of nerve fibres in the brain which transmit impulses from peripheral sense organs to higher centres of brain; *alt.* reticular formation.

reticular cells mesenchymal cells of bone marrow, lymph glands, and spleen, giving rise to granulocytes, lymphocytes, and monocytes.

reticular theory a modification of the fibrillar theory where the fibrils did form networks.

reticulate a. [L. *reticulatus*, latticed] like network; *appl.* nervation of leaf or insect wing; *appl.* thickening of cell wall; *appl.* species formation due to intercrossing between several lines; *alt.* netted.

reticule reticulum, *q.v.*

reticulin n. [L. *reticulum*, small net] a scleroprotein resembling collagen, occurring in fibres of reticular tissue.

reticulocyte n. [L. *reticulum*, small net; *kytos*, hollow] a small erythrocyte or proerythrocyte of reticular appearance when stained.

reticulo-endothelial a. [L. *reticulum*, small net; Gk. *endon*, within; *thēlē*, nipple] *appl.* cells, or stationary histiocytes of various organs, functioning as phagocytes in the production of antibodies, or in destroying erythrocytes; *appl.* system, or metabolic apparatus, consisting of reticulum and endothelial cells and of wandering histiocytes.

reticulopodia n.plu. [L. *reticulum*, small net; Gk. *pous*, foot] anastomosing thread-like pseudopodia, as of Foraminifera; *cf.* filopodia.

reticulose a. [L. *reticulum*, small net] of network formation.

reticulospinal a. [L. *reticulum*, small net; *spina*, spine] connecting reticular formation of the brain with spinal cord, *appl.* nerve fibres.

reticulum n. [L. *reticulum*, small net] delicate network of cell protoplasm; cross-fibres about base of petioles in palms; in ruminants, the 2nd chamber of the stomach in which water is stored, *alt.* honeycomb bag, honeycomb stomach; the framework of reticular tissue in many organs; *alt.* reticle, reticule.

retiform a. [L. *rete*, net; *forma*, shape] in form of a network; *alt.* reteform, reticulate, retiary.

retina n. [L. *rete*, net] the inner light-sensitive layer of eye, consisting of rods, or rods and cones.

retinaculum n. [L. *retinaculum*, tether] a small glandular mass to which an orchid pollinium adheres at dehiscence; a fibrous band which holds parts closely together; a minute hooked prominence holding egg sac in position in cirripedes; a structure linking together fore-and hind-wings of some insects; appendages modified to hold furcula beneath abdomen in springtails; *plu.* retinacula.

retinaculum tendinum annular ligament of wrist or ankle.

retinal a. [L. *rete*, net] *pert.* the retina. n. Retinene, *q.v.*

retine a natural constituent of body cells that retards the growth of cells in multicellular organisms.

retinella n. [*Dim.* of L. *rete*, net] neurofibrillar network of phaosome.

retinene n. [L. *retina*, retina] a carotenoid retinal pigment formed from visual yellow in dark-adapted eye, retinene$_1$ being a constituent of rhodopsin, retinene$_2$ of porphyropsin; *alt.* vitamin A aldehyde, retinal.

retinerved a. [L. *rete*, net; *nervus*, sinew] having reticulate veins or nerves.

retinoblasts n.plu. [L. *rete*, net; Gk. *blastos*, bud] retinal epithelial cells which give rise to neuroblasts and spongioblasts.

retinochrome n. [L. *rete*, net; Gk. *chrōma*, colour] a photopigment extracted from the rhabdome-free retina in squids.

retinol n. [L. *rete*, net] *see* vitamin A.

retinophore n. [L. *rete*, net; Gk. *pherein*, to bear] vitrella, *q.v.*

retinula n. [L. *rete*, net] group of elongated pigmented cells, innermost element of an ommatidium.

retisolution n. [L. *rete*, net; *solutio*, solution] dissolution of the Golgi apparatus.

retispersion n. [L. *rete*, net; *dispersio*, dispersion] peripheral distribution of Golgi apparatus in a cell.

retort-shaped organs glandular tissue at proximal ends of maxillary stylets, in Hemiptera.

retractile a. [L. *retractus*, withdrawn] *appl.* a part or organ that may be drawn inwards, as feelers, claws, etc.

retractor n. [L. *retrahere*, to draw back] a muscle which by contraction withdraws the part attached to it, bringing it towards the body; *cf.* protractor.

retrahens n. [L. *retrahere*, to draw back] a muscle which draws a part backwards, as the auricularis posterior.

retral a. [L. *retro*, backward] backward; posterior.

retroarcuate a. [L. *retro*, backwards; *arcuatus*, curved] curving backwards.

retrobulbar a. [L. *retro*, backwards; *bulbus*, bulb] posterior to eyeball; on dorsal side of medulla oblongata.

retrocaecal, retrocecal a. [L. *retro*, backwards; *caecus*, blind] behind caecum, *appl.* fossae.

retrocerebral a. [L. *retro*, behind; *cerebrum*, brain] situated behind the cerebral ganglion.

retrocurved a. [L. *retro*, backwards; *curvus*, bent] recurved, *q.v.*

retrofract a. [L. *retro*, backwards; *fractus*, broken] bent backwards at an angle.

retrogression n. [L. *retrogressus*, going back] a step from higher to lower type in individual or race; degeneration; a. retrogressive.

retrolingual a. [L. retro, backwards; lingua, tongue] behind the tongue, appl. a gland.

retromandibular a. [L. retro, behind; mandibula, jaw] appl. posterior facial or temporomaxillary vein.

retromorphosis n. [L. retro, backwards; Gk. morphōsis, form] development with degenerating tendency.

retroperitoneal a. [L. retro, backwards; Gk. peri, round; teinein, to stretch] behind peritoneum; appl. space between peritoneum and spinal column.

retropharyngeal a. [L. retro, backwards; Gk. pharyngx, pharynx] behind the pharynx, appl. a space, lymph glands.

retropubic a. [L. retro, backward; pubes, mature] appl. a pad or mass of fatty tissue behind pubic symphysis.

retrorse a. [L. retrorsum, backwards] turned or directed backwards; alt. retroverse; cf. antrorse.

retroserrate a. [L. retro, backwards; serra, saw] toothed, with teeth directed backwards; alt. runcinate.

retroserrulate a. [L. retro, backwards; serrula, small saw] with small retrorse teeth.

retrosiphonate a. [L. retro, backwards; Gk. siphōn, tube] with septal necks directed backwards.

retro-uterine a. [L. retro, backwards; uterus, womb] behind the uterus.

retroverse a. [L. retroversus, turned backwards] retrorse, q.v.

retroversion n. [L. retroversus, turned backwards] state of being reversed or turned backwards.

retuse a. [L. retusus, blunted] obtuse with a broad shallow notch in middle, appl. leaves, molluscan shells.

revehent a. [L. revehens, carrying back] in renal portal system, appl. vessels carrying blood back from excretory organs.

reverse mutation back mutation, q.v.

reverse transcriptase the enzyme that allows synthesis of DNA on an RNA template; alt. revertase, RNA-dependent DNA polymerase.

reverse transcription the synthesis of DNA from a messenger RNA template by matching the sequence of bases.

reversed a. [L. reversus, turned back] inverted; appl. a spiral shell whose turns are directed sinistrally; appl. barbs united to rachis by their apices.

reversion n. [L. reversio, turning back] atavism, i.e. a return in a greater or lesser degree to some ancestral type, as a return from cultivation or domestication to the wild state; back mutation, q.v.

revertant an allele, or the organism bearing it, that undergoes reverse mutation.

revertase reverse transcriptase, q.v.

revolute a. [L. revolvere, to roll back] rolled backwards from margin upon under surface, as some leaves; cf. involute.

reward positive reinforcement, q.v.

R-factor resistance factor, q.v.

Rh rhesus factor, q.v.

rhabdacanth n. [Gk. rhabdos, rod; akantha, thorn] in Rugosa, a compound trabecula consisting of small rod-like trabeculae wrapped around with lamellar tissue.

rhabdi plu. of rhabdus.

rhabdions n.plu. [Gk. rhabdos, rod] sclerotized ridges or rods in the cuticle lining the buccal cavity in some nematodes.

rhabdite n. [Gk. rhabdos, rod] one of short rod-like bodies in epidermal cells in Turbellaria and Temnocephaloidea; one of the gonapophyses, q.v.

Rhabditida, Rhabditata n. [Gk. rhabdos, rod] a large order of Phasmidia, many of which are parasites, some semiparasites, and some free living, which have an oesophagus with a large basal bulb.

rhabditiform a. [Gk. rhabdos, rod; L. forma, shape] appl. larvae of nematodes with short straight oesophagus, with double bulb.

rhabditis n. [Gk. rhabdos, rod] larva of certain nematodes.

Rhabdocoela n. [Gk. rhabdos, rod; koilos, hollow] an order of turbellarians whose intestine, if present, is simple and sac-like; alt. Rhabdocoelida.

rhabdocrepid a. [Gk. rhabdos, rod; krēpis, foundation] appl. a desma with uniaxial crepis, in sponge spicules.

rhabdoid a. [Gk. rhabdos, rod; eidos, form] rod-like. n. Any rod-shaped body.

rhabdolith n. [Gk. rhabdos, rod; lithos, stone] a calcareous rod found in some Protozoa, strengthening the walls.

rhabdom(e) n. [Gk. rhabdos, rod] a refractive rod composed of rhabdomeres enclosed by retinula cells of ommatidium, alt. optic rod; a rhabdus bearing a cladome, in sponges.

rhabdomere n. [Gk. rhabdos, rod; meros, part] the refracting element in a retinula.

Rhabdomonadales n. [Rhabdomonas, generic name] an order of Euglenophyceae which are colourless and saprophytic.

Rhabdopleurida n. [Gk. rhabdos, rod; pleura, side] an order of hemichordates of the class Pterobranchia which form true colonies with the individuals in organic union with each other.

rhabdopod n. [Gk. rhabdos, rod; pous, foot] an element of clasper of some male insects.

rhabdosome n. [Gk. rhabdos, rod; sōma, body] in graptolites, a colonial form produced from a single individual.

rhabdosphere n. [Gk. rhabdos, rod; sphaira, globe] aggregated rhabdoliths found in deep-sea calcareous oozes.

rhabdus n. [Gk. rhabdos, rod] a rod-like spicule; a stipe of certain fungi; plu. rhabdi.

rhachi- rachi-, q.v.

Rhachitomi, rhachitomes n., n.plu. [Gk. rhachis, spine; tomē, cutting] a group of temnospondyl labyrinthodont amphibians.

Rhaetic a. [L. Rhaetia, Grisons and Tirol] appl. fossils found in marls, shales, and limestone between Trias and Lias; alt. Rhaetian.

rhagiocrine a. [Gk. rhax, grape; krinein, to separate] appl. cells: histiocytes, q.v.

rhagon n. [Gk. rhax, grape] a bun-shaped type of sponge with apical osculum and large gastral cavity.

Rhamnales n. [Rhamnus, generic name] an order of dicots of the Rosidae or Archichlamydeae including the families Rhamnaceae, e.g. buckthorns, and Vitaceae, e.g. grape vines.

rhamphoid a. [Gk. rhamphos, beak; eidos, form] beak-shaped.

rhamphotheca n. [Gk. rhamphos, beak; thēkē, case] the horny sheath of a bird's beak.

rhapidosome n. [Gk. rhapis, rod; sōma, body] a group of rod-shaped ribonucleoprotein particles normally present in cells and resembling viruses but not known to be transmissible.

Rheiformes n. [Rhea, generic name; L. forma, shape] an order of birds of the subclass Neornithes, including the rheas.

Rhenanidiformes n. [Rhenanidus, generic name; L. forma, shape] in some classifications, an order of Devonian placoderms.

rheobase n. [Gk. rheein, to flow; basis, ground] the minimal or liminal electric stimulus that will produce a response; alt. rheobasis.

rheogameon n. [Gk. rheein, to flow; gamos, marriage; on, being] rassenkreis, q.v.

rheology n. [Gk. rheein, to flow; logos, discourse] the study of flow, e.g. of running waters, circulation of blood, etc.

rheophile, rheophilic a. [Gk. rheein, to flow; philos, loving] preferring to live in flowing water; n. rheophily.

rheoplankton n. [Gk. rheein, to flow; plangktos, wandering] the plankton of running waters.

rheoreceptors n.plu. [Gk. rheein, to flow; L. recipere, to receive] cutaneous sense organs of fishes and certain amphibians, receiving stimulus of water current, as pit organs, lateral-line organs, ampullae of Lorenzini, vesicles of Savi.

rheotaxis n. [Gk. rheein, to flow; taxis, arrangement] a taxis in response to the stimulus of a current, usually a water current; a. rheotactic.

rheotropism n. [Gk. rheein, to flow; tropē, turn] a growth curvature in response to a water or air current; a. rheotropic.

rhesus factor Rh factor, antigen in blood of rhesus monkey and man, and agglutinated by an (rh) antibody in individuals lacking the factor, which is inherited as a Mendelian dominant.

rhexigenous a. [Gk. rhēxis, a breaking; -genēs, born] resulting from rupture or tearing; alt. rhexogenous, lysigenous.

rhexilysis n. [Gk. rhēxis, a breaking; lysis, loosing] the separation of parts, or production of openings or cavities, by rupture of tissues; alt. rhexolysis.

rhexis n. [Gk. rhexis, a breaking] fragmentation of chromosomes, caused by physical or chemical agents.

rhexogenous rhexigenous, q.v.

rhexolysis rhexilysis, q.v.

rhigosis n. [Gk. rhigos, cold] sensation of cold.

rhinal a. [Gk. rhis, nose] of or pert. the nose; appl. fissure separating rhinencephalon, or olfactory lobe and tract, and cerebral hemisphere.

rhinarium n. [Gk. rhis, nose] the muzzle or external nasal area of mammals; nostril area; part of nasus of some insects.

rhinencephalon n. [Gk. rhis, nose; engkephalos, brain] the part of the fore-brain forming most of the hemispheres in fishes, amphibians, and reptiles, and comprising in man the olfactory lobe, uncus, the supracallosal, subcallosal and dentate gyri, fornix, and hippocampus.

rhinion n. [Gk. rhis, nose] most prominent point at which nasal bones touch.

rhinocaul n. [Gk. rhis, nose; kaulos, stalk] narrowed portion of brain which bears the olfactory lobe; alt. olfactory peduncle.

rhinocoel(e) n. [Gk. rhis, nose; koilos, hollow] cavity in olfactory lobe of brain.

rhinopharynx nasopharynx, q.v.

rhinophore n. [Gk. rhis, nose; pherein, to bear] in some molluscs, an organ made of sensory epithelium sometimes in a pit, usually found on the tentacles and thought to have an olfactory function.

rhinotheca n. [Gk. rhis, nose; thēkē, case] the sheath of upper jaw of a bird.

rhipidate a. [Gk. rhipis, fan] fan-shaped; alt. flabelliform.

Rhipidistia n. [Gk. rhipis, fan] an order of choanate crossopterygians existing from Devonian to Permian times and including ancestors of land vertebrates.

rhipidium n. [Gk. rhipis, fan; idion, dim.] a fan-shaped cymose inflorescence; a fan-shaped colony of zooids.

rhipidoglossate a. [Gk. rhipis, fan; glōssa, tongue] having a radula with numerous teeth in a fan-like arrangement, as ear shells.

rhipidostichous a. [Gk. rhipis, fan; stichos, row] appl. fan-shaped fins.

rhiptoglossate a. [Gk. rhiptos, thrown; glōssa, tongue] having a long, prehensile tongue, e.g. chameleon.

rhizanthous a. [Gk. rhiza, root; anthos, flower] radiciflorous, q.v.

rhizautoicous a. [Gk. rhiza, root; autos, self; oikos, house] with antheridial and archegonial branches coherent.

rhizine n. [Gk. rhiza, root] a rhizoid of a lichen made of hyphae either arranged singly or a number of parallel hyphae lying side by side.

rhizobia n.plu. [Gk. rhiza, root; bios, life] bacteria of the genus Rhizobium forming root nodules in leguminous plants.

rhizoblasts rhizoplasts, q.v.

rhizocaline n. [Gk. rhiza, root; kalein, to summon] a hormone, not an auxin, which may play a part in development of roots; cf. caulocaline.

rhizocarp n. [Gk. rhiza, root; karpos, fruit] a perennial herbaceous plant whose stems die down in winter so that it persists by underground organs only; a plant producing underground flowers; a. rhizocarpic, rhizocarpous.

rhizocaul n. [Gk. rhiza, root; caulos, stem] hydrorhiza, q.v.

Rhizocephala n. [Gk. rhiza, root; kephalē, head] an order of barnacles which are parasitic, mainly in decapod crustaceans, and have the appearance of absorptive roots.

Rhizochloridales n. [Rhizochloris, generic name] an order of Xanthophyceae having an amoeboid vegetative form.

Rhizochrysidales n. [Gk. rhiza, root; chrysos, gold] an order of Chrysophyta having rhizopodial and dendroid forms.

rhizocorm n. [Gk. rhiza, root; kormos, trunk] an underground stem like a single-jointed rhizome; a bulb or corm.

rhizodermis n. [Gk. rhiza, root; derma, skin] epiblema, q.v.

Rhizodiniales n. [Gk. rhiza, root; dinos, rotation] an order of Dinophyceae having amoeboid forms; alt. Dinamoebidiales.

rhizogenesis n. [Gk. rhiza, root; genesis, origin] differentiation and development of roots.

rhizogenic, rhizogenous a. [Gk. rhiza, root;

genos, descent] root-producing; arising from endodermic cells, not developed from pericycle; *pert.*, or stimulating, root formation.

rhizoid *n*. [Gk. *rhiza*, root; *eidos*, form] a unicellular or multicellular filamentous outgrowth from a thallus functioning like a root as in algae, bryophytes, and pteridophyte prothalli; a hypha functioning within a substrate. *a*. Root-like, *appl*. form of bacterial colony.

Rhizomastigina *n*. [Gk. *rhiza*, root; *mastix*, whip] an order of Zoomastigina having 1 or 2 flagella and whose whole body can be amoeboid; sometimes placed in the Rhizopoda.

rhizomatous *a*. [Gk. *rhizoda*, root] like a rhizome, *appl*. mycelium within a substratum or host.

rhizome *n*. [Gk. *rhizōma*, root] a thick horizontal stem usually underground, sending out shoots above and roots below; *cf*. stolon, rootstock.

rhizomorph *n*. [Gk. *rhiza*, root; *morphē*, form] a root-like or bootlace-like structure formed from interwoven hyphae in certain fungi.

rhizomorphic rhizomorphous, *q.v.*

rhizomorphoid *a*. [Gk. *rhiza*, root; *morphē*, form; *eidos*, particular kind] resembling a rhizomorph; brancing like a root.

rhizomorphous *a*. [Gk. *rhiza*, root; *morphē*, form] in form of a root; root-like; *pert.* a rhizomorph; *alt*. rhizomorphic.

rhizomycelium *n*. [Gk. *rhiza*, root; *mykēs*, fungus] a rhizoid mycelium connecting reproductive bodies in certain Phycomycetes.

rhizophagous *a*. [Gk. *rhiza*, root; *phagein*, to eat] root-eating.

rhizophore *n*. [Gk. *rhiza*, root; *pherein*, to bear] a naked branch which grows down into soil and develops roots from apex, as in club mosses.

rhizophorous *a*. [Gk. *rhiza*, root; *pherein*, to bear] root-bearing.

rhizopin *n*. [*Rhizopus*, a genus of Mucorales] a plant growth-promoting substance extracted from substrate of *Rhizopus* and probably identical with heterouxin.

rhizoplane *n*. [Gk. *rhiza*, root; L. *planus*, flat] the root surface.

rhizoplasts *n.plu.* [Gk. *rhiza*, root; *plastos*, moulded] fibrillae connecting parabasal body or blepharoplast and nucleus in Flagellata; intracytoplasmic portions of axonemes; *alt*. rhizoblasts.

Rhizopoda *n*. [Gk. *rhiza*, root; *pous*, foot] Sarcodina, *q.v.*; in some classifications, a subclass of Sarcodina.

rhizopodium *n*. [Gk. *rhiza*, root; *pous*, foot] a branching and anastomosing filamentous pseudopodium.

rhizosphere *n*. [Gk. *rhiza*, root; *sphaira*, ball] the soil immediately surrounding the root system of a plant.

Rhizostomeae *n*. [Gk. *rhiza*, root; *stoma*, mouth] an order of scyphozoans having medusae with no tentacles and oral lobes fused together sealing up the mouth, but with many small mouths on the oral lobes.

rhizotaxis *n*. [Gk. *rhiza*, root; *taxis*, arrangement] root arrangement.

Rhodesian man a subspecies of *Homo sapiens* known from fossils in southern Africa, living in the middle and upper Pleistocene.

Rhodochaetales *n*. [Gk. *rhodon*, rose; *chaitē*, hair] an order of red algae having tiny branched filaments.

rhodocyte erythrocyte, *q.v.*

rhodogenesis *n*. [Gk. *rhodon*, rose; *genesis*, origin] formation, or reconstitution after bleaching, of rhodopsin.

rhodophane *n*. [Gk. *rhodon*, rose; *phainein*, to appear] a red chromophane in retinal cones of some fishes, reptiles, and birds.

Rhodophyceae *n*. [Gk. *rhodon*, rose; *phykos*, seaweed] a class of algae, commonly called red algae, which have no flagella at any stages, whose photosynthetic pigments include phycocyanin and phycoerythrin, chlorophyll d and taraxanthin, and whose food storage products include floridean starch and floridoside; in some classifications raised to division status known as Rhodophyta.

rhodophyll *n*. [Gk. *rhodon*, rose; *phyllon*, leaf] the red colouring matter of red algae.

Rhodophyta *n*. [Gk. *rhodon*, rose; *phyton*, plant] in some modern classifications, the Rhodophyceae, *q.v.*, raised to division status.

rhodopin *n*. [Gk. *rhodon*, rose; *piein*, to absorb] a carotenoid pigment of certain bacteria.

rhodoplast *n*. [Gk. *rhodon*, rose; *plastos*, formed] a reddish plastid in red algae.

rhodopsin *n*. [Gk. *rhodon*, rose; *opsis*, sight] a reddish-purple pigment in retinal rods; *alt*. visual purple, erythropsin.

rhodosporous *a*. [Gk. *rhodon*, rose; *sporos*, seed] with pink spores.

rhodoxanthin *n*. [Gk. *rhodon*, rose; *xanthos*, yellow] a carotenoid pigment, found in aril of yew.

Rhodymeniales *n*. [*Rhodymenia*, generic name] an order of red algae in which the auxiliary cell does not differentiate as such until after fertilization.

Rhoeadales *n*. [*Rhoeo*, generic name] formerly an order of dicots of the Dilleniidae or Archichlamydeae, now usually divided into the Capparales and Papaverales due to differences in their cytology, serology, embryology, etc.

rhombencephalon *n*. [Gk. *rhombos*, rhomb; *engkephalos*, brain] hind-brain, consisting of the isthmus rhombencephali, metencephalon, and myelencephalon, derived from the 3rd embryonic or primary vesicle, *alt*. metepencephalon; the 3rd primary vesicle.

rhombic *a*. [Gk. *rhombos*, rhomb] *appl*. lip and grooves of brain at rhomboid fossa.

Rhombifera *n*. [Gk. *rhombos*, rhombus; L. *ferre*, to carry] an order of echinoderms of the subclass Hydrophoridea, having a diamond-shaped pattern of thecal pores.

rhombocoele *n*. [Gk. *rhombos*, rhombus; *koilos*, hollow] dilatation of the central canal of the medulla spinalis near its posterior end, the terminal ventricle.

rhombogen(e) *n*. [Gk. *rhombos*, rhomb; *-genēs*, producing] phase of parent form in life cycle of some Mesozoa, involving production of infusoriform embryos, or makes; *cf*. nematogene.

rhomboid *a*. [Gk. *rhombos*, rhombus; *eidos*, form] having the shape of a rhombus, i.e. a diamond in a pack of playing cards, *appl*. fossa, sinus, ligament, scales, etc.

rhomboidal rhomboid, *appl*. an apical plate in certain Dinoflagellata.

rhomboideum the rhomboid or costoclavicular ligament.

rhomboideus, major and minor parallel muscles connecting scapula with thoracic vertebrae.

rhomboid-ovate between rhomboid and oval in shape.

rhombopore *n.* [Gk. *rhombos*, rhombus; *poros*, channel] any of the pores making up a pore rhomb.

Rhombozoa *n.* [Gk. *rhombos*, rhombus; *zōon*, animal] *see* Dicyemida.

rhopalium *n.* [Gk. *rhopalon*, club] a marginal sense organ of Discomedusae.

rhopheocytosis *n.* [Gk. *rhophein*, to gulp; *kytos*, hollow] the process of taking large molecules into the interior of the cell, by the molecule being attached to the cell membrane which is then stimulated in some way to invaginate it.

Rhynchobdellae *n.* [Gk. *rhyngchos*, snout; *bdella*, leech] an order of leeches that are parasitic on fishes, amphibians, and reptiles, and have an eversible proboscis and no jaws.

Rhynchocephalia *n.* [Gk. *rhyngchos*, snout; *kephalē*, head] an order of lepidosaur reptiles from Triassic to present day with the tuatara as the only living genus.

rhynchocoel *n.* [Gk. *rhyngchos*, snout; *koilos*, hollow] in nemertines, a tubular cavity with muscular walls serving to evert the proboscis.

Rhynchocoela *n.* [Gk. *rhyngchos*, snout; *koilos*, hollow] Nemertini, *q.v.*

rhynchodaeum *n.* [Gk. *rhyngchos*, snout; *hodaios*, pert. a way] the precerebral region of a nemertine.

rhynchodont *a.* [Gk. *rhyngchos*, snout; *odous*, tooth] with a toothed beak.

rhynchoporous *a.* [Gk. *rhyngchos*, snout; *pherein*, to bear] beaked, *appl.* weevils.

rhynchostome *n.* [Gk. *rhyngchos*, snout; *stoma*, mouth] anterior terminal pore through which proboscis is everted, in nemertines.

Rhynchota *n.* [Gk. *rhynchos*, snout] Hemiptera, *q.v.*

rhythm *n.* [Gk. *rhythmos*, measured motion] regularity of movement as seen in heart beat, etc.; a cyclic variation in the intensity of a metabolic activity or some aspect of behaviour in an organism, the cycle being of a day, month, year, etc., *alt.* periodicity.

rhytidome *n.* [Gk. *rhytis*, wrinkle; *domos*, layer] the outer bark consisting of a periderm and tissues isolated by it.

rib *n.* [A.S. *ribb*, rib] an elongated protrusion; a curved bone of thorax articulating with spine and either free at other end or connected with sternum; primary or central veing of a leaf; *alt.* costa.

Ribaga's organ abdominal opening leading to Berlese's organ, *q.v.*

rib meristem one in which cells divide perpendicular to the longitudinal axis, producing a complex of parallel files or ribs of cells; *alt.* file meristem.

riboflavin *n.* [L. *ribes*, currant; *flavus*, yellow] a flavine derived from ribose, and a member of the vitamin B complex, vitamin B_2, occurring free in milk, etc., and combined as a flavoprotein enzyme and nucleotide in liver, yeast, green vegetables, etc., and which has the deficiency symptom of skin cracking and lesions, a condition known as ariboflavinosis, and is important in cell meta-

bolism as a coenzyme; *alt.* growth factor G, vitamin G, lactoflavin, lactochrome.

riboflavin phosphate FMN, *q.v.*

ribonuclease RNAse, *q.v.*

ribonucleic acid (RNA) a nucleic acid consisting of a polymeric chain of a backbone of alternating ribose sugar units and phosphate groups, with an organic base, adenine, guanine, uracil, or cytosine attached to each sugar unit; *see* messenger RNA, transfer RNA, ribosomal RNA; *cf.* deoxyribonucleic acid.

ribonucleoprotein (RNP) a macromolecule containing both RNA and protein.

ribonucleotide a nucleotide composed of ribose, a phosphate, and a purine or pyrimidine base.

ribose *n.* [Altered form of *arabinose*] a pentose sugar found in RNA and often combined with phosphate, acting as an intermediate in various metabolic cycles.

ribosomal RNA (rRNA) the kind of RNA which, together with protein, makes up the ribosomes.

ribosome *n.plu.* [*ribo*nucleic acid; Gk. *sōma*, body] particles made of RNA and protein which are associated with the endoplasmic reticulum and are the site of protein synthesis in the cell; *alt.* Palade granules, microsomes.

ribovirus *n.* [*ribo*nucleic acid; L. *virus*, poison] a virus containing RNA; *alt.* RNA virus.

ribulose a ketose pentose sugar, important in intermediate metabolism, especially as ribulose diphosphate (bisphosphate) in photosynthesis.

ribulose diphosphate (RDP) a 5-carbon compound which is thought to be the carbon dioxide acceptor in photosynthesis; now more properly called ribulose bisphosphate.

Ricinulei *n.* [*Ricinoides*, generic name] an order of arachnids having a thick warty cuticle and a hood on the prosoma.

rickettsiae *n.plu.* [*Rickettsia*, generic name] a group of intracellular parasites found on ticks, lice, mites, and fleas, which are prokaryotic and somewhat intermediate between bacteria and viruses; *alt.* rickettsias.

Rickettsiales rickettsiae, *q.v.*, when considered to be an order of bacteria; sometimes considered as both the rickettsiae and rickettsia-like organisms.

rickettsia-like organism (RLO) any of a group of organisms which resemble rickettsiae but are found in plants.

rictal *a.* [L. *rictus*, mouth aperture] *pert.* mouth gape of birds.

rictus *n.* [L. *rictus*, a gaping] opening or throat of calyx; gape of a bird's beak.

rigor *n.* [L. *rigor*, stiffness] the rigid state of plants when not sensitive to stimuli; contraction and loss of irritability of muscle on heating, due to coagulation of proteins.

rigor mortis stiffening of body after death, due to myosin formation, and lasting until commencement of decomposition.

rima *n.* [L. *rima*, cleft] a cleft or fissure; orifice of mouth; a slit-like ostiole.

rimate *a.* [L. *rima*, cleft] having fissures.

rimiform *a.* [L. *rima*, cleft; *forma*, shape] in shape of a narrow fissure.

rimose *a.* [L. *rima*, cleft] having many clefts or fissures.

rimulose *a.* [L.L. *rimula*, small cleft] having many small clefts.

rind *n.* [A.S. *rinde*, bark] the outer layer, tissue, or cortex.

ring *see also* annulus.

ring bark bark of a tree where formations of phellogen are cylindrical, and so which splits off in rings; *cf.* scale bark.

ring canal a circular canal running close to, and parallel with, umbrella margin in medusae; circular vessel around gullet in echinoderms.

ring cell a thick-walled cell of sporangium annulus of ferns.

ring centriole disc at end of body or middle portion of spermatozoon, perforated for axial filament; *alt.* end-ring, end-disc, terminal disc.

ring chromosomes chromosomes formed by fusion of ends of the centric fragment after breaks on opposite sides of centromere in the same chromosome; chromosomes with no ends in mitosis, attached end to end in meiosis.

ringed worms the common name for the Annelida, *q.v.*

ringent *a.* [L. *ringi*, to open mouth] having lips, as of a corolla, or valves, separated by a distinct gap; with upper lip arched; gaping.

ring gland glandular structure around aorta, with elements representing corpus allatum, corpus cardiacum, pericardial gland, and hypocerebral ganglion, secreting the metamorphosis-producing hormone in Diptera; *alt.* Weismann's gland or ring.

ringless *appl.* ferns without an annulus.

ring-porous *appl.* wood in which the vessels tend to be larger and have thinner walls than those in diffuse-porous wood; *appl.* wood in which the vessels of early wood are clearly larger than those of late wood, so forming a clear ring in cross-section.

ring species two species which overlap in range and behave as true species with no interbreeding between them, but are connected by a ring of subspecies so that no true specific separation can be made.

ring vessel a structure in head of cestodes, which unites the 4 longitudinal excretory trunks.

ripa *n.* [L. *ripa*, river bank] a line of ependymal fold over a plexus or a tela.

riparium, riparial, riparious *a.* [L. *ripa*, river bank] frequenting, growing on, or living on the banks of streams or rivers; *pert.* ripa.

risorius *n.* [L. *risus*, laughter] a cheek muscle stretching from over masseter muscle to corner of mouth; *alt.* Santorini's muscle.

ritualization *n.* [L. *ritus*, rite] the process in the course of evolution by which the activities of one animal (often displacement activities) come to cause certain responses in another animal of the same species, especially during courtship.

rivinian *a.* [A. Q. *Rivinus*, German anatomist] *appl.* sublingual glands and ducts; *appl.* notch in ring of bone surrounding tympanic membrane.

rivose *a.* [L. *rivus*, stream] marked with irregularly winding furrows or channels.

rivulose *a.* [L. *rivulus*, rivulet] marked with sinuate narrow lines or furrows.

RLO rickettsia-like organism, *q.v.*

RNA ribonucleic acid, *q.v.*

RNAase RNAse, *q.v.*

RNA-dependent DNA polymerase reverse transcriptase, *q.v.*

RNA polymerase transcriptase, *q.v.*

RNase any enzyme which hydrolyses RNA; *alt.* ribonuclease, RNAase.

RNA viruses riboviruses, *q.v.*

RNP ribonucleoprotein, *q.v.*

rod *see* rods.

Rodentia, rodents *n., n.plu.* [L. *rodere*, to gnaw] the largest order of placental mammals known from the Palaeocene and including the rats, mice, porcupines, beavers, and squirrels, having chisel-like incisors for gnawing and growing from persistent pulp, no canines, and being herbivores or omnivores with a large caecum.

rod epithelium epithelium consisting of apparently striated cells.

rod fibre fibre with which a rod of retina is connected internally.

rod fructification fructification occurring in Basidiomycetes by means of rod-like gonidia from a hyphal branch.

rod granule nucleus of rod fibre.

roding *n.* [A.S. *rode*, raid] patrolling flight of birds defending territory.

rods bacteria shaped as straight or slightly curved cylinders found singly or in chains of various lengths; rod-shaped, light-sensitive cells of retina responsible for non-colour vision and vision in poor light.

rods and cones nerve epithelium layer of retina.

rods of Corti Corti's rods, *q.v.*

rod vision dark-adapted or scotopic vision.

rogue sport, *q.v.*

rolandic *a.* [L. *Rolando*, Italian anatomist] *appl.* fissure or central sulcus of cerebral hemispheres; *appl.* tubercle or tuberculum cinereum of posterior region of medulla oblongata, and gelatinous substance of dorsal horn of spinal medulla.

root *n.* [A.S. *wyrt*, root] descending portion of plant, fixing it in soil, and absorbing water and minerals, and having a characteristic arrangement of vascular tissue; radix, *q.v.*; embedded part of hair, nail, tooth, or other structure; pulmonary veins and artery joining lung to heart and trachea; pedicle of vertebra; efferent and afferent fibres of a spinal nerve, leaving or entering the spinal cord.

root borer a larval form or insect which bores into the roots of plants.

root cap a protective cap of tissue at apex of root; *alt.* calyptra, pileorhiza.

root cell clear colourless base of an alga, attaching thallus to substratum.

root climber a plant which climbs by roots developed from stem.

root hairs unicellular epidermal outgrowths from roots, concerned with absorption.

rootlet an ultimate branch of a root; *alt.* radicel, radicule.

root nodules small swellings on roots of leguminous and other plants, containing nitrogen-fixing bacteria; *alt.* root tubercles, mycodomatia, bacteriorrhizae.

root parasitism a condition exhibited by semiparasitic plants, roots of which penetrate roots of neighbouring plants and draw from them elaborated food material.

root pocket a sheath containing a root, especially of aquatic plants.

root pressure a pressure developed in the roots of some plants which, if the shoot is cut off, causes fluid to exude from the root stump by way of the stele, but the tissues involved and the cause of the pressures are uncertain; *alt.* bleeding pressure.

root process a branched structure fixing an algal thallus to substratum.

root sheath coleorhiza, *q.v.*; an orchid velamen; that part of a hair follicle continuous with epidermis.

root stalk a rootstock or rhizome; root-like horizontal portion of Hydrozoa.

rootstock more-or-less erect underground part of stem; rhizome, *q.v.*

root tubercles root nodules, *q.v.*

root tubers swollen food-storing roots of certain plants such as lesser celandine and some orchids.

roridous *a.* [L. *ros*, dew] like dew; covered with droplets.

rosaceous *a.* [L. *rosa*, rose] resembling a rose; *pert.* dicot family Rosaceae which includes the roses.

Rosales *n.* [*Rosa*, generic name] an order of dicots of the Rosidae, Archichlamydeae, or Polypetalae, used in different ways by different authorities but always including the Rosaceae (e.g. roses).

rosellate *a.* [L. *rosa*, rose] arranged in rosettes; *alt.* rosular, rosulate.

Rosenmüller's organ [*J. C. Rosenmüller,* German anatomist] epoophoron, *q.v.*

rosette *n.* [F. from L. *rosa*, rose] a cluster of leaves arising in close circles from a central axis; a group of cells between embryo and proembryonic remains, also arrangement of embryos, as in pine; a plant disease due to deficiency of boron or of zinc; a cluster of crystals, as in certain plant cells; a swirl or vortex of hair in pelage; a small cluster of blood cells; group of spiracular channels in exocuticle of some aquatic insects; a thin plate formed by coalescence of interradial basals of larval crinoid; pellions of echinoids; a large ciliated funnel leading out of anterior sperm reservoir of earthworm; 2 circles of ciliated cells forming excretory organ in Ctenophora; *alt.* rosula.

rosette organ in certain ascidians, ventral complex stolon from which buds are constricted off.

Rosidae *n.* [*Rosa*, generic name] a subclass of dicots having cyclic polypetalous flowers with cup-shaped or disc-like receptacles; *alt.* Rosiflorae.

Rosiflorae *n.* [*Rosa*, generic name; L. *flos*, flower] Rosidae, *q.v.*

rosin *n.* [M.E. variant of resin] a resin obtained from pine exudates.

rostel *n.* [L. *rostellum*, dim. of *rostrum*, beak] rostellum, *q.v.*

rostellar *a.* [L. *rostellum*, small beak] *pert.* a rostellum.

rostellate *a.* [L. *rostellum*, small beak] furnished with a rostellum.

rostelliform *a.* [L. *rostellum*, small beak; *forma,* shape] shaped like a small beak.

rostellum *n.* [L. *rostellum*, small beak] a small rostrum; projecting structure developed from a stigmatic surface of orchid flowers; rounded prominence, furnished with hooks, on scolex of tapeworm; tubular mouth-parts of certain apterous insects; a beaked-shaped process; *alt.* rostel.

rostrad *adv.* [L. *rostrum*, beak; *ad,* toward] towards anterior end of body; *cf.* caudad.

rostral *a.* [L. *rostrum*, beak] *pert.* a rostrum.

rostral gland premaxillary part of labial gland, as in snakes; labral gland of spiders.

rostrate *a.* [L. *rostrum*, beak] beaked.

rostriform, rostroid *a.* [L. *rostrum*, beak; *forma,* shape; Gk. *eidos,* form] beak-shaped.

rostrulate *a.* [L.L. *rostrulum*, small beak] with, or like, a rostrulum.

rostrulum *n.* [L.L. *rostrulum*, small beak] a small rostrum.

rostrum *n.* [L. *rostrum*, beak] beak of birds; beak-like process; anterior end of gregarine, which forms epimerite; process projecting between eyes of crayfish; a median ventral plate at base of carapace of Cirripedia; labrum of spiders; prominence or mucro at posterior end of sepion; prenasal region; anterior continuation of basisphenoid; backward prolongation of anterior end of corpus callosum.

rosula *n.* [L. *rosa*, rose] rosette, *q.v.*

rosular, rosulate *a.* [L. *rosa*, rose] rosellate, *q.v.*

rotate *a.* [L. *rota*, wheel] shaped like a wheel; *alt.* rotiform.

rotation *n.* [L. *rota*, wheel] turning as on a pivot, as limbs; circulation, as of cell sap.

rotator *n.* [L. *rota*, wheel] a muscle which allows of circular motion.

rotatores spinae paired muscles, one on either side of thoracic vertebra, each arising from transverse process and inserted into vertebra next above.

rotatorium trochoid articulation or pivot joint.

Rotifera, rotifers *n., n.plu.* [L. *rota*, wheel; *ferre,* to carry] a phylum of aquatic microscopic pseudocoelomate animals, commonly called wheel animals, having the anterior end surmounted by a ciliated organ called a corona, which looks like a rotating wheel; the group in some classifications is considered a class of the phylum Aschelminthes.

rotiform *a.* [L. *rota*, wheel; *forma,* shape] rotate, *q.v.*; circular.

rotula *n.* [L. *rotula*, small wheel] one of 5 radially-directed bars bounding circular aperture of oesophagus of a sea urchin; patella or knee cap; *a.* rotular.

rotuliform *a.* [L. *rotula*, small wheel; *forma,* shape] shaped like a small wheel.

rotundifolious *a.* [L. *rotundus*, round; *folium,* leaf] with rounded leaves.

Rouget cells [*A. D. Rouget,* French physiologist] branched cells lying external to walls of capillaries and formerly thought to be responsible for their contraction; *cf.* pericytes.

rouleaux *n.plu.* [F. *rouleau,* roll] formations like piles of coins into which red blood corpuscles tend to aggregate.

round dance a dance of bees in which the bee dances in a circular motion which indicates that the new food source is close to the hive; *cf.* waggle dance.

round window fenestra rotunda, *q.v.*

roundworms the common name for the Nematoda, *q.v.*

royal jelly in bees, a secretion produced by special glands on the worker and used to feed new larvae that are to be queens.

RQ respiratory quotient, *q.v.*

rRNA ribosomal RNA, *q.v.*

rubber *n.* [M.E. *rubben,* to rub] coagulated latex

of several trees, mainly *Hevea* species, being hydrocarbons and polymers of isoprene.

rubiaceous *a.* [*Rubia*, generic name] paracytic, *q.v.*; *pert.* a member of the dicot flower family Rubiaceae.

Rubiales *n.* [*Rubid*, generic name] formerly an order of dicots of the Gamopetalae or Sympetalae, now usually divided into the Gentianales and Dipsacales.

rubiginose, rubiginous *a.* [L. *rubigo*, rust] of a brownish-red tint, rust-coloured; affected by rust parasites.

rubriblast *n.* [L. *ruber*, red; Gk. *blastos*, bud] immature proerythrocyte, *alt.* proerythroblast; promegaloblast, *q.v.*

rubricyte *n.* [L. *ruber*, red; Gk. *kytos*, hollow vessel] a polychromatophil erythroblast.

rubrospinal *a.* [L. *ruber*, red; *spina*, spine] *appl.* descending tract or fasciculus of axons of red nucleus, in ventrolateral column of spinal cord; *alt.* prepyramidal tract.

ruderal *a.* [L. *rudus*, debris] growing among rubbish or debris.

rudiment *n.* [L. *rudimentum*, first beginning] an initial or primordial group of cells which gives rise to a structure; a vestige (certain authors).

rudimentary *a.* [L. *rudimentum*, first attempt] in an imperfectly developed condition; at an early stage of development; arrested at an early stage; vestigial (certain authors).

ruff *n.* [A.S. *ruh*, rough] a neck fringe of hair or feathers.

Ruffini's organs [*A. Ruffini*, Italian anatomist] cylindrical end-bulbs containing interlaced branches of nerve endings, warmth receptors in subcutaneous tissue of finger; *alt.* corpuscles of Ruffini.

rufine *n.* [L. *rufus*, reddish] a red pigment in mucous glands of slugs.

rufinism *n.* [L. *rufus*, reddish] red pigmentation due to inhibition of formation of dark pigment; *alt.* rutilism; *cf.* erythrism.

ruga *n.* [L. *ruga*, wrinkle] a fold or wrinkle, as of skin, or of mucosa of certain organs.

rugate *a.* [L. *rugare*, to wrinkle] wrinkled; ridged.

rugose, rugous with many wrinkles or ridges.

rugulose *a.* [L. *ruga*, wrinkle] finely wrinkled.

ruling reptiles archosaurs, *q.v.*

rumen *n.* [L. *rumen*, cud] in ruminants, the 1st stomach in which food undergoes bacterial digestion and from which it can be regurgitated into the mouth for further chewing; *alt.* ingluvies, paunch.

ruminant *n.* [L. *ruminare*, to chew the cud] an animal which chews the cud.

ruminate *a.* [L. *ruminare*, to chew the cud] appearing as if chewed, *appl.* endosperm with infolding of testa or of perisperm, appearing mottled in section, *appl.* seeds having such endosperm, as betel nut and nutmeg. *v.* To chew the cud.

rumination *n.* [L. *ruminatio*, chewing of cud] the act of ruminant animals in returning food from 1st stomach to mouth in small quantities for thorough mastication and insalivation; *alt.* chewing the cud.

runcinate *a.* [L. *runcina*, plane] *appl.* a pinnatifid leaf when divisions point towards base, as in dandelion; *alt.* retroserrate.

runner *n.* [A.S. *rinnan*, to run] a specialized stolon consisting of a prostrate stem rooting at the node

and forming a new plant which eventually becomes detached from the parent, as in strawberry.

runoff the drainage of water from waterlogged or impermeable soil.

rupestrine, rupicoline, rupicolous *a.* [L. *rupes*, rock; *colere*, to inhabit] growing or living among rocks.

ruptile *a.* [L. *rumpere*, to break] bursting in an irregular manner.

rust *n.* [A.S. *rust*, redness] a disease of grasses and other plants caused by Uredinales, parasitic fungi which produce uredospores in summer, teleutospores in winter.

rust fungi the common name for the Uredinales, *q.v.*

rut *n.* [O.F. *ruit*, rut, from L. *rugire*, to roar] period of heat in male animals; *cf.* oestrus.

Rutales *n.* [*Ruta*, generic name] an order of dicots (Rosidae), sometimes considered a group of Terebinthales being mainly tropical woody plants secreting oils, resins, or balsams.

rutilant *n.* [L. *rutilus*, red] of a bright bronze-red colour.

rutilism *n.* [L. *rutilus*, red] rufinism, *q.v.*

Ruysch's penicilli *see* penicillus.

S

sabuline *a.* [L. *sabulum*, sand] sandy, *alt.* sabulose, sabulous; growing in sand, especially coarse sand.

sac *n.* [L. *saccus*, sack] a sack, bag, or pouch.

saccadic *a.* [F. *saccader*, to jerk] *appl.* brief movement of the eyes when suddenly looking at a different fixation point.

saccate *a.* [L. *saccus*, sack] pouched; gibbous, *q.v.*

saccharase *n.* [Gk. *sakchar*, sugar] invertase, *q.v.*

saccharide *n.* [Gk. *sakchar*, sugar] *see* monosaccharide, disaccharide, oligosaccharide, polysaccharide.

saccharobiose *n.* [Gk. *sakchar*, sugar; L. *bis*, twice] sucrose, *q.v.*

Saccharomycetales, Saccharomycetaceae *n.* [Gk. *sakchar*, sugar; *mykēs*, fungus] a group of fungi including the yeasts, sometimes given order status and also called Endomycetales, or sometimes reduced to a family and included in the Endomycetales, *q.v.*

saccharose *n.* [Gk. *sakchar*, sugar] sucrose, *q.v.*

sacciferous *a.* [L. *saccus*, sack; *ferre*, to bear] furnished with a sac.

sacciform *a.* [L. *saccus*, sack; *forma*, shape] like a sac or pouch; *alt.* saccular.

Saccopharyngiformes *n.* [L. *saccus*, sack; Gk. *pharynx*, gullet; L. *forma*, shape] an order or suborder of deep-sea teleosts, including the gulper eels, having a large mouth and capable of swallowing prey larger than themselves; *alt.* Lyomeri.

sacculate *a.* [L. *sacculus*, small bag] provided with sacculi.

sacculation *n.* [L. *sacculus*, small bag] the formation of sacs or saccules; a series of sacs, as of haustra of colon.

saccule, sacculus *n.* [L. *sacculus*, small bag] a

small sac; peridium, *q.v.*; lower part of vestibule of ear; appendix of laryngeal ventricle; lower portion of harpe.

sacculus rotundus dilatation between ileum and caecum, with chyle-retaining lymphoid tissue, in Lagomorpha.

saccus *n.* [L. *saccus*, sack] a sac-like structure, e.g. saccus vasculosus, saccus endolymphaticus, saccus lacrimalis (dacryocyst); 9th abdominal sternite of certain male insects; median invagination of vinculum in Lepidoptera.

Sacoglossa *n.* [Gk. *sakos*, shield; *glōssa*, tongue] an order of opisthobranchs that are herbivorous and suctorial with a modified radula.

sacral *a.* [L. *sacer*, sacred] *pert.* the sacrum.

sacral index one hundred times the breadth of sacrum at base, divided by anterior length.

sacralization *n.* [L. *sacer*, scared] fusion of lumbar and sacral vertebrae.

sacral ribs elements of sacrum joining true sacral vertebrae to pelvis.

sacrocaudal *a.* [L. *sacer*, sacred; *cauda*, tail] *pert.* sacrum and tail region.

sacrococcygeal *a.* [L. *sacer*, sacred; Gk. *kokkyx*, cuckoo] *pert.* sacrum and coccyx.

sacro-iliac *a.* [L. *sacer*, sacred; *ilia*, flanks] *pert.* sacrum and ilium, *appl.* joint, ligaments.

sacrolumbar *a.* [L. *sacer*, sacred; *lumbus*, loin] *pert.* sacral and lumbar regions.

sacrospinal *a.* [L. *sacer*, sacred; *spina*, spine] *pert.* sacral region and spine; *appl.* muscle: erector spinae; *appl.* ligament between sacrum and spine of ischium: sacrosciatic ligament.

sacrovertebral *a.* [L. *sacer*, sacred; *vertebra*, joint] *pert.* sacrum and vertebrae.

sacrum *n.* [L. *sacer*, sacred] the bone forming termination of vertebral column, usually of several fused vertebrae, *alt.* os sacrum; vertebra or vertebrae to which pelvic girdle is attached.

saddle clitellum, *q.v.*

Saefftigen's pouch in many acanthocephalans, an elongated pouch inside the genital sheath which injects fluid into the bursal cap to help eversion of the bursa.

sagitta *n.* [L. *sagitta*, arrow] an elongated otolith in sacculus of teleosts; a genus of arrow worms.

sagittae *n.plu.* [L. *sagitta*, arrow] the inner genital valves in Hymenoptera.

sagittal *appl.* the suture between parietals; *appl.* section or division in median longitudinal plane.

sagittate, sagittiform *a.* [L. *sagitta*, arrow] shaped like head of an arrow, *appl.* leaf.

sagittocyst *n.* [L. *sagitta*, arrow; Gk. *kystis*, bladder] a cyst or capsule in turbellarians, containing a single spindle.

salamanders *see* Urodela.

Salicales *n.* [*Salix*, generic name] an order of dicots of the Dilleniidae or Archichlamydeae, formerly placed in the Amentiferae, having unisexual dioecious flowers arranged in catkins and comprising the family Salicaceae, e.g. poplars and willows.

Salient(i)a *n.* [L. *saliens*, leaping] in some classifications, a superorder of Apsidospondyli amphibians, including the frogs and toads and containing one order, the Anura, *q.v.*

salinization *n.* [L. *salinus*, salty] the enrichment of the soil with salt.

saliva *n.* [L. *saliva*, spittle] the secretion produced by salivary glands, in mammals containing mucin and sometimes ptyalin, in insects containing various enzymes depending on diet, and in blood-sucking insects containing anticoagulants.

salivarium *n.* [L. *saliva*, spittle] recess of preoral food cavity, with opening of the salivary duct, in insects.

salivary *a.* [L. *saliva*, spittle] *pert.* saliva; *appl.* giant chromosomes conspicuous in salivary gland cells of Diptera; *appl.* amylase: ptyalin, *q.v.*

salivary glands glands opening into or near the mouth in various animals which secrete saliva; *alt.* sialadens.

Salmoniformes *n.* [*Salmo*, generic name of salmon; L. *forma*, shape] a large order of fairly primitive marine and freshwater teleosts, including salmon, trout, pike, etc.

Salpida(e) *n.* [*Salpa*, generic name] an order of salps having no larval stage, and strong often complex muscles forming incomplete rings around body; *alt.* Hemimyaria.

salpingian *a.* [Gk. *salpingx*, trumpet] *pert.* Eustachian or to Fallopian tube.

salpingopalatine *a.* [Gk. *salpingx*, trumpet; L. *palatum*, palate] *pert.* Eustachian tubes and palate.

salpinx *n.* [Gk. *salpingx*, trumpet] Eustachian tube, *q.v.*; Fallopian tube, *q.v.*; any of various trumpet-shaped structures.

salps the common name for the Thaliacea, *q.v.*

salsuginous *a.* [L. *salsugo*, saltiness] growing in soil impregnated with salts, as in a salt marsh.

saltant *n.* [L. *saltare*, to leap] an organism showing saltation.

saltation *n.* [L. *saltare*, to leap] an irreversible and inherited change in a cell or organism, which may be due to mutation or to some other mechanism.

saltatorial *a.* [L. *saltare*, to leap] adapted for, or used in, leaping, *appl.* limbs of jumping insects; *alt.* saltatory.

saltatory *a.* [L. *saltare*, to leap] saltatorial, *q.v.*; moving across a gap, as conduction at nodes of Ranvier in myelinated nerve fibres.

salted animals those which have survived certain diseases but remain infective and provide a source of material for preventive inoculation.

salt gland organ near eye in marine reptiles and birds for excretion of excess sodium chloride; a similar structure in gills of fishes; a hydathode from which salts are exuded from certain leaves.

saltigrade *a.* [L. *saltare*, to leap; *gradus*, step] moving by leaps, as some insects and spiders.

salt marsh the intertidal area on sandy mud in sheltered coastal areas and in estuaries, supporting characteristic plant communities.

Salviniales *n.* [*Salvinia*, generic name] an order of heterosporous leptosporangiate Pteropsida (Filicopsida) which are free-floating plants with or without roots and are not fern-like.

samara *n.* [L. *samara*, seed of elm] a winged indehiscent fruit, as of elm, ash, maple; *a.* samaroid, samariform.

SAN sinu-atrial node, *q.v.*

sand *n.* [M.E. *sond*, sand] a soil having most particles between 2 mm and 0·02 mm in size, made usually of silica, and being well drained and aerated.

sand dollars the common name for the Clypeasteroida, *q.v.*

sanguicolous *a.* [L. *sanguis*, blood; *colere*, to inhabit] living in blood.

sanguiferous *a.* [L. *sanguis*, blood; *ferre*, to carry] conveying blood, as arteries, veins.

sanguimotor *a.* [L. *sanguis*, blood; *movere*, to move] *pert.* circulation of blood.

sanguivorous *a.* [L. *sanguis*, blood; *vorare*, to devour] feeding on blood.

sanidaster *n.* [Gk. *sanidion*, panel; *astēr*, star] a slender rod-like spicule with spines at intervals.

Santalales *n.* [*Santalum*, generic name] an order of dicots of the Rosidae or Archichlamydeae, showing varying degrees of parasitism.

Santorini's cartilages [*G. D. Santorini*, Italian anatomist] corniculate cartilages, *q.v.*

Santorini's duct the accessory pancreatic duct; *cf.* Wirsung's duct.

Santorini's muscle risorius, *q.v.*

sap cavity vacuole in a plant cell.

saphena *n.* [Gk. *saphēnēs*, clear] a conspicuous vein of leg, extending from foot to femoral vein.

saphenous *a.* [Gk. *saphēnēs*, clear] *pert.* internal or external saphena, *appl.* a branch of femoral nerve.

sap hypha a laticiferous hypha.

sapients *n.plu.* [L. *sapere*, to be wise] the modern human phase of hominids, being of the species *Homo sapiens*, such as Neanderthal man, Cro-Magnon man, and modern living man.

Sapindales *n.* [*Sapindus*, generic name] an order of dicots of the Rosidae, Archichlamydeae, or Polypetalae, sometimes considered to be a group of the Terebinthales, being woody plants, and not secreting oils, balsams, and resins.

sapogenin *n.* [L. *sapo*, soap; Gk. *genos*, birth] the non-sugar part of a saponin sometimes found free in plants, but usually obtained from a saponin by hydrolysis.

saponin *n.* [L. *sapo*, soap] any of various steroid glycosides found in many plants, such as soap-bark and soapwort, which produce soapy foaming solutions in water, and on hydrolysis yield sugars and sapogenins.

saporiphore *n.* [L. *sapor*, taste; Gk. *pherein*, to bear] a radical or group of atoms in a compound, producing sensation of taste.

saprobe a saprobic organism.

saprobic *a.* [Gk. *sapros*, rotten; *bios*, life] living on decaying organic matter, *appl.* certain Protista; saprophytic or saprozoic, *q.q.v.; cf.* katharobic.

saprobiont *n.* [Gk. *sapros*, rotten; *biōnai*, to live] saprophage, *q.v.*

saprogenic *a.* [Gk. *sapros*, rotten; *-genēs*, producing] causing decay; resulting from decay.

Saprolegniales *n.* [*Saprolegnia*, generic name] an order of Phycomycetes having an extensive mycelial thallus without holdfast, asexual reproduction by biflagellate zoospores, and oogamous sexual reproduction.

sapropelic *a.* [Gk. *sapros*, rotten; *pēlos*, mud] living among debris of bottom ooze.

saprophage *n.* [Gk. *sapros*, rotten; *phagein*, to eat] a heterotrophic organism which feeds on dead organic matter, being a saprophyte or saprozoite; *alt.* saprotroph, saprobiont; *cf.* biophage; *a.* saprophagic.

saprophyte *n.* [Gk. *sapros*, rotten; *phyton*, plant] a plant which lives on dead and decaying organic matter; *alt.* humus plant, hysterophyte; *a.* saprophytic; *cf.* autophyte.

saprophyte chain a food chain starting with dead organic matter and passing to saprophytic micro-organisms; *cf.* parasite chain, predator chain.

saprophytic *pert.* saprophytes.

saprotroph *n.* [Gk. *sapros*, rotten; *trophē*, nourishment] saprophage, *q.v.; a.* saprotrophic.

saprozoic *pert.* saprozoites.

saprozoite *n.* [Gk. *sapros*, rotten; *zōon*, animal] an animal which lives on dead or decaying organic matter; *a.* saprozoic.

sapwood the more superficial, paler softer wood of trees that is water-conducting and contains living cells; *alt.* alburnum, splintwood; *cf.* heartwood.

sarcenchyma *n.* [Gk. *sarx*, flesh; *engchyma*, infusion] parenchyma whose ground substance is granular and not abundant.

sarcinaeform, sarciniform *a.* [L. *sarcina*, package; *forma*, shape] arranged in more-or-less cubical clumps, *appl.* cocci.

sarcocarp *n.* [Gk. *sarx*, flesh; *karpos*, fruit] the fleshy or pulpy part of a fruit; *cf.* mesocarp.

sarcocystin *n.* [Gk. *sarx*, flesh; *kystis*, bladder] a toxin derived from Sarcosporidia, a suborder of Neosporidia.

sarcocyte *n.* [Gk. *sarx*, flesh; *kytos*, hollow] the middle layer of ectoplasm in some Protozoa such as gregarines.

sarcode *n.* [Gk. *sarkōdēs*, like flesh] the body protoplasm of Protista; protoplasm in general.

sarcoderm *n.* [Gk. *sarx*, flesh; *derma*, skin] the fleshy layer between a seed and external covering; *alt.* sarcosperm.

sarcodic *a.* [Gk. *sarkōdēs*, fleshy] *pert.* or resembling protoplasm.

sarcodictyum *n.* [Gk. *sarx*, flesh; *diktyon*, net] the 2nd or network protoplasmic zone of Radiolaria.

Sarcodina *n.* [Gk. *sarkōdēs*, like flesh] a class of Protozoa having pseudopodia for at least part of their life history, usually free living, with little differentiation of body; *alt.* Rhizopoda.

sarcogenic *a.* [Gk. *sarx*, flesh; *-genēs*, producing] flesh-producing.

sarcoid *a.* [Gk. *sarx*, flesh; *eidos*, form] fleshy, as sponge tissue.

sarcolactic *a.* [Gk. *sarx*, flesh; L. *lac*, milk] *appl.* an acid in muscle, an isomer of lactic acid; *alt.* paralactic.

sarcolemma *n.* [Gk. *sarx*, flesh; *lemma*, skin] the tubular membranous sheath of a muscle fibre; *alt.* myolemma.

sarcolyte *n.* [Gk. *sarx*, flesh; *lytērios*, loosing] a non-nucleared muscle fragment undergoing phagocytosis in development of insects; a transient striated cell in thymus, *alt.* myoid cell.

sarcoma *n.* [Gk. *sarx*, flesh] a fleshy excrescence or tumour, usually malignant; sarcosoma, *q.v.*

sarcomatrix *n.* [Gk. *sarx*, flesh; L. *matrix*, dam] the 4th protoplasmic zone of a radiolarian, the seat of digestion and assimilation.

sarcomere *n.* [Gk. *sarx*, flesh; *meros*, part] a transverse portion of a myofibril between 2 Z-discs; *alt.* inocomma, comma, komma.

sarcophagous *a.* [Gk. *sarx*, flesh; *phagein*, to eat] flesh-eating.

sarcoplasm *n.* [Gk. *sarx*, flesh; *plasma*, mould]

the longitudinal interstitial substance between fibrils of muscular tissue; *cf.* myoplasm.

sarcoplasmic reticulum endoplasmic reticulum in striated muscle, where it surrounds myofibrils and is intimately connected with the process of contraction.

Sarcopterygii *n.* [Gk. *sarkōdēs*, fleshy; *pterygion*, little wing or fin] a subclass of the Osteichthyes including the Crossopterygii and Dipnoi, having fleshy fins and internal nostrils in some groups; *alt.* Choanichthyes.

sarcosoma *n.* [Gk. *sarx*, flesh; *sōma*, body] the fleshy as opposed to the skeletal portion of an animal's body; *alt.* sarcoma.

sarcosomes mitochondria in muscle cells.

sarcosperm *n.* [Gk. *sarx*, flesh; *sperma*, seed] sarcoderm, *q.v.*

sarcostyle *n.* [Gk. *sarx*, flesh; *stylos*, pillar] myofibril, *q.v.*; a dactylozooid column; *alt.* muscle column.

sarcotesta *n.* [Gk. *sarx*, flesh; L. *testa*, shell] softer fleshy outer portion of a testa; *cf.* sclerotesta.

sarcotheca *n.* [Gk. *sarx*, flesh; *thēkē*, box] the sheath of a hydrozoan sarcostyle.

sarcous *a.* [Gk. *sarx*, flesh] *pert.* flesh or muscle tissue.

sarcous disc A-band, *q.v.*

sargasterol *n.* [*Sargassum*, generic name of a brown seaweed; Gk. *stereos*, solid; alcoho*l*] a sterol found in some species of algae of the classes Phaeophyceae, Chlorophyceae, Rhodophyceae, Chrysophyceae, and Bacillariophyceae.

sarmentaceous *a.* [L. *sarmentum*, twig] having slender prostrate stems or runners; *alt.* sarmentose, sarmentous.

sarmentum *n.* [L. *sarmentum*, twig] the slender stem of a climber or runner.

sarothrum *n.* [Gk. *saron*, broom; *throna*, flowers] pollen brush, *q.v.*

Sarraceniales *n.* [*Sarracenia*, generic name] an order of dicots of the Rosidae or Archichlamydeae having many carnivorous members, such as pitcher plants, sundew, Venus' fly trap.

sartorius *n.* [L. *sartor*, tailor] a thigh muscle which enables legs to be moved inwards.

satellite *n.* [L. *satelles*, attendant] the 2nd of any pair of individuals of a catenoid colony in pseudo-conjugation of Gregarinida, *cf.* primite; trabant, *q.v.*; *appl.* cells closely applied to others, as Schwann's sheath to medullary sheath or as astrocytes and oligodendrocytes to nerve fibres; *appl.* a minute body adjacent to nucleolus and containing DNA, as in nerve cells.

saturnine *a.* [L. *Saturnus*, planet Saturn] forming, having, or *pert.* an equatorial ring, *appl.* arrangement of chromosomes, *appl.* rim on spores.

Sauria, saurians *n.*, *n.plu.* [Gk. *sauros*, lizard] in some earlier classifications, a group of reptiles including lizards, crocodiles, and some extinct forms which superficially resembled lizards; now used in some classifications as a suborder of Squamata coextensive with the Lacertilia (lizards).

saurian *a.* [Gk. *sauros*, lizard] *pert.* or resembling a lizard.

Saurichthyformes *n.* [Gk. *sauros*, lizard; *ichthys*, fish; L. *forma*, shape] an order of Triassic to Jurassic predaceous Chondrostei which were elongated and had a long snout.

Saurischia *n.* [Gk. *sauros*, lizard; *ischion*, hip] an order of Mesozoic archosaurs, commonly called lizard-hipped dinosaurs, including both bipedal carnivores and very large quadrupedal herbivores.

saurognathous *a.* [Gk. *sauros*, lizard; *gnathos*, jaw] with a saurian arrangement of jaw bones.

sauroid *a.* [Gk. *sauros*, lizard; *eidos*, form] resembling a saurian or part of a saurian; *appl.* cells: normoblasts, *q.v.*

Sauropsida *n.* [Gk. *sauros*, lizard; *opsis*, appearance] collectively, reptiles and birds; *cf.* Ichthyopsida.

Sauropterygia, sauropterygians *n.*, *n.plu.* [Gk. *sauros*, lizard; *pterygion*, little fin or wing] a group of extinct Mesozoic reptiles, sometimes included in the subclass Synaptosauria, that were amphibious or marine, and includes the nothosaurs, plesiosaurs, and placodonts.

savanna *n.* [Sp. *sabana*] subtropical or tropical grassland with drought-resistant vegetation and scattered trees; transitional zone between grasslands and tropical rain forests.

saxatile *a.* [L. *saxatilis*, found among rocks] lithophilous, *q.v.*

saxicavous *a.* [L. *saxum*, rock; *cavus*, hollow] rock-boring, as some molluscs; *alt.* lithophagous.

saxicoline, saxicolous *a.* [L. *saxum*, rock; *colere*, to inhabit] lithophilous, *q.v.*

Saxifragales *n.* [*Saxifraga*, generic name] an order of dictos (Rosidae) having endospermous seeds.

scaberulous *a.* [L. *scaber*, rough] somewhat rough.

scabrate, scabrous *a.* [L. *scaber*, rough] rough with a covering of stiff hairs, scales, or points.

scala *n.* [L. *scala*, ladder] any of 3 canals in cochlea of ear: vestibuli, tympani, media.

scalariform *a.* [L. *scala*, ladder; *forma*, shape] ladder-shaped; *appl.* vessels or tissues having bars like a ladder; *appl.* series of pits in cell walls; *appl.* conjugation between opposite cells of parallel filaments, as in *Spirogyra*; *alt.* scaliform.

scale *n.* [A.S. *sceala*, shell, husk] a flat, small, plate-like external structure, dermal or epidermal; a bony, horny, or chitinous outgrowth; bract of a catkin; ligule of certain flowers; modification of a stellate hair on certain leaves. [L. *scala*, ladder] a graduated measure; range of frequency, as of audible wavelengths.

scale bark bark in irregular sheets or patches, due to irregular or dipping formation of phellogen; *cf.* ring bark.

scale leaf a bud-protecting cataphyllary leaf; any reduced membranous leaf.

scalene *a.* [Gk. *skalēnos*, uneven] *pert.* scalenus muscles; *appl.* tubercle on 1st rib, for attachment of scalenus anticus or anterior.

scalenus *n.* [Gk. *skalēnos*, uneven] one of 3 neck muscles; scalenus posticus, medius, anticus.

scalids *n.plu.* [Gk. *skaleuein*, to hoe; *idion*, dim.] spines arranged in a series of rings around mouth cones in Kinorhyncha.

scaliform *a.* [L. *scala*, ladder; *forma*, shape] scalariform, *q.v.*

scalpella *n.plu.* [L. *scalpellum*, small knife] paired pointed processes, parts of maxillae of Diptera.

scalpriform *a.* [L. *scalprum*, chisel; *forma*, shape] chisel-shaped, *appl.* incisors of rodents.

scalprum *n.* [L. *scalprum*, chisel] the cutting edge of an incisor.

scandent *a.* [L. *scandere*, to climb] climbing by stem roots or tendrils; trailing, as grasses over shrubs.

scansorial *a.* [L. *scandere*, to climb] formed or adapted for climbing; habitually climbing.

scape *n.* [Gk. *skapos*, stalk] a flower stalk arising at or under ground; a radical peduncle, as cowslip; a structure formed by 2 basal segments of antennae of some insects as Diptera; an epignal structure protecting vulva in spiders; scapus, *q.v.*

scapha *n.* [L. *scapha*, boat] narrow curved groove between helix and antihelix of ear.

scaphium *n.* [Gk. *skaphion*, small boat] process of 9th (copulatory) segment of male Lepidoptera; anterior Weberian ossicle; keel of leguminous flower; a boat-shaped structure.

scaphocephalic *a.* [Gk. *skaphē*, boat; *kephalē*, head] with narrow, elongated skull.

scaphocerite *n.* [Gk. *skaphē*, boat; *keras*, horn] boat-shaped exopodite of 2nd antenna of Decapoda.

scaphognathite *n.* [Gk. *skaphē*, boat; *gnathos*, jaw] epipodite of 2nd maxilla of Decapoda, regulating flow of water through respiratory chamber; *alt.* bailer, baler.

scaphoid *a.* [Gk. *skaphē*, boat; *eidos*, form] shaped like a boat, *appl.* carpal and tarsal bones, *appl.* fossa above pterygoid fossa; *alt.* navicular, cymbiform. *n.* Os naviculare.

scapholunar *a.* [Gk. *skaphē*, boat; L. *luna*, moon] *pert.* scaphoid and lunar carpal bones, or those bones fused; *alt.* scapholunatum.

Scaphopoda *n.* [Gk. *skaphē*, boat; *pous*, foot] a class of marine molluscs, commonly called tusk shells or elephant-tooth shells, which have a tubular shell, a reduced foot, and no ctenidia.

scapiform, scapigerous, scapoid, scapose *a.* [Gk. *skapos*, stalk; L. *forma*, shape] resembling or consisting of a scape.

scapula *n.* [L. *scapula*, shoulder blade] the shoulder blade, i.e. dorsal part of pectoral girdle; name given to various structures suggestive of a shoulder blade, as tegula, patagium, mesothoracic pleuron, foreleg trochanter of certain insects; in Crinoidea, proximal plate of ray that has an articular facet for arms.

scapular *a.* [L. *scapula*, shoulder blade] *pert.* scapula. *n.* A feather growing from shoulder and lying laterally along back.

scapulus *n.* [L. *dim.* of *scapus*, stem] modified submarginal region in certain sea anemones.

scapus *n.* [L. *scapus*, stem, stalk] scape, *q.v.*; stem of feather; hair shaft; part of column below, and including, parapet in sea anemones.

scarab(a)eiform *a.* [L. *Scarabaeus*, a genus of beetles; *forma*, form] *appl.* C-shaped larval type of certain beetles.

scarious *a.* [F. *scarieux*, membranous] thin, dry, membranous; scaly or scurfy.

Scarpa's fascia [*A.* *Scarpa*, Italian anatomist] deep layer of superficial abdominal fascia.

Scarpa's foramina two openings, for nasopalatine nerves, in middle line of palatine process of maxilla.

Scarpa's ganglion vestibular ganglion in internal ear.

Scarpa's triangle the femoral triangle formed by adductor longus, sartorius, and inguinal ligament.

scatophagous *a.* [Gk. *skatophagous*, dung-eating] coprophagous, *q.v.*

scent scales androconia, *q.v.*

schemochromes *n.plu.* [Gk. *schēma*, shape; *chrōma*, colour] physical or structural colours evoked by material devoid of pigment coloration but arranged in very small ultramicroscopic states of subdivision; *a.* schemochromic.

schindylesis *n.* [Gk. *schindylein*, to cleave] articulation in which a thin plate of bone fits into a cleft or fissure, as that between vomer and palatines; *alt.* wedge-and-groove suture.

schistocytes *n.plu.* [Gk. *schizein*, to cleave; *kytos*, hollow] fragments of erythrocytes; blood corpuscles undergoing fragmentation; microcytes, *q.v.*; poikilocytes, *q.v.*

Schistostegales *n.* [*Schistostega*, generic name] an order of mosses of subclass Bryidae, including the luminous moss *Schistostega*.

schizocarp *n.* [Gk. *schizein*, to cleave; *karpos*, fruit] a dry fruit formed from a syncarpous ovary which dehisces by splitting into 1-seeded portions (mericarps).

schizocarpic *appl.* dry fruits which split into 2 or more mericarps, as carcerulus, cremocarp, lomentum, regma, compound samara.

schizocele schizocoel, *q.v.*

schizochroal *a.* [Gk. *schizein*, to cleave; *chrōs*, body surface] with lenses separate and cornea not continuous, *appl.* certain trilobite eyes.

schizocoel(e) *n.* [Gk. *schizein*, to cleave; *koilos*, hollow] coelom formed by splitting of mesoblast into layers; *alt.* schizocele.

Schizodonta *n.* [Gk. *schizein*, to cleave; *odous*, tooth] an order of lamellibranchs with eulamellibranch gills and divided hinge teeth; *cf.* Eulamellibranchiata.

schizogamy *n.* [Gk. *schizein*, to cleave; *gamos*, marriage] fission into a sexual and a non-sexual zooid in some Polychaeta.

schizogenesis *n.* [Gk. *schizein*, to cleave; *genesis*, descent] reproduction by fission; *alt.* scissiparity.

schizogenetic, schizogenic, schizogenous *a.* [Gk. *schizein*, to cleave; *genesis*, descent] reproducing or formed by fission; *appl.* intercellular spaces or glands formed by separation of cell walls along middle lamella; *n.* schizogeny.

schizognathous *a.* [Gk. *schizein*, to cleave; *gnathos*, jaw] having vomer small and pointed in front and maxillopalatines not united with each other and vomer, *appl.* a type of palate found in some Carinatae such as pigeon.

Schizogoniales *n.* [Gk. *schizein*, to cleave; *gonos*, offspring] Prasiolales, *q.v.*

schizogony *n.* [Gk. *schizein*, to cleave; *gonos*, offspring] reproduction by multiple fission in Protozoa; *alt.* agamogony, merogony.

schizokinete *n.* [Gk. *schizein*, to cleave; *kinētēs*, mover] motile vermicule stage in life history of Haemosporidia.

schizolysigenous *a.* [Gk. *schizein*, to cleave; *lysis*, loosing; *gennaein*, to produce] formed schizogenously, by separation, and enlarged lysigenously, by breakdown, such as glands and cavities in pericarp, e.g. of citrus fruits.

schizolysis *n.* [Gk. *schizein*, to cleave; *lysis*, loosing] fragmentation; disjunction at septa, as of hyphae.

Schizomeri *n.* [Gk. *schizein*, to cleave; *meros*, part] an order of probably aquatic fossil Mississippian amphibians.

Schizomycetes *n.* [Gk. *schizein*, to cleave; *mykēs*, fungus] alternative name for the bacteria when placed in the Schizophyta; *alt.* fission fungi, Schizomycophyta.

Schizomycophyta

Schizomycophyta n. [Gk. *schizein*, to cleave; *mykēs*, fungus; *phyton*, plant] Schizomycetes, *q.v.*

schizont n. [Gk. *schizein*, to cleave; *onta*, beings] in some Protozoa especially Sporozoa, the stage following the trophozoite and reproducing in the host by multiple fission; *alt.* agamont, merocyte.

schizontoblast n. [Gk. *schizein*, to cleave; *onta*, beings; *blastos*, bud] agametoblast, *q.v.*

schizontocytes n.*plu.* [Gk. *schizein*, to cleave; *on*, being; *kytos*, hollow] cytomeres into which a schizont divides, and which themselves divide into clusters of merozoites.

schizopelmous a. [Gk. *schizein*, to cleave; *pelma*, sole of foot] with 2 separate flexor tendons connected with toes, as some birds.

Schizopeltida n. [Gk. *schizein*, to cleave; *peltē*, shield] a group of very small arachnids sometimes considered to be of order status or included with the whip scorpions, having a prosoma covered with thin and separate plates of chitin.

Schizophyceae n. [Gk. *schizein*, to cleave; *phykos*, seaweed] alternative name for the blue–green algae when they are placed in the Schizophyta.

Schizophyta n. [Gk. *schizein*, to cleave; *phyton*, plant] plant group including the bacteria, known as the Schizomycetes, and blue–green algae known as the Schizophyceae.

schizophyte n. [Gk. *schizein*, to cleave; *phyton*, plant] a plant which reproduces solely by fission, as bacteria, yeasts, blue–green algae; a member of the Schizophyta, *q.v.*

schizopod stage that stage in development of a decapod crustacean larva when it resembles an adult *Mysis* in having exopodite and endopodite to all thoracic limbs.

schizorhinal a. [Gk. *schizein*, to cleave; *rhis*, nose] having external narial opening elongated and posterior border angular or slit-like; *cf.* holorhinal.

schizostele n. [Gk. *schizein*, to cleave; *stēlē*, post] one of a number of strands formed by division of plerome of stem.

schizostely n. [Gk. *schizein*, to cleave; *stēlē*, post] condition of stem in which plerome gives rise to a number of strands, each composed of one vascular bundle; *alt.* astely.

schizothecal a. [Gk. *schizein*, to cleave; *thēkē*, case] having scale-like horny tarsal plates.

schizozoite n. [Gk. *schizein*, to cleave; *zōon*, animal] in Sporozoa the stage in the life cycle produced by schizogony; *alt.* agamete, merozoite.

Schlemm, canal of [*F. S. Schlemm*, German anatomist] circular canal near sclerocorneal junction and joining with anterior chamber of eye and anterial ciliary veins; *alt.* sinus venosus sclerae.

Schwann cell [*Th. Schwann*, German anatomist] a uninucleate cell, rich in fatty substances, which surrounds a medullated nerve fibre and forms the myelin sheath, between the nodes of Ranvier; *alt.* neurilemmal cell.

Schwann sheath neurilemma, *q.v.*

sciaphyte skiaphyte, *q.v.*

sciatic a. [Gk. *ischion*, hip joint] *pert.* hip region, *appl.* artery, nerve, veins, etc.; *alt.* ischiatic.

scion n. [F. *scion*, shoot] a branch or shoot for grafting purposes; *alt.* cion.

sciophilous skiophilous, *q.v.*

sciophyll skiophyll, *q.v.*

scissile a. [L. *scissilis*, cleavable] cleavable; splitting, as into layers.

scissiparity n. [L. *scissio*, cleaving; *parere*, to beget] schizogenesis, *q.v.*

Scitamineae n. [L. *scitamenta*, delicacies] Zingiberales, *q.v.*

sclera n. [Gk. *sklēros*, hard] the tough, opaque, fibrous tunic of the eyeball; *alt.* sclerotic coat, sclerotica.

Scleractinia n. [Gk. *sklēros*, hard; *aktis*, ray] an order of usually colonial Zoantharia known as true corals, having a compact calcareous skeleton and no siphonoglyph; *alt.* Madreporaria.

scleratogenous layer strand of the fused sclerotomes formed along the neural tube, later surrounding the notochord.

sclere n. [Gk. *sklēros*, hard] a small skeletal structure; sponge spicule.

sclereid(e) n. [Gk. *sklēros*, hard; *eidos*, form] any cell with a thick lignified wall; a sclerenchymatous cell; stone cell, *q.v.*; *alt.* sclerid.

sclerenchyma n. [Gk. *sklēros*, hard; *engchyma*, infusion] plant tissue with thickened, usually lignified cell walls, which acts as supporting tissue; hard tissue of coral; *a.* sclerenchymatous.

sclerid sclereid, *q.v.*

sclerification n. [Gk. *sklēros*, hard; L. *facere*, to make] the process of becoming sclerenchyma.

sclerins scleroproteins, *q.v.*

sclerite n. [Gk. *sklēros*, hard] calcareous plate or spicule; chitinous plate, part of exoskeleton; *a.* scleritic.

sclerobase n. [Gk. *skleros*, hard; *basis*, base] the calcareous axis of Alcyonaria.

sclerobasidium n. [Gk. *sklēros*, hard; *basis*, base; *idion*, dim.] a thick-walled resting body or encysted probasidium of rust and smut fungi; *alt.* hypnobasidium.

scleroblast n. [Gk. *sklēros*, hard; *blastos*, bud] a sponge cell from which a sclere develops; an immature sclereid.

scleroblastema n. [Gk. *sklēros*, hard; *blastēma*, bud] embryonic tissue involved in development of skeleton.

scleroblastic a. [Gk. *sklēros*, hard; *blastos*, bud] *appl.* skeletal-forming tissue.

sclerocarp n. [Gk. *sklēros*, hard; *karpos*, fruit] the hard seed coat or stone, usually the endocarp, of succulent fruit.

sclerocauly n. [Gk. *sklēros*, hard; *kaulos*, stalk] condition of excessive skeletal structure in a stem.

sclerocorneal a. [Gk. *sklēros*, hard; L. *cornea*, cornea] corneosclerotic, *q.v.*

scleroderm n. [Gk. *sklēros*, hard; *derma*, skin] a hard integument; skeletal part of corals.

Sclerodermatales n. [Gk. *sklēros*, hard; *derma*, skin] an order of Gasteromycetes, commonly called earth balls, having a thick hard periderm and a gleba the inside of which is usually dark coloured.

sclerodermatous a. [Gk. *sklēros*, hard; *derma*, skin] with an external skeletal structure.

sclerodermite n. [Gk. *sklēros*, hard; *derma*, skin] the part of the exoskeleton over one arthropod segment.

sclerogen n. [Gk. *sklēros*, hard; *-genēs*, producing] wood-producing cells.

sclerogenic, sclerogenous a. [Gk. *sklēros*, hard; *-genēs*, producing] producing lignin.

scleroid a. [Gk. *sklēros*, hard; *eidos*, form] hard; skeletal; *alt.* sclerous.

scleromeninx n. [Gk. *sklēros*, hard; *meningx*, membrane] the dura mater, *q.v.*

Scleroparei n. [Gk. *sklēros*, hard; *pareia*, cheek] a large order of teleosts including the gurnards, greenlings, and scorpion fish, having one of the suborbital bones extending towards the preoperculum.

sclerophyll n. [Gk. *sklēros*, hard; *phyllon*, leaf] a sclerophyllous plant; one of the leaves of such a plant.

sclerophyllous a. [Gk. *sklēros*, hard; *phyllon*, leaf] hard-leaved, *appl.* leaves resistant to drought through having a thick cuticle, much sclerenchymatous tissue, and reduced intercellular spaces.

sclerophylly n. [Gk. *sklēros*, hard; *phyllon*, leaf] condition of excessive skeletal structure in leaves.

scleroproteins n.plu. [Gk. *sklēros*, hard; *prōteion*, first] a group of proteins occurring in connective, skeletal, and epidermal tissues, as ossein, collagen, gelatin, chondrin, elastin, keratin, etc.; *alt.* albuminoids, sclerins.

scleroseptum n. [Gk. *sklēros*, hard; L. *septum*, division] a radial vertical wall of calcium carbonate in corals of the order Scleractinia.

sclerosis n. [Gk. *skleros*, hard] hardening by increase of connective tissue or of lignin.

sclerospermous a. [Gk. *sklēros*, hard; *sperma*, seed] having the seeds covered by a hard coat.

sclerotal a. [Gk. *sklēros*, hard] sclerotic, *q.v.*

sclerotesta n. [Gk. *sklēros*, hard; L. *testa*, shell] the hard lignified inner layer of a testa; *cf.* sarcotesta.

sclerotic n. [Gk. *sklēros*, hard] the sclera, *q.v. a.* Indurated; containing lignin; *pert.* sclerosis; *pert.* sclera; *alt.* sclerotal, sclerous.

sclerotica sclera, *q.v.*

sclerotic ossicles ring of small bones around sclera of birds; plates surrounding the eye of certain fishes.

sclerotioid, sclerotiform a. [Gk. *sklēros*, hard; *eidos*, form; L.L. *forma*, shape] like or *pert.* a sclerotium.

sclerotin n. [Gk. *sklēros*, hard] a highly resistant, stable, quinone-tanned protein occurring in insect cuticle and in structural proteins in widely different groups of vertebrates and invertebrates.

sclerotium n. [Gk. *sklēros*, hard] resting, dormant, or winter stage of some fungi when they become a mass of hardened mycelium or of waxy protoplasm, *alt.* hypothallus; in slime moulds, a collection of multinucleate cellulose cysts formed in times of drought.

sclerotization n. [Gk. *sklēros*, hard] the process of hardening and darkening the exoskeleton which occurs in insects after ecdysis.

sclerotome n. [Gk. *sklēros*, hard; *tomē*, cutting] a partition of connective tissue between 2 myomeres; mesenchymatous tissue destined to form a vertebra.

sclerous a. [Gk. *sklēros*, hard] hard; indurated; sclerotic, *q.v.*; scleroid, *q.v.*

scobiscular, scobisculate, scobiform a. [*Dim.* of L. *scobis*, sawdust] granulated; resembling sawdust.

scobina n. [L. *scobina*, file] pedicel of a spikelet of grasses.

scobinate a. [L. *scobina*, file] having a rasp-like surface.

scoleces *plu.* of scolex.

scolecid a. [Gk. *skōlēx*, worm] *pert.* a scolex; *alt.* scolecoid.

scoleciform a. [Gk. *skōlēx*, worm; L. *forma*, shape] like a scolex; *alt.* scolecoid.

scolecite n. [Gk. *skōlēx*, worm] vermiform body branching from mycelium of Discomycetes; Woronin hypha, *q.v.*

scolecoid scolecid, *q.v.*; scoleciform, *q.v.*

scolecospore n. [Gk. *skōlēx*, worm; *sporos*, seed] a worm-like or thread-like spore.

scolex n. [Gk. *skōlēx*, worm] the head or anterior end of a tapeworm; *plu.* scoleces, scolices.

scolite n. [Gk. *skōlēx*, worm; *lithos*, stone] a fossil worm burrow.

scolopale n. [Gk. *skōlos*, stake; *palē*, struggle] vibratile central peg-like portion of a scolophore.

Scolopendromorpha n. [*Scolopendra*, generic name; Gk. *morphē*, form] an order of centipedes of the subclass Epimorpha having spiracles not on every segment.

scolophore, scolopidium n. [Gk. *skōlos*, stake; *pherein*, to bear] chordotonal sensilla, *q.v.*

scolus n. [Gk. *skōlos*, thorn] a thorny process of some insect larva.

scopa n. [L. *scopa*, brush] pollen brush, *q.v.*

scopate a. [L. *scopa*, brush] having a tuft of hairs like a brush; *alt.* scopiferous.

Scopeliformes n. [*Scopelus*, generic name; L. *forma*, shape] an order or suborder of marine teleosts with many deep-sea members and including the lantern fish and lizard fish; *alt.* Iniomi.

Scopelomorpha n. [*Scopelus*, generic name; Gk. *morphē*, form] in some classifications, Scopeliformes, *q.v.*, when considered as a superorder.

scopiferous a. [L. *scopa*, brush; *ferre*, to bear] scopate, *q.v.*

scopiform a. [L. *scopa*, brush; *forma*, shape] brush-like; *alt.* scopulate.

scopula n. [L. *scopula*, small brush] a small tuft of hairs; brush-like adhesive organ formed by cilia in certain peritrichous ciliates; a needle-like sponge spicule with brush-like head; in climbing spiders an adhesive tuft of club-like hairs on each foot, replacing 3rd claw.

scopulate a. [L. *scopula*, small brush] scopiform, *q.v.*

scopuliferous a. [L. *scopula*, small brush; *ferre*, to carry] having a small brush-like structure.

scopuliform a. [L. *scopula*, small brush; *forma*, shape] resembling a small brush.

Scorpaeniformes n. [Gk. *skorpios*, scorpion; L. *forma*, shape] a widely divergent order of teleosts including the scorpion fish, gurnards, and bullheads.

scorpamins scorpion toxins.

scorpioid a. [Gk. *skorpios*, scorpion; *eidos*, form] circinate, *appl.* inflorescence; resembling a scorpion.

scorpioid cyme a uniparous cymose inflorescence in which daughter axes are developed right and left alternately; *alt.* cincinnus.

Scorpiones, Scorpionoidea n. [Gk. *skorpios*, scorpion; *eidos*, form] an order of arachnids including the scorpions, having a dorsal carapace on the prosoma, the opisthosoma divided into 2 regions, and a telson with a poisonous sting.

scorpion flies the common name for the Mecoptera, *q.v.*

scorteal *a.* [L. *scorteus*, leathern] *appl.* or *pert.* a tough cortex, as of certain fungi.

scotochromogens *n.plu.* [Gk. *skotos*, darkness; *chrōma*, colour; *gennaein*, to produce] bacteria that produce pigments in the dark.

scotoma *n.* [Gk. *skotos*, darkness] a spot where vision is absent within the visual field, a blind spot.

scotophobin *n.* [Gk. *skotos*, darkness; *phobos*, fear] a chemical which is claimed to instil fear of the dark into rats and is transmissible by extracts of brain tissue.

scotopia *n.* [Gk. *skotos*, darkness; *ōps*, eye] adaptation of the eye to darkness; *cf.* photopia; *a.* scotopic.

scotopsin *n.* [Gk. *skotos*, darkness; *opsis*, sight] the protein component of rhodopsin.

scrobe *n.* [L. *scrobis*, ditch] a groove on either side of beetle rostrum.

scrobicula *n.* [L.L. *dim.* of *scrobis*, ditch] the smooth area around boss of echinoid test; *alt.* scrobicule, areola.

scrobicular in region of scrobicula.

scrobiculate *a.* [L.L. *dim.* of *scrobis*, ditch] marked with little pits or depressions.

scrobicule scrobicula, *q.v.*; scrobiculus, *q.v.*

scrobiculus *n.* [L.L. *dim.* of *scrobis*, ditch] a pit or depression; *alt.* scrobicule.

scrobiculus cordis pit of stomach.

Scrophulariales *n.* [*Scrophularia*, generic name] an order of dicots (Asteridae) having zygomorphic flowers with bicarpellary ovaries; *alt.* Personales, Personatae.

scrotal *a.* [L. *scrotum*, scrotum] *pert.* or in region of scrotum.

scrotum *n.* [L. *scrotum*, scrotum] external sac or sacs containing testicles in mammals; covering of testis in insects.

scrub *n.* [M.E. *shrobbe*, shrub] a plant community dominated by shrubs.

scurf *n.* [A.S. *scurf*] scaly skin; dried outer skin peeling off in scales; scaly epidermal covering of some leaves.

scuta *plu.* of scutum.

scutal *a.* [L. *scutum*, shield] *pert.* a scutum.

scutate *a.* [L. *scutum*, shield] protected by large scales or horny plates.

scute *n.* [L. *scutum*, shield] an external scale, as of reptile, fish, or scaly insect; a scale-like structure; bony plate separating sinuses of mastoid bone from tympanic cavity; scutum, *q.v.*

scutella *n.* [L. *scutellum*, small shield] a scutellum or shield-like structure; *plu.* of scutellum.

scutellar *pert.* a scutellum.

scutellate *a.* [L. *scutellum*, small shield] shaped like a small shield; *alt.* scutelliform.

scutellation *n.* [L. *scutellum*, small shield] arrangement of scales, as on tarsus of bird; *cf.* pholidosis.

scutelliform scutellate, *q.v.*

scutelligerous *a.* [L. *scutellum*, small shield; *gerere*, to bear] furnished with scutella or a scutellum.

scutelliplantar *a.* [L. *scutellum*, small shield; *planta*, sole of foot] having tarsus covered with small plates or scutella.

scutellum *n.* [L. *scutellum*, small shield] a tarsal scale of birds; posterior part of insect notum; development of part of cotyledon which separates embryo from endosperm in seed of grasses; a shield-shaped structure; *plu.* scutella.

scutiferous scutigerous, *q.v.*

scutiform *a.* [L. *scutum*, shield; *forma*, shape] shaped like a shield, as floating leaf of the fern *Salvinia*.

Scutigeromorpha *n.* [*Scutigera*, generic name; Gk. *morphē*, form] an order of fast-running centipedes of the subclass Anamorpha having long legs and unpaired spiracles.

scutigerous *a.* [L. *scutum*, shield; *gerere*, to bear] bearing a shield-like structure; *alt.* scutiferous.

scutiped *a.* [L. *scutum*, shield; *pes*, foot] having foot or part of it covered by scutella.

scutum *n.* [L. *scutum*, shield] broad apex of style, as in dicots of family Asclepiadaceae; 1 of 8 plates surrounding antheridium of Charales; a shield-like plate, horny, bony, or chitinous, developed in integument; fornix or modified spine overhanging aperture in some Cheilostomata; middle sclerite of insect notum; dorsal shield of ticks; one of the pair of anterior valves of *Lepas*, the goose barnacle; *alt.* scute, shield; *plu.* scuta.

scyllitol *n.* [*Scyliorhinus*, generic name of dogfish] a sweet alcohol, $C_6H_6(OH)_6$, one of the inositol group, found in dogfish and in various plants such as coconut palm leaves and acorns.

scyphi *plu.* of scyphus.

scyphiferous *a.* [L. *scyphus*, cup; *ferre*, to bear] bearing scyphi, as some lichens.

scyphiform *a.* [L. *scyphus*, cup; *forma*, shape] shaped like a cup; *alt.* scyphoid, scyphose.

scyphistoma *n.* [Gk. *skyphos*, cup; *stoma*, mouth] the segmenting polyp stage in development of scyphozoans; *alt.* scyphula; *cf.* hydratuba.

scyphoid *a.* [Gk. *skyphos*, cup; *eidos*, form] scyphiform, *q.v.*

Scyphomedusae *n.* [Gk. *skyphos*, cup; *Medousa*, Medusa] Scyphozoa, *q.v.*

scyphose *a.* [L. *scyphus*, cup] having scyphi; scyphiform, *q.v.*

Scyphozoa, scyphozoans *n.*, *n.plu.* [Gk. *skyphos*, cup; *zōon*, animal] a class of cnidarian coelenterates, commonly called jellyfish, having a dominant medusoid phase which is free swimming or attached by an aboral stalk, and no velum; *alt.* Scyphomedusae.

scyphula *n.* [L.L. *dim.* of *scyphus*, cup] scyphistoma, *q.v.*

scyphulus *n.* [*Dim.* of L. *scyphus*, cup] a small cup-shaped structure.

scyphus *n.* [L. *scyphus*, Gk. *skyphos*, cup] funnel-shaped corolla as of daffodil; cup-shaped expansion of podetium in some lichens; *plu.* scyphi.

sea anemones the common name for the Actiniaria, *q.v.*

sea butterflies the common name for the Pteropoda, *q.v.*

sea combs a common name for the Ctenophora, *q.v.*

sea cows the common name for the Sirenia, *q.v.*

sea cucumbers the common name for the Holothuroidea, *q.v.*

sea gooseberries a common name for the Ctenophora, *q.v.*

sea lilies the common name for the Crinoidea, *q.v.*

sea pens the common name for the Pennatulacea, *q.v.*

sea squirts the common name for the ascidians, *q.v.*

sea stars a common name for the Asteroidea, *q.v.*

sea urchins the common name for the Echinoidea, *q.v.*

sebaceous *a.* [L. *sebum*, tallow] containing or secreting fatty matter, *appl.* glands; *alt.* sebific, sebiparous.

sebiferous *a.* [L. *sebum*, tallow; *ferre*, to carry] conveying fatty matter.

sebific *a.* [L. *sebum*, tallow; *facere*, to make] sebaceous, *q.v.*; *appl.* gland: colleterium, *q.v.*

sebiparous *a.* [L. *sebum*, tallow; *parere*, to beget] sebaceous, *q.v.*

sebum *n.* [L. *sebum*, tallow] the secretion of sebaceous glands, consisting of fat and fatty materials.

Secernentea *n.* [L. *secernere*, to separate] Phasmidia, *q.v.*

secodont *a.* [L. *secare*, to cut; Gk *odous*, tooth] furnished with teeth adapted for cutting.

secondary *a.* [L. *secundus*, second] second in importance or in position; arising, not from growing point, but from other tissue; Mesozoic. *n.* A forearm quill feather of bird's wing, *alt.* cubital; an insect hind-wing.

secondary bud an axillary bud, accessory to normal one.

secondary capitula six small cells arising from each capitulum of Charales.

secondary cell wall under the light microscope the cell wall laid down on top of the primary wall and consisting of various materials including more cellulose, lignin, and cutin; under the electron microscope, the cell wall in which the microfibrils show a parallel orientation.

secondary constriction a non-staining region of the chromosome, other than the centromere, not showing attraction to the equator of the spindle at metaphase.

secondary consumer carnivore which eats herbivores; *cf.* tertiary consumer.

secondary cortex phelloderm, *q.v.*

secondary growth growth of a plant body by the activity of the lateral meristems, the vascular cambium and phellogen; *alt.* secondary thickening.

secondary host intermediate host, *q.v.*

secondary meristem phellogen, *q.v.*

secondary nucleus fusion nucleus, *q.v.*

secondary plant body the plant body formed by growth from lateral meristems, the vascular cambium and phellogen.

secondary phloem phloem tissue formed from the vascular cambium during secondary growth, sometimes also called inner bark.

secondary prothallium a tissue produced in megaspore of the club moss *Selaginella* after true prothallium is formed.

secondary roots branches of primary root, arising within its tissue, and in turn giving rise to tertiary roots; roots arising at other than normal points of origin.

secondary sexual characteristic a characteristic found in one sex, usually developing under the influence of androgens and oestrogens, including growth of beard in men, antlers in stags, and enlarged breasts in women.

secondary spore a small or abjointed spore; a mycelial spore.

secondary succession a sere or plant succession following the interruption of the normal or primary succession.

secondary thickening secondary growth, *q.v.*; the production of secondary cell wall.

secondary tissue tissue formed from phellogen, externally cork, and internally phelloderm.

secondary vascular tissue xylem and phloem formed by vascular cambium during secondary growth.

secondary wood secondary xylem, *q.v.*

secondary xylem xylem formed by vascular cambium during secondary growth; *alt.* secondary wood, wood.

secreta *n.plu.* [L. *secretum*, separated] any products of a secretory process; all the secretions.

secretin *n.* [L. *secernere*, to separate] a hormone produced by the duodenum as a result of the presence of acid chyme, which causes the pancreas to secrete pancreatic juice; *cf.* gastric secretin.

secretion *n.* [L. *secretio*, separation] substance or fluid which is separated and elaborated by cells or glands; process of such separation.

secretitious *a.* [L. *secernere*, to separate] produced by secretion, *appl.* substance or fluid.

secretory *a.* [L. *secernere*, to separate] effecting or *pert.* the secretion; secreting.

sectile *a.* [L. *secare*, to cut] cut into small partitions or compartments.

section a taxonomic group, often used as a subdivision of a genus, but used in different ways by different authors and never precisely defined.

sectorial *a.* [L. *sector*, cutter] formed or adapted for cutting, as certain teeth; *appl.* chimaera when 2 different tissues extend from centre to periphery, a wedge of one tissue inserted in the other.

secund *a.* [L. *secundus*, following] arranged on one side, *appl.* flowers or leaves on stem.

secundiflorous *a.* [L. *secundus*, following; *flos*, flower] having flowers on one side of stem only.

secundine *n.* [L. *secundus*, following] the internal integument of ovule; *cf.* primine.

secundines foetal membranes collectively; afterbirth, *q.v.*

secundly *adv.* [L. *secundus*, following] on one side of a stem or axis.

sedentaria *n.* [L. *sedere*, to sit] sedentary organisms, the term sometimes being used as a taxonomic group; *cf.* errantia.

sedentary *a.* [L. *sedere*, to sit] not free living, *appl.* animals attached by a base to some substratum; not migratory.

sedoheptulose *n.* [*Sedum*, generic name; Gk. *hepta*, seven] a heptose sugar important in plants as a free sugar in the leaves of succulent Crassulaceae and as sedoheptulose-1,7-diphosphate (bisphosphate) in the carbon assimilation cycle of photosynthesis.

seed *n.* [A.S. *saed*, seed] a reproductive unit formed from a fertilized ovule consisting of an embryo, food store, and protective coat; semen, *q.v.* *v.* To introduce micro-organisms into a culture medium.

seed bank a place in which seeds of rare plants or obsolete varieties are stored, usually under cold, vacuum-packed conditions, to prolong their viability.

seed bud ovule, *q.v.*

seed coat *see* testa.

seed ferns Pteridospermae, *q.v.*

seed leaf cotyledon, *q.v.*
seed plants Spermatophyta, *q.v.*
seed stalk *see* funicle.
seed vessel fruit, *q.v.*, especially a dry fruit.
Seessel's pouch [*A. Seessel*, American embryologist] a dorsal endodermal diverticulum from anterior end of fore-gut, behind buccopharyngeal membrane in vertebrate embryos.
segment *n.* [L. *segmentum*, piece] a division formed by cleavage of an ovum; part of an animal or of a jointed appendage; metamere, *q.v.*; division of leaf if cleft nearly to base; portion of a chromosome.
segmental *a.* [L. *segmentum*, piece] of the nature of a segment; *pert.* a segment.
segmental arteries diverticula from dorsal aortae arising in spaces between successive somites.
segmental duct an embryonic nephridial duct which gives rise to Wolffian or Müllerian duct.
segmental interchange exchange of non-homologous segments as between 2 chromosomes; mutual translocation.
segmental organ an embryonic excretory organ; *alt.* nephridium.
segmental papillae conspicuous pigment spots by which true segments may be recognized in leeches.
segmental reflex a reflex involving a single region of the spinal cord.
segmentation *n.* [L. *segmentum*, piece] the division or splitting into segments or portions; cleavage of an ovum.
segmentation cavity central cavity of a blastula formed at an early stage of egg cleavage; *alt.* blastocoel.
segmentation nucleus body formed by union of male and female pronuclei in fertilization of ovum.
segregation *n.* [L. *segregare*, to separate] separation of parental chromosomes at meiosis and dissociation of paternal and maternal characters; separation of allelic genes.
seiospore *n.* [Gk. *seiein*, to shake; *sporos*, seed] a spore shaken from a sporophore and becoming air-borne.
seiro- *alt.* siro-.
seiroderm *n.* [Gk. *seira*, chain; *derma*, skin] dense outer tissue composed of parallel chains of hyphal cells, in certain fungi.
seirospore *n.* [Gk. *seira*, chain; *sporos*, seed] one of spores arranged like a chain; formerly, a catenulate spore of certain red algae.
seismaesthesia *n.* [Gk. *seismos*, a shaking; *aisthēsis*, perception] perception of mechanical vibrations.
seismonasty *n.* [Gk. *seismos*, a shaking; *nastos*, pressed close] plant movements in response to the stimulus of mechanical shock or vibration; *a.* seismonastic.
seismotaxis *n.* [Gk. *seismos*, a shaking; *taxis*, arrangement] a taxis in response to mechanical vibrations; *alt.* vibrotaxis.
Seisonidea *n.* [*Seison*, generic name] an order of aberrant marine rotifers with a poorly developed corona, that are epizootic on crustaceans.
sejugate, sejugous *a.* [L. *sex*, six; *jugum*, yoke] with 6 pairs of leaflets.
Selachii, selachians *n.*, *n.plu.* [Gk. *selachos*, shark] an order of elasmobranch fish existing from Devonian times to the present day, including the sharks and dogfish, having claspers, and fins with a constricted base.
selachine *n.* [Gk. *selachos*, shark] a neurohumor of selachians which induces blanching of skin.
Selaginellales *n.* [*Selaginella*, generic name] an order of heterosporous Lycopsida having ligulate leaves, sporophylls in cones and no secondary thickening; commonly called club mosses, although this name is sometimes restricted to the Lycopodiales, or used for the whole of the Lycopsida.
selection pressure the force of natural selection acting on a population.
selective advantage *pert.* any characteristic which gives an organism an increased chance of surviving and breeding.
selenodont *a.* [Gk. *selēnē*, moon; *odous*, tooth] *appl.* molars lengthened out anteroposteriorly and curved; *appl.* or having molar teeth with crescent-shaped ridges on the grinding surface as in artiodactyls.
selenoid *a.* [Gk. *selēnē*, moon; *eidos*, form] crescentic.
selenotropism *n.* [Gk. *selēnē*, moon; *tropē*, turn] tendency to turn towards moon's rays.
selenozone *n.* [Gk. *selēnē*, moon; *zōnē*, girdle] lateral stripe of crescentic growth lines on whorl of a gastropod shell.
self-fertile, self-sterile capable, incapable, of being fertilized by its own male gametes, *appl.* hermaphrodite plants and animals.
selfing self-pollination or self-fertilization.
self-fertilization the fusion of male and female gametes from the same individual; *alt.* autogamy, automixis.
self-pollination transference of pollen grains from anthers to stigma of same flower; *alt.* selfing, endogamy.
self-sterility the inability of a hermaphrodite to form viable offspring by self-fertilization.
sellaeform *a.* [L. *sella*, saddle; *forma*, shape] saddle-shaped; *alt.* selliform.
sellar *a.* [L. *sella*, saddle] *pert.* pituitary fossa or sella turcica.
sella turcica *n.* [L. *sella*, saddle; *turcicus*, Turkish] deep depression on superior surface of sphenoidal bone behind tuberculum sellae, the deepest part, fossa hypophyseos, lodging the pituitary body; transverse bar formed by union of apodemes of posterior somites of certain Decapoda; *alt.* Turkish saddle.
selliform sellaeform, *q.v.*
selva *n.* [Sp. *selva*, from L. *silva*, forest] tropical rain forest; *alt.* silva.
Semaeostomeae *n.* [Gk. *sēmeia*, token; *sōma*, body] an order of scyphozoans having no tentacles and with the corners of the mouth extended into 4 frilly lobes.
semantide *n.* [Gk. *sēmantikos*, significant] a molecule which carries information, such as DNA, or a transcript of the information, such as RNA.
sematic *a.* [Gk. *sēma*, sign] functioning as a danger signal, as warning colours or odours, *appl.* warning and recognition markings; *cf.* aposematic, episematic, parasematic.
semen *n.* [L. *semen*, seed] fluid composed of secretions of testes and accessory glands, and containing spermatozoa; *alt.* seed, sperm.

semiamplexicaul *a.* [L. *semi*, half; *amplecti*, to embrace; *caulis*, stem] partially surrounding stem.

semianatropous *a.* [L. *semi*, half; Gk. *ana*, up; *trope̅*, turn] with half-inverted ovule.

semicaudate *a.* [L. *semi*, half; *cauda*, tail] with tail rudimentary.

semicells [L. *semi*, half; *cellula*, small cell] the 2 halves of a cell, which are interconnected by an isthmus, as in certain green algae.

semicircular *a.* [L. *semi*, half; *circulus*, circle] describing a half-circle, *appl.* canals and ducts of ear labyrinth.

semiclasp *n.* [L. *semi*, half; A.S. *clyppan*, to embrace] one of 2 apophyses which may combine to form the clasper in certain male insects.

semicomplete *a.* [L. *semi*, half; *completus*, filled] incomplete, *appl.* metamorphosis.

semiconservative replication replication in DNA, where the molecule unwinds and each half acts as a template for a new strand.

semicylindrical *a.* [L. *semi*, half; *cylindrus*, cylinder] round on one side, flat on the other, *appl.* leaves.

semidominant codominant, *q.v.*

semifloret, semifloscule *n.* [L. *semi*, half; *flos*, flower] a ligulate floret of Compositae.

semiflosculous *a.* [L. *semi*, half; *flosculus*, small flower] having ligulate florets.

semigamy hemigamy, *q.v.*

semiherbaceous *a.* [L. *semi*, half; *herbaceus*, grassy] having lower part of stem woody and upper part herbaceous.

semilethal *a.* [L. *semi*, half; *lethalis*, deadly] not wholly lethal; *appl.* genes causing a mortality of more than 50%, or permitting survival until reproduction has been effected; *cf.* subvital.

semiligneous *a.* [L. *semi*, half; *ligneus*, wooden] partially lignified; with stem woody only near base.

semilocular *a.* [L. *semi*, half; *loculus*, compartment] *appl.* ovary with incomplete loculi.

semilunar *a.* [L. *semi*, half; *luna*, moon] half-moon shaped, *appl.* branches of internal carotid artery, fibrocartilages of knee, ganglia, fascia, lobules of cerebellum, valves, *appl.* notch, greater sigmoid cavity between olecranon and coronoid process of ulna. *n.* A carpal bone: lunar bone, *q.v.*

semimembranosus *n.* [L. *semi*, half; *membranosus*, membranous] a thigh muscle with flat membrane-like tendon at upper extremity.

semimetamorphosis *n.* [L. *semi*, half; Gk. *metamorphōsis*, transformation] partial, or semicomplete metamorphosis.

seminal *a.* [L. *semen*, seed] *pert.* semen, *appl.* fluid, duct, vesicle.

seminal funnel internal opening of vasa deferentia in Oligochaeta.

seminal receptacle spermatheca, *q.v.*; vesicular seminalis, *q.v.*

seminal root the first-formed root, developed from the radicle of the seed; *cf.* coronal.

seminal vesicle one of a pair of glands of male mammals, which secrete the alkaline fluid component of semen; an organ for storing sperm in some invertebrates and lower vertebrates; *alt.* vesicula seminalis.

semination *n.* [L. *seminatio*, sowing] dispersal of seeds; discharge of spermatozoa; *cf.* insemination.

seminiferous *a.* [L. *semen*, seed; *ferre*, to carry] secreting or conveying semen; bearing seed.

seminude *a.* [L. *semi*, half; *nudus*, naked] with ovules or seeds exposed.

seminymph *n.* [L. *semi*, half; *nympha*, nymph] stage in development of insects approaching complete metamorphosis.

Semionotiformes *n.* [*Semionotus*, generic name; L. *forma*, shape] an order of holosteans existing from Permian times to the present day and including the modern gar pikes; *alt.* Semionotoidea.

semiorbicular *a.* [L. *semi*, half; *orbis*, orb] half-rounded, hemispherical.

semiovate *a.* [L. *semi*, half; *ovum*, egg] half-oval; somewhat oval.

semioviparous *a.* [L. *semi*, half; *ovum*, egg; *parere*, to beget] between oviparous and viviparous, as a marsupial whose young are imperfectly developed when born.

semiovoid *a.* [L. *semi*, half; *ovum*, egg; Gk. *eidos*, form] somewhat ovoid in shape.

semipalmate *a.* [L. *semi*, half; *palma*, palm of hand] having toes webbed halfway down.

semiparasite *n.* [L. *semi*, half; Gk. *parasitos*, eating beside another] a plant which derives only part of its nutrient from host and has some photosynthetic capacity; *alt.* partial parasite.

semipenniform *a.* [L. *semi*, half; *penna*, feather; *forma*, shape] *appl.* certain muscles bearing some resemblance to the lateral half of a feather.

semipermeable *a.* [L. *semi*, half; *per*, through; *meare*, to pass] partially permeable; *appl.* membrane permeable to a solvent, especially water, but not to solutes.

semiplacenta *n.* [L. *semi*, half; *placenta*, flat cake] a non-deciduate placenta.

semiplume *n.* [L. *semi*, half; *pluma*, feather] a feather with ordinary shaft but downy web.

semipupa *n.* [L. *semi*, half; *pupa*, puppet] coarctate pupa, *q.v.*

semirecondite *a.* [L. *semi*, half; *recondere*, to conceal] half-concealed as insect head by thorax.

semisagittate *a.* [L. *semi*, half; *sagitta*, arrow] shaped like a half arrowhead.

semisaprophyte *n.* [L. *semi*, half; Gk. *sapros*, rotten; *phyton*, plant] a plant partially saprophytic.

semispecies *n.* [L. *semi*, half; *species*, particular kind] a taxonomic group intermediate between a species and subspecies especially as a result of geographical isolation.

semispinalis *n.* [L. *semi*, half; *spinalis*, spinal] a muscle of back, also of neck, on each side of spinal column, arising from transverse and inserted into spinous processes.

semistreptostylic between monimostylic and streptostylic; with slightly moveable quadrate.

semitendinosus *n.* [L. *semi*, half; *tendo*, sinew] a dorsal muscle of the thigh stretching from the tuberosity of the ischium to the tibia.

semitendinous *a.* [L. *semi*, half; *tendere*, to stretch] half-tendinous.

semituberous *a.* [L. *semi*, half; *tuber*, hump] having somewhat tuberous roots.

senescence *n.* [L. *senescere*, to grow old] advancing age; the complex of ageing processes that eventually lead to death; *a.* senescent.

senility *n.* [L. *senilis*, senile] degeneration due to old age; vital exhaustion of Protozoa.

sense hillock neuromast, *q.v.*

sense organ an organ functional in receiving external stimulation; *alt.* receptor.

sensiferous, sensigerous *a.* [L. *sensus,* sense; *ferre, gerere,* to carry] receiving or conveying sense impressions.

sensile *a.* [L. *sensilis,* sensitive] capable of affecting a sense.

sensilla *n.* [L. *sensus,* sense] a small sense organ.

sensitive *a.* [L. *sensus,* sense] capable of receiving impressions from external objects; reacting to a stimulus.

sensitization the process of increasing an animal's reaction to an antigen.

sensomobile *a.* [L. *sensus,* sense; *mobilis,* movable] responding to irritation with movement.

sensorium *n.* [L. *sensus,* sense] seat of sensation or consciousness; entire nervous system with sense organs; the sensory, neuromuscular, and glandular system; *a.* sensorial; *cf.* motorium.

sensory *a.* [L. *sensus,* sense] having direct connection with any part of sensorium.

sensory neurone nerve cell concerned with carrying impulses to the CNS from a sense organ; *alt.* afferent neurone; *cf.* motor neurone.

sensu lato in a broad sense.

sensu stricto in a restricted sense.

sentient *a.* [L. *sentire,* to feel] *appl.* cells which are sensitive and perceptive.

sepal *n.* [F. *sépale*; L. *separare,* to separate] one of the outer series of perianth segments (the calyx) particularly if green and leaf-like.

sepaled *a.* [L. *separare,* to separate] having sepals; *alt.* sepalous.

sepaline *a.* [L. *separare,* to separate] like a sepal; *alt.* sepaloid.

sepalody *n.* [L. *separare,* to separate; Gk. *eidos,* form] conversion of petals or other parts of a flower into sepals.

sepaloid *a.* [L. *separare,* to separate; Gk. *eidos,* form] sepaline, *q.v.*

sepalous sepaled, *q.v.*

separation layer abscission layer, *q.v.*

sepiapterin [Gk. *sepia,* cuttlefish; *pteron,* wing] a pteridine pigment which is a precursor of the drosopterins, and gives the fruitfly, *Drosophila,* testis its yellow colour.

sepicolous *a.* [L. *sepes,* hedge; *colere,* to inhabit] living in hedges.

sepiment *n.* [L. *sepimentum,* fence] a partition; dissepiment, *q.v.*

sepion *n.* [Gk. *sēpion,* cuttlebone] calcareous shell of cuttlefish, in the form of a broad oval plate, solid in the upper and lamellate on the lower side; *alt.* sepia bone, sepiost, sepiostaire, sepium.

septa *plu.* of septum.

septal *a.* [L. *septum,* partition] *pert.* a septum; *pert.* hedgerows, *appl.* flora.

septal fossula a small primary septum which appears to lie in a pit in some fossil corals.

septal neck in the cephalopod, *Nautilus,* a shelly tube continuous for some distance beyond each septum as support to siphuncle.

septate *a.* [L. *septum,* partition] *pert.* a septum; divided by partitions; *alt.* polycystid.

septempartite *a.* [L. *septum,* seven: *pars,* part] *appl.* leaf with 7 divisions extending nearly to base.

septenate *a.* [L. *septeni,* seven each] with parts in 7s; *appl.* 7 leaflets of a leaf.

septibranch *a.* [L. *septum,* partition; *branchiae,* gills] *appl.* gills of bivalve molluscs which are small and transformed into a transverse muscular pumping septum.

Septibranchia, septibranchs *n., n.plu.* [L. *septum,* partition; *branchiae,* gills] an order of lamellibranchs having small gills which form a muscular septum and act as a pumping system, and which feed on large particles of animal remains; in some classifications, raised to subclass status.

septicidal *a.* [L. *septum,* division; *caedere,* to cut] dividing through middle of ovary septa; dehiscing at septum.

septiferous *a.* [L. *septum,* partition; *ferre,* to bear] having septa.

septifolious *a.* [L. *septum,* seven; *folium,* leaf] with 7 leaves or leaflets.

septiform *a.* [L. *septum,* partition; *forma,* shape] in form of a septum.

septifragal *a.* [L. *septum,* partition; *frangere,* to break] with slits as in septicidal dehiscence, but with septa broken and placentae and seeds left in middle.

Septobasidiales *n.* [L. *septum,* partition; *basis,* base; *idion,* dim.] an order of Basidiomycetes having basidia in which development stops for a while and is continued at a later stage.

septomaxillary *a.* [L. *septum,* partition; *maxilla,* jaw] *pert.* maxilla and nasal septum; *appl.* a small bone in many amphibians and reptiles and in certain birds.

septonasal *a.* [L. *septum,* partition; *nasus,* nose] *pert.* nasal, or internarial, septum.

septulum *n.* [L. *septulum,* small septum] a small or secondary septum; *a.* septulate.

septum *n.* [L. *septum,* partition] a partition separating 2 cavities or masses of tissue, as in fruits, chambered shells, corals, heart, nose, tongue, etc.; *plu.* septa; *a.* septate, septal.

septum lucidum thin inner walls of cerebral hemispheres, between corpus callosum and fornix; *alt.* septum pellucidum.

septum narium partition betweeen nostrils; *alt.* septum mobile nasi.

septum transversum foetal diaphragm; ridge within ampulla of semicircular canal.

ser serine, *q.v.*

sera *plu.* of serum.

seral *a.* [L. *serere,* to put in a row] *pert.* a sere; *appl.* a plant community before reaching equilibrium or climax.

sere *n.* [L. *serere,* to put in a row] a successional series of plant communities, as from prisere to climax; a state in a succession.

seriate *a.* [L. *serere,* to put in a row] arranged in a row or series.

sericate, sericeous *a.* [L. *sericus,* silken] covered with fine close-pressed silky hairs; silky.

serific *a.* [L. *sericum,* silk; *facere,* to make] silk-producing.

serine *n.* [L. *serum,* whey] a non-essential amino acid, β-hydroxyalanine, obtained by the hydrolysis of various proteins, and also important in the synthesis of phosphatides; abbreviated to ser.

serodeme *n.* [L. *serum,* whey; Gk. *dēmos,* people] a deme, e.g. of parasites, differing from others in immunological characteristics.

serology n. [L. *serum*, whey; Gk. *logos*, discourse] the study of sera.

serophyte xerophyte, *q.v.*

serosa n. [L. *serum*, whey] serous membrane, *q.v.*; visceral peritoneum; false amnion or outer layer of amniotic fold; outer larval membrane of insects.

serosity n. [L. *serum*, whey] watery part of animal fluid; condition of being serous.

serotinal, serotinous a. [L. *serus*, late] appearing or blooming late in the season; *pert.* late summer, *cf.* aestival; flying late in the evening, as bats.

serotonin n. [L. *serum*, whey; Gk. *tonos*, tightening] a vasoconstrictor compound, chemically 5-hydroxytryptamine, found in blood platelets, brain cells, mast cells, and intestinal tissue, which causes contraction of smooth muscle.

serous a. [L. *serum*, serum] watery; *pert.* serum or other watery fluid, *appl.* fluid, tissue, cells, glands; *alt.* orrhoid.

serous alveoli alveoli which secrete a watery non-viscid saliva; *cf.* mucous alveoli.

serous membrane a thin membrane of connective tissue, lining some closed cavity of body, and reflected over viscera, as mesentery; *alt.* serosa, tunica serosa.

serozyme n. [L. *serum*, serum; Gk. *zymē*, leaven] prothrombin, *q.v.*

serozymogenic a. [L. *serum*, serum; Gk. *zymē*, leaven; *gennaein*, to produce] *appl.* cells of serous alveoli when containing zymogen granules.

Serpentes n. [L. *serpens*, serpent] Ophidia, *q.v.*

Serpulimorpha n. [L. *serpula*, small snake; Gk. *morphē*, form] an order of polychaetes having an anterior cone-shaped set of gills for filtering food material from the water.

serpulite n. [L. *serpula*, small snake; Gk. *lithos*, stone] the fossil tube of *Serpulites* thought to be an annelid or nautiloid. a. *Appl.* grit containing these tubes.

serra n. [L. *serra*, saw] any saw-like structure.

serrate a. [L. *serra*, saw] notched on edge like a saw, *appl.* leaves and other structures.

serrate-ciliate with hairs fringing toothed edges.

serrate-dentate with serrate edges themselves toothed.

serratiform a. [L. *serra*, saw; *forma*, shape] like a saw.

serration n. [L. *serra*, saw] saw-like formation; *alt.* serrature.

serratirostral a. [L. *serra*, saw; *rostrum*, beak] with serrate bill, *appl.* birds.

serratodenticulate a. [L. *serra*, saw; *dens*, tooth] with many-toothed serrations.

serratulate serrulate, *q.v.*

serrature n. [L. *serra*, saw] a saw-like notch; serration, *q.v.*

serratus magnus a muscle stretching from upper ribs to scapula; *alt.* serratus anterior.

serratus posterior superior and inferior: 2 thin thoracic muscles aiding in respiration, spreading respectively backward to anterior ribs, and forward to posterior ribs.

serriferous a. [L. *serra*, saw; *ferre*, to carry] furnished with a saw-like organ or part.

serriform a. [L. *serra*, saw; *forma*, shape] like a saw.

serriped a. [L. *serra*, saw; *pes*, foot] with notched feet.

serrula n. [L. *serrula*, small saw] a comb-like ridge on chelicerae of some Arachnida.

serrulate a. [L. *serrula*, small saw] finely notched; *alt.* serratulate.

serrulation n. [L. *serrula*, small saw] small notch; condition of being finely notched.

Sertoli cells [E. *Sertoli*, Italian histologist] enlarged lining epithelium cells connected with groups of developing spermatozoa in testes; *alt.* supporting cells, spermatoblasts.

serule n. [L. *serere*, to put in a row; *dim.*] a minor sere; succession of minor life forms.

serum n. [L. *serum*, whey] watery fluid which separates from blood on coagulation; the secretion of a serous membrane; whey; *plu.* sera.

serum albumin, serum globulin two of the proteins of serum.

sesamoid a. [Gk. *sēsamon*, sesame; *eidos*, form] *appl.* a bone developed within a tendon and near a joint, as patella, radial or ulnar sesamoid, fabella. n. A sesamoid bone.

sesamoidal *pert.* a sesamoid bone.

sessile a. [L. *sedēre*, to sit] sitting directly on base without support, stalk, pedicel, or peduncle; attached or stationary, as opposed to free living or motile.

seston n. [Gk. *sēsis*, sifting] microplankton, *q.v.*; all bodies, living and non-living, floating or swimming in water; *cf.* nekton, neuston, plankton, tripton.

seta n. [L. *seta*, bristle] a bristle-like structure; the stalk bearing the capsule in bryophytes; chaeta of Chaetopoda; extension of exocuticle, produced by trichogen: a hair, bristle, or scale of insects.

setaceous a. [L. *seta*, bristle] bristle-like; set with bristles.

setiferous a. [L. *seta*, bristle; *ferre*, to carry] bearing setae or bristles; *alt.* setigerous, chaetiferous.

setiform a. [L. *seta*, bristle; *forma*, shape] bristle-shaped; *appl.* teeth when very fine and closely set.

setigerous a. [L. *seta*, bristle; *gerere*, to bear] setiferous, *q.v.*

setigerous sac a sac, in which is lodged a bundle of chaeta, formed by invagination of epidermis in parapodium of Chaetopoda; *alt.* chaeta sac.

setiparous a. [L. *seta*, bristle; *parere*, to produce] producing setae or bristles.

setirostral a. [L. *seta*, bristle; *rostrum*, beak] *appl.* birds with beak bristles.

setobranchia n. [L. *seta*, bristle; *branchiae*, gills] a tuft of setae attached to gills of certain decapods being coxopoditic setae.

set of chromosomes the characteristic group of chromosomes of a typical gamete of an organism.

setose a. [L. *seta*, bristle] set with bristles, bristly.

setula, setule n. [*Dim.* from L. *seta*, bristle] a thread-like or hair-like bristle.

setuliform a. [*Dim.* from L. *seta*, bristle; *forma*, shape] thread-like; like a setula or fine bristle.

setulose a. [*Dim.* from L. *seta*, bristle] set with small bristles.

Sewall Wright effect [*Sewall Wright*, American geneticist] genetic drift, *q.v.*

sex n. [L. *sexus*, sex] the sum of characteristics, structures, functions, by which an animal or plant is classed as male or female.

sex cells gametes, *q.v.*

sex chromatin chromatin forming a condensed mass representing an X-chromosome, for each X-chromosome in excess of one in the mammalian nucleus.

sex chromosomes the chromosomes whose presence, absence, or particular form may determine sex, being the X-, Y-, W-, or Z-chromosomes; *alt.* allosomes, idiochromosomes, heterochromosomes, monosomes, accessory chromosomes; *cf.* autosomes.

sex-conditioned character a phenotype controlled by a gene which may be dominant in one sex and recessive in the other, such as human baldness; *alt.* sex-influenced character.

sex-controlled genes genes encountered in both sexes, but manifesting themselves differently in males and females.

sex cords proliferations from germinal epithelium which give rise either to seminiferous tubules or to medullary cords of ovary.

sex determination any of various methods by which the sex of an individual is established.

sex differentiation differentiation of gametes; differentiation of organisms into kinds with different sexual organs.

sexdigitate *a.* [L. *sex*, six; *digitus*, finger] with 6 fingers or toes.

sexduction *n.* [L. *sexus*, sex; *ducere*, to lead] the process in which a small piece of genetic material from one bacterium is carried with the fertility (F) factor to another bacterium; *alt.* F-duction, F-mediated transduction.

sex factor F-episome, *q.v.*

sexfid *a.* [L. *sex*, six; *findere*, to cleave] cleft into 6, as a calyx.

sexfoil *n.* [L. *sex*, six; *folium*, leaf] a group of 6 leaves or leaflets around one axis.

sex gland gonad, *q.v.*

sex hormones gonad hormones and gonadotropic hormones.

sex-influenced character sex-conditioned character, *q.v.*

sex-limited inheritance inheritance of characters whose genes have a phenotypic effect in one sex only.

sex-linked inheritance transmission of characters and the genes determining them which are borne on the sex chromosomes; *alt.* sex linkage.

sex mosaic an intersex, *q.v.*; gynandromorph, *q.v.*

sex pilus F-pilus, *q.v.*

sexradiate *a.* [L. *sex*, six; *radiatus*, rayed] hexactinal, *q.v.*

sex ratio number of males per 100 females, or per 100 births; percentage of males in a population.

sex reversal, sex transformation a changeover from one sex to the other, natural, pathological, or artificially induced.

sexual *a.* [L. *sexus*, sex] *pert.* sex, *appl.* reproduction, etc.

sexual cell gamete, *q.v.*

sexual coloration the colours displayed during the breeding season, often different in the 2 sexes.

sexual dimorphism marked differences, in shape, size, structure, colour, etc., between male and female of the same species; *alt.* antigeny.

sexual reproduction any kind of reproduction involving fusion of gametes to form a zygote.

sexual selection a mechanism causing evolution by which the females (or males) in a population select makes (or females) with certain characteristics which are then inherited by their progeny.

sexuparous *a.* [L. *sexus*, sex; *parere*, to bear] producing sexual offspring, as after bearing parthenogenetic females in Pterygota.

Seymouriamorpha *n.* [*Seymouria*, generic name; Gk. *morphē*, form] an order of labyrinthodont amphibians, in some classifications placed as reptiles in the Cotylosauria.

shaft *n.* [A.S. *sceaft*, spear shaft] rachis; distal part of stem of feather; stem of hair; scapus, *q.v.*; straight cylindrical part of long bone.

Sharpey's fibres [*W. Sharpey*, Scottish surgeon] calcified bundles of white fibres and elastic fibres perforating and holding together periosteal bone lamellae; *alt.* perforating fibres.

sheath *n.* [A.S. *sceth*, shell or pod] a protective covering; theca, *q.v.*; lower part of leaf enveloping stem or culm; insect wing cover, especially elytron.

sheet erosion the removal of the topsoil over a wide area during heavy rain.

shell *n.* [A.S. *scell*, shell] the hard outer covering of animal or fruit; a calcareous, siliceous, bony, horny, or chitinous covering.

shell gland, shell sac organ in which material for forming a shell is secreted.

shield *n.* [A.S. *scyld*, shield] carapace, *q.v.*; clypeus, *q.v.*; scutellum, *q.v.*; scutum, *q.v.*; dorsal cover, as of Entomostraca and Palaeostraca; disc-like ascocarp or apothecium borne on thallus of lichens.

shift *n.* [A.S. *sciftan*, to divide] *appl.* translocation in which the portion between 2 breaks is transferred to a gap left by a 3rd break in the same chromosome; *cf.* insertional.

shoot *n.* [A.S. *sceótan*, to dart] the part of a vascular plant derived from the plumule, being the stem and usually the leaves; a sprouted part, branch, or offshoot of a plant.

short-day *appl.* plants in which the flowering period is hastened by a relatively short photoperiod, ordinarily less than 12 hours and a correspondingly long dark period; *cf.* long-day, day-neutral.

short shoot brachyplast, *q.v.*

shoulder girdle pectoral girdle, *q.v.*

Shrapnell's membrane [*H. J. Shrapnell*, English anatomist] small, flaccid part of the tympanic membrane above malleolar folds.

shrub *n.* [M.E. *shrubbe*, brushwood] a woody plant which branches from the base and does not have a main trunk.

sialaden *n.* [Gk. *sialon*, saliva; *adēn*, gland] salivary gland, *q.v.*

sialic *a* [Gk *sialon*, saliva] *pert.* saliva.

sialic acids a group of amino sugars derived from neuramic acid and found in glycolipids and glycoproteins.

sialoid *a.* [Gk. *sialon*, saliva; *eidos*, form] like saliva.

siblings *n.plu.* [A.S. *sibb*, kin] offspring of the same parents but not at the same birth, i.e. brothers and sisters; *alt.* sibs.

sibling species true species which do not interbreed but are difficult to separate on morphological evidence alone.

sibmating mating of siblings.

sibship collectively, the siblings of one family.

siccous a. [L. siccus, dry] dry; with little or no juice.

sickle dance a dance of bees where the bee performs a sickle-like semicircular movement and the axis of the semicircle indicates the direction of the food.

sicula n. [L. sicula, small dagger] a small dagger-shaped body at end of graptolite, supposed to be skeleton of primary zooid of colony.

sicyoid a. [Gk. sikyos, gourd; eidos, form] gourd-shaped.

side-chain theory Ehrlich's theory of phenomena of immunity, i.e. that toxins unite with living protoplasm by possessing the same property as that by which nutritive proteins are normally assimilated; alt. lateral-chain theory.

siderocyte n. [Gk. sidēros, iron; kytos, hollow] an erythrocyte containing free iron not utilized in haemoglobin formation.

siderophil a. [Gk. sidēros, iron; philos, loving] staining deeply with iron-coating stains; tending to absorb iron; alt. siderophilous. n. An organism which thrives in the presence of iron.

sierozem n. [Russ. seryi, grey; zemlya, soil] grey soil, containing little humus, of middle latitude continental desert regions.

sieve area a region in the cell wall of a sieve tube element, sieve cell, or parenchyma cell, having pores through which cytoplasmic connections pass to adjacent cells, in sieve tube elements being mostly on end walls between adjoining elements where they form sieve plates; alt. sieve field.

sieve cell a cell of the phloem of gymnosperms and pteridophytes, being elongated and tapering and having sieve areas not aggregated into sieve plates, but being similar in function to a sieve tube element.

sieve disc sieve plate, q.v.

sieve elements the conducting parts of the phloem, consisting of sieve cells or sieve tube elements.

sieve field sieve area, q.v.

sieve pit a primary pit giving rise to a sieve pore.

sieve plate the wall between adjoining sieve tube elements containing sieve areas, alt. sieve disc; madrepore, q.v.; area of coxal lobe of pedipalp, with openings of salivary ducts, in spiders.

sieve pore one of the perforations in a sieve area or sieve plate.

sieve tissue phloem, especially the sieve elements.

sieve tube element the basic unit of a sieve tube consisting of an elongated cell with no nucleus at maturity, joined end to end with similar cells by a sieve plate, and in seed plants connected with a companion cell.

sieve tubes tubes in phloem tissue made of end-to-end sieve cells or sieve tube elements, whose function is the translocation of organic compounds, especially sugars.

sigillate a. [L. sigillum, seal] having seal-like markings, as certain rhizomes and roots.

sigma n. [Gk. ⁶, sigma] a C-shaped or S-shaped sponge spicule; symbol for 0·001 second, or for standard deviation.

sigmaspire n. [Gk. Σ, sigma; L. spira, coil] a sigma with an additional twist.

sigmoid a. [Gk. Σ, sigma; eidos, form]curved like a sigma, in 2 directions, appl. arteries, cavities, valves.

sigmoid flexure an S-shaped double curve as in a bird's neck; S-shaped curve of colon; any S-shaped bend.

sign stimulus an environmental stimulus which acts as a releaser of species-specific behaviour.

silicified a. [L. silex, flint] being impregnated with silica, as the walls of diatoms.

silicle silicula, q.v.

silicoflagellates n.plu. [L. silex, flint; flagellum, whip] a group of Chrysomonadina, q.v.

silicole n. [L. silex, flint; colere, to inhabit] a plant thriving in markedly siliceous soil; cf. calcifuge.

silicular, silicule n. [L. silicula, little pod] a specialized broad flat capsule divided into 2 by a false septum and found in members of the Cruciferae such as honesty; alt. silicle; cf. siliqua.

silicular, siliculose, siliculous a. [L. silicula, little pod] like, pert. or having a silicula.

siliqua n. [L. siliqua, pod] a specialized long thin capsule divided into 2 by a false septum, found in members of the Cruciferae such as wallflower, alt. silique, cf. silicula; siliqua olivae: a pod-shaped structure in mammalian brain, being a group of fibres around the olive.

siliquiform a. [L. siliqua, pod; forma, shape] formed like a siliqua.

siliquose, siliquous a. [L. siliqua, pod] bearing siliquas.

silt n. [M.E. cylte, gravel] a soil intermediate between sands and clays in size of the particles.

Silurian n. [L. Silures, a people of South Wales] pert. or appl. period of Palaeozoic era, between Ordovician and Devonian.

Siluriformes n. [Silurus, generic name of catfish; L. forma, shape] an order of teleosts including the catfish, without true scales but often with dermal plates or bony spines in the skin.

silva selva, q.v.; sylva, q.v.

silverfish a common name for many members of the Thysanura, q.v.

silvicolous a. [L. silvicola, forest inhabitant] inhabiting or growing in woodlands, appl. plant formations.

simblospore n. [Gk. simblos, beehive; sporos, seed] zoospore, q.v.

simian a. [L. simia, ape] possessing characteristics of, or pert., anthropoid apes.

simple not compound.

simple eyes ocelli which occur with or without compound eyes in adults of many insects, and usually the only eyes possessed by larvae; eyes with only one lens.

simple viruses those which have a regular particle structure and whose main chemical components are nucleic acid and protein only.

simplex a. [L. simplex, simple] having one dominant gene, in polyploidy.

simulation n. [L. simulare, to simulate] assumption of features or structures intended to deceive enemies, as forms of leaf and stick insects, and all varieties of protective coloration, i.e. mimicry.

Sinanthropus n. [L.L. Sinae, Chinese; Gk. anthrōpos, man] the old generic name for Peking man, q.v.

sinciput n. [L. semi, half; caput, head] upper or fore part of head; a. sincipital.

sinigrin n. [Sinapsis nigra, black mustard] myrosin, q.v.

sinigrinase myrosinase, *q.v.*

sinistral *a.* [L. *sinister*, left] on or *pert.* the left, *cf.* dextral; sinistrorse, *q.v.*

sinistrorse *a.* [L. *sinister*, left; *vertere*, to turn] growing in a spiral which twines from right to left, anticlockwise, as in most gastropod shells; *alt.* sinistral; *cf.* dextrorse.

sinuate, sinous *a.* [L. *sinus*, curve] winding; tortuous; having a wavy indented margin as leaves or gills of an agaric.

sinu-atrial, sinu-auricular *a.* [L. *sinus*, gulf; *atrium*, central room; *auricula*, small ear] *appl.* node: group of specialized cells in the wall of right atrium near opening of superior or anterior vena cava, in which heart beat is initiated, *alt.* SAN, Keith and Flack's node, pacemaker; *appl.* valves between sinus venosus and atrium.

sinupalliate *a.* [L. *sinus*, curve; *pallium*, mantle] in molluscs, having well-developed siphon, and so an indented pallial line; *cf.* integripalliate.

sinus *n.* [L. *sinus*, curve or gulf] a cavity, depression, recess, or dilatation; a groove or indentation; *alt.* lacuna.

sinuses of Valsalva [*A. M. Valsalva*, Italian anatomist] dilatations of pulmonary artery and of aorta, opposite pulmonary and aortic semilunar valves of heart.

sinus glands endocrine glands in eye-stalks of decapod crustaceans.

sinusoid *n.* [L. *sinus*, gulf; Gk. *eidos*, form] a minute blood space in organ tissue formed from intercrescence of endodermal cells and vascular endothelium, as in liver; blood space with irregular lumen connecting arterial and venous capillaries.

sinus pocularis uterus masculinus, *q.v.*

sinus rhomboidalis in vertebrate embryos, posterior incompletely closed part of medullary canal; later, a dilatation of canal in sacral region, formed from it.

sinus venosus posterior chamber of tubular heart of embryo; in lower vertebrates, a corresponding structure receiving venous blood and opening into auricle; cavity of auricle.

siphon *n.* [Gk. *siphōn*, reed or tube] a tubular or siphon-like structure of various organisms, subserving various purposes.

siphonaceous *a.* [Gk. *siphōn*, tube] *see* siphoneous.

Siphonales *n.* [Gk. *siphōn*, tube] Caulerpales, *q.v.*; in some classifications considered a separate order of Chlorophyceae.

Siphonaptera *n.* [Gk. *siphōn*, tube; *apteros*, wingless] an order of insects, commonly called fleas, that are parasitic and blood-sucking on birds and mammals, have complete metamorphosis, and are eyeless and wingless as adults; *alt.* Aphaniptera.

siphonate *a.* [Gk. *siphōn*, tube] furnished with a siphon or siphons.

siphoneous *a.* [Gk. *siphōn*, tube] tubular; in algae, a form consisting of a more-or-less elaborate multinucleate thallus, not possessing septa, *alt.* siphonaceous.

siphonet *n.* [Gk. *siphōn*, tube] the honey dew tube of an aphid; *alt.* siphuncle.

siphoneum siphonium, *q.v.*

siphonium *n.* [Gk. *siphōn*, tube] membranous tube connecting air passages of quadrate with air space in mandible; *alt.* siphoneum; *a.* siphonial.

Siphonocladales *n.* [*Siphonocladus*, generic name] an order of green algae having multicellular thalli attached to the substratum by rhizoids.

siphonocladial *a.* [Gk. *siphōn*, tube; *kladion*, twig] *appl.* filaments with tubular segments, as in certain green algae.

siphonogam *n.* [Gk. *siphōn*, tube; *gamos*, marriage] a plant securing fertilization through siphonogamy; *a.* siphonogamic, siphonogamous.

siphonogamy *n.* [Gk. *siphōn*, tube; *gamos*, marriage] fertilization by means of a pollen tube through which contents of pollen grain pass to the embryo sac.

siphonoglyph *n.* [Gk. *siphōn*, tube; *glyphein*, to engrave] one of 2 longitudinal grooves or sulci of gullet of some anthozoans.

Siphonophora, siphonophores *n.*, *n.plu.* [Gk. *siphōn*, tubes; *phora*, producing] an order of pelagic hydrozoans with polymorphic colonies of both polyps and medusoid forms on the same colony, some individuals modified as a float or for swimming; the Chondrophora were formerly included in this order.

siphonoplax *n.* [Gk. *siphōn*, tube; *plax*, tablet] a calcareous plate connected with siphon of certain molluscs.

Siphonopoda *n.* [Gk. *siphōn*, tube; *pous*, foot] Cephalopoda, *q.v.*

siphonostele *n.* [Gk. *siphōn*, tube; *stēlē*, post] the hollow vascular cylinder of a stem, which may contain pith; *a.* siphonostelic.

siphonostomatous *a.* [Gk. *siphōn*, tube; *stoma*, mouth] with tubular mouth; having front margin of shell notched for emission of siphon.

siphonozoid, siphonozooid *n.* [Gk. *siphōn*, tube; *zōon*, animal; *eidos*, form] small modified polyp without tentacles and serving to propel water through canal system of certain alcyonarian colonies.

siphorhinal *a.* [Gk. *siphōn*, tube; *rhines*, nostrils] with tubular nostrils.

siphuncle *n.* [L. *siphunculus*, small tube] siphonet, *q.v.*; a median tube of skin, partly calcareous, connecting up all compartments of a nautiloid shell.

Siphunculata *n.* [L. *siphunculus*, small tube] Anoplura, *q.v.*

siphunculate *a.* [L. *siphunculus*, small tube] having a siphuncle; having mouth-parts modified for sucking, as certain lice.

Sipuncula, sipunculids *n.*, *n.plu.* [*Sipunculus*, generic name] a phylum of marine, mainly sedentary unsegmented coelomate worms that were formerly classed as annelids of the class Sipunculoidea, which have the anterior end of the body introverted and used as a proboscis, and tentacles around the mouth.

Sipunculoidea *see* Sipuncula.

Sirenia, sirenians *n.*, *n.plu.* [Gk. *Seirēn*, a Siren] an order of placental mammals commonly called sea cows, including the dugong and manatee, which are highly modified for aquatic life, with a naked body and front limbs as paddles.

sirenin *n.* [Gk. *Seirēn*, a Siren] a substance secreted by certain Phycomycetes during development of female gametes and facilitating fertilization.

siro- seiro-, *q.v.*

sister cell or nucleus one of a pair of cells or nuclei produced by division of an existing cell.

sitology n. [Gk. sitos, food; logos, discourse] science of food, diet, and nutrition.

sitophore n. [Gk. sitos, food; pherein, to bear] trough of hypopharynx between arms of suspensorium.

sitosterol n. [Gk. sitos, food; stereos, solid; alcohol] complex mixtures of phytosterols occurring in fatty or oily tissue of higher plants, especially in corn and wheat-germ oil and in certain algae, particularly the Chlorophyceae.

sitotoxin n. [Gk. sitos, food; toxikon, poison] food poison.

sitotropism n. [Gk. sitos; food; trope, turn] tendency to turn in direction of food; reaction towards stimulating influences of food.

skein nuclear reticulum, q.v.

skeletal a. [Gk. skeletos, dried] pert. the skeleton.

skeletal muscle voluntary muscle, q.v.

skeletogenous a. [Gk. skeletos, hard; gennaein, to produce] appl. embryonic structures or parts which later become parts of skeleton; in sponges the part of the dermal layer made of scattered cells and the jelly in which it is embedded.

skeleton n. [Gk. skeletos, dried, hard] hard framework, internal or external, which supports and protects softer parts of plant or animal, and usually to which muscles attach in animals.

skeletoplasm n. [Gk. skeletos, hard; plasma, mould] formative material destined to form supporting structures.

Skene's glands [A. J. C. Skene, Scottish gynaecologist] racemose mucous glands of the female urethra; alt. para-urethral glands, Guerin's glands.

skiaphyte n. [Gk. skia, shade; phyton, plant] a plant growing in the shade, as algae under rocks, or as undergrowth in the forest; alt. sciaphyte, skiophyte, skiarophyte; cf. heliophyte.

skin n. [A.S. scinn, skin] the external covering of an animal, plant, fruit, or seed.

skin gills see papulae.

skin rings annular markings on body of annelids.

skiophil(ous) a. [Gk. skia, shade; philein, to love] shade-loving, growing in shade; alt. sciophyllous, heliophobic, umbracticolous.

skiophyll n. [Gk. skia, shade; phyllon, leaf] a plant having dorsiventral leaves; alt. sciophyll; cf. heliophyll.

skiophyte skiaphyte, q.v.

skotoplankton n. [Gk. skotos, darkness; plangktos, wandering] plankton living at depths of 500 m; cf. knephoplankton, phaoplankton.

skototaxis n. [Gk. skotos, darkness; taxis, arrangement] positive orientation towards darkness, not negative phototaxis.

skull n. [M.E. skulle, cranium] hard cartilaginous or bony part of head of vertebrate, containing brain, and including the jaws; alt. cranium.

slavery dulosis, q.v.

sleep movements change in position of leaves, leaflets, petals, etc., at night, which may be brought about by external stimuli of light or temperature (nyctinasty), or may be an endogenous circadian rhythm.

sliding growth of cells, when new part of cell wall slides over walls of cells with which it comes in contact; alt. gliding growth; cf. interpositional growth.

slime a viscous substance, usually of protein.

slime bodies cytoplasmic bodies elaborating a viscid proteid, as in sieve tube cells.

slime layer sheath of certain bacterial cells, usually of polysaccharide or polypeptide but of variable composition, called a capsule when thickened.

slime moulds the common name for the Myxomycophyta, q.v.

slime pits in Anthocerotales, cavities in the thallus filled with mucilage and sometimes colonized by algae; alt. slime pores.

slime spore myxospore, q.v.

slime tubes Cuvierian organs, q.v.

slough n. [M.E. slouh, skin of snake] the dead outer skin cast off periodically by snakes. v. To shed the skin.

small intestine collectively, the duodenum, jejunum, and ileum.

smegma n. [Gk. smegma, unguent] secretion of preputial glands, or of clitoris glands; alt. sebum praeputiale.

smooth muscle involuntary muscle, q.v.

smut n. [A.S. smitta, spot] a disease of grasses and other plants, caused by Ustilaginales, fungi producing numerous black spores; any smut fungus.

smut fungi a common name for the Ustilaginales, q.v.

snakes see Squamata.

sobole(s) n. [L. soboles, offshoot] see sucker; an underground creeping stem developing roots and leaves at intervals.

soboliferous a. [L. soboles, offshoot; ferre, to carry] having soboles.

social a. [L. sociare, to associate] gregarious; living in organized groups or colonies.

social facilitation the effect of the actions of one animal of a group on the activity of the others; alt. mimesis.

sociation n. [L. sociare, to associate] a minor unit of vegetation, or micro-association.

socies n. [L. sociare, to associate] a society of plants representing a stage in the process of succession.

society n. [L. societas, company] a number of organisms forming a community; a community of plants other than dominants within an association or consociation.

sociohormone n. [L. sociare, to associate; hormaein, to excite] pheromone, q.v.

sodium pump a mechanism in the membrane of neurones (and some other cells) which actively moves sodium ions out of the cell and maintains the resting potential across the membrane.

soft corals the common name for the Alcyonacea, q.v.

soft-rayed having jointed fin-rays.

softwoods woods from conifers; conifers themselves.

soil profile a section through the soil showing the different layers from the surface soil to underlying bedrock.

soil structure the extent to which soil aggregates are developed, and their size and shape.

soil texture the property of the soil which depends on the size of the mineral particles and which controls aeration and drainage.

sola plu. of solum.

solaeus soleus, q.v.

solar a. [L. sol, sun] having branches or filaments like rays of sun; dextrorse, q.v.

solarization n. [L. *solaris*, solar] retardation or inhibition of photosynthesis due to prolonged exposure to intense light; *alt*. heliosis.

solar plexus a network of sympathetic nerves with some ganglia, situated behind stomach and supplying abdominal viscera; *alt*. coeliac plexus.

soleaform a. [L. *solea*, sandal; *forma*, shape] slipper-shaped.

solenia n.plu. [Gk. *sōlēn*, channel] endoderm-lined canals, diverticula from coelenteron of zooid colony.

Solenichthyes n. [Gk. *sōlēn*, pipe; *ichthys*, fish] an order or superorder of teleosts comprising the Gasterosteiformes and Syngnathiformes.

solenidion n. [Gk. *sōlēn*, pipe; *idion*, dim.] a modified blunt seta associated with a sensory cell, on legs of Acarina; *plu*. solenidia.

solenocytes n.plu. [Gk. *sōlēn*, channel; *kytos*, hollow] in some invertebrates and lower chordates an excretory organ similar to a flame cell in action but supplied with blood vessels, to help filtration and absorption from blood, *alt*. archinephridium; flame cell, *q.v.*

solenophage n. [Gk. *sōlēn*, pipe; *phagein*, to eat] a blood-sucking insect that feeds from the lumen of a blood vessel; *alt*. vessel feeder.

solenostele n. [Gk. *sōlēn*, channel; *stēlē*, column] a stage after siphonostele in fern development, having phloem both internal and external to xylem.

soleus n. [L. *solea*, sole of foot] a flat calf muscle beneath gastrocnemius; *alt*. solaeus.

Solifuga n. [L.L. *solifuga*, a venomous ant or spider] Solpugida, *q.v.*

soliped, solidungulate a. [L. *solus*, single; *pes*, foot] single-hoofed, as horse.

solitaria phase in locusts, the phase during which locusts lack gregarious tendencies; *cf*. gregaria phase.

solitary cells large pyramidal cells of brain, with axons terminating in superior colliculus or in mid-brain.

solitary glands or follicles lymphoid nodules occurring singly on intestines, and constituting Peyer's patches when aggregated.

Solo man a fossil hominid known from remains found near the Solo river in Java, and considered to be either a subspecies of *Homo erectus, H. erectus soloensis*, or a member of *Homo sapiens*.

solonchak n. [Russ. *solonchak*, salt marsh] any of a group of pale saline soils typical of certain poorly drained semiarid regions.

solonets n. [Russ. *solonet'*, to become salty] any of a group of dark alkaline soils formed from solonchak by leaching.

Solpugida n. [*Solpuga*, generic name] an order of arachnids, commonly called false spiders or sun spiders, having very hairy bodies, large chelicerae, and a segmented prosoma and opisthosoma; *alt*. Solifuga.

soluble RNA (sRNA) transfer RNA, *q.v.*

solum n. [L. *solum*, ground, soil] floor, as of cavity; soil between source material and topsoil, *alt*. true soil; *plu*. sola.

soma n. [Gk. *sōma*, body] the animal or plant body as a whole with exception of germinal cells.

somactids n.plu. [Gk. *sōma*, body; *aktis*, ray] endoskeletal supports of dermal fin-rays; *alt*. radials.

somacule n. [Gk. *sōma*, body] a hypothetical unit, *q.v.*

somaesthesis n. [Gk. *sōma*, body; *aisthēsis*, sensation] sensation due to stimuli from skin, muscle, or internal organs.

somaplasm somatoplasm, *q.v.*

Somasteroidea n. [Gk. *sōma*, body; *astēr*, star; *eidos*, form] an extinct class of lower Ordovician echinoderms of the subphylum Eleutherozoa.

somatic a. [Gk. *sōma*, body] *pert*. purely bodily part of animal or plant, *cf*. germinal; *appl*. mutation occurring in a body cell; *appl*. number: basic number of chromosomes in somatic cells.

somatic cell any cell of the body other than a reproductive cell; *cf*. germ cell.

somatoblast n. [Gk. *sōma*, body; *blastos*, bud] a cell which gives rise to somatic cells; a specialized micromere in oosperm division of Annulates.

somatocoels n.plu. [Gk. *sōma*, body; *koilos*, hollow] a pair of sacs which bud off from the primary coelomic sacs in the larval development of echinoderms and later form the main coelom of the adult.

somatocyst n. [Gk. *sōma*, body; *kystis*, bladder] a cavity in pneumatophore of Siphonophora containing air or an oil droplet.

somatoderm n. [Gk. *sōma*, body; *derma*, skin] the outer cells in Mesozoa.

somatogamy n. [Gk. *sōma*, body; *gamos*, marriage] pseudomixis, *q.v.*

somatogenic a. [Gk. *sōma*, body; *gennaein*, to produce] developing from somatic cells, *alt*. somatogenetic; *appl*. variation or adaptations arising from external stimuli.

somatology n. [Gk. *sōma*, body; *logos* discourse] the scientific study of the constitution of the body; study of human constitutional types.

somatome n. [Gk. *sōma*, body; *tomē*, cutting] somite, *q.v.*

somatomedin n. [Gk. *sōma*, body; L. *medius*, middle] in mammals, a polypeptide which is released from the liver, and possibly also from the kidneys, into the blood by somatotrophin; *alt*. plasma growth factor.

somatophyte n. [Gk. *sōma*, body; *phyton*, plant] a plant whose cells develop mainly into adult body tissue.

somatoplasm n. [Gk. *sōma*, body; *plasma*, mould] the substance of a somatic cell; *alt*. somaplasm.

somatopleure n. [Gk. *sōma*, body; *pleura*, side] the body wall formed by somatic layer of mesoblast becoming closely connected with surface epiblast; *a*. somatopleural.

somatotrophic a. [Gk. *sōma*, body; *trephein*, to increase] stimulating nutrition and growth; *appl*. hormone (STH): somatotrophin, *q.v.*; *alt*. somatotropic, somatropic.

somatotrophin n. [Gk. *sōma*, body; *trephein*, to increase] a protein hormone which is produced by the anterior lobe of the pituitary gland and promotes general body growth; *alt*. somatotrophic or somatotropic hormone, STH, growth hormone, phyone, tethelin.

somatotype n. [Gk. *sōma*, body; *typos*, pattern] body type or conformation as rated by measurements.

somatropic a. [Gk. *sōma*, body; *tropikos*, turning] somatotrophic, *q.v.*

somite n. [Gk. *sōma*, body] a mesoblastic segment

or compartment; a body segment of a metamerically segmented animal; *alt.* somatome.

somitic *a.* [Gk. *sōma*, body] *pert.* or giving rise to somites, *appl.* mesoderm, *appl.* paraxial plate.

sonic *a.* [L. *sonare*, to sound] *pert.* or produced by sound.

sonochemical *a.* [L. *sonus*, sound; Gk. *chēmeia*, transmutation] *appl.* or *pert.* biochemical reactions, as disruption of cells, induced by sound waves.

Sonoran *a.* [*Sonora*, Mexican State] *appl.* or *pert.* zoogeographical region of southern North America, including northern Mexico, between nearctic and neotropical regions; *alt.* Medio-Columbian.

soral *a.* [Gk. *sōros*, heap] *pert.* a sorus.

soralium *n.* [Gk. *sōros*, heap] a well-defined group of soredia surrounded by a distinct margin on the thallus of a lichen.

sorbitol *n.* [*Sorbus*, generic name of mountain ash] a faintly sweet alcohol found in the fruits of Rosaceae, especially mountain ash, and which may be obtained from the hexose sugar sorbose or from glucose, and is isomeric with mannitol.

sorede soredium, *q.v.*

soredia *plu.* of soredium.

soredial *a.* [Gk. *sōros*, heap] *pert.* or resembling a soredium.

sorediate *a.* [Gk. *sōros*, heap] bearing soredia; with patches on the surface.

soredium *n.* [Gk. *sōros*, heap] a scale-like or globular body consisting of fungal hyphae with some algal cells, on thallus of some lichens, and serving for propagation; *alt.* brood bud, sorede, soreuma; *plu.* soredia.

soreuma *n.* [Gk. *sōreuma*, heap] soredium, *q.v.*

sori *plu.* or sorus.

soriferous *a.* [Gk. *sōros*, heap; L. *ferre*, to carry] bearing sori; *alt.* sorose.

sorocarp *n.* [Gk. *sōros*, heap; *karpos*, fruit] the unenclosed simple fruit body of certain slime moulds.

sorogen *n.* [Gk. *sōros*, heap; *gennaein*, to produce] the cell or tissue that develops into a sorus.

sorophore *n.* [Gk. *sōros*, heap; *pherein*, to bear] base or stalk bearing a sorus or sorocarp.

sorose *a.* [Gk. *sōros*, heap] soriferous, *q.v.*

sorosis *n.* [Gk. *sōros*, heap] a composite fruit (anthocarp) formed by fusion of fleshy axis and flowers, as pineapple.

sorotrochous *a.* [Gk. *sōros*, heap; *trochos*, wheel] having a compound wheel organ or trochal disc, as certain Rotifera.

sorption *n.* [L. *sorbere*, to suck in] retention of material at a surface, by absorption or by adsorption.

sorus *n.* [Gk. *sōros*, heap] a collection of small stalked sporangia on under surface of a fern pinnule, *alt.* fruit spot; a group of sporangia in fungi; a group of antheridia on frond of seaweeds; a mass of soredia in lichens; clusters of spores in some Sarcodina and slime moulds; *plu.* sori; *a.* soral.

soup the prebiotic solution or suspension of organic molecules thought to have been formed before the origin of life on earth.

SP suction pressure, *q.v.*

spadiceous *a.* [L. *spadix*, palm branch] arranged like a spadix; *alt.* spadicifloral.

Spadiciflorae *n.* [L. *spadix*, palm branch; *flos*, flower] Arecidae, *q.v.*

spadiciform, spadicose *a.* [L. *spadix*, palm branch; *forma*, shape] resembling a spadix.

spadix *n.* [L. *spadix*, palm branch] a racemose inflorescence with elongated axis, sessile flowers, and an enveloping spathe; a succulent spike; endodermal rudiment of developing manubrium of certain coelenterates; conoid amalgamation of internal lateral lobes of tentacles in *Nautilus*, a cephalopod.

spanandry *n.* [Gk. *spanos*, scarce; *anēr*, male] a scarcity of males; progressive decrease in number of males, as in some insects.

spanogamy *n.* [Gk. *spanos*, scarce; *gamos*, marriage] progressive decrease in number of females.

spasm *n.* [Gk. *spasmos*, tension] involuntary muscular contraction; spastic or spasmodic contraction of muscle fibres.

spasmoneme *n.* [Gk. *spasmos*, tension; *nēma*, thread] in certain stalked Ciliphora, a stalk muscle formed by union of longitudinal myonemes.

spat *n.* [A.S. *spaetan*, to spit] the spawn or young of bivalve molluscs.

Spatangoida *n.* [*Spatangus*, generic name; Gk. *eidos*, form] an order of heart- or cushion-shaped sea urchins of the superorder Atelostomata.

spathaceous, spathal *a.* [L. *spatha*, broad blade] resembling or bearing a spathe.

spathe *n.* [Gk. *spathē*, broad blade] a large enveloping leaf-like structure, green or petaloid, protecting a spadix.

spathed *a.* [Gk. *spathē*, broad blade] furnished with a spathe.

spathella *n.* [L. *spatha*, broad blade] small spathe surrounding division of palm spadix.

Spathiflorae *n.* [L. *spatha*, broad blade; *flos*, flower] Arales, *q.v.*

spathose *a.* [L. *spatha*, broad blade] with or like a spathe.

spatia *n.plu.* [L. *spatium*, space] spaces, e.g. intercostal spaces.

spatia zonularia a canal surrounding marginal circumference of lens of eye; *alt.* Petit's canal.

spatula *n.* [L. *spatula*, spoon] a breast bone or anchor process of certain dipterous larvae; a spoon-shaped structure.

spatulate *a.* [L. *spatula*, spoon] spoon-shaped, *appl.* a leaf with broad, rounded apex, thence tapering to base.

spawn *n.* [O.F. *espandre*, to shed] collection of eggs deposited by bivalve molluscs, fishes, frogs, etc.; mycelium of certain fungi. *v.* To deposit eggs, as by fishes, etc.

spay *v.* [L. *spado*, eunuch] to deprive of ovaries.

Special Creation the doctrine that organisms did not evolve but that each was created in its present form.

specialization *n.* [L. *specialis*, special] adaptation to a particular mode of life or habitat in the course of evolution; specificity, *q.v.*

speciation *n.* [L. *species*, particular kind] the evolution of species; development of a specific quality; species formation.

species *n.* [L. *species*, particular kind] a group of interbreeding individuals not interbreeding with another such group, being a taxonomic unit including geographical races and varieties and having 2 names in binomial nomenclature, the generic name and specific epithet, similar and related species being grouped into a genus.

species aggregate a group of very closely related species which have more in common with each other than with other species.

species-specific behaviour the behaviour patterns that are inborn in a species and performed by all members of that species under the same conditions and which are not modified by learning; *alt*. inborn behaviour, innate behaviour, instinct.

specific *a*. [L. *species*, particular kind; *facere*, to make] peculiar to; *pert*. a species, *appl*. characteristics distinguishing a species.

specific dynamic action *see* dynamic.

specific epithet the 2nd name in binomial nomenclature; *alt*. specific name.

specificity *n*. [L. *species*, particular kind; *facere*, to make] condition of being specific; being limited to a species; restriction of parasites to particular hosts; restriction of enzymes to certain substrates; *alt*. specialization.

spectrum *n*. [L. *spectrum*, appearance] a statistical survey of the distribution of species for determination and comparison of biogeographical regions; the series of colours resulting from dispersal of white light; the range of audible wavelengths.

speculum *n*. [L. *speculum*, mirror] ocellus, *q.v.*; a wing bar having a metallic sheen, as in drakes.

Spelaeogriphacea *n*. [*Spelaeogriphus*, generic name] an order of blind malacostracans that live in caves and have an elongated body and small carapace.

spel(a)eology *n*. [Gk. *spēlaion*, cave; *logos*, discourse] the study of caves and cave life.

sperm *n*. [Gk. *sperma*, seed] any male gamete; spermatozoid, *q.v.*; spermatozoon, *q.v.*; semen, *q.v.*

spermaduct *n*. [Gk. *sperma*, seed; L. *ducere*, to lead] duct for conveying spermatozoa from testis to exterior; *alt*. spermiduct, spermoduct.

spermagglutination agglutination of spermatozoa such as is brought about by members of the myxovirus group in some mammals and birds.

spermagone, spermagonium spermatogonium, *q.v.*; spermogonium, *q.v.*

spermangium *n*. [Gk. *sperma*, seed; *anggeion*, vessel] an organ producing male spore-like cells, in Ascomycetes; *alt*. andrangium.

spermaphore *n*. [Gk. *sperma*, seed; *pherein*, to bear] placenta of plants.

Spermaphyta, spermaphytes *n*., *n.plu*. [Gk. *sperma*, seed; *phyton*, plant] Spermatophyta, *q.v.*

spermaphytic *a*. [Gk. *sperma*, seed; *phyton*, plant] seed-bearing.

spermary, spermarium *n*. [Gk. *sperma*, seed] an organ in which male gametes are produced, such as the testis.

spermatangium *n*. [Gk. *sperma*, seed; *anggeion*, vessel] male sex organ in red algae which liberates a single non-motile spermatium.

spermateleosis *n*. [Gk. *sperma*, seed; *teleiōsis*, completion] development of spermatozoon from spermatid in spermatogenesis, *q.v.*; the whole of spermatogenesis, *q.v.*; *alt*. spermioteleois, spermiogenesis.

spermatheca *n*. [Gk. *sperma*, seed; *thēkē*, case] in female or hermaphrodite invertebrates, a sac for storing spermatozoa received in copulation; *alt*. receptaculum seminis, seminal receptacle, spermotheca.

spermatia *plu*. of spermatium.

spermatic *a*. [Gk. *sperma*, seed] *pert*. spermatozoa; *pert*. testis.

spermatid *n*. [Gk. *sperma*, seed] a haploid cell arising by division of secondary spermatocyte, and becoming a spermatozoon; *alt*. spermatoblast.

spermatiferous *a*. [Gk. *sperma*, seed; L. *ferre*, to carry] bearing spermatia.

spermatiform *a*. [Gk. *sperma*, seed; L. *forma*, shape] resembling a spermatium.

spermatiophore *n*. [Gk. *sperma*, seed; *pherein*, to bear] a spermatia-producing sporophore; *alt*. spermatophore.

spermatism spermism, *q.v.*

spermatium *n*. [Gk. *sperma*, seed] a non-motile gamete of red algae; pycnidiospore in rust fungi; oidium in toadstools and mushrooms; small conidium in cup fungi; any spore-like structure in fungi or lichens which acts as a gamete; *alt*. pollinoid; *plu*. spermatia.

spermatize *v*. [Gk. *sperma*, seed] to impregnate.

spermatoblast *n*. [Gk. *sperma*, seed; *blastos*, bud] spermatid, *q.v.*; Sertoli cell, *q.v.*; *alt*. spermoblast.

spermatoblastic *a*. [Gk. *sperma*, seed; *blastos*, bud] sperm-producing.

spermatocyst *n*. [Gk. *sperma*, seed; *kystis*, bladder] a seminal sac.

spermatocyte *n*. [Gk. *sperma*, seed; *kytos*, hollow] a cell arising by growth from a spermatogonium: a primary spermatocyte divides to form 2 secondary spermatocytes, each of which gives rise to 2 spermatids.

spermatocytogenesis *n*. [Gk. *sperma*, seed; *kytos*, hollow; *genesis*, descent] first phase of spermatogenesis, preceding spermateliosis.

spermatogenesis *n*. [Gk. *sperma*, seed; *genesis*, origin] sperm formation, from spermatogonium, through primary and secondary spermatocytes and spermatid, to spermatozoon; *alt*. spermogenesis; *cf*. spermateleosis.

spermatogenetic *a*. [Gk. *sperma*, seed; *genesis*, descent] *pert*. sperm formation; sperm-producing; *alt*. spermatogenic, spermatogenous.

spermatogonium *n*. [Gk. *sperma*, seed; *gonos*, offspring] primordial male germ cells; sperm mother cell; spermogonium, *q.v.*; *alt*. spermagonium; *a*. spermatogonial.

spermatoid *a*. [Gk. *sperma*, seed; *eidos*, form] like a sperm.

spermatomerites *n.plu*. [Gk. *sperma*, seed; *meros*, part] chromatin granules formed from sperm nucleus.

spermatophore *n*. [Gk. *sperma*, seed; *pherein*, to bear] a capsule of albuminous matter containing a number of sperms; spermatiophore, *q.v.*

Spermatophyta, spermatophytes *n*., *n.plu*. [Gk. *sperma*, seed; *phyton*, plant] a major division of the plant kingdom, characterized by reproducing by seed, and subdivided into the Gymnospermae and Angiospermae and sometimes the Pteridospermae, or in some classifications into the Coniferophytina, Cycadophytina, and Magnoliophytina; *alt*. seed plants, Spermaphyta, Spermophyta, Anthophyta, Phanerogamia, Phaenogamia.

spermatoplasm *n*. [Gk. *sperma*, seed; *plasma*, mould] protoplasm of sperm cells.

spermatoplast *n*. [Gk. *sperma*, seed; *plastos*, moulded] a male gamete.

spermatosome n. [Gk. *sperma*, seed; *sōma*, body] spermatozoon, *q.v.*

spermatostrate a. [Gk. *sperma*, seed; L. *stratus*, strewn] spread by means of seeds.

spermatoxin n. [Gk. *sperma*, seed; *toxikon*, poison] antibodies causing sterility, formed after injection of spermatozoa in serum.

spermatozeugma n. [Gk. *sperma*, seed; *zeugma*, bond] union by conjugation of 2 or more spermatozoa, as in vas deferens of some insects.

spermatozoa *plu.* of spermatozoon.

spermatozoid, spermatozooid n. [Gk. *sperma*, seed; *zōon*, animal; *idion*, *dim.*] a free-swimming male gamete in lower plants; *alt.* antherozoid, sperm, zoosperm.

spermatozoon n. [Gk. *sperma*, seed; *zōon*, animal] a male gamete consisting usually of a head containing the nucleus, middle piece containing mitochondria, and tail piece of locomotory flagellum; *alt.* spermium, sperm, spermatosome; *plu.* spermatozoa.

sperm centrosome end-knob of axial filament of spermatozoon, situated on middle piece just at base of head; the small body at apex of head.

spermiducal a. [Gk. *sperma*, seed; L. *ducere*, to lead] *appl.* glands into or near which sperm ducts open, in many vertebrates; *appl.* glands associated with male ducts, or prostates, in Oligochaeta.

spermiduct spermaduct, *q.v.*

spermine n. [Gk. *sperma*, seed] a polyamine found as its phosphate in semen, in body tissue, pancreas, blood serum, and yeast.

spermiocalyptrotheca n. [Gk. *sperma*, seed; *kalyptra*, covering; *thēkē*, case] the head-cap of a spermatozoon.

spermiocyte n. [Gk. *sperma*, seed; *kytos*, hollow] primary spermatocyte.

spermiogenesis, spermioteleosis n. [Gk. *sperma*, seed; *genesis*, origin] spermateleosis, *q.v.*

spermism n. [Gk. *sperma*, seed] theory held by spermists or animalculists that embryo is derived from spermatozoon alone; *alt.* spermatism; *cf.* ovism.

spermium n. [Gk. *sperma*, seed] spermatozoon, *q.v.*

sperm nucleus male pronucleus, *q.v.*

spermoblast n. [Gk. *sperma*, seed; *blastos*, bud] spermatoblast, *q.v.*

spermocarp n. [Gk. *sperma*, seed; *karpos*, fruit] an oogonium after fertilization.

spermocentre n. [Gk. *sperma*, seed; L. *centrum*, a centre] the male centrosome during fertilization.

spermoderm n. [Gk. *sperma*, seed; *derma*, skin] the seed coat, consisting of inner tegmen and outer testa.

spermodochium n. [Gk. *sperma*, seed; *docheion*, holder] a group of spermatiophores derived from a single cell and lacking a capsule; *cf.* spermogonium.

spermoduct spermaduct, *q.v.*

spermogenesis spermatogenesis, *q.v.*

spermogoniferous a. [Gk. *sperma*, seed; *gonos*, offspring; L. *ferre*, to carry] having spermogonia.

spermogonium n. [Gk. *sperma*, seed; *gonos*, generation] a capsule containing spermatia in certain fungi and lichens, *cf.* spermodochium; pycnidium, *q.v.*; *alt.* spermagonium, spermatogonium, pycnoconidangium; *plu.* spermogonia.

spermogonous a. [Gk. *sperma*, seed; *gonos*, offspring] like or *pert.* a spermogonium.

spermology n. [Gk. *sperma*, seed; *logos*, discourse] the study of seeds.

Spermophyta, spermophytes Spermatophyta, *q.v.*

spermospore n. [Gk. *sperma*, seed; *sporos*, seed] a male gamete produced in a spermangium.

spermotheca n. [Gk. *sperma*, seed; *thēkē*, case] spermatheca, *q.v.*

spermotype n. [Gk. *sperma*, seed; *typos*, pattern] a plant specimen grown from seed of a type plant.

spermozeugma n. [Gk. *sperma*, seed; *zeugma*, bond] a mass of regularly aggregated spermatozoa, for delivery into a spermatheca.

sperm pump pompetta, *q.v.*

Sphacelariales n. [*Sphacelaria*, generic name] an order of brown algae having isomorphic alternation of generations and thalli produced by growth of a single apical cell.

sphacelate a. [Gk. *sphakelos*, gangrene] decayed; withered; looking as though decayed and withered.

sphacelia n. [Gk. *sphakelos*, gangrene] conidial or honey-dew stage in development of fungus, producing sclerotium or ergot.

sphaer- *alt.* spher-, *q.v.*

sphaeraphides n.plu [Gk. *sphaira*, globe; *rhaphis*, needle] globular clusters of minute crystals in plant cells; *alt.* cluster crystals, sphere crystals, sphaerraphides, sphaerites.

sphaerenchyma n. [Gk. *sphaira*, globe; *engchyma*, juice] tissue of spherical cells.

Sphaeriales n. [Gk. *sphaira*, globe] an order of Pyrenomycetes having globose sessile perithecia standing free from the remainder of the mycelium.

sphaeridia n.plu. [Gk. *sphaira*, globe; *idion*, *dim.*] small rounded bodies, possibly balancing organs or some other kind of sense organ, found on the surface of some echinoderms.

sphaerite n. [Gk. *sphaira*, globe] a sphaeraphide, especially when made of calcium oxalate or starch.

Sphaerocarpales n. [Gk. *sphaira*, globe; *karpos*, fruit] an order of liverworts, having delicate simple thalli and large involucres surrounding spherical capsules during development.

sphaeroid a. [Gk. *sphaira*, globe; *eidos*, form] globular, ellipsoidal, or cylindrical; *appl.* an aggregate of individual Protozoa; *appl.* a dilated hyphal cell containing oil droplets, in lichens.

sphaeroplast n. [Gk. *sphaira*, globe; *plastos*, formed] spheroplast, *q.v.*

Sphaeropleales n. [*Sphaeroplea*, generic name] an order of green algae having filaments of multinucleate cells.

Sphaeropsidales n. [*Sphaeropsis*, generic name] an order of Fungi Imperfecti which are plant parasites and have conidia borne in cavities within the host or in rounded fruit bodies.

sphaerraphides sphaeraphides, *q.v.*

Sphagnales n. [*Sphagnum*, generic name] an order or in some classifications a superorder of mosses of subclass Sphagnidae.

sphagnicolous a. [Gk. *sphagnos*, moss; L. *colere*, to inhabit] inhabiting peat moss.

Sphagnidae n. [*Sphagnum*, generic name] a subclass of mosses with one order Sphagnales, having the gametophore with branches in whorls, many dead water-absorbing cells, and comprising one genus *Sphagnum*, commonly called bog moss or peat moss.

sphagnous *a.* [Gk. *sphagnos*, moss] *pert.* peat moss.

sphalerocarp *n.* [Gk. *sphaleros*, ready to fall; *karpos*, fruit] a multiple fruit or anthocarp formed from an enlarged and fleshy perianth; *cf.* diclesium.

S-phase the phase of the cell cycle in which the amount of DNA is doubled and the chromosomes replicate.

sphenethmoid *n.* [Gk. *sphēn*, wedge; *ēthmos*, sieve; *eidos*, form] single bone replacing the orbitosphenoids in anurans.

sphenic *a.* [Gk. *sphēn*, wedge] like a wedge.

Sphenisciformes *n.* [*Spheniscus*, generic name of penguin; L. *forma*, shape] an order of birds of the subclass Neornithes, including the penguins.

spheno-ethmoidal *a.* [Gk. *sphēn*, wedge; *ēthmos*, sieve; *eidos*, form] *pert.* or in region of sphenoid and ethmoid, *appl.* a recess above superior nasal concha, and a suture.

sphenofrontal *a.* [Gk. *sphēn*, wedge; L. *frons*, forehead] *pert.* sphenoid and frontal bones, *appl.* a suture.

sphenoid *n.* [Gk. *sphēn*, wedge; *eidos*, form] a basal compound skull bone of some vertebrates, *alt.* butterfly bone. *a.* Wedge-shaped, *alt.* cuneate, cuneiform.

sphenoidal *a.* [Gk. *sphēn*, wedge; *eidos*, form] wedge-shaped; *pert.* or in region of sphenoid; *appl.* fissure, processes, nostrum, sinus; *cf.* cuneiform.

sphenolateral *n.* [Gk. *sphēn*, wedge; L. *latus*, side] one of a dorsal pair of cartilages parallel to trabeculae; *alt.* pleurosphenoid.

sphenomandibular *a.* [Gk. *sphēn*, wedge; L. *mandibulum*, jaw] *pert.* sphenoid and mandible, *appl.* ligament.

sphenomaxillary *a.* [Gk. *sphēn*, wedge; *maxilla*, jaw] *pert.* sphenoid and maxilla, *appl.* fissure and (pterygopalatine) fossa.

Sphenomonadales *n.* [*Sphenomonas*, generic name] an order of saprophytic Euglenophyceae.

sphenopalatine *a.* [Gk. *sphēn*, wedge; L. *palatus*, palate] *pert.* sphenoid and palatine, *appl.* artery, foramen, nerves.

sphenopalatine ganglion an autonomic ganglion on the maxillary nerve in pterygopalatine fossa; *alt.* Meckel's ganglion, pterygopalatine ganglion.

sphenoparietal *a.* [Gk. *sphēn*, wedge; L. *paries*, wall] *pert.* sphenoid and parietal, *appl.* a cranial suture.

Sphenophyllales *n.* [Gk. *sphēn*, wedge; *phyllon*, leaf] an order of fossil Sphenopsida having weak stems with secondary thickening.

Sphenophyta *n.* [Gk. *sphēn*, wedge; *phyton*, plant] in some classifications a division of the Tracheophyta equivalent to the Sphenopsida, *q.v.*

Sphenopsida *n.* [Gk. *sphēn*, wedge; *opsis*, appearance] a class of pteridophytes (in some classifications a subdivision of the Tracheophyta), commonly called horsetails, having a sporophyte with roots, jointed stems, and whorled leaves, and reflexed sporangia borne on sporangiophores; *alt.*

Arthrophyta, Articulae, Articulatae, Equisetatae, Calamophyta.

sphenopterygoid *a.* [Gk. *sphēn*, wedge; *pteryx*, wing; *eidos*, form] *pert.* sphenoid and pterygoid; *appl.* mucous pharyngeal glands, near openings of Eustachian tubes, as in birds.

sphenosquamosal *a.* [Gk. *sphēn*, wedge; L. *squama*, scale] *appl.* cranial suture between sphenoid and squamosal.

sphenotic *n.* [Gk. *sphēn*, wedge; *ous*, ear] postfrontal cranial bone of many fishes.

sphenoturbinal *n.* [Gk. *sphēn*, wedge; L. *turbo*, whirl] laminar process of sphenoid.

sphenozygomatic *a.* [Gk. *sphēn*, wedge; *zygōma*, *zygon*, cross-bar] *appl.* cranial suture between sphenoid and zygomatic.

spher- *alt.* sphaer-, *q.v.*

spheraster *n.* [Gk. *sphaira*, globe; *astēr*, star] a many-rayed globular spicule.

sphere crystals sphaeraphides, *q.v.*

spheridium *n.* [Gk. *sphaira*, globe; *idion*, dim.] a spherical apothecium or capitulum in certain lichens; *plu.* spheridia.

spheroidal *a.* [Gk. *sphaira*, globe; *eidos*, form] globular but not perfectly spherical, *appl.* glandular epithelium.

spheroidocyte *n.* [Gk. *sphaira*, globe; *eidos*, form; *kytos*, hollow] a type of haemocyte in insects.

spherome *n.* [Gk. *esphairōmēn*, made globular] cell inclusions producing oil or fat globules; intracellular fatty globules as a whole.

spheromere *n.* [Gk. *sphaira*, globe; *meros*, part] a segment of a radiate animal.

spheroplast *n.* [Gk. *sphaira*, globe; *plastos*, formed] a plant or bacterial cell from which the cell wall has been artificially removed, but which still retains a number of cellular functions; mitochondrion, *q.v.*; hypothetical unit, *q.v.*; any of various organelles in the cytoplasm; *alt.* sphaeroplast.

spherosome *n.* [Gk. *sphaira*, sphere; *sōma*, body] in eukaryotes, an organelle derived from endoplasmic reticulum and bounded by a single membrane, that synthesizes lipids.

spherula *n.* [L. *sphaerula*, small globe] a spherule or small sphere; a small spherical spicule.

spherulate *a.* [L. *sphaerula*, small globe] covered with small spheres.

sphincter *n.* [Gk. *sphinggein*, to bind tight] a muscle which contracts or closes an orifice, as that of bladder, mouth, anus, vagina, etc.

sphingolipid(e)s *n.plu.* [Gk. *sphingein*, to draw tight; *lipos*, fat] complex lipids which on hydrolysis yield lipid and sphingosine or dihydrosphingosine.

sphingomyelins *n.plu.* [Gk. *sphingein*, to draw tight; *myelos*, marrow] compounds containing sphingosine, phosphorylcholine, and a fatty acid group, found in the myelin sheath of nerves.

sphingosine *n.* [Gk. *sphingein*, to draw tight] a nitrogenous base found in gangliosides, glycosphingolipids, sphingolipids, and sphingomyelins.

sphragidal *a.* [Gk. *sphragis*, seal] *appl.* plastic fluid secreted by tubular glands opening into vesiculae seminales in male Lepidoptera and forming a sphragis.

sphragis *n.* [Gk. *sphragis*, seal] a structure sealing bursa copulatrix on female abdomen of certain Lepidoptera after pairing, and consisting of hardened sphragidal fluid.

sphygmic *a.* [Gk. *sphygmos*, pulse] *pert.* the pulse; *appl.* 2nd phase of systole.

sphygmoid *a.* [Gk. *sphygmos*, pulse; *eidos*, form] pulsating; like a pulse.

sphygmus *n.* [Gk. *sphygmos*, pulse] the pulse.

spica *n.* [L. *spica*, spike] spike, *q.v.*; calcar, *q.v.*, of birds.

spicate, spiciferous *a.* [L. *spica*, spike] spiked; arranged in spikes, as an inflorescence; bearing spikes; with spur-like prominence; *alt.* spicigerous.

spiciform *a.* [L. *spica*, spike; *forma*, form] spike-shaped.

spicigerous spicate, *q.v.*

spicose *a.* [L. *spica*, spike] with spikes or ears, as corn.

spicula *n.* [L. *spicula*, small spike] a small spike; a needle-like body; spicule, *q.v.*; *plu.* of spiculum.

spicular *pert.* or like a spicule.

spiculate *a.* [L. *spicula*, small spike] set with spicules; divided into small spikes.

spicule *n.* [L. *spicula*, small spike] a minute needle-like body, siliceous or calcareous, found in invertebrates; a minute pointed process; formerly, a sterigma; *alt.* spicula.

spiculiferous *a.* [L. *spicula*, small spike; *ferre*, to carry] furnished with or protected by spicules; *alt.* spiculigerous, spiculose.

spiculiform *a.* [L. *spicula*, small spike; *forma*, shape] spicule-shaped.

spiculum *n.* [L. *spiculum*, a dart] a spicular structure; the dart of a snail; the pointed tip of a sterigma.

spider cells fibrous astrocytes, *q.v.*

spiders the common name for the Araneida, *q.v.*

Spigelian *a.* [*A. van den Spieghel* or *Spigelius*, Flemish anatomist] *appl.* a small lobe of liver, originally named lobus exiguus, in mammals.

spigots *n.plu.* [L. *spica*, spike] conical spinning tubes, in spiders.

spike *n.* [L. *spica*, spike, ear of corn] a simple racemose inflorescence with sessile flowers or spikelets along the axis; *alt.* spica; *a.* spikate.

spikelet *n.* [L. *spica*, spike] one of the units of which the inflorescence is made in grasses, consisting of several florets on a thin axis, at the base of which are 2 bracts (glumes) marking the end of the spikelet; one of the units of the inflorescence in sedges; *alt.* locusta.

spina *n.* [L. *spina*, spine] a spine; median apodeme behind furca, as in many Orthoptera.

spinal *a.* [L. *spina*, spine] *pert.* backbone; *pert.* spinal cord; *appl.* foramen, ganglion, nerves, etc.

spinal canal vertebral canal containing spinal cord.

spinal column backbone, vertebral column.

spinal cord dorsal nerve cord posterior to brain and contained in spinal canal; *alt.* medulla spinalis, neural axis.

spinalis *n.* [L. *spina*, spine] name given to muscles connecting vertebrae; *plu.* spinales.

spinal segment *see* neuromere.

spinasternum *n.* [L. *spina*, thorn; *sternum*, breast bone] an intersegmental sternal sclerite or post-sternellum with an internal spine, in certain insects.

spinate *a.* [L. *spina*, thorn] spine-shaped; spine-bearing; *alt.* spiniferous, spinigerous.

spination *n.* [L. *spina*, thorn] the occurrence, development, or arrangement of spines.

spindle *n.* [A.S. *spinnan*, to spin] a structure resembling a spinning-machine spindle; an elongated peduncle bearing sessile flowers; a structure formed of achromatin fibres during mitosis; a muscle spindle, *q.v.*; fuseau, *q.v.*

spindle cells spindle-shaped type of coelomocytes.

spindle-fibre locus centrosome, *q.v.*

spindle organ muscle spindle, *q.v.*

spine *n.* [L. *spina*, a spine] a sharp-pointed process as on leaves, bones, echinoids, porcupines; the backbone or vertebral column; pointed process of vertebra; scapular ridge; fin-ray.

spinescent *a.* [L. *spinescere*, to become spiny] tapering; tending to become spiny.

spiniferous *a.* [L. *spina*, spine; *ferre*, to carry] spine-bearing, *appl.* pads on ventral side of distal end of leg in the onychophoran, *Peripatus*; *alt.* spinate, spinigerous.

spiniform *a.* [L. *spina*, spine; *forma*, shape] acanthoid, *q.v.*

spinigerous *a.* [L. *spina*, spine; *gerere*, to bear] spine-bearing, *appl.* hedgehog; *alt.* spinate, spiniferous.

spinisternite *n.* [L. *spina*, spine; *sternum*, breast bone] a small sternite with spiniform apodema, between thoracic segments of insects.

spinneret *n.* [A.S. *spinnan*, to spin] one of organs perforated by tubes connected with glands secreting liquid silk, in spiders; one of organs preparing material for puparia, as in Coccidae bugs.

spinnerule *n.* [A.S. *spinnan*, to spin] a tube discharging silk secretion of spiders.

spinning glands glands which secrete silky material in arthropods, such as for webs in spiders and cocoons in caterpillars.

spinocaudal *a.* [L. *spina*, spine; *cauda*, tail] *pert.* trunk of vertebrates; *appl.* inductor: posterior roof of archenteron, *alt.* trunk inductor.

spino-occipital *a.* [L. *spina*, spine; *occiput*, back of head] *appl.* nerves arising in trunk somites which later form part of the skull; *appl.* nerve roots from medulla oblongata which join to form the occipital nerve in Selachii.

spinose *a.* [L. *spinosus*, prickly] bearing many spines.

spinous *a.* [L. *spina*, spine] spiny; spine-like.

spinous process median dorsal spine-like process of vertebra; a process of sphenoid; a process between articular surfaces of proximal end of tibia.

spinulate *a.* [L. *spinula*, small spine] covered with small spines; *alt.* spinulose, spinulous.

spinulation *n.* [L. *spinula*, small spine] a defensive spiny covering; state of being spinulate.

spinule *n.* [L. *spinula*, small spine] a small spine.

spinulescent *a.* [L. *spinula*, small spine] tending to be spinulate.

spinuliferous *a.* [L. *spinula*, small spine; *ferre*, to bear] bearing small spines.

Spinulosa *n.* [L. *spinula*, small spine] an order of starfish that often do not have marginal plates.

spinulose, spinulous *a.* [L. *spinula*, small spine] spinulate, *q.v.*

spiny-finned bearing fins with spiny rays for support.

spiny-rayed *appl.* fins supported by spiny rays.

Spiomorpha *n.* [*Spio*, generic name; Gk. *morphē*, form] an order of polychaetes having segments all fairly similar and the prostomium with long tentacles for food collecting.

spiracle n. [L. spiraculum, air hole] first pharyngeal aperture or visceral cleft; branchial passage between mandibular and hyoid arches in fishes; lateral branchial opening in tadpoles; nasal aperture of Cetacea; respiratory aperture behind eye of elasmobranchs; breathing aperture of insects and myriapods; aperture of book lungs; any of 5 openings around mouth of Blastoidea; alt. spiraculum; a. spiracular.

spiraculate, spiraculiferous a. [L. spiraculum, air hole] having spiracles.

spiraculiform a. [L. spiraculum, air hole; forma, shape] spiracle-shaped.

spiraculum spiracle, q.v.

spiral a. [L. spira, coil] winding like a screw, appl. leaves alternately placed, appl. flowers with spirally inserted parts, appl. thickening of cell wall in vessels, cells, and tracheids, appl. chromatids and chromosomes. n. A coiled structure; coil of the chromosome thread in mitosis and meiosis; cf. internal, relational, relic spiral.

spiral cleavage cleavage into unequal parts, arranged in mosaic fashion and interlocking, upper cells rotating to right to alternate with lower; alt. oblique or alternating cleavage.

spiral filament spiral fold of the inner wall of tracheal tubes in insects.

spiralia n.plu. [L. spira, coil] coiled structures supported by crura, in certain brachiopods.

spiral organ Corti's organ, q.v.

spiral valve in certain more primitive fishes, such as elasmobranchs, ganoids, and dipnoans, a spiral infolding of intestine wall; of Heister, folds of mucous membrane in neck of gall bladder.

spiral vessels first xylem elements of a stele, spiral fibres coiled up inside tubes and so adapted for rapid elongation.

spiranthy n. [Gk. speira, coil; anthos, flower] displacement of flower parts through twisting.

spiraster n. [L. spira, coil; astēr, star] a spiral and rayed sponge spicule.

spire n. [L. spira, coil] totality of whorls of a spiral shell.

spireme n. [Gk. speirēma, coil] thread-like appearance of nuclear chromatin during prophase of mitosis.

spiricles n.plu. [L. spira, coil] thin, coiled, thread-like outgrowths of some seed coats.

spiriferous a. [L. spira, coil; ferre, to bear] having a spiral structure.

spirilla a. [L. spirillum, small coil] pert. or resembling a spirillum.

spirillum n. [L. spirillum, small coil] a thread-like wavy or coiled bacterium; former term for antherozooid.

spirivalve n. [L. spira, coil; valvae, folding doors] a gastropod with spiral shell.

Spiroch(a)etales n. [Gk. speira, coil; chaitē, hair] an order of slender spiral bacteria with a flexible body and no rigid wall; alt. spirochaetes.

spirocyst n. [Gk. speira, coil; kystis, bladder] a nematocyst in which the thread is coiled spirally.

spiroid a. [Gk. speira, coil; eidos, form] spirally formed.

spirolophe n. [Gk. speira, coil; lophē, crest] a stage in the development of the branchiopod lophophore in which 2 spirally twisted arms are formed with cirri on one side.

spironeme n. [Gk. speira, coil; nēma, thread] coiling contractile thread in stalk of certain ciliates.

spiroplasmas n.plu. [Gk. speira, coil; plasma, form] spiral organisms found in plants, which otherwise resemble mycoplasma-like organisms and may be interconvertible with them.

Spirotricha n. [L. spira, coil; thrix, hair] an order of Ciliata having a gullet with an undulating membrane; in some classifications considered a subclass of Ciliphora.

spirulate a. [L. spira, coil] appl. any spiral structure or coiled arrangement.

Spirurida, Spirurata n. [L. spira, coil] an order of Phasmidia, mostly parasitic on man and large mammals with an invertebrate intermediate host, having the oesophagus partly muscular and partly glandular.

splanchnic a. [Gk. splangchnon, entrail] pert. viscera.

splanchnocoel n. [Gk. splangchnon, entrail; koilos, hollow] the cavity of lateral plates of embryo, persisting as visceral cavity of adult.

splanchnocranium viscerocranium, q.v.

splanchnology n. [Gk. splangchnon, entrail; logos, discourse] the branch of anatomy dealing with viscera.

splanchnopleure n. [Gk. splangchnon, entrail; pleura, side] inner layer of mesoblast, applied to viscera.

spleen n. [Gk. splēn, spleen] a vascular organ in which lymphocytes are produced and red blood corpuscles destroyed, in vertebrates; alt. lien, milt.

splenetic splenic, q.v.

splenial a. [L. splenium, a patch] pert. splenius muscle, or splenial bone, or splenium.

splenial bone membrane bone in lower jaw of some vertebrates.

splenic a. [Gk. splēn, spleen] pert. the spleen; alt. splenetic, lienal.

splenic nodule or corpuscle see Malpighian body.

splenium n. [L. splenium, patch] posterior border of corpus callosum.

splenius n. [L. splenium, patch] muscle of upper dorsal region and back of neck.

splenocyte n. [Gk. splēn, spleen; kytos, hollow] a large monocyte believed to originate in spleen.

splenophrenic a. [Gk. splēn, spleen; phrēn, midriff] pert. spleen and diaphragm.

splintwood alburnum, q.v.

spondyl(e), spondylus n. [Gk. sphondylos, vertebra] vertebra, q.v.; a. spondylous.

sponges the common name for the Porifera, q.v.

spongicolous a. [L. spongia, sponge; colere, to inhabit] living in sponges.

spongin n. [L. spongia, sponge] material of skeletal fibres of sponges such as horny sponges.

sponginblast n. [Gk. sponggia, sponge; blastos, bud] a spongin-producing cell; alt. spongoblast.

spongioblasts n.plu. [Gk. sponggia, sponge; blastos, bud] embryonic epithelial cells which give rise to neuroglia cells and fibres radiating to periphery of spinal cord.

spongiocoel n. [Gk. sponggia, sponge; koilos, hollow] the cavity, or system of cavities in sponges; alt. spongocoel.

spongiocyte n. [Gk. sponggia, sponge; kytos, hollow] a vacuolated cell of zona fasciculata of adrenal cortex.

spongioplasm n. [Gk. *sponggia*, sponge; *plasma*, mould] cytoplasmic threadwork of a cell; *alt.* cytoreticulum, mitome, polioplasm.

spongiose a. [L. *spongia*, sponge] of a spongy texture, spongoid; full of small cavities.

spongoblast sponginblast, q.v.

spongocoel spongiocoel, q.v.

spongophare n. [Gk. *sponggos*, sponge; *pherein*, to bear] the upper part of a sponge containing the flagellated chambers; *cf.* hypophare.

spongophyll n. [Gk. *sponggos*, sponge; *phyllon*, leaf] a leaf having spongy parenchyma, without palisade tissue, between upper and lower epidermis, as in certain aquatics.

spongy a. [L. *spongia*, sponge] of open texture, lacunar, *appl.* parenchyma of mesophyll, *appl.* tissue surrounding embryo sac, as in gymnosperms.

spontaneous generation see abiogenesis.

spools fusulae, q.v.

spoon small sclerite at base of balancers in Diptera; pinion of tegula.

sporabola n. [Gk. *sporos*, seed; *bolos*, a throw] the trajectory of a spore discharged from a sterigma.

sporadic a. [Gk. *sporadikos*, scattered] *appl.* plants confined to limited localities.

sporadin n. [Gk. *sporadēn*, scattered about] trophozoite of gregarines moving about in lumen of gut.

sporange sporangium, q.v.

sporangia plu. of sporangium.

sporangial a. [Gk. *sporos*, seed; *anggeion*, vessel] *pert.* a sporangium.

sporangiferous n. [Gk. *sporos*, seed; *anggeion*, vessel; L. *ferre*, to bear] sporangia-bearing.

sporangiform, sporangioid a. [Gk. *sporos*, seed; *anggeion*, vessel; L. *forma*, shape; Gk. *eidos*, form] like a sporangium.

sporangiocarp n. [Gk. *sporos*, seed; *anggeion*, vessel; *karpos*, fruit] an enclosed collection of sporangia; a structure of asci and sterile hyphae surrounded by a peridium; ascocarp, q.v.

sporangiocyst n. [Gk. *sporos*, seed; *anggeion*, vessel; *kystis*, pouch] a membrane enclosing a sporangium; a thick-walled resistant sporangium.

sporangiolum, sporangiole n. [Gk. *sporos*, seed; *anggeion*, vessel] a secondary or small few-spored sporangium; ptyosome, q.v.

sporangiophore n. [Gk. *sporos*, seed; *anggeion*, vessel; *pherein*, to bear] a stalk-like structure bearing sporangia.

sporangiosorus n. [Gk. *sporos*, seed; *anggeion*, vessel; *sōros*, heap] a compact group of sporangia.

sporangiospore n. [Gk. *sporos*, seed; *anggeion*, vessel; *sporos*] a spore, particularly if non-motile, formed in a sporangium.

sporangium n. [Gk. *sporos*, seed; *anggeion*, vessel] a spore case, capsule, or cell in which spores are produced; *alt.* sporange, ootheca; *plu.* sporangia.

spore n. [Gk. *sporos*, seed] a 1-celled, or sometimes several-celled, very small reproductive body produced in plants, bacteria, and Protozoa.

spore ball cephalosporium, q.v.

spore case see theca.

spore coat envelope of a bacterial spore, external to cortex, and surrounded by exosporium.

spore formation reproduction by encystation

followed by division and free cell liberation, *alt.* endogenous multiplication; any other method by which spores are formed; *alt.* sporogony, sporogenesis, sporulation.

spore group sporodesm, q.v.

spore mother cell a diploid cell which by meiosis gives rise to 4 haploid cells; sporoblast, q.v.; *alt.* sporocyte.

sporetia n.plu. [Gk. *sporos*, seed] idiochromidia, q.v.

sporidesm sporodesm, q.v.

sporidiferous a. [Gk. *sporos*, seed; L. *ferre*, to bear] sporidia-bearing.

sporidiole n. [Gk. *sporos*, seed] a sporidium arising from a promycelium, such as a protobasidium.

sporidium n. [Gk. *sporos*, seed; *idion*, *dim.*] a small spore particularly one produced from a promycelium and/or by abstriction.

sporidochium n. [Gk. *sporos*, seed; *docheion*, holder] receptacle of certain fungi; *cf.* sporodochium.

sporiferous a. [Gk. *sporos*, seed; L. *ferre*, to bear] spore-bearing.

sporification n. [Gk. *sporos*, seed; L. *facere*, to make] formation of spores.

sporiparity n. [Gk. *sporos*, seed; L. *parere*, to beget] reproduction by spore formation.

sporiparous a. [Gk. *sporos*, seed; L. *parere*, to beget] sporogenous, q.v.

sporoblast n. [Gk. *sporos*, seed; *blastos*, bud] an archespore, q.v.; a stage in spore formation, a sporoblast giving rise to spores, and these to sporozoites; *alt.* spore mother cells.

sporocarp n. [Gk. *sporos*, seed; *karpos*, fruit] an ascocarp, q.v.; a sorus covered by indusium; a structure inside which spores are formed; *cf.* carpophore.

Sporochnales n. [*Sporochnus*, generic name] an order of brown algae having a sporophyte in which each branch terminates in a tuft of hairs.

sporocladium n. [Gk. *sporos*, seed; *kladion*, small young branch] branch of a conidiophore, bearing sporangia or conidia; *cf.* prophialide.

sporocyst n. [Gk. *sporos*, seed; *kystis*, bladder] a stage in spore formation preceding liberation of spores; protective envelope of a spore, in Protozoa; encysted embryo stage of trematode after degeneration following entry into intermediate host; *alt.* gamocyst, zoocyst.

sporocystid a. [Gk. *sporos*, seed; *kystis*, bladder; *eidos*, form] *appl.* oocyst of Sporozoa when the zygote forms sporocysts.

sporocyte n. [Gk. *sporos*, seed; *kytos*, hollow] spore mother cell, q.v.

sporoderm n. [Gk. *sporos*, seed; *derma*, skin] the pollen grain wall.

sporodesm n. [Gk. *sporos*, seed; *desmos*, bond] a compound or multicellular spore in which each cell can germinate independently; *alt.* multilocular or septate or pluricellular spore, spore group, sporidesm, compound spore.

sporodochium n. [Gk. *sporos*, seed; *docheion*, holder] a hemispherical aggregate of conidiophores; *cf.* sporidochium.

sporoduct n. [Gk. *sporos*, seed; L. *ducere*, to lead] a special apparatus for dissemination of spores of Sporozoa and of some fungi.

sporogenesis spore formation, q.v.

sporogenous

sporogenous *a.* [Gk. *sporos*, seed; *gennaein*, to produce] spore-producing; *alt.* sporiparous.

sporogonium *n.* [Gk. *sporos*, seed; *gonos*, off-spring] the sporophyte generation in bryophytes consisting of a capsule and seta, developing from the fertilized ovum in the archegonium, and giving rise to spores; *a.* sporogonial.

sporogony *n.* [Gk. *sporos*, seed; *gonos*, birth] spore formation, *q.v.*; the formation of sporozoites or spores from a sporont in Protozoa; the formation of gametes from a gamont, their fusion and subsequent formation of spores and sporozoites from the zygote (sporont), *alt.* gamogony in Protozoa.

sporoid *a.* [Gk. *sporos*, seed; *eidos*, like] like a spore.

sporokinete *n.* [Gk. *sporos*, seed; *kinein*, to move] a motile spore from the oocyst of certain Haemosporidia.

sporonine *n.* [Gk. *sporos*, seed] a substance related to the terpenes, found in the walls of spores and pollen grains.

sporont *n.* [Gk. *sporos*, seed; *on*, being] in some Protozoa, the individual or generation which gives rise to a generation of sporozoites; *alt.* amphiont, gamont.

sporophore *n.* [Gk. *sporos*, seed; *pherein*, to bear] a spore-bearing structure in fungi, may be simple as in sporangiophore or complex as in agarics; process of plasmodium producing spores on free surface, in slime moulds; an inflorescence.

sporophydium *n.* [Gk. *sporos*, seed; *phyas*, shoot; *idion*, dim.] the sporangium of certain thallophytes.

sporophyll *n.* [Gk. *sporos*, seed; *phyllon*, leaf] a leaf or leaf-like structure which bears or subtends a sporangium; *cf.* trophophyll.

sporophyte *n.* [Gk. *sporos*, seed; *phyton*, plant] the diploid spore-producing phase in plants with alternation of generations; *alt.* synkaryophyte; *cf.* gametophyte.

sporoplasm *n.* [Gk. *sporos*, seed; *plasma*, mould] the protoplasm which gives rise to a spore; in some Sporozoa, the protoplasmic body released from a cyst and forming an amoebula; *cf.* epiplasm.

sporopollenin *n.* [Gk. *sporos*, seed; L. *pollen*, fine flour] an alcohol which is found in the wall of spores and is related to suberin and cutin but is more durable, resulting in spores surviving for millions of years.

sporosac *n.* [Gk. *sporos*, seed; L. *saccus*, sack] an ovoid pouch-like body, consisting of a gonad, a degraded reproductive zooid of a siphonophore.

sporotamium *n.* [Gk. *sporos*, seed; *tamieion*, store] cell layer beneath apothecium, as in lichens.

sporothallus *n.* [Gk. *sporos*, seed; *thallos*, young shoot] a thallus which produces spores; *cf.* gametothallus.

sporotheca *n.* [Gk. *sporos*, seed; *thēķē*, case] a membrane enclosing sporozoites.

Sporozoa *n.* [Gk. *sporos*, seed; *zōon*, animal] a class of parasitic Protozoa reproducing by spores and usually having no feeding or locomotion organelles.

sporozoid *n.* [Gk. *sporos*, seed; *zōon*, animal; *eidos*, form] a motile spore; zoospore, *q.v.*

sporozoite *n.* [Gk. *sporos*, seed; *zōon*, animal] spore liberated through dissolving of membrane of sporocyst, a phase in life history of Sporozoa; *alt.* exotospore.

sport a somatic mutation; *alt.* rogue.

sporula, sporule *n.* [Gk. *sporos*, seed] a small spore; formerly a spore.

sporulation *n.* [L. *sporula*, small seed] brood formation by multiple cell fission; spore formation, *q.v.*; liberation of spores.

spot fruit sorus, as of ferns.

spreading factor hyaluronidase, *q.v.*

springtails the common name for the Collembola, *q.v.*

spring wood early wood, *q.v.*

sprout mycelium pseudomycelium, *q.v.*

spur *n.* [A.S. *spora*, spur] a calcar, *q.v.*; cog tooth of malleus; rim of sclera outside iridial angle; cuticular outgrowth on legs of certain insects; a process of a petal or of a sepal, functioning as a nectar receptacle; small reproductive shoot; a brachyplast, *q.v.*

spuriae *n.plu.* [L. *spurius*, false] feathers of bastard wing, *q.v.*

spurious *a.* [L. *spurius*, false] seemingly true but morphologically false, *appl.* fruit, teeth, vein, wing.

spurious dissepiment false dissepiment, *q.v.*

squama *n.* [L. *squama*, scale] a scale; a part arranged like a scale; vertical part of frontal bone; part of occipital bone above and behind foramen magnum; anterior and upper part of temporal bone; a scale below wing base of Diptera, *alt.* antitegula, calyptron, prehaltere; a scale-like body attached to 2nd podomere of antenna of some Crustacea; *alt.* squame.

Squamata *n.* [L. *squama*, scale] an order of lepidosaur reptiles including the lizards and snakes.

squamate *a.* [L. *squama*, scale] scaly.

squamation *n.* [L. *squama*, scale] pholidosis, *q.v.*

squame squama, *q.v.*

squamella *n.* [Dim. of L. *squama*, scale] a small scale or bract; palea, *q.v.*

squamellate, squamelliferous *a.* [L. *squama*, a scale] having small scales or bracts.

squamelliform *a.* [L. *squama*, scale; *forma*, shape] resembling a squamella.

squamiferous *a.* [L. *squama*, scale; *ferre*, to bear] bearing scales; *alt.* squamigerous.

squamiform *a.* [L. *squama*, scale; *forma*, shape] scale-like.

squamosal *n.* [L. *squama*, scale] a membrane bone of vertebrate skull forming part of posterior side wall; *alt.* paraquadrate.

squamose *a.* [L. *squama*, scale] covered with scales; *alt.* squamous.

squamous *a.* [L. *squama*, scale] consisting of scales; *appl.* simple epithelium of flat nucleated cells, scaly or pavement epithelium; squamose, *q.v.*

squamula *n.* [L. *squama*, scale; *dim.*] a small scale; lodicule, *q.v.*; tegula of some insects; one of small circular areas into which pouch scales of Gymnophiona are divided; *alt.* squamule.

squamulate, squamulose *a.* [L. *squama*, scale; *dim.*] having minute scales.

squamule squamula, *q.v.*

squarrose *a.* [L.L. *squarrosus*, scurfy] rough with projecting scales or rigid leaves.

squarrulose *a.* [L.L. *squarrosus*, scurfy, *dim.*] tending to become squarrose.

sRNA soluble RNA, *q.v.*

stabilate n. [L. stabilis, firm] a stable population of an organism similar to a strain.

stachyose n. [Stachys, generic name of artichoke] a tetrasaccharide found especially in the roots of the white dead-nettle and in tubers of Stachys tubifera (Chinese artichoke), hydrolysis of which gives 2 molecules of galactose, one of glucose, and one of fructose.

stachyosporous a. [Gk. stachys, ear of corn; sporos, seed] bearing sporangia on the axis.

stadium n. [L. stare, to stand] a stage in development or life history of plant or animal; interval between 2 successive ecdyses; alt. stade.

stag-horned a. [Icel. stiga, to mount; A.S. horn] having large branched mandibles, as a stag beetle.

stagnicolous a. [L. stagnum, standing water; colere, to inhabit] living in stagnant water.

stalk n. [A.S. stel, stalk] a supporting structure, as a caudicle, caulicle, caulis, filament, haulm, pedicel, pedicle, peduncle, petiole, stem, stipe, stipule, etc.

stalk cell the barren cell of 2 into which the antheridial cell of gymnosperms divides; basal cell of crozier in Discomycetes.

stamen n. [L. stamen, warp] one of the male reproductive organs of a flower, consisting of stalk or filament with anther containing pollen; alt. microsporophyll.

staminal a. [L. stamen, warp] pert. or derived from a stamen.

staminate a. [L. stamen, warp] producing, or consisting of, stamens; appl. flower having stamens but not pistils; cf. pistillate.

staminiferous, staminigerous a. [L. stamen, warp; ferre, gerere, to bear] stamen-bearing.

staminode, staminodium n. [L. stamen, warp; Gk. eidos, form] a foliaceous scale-like body in some flowers, derived from a metamorphosed stamen; a rudimentary, imperfect, or sterile stamen; alt. parastamen, parastemon.

staminody n. [L. stamen, warp; Gk. eidos, form] the conversion of any organs in a flower into stamens.

staminose a. [L. stamen, warp] having very obvious stamens in the flower.

standard n. [O.F. estandart, from L. stare, to stand] the vexillum, q.v., of a papilionaceous flower; a tree or shrub not supported by a wall; a unit of measurement or solution used in calibration.

standard nutritional unit the unit expressing the energy available at a certain trophic level for the next level in the food chain, usually expressed as 10^6 kilocalories per hectare per year.

standing crop the biomass at a certain time.

stapedius n. [L.L. stapes, stirrup] a muscle pulling the head of the stapes.

stapes n. [L.L. stapes, stirrup] stirrup-shaped innermost bone of middle ear; operculum or internal end of columella auris, fitting into and filling fenestra ovalis in amphibians.

staphyle n. [Gk. staphylē, bunch of grapes] uvula, q.v.

star cells Kupffer cells, q.v.

starch n. [A.S. stearc, stiff] a hexosan polysaccharide composed of a core of amylose surrounded by amylopectin which can be hydrolysed via dextrins to maltose by amylase and then to glucose

by maltase and which occurs in plants in granules as storage molecules; alt. amylum.

starch gums dextrins, q.v.

starch sheath endodermis with starch grains.

starch sugar glucose, q.v.

starfish a common name for the Asteroidea, q.v.

start codon the DNA triplet signalling the start of the transcription of mRNA; the RNA triplet signalling the start of translation of a polypeptide chain.

stasimorphy n. [Gk. stasis, standing; morphē, form] a deviation in form due to arrested development.

stasis n. [Gk. stasis, standing] stoppage, or retardation, as of growth, or of movement of animal fluids.

stathmokinesis n. [Gk. stathmos, station; kinēsis, movement] inhibition of cell division, as by colchicine or other agent.

static a. [Gk. statikos, causing to stand] pert. system at rest or in equilibrium, appl. postural reactions, cf. kinetic; appl. proprioceptors, as otoliths and semicircular canals; cf. dynamic.

stationary phase third stage in bacterial multiplication during which the viable count remains steady.

stato-acoustic a. [Gk. statos, standing; akouein, to hear] pert. sense of balance and of hearing, appl. 8th cranial or acoustic nerve, dividing into vestibular and cochlear nerves.

statoblast n. [Gk. statos, stationary; blastos, bud] a specialized bud or 'winter egg' of some Polyzoa, developed on funiculus and set free on death of parent organism; a gemmule of certain sponges.

statocone n. [Gk. statos, stationary; konis, dust] a minute structure contained in a statocyst.

statocyst n. [Gk. statos, stationary; kystis, bladder] a vesicle lined with sensory cells and containing bodies falling under gravity, found in many invertebrates, which is concerned with perception of gravity, cf. otocyst; statocyte, q.v.; alt. lithocyst.

statocyte n. [Gk. statos, stationary; kytos, hollow] a cell containing statoliths and probably acting as a georeceptor, such as a root cap cell containing granules such as starch grains; alt. statocyst.

statokinetic a. [Gk. statos, standing; kinētikos, putting in motion] pert. maintenance of equilibrium and associated movements, appl. reflexes.

statolith n. [Gk. statos, stationary; lithos, stone] a structure of calcium carbonate, sand grain, or secreted substance contained in a statocyst, alt. lithite, otolith; a cell inclusion, as oil droplet, starch grain, or crystal, which changes its intracellular position under the influence of gravity in a statocyte; cf. amylostatolith, chlorostatolith.

statorhabd n. [Gk. statos, stationary; rhabdos, rod] a short tentacular process carrying the statolith in Trachymedusae.

statospore n. [Gk. statos, stationary; sporos, seed] a resting spore.

status quo hormone juvenile hormone, q.v.

Stauromedusae n. [Gk. stauros, cross; Medousa, Medusa] an order of scyphozoans which are sessile and develop directly from the polyp stage; in some classifications considered as a subclass.

staurophyll n. [Gk. stauros, palisade; phyllon,

leaf] a leaf having palisade or other compact tissue throughout.

staurospore *n.* [Gk. *stauros*, cross; *sporos*, seed] a cross-shaped or a triquetrous spore.

steapsin *n.* [Gk. *stear*, tallow; *pepsis*, digestion] a lipolytic enzyme occurring in digestive juice of certain animals, such as in pancreatic juice of vertebrates; EC 3.1.1.3; *r.n.* triacylglycerol lipase.

stearic acid a widely distributed fatty acid.

stearin *n.* [Gk. *stear*, tallow] the solid part of fat, held dissolved by olein at body temperature; a component of many animal and vegetable fats.

steatogenesis *n.* [Gk. *stear*, tallow; *genesis*, origin] the production of lipid substance, as in mammalian seminiferous tubules.

steganopodous *a.* [Gk. *steganos*, covered; *pous*, foot] having feet completely webbed; *alt.* totipalmate.

stege *n.* [Gk. *stege*, roof] the inner layer of rods of Corti.

stegocarpic, stegocarpous *a.* [Gk. *stegein*, to cover; *karpos*, fruit] having a capsule with operculum and peristome, *appl.* mosses; *cf.* cleistocarpous.

Stegocephalia *n.* [Gk. *stege*, roof; *kephale*, head] in some classifications, a group of Amphibia comprising all pre-Jurassic and some later extinct forms with a salamander-like body and large size.

stegocrotaphic *a.* [Gk. *stege*, roof; *krotaphos*, the temples] anapsid, *q.v.*

Steinheim skull a fossil hominid skull found near Stuttgart, W. Germany, which is intermediate between *Homo erectus* and *Homo sapiens*.

stelar parenchyma pith, *q.v.*

stelar system of plants, vascular and associated conjunctive tissue.

stele *n.* [Gk. *stele*, pillar] a bulky strand or cylinder of vascular tissue and associated ground tissue contained in stem and root of plants, and developed from plerome in roots; *alt.* central cylinder.

stellar stellate, *q.v.*

stellate *a.* [L. *stella*, star] star-shaped, radiating, *appl.* cells: Kupffer cells, *q.v.*; *appl.* leaf, hair, spicule, ganglion of sympathetic system, ligament of rib, veins beneath fibrous tunic of kidney, etc.; *alt.* stellar, stelliform, arachniform, asteroid.

stellate reticulum enamel pulp of dental germ.

Stelleroidea *n., n.plu.* [L. *stella*, star; Gk. *eidos*, form] in some classifications, a class of echinoderms including the Asteroidea and Ophiuroidea which are reduced to subclass status.

stelliform stellate, *q.v.*

stelocyttarous *a.* [Gk. *stele*, pillar; *kyttaros*, honeycomb cell] building, or *pert.*, stalked combs, as of certain wasps; *cf.* phragmocyttarous.

stem *n.* [A.S. *stemn*, tree stem] main axis of a plant, having leaves or scale leaves and a characteristic arrangement of vascular tissue.

stem body equatorial part of the spindle, as between 2 nuclei at telophase.

stem cell a primordial germ cell; haemocytoblast, *q.v.*

stemma *n.* [Gk. *stemma*, garland] ocellus of arthropods; ommatidium, *q.v.*; a tubercle bearing an antenna; *plu.* stemmata, stemmas.

stem nematogen a young nematogen with 2 or 3 axial cells, in certain Dicyemida.

Stemonitales *n.* [*Stemonitis*, generic name] an order of usually stalked slime moulds whose massed spores are black, violet, or rust coloured.

stenobaric *a.* [Gk. *stenos*, narrow; *baros*, weight] *appl.* animals adaptable only to small differences in pressure or altitude; *cf.* eurybaric.

stenobathic *a.* [Gk. *stenos*, narrow; *bathys*, deep] having a narrow vertical range of distribution; *cf.* eurybathic.

stenobenthic *a.* [Gk. *stenos*, narrow; *benthos*, depth of the sea] *pert.*, or living within a narrow range of depth of the sea bottom; *cf.* eurybenthic.

stenochoric *a.* [Gk. *stenos*, narrow; *choros*, place] having a narrow range of distribution; *cf.* eurychoric.

stenocyst *n.* [Gk. *stenos*, narrow; *kystis*, bladder] one of the auxillary cells in leaves of certain mosses; *cf.* eurycyst.

stenoecious *a.* [Gk. *stenos*, narrow; *oikos*, abode] having a narrow range of habitat selection; *cf.* euryoecious.

Stenoglossa *n.* [Gk. *stenos*, narrow; *glossa*, tongue] Neogastropoda, *q.v.*

stenohaline *a.* [Gk. *stenos*, narrow; *halinos*, saline] *appl.* organisms adaptable to a narrow range of salinity; *cf.* euryhaline.

stenohygric *a.* [Gk. *stenos*, narrow; *hygros*, wet] *appl.* organisms adaptable to a narrow variation in atmospheric humidty; *cf.* euryhygric.

stenomorphic *a.* [Gk. *stenos*, narrow; *morphe*, form] dwarfed; smaller than typical form, owing to cramped habitat.

stenonian duct Stensen's duct, *q.v.*

stenonatal *a.* [Gk. *stenos*, narrow; *noton*, back] with very small thorax, as worker insect.

stenopetalous *a.* [Gk. *stenos*, narrow; *petalon*, leaf] with narrow petals.

stenophagous, stenophagic *a.* [Gk. *stenos*, narrow; *phagein*, to eat] subsisting on a limited variety of food; *cf.* euryphagous, monophagous, polyphagous.

stenophyllous *a.* [Gk. *stenos*, narrow; *phyllon*, leaf] narrow-leaved.

stenopodium *n.* [Gk. *stenos*, narrow; *pous*, foot] a crustacean limb in which the protopodite bears distally both endopodite and exopodite.

stenosepalous *a.* [Gk. *stenos*, narrow; F. *sepale*, sepal] with narrow sepals.

stenosis *n.* [Gk. *stenos*, narrow] narrowing or constriction of a tubular structure, as of a pore, duct, or vessel.

stenostomatous *a.* [Gk. *stenos*, narrow; *stoma*, mouth] narrow-mouthed.

stenotele *n.* [Gk. *stenos*, narrow; *telos*, end] *appl.* type of nematocyst having distal end narrower than base, as a stinging thread; *alt.* penetrant.

stenothermic, -al, -ous *a.* [Gk. *stenos*, narrow; *therme*, heat] *appl.* organisms adaptable only to slight variations in temperature; *cf.* eurythermic.

stenotopic *a.* [Gk. *stenos*, narrow; *topos*, place] having a restricted range of geographical distribution; *cf.* eurytopic.

stenotropic *a.* [Gk. *stenos*, narrow; *trope*, turn] having a very limited adaptation to varied conditions.

Stensen's duct [*N. Stensen*, Danish physiologist] duct of parotid gland which opens into mouth; *alt.* stenonian duct.

Stensioelliformes *n.* [*Stensioellus*, generic name; L. *forma*, shape] an order of lower Devonian placoderms that seem to lack bone development.

stephanion *n.* [Gk. *stephanos*, crown] the point where superior temporal ridge is crossed by coronal suture.

stephanokont *a.* [Gk. *stephanos*, crown; *kontos*, punting pole] having a ring of flagella or cilia around anterior end, *appl.* zoospores.

steppe *n.* [Russ. *step'*, lowland] xerophilous and generally treeless grassland; short-grass plains.

stercobilin *n.* [L. *stercus*, dung; *bilis*, bile] hydrobilirubin, *q.v.*; urobilin, *q.v.*

stercobilinogen hydrobilirubinogen, *q.v.*

stercomarium *n.* [L. *stercus*, dung] the system of stercome-containing tubes of certain Sarcodina.

stercome *n.* [L. *stercus*, dung] faecal matter of Sarcodina, in masses of brown granules.

stercoral *a.* [L. *stercus*, dung] *pert.* faeces; *appl.* a dorsal pocket or sac of proctodaeum in spiders.

stereid(e) *n.* [Gk. *stereos*, solid; *eidos*, form] a lignified parenchyma cell with pit canals; stone cell, *q.v.*

stereid(e) bundles bands or bundles of sclerenchymatous fibres.

stereoblastula *n.* [Gk. *stereos*, solid; *blastos*, bud] abnormal form of echinoid larva unable to gastrulate.

stereocilia *n.plu.* [Gk. *stereos*, rigid; L. *cilium*, eyelash] non-motile secretory projections on epithelium, as of duct of epididymis.

stereognostic *a.* [Gk. *stereos*, solid; *gnōstos*, to be known] *appl.* sense which appreciates size, shape, weight.

stereoisomers *n.plu.* [Gk. *stereos*, solid; *isos*, equal; *meros*, part] isomers having different 3-dimensional configurations.

stereokinesis *n.* [Gk. *stereos*, solid; *kinēsis*, movement] thigmokinesis, *q.v.*

stereom(e) *n.* [Gk. *stereōma*, solid body] sclerenchymatous and collenchymatous masses along with hardened parts of vascular bundles forming supporting tissue in plants; the thick-walled elongated cells of the central cylinder in mosses.

stereoplasm *n.* [Gk. *stereos*, solid; *plasma*, mould] the more solid part of the cytoplasm, *cf.* hygroplasm; a vesicular substance filling interseptal spaces of certain corals.

Stereospondyli, stereospondyls *n., n.plu.* [Gk. *stereos*, solid; *sphondylos*, vertebra] a group of Triassic degenerate labyrinthodont amphibians.

stereospondylous *a.* [Gk. *stereos*, solid; *sphondylos*, vertebra] having vertebrae each fused into one piece; *cf.* temnospondylous; *n.* stereospondyly.

stereotaxis, stereotaxy *n.* [Gk. *stereos*, solid; *taxis*, arrangement] the response of an organism to the stimulus of contact with a solid, such as the tendency of certain organisms to attach themselves to solid objects or to live in crannies or tunnels in total contact with solids; *alt.* stereotropism, thigmotaxis.

stereotropism *n.* [Gk. *stereos*, solid; *tropē*, turn] a tropism in relation to contact, especially with a solid or rigid structure; stereotaxis, *q.v.*; *cf.* thigmotropism.

stereotyped behaviour *see* fixation.

stereozone *n.* [Gk. *stereos*, solid; *zōnē*, girdle] the dense calcareous region of epitheca in certain corallites.

sterigma *n.* [Gk. *stērigma*, support] a short hypha arising from basidium or conidiophore and giving rise to spores by abstriction, *alt.* phialide, trichidium; flange- or rib-like part of a decurrent leaf, lying along the stem; *plu.* sterigmata.

sterile *a.* [L. *sterilis*, barren] incapable of propagation; aseptic; axenic, *q.v.*

sterile glume *see* glume.

sterilize *v.* [L. *sterilis*, barren] to render incapable of reproduction or of conveying infection.

sternal *a.* [Gk. *sternon*, chest] *pert.* sternum, or sternite; *appl.* angle: angulus, *q.v.*

Sternaspimorpha *n.* [Gk. *sternon*, chest; *aspis*, shield; *morphē*, form] an order of polychaetes with a globular, almost unsegmented body.

sternebrae *n.plu.* [L. *sternum*, breast bone; *-ebra*, on analogy of vert*ebra*] divisions of a segmented sternum or breast bone.

sternellum *n.* [*Dim.* of L. *sternum*, breast bone] a sternal sclerite of insects, especially sclerite behind eusternum; *alt.* poststernite.

sternite *n.* [Gk. *sternon*, chest] a ventral plate of an arthropod segment; *alt.* sternal sclerite.

sternobranchial *a.* [L. *sternum*, breast bone; *branchiae*, gills] *appl.* vessel conveying blood to gills, in certain Crustacea.

sternoclavicular *a.* [L. *sternum*, breast bone; *clavicula*, small key] *appl.* and *pert.* articulation between sternum and clavicle.

sternocostal *a.* [L. *sternum*, breast bone; *costa*, rib] *pert.* sternum and ribs, *appl.* ligament, *appl.* surface of heart.

sternohyoid *a.* [Gk. *sternon*, chest; *hyoeidēs*, Y-shaped] *appl.* muscle between back of manubrium of sternum and hyoid.

sternokleidomastoid *a.* [Gk. *sternon*, chest; *kleis*, key; *mastos*, breast; *eidos*, form] *appl.* an oblique neck muscle stretching from sternum to mastoid process.

sternopericardial *a.* [Gk. *sternon*, chest; *peri*, around; *kardia*, heart] *appl.* ligament connecting dorsal surface of sternum and fibrous pericardium.

sternopleurite *n.* [Gk. *sternon*, chest; *pleura*, side] thoracic sclerite formed by union of episternum and sternum in insects, *alt.* sternopleuron; meskatepisternum of Diptera.

sternoscapular *a.* [L. *sternum*, breast bone; *scapula*, shoulder blade] *appl.* a muscle connecting sternum and scapula.

sternothyroid *a.* [Gk. *sternon*, chest; *thyra*, door; *eidos*, form] *appl.* muscle connecting manubrium of sternum and thyroid cartilage.

sternotribe *a.* [Gk. *sternon*, chest; *tribein*, to rub] *appl.* flowers with fertilizing elements so placed as to be brushed by sternites of visiting insects.

sternoxiphoid *a.* [Gk. *sternon*, chest; *xiphos*, sword; *eidos*, form] *appl.* plane through junction of sternum and xiphoid cartilage.

sternum *n.* [L. *sternum*, breast bone] breast bone of vertebrates to which ribs are attached in higher forms; ventral plate of typical arthropod segment; all the ventral sclerites of a thoracic segment in insects.

steroids *n.plu.* [Gk. *stereos*, solid; *eidos*, form] complex polycyclic lipids with a hydrocarbon nucleus, including the sterols, bile acids, various hormones, cardiac glycosides, and saponins.

sterols *n.plu.* [Gk. *stereos*, solid; alcoho*l*] steroid alcohols found in plants and animals, comprising the mycosterols, phytosterols, and 'zoosterols.

sterraster *n.* [Gk. *sterros*, solid; *astēr*, star] aster with actines soldered together by silica.

sterroblastula *n.* [Gk. *sterros*, solid; L. *blastula*, little bud] more-or-less solid blastula formed in yolk-rich eggs of some molluscan species.

sterrula *n.* [Gk. *sterros*, solid] solid free-swimming larva of Alcyonaria, preceding planula.

Stewart's organs five vesicles of coelom of lantern protruding into the perivisceral space and acting as internal gills in some Echinoidea.

STH somatotrophic hormone: somatotrophin, *q.v.*

stichic *a.* [Gk. *stichos*, row] in a row parallel to long axis.

stichidium *n.* [Gk. *stichos*, row; *idion*, dim.] a tetraspore receptacle of some red algae.

stichobasidium *n.* [Gk. *stichos*, row; *basis*, base; *idion*, dim.] a cylindrical non-septate basidium having a longitudinal series of nuclear spindles.

stichochrome *a.* [Gk. *stichos*, row; *chrōma*, colour] with Nissl granules arranged in rows, as in motor neurones.

stichosome *n.* [Gk. *stichos*, row; *sōma*, body] an oesophageal structure in Trichurata consisting of a capillary-like oesophagus with many large glands.

stick insects the common name for many members of the Phasmida, *q.v.*

stigma *n.* [Gk. *stigma*, mark] portion of pistil which receives pollen; eye-spot of some Protista; an arthropod spiracle; coloured wing spot of certain butterflies and other insects; thickened area near apex of wing membrane in dragonflies; gill slit of tunicates; spot or stoma formed as an artefact in walls of capillaries; *alt.* stigmata, stigmas.

stigmasterol *n.* [Gk. *stigma*, mark; *stereos*, solid; alcoho*l*] a plant sterol, also present in milk, when deficient in diet causing muscular atrophy and calcium phosphate deposits in muscles and joints; *alt.* antistiffness factor.

stigmata *plu.* of stigma.

stigmatic *a.* [Gk. *stigma*, mark] *appl.* lid cell of an archegonium; *pert.* a stigma.

stigmatiferous *a.* [Gk. *stigma*, mark; L. *ferre*, to carry] stigma-bearing.

stigmatiform, stigmatoid *a.* [Gk. *stigma*, mark; L. *ferre*, to carry] stigma-bearing.

Stigonematales *n.* [*Stigonema*, generic name] an order of blue–green algae which have a filamentous plant body with true branching, and produce hormogonia.

stile(t) style(t), *q.v.*

stilliform *a.* [L. *stilla*, a drop; *forma*, shape] guttiform, *q.v.*

stilt roots buttress roots, *q.v.*

stimulation *n.* [L. *stimulare*, to incite] excitation or irritation of an organism or part by external or internal influences.

stimulose *a.* [L. *stimulare*, to incite] furnished with stinging hairs or cells.

stimulus *n.* [L. *stimulus*, goad] an agent which causes a reaction or change in an organism or in any of its parts, *cf.* response; a stinging hair.

sting *n.* [A.S. *stingan*, to sting] stinging hair or cell; spine of sting-ray; offensive and defensive organ for piercing, also for inoculating with poison.

stinkhorn the common name for a member of the Phallales, *q.v.*

stipe *n.* [L. *stipes*, stalk] the stem bearing pileus in mushrooms and toadstools; stalk of seaweeds; stem or caudex of palms and tree ferns; stem of fern fronds; stipes, *q.v.*; series of thecae in graptolites.

stipel *n.* [L. *stipes*, stalk] an outgrowth of leaflets resembling the stipule of a leaf base.

stipella *n.* [*Dim.* from L. *stipes*, stalk] stipule of a leaflet in a compound leaf.

stipellate *a.* [L. *stipes*, stalk] bearing stipels.

stipes *n.* [L. *stipes*, stalk] peduncle of a stalked eye; distal part of protopodite of 1st maxilla of insects, itself divided into eustipes and parastipes, and the eustipes further into dististipes, proxistipes, and basistipes; distal portion of embolus in spiders; *alt.* stipe; *plu.* stipites.

stipiform *a.* [L. *stipes*, stalk; *forma*, shape] resembling a stalk or stem; *alt.* stipitiform.

stipitate *a.* [L. *stipes*, stalk] stalked.

stipites *n.plu.* [L. *stipes*, stalk] *plu.* of stipes; paired part, anterior to mentum, of gnathochilarium.

stipitiform stipiform, *q.v.*

stipular *a.* [L. *stipula*, small stalk] like, *pert.*, or growing in place of, stipules.

stipulate *a.* [L. *stipula*, small stalk] stipuliferous, *q.v.*

stipule *n.* [L. *stipula*, small stalk] one of 2 foliaceous or membranaceous processes developed at base of a leaf petiole, sometimes in tendril or spine form; paraphyll, *q.v.*; pin feather, *q.v.*

stipuliferous *a.* [L. *stipula*, small stalk; *ferre*, to carry] bearing stipules; *alt.* stipulate.

stipuliform *a.* [L. *stipula*, small stalk; *forma*, shape] in the form of a stipule.

stipuloid *n.* [L. *stipula*, small stalk; Gk. *eidos*, form] a unicellular outgrowth from basal node of branches in Charophyta.

stirps *n.* [L. *stirps*, stock] the sum total of determinants to be found in a newly fertilized ovum; a group of organisms descended from a common ancestor; a variety of plants with stable characteristics which are retained under cultivation; a group of animals equivalent to a superfamily; *plu.* stirpes.

stock *n.* [A.S. *stocc*, post] stem of tree or bush receiving bud, or scion, in grafting; the perennial part of a herbaceous perennial plant; an asexual zooid which produces sexual zooids of one sex by gemmation, as in Polychaeta; livestock; one or a group of individuals initiating a line of descent.

stolon *n.* [L. *stolo*, shoot] a creeping stem or runner capable of developing rootlets and stem, and ultimately forming a new individual, *cf.* rhizome; a creeping hypha which can form aerial mycelium and rhizoids or haustoria; a cylindrical stem of some Polyzoa from which individuals grow out at intervals; a horizontal tubular branch of some coelenterates from which new zooids arise by budding; the cadophore and bud-forming ventral outgrowth of tunicates.

stolonate *a.* [L. *stolo*, shoot] having stolons; resembling a stolon; developing from a stolon; *appl.* plants and animals which develop by means of stolons; *alt.* stoloniferous.

stolonial *a.* [L. *stolo*, shoot] *pert.* a stolon or stolons, *appl.* a mesodermal septum in certain tunicates.

Stolonifera *n.* [L. *stolo*, shoot; *ferre*, to carry] an order of Octocorallia in which the polyps are connected together by basal stolons; sometimes included in the Alcyonacea.

stoloniferous *a.* [L. *stolo*, shoot; *ferre*, to carry] bearing a stolon or stolons; *alt.* stolonate.

Stolonoidea *n.* [L. *stolo*, shoot; Gk. *eidos*, form] an order of graptolites with irregular branching, the branches being made of complete rings.

stolotheca *n.* [L. *stolo*, shoot; *theca*, case] theca budded from side of metasicula of graptolites, and producing buds of autotheca, bitheca, and a 2nd stolotheca.

stoma *n.* [Gk. *stoma*, mouth] a small orifice; minute openings, with guard cells, in epidermis of plants, especially on undersurface of leaves, or, the stomatic pores only, *alt.* air pore; apertures in endothelium of serous membranes; part of alimentary canal between mouth opening and oesophagus, in nematodes; *alt.* stomate; *plu.* stomata.

stomach *n.* [Gk. *stomachos*, throat, gullet] saclike portion of gut between oesophagus and intestines in vertebrates; corresponding part, or entire digestive cavity of invertebrates; *alt.* ventriculus; *a.* stomachic.

stomal stomatal, *q.v.*

stomata *plu.* of stoma.

stomatal *a.* [Gk. *stoma*, mouth] *pert.* or like a stoma; *appl.* index: the ratio between number of stomata and number of epidermal cells per unit area; *alt.* stomal, stomatic.

stomate *a.* [Gk. *stoma*, mouth] with stoma or stomata, *alt.* stomatose, stomatous; *n.* stoma, *q.v.*

stomatic *a.* [Gk. *stoma*, mouth] stomatal, *q.v.*

stomatiferous *a.* [Gk. *stoma*, mouth; L. *ferre*, to carry] bearing stomata.

stomatogastric *a.* [Gk. *stoma*, mouth; *gastēr*, stomach] *pert.* mouth and stomach; *appl.* visceral system of nerves supplying anterior part of alimentary canal; *appl.* recurrent nerve from frontal to stomachic ganglion, in insects.

stomatogenesis *n.* [Gk. *stoma*, mouth; *genesis*, origin] the formation of a mouth, as in Ciliata.

Stomatopoda, stomatopods *n., n.plu.* [Gk. *stoma*, mouth; *pous*, foot] an order of malacostracans having at least 4 thoracic segments not fused to the carapace.

stomatose, stomatous *see* stomate.

stomidium *n.* [Gk. *stoma*, mouth; *idion*, *dim.*] aperture representing terminal pore of degenerated tentacles of Actiniaria.

stomions *n.plu.* [Gk. *stomion*, small mouth] dermal pores or ostia perforating dermal membrane of developing sponge.

stomium *n.* [Gk. *stomion*, small mouth] group of thin-walled cells in fern sporangium where rupture of mature capsule takes place; slit of dehiscing anther.

stomocoel *n.* [Gk. *stoma*, mouth; *koilos*, hollow] system of cavities in lips.

stomodaeal apparatus masticatory stomach, *q.v.*

stomodaeal canal in Ctenophora, a canal given off by each perradial canal, and situated parallel to stomodaeum.

stomodaeum *n.* [Gk. *stoma*, mouth; *hodaios*, *pert.* way] anterior ectoderm-lined portion of gut;

anterior portion of embryonic gut, arising as an invagination of the ectoderm; *alt.* fore-gut.

stone canal an S-shaped cylinder extending from madreporite to near mouth border in echinoderms; *alt.* hydrophoric canal, madreporic canal.

stone cells short isodiametric sclereids as found in pear; *alt.* grit cells, stereids, brachysclereids, sclereids.

stoneflies the common name for the Plecoptera, *q.v.*

stone fruit drupe, *q.v.*

stonewort the common name for a member of the Charophyta, *q.v.*

stop codon the DNA triplet signalling the end of transcription of mRNA; the RNA triplet signalling the end of translation of a polypeptide chain.

storage trunk pseudorhiza, *q.v.*

storied arranged in horizontal rows on tangential surfaces, *appl.* axial cells and ray cells of wood cambium, *cf.* non-storied; *appl.* cork in monocotyledons with suberized cells in radial rows.

strangulated *a.* [L. *strangulare*, to throttle] constricted in places; contracted and expanded irregularly.

strata *plu.* of stratum.

stratification *n.* [L. *stratum*, layer; *facere*, to make] arrangement in layers; superimposition of layers of epithelium cells; vertical grouping within a community or ecosystem; differentiation of horizons of soil.

stratified epithelium epithelium cells arranged in many superimposed layers.

stratiform *a.* [L. *stratum*, layer; *forma*, shape] layered, *appl.* fibrocartilage coating osseous grooves, or developed in some tendons.

stratose *a.* [L. *stratum*, layer] arranged in layers.

stratum *n.* [L. *stratum*, layer] a layer, as of cells, or of tissue; a group of organisms inhabiting a vertical division of an area; vegetation of similar height in a plant community, as trees, shrubs, herbs, and mosses; a layer of rock; *plu.* strata.

stratum compactum surface layer of decidua vera.

stratum corneum horny external layer of epidermis.

stratum cylindricum inner ectodermal layer surrounding mesodermal pulp of feather.

stratum fibrosum external fibrous tissue of articular capsule.

stratum germinativum Malpighian layer, *q.v.*

stratum granulosum superficial layer of rete mucosum of skin.

stratum lucidum layer of cells between stratum corneum and stratum granulosum of skin.

stratum opticum layer of nerve fibres constituting innermost layer of retina; layer of multipolar nerve cells of anterior corpora quadrigemina, *cf.* stratum zonale.

stratum spinosum layer of prickle cells in epidermis.

stratum spongiosum deeper three-fourths of decidua vera.

stratum synoviale synovial membrane, *q.v.*

stratum transect a profile of vegetation, drawn to scale and intended to show the heights of plant shoots; *alt.* profile transect; *cf.* bisect.

stratum zonale, cinereum, opticum, lem-

nisci strata of anterior corpora quadrigemina, from surface inwards.

strepsinema *n.* [Gk. *strepsis*, twisting; *nēma*, thread] chromosome thread at the strepsitene stage.

Strepsiptera *n.* [Gk. *strepsis*, twisting; *pteron*, wing] an order of insects with complete metamorphosis, commonly called stylopids, whose females are endoparasites of other insects, and the males free living.

strepsitene *a.* [Gk. *strepsis*, twisting; *tainia*, band] *appl.* stage in meiosis where the diplotene threads appear to be twisted and where crossing-over occurs.

streptomycin *n.* [*Streptomyces*, generic name] an antibiotic synthesized by the actinomycete *Streptomyces griseus*, which inhibits protein synthesis by causing faulty translation of mRNA to protein.

Streptoneura *n.* [Gk. *streptos*, twisted; *neuron*, nerve] Prosobranchia, *q.v.*

streptoneurous *a.* [Gk. *streptos*, twisted; *neuron*, nerve] having visceral cord twisted, forming a figure of 8, as certain gastropods; *cf.* euthyneurous.

streptonigrin *n.* [*Streptomyces*, generic name; L. *niger*, black] an antibiotic synthesized by the actinomycete *Streptomyces flocculus*, which causes chromosome breakage.

streptostylic *a.* [Gk. *streptos*, pliant; *stylos*, column] having quadrate in moveable articulation with squamosal; *cf.* monimostylic; *n.* streptostyly.

stria *n.* [L. *stria*, groove, channel] a narrow line, streak, band, groove, or channel.

striate(d) *a.* [L. *striatus*, grooved] marked by narrow lines or grooves, usually parallel.

striated muscle voluntary muscle, *q.v.*

striatum *a.* [L. *striatus*, grooved] corpus striatum, *q.v.*

stridulating organs a special apparatus on metathoracic and anterior abdominal segments for producing song of cicadas; sound-producing organs of various other Arthropoda.

stridulation *n.* [L. *stridor*, grating] in certain insects and some other arthropods, sound produced by rubbing one part of the body against another.

striga *n.* [L. *striga*, ridge, furrow] a band of upright, stiff pointed hairs or bristles; a bristle-like scale; *plu.* strigae.

strigate *a.* [L. *striga*, ridge] bearing strigae.

Strigiformes *n.* [*Strix*, generic name of some owls; L. *forma*, shape] an order of birds of the subclass Neornithes, including the owls.

strigil(is) *n.* [L. *strigilis*, currycomb] a mechanism for cleaning antennae, at junction of tibia and tarsus on 1st leg of certain insects such as bees.

strigillose *a.* [L. *strigilla*, small ridge] minutely strigose.

strigose *a.* [L. *striga*, ridge] covered with stiff hairs; ridged; marked by small furrows.

striola *n.* [L. *striola*, small channel] fine narrow line or streak.

striolate *a.* [L. *striola*, small channel] finely striate.

striped muscle voluntary muscle, *q.v.*

stripe of Hensen Hensen's stripe, *q.v.*

strobila *n.* [Gk. *strobilos*, fir cone] stage in development of some Scyphozoa, where from a suc-

cession of annular discs embryos take form of a pile of discs separated off in turn; chain of proglottides of tapeworms; strobile, *q.v.*

strobilaceous *a.* [Gk. *strobilos*, fir cone] cone-shaped; *pert.* or having strobiles.

strobilation *n.* [Gk. *strobilos*, fir cone] reproduction by body segmentation into zooids, as in coelenterates, or into proglottides, as in tapeworms; *alt.* strobilization.

strobile *n.* [Gk. *strobilos*, fir cone] an assemblage of sporophylls arranged in a cone-shaped structure in horsetails, club mosses, and conifers, *alt.* cone; a spike formed by persistant membranous bracts, each having a pistillate flower; strobila, *q.v.*; *alt.* strobilus.

strobiliferous *a.* [Gk. *strobilos*, fir cone; L. *ferre*, to carry] producing strobiles.

strobiliform, strobiloid *a.* [Gk. *strobilos*, fir cone; *eidos*, form; L. *forma*, form] resembling or shaped like a strobilus or cone.

strobilization strobilation, *q.v.*

strobilus *n.* [Gk. *strobilos*, fir cone] strobile, *q.v.*

stroma *n.* [Gk. *strōma*, bedding] transparent filmy framework of red blood corpuscles; in chloroplasts, the colourless regions where grana are absent, containing enzymes concerned with carbon dioxide fixation; the non-pigmented part of other plastids; connective tissue binding and supporting an organ; in ovary, a soft, vascular, reticular framework in meshes of which ovarian follicles are embedded; tissue of hyphae, or of fungous cells with host tissue, in or upon which spore-bearing structure may be produced; *plu.* stromata.

stromata *plu.* of stroma; short protrusions from a sclerotium, each composed of hyphae, in which perithecia are developed in some thallophytes.

stromate *a.* [Gk. *strōma*, bedding] having, or being within or upon, a stroma, *appl.* fruit bodies of fungi.

stromatic *a.* [Gk. *strōma*, bedding] *pert.*, like, in form or nature of, a stroma; *alt.* stromatiform, stromatous, stromatoid, stromoid.

stromatin *n.* [Gk. *strōma*, bedding] the protein constituent of the plasma membrane, as of red blood corpuscles.

stromatolites *n.plu.* [Gk. *strōma*, bedding; *lithos*, stone] structures formed by blue–green algae in warm waters and made of layers of calcareous material, and similar structures found in Pre-Cambrian rocks, indicating a form of life at that time.

stromatolysis *n.* [Gk. *strōma*, bedding; *lysis*, loosing] continued action of a haemolysin on cell stroma after haemoglobin has been liberated.

strombuliferous *a.* [*Dim.* of L. *strombus*, spiral shell; *ferre*, to carry] having spirally coiled organs or structures.

strombuliform *a.* [*Dim.* of L. *strombus*, spiral shell; *forma*, shape] spirally coiled.

stromoid stromatic, *q.v.*

Strongylata *n.* [*Strongylus*, generic name] in some classifications an order of Phasmidia or a subdivision of Rhabditida, having a simple mouth and no papillae.

strongyle, strongylon *n.* [Gk. *stronggylos*, rounded] a 2-rayed rod sponge spicule rounded at both ends; a type of nematode larva.

strophanthin *n.* [*Strophanthus*, generic name] a saponin glycoside obtained from various plants of the family Apocynaceae, used as a tropical arrow poison.

strophiolate *a.* [L. *strophiolum*, small garland] having excrescences around hilum; *alt.* carunculate.

strophiole *n.* [L. *strophiolum*, small garland] one of the small excrescences arising from various parts of a seed testa, never developed before fertilization; *alt.* caruncle.

strophotaxis *n.* [Gk. *strophos*, twisted; *taxis*, arrangement] twisting movement or tendency, in response to an external stimulus.

structural colours those due to surface structure and not to pigment.

structural gene a gene in an operon which produces mRNA to make a polypeptide chain; *alt.* cistron.

struma *n.* [L. *struma*, scrofulous tumour] a swelling on a plant organ.

strumiferous *a.* [L. *struma*, wen; *ferre*, to carry] having a struma or strumae.

strumiform *a.* [L. *struma*, wen; *forma*, shape] cushion-like.

strumose, strumulose *a.* [L. *struma*, wen] having small cushion-like swellings.

Struthioniformes *n.* [*Struthio*, generic name of ostrich; L. *forma*, shape] an order of birds of the subclass Neornithes, including the ostriches.

strut roots buttress roots, *q.v.*

strychnine *n.* [*Strychnos*, generic name] an alkaloid produced from seeds of *Strychnos* species, and some other plants.

stupeous, stupose *a.* [L. *stupa*, tow] tow-like; having a tuft of matted filaments.

stupulose *a.* [L. *stupa*, tow] covered with short filaments.

stylar *a.* [L. *stylus*, pricker] *pert.* style.

Stylasterina *n.* [*Stylaster*, generic name] an order of hydrozoans that are very similar to the Milleporina but lack a medusoid stage.

stylate *a.* [L. *stylus*, pricker] having a style or styles.

style *n.* [Gk. *stylos*, pillar; L. *stylus*, pricker] slender upper part of pistil, supporting stigma; a rod-like sponge spicule, pointed at one end; a calcareous projection from pore tabula in some Millepora; abdominal bristle-like process on male insects; arista, *q.v.*; embolus of spiders; any of the small projections of cingulum of a molar tooth; crystalline style, *q.v.*; *alt.* stylus.

style sac a tubular gland in some molluscs, which secretes the crystalline style.

stylet *n.* [L. *stylus*, pricker] small, pointed bristle-like appendage; unpaired part of terebra or sting, held in position by stylet sheath; needle-like digit of chelicerae in certain parasitic Acarina; sharp projection at the tip of eversible part of proboscis in Metanemertea; *alt.* stilet, stylus.

stylifer *n.* [L. *stylus*, pricker; *ferre*, to carry] portion of clasper which carries style.

styliferous *a.* [L. *stylus*, pricker; *ferre*, to carry] bearing a style; having bristly appendages.

styliform *a.* [L. *stylus*, pricker; *forma*, shape] prickle- or bristle-shaped.

styloconic *a.* [Gk. *stylos*, pillar; *kōnos*, cone] having terminal peg on conical base, *appl.* type of olfactory sensilla in insects.

styloglossus *n.* [Gk. *stylos*, pillar; *glōssa*, tongue] a muscle connecting styloid process and side of tongue; *a.* styloglossal.

stylogonidium conidium, *q.v.*

stylohyal *n.* [Gk. *stylos*, pillar; *hyoeidēs*, shaped] distal part of styloid process of temporal bone; a small interhyal between hyal and hyomandibular.

stylohyoid *a.* [Gk. *stylos*, pillar; *hyoeidēs*, shaped] *appl.* a ligament attached to styloid process and lesser cornu of hyoid; *appl.* a muscle; *appl.* a branch of facial nerve.

styloid *a.* [Gk. *stylos*, pillar; *eidos*, form] pillar-like, *appl.* processes of temporal bone, fibula, radius, ulna. *a.* A columnar crystal.

stylomandibular *a.* [Gk. *stylos*, pillar; L. *mandibulum*, jaw] *appl.* ligamentous band extending from styloid process of temporal bone to angle of lower jaw.

stylomastoid *a.* [Gk. *stylos*, pillar; *mastos*, breast; *eidos*, like] *appl.* foramen between styloid and mastoid processes, also an artery entering that foramen.

Stylommatophora *n.* [Gk. *stylos*, pillar; *omma*, eye; *phora*, bearing] an order of mainly terrestrial pulmonates, including the land snails and slugs.

stylopharyngeus *n.* [Gk. *stylos*, pillar; *pharyngx*, pharynx] a muscle extending from the base of styloid process downwards along side of pharynx to thyroid cartilage.

stylopids the common name for the Strepsiptera, *q.v.*

stylopodium *n.* [Gk. *stylos*, pillar; *pous*, food] a conical swelling surrounding bases of divaricating styles of the dicot family Umbelliferae; structure attaching mericarps to carpophore; upper arm, or thigh.

stylospore *n.* [Gk. *stylos*, pillar; *sporos*, seed] a spore borne on a stalk such as pycnidiospore or conidium; *alt.* myceloconidium; *a.* stylosporous.

stylostegium *n.* [Gk. *stylos*, pillar; *stegē*, roof] inner corona of milkweed plants.

stylostome *n.* [L. *stylus*, pricker; Gk. *stoma*, mouth] a tube in skin, produced by tissue reaction of host to insertion of chelicerae of a mite.

stylus *n.* [L. *stylus*, pricker] style, *q.v.*; stylet, *q.v.*; simple pointed spicule; molar cusp; pointed process.

sub- [L. *sub*, under] prefix meaning not quite, somewhat, nearly; below; in classification, a group just below the status of the taxon following it as in subclass, etc.

subabdominal *a.* [L. *sub*, under; *abdomen*, belly] nearly in abdominal region.

subacuminate *a.* [L. *sub*, under; *acumen*, point] somewhat tapering.

subaduncate *a.* [L. *sub*, under; *aduncus*, hooked] somewhat crooked.

subaerial *a.* [L. *sub*, under; *aer*, air] growing just above surface of ground.

subalpine *a.* [L. *sub*, under; *alpinus*, alpine] *appl.* zone below timber line, or to plants or animals growing or living there.

subalternate *a.* [L. *sub*, under; *alternus*, one after another] tending to change from alternate to opposite.

subanconeous *n.* [L. *sub*, under; Gk. *angkōn*, elbow] small muscle extending from triceps to elbow.

subapical *a.* [L. *sub*, under; *apex*, extremity] nearly at the apex.

subarachnoid *a.* [L. *sub*, under; Gk. *arachnē*, spider's web; *eidos*, form] *appl.* a cavity filled with cerebrospinal fluid between arachnoid and pia mater; *appl.* cisternae of brain, and longitudinal septum in region of spinal medulla.

subarborescent *a.* [L. *sub*, under; *arborescens*, growing into a tree] somewhat like a tree.

subarcuate *a.* [L. *sub*, under; *arcus*, bow] *appl.* a blind fossa which extends backwards under superior semicircular canal, in infant skull.

subatrial *a.* [L. *sub*, under; *atrium*, hall] below the atrium, *appl.* longitudinal ridges on inner side of metapleural folds, uniting to form ventral part of atrium, in development of lancelet, *Amphioxus.*

subauricular *a.* [L. *sub*, under; *auricula*, external ear] below the ear.

subaxillary *a.* [L. *sub*, under; *axilla*, armpit] *appl.* outgrowths just beneath the axil.

sub-basal *a.* [L. *sub*, under; Gk. *basis*, foundation] situated near the base.

sub-branchial *a.* [L. *sub*, under; *branchiae*, gills] under the gills.

sub-bronchial *a.* [L. *sub*, under; Gk. *brongchos*, windpipe] below the bronchials.

subcalcareous *a.* [L. *sub*, under; *calx*, lime] somewhat limy.

subcalcarine *a.* [L. *sub*, under; *calcar*, spur] under the calcarine fissure, *appl.* lingual gyrus of brain.

subcallosal *a.* [L. *sub*, under; *callus*, hard skin] *appl.* a gyrus below corpus callosum.

subcampanulate *a.* [L. *sub*, under; *campanula*, little bell] somewhat bell-shaped.

subcapsular *a.* [L. *sub*, under; *capsula*, little chest] inside a capsule.

subcardinal *a.* [L. *sub*, under; *cardo*, hinge] *appl.* pair of veins between mesonephroi.

subcarinate *a.* [L. *sub*, under; *carina*, keel] somewhat keel-shaped.

subcartilaginous *a.* [L. *sub*, under; *cartilago*, gristle] not entirely cartilaginous.

subcaudal *a.* [L. *sub*, under; *cauda*, tail] situated under tail, as a shield or plate.

subcaudate *a.* [L. *sub*, under; *cauda*, tail] having a tail-like process.

subcaulescent *a.* [L. *sub*, under; *caulis*, stalk] borne on a very short stem.

subcellular *a.* [L. *sub*, under; *cellula*, small cell] *appl.* functional units or organelles within the cell, as chloroplasts, chromosomes, etc.

subcentral *a.* [L. *sub*, under; *centrum*, centre] nearly central.

subchela *n.* [L. *sub*, under; Gk. *chēlē*, claw] in some arthropods, a prehensile claw of which the last joint folds back on the preceding one.

subchelate *a.* [L. *sub*, under; Gk. *chēlē*, claw] having subchelae; having imperfect chelae.

subcheliceral *a.* [L. *sub*, under; Gk. *chēlē*, claw; *keras*, horn] beneath the chelicerae; *appl.* plate or epistome, for attachment of pharyngeal dilators in certain Acarina.

subchordal *a.* [L. *sub*, under; *chorda*, cord] under the notochord.

subcingulum *n.* [L. *sub*, under; *cingulum*, girdle] the lower lip part of a cingulum or girdle of rotifers.

subclavate *a.* [L. *sub*, under; *clavus*, club] somewhat club-shaped.

subclavian *a.* [L. *sub*, under; *clavis*, key] below clavicle, *appl.* artery, vein, nerve.

subclavius a small muscle connecting 1st rib and clavicle.

subclimax *n.* [L. *sub*, under; Gk. *klimax*, ladder] stage in plant succession preceding the climax which persists because of some arresting factor as in biotic climax, fire climax, plagioclimax; proclimax, *q.v.*; *cf.* preclimax.

subcoracoid *a.* [L. *sub*, under; Gk. *korax*, crow; *eidos*, like] below the coracoid.

subcordate *a.* [L. *sub*, under; *cor*, heart] tending to be heart-shaped.

subcorneous *a.* [L. *sub*, under; *cornu*, horn] under a horny layer; slightly horny.

subcortical *a.* [L. *sub*, under; *cortex*, bark] under cortex, or cortical layer; *appl.* cavities under dermal cortex of sponges.

subcosta *n.* [L. *sub*, under; *costa*, rib] an auxiliary vein joining costa of insect wing.

subcostal *a.* [L. *sub*, under; *costa*, rib] below ribs, *appl.* zone, muscles, arteries, nerve, plane; *pert.* subcosta.

subcoxa *n.* [L. *sub*, under; *coxa*, hip] basal ring, or segment, articulated distally with coxa of arthropod leg; *alt.* praecoxa, precoxa, pretrochantin.

subcrenate *a.* [L. *sub*, under; L.L. *crena*, notch] tending to have rounded scallops, as a leaf margin.

subcrureus *n.* [L. *sub*, under; *crus*, leg] muscle extending from lower femur to knee; *alt.* articularis genus.

subcubical *a.* [L. *sub*, under; *cubus*, cube] *appl.* cells not quite so long as broad, as those lining alveoli of thyroid.

subcutaneous *a.* [L. *sub*, under; *cutis*, skin] under the skin; *appl.* parasites living just under skin; *appl.* inguinal or external abdominal ring; *appl.* fat under the skin.

subcuticula *n.* [L. *sub*, under; *cuticula*, cuticle] epidermis beneath cuticle, as in nematodes.

subcuticular *a.* [L. *sub*, under; *cuticula*, cuticle] under the cuticle, epidermis, or outer skin.

subcutis *n.* [L. *sub*, under; *cutis*, skin] a loose layer of connective tissue between corium and deeper tissues of skin; inner layer of cutis of agarics, under the epicutis.

subdentate *a.* [L. *sub*, under; *dens*, a tooth] slightly toothed or notched.

subdermal *a.* [L. *sub*, under; Gk. *derma*, skin] beneath the skin; beneath derma.

subdorsal *a.* [L. *sub*, under; *dorsum*, back] situated almost on dorsal surface.

subdural *a.* [L. *sub*, under; *durus*, hard] *appl.* the space separating spinal dura mater from arachnoid.

subepicardial *a.* [L. *sub*, under; Gk. *epi*, upon; *kardia*, heart] *appl.* areolar tissue attaching visceral layer of pericardium to muscular wall of heart.

subepiglottic *a.* [L. *sub*, under; Gk. *epi*, upon; *glōtta*, tongue] beneath epiglottis.

subepithelial *a.* [L. *sub*, under; Gk. *epi*, upon; *thēlē*, nipple] below epithelium, *appl.* plexus of cornea; *appl.* endothelium: Débove's membrane, *q.v.*

suber *n.* [L. *suber*, cork tree] cork tissue.

subereous *a.* [L. *suber*, cork tree] of corky texture.

suberic *a.* [L. *suber*, cork tree] *pert.* or derived from cork.

suberiferous *a.* [L. *suber*, cork tree; *ferre*, to bear] cork-producing.

suberification n. [L. *suber*, cork tree; *facere*, to make] conversion into cork tissue.

suberin n. [L. *suber*, cork tree] the waxy substance developed in a thickened cell wall, characteristic of cork tissues; *cf.* cutin.

suberization n. [L. *suber*, cork tree] modification of cell walls due to suberin formation.

suberose a. [L. *suber*, cork tree] with corky, waterproof texture. [L. *sub*, under; *erosus*, gnawed] as if somewhat gnawed.

subesophageal suboesophageal, *q.v.*

subfusiform a. [L. *sub*, under; *fusus*, spindle; *forma*, shape] somewhat spindle-shaped, elliptic-fusiform; *alt.* boletiform.

subgalea n. [L. *sub*, under; *galea*, helmet] part of maxilla, at base of stipes, of insects; *alt.* parastipes.

subgeniculate a. [L. *sub*, under; *geniculum*, little knee] somewhat bent.

subgenital a. [L. *sub*, under; *genitalis*, genital] below reproductive organs; *appl.* shallow pit or pouch beneath gonad in the jellyfish, *Aurelia*; *appl.* portico formed by fusion of subgenital pouches of Discomedusae; *appl.* plate formed by 9th abdominal sternite and coxites, hypandrium of certain insects.

subgerminal a. [L. *sub*, under; *germen*, bud] beneath the germinal disc, *appl.* cavity.

subglenoid a. [L. *sub*, under; Gk. *glēnē*, socket; *eidos*, form] beneath glenoid cavity.

subglossal a. [L. *sub*, under; Gk. *glōssa*, tongue] beneath the tongue.

subharpal a. [L. *sub*, under; Gk. *harpē*, sickle] *appl.* plate in area below harpe in insects.

subhyaloid a. [L. *sub*, under; Gk. *hyalos*, glass; *eidos*, like] beneath hyaloid membrane or fossa of eye.

subhymenium n. [L. *sub*, under; Gk. *hymēn*, membrane] layer of small cells between trama and hymenium in gill of agarics, *alt.* hymenopodium, subhymenial layer; hypothecium, *q.v.*

subhyoid a. [L. *sub*, under; Gk. *hyoeidēs*, Υ-shaped] below hyoid at base of tongue.

subicle subiculum of fungi.

subiculum n. [L. *subiculum*, under layer] a mycelial covering of substrate, *alt.* subicle; part of the hippocampus bordering the hippocampal fissure; bony ridge bounding oval opening in interior wall of middle ear.

subimago n. [L. *sub*, under; *imago*, likeness] a stage between pupa and imago in life history of some insects; *alt.* pseudimago.

subinguinal a. [L. *sub*, under; *inguen*, groin] situated below a horizontal line at level of great saphenous vein termination, *appl.* lymph glands.

subjugal a. [L. *sub*, under; *jugum*, yoke] below jugal or cheek bone.

subjugular a. [L. *sub*, under; *jugulum*, collar bone] *appl.* a ventral fish fin nearly far enough forward to be jugular.

sublanceolate a. [L. *sub*, under; *lanceolatus*, speared] tending to be narrow and to taper towards both ends.

sublaryngeal a. [L. *sub*, under; Gk. *larynx*, larynx] situated below larynx.

sublenticular a. [L. *sub*, under; *lenticula*, small lentil] somewhat lens-shaped.

subliminal a. [L. *sub*, under; *limen*, threshold] inadequate for perceptible response, *appl.* stimuli, or to differences between stimuli; *cf.* limen.

sublingua n. [L. *sub*, under; *lingua*, tongue] a single or double projection or fold beneath tongue, in some mammals.

sublingual [L. *sub*, under; *lingua*, tongue] beneath tongue, *appl.* gland, artery, etc., *appl.* ventral pharyngeal gland, in Hymenoptera.

sublitoral a. [L. *sub*, under; *litus*, seashore] below litoral, *appl.* shallow water zone to about 200 m; *alt.* sublittoral.

sublobular a. [L. *sub*, under; *lobus*, lobe] *appl.* veins at base of lobules of liver.

sublocular a. [L. *sub*, under; *loculus*, compartment] somewhat locular or cellular.

submalleate a. [L. *sub*, under; *malleus*, hammer] somewhat hammer-shaped, *appl.* trophi of rotifer mastax.

submandibular a. [L. *sub*, under; *mandibulum*, jaw] submaxillary, *q.v.*

submarginal a. [L. *sub*, under; *margo*, margin] placed nearly at margin.

submarginate a. [L. *sub*, under; *margo*, margin] *appl.* a bordering structure near a margin.

submaxilla n. [L. *sub*, under; *maxilla*, jaw] mandible, *q.v.*

submaxillary a. [L. *sub*, under; *maxilla*, jaw] beneath lower jaw, *appl.* duct, ganglion, gland, triangle; *alt.* mandibular, submandibular.

submedian a. [L. *sub*, under; *medius*, middle] *appl.* tooth or vein next median.

submental a. [L. *sub*, under; *mentum*, chin] beneath chin, *appl.* artery, glands, triangle, vibrissae; *pert.* submentum.

submentum n. [L. *sub*, under; *mentum*, chin] basal part of labium of insects.

submersed a. [L. *submergere*, to submerge] *appl.* plants growing entirely under water.

submetacentric a. [L. *sub*, under; Gk. *meta*, among; *kentron*, centre] *appl.* chromosome that is shaped like a J at anaphase because the centromere is nearer one end than the other.

submicron n. [L. *sub*, under; Gk. *mikros*, small] a small particle that is only visible with the ultramicroscope, particularly one less than 1 μm in diameter; *cf.* amicron.

submission n. [L. *submittere*, to let down] the behaviour of a losing animal in a conflict situation where it takes a certain submissive posture to prevent further attack.

submucosa n. [L. *sub*, under; *mucosus*, mucous] a layer of the gut wall between the mucosa and muscularis externa made of yellow and white fibres and accommodating blood vessels, nerves, Meissner's plexus, and some glands.

subnasal a. [L. *sub*, under; *nasus*, nose] beneath the nose.

subneural a. [L. *sub*, under; Gk. *neuron*, nerve] *appl.* blood vessel in annelids; *appl.* gland and ganglion of nervous system of tunicates; *appl.* sarcoplasm in motor end plates.

subnotochord n. [L. *sub*, under; Gk. *nōton*, back; *chordē*, cord] a rod ventral to true notochord; *alt.* hypochord.

suboccipital a. [L. *sub*, under; *occiput*, back of head] *appl.* muscles, nerve, triangle, under occipitals of skull.

subocular shelf ingrowth from suborbitals supporting eyeball of certain fishes.

suboesophageal a. [L. *sub*, under; Gk. *oisophagos*, gullet] below the gullet; *appl.* anterior ganglion of ventral nerve cord; *alt.* subesophageal.

subopercle suboperculum, *q.v.*

subopercular *a.* [L. *sub*, under; *operculum*, cover] under operculum of fishes, or shell lid of molluscs.

suboperculum *n.* [L. *sub*, under; *operculum*, cover] a membrane bone of operculum of fishes; *alt.* subopercle.

suboptic *a.* [L. *sub*, under; Gk. *optikos*, relating to sight] below the eye.

suboral *a.* [L. *sub*, under; *os*, mouth] below or near mouth.

suborbital *a.* [L. *sub*, under; *orbis*, eye socket] *appl.* structures below orbit.

subovate *a.* [L. *sub*, under; *ovum*, egg] somewhat oval or egg-shaped; *alt.* suboval, subovoid.

subpalmate *a.* [L. *sub*, under; *palma*, palm] tending to become palmate, *appl.* leaves.

subparietal *a.* [L. *sub*, under; *paries*, wall] beneath parietals, *appl.* sulcus which is lower boundary of parietal lobe.

subpectinate *a.* [L. *sub*, under; *pecten*, comb] tending to be comb-like in structure.

subpedunculate *a.* [L. *sub*, under; L.L. *pedunculus*, little foot] resting on very short stalk.

subpericardial *a.* [L. *sub*, under; Gk. *peri*, around; *kardia*, heart] under pericardium.

subperitoneal *a* [L. *sub*, under; Gk. *peritoneion*, something stretched around] *appl.* connective tissue under peritoneum; *alt.* extraperitoneal.

subpessular *a.* [L. *sub*, under; *pessulus*, bolt] below the pessulus of syrinx, *appl.* air sac.

subpetiolar, subpetiolate *a.* [L. *sub*, under; *petiolus*, little foot] within petiole, *appl.* bud so concealed; almost sessile.

subpharyngeal *a.* [L. *sub*, under; Gk. *pharyngx*, throat] below the throat; *appl.* gland or endostyle beneath pharynx, with cells containing iodine in ammocoetes.

subphrenic *a.* [L. *sub*, under; Gk. *phrēn*, midriff] below the diaphragm.

subpial *a.* [L. *sub*, under; *pia*, kind] under the pia mater.

subpleural *a.* [L. *sub*, under; Gk. *pleura*, side] beneath inner lining of thoracic wall.

subpolar *a.* [L. *sub*, close to; Gk. *polos*, pole] situated near a pole, as in bacterial flagellum.

subpubic *a.* [L. *sub*, under; *pubes*, adult] below the pubic region, *appl.* arcuate ligament.

subpulmonary *a.* [L. *sub*, under; *pulmo*, lung] beneath the lungs.

subradicate *v.* [L. *sub*, slightly; *radicari*, to take root] to have a slight downward extension of base, as of stipe.

subradius *n.* [L. *sub*, under; *radius*, ray] in radiate animals, a radius of 4th order, that between adradius and perradius, or between adradius and interradius.

subradular *a.* [L. *sub*, under; *radere*, to scrape] *appl.* organ containing nerve endings, situated at anterior end of odontophore.

subramose *a.* [L. *sub*, under; *ramus*, branch] slightly branching.

subreniform *a.* [L. *sub*, under; *renes*, kidneys; *forma*, shape] slightly kidney-shaped.

subretinal *a.* [L. *sub*, under; *rete*, net] beneath retina.

subrostral *a.* [L. *sub*, under; *rostrum*, beak] below the beak or rostrum, *appl.* a cerebral fissure.

subsacral *a.* [L. *sub*, under; *sacrum*, sacred] below the sacrum.

subsartorial *a.* [L. *sub*, under; *sartor*, tailor] *appl.* plexus under sartorius of thigh.

subscapular *a.* [L. *sub*, under; *scapula*, shoulder blade] beneath the scapula, *appl.* artery, muscles, nerves, etc.

subsclerotic *a.* [L. *sub*, under; Gk. *sklēros*, hard] beneath sclera; between sclerotic and choroid layers of eye.

subscutal *a.* [L. *sub*, under; *scutum*, shield] under a scutum; *appl.* Gené's organ, *q.v.*

subsere *n.* [L. *sub*, under; *serere*, to put in a row] plant succession on denuded area, a secondary succession.

subserous *a.* [L. *sub*, under; *serum*, whey] beneath a serous membrane, *appl.* areolar tissue.

subserrate *a.* [L. *sub*, under; *serra*, saw] somewhat notched or saw-toothed.

subsessile *a.* [L. *sub*, under; *sedere*, to sit] nearly sessile, with almost no stalk.

subsidiary cells auxiliary cells, *q.v.*

subspatulate *a.* [L. *sub*, under; *spatula*, spoon] somewhat spoon-shaped.

subspecies *n.* [L. *sub*, under; *species*, particular kind] a taxonomic term used in slightly different ways by specialists but usually meaning a group consisting of individuals within a species having certain distinguishing characteristics separating them from other members of the species, and forming a breeding group; variety, *q.v.*

subspinous *a.* [L. *sub*, under; *spina*, spine] tending to become spiny.

substantia *n.* [L. *substantia*, substance] substance or matter.

substantia adamantina enamel of teeth.

substantia alba white matter of brain and spinal cord.

substantia eburnea dentine, *q.v.*

substantia gelatinosa gelatinous neuroglia, with some nerve cells, in spinal cord.

substantia grisea grey matter of brain and spinal cord.

substantia nigra a semilunar layer of grey cells of mid-brain.

substantia ossea cement of teeth; *alt.* crusta petrosa.

substantia reticularis anterior and lateral reticular formations in medulla oblongata.

substantia spongiosa cancellous tissue of bone.

substantive variation changes in actual constitution or substance of parts; *cf.* meristic variation.

substernal *a.* [L. *sub*, under; *sternum*, breast bone] below the sternum.

substipitate *a.* [L. *sub*, under; *stipes*, stalk] having an extremely short stem.

substomatal hypostomatic, *q.v.*

substrate *n.* [L. *sub*, under; *stratum*, layer] the substance upon which an enzyme acts, *alt.* zymolyte; a substance undergoing oxidation utilized in respiration, a respiratory substrate; inert substance containing or receiving a nutrient solution, *alt.* substratum.

substratose *a.* [L. *sub*, under; *stratum*, layer] slightly or indistinctly stratified.

substratum *n.* [L. *sub*, under; *stratum*, layer] the base to which a stationary animal or a plant is fixed; *see* substrate.

subtalar *a.* [L. *sub*, under; *talus*, ankle] *appl.* joint: talocalcaneal articulation.

subtectal *a.* [L. *sub*, under; *tectum*, roof] lying

under a roof, especially of skull; *pert.* alisphenoid of fish skull.

subtegminal *a.* [L. *sub*, under; *tegmen*, covering] under a tegmen or inner coat of a seed.

subtegulum *n.* [L. *sub*, under; *tegula*, tile] a chitinous structure protecting the haematodocha in certain spiders.

subtentacular canals two prolongations of echinoderm coelom.

subterminal *a.* [L. *sub*, close to; *terminus*, end] situated near the end.

subthalamus hypothalamus, *q.v.*; part of hypothalamus excluding optic chiasma and region of mamillary bodies.

subthoracic *a.* [L. *sub*, under; Gk. *thōrax*, chest] not so far forward as to be called thoracic, *appl.* certain fish fins.

subtrapezoidal *a.* [L. *sub*, under; Gk. *trapezion*, small table; *eidos*, form] somewhat trapezoid-shaped.

subtruncate *a.* [L. *sub*, under; *truncatus*, maimed] terminating rather abruptly.

subtypical *a.* [L. *sub*, under; *typus*, image] deviating slightly from type.

subulate *a.* [L. *subula*, awl] awl-shaped, i.e. narrow and tapering from base to a fine point, *appl.* leaves, as of onion.

subumbellate *a.* [L. *sub*, under; *umbella*, small shade] tending to an umbellate arrangement with peduncles arising from a common centre.

subumbonal *a.* [L. *sub*, under; *umbo*, boss] beneath or anterior to umbo of bivalve shell.

subumbonate *a.* [L. *sub*, under; *umbo*, boss] slightly convex; having a low rounded protuberance.

subumbrella *n.* [L. *sub*, under; *umbra*, shade] concave inner surface of medusoid bell.

subuncinate *a.* [L. *sub*, under; *uncus*, hook] having a somewhat hooked process; somewhat hook-shaped.

subungual *a.* [L. *sub*, under; *unguis*, nail] under a nail, claw, or hoof; *alt.* hyponychial.

subunguis *n.* [L. *sub*, under; *unguis*, nail] the ventral scale of a claw or nail.

subuniversal veil protoblema, *q.v.*

subvaginal *a.* [L. *sub*, under; *vagina*, sheath] within or under a sheath.

subvertebral *a.* [L. *sub*, under; *vertebra*, a joint] under the vertebral column.

subvital *a.* [L. *sub*, under; *vitalis*, vital] deficient in vitality; *appl.* genes causing a mortality of less than 50%; *cf.* semilethal.

subzonal *a.* [L. *sub*, under; *zona*, belt] *appl.* layer of cells internal to zona radiata.

subzygomatic *a.* [L. *sub*, under; Gk. *zygon*, yoke] under the cheek bone.

succate *a.* [L. *succus*, sap] containing juice, juicy; *alt.* succose, succous.

succession *n.* [L. *successio*, succession] a geological, ecological, or seasonal sequence of species; the development of plant communities leading to a climax; chronological distribution of organisms in a given area; lagging of sex chromosomes behind autosomes in moving to the poles after meiosis.

succiferous *a.* [L. *succus*, sap; *ferre*, to carry] sap-conveying.

succinic acid an acid of the Krebs' cycle which is oxidized to fumaric acid by succinic dehydrogenase.

succinic dehydrogenase an enzyme which is a flavoprotein (FAD) containing iron, which oxidizes succinate (succinic acid) to fumarate (fumaric acid) using an acceptor; EC 1.3.99.1; *r.n.* succinate dehydrogenase.

succiput *n.* [L. *sub*, under; *caput*, head] area below foramen of neck in insects.

succise *a.* [L. *succisus*, lopped off] abrupt; appearing as if a part were cut off.

succubous *a.* [L. *sub*, under; *cubare*, to lie down] with each leaf covering part of that under it; *cf.* incubous.

succulent *a.* [L. *succus*, sap] full of juice or sap; *appl.* fruit having a fleshy pericarp as berries, drupes, *cf.* dry.

succus *n.* [L. *succus*, juice, sap] the juice of a plant; fluid secreted by glands.

succus entericus the digestive juice secreted by the small intestine containing various digestive enzymes; *alt.* intestinal juice.

sucker *n.* [A.S. *sucan*, to suck] a stem branch, first subterranean and then aerial, which may ultimately form an independent plant, *alt.* sobole, surculus; haustorium, *q.v.*; an organ adapted for creating a vacuum, in some animals for purposes of ingestion, in others to assist in locomotion or attachment.

sucking disc a disc assisting in attachment, as at end of echinoderm tube-foot.

sucking lice a common name for the Anoplura, *q.v.*

sucrase *n.* [F. *sucre*, sugar; *-ase*] an enzyme which catalyses the hydrolysis of sucrose and maltose by an α-D-glucoside-type action; EC 3.2.1.48; *r.n.* sucrose α-D-glycohydrolase.

sucrose *n.* [F. *sucre*, sugar] a non-reducing disaccharide present in many green plants and hydrolysed by the enzymes invertase or sucrase or by dilute acids to produce glucose and fructose; *alt.* cane sugar, beet sugar, invert sugar, saccharose, saccharobiose.

suction pressure (SP) the capacity of a plant cell to take up water by osmosis, being the difference between the osmotic pressure of the vacuolar cell sap causing water to enter and the back pressure exerted by the cell wall (the turgor pressure) tending to prevent water entering: when the cell is turgid it has no suction pressure; *alt.* diffusion pressure deficit, water potential.

Suctoria *n.* [L. *sugere*, to suck] a subclass of Ciliophora which usually lose their cilia in the adult stages and possess one or more suctorial tentacles; in some classifications considered an order of Holotricha and called the Suctorida; *alt.* Acinetae.

suctorial *a.* [L. *sugere*, to suck] adapted for sucking; *pert.* sucker; furnished with suckers; *appl.* a pad of fat in relation with buccinator, supposed to assist in sucking; *alt.* sugent, sugescent.

sudation *n.* [L. *sudatio*, perspiration] discharge of water and substances in solution, as through pores; sweating.

sudor *n.* [L. *sudor*, sweat] perspiration.

sudoriferous *a.* [L. *sudor*, sweat; *ferre*, to carry] conveying, producing, or secreting sweat, *appl.* glands and their ducts; *alt.* sudoriparous.

sudorific *a.* [L. *sudor*, sweat; *facere*, to make] causing or *pert.* secretion of sweat.

sudoriparous sudoriferous, *q.v.*

sufflaminal *a.* [L. *sufflamen*, blast] *appl.* a plate

suffrutex

partly forming gill chamber in certain extinct fishes.

suffrutex *n.* [L. *sub*, under; *frutex*, shrub] an under-shrub; *plu.* suffrutices.

suffruticose *a.* [L. *sub*, under; *frutex*, shrub] somewhat shrubby.

sugar *n.* [M.E. *sugre*, sugar] the common name for any mono-, di-, or trisaccharide, particularly sucrose.

sugent, sugescent *a.* [L. *sugere*, to suck] suctorial, *q.v.*

sulcate *a.* [L. *sulcus*, furrow] furrowed or grooved.

sulcation *n.* [L. *sulcatio*, ploughing] fluting; formation of ridges and furrows, as in elytra.

sulcus *n.* [L. *sulcus*, furrow] a groove; *appl.* cerebral grooves; those of heart, tongue, cornea, bones, etc.; stomodaeal groove of Anthozoa; longitudinal flagellum groove of Dinoflagellata; *alt.* sulculus, fissure.

sulphatidates a group of cerebrosides containing an esterified sulphate group and found in the white matter of the brain.

summation *n.* [L. *summa*, sum total] combined action of either simultaneous or successive subliminal stimuli or impulses which produces an excitatory or inhibitory response.

summer egg thin-shelled, quickly developing egg of some freshwater forms laid in spring or summer; *cf.* winter egg.

summer wood late wood, *q.v.*

sun spiders a common name for the Solpugida, *q.v.*

super- [L. *super*, over] in classification, a group just above the status of the taxon following it, as in superclass.

supercarpal *a.* [L. *super*, over; *carpus*, wrist] upper carpal or above the carpus.

supercilia *n.plu.* [L. *supercilia*, eyebrows] the eyebrows.

superciliary *a.* [L. *super*, over; *cilia*, eyelids] *pert.* eyebrows; above orbit.

superciliary arches two arched elevations below frontal eminences.

superdominance *n.* [L. *super*, over; *dominans*, ruling] overdominance, *q.v.*

superfemale *n.* [L. *super*, over; *femina*, women] metafemale, *q.v.*

superfetation superfoetation, *q.v.*

superficial *a.* [L. *super*, over; *facies*, face] on, or near, the surface, *appl.* arteries, veins, etc.; *appl.* placentation where ovules are scattered over inner surface of ovary wall.

superfoetation *n.* [L. *super*, over; *foetus*, big with] fertilization of ovules of an ovary by more than one kind of pollen; successive fertilization, of 2 ova of different oestrous periods, in the same uterus, *alt.* hypercyesis; *alt.* superfetation.

supergene *n.* [L. *super*, over; Gk. *genos*, descent] a region of chromosome which contains a number of genes, but where recombination does not occur, so the genes are transmitted together from generation to generation.

superglottal *a.* [L. *super*, over; Gk. *glôtta*, tongue] above the glottis.

superior *a.* [L. *superior*, upper] upper; higher; anterior; growing or arising above another organ; *appl.* ovary having perianth inserted around the base; *appl.* vena cava: precava, *q.v.*; *cf.* inferior.

superlinguae *n.plu.* [L. *super*, over; *lingua*, tongue] paired lobes of hypopharynx in certain insects.

supermale *n.* [L. *super*, over; *mas*, male] metamale, *q.v.*

supernormal *appl.* stimulus, often artificial, which is stronger than normal stimulus and more successful at producing a species-specific response.

supernumerary chromosomes extra heterochromatic chromosomes found in some plant cells above the normal number for that species; *alt.* B-chromosomes, accessory chromosomes.

superovulation *n.* [L. *super*, over; *ovum*, egg; *latum*, borne away] the production of an unusually large number of eggs at any one time.

superparasite hyperparasite, *q.v.*

super-regeneration the development of additional or superfluous parts in the process of regeneration.

supersacral *a.* [L. *super*, over; *sacrum*, sacred] above the sacrum.

supersonic *a.* [L. *super*, over; *sonare*, to sound] ultrasonic, *q.v.*

superspecies *n.* [L. *super*, over; *species*, particular kind] a group of closely related species having many morphological similarities.

supersphenoidal *a.* [L. *super*, over; Gk. *sphēn*, wedge; *eidos*, form] above sphenoid bone.

supervolute *a.* [L. *super*, over; *volvere*, to roll] having a plaited and rolled arrangement in the bud.

supinate *a.* [L. *supinus*, bent backwards] inclining or leaning backwards.

supination *n.* [L. *supinus*, bent backward] movement of arm by which palm of hand is turned upwards; *cf.* pronation.

supinator brevis and longus two arm muscles used in supination.

supplemental air volume of air which can be expelled from the lungs after normal breathing out.

supplementary type hypotype, *q.v.*

supporting tissue in plants, tissue made of cells with thickened walls as collenchyma and sclerenchyma, adding strength to plant body, *alt.* mechanical tissue; in animals, skeletal tissue forming endo- or exoskeleton.

suppression *n.* [L. *suppressio*, a keeping back] non-development of an organ or part.

suppressor *appl.* genes which nullify the phenotypic effect of another gene.

supra-acromial *a.* [L. *supra*, above; Gk. *akros*, summit; *ōmos*, shoulder] above the acromion of the shoulder blade.

supra-anal *a.* [L. *supra*, above; *anus*, anus] above anus or anal region; *alt.* suranal.

supra-angular surangular(e) *q.v.*

supra-auricular *a.* [L. *supra*, above; *auricula*, external ear] above the auricle or ear, *appl.* feathers.

suprabranchial *a.* [L. *supra*, above; *branchiae*, gills] above the gills.

suprabuccal *a.* [L. *supra*, above; *bucca*, cheek] above cheek and mouth.

suprabulbar *a.* [L. *supra*, above; *bulbus*, bulb] *appl.* region between hair bulb and fibrillar region of hair.

supracallosal *a.* [L. *supra*, above; *callosus*, hard] *appl.* a gyrus on upper surface of corpus callosum of brain.

supracaudal *a.* [L. *supra*, above; *cauda*, tail] above the tail or caudal region.

supracellular *a.* [L. *supra*, above; *cellula*, small cell] *appl.* structures, fibrous or laminar, originating from many cells.

supracerebral *a.* [L. *supra*, above; *cerebrum*, brain] *appl.* lateral pharyngeal glands, as in Hymenoptera.

suprachoroid *a.* [L. *supra*, above; Gk. *chorion*, skin] over the choroid; between choroid and sclera; *alt.* suprachorioid.

supraclavicle *n.* [L. *supra*, above; *clavicula*, small key] a bone of pectoral girdle of fishes; *alt.* supracleithrum.

supraclavicular *a.* [L. *supra*, above; *clavicula*, small key] above or over the clavicle, *appl.* nerves.

supracleithrum *n.* [L. *supra*, above; Gk. *kleithron*, bolt] supraclavicle, *q.v.*

supracondylar *a.* [L. *supra*, above; Gk. *kondylos*, knuckle] above a condyle, *appl.* ridge and process.

supracostal *a.* [L. *supra*, above; *costa*, rib] over or externally to the ribs.

supracranial *a.* [L. *supra*, above; Gk. *kranion*, skull] over or above the skull.

supradorsal *a.* [L. *supra*, above; *dorsum*, back] on or over the dorsal surface; *appl.* small cartilaginous elements in connection with primitive vertebral column.

supra-episternum *n.* [L. *supra*, above; Gk. *epi*, upon; L. *sternum*, breast bone] upper sclerite of episternum in some insects.

supraesophageal supraoesophageal, *q.v.*

supraethmoid *n.* [L. *supra*, above; Gk. *ēthmos*, sieve; *eidos*, form] a bone external to mesethmoid; *alt.* dermethmoid.

supraglenoid *a.* [L. *supra*, above; Gk. *glēnē*, socket] above the glenoid cavity; *appl.* tuberosity at apex of glenoid cavity.

suprahyoid *a.* [L. *supra*, above; Gk. *hyoeidēs*, Υ-shaped] over the hyoid bone, *appl.* aponeurosis, glands, muscles.

supralabial *a.* [L. *supra*, above; *labium*, lip] on the lip, *appl.* scutes or scales.

supraliminal *a.* [L. *supra*, above; *limen*, threshold] above the threshold of sensation, *appl.* stimuli.

supralitoral *a.* [L. *supra*, above; *litus*, seashore] *pert.* seashore above high-water mark, or spray zone; *alt.* supralittoral.

supraloral *a.* [L. *supra*, above; *lorum*, thong] above the loral region, as in birds, snakes.

supramastoid crest ridge at upper boundary of mastoid region of temporal bone; *alt.* temporal line.

supramaxillary *a.* [L. *supra*, above; *maxilla*, jaw] *pert.* upper jaw.

suprameatal *a.* [L. *supra*, above; *meatus*, passage] *appl.* triangle and spine over external acoustic meatus.

supranasal *a.* [L. *supra*, above; *nasus*, nose] over nasal bone or nose.

supraoccipital *n.* [L. *supra*, above; *occiput*, back of head] a large median bone of upper occipital region.

supraocular *a.* [L. *supra*, above; *oculus*, eye] over or above the eye.

supraoesophageal *a.* [L. *supra*, above; Gk. *oisophagos*, gullet] above or over the gullet; *alt.* supraesophageal.

supraorbital *a.* [L. *supra*, above; *orbis*, eye socket] above orbital cavities, *appl.* process, artery, foramen, nerve, vein, etc. *n.* A skull bone in certain fishes.

suprapatellar *a.* [L. *supra*, above; *patella*, knee cap] *appl.* bursa between upper part of patella and femur.

suprapericardial *see* ultimobranchial.

suprapharyngeal *a.* [L. *supra*, above; Gk. *pharyngx*, pharynx] above or over pharynx.

suprapubic *a.* [L. *supra*, above; *pubes*, adult] above the pubic bone.

suprapygal *a.* [L. *supra*, above; Gk. *pygē*, rump] above the pygal bone.

suprarenal *a.* [L. *supra*, above; *renes*, kidneys] situated above kidneys; *alt.* adrenal, surrenal.

suprarenal bodies, capsules or glands *see* adrenal body.

suprarenal medulla the outer part of the adrenal bodies; *alt.* chromaffin tissue.

suprarenin *n.* [L. *supra*, above; *renes*, kidneys] synthetic adrenaline.

suprarostral *a.* [L. *supra*, above; *rostrum*, beak] *appl.* a cartilaginous plate anterior to trabeculae in Amphibia.

suprascapula *n.* [L. *supra*, above; *scapula*, shoulder blade] a cartilage of dorsal part of pectoral girdle in certain cartilaginous fishes; an incompletely ossified extension of scapula of amphibians and certain reptiles.

suprascapular above the shoulder blade, *appl.* artery, ligament, nerve.

supraseptal *a.* [L. *supra*, above; *septum*, partition] *appl.* 2 plates diverging from interorbital septum.

suprasphenoid *n.* [L. *supra*, above; Gk. *sphēn*, wedge; *eidos*, form] membrane bone dorsal to sphenoid cartilage.

suprasphenoidal *a.* [L. *supra*, above; Gk. *sphēn*, wedge] above sphenoid bone of skull.

supraspinal *a.* [L. *supra*, above; *spina*, spine] above or over spinal column; above ventral nerve cord, in insects.

supraspinatous *a.* [L. *supra*, above; *spina*, spine] *appl.* scapular fossa and fascia for origin of supraspinatus.

supraspinatus shoulder muscle inserted into proximal part of greater tubercle of humerus.

suprastapedial *n.* [L. *supra*, above; *stapes*, stirrup] the part of columella of ear above stapes, homologous with mammalian incus.

suprasternal *a.* [L. *supra*, above; *sternum*, breast bone] over or above breast bone, *appl.* a slit-like space in cervical muscle, *appl.* supernumerary sternal elements in some mammals, *appl.* body plane.

suprastigmal *a.* [L. *supra*, above; *stigma*, mark] above a stigma or breathing pore of insects.

supratemporal *a.* [L. *supra*, above; *tempora*, temples] *pert.* upper temporal region of skull; *appl.* bone, arch, fossa; pterotic of teleosts.

suprathoracic *a.* [L. *supra*, above; Gk. *thōrax*, chest] above thoracic region.

supratidal *a.* [L. *supra*, above; A.S. *tid*, time] above high-tide mark, *appl.* spray zone, or to organisms living there.

supratonsillar *a.* [L. *supra*, above; *tonsillae*, tonsils] *appl.* a small depression in lymphoid mass of palatine tonsil.

supratrochlear *a.* [L. *supra*, above; *trochlea*, pulley] over trochlear surface, *appl.* nerve, foramen, lymph glands.

supratympanic *a.* [L. *supra*, above; *tympanum*, drum] above the ear drum.

sural *a.* [L. *sura*, calf of leg] *pert.* calf of leg, *appl.* arteries and nerves.

suranal supra-anal, *q.v.*

surangular(e) *n.* [L. *supra*, above; *angulus*, angle] a bone of lower jaw of some fishes, reptiles, and birds, *alt.* supra-angular.

surculose *a.* [L. *surculus*, shoot] bearing suckers in plants; *alt.* surculous, surculigerous.

surculus *n.* [L. *surculus*, shoot] sucker, in plants.

surcurrent *a.* [L. *supra*, above; *currere*, to run] proceeding or prolonged up a stem; *cf.* decurrent.

surrenal suprarenal, *q.v.*

suscept *n.* [L. *suscipere*, to undergo] a plant or animal susceptible to disease; a species harbouring a virus.

suspensor *n.* [L. *suspendere*, to hang up] a modified portion of a hypha from which a gametangium or a zygospore is suspended, *alt.* zygosporophore; a chain of cells developed from hypobasal segment of angiosperm zygote, attaching embryo to embryo sac, occurring in modified form in other plants; terminal filament of ovariole.

suspensorium *n.* [L. *suspendere*, to hang up] the upper part of hyoid arch from which lower jaw is suspended; suspensory structure of hypopharynx; the skeletal support of a gonopodium.

suspensory *a.* [L. *suspendere*, to hang up] *pert.* a suspensorium; serving for suspension, *appl.* various ligaments.

sustentacular *a.* [L. *sustentaculum*, prop, support] supporting, *appl.* cells; *appl.* connective tissue acting as a supporting framework for an organ; *appl.* Müller's fibres, *q.v.*

sustentaculum lienis fold of peritoneum supporting spleen.

sustentaculum tali projection of calcaneus supporting middle articular surface for ankle bone.

sustentator, sustentor *n.* [L. *sustinere*, to sustain] hooked cremaster of Lepidoptera.

sutural *a.* [L. *sutura*, seam] *pert.* a suture; *appl.* dehiscence taking place at a suture.

sutural bones irregular isolated bones occurring in the course of sutures, especially in lambdoidal suture and posterior fontanelle; *alt.* ossa suturara, ossa triquetra, Wormian bones.

suture *n.* [L. *sutura*, seam] line of junction of 2 parts immovably connected; line of union of shell wall and edge of septum, as in ammonites; line of junction between sclerites; an immovable articulation of bone as in skull; line along which dehiscence occurs.

Svedberg unit [*T. Svedberg*, Swedish chemist] a measurement of the rate of sedimentation of a particle in an ultracentrifuge, related to its weight and shape, which is used, e.g., for measuring size of ribosome subunits; symbol S.

Swammerdam's glands [*J. Swammerdam*, Dutch naturalist] paired outgrowths of prolonged saccus endolymphaticus on each side of vertebral column, as in frog; *alt.* periganglionic glands, calcigerous glands, calcareous bodies.

Swammerdam's vesicle the spermatheca of gastropods.

Swanscombe skull a fossil hominid skull found

at Swanscombe in Kent, which is intermediate between *Homo erectus* and *Homo sapiens*.

swarm *n.* [A.S. *swearm*, swarm] a large number of small motile organisms viewed collectively; *alt.* synhesma; departure of a number of bees from one hive to form another.

swarm cell, swarm spore any of various sexual or non-sexual motile spores such as zoospore or planogamete, especially the motile isogamete of certain fungi.

swimmerets paired abdominal appendages of crustaceans, functional partly for swimming.

swimming bell nectocalyx, *q.v.*

swimming or swim bladder air bladder of fishes, developed as a diverticulum of the alimentary canal.

swimming funnel tube of Dibranchiata through which water is expelled from mantle cavity, expulsion providing means of propulsion.

swimming ovaries groups of ripe ova of Acanthocephala, detached from ovary and floating in body cavity.

swimming plates in Ctenophora, ciliated comblike plates, arranged in 8 equidistant bands, propellers of the organism; *alt.* ctene, comb-rib, costa.

switch gene a gene that causes development to switch from one pathway to another.

switch plant a xerophyte which produces normal leaves when it is young, then sheds them, and photosynthesis is taken over by another structure such as a cladode or phyllode.

Sycettida *n.* [*Sycetta*, generic name] an order of Calcaronea having typically a continuous dermal membrane or cortex and the choanocytes in flagellated chambers, not lining the central cavity.

syconium, syconus *n.* [Gk. *sykon*, fig] a composite fruit consisting mainly of enlarged succulent receptacle, as in the fig; *alt.* synconium.

sylva *n.* [L. *sylva*, forest] forest of a region; forest trees collectively; *alt.* silva.

sylvestral *a.* [L. *sylvestris*, *pert.* forest] *appl.* flora of woodlands and forest.

sylvian *a.* [*F. Sylvius*, or *de la Boe*, French anatomist] *appl.* structures described by Sylvius, as aqueduct, *q.v.*, fissure (lateral cerebral fissure), fossa, veins, etc.

symbasis *n.* [Gk. *symbasis*, agreement] the common evolutionary trend in an interbreeding association of organisms.

symbion(t) *n.* [Gk. *symbiōnai*, to live with] one of the partners in symbiosis; *alt.* symbiote.

symbiosis *n.* [Gk. *symbiōnai*, to live together] living together of different species not necessarily only those of mutual benefit; the term is often used for an association of mutual benefit which is more properly called mutualism; *a.* symbiotic.

symbiote symbiont, *q.v.*

symbiotic *a.* [Gk. *symbiōnai*, to live together] *pert.* symbiosis.

Symbranchiformes, Symbranchii *n.* [*Symbranchus*, generic name; L. *forma*, shape] an order of teleosts which lack scales and are usually air breathing using pharyngeal sacs or intestine; *alt.* Synbranchiformes.

symmetrical *a.* [Gk. *syn*, with; *metron*, measure] *pert.* symmetry.

Symmetrodonta, symmetrodonts *n.*, *n.plu.* [Gk. *symmetria*, due proportion; *odous*, tooth] an order of Mesozoic trituberculate mammals having molar teeth with 3 or more cusps in a triangle.

symmetry *n.* [Gk. *symmetria*, due proportion] state of divisibility into similar halves; regularity of form; similarity of structure on each side of an axis, central, dorsoventral, or anteroposterior; *see* bilateral and radial symmetry.

symparasitism *n.* [Gk. *syn*, together; *parasitos*, parasite] the development of several competing species of parasite within or on one host individual.

sympathetic *a.* [Gk. *syn*, with; *pathos*, feeling] *appl.* system of nerves supplying viscera and blood vessels, and intimately connected with spinal and some cerebral nerves and having adrenergic nerve endings; *appl.* segmental nerves supplying spiracles in insects; *appl.* coloration in imitation of surroundings.

sympathin *n.* [Gk. *syn*, with; *pathos*, feeling] adrenalin or noradrenalin when they are released from the ends of adrenergic nerves and are acting as neurotransmitters.

sympathoblast *n.* [Gk. *syn*, with; *pathos*, feeling; *blastos*, bud] a cell which develops into a neurone of sympathetic ganglia.

sympathochromaffin *a.* [Gk. *syn*, with; *pathos*, feeling; *chrōma*, colour; L. *affinis*, related] *appl.* cells forming sympathoblasts and chromaffin bodies.

sympathomimetic *a.* [Gk. *syn*, with; *pathos*, feeling; *mimētikos*, imitative] *appl.* substances which produce effects like those produced by sympathetic stimulation.

sympatric *a.* [Gk. *syn*, with; *patra*, native land] having the same, or overlapping, areas of geographical distribution; *cf.* allopatric.

Sympetalae *n.* [Gk. *syn*, with; *petalon*, leaf] a subclass of dicots, the flowers of which have corollas with united petals; *alt.* Gamopetalae, Metachlamydeae.

sympetalous *a.* [Gk. *syn*, with; *petalon*, leaf] gamopetalous, *q.v.*

symphile *n.* [Gk. *syn*, with; *philein*, to love] an insect living as a guest in the nest of a social insect which feeds and protects it in return for its secretions which are used as food, e.g. certain beetles in the nests of ants or termites.

symphily *n.* [Gk. *syn*, with; *philein*, to love] commensalism of symphiles with mutual benefit or attraction.

symphoresis *n.* [Gk. *symphorēsis*, a bringing together] conveyance collectively, as movement of spermatid group to a Sertoli cell.

symphyantherous synantherous, *q.v.*

Symphyla *n.* [Gk. *syn*, with; *phylon*, race] a class of arthropods, sometimes considered myriapods, having 14 body segments and 6 pairs of walking legs.

symphyllodium *n.* [Gk. *syn*, with; *phyllon*, leaf; *eidos*, form] a structure formed by coalescence of external coats of 2 or more ovules; a compound ovuliferous scale.

symphyllous gamophyllous, *q.v.*

symphogenesis *n.* [Gk. *symphyein*, to grow together; *genesis*, descent] development of an organ from union of 2 others.

symphysis *n.* [Gk. *symphysis*, a growing together] the coalescence of parts; the line of junction of 2 pieces of bone separate in early life, as pubic symphysis; slightly moveable articulation

with bony surfaces connected by fibrocartilage; *a.* symphysial, symphyseal, symphysian.

symplast *n.* [Gk. *syn*, with; *plastos*, formed] multinucleate body formed by nuclear fragmentation of a single energid; coenocyte, *q.v.*

symplastic *a.* [Gk. *symplassein*, to mould together] being formed with coordinated development of parts; *appl.* growth of contiguous cells without displacement of cell walls; *alt.* coordinated growth.

symplectic(um) *n.* [Gk. *symplektos*, plaited] mesotympanic, *q.v.*

symplex *n.* [Gk. *symplektos*, plaited] the combination of the active substance and protein which constitutes an enzyme; *cf.* agon, pheron.

symplocium *n.* [Gk. *symplokē*, an intertwining] annulus, *q.v.*, in fern sporangium.

sympodial *a.* [Gk. *syn*, with; *pous*, foot] *pert.* or resembling a sympodium in principle; *appl.* branching, growth of axillary shoots when apical budding has ceased.

sympodite *n.* [Gk. *syn*, with; *pous*, foot] protopodite, *q.v.*

sympodium *n.* [Gk. *syn*, with; *pous*, foot] an apparent main axis, made up not from a terminal bud but from a lateral branch or branches which also stop growing after a while and are replaced by others; *alt.* pseudaxis; *cf.* monopodium.

synacme, synacmy *n.* [Gk. *syn*, with; *akmē*, prime] condition when stamens and pistils mature simultaneously; *alt.* homogamy, synanthesis; *cf.* dichogamy.

synaesthesia *n.* [Gk. *syn*, with; *aisthēsis*, sensation] the accompaniment of a sensation due to stimulation of the appropriate receptor, as sound, by a sensation characteristic of another sense, as colour; *alt.* synesthesis.

synandrium *n.* [Gk. *syn*, with; *anēr*, male] the cohesion of anthers in male flowers of some aroids.

synandry *n.* [Gk. *syn*, with; *anēr*, male] the condition where stamens normally separated are united.

synangium *n.* [Gk. *syn*, with; *anggeion*, vessel] a compound sporangium in which sporangia are coherent, as in some ferns; anterior portion of truncus arteriosus; any arterial trunk from which arteries arise.

Synanthae *n.* [Gk. *syn*, with; *anthos*, flower] an order of monocots which are often palm-like, climbers, or herbaceous plants with male and female flowers alternating over the surface of a spike.

synantherous *a.* [Gk. *syn*, with; *anthēros*, flowery] having anthers united to form a tube; *alt.* symphyantherous.

synanthesis synacme, *q.v.*

synanthous *a.* [Gk. *syn*, with; *anthos*, flower] having flowers and leaves appearing simultaneously; having flowers united together as in Compositae.

synanthropic *a.* [Gk. *syn*, together; *anthrōpos*, man] associated with man or human dwellings.

synanthy *n.* [Gk. *syn*, with; *anthos*, flower] adhesion of flowers usually separate.

synaporium *n.* [Gk. *syn*, with; *aporia*, want] an animal association formed owing to unfavourable environmental conditions or disease.

synaposematic *a.* [Gk. *syn*, with; *apo*, from; *sēma*, sign] having warning colours in common, *appl.* mimicry of a more powerful species as means of defence.

synapse *n.* [Gk. *synapsis*, union] the region at which nervous impulses pass from one nerve fibre to another; *alt.* neurosynapse, synapsis.

synapsid *a.* [Gk. *synapsis*, union] *appl.* skulls with supra- and infratemporal fossae united in a single fossa, or with the infratemporal fossa only; *cf.* diapsid.

Synapsida, synapsids *n., n.plu.* [Gk. *synapsis*, union] a subclass of Carboniferous to Triassic reptiles, the mammal-like reptiles, some forms of which gave rise to the mammals, and including the pelycosaurs and therapsids.

synapsis *n.* [Gk. *synapsis*, union] stage or period from contraction of nucleus to segmentation of spireme into chromosomes; syndesis, *q.v.*; synapse, *q.v.*

synaptene *a.* [Gk. *synapsis*, union; *tainia*, band] zygotene, *q.v.*

synaptic knobs knob-like swellings on the end of nerve axons, making contact with the dendrites of another neuron.

synaptic mates homologous chromosomes undergoing synapsis.

synaptic membrane a membrane intervening between nerve ending and muscle fibre supplied by it, also between processes of one neurone and those of another.

synapticula *n.* [Gk. *synaptos*, joined] one of the small rods connecting septa as in certain corals and gill bars in *Amphioxus*, the lancelet.

Synaptida *n.* [Gk. *synaptos*, joined] Apoda (sea cucumbers), *q.v.*

synaptinemal complexes linear bodies seen by the electron microscope which lie between paired homologous chromosomes during meiosis.

synaptomeres *n.plu.* [Gk. *synaptos*, joined; *meros*, part] hypothetical segments of chromosome which cause pairing at synapsis.

Synaptosauria *n.* [Gk. *synaptos*, joined; *sauros*, lizard] a subclass of extinct Mesozoic reptiles having a single temporal opening high up in the cheek, which includes the groups Sauropterygia and Protorosauria that are sometimes raised to subclass status; *alt.* Euryapsida.

synaptospermous *a.* [Gk. *synaptos*, joined; *sperma*, seed] having seeds germinating close to the parent plant.

synaptospore *n.* [Gk. *synaptos*, joined; *sporos*, seed] aggregate spore; clinospores joined together.

synaptotene *a.* [Gk. *synaptos*, joined; *tainia*, band] zygotene, *q.v.*

synaptychus *n.* [Gk. *syn*, with; *a*, together; *ptychē*, plate] aptychus in which paired plates are permanently united; *cf.* anaptychus.

synarthrosis *n.* [Gk. *syn*, with; *arthron*, joint] an articulation in which bone surfaces are in almost direct contact, fastened together by connective tissue or hyaline cartilage, with no appreciable motion; *cf.* synchondrosis, syndesmosis.

synascus *n.* [Gk. *syn*, together; *askos*, bag] an ascogonium containing a number of asci.

Synbranchiformes *n.* [Gk. *syn*, with; *brangchia*, gills; L. *forma*, shape] Symbranchiformes, *q.v.*

Syncarida *n.* [Gk. *syn*, with; L. *caris*, a kind of crab] a group of freshwater malacostracans having no carapace, considered an order or a superorder subdivided into the Anaspidacea, *q.v.*, and Bathynellacea, *q.v.*

syncarp(ium) *n.* [Gk. *syn*, with; *karpos*, fruit] an aggregate fruit with united carpels; *cf.* etaerio.

syncarpous *a.* [Gk. *syn*, with; *karpos*, fruit] with carpels united, *cf.* apocarpous; bearing a syncarp.

syncarpy *n.* [Gk. *syn*, with; *karpos*, fruit] condition of having carpels united to form a compound gynaecium; *cf.* apocarpy.

syncaryo- synkaryo-, *q.v.*

syncerebrum *n.* [Gk. *syn*, with; L. *cerebrum*, brain] a secondary brain formed by union with brain of one or more of ventral cord ganglia, in some arthropods.

syncheimadia *n.plu.* [Gk. *syn*, with; *cheimadion*, winter dwelling] societies overwintering together.

synchondrosis *n.* [Gk. *syn*, with; *chondros*, cartilage] a synarthrosis in which the connecting medium is cartilage.

synchorology *n.* [Gk. *syn*, with; *chōros*, place; *logos*, discourse] study of the distribution of plant or animal associations; geographical distribution of communities.

synchronic *a.* [Gk. *syn*, with; *chronos*, time] contemporary; existing at the same time, *appl.* species, etc.; *cf.* allochronic.

synchronizer *n.* [Gk. *syn*, with; *chronos*, time] a factor of the environment such as light intensity or temperature that acts as a stimulus to an endogenous circadian rhythm causing it to be precisely synchronized (entrained) to a 24-hour cycle, rather than free running; *alt.* zeitgeber.

syncladous *a.* [Gk. *syn*, together; *klados*, branch] with offshoots or branchlets in tufts, *appl.* certain mosses.

synconium, synconus *n.* [Gk. *syn*, with; *kōnos*, cone] syconium, *q.v.*

syncraniate *a.* [Gk. *syn*, with; *kranion*, skull] having vertebral elements fused with skull.

syncranterian *a.* [Gk. *syn*, with; *krantērēs*, wisdom teeth] with teeth in a continuous row.

syncryptic *a.* [Gk. *syn*, with; *kryptos*, hidden] *appl.* animals which appear alike, though unrelated, due to common protective resemblance to surroundings.

syncytiotrophoblast *see* syncytium.

syncytium *n.* [Gk. *syn*, with; *kytos*, hollow] a multinucleated mass of protoplasm without differentiation into cells; outer stratum of trophoblast of mammalian ovum, *alt.* syncytiotrophoblast; plasmodium, *q.v.*; *cf.* coenocyte; *a.* syncytial.

syndactyl *a.* [Gk. *syn*, with; *daktylos*, digit] with fused digits, as in many birds.

syndactylism *n.* [Gk. *syn*, with; *daktylos*, digit] whole or part fusion of 2 or more digits.

syndesis *n.* [Gk. *syndēsai*, to bind together] coming together of homologous chromosomes in meiosis; *alt.* synapsis.

syndesmology *n.* [Gk. *syndesmos*, ligament; *logos*, discourse] the branch of anatomy dealing with ligaments and articulations; *cf.* desmology.

syndesmosis *n.* [Gk. *syndesmos*, ligament] a slightly moveable articulation, with bone surfaces connected by an interosseous ligament; *cf.* synarthrosis.

syndetocheilic *a.* [Gk. *syndetos*, bound together; *cheilos*, edge] *appl.* type of stomata found in gymnosperms in which the subsidiary cells are derived from the same cell as the guard mother cell.

syndrome *n.* [Gk. *syn*, together; *dramein*, to run] a group of concomitant symptoms.

synecete synoekete, *q.v.*

synechthrans *n.plu.* [Gk. *synechthairein*, to join in hating] unwelcome ant intruders in the nest of other ants.

synecology *n.* [Gk. *syn*, together; *oikos*, household; *logos*, discourse] ecology of plant or of animal communities; *cf.* autecology.

synecthry *n.* [Gk. *syn*, with; *echthros*, hatred] commensalism of synecthrans with mutual dislike.

synema synnema, *q.v.*

synencephalon *n.* [Gk. *syn*, with; *engkephalos*, brain] the part of the embryonic brain between diencephalon and mesencephalon.

synenchyma *n.* [Gk. *syn*, together; *engchyma*, infusion] fungous tissue composed of laterally closed joined hyphae.

Synentognathi *n.* [Gk. *syn*, with; *entos*, within; *gnathos*, jaw] Beloniformes, *q.v.*

syneresis *n.* [Gk. *syn*, together; *ereidein*, to press] contraction of a gel with expression of liquid; contraction of clotting blood and separation of serum.

synergic, synergetic *a.* [Gk. *synergos*, cooperator] operating together; *appl.* muscles which combine with prime movers and fixation muscles in movement; *appl.* system of muscles and nerves affecting a particular movement; *appl.* certain hormones; *alt.* metaleptic.

synergid *n.* [Gk. *synergos*, cooperator] each of 2 cells without cell walls lying beside ovum at micropylar end of embryo sac of an ovule; *alt.* help cell; *plu.* synergidae.

synergism *n.* [Gk. *synergos*, cooperator] the phenomenon where 2 substances such as hormones, interact to produce an effect greater than the sum of their individual effects.

synesthesis synaesthesia, *q.v.*

synethogametism *n.* [Gk. *synēthēs*, well suited; *gametēs*, spouse] ability of gametes to fuse; gametal compatibility; *cf.* asynethogametism.

syngametic *a.* [Gk. *syn*, together; *gametēs*, spouse] *pert.* union of morphologically similar cells.

syngamodeme *n.* [Gk. *syn*, with; *gamos*, marriage; *dēmos*, people] a unit made up of coenogamodemes whose members can form sterile hybrids.

syngamy *n.* [Gk. *syn*, with; *gamos*, marriage] sexual reproduction, consisting of the fusion of gametes or of mating types of unicellular organisms; *alt.* gametogamy, hylogamy; *cf.* pseudomixis.

syngen *n.* [Gk. *syn*, with; *genos*, birth] a taxonomic term equivalent to variety, and used in protozoan genetics.

syngeneic, syngenic *a.* [Gk. *syn*, with; *genos*, birth] isogenic, *q.v.*

syngenesious *a.* [Gk. *syn*, with; *genesis*, descent] having stamens united in cylindrical form by anthers; with anthers united.

syngenesis *n.* [Gk. *syn*, with; *genesis*, descent] sexual reproduction; theory that germs of all human beings, past, present, and future, were created simultaneously, and that there are germs within germs *ad infinitum*; coenogenesis, *q.v.*

syngenetic *a.* [Gk. *syn*, with; *genesis*, descent] sexually reproduced; descended from the same ancestors.

Syngnathiformes *n.* [*Syngnathus*, generic name; L. *forma*, shape] an order of teleosts including the seahorses and pipefish, sometimes considered as part of the Solenichthyes.

syngnaths *n.plu.* [Gk. *syn*, with; *gnathos*, jaw] paired jaws or mouth-plates of Stelleroidea.

syngonic *a.* [Gk. *syn*, with; *gonē*, seed] producing male and female gametes in the same gone; *cf.* amphigonic, digonic.

syngraft *n.* [Gk. *syn*, with; O.F. *graffe*, graft] isograft, *q.v.*

syngynous epigynous, *q.v.*

synhesma *n.* [Gk. *syn*, with; *hesmos*, a swarm] a swarm; a swarming society.

synizesis *n.* [Gk. *synizēsis*, contraction] the contraction of chromatin in nucleus during synapsis; contraction of pupil.

synkaryo- *alt.* syncaryo-.

synkaryon *n.* [Gk. *syn*, with; *karyon*, nucleus] zygote nucleus resulting from fusion of gamete nuclei.

synkaryophyte *n.* [Gk. *syn*, with; *karyon*, nucleus; *phyton*, plant] diploid plant; sporophyte, *q.v.*

synkaryotic *a.* [Gk. *syn*, together; *karyon*, nucleus] diploid, *appl.* nucleus.

synnema *n.* [Gk. *syn*, with; *nēma*, thread] coremium, *q.v.*; the united stamen filaments of a monadelphous flower; *alt.* synema.

synochreate, synocreate *a.* [Gk. *syn*, with; L. *ocrea*, legging] with stipules united, enclosing stem in a sheath.

synoecete synoekete, *q.v.*

synoecious, synoicous *a.* [Gk. *syn*, together; *oikos*, house] having antheridia and archegonia on same receptacle, or stamens and pistils on same flower, or male and female flowers on same capitulum; *alt.* synoikous.

synoekete *n.* [Gk. *syn*, with; *oikētēs*, dweller] a tolerated guest in a colony; *alt.* synecete, synoecete.

synoikous synoecious, *q.v.*

synosteosis, synostosis *n.* [Gk. *syn*, with; *osteon*, bone] ossification from 2 or more centres in the same bone, as from diaphysis and epiphyses in long bones; anchylosis, *q.v.*

synotic tectum in higher vertebrates, a cartilaginous arch between otic capsules representing cartilaginous roof or tegmen of cranium in lower vertebrates.

synovia *n.* [Gk. *syn*, with; L. *ovum*, egg] viscid, glairy secretion of synovial membrane.

synovial membrane inner stratum of articular capsule around a moveable joint, made of connective tissue and secreting synovia to lubricate the joints; *alt.* stratum synoviale.

synovial villi Haversian fringes, *q.v.*

synovin synovial mucin.

synoviparous *a.* [Gk. *syn*, with; L. *ovum*, egg; *parere*, to beget] secreting synovia.

synpelmous *a.* [Gk. *syn*, with; *pelma*, sole] having 2 tendons united before they go to separate digits.

synpolydesmic *a.* [Gk. *syn*, with; *polys*, many; *desmos*, bond] *appl.* cyclomorial scales made up of fused monodesmic scales with continuous dentine layer; *cf.* polydesmic.

synsacrum n. [Gk. syn, with; L. sacrum, sacred] a mass of fused vertebra supporting the pelvic girdle of birds and of certain extinct saurians.

synsepalous a. [Gk. syn, with; F. sépale, sepal] with calyx composed of fused or united sepals.

synspermous a. [Gk. syn, with; sperma, seed] having seeds united.

synsporous a. [Gk. syn, with; sporos, seed] propagating by cell conjugation, as in algae.

syntagma n. [Gk. syn, together; tagma, corps] see tagma; plu. syntagmata.

syntechnic n. [Gk. syn, with; technē, art] resemblance in unrelated animals, due to environment; alt. convergence.

syntelome n. [Gk. syn, with; telos, end] a compound telome.

syntenosis n. [Gk. syn, with; tenōn, sinew] articulation of bones by means of tendons.

synthetases n.plu. [Gk. syntithenai, to put together] a group of enzymes that catalyse the synthesis of a molecule by the condensation of 2 other molecules using energy from the breakdown of ATP; now known as ligases (EC group 6).

syntonin n. [Gk. syntonos, stretched] muscle fibrin.

syntrophoblast n. [Gk. syn, together; trephein, to nourish; blastos, bud] trophoblastic syncytium; alt. plasmoditrophoblast.

syntrophy n. [Gk. syn, together; trophē, nourishment] nutritional interdependence.

syntropic a. [Gk. syn, together; tropē, turn] turning or arranged in the same direction, as ribs on one side; cf. antitropic.

syntype n. [Gk. syn, with; typos, pattern] any one specimen of a series used to designate a species when holotype and paratypes have not been selected; alt. cotype.

synusia n. [Gk. synousia, a living together] a plant community of relatively uniform composition, living in a particular environment, and forming part of a phytocoenosis, q.v.

synzoospore n. [Gk. syn, with; zōon, animal; sporos, seed] a group of zoospores which do not separate; cf. zoocoenocyte.

syringeal a. [Gk. syringx, pipe] pert. the syrinx.

syringium n. [Gk. syringx, pipe] a syringe-like organ for ejection of disagreeable fluid of some insects.

syringograde a. [Gk. syringx, pipe; L. gradus, step] jet-propelled, moving by alternate suction and ejection of water through siphons as squids and salps.

syrinx n. [Gk. syringx, pipe] vocal organ of birds, at base of trachea.

systaltic a. [Gk. systellein, to draw in] contractile; alternately contracting and dilating.

systematics n. [Gk. systema, whole made of several parts] taxonomy, q.v.

systemic circulation course of blood from left ventricle through the body to right auricle, as compared with pulmonary or lesser circulation.

systemic heart heart of invertebrates, and auricle and ventricle of left side of heart of higher vertebrates; cf. respiratory heart.

systilius systylius, q.v.

systole n. [Gk. systolē, drawing together] contraction of heart causing circulation of blood; contraction of any contractile cavity; cf. diastole.

systrophe n. [Gk. systrophē, a gathering] an aggregation of starch grains in chloroplasts, induced by illumination.

systylius n. [Gk. syn, with; stylos, column] the columella lid of some mosses; alt. systilius.

systylous a. [Gk. syn, with; stylos, column] with coherent styles; with fixed columella lid, as in mosses.

syzygium n. [Gk. syzygia, union] group of associated gregarines.

syzygy n. [Gk. syzygia, union] a close suture of 2 adjacent arms, found in crinoids; a number of individuals, from 2 to 5, adhering in strings in association of gregarines; reunion of chromosome fragments at meiosis.

T

tabellae n.plu. [L. tabella, tablet] small tabulae or horizontal plates around axis of a corallite.

tables n.plu. [L. tabula, board] outer and inner layers of flat compact bones, especially of skull.

tabula n. [L. tabula, board] horizontal partitions traversing vertical canals of Hydrocorallina and of tabulate corals; plu. tabulae.

tabular arranged in a flat surface or table; flattened, as certain cells.

tabulare n. [L. tabula, board] skull bone posterior to parietal in some vertebrates.

tabularium all the tabulae surrounding the axis of a corallite.

tachyauxesis n. [Gk. tachys, quick; auxēsis, growth] relatively quick growth; growth of a part at a faster rate than that of the whole; cf. bradyauxesis, isauxesis.

tachyblastic a. [Gk. tachys, quick; blastos, growth] with cleavage immediately following oviposition, appl. quickly hatching eggs; cf. opsiblastic.

tachygen n. [Gk. tachys, quick; gennaein, to produce] a structure originating abruptly in evolution.

tachygenesis n. [Gk. tachys, quick; genesis, descent] development with omission of certain embryonic stages as in some crustaceans, or of nymphal stages as in some insects, or of tadpole stages in some amphibians, resulting in accelerated development in phylogeny; cf. bradygenesis.

tachysporous a. [Gk. tachys, quick; sporos, seed] dispersing seeds quickly.

tachytelic a. [Gk. tachys, quick; telos, fulfilment] evolving at a rate faster than the standard rate; n. tachytely; cf. bradytelic, horotelic.

tactic a. [Gk. taktos, arranged] pert. taxis, q.v.

tactile a. [L. tactilis, that may be touched] serving the sense of touch, as special end organs or tangoreceptors, appl. cells, cones, corpuscles, discs, hairs, etc.

tactism n. [Gk. taktos, arranged] taxis, q.v.

tactor n. [L. tactus, touch] tactile end organ; tangoreceptor, q.v.

tactual a. [L. tactus, touch] pert. sense of touch.

taenia n. [L. taenia, ribbon] a band, as of nerve or of muscle; ligula, q.v.; alt. tenia.

taeniate a. [L. taenia, ribbon] ribbon-like; striped.

taenidium n. [Gk. tainia, ribbon; idion, dim.] spiral ridge of cuticle strengthening the chitinous layer of insect tracheae and tracheoles; *plu.* taenidia.

Taeniidea *n.* [L. *taenia*, ribbon] Cyclophyllidea, *q.v.*

Taeniodonta *n.* [Gk. *tainia*, ribbon; *odous*, tooth] an order of extinct placental mammals of the Palaeocene and Eocene.

taenioid *a.* [Gk. *tainia*, ribbon; *eidos*, form] ribbon-shaped; like a tapeworm.

taenioles *n.plu.* [L. *taeniola*, small ribbon] four longitudinal ridges of a scyphistoma.

tagma *n.* [Gk. *tagma*, corps] a segment of the body of a metamerically segmented animal formed by fusion of somites, e.g. head, thorax, abdomen of insects; a unit or part; a molecular group; *plu.* tagmata; *alt.* syntagma.

tagmosis *n.* [Gk. *tagma*, corps] the fusion or grouping of somites to form tagmata in a metamerically segmented animal.

taiga *n.* [Russ. rocky] northern coniferous forest zone, especially in Siberia.

tail fan in decapod crustaceans, the uropods and telson.

Takakiales *n.* [*Takakia*, generic name] an order of unusual liverworts that superficially resemble the stoneworts (Charales).

talandic *a.* [Gk. *talanteain*, to swing to and fro] *appl.* rhythmic changes occurring in a cell, possibly concerned with control of metabolism.

tali *plu.* of talus.

talocalcaneal *a.* [L. *talus*, ankle bone; *calcaneum*, heel] *pert.* talus and calcaneus, *appl.* articulation, ligaments.

talocrural *a.* [L. *talus*, ankle; *crus*, leg] *pert.* ankle and shank bones; *appl.* articulation: the ankle joint.

talon *n.* [F., from L. *talus*, ankle] claw of bird of prey; posterior heel of molar tooth.

taloscaphoid *a.* [L. *talus*, ankle; Gk. *skaphē*, boat; *eidos*, form] *pert.* talus and scaphoid bone.

talus *n.* [L. *talus*, ankle] astragalus, *q.v.*; *plu.* tali.

Tanaidacea *n.* [*Tanais*, generic name] an order of marine Peracarida in which the carapace covers only 2 thoracic segments.

tandem *appl.* satellites separated from each other by a constriction.

tangoreceptor *n.* [L. *tangere*, to touch; *receptor*, receiver] a receptor sensitive to slight pressure differences; *alt.* tactor.

tannase an enzyme which hydrolyses the tannin digallate to gallate, but will also hydrolyse certain ester linkages in other tannins; EC 3.1.1.20; *r.n.* tannase.

tannins *n.plu.* [L. *tannum*, oak bark] complex aromatic compounds some of which are glucosides, occurring in the bark of various trees, possibly giving protection or concerned with pigment formation, and used in tanning and dyeing.

T-antigen an antigen which is found in cells infected with oncogenic virus, and is probably a protein coded by the virus.

tanyblastic *a.* [Gk. *tanyein*, to stretch; *blastos*, bud] with a long germ band; *cf.* brachyblastic.

tapesium *n.* [Gk. *tapēs*, rug] a dense outer mycelium bearing ascus-producing hyphae.

tapetum *n.* [L. *tapete*, carpet] outer and posterior part of choroid; pigment layer of retina; main body of fibres of corpus callosum; special nutritive layer investing sporogenous tissue of sporangium; in anther, a layer of cells lining the cavity and absorbed as pollen grains mature; *a.* tapetal.

tapeworms the common name for the Cestoda, *q.v.*

Taphrinales *n.* [*Taphrina*, generic name] a group of lower (hemi) Ascomycetes having a mycelium producing asci that lie parallel to each other in a palisade layer without any enclosing peridium; *alt.* Exoascales.

taphrophyte *n.* [Gk. *taphros*, ditch; *phyton*, plant] ditch-dwelling plant.

tap root *n.* [M.E. *tappe*, short pipe; A.S. *wyrt*, root] an elongated parent root with secondary roots in acropetal succession; persistent primary root formed from radicle of embryo.

taraxanthin a xanthophyll carotinoid pigment found in red algae.

Tardigrada, tardigrades *n., n.plu.* [L. *tardigradus*, slow paced] a phylum of small animals, commonly called water bears, that have some features similar to arthropods and have been classed as arachnids; an infraorder of Edentata, including the tree sloths.

target organ the organ upon which a hormone acts.

Tarrasiiformes *n.* [*Tarrasia*, generic name; L. *forma*, shape] an order of Carboniferous palaeoniscid-like Chondrostei.

tarsal *a.* [Gk. *tarsos*, sole of foot] *pert.* tarsus, of foot and eyelid, *appl.* arteries, bones.

tarsal(e) *n.* [Gk. *tarsos*, sole of foot] ankle bone; *plu.* tarsalia.

tarsal glands Meibomian glands, *q.v.*

tarsi *n.plu.* [Gk. *tarsos*, sole of foot] *plu.* of tarsus; 2 thin elongated plates of dense connective tissue helping to support the eyelid.

tarsomeres *n.plu.* [Gk. *tarsos*, sole of foot; *meros*, part] the 2 parts of dactylopodite in spiders, basitarsus and telotarsus.

tarsometatarsal *a.* [Gk. *tarsos*, sole of foot; *meta*, beyond] *pert.* an articulation of tarsus with metatarsus.

tarsometatarsus *n.* [Gk. *tarsos*, sole of foot; *meta*, beyond] a short straight bone of bird's leg formed by fusion of distal row of tarsals with 2nd to 5th metatarsals; *alt.* cannon bone.

tarsophalangeal *a.* [Gk. *tarsos*, sole of foot; *phalangx*, line of battle] *pert.* tarsus and phalanges.

tarsus *n.* [Gk. *tarsos*, sole of foot] ankle bones, usually consisting of 2 rows; segment of leg distal to tibia, in insects and myriapods; telotarsus, *q.v.*; fibrous connective tissue plate of eyelid.

tartareous *a.* [L.L. *tartarum*, an acid salt] having a rough and crumbling surface.

tartaric acid an organic acid which gives acid taste to grapes.

tassel *n.* [O.F. *tasel*, a clasp] male inflorescence of maize plant; appendix colli of goat, sheep, pig, etc.

tassel-finned fishes a common name for the Crossopterygii, *q.v.*

tassement polaire the deeply staining mass of chromosomes at the poles of the spindle at the end of anaphase and beginning of telophase.

taste bud an ovoid buccal sense organ of taste consisting of a flask-shaped group of sensory gustatory cells together with supporting cells, found on tongue and adjacent parts; *alt.* calyculus gustatorius, taste bulb.

taste pore orifice in epithelium, leading to terminal hairs of sensory cells in a taste bud.

tauidion *n.* [Gk. *tau*, T; *idion*, *dim.*] part of cranial floor of *Palaeospondylus*, an extinct primitive Devonian vertebrate.

taurine *n.* [L. *taurus*, bull] an amino acid containing sulphonic acid which is a constituent of taurocholic acid and causes hyperpolarization of some neurones.

taurocholic *a.* [L. *taurus*, bull; Gk. *cholē*, bile] *appl.* a bile acid, hydrolysed to taurine and cholic acid.

tautomeric *a.* [Gk. *tautos*, the same; *meros*, part] *pert.* the same part; *appl.* cells: neurones with axis cylinders passing into white matter of same side of spinal cord; *appl.* organic compounds of the same composition but differing in structure.

tautonym *n.* [Gk. *tautos*, the same; *onyma*, name] the same name given to a genus and one of its species or subspecies.

tautotype *n.* [Gk. *tautos*, the same; *typos*, pattern] a genotype by virtue of tautonymy.

Tawara's node [*S. Tawara*, Japanese pathologist] the atrioventricular node, *q.v.*

Taxales *n.* [*Taxus*, generic name] an order of Coniferopsida, being evergreen shrubs or small trees with small linear leaves and ovules solitary and surrounded by an aril.

taxeopodous *a.* [Gk. *taxis*, arrangement; *pous*, foot] having proximal and distal tarsal bone in straight lines parallel to limb axis.

taxis *n.* [Gk. *taxis*, arrangement] a movement of a freely motile, usually simple organism, especially Protista, or part of an organism, towards (positive), or away from (negative), a source of stimulation, such as light, temperature, chemicals; an orientation behaviour related to a directional stimulus; *alt.* taxy, tactic movement, tactism; *plu.* taxes; *a.* tactic; *cf.* kinesis, tropism.

Taxodonta *n.* [Gk. *taxis*, arrangement; *odous*, tooth] an order of lamellibranchs having free gill filaments and many similar teeth to the shell hinge.

taxometrics *n.* [Gk. *taxis*, arrangement; *metron*, measure] quantitative estimates of similarities between organisms leading to their ordering into taxa; *alt.* numerical taxonomy.

taxon *n.* [Gk. *taxis*, arrangement] any definite unit in classification of plants and animals; *alt.* taxonomic unit.

taxonomic species linneon, *q.v.*

taxonomic unit taxon, *q.v.*

taxonomy *n.* [Gk. *taxis*, arrangement; *nomos*, law] the study of the rules, principles, and practice of classifying, especially living organisms; *alt.* systematics, biosystematics, genonomy.

taxy taxis, *q.v.*

TCA cycle tricarboxylic acid cycle, *q.v.*

tear pit dacryocyst, *q.v.*

tectal *a.* [L. *tectum*, roof] of or *pert.* tectum.

tectin *n.* [L. *tectus*, covered] a pseudochitin, organic compound in shell of Foraminifera.

tectology *n.* [Gk. *tektōn*, builder; *logos*, discourse] morphology in which an organism is considered as a group of morphological as distinct from physiological units or individuals.

tectorial *a.* [L. *tectorius*, *pert.* cover] covering; *appl.* membrane covering the spiral organ of Corti.

tectorium *n.* [L. *tectorium*, cover] membrane of Corti; the coverts of birds.

tectospondylic *a.* [L. *tectus*, covered; Gk. *sphondylos*, vertebra] having vertebrae with several concentric rings of calcification, as in some elasmobranchs; *alt.* tectospondylous.

tectostracum *a.* [L. *tectum*, cover; Gk. *ostrakon*, shell] thin, waxy outer covering of exoskeleton, as of Acarina.

tectotype *n.* [Gk. *tektōn*, builder; *typos*, pattern] description of a species, based on microscopical examination of a prepared section; the section used.

tetrices *n.plu.* [L. *tectus*, covered] small feathers covering bases of remiges; *alt.* wing coverts.

tectum *n.* [L. *tectum*, roof] a roof-like structure, as corpora quadrigemina, forming roof of mesencephalon; dorsal wall of capitulum in Acarina.

teeth *n.plu.* [A.S. *toth*, tooth] hard bony growths on maxillae, premaxillae, and mandibles of mammals; growths of similar, of chitinous, or of horny formation borne on jaws, tongue, or pharynx; pointed outgrowths on margin of leaf, calyx, or corolla.

tegmen *n.* [L. *tegmen*, covering] the integument, endopleura, or inner seed coat; calyx covers of Crinoidea; 9th abdominal tergite of male insects, *alt.* tegumen; thin hardened fore-wing of Orthoptera, Phasmida, and Dictyoptera; plate of bone over tympanic antrum; *alt.* tegmentum; *plu.* tegmina.

tegmen cranii roof of chondrocranium.

tegmenta *plu.* of tegmentum.

tegmentum *n.* [L. *tegmen*, covering] a protective bud scale; dorsal part of cerebral peduncles; in chitons, the upper part of each of the body plates, *cf.* articulamentum; tegmen, *q.v.*

tegmina *plu.* of tegmen.

tegula *n.* [L. *tegula*, tile] a small sclerite on mesothorax overhanging articulation of wings in Lepidoptera and Hymenoptera; a small lobe or alula at wing base of Diptera; a tile-shaped structure; *cf.* parapteron.

tegular *a.* [L. *tegula*, tile] *pert.* a tegula; consisting of a tile-like structure.

tegumen *n.* [L. *tegumen*, cover] *see* tegmen.

tegument *n.* [L. *tegumentum*, covering] integument, *q.v.*; cuticle of trematodes.

tegumental *a.* [L. *tegumentum*, covering] *pert.* an integument; *appl.* gland cells of epidermis which secrete epicuticle in various arthropods.

teichoic acids substances found in cell wall or cell membranes of certain bacteria, being phosphodiester linked polymers of glycerol or ribitol, often containing N-acetylglucosamine, N-acetylgalactosamine, alanine, and gentiobiose.

tela *n.* [L. *tela*, web] a web-like tissue; *appl.* chorioidea, folds of the pia mater forming membranous roof of 3rd and 4th ventricles; *appl.* interlacing fibrilliform or hyphal tissue of fungi, tela contexta.

telamon *n.* [Gk. *telamōn*, supporting strap] chitinized curved plate in lateral wall of cloaca in male nematodes.

telarian *a.* [L. *tela*, web] web-spinning.

teleblem *n.* [Gk. *teleos*, complete; *blēma*, coverlet] volva, *q.v.*

teleceptor *n.* [Gk. *tēle*, far; L. *capere*, to take] distance receptor, *q.v.*

telegamic *a.* [Gk. *tēle*, far; *gamos*, marriage] attracting females from a distance, *appl.* scent apparatus of Lepidoptera.

telegenesis *n.* [Gk. *tēle*, far; *genesis*, descent] artificial insemination, *q.v.*

telegony *n.* [Gk. *tēle*, far; *gonos*, offspring] the supposed influence of a male parent on offspring, subsequent to his own, of the same female parent by another sire.

teleianthous *a.* [Gk. *teleios*, complete; *anthos*, flower] *appl.* a flower having both gynaecium and androecium.

teleiochrysalis *n.* [Gk. *teleios*, complete; *chrysallis*, from *chrysos*, gold] nymph during the resting stage preceding the adult form of certain mites.

telemetacarpal *a.* [Gk. *tēle*, far; *meta*, after; *karpos*, wrist] *appl.* condition of retaining distal elements of metacarpals, as in some Cervidae (deer); *cf.* plesiometacarpal.

telemorphosis *n.* [Gk. *tēle*, far; *morphōsis*, a shaping] alteration of form in response to a distant stimulus, as of hypha or zygophore in response to another hypha or zygophore.

telencephalon *n.* [Gk. *telos*, end; *engkephalos*, brain] the anterior part of fore-brain, including the cerebral hemispheres, lateral ventricles, optic part of hypothalamus, and anterior portion of 3rd ventricle; *alt.* end-brain; *a.* telencephalic.

teleneurite *n.* [Gk. *telein*, to end; *neuron*, nerve] telodendrion, *q.v.*

teleoblem(a) *n.* [Gk. *teleos*, complete; *blēma*, coverlet] volva, *q.v.*

teleodont *a.* [Gk. *teleos*, complete; *odous*, tooth] *appl.* forms of stag beetles with largest mandible development.

teleology *n.* [Gk. *teleos*, complete; *logos*, discourse] the doctrine of adaptation to a definite purpose, and that evolution is purposive; *a.* teleological.

teleonomy *n.* [Gk. *teleos*, complete; *nomos*, law] the idea that if a structure or process exists in an organism, it must have conferred evolutionary advantage.

teleophore *n.* [Gk. *teleos*, complete; *pherein*, to bear] gonotheca, *q.v.*

teleoptile *n.* [Gk. *teleos*, complete; *ptilon*, feather] a feather of definitive plumage, *cf.* neossoptile; a pennaceous feather, *cf.* mesoptile, metaptile.

teleorganic *a.* [Gk. *telein*, to fulfil; *organon*, instrument] *appl.* functions vital to an organism.

teleosis *n.* [Gk. *teleiōsis*, completion] purposive development or evolution; *a.* telic.

Teleostei, teleosts *n., n.plu.* [Gk. *telos*, end; *osteon*, bone] a group of advanced actinopterygians including the majority of modern fishes, existing from the Jurassic to the present day, having scales of thin bony plates covered by epidermis, a homocercal tail, a hydrostatic air bladder, and no spiracle or spiral valve in the gut; *alt.* Teleostomi.

Teleostomi *n.* [Gk. *telos*, end; *osteon*, bone; *ōmos*, shoulder] Teleostei, *q.v.*

teleotrocha trochosphere, *q.v.*

telereceptor *n.* [Gk. *tēle*, far; L. *receptor*, receiver] distance receptor, *q.v.*

telescopiform *a.* [Gk. *tēle*, far; *skopein*, to view; L. *forma*, shape] having joints that telescope into each other.

Telestacea *n.* [*Telesto*, generic name] an order of Octocorallia having a colony consisting of a cluster of long polyps; sometimes included in the Alcyonacea.

telethmoid prenasal, *q.v.*

teleutobud teleutospore, *q.v.*

teleutoform stage teleutostage, *q.v.*

teleutogonidium teleutospore, *q.v.*

teleutosorus *n.* [Gk. *teleutē*, completion; *sōros*, heap] in rust fungi, a group of developing teleutospores forming a pustule on the host; *alt.* telium, teliosorus.

teleutospore *n.* [Gk. *teleutē*, completion; *sporos*, seed] in rust fungi, a winter spore formed in autumn which remains dormant all winter and germinates the following spring; *alt.* teliospore, teleutobud, teleutogonidium, brand spore, winter bud.

teleutosporiferous *a.* [Gk. *teleutē*, completion; *sporos*, seed; L. *ferre*, to carry] *appl.* rusts bearing teleutospores; *alt.* teliosporiferous

teleutostage *n.* [Gk. *teleutē*, completion; L. *stare*, to stand] the stage when teleutospores are produced in the life cycle of a rust fungus; *alt.* telial stage, teliostage, teleutoform stage.

telia *plu.* of telium.

telial *pert.*, or having, telia.

telial stage teleutostage, *q.v.*

telic *a.* [Gk. *telos*, end] purposive; *pert.* teleosis.

Teliomycetes *n.* [Gk. *telos*, end; *mykēs*, fungus] a subdivision of Basidiomycetes including the rusts and smuts, having a compound structure consisting of a teleutospore and its promycelia taking the place of basidia in the rest of the Basidiomycetes.

teliosorus teleutosorus, *q.v.*

teliospore teleutospore, *q.v.*

teliostage *n.* [Gk. *telos*, end; F. *étage*, stage, from L. *stare*, to stand] teleutostage, *q.v.*

telium *n.* [Gk. *telos*, end] teleutosorus, *q.v.*

telmophage *n.* [Gk. *telma*, pool; *phagein*, to eat] a blood-sucking insect which feeds from a blood pool produced by laceration of blood vessels; *alt.* pool feeders.

teloblast *n.* [Gk. *telos*, end; *blastos*, bud] a stage derived from tritoblast and dividing into sporoblasts, in Neosporidia; a large cell which buds forth rows of smaller cells, as in annelid and mollusc embryos, *alt.* pole cell.

telocentric *a.* [Gk. *telos*, end; *kentron*, centre] with terminal centromere, *appl.* chromosomes; *cf.* acrocentric, metacentric.

telocoele *n.* [Gk. *telos*, end; *koilos*, hollow] first, or 2nd ventricle of brain; lateral ventricle; telencephalic vesicle.

telodendrion *n.* [Gk. *telos*, end; *dendrion*, dim. of *dendron*, tree] the terminal arborization of an axon; *alt.* end-brush, teleneurite; *plu.* telodendria.

telofemur *n.* [Gk. *telos*, end; L. *femur*, thigh] distal segment of femur, between basifemur and genu, in certain Acarina.

telogen *n.* [Gk. *telos*, end; *gennaein*, to produce] resting stage in the hair growth cycle.

teloglia *n.plu.* [Gk. *telos*, end; *glia*, glue] cells around endings of axon at a neuromuscular junction; terminal Schwann cells.

telokinesis *n.* [Gk. *telos*, end; *kinēsis*, movement] telophase, *q.v.*; changes in cell after telophase, where this is considered only as a cytoplasmic division, resulting in the reconstitution of daughter nuclei.

telolecithal *a.* [Gk. *telos*, end; *lekithos*, yolk] having yolk accumulated in one hemisphere, as in mesolecithal and polylecithal eggs; *cf.* oligolecithal.

telolemma *n.* [Gk. *telos*, end; *lemma*, skin] a capsule containing a nerve fibre termination, in neuromuscular spindles; *alt.* end-sheath.

telome *n.* [Gk. *telos*, end] a morphological unit in vascular plants being either a terminal branch bearing a sporangium and a vascular supply, or the simplest unit of a plant body, whether terminal or not.

telomere *n.* [Gk. *telos*, end; *meros*, part] end of each chromosome arm distal to centromere.

telome theory a theory of land plant evolution which suggests that all vascular plants evolved from a simple leafless ancestral type such as *Rhynia* made of sterile and fertile axes called telomes which become modified in different taxonomic groups.

telomitic *a.* [Gk. *telos*, end; *mitos*, thread] having chromosomes attached by their end to spindle fibres due to having centromeres terminal.

telophase *n.* [Gk. *telos*, end; *phasis*, aspect] final stage of mitosis and meiosis with cytoplasm division and reconstitution of nuclei, *alt.* telokinesis; sometimes considered only as cytoplasmic division, and the reconstitution of nuclei is telokinesis.

telophragma *n.* [Gk. *telos*, end; *phragma*, fence] Z-disc, *q.v.*

telopod *n.* [Gk. *telos*, end; *pous*, foot] a special male copulatory appendage in some millipedes.

teloreceptor *n.* [Gk. *telos*, end; L. *receptor*, receiver] distance receptor, *q.v.*

Telosporidia *n.* [Gk. *telos*, end; *sporos*, seed] a subclass of Sporozoa having a trophic phase (trophozoite) quite distinct from the reproductive phase, and adults having only one nucleus.

telosynapsis telosynedesis, *q.v.*

telosynedesis *n.* [Gk. *telos*, end; *syndēsai*, to bind together] end-to-end union of chromosome halves in meiosis, now thought to be an artefact; *alt.* telosynapsis, metasyndesis, acrosyndesis; *cf.* parasyndesis.

telotarsus *n.* [Gk. *telos*, end; *tarsos*, sole of foot] distal tarsomere of dactylopodite of spiders; *alt.* tarsus; *cf.* basitarsus.

telotaxis *n.* [Gk. *telos*, end; *taxis*, arrangement] movement along line between animal and source of stimulus; *alt.* goal orientation; *cf.* kinotaxis, tropotaxis.

telotroch *n.* [Gk. *telos*, end; *trochos*, wheel] preanal tuft or circlet of cilia of a trochophore.

telotrocha trochosphere, *q.v.*

telotrophic *a.* [Gk. *telos*, end; *trophē*, nourishment] acrotrophic, *q.v.*

telson *n.* [Gk. *telson*, extremity] the unpaired terminal abdominal segment of Crustacea and the king crab, *Limulus*; curved caudal spine or sting in scorpions; 12th abdominal segment in Protura and in some insect embryos.

telum *n.* [Gk. *telos*, end] last abdominal segment of insects.

Temnocephaloidea *n.* [Gk. *temnein*, to cut; *kephalē*, head; *eidos*, form] a class of flatworms being ectocommensals living in freshwater crustaceans or chelonians or occasionally molluscs, and containing one order, the Temnocephalidea; formerly included in the Rhabdocoela or as an order of turbellarians, called the Temnocephalea.

Temnopleuroida *n.* [Gk. *temnein*, to cut; *pleura*,

side; *eidos*, form] an order of Echinacea usually having a pitted test.

Temnospondyli, temnospondyls *n., n.plu.* [Gk. *temnein*, to cut; *sphondylos*, vertebra] an order of labyrinthodonts existing from Carboniferous to Triassic times, having 4 fingers and a steadily flattening skull over evolutionary time.

temnospondylous *a.* [Gk. *temnein*, to cut; *sphondylos*, vertebra] with vertebrae not fused but in articulated pieces; *alt.* rachitomous; *cf.* stereospondylous; *n.* temnospondyly.

temperate bacteriophage (phage) a bacteriophage (phage) in which the virus lives in the bacterium but does not cause lysis; *cf.* prophage, virulent phage.

temperature coefficient, Q_{10} quotient of 2 reaction rates at temperatures differing by $10°C$.

template a blueprint or pattern from which objects can be made, *appl.* DNA acting as template for RNA and protein synthesis.

template RNA messenger RNA, *q.v.*

temporal *a.* [L. *tempora*, temples] *pert.* or in region of temples; *appl.* compound bone on side of mammalian skull whose formation includes the fusion of petrosal and squamosal.

temporalis broad radiating muscle arising from whole of temporal fossa and extending to coronoid process of mandible.

temporal isolation the prevention of interbreeding between species of plants and animals as a result of time differences, such as of shedding pollen or mating.

temporomalar *a.* [L. *tempora*, temples; *mala*, cheek] *appl.* branch of maxillary nerve supplying temple and cheek, *alt.* zygomatic nerve.

temporomandibular *a.* [L. *tempora*, temples; *mandibula*, jaw] *appl.* articulation: the hinge of the jaws; *appl.* external lateral ligament between zygomatic process of temporal bone and neck of mandible.

temporomaxillary *a.* [L. *tempora*, temples; *maxilla*, jaw] *pert.* temporal and maxillary region, *appl.* posterior facial vein.

tenacle, tenaculum *n.* [L. *tenax*, holding] holdfast of algae; filaments surrounding ostiole of ascus and containing the spore mass in Haerangiomycetes; an ectodermal area modified for adhesion of sand grains, in certain sea anemones; in Collembola, paired appendages of 3rd abdominal segment, modified to retain furcula; in teleosts, fibrous band extending from eyeball to skull.

tendines tendons, *plu.* of tendo.

tendinous *a.* [L. *tendere*, to stretch] of the nature of a tendon; having tendons.

tendo calcaneus, tendo Achillis Achilles tendon, *q.v.*

tendon *n.* [L. *tendo*, tendon, from *tendere*, to stretch] a white fibrous band or cord connecting a muscle with a moveable structure such as a bone; *alt.* tendo.

tendon cells cells in white fibrous connective tissue, with wing-like processes extending between bundles of fibres.

tendon reflex contraction of muscles in a state of slight tension by a tap on their tendons.

tendril *n.* [O.F. *tendrillon*, tender sprig] a specialized twining stem, leaf, petiole, or inflorescence by which climbing plants support themselves.

tendril fibres cerebellar fibres with branches adhering to dendrites of Purkinje's cells; *alt.* clinging fibres; *cf.* basket cells.

tendrillar *a.* [O.F. *tendrillon*, tender sprig] acting as a tendril; twining.

tenent *a.* [L. *tenere*, to hold] holding; *appl.* tubular hairs with expanded tips, of arolium; *appl.* hairs secreting an adhesive fluid, on tarsus of spiders.

teneral *a.* [L. *tener*, tender] immature; *appl.* stage on emergence from nymphal integument.

tenia taenia, *q.v.*

tenofibrils *n.plu.* [L. *tenere*, to hold; *fibrilla*, small fibre] delicate fibrils connecting epithelial cells and passing through intercellular bridges.

Tenon, capsule of [*J. R. Tenon*, French anatomist] the fibroelastic membrane surrounding the eyeball from optic nerve to ciliary region; *alt.* fascia bulbi.

tenoreceptor *n.* [Gk. *tenōn*, tendon; L. *recipere*, to receive] a proprioceptor in tendon reacting to contraction.

tension wood reaction wood of dicotyledons, having little lignification and many gelatinous fibres, and produced on the upper side of bent branches.

tensor *a.* [L. *tendere*, to stretch] *appl.* muscles which stretch parts of body.

tentacles *n.plu.* [L.L. *tentaculum*, feeler] slender flexible organs on head of many invertebrate animals, used for feeling, exploration, prehension, or attachment; adhesive structure of insectivorous plants, as of sundew; *alt.* tentacula.

tentacular *a.* [L.L. *tentaculum*, feeler] *pert.* tentacles; *appl.* a canal branching from perradial canal to tentacle base in ctenophores.

Tentaculata *n.* [L.L. *tentaculum*, feeler] a class of Ctenophora having 2 tentacles.

tentaculiferous *a.* [L.L. *tentaculum*, feeler; L. *ferre*, to carry] bearing tentacles.

tentaculiform *a.* [L.L. *tentaculum*, feeler; L. *forma*, shape] like a tentacle in shape or structure.

tentaculocyst *n.* [L.L. *tentaculum*, feeler; Gk. *kystis*, bladder] in some coelenterate medusae, a sense organ consisting of a modified club-shaped tentacle on the umbrella margin, containing one or more lithites.

tentaculozooids *n.plu.* [L.L. *tentaculum*, feeler; Gk. *zōon*, animal; *eidos*, form] long slender tentacular individuals at outskirts of hydrozoan colony.

tentaculum *n.* [L.L. *tentaculum*, feeler] tentacle, *q.v.*; *plu.* tentacula.

tentilla, tentillum *n.* [L. *tentare*, to feel] a tentacle branch.

tentorium *n.* [L. *tentorium*, tent] a chitinous framework supporting brain of insects; a transverse fold of dura mater, ossified in some mammals, between cerebellum and occipital lobes of brain; *alt.* corpus tentorii.

tenuissimus *a.* [L. *tenuis*, slight] *appl.* a slender muscle beneath biceps femoris in certain mammals.

tepal *n.* [F. *tépale*, from *pétale*] a perianth segment which is not differentiated into petal or sepal.

tephrous *a.* [Gk. *tephra*, ashes] cinerous, *q.v.*

teratogen *n.* [Gk. *teras*, monster; *genos*, birth] any substance causing teratogenesis.

teratogenesis *n.* [Gk. *teras*, monster; *genesis*, descent] the production of monstrous foetuses or growths.

teratology *n.* [Gk. *teras*, monster; *logos*, discourse] science studying and treating of malformations and monstrosities of plants and animals.

tercine *n.* [L. *tertius*, third] the 3rd coat of an ovule or a layer of the 2nd.

Terebellomorpha *n.* [*Terebella*, generic name; Gk. *morphē*, form] an order of sedentary polychaetes that live in burrows or tubes and have tentacles for gathering food at the anterior end.

Terebinthales *n.* [Gk. *terebinthos*, turpentine tree] an order of dicots, now usually divided into the Rutales and Sapindales.

terebra *n.* [L. *terebra*, borer] an ovipositor modified for boring, sawing, or stinging, as in certain Hymenoptera.

terebrate *a.* [L. *terebra*, borer] furnished with a boring organ; adapted for boring.

terebrator a boring organ; trichogyne, *q.v.*, of lichens.

teres *n.* [L. *teres*, rounded] the round ligament of liver; 2 muscles, teres major and minor, extending from scapula to humerus.

terete, teretial *a.* [L. *teres*, rounded] nearly cylindrical in section, as stems.

terga *plu.* of tergum.

tergal *a.* [L. *tergum*, back] *pert.* tergum; situated at the back, *cf.* notal.

tergeminate *a.* [L. *ter*, thrice; *gemini*, twins] thrice forked with twin leaflets.

tergite *n.* [L. *tergum*, back] dorsal chitinous plate of each segment of most Arthropoda; *alt.* tergal sclerite.

tergosternal *a.* [L. *tergum*, back; *sternum*, breast bone] connecting tergite and corresponding sternite, *appl.* muscles in insects.

tergum *n.* [L. *tergum*, back] the back generally; dorsal part of arthropod segment; notum, *q.v.*; dorsal plate of barnacles; *plu.* terga.

terminal *a.* [L. *terminus*, end] *pert.*, or situated at, the end, as terminal bud at end of twig; *appl.* nerve: nervus terminalis, *q.v.*; *appl.* filament, slender prolongation of ovariole; *appl.* chiasma at extreme end of chromatid; *appl.* gene at end of telomere; *appl.* organs: receptor and effector; *appl.* disc: ring centriole, *q.v.*

terminalia *n.plu.* [L. *terminus*, end] external genitalia, or hypopygium, in Diptera.

terminalization *n.* [L. *terminus*, end] movement of chiasmata towards chromosome ends during diplotene and diakinesis.

terminal oxidase an oxidase which reacts with molecular oxygen to form water at the end of electron transfer chains, such as cytochrome oxidase, ascorbic oxidase, polyphenol oxidase.

termitarium *n.* [L. *termes*, woodworm] an elaborately constructed nest of a termite colony.

termites the common name for the Isoptera, *q.v.*

termitophil *a.* [L. *termes*, woodworm; Gk. *philein*, to love] living in termite nest, *appl.* certain fungi and insects.

termones *n.plu.* [Gk. *termōn*, limit] sex-determining substances as in certain protozoans, algae, and fungi.

ternary, ternate *a.* [L. *terni*, three each] arranged in 3s; having 3 leaflets to a leaf, *alt.* trifoliolate; trilateral, *appl.* symmetry.

ternatopinnate *a.* [L. *terni*, three each; *pinna*,

feather] having 3 pinnate leaflets to each compound leaf.

terpenes *n.plu.* [Gk. *terebinthos*, turpentine tree] hydrocarbons present in parts of many higher plants, and constituents of fragrant or essential oils.

terraneous *a.* [L. *terra*, earth] *appl.* land vegetation.

terrestrial *a.* [L. *terra*, earth] *appl.* organisms living on land; *alt.* terricolous.

terricolous *a.* [L. *terra*, earth; *colere*, to inhabit] inhabiting the soil; terrestrial, *q.v.*; *alt.* terricoline.

terrigenous *a.* [L. *terra*, earth; *gignere*, to produce] derived from land, *appl.* deposits.

territorality a social system in which an animal establishes a territory which it defends against other members of the same species.

territory *n.* [L. *territorium*, domain] an area defended by an animal shortly before and during the breeding season; an area sufficient for food requirements of an animal or aggregation of animals; foraging area; *alt.* home range.

tertial *n.* [L. *tertius*, third] flight wing feather of 3rd row, attached to humerus; *alt.* scapular, tertiary feather.

tertiary *a.* [L. *tertius*, third] *appl.* roots produced by secondary roots; *appl.* inner wall of some wood fibres; *appl.* feather: tertial, *q.v.*; *appl.* consumer: a carnivore that eats other carnivores.

Tertiary *appl.* era following the Mesozoic and preceding Quaternary; earlier period of Caenozoic era, Eocene to Pliocene epochs.

tertiary parasite an organism parasitic on a hyperparasite.

tessellate(d) *a.* [L. *tessella*, small stone cube] chequered, *appl.* markings or colours arranged in squares, *appl.* epithelium.

tesserae *n.plu.* [L. *tessera*, square block] prisms of lime, in calcification of cartilage; patches of endoplasmic reticulum of different function, each of which carries a different set of enzymes.

test *n.* [L. *testa*, shell] a shell or hardened outer covering; testa, *q.v.*

testa *n.* [L. *testa*, shell] test, *q.v.*; outer coat of seed, *alt.* episperm; the whole of the outer covering of a seed, made from the integuments of the ovule, *alt.* seed coat.

Testacea *n.* [L. *testa*, shell] an order of Sarcodina, sometimes included in the Foraminifera as the suborder Monothalamia, having shells made of protein impregnated with silica.

testaceous *a.* [L. *testa*, shell] protected by a shell-like outer covering; made of shell or shell-like material; red-brick in colour.

test cross the mating of an organism to a double recessive in order to determine whether it is homozygous or heterozygous for a character under consideration; *cf.* back cross.

testes *plu.* of testis.

Testicardines *n.* [L. *testis*, testicles; *cardo*, hinge] a class of Brachiopoda having shells with a hinge and an internal skeleton.

testicle *n.* [L. *dim.* of *testis*, testicle] testis, *q.v.*

testicular, **testiculate** *a.* [L. *dim.* of *testis*, testicle] having 2 oblong tubercles, as in some orchids; testicle-shaped; *pert.* testis; *alt.* orchitic.

testis *n.* [L. *testis*, testicle] male reproductive gland. producing spermatozoa; *alt.* testicle; *plu.* testes.

testosterone *n.* [L. *testis*, testicle; Gk. *stear*, suet] an androgen produced mainly by the testis in vertebrates.

Testudinata *n.* [L. *testudo*, tortoise] Chelonia, *q.v.*

testudinate *a.* [L. *testudo*, tortoise] having a hard protective shell, as of tortoise.

tetaniform *a.* [Gk. *tetanos*, stretched; L. *forma*, shape] like tetanus; *alt.* tetanoid.

tetanus *n.* [Gk. *tetanos*, stretched] state of a muscle undergoing a continuous fused series of contractions due to electrical stimulation; a rigid state of plant tissue caused by continued stimulus; *v.* tetanize.

tethelin *n.* [Gk. *tethēlōs*, swelling] somatotrophin, *q.v.*

Tethys (Tethyian) sea [Gk. *Tethys*, a titaness, wife of Oceanus] the sea between Laurasia and Gondwanaland.

tetra-allelic *appl.* polyploid with 4 different alleles at a locus.

Tetrabothrioidea *n.* [Gk. *tetras*, four; *bothros*, trench; *eidos*, form] an order of segmented tapeworms having a scolex with 4 suckers, sometimes included in the Cyclophyllidea.

Tetrabranchiata, **tetrabranchs** *n.*, *n.plu.* [Gk. *tetras*, four; *brangchia*, gills] in some classifications, a group of cephalopod molluscs with 2 pairs of gills and an external shell, and including the Nautiloidea and Ammonoidea.

tetrabranchiate *a.* [Gk. *tetras*, four; *brangchia*, gills] having 4 gills.

tetracarpellary *a.* [Gk. *tetras*, four; *karpos*, fruit] having 4 carpels.

tetracerous *a.* [Gk. *tetras*, four; *keras*, horn] four-horned.

tetrachaenium *n.* [Gk. *tetras*, four; *a*, not; *chainein*, to gape] four adherent achenes, as constituting fruit of the dicot family Labiatae.

tetrachotomous *a.* [Gk. *tetracha*, fourfold; *tomē*, cutting] divided up into 4s.

tetracoccus *n.* [Gk. *tetras*, four; *kokkos*, kernel] any micro-organism found in groups of 4.

tetracont tetrakont, *q.v.*

tetracotyledonous *a.* [Gk. *tetras*, four; *kotylēdōn*, cup-like hollow] with 4 cotyledons.

tetracrepid *a.* [Gk. *tetras*, four; *krēpis*, edge] *appl.* a minute caltrop or 4-rayed spicule.

tetract *n.* [Gk. *tetras*, four; *aktis*, ray] a 4-rayed spicule.

tetractinal *a.* [Gk. *tetras*, four; *aktis*, ray] quadriradiate, *q.v.*

tetractine *n.* [Gk. *tetras*, four; *aktis*, ray] a spicule of 4 equal and similar rays meeting at equal angles; *alt.* tetraxon; *a.* tetractinal.

Tetractinellida *n.* [Gk. *tetras*, four; *aktis*, ray] a subclass or in some classifications an order of Demospongea having 4-rayed spicules or no spicules, and no spongin.

tetracyclic *a.* [Gk. *tetras*, four; *kyklos*, circle] with 4 whorls.

tetracycline *n.* [Gk. *tetras*, four; *kyklos*, circle] any of various antibiotics produced by the actinomycete *Streptomyces*, which inhibit the binding of tRNA to ribosome.

tetracyte *n.* [Gk. *tetras*, four; *kytos*, hollow] one of 4 daughter cells formed from a mother cell by meiosis.

tetracytic *a.* [Gk. *tetra*, four; *kytos*, hollow] *appl.* stoma accompanied by 4 subsidiary cells.

tetrad *n.* [Gk. *tetras*, four] a group of 4; *appl.* 4 spores formed by 1st and 2nd meiotic divisions of spore mother cell; 4-cell stage in development of bryophytes and pteridophytes; a quadruple group of chromatids at meiosis; a quadrangular mass or loop of chromosomes in a stage of mitosis; *cf.* quartet.

tetradactyl *a.* [Gk. *tetras*, four; *daktylos*, finger] having 4 digits.

Tetradesmida *n.* [Gk. *tetras*, four; *desmos*, bond] a subclass of tapeworms having typically 4 suckers or proboscides on the scolex.

tetradidymous *a.* [Gk. *tetras*, four; *didymos*, double] having or *pert.* 4 pairs.

tetradymous *a.* [Gk. *tetradymos*, fourfold] having 4 cells, *appl.* spores.

tetradynamous *a.* [Gk. *tetras*, four; *dynamis*, power] having 4 long stamens and 2 short.

tetragenic *a.* [Gk. *tetras*, four; *genos*, descent] controlled by 4 genes.

tetragonal *a.* [Gk. *tetras*, four; *gōnia*, angle] having 4 angles; *alt.* quaternary.

tetragynous *a.* [Gk. *tetras*, four; *gynē*, female] having 4 pistils; having 4 carpels to a gynaecium.

tetrahedral *a.* [Gk. *tetras*, four; *hedra*, base] having 4 triangular sides; *appl.* apical cell in plants having a unicellular growing point.

tetrahydrofolic acid a coenzyme concerned in purine and pyrimidine synthesis.

tetraiodothyronine thyroxine, *q.v.*

tetrakon(an) *a.* [Gk. *tetras*, four; *kontos*, punting pole] having 4 flagella; *alt.* tetracont.

tetralophodont *a.* [Gk. *tetras*, four; *lophos*, crest; *odous*, tooth] *appl.* molar teeth with 4 ridges.

tetralophous *a.* [Gk. *tetras*, four; *lophos*, crest] *appl.* a spicule with 4 rays branched or crested.

tetramerous *a.* [Gk. *tetras*, four; *meros*, part] composed of 4 parts; in multiples of 4; *alt.* tetrameral, tetrameric.

tetramite *n.* [Gk. *tetras*, four; *mitos*, thread] a tetrad formed by 4 parallel chromatids prior to diakinesis.

tetramorphic *a.* [Gk. *tetras*, four; *morphē*, form] having 4 forms; of 4 different lengths, as basidia.

tetrandrous *a.* [Gk. *tetras*, four; *anēr*, man] having 4 stamens.

Tetraodontiformes *n.* [Gk. *tetras*, four; *odous*, tooth; L. *forma*, shape] an order of teleosts having a beak-like snout, small mouth, and heavy teeth, and including the trigger fish, puffers, and porcupine fish; *alt.* Plectognathi, Tetrodontiformes.

tetrapetalous *a.* [Gk. *tetras*, four; *petalon*, leaf] having 4 petals.

Tetraphidales *n.* [*Tetraphis*, generic name] an order of mosses of subclass Bryidae.

Tetraphyllidea *n.* [Gk. *tetras*, four; *phyllon*, leaf] an order of segmented tapeworms having a scolex with 4 suckers (bothria) which are often hooked.

tetraphyllous *a.* [Gk. *tetras*, four; *phyllon*, leaf] quadrifoliate, *q.v.*

tetraploid *a.* [Gk. *tetraplē*, fourfold] with 4 times the normal haploid number of chromosomes. *n.* A tetraploid organism or cell.

tetrapneumonous *a.* [Gk. *tetras*, four; *pneumōn*, lung] having 4 lung books, as certain spiders.

tetrapod *n.* [Gk. *tetras*, four; *pous*, foot] a 4-footed animal; *alt.* quadruped.

tetrapterous *a.* [Gk. *tetras*, four; *pteron*, wing] having 4 wings or wing-like processes; *alt.* quadripennate.

tetrapyrenous *a.* [Gk. *tetras*, four; *pyrēn*, fruit stone] having 4 fruit stones; being a 4-stoned fruit.

tetraquetrous *a.* [Gk. *tetras*, four; L. *quadratus*, squared] having 4 angles, as some stems.

tetraradiate *a.* [Gk. *tetras*, four; L. *radius*, ray] *appl.* pelvic girdle consisting of pubis, prepubis, ilium, and ischium; *cf.* triradiate.

tetrarch *a.* [Gk. *tetras*, four; *archē*, element] with 4 alternating xylem and phloem groups; with 4 protoxylem bundles.

Tetrarhynchoidea *n.* [Gk. *tetras*, four; *rhyngchos*, snout; *eidos*, form] an order of segmented tapeworms having a scolex with 4 proboscides and 4 suckers (bothria); *alt.* Trypanorhyncha.

tetrasaccharide *n.* [Gk. *tetras*, four; *sakchar*, sugar] any of a group of carbohydrates, such as stachyose, made up of 4 units of monosaccharides.

tetraselenodont *a.* [Gk. *tetras*, four; *selēnē*, moon; *odous*, tooth] having 4 crescentic ridges on molar teeth.

tetrasepalous *a.* [Gk. *tetras*, four; F. *sépale*, sepal] having 4 sepals.

tetraseriate *a.* [Gk. *tetras*, four; L. *serere*, to put in a row] tetrastichous, *q.v.*; quadriserial, *q.v.*

tetrasome *n.* [Gk. *tetras*, four; *sōma*, body] association of 4 homologous chromosomes in meiosis.

tetrasomic *a.* [Gk. *tetras*, four; *sōma*, body] *pert.* or having 4 homologous chromosomes. *n.* An organism with 4 chromosomes of one type.

tetraspermous *a.* [Gk. *tetras*, four; *sperma*, seed] having 4 seeds.

Tetrasporales *n.* [Gk. *tetras*, four; *sporos*, seed] an order of green algae having non-motile colonial vegetative forms with vegetative cell division.

tetrasporangium *n.* [Gk. *tetras*, four; *sporos*, seed; *anggeion*, vessel] sporangium producing tetraspores, as in red algae.

tetraspore *n.* [Gk. *tetras*, four; *sporos*, seed] one of a group of 4 non-motile spores produced by sporangium of certain algae; one of 4 basidial spores, as in Hymenomycetes.

tetrasporic, tetrasporous four-spored.

tetrasporocystid *a.* [Gk. *tetras*, four; *sporos*, seed; *kystis*, bladder] *appl.* oocyst of Sporozoa when 4 sporocysts are present.

tetrasporophyte *n.* [Gk. *tetras*, four; *sporos*, seed; *phyton*, plant] in red algae, an asexual generation which produces tetrasporangia.

tetraster *n.* [Gk. *tetras*, four; *astēr*, star] a mitotic figure having 4 astral poles rather than 2, and found in zygotes or embryos produced by polyspermy.

tetrasternum *n.* [Gk. *tetras*, four; *sternon*, chest] sternite of 4th segment of prosoma or 2nd segment of podosoma in Acarina.

tetrastichous *a.* [Gk. *tetras*, four; *stichos*, row] arranged in 4 rows; *alt.* tetraseriate.

tetrathecal *a.* [Gk. *tetras*, four; *thēkē*, case] having 4 loculi; *alt.* quadrilocular.

tetraxon *n.* [Gk. *tetras*, four; *axōn*, axis] tetractine, *q.v.*

tetrazoic *a.* [Gk. *tetras*, four; *zōon*, animal] having 4 sporozoites.

tetrazooid *n.* [Gk. *tetras*, four; *zōon*, animal; *eidos*, form] zooid developed from each of 4 parts

Tetrodontiformes

constricted from stolon process of embryonic ascidian.

Tetrodontiformes Tetraodontiformes, *q.v.*

tetroses *n.* [Gk. *tetras*, four] monosaccharides having the formula $C_4H_4O_4$ such as erythrose.

textura *n.* [L. *textura*, fabric] tissue.

thalamencephalon *n.* [Gk. *thalamos*, chamber; *engkephalos*, brain] the part of the fore-brain comprising thalamus, corpora geniculata, and epithalamus.

Thalammophora *n.* [Gk. *thalamos*, chamber; *phora*, producing] an order of Sarcodina sometimes included in the Foraminifera as the suborder Polythalamia, which are usually marine and have many-chambered, calcareous shells.

thalamomamillary *a.* [Gk. *thalamos*, chamber; L. *mamilla*, nipple] *appl.* fasciculus: Vicq-d'Azyr bundle, *q.v.*

thalamus *n.* [Gk. *thalamos*, chamber] the receptacle of a flower; ovoid ganglionic mass on either side of 3rd ventricle of brain.

thalassin *n.* [Gk. *thalassa*, sea] a toxin of sea anemone tentacles.

thalassoid *a.* [Gk. *thalassa*, sea; *eidos*, form] *pert.* freshwater organisms resembling, or originally, marine forms; halolimnic, *q.v.*

thalassophyte *n.* [Gk. *thalassa*, sea; *phyton*, plant] any marine alga especially a seaweed.

thalassoplankton *n.* [Gk. *thalassa*, sea; *plangktos*, wandering] marine plankton.

Thaliacea *n.* [*Thalia*, generic name] a class of free-swimming urochordates, commonly called salps, having a transparent gelatinous test and 2 distinct phases in the life history, one reproducing sexually and the other asexually.

thalliform thalloid, *q.v.*

thalline *a.* [Gk. *thallos*, young shoot] consisting of a thallus; thalloid, *q.v.*

thallodal thalloid, *q.v.*

thallogen thallophyte, *q.v.*

thalloid *a.* [Gk. *thallos*, young shoot; *eidos*, form] resembling a thallus; *alt.* thalliform, thallodal, thalline, thallose.

thallome *n.* [Gk. *thallos*, young shoot] a thallus-like structure; a thallus, *q.v.*

Thallophyta, thallophytes *n., n.plu.* [Gk. *thallos*, young shoot; *phyton*, plant] a major division of the plant kingdom in which the plant body is not differentiated into root, stem, and leaves, but varies widely in form and which includes Algae, Fungi, and in some classifications Lichens, Bacteria, and Myxomycophyta; *alt.* thallogens; *cf.* Cormophyta.

thallose thalloid, *q.v.*

thallospore *n.* [Gk. *thallos*, young shoot; *sporos*, seed] spore cell in vegetative part of a fungus.

thallus *n.* [Gk. *thallos*, young shoot] a plant body not differentiated into leaf and stem, as vegetative body of thallophytes and some liverworts; *alt.* thallome; *cf.* cormus.

thalposis *n.* [Gk. *thalpos*, warmth] sensation of warmth.

thamniscophagy *n.* [Gk. *thamnos*, bush; *dim.*; *phagein*, to eat] disintegration and absorption of arbusculae and sporangioles in mycorrhiza.

thamniscophysalidophagy *n.* [Gk. *thamnos*, bush; *dim.*; *physalis*, bubble; *phagein*, to eat] the joint disintegration of arbusculae and vesicles in certain mycorrhizae.

thamnium *n.* [Gk. *thamnos*, bush] a branched or fruticose thallus of certain lichens.

thanatoid *a.* [Gk. *thanatos*, death; *eidos*, form] deadly poisonous; resembling death.

thanatology *n.* [Gk. *thanatos*, death; *logos*, discourse] theories concerning death.

thanatosis *n.* [Gk. *thanatos*, death] habit or act of feigning death; death of a part.

Theales *n.* [*Thea*, generic name] an order of dicots (Dilleniidae) which are usually woody plants with a superior syncarpous ovary containing seeds with little endosperm and a large embryo; *alt.* Guttiferae.

thebesian *a.* [*A. C. Thebesius*, German anatomist] *appl.* valve of coronary sinus.

theca *n.* [Gk. *thēkē*, case] a spore- or pollen-case; a sporangium; a capsule; a structure serving as protective covering for organ or organism, as of spinal cord, follicle, pupa, proboscis, tube animals; the calcareous shell (test) in echinoderms; ascus, *q.v.*; *alt.* sheath.

thecacyst *n.* [Gk. *thēkē*, case; *kystis*, bladder] sperm envelope or spermatophore formed by spermatheca.

thecal *a.* [Gk. *thēkē*, case] surrounded by a protective membrane or tissue; *pert.* a theca; *pert.* an ascus.

theca lutein paralutein, *q.v.*

Thecanephria *n.* [Gk. *thēkē*, case; *nephros*, kidney] an order of Pogonophora often having tentacles jointed together by a membrane.

thecaphore *n.* [Gk. *thēkē*, case; *pherein*, to bear] a structure on which a theca is borne.

thecaspore ascospore, *q.v.*

thecasporous *a.* [Gk. *thēkē*, case; *sporos*, a seed] having spores enclosed.

Thecata *n.* [Gk. *thēkē*, case] an order of hydrozoans having hemispherical or flattened medusae and polyps usually with a hydrotheca into which they can withdraw; *alt.* Calyptoblastea, Leptomedusae.

thecate *a.* [Gk. *thēkē*, case] covered or protected by theca; *alt.* theciferous, thecigerous.

thecial *a.* [Gk. *thēkē*, case] within or *pert.* a thecium.

thecium *n.* [Gk. *thēkē*, case] that part of a fungus or lichen containing sporules; ascus hymenium.

thecodont *a.* [Gk. *thēkē*, case; *odous*, tooth] having teeth in sockets.

Thecodontia, thecodonts *n., n.plu.* [Gk. *thēkē*, case; *odous*, tooth] an order of primitive archosaurs of Permian to Triassic age, including bipedal or crocodile-like forms and thought to be ancestral to several other groups.

Thecosomata *n.* [Gk. *thēkē*, case; *sōma*, body] an order of planktonic pteropod opisthobranchs which have a shell.

theelin *n.* [Gk. *thēlys*, female] oestrone, *q.v.*

theelol *n.* [Gk. *thēlys*, female] oestriol, *q.v.*

thelephorous *a.* [Gk. *thēlē*, teat; *pherein*, to bear] having nipples or nipple-like projections; with a closely nippled surface.

Thelodonti *n.* [*Thelodus*, generic name] Coelolepida, *q.v.*

thelyblast *n.* [Gk. *thēlys*, female; *blastos*, bud] a matured female gamete.

thelygenic *a.* [Gk. *thēlys*, female; *-genēs*, producing] producing offspring preponderantly or entirely female; *n.* thelygeny, *cf.* thelyotoky, monogeny, arrhenogeny.

thelyotoky *n.* [Gk. *thēlys*, female; *tokos*, offspring] parthenogenesis in case where females only are produced; *alt.* thelytoky; *a.* thelyotokous, *cf.* thelygenic, arrhenotoky, deuterotoky.

thelyplasm *n.* [Gk. *thēlys*, female; *plasma*, mould] female plasm; *cf.* arrhenoplasm.

thenal *a.* [Gk. *thenar*, palm of hand] palmar, *q.v.*

thenar *n.* [Gk. *thenar*, palm of hand] the muscular mass forming ball of thumb.

theobromine *n.* [*Theobroma*, generic name of cocoa tree] a bitter compound found in cocoa beans and in some other plants, which is similar to caffeine acting as a stimulant and diuretic.

Therapsida, therapsids *n.*, *n.plu.* [Gk. *theraps*, attendant] an order of Permian to Triassic mammal-like reptiles with many mammalian features and probably including mammalian ancestors.

Theria, therians *n.*, *n.plu.* [Gk. *thērion*, small animal] a subclass of mammals whose living members are viviparous and have a molar tooth pattern based on a triangle of cusps, cervical ribs fused to vertebrae, and a spiral cochlea.

thermaesthesia *n.* [Gk. *thermē*, heat; *aisthēsis*, perception] sensitivity to temperature stimuli.

thermium *n.* [Gk. *thermai*, hot springs] plant community in warm or hot springs.

thermobiology *n.* [Gk. *thermē*, heat; *bios*, life; *logos*, discourse] study of the effects of thermal energy on all types of living organisms and biological molecules.

thermocleistogamy *n.* [Gk. *thermē*, heat; *kleistos*, closed; *gamos*, marriage] self-pollination of flowers when unopened owing to unfavourable temperature.

thermocline *n.* [Gk. *thermē*, heat; *klinein*, to swerve] the layer of water of rapidly changing temperature in lakes and seas in summer, above which is the epilimnion and below is the hypolimnion; *alt.* discontinuity layer.

thermoduric *a.* [Gk. *thermē*, heat; L. *durus*, hardy] resistant to relatively high temperatures, *appl.* micro-organisms.

thermogenesis *n.* [Gk. *thermē*, heat; *genesis*, production] body-heat production by oxidation; heat production by bacteria.

thermolysis *n.* [Gk. *thermē*, heat; *lysis*, loosing] loss of body heat; chemical dissociation owing to heat.

thermonasty *n.* [Gk. *thermē*, heat; *nastos*, close pressed] a nastic movement in plants in response to variations of temperature; *a.* thermonastic.

thermoperiodicity, thermoperiodism *n.* [Gk. *thermē*, heat; *periodos*, period] the response of living organisms to regular changes of temperature, either with day and night or season to season; *cf.* photoperiodism.

thermophase *n.* [Gk. *thermē*, heat; *phainein*, to appear] first developmental stage in some annual and perennial plants, and which can be partly or entirely completed during seed ripening if temperature and humidity are favourable; *alt.* vernalization phase.

thermophil *a.* [Gk. *thermē*, heat; *philos*, loving] thriving at relatively high temperatures, above 40°C, *appl.* certain bacteria, *alt.* thermophilic. *n.* thermophile.

thermophobic *a.* [Gk. *thermē*, heat; *phobein*, to fear] able to live or thrive only at relatively low temperatures.

thermophylatic *a.* [Gk. *thermē*, heat; *phylaktikos*, fit for preserving] heat-resistant; tolerating heat, as certain bacteria.

thermophyte *n.* [Gk. *thermē*, heat; *phyton*, plant; a heat-tolerant plant; therophyte, *q.v.*

thermoreceptor *n.* [Gk. *thermē*, heat; L. *recipere*, to receive] an organ which reacts to temperature stimuli.

Thermosbaenacea *n.* [*Thermosbaena*, generic name] an order of malacostracans that live in hot springs and have a cylindrical body and a dorsal brood pouch.

thermoscopic *a.* [Gk. *thermē*, heat; *skopein*, to view] adapted for recognizing changes of temperature, as special sense organs of certain cephalopods.

thermotactic *a.* [Gk. *thermē*, heat; *taxis*, arrangement] *pert.* thermotaxis; *appl.* optimum: the range of temperature preferred by an organism.

thermotaxis *n.* [Gk. *thermē*, heat; *taxis*, arrangement] locomotor reaction to temperature stimulus; regulation of body temperature.

thermotropism *n.* [Gk. *thermē*, heat; *tropē*, turn] curvature in plants in response to temperature stimulus.

Theromorpha *n.* [Gk. *theros*, summer; *morphē*, form] Pelycosauria, *q.v.*

therophyllous *a.* [Gk. *theros*, summer; *phyllon*, leaf] having leaves in summer; with deciduous leaves.

therophyte *n.* [Gk. *theros*, summer; *phyton*, plant] a plant which completes life cycle within a single season, being dormant as seed during unfavourable period, i.e. an annual; *alt.* thermophyte.

thesocytes *n.plu.* [Gk. *thesis*, deposit; *kytos*, hollow] sponge cells storing reserve material.

theta (θ) factor thyrotropic hormone, *q.v.*

thiamin(e) a water-soluble vitamin, a member of the vitamin B complex, found especially in seed embryos and yeast, absence causing beri-beri in man, or polyneuritis, and which is a precursor of the coenzyme thiamine pyrophosphate; *alt.* vitamin B₁, aneurine, antineuritic vitamin.

thiamin(e) pyrophosphate (TPP) a coenzyme found in most living cells, which is involved in certain decarboxylation reactions.

thigmaesthesia *n.* [Gk. *thigēma*, touch; *aisthēsis*, sensation] the sense of touch; *alt.* thigmesthesis.

thigmocyte *n.* [Gk. *thigēma*, touch; *kytos*, hollow] a blood corpuscle which undergoes cytolysis on contact with a foreign substance, as a thrombocyte.

thigmokinesis *n.* [Gk. *thigēma*, touch; *kinēsis*, movement] movement, or inhibition of movement, in response to contact stimuli; *alt.* stereokinesis.

thigmomorphosis [Gk. *thigēma*, touch; *morphōsis*, form] structural change due to contact, as swelling at ends of contacting zygophores, formation of haustoria, etc.

thigmotaxis *n.* [Gk. *thigēma*, touch; *taxis*, arrangement] stereotaxis, *q.v.*; taxis in response to touch stimulus; *a.* thigmotactic.

thigmotropism *n.* [Gk. *thigēma*, touch; *tropē*, turn] growth curvature in response to a contact stimulus found in clinging plant organs such as stems, tendrils; response of sessile organisms to stimulus of contact; *alt.* haptotropism; *a.* thigmotropic; *cf.* stereotropism.

thinophyte

thinophyte n. [Gk. *this*, sand heap; *phyton*, plant] dune plant.

Thiobacteriales n. [Gk. *theion*, sulphur; *baktērion*, small rod] an order of bacteria that grow well in the presence of hydrogen sulphide and contain bacteriopurpurin and/or sulphur.

thioester an ester with sulphur instead of oxygen.

thiogenic a. [Gk. *theion*, sulphur; *gennaein*, to produce] sulphur-producing, *appl.* bacteria utilizing sulphur compounds.

thiophil n. [Gk. *theion*, sulphur; *philein*, to love] an organism thriving in the presence of sulphur compounds as certain bacteria; a. thiophilic.

thoracic a. [Gk. *thōrax*, chest] *pert.*, or in region of, thorax.

Thoracica n. [Gk. *thōrax*, chest] an order of free-living barnacles having 6 pairs of plumose thoracic limbs.

thoracic duct vessel conveying lymph and chyle from abdomen to left subclavian vein; *cf.* receptaculum chyli.

thoracic index one hundred times depth of thorax at nipple level divided by breadth.

thoracic ring the ring formed by notum, pleura, and sternum in insects.

thoracolumbar a. [Gk. *thōrax*, chest; L. *lumbus*, loin] *pert.* thoracic and lumbar part of spine; *appl.* nerves, the sympathetic system.

thoracopod n. [Gk. *thōrax*, chest; *pous*, foot] any thoracic leg of Malacostraca.

thorax n. [Gk. *thōrax*, chest] in higher vertebrates, that part of body between neck and abdomen containing heart, lungs, etc.; body region behind head of other animals.

thorny-headed worms the common name for the Acanthocephala, *q.v.*

thr threonine, *q.v.*

thread capsule nematocyst, *q.v.*

thread cells stinging cells of cnidoblasts in coelenterates; in skin of myxinoids, cells whose long threads form a network in which mucous secretion of ordinary gland cells is entangled.

thread press the muscular portion of a spinning tube.

thremmatology n. [Gk. *thremma*, nursling; *logos*, discourse] the science of breeding animals and plants under domestic conditions.

threonine n. [Altered form of Gk. *erythros*, red] an essential amino acid, aminohydroxybutyric acid, obtained by hydrolysis of various proteins such as egg proteins and casein; abbreviated to thr.

threpsology n. [Gk. *trephein*, to nourish; *logos*, discourse] the science of nutrition.

threshold *see* limen.

thrips the common name for the Thysanoptera, *q.v.*

thrombase thrombin, *q.v.*

thrombin n. [Gk. *thrombos*, clot] a proteinase produced from prothrombin at a wound, which converts the soluble protein fibrinogen to insoluble fibrin; *alt.* thrombase; EC 3.4.21.5; *r.n.* thrombin.

thrombinogen n. [Gk. *thrombos*, clot; *-genēs*, producing] prothrombin, *q.v.*

thrombocytes n.plu. [Gk. *thrombos*, clot; *kytos*, hollow] blood platelets, *q.v.*; in non-mammalian vertebrates, nucleated spindle-shaped cells concerned with clotting of blood; *alt.* haematoblasts, thigmocytes.

thrombogen n. [Gk. *thrombos*, clot; *-genēs*, producing] prothrombin, *q.v.*

thrombokinase n. [Gk. *thrombos*, clot; *kinein*, to move] a proteinase which converts prothrombin to thrombin in the presence of calcium ions; *alt.* cytozyme, thromboplastin, thrombozyme; EC 3.4.21.6; *r.n.* coagulation factor Xa.

thrombokinesis n. [Gk. *thrombos*, clot; *kinēsis*, movement] the process of blood clotting.

thromboplastid n. [Gk. *thrombos*, clot; *plastos*, moulded] blood platelet, *q.v.*

thromboplastin n. [Gk. *thrombos*, clot; *plastos*, moulded] thrombokinase, *q.v.*; *alt.* thromboplastic factor.

thrombosis n. [Gk. *thrombos*, clot] clotting, as of blood.

thrombozyme n. [Gk. *thrombos*, clot; *zymē*, leaven] thrombokinase, *q.v.*

thrum-eyed short-styled, with long stamens extending to mouth of tubular corolla; *cf.* pin-eyed.

thryptophyte n. [Gk. *thryptein*, to enfeeble; *phyton*, plant] any fungus that modifes host tissue without any direct lethal effect.

thylacogens n.plu. [Gk. *thylax*, bag; *gennaein*, to produce] substances produced by parasites and causing reactive hypertrophy of the host's tissue at the site of infection.

thylakentrin n. [Gk. *thylakē*, scrotum; *kentron*, sharp point] follicle-stimulating hormone, *q.v.*

thylakoids n.plu. [Gk. *thylakos*, pouch; *eidos*, form] the vesicles making up the grana of chloroplasts, bearing the photosynthetic pigments on their walls; invaginations of the cytoplasmic membranes in cells of blue-green algae, forming lamellae on which photosynthetic pigments are borne.

thylosis n. [Gk. *thylakos*, pouch] tylosis, *q.v.*; *plu.* thyloses.

Thymelaeales n. [*Thymelaea*, generic name] an order of dictos of uncertain position, now usually placed in the Dilleniidae, comprising the single family Thymelaeaceae, e.g. mezereon.

thymic a. [Gk. *thymos*, thymus] *pert.* the thymus; *appl.* corpuscles: the concentric corpuscles of Hassall.

thymic acid a nucleotide formed from thymidine and phosphoric acid.

thymidine a nucleoside with thymine as its base.

thymine n. [Gk. *thymos*, thymus] a pyrimidine base which is part of the genetic code of DNA, where it pairs with adenine, and in RNA is replaced by uracil.

thymocyte n. [Gk. *thymos*, thymus; *kytos*, hollow] a small lymphocyte in cortex of thymus.

thymonucleic acid [Gk. *thymos*, thymus; L. *nucleus*, kernel] the DNA found in the thymus.

thymovidin n. [Gk. *thymos*, thymus; L. *ovum*, egg] a thymus hormone of birds, which influences egg albumin and shell formation.

thymus n. [Gk. *thymos*, thymus] a gland in lower anterior part of neck, or surrounding heart, in man regressing after maximum development at puberty, and which controls many of the immune responses of the body.

thyreo- *see also* thryo-.

thyreoid thyroid, *q.v.*

thyreothecium n. [Gk. *thyreos*, oblong shield; *thēkē*, case] a shield-like fruit body of certain ectoparasitic fungi.

thyridium n. [Gk. thyra, door; idion, dim.] hairless whitish area on certain insect wings.

thyro- see also thyreo-.

thyro-arytaenoid n. [Gk. thyra, door; arytaina, pitcher; eidos, form] a muscle of larynx.

thyrocalcitonin n. [Gk. thyra, door; L. calyx, lime; Gk. tonos, tension] calcitonin, q.v.

thyroepiglottic a. [Gk. thyra, door; epi, upon; glotta, tongue] appl. ligament connecting epiglottis stem and angle of thyroid cartilage.

thyroglobulin n. [Gk. thyra, door; L. globus, globe] a glycoprotein containing iodine and mannose or N-acetylglucosamine, the iodine of the thyroid gland being stored chiefly as this protein.

thyroglossal a. [Gk. thyra, door; glossa, tongue] pert. thyroid and tongue; appl. an embryonic duct, the ductus thyreoglossus from which the thyroid gland develops.

thyrohyals n.plu. [Gk. thyra, door; hyoeidēs, Υ-shaped] greater cornua of hyoid bone.

thyrohyoid a. [Gk. thyra, door; hyoeidēs, Υ-shaped] appl. muscle extending from thyroid cartilage to hyoid cornu.

thyroid a. [Gk. thyra, door; eidos, form] shield-shaped, peltate, appl. gland, arteries, cartilage, and veins; alt. thyreoid.

thyroid gland an endocrine gland in the neck of vertebrates which secretes the iodine-containing hormones thyroxine and tri-iodothyronine, whose effects are to increase the rates of oxidation in the body and initiate metamorphosis in tadpoles, and which in mammals also secretes calcitonin.

thyroid hormones the hormones secreted by the thyroid gland, q.v.

thyroid-stimulating hormone (TSH) thyrotrophin, q.v.

thyroid-stimulating prepituitary hormone (TSP) thyrotrophin, q.v.

thyrotrophic, thyrotropic a. [Gk. thyra, door; trophē, nourishment; L. tropē, change] influencing the activity of the thyroid gland; appl. hormone: thyrotrophin, q.v., alt. theta factor.

thyrotrophin, thyrotropin n. [Gk. thyra, door; trophē, nourishment; tropē, a change] a glycoprotein hormone secreted by the anterior lobe of the pituitary gland which regulates the growth of the thyroid gland and the formation and secretion of its hormones; alt. thyrotrophic or thyrotropic hormone, thyroid-stimulating hormone (TSH), thyroid-stimulating prepituitary hormone (TSP), theta factor.

thyroxin(e) n. [Gk. thyra, door; oxys, sharp] an iodine-containing hormone, tetraiodothyronine, derived from the amino acid tyrosine and produced by the thyroid gland.

thyrse thyrsus, q.v.

thyrsoid a. [Gk. thyrsos, wand; eidos, form] resembling a thyrsus in shape.

thyrsus n. [Gk. thyrsos, wand] a mixed inflorescence with main axis racemose, later axes cymose, with cluster almost double-cone shaped; hypha-bearing lateral chlamydospores; penis; alt. thyrse.

Thysanoptera n. [Gk. thysanos, fringe; pteron, wing] an order of small slender insects, commonly called thrips, having incomplete metamorphosis and piercing mouth-parts, and being wingless or with very narrow wings.

Thysanura n. [Gk. thysanos, fringe; oura, tail] an order of wingless insects including the bristletails

and silverfish, having biting mouth-parts, a pair of cerci, and a median process on the last segment.

thysanuriform a. [Gk. thysanos, fringe; oura, tail; L. forma, form] campodeiform, q.v.

tibia n. [L. tibia, shin] shin bone, inner and larger of leg bones between knee and ankle; 4th joint of insect, arachnid, and some myriapod legs.

tibial a. [L. tibia, shin] pert. or in region of tibia.

tibiale n. [L. tibia, shin] embryonic structure partly represented by astragalus; a sesamoid bone in tendon of posterior tibial muscle.

tibialis anterior and posterior: tibial muscles acting on ankle and intertarsal joints.

tibiofibula n. [L. tibia, shin; fibula, buckle] bone formed of fused tibia and fibula.

tibiofibular pert. tibia and fibula, appl. articulation, syndesmosis; pert. tibiofibula.

tibiotarsal a. [L. tibia, shin; Gk. tarsos, sole of foot] pert. tibia and tarsus; pert. or in region of tibiotarsus.

tibiotarsus n. [L. tibia, shin; Gk. tarsos, sole of foot] tibial bone to which proximal tarsals are fused, in birds.

ticks a common name for many members of the Acari, q.v.

tidal a. [A.S. tid, time] pert. tides; ebbing and flowing; appl. air: volume of air normally inhaled and exhaled at each breath; appl. wave, main flow of blood during systole.

Tiedmann's vesicles [F. Tiedmann, German anatomist] small rounded glandular chambered bodies at neck of Polian vesicles, being racemose vesicles of Asteroidea.

tige n. [F. tige, stem] paturon, q.v.; stem.

tigellum n. [F. tigelle, dim. of tige, stem] the central embryonic axis, consisting of radicle and plumule.

tigroid a. [Gk. tigroeidēs, spotted] appl. granules or bodies: Nissl granules, q.v.

tigrolysis n. [Gk. tigroeidēs, spotted; lysis, loosing] chromatolysis of tigroid granules.

Tillodont(i)a n. [Gk. tillein, to tear; odous, tooth] an order of extinct placental mammals of the Palaeocene and Eocene.

Tilopteridales n. [Tilopteris, generic name] an order of brown algae having a freely branching plant body and trichothallic growth.

timbal n. [F. timbale, kettle drum] sound-producing organ in cicadas; alt. tymbal.

timberline the latitude above which trees are unable to grow.

Timofeev's corpuscles [D. A. Timofeev, Russian anatomist] specialized sensory nerve endings in submucosa of urethra and in prostatic capsule.

Tinamiformes n. [Tinamus, generic name of tinamou; L. forma, shape] an order of birds of the subclass Neornithes, including tinamous.

tinctorial a. [L. tinctorius, pert. dyeing] producing dyestuff.

tinsel appl. flagella which are branched and feathery; alt. flimmer; cf. whiplash.

Tintinnida n. [Tintinnidium, generic name] an order or suborder of mainly marine and planktonic Spirotricha having body in a vase-like case (lorica); sometimes included in the Oligotrichida.

tip cell the uninucleate ultimate cell of a hyphal crozier, distal to the dome cell, and directed towards the basal cell; alt. ultimate cell.

tiphophyte n. [Gk. *tiphos*, pool; *phyton*, plant] pond plant.

tissue n. [F. *tissu*, woven] the fundamental structure of which animal and plant organs are composed; an organization of like cells.

tmema n. [Gk. *tmētos*, cut] an intercalary cell which separates aecidiospores of certain rust fungi.

TMV tobacco mosaic virus, *q.v.*

toad an amphibian of the order Anura.

tobacco mosaic virus (TMV) a virus used in early work on nucleic acid structure and consisting of RNA and a protein coat.

tocopherol n. [Gk. *tokos*, birth; *pherein*, to carry] vitamin E, *q.v.*

token stimulus a stimulus which operates indirectly by having become linked through experience with an action or object, e.g. colour is the token stimulus attracting bees to some flowers, although they actually want the nectar.

tokocytes n.plu. [Gk. *tokos*, offspring; *kytos*, hollow] reproductive cells of sponges.

tokostome n. [Gk. *tokos*, birth; *stoma*, mouth] female genital aperture, as in mites, etc.

tolypophagy n. [Gk. *tolypē*, ball of wool; *phagein*, to eat] disintegration and absorption by the host of hyphal coils in mycorrhizae.

tomentose a. [L. *tomentum*, stuffing] covered closely with matted hairs or fibrils.

tomentum n. [L. *tomentum*, stuffing] the closely matted hair on leaves or stems; or filaments on pileus and stipe; in lichens, a structure like rhizines but lacking compaction, and being a felty mass of multicellular rhizoids.

Tomes' fibres (*Sir J. Tomes*, English dentist] dentinal fibres, processes of odontoblasts in dentinal tubules.

Tomes' granular layer a layer of interglobular spaces in dentine.

tomite n. [Gk. *tomē*, cutting; *mitos*, thread] free-swimming, non-feeding stage following protomite stage in life cycle of Holotricha.

tomium n. [Gk. *tomos*, cutting] the sharp edge of a bird's beak.

tomont n. [Gk. *tomē*, cutting; *onta*, beings] stage in life cycle of Holotricha when body divides, usually in a cyst.

tone n. [Gk. *tonos*, tension, tone] tonus, *q.v.*; the condition of tension found in living animal tissue especially muscle, *alt.* tonicity, tonus; quality of sensation due to particular audible wavelengths.

tongue n. [A.S. *tunge*, tongue] an organ on floor of mouth, usually movable and protrusible; any tongue-like structure, as radula, ligula; hypopharynx, in some insects; *alt.* lingua.

tonicity n. [Gk. *tonos*, tension] see tone, tonus.

tonofibrillae n.plu. [Gk. *tonos*, tension; L. *fibrilla*, small fibre] epitheliofibrillae, *q.v.*, regarded as skeletal or supporting structures rather than as myofibrillae; supporting fibrils, as of cilia.

tonoplast n. [Gk. *tonos*, tension; *plastos*, modelled] a vacuolar membrane; a plastid with distinct vacuole walls; a special form of vacuole-producing plastid.

tonotaxis n. [Gk. *tonos*, tension; *taxis*, arrangement] a taxis in response to change in density of surrounding medium; *cf.* osmotaxis.

tonsil n. [L. *tonsilla*, tonsil] one of aggregations of lymphoid tissue in pharynx or near tongue base; *alt.* tonsilla.

tonsilla n. [L. *tonsilla*, tonsil] tonsil, *q.v.*; posterior lobule side of cerebellar hemisphere on either side of uvula of inferior vermis.

tonsillar ring partial ring of lymphoid tissue formed by the palatine, pharyngeal, and lingual tonsils; *alt.* Waldeyer's tonsillar ring.

tonus n. [Gk. *tonos*, tension] a condition of persistent partial excitation, which in muscles results in a state of partial contraction; in certain nerve centres, the state in which motor impulses are given out continuously without any sensory impulses from the receptors; *alt.* tonicity, tone; *a.* tonic.

tool using the manipulation of an object by an animal to achieve some end which it could not achieve without it.

topaesthesia, topesthesis n. [Gk. *topos*, place; *aisthēsis*, sensation] appreciation of locus of a tactile sensation.

topochemical a. [Gk. *topos*, place; *chēmeia*, transmutation] *appl.* sense, the perception of odours in relation to track or place, as in ants.

topocline n. [Gk. *topos*, place; *klinein*, to slant] a geographical variation, not always related to an ecological gradient, but to other factors such as topography and climate.

topodeme n. [Gk. *topos*, place; *dēmos*, people] deme occupying a particular geographical area.

topogamodeme n. [Gk. *topos*, place; *gamos*, marriage; *dēmos*, people] individuals occupying a precise locality which form a reproductive or breeding unit.

topoinhibition n. [Gk. *topos*, place; L. *inhibere*, to restrain] inhibition of a cellular process as a result of the proximity of other cells.

toponym n. [Gk. *topos*, place; *onyma*, name] the name of a place or of a region; a name designating the place of origin of a plant or animal.

topophysis n. [Gk. *topos*, place; *physis*, constitution] persistent growth and differentiation, without genetic change, of a plant cutting, depending on the tissue of source.

topotaxis n. [Gk. *topos*, place; *taxis*, arrangement] movement induced by spatial differences in stimulation intensity, and orientation in relation to sources of stimuli, as telotaxis, tropotaxis, menotaxis, mnemotaxis; tropism, *q.v.*

topotype n. [Gk. *topos*, place; *typos*, pattern] a specimen from locality of original type.

toral a. [L. *torus*, a swelling] of or *pert.* a torus.

torcular n. [L. *torcular*, wine press] occipital junction of venous sinuses of dura mater; *alt.* confluens sinuum, torcular Herophili, confluence.

tori *plu.* of torus.

torma n. [Gk. *tormos*, socket] a thickening at junction of labrum and clypeus.

tormogen n. [Gk. *tormos*, socket; *-genēs*, producing] a cell secreting the socket of a bristle, in insects.

tornaria n. [L. *tornare*, to turn] the free larval stage in development of some Enteropneusta.

tornate a. [L. *tornare*, to turn] with blunt extremities, as a spicule; rounded off.

torose a. [L. *torus*, swelling] having fleshy swellings; knobbed.

torques n. [L. *torques*, necklace] a necklace-like arrangement of fur, feathers, or the like.

torsion n. [L. *torquere*, to twist] spiral bending; the twisting around of a gastropod body as it develops; *a.* torsive.

torticone n. [L. *torquere*, to twist; *conus*, cone] a turreted, spirally twisted shell.

tortoises see Chelonia.

torula n. [L. *torulus*, small swelling] a small torus; torulus, q.v.; a small round protuberance; a yeast plant.

torulaceous torulose, q.v.

torula condition yeast-like isolated cells resulting from growth of blue mould conidia in sugar solution.

toruloid a. [L. *torulus*, small swelling; Gk. *eidos*, form] appl. a structure, plasmatoönkosis, storage organ of zoosporangium, as in Peronosporales.

torulose a. [L. *torulus*, small swelling] with small swellings; beaded; alt. moniliform, moniloid, torulose.

torulus n. [L. *torulus*, small swelling] antennifer, q.v.; alt. torula.

torus n. [L. *torus*, swelling] axis bearing floral leaves; receptacle of a flower; thickened centre of a bordered pit membrane; firm prominence, or marginal fold or ridge; ridge bearing uncini in Polychaeta; pedicel in Diptera; any of the pads on feet of various animals, as of cat; plu. tori.

totipalmate a. [L. *totus*, all; *palma*, palm of hand] steganopodous, q.v.

totipotent(ial) a. [L. *totus*, all; *potens*, powerful] appl. blastomeres which can develop into complete embryos when separated from aggregate of blastomeres; appl. meristematic cells capable of specialization in response to hormones from growth centres; alt. equipotent; cf. unipotent.

toxa toxon, q.v.

toxaspire n. [Gk. *toxon*, bow; L. *spira*, coil] a spiral spicule of rather more than one revolution.

toxic a. [Gk. *toxikon*, poison] pert., caused by, or of the nature of a poison; poisonous.

toxicant any poison or toxic agent.

toxicity n. [Gk. *toxikon*, poison] the nature of a poison; the virulence of a poison.

toxicogenic toxigenic, q.v.

toxicology n. [Gk. *toxikon*, poison; *logos*, discourse] the science treating of poisons and their effects.

toxicophorous, toxiferous a. [Gk. *toxikon*, poison; *phora*, producing; L. *ferre*, to carry] holding or carrying poison.

toxigenic a. [Gk. *toxikon*, poison; *-genēs*, producing] producing a poison; alt. toxicogenic.

toxiglossate a. [Gk. *toxikon*, poison; *glōssa*, tongue] having hollow lateral radula teeth conveying poisonous secretion of salivary glands, as certain carnivorous marine gastropods.

toxin n. [Gk. *toxikon*, poison] any poison derived from a plant or animal; phytotoxin or zootoxin.

toxognaths n.plu. [Gk. *toxikon*, poison; *gnathos*, jaw] first pair of limbs, with opening of poison duct, in Chilopoda.

toxoid n. [Gk. *toxikon*, poison; *eidos*, form] a toxin deprived of its toxic but not of its antigenic capacity; alt. anatoxin.

toxon n. [Gk. *toxon*, bow] a bow-shaped spicule; alt. toxa.

toxophores n.plu. [Gk. *toxikon*, poison; *pherein*, to carry] the poisoning qualities of toxin molecules; cf. haptophores.

TP turgor pressure, q.v.

TPN triphosphopyridine nucleotide now known as NADP, formerly coenzyme II.

TPP thiamine pyrophosphate, q.v.

trabant n. [Ger. *Trabant*, satellite] short segment of chromosome constricted from the rest; alt. satellite.

trabeculae n.plu. [L. *trabecula*, little beam] primordial lamellae of agarics; plates of sterile cells extending across sporangium of pteridophytes; a rod-like part of a cell wall extending across lumen of cell; rows of cells bridging a cavity; 2 curved bars of cartilage embracing hypophysis cerebri of embryo; small fibrous bands forming imperfect septa or framework of organs; muscular columns projecting from inner surface of ventricles of heart; sing. trabecula.

trabecular a. [L. *trabecula*, little beam] pert. or of nature of a trabecula; having a cross-barred framework; alt. trabeculate.

trabs cerebri corpus callosum, q.v.

trace elements elements occurring in minute quantities as natural constituents of living organisms; alt. micronutrient, microelement, minor element.

tracer elements isotopes used for tracing chemical elements and compounds in living tissue; alt. tracers.

trachea n. [L.L. *trachia*, windpipe] the windpipe; a respiratory tubule of insects and other arthropods; spiral or annular xylem tissue of plants; alt. wood vessel, fistula.

tracheal pert., resembling, or having trachea; alt. tracheate, tracheary.

tracheal gills small wing-like respiratory outgrowths from the abdomen of aquatic larvae of insects.

trachean tracheate, q.v.

tracheary tracheal, q.v.; tracheate, q.v.

tracheate a. [L.L. *trachia*, windpipe] having tracheae; alt. trachean, tracheal, tracheary.

tracheid(e) n. [L.L. *trachia*, windpipe] one of the cells with spiral thickening or bordered pits, conducting water and some solutes, and forming woody tissue; alt. hydroid.

tracheidal cells pericycle cells resembling tracheids.

trachein n. [L.L. *trachia*, windpipe] colloid substance of tracheal air sacs, contracting or expanding according to degree of moisture, in certain buoyant insect larvae.

trachelate a. [Gk. *trachēlos*, neck] narrowed, as in neck-formation.

trachelomastoid a. [Gk. *trachēlos*, neck; *mastos*, breast; *eidos*, form] pert. neck region and mastoid process, appl. muscle, longissimus capitis.

trachenchyma n. [L.L. *trachia*, windpipe; Gk. *engchyma*, infusion] tracheal vascular tissue.

tracheobronchial a. [L.L. *trachia*, windpipe; Gk. *brongchos*, bronchial tube] pert. trachea and bronchi, appl. lymph glands; appl. a syrinx formed of lower end of trachea and upper bronchi.

tracheole n. [L.L. *trachia*, windpipe] an ultimate branch of tracheal system.

Tracheophyta, tracheophytes n., n.plu. [L.L. *trachia*, windpipe; Gk. *phyton*, plant] in some classifications, a major division of plants comprising those with vascular tissue, being the Pteridophyta and Spermatophyta, commonly called vascular plants.

Trachomedusae Trachymedusae, q.v.

trachychromatic n. [Gk. *trachys*, rugged; *chrōma*, colour] staining or stained deeply; *cf.* amblychromatic.

trachyglossate a. [Gk. *trachys*, rough; *glōssa*, tongue] with rasping or toothed tongue.

Trachylina n. (Gk. *trachys*, rough] Trachymedusae, *q.v.*, or the Trachymedusae and Narcomedusae.

Trachymedusae n. [Gk. *trachys*, rough; *Medousa*, Medusa] an order of hydrozoans having no polyp phase and the medusa with solid and/or hollow tentacles; *alt.* Trachomedusae and Trachylina, or in some classifications included with the Narcomedusae in the Trachylina.

Trachysomata n. [Gk. *trachys*, rough; *sōma*, body] in some classifications, an order of eel-like lepospondyl amphibians including the sirens.

tract n. [L. *tractus*, region] a region or area or system considered as a whole, as alimentary tract; a band, bundle, or system of nerve fibres; pteryla, *q.v.*

tractellum n. [L. *trahere*, to draw] a flagellum of forward end of Mastigophora, or of zoospores, with circumductory motion.

tragus n. [Gk. *tragos*, goat] a small pointed eminence in front of concha of ear; its hair; *alt.* antilobium.

trama n. [L. *trama*, woof] a central core of interwoven hyphae of a fungal gill; *alt.* dissepiment.

tramal in, from, or *pert.* trama.

transacetylase any of the enzymes catalysing the transfer of acetyl groups, EC sub-subgroup 2.3.1, particularly one in bacteria concerning the conversion of acetyl CoA to acetyl phosphate, EC 2.3.1.8, *r.n.* phosphate acetyltransferase.

transad adv. [L. *trans*, across; *ad*, to] *appl.* organisms of the same or closely related species which have become separated by an environmental barrier, as European and American reindeer.

transaminases a group of enzymes that catalyse transamination; *alt.* aminopherases, aminotransferases; EC sub-subgroup 2.6.1.

transamination n. [L. *trans*, across; Gk. *ammōniakon*, resinous gum] transfer of amino (NH_2) groups to another molecule.

transapical a. [L. *trans*, across; *apex*, summit] *appl.* transverse axis and plane of diatom valve.

trans-configuration the situation where 2 mutants at different sites in the same cistron are on different chromosomes; one of 2 configurations of a molecule caused by the limitation of rotation around a double bond, the other configuration being the *cis*-configuration; *cf.* *cis*-configuration.

transcriptase an enzyme catalysing transcription; *alt.* RNA polymerase, DNA-dependent RNA polymerase.

transcription n. [L. *transcribere*, to copy out] the synthesis of mRNA from DNA in the nucleus by matching the sequences of bases; *cf.* reverse transcription, translation.

transduction n. [L. *transducere*, to transfer] the process of conveying or carrying over; *appl.* the carrying of a gene from one micro-organism to another by a virus, especially a bacteriophage.

transect n. [L. *trans*, across; *secare*, to cut] a line, strip, or profile, as of vegetation, chosen for studying and charting.

transection n. [L. *trans*, across; *sectio*, to cut]

section across a longitudinal axis; *alt.* cross-section, transverse section; *cf.* longisection.

transeptate a. [L. *trans*, across; *septum*, partition] having transverse septa.

transferase n. [L. *transferre*, to carry across; *-ase*] any enzyme which catalyses the transfer of a radical or group of atoms from one molecule to another as in various biosyntheses; EC group 2.

transfer RNA (tRNA) a small RNA molecule which places an amino acid in the correct place along a molecule of mRNA, there being a different kind of tRNA for each of the fundamental amino acids; *alt.* soluble RNA, acceptor RNA, adaptor RNA.

transformation n. [L. *transformare*, to change in shape] the inherited modification of one strain of bacterium by DNA isolated from another strain; change in form as in metamorphosis; metabolism, *q.v.*

transfusion tissue tissue of gymnosperm leaves, consisting of parenchymatous and tracheidal cells.

transgenation point mutation, *q.v.*

transhydrogenase formerly, one of a group of enzymes which catalyse the transfer of hydrogen from one molecule to another, now placed in EC group 1.

transient a. [L. *transire*, to pass by] passing; of short duration; *appl.* orange: the unstable product of bleaching rhodopsin, and transformed into indicator yellow.

transient polymorphism the coexistence of 2 or more distinct types of individuals in the same breeding population only for a short while, then one type replacing the other; *cf.* balanced polymorphism.

transilient a. [L. *transilire*, to leap over] *appl.* nerve fibres connecting brain convolutions not adjacent. *n.* A mutation.

transitional a. [L. *transire*, to go across] *appl.* epithelium occurring in ureters and urinary bladder renewing itself by mitotic division of 3rd and innermost layer of cells; *appl.* cell: endotheliocyte, *q.v.*; *appl.* inflorescence intermediate between racemose and cymose.

translation n. [L. *translatio*, transferring] the synthesis of a polypeptide chain from mRNA at the ribosomes by matching the codons on mRNA with the anticodons on tRNA; *cf.* transcription.

translocation n. [L. *trans*, across; *locus*, place] removal to a different place or habitat; diffusion, as of food material; movement of material in solution, especially in phloem of a plant; change in position of a chromosome segment to another part of the same chromosome or of a different chromosome.

translocation quotient ratio of chemical content of shoot to that of root, a measure of mobility or relative translocation, e.g. of manganese.

transmedian a. [L. *trans*, across; *medius*, middle] *pert.* or crossing the middle plane.

transmitter neurotransmitter, *q.v.*

transmutation theory theory that one species can evolve from another.

transpalatine n. [L. *trans*, across; *palatus*, the palate] a cranial bone of crocodiles, connecting pterygoid with jugal and maxilla.

transpinalis n. [L. *trans*, across; *spina*, spine] a muscle connecting transverse processes of vertebrae.

transpiration n. [L. trans, across; spirare, to breathe] the evaporation of water through stomata.

transpiration stream or current the flow of water through the xylem of a plant from roots to the stem and leaves and evaporation off stomata of leaves, thought to be important in causing rise of water in the xylem.

transplant v. [L. trans, across; plantare, to plant] to transfer tissue from one part to another part of the body of the same or that of another individual. n. Tissue transferred to another part. Alt. graft.

transplantation antigen histocompatibility antigen, q.v.

transposon specialized DNA sequences which can jump or transpose on to any DNA molecule in the same cell.

transpyloric plane upper of imaginary horizontal planes dividing abdomen into artificial regions.

transtubercular a. [L. trans, across; tuberculum, shall hump] appl. plane of body through tubercles of iliac crests; alt. intertubercular.

transudate n. [L. trans, beyond; sudare, to sweat] any substance which has oozed through a membrane or pores.

transversal a. [L. transversus, across] lying across or between, as a transversal wall.

transverse a. [L. transversus, across] lying across or between, as artery, colon, ligament, process.

transversion n. [L. transversus, across] in DNA or RNA, the substitution of purine for pyrimidine or vice versa, in DNA resulting in a mutation.

transversum n. [L. transversus, across] in many reptiles, a cranial bone extending from pterygoid to maxilla.

transversus n. [L. transversus, across] a transverse muscle, as of abdomen, thorax, pinna, tongue, foot, perinaeum.

trapeziform a. [Gk. trapezion, small table; L. forma, shape] trapezium-shaped.

trapezium n. [Gk. trapezion, small table] the 1st carpal bone, at base of 1st metacarpal, alt. greater multangular bone; portion of pons Varolii.

trapezius a broad, flat, triangular muscle of neck and shoulders.

trapezoid a. [Gk. trapezion, small table; eidos, form] trapezium-shaped, appl. ligament, nucleus, ridge. n. The 2nd carpal bone at base of 1st metacarpal, alt. lesser multangular bone.

traumatic a. [Gk. trauma, wound] pert. or caused by, a wound or other injury; appl. an acid which stimulates healing of plant wounds.

traumatin n. [Gk. trauma, wound] a substance obtained from wounded bean pods and considered to be a wound hormone, which is effective in inducing cell division.

traumato- see also trauma-, traumo-.

traumatonasty n. [Gk. trauma, wound; nastos, close pressed] a nastic movement in response to wounding.

traumatropism n. [Gk. trauma, wound; trope, turn] a growth curvature in plants in response to wounds; a. traumatropic.

traumo- see also traumato-.

traumotaxis n. [Gk. trauma, wound; taxis, arrangement] reaction after wounding, as in nuclei and protoplasts.

tree n. [M.E. tre, tree] a woody plant with a single main trunk bearing lateral branches.

trefoil n. [L. trifolius, three-leaved] flower or leaf with 3 lobes.

trehalose n. [L. trehala, sugary substance secreted by beetles] a disaccharide as found in lichens and in insect blood; alt. mycose.

Trematoda n. [Gk. trematodes, having holes] a class of endoparasitic flatworms commonly called flukes, having a leaf-like body covered by a tough cuticle and with hooks, suckers, or both.

Tremellales n. [Tremella, generic name] an order of jelly fungi having a basidium in which the hypobasidium becomes vertically divided into cells, each of which develops an epibasidium; in some classifications includes all jelly fungi.

tremelloid, tremellose a. [L. tremere, to tremble] gelatinous in substance or appearance.

trephocyte n. [Gk. trephein, to nourish; kytos, hollow] a cell which forms and stores substances for nourishing adjacent cells, as nurse cell, mast cell, plasmacyte, sustentacular cell.

trephones n.plu. [Gk. trephein, to nourish] nutritive substances formed on breaking down of cells and which stimulate cell division.

treption n. [Gk. treptos, turned] an adaptation in a cell in response to a change in its environment.

TRH TSH-releasing hormone, q.v.

triactinal a. [Gk. tria, three; aktis, ray] three-rayed.

triadelphous a. [Gk. tria, three; adelphos, brother] having stamens united by their filaments into 3 bundles.

triaene n. [Gk. triaina, trident] a somewhat trident-shaped spicule.

trial-and-error conditioning or learning a kind of learning in which a random and spontaneous response becomes associated with a particular stimulus, because that response has always produced a reward whereas other responses have not done so; alt. habit formation, operant conditioning, instrumental conditioning.

triallelic a. [Gk. tria, three; allelon, one another] appl. polyploid with 3 different alleles at a locus.

triandrous a. [Gk. tria, three; aner, man] having 3 stamens.

triangularis n. [L. triangularis, three-cornerd] muscle from mandible to lower lip, which pulls down corner of mouth, depressor anguli oris; muscle and tendinous fibres between dorsal surface of sternum and costal cartilages, transversus thoracis, which assists expiration.

trianthous a. [Gk. tria, three; anthos, flower] having 3 flowers.

triarch n. [Gk. tria, three; arche, elements] with 3 alternating xylem and phloem groups; with 3 vascular bundles.

triarticulate a. [L. tres, three; articulus, joint] three-jointed.

Triassic a. [Gk. tria, three] appl. the early period of the Mesozoic era; n. Trias.

triaster n. [Gk. tria, three; aster, star] three chromatin masses resulting from tripolar mitosis, as in cancer cells.

triaxon n. [Gk. tria, three; axon, axle] a sponge spicule with 3 axes.

tribe n. [L. tribus, tribe] in classification, a subdivision of a family and differing in minor characters from other tribes.

triboloid a. [Gk. tribolos, burr; eidos, form] like a burr; prickly; echinulate, q.v.

triboluminescence

triboluminescence *n.* [Gk. *tribein*, to rub; L. *luminescere*, to grow light] luminescence produced by friction.

Tribonematales *n.* [*Tribonema*, generic name] Heterotrichales, *q.v.*

tribracteate *a.* [L. *tres*, three; *bractea*, thin plate of metal] with 3 bracts.

trica *n.* [F. *tricoter*, to knit] a lichen apothecium with ridged spherical surface.

tricarboxylic acid cycle (TCA cycle) Krebs' cycle, *q.v.*

tricarpellary *a.* [Gk. *tria*, three; *karpos*, fruit] with 3 carpels.

tricentric *a.* [Gk. *tria*, three; *kentron*, centre] having 3 centromeres, *appl.* chromosomes.

triceps *n.* [L. *tres*, three; *caput*, head] *appl.* a muscle with 3 heads or insertions; *a.* tricipital.

Trichiales *n.* [*Trichia*, generic name] an order of stalked or sessile slime moulds.

trichidium *n.* [Gk. *thrix*, hair; *idion*, dim.] *see* sterigma.

trichilium *n.* [Gk. *thrix*, hair; *ilē*, crowd] a pad of matted hairs at base of certain leaf petioles.

trichimaera *n.* [Gk. *tria*, three; L. *chimaera*, monster] a periclinal chimaera in which all 3 tissue layers are genetically different from one another.

trichites *n.plu.* [Gk. *thrix*, hair] fine rod-like extrusible structures in mouth region of certain ciliates; siliceous spicules in certain sponges; hypothetical amylose crystals constituting a starch granule.

trichoblast *n.* [Gk. *thrix*, hair; *blastos*, bud] a cell of plant epidermis, which develops into a root hair.

trichobothrium *n.* [Gk. *thrix*, hair; *bothros*, pit] a conical protuberance with sense hair, on each side of anal segment in certain myriapods; a vibratory sense hair or setula in spiders.

trichocarpous *a.* [Gk. *thrix*, hair; *karpos*, fruit] with hairy fruits.

trichocutis *n.* [Gk. *thrix*, hair; L. *cutis*, skin] cutis of a stipe, formed by coherent hairs or filaments of trichoderm.

trichocyst *n.* [Gk. *thrix*, hair; *kystis*, bladder] an oval or spindle-shaped protrusible body found in ectoplasm of ciliates and dinoflagellates.

trichoderm *n.* [Gk. *thrix*, hair; *derma*, skin] a filamentous outer layer of pileus and stipe of agarics; *cf.* epitrichoderm.

trichodragmata *n.plu.* [Gk. *thrix*, hair; *dragma*, sheaf] straight, fine hair-like spicules in bundles; *sing.* trichodragma.

trichogen *n.* [Gk. *thrix*, hair; *-genēs*, producing] a seta-producing cell in some arthropods.

trichogyne *n.* [Gk *thrix*, hair; *gynē*, woman] an elongated hair-like receptive cell at the end of the female sex organ in some fungi, lichens, and in some red and green algae, which in some cases may receive the male gamete.

trichohyalin *n.* [Gk. *thrix*, hair; *hyalos*, glass] a substance resembling eleidin, in granules in Huxley's layer of hair follicle.

trichoid *a.* [Gk. *thrix*, hair; *eidos*, form] hair-like; *appl.* a type of tactile sensilla in insects; *alt.* piliform.

trichome *n.* [Gk. *trichōma*, growth of hair] an outgrowth of plant epidermis, either hairs or scales; a hair tuft; a line of cells and their surrounding sheath forming the filament of blue-green algae; *alt.* trichoma.

Trichomycetes *n.* [Gk. *thrix*, hair; *mykēs*, fungus] a class of fungi which live as commensals on arthropods, having a filamentous thallus, and asexual reproduction by trichospores, arthrospores, or amoeboid cells.

trichophore *n.* [Gk. *thrix*, hair; *pherein*, to bear] a group of cells bearing a trichogyne; chaetigerous sac of annelids.

trichopore *n.* [Gk. *thrix*, hair; *poros*, channel] opening for an emerging hair or bristle, as in spiders.

Trichoptera *n.* [Gk. *thrix*, hair; *pteron*, wing] an order of insects commonly called caddis flies, that have complete metamorphosis and aquatic larvae which live in cases.

trichosclereid *n.* [Gk. *thrix*, hair; *sklēros*, hard] a sclereid with thin hair-like branches extending into intercellular spaces.

trichosiderin *n.* [Gk. *thrix*, hair; *sidēros*, iron] iron-containing red pigment isolated from human red hair.

trichosis *n.* [Gk. *thrix*, hair] distribution of hair; abnormal hair growth.

trichospore *n.* [Gk. *thrix*, hair; *sporos*, seed] zoospore, *q.v.*

Trichostomatida, Trichostomata *n.* [Gk. *thrix*, hair; *stoma*, mouth] an order of Holotricha having the mouth at the base of a funnel-like depression.

Trichosyringida *n.* [Gk. *thrix*, hair; *syringx*, pipe] an order of Aphasmidia having adults free living in the soil and larvae parasitic in arthropods; *alt.* Mermithoidea.

trichothallic *a.* [Gk. *thrix*, hair; *thallos*, young shoot] having a filamentous thallus; *appl.* growth of a filamentous thallus in certain algae by division of intercalary meristematic cells at the base of a terminal hair.

trichotomous *a.* [Gk. *tricha*, threefold; *tomē*, cutting] divided into 3 branches.

trichroic *a.* [Gk. *tria*, three; *chrōs*, colour] showing 3 different colours when seen in 3 different aspects.

trichromatic, trichromic *a.* [Gk. *tria*, three; *chrōma*, colour] *pert.* or able to perceive, the 3 primary colours.

Trichurata *n.* [*Trichuris*, generic name] an order of Aphasmidia which are parasites of vertebrates and have a stichosome.

tricipital *a.* [L. *tres*, three; *caput*, head] having 3 heads or insertions, as triceps; *pert.* triceps.

Tricladida, triclads *n.*, *n.plu.* [Gk. *tria*, three; *klados*, sprout] an order of elongated turbellarians commonly called planarians, having an intestine with 3 main branches and well-developed sense organs.

Tricoccae *n.* [Gk. *tria*, three; *kokkos*, kernel] Euphorbiales, *q.v.*

tricoccous *a.* [Gk. *tria*, three; *kokkos*, kernel] *appl.* a 3-carpel fruit.

triconodont *a.* [Gk. *tria*, three; *kōnos*, cone; *odous*, tooth] *appl.* tooth with 3 crown prominences in a line parallel to jaw axis.

Triconodonta, triconodonts *n.*, *n.plu.* [Gk. *tria*, three; *kōnos*, cone; *odous*, tooth] an order of Jurassic and Cretaceous eotherians having teeth with 3 cusps in a straight line, one of the subgroups of which may be true therians; in some classifications considered to be prototherians.

tricostate *a.* [L. *tres*, three; *costa*, rib] with 3 ribs.

tricotyledonous *a.* [Gk. *tria*, three; *kotylēdōn*, cup-like hollow] with 3 cotyledons.

tricrotic *a.* [Gk. *tria*, three; *krotein*, to beat] having a triple beat in the arterial pulse.

tricrural *a.* [L. *tres*, three; *crus*, leg] with 3 branches.

tricuspid *a.* [L. *tres*, three; *cuspis*, point] three-pointed; *appl.* triangular valve of heart; trituber-cular, *q.v.*

tricuspidate *a.* [L. *tres*, three; *cuspis*, point] having 3 points, *appl.* leaf.

tridactyl(e) *a.* [Gk. *tria*, three; *daktylos*, finger] having 3 digits; with 3 jaws, *appl.* pedicellariae.

tridentate *a.* [L. *tridens*, three-pronged] having 3 tooth-like divisions.

tridynamous *a.* [Gk. *tria*, three; *dynamis*, power] with 3 long and 3 short stamens.

trifacial *a.* [L. *tres*, three; *facies*, *face*] *appl.* 5th cranial nerve, the trigeminal.

trifarious *a.* [L. *trifarius*, of three sorts] in groups of 3; of 3 kinds; in 3 rows, *alt.* triserial; having 3 surfaces.

trifid *a.* [L. *trifidus*, three-forked] cleft to form 3 lobes.

triflagellate *a.* [L. *tres*, three; *flagellum*, whip] having 3 flagella.

trifoliate *a.* [L. *tres*, three; *folium*, leaf] having 3 leaves growing from same point; *cf.* triphyllous.

trifoliolate *a.* [L. *tres*, three; *dim.* of *folium*, leaf] with 3 leaflets growing from same point; *alt.* ternate.

trifurcate *a.* [L. *trifurcatus*, three-forked] with 3 forks or branches.

trigamma *n.* [Gk. *tria*, three; γ, gamma] three-pronged forked wing venation in Lepidoptera.

trigamous *a.* [Gk. *tria*, three; *gamos*, marriage] *appl.* flower head with staminate, pistillate, and hermaphrodite flowers.

trigeminal *a.* [L. *trigeminus*, triplet] consisting of, or *pert.*, 3 structures; *appl.* 5th cranial nerve, with ophthalmic, maxillary, and mandibular divisions, *alt.* facial; *appl.* arrangement of pairs of pores in 3 rows in ambulacra of some echinoids.

trigeneric *a.* [L. *tres*, three; *genus*, race] *pert.* or derived from 3 genera.

trigenetic *a.* [Gk. *tria*, three; *genesis*, origin] requiring 3 different hosts in the course of a life cycle.

trigenic *a.* [Gk. *tria*, three; *genos*, descent] *pert.* or controlled by 3 genes.

triglyceride *n.* [Gk. *tria*, three; *glykys*, sweet] any of various esters of glycerol and fatty acids, which are important components of animal and plant fats and oils.

trigon *n.* [Gk. *tria*, three; *gōnia*, angle] triangle of cusps of upper jaw molar teeth.

trigonal *a.* [Gk. *trigōnos*, triangular] ternary or triangular when *appl.* symmetry with 3 parts to a whorl; triangular in section, *appl.* stems; *alt.* trigonous.

trigone *n.* [Gk. *trigōnon*, triangle] a small triangular space, as olfactory trigone, trigonum vesicae, etc.; *alt.* trigonum.

trigonelline *n.* [*Trigonella*, generic name of fenugreek] a betaine found in legume seeds, potato, and dahlia tubers.

trigonid *n.* [Gk. *trigōnon*, triangle] triangle of cusps of lower molar teeth.

trigonous trigonal, *q.v.*

trigonum *n.* [Gk. *trigōnon*, triangle] trigone, *q.v.*; os trigonum: posterior process of talus forming a separate ossicle.

trigynous *a.* [Gk. *tria*, three; *gynē*, woman] having 3 pistils or styles; having 3 carpels to the gynaecium.

triheterozygote *n.* [Gk. *tria*, three; *heteros*, other; *zygōtos*, yoked together] an organism heterozygous for 3 genes.

trihybrid *n.* [L. *tres*, three; *hibrida*, mixed offspring] a cross whose parents differ in 3 distinct characters. *a.* Heterozygous for 3 pairs of alleles.

tri-iodothyronine a iodine-containing hormone derived from the amino acid tyrosine produced by the thyroid gland.

trijugate *a.* [L. *tres*, three; *jugum*, yoke] having 3 pairs of leaflets.

trilabiate *a.* [L. *tres*, three; *labium*, lip] with 3 lips.

trilacunar *a.* [L. *tres*, three; *lacuna*, cavity] with 3 lacunae; having 3 leaf gaps, *appl.* nodes.

trilobate *a.* [Gk. *tria*, three; *lobos*, lobe] three-lobed.

Trilobita, Trilobitomorpha, trilobites *n.*, *n.plu.* [*Trilobites*, generic name; Gk. *morphē*, form] a subphylum, or in some classifications a class of arthropods, which are known only as fossils and have a body divided into 3 regions or lobes.

trilocular *a.* [L. *tres*, three; *loculus*, compartment] having 3 compartments or loculi.

trilophodont *a.* [Gk. *tria*, three; *lophos*, crest; *odous*, tooth] having 3-crested teeth.

trilophous *a.* [Gk. *tria*, three; *lophos*, crest] *appl.* rayed spicule with 3 rays branched or ridged.

trimerous *a.* [Gk. *tria*, three; *meros*, part] composed of 3 or multiples of 3, as parts of flower; *appl.* tarsi of certain beetles.

trimethylamine oxide the excretory product of some marine fishes.

trimitic *a.* [Gk. *tria*, three; *mitos*, thread] having 3 kinds of hyphae: supporting, connective, and reproductive; *cf.* dimitic, monomitic.

trimonoecious *a.* [Gk. *tria*, three; *monos*, alone; *oikos*, house] with male, female, and hermaphrodite flowers on the same plant.

trimorphic *a.* [Gk. *tria*, three; *morphē*, form] having 3 different forms; *appl.* a polymorphic species with 3 different types of individual; with stamens and pistils of 3 different lengths.

trimorphism *n.* [Gk. *tria*, three; *morphē*, form] occurrence of 3 distinct forms or forms of organs in one life cycle or in one species; trimorphous condition.

trinervate *a.* [L. *tres*, three; *nervus*, sinew] having 3 veins or ribs running from base to margin of leaf.

trinomial *a.* [L. *tres*, three; *nomen*, name] consisting of 3 terms, as names of subspecies; *cf.* binomial, trionym, multinomial.

triod *n.* [Gk. *triodos*, meeting of three roads] a 3-rayed or triactinal spicule in sponges.

trioecious *a.* [Gk *tria*, three; *oikos*, house] producing male, female, and hermaphrodite forms on different plants; *alt.* trioikous.

trioikous trioecious, *q.v.*

trionym *n.* [Gk. *tria*, three; *onyma*, name] a name with 3 terms; *alt.* trinomial name.

trioses

trioses *n.plu.* [Gk. *tria*, three] monosaccharides having the formula $C_3H_6O_3$.

triosseum *a.* [L. *tres*, three; *ossa*, bones] *appl.* foramen, the opening between coracoid, clavicle and scapula.

triovulate *a.* [L. *tres*, three; *ovum*, egg] having 3 ovules.

tripartite *a.* [L. *tres*, three; *partitus*, separated] divided into 3 parts, as a leaf.

tripetalous *a.* [Gk. *tria*, three; *petalon*, leaf] having 3 petals.

triphosphopyridine nucleotide TPN, *q.v.*

triphyllous *a.* [Gk. *tria*, three; *phyllon*, leaf] three-leaved; *alt.* trifoliate.

tripinnate *a.* [L. *tres*, three; *pinna*, feather] divided pinnately 3 times.

tripinnatifid *a.* [L. *tres*, three; *pinna*, feather; *findere*, to cleave] divided 3 times in a pinnatifid manner.

tripinnatisect *a.* [L. *tres*, three; *pinna*, feather; *secare*, to cut] thrice pinnatisect, i.e. 3 times lobed with divisions nearly to midrib.

triplechinoid *see* diadematoid.

triple-nerved *appl.* a leaf with 3 prominent veins; *alt.* triplinerved.

triplet in DNA or RNA, a codon made of 3 bases which code for a particular amino acid.

triplet code the genetic code where each codon is a triplet.

triplex *a.* [L. *triplex*, threefold] having 3 dominant genes, in polyploidy.

triplicostate *a.* [L. *triplus*, triple; *costa*, rib] having 3 ribs.

triplinerved triple nerved, *q.v.*

triploblastic *a.* [Gk. *triploos*, triple; *blastos*, bud] with 3 primary germinal layers: epiblast, mesoblast, hypoblast.

triplocaulescent *a.* [L. *triplus*, triple; *caulis*, stalk] having axes of the 3rd order, i.e. having a main stem with branches which are branched.

triploid *a.* [Gk. *triploos*, threefold] with 3 times the normal haploid number of chromosomes. *n.* A triploid organism or cell.

triplostichous *a.* [Gk. *triploos*, threefold; *stichos*, row] arranged in 3 rows, as of cortical cells on small branches of *Chara* (Charales); *appl.* eyes with preretinal, retinal, and postretinal layers, as of larval scorpion.

tripod *n.* [Gk. *tria*, three; *pous*, foot] a tripod-shaped or 3-legged spicule.

tripolar *a.* [Gk. *tria*, three; *polos*, axis] *appl.* division of chromatin to 3 poles in diseased cells, instead of normal 2 poles in mitosis.

tripolite *n.* [*Tripoli* in North Africa; Gk. *lithos*, stone] siliceous deposit formed mainly of frustules of diatoms; *alt.* diatomaceous earth, infusorial earth, kieselguhr.

tripton *n.* [Gk. *triptos*, pounded] non-living seston; *alt.* abioseston, trypton.

tripus *n.* [L. *tripus*, tripod] posterior Weberian ossicle, adjoining air bladder; trifurcation of coeliac artery into left gastric or coronary, hepatic, and splenic arteries, tripus Halleri.

triquetral, triquetrous *a.* [L. *triquetrus*, three-cornered] *appl.* stem with 3 angles and 3 concave faces; *appl.* 3-cornered or wedge-shaped bone: triquetrum, *q.v.*

triquetrum *n.* [L. *triquetrum*, triangle] the cuneiform carpal bone, *alt.* triquetral or Wormian bone.

triquinate *a.* [L. *tres*, three; *quini*, five each] divided into 3, with each lobe again divided into 5.

triradial *a.* [L. *tres*, three; *radius*, ray] having 3 branches as radii from one centre; triradiate, *appl.* orbital sulcus; triactinal, *appl.* spicules.

triradiate *appl.* pelvic girdle consisting of pubis, ilium, and ischium; *cf.* tetraradiate.

triramous, triramose *a.* [L. *tres*, three; *ramus*, branch] divided into 3 branches.

trisaccharide *n.* [Gk. *tria*, three; *sakchar*, sugar] a carbohydrate made of 3 molecules of monosaccharide with the elimination of water, and including raffinose.

trisepalous *a.* [Gk. *tria*, three; F. *sépale*, sepal] having 3 sepals.

triseptate *a.* [L. *tres*, three; *septum*, partition] having 3 septa.

triserial *a.* [L. *tres*, three; *series*, row] arranged in 3 rows, *alt.* trifarious; having 3 whorls.

trisomic *a.* [Gk. *tria*, three; *sōma*, body] *pert.*, or having, 3 homologous chromosomes. *n.* An organism with 3 chromosomes of one type.

trisomy the trisomic condition.

trispermous *a.* [Gk. *tria*, three; *sperma*, seed] having 3 seeds.

trisporic, trisporous *a.* [Gk. *tria*, three; *sporos*, seed] having 3 spores.

tristachyous *a.* [Gk. *tria*, three; *stachys*, ear of corn] with 3 spikes.

tristichous *a.* [Gk. *tria*, three; *stichos*, row] arranged in 3 vertical rows.

tristyly *n.* [Gk. *tria*, three; *stylos*, pillar] the condition of having short, medium-length, and long styles.

triternate *a.* [L. *tres*, three; *terni*, three each] thrice ternately divided.

tritibial *n.* [L. *tres*, three; *tibia*, shin] compound ankle bone formed when centrale unites with talus.

tritoblasts *n.plu.* [Gk. *tritos*, third; *blastos*, bud] a generation of Neosporidia produced by deutoblasts and in turn giving rise to teloblasts.

tritocerebrum *n.* [Gk. *tritos*, third; L. *cerebrum*, brain] third lobe of insect brain indicated during development; part of brain of higher Crustacea, consisting of antennal nerve centres; *alt.* tritocerebron, metacerebrum.

tritocone *n.* [Gk. *tritos*, third; *kōnos*, cone] premolar cusp.

tritonymph *n.* [Gk. *tritos*, third; *nymphē*, chrysalis] developmental stage or instar following the deutonymph in Acarina.

tritor *n.* [L. *tritor*, grinder] grinding surface of a tooth.

tritosternum *n.* [Gk. *tritos*, third; *sternon*, chest] sternite of 3rd segment of prosoma or 1st segment of podosoma in Acarina.

tritozooid *n.* [Gk. *tritos*, third; *zōon*, animal; *eidos*, form] a zooid of 3rd generation.

tritubercular *a.* [L. *tres*, three; *tuberculum*, small hump] *appl.* molar teeth with 3 cusps; *alt.* tricuspid.

Trituberculata, trituberculates *n.*, *n.plu.* [L. *tres*, three; *tuberculum*, small hump] an infraclass of Mesozoic therian mammals known mainly from remains of jaws and teeth, that are forerunners of the living therians.

trituberculy *n.* [L. *tres*, three; *tuberculum*, small hump] theory of molar tooth development.

triungulin *n.* [L. *tres*, three; *ungula*, claw] small,

6-legged larva of Strepsiptera and Meloidae (blister and oil beetles); *alt.* triungulus.

Triuridales *n.* [*Triuris,* generic name] an order of saprophytic monocots having scale leaves and small, long-stalked flowers.

trivalent *n.* [L. *tres,* three; *valere,* to be strong] association of 3 chromosomes held together by chiasmata between dipolotene and metaphase of 1st division in meiosis. *a. Appl.* amboceptor which can bind 3 different complements.

trivium *n.* [L. *trivium,* crossroad] the 3 rays of certain echinoderms farthest from madreporite; *cf.* bivium.

trixenous *a.* [Gk. *tria,* three; *xenos,* host] of a parasite, having 3 hosts.

trizoic *a.* [Gk. *tria,* three; *zōon,* animal] *appl.* protozoan spore containing 3 sporozoites.

tRNA transfer RNA, *q.v.*

troch *n.* [Gk. *trochos,* wheel] a circlet or segmental band of cilia of a trochophore.

trochal *a.* [Gk. *trochos,* wheel] wheel-shaped; *appl.* anterior disc of Rotifera.

trochantellus *n.* [Gk. *trochantēr,* runner] a segment of leg between trochanter and femur, in some insects.

trochanter *n.* [Gk. *trochantēr,* runner] processes or prominences at upper end of thigh bone— greater (major), lesser (minor) and third (tertius); small segment of leg between coxa and femur in insects and spiders.

trochanteric fossa a deep depression on medial surface of neck of femur.

trochantin *n.* [Gk. *trochantēr,* runner] a small sclerite at base of coxa of insect leg; sclerite for articulation of mandible in Orthoptera; lesser trochanter.

trochate *a.* [Gk. *trochos,* wheel] having a wheel-like structure; wheel-shaped; *alt.* trochiferous, trochiform.

Trochelminthes, trochelminths *n., n.plu.* [Gk. *trochos,* wheel; *helmins,* worm] in some classifications, a phylum including Rotifera, Gastrotricha, and a few obscure types.

trochite *n.* [Gk. *trochos,* wheel] segment or joint of stem of Crinoidea.

trochlea *n.* [Gk. *trochilia,* pulley] a pulley-like structure, especially one through which a tendon passes, as of humerus, femur, orbit.

trochlear shaped like a pulley; *pert.* trochlea; *appl.* nerve: 4th cranial nerve to superior oblique muscle of eye, *alt.* pathetic nerve.

trochoblasts *n.plu.* [Gk. *trochos,* wheel; *blastos,* bud] portions of segmenting egg destined to become prototroch of a trochosphere.

Trochodendrales *n.* [*Trochodendron,* generic name] a very small order of dicots (Hamamelididae) in which vessels are usually absent.

trochoid *a.* [Gk. *trochos,* wheel; *eidos,* form] wheel-shaped; capable of rotating motion, as a pivot joint.

trochophore, trochosphere *n.* [Gk. *trochos,* wheel; *phora,* bearing; *sphaira,* globe] free-swimming pelagic larval stage of annelids, polyzoans, and some molluscs having a preoral whorl of cilia; *alt.* telotrocha, teleotrocha.

trochus *n.* [Gk. *trochos,* wheel] inner, anterior, coarser ciliary zone of rotifer discs; *cf.* cingulum.

troglobiont *n.* [Gk. *trōglē,* hole; *biōnai,* to live] any organism living in caves only.

Trogoniformes *n.* [*Trogon,* generic name; L.

forma, shape] an order of Neornithes, including the trogons, being very soft-plumaged arboreal birds.

tropeic *a.* [Gk. *tropis,* keel] cariniform, *q.v.*

trophagones *n.plu.* [Gk. *trophē,* nourishment; *agein,* to lead] xenagones whose action on the host results in the carriage of food substances towards the parasite's habitat.

trophallaxis *n.* [Gk. *trophē,* nourishment; *allaxis,* interchange] interchange of food between larvae and adults in certain insects, especially social insects, where it bonds insect societies together; *alt.* reciprocal feeding, oecotrophobiosis.

trophamnion *n.* [Gk. *trophē,* nourishment; *amnion,* foetal membrane] sheath around developing egg of some insects, and passing nourishment to the embryo.

trophectoderm *n.* [Gk. *trophē,* nourishment; *ektos,* outside; *derma,* skin] outer layer of mammalian blastocyst; trophoblast, *q.v.*

trophi *n.plu.* [Gk. *trophē,* nourishment] hard chitinous chewing organs of rotifers; mouth-parts of arthropods, especially insects; mandibles and maxillae collectively.

trophic *a.* [Gk. *trophē,* nourishment] *pert.,* or connected with, nutrition, *appl.* nerves, stimuli, enlargement, etc.; *appl.* hormones influencing activity of endocrine glands and growth, as those secreted by the anterior lobe of the hypophysis; *appl.* nucleus: trophonucleus, *q.v.*

trophic level a division of a food chain defined by the method of obtaining food, as primary producer, primary, secondary, or tertiary consumer.

trophidium *n.* [Gk. *trophē,* brood; *idion, dim.*] the 1st larval stage of certain ants.

trophifer, trophiger *n.* [Gk. *trophē,* nourishment; L. *ferre, gerere,* to carry] posterolateral region of insect head with which mouthparts articulate.

trophobiosis *n.* [Gk. *trophē,* nourishment; *biōsis,* a living] the life of ants in relation to their nutritive organisms, as to certain fungi and insects.

trophoblast *n.* [Gk. *trophē,* nourishment; *blastos,* bud] the outer layer of cells of epiblast, or of morula; trophoderm, *q.v.; alt.* trophectoderm; *cf.* formative cell.

trophochromatin *n.* [Gk. *trophē,* nourishment; *chrōma,* colour] vegetative chromatin, that which regulates metabolism of the cell; *cf.* idiochromatin.

trophochrome *a.* [Gk. *trophē,* nourishment; *chrōma,* colour] *appl.* cells with secretory granules giving staining reaction for mucus.

trophochromidia *n.plu.* [Gk. *trophē,* nourishment; *chrōma,* colour] vegetative chromidia, especially those concerned with nutritive process; *cf.* idiochromidia.

trophocyst *n.* [Gk. *trophē,* nourishment; *kystis,* bag] primordial structure giving rise to a sporangiophore, as in the fungus *Pilobolus.*

trophocytes *n.plu.* [Gk. *trophē,* nourishment; *kytos,* hollow] cells providing nutritive material for other cells, e.g. for archaeocytes, or fat-cells used in insect development; *alt.* nurse cells.

trophoderm *n.* [Gk. *trophē,* nourishment; *derma,* skin] outer layer of chorion; trophectoderm with a mesodermal cell layer; *alt.* trophoblast.

trophodisc *n.* [Gk. *trophē,* nourishment; *diskos,* plate] female gonophore of certain Hydrozoa.

trophogenic *a.* [Gk. *trophē,* nourishment; *-genēs,*

trophogone

producing] due to food or feeding, *appl.* characters in social Hymenoptera.

trophogone *n.* [Gk. *trophē*, nourishment; *gonē*, seed] a nutritive organ in Ascomycetes, considered as an antheridium which has lost its normal function.

trophology *n.* [Gk. *trophē*, nourishment; *logos*, discourse] the science of nutrition.

trophonemata *n.plu.* [Gk. *trophē*, nourishment; *nēma*, thread] uterine villi or hair-like projections which transfer nourishment to embryo through spiracle of elasmobranchs; villi, *q.v.*; *sing.* trophonema.

trophont *n.* [Gk. *trephein*, to feed; *on*, being] growth stage in Holotricha.

trophonucleus *n.* [Gk. *trophē*, nourishment; L. *nucleus*, kernel] meganucleus, *q.v.*; the nucleus of a haemoflagellate; *alt.* trophic nucleus; *cf.* kinetonucleus.

trophophore *n.* [Gk. *trophē*, nourishment; *pherein*, to bear] in sponges, an internal bud or group of cells destined to become a gemmule.

trophophyll *n.* [Gk. *trophē*, nourishment; *phyllon*, leaf] a sterile or foliage leaf of certain pteridophytes; *cf.* sporophyll.

trophoplasm *n.* [Gk. *trophē*, nourishment; *plasma*, mould] vegetative or nutritive part of protoplasm; *cf.* idioplasm.

trophoplast *n.* [Gk. *trophē*, nourishment; *plastos*, moulded] a cell, nucleated or not; plastid, *q.v.*

trophosome *n.* [Gk. *trophē*, nourishment; *sōma*, body] the nutritive polypoid persons of a hydroid colony.

trophospongia *n.* [Gk. *trophē*, nourishment; *sponggia*, sponge] spongy vascular layer of mucous membrane between uterine wall and trophoblast.

trophospongium *n.* [Gk. *trophē*, nourishment; *sponggia*, sponge] canalization of nerve cells, canaliculi occupied by branching processes of neuroglia cells.

trophotaeniae *n.plu.* [Gk. *trophē*, nourishment; *tainia*, ribbon] embryonic rectal processes, for absorption of nutritive substances from ovarian fluid, in Goodeidae and certain other fishes.

trophotaxis *n.* [Gk. *trophē*, nourishment; *taxis*, arrangement] response to stimulation by an agent which may serve as food.

trophothylax *n.* [Gk. *trophē*, nourishment; *thylax*, sack] food pocket on 1st abdominal segment of certain ant larvae.

trophotropism *n.* [Gk. *trophē*, nourishment; *tropē*, turn] tendency of a plant organ to turn towards food, or of an organism to turn towards a food supply.

trophozoite *n.* [Gk. *trophē*, nourishment; *zōon*, animal] the adult stage of a sporozoan.

trophozooid *n.* [Gk. *trophē*, nourishment; *zōon*, animal; *eidos*, form] gastrozooid, *q.v.*, of tunicates.

tropibasic *a.* [Gk. *tropē*, turn; *basis*, base] *appl.* chondrocranium with small hypophysial fenestra and common trabecula; *alt.* tropitrabic; *cf.* platybasic.

tropic *a.* [Gk. *tropikos*, pert. turn] *pert.* tropism, *appl.* movement of curvature in response to a directional or unilateral stimulus; having or *pert.* a directive influence, *appl.* hormones, as tropins; tropical, *appl.* regions.

tropine *n.* [Gk. *tropē*, turn] opsonin, *q.v.*; the base, $C_8H_{15}NO$, of atropine.

tropins pituitary hormones which have a tropic or trophic influence on other endocrine organs, melanophores, etc.

tropis *n.* [Gk. *tropē*, turn] in ostracods, a heavy chitinous rod connecting the zygum to the pastinum.

tropism *n.* [Gk. *tropē*, turn] a growth movement, usually curvature towards (positive), or away from (negative), the source of a stimulus in sessile animals and plants, *alt.* orienting curvature; in animal behaviour, formerly used for directional movements of animals when these movements were thought to be similar to plants; *cf.* taxis, kinesis.

tropitrabic *a.* [Gk. *tropē*, turn; L. *trabs*, beam] tropibasic, *q.v.*

tropocollagen *n.* [Gk. *tropos*, mode; *kolla*, glue; *gennaein*, to produce] a long molecule secreted in a fibrocyte, which, outside the cell, unites with others to form a collagen.

troponin and tropomyosin *n.* [Gk. *tropos*, mode; *mys*, muscle] proteins in striated muscle which inhibit contraction in the absence of calcium, tropomyosin consisting of 2 proteins, tropomyosin A being insoluble and B soluble in water.

tropophil(ous) *a.* [Gk. *tropos*, turn; *philos*, loving] tolerating alternating periods of cold and warmth, or of moisture and dryness; adapted to seasonal changes.

tropophyte *n.* [Gk. *tropos*, turn; *phyton*, plant] a changing plant, or one which is more-or-less hygrophilous in summer and xerophilous in winter; a plant growing in the tropics.

tropotaxis *n.* [Gk. *tropos*, turn; *taxis*, arrangement] a taxis in which an animal orients itself in relation to a source of stimulation by simultaneously comparing the amount of stimulation on either side of it by symmetrically placed sense organs; *cf.* klinotaxis, telotaxis.

trp tryptophane, *q.v.*

true corals the common name for the Scleractinia, *q.v.*

true flies the common name for the Diptera, *q.v.*

true ribs ribs which are directly connected with sternum.

true soil solum, *q.v.*

trumpet hyphae elongated cells with enlarged ends in contact with those of adjoining cells, and comparable to sieve tubes, as in medulla of thallus in the brown alga *Laminaria*.

truncate *a.* [L. *truncatus*, cut off] terminating abruptly, as if tapering end were cut off; *alt.* abrupt.

truncus trunk, *q.v.*

truncus arteriosus most anterior region of amphibian, or foetal, heart; through which blood is driven from ventricle.

trunk *n.* [F. *tronc*, from L. *truncus*, stem of tree] main stem of tree; body exclusive of head and extremities; main stem of a vessel or nerve; proboscis, as of elephant; *alt.* truncus.

trunk inductor posterior roof of archenteron in vertebrates; *alt.* spinocaudal inductor.

trunk legs pereiopods, *q.v.*

try tryptophan, *q.v.*

Tryblidioidea *n.* [*Tryblidium*, generic name; Gk.

eidos, form] the only order of the Monoplacophora, *q.v.*

tryma *n.* [Gk. *trymē*, hole] a drupe with separable rind and 2-halved endocarp with spurious dissepiments, as walnut.

trypanomonad *a.* [Gk. *trypan*, to bore; *monas*, unit] *appl.* phase in development of trypanosome while in its invertebrate host; *alt.* crithidial.

Trypanorhyncha *n.* [Gk. *trypan*, to bore; *rhyngchos*, snout] Tetrarhynchoidea, *q.v.*

trypanorhynchus *n.* [Gk. *trypan*, to bore; *rhyngchos*, snout] a spiniferous protrusible proboscis accompanying each bothridium in certain cestodes.

trypsin *n.* [Gk. *tryein*, to rub down; *pepsis*, digestion] a proteinase found in pancreatic juice of mammals, produced by the action of enterokinase on trypsinogen, EC 3.4.21.4, *r.n.* trypsin; a similar enzyme found in other animals and in plants; *a.* tryptic.

trypsinogen *n.* [Gk. *tryein*, to rub down; *pepsis*, digestion; *-genēs*, producing] enzyme precursor (zymogen) secreted by cells of pancreas and converted to trypsin by enterokinase of succus entericus.

tryptic *a.* [Gk. *tryein*, to rub down; *pepsis*, digestion] produced by, of *pert.*, trypsin.

trypton *n.* [Gk. *tripsai*, to grind down] tripton, *q.v.*

tryptophanase an enzyme found especially in colon bacilli, which catalyses the breakdown of tryptophan into ammonia, pyruvic acid, and indole; EC 4.1.99.1; *r.n.* tryptophanase.

tryptophan(e) *n.* [Gk. *tryein*, to rub down; *pepsis*, digestion; *phainein*, to appear] an essential amino acid, ß-indolealanine, obtained by the hydrolysis of various proteins such as casein and fibrin, that is important in the synthesis of haemoglobin and plasma proteins and possibly in the synthesis of nicotinic acid in some organisms, and from which the auxin IAA is derived; abbreviated to try or trp.

tryptophan(e) synthetase the enzyme that catalyses the synthesis of tryptophan from indoleglycerol phosphate and serine; EC 4.2.1.20; *r.n.* tryptophan synthase.

TSH thyroid-stimulating hormone, *q.v.*

TSH-releasing hormone (TRH) a tripeptide hormone secreted by the hypothalamus which causes the secretion and release of TSH from the pituitary gland.

TSP thyroid-stimulating prepituitary hormone, *q.v.*

tuba *n.* [L. *tuba*, trumpet] a salpinx or tube, as tuba acustica or auditiva, the Eustachian tube; tuba uterina, Fallopian tube; *a.* tubal.

tubar consisting of an arrangement of tubes, or forming a tube, as *appl.* system and skeleton in sponges.

tubate *a.* [L. *tubus*, pipe] tube-shaped, tubular; *alt.* tubiform.

tube *n.* [L. *tubus*, pipe] any tubular structure; cylindrical structure, as protective enveloping case of many animals; a mollusc siphon.

tube-feet organs connected with the water vascular system in various echinoderms, for locomotion, also modified for sensory, food-catching, and respiratory functions; *alt.* ambulacra, podia.

tuber *n.* [L. *tuber*, hump] thickened fleshy food-

storing underground root, or stem with surface buds; rounded protuberance.

Tuberales *n.* [*Tuber*, generic name] an order of Discomycetes including the truffles and other mycorrhizal species, having an underground closed ascocarp which does not dehisce.

tuber cinereum hollow protuberance of grey matter between optic chiasma and corpora mammillaria of hypothalamus; *alt.* tuber anterius.

tubercle *n.* [L. *tuberculum*, small hump] a small rounded protuberance; root swelling or nodule; bulbil, *q.v.*; a dorsal articular knob on a rib; a cusp of a tooth; *alt.* tuberculum.

tubercular, tuberculate *a.* [L. *tuberculum*, small hump] *pert.*, resembling or having tubercles.

tuberculose *a.* [L. *tuberculum*, small hump] having many tubercles.

tuberculum tubercle, *q.v.*

tuberiferous *a.* [L. *tuber*, hump; *ferre*, to bear] bearing or producing tubers.

tuberiform, tuberoid *a.* [L. *tuber*, hump; *forma*, shape; Gk. *eidos*, shape] resembling or shaped like a tuber.

tuberosity *n.* [L. *tuber*, hump] rounded eminence on a bone, usually for muscle attachment.

tuberose, tuberous *a.* [L. *tuber*, hump] covered with or having many tubers.

tuber vermis part of superior vermis of cerebellum, continuous laterally with inferior semilunar lobules.

tube tonsil lymphoid tissue near pharyngeal opening of auditory tube.

tubicolous *a.* [L. *tubus*, tube; *colere*, to inhabit] inhabiting a tube.

tubicorn *a.* [L. *tubus*, tube; *cornu*, horn] with hollow horns.

tubifacient *a.* [L. *tubus*, tube; *faciens*, making] tube-making, as some polychaetes.

Tubiflorae *n.* [L. *tubus*, tube; *flos*, flower] an order of dicots (Sympetalae) which are typically herbaceous plants with epipetalous stamens and ovules with only one integument.

tubiflorous tubuliflorous, *q.v.*

tubiform *a.* [L. *tubus*, tube; *forma*, shape] tubate, *q.v.*

tubilingual *a.* [L. *tubus*, tube; *lingua*, tongue] having a tubular tongue, adapted for sucking.

tubiparous *a.* [L. *tubus*, tube; *parere*, to beget] secreting tube-forming material, *appl.* glands.

Tuboidea *n.* [L. *tubus*, tube; Gk. *eidos*, form] an order of graptolites having the thecae arranged on one side of the stipe but having varied branching.

tubo-ovarian *a.* [L. *tubus*, pipe; *ovarium*, ovary] of or *pert.* oviduct and ovary.

tubotympanic *a.* [L. *tubus*, pipe; *tympanum*, drum] *appl.* recess between 1st and 3rd visceral arches, from which are derived the tympanic cavity and Eustachian tube.

tubular *a.* [L. *tubulus*, small tube] having the form of a tube or tubule, *alt.* tubate, tubiform; containing tubules; *appl.* dentine: orthodentine, *q.v.*

tubulate *a.* [L. *tubulus*, small tube] tubiform; tubular; tubuliferous.

tubule *n.* [L. *tubulus*, small tube] any small, hollow, cylindrical structure; *alt.* tubulus.

tubili *plu.* of tubulus.

tubuli contorti the convoluted seminiferous tubules.

Tubulidentata *n.* [L. *tubulus*, small tube; *dens*,

tooth] an order of placental mammals known from the Miocene, or possibly Eocene, whose only living relative is the aardvark, which is ant-eating with digging forelimbs.

tubuliferous *a.* [L. *tubulus*, small tube; *ferre*, to carry] having a tubule or tubules.

tubuliflorous *a.* [L. *tubulus*, small tube; *flos*, flower] having florets with tubular corolla; *alt.* tubiflorous.

tubuliform *a.* [L. *tubulus*, small tube; *forma*, shape] tube-shaped, *appl.* type of spinning glands; *alt.* cylindrical.

tubuli recti straight tubules connecting semini-ferous tubules and rete testis.

tubulose *a.* [L. *tubulus*, small tube] having, or composed of, tubular structures, *appl.* aster head, *appl.* the coral *Tubipora*, organ pipe coral; hollow and cylindrical.

tubulus *n.* [L. *tubulus*, small tube] tubule, *q.v.*; a hymenial pore; a cylindrical ovipositor. *Plu.* tubuli: any small tubular structure as tubuli lac-tiferi, recti, seminiferi.

tumescence *n.* [L. *tumescere*, to swell] a swelling; a tumid state; *cf.* intumescence, detumescence.

tumid *a.* [L. *tumidus*, swollen] swollen, turgid.

tundra *n.* [Russ. arctic hill] treeless region with permanently frozen subsoil.

tunic *n.* [L. *tunica*, coating] an investing mem-brane or tissue, as those of bulbs, eye, kidney, ovary, testis, arteries, etc.; body wall or test, as of tunicates; *alt.* tunica.

tunica tunic, *q.v.*; apical meristematic cells divid-ing anticlinally and giving surface growth, form-ing a mantle over the corpus.

tunica adventitia *see* adventitia.

tunica albuginea *see* albuginea.

tunica-corpus theory the concept that the apical meristem of a shoot is differentiated into 2 regions which can be distinguished by their method of growth, the peripheral tunica having mainly anticlinal divisions and the inner corpus having divisions in various planes; *cf.* histogens.

tunica propria in an insect ovariole, a membrane attached to the surface of the vitellarium, germar-ium, and terminal filament.

Tunicata, tunicates *n., n.plu.* [L. *tunica*, coat-ing] Urochordata, *q.v.*

tunicate *a.* [L. *tunica*, coating] provided with a tunic or test; *appl.* bulbs with numerous concen-tric broad layers; enveloped in tough test or mantle. *n.* A member of the Tunicata, *q.v.*

tunicin(e) *n.* [L. *tunica*, coating] a polysaccharide related to cellulose, in tunic of ascidians; *alt.* animal cellulose.

tunicle *n.* [L. *tunicula*, little coat] a natural cover-ing or integument.

tunnel of Corti [*A. Corti*, Italian histologist] triangular tunnel enclosed by 2 rows of pillars of Corti and basilar membrane.

turacin *n.* [*Turaco*, an African bird] a water-soluble red plumage pigment containing copper, in turaco and other Musophagidae (plantain-eaters).

turacoverdin *n.* [*Turaco*, F. *vert*, green] a green feather pigment containing iron, in certain Musophagidae (plantain-eaters).

Turbellaria, turbellarians *n., n.plu.* [L. *turbellae*, stir, row] a class of free-living flatworms with a leaf-like shape and a glandular, ciliated epidermis.

turbinal *a.* [L. *turbo*, whirl] spirally rolled or coiled, as bone or cartilage, *n.* One of the bones in nose of vertebrates, supporting the olfactory membrane, *alt.* turbinate bone.

turbinate *a.* [L. *turbo*, whirl] top-shaped, *appl.* pileus, *appl.* shells; *appl.* certain nasal bones, *alt.* conchae nasales, turbinals.

turbinulate *a.* [*Dim.* of L. *turbo*, whirl] shaped like a small top, *appl.* certain apothecia.

turgescence, turgidity, turgor *n.* [L. *turge-scere*, to swell] the distention of plant cells due to internal pressure of vacuolar contents; distention of living tissues due to internal pressures; *a.* turgid, turgescent.

turgor pressure (TP) the pressure set up inside a plant cell due to the hydrostatic pressure of the vacuole contents pressing against the cell wall; *alt.* wall pressure.

turio(n) *n.* [L. *turia*, shoot] young scaly shoot budded off from underground stem; detachable winter bud used for perennation in many water plants.

Turkish saddle sella turcica, *q.v.*

turnip yellow mosaic virus (TYMV) a virus used on early studies of nucleic acid and ultra-structure.

turtles *see* Chelonia.

tusk shells a common name for the Scaphopoda, *q.v.*

tutamen *n.* [L. *tutamen*, protection] means of protection; a protective structure, as eyelid; *plu.* tutamina.

tween-brain diencephalon, *q.v.*

tychocoen *n.* [Gk. *tychē*, chance; *koinos*, com-mon] those members of a biocoenosis which thrive under different habitat conditions; *cf.* eucoen.

tycholimnetic *a.* [Gk. *tychē*, chance; *limnē*, marshy lake] temporarily attached to the bed of a lake and at other times floating.

tychoplankton *n.* [Gk. *tychē*, chance; *plangktos*, wandering] drifting or floating organisms which have been detached from their previous habitat, as in plankton of the Sargasso Sea; inshore plank-ton; *cf.* euplankton.

tychopotamic *a.* [Gk. *tychē*, chance; *potamos*, river] thriving only in backwaters, *appl.* potamo-plankton.

Tylenchida *n.* [*Tylenchus*, generic name] an order of Phasmidia, some of which are plant parasites and have a piercing spear on the anterior of the gut.

tylhexactine *n.* [Gk. *tylos*, knob; *hex*, six; *aktis*, ray] a hexactine spicule with rays ending in knobs.

tylosis *n.* [Gk. *tylos*, callus] development of ir-regular cells in a cell cavity; an intrusion of parenchyma cells into xylem vessel, especially of secondary xylem, through pits; a callosity; callus formation; *plu.* tyloses; *alt.* thylosis, tylose.

tylosoid *n.* [Gk. *tylos*, knob; *eidos*, form] resem-bling a tylosis, as a resin duct filled with paren-chymatous cells.

tylostyle *n.* [Gk. *tylos*, knob; *stylos*, pillar] spicule pointed at one end, knobbed at other.

tylotate *a.* [Gk. *tylōtos*, knobbed] with a knob at each end.

tylote *n.* [Gk. *tylōtos*, knobbed] a slender dumbell-shaped spicule.

tylotic *a.* [Gk. *tylos*, callus] affected by tylosis.

tylotoxea *n.* [Gk. *tylos*, knob; *oxys*, sharp] a tylote with one sharp end, directed towards surface of sponge.

tylotrich *a.* [Gk. *tylos*, knob; *thrix*, hair] *appl.* a type of hair follicle found in many mammals.

tylus *n.* [Gk. *tylos*, knob] a medial protuberance on head of certain Hemiptera.

tymbal timbal, *q.v.*

tympanic *a.* [Gk. *tympanon*, drum] *pert.* tympanum.

tympanohyal *a.* [Gk. *tympanon*, drum; *hyoeidēs*, Υ-shaped] *pert.* tympanum and hyoid. *n.* Part of hyoid arch embedded in petromastoid.

tympanoid *a.* [Gk. *tympanon*, drum; *eidos*, form] shaped like a flat drum, *appl.* certain diatoms.

tympanum *n.* [Gk. *tympanon*, drum] the epiphragm of mosses; the drum-like cavity constituting middle ear; eardrum; membrane of auditory organ on tibia, metathorax, or abdomen of insect; inflatable air sac on neck of some Tetraoninae, grouse family; any of various other drum-like structures.

TYMV turnip yellow mosaic virus, *q.v.*

type *n.* [L. *typus*, pattern] sum of characteristics common to a large number of individuals, serving as a ground for classification; a primary model; the actual specimen described as the original of a new genus or species, *alt.* holotype.

type locality the locality in which the holotype or other type used for designation of a species was found.

typembryo *n.* [Gk. *typos*, pattern; *embryon*, embryo] a larval stage in Brachiopoda, attached to substrate by terminal segment.

type number the most frequently occurring chromosome number in a taxonomic group; *alt.* modal number.

type specimen holotype, *q.v.*

Typhales *n.* [*Typha*, generic name] an order of monocots (Arecidae) which are herbaceous marsh or aquatic plants with rhizomes, and unisexual flowers densely crowded in heads or on spikes.

typhlosole *n.* [Gk. *typhlos*, blind; *sōlēn*, channel] median dorsal longitudinal fold of intestine projecting into lumen of gut in some invertebrates and in cyclostomes.

typical *a.* [Gk. *typos*, pattern] *appl.* specimen conforming to type or primary example; exhibiting in marked degree the essential characteristics of genus or species.

typogenesis *n.* [Gk. *typos*, pattern; *genesis*, descent] phase of rapid type formation in phylogenesis; quantitative or 'explosive' evolution.

typology *n.* [Gk. *typos*, pattern; *logos*, discourse] the study of types, as of constitutional types.

typolysis *n.* [Gk. *typos*, pattern; *lysis*, loosing] phase preceding extinction of type; *alt.* phylogerontic stage.

typonym *n.* [Gk. *typos*, pattern; *onyma*, name] a name designating or based on a type specimen or type species.

typostasis *n.* [Gk. *typos*, pattern; *stasis*, halt] relative absence of type formation, a static phase in phylogenesis.

tyr tyrosine, *q.v.*

tyramine *n.* [Gk. *tyros*, cheese; *ammoniakon*, resinous gum] a phenolic amine produced by bacterial action on tyrosine and also secreted by cephalopods and found in various plants such as mistletoe and ergot, which causes a rise in arterial pressure.

tyrosinase polyphenol oxidase, *q.v.*; formerly, one of a group of enzymes which convert tyrosine to melanin.

tyrosine *n.* [Gk. *tyros*, cheese] an amino acid which is important as one of the precursors of adrenalin and noradrenalin and of thyroxine and melanin; abbreviated to tyr.

Tyson's glands [*E. Tyson*, English anatomist] sebaceous glands around the corona of the glans penis.

U

ubiquinone a substituted benzoquinone, which receives electrons from a flavoprotein in electron transfer chains; *alt.* coenzyme Q.

UCR unconditioned response or reflex, *q.v.*

UCS unconditioned stimulus, *q.v.*

Udonelloidea *n.* [*Udonella*, generic name; Gk. *eidos*, form] an order of Monopisthocotylea which are cylindrical with an unarmed opisthaptor.

UDP uridine diphosphate, *q.v.*

UDPG uridine diphosphate glucose, *q.v.*

uintatheres *n.plu.* [*Uintatherium*, generic name] the common name for the Dinocerata, *q.v.*

ula *n.plu.* [Gk. *oula*, the gums] the gums; *alt.* gingivae; *sing.* ulon.

uletic *a.* [Gk. *oulon*, gum] *pert.* the gums; *alt.* gingival.

uliginose, uliginous *a.* [L. *uliginosus*, oozy] swampy; growing in mud or swampy soil; *alt.* paludal.

ulna *n.* [L. *ulna*, elbow] a long bone on medial side of forearm parallel with radius; *alt.* cubitus.

ulnar *pert.* ulna, *appl.* artery, nerve, veins, bone, ligaments.

ulnare *n.* [L. *ulna*, elbow] bone, in proximal row of carpals, lying at distal end of ulna.

ulnar nervure radiating or cross nervure in wing of insects.

ulnocarpal *a.* [L. *ulna*, elbow; *carpus*, wrist] *pert.* ulna and carpus.

ulnoradial *a.* [L. *ulna*, elbow; *radius*, radius] *pert.* ulna and radius.

uloid *a.* [Gk. *oulē*, scar; *eidos*, form] resembling a scar.

ulon *sing.* of ula.

Ulotrichales *n.* [*Ulothrix*, generic name] an order of green algae having the plant body as an unbranched simple filament with diffuse growth.

ulotrichous *a.* [Gk. *oulos*, woolly; *thrix*, hair] having woolly or curly hair.

ultimate cell tip cell, *q.v.*

ultimobranchial bodies pair of gland rudiments derived from 5th pharyngeal pouches, which secrete calcitonin and later degenerate and disappear; *alt.* postbranchial or suprapericardial bodies.

ultra-abyssal hadral, *q.v.*

ultramicroscopic *a.* [L. *ultra*, beyond; Gk. *mikros*, small; *skopein*, to look at] *appl.* structures or organisms that are too small to be seen with the light microscope, but which can be made visible by other means, such as with the electron microscope.

ultrasonic a. [L. *ultra*, beyond; *sonus*, sound] *appl.* sounds of high frequency inaudible to the human ear, as emitted by certain animals; *alt.* supersonic.

ultrastructure n. [L. *ultra*, beyond; *structura*, structure] the fine structure of cells as seen with the ultramicroscope.

ultravirus n. [L. *ultra*, beyond; *virus*, poison] a former term for a virus, i.e. an organism which could only be seen with the electron microscope, and equivalent to a filterable virus.

Ulvales n. [*Ulva*, generic name] an order of green algae having a thallus of uninucleate cells and isomorphic alternation of generations.

umbel n. [L. *umbella*, dim. of *umbra*, shade] an arrangement of flowers or of polyps springing from a common centre and forming a flat or rounded cluster; *alt.* umbella.

umbella n. [L. *umbella*, sunshade] an umbel, *q.v.*; umbrella of jellyfish.

umbellate a. [L. *umbella*, sunshade] arranged in umbels.

umbellet umbellule, *q.v.*

umbelliferous a. [L. *umbella*, sunshade; *ferre*, to carry] producing umbels; n. umbellifer.

Umbelliflorae n. [L. *umbella*, sunshade; *flos*, flower] an order of dicots of the Archichlamydeae or Polypetalae, or a superorder of Rosidae, usually having flowers in umbels.

umbelliform a. [L. *umbella*, sunshade; *forma*, shape] shaped like an umbel.

umbelligerous a. [L. *umbella*, sunshade; *gerere*, to carry] bearing flowers or polyps in umbellate clusters.

umbellula n. [L.L. *umbellula*, dim. of *umbella*, sunshade] a large cluster of polyps at tip of elongated stalk of rachis: umbellule, *q.v.*

umbellulate a. [L.L. *umbellula*, small umbel] arranged in umbels and umbellules.

umbellule n. [L.L. *umbellula*, small umbel] a small or secondary umbel; *alt.* umbellet, umbellula.

umbilical a. [L. *umbilicus*, navel] *pert.* navel, or umbilical cord, *appl.* arteries, veins, tissues, vesicle, plane, etc.; *alt.* omphalic.

umbilical cord navel cord connecting embryo with placenta; funicle or prolongation by which ovule is attached to placenta.

umbilicate(d) a. [L. *umbilicus*, navel] having a central depression; navel-like; *alt.* omphaloid.

umbilicus n. [L. *umbilicus*, navel] the navel, central abdominal depression at place of attachment of umbilical cord; hilum, *q.v.*; basal depression of certain spiral shells; an opening near base of feather; a structure for attachment of thallus in certain lichens.

umbo n. [L. *umbo*, shield boss] a protuberance like boss of a shield; swollen point of a cone scale; convexity of tympanic membrane at point of attachment of manubrium mallei; beak or older part of bivalve shell; a prothoracic projection in certain insects; *plu.* umbones.

umbonal a. [L. *umbo*, shield boss] *pert.* an umbo.

umbonate a. [L. *umbo*, shield boss] having, *pert.*, or resembling an umbo; *alt.* bossed.

umbones *plu.* of umbo.

umbraculiferous a. [L. *umbraculum*, sunshade] bearing an umbrella-like organ or structure.

umbraculiform a. [L. *umbraculum*, sunshade; *forma*, shape] shaped like an expanded umbrella.

umbraculum n. [L *umbraculum*, sunshade] any umbrella-like structure; pigmented fringe of iris, in certain ungulates; pupillary appendage, in amphibians.

umbraticolous a. [L. *umbraticola*, one who likes the shade] growing in a shaded habitat; *alt.* skiophilous.

umbrella n. [L. *umbella*, sunshade] an umbrella-like structure; the contractile disc of a jellyfish; web between arms of certain Octopoda; *alt.* bell.

umbrella cells wing cells, *q.v.*

UMP uridine monophosphate, *q.v.*

uncate a. [L. *uncus*, hook] hooked; *alt.* hamate.

unciferous a. [L. *uncus*, hook; *ferre*, to carry] bearing hooks or hook-like processes.

unciform a. [L. *uncus*, hook; *forma*, shape] shaped like a hook or barb, *alt.* hamiform; *appl.* process of ethmoid bone. n. Unciform bone of wrist, *alt.* os hamatum, uncinatum.

uncinate a. [L. *uncinus*, hook] hook-like, *alt.* unciform, hamate; *appl.* fasciculus associating temporal and frontal lobes of brain; *appl.* process of ribs of birds; *appl.* process of ethmoid, of head of pancreas; *appl.* decurrent lamellae of agarics; *appl.* apex, as of a leaf.

uncinus n. [L. *uncinus*, hook] small hooked, or hook-like, structure; crotchet, *q.v.*; one of small hooks found on segments of many annelids; a hook-like structure found in certain Protozoa; a marginal tooth of radula in gastropods.

unconditioned *appl.* inborn reflex (UCR), *cf.* conditioned or acquired reflex; *appl.* stimulus (UCS) producing a simple reflex response.

uncus n. [L. *uncus*, hook] hook-shaped anterior extremity of hippocampal gyrus; hooked head of malleus of rotifers; hook-like or bifid process on dorsal portion of 9th abdominal segment of male Lepidoptera; uncinate hair.

undate a. [L. *undare*, to rise in waves] wavy, undulating; undose, *q.v.*; undulate, *q.v.*

under-wing one of posterior wings of any insect.

undifferentiated in immature form, still in embryonic or meristematic state; in mature form, relatively unspecialized.

undose a. [L. *undosus*, billowy] having undulating and nearly parallel depressions which run into one another and resemble ripple marks; *alt.* undate.

undulate a. [L.L. *undulatus*, risen like waves] having wave-like elevations, *appl.* leaves; *alt.* undate.

undulating membrane a cytoplasmic membrane between body and part of flagellum; a similar structure in tail of certain spermatozoa; sometimes used for vibratile membrane, *q.v.*

unequally pinnate imparipinnate, *q.v.*

ungual a. [L. *unguis*, nail] *pert.* or having a nail or claw; *appl.* phalanges bearing claws or nails.

unguicorn dertrotheca, *q.v.*

Unguiculata [L. *unguiculus*, little nail] an infraclass or cohort of eutherians, including the living orders Insectivora, Dermoptera, Pholidota, Edentata, Chiroptera, and Primates, and some fossil orders.

unguiculate a. [L. *unguiculus*, little nail] clawed; *appl.* petals with narrowed stalk-like portion below.

unguiculus *n.* [L. *unguiculus*, little nail] a small nail, or claw, as on tibiotarsus of Collembola.

unguis *n.* [L. *unguis*, claw] a nail or claw; narrow stalk-like portion of some petals, *alt.* ungula; a chitinous hook on foot of insect; distal joint, the crochet or fang, of arachnid chelicerae; lacrimal bone, *q.v.*; the calcar avis, *q.v.*

unguitractor *n.* [L. *unguis*, claw; *tractus*, pull] a median plate of pretarsus for attachment of retractor or flexor muscle of claw, in insects.

ungula *n.* [L. *ungula*, hoof] hoof; unguis of petal.

Ungulata, ungulates *n., n.plu.* [L. *ungula*, hoof] a group of hoofed grazing placental mammals which are now thought not to be closely related and contains 4 superorders, the Protoungulata, Paenungulata, Mesaxonia, and Paraxonia.

ungulate *a.* [L. *ungula*, hoof] hoofed; hoof-like.

unguliform *a.* [L. *ungula*, hoof; *forma*, form] hoof-shaped.

unguligrade *a.* [L. *ungula*, hoof; *gradus*, step] walking upon hoofs.

uniascal *a.* [L. *unus*, one; Gk. *askos*, bag] containing a single ascus, *appl.* locules.

uniaxial *a.* [L. *unus*, one; *axis*, axis] with one axis, *alt.* monaxial, uniradiate; *appl.* movement only in one plane, as of hinge joint.

unibranchiate *a.* [L. *unus*, one; *branchiae*, gills] having one gill.

unicamerate *a.* [L. *unus*, one; *camera*, vault] one-chambered; *alt.* unilocular.

unicapsular *a.* [L. *unus*, one; *capsula*, small case] having only one capsule.

unicell *n.* [L. *unus*, one; *cellula*, cell] a unicellular organism, being a protophyton or protozoon.

unicellular *a.* [L. *unus*, one; *cellula*, cell] having only one cell, or consisting of one cell; *alt.* monocellular.

uniciliate *a.* [L. *unus*, one; *cilium*, eyelash] having one cilium or flagellum.

unicolour *a.* [L. *unicolor*, of one colour] having only one colour; of the same colour throughout; *alt.* unicoloured, unicolorate, unicolorous, monochromatic.

unicorn *a.* [L. *unus*, one; *cornu*, horn] having a single horn-like spine.

unicostate *a.* [L. *unus*, one; *costa*, rib] having a single prominent midrib, as certain leaves.

unicotyledonous *a.* [L. *unus*, one; Gk. *kotylēdōn*, cup] *see* monocotyledonous.

unicuspid *a.* [L. *unus*, one; *cuspis*, point of spear] having one tapering point, as a tooth.

unidactyl *a.* [L. *unus*, one; Gk. *daktylos*, finger] monodactylous, *q.v.*

uniembryonate *a.* [L. *unus*, one; Gk. *embryon*, foetus] having one embryo only.

unifacial *a.* [L. *unus*, one; *facies*, face] having one face or chief surface; having similar structure on both sides.

unifactorial *a.* [L. *unus*, one; *facere*, to make] *pert.* or controlled by a single gene; *alt.* monogenic, monofactorial.

uniflagellate *a.* [L. *unus*, one; *flagellum*, whip] having only one flagellum; *alt.* monocont, monokont, monociliated, monocerous.

uniflorous *a.* [L. *unus*, one; *flos*, flower] bearing only one flower.

unifoliate *a.* [L. *unus*, one; *folium*, leaf] with one leaf, *alt.* monophyllous; with a single layer of zooecia, *appl.* polyzoan colony.

unifoliolate *a.* [L. *unus*, one; *foliolum*, dim. of *folium*, leaf] having one leaflet only.

uniforate *a.* [L. *unus*, one; *foratus*, pierced] having only one opening.

unigeminal *a.* [L. *unus*, one; *geminus*, twin-born] *appl.* arrangement of pore pairs in one row, in ambulacra of some echinoids.

unigenesis *n.* [L. *unus*, one; Gk. *genesis*, descent] monogenesis, *q.v.*

unihumoral *a.* [L. *unus*, one; *humor*, fluid] activated by only one neurohumor, *appl.* certain chromatophores.

unijugate *a.* [L. *unus*, one; *jugum*, yoke] *appl.* pinnate leaf having one pair of leaflets.

unilabiate *a.* [L. *unus*, one; *labium*, lip] with one lip or labium.

unilacunar *a.* [L. *unus*, one; *lacuna*, cavity] with one lacuna; having one leaf gap, *appl.* nodes.

unilaminate *a.* [L. *unus*, one; *lamina*, layer] having one layer only, *appl.* tissues.

unilateral *a.* [L. *unus*, one; *latus*, side] arranged on one side only.

unilocular *a.* [L. *unus*, one; *loculus*, compartment] having a single small compartment or cell; containing a single oil droplet, as cells in white fat; *alt.* monolocular, monothalamous, unicamerate; *cf.* multilocular.

unimodel *a.* [L. *unus*, one; *modus*, measure] having only one mode; *appl.* frequency distribution with a single maximum.

unimucronate *a.* [L. *unus*, one; *mucro*, sharp point] having a single sharp point or tip.

unineme *n.* [L. *unus*, one; *nēma*, thread] mononeme, *q.v.*

uninomial, uninominal, unitary mononomial, *q.v.*

uninuclear, uninucleate *a.* [L. *unus*, one; *nucleus*, nucleus] having one nucleus; *alt.* mononuclear.

uniovular *a.* [L. *unus*, one; *ovum*, egg] *pert.* a single ovum; monozygotic, *q.v.*, *appl.* twinning; *alt.* monovular.

uniparous *a.* [L. *unus*, one; *parere*, to beget] producing one offspring at a birth; having a cymose inflorescence with one axis at each branching.

unipennate *a.* [L. *unus*, one; *penna*, feather] *appl.* muscle having its tendon of insertion extending along one side.

unipetalous *a.* [L. *unus*, one; Gk. *petalon*, leaf] having one petal; *alt.* monopetalous.

unipolar *a.* [L. *unus*, one; *polus*, pole] having one pole only; *appl.* some nerve cells; *appl.* spindle when cone-shaped at 1st meiotic division, as in the aberrant type of meiosis in certain families of Diptera; *alt.* monopolar.

unipotent(ial) *a.* [L. *unus*, one; *potens*, powerful] *appl.* cells which can develop into cells of one kind only; *cf.* totipotent.

uniradiate *a.* [L. *unus*, one; *radius*, ray] one-rayed; uniaxial, *q.v.*; monactinal, *appl.* spicule.

uniramous *a.* [L. *unus*, one; *ramus*, branch] having one branch; *appl.* crustacean appendage lacking an exopodite; *appl.* antennule.

unisegmental *a.* [L. *unus*, one; *segmentum*, a slice] *pert.*, or involving a single segment.

unisepalous *see* monosepalous.

uniseptate *a.* [L. *unus*, one; *septum*, hedge] having one septum or dividing partition.

uniserial *a.* [L. *unus*, one; *series*, rank] arranged

453

in one row or series; *appl.* certain ascospores; *appl.* fins with radials on one side of basalia; *appl.* medullary rays; *appl.* thecae of graptolites; *appl.* gills of some molluscs; *alt.* uniseriate.

uniserrate *a.* [L. *unus*, one; *serra*, saw] having only one row of serrations on edge.

uniserrulate *a.* [L. *unus*, one; *serra*, saw] having one row of small serrations on edge.

unisetose *a.* [L. *unus*, one; *seta*, bristle] bearing one bristle.

Unisexales *n.* [L. *unus*, one; *sexus*, sex] Casuarinales, *q.v.*

unisexual *a.* [L. *unus*, one; *sexus*, sex] of one or other sex, distinctly male or female; *alt.* diclinous, dioecious, gonochoristic.

unispiral *a.* [L. *unus*, one; *spira*, coil] having one spiral only.

unistrate *a.* [L. *unus*, one; *stratum*, layer] having only one layer; *alt.* unistratose.

unitubercular *a.* [L. *unus*, one; *tuberculum*, small swelling] having a single small prominence, tubercle, or cusp.

univalent *a.* [L. *unus*, one; *valere*, to be strong] *appl.* a single unpaired chromosome at meiosis; *alt.* monovalent.

univalve *n.* [L. *unus*, one; *valvae*, folding doors] a shell consisting of one piece or valve, as a gastropod shell.

universal donor person with blood of group O, or 4, whose blood may be transfused into, or whose skin may be grafted on to, a member of any other group, without harmful reaction.

universal recipient person with blood of group AB, or one, into whom blood may be transfused from a member of any other group, without harmful reaction.

universal veil volva, *q.v.*; *alt.* velum universale.

univoltine *a.* [L. *unus*, one; It. *volta*, time] producing only one brood in a season; *alt.* monovoltine.

unpaired *a.* [A.S. *un-*, not; L. *par*, equal] situated in median line of body, consequently single.

unstriped *appl.* muscle: involuntary muscle, *q.v.*

urachus *n.* [Gk. *ouron*, urine; *echein*, to hold] the median umbilical ligament; fibrous cord extending from apex of bladder to umbilicus.

uracil *n.* [Gk. *ouron*, urine] a pyrimidine base which is part of the genetic code of RNA where it pairs with adenine and is replaced by thymine in DNA, and also forms part of the molecule of uridine.

urate *n.* [Gk. *ouron*, urine] a salt of uric acid; *appl.* excretory cells in fat-body of insects lacking Malpighian tubles.

urceolate *a.* [L. *urceolus*, small pitcher] urn- or pitcher-shaped, *appl.* apothecium, *appl.* calyx or corolla, *appl.* shells of various Protozoa; having an urceolus.

urceolus *n.* [L. *urceolus*, small pitcher] any pitcher-shaped structure especially the external tube of certain rotifers.

urea *n.* [Gk. *ouron*, urine] carbamide, an amide derived from protein in the diet, the chief nitrogenous constituent of urine of mammals and some other animals, and also found in some fungi and seed plants.

urea cycle a biochemical cycle concerned with urea synthesis from bound ammonia, chiefly from protein metabolism, via a cyclic series of reactions driven by ATP hydrolysis; *alt.* Krebs'-

Henseleit cycle, arginine-urea cycle, ornithine cycle.

urease *n.* [Gk. *ouron*, urine; *-ase*] an enzyme which catalyses hydrolysis of urea into ammonia and carbon dioxide; EC 3.5.1.5; *r.n.* urease.

uredia *plu.* of uredium.

uredial *a.* [L. *uredo*, blight] *appl.* or *pert.* the summer stage of rust fungi; *alt.* uredinial.

Uredinales *n.* [L. *uredo*, blight] an order of parasitic Basidiomycetes, commonly called rust fungi, with some species being economically important, having a mycelium living inside the plant host and producing spores in pustules, and often having a complex life cycle with 2 hosts.

urediniospore uredospore, *q.v.*

uredinium *n.* [L. *uredo*, blight] uredosorus, *q.v.*

urediospore uredospore, *q.v.*

uredium *n.* [L. *uredo*, blight] uredosorus, *q.v.*

uredo *n.* [L. *uredo*, blight] summer stage of rust fungi.

uredobud uredospore, *q.v.*

uredogonidium uredospore, *q.v.*

uredosorus *n.* [L. *uredo*, blight; Gk. *sōros*, heap] in rust fungi, a group of developing uredospores forming a pustule on the host; *alt.* uredinium, uredium.

uredospore *n.* [L. *uredo*, blight; Gk. *sporos*, seed] in rust fungi, one of the reddish spores produced on sporophore in summer; *alt.* urediniospores forming a pustule on the host; *alt.* uredinium, uredium.

uredostage *n.* [L. *uredo*, blight; *stare*, to stand] the stage in a rust fungus when uredospores are formed.

ureotelic *a.* [Gk. *ouron*, urine; *telos*, end] excreting nitrogen as urea, e.g. adult amphibia, elasmobranchs, mammals; *cf.* uricotelic.

ureter *n.* [Gk. *ourētēr*, ureter] duct conveying urine from kidney to bladder or cloaca.

ureteric *pert.* ureters; *appl.* bud: embryonic diverticulum of metanephros giving rise ultimately to ureters.

urethra *n.* [Gk. *ourēthra*, from *ouron*, urine] duct leading off urine from bladder, and in male conveying semen in addition.

uric acid trioxypurine, the end-product of purine metabolism in certain mammals, and the main nitrogenous excretory product in birds, reptiles, and some invertebrates, especially insects.

uricase an enzyme of kidney and liver, also of some fungi, involved in the oxidation of uric acid to allantoin and carbon dioxide; *alt.* uric acid oxidase, urico-oxidase; EC 1.7.3.3; *r.n.* urate oxidase.

uricolytic *a.* [Gk. *ouron*, urine; *lyein*, to loose] decomposing uric acid; *appl.* index: the ratio between nitrogen excreted as allantoin to that present in urine as uric acid.

urico-oxidase uricase, *q.v.*

uricosuric *appl.* a substance which reduces blood levels of uric acid and eventually the amount of uric acid in urine.

uricotelic *a.* [Gk. *ouron*, urine; *telos*, end] excreting nitrogen as uric acid, e.g. insects, reptiles, birds; *cf.* ureotelic.

uridine a nucleoside with uracil as its base.

uridine diphosphate (UDP) a nucleotide cofactor analogous to ADP, and used to make UTP.

uridine diphosphate glucose (UDPG) a compound which acts as a coenzyme to various enzymes and is important in carbohydrate metabolism.

uridine monophosphate (UMP) uridylic acid, *q.v.*, a nucleotide cofactor analogous to AMP.

uridine triphosphate (UTP) a nucleotide cofactor analogous to ATP but which cannot be directly phosphorylated except by reaction with ATP itself.

uridylic acid a nucleotide formed from uridine and phosphoric acid; *alt.* uridine monophosphate.

urinary *a.* [L. *urina*, urine] *pert.* urine, *appl.* organs including kidneys, ureters, bladder, and urethra.

urine *n.* [L. *urina*, urine] a fluid excretion from kidneys in mammals, a solid or semisolid excretion in birds and reptiles.

uriniferous, uriniparous *a.* [L. *urina*, urine; *ferre*, to bear; *parere*, to bring forth] urine-producing; *appl.* tubules of nephron leading from Bowman's capsules to collecting ducts.

urinogenital *a.* [L. *urina*, urine; *gignere*, to beget] *pert.* urinary and genital systems; *alt.* urogenital, genito-urinary.

urinogenital ridge a paired ridge from which urinary and genital systems are developed.

urinogenital sinus bladder or pouch in connection with urinary and genital systems in many animals.

urite *n.* [Gk. *oura*, tail] uromere, *q.v.*; anal cirrus in polychaetes.

urn *n.* [L. *urna*, jar] an urn-shaped structure; the base of a pyxis in lichens; theca or capsule of mosses; one of the ciliate bodies floating in coelomic fluid of annulates.

urobilin *n.* [Gk. *ouron*, urine; L. *bilis*, bile] a brown pigment of urine; *alt.* stercobilin.

urobilinogen *n.* [Gk. *ouron*, urine; L. *bilis*, bile; Gk. *genēs*, produced] a colourless compound derived from bilirubin, oxidized to urobilin, and excreted in urine.

urocardiac ossicle a short stout bar forming part of gastric mill in certain Crustacea.

urochord *n.* [Gk. *oura*, tail; *chordē*, cord] the notochord when confined to caudal region, as in tunicates.

Urochorda(ta), urochordates *n., n.plu.* [Gk. *oura*, tail; *chordē*, cord] a subphylum of chordates, sometimes considered a group of protochordates, in which the chordate features are found only in the larval stages and are mostly lost in the adult; *alt.* Tunicata.

urochrome *n.* [Gk. *ouron*, urine; *chrōma*, colour] a yellowish pigment to which ordinary colour of urine is due.

urocoel *n.* [Gk. *ouron*, urine; *koilos*, hollow] an excretory organ in Mollusca.

urocyst *n.* [Gk. *ouron*, urine; *kystis*, bladder] the urinary bladder; *alt.* vesica urinaria.

urodaeum *n.* [Gk. *ouron*, urine; *hodaios*, way] the part or chamber of cloaca into which ureters and genital ducts open; *alt.* urodeum.

Urodela, urodeles *n., n.plu.* [Gk. *oura*, tail; *dēlos*, visible] an order of amphibians, placed in the subclass Lepospondyli in some classifications and in the Lissamphibia in others, and including the newts and salamanders; also called Caudata in some classifications.

urodelous *a.* [Gk. *oura*, tail; *dēlos*, visible] with persistent tail.

urodeum urodaeum, *q.v.*

urogastric *a.* [Gk. *oura*, tail; *gastēr*, stomach] *pert.* the posterior portion of the gastric region in certain crustaceans.

urogastrone a substance resembling enterogastrone and inhibiting gastric secretion, found in mammalian urine.

urogenital urinogenital, *q.v.*

urogomphi *n.plu.* [Gk. *oura*, tail; *gomphos*, nail] processes found on the terminal segments of certain beetle larvae.

urohyal *n.* [Gk. *oura*, tail; *hyoiedēs*, Y-shaped] a median bony element in hyoid arch below hypohyals; *alt.* basibranchiostegal.

uroid *n.* [Gk. *oura*, tail; *eidos*, shape] in some amoebae, a region of posterior gelation, usually sticky and trailing material behind it; *alt.* urosphere.

urokinase *n.* [Gk. *ouron*, urine; *kinein*, to move] a proteolytic enzyme found in human urine.

uromere *n.* [Gk. *oura*, tail; *meros*, part] an abdominal segment in Arthropoda; *alt.* urite.

uromorphic *a.* [Gk. *oura*, tail; *morphē*, shape] like a tail; *alt.* uromorphous.

uroneme *n.* [Gk. *oura*, tail; *nēma*, thread] a tail-like structure of some ciliate Protozoa.

uronic acid any of a group of acids formed as oxidation products of sugars, and found in urine and as part of the molecule of certain polysaccharides.

uropatagium *n.* [Gk. *oura*, tail; L. *patagium*, border] membrane stretching from one femur to the other in bats; part of patagium extending between hind-feet and tail in flying lemurs; podical plate of insects.

urophan *n.* [Gk. *ouron*, urine; *phainein*, to show] any ingested substance found chemically unchanged in urine; *a.* urophanic.

urophysis *n.* [Gk. *oura*, tail; *physis*, growth] in teleosts and elasmobranchs, a concentration of neurosecretory nerve endings at the end of the spinal cord, resembling the neurohypophysis of mammals.

uropod *n.* [Gk. *oura*, tail; *pous*, foot] an abdominal appendage of Crustacea in front of telson.

uropolar *a.* [Gk. *oura*, tail; *polos*, axis] *appl.* cells at posterior end of rhombogen in Dicyemida.

uropore *n.* [Gk. *ouron*, urine; *poros*, passage] opening of excretory duct in Acarina.

uroporphyrin *n.* [Gk. *ouron*, urine; *porphyra*, purple] a brownish-red iron-free product of haem metabolism, a pigment of urine.

uropterine *n.* [Gk. *ouron*, urine; *pteron*, wing] xanthopterine, *q.v.*, especially when derived from urine.

uropyge uropygium, *q.v.*

Uropygi *n.* [G. *oura*, tail; *pygē*, rump] an order of arachnids, commonly called whip scorpions, having the last segment bearing a long jointed flagellum.

uropygial *a.* [Gk. *oura*, tail; *pygē*, rump] *pert.* uropygium, *appl.* oil gland.

uropygium *n.* [Gk. *oura*, tail; *pygē*, rump] the hump at end of bird's trunk, containing caudal vertebrae, and supporting tail feathers; *alt.* uropyge.

uropyloric *a.* [Gk. *oura*, tail; *pylōros*, gatekeeper] *pert.* posterior portion of crustacean stomach.

urorectal *a.* [L. *urina*, urine; *rectus*, straight] *appl.*

urorubin

embryonic septum, which ultimately divides intestine into anal and urinogenital parts.

urorubin *n*. [Gk. *ouron*, urine; L. *ruber*, red] the red pigment of urine.

urosacral *a*. [Gk. *oura*, tail; *sacrum*, sacred] *pert*. caudal and sacral regions of the vertebral column.

urosome *n*. [Gk. *oura*, tail; *sōma*, body] tail region of fish; abdomen of some arthropods; *alt*. urosoma.

urosphere *n*. [Gk. *oura*, tail; *sphaira*, globe] uroid, *q.v.*

urostege, urostegite *n*. [Gk. *oura*, tail; *stegē*, roof] ventral tail plate of snakes.

urosteon *n*. [Gk. *oura*, tail; *osteon*, bone] median ossification on the back portion of the keel-bearing part of the sternum in birds.

urosternite *n*. [Gk. *oura*, tail; *sternon*, chest] ventral plate of arthropodan abdominal segment.

urosthenic *a*. [Gk. *oura*, tail; *sthenos*, strength] having tail strongly developed for propulsion.

urostyle *n*. [Gk. *oura*, tail; *stylos*, pillar] an unsegmented bone, posterior part of vertebral column of anurous amphibians; hypural bone in fishes.

uroxanthin *n*. [Gk. *ouron*, urine; *xanthos*, yellow] a yellow pigment of normal urine.

Urticales, Urticiflorae *n*. [*Urtica*, generic name; L. *flos*, flower] an order of dicots of the Hamamelididae or Archichlamydeae being usually woody plants, although sometimes herbaceous, with simple leaves which re often lobed and have stipules.

urticant *a*. [L. *urtica*, nettle] nettling, stinging, irritating.

urticarial *a*. [L. *urtica*, nettle] nettling, urticant, *appl*. hairs, as of some caterpillars.

urticator *n*. [L. *urtica*, nettle] a nettling or stinging cell; nematocyst, *q.v.*

use inheritance transmission of acquired characteristics.

Ustilaginales *n*. [*Ustilago*, generic name] an order of Basidiomycetes, commonly known as smut, brand, or bunt fungi, some species of which are economically important, having a mycelium living inside the plant host and frequently replacing the flowers by soot-like spores.

uterine *a*. [L. *uterus*, womb] *pert*. uterus, *appl*. artery, vein, plexus, glands, etc., of mammals.

uterine bell muscular bell-like structure in female of certain nematodes, communicating with coelom and uterus.

uterine crypts depressions in uterine mucosa, for accommodation of chorionic villi.

uterine tube Fallopian tube, *q.v.*

uteroabdominal *a*. [L. *uterus*, womb; *abdomen*, stomach] *pert*. uterus and abdominal region.

uterosacral *a*. [L. *uterus*, womb; *sacrum*, sacred] *appl*. 2 ligaments of sacrogenital folds attached to sacrum.

uterovaginal *a*. [L. *uterus*, womb; *vagina*, sheath] *pert*. uterus and vagina.

uteroverdin *n*. [L. *uterus*, womb; O.F. *verd*, from L. *viridis*, green] green placental pigment in Canidae (dogs, etc.), oxidized biliverdin, of biliverdin itself.

uterovesical *a*. [L. *uterus*, womb; *vesicula*, vesicle] *pert*. uterus and urinary bladder.

uterus *n*. [L. *uterus*, womb] the organ in female mammals in which the embryo develops and is nourished before birth; an enlarged portion of

oviduct modified to serve as a place for development of young or of eggs; *alt*. metra, matrix.

uterus masculinus median sac, vestigial Müllerian duct in male attached to dorsal surface of urinogenital canal; *alt*. utriculus prostaticus, vesica prostatica, sinus pocularis, Weber's organ, utricle.

utilization time interval between a liminal stimulus and reaction, as between rheobase and excitation.

UTP uridine triphosphate, *q.v.*

utricle *n*. [L. *utriculus*, small bag] former term for ascus; bladder-like sporocarp of certain fungi; an air bladder of aquatic plants; membranous indehiscent 1-celled fruit; protoplasm enveloping a vacuole; membranous sac of ear labyrinth; uterus masculinus, *q.v.*; *alt*. utriculus.

utricular, utriculate *a*. [L. *utriculus*, small bag] *pert*. utricle; utriculiform, *q.v.*; containing vessels like small bags; *appl*. modification of laticiferous tissue.

utriculiform *a*. [L. *utriculus*, small bag; *forma*, shape] shaped like a utricle or small bladder; *alt*. utricular, utriculate.

utriculus utricle, *q.v.*

utriform *a*. [L. *uter*, leather bottle; *forma*, shape] bladder-shaped, with a shallow constriction.

uva *n*. [L. *uva*, grape] a berry formed from a superior ovary and with a central placenta, such as a grape; *cf*. nuculanium.

uvea *n*. [L. *uva*, grape] pigmented epithelium covering posterior surface of iris: pars iridica retinae; or the uveal tract: choroid, ciliary body, and iris.

uvette *n*. [F. from L. *uva*, grape] the glandular junction of the 2 demanian vessels whence duct passes to exterior.

uvula *n*. [L.L. *dim*. of L. *uva*, grape] part of inferior vermis of cerebellum; conical pendulous process from soft palate; small elevation in mucous membrane of urinary bladder, caused by prostate; *alt*. staphyle.

V

V–A vesicular–arbuscular, *appl*. mycorrhiza.

vaccine *n*. [L. *vacca*, cow] a preparation of a pathogenic micro-organism which is treated to destroy its virulence then introduced into the body, causing the body to produce antibodies and so conferring immunity.

vacuolar *a*. [L. *vacuus*, empty] *pert*. or like a vacuole.

vacuolate(d) *a*. [L. *vacuus*, empty] containing vacuoles.

vacuolation vacuolization, *q.v.*

vacuole *n*. [L. *vacuus*, empty] one of the spaces in cell cytoplasm containing air, water, sap, partially digested food, or other materials.

vacuolization *n*. [L. *vacuus*, empty] the formation of vacuoles; appearance or formation of drops of clear fluid in growing or ageing cells; *alt*. vacuolation.

vacuome *n*. [L. *vacuus*, empty] the vacuolar system of a single cell.

vacuum activities actions that take place without apparent need or stimuli.

vagal *a.* [L. *vagus*, wandering] *pert.* the vagus.

vagiform *a.* [L. *vagus*, indefinite; *forma*, shape] having an indeterminate form; amorphous.

vagile *a.* [L. *vagus*, wandering] freely motile; able to migrate; *n.* vagility.

vagina *n.* [L. *vagina*, sheath] a sheath or sheathlike tube; expanded sheath-like portion of leaf base; canal leading from uterus to external opening of genital canal.

vaginae mucosae mucous sheaths lessening friction of tendons gliding in fibro-osseous canals, as in hand or foot.

vaginal *a.* [L. *vagina*, sheath] *pert.* or supplying vagina, *appl.* arteries, nerves, etc.

vaginal process projecting lamina on inferior surface of petrous portion of temporal; a lamina on sphenoid.

vaginate *a.* [L. *vagina*, sheath] invested by a sheath; *alt.* vaginiferous.

vaginervose *a.* [L. *vagus*, wandering; *nervus*, sinew] with irregularly arranged veins.

vaginicolous *a.* [L. *vagina*, sheath; *colere*, to inhabit] *appl.* certain Protozoa which build and inhabit sheaths or cases.

vaginiferous *a.* [L. *vagina*, sheath; *ferre*, to carry] vaginate, *q.v.*

vaginipennate *a.* [L. *vagina*, sheath; *penna*, wing] having wings protected by a sheath.

vaginula *n.* [L. *vaginula*, *dim.* of *vagina*, sheath] a small sheath; sheath surrounding basal portion of sporogonium in bryophytes.

vagus *n.* [L. *vagus*, wandering] the pneumogastric or 10th cranial nerve; visceral accessory nervous system in insects.

val valine, *q.v.*

valency *n.* [L. *valentia*, strength] the way in which chromosomes are paired in a nucleus in meiosis as in univalent and bivalent.

valine an amino acid essential in the diet of man and some other animals, obtained by the hydrolysis of proteins such as casein and zein; abbreviated to val.

vallate *a.* [L. *vallatus*, surrounded by a rampart] with a rim surrounding a depression; *appl.* papillae with taste buds on back part of tongue; *alt.* circumvallate.

vallecula *n.* [L.L. *dim.* of L. *vallis*, valley] a depression or groove.

vallecular canal one of canals in cortical tissue of stem of horsetails.

valleculate *a.* [L.L. *dim.* of L. *vallis*, valley] grooved.

Valsalva *see* sinuses of Valsalva.

valval, valvar *a.* [L. *valva*, fold] *appl.* view of diatom when one whole valve is next the observer.

valvate *a.* [L. *valva*, fold] hinged at margin only; meeting at edges; opening by or furnished with valves; *pert.* valves.

valve *n.* [L. *valva*, fold] any of various structures which permit flow in one direction, but are capable of closing tube or vessel and preventing backward flow; any of pieces formed by a capsule on dehiscence; lid-like structure of certain anthers; lemma, *q.v.*; one of pieces forming shell of diatom; any of pieces which form shell in certain molluscs, barnacles, etc.; one of pieces forming sheath of ovipositor or of clasper in certain insects.

valvelet *n.* [L. *valvula*, *dim.* of *valva*, fold] a small fold or valve; *alt.* valvula.

valve of Thebesius [*A. C. Thebesius*, Germany anatomist] valve of the coronary sinus in right atrium; *alt.* thebesian valve, valvula sinus, coronarii cordis.

valve of Vieussens [*R. Vieussens*, French anatomist] thin layer of white matter extending between superior peduncles of cerebellum; *alt.* anterior medullary velum, Willis' valve.

valvifer *n.* [L. *valva*, fold; *ferre*, to bear] one of the sclerites or coxites at base of valves of ovipositor in certain insects.

valvula *n.* [L. *dim.* of *valva*, fold] a small valve; *alt.* valvelet, valvule.

valvulae conniventes circular, spiral, or bifurcated folds of mucous membrane found in alimentary canal from duodenum to ileum, affording increased area for secretion and absorption; *alt.* Kerckring's valves, plicae circulares.

valvular *pert.* or like, a valve or valvula; *appl.* dehiscence of certain capsules and anthers.

valvule *n.* [L. *dim,* of *valva*, fold] valvula, *q.v.*; upper palea of grasses.

Vampyromorpha *n.* [L. *vampyrus*, vampire; Gk. *morphē*, form] an order of cephalopods of the subclass Coleoidea, known as vampire squids, having 8 long webbed arms and 2 small retractile tentacles.

vanadocyte *n.* [*Vanad*ium; Gk. *kytos*, hollow] a blood corpuscle containing a vanadium compound, in certain ascidians.

vancomycin an antibiotic which blocks steps of cell wall synthesis in bacteria.

vane *n.* [A.S. *fana*, small flag] the web of a feather, consisting of barbs, etc.; *alt.* vexillum.

vannal *a.* [L. *vannus*, fan] *pert.* vannus, *appl.* veins.

vannus *n.* [L. *vannus*, fan] fan-like posterior lobe of hind-wing in some insects; *alt.* anal lobe, neala.

variance *n.* [L. *variare*, to change) the condition of being varied; the mean of the squares of individual deviations from the mean.

variant *n.* [L. *varians*, changing] an individual or species deviating in some character or characters from type.

variate *n.* [L. *variare*, to change] the variable quantity in variation; a character variable in quality or magnitude.

variation *n.* [L. *variare*, to change] divergence from type in certain characteristics; deviation from the mean.

varicellate *a.* [L. *varix*, dilatation] *appl.* shells with small or indistinct ridges.

varices *n.plu.* [L. *varix*, dilatation] prominent ridges across whorls of various univalve shells, showing previous position of outer lip; *sing.* varix.

variegation *n.* [L. *variegare*, to make various] variation of pigmentation of leaves or other plant organs, caused by genetic or viral interference with their normal coloration.

variety *n.* [L. *varietas*, variety] a taxonomic group below the species used in different senses by different specialists, including a race, stock, strain, breed, subspecies, geographical race, or mutant.

variole *n.* [L.L. *variola*, smallpox] a small pit-like marking found on various parts in insects.

varix *sing.* of varices.

vas *n.* [L. *vas*, vessel] a small vessel, duct, canal, or blind tube; *plu.* vasa, *q.v.*

vasa afferentia lymphatic vessels entering lymph nodes.

vasa deferentia ducts leading from testes to penis, exterior, urinogenital canal, or cloaca; *alt.* deferent ducts, ducti deferentia; *sing.* vas deferens.

vasa efferentia ductules leading from testis to vas deferens; lymphatic vessels leading from lymph nodes; *sing.* vas efferens.

vasal *a.* [L. *vas*, vessel] *pert.* or connected with a vessel.

vasa vasorum blood vessels supplying the larger arteries and veins and found in their coats.

vascular *a.* [L. *vasculum*, small vessel] *pert.*, consisting of, or containing vessels adapted for transmission or circulation of fluid.

vascular areas scattered areas developed between endoderm and mesoderm of yolk sac, beginnings of primitive blood vessels; *alt.* area vasculosa.

vascular bundle a group of special cells consisting of 2 parts, xylem and phloem, sometimes separated by a strip of cambium; *alt.* vascular strand.

vascular cylinder vascular tissues and associated ground tissue in stem or root, equivalent to the stele but not implying the stelar concept.

vascular plants the common name for the Tracheophyta, *q.v.*

vascular tissue in plants, xylem and/or phloem; *alt.* angienchyma.

vascular tunic of eye: choroid, ciliary body, and iris.

vasculum *n.* [L. *vasculum*, small vessel] a pitcher-shaped leaf or ascidium; a small blood vessel.

vasifactive *a.* [L. *vas*, vessel; *facere*, to make] producing new blood vessels; *appl.* cell: angioblast, *q.v.*; *alt.* vasoformative.

vasiform *a.* [L. *vas*, vessel; *forma*, shape] functioning as or resembling a duct; vascular.

vasochorial *a.* [L. *vas*, vessel; *chorion*, skin] endotheliochorial, *q.v.*

vasoconstrictor *a.* [L. *vas*, vessel; *constringere*, to draw tight] causing constriction of blood vessels; *alt.* vasohypertonic.

vasodentine *n.* [L. *vas*, vessel; *dens*, tooth] a variety of dentine permeated by blood vessels.

vasodilatin *n.* [L. *vas*, vessel; *dilatus*, separated] product of protein disintegration corresponding in properties with histamine.

vasodilator *a.* [L. *vas*, vessel; *dilatus*, separated] relaxing or enlarging the vessels; *alt.* vasohypotonic, vasoinhibitory.

vasoformative vasifactive, *q.v.*

vasoganglion *n.* [L. *vas*, vessel; Gk. *gangglion*, little tumour] a dense network of blood vessels such as a rete mirabile.

vasohypertonic vasoconstrictor, *q.v.*

vasohypotonic vasodilator, *q.v.*

vasoinhibitory vasodilator, *q.v.*

vasomotion *n.* [L. *vas*, vessel; *movere*, to move] a change in calibre of blood vessel.

vasomotor *a.* [L. *vas*, vessel; *movere*, to move] *appl.* nerves supplying muscles in walls of blood vessels and regulating calibre of blood vessels, through containing both vasoconstrictor and vasodilator fibres.

vasoperitoneal *a.* [L. *vas*, vessel; Gk. *periteinein*, to stretch around] *appl.* vesicles of archenteron which give rise to body cavity and to the primordial water vascular system in echinoderms.

vasopressin *n.* [L. *vas*, vessel; *pressus*, pressure] a polypeptide hormone secreted by posterior lobe of pituitary gland that stimulates plain muscle, causes constriction of arteries, and raises blood pressure, and has an antidiuretic effect; *alt.* β-hypophamine, pitressin, antidiuretic hormone.

vasotocin *n.* [L. *vas*, vessel; Gk. *tokos*, birth] a peptide hormone having properties similar to vasopressin and oxytocin, and increasing the permeability of skin in amphibians, secreted by the posterior lobe of pituitary gland in lower vertebrates and by mammalian foetus and pineal gland.

vastus *n.* [L. *vastus*, immense] a division of quadriceps muscle of thigh.

Vater's ampulla [*A. Vater*, German anatomist] dilatation of the united common bile duct and pancreatic duct.

Vater's corpuscles Pacinian corpuscles, *q.v.*

Vaucheriales *n.* [*Vaucheria*, generic name] Heterosiphonales, *q.v.*

V-chromosomes mediocentric chromosomes, *q.v.*

vector *n.* [L. *vector*, bearer] a carrier, as many invertebrate hosts, of pathogenic organisms; any agent transferring a parasite to a host.

vegetable base alkaloid, *q.v.*

vegetable casein legumin, *q.v.*

vegetable insulin glucokinin, *q.v.*

vegetable starch starch, as compared with glycogen; *alt.* amylum.

vegetal pole that side of a blastula at which megameres collect; the lower more slowly segmenting portion of a telolecithal egg due to presence of yolk; *alt.* vegetative pole; *cf.* animal pole.

vegetative *a.* [L. *vegetare*, to enliven] *appl.* stage of growth in plants when reproduction does not occur; assimilative, *appl.* fungi; *appl.* foliage shoots on which flowers are not formed; *appl.* reproduction by bud formation or other asexual method in plants and animals; *appl.* nervous system, the autonomic nervous system.

vegetative cone the apical meristem, *q.v.*

vegetative nucleus meganucleus, *q.v.*; pollen tube nucleus.

vegetative pole vegetal pole, *q.v.*

veil *n.* [L. *velum*, covering] velum, *q.v.*; calyptra, *q.v.*; indusium, *q.v.*

veins *n.plu.* [L. *vena*, vein] branched vessels which convey blood to heart; ribs or nervures of insect wing; ridges between lamellae of agarics; branching ribs or strands of vascular tissue of leaf; *alt.* venae.

vela *plu.* of velum.

velamen *n.* [L. *velamen*, covering] a membrane; a multiple epidermal sheath covering aerial roots of some orchids and aroids, acting as specialized moisture-absorbing tissue; *alt.* velamentum; *plu.* velamina.

velaminous *a.* [L. *velamen*, covering] having a velamen.

velangiocarpy *n.* [L. *velum*, covering; Gk. *anggeion*, vessel; *karpos*, fruit] the enclosure of a fungal fruit body by an early formed veil or velum.

velar *a.* [L. *velum*, covering] *pert.* or situated near a velum.

velarium *n.* [L. *velarium*, awning] velum of certain Cubomedusae, which differs from a true velum in containing endodermic canals; margin of umbrella, including tentacles, in Scyphozoa.

velate *a.* [L. *velum*, covering] veiled; covered by a velum.

veliger *n.* [L. *velum*, covering; *gerere*, to carry] second stage in larval life of certain molluscs, developed from a trochophore, where the head bears a velum.

vellus *n.* [L. *vellus*, fleece] the stipe of certain fungi; hair replacing primary hair (lanugo).

velum *n.* [L. *velum*, covering] a membrane or structure similar to a veil; in Hydrozoa and certain jellyfishes, the annular membrane projecting inwards from margin of bell; membrane in connection with buccal cavity in lancelet, *Amphioxus*; flap-like structure for closing off choanae from mouth cavity in Crocodilia; membrane-like structure bordering oral cavity of certain ciliates; ciliated swimming organ of veliger larva; mass of tissue stretching from stipe of pileus in certain Basidiomycetes; membrane partly covering opening of fovea in the quillwort, *Isoetes*; *alt.* veil; *plu.* vela.

velum partiale partial veil, *q.v.*

velum universale universal veil, *q.v.*

velutinous *a.* [It. *velluto*, velvet] velvety; covered with very fine, dense, short upright hairs.

velvet *n.* [L.L. *velluetum*, velvet] soft vascular skin which covers antlers of deer during growth.

vena *n.* [L. *vena*, vein] a vein, *q.v.*, especially a large blood vessel carrying blood to heart; *plu.* venae.

vena cava one of the main veins that carries blood to the right auricle of the heart.

vena comitantes veins accompanying or alongside an artery or nerve.

venae *plu.* of vena.

venation *n.* [L. *vena*, vein] the whole system of veins; nervation, *q.v.*

venin *n.* [L. *venenum*,.poison] a toxic substance of snake venom.

veniplex *n.* [L. *vena*, vein; *plexus*, interwoven] a plexus of veins.

venomosalivary *a.* [L. *venenum*, poison; *salivare*, to salivate] *pert.* salivary glands of which the secretion is poisonous.

venomous *a.* [L. *venenum*, poison] having poison glands; able to inflict a poisonous wound.

venose *a.* [L. *vena*, vein] with many and prominent veins.

venous *a.* [L. *vena*, vein] *pert.* veins; *appl.* blood returning to heart after circulation in body.

vent *n.* [L. *findere*, to cleave] the anus; cloacal or anal aperture in lower vertebrates; *appl.* feather: an under-tail covert.

venter *n.* [L. *venter*, belly] the abdomen; lower abdominal surface; protuberance, as of muscle; smooth concave surface; swollen basal portion of an archegonium containing an ovum.

venter canal cell ventral canal cell, *q.v.*

ventrad *adv.* [L. *venter*, belly; *ad*, to] towards lower or abdominal surface; *cf.* dorsad.

ventral *a.* [L. *venter*, belly] *pert.* or situated on lower or abdominal surface; *pert.* or designating that surface of a petal, etc., that faces centre or axis of flower; *appl.* lower surface of flattened ribbon-like thalli; *pert.* a venter; *cf.* dorsal.

ventral canal cell a small cell without a wall at the base of the neck canal cells and above the ovum in an archegonium; *alt.* venter canal cell.

ventral root a cranial nerve root with some sensory fibres; a spinal nerve root with motor fibres.

ventrianal *a.* [L. *venter*, belly; *anus*, anus] *appl.* plate formed by fused ventral and anal sclerites, in certain Acarina.

ventricle *n.* [L. *ventriculus*, *dim.* of *venter*, belly] a cavity or chamber, as in heart or brain; *appl.* fusiform fossa of larynx; gizzard of birds; chylific ventricle: mid-gut of insects; *alt.* ventriculus.

ventricose *a.* [L. *venter*, belly] swelling out in the middle, or unequally, *appl.* corolla, spores, stipe, shells.

ventricular *a.* [L. *ventriculus*, ventricle] *pert.* a ventricle, *appl.* ligaments and folds of larynx, *appl.* septum and valves in heart.

ventricula stomach, *q.v.*; ventricle, *q.v.*

ventrodorsal *a.* [L. *venter*, belly; *dorsum*, back] extending from ventral to dorsal surface.

ventrolateral *a.* [L. *venter*, belly; *latus*, side] at side of ventral region; central and lateral; *alt.* anterolateral.

venule *n.* [L. *venula*, *dim.* of *vena*, vein] small vein of leaf or a vent wing; small vessel conducting venous blood from capillaries to vein.

venulose *a.* [L. *venula*, veinlet] having numerous small veins.

Vermes *n.* [L. *vermis*, worm] an old name for a group of animals that looked like worms and included the Platyhelminthes, Nemertini, Nematoda, and Annelida.

vermian *a.* [L. *vermis*, worm] worm-like; *pert.* vermis.

vermicular *a.* [*Dim.* of L. *vermis*, worm] resembling a worm in appearance or movement.

vermiculate *a.* [*Dim.* of L. *vermis*, worm] marked with numerous sinuate fine lines or bands of colour or by irregular depressed lines.

vermiculation *n.* [*Dim.* of L. *vermis*, worm] worm-like or peristaltic movement; fine wavy markings.

vermicule *n.* [*Dim.* of L. *vermis*, worm] a small worm-like structure especially motile or ookinete stage of some Sporozoa.

vermiform *a.* [L. *vermis*, worm; *forma*, shape] shaped like a worm, *alt.* helminthoid; *appl.* certain Protista and numerous structures, especially appendix; *appl.* body, a scolecite, *q.v.*; *appl.* cells: plasmatocyte-like blood cells in insects.

vermiform appendix a remnant of the caecum found in some mammals, in man being a worm-like blind tube extending from the gut; *alt.* epityphlon.

vermis *n.* [L. *vermis*, worm] annulated median portion of cerebellum; central portion of cerebellum in birds and reptiles.

vernal *a.* [L. *vernalis*, of the spring] *pert.* or appearing in mid or late spring.

vernalin *n.* [L. *vernalis*, of the spring] a hypothetical hormone believed to control the temperature effect in vernalization, and possibly the formation of florigen.

vernalization *n.* [L. *vernalis*, of the spring] the exposure of certain plants or their seeds to a period of cold which is necessary either to cause them to flower at all, or to make them flower earlier than usual, and is used especially on cereals such as winter varieties of wheat, oats, and rye; *alt.* jarovization, yarovization.

vernation *n.* [L. *vernatio*, sloughing] the arrangement of leaves within a bud; *alt.* foliation; *cf.* prefoliation.

vernicose *a.* [F. *vernis*, varnished] having a varnished appearance, glossy.

vernix caseosa shed flakes of epidermis mixed with sebaceous secretions gradually coating the skin during 2nd half of human foetal life.

verruca *n.* [L. *verruca*, wart] a wart-like projection; a wart-like apothecium; one of small wart-like projections surrounding base of polyps in many Alcyonaria; one of the blister-like evaginations of body wall in some sea anemones; a cuticular protuberance tufted with bristles, as in certain larval insects.

verruciform *a.* [L. *verruca*, wart; *forma*, shape] wart-shaped.

verrucose *a.* [L. *verrucosus*, warty] covered with wart-like projections.

verruculose *a.* [L. *verrucula*, small wart] covered with minute wart-like excrescences.

versatile *a.* [L. *versatilis*, turning around] swinging freely, *appl.* anthers; capable of turning backwards and forwards, *appl.* bird's toe.

versicoloured *a.* [L. *versicolor*, changing colour] variegated in colour; capable of changing colour.

versiform *a.* [L. *versare*, to turn; *forma*, form] changing shape; having different forms.

Verson's glands ecdysial glands, *q.v.*

vertebra *n.* [L. *vertebra*, turning joint] any of the bony or cartilaginous segments that make up the backbone; one of the ossicles in an ophiuroid arm; *alt.* spondyl, spondylus.

vertebral *a.* [L. *vertebra*] *pert.* backbone; *appl.* various structures situated near or connected with backbone, or with any structure like a backbone; *alt.* spondylous.

vertebral column the series of vertebrae running from head to tail dorsally in the body of vertebrates, and protecting the spinal cord; *alt.* backbone, spinal column.

vertebra prominens seventh cervical vertebra.

vertebrarterial canal canal formed by foramina in transverse processes of cervical vertebrae or between cervical rib and vertebra.

Vertebrata, vertebrates *n., n.plu.* [L. *vertebra*, vertebra] Craniata, *q.v.*

vertebration *n.* [L. *vertebra*, vertebra] division into segments or parts resembling vertebrae.

vertebropelvic *a.* [L. *vertebra*, vertebra; *pelvis*, basin] *appl.* ligaments: the iliolumbar, sacrospinous, and sacrotuberous ligaments.

vertex *n.* [L. *vertex*, top] top of head; highest point of skull; region between compound eyes in insects.

vertical *a.* [L. *vertex*, top] standing upright; lengthwise, in direction of axis; *pert.* vertex of head.

vertical margin limit between frons and occiput in Diptera.

verticil *n.* [L. *verticillus*, *dim.* of *vertex*, whorl] an arrangement of flowers, inflorescences, or other structures about the same point on the axis; *alt.* whorl.

verticillaster *n.* [L. *verticillus*, small whorl; *aster*, star] a much condensed cyme with appearance of whorl, but in reality arising in axils of opposite leaves.

Verticillatae *n.* [L. *verticillus*, small whorl] Casuarinales, *q.v.*

verticillate *a.* [L. *verticillus*, small whorl] arranged in verticils, whorled; *appl.* antennae whose joints are surrounded, at equal distances, by stiff hairs.

veruculate *a.* [L. *veruculum*, skewer] rod-shaped and pointed.

verumontanum *n.* [L. *veru*, spit; *montanum*, mountainous] ridge on floor of urethra, with small elevation where seminal ducts enter the colliculus seminalis; *alt.* urethral crest, crista urethralis.

vesica *n.* [L. *vesica*, bladder] bladder, especially urinary bladder.

vesica fellea gall bladder, *q.v.*

vesical *a.* [L. *vesica*, bladder] *pert.* or in relation with bladder, *appl.* arteries, etc.

vesica prostatica uterus masculinus, *q.v.*

vesica urinaria urocyst, *q.v.*

vesicle *a.* [L. *vesicula*, *dim.* of *vesica*, bladder] small globular or bladder-like air space in tissues; small cavity or sac usually containing fluid; a hyphal swelling as in mycorrhiza; hollow prominence on shell or coral; base of postanal segment in scorpions; one of 3 primary cavities of brain; *alt.* vesicula.

vesicula *n.* [L. *vesicula*, small bladder] a small bladder-like cyst or sac; vesicle, *q.v.*

vesicular *a.* [L. *vesicula*, small bladder] composed of or marked by presence of vesicle-like cavities; bladder-like.

vesicular–arbuscular (V–A) *appl.* a kind of mycorrhiza in which the fungus is a phycomycete and the roots do not show morphological peculiarities, the hyphae bear globular swellings (vesicles) and minute processes (arbuscles) which are finely branched and introduced into the host cells.

vesicular gland a gland in tissue underlying epidermis in plants and containing essential oils.

vesicular ovarian follicle Graafian follicle, *q.v.*

vesiculase *n.* [L. *vesicula*, small bladder] an enzyme from secretion of prostate gland, capable of coagulating contents of seminal vesicles.

vesicula seminalis a sac in which spermatozoa complete their development and are stored; *alt.* seminal receptacle, seminal vesicle, glandula vesiculosa.

vespertine *a.* [L. *vespertinus*, of the evening] blossoming or active in the evening; *alt.* crepuscular.

vespoid *a.* [L. *vespa*, wasp; Gk. *eidos*, like] wasp-like; *alt.* vespiform.

vessel *n.* [L. *vascellum*, *dim.* of *vas*, vessel] any tube or canal with properly defined walls in which fluids as blood, lymph, sap, etc., move; a continuous tube formed by superposition of numerous cells whose common walls have perforations or have completely broken down, as in xylem vessel.

vessel element a unit of a vessel consisting of a cell whose cross walls have broken down, as in xylem.

vessel feeder solenophage, *q.v.*

vestibular *a.* [L. *vestibulum*, porch] *pert.* a vestibule, *appl.* artery, bulb, fissure, gland, nerve, etc.

vestibulate *a.* [L. *vestibulum*, porch] in the form of a passage betweenn 2 channels; resembling, or having, a vestibule.

vestibule, vestibulum *n.* [L. *vestibulum*, porch] a cavity leading into another cavity or passage, as cavity of ear labyrinth; space between labia minora containing opening of urethra; portion of ventricle directly below opening of aortic arch; cavity leading to larynx; nasal cavity; posterior chamber of bird's cloaca; small tubular or grooved depression leading to mouth in most ciliates; space within circle of tentacles in Endoprocta; pit leading to stoma of leaf.

vestibulocochlear *a.* [L. *vestibulum*, porch; Gk. *kochlias*, snail] *appl.* nerve: the auditory nerve, which innervates the inner ear.

vestige *n.* [L *vestigium*, trace] a small degenerate or imperfectly developed organ or part which may have been complete and functional in some ancestor.

vestigial *a.* [L. *vestigium*, trace] small and imperfectly developed; *pert.* vestige.

vestiture *n.* [L. *vestitus*, garment] a body covering, as of scales, hairs, feathers, etc.

vexilla *plu.* of vexillum.

vexillar(y) *a.* [L. *vexillum*, standard] *pert.* a vexillum; *appl.* type of imbricate aestivation in which upper petal is folded over others.

vexillate *a.* [L. *vexillum*, standard] bearing a vexillum.

vexillum *n.* [L. *vexillum*, standard] the upper petal standing at the back of a papilionaceous flower such as pea, bean, etc., which helps to make the flower conspicuous, *alt.* banner, standard; the vane of a feather, *alt.* web.

via *n.* [L. *via*, way] a way or passage.

viable *a.* [F. *vie*, life] capable of living; capable of developing or of surviving parturition.

viatical *a.* [L. *via*, way] *appl.* plants growing by the roadside.

vibracularium *n.* [L. *vibrare*, to quiver] the vibracula collectively.

vibraculum *n.* [L. *vibrare*, to quiver] modified whip-like avicularium for defence purposes in ectoproct polyzoans; *plu.* vibracula.

vibratile *a.* [L. *vibrare*, to quiver] oscillating; *appl.* antennae of insects.

vibratile corpuscles corpuscles closely resembling sperms found in coelomic fluid of starfish.

vibratile membrane a membrane formed by fusion of cilia for wafting food to the mouth in ciliates, sometimes called undulating membrane.

vibrioid *a.* [L. *vibrare*, to quiver; Gk. *eidos*, like] like a vibrio, a bacterium with thread-like appendages and a vibratory motion.

vibrioid body a slender cylindrical body found in superficial cytoplasmic layer of certain algae and fungi.

vibrissa *n.* [L. *vibrissa*, nostril hair] a hair growing on nostril or face of animals, as whiskers of cat, acting often as tactile organ; a feather at base of bill or around eye; one of paired bristles near upper angles of mouth cavity in Diptera; one of the sensitive hairs of an insectivorous plant, as of *Dionaea*, Venus' fly trap.

vibrotaxis *n.* [L. *vibrare*, to quiver; Gk. *taxis*, arrangement] seismotaxis, *q.v.*

vicariad *n.* [L. *vicarius*, deputy] a number of species inhabiting different areas and produced from a common ancestor.

vicariation *n.* [L. *vicarius*, deputy] the separate occurrence of corresponding species, as reindeer and caribou, in corresponding but separate environments.

vicinism *n.* [L. *vicinus*, neighbour] tendency to variation due to proximity of related forms.

Vicq-d'Azyr, bundles of [F. *Vicq-d'Azyr*, French anatomist] a bundle of nerve fibres running from the corpora mamillaria to the thalamus; *alt.* thalamomamillary fasciculus, mamillothalamic tract.

villi *plu.* of villus.

villiform *a.* [L. *villus*, shaggy hair; *forma*, shape] having form or appearance of velvet, *appl.* dentition.

villikinin *n.* [L. *villus*, shaggy hair; Gk. *kinein*, to move] a factor, in yeast and duodenal mucosa, which stimulates contractility of intestinal villi.

villose, villous *a.* [L. *villus*, shaggy hair] pubescent; shaggy; having villi or covered with villi.

villus *n.* [L. *villus*, shaggy hair] one of the minute vascular processes on small intestine lining, *alt.* trophonema; one of processes on chorion through which nourishment passes to embryo; Pacchionian body, *q.v.*, of arachnoid; invagination, into joint cavity, of a synovial membrane; fine straight process on epidermis of plants; *plu.* villi.

vimen *n.* [L. *vimen*, osier] long slender shoot or branch; *plu.* vimina.

vinculum *n.* [L. *vinculum*, bond] slender tendinous bands; accessory connecting bands of fibres, as vincula brevia; band uniting 2 main tendons of foot in birds; part of sternum bearing claspers of male insects; sternal region of 9th segment in Lepidoptera; *alt.* coxosternum; *plu.* vincula.

Violales *n.* [*Viola*, generic name] an order of dicots (Dilleniidae) having oily endosperm.

violaxanthin *n.* [*Viola*, generic name; Gk. *xanthos*, yellow] a xanthophyll carotenoid pigment found in algae, especially of the classes Chlorophyceae and Phaeophyceae, in pansies and in some other plants.

viosterol *n.* [ultra*vio*let; ergo*sterol*] vitamin D_2, calciferol, especially a preparation of it dissolved in edible vegetable oil and given as a dietary supplement to young children.

viral *a.* [L. *virus*, poison] *pert.*, consisting of, or due to, a virus.

Virales *n.* [L. *virus*, poison] the viruses, when considered an order of Schizophyta.

virescence *n.* [L. *virescere*, to grow green] production of green colouring matter in petals instead of usual pigment.

virescent turning greenish or green.

virgalium *n.* [L. *virga*, rod] a series of rod-like elements forming petaloid rays of an ambulacral plate, as in some Asteroidea.

virgate *a.* [L. *virga*, rod] rod-shaped; striped.

virgula *n.* [L. *dim* of *virga*, rod] a small rod, axis of graptolite; a paired or bilobed structure or organ at oral sucker in certain trematodes.

virgulate *a.* [L. *virgula*, little rod] with or like a small rod or twig; having minute stripes.

viridant *a.* [L. *viridare*, to make green] becoming or being green.

virion *n.* [L. *virus*, poison] a complete virus consisting of a protein coat and nucleic acid core.

viroids *n.plu.* [L. *virus*, poison; Gk. *eidos*, form] hypothetical ultramicroscopic virus-like particles which live symbiotically with the host but can give rise to viruses by mutation, *cf.* neovirus, palaeovirus; a virus-like potential pathogen not producing nucleoprotein particles.

virology *n.* [L. *virus*. poison; Gk. *logos*, discourse] the study of viruses; *alt.* virusology.

virose, virous *a.* [L. *virosus*, poisonous] containing a virus.

virulence *n.* [L. *virus*, poison] the ability to cause disease.

virulent bacteriophage (phage) a phage which causes lysis of the host; *alt.* intemperate phage; *cf.* temperate phage.

virulin aggressin, *q.v.*

virus *n.* [L. *virus*, poisonous liquid] an intracellular obligate parasite, which is inert outside the host cell, whose nuclear material is DNA or RNA, visible only under the electron microscope and causing many diseases in man, animals, and plants, and having a simple structure, often of protein coat surrounding nucleic acid core, although the protein coat may be absent; formerly used for any disease-causing micro-organism, when the present use of virus was known as an ultravirus or filterable virus.

virusology virology, *q.v.*

virus receptors the region on the cell membrane containing neuraminic acid at which viruses attach.

viscera *n.plu.* [L. *viscera*, bowels] the internal organs contained in various cavities of body; *sing.* viscus.

visceral *a.* [L. *viscera*, bowels] *pert.* viscera, *appl.* to numerous structures and organs.

visceral arches a series of arches developed in connection with the mouth and pharynx including the gill arches and gill bars; also used for gill arch, *q.v.*

visceral clefts a series of furrows or clefts in neck region between successive visceral arches; *alt.* gill clefts, branchial grooves.

visceral hump or mass in molluscs, a central concentration of the viscera covered by a soft skin, the mantle.

viscerocranium *n.* [L. *viscera*, bowels; L.L. *cranium*, skull] jaws and visceral arches; *alt.* splanchnocranium; *cf.* neurocranium.

visceromotor *a.* [L. *viscera*, bowels; *movere*, to move] carrying motor impulses to viscera.

viscid *a.* [L. *viscum*, mistletoe] sticky.

viscin *n.* [L. *viscum*, mistletoe] sticky substance obtained from various plants, especially from berries of mistletoe.

viscidium *n.* [L. *viscum*, mistletoe] in orchid flower, a sticky disc at the end of the stalk of the pollinium, by which it is attached to an insect's head.

viscosity *n.* [L. *viscosus*, viscous] internal friction in fluids due to adherence of particles to one another.

viscus *sing.* of viscera.

visual axis the straight line between the point to which the focused eye is directed and the fovea centralis; *alt.* visual line.

visual purple porphyropsin, *q.v.*; rhodopsin, *q.v.*

visual violet iodopsin, *q.v.*

visual white the product of visual yellow irradiated by ultraviolet rays, being vitamin A alcohol; *alt.* leucopsin.

visual yellow a pigment formed by the action of light upon visual purple; a retinal pigment in certain fish; *alt.* xanthopsin.

vital capacity of lungs, the sum of complemental, tidal, and supplemental air.

vital force form of energy manifested in living phenomena when considered distinct from chemical, physical, and mechanical forces; *alt.* élan vital, bionergy; *cf.* horme.

vital functions functions of body on which life depends.

vitalism *n.* [L. *vita*, life] belief of vitalists, that phenomena exhibited in living organisms are due to a special force distinct from physical and chemical forces.

vital staining staining of living cells or tissues with non-toxic dyes.

vitamers *n.plu.* [L. *vita*, life; Gk. *meros*, part] compounds having a chemical structure and physiological effects similar to those of natural vitamins.

vitamin *n.* [L. *vita*, life; *ammoniacum*, resinous gum] any of various organic compounds needed in small quantities for various metabolic processes and synthesized by plants and some lower animals, but necessary in the diet of higher animals and some micro-organisms, lack of the appropriate vitamin in animals causing a deficiency disease; *alt.* accessory food factor, nutramin.

vitamin A a fat-soluble vitamin derived from carotenes, found in liver oils of certain fish and in milk and eggs, which is a precursor of the light-sensitive pigments of rods and cones of the eye, and whose deficiency retards growth, causes night blindness and keratinization of epithelia; it occurs as vitamin A_1, known as axerophthol, antixerophthalmic vitamin, retinol, and as vitamin A_2, dehydroretinol.

vitamin B complex a group of water-soluble vitamins obtained from similar sources such as yeast, cereal germ, liver, and given separate B numbers, now usually replaced by specific names; *alt.* bios.

vitamin Bc folic acid, *q.v.*

vitamin B_1 thiamine, *q.v.*

vitamin B_2 riboflavin, *q.v.*

vitamin B_3 pantothenic acid, *q.v.*

vitamin B_4 biotin, *q.v.*

vitamin B_5 pantothenic acid, *q.v.*

vitamin B_6 any or all of 3 interconvertible compounds, pyridoxine, pyridoxal, and pyridoxamine (especially pyridoxine) found in eggs, milk, meat, whole grain, fresh vegetables, and yeast, which were first defined as the vitamin which cured dermatitis in rats, and are now considered essential in the nutrition of vertebrates including man, for some insects, and as a growth factor for micro-organisms; *alt.* adermin.

vitamin B_7 nicotinamide or nicotinic acid, *q.v.*

vitamin B_8 *see* adenylic acid.

vitamin B_{12} *see* cobalamine.

vitamin B_{17} laetrile, *q.v.*

vitamin C a water-soluble vitamin, ascorbic acid, found in fresh fruit and vegetables, especially citrus fruit and blackberries, required in the diet of primates and some other animals, deficiency causing scurvy, but its precise role in metabolism is not known, although it is thought to act as a cofactor in oxidation of tyrosine and possibly other oxidation–reduction reactions, and is involved in synthesis of bone, cartilage, and dentine; *alt.* cevitamic acid, antiscorbutic vitamin, hexuronic acid.

vitamin D any or all of several fat-soluble antirachitic steroids, found especially in fish liver oils, egg yolk, and milk, or by ultraviolet irradiation, and being necessary for normal bone and teeth structure, increasing calcium and phosphate absorption from the gut; vitamin D_2 is calciferol, *q.v.*, vitamin D_3 is cholecalciferol, *q.v.*, vitamin

D₄ is dihydrotachysterol, *q.v.*; *alt.* antirachitic vitamin.

vitamin E any of several fat-soluble vitamins, the most common being α-tocopherol (vitamin E or E₁), β-tocopherol (vitamin E₂), and γ-tocopherol (vitamin E₃), which occur in leaves of various plants and in oils of some seed germs, their absence leading to sterility in some animals, and possibly necessary for reproduction in all mammals, having strong antioxidant properties, and may be necessary for stabilizing membranes and preventing oxidation in cells; *alt.* antisterility factor, fertility vitamin.

vitamin G riboflavin, *q.v.*

vitamin H biotin, *q.v.*

vitamin K a group of vitamins necessary for blood clotting as they are concerned in production of prothrombin in the liver, and which are obtained from green leaves, putrefied fish, or synthesized by bacteria in the intestines; vitamin K₁ is α-phylloquinone or phytonadione), vitamin K₂ is β-phylloquinone (farnoquinone), vitamin K₃ is menadione; *alt.* antihaemorrhagic factor, blood-coagulation factor.

vitamin L *see* lactation vitamins.

vitamin M folic acid, *q.v.*

vitamin P citrin, with its active constituent hesperidin, which affects the permeability and fragility of blood capillaries; *alt.* permeability factor or vitamin, bioflavonoid.

vitamin P-P pellagra-preventive factor, *q.v.*

vitazyme *n.* [L. *vita*, life; *zymē*, leaven] an enzyme having vitamins as part of its chemical structure.

vitellarium *a.* [L. *vitellus*, yolk] yolk gland, *q.v.*

vitelligenous *a.* [L. *vitellus*, yolk; *gignere*, to beget] producing yolk; *alt.* vitellogene, vitellogenous.

vitellin *n.* [L. *vitellus*, yolk] the phosphoprotein of egg yolk, *alt.* ovovitellin; similar or related substance in seeds.

vitelline *a.* [L. *vitellus*, yolk] *pert.* yolk or yolk-producing organ, *app.* artery, vein, duct, gland; *appl.* membrane: zona pellucida, *q.v.*; yolk-coloured.

vitelline body yolk nucleus, *q.v.*

vitelloduct *n.* [L. *vitellus*, yolk; *ductus*, led] albuminiferous canal, duct conveying vitellus from yolk gland into oviduct.

vitellogen *n.* [L. *vitellus*, yolk; *gignere*, to produce] yolk gland, *q.v.*

vitellogene, vitellogenous vitelligenous, *q.v.*

vitellogenesis *n.* [L. *vitellus*, yolk; Gk. *genesis*, descent] yolk formation.

vitellogenic hormone juvenile hormone, *q.v.*

vitellogenin *n.* [L. *vitellus*, yolk; *gignere*, to produce] a protein which is produced in the liver of certain female amphibians, such as the toad *Xenopus*, and is transformed into yolk protein.

vitellophags *n.plu.* [L. *vitellus*, yolk; Gk. *phagein*, to eat] isolated cells forming hypoblast of crustacean and insect egg.

vitellose *n.* [L. *vitellus*, yolk] a substance formed in digestion of yolk.

vitellus *n.* [L. *vitellus*, yolk] yolk of ovum or egg.

vitrella *n.* [L. *vitrum*, glass] a cell of an ommatidium which secretes the crystalline cone; *alt.* retinophore.

vitreodentine *n.* [L. *vitreus*, glassy; *dens*, tooth] a very hard variety of dentine; *alt.* vitrodentine.

vitreous *a.* [L. *vitreus*, glassy] hyaline; transpar-

ent; *appl.* humor or body, the clear jelly-like substance in inner chamber of eye; *appl.* membrane: the innermost layer of dermic coat of hair follicle, and posterior elastic lamina of cornea.

vitreum *n.* [L. *vitreus*, glassy] vitreous humor of the eye; *alt.* vitrina.

vitrification *n.* [L. *vitrum*, glass; *facere*, to make] condition of cells or organisms instantaneously frozen but able to resume all vital activities on being thawed out.

vitrina vitreum, *q.v.*

vitrodentine vitreodentine, *q.v.*

vitta *n.* [L. *vitta*, band or fillet] one of the resinous canals in pericarp of dicots of the Umbelliferae and some other families; a longitudinal ridge of diatoms; a band of colour; *plu.* vittae.

vittate *a.* [L. *vittatus*, with a fillet] having ridges, stipes, or bands lengthwise.

vivification *n.* [L. *vivus*, living; *facere*, to make] one of series of changes in assimilation by which material which has been taken up by cell is able to exhibit phenomena of living protoplasm.

viviparous *a.* [L. *vivus*, living; *parere*, to beget] producing young alive rather than in eggs, as in most mammals, *alt.* zoogonous, *cf.* oviparous, ovoviviparous; germinating while still attached to the parent plant as some seeds, e.g. mangrove; multiplying by vegetative means such as buds or bulbils in the position of flowers; *n.* viviparity, vivipary.

vocal cords folds of mucous membrane projecting into larynx by vibration of which sound is produced.

voice box larynx, *q.v.*

volar *a.* [L. *vola*, palm of hand] *pert.* palm of hand or sole of foot.

Volkmann's canals [*A. W. Volkmann*, German physiologist] simple canals piercing circumferential or periosteal lamellae of bone, for blood vessels, and joining Haversian canal system.

voltine *a.* [It. *volta*, time] *pert.* number of broods in a year.

voltinism *n.* [It. *volta*, time] polymorphism in insects where some enter diapause and some do not.

Voltziales *n.* [*Voltzia*, generic name] an order of conifers in some classifications.

voluble *a.* [L. *volvere*, to roll] twining spirally.

voluntary *a.* [L. *voluntas*, will] subject to or regulated by the will.

voluntary muscle striated muscle fibres, i.e. fibres having transverse striations, and under control of the will, such as those attached to the skeleton; *alt.* skeletal, striated, striped muscle; *cf.* involuntary muscle.

volute *a.* [L. *volvere*, to roll] rolled up; spirally twisted.

volutin *n.* [L. *volvere*, to roll] metachromatin found as granules in the cytoplasm, especially of micro-organisms.

volution *n.* [L. *volvere*, to roll] spiral twist of a shell or of cochlea.

volva *n.* [L. *volva*, wrapper] tissue enveloping the pileus and stipe of some Agaricales and Boletales, *alt.* universal veil; universal veil after becoming detached from pileus and limited to the lower part of the stipe forming cup or pouch; *alt.* velum universale, teleoblema, teleoblem, teleblem.

volvate *a.* [L. *volva*, wrapper] provided with a volva.

volvent *n.* [L. *volvere*, to roll] desmoneme, *q.v.*

Volvocales *n.* [*Volvox*, generic name] an order of green algae having unicellular and colonial forms with motile vegetative stages.

Volvocina *n.* [*Volvox*, generic name] an order of Phytomastigina, also classified as algae of the group Volvocales, being plant-like flagellates each usually with chloroplast and cellulose cell wall, and many being colonial; *alt.* Phytomonadina.

vomer *n.* [L. *vomer*, ploughshare] a bone in nasal region; *alt.* ploughshare bone.

vomerine *a.* [L. *vomer*, ploughshare] *pert.* vomer; *appl.* teeth borne on vomers.

vomeronasal *a.* [L. *vomer*, ploughshare; *nasus*, nose] *appl.* cartilage and organ in region of vomer and nasal cavity; *cf.* Jacobson's cartilage and organ.

vomeropalatine *n.* [L. *vomer*, ploughshare; *palatum*, palate] fused vomer and palatine, in some ganoids and amphibians.

von Baer's law [*K. E. von Baer*, German biologist] recapitulation theory, *q.v.*

vortex *n.* [L. *vortex*, vortex] spiral arrangement of muscle fibres at apex of heart; spiral arrangement of hairs.

vulva *n.* [L. *vulva*, vulva] the external female genitalia, *alt.* pudendum; recess of 3rd ventricle, between columns of fornix; epigynum, *q.v.*

vulviform *a.* [L. *vulva*, vulva; *forma*, shape] like a cleft with projecting lips; shaped like a vulva.

vulvouterine *a.* [L. *vulva*, vulva; *uterus*, womb] *pert.* vulva and uterus.

vulvovaginal *a.* [L. *vulva*, vulva; *vagina*, sheath] *pert.* vulva and vagina.

W

waggle dance a dance of bees consisting of a figure of 8 in which the bee waggles her abdomen from side to side giving information about the distance away of the food source and its direction in relation to the sun; *cf.* round dance.

Wagner's corpuscles [*R. Wagner*, German physiologist] Meissner's corpuscles, *q.v.*

Waldeyer's tonsillar ring [*H. W. G. von Waldeyer*, German anatomist] tonsillar ring, *q.v.*

Wallace's Line [*A. R. Wallace*, English naturalist] imaginary line, separating Australian and Oriental zoogeographical regions, between Bali and Lombok, between Celebes and Borneo, and then eastward of Philippines.

Wallerian degeneration [*A. V. Waller*, English physiologist] degeneration of nerve fibres following injury, produced distally to the injury.

wall pressure (WP) turgor pressure, *q.v.*

wandering cells amoeboid cells of mesogloea; cercids, *q.v.*; migratory leucocytes of areolar tissue; planocytes, *q.v.*; *alt.* aletocytes.

wandering resting cells macrophages in connective tissue, being clasmatocytes and histiocytes.

wandernymph deutonymph, *q.v.*

Warburg–Dickens pathway pentose phosphate pathway, *q.v.*

Warburg's factor [*O. H. Warburg*, German physiologist] cytochrome oxidase, *q.v.*; *alt.* Warburg's respiratory enzyme.

Warburg's yellow enzyme yellow enzyme, *q.v.*

warm-blooded homoiothermal, *q.v.*

warning coloration aposematic coloration, *q.v.*

wart *n.* [A.S. *wearte*, wart] a dry excrescence formed on skin; firm glandular protuberance; veruca, *q.v.*

water bears the common name for the tardigrades, *q.v.*

water culture experimental raising of plants in water to see effects of different nutrient solutions; *cf.* hydroponics.

water gland structure in mesophyll of leaves regulating secretion of water through hydathode.

water pore in various invertebrates, a pore by which water tubes connect to the exterior, especially a pore in echinoderms from which the madreporite is derived; hydathode, *q.v.*

water potential (WP) suction pressure, *q.v.*

water stoma hydathode, *q.v.*; *plu.* water stomata.

water vascular system system of canals circulating watery fluid throughout body of Echinodermata; also applied to excretory system of Platyhelminthes; *alt.* hydrocoel.

water vesicle an enlarged epidermal plant cell with a high water content.

Watson–Crick DNA model [*J. D. Watson*, American geneticist; *F. H. C. Crick*, English biophysicist] the structure of DNA suggested by Watson and Crick, consisting of 2 polynucleotide chains arranged as a double helix, each consisting of a sugar-phosphate backbone and purine and pyrimidine bases on the inside of the chain, and paired one purine with one pyrimidine forming cross-linkages, held together with hydrogen bonds; *alt.* duplex DNA.

wattle *n.* [M.E. *watel*, bag] fleshy process under throat of cock or turkey, and of certain reptiles; appendix colli, *q.v.*; barbel, *q.v.*

wax *n.* [A.S. *weax*, wax] esters of alcohol higher than glycerol which are insoluble in water and difficult to hydrolyse and so form protective waterproof layers on leaves, stems, fruits, animal fur, and integuments of insects, and including lanolin, beeswax, carnauba wax.

wax cells modified leucocytes charged with wax, as in certain insects.

wax hair a filament of wax extruded through pore of a wax gland, as in certain scale insects.

wax pocket one of the paired wax-secreting glands on abdomen of worker bee.

W-chromosome the X-chromosome when female is the heterozygous sex.

weathering the action of external factors such as rain, frost, snow, sun, or wind on rocks, altering their texture and composition and converting them to soil.

web *n.* [A.S. *webbe*, web] membrane stretching from toe to toe, as in frog and swimming birds; *see* vexillum; network of threads spun by spiders.

Weber–Fechner Law [*E. H. Weber*, German physiologist; *G. T. Fechner*, German psychologist] Fechner's Law, *q.v.*

Weberian apparatus [*E. H. Weber*, German physiologist] an apparatus found in Cypriniformes, and including Weberian ossicles, a chain of 4 small bones stretching on each side from a membranous fenestra of atrium to air bladder.

Weberian ossicles *see* Weberian apparatus.

Weber's law inference that, within limits, equal relative differences between 2 stimuli of the same kind are equally perceptible.

Weber's line [*M. Weber*, Dutch zoologist] imaginary line separating islands with a preponderant Indo-Malayan fauna from those with a preponderant Papuan fauna.

Weber's organ [*M. I. Weber*, German anatomist] uterus masculinus, *q.v.*

wedge-and-groove suture schindylesis, *q.v.*

wedge bones small infravertebral ossifications at junction of 2 vertebrae, often present in lizards.

weed *n.* [M.E. *wede*, weed] a plant in the wrong place growing where it is not wanted.

Weismannism *n.* [*A. F. L. Weismann*, German biologist] the teaching of Weismann in connection with evolution and heredity, dealing chiefly with continuity of germ plasm, and non-transmissibility of acquired characters.

Weismann's gland or ring ring gland, *q.v.*

Welwitschiales *n.* [*Welwitschia*, generic name] an order of Gnetopsida including the single living genus *Welwitschia*, having a mainly subterranean stem and 2 thick long leaves surviving throughout the plant's life.

wetlands areas of shallow water containing much vegetation.

whales the common name for the Cetacea, *q.v.*

Wharton's duct [*T. Wharton*, English anatomist] the duct of the submaxillary gland; *alt.* submaxillary duct.

Wharton's jelly the gelatinous core of the umbilical cord.

wheel animals the common name for the Rotifera, *q.v.*

wheel organ locomotory ciliated ring or trochal disc of Rotifera; specialized ciliated epithelial structure in buccal cavity of Cephalochorda.

whiplash *appl.* flagella that are unbranched and not feathery; *cf.* tinsel.

whip scorpions the common name for the Uropygi, *q.v.*

whirl whorl, *q.v.*

white blood cell leucocyte, *q.v.*

white body so-called optic gland of molluscs, a large soft body of unknown function.

white commissure anterior commissure, a transverse band of white fibres forming floor of median ventral fissue of spinal cord.

white fibres unbranched inelastic fibres in connective tissue made of collagen and occurring in wavy bundles; *cf.* yellow fibres.

white matter tracts of medullated nerve fibres in brain and spinal cord, and being paler than grey matter.

white yolk spheres minute vesicles forming a flask-shaped plug in centre of egg yolk, and fine layers alternating with yellow yolk.

whorl *n.* [A.S. *hweorfan*, to turn] the spiral turn or volution of a univalve shell; circle of flowers, parts of a flower, or leaves, arising from one point; verticil, *q.v.*; concentric arrangement of papillary ridges on fingers; *alt.* whirl.

wild type the typical form or genotype of an organism as found in nature.

Willis's circle [*T. Willis*, English anatomist] arterial circle, an anastomosis in subarachnoid space at base of brain.

Willis's valve valve of Vieussens, *q.v.*

wilting loss of turgidity in plant cells, due to inadequate water absorption.

wilting coefficient percentage of moisture in soil when wilting takes place.

wing *n.* [M.E. *winge*, wing] one of 2 lateral petals in a papilionaceous flower; lateral expansion on many fruits or seeds; any broad membranous expansion; large lateral process of sphenoid; forelimb modified for flying, in pterodactyls, birds, and bats; flight organ of insects; *alt.* ala.

wing cells distally rounded polyhedral cells in epithelium of cornea, proximally with extensions between heads of basal cells; *alt.* umbrella cells.

wing coverts tectrices, *q.v.*

winged insects the common name for the Pterygota, *q.v.*

winged stem stem having photosynthetic expansions.

wingless insects the common name for the Apterygota, *q.v.*

wing pad undeveloped wing of insect pupae.

wing quill remex, *q.v.*

wing sheath elytrum of insects.

Winslow's foramen [*J. B. Winslow*, Danish anatomist] epiploic foramen, *q.v.*

winter bud dormant bud, protected by hard scales during winter; teleutospore, *q.v.*

winter egg egg of many freshwater forms, provided with thick shell which preserves it as it lies quiescent during winter; *cf.* summer egg.

Wirsung's duct [*J. G. Wirsung*, Bavarian surgeon] the main pancreatic duct; *alt.* Santorini's duct.

wisdom teeth four molar teeth which complete permanent set in man, erupting late.

wobble hypothesis an hypothesis to explain how one tRNA may recognize more than one codon, because there is a certain amount of 'wobble' around the 3rd base of an anticodon so that it can pair with more than one base of a codon.

Wolffian *a.* [*C. F. Wolff*, German embryologist] *appl.* certain structures first discovered by Wolff.

Wolffian body mesonephros, *q.v.*

Wolffian duct duct of the mesonephros; *alt.* Leydig's duct, mesonephric duct.

Wolffian ridges ridges which appear on either side of middle line of early embryo, and upon which limb buds are formed.

Wolfring's glands [*E. F. Wolfring*, Polish ophthalmologist] tubulo-alveolar glands near proximal end of tarsi of eyelids, with ducts opening on conjunctiva.

wolf tooth a small premolar tooth at front of premolar series, occasionally present in horses.

wood *n.* [A.S. *wudu*, wood] secondary xylem; the hard substance of a tree stem; formerly used for xylem of vascular bundle.

woodlice *see* Isopoda.

wood vessel a xylem vessel; *alt.* trachea.

Woolner's tubercle [*T. Woolner*, British sculptor] Darwinian tubercle, *q.v.*

worker non-fertile female in a colony of social insects.

worm *n.* [A.S. *wyrm*, worm] a member of the Vermes, *q.v.*; lytta as of dog.

Wormian bones [*O. Worm*, or *Wormius*, Danish anatomist] sutural bones, *q.v.*

Woronin bodies [*M. S. Woronin*, Russian mycologist] metachromatic bodies in protoplasm of certain hyphal cells, as in Discomycetes.

Woronin hypha a hypha inside coil of perithecial hyphae and giving rise to ascogonia, as in Sphaeriales; *alt.* scolecite.

wound cambium cambium forming protective tissue at site of an injury.

wound hormones substances produced in wounded cells, which stimulate renewed growth near the wounds; *cf.* traumatin.

WP wall pressure, *q.v.*; water potential, *q.v.*

w-substance a hormone secreted by the pituitary gland and inducing contraction of chromatophores and lightening of the animal.

X

xanthein *n.* [Gk. *xanthos*, yellow] a water-soluble yellow colouring matter of cell sap.

xanthin *n.* [Gk. *xanthos*, yellow] a yellow carotinoid pigment occurring in flowers.

xanthine *n.* [Gk. *xanthos*, yellow] a nitrogenous compound, dihydroxypurine, or any of its derivatives, found especially in animal tissues such as muscle, liver, pancreas, spleen, and urine, and also in certain plants, which is made by hydrolysis of guanine and produces uric acid on oxidation.

xanthine oxidase an enzyme transforming xanthine to uric acid, using molecular oxygen; EC 1.2.3.2; *r.n.* xanthine oxidase.

xanthocarpous *a.* [Gk. *xanthos*, yellow; *karpos*, fruit] having yellow fruits.

xanthochroic *a.* [Gk. *xanthos*, yellow; *chrōs*, skin colour] having a yellow or yellowish skin, *appl.* a human ethnological group; *n.* xanthochroism.

xanthodermic *a.* [Gk. *xanthos*, yellow; *derma*, skin] having a yellowish skin.

xanthodont *a.* [Gk. *xanthos*, yellow; *odous*, tooth] having yellow-coloured incisors, *appl.* certain rodents.

xantholeucite *n.* [Gk. *xanthos*, yellow; *leukos*, white] leucoplast of an etiolated plant.

xantholeucophore *n.* [Gk. *xanthos*, yellow; *leukos*, white; *pherein*, to bear] xanthophore, *q.v.*

xanthommatin *n.* [Gk. *xanthos*, yellow; *omma*, eye] a brown ommochrome pigment made by condensation of 2 molecules of hydroxykynurenine and found in *Drosophila* (fruit fly) eye.

xanthophane *n.* [Gk. *xanthos*, yellow; *phainein*, to appear] a yellow chromophane.

xanthophore *n.* [Gk. *xanthos*, yellow; *pherein*, to bear] a yellow chromatophore; *alt.* guanophore, lipophore, ochrophore, xantholeucophore.

Xanthophyceae *n.* [Gk. *xanthos*, yellow; *phykos*, seaweed] a class of algae having an excess of carotinoid over chlorophyll and commonly called golden-brown algae, whose food store is fat or oil, and whose cell wall is absent or containing pectic material and often silica; in some classifications raised to division status called Xanthophyta, in other classifications, considered to be a class of Chrysophyta, formerly included in the Chlorophyceae.

xanthophylls *n.plu.* [Gk. *xanthos*, yellow; *phyllon*, leaf] a group of neutral yellow or brown carotinoid pigments that are oxygenated derivatives of carotenes, and are widely distributed in plants; *alt.* phylloxanthin; *see also* lutein.

Xanthophyta *n.* [Gk. *xanthos*, yellow; *phyton*, plant] in some classifications the Xanthophyceae, *q.v.*, raised to division (phylum) status.

xanthoplast *n.* [Gk. *xanthos*, yellow; *plastos*, formed] a yellow plastid or chromatophore.

xanthopous *a.* [Gk. *xanthos*, yellow; *pous*, foot] having a yellow stem.

xanthopsin *n.* [Gk. *xanthos*, yellow; *opsis*, sight] yellow pigment of insect eyes; visual yellow, *q.v.*

xanthopterin(e) *n.* [Gk. *xanthos*, yellow; *pteron*, wing] a yellow pteridine pigment found especially in the wings of yellow butterflies and in the yellow bands of wasps and other insects, and also in mammalian urine, which can be oxidized to leucopterine and converted to folic acid by microorganisms; *alt.* uropterine.

xanthosomes *n.plu.* [Gk. *xanthos*, yellow; *sōma*, body] amber-coloured excretory granules in Foraminifera.

xanthospermous *a.* [Gk. *xanthos*, yellow; *sperma*, seed] having yellow seeds.

X-bodies protein-like inclusions in cells affected by a virus.

X-chromosome sex chromosome, single in the heterogametic sex, paired in the homogametic sex.

Xenacanthodii *n.* [Gk. *xenos*, strange; *akantha*, thorn] an order of Devonian to Triassic elasmobranch fish having long head spines; *alt.* Xenacanthiformes.

xenagones *n.plu.* [Gk. *xenos*, host; *agein*, to lead towards] specific substances produced by a live parasite and acting on the host.

xenarthral *a.* [Gk. *xenos*, strange; *arthron*, joint] having additional articular facets on dorsolumbar vertebrae.

xenia *n.* [Gk. *xenios*, hospitable] the effect of genes of the male gamete, especially the pollen grain, on female structures, especially structures in a seed plant other than the embryo, such as the endosperm or fruit.

xeniobiosis *n.* [Gk. *xenios*, hospitable; *biōsis*, living] hospitality in ant colonies.

xenodeme *n.* [Gk. *xenos*, host; *dēmos*, people] a deme of parasites differing from others in host specificity.

xenoecic *a.* [Gk. *xenos*, host; *oikos*, house] living in the empty shell of another organism.

xenogamy *n.* [Gk. *xenos*, strange; *gamos*, marriage] cross-fertilization, *q.v.*

xenogenesis *n.* [Gk. *xenos*, strange; *genesis*, descent] heterogenesis, *q.v.*

xenogenous *a.* [Gk. *xenos*, strange; *genos*, descent] originating outside the organism; caused by external stimuli; *alt.* exogenous.

xenograft *n.* [Gk. *xenos*, strange; O.F. *graffe*, graft] heterograft, *q.v.*

xenology *n.* [Gk. *xenos*, host; *logos*, discourse] the study of hosts in relation to the life history of parasites; *cf.* definitive host, intermediate host.

xenomixis *n.* [Gk. *xenos*, strange; *mixis*, mingling] exomixis, *q.v.*

xenomorphosis *n.* [Gk. *xenos*, strange; *morphōsis*, a shaping] heteromorphosis, *q.v.*

xenophya *n.plu.* [Gk. *xenos*, stranger; *phyein*, to grow] foreign bodies deposited in interspaces of certain Sarcodina, or used in formation of shells of certain Protozoa; *cf.* autophya.

xenoplastic *a.* [Gk. *xenos*, stranger; *plastos*, formed] *appl.* graft established in a different host; *cf.* heteroplastic.

Xenopterygii *n.* [Gk. *xenos*, host; *pterygion*, little wing or fin] a small order of marine teleosts which cling to stones in the intertidal zone by a sucking disc between the pelvic fins.

xerantic *a.* [Gk. *xēransis*, parching] drying up; withering, parched, exsiccant.

xerarch *a.* [Gk. *xēros*, dry; *archē*, beginning] *appl.* seres progressing from xeric towards mesic conditions; *cf.* mesarch, hydrarch.

xeric *a.* [Gk. *xēros*, dry] characterized by a scanty supply of moisture; tolerating, or adapted to, arid conditions; *cf.* hygric.

xerochasy *n.* [Gk. *xēros*, dry; *chasis*, separation] dehiscence of fruits when induced by drying; *cf.* hydrochasy; *a.* xerochasis.

xerocle *n.* [Gk. *xēros*, dry; *colere*, to dwell] an animal which lives in a dry place.

xeromorphic *a.* [Gk. *xēros*, dry; *morphē*, form] structurally modified so as to retard transpiration, *appl.* characters of xerophytes; *n.* xeromorphy.

xerophilous *a.* [Gk. *xēros*, dry; *philein*, to love] able to withstand drought; *appl.* plants adapted to a limited water supply; *n.* xerophil, *alt.* xerophyte.

xerophobous *a.* [Gk. *xēros*, dry; *phobos*, fear] not tolerating drought.

xerophyte *n.* [Gk. *xēros*, dry; *phyton*, plant] a plant adapted to dry conditions, either having xeromorphic characteristics or being a mesophyte growing only during the wet period; *alt.* xerophil, serophyte.

xerophyton *n.* [Gk. *xēros*, dry; *phyton*, plant] a plant inhabiting dry land.

xeropoium *n.* [Gk. *xēros*, dry; *poia*, grass] steppe vegetation.

xerosere *n.* [Gk. *xēros*, dry; L. *serere*, to put in a row] a plant succession originating on dry soil.

xerotherm *n.* [Gk. *xēros*, dry; *thermē*, heat] a plant surviving in conditions of drought and heat.

x-generatiom gametophyte; 2*x*, sporophyte generation.

xiphihumeralis *n.* [Gk. *xiphos*, sword; L. *humerus*, shoulder] a muscle extending from xiphoid cartilage to humerus.

xiphioid xiphoid, *q.v.*

xiphiplastron *n.* [Gk. *xiphos*, sword; F. *plastron*, breast plate] fourth lateral plate in plastron of Chelonia.

xiphisternum *n.* [Gk. *xiphos*, sword; L. *sternum*, breast bone] the posterior segment of the sternum, usually cartilaginous; *alt.* ensiform process, metasternum, xiphoid process.

xiphoid *a.* [Gk. *xiphos*, sword; *eidos*, shape] sword-shaped; *alt.* ensiform, xiphioid.

xiphoid process xiphisternum, *q.v.*; tail or telson of the king crab, *Limulus.*

xiphophyllous *a.* [Gk. *xiphos*, sword; *phyllon*, leaf] having sword-shaped leaves.

Xiphosura *n.* [Gk. *xiphos*, sword; *oura*, tail] an order of aquatic arthropods, commonly called king or horseshoe crabs, formerly included in the Arachnida but now placed in the Merostomata, having a heavily chitinized body with the prosoma covered with a horseshoe-shaped carapace, *Limulus* being a living example, but the group is known from the Palaeozoic.

X-organ small compact or sac-like neurosecretory organ in eye-stalk of certain Crustacea.

xylan *n.* [Gk. *xylon*, wood] a pentosan found in straw, bran, canes, and especially in wood of de-ciduous trees, which on hydrolysis yields the pentose, xylose.

Xylariales *n.* [Gk. *xylon*, wood] an order of Ascomycetes whose perithecia are dark, resembling black, burned wood.

xylary *a.* [Gk. *xylon*, wood] *pert.* xylem, *appl.* fibres, etc.; *appl.* procambium which gives rise to xylem; *alt.* xyloic.

xylem *n.* [Gk. *xylon*, wood] the main water-conducting tissue in a plant, consisting of lignified tracheids or vessels, and which also acts as a supporting tissue; secondary xylem: wood.

xylem canal narrow tubular space replacing central xylem in demersed stem of some aquatic plants.

xylem parenchyma short lignified cells surrounding vascular cells or produced with other xylem cells toward the end of the growing season.

xylem ray ray or plate of xylem between 2 medullary rays; part of a ray found in secondary xylem.

xylocarp *n.* [Gk. *xylon*, wood; *karpos*, fruit] a hard woody fruit.

xylochrome *n.* [Gk. *xylon*, wood; *chrōma*, colour] pigment of tannin, produced before death of xylem cells and giving colour to heartwood.

xylogen *n.* [Gk. *xylon*, wood; *-genēs*, producing] the forming xylem in a bundle; lignin, *q.v.*

xyloic *a.* [Gk. *xylon*, wood] xylary, *q.v.*

xyloid *a.* [Gk. *xylon*, wood; *eidos*, shape] ligneous, *q.v.*

xyloma *n.* [Gk. *xylon*, wood] a hardened mass of mycelium which gives rise to spore-bearing structures in certain fungi; a tumour of woody plants.

xylophagous *a.* [Gk. *xylon*, wood; *phagein*, to eat] wood-eating; *appl.* certain molluscs, insects, fungi.

xylophilous *a.* [Gk. *xylon*, wood; *philein*, to love] preferring wood; growing on wood.

xylophyte *n.* [Gk. *xylon*, wood; *phyton*, plant] a woody plant.

xylose *n.* [Gk. *xylon*, wood] an aldose pentose sugar found in straw, canes, and husks of seeds.

xylostroma *n.* [Gk. *xylon*, wood; *strōma*, bedding] the felt-like mycelium of certain wood-destroying fungi.

xylotomous *a.* [Gk. *xylotomos*, wood-cutting] able to bore or cut wood.

xylotomy *n.* [Gk. *xylotomos*, wood-cutting] the anatomy of wood or xylem.

xylulose *n.* [Gk. *xylon*, wood] a ketose pentose sugar.

X-zone transitory region of inner adrenal cortex.

Y

yarovization *n.* [Russ. *yarovizatsya*, from *yarovoi*, vernal] vernalization, *q.v.*

Y-cartilage cartilage joining ilium, ischium, and os pubis in the acetabulum.

Y-chromosome the sex chromosome which pairs with the X-chromosome in the heterogametic sex.

yelk yolk, *q.v.*

yellow body corpus luteum, *q.v.*

yellow cartilage a cartilage with matrix pervaded by yellow or elastic connective tissue fibres.

yellow cells chloragogen cells surrounding gut of

Annelida; cells occurring in intestine of Turbellaria; in Radiolaria, symbiotic algae or zoochlorellae; zooxanthellae, *q.v.*; chromo-argentaffin cells, *q.v.*

yellow enzyme a flavoprotein enzyme consisting of FMN firmly bound to a protein, concerned in cellular respiration, *alt.* yellow oxidation catalyst, cytoflavin, Warburg's yellow enzyme, old yellow enzyme; now used for an enzyme which oxidizes NADPH using any hydrogen acceptor; EC 1.6.99.1; *r.n.* NADPH dehydrogenase.

yellow fibres branched elastic fibres in connective tissue, made of elastin and occurring singly; *cf.* white fibres.

yellow spot macula lutea, *q.v.*

Y-granules granules, chemically allied to yolk, found in male germ cells; yolk granules.

Y-ligament iliofemoral ligament, *q.v.*

yolk *n.* [A.S. *geoloca*, yellow part] inert, or nonformative nutrient material in ovum, *alt.* vitellus, yelk; suint or greasy substance of fleece.

yolk duct vitelline duct.

yolk epithelium epithelium surrounding yolk sac.

yolk gland a gland in connection with reproductive system by which egg is furnished with a supply of food material; *alt.* vitellarium, vitellogen.

yolk nucleus cytoplasmic body appearing temporarily in oocyte, consisting mainly of Golgi bodies and mitochondria surrounding the centrosome before formation of yolk platelets; *alt.* vitelline body, Balbiani's body or nucleus.

yolk plates parallel lamellae into which deutoplasm may be split up in amphibians and many fishes.

yolk plug mass of yolk cells filling up blastopore, as in some amphibians such as frog.

yolk pyramids certain cells formed in segmenting egg of crayfish.

yolk sac membranous sac attached to embryo and containing yolk which passes to intestine through vitelline duct and acts as food for developing embryo.

yolk spherules remains of neighbouring cells found in ovum.

yolk stalk a short stalk or strand containing ducts and connecting yolk sac with embryo.

Y-organs in malacostracans, a pair of glands in the antennary or maxillary segment, which resemble the prothoracic glands of insects and secrete ecdysone.

ypsiliform *a.* [Gk. Υ, upsilon; L. *forma*, shape· Υ-shaped, *appl.* germinal spot at a certain stage in its development; *alt.* ypsiloid.

ypsiloid *a.* [Gk. Υ, upsilon; *eidos*, form] Υ-shaped, *appl.* cartilage anterior to pubis in salamanders for attachment of muscles used in breathing; *alt.* hypsiloid, ypsiliform.

Y-shaped ligament of Bigelow [*J. Bigelow*, American physician] the iliofemoral ligament, *q.v.*

Z

zalambdodont *a.* [Gk. *za*, very; *lambda*, λ; *odous*, tooth] *appl.* insectivores with narrow molar teeth with V-shaped transverse ridges; *cf.* dilambdodont.

Z-chromosome the Y-chromosome when female is the heterozygous sex.

Z-disc *n.* [Ger. *zwischenscheibe*, intermediate] a disc or membrane separating sarcomeres of muscle myofibrils; *alt.* Z-line, intermediate disc, telophragma, Krause's membrane, plasmophore, Dobie's line.

zeatin *n.* [*Zea mays*, maize] a natural cytokinin isolated from maize, and a derivative of adenine.

zeaxanthin *n.* [L.L. *zea*, corn; Gk. *xanthos*, yellow] a xanthophyll carotinoid pigment found in many plant cells including maize, in some classes of algae and in egg yolk, and which is an isomer of lutein.

Zeiformes *n.* [*Zeus*, generic name; L. *forma*, shape] an order of teleosts including the John Dory and boarfish; *alt.* Zeomorphi.

zein *n.* [L.L. *zea*, corn] a prolamine, lacking tryptophane and lysine, in seeds of maize.

Zeis, glands of sebaceous glands associated with eyelashes.

zeitgeber *n.* [Ger. *Zeit*, time; *Geber*, giver] synchronizer, *q.v.*

Zeomorphi *n.* [*Zeus*, generic name; Gk. *morphē*, form] Zeiformes, *q.v.*

Zeugloptera *n.* [Gk. *zeugle*, loop;·*pteron*, wing] a group of insects considered as a suborder of the Lepidoptera or as a separate order, as they have functional mandibles.

zeugopodium *n.* [Gk. *zeugos*, joined; *pous*, foot] forearm; shank, *alt.* crus.

Zingiberales *n.* [*Zingiber*, generic name] an order of monocots (Liliidae) having zygomorphic flowers and being typically· herbaceous but sometimes large and woody; *alt.* Scitamineae.

Zinjanthropus boisei [*Zing*, area of E. Africa; Gk. *anthropōs*, man] a man-like species of ape known from fossils from Olduvai gorge and which made tools and existed 1·7 million years ago, and which is now renamed *Australopithecus boisei.*

Zinn, zonule of [*J. G. Zinn*,. German anatomist] zonula ciliaris, *q.v.*

Z-line Z-disc, *q.v.*

zoae- *see also* zoe-.

zoaea zoëa, *q.v.*

Zoantharia *n.* [Gk. *zōon*, animal; *anthos*, flower] a subclass of anthozoans which are solitary or colonial with paired septa usually in multiples of 6, and have the tentacles, if present, external and not made of spicules; *alt.* Hexactinia, Hexaradiata.

zoanthella *n.* [Gk. *zōon*, animal; *anthos*, flower] type of zoanthid larva with transverse girdle of cilia.

Zoanthidia, zoanthids Zoanthiniaria, *q.v.*; *alt.* Zoanthidea.

zoanthina *n.* [Gk. *zōon*, animal; *anthinos*, blooming] type of zoanthid larva with longitudinal band of cilia.

Zoanthiniaria *n.* [Gk. *zōon*, animal; *anthinos*, blooming] an order of Zoantharia which are small and solitary or colonial without a skeleton, many · being epizoic on invertebrates; *alt.* Zoanthidea, Zoanthidia.

zoanthodeme *n.* [Gk. *zōon*, animal; *anthos*, flower; *demas*, body] a compound animal organism formed by zooids; a coherent colony of polyps.

zoarium *n.* [Gk. *zōarion*, animalcule] all the individuals of a polyzoan colony; polypary, *q.v.*

zodiophilous zoophilous, *q.v.*

zoëa *n.* [Gk. *zōē*, life] early larval form of certain decapod crustaceans; *alt.* zoaea.

zoëaform *a.* [Gk. *zōē*, life; L. *forma*, shape] shaped like a zoea; *alt.* zoaeaform.

zoecial, zoecium zooecial, zooecium, *q.q.v.*

zoetic *a.* [Gk. *zōē*, life] of or *pert.* life.

zoic *a.* [Gk. *zōikos, pert.* life] containing remains of organisms and their products, *cf.* azoic. [Gk. *zōon,* animal] *pert.* animals or animal life.

zoid *n.* [Gk. *zōon,* animal; *idion,* dim.] zoospore, *q.v.*; a sporozoite formed by division of sporoblasts of Haemosporidia.

zoidiogamic *a.* [Gk. *zōon,* animal; *idion,* dim.; *gamos,* marriage] *appl.* plants fertilized by spermatozoids carried by water.

zoidiogamy *n.* [Gk. *zōon,* animal; *idion,* dim.; *gamos,* marriage] fertilization by motile spermatozoids or antherozoids.

zoidophore *n.* [Gk. *zōon,* animal; *idion,* dim.; *pherein,* to bear] a spore mother cell or sporoblast formed by segmentation of oocyte in Haemosporidia.

zona *n.* [L. *zona,* girdle] a zone, band, or area.

zona arcuata inner part of basilar membrane, supporting spiral organ of Corti; *alt.* zona tecta.

zona fasciculata radially arranged columnar cells in suprarenal cortex below zona glomerulosa.

zona glomerulosa rounded groups of cells forming external layer of suprarenal cortex beneath capsule.

zona granulosa discus proligerus, *q.v.*

zonal *a.* [L. *zonalis, pert.* zone] of or *pert.* a zone.

zonality *n.* [L. *zona,* girdle] zonal distribution; zonal character.

zonal symmetry metamerism, *q.v.*

zonal view view of diatom when the girdle is seen.

zona orbicularis circular fibres of capsule of hip joint, around neck of femur.

zona pectinata outer division of basilar membrane of cochlea.

zona pellucida thick transparent membrane surrounding ovum; *alt.* zona striata, oolemma, vitelline membrane.

zona radiata radially striated inner egg envelope, as in Polychaeta; membrane with radially arranged pores receiving cell processes from corona radiata, *q.v.*

zona reticularis or reticulata inner layer of suprarenal cortex.

zonary *a.* [L. *zona,* girdle] *appl.* placenta with villi arranged in a band or girdle.

zona striata zona pellucida, *q.v.*

zonate *a.* [L. *zona,* girdle] zoned or marked with rings; arranged in a single row, as some tetraspores.

zona tecta zona arcuata, *q.v.*

zonation *n.* [L. *zona,* girdle] arrangement or distribution in zones.

zone *n.* [Gk. *zōnē,* girdle] an area characterized by similar fauna or flora; a belt or area to which certain species are limited; stratum or set of beds characterized by typical fossil or set of fossils; an area or region of the body; *alt.* zona.

zonite *n.* [Gk. *zōnē,* girdle] a body segment of Diplopoda.

zonociliate *a.* [Gk. *zōnē,* girdle; L. *cilium,* eyelash] banded with cilia, as certain annelid larvae.

zonoid *a.* [Gk. *zōnē,* girdle; *eidos,* form] like a zone.

zonolimnetic *a.* [Gk. *zōnē,* girdle; *limnē,* lake] of or *pert.* a certain zone in depth.

zonoplacental *a.* [L. *zona,* girdle; *placenta,* cake] having a zonary placenta.

zonula zonule, *q.v.*

zonula ciliaris *n.* [L. *zonula,* dim. of *zona,* girdle; *cilium,* eyelash] the hyaloid membrane forming suspensory ligament of lens of eye; *alt.* zonule of Zinn.

zonule *n.* [L. *zonula,* dim. of *zona,* girdle] a little zone, belt, or girdle; *alt.* zonula.

zooamylon *n.* [Gk. *zōon,* animal; *amylon,* starch] food reserve in refractile bodies of cytoplasm, such as paramylon or paraglycogen in Protozoa.

zooanthellae *n.plu.* [Gk. *zōon,* animal; *anthos,* flower] cryptomonads symbiotic with certain marine Protozoa.

zooapocrisis *n.* [Gk. *zōon,* animal; *apokrisis,* answer] the response of animals to their environmental conditions as a whole.

zoobenthos *n.* [Gk. *zōon,* animal; *benthos,* depths of sea] the fauna of the sea bottom, or of the bottom of inland waters.

zoobiotic *a.* [Gk. *zōon,* animal; *biōtikos, pert.* life] parasitic with, or living on an animal.

zooblast *n.* [Gk. *zōon,* animal; *blastos,* bud] an animal cell.

zoocaulon *n.* [Gk. *zōon,* animal; *kaulos,* stalk] zoodendrium, *q.v.*

zoochlorellae *n.plu.* [Gk. *zōon,* animal; *chlōros,* green] symbiotic green algae living in various animals, e.g. in Sarcodina, Radiolaria, hydra.

zoochoric *a.* [Gk. *zōon,* animal; *chōrein,* to spread] dispersed by animals, *appl.* plants; *n.* zoochory.

zoocoenocyte *n.* [Gk. *zōon,* animal; *koinos,* common; *kytos,* hollow] a coenocyte bearing cilia, in certain algae; *cf.* synzoospore.

zoocyst *n.* [Gk. *zōon,* animal; *kystis,* sac] sporocyst, *q.v.*

zoocytium *n.* [Gk. *zōon,* animal; *kytos,* hollow] in certain ciliates, the common gelatinous and often branched matrix; *alt.* zoothecium.

zoodendrium *n.* [Gk. *zōon,* animal; *dendron,* tree] the tree-like branched stalk of certain colonial ciliates; *alt.* zoocaulon.

zoodynamics *n.* [Gk. *zōon,* animal; *dynamis,* power] the physiology of animals; *alt.* zoonomy.

zooecial *a.* [Gk. *zōon,* animal; *oikos,* house] *pert.* or resembling a zooecium; *alt.* zoecial.

zooecium *n.* [Gk. *zōon,* animal; *oikos,* house] a chamber or sac enclosing a polyzoan nutritive zooid; *alt.* zoecium.

zooerythrin *n.* [Gk. *zōon,* animal; *erythros,* red] red pigment found in plumage of various birds; *alt.* zoonerythrin.

zoofulvin *n.* [Gk. *zōon,* animal; L. *fulvus,* yellow] yellow pigment found in plumage of various birds.

zoogamete *n.* [Gk. *zōon,* animal; *gametēs,* spouse] a motile gamete; *alt.* planogamete.

zoogamy *n.* [Gk. *zōon,* animal; *gamos,* marriage] sexual reproduction in animals.

zoogenesis *n.* [Gk. *zōon,* animal; *genesis,* descent] the origin of animals; ontogeny and phylogeny of animals.

zoogenetics *n.* [Gk. *zōon,* animal; *genesis,* descent] animal genetics.

zoogenous *a.* [Gk. *zōon*, animal; *gennaein*, to produce] produced or caused by animals.

zoogeography *n.* [Gk. *zōon*, animal; *gē*, earth; *graphein*, to write] the science of distribution of animals on the earth.

zoogloea *n.* [Gk. *zōon*, animal; *gloia*, glue] a mass of bacteria embedded in a mucilaginous matrix frequently forming an iridescent film; *alt.* zooglea.

zoogonidangium *n.* [Gk. *zōon*, animal; *gonos*, offspring; *idion*, dim.; *anggeion*, vessel] a cell which produces zoospores or zoogonidia, in algae.

zoogonid(ium) *n.* [Gk. *zōon*, animal; *gonos*, offspring; *idion*, dim.] one of motile spores formed in gonidangium of algae.

zoogonous *a.* [Gk. *zōon*, animal; *gonos*, offspring] *see* viviparous.

zooid *n.* [Gk. *zōon*, animal; *eidos*, like] a member of a compound animal organism; an individual or person in a coelenterate or polyzoan colony, *alt.* polyp; posterior genital and non-sexual region formed in many polychaetes; *cf.* phytoid.

zoolith, zoolite *n.* [Gk. *zōon*, animal; *lithos*, stone] any fossil animal.

zoology *n.* [Gk. *zōon*, animal; *logos*, discourse] the science dealing with structure, functions, behaviour, history, classification, and distribution of animals.

Zoomastigina, Zoomastigophora *n.* [Gk. *zōon*, animal; *mastix*, whip; *pherein*, to bear] a subclass of Flagellata having definitely holozoic nutrition with no chloroplasts, and sometimes more than 4 flagella, some being important parasites.

zoöme *n.* [Gk. *zōon*, animal] animals considered as an ecological unit.

zoomorphosis *n.* [Gk. *zōon*, animal; *morphōsis*, a forming] formation of structures in plants owing to animal agents, as production of galls.

zoon *n.* [Gk. *zōon*, animal] an individual developed from an egg.

zooerythrin *n.* [Gk. *zōon*, animal; *erythros*, red] red lipochrome pigment found in various animals; zooerythrin, *q.v.*

zoonite *n.* [Gk. *zōon*, animal] a body segment of an articulated animal.

zoonomy *n.* [Gkl *zōon*, animal; *nomos*, law] physiology, especially animal physiology; *alt.* zoodynamics.

zoonosis *n.* [Gk. *zōon*, animal; *nosos*, disease] disease of animals; animal disease transmitted to man; *cf.* zoosis.

Zoopagales *n.* [*Zoopage*, generic name] an order of Phycomycetes which are predaceous and capture amoebae, eelworms, and other small animals.

zooparasite *n.* [Gk. *zōon*, animal; *parasitos*, parasite] any parasitic animal.

zoopherin a former name for cobalamine, *q.v.*

zoophilous *a.* [Gk. *zōon*, animal; *philein*, to love] *appl.* plants adapted for pollination by animals other than insects; *alt.* zodiophilous.

zoophobic *a.* [Gk. *zōon*, animal; *phobos*, fear] shunning or shunned by animals, *appl.* plants protected by spines, hairs, secretions, etc.

Zoophyta *n.* [Gk. *zōon*, animal; *phyton*, plant] an artificial group of invertebrates consisting of zoophytes, especially sponges and coelenterates, and particularly anthozoans.

zoophyte *n.* [Gk. *zōon*, animal; *phyton*, plant] an animal resembling a plant in appearance or growth.

zooplankton *n.* [Gk. *zōon*, animal; *plangktos*, wandering] animal plankton.

zooplasm *n.* [Gk. *zōon*, animal; *plasma*, mould] living substance which depends on the products of other living organisms for nutritive material.

zoosis *n.* [Gk. *zōon*, animal] any disease produced by animals; *cf.* zoonosis.

zoosperm *n.* [Gk. *zōon*, animal; *sperma*, seed] spermatozoid, *q.v.*; zoospore, *q.v.*

zoosphere *n.* [Gk. *zōon*, animal; *sphaira*, globe] biciliate zoospore of algae.

zoosporangiophore *n.* [Gk. *zōon*, animal; *sporos*, seed; *anggeion*, vessel; *phoros*, bearing] structure bearing zoosporangia.

zoosporangium *n.* [Gk. *zōon*, animal; *sporos*, seed; *anggeion*, vessel] a sporangium in which zoospores develop; *alt.* swarm sporangium.

zoospore *n.* [Gk. *zōon*, animal; *sporos*, seed] a flagellated cell for asexual reproduction, in algae and fungi; a flagellated or amoeboid cell in protozoans; *alt.* swarm cell or spore, sporozoid, simbospore, zooid, trichospore, zoosperm, planospore.

zoosporocyst *n.* [Gk. *zōon*, animal; *sporos*, seed; *kystis*, bladder] zoosporangium of certain saprophytic Phycomycetes.

zoosterols *n.plu.* [Gk. *zōon*, animal; *stereos*, solid; *alcohol*] sterols originally discovered in animals but now found to be more widely distributed such as cholesterol; *cf.* mycosterols, phytosterols.

zootaxy *n.* [Gk. *zōon*, animal; *taxis*, arrangement] the classification of animals.

zootechnics *n.* [Gk. *zōon*, animal; *technē*, craft] science applied to the art of breeding, rearing, and utilizing animals; *alt.* zootechny.

zoothecium *n.* [Gk. *zōon*, animal; *thēkē*, case] zoocytium, *q.v.*

zoothome *n.* [Gk. *zōon*, animal; *thōmos*, heap] any group of individuals in a living coral.

zootomy *n.* [Gk. *zōon*, animal; *temnein*, to cut] dissection or anatomy of animals other than man.

zootoxin *n.* [Gk. *zōon*, animal; *toxikon*, poison] any toxin produced by animals.

zootrophic *a.* [Gk. *zōon*, animal; *trephein*, to nourish] heterotrophic, *q.v.*; holozoic, *q.v.*

zootype *n.* [Gk. *zōon*, animal; *typos*, pattern] representative type of animal.

zooxanthellae *n.plu.* [Gk. *zōon*, animal; *xanthos*, yellow] yellow or brown cells or symbiotic unicellular algae living in various animals.

zooxanthin *n.* [Gk. *zōon*, animal; *xanthos*, yellow] yellow pigment found in plumage of certain birds.

zoozygosphere planogamete, *q.v.*

zoozygospore *n.* [Gk. *zōon*, animal; *zygon*, yoke; *sporos*, seed] a motile zygospore.

Zoraptera *n.* [Gk. *zoros*, pure; *pteron*, wing] an order of small carnivorous insects with slight metamorphosis, the females having wings and eyes, and the males having neither.

Zuckerkandl's bodies or organs [*E. Zuckerkandl*, Austrian anatomist] aortic bodies, *q.v.*

zwitterion *n.* [Ger. *Zwitter*, half breed; Gk. *iōn*, going] an ion with both positive and negative charges, as all amino acids.

zygantrum *n.* [Gk. *zygon*, yoke; *antron*, cave] a fossa on posterior surface of neural arch of vertebrae of snakes and certain lizards; *cf.* zygosphene.

zygapophysis *n.* [Gk. *zygon*, yoke; *apophysis*, process of a bone] one of processes of a vertebra by which it articulates with adjacent vertebrae.

Zygnematales *n.* [*Zygnema*, generic name] an order of green algae having isogamous sexual reproduction, with amoeboid gametes without flagella; *alt.* Conjugales, Conjugatae, Acontae.

zygobranchiate *a.* [Gk. *zygon*, yoke; *brangchia*, gills] having gills symmetrically placed and renal organs paired, *appl.* certain Gastropoda.

zygocardiac ossicles paired lateral ossicles in gastric mill of Crustacea.

zygodactyl(ous) *a.* [Gk. *zygon*, yoke; *daktylos*, digit] having 2 toes pointing forward, 2 backward, as in parrots.

zygodont *a.* [Gk. *zygon*, yoke; *odous*, tooth] having molar teeth in which the 4 tubercles are united in pairs.

zygogamy *n.* [Gk. *zygon*, yoke; *gamos*, marriage] isogamy, *q.v.*

zygogenetic, zygogenic *a.* [Gk. *zygon*, yoke; *genesis*, origin] produced by fertilization; *cf.* parthenogenetic.

zygoid *a.* [Gk. *zygon*, yoke; *eidos*, form] diploid, *appl.* parthenogenesis.

zygolysis *n.* [Gk. *zygon*, yoke; *lysis*, loosing] separation of a pair, as of alleles.

zygoma *n.* [Gk. *zygōma*, yoke] the bony arch of the cheek, formed by temporal process of zygomatic bone and zygomatic process of temporal bone; *alt.* arcus zygomaticus.

zygomatic *a.* [Gk. *zygōma*, yoke] malar, *q.v.*; *pert.* zygoma, *appl.* arch, bone, fossa, processes, muscle; *appl.* nerve: temporomalar nerve, *q.v.*

zygomatic gland the infraorbital salivary gland.

zygomaticofacial *a.* [Gk. *zygōma*, yoke; L. *facies*, face] *appl.* foramen on malar surface of zygomatic bone for passage of nerve and vessels; *appl.* branch of zygomatic or temporomalar nerve.

zygomaticotemporal *a.* [Gk. *zygōma*, yoke; L. *tempora*, temples] *appl.* suture, foramen, nerve, etc., at temporal surface of zygomatic bone.

zygomaticus muscle from zygomatic bone to angle of mouth.

zygomelous *a.* [Gk. *zygon*, yoke; *melos*, limb] having paired appendages, *appl.* fins; *cf.* azygomelous.

zygomite *n.* [Gk. *zygon*, yoke; *mitos*, thread] one of a pair of conjugated filaments.

zygomorphic *a.* [Gk. *zygon*, yoke; *morphē*, shape] bilaterally symmetrical, with only one plane of symmetry; *alt.* zygomorphous, irregular, monosymmetrical; *cf.* actinomorphic.

Zygomycetales *n.* [Gk. *zygon*, yoke; *mykēs*, fungus] the Zygomycetes when considered as an order of Phycomycetes.

Zygomycetes *n.* [Gk. *zygos*, yoke; *mykēs*, fungus] a group of fungi, sometimes placed in the Phycomycetes or sometimes raised to class status, having no motile spores, and sexual processes resulting in zygospores.

zygonema *n.* [Gk. *zygon*, yoke; *nēma*, thread] chromosome thread during zygotene.

zygoneure *n.* [Gk. *zygon*, yoke; *neuron*, nerve] a nerve cell connected with other nerve cells.

zygoneury *n.* [Gk. *zygon*, yoke; *neuron*, nerve] in certain Gastropoda, having a connective between

pleural ganglion and ganglion on visceral branch of opposite side; *cf.* dialyneury.

zygophase *n.* [Gk. *zygon*, yoke; *phasis*, aspect] diplophase, *q.v.*; *cf.* gamophase.

zygophore *n.* [Gk. *zygon*, yoke; *pherein*, to bear] a conjugating hypha in certain fungi; *alt.* zygosporophore.

zygophyte *n.* [Gk. *zygon*, yoke; *phyton*, plant] a plant with 2 similar reproductive cells which unite in fertilization.

zygopleural *a.* [Gk. *zygon*, yoke; *pleuron*, side] bilaterally symmetrical; *n.* zygopleury.

zygopodium *n.* [Gk. *zygon*, yoke; *pous*, foot] forearm; shank.

zygosis *n.* [Gk. *zygōsis*, a joining] conjugation; union of gametes.

zygosome *n.* [Gk. *zygon*, yoke; *sōma*, body] mixochromosome, *q.v.*

zygosperm *n.* [Gk. *zygon*, yoke; *sperma*, seed] zygospore, *q.v.*

zygosphene *n.* [Gk. *zygon*, yoke; *sphēn*, wedge] an articular process on anterior surface of neural arch of vertebrae of snakes and certain lizards, which fits into zygantrum.

zygosphere *n.* [Gk. *zygon*, yoke; *sphaira*, globe] a gamete which conjugates with a similar one to form a zygospore.

zygosporangium *n.* [Gk. *zygon*, yoke; *sporos*, seed; *anggeion*, vessel] a sporangium in which zygospores are formed.

zygospore *n.* [Gk. *zygon*, yoke; *sporos*, seed] a cell, or resting spore, formed by conjugation of similar reproductive cells, as in Conjugales and Zygomycetes; *alt.* zygosperm.

zygosporocarp *n.* [Gk. *zygon*, yoke; *sporos*, seed; *karpos*, fruit] a fruit body in which zygospores are produced.

zygosporophore *n.* [Gk. *zygon*, yoke; *sporos*, seed; *pherein*, to bear] zygophore, *q.v.*

zygotaxis *n.* [Gk. *zygon*, yoke; *taxis*, arrangement] tendency towards conjugation between 2 specialized hyphae in certain fungi; mutual attraction between gametes of the opposite sex; *alt.* zygotactism.

zygote *n.* [Gk. *zygōtos*, yoked] cell formed by union of 2 gametes or reproductive cells; *alt.* fertilized ovum, mixote, amphiont.

zygotene *n.* [Gk. *zygon*, yoke; *tainia*, band] a stage of prophase of meiosis where spireme chromosome threads are uniting in pairs; *alt.* amphitene, synaptene, zygotene.

zygotic *a.* [Gk. *zygōtos*, yoked] *pert.* a zygote; *appl.* mutation occurring immediately after fertilization; *appl.* number: diploid or somatic number, 2*n*; *cf.* gametic.

zygotoblast *n.* [Gk. *zygōtos*, yoked; *blastos*, bud] a sporozoite produced by segmentation of zygotomere in haemamoebae.

zygotoid *n.* [Gk. *zygōtos*, yoked; *eidos*, form] product of union of 2 gametoids, as in mucorine fungi; *alt.* mixote.

zygotomere *n.* [Gk. *zygōtos*, yoked; *meros*, part] a cell formed by segmentation of zygote in haemamoebae.

zygotonucleus *n.* [Gk. *zygōtos*, yoked; L. *nucleus*, kernel] a nucleus formed by fusion of 2 gametonuclei.

zygotropism *n.* [Gk. *zygon*, yoke; *tropē*, turn] the growth of zygophores towards each other.

zygozoospore *n.* [Gk. *zygon*, yoke; *zōon*, animal;

sporos, seed] a motile cell formed by union of 2 similar cells.

zygum *n.* [Gk. *zygon*, yoke] in ostracods, a single chitinous piece suspended in the posterior shell region by a system of chitinous rods.

zymase *n.* [Gk. *zymē*, leaven] a complex of enzymes which catalyses the breakdown of glucose to ethyl alcohol and carbon dioxide in fermentation and requires many coenzymes for its activity.

zymin *n.* [Gk. *zymē*, leaven] formerly, enzyme, *q.v.*; yeast extract.

zymocont *n.* [Gk. *zymē*, leaven; *kontos*, punting pole] rod-shaped mitochondrion of a pancreatic cell.

zymo-excitor a substance activating a zymogen, e.g. hydrochloric acid, which activates pepsin.

zymogen *n.* [Gk. *zymē*, leaven; -*genēs*, producing] formerly a substance capable of being transformed into an enzyme, i.e. a precursor of an enzyme, *alt.* proenzyme, proferment; a zymogenic organism.

zymogenesis *n.* [Gk. *zymē*, leaven; *genesis*, origin] the production of an enzyme by a zymogen activated by a kinase.

zymogenic *a.* [Gk. *zymē*, leaven; -*genēs*, producing] enzyme-producing, *appl.* certain cells of gas-

tric gland tubule, *appl.* micro-organisms, as bacteria; *pert.* a zymogen.

zymogenous *a.* [Gk. *zymē*, leaven; -*genēs*, producing] *appl.* microflora in soils normally present in resting stage and only becoming active when fresh organic materials are added.

zymohydrolysis *n.* [Gk. *zymē*, leaven; *hydōr*, water; *lysis*, breaking down] formerly, hydrolysis due to the action of an enzyme.

zymolysis *n.* [Gk. *zymē*, leaven; *lysis*, loosing] formerly, decomposition by the action of enzymes.

zymolyte *n.* [Gk. *zymē*, leaven; *lysis*, loosing] formerly, substrate on which enzyme acts.

zymophore *n.* [Gk. *zymē*, leaven; *phoros*, bearing] formerly, the active portion of an enzyme.

zymoprotein *n.* [Gk. *zymē*, leaven; *prōteion*, first] formerly, proteins that are enzymes.

zymosis *n.* [Gk. *zymē*, fermentation] formerly, fermentation, *q.v.*; any reactions catalysed by enzymes.

zymosthenic *a.* [Gk. *zymē*, leaven; *sthenein*, to be strong] formerly, enhancing the activity of an enzyme.

zymotic *a.* [Gk. *zymōtikos*, causing fermentation] formerly, *pert.* or caused by fermentation; *appl.* diseases induced by infection.

APPENDICES

A. CLASSIFICATION OF THE PLANT KINGDOM

division [1] Schizophyta [2]
   class Schizomycetes (Bacteria)
      order Pseudomonadales
           Chlamydobacteriales
           Hyphomicrobiales
           Eubacteriales
           Caryophanales
           Actinomycetales [3]
           Myxobacteriales [3]
           Spirochaetales
           Beggiatoales
           Thiobacteriales
   class Mollicutes [4]
      order Mycoplasmatales
   class Microtabiotes (Microtatobiotes)
      order Rickettsiales [4]
           Virales (Viruses) [5]

   class Schizophyceae [6]
      order Chroococcales
           Chamaesiphonales [7]
           Pleurocapsales ⎫
           Nostocales   ⎬ Hormogonales
           Stigonematales ⎭

Algae [8] (Phycophyta)

1. Divisions are also known as phyla.
2. The Schizophyta is placed in the Prokaryota and all other organisms in the Eukaryota.
3. May be placed in separate classes.
4. May be placed in the Schizomycetes.
5. May be placed in a separate division.
6. The Schizophyceae may be considered as a class of Algae, the Cyanophyceae, or as a division, the Cyanophyta.
7. May be included in the Chroococcales.
8. The divisions of Algae are not recognized in all classifications. Then Algae is the division and the divisions are reduced to class status.

division Cyanophyta (Myxophyta)
    class Cyanophyceae [9] (Myxophyceae)

division Chlorophyta
    class Chlorophyceae
        order Volvocales
            Tetrasporales
            Chlorococcales (Protococcales)
            Ulotrichales
            Sphaeropleales
            Chaetophorales
            Ulvales
            Prasiolales (Schizogoniales)
            Oedogoniales
            Zygnematales (Conjugales)
            Cladophorales
            Acrosiphonales
            Siphonocladales
            Dasycladales
            Derbesiales
            Dichotomosiphonales
            Caulerpales
            (Siphonales) [10]
    class Prasinophyceae [11]
        order Halosphaerales

division Charophyta [12]
    class Charophyceae
        order Charales

division Xanthophyta
    class Xanthophyceae
        order Heterochloridales (Chloramoebales)
            Heterocapsales (Heterogloeales)
            Heterococcales (Mischococcales)
            Tribonematales (Heterotrichales)
            Vaucheriales (Heterosiphonales)
            Rhizochloridales

9. The Cyanophyceae may be considered to be a class of Schizomycetes, the Schizophyceae. The orders are as above, under Schizophyceae, and the Cyanophyta then does not exist.

10. This order is used in different ways in different systems, often including some of the orders above it.

11. Of uncertain position, placed here by some authorities.

12. This division may not be recognized, and the Charophyceae is then placed in the Chlorophyta, or reduced to order status, the Charales, in the Chlorophyceae.

division Chrysophyta
    class Chrysophyceae
        order Chrysomonadales
             Rhizochrysidales
             Chrysocapsales
             Chrysosphaerales
             Chrysotrichales (Phaeothamniales)
             (Bacillariales) [13]
    class Haptophyceae
        order Isochrysidales
             Prymnesiales

division Bacillariophyta [13]
    class Bacillariophyceae
        order Centrales
             Pennales

division Pyrrophyta [14]
    class Desmophyceae
        order Prorocentrales
    class Dinophyceae
        order Gymnodiniales
             Peridiniales
             Dinophysidales
             Dinocapsales
             Dinotrichales
             Rhizodiniales (Dinamoebidiales)
             Phytodiniales (Dinococcales)

division Cryptophyta [15]
    class Cryptophyceae
        order Cryptomonadales
             Cryptococcales

division Chloromonadophyta [16]
    class Chloromonadophyceae

division Euglenophyta
    class Euglenophyceae
        order Euglenales

13. The Bacillariophyta may be reduced to class status (Bacillariophyceae) within the Chrysophyta or to order status (Bacillariales) within the Chrysophyceae.
14. In classifications where the Pyrrophyta is reduced to class status it is known as the Pyrrophyceae, and its classes are considered to be orders.
15. May be considered to be a class of Pyrrophyta.
16. A group of uncertain position; may be considered a class of Pyrrophyta.

order Rhabdomonadales
Heteronematales
Sphenomonadales
Euglenomorphales

division Phaeophyta
class Phaeophyceae [17]
order Ectocarpales
Tilopteridales
Sphacelariales
Cutleriales
Dictyotales
Chordariales
Sporochnales
Desmarestiales
Dictyosiphonales
Laminariales
Fucales
Ascoseirales [18]
Durvilleales [18]

division Rhodophyta
class Rhodophyceae
subclass Bangiophycidae (Bangioideae)
order Porphyridiales
Bangiales
Goniotrichales
Compsopogonales
Rhodochaetales
subclass Florideophycidae (Florideae)
order Nemalionales
Bonnemaisoniales
Gelidiales
Cryptonemiales
Gigartinales
Rhodymeniales
Ceramiales

division Fungi (Mycophyta)
subdivision Myxomycophyta [19]

17. The class Phaeophyceae may be divided into three classes, the Isogeneratae, orders 1–5, the Heterogeneratae, orders 6–10, and the Cyclosporae, order 11.

18. These orders may not be recognized.

19. May be considered as full divisions, sometimes with the Eumycophyta being equivalent to the Fungi.

    class Myxomycetes
      subclass Ceratiomyxomycetidae
        order Ceratiomyxales
      subclass Myxogastromycetidae
        order Liceales
            Echinosteliales
            Trichiales
            Stemonitales
            Physarales
    Alternative classification of Myxomycetes
    class Myxomycetes
        order Myxomycetales
            Acrasiales
            Plasmodiophorales * [20]
Alternative classification of Myxomycophyta
subdivision Myxomycophyta
    class Myxomycetae
        order Exosporeae
            Endosporeae (Myxogastres)
    class Acrasieae
    class Plasmodiophoreae * [20]

subdivision Eumycophyta [19]
    class Phycomycetes
        order Plasmodiophorales [20]
            Chytridiales
            Blastocladiales
            Monoblepharidales
            Hyphochytriales (Anisochytridiales)
            Oomycetes (Oomycetales)
            Zygomycetes (Zygomycetales)
    Alternative classification of Phycomycetes [21]
    class Chytridiomycetes
        order Chytridiales
            Blastocladiales
            Monoblepharidales
    class Hyphochytridiomycetes
        order Hyphochytriales
    class Plasmodiophoromycetes
        order Plasmodiophorales
    class Oomycetes
        order Lagenidiales

20. May be considered as a class or order of Myxomycetes or of Phycomycetes.
* End of alternative classification.
21. May not be recognized in this classification, or may be considered as a superclass.

        order Saprolegniales
             Leptomitales
             Peronosporales
class Zygomycetes
      order Mucorales
             Entomophthorales
             Zoopagales
class Trichomycetes
      order Amoebidiales
             Eccrinales
             Asellariales
             Harpellales *
class Ascomycetes
  subclass Hemiascomycetes (Hemiascomycetidae)
      order Protomycetales
             Endomycetales (Saccharomycetales)
             Taphrinales (Exoascales)
  subclass Euascomycetes (Euascomycetidae)
      order Eurotiales (Plectomycetes)
             Microascales
             Onygenales
             Erysiphales (Perisporiales)
             Meliolales
             Chaetomiales
             Xylariales
             Diaporthales
             Hypocreales
             Clavicipitales
             Coryneliales
             Coronophorales
             Laboulbeniales
             Ostropales
             Phacidiales
             Helotiales
             Pezizales
             Tuberales
Alternative classification of Euascomycetes
subclass Plectomycetes (Plectomycetidae)
      order Plectascales (Aspergillales)
             Erysiphales (Perisporiales)
subclass Pyrenomycetes (Pyrenomycetidae)
      order Sphaeriales
                 Clavicipitales
                 Laboulbeniales
                 Dothideales (Dothiorales)

* End of alternative classification.

subclass Discomycetes (Discomycetidae)
    order Helvellales
        Helotiales
        Pezizales
        Phacidiales
        Tuberales *
subclass Loculomycetes (Loculomycetidae)
    order Myriangiales
        Dothideales (Dothiorales)
        Pleosporales
        Pseudosphaeriales
        Microthyriales (Hemisphaeriales)
        Hysteriales
class Basidiomycetes
  subclass Heterobasidiomycetes (Heterobasidiomycetidae)
    order Auriculariales ⎫
        Tremellales ⎬ Tremellales
        Dacromycetales ⎭
        Septobasidiales
        Uredinales
        Ustilaginales
  subclass Homobasidiomycetes (Homobasidiomycetidae)
    Hymenomycetes
      order Exobasidiales
          Poriales (Polyporales)
          Agaricales
    Gasteromycetes
      order Lycoperdales
          Sclerodermatales
          Phallales
          Nidulariales
          Hymenogastrales
class Deuteromycetes (Fungi Imperfecti)
    order Sphaeropsidales
        Melanconiales
        Moniliales (Hyphomycetes)
        Mycelia Sterila

division Lichenes
  class Basidiolichenes (Hymenolichenes)
  class Ascolichenes
      order Caliciales
        Graphidales
        Cyanophilales

* End of alternative classification.

order Lecanorales
Caloplacales
Alternative classification of Ascolichenes I
class Ascolichenes
subclass Gymnocarpeae
subclass Pyrenocarpeae *
Alternative classification of Ascolichenes II
subclass Ascomycetidae
order Lecanorales
Sphaeriales
Caliciales
subclass Loculoascomycetidae
order Myrangiales
Pleosporales
Hysteriales *

division Bryophyta [22]
class Hepaticae (Hepaticopsida)
subclass Hepaticidae
order Sphaerocarpales
Marchantiales
Calobryales
Jungermanniales
subclass Anthocerotidae [23]
order Anthocerotales
Takakiales
class Musci (Bryopsida)
subclass Sphagnidae
order Sphagnales
subclass Andreaeidae
order Andreaeales
subclass Bryidae
order Archidiales
Dicranales
Fissidentales
Pottiales
Grimmiales
Funariales
Eubryales [24]
Schistostegales
Tetraphidales

22. The Bryophyta, Pteridophyta, and Spermatophyta may be placed in the Embryophyta; the Bryophyta and Pteridophyta may be placed in the Archegoniatae.

23. May be given equal status with the Hepaticopsida and Bryopsida, and be known as Anthocerotopsida.

24. May be considered a subclass (Eubrya).

* End of alternative classification.

                    Hookeriales
                    Buxbaumiales
                    Polytrichales[25]
                    Isobryales
                    Hypnobryales
    Alternative classification of Musci
            superorder or
                order Sphagnales
                      Andreaeales
                      Bryales*

division Pteridophyta[22, 26]
    class Psilophytopsida[27]
                order Psilophytales
    class Psilotopsida[27]
                order Psilotales
    class Lycopsida
                order Protolepidodendrales
                      Lycopodiales
                      Lepidodendrales
                      Isoetales
                      Selaginellales
    class Sphenopsida
                order Hyeniales
                      Sphenophyllales
                      Calamitales
                      Equisetales
                      Pseudoborniales
    class Filicopsida[28]
        subclass Primofilices (Primofilicidae)
                order Cladoxylales
                      Coenopteridales
                      Protopteridales[29]
                      Archaeopteridales[29]
        subclass Eusporangiatae (Eusporangidae)
                order Ophioglossales
                      Marattiales
        subclass Leptosporangiatae (Leptosporangidae)

25. May be considered a superorder.
* End of alternative classification.
26. The Pteridophyta and Spermatophyta may be placed in the Tracheophyta.
27. The Psilophytopsida and Psilotopsida may be considered subclasses of the class Psilophytatae (Psilopsida, Psilophyta).
28. Also called Pteropsida, or this term may be used to include the Filicopsida, Gymnospermae, and Angiospermae.
29. These groups may be considered to be intermediate between Pteridophyta and Gymnospermae, or may be placed in the Progymnospermopsida in some classifications.

     order Filicales
       Marsileales
       Salviniales
   subclass Osmundidae
     order Osmundales
  Alternative classification of Filicopsida
  subclass Primofilices (as above)
  subclass Eusporangiatae (as above)
  subclass Leptosporangiatae
    order Filicales
      Osmundales
  subclass Hydropterides
    order Marsileales
      Salviniales *
Alternative classification of Pteridophyta
Each class may be raised to division status
Psilophyta = Psilopsida
Lycophyta = Lycopsida
Sphenophyta = Sphenopsida
Filicophyta = Filicopsida *

division Spermatophyta[22, 26, 28]
 Gymnospermae
  subdivision Cycadophytina (Cycadicae)[30]
   class Lyginopteridatae (Pteridospermae)
      order Lyginopteridales (Cycadofilicales,
        Pteridospermales)
       Caytoniales
   class Cycadatae
     order Nilssonniales
      Cycadales
   class Bennettitatae
    subclass Bennettitidae
     order Bennettitales
    subclass Pentoxylidae
     order Pentoxylales
   class Gnetatae (Chlamydospermae)
    subclass Welwitschiidae
     order Welwitschiales
    subclass Gnetidae
     order Gnetales
    subclass Ephedridae
     order Ephedrales
  subdivision Coniferophytina (Pinicae)[30]

* End of alternative classification
30. These groups may be considered equivalent to the Angiospermae in weight, and the Gymnospermae is not recognized.

     class Ginkgoatae
          order Ginkgoales
     class Pinatae
       subclass Cordaitidae
          order Cordaitales
       subclass Pinidae (Coniferae)
          order Voltziales
               Pinales
       subclass Taxidae

Alternative classification of Gymnospermae

subdivision Gymnospermae (Pinophyta)
   class Progymnospermopsida [31]
   class Cycadopsida
         order Pteridospermales
             Bennettitales (Cycadeoidales)
             Pentoxylales
             Caytoniales [32]
   class Coniferopsida
         order Cordaitales
             Coniferales
             Taxales
             Ginkgoales
   class Gnetopsida (Gneticae)
         order Welwitschiales ⎤
             Gnetales   ⎬Gnetales
             Ephedrales * ⎦

subdivision Angiospermae (Magnoliophytina)
   class Dicotyledoneae (Magnoliatae, Magnoliopsida)
       subclass Magnoliidae (Polycarpicae)
          superorder Magnolianae
            order Magnoliales
               Piperales
               Aristolochiales
               Rafflesiales
               Nymphaeales
          superorder Ranunculanae
            order Ranunculales
               Papaverales [33]
       subclass Hamamelididae (Amentiflorae, Amentiferae)
          order Trochodendrales
              Hamamelidales

---

31. May be considered as Pteridophyta; see note 29.
32. May be considered to be a family of Pteridospermales.
* End of alternative classification.
33. The Papaverales and Capparales (Dilleniidae) were formerly placed together in the Rhoeadales.

                order Fagales
                        Casuarinales
                        Urticales
                        Myricales
                        Juglandales
    subclass Rosidae (Rosiflorae)
      superorder Rosanae
        order Saxifragales
                Rosales
                Fabales (Leguminosae)
                Sarraceniales
                Podostemonales
      superorder Myrtanae
        order Myrtales
                Haloragales
                Elaeagnales
      superorder Rutanae
        order Rutales   ⎫
                Sapindales ⎬ Terebinthales
                       ⎭
                Geraniales (Gruinales)
                Polygalales
      superorder Celastranae
        order Celastrales
                Rhamnales
                Euphorbiales (Tricoccae)
                Santalales
      superorder Proteanae
        order Proteales
      superorder Aralianae
        order Cornales
                Araliales
    subclass Dilleniidae
      superorder Dillenianae
        order Dilleniales
                Theales (Guttiferae)
                Violales
                Capparales [34]
                Salicales [35]
                Begoniales
                Cucurbitales [36]
      superorder Malvanae
        order Malvales (Columniferae)
                Thymelaeales

34. See note 33.
35. Formerly placed in the Amentiferae.
36. Formerly placed in the Sympetalae.

superorder Ericanae
 order Ericales (Bicornes)
  Ebenales
  Primulales

subclass Caryophyllidae
 order Caryophyllales (Centrospermae)
  Polygonales
  Plumbaginales

subclass Asteridae
 superorder Lamianae
  order Gentianales (Contortae and part of Rubiales)
   Dipsacales (Rubiales in part)
   Oleales (Ligustrales)
 superorder Tubiflorae (broad sense)
  order Polemoniales (Tubiflorae in part)
   Scrophulariales (Personatae)
   Lamiales
 superorder Asteranae
  order Campanulales
   Asterales (Compositae)

class Monocotyledoneae (Liliatae, Liliopsida)
subclass Alismatidae (Helobiae)
 order Alismatales
  Hydrocharitales
  Naiadales

subclass Liliidae
 superorder Lilianae
  order Liliales (Liliiflorae)
   Orchidales (Gynandrae, Microspermae)
 superorder Bromelianae
  order Bromeliales
   Zingiberales (Scitamineae)
 superorder Juncanae (Junciflorae)
  order Juncales
   Cyperales
 superorder Commelinanae (Farinosae, broad sense)
  order Commelinales
   Eriocaulales
   Restionales
   Poales (Glumiflorae)

subclass Arecidae (Spadiciflorae)
 order Arecales (Principes)
  Cyclanthales
  Arales
  Pandanales
  Typhales

Alternative classification of Angiospermae I
subdivision Angiospermae
  class Dicotyledoneae
    subclass Archichlamydeae (Choripetalae)
      order Verticillatae
        Piperales
        Salicales
        Garryales
        Myricales
        Balanopsidales
        Leitneriales
        Juglandales
        Batidales
        Julianiales
        Fagales
        Urticales
        Proteales
        Santalales
        Aristolochiales
        Polygonales
        Centrospermae
        Ranales
        Rhoeadales
        Sarraceniales
        Rosales
        Pandales
        Geraniales
        Sapindales
        Rhamnales
        Malvales
        Parietales
        Opuntiales
        Myrtiflorae
        Umbelliflorae
    subclass Sympetalae
      order Ericales
        Primulales
        Plumbaginales
        Ebenales
        Contortae
        Tubiflorae
        Plantaginales
        Rubiales
        Cucurbitales
        Campanulales (Campanulatae)
  class Monocotyledoneae
      order Pandanales

Helobiae
Triuridales
Glumiflorae
Principes
Synanthae
Spathiflorae
Farinosae
Liliiflorae
Scitamineae
Microspermae *

Alternative classification of Angiospermae II
subdivision Angiospermae
class Dicotyledones
subclass Polypetalae
order Ranales
Parietales
Polygalinae
Caryophyllinae
Guttiferales
Malvales
Geraniales
Olacales
Celastrales
Sapindales
Rosales
Myrtales
Passiflorales
Ficoidales
Umbellales (Umbelliflorae)
subclass Gamopetallae
order Rubiales
Asterales
Campanulales
Ericales
Primulales
Ebenales
Gentianales
Polemoniales
Personales (Personatae)
Lamiales
subclass Monochlamydeae [37] (Incompletae, Apetalae)
class Monocotyledones [37]

* End of alternative classification.
37. No orders listed, only series and families.

B. CLASSIFICATION OF THE ANIMAL KINGDOM

subkingdom Protozoa
  phylum Protozoa
      class Flagellata (Mastigophora)
       subclass Phytomastigina
         order Chrysomonadina
             Cryptomonadina
             Euglenoidina
             Chloromonadina
             Dinoflagellata
             Volvocina (Phytomonadina)
       subclass Zoomastigina
           order Rhizomastigina [1]
               Holomastigina
               Protomonadina
               Polymastigina ⎫
               Hypermastigina ⎬ Metamonadina
               Opalina [2]
      class Sarcodina (Rhizopoda)
          order Amoebina
              Thalammophora ⎫
              Testacea ⎬ Foraminifera
              Radiolaria
              Acantharia [3]
              Heliozoa
              Mycetozoa
      class Sporozoa
       subclass Telosporidia
           order Coccidiomorpha
              Gregarinidea
       subclass Neosporidia
           order Cnidosporidia
              Acnidosporidia
      class Ciliophora
       subclass Ciliata
          order Holotricha
              Spirotricha
              Peritricha
          order Chonotricha
       subclass Suctoria (Acinetae)

1. May be placed in the Rhizopoda.
2. May be placed in a separate class, Opalinata.
3. Formerly placed in the Radiolaria.

Alternative classification of Ciliophora
    subclass Holotricha
            order Gymnostomatida
                Suctorida
                Trichostomatida
                Hymenostomatida
                Peritrichida
                Astomatida
                Chonotricha
    subclass Spirotricha
            order Heterotrichida
                Oligotrichida
                Entodiniomorphida
                Tintinnida
                Hypotrichida *

subkingdom Parazoa [4]
 phylum Porifera
  subphylum Nuda
    class Hexactinellida
        order Hexasterophora
            Amphidiscophora
  subphylum Gelatinosa
    class Calcarea
        order Homocoela
            Heterocoela
    Alternative classification of Calcarea
    subclass Calcinea
        order Clathrinida
            Leucettida
            Pharetronida
    subclass Calcaronea
        order Leucosoleniidae
            Sycettida *
    class Demospongea
      subclass Tetractinellida [5]
          order Myxospongida
             Carnosa
             Choristida
      subclass Monaxonida [5]
          order Hadromerina
          order Halichondrina
             Poecilosclerina
             Haplosclerina

---

* End of alternative classification.
4. The Parazoa and Mesozoa may be considered as Metazoa in the broad sense. All animals from Coelenterata upwards are then Eumetazoa.
5. May be considered as orders.

subclass Keratosa [5]

subkingdom Mesozoa [6]
  phylum Mesozoa
    class Orthonectida
    class Dicyemida (Rhombozoa)
        order Dicyemida
             Heterocyemida

subkingdom Metazoa—Eumetazoa [6]
  phylum Coelenterata [7]
   subphylum Cnidaria
    class Hydrozoa
        order Athecata (Anthomedusae, Gymnoblastea)
             Milleporina ⎫ Hydrocorallina
             Stylasterina ⎭
             Chondrophora
             Thecata (Leptomedusae, Calyptoblastea)
             Limnomedusae
       These 6 orders are sometimes placed in the single order Hydroida (Hydrida)
             Trachymedusae (Trachylina) ⎫ Trachylina
             Narcomedusae ⎭
             Siphonophora
             Actinulida
    class Scyphozoa (Scyphomedusae)
     subclass Stauromedusae
        order Stauromedusae
     subclass Discomedusae
        order Cubomedusae
             Coronatae
             Semaeostomeae
             Rhizostomeae
    class Anthozoa (Actinozoa)
     subclass Ceriantipatharia
        order Antipatharia
             Ceriantharia
     subclass Octocorallia (Octoradiata, Alcyonaria)
        order Alcyonacea
             Stolonifera
             Telestacea

6. See note 4.
7. The Coelenterata may be considered a superphylum and the Cnidaria and Ctenophora are then phyla.

order Coenothecalia
Gorgonacea
Pennatulacea
subclass Zoantharia (Hexactinia, Hexaradiata)
order Zoanthiniaria (Zoanthidea)
Corallimorpharia
Actiniaria
Scleractinia (Madreporaria)
subphylum Ctenophora
class Tentaculata
order Cydippida
Lobata
Cestida
Platyctenia
class Nuda
order Beroida

Acoelomates
phylum Platyhelminthes
class Turbellaria
order Acoela
Rhabdocoela
Alloeocoela (Alloiocoela)
Tricladida
Polycladida
class Temnocephaloidea [8]
order Temnocephalidea (Temnocephalea)
class Monogenea (Heterocotylea) [9]
subclass Monopisthocotylea
order Capsaloidea
Gyrodactyloidea
Acanthocotyloidea
Protogyrodactyloidea
Udonelloidea
subclass Polyopisthocotylea
order Chimaericoloidea
Diclidophoroidea
Diclybothrioidea
Polystomatoidea
class Cestodaria [10]
order Gyrocotyloidea
Amphilinoidea
class Cestoda
order Pseudophyllidea

8. May be considered an order of Turbellaria.
9. May be considered an order of Trematoda.
10. May be considered an order of Cestoda.

      order Haplobothrioidea
        Tetrarhynchoidea (Trypanorhyncha)
        Diphyllidea
        Tetraphyllidea
        Lecanicephaloidea
        Tetrabothrioidea
        Proteocephaloidea
        Nippotaeniidea
        Cyclophyllidea (Taeniidea)
   Alternative classification of Cestoda
   class Cestoda
      order Cestodaria
        Eucestoda*
  class Trematoda
   subclass Aspidogastrea (Aspidobothria)
   subclass Digenea (Malacocotylea)
   Alternative classification of Trematoda
   class Trematoda
      order Heterocotylea (Monogenea)
        Malacocotylea (Digenea)*
 phylum Nemertini (Nemertea, Rhynchocoela)
  class Anopla
      order Palaeonemertini
        Heteronemertini
  class Enopla
      order Hoplonemertini
        Bdellomorpha

Pseudocoelomates
 phylum Rotifera [11]
      order Seisonidea
        Bdelloidea
        Monogononta
 phylum Gastrotricha [11]
      order Macrodasyoidea
        Chaetonotoidea
 phylum Kinorhyncha [11]
 phylum Nematoda [11]
  class Aphasmidia (Adenophora)
      order Monhysterida
        Dorylaimida
        Chromadorida
        Enoplida
        Trichosyringida (Mermithoidea)
        Trichurata

* End of alternative classification.
11. May be considered classes of the phylum Aschelminthes.

order Dioctophymatida
class Phasmidia (Secernentea)
order Tylenchida
Rhabditida
Spirurida
Camallanida
Alternative classification of Phasmidia
order Ascaridata
Strongylata
Rhabditata
Spirurata
Camallanata *

phylum Nematomorpha [11]
order Gordioidea
Nectonematoidea

phylum Acanthocephala
order Archiacanthocephala
Palaeacanthocephala
Eoacanthocephala

phylum Endoprocta [12]
phylum Priapulida

Protostome coelomates
phylum Mollusca
class Monoplacophora [13]
order Triblidioidea
class Amphineura
subclass Polyplacophora [13]
order Lepidopleurida
Chitonida
subclass Aplacophora [13]
order Neomeniomorpha
Chaetodermomorpha
class Gastropoda
subclass Prosobranchia [14]
order Archaeogastropoda
Neritacea
Mesogastropoda
Neogastropoda (Stenoglossa)
subclass Opisthobranchia [15]
order Cephalaspidea (Bullomorpha)

* End of alternative classification.
12. Formerly, the Endoprocta and Ectoprocta were placed together in the phylum Bryozoa (Polyzoa).
13. May be considered as orders of the Amphineura.
14. May be considered an order of Gastropoda and also called Streptoneura.
15. May be considered an order of Gastropoda and also known as Euthyneura A.

order Anaspidea (Aplysiomorpha)
Thecosomata
Gymnosomata
Notaspidea (Pleurobranchomorpha)
Acochlidiacea
Sacoglossa
Acoela (Nudibranchia)
subclass Pulmonata [16]
order Basommatophora
Stylommatophora
class Scaphopoda
class Bivalvia (Lamellibranchiata, Pelecypoda)
subclass Protobranchia [17]
subclass Lamellibranchia [18]
order Taxodonta ⎫ Filibranchiata
Anisomyaria ⎭
Heterodonta ⎫
Schizodonta ⎪
Adapedonta ⎬ Eulamellibranchiata
Anomalodesmata ⎭
Septibranchia
class Cephalopoda (Siphonopoda)
subclass Nautiloidea
subclass Ammonoidea
subclass Coleoidea
order Decapoda
Octopoda
Vampyromorpha
Alternative classification of Cephalopoda
class Cephalopoda
order Dibranchiata
suborder Decapoda
Octopoda
order Tetrabranchiata
suborder Nautiloidea
Ammonoidea *
phylum Sipuncula [19]
phylum Echiura [19]
phylum Annelida
class Polychaeta [20]

16. May be considered an order of Gastropoda and also known as Euthyneura B.
17. May be considered an order of Lamellibranchiata and called Protobranchiata.
18. Not recognized in all classifications.
* End of alternative classification.
19. May be considered classes of Annelida and known as the Sipunculoidea and Echiuroidea respectively.
20. The Polychaeta and Oligochaeta may be considered as subclasses of the class Chaetopoda.

        order Amphinomorpha
               Eunicemorpha
               Phyllodocemorpha
               Spiomorpha
               Drilomorpha
               Oweniimorpha
               Sternaspimorpha
               Flabelligerimorpha
               Terebellomorpha
               Serpulimorpha
               Psammodrilomorpha
               Archiannelida [21]
   class Myzostomaria
   class Oligochaeta [20, 22]
        order Prosopora
               Plesiopora
               Opisthopora
   class Hirudinea [22]
        order Acanthobdellae
               Rhynchobdellae
               Gnathobdellae
               Pharyngobdellae
phylum Tardigrada [23]
        order Heterotardigrada
               Eutardigrada
phylum Pentastomida [23] (Linguatulida)
        order Cephalobaenida
               Porocephalida
phylum Onychophora [24]
phylum Arthropoda
  subphylum Trilobitomorpha
   class Trilobita
  subphylum Chelicerata
   class Merostomata [25]
        order Eurypterida
               Xiphosura
   class Arachnida
        order Scorpiones (Scorpionoidea)
               Pseudoscorpiones
               Solpugida (Solifuga)
               Palpigradi (Palpigrada)

21. May be considered a separate class of Annelida.
22. The Oligochaeta and Hirudinea may be considered as subclasses of the class Clitellata.
23. May be considered as classes or orders of Arachnida.
24. May be considered a subphylum or class of Arthropoda.
25. May not exist, and its orders may be considered orders of Arachnida.

       order Uropygi
          Amblypygi
          Araneida
          Ricinulei
          Opiliones (Phalangida)
          Acari (Acarina)
   class Pycnogonida (Pantopoda)[26]
       order Colossendeomorpha
          Nymphonomorpha
          Ascorhynchomorpha
          Pycnogonomorpha
 subphylum Mandibulata
  class Crustacea
   subclass Cephalocarida[27]
   subclass Branchiopoda
      order Anostraca
         Lipostraca
         Notostraca
         Conchostraca ⎫
         Cladocera ⎬ Diplostraca
   subclass Ostracoda
      order Myodocopa
         Cladocopa
         Podocopa
         Platycopa
   subclass Copepoda
      order Calanoida
         Harpacticoida
         Cyclopoida
         Notodelphyoida
         Monstrilloida
         Caligoida
         Lernaeopodoida
   subclass Mystacocarida
      order Derocheilocarida
   subclass Branchiura
   subclass Cirripedia
      order Thoracica
         Acrothoracica
         Ascothoracica
         Apoda
         Rhizocephala
   subclass Malacostraca

26. May be considered an order of Arachnida.
27. May be considered an order of Branchiopoda.

series Leptostraca [28]
  superorder Phyllocarida
    order Nebaliacea
series Eumalacostraca
  superorder Syncarida [28]
    order Anaspidacea
       Bathynellacea
  superorder Pancarida [28]
    order Thermosbaenacea
  superorder Peracarida
    order Mysidacea
       Cumacea
       Tanaidacea
       Gnathiidea
       Isopoda
       Spelaeogriphacea
       Amphipoda
  superorder Eucarida [28]
    order Euphausiacea
       Decapoda
  superorder Hoplocarida [28]
    order Stomatopoda
class Diplopoda [29]
  subclass Pselaphognatha
    order Polyxenida
  subclass Chilognatha
    superorder Pentazonia
      order Glomerida (Oniscomorpha)
         Glomeridesmida (Limacomorpha)
    superorder Helminthomorpha
      order Chordeumida
         Polydesmida
         Juliformia
    superorder Colobognatha
class Chilopoda [29]
  subclass Epimorpha
    order Geophilomorpha
       Scolopendromorpha
  subclass Anamorpha
    order Lithobiomorpha
       Scutigeromorpha
class Symphyla [29]
class Pauropoda [29]
class Insecta (Hexapoda)

---

28. May be considered as orders of Malacostraca.
29. These groups may be considered as subclasses of the class Myriapoda.

subclass Apterygota (Ametabola)
 superorder Entotropha
  order Collembola
   Protura
   Diplura
 superorder Ectotropha
  order Thysanura
subclass Pterygota (Metabola)
 section Palaeoptera (Exopterygota, Hemimetabola)
  superorder Ephemeropteroidea
   order Ephemeroptera
  superorder Odonatopteroidea
   order Odonata
 section Polyneoptera[30] (Polyneuroptera, Exopterygota, Heterometabola)
  superorder Blattopteroidea
   order Dictyoptera
    Isoptera
    Zoraptera
  superorder Orthopteroidea
   order Plectoptera
    Grylloblattodea (Notoptera)
    Phasmida (Cheleutoptera)
    Orthoptera
    Embioptera
  superorder Dermapteroidea
   order Dermaptera
 section Paraneoptera[30] (Paraneuroptera, Exopterygota, Heterometabola)
  superorder Psocopteroidea
   order Psocoptera
    Mallophaga
    Anoplura (Siphunculata)
  superorder Thysanopteroidea
   order Thysanoptera
  superorder Rhynchota (Hemiptera)
   order Homoptera ⎱ Hemiptera
     Heteroptera ⎰
 section Oligoneoptera[30] (Oligoneuroptera, Endopterygota, Holometabola)
  superorder Coleopteroidea
   order Coleoptera
  superorder Neuropteroidea
   order Megaloptera ⎱ Neuroptera
     Planipennia ⎰

30. May be placed together in the section Neoptera.

  order Raphidioptera
  superorder Mecopteroidea
   order Mecoptera
    Trichoptera
    Lepidoptera
    Zeugloptera
    Diptera
  superorder Siphonapteroidea
   order Siphonaptera
  superorder Hymenopteroidea
   order Hymenoptera
    Strepsiptera

Lophophorate coelomates
 phylum Phoronida (Phoronidea)
 phylum Ectoprocta[31]
  class Phylactolaemata
  class Gymnolaemata
    order Ctenostomata
     Cheilostomata
     Cyclostomata
 phylum Brachiopoda
  class Inarticulata
    order Atremata
     Neotremata
  class Articulata

Deuterostome ceolomates
 phylum Chaetognatha
 phylum Pogonophora[32]
    order Athecanephria
     Thecanephria
 phylum Hemichorda(ta)[33]
  class Pterobranchia
    order Rhabdopleurida
     Cephalodiscida
  class Enteropneusta
  class Planktosphaeroidea
  class Graptolita
    order Dendroidea
     Tuboidea
     Camaroidea
     Stolonoidea
     Graptoloidea

31. See note 12.
32. Formerly considered members of the Pterobranchiata.
33. May be considered a subphylum of Chordata, or with the Urochordata and Cephalochordata as subphyla of the phylum Protochordata.

phylum Echinodermata
  subphylum Pelmatozoa
    class Crinoidea
        order Articulata
           Comerata
           Inadunata
           Flexibilia
    class Cystoidea (Cystidea)
      subclass Hydrophoridea
        order Diploporita
           Rhombifera
      subclass Blastoidea
    class Eocrinoidea
    class Paracrinoidea
    class Edrioasteroidea
    class Carpoidea
    class Machaeridea
    class Cyamoidea
    class Cycloidea
  subphylum Eleutherozoa
    class Holothuroidea
        order Aspidochirota(e)
           Elasipoda
           Pelagothurida [34]
           Dendrochirota(e)
           Molpadonia
           Apoda (Synaptida, Paractinopoda)
    class Echinoidea
      subclass Perischoechinoidea
        order Cidaroidea
      subclass Euechinoidea
        superorder Diadematacea
        order Diadematoida
          Echinothuroida
          Pedinoida
        superorder Echinacea
        order Hemicidaroida
          Phymosomatoida
          Arbacoida
          Temnopleuroida
          Echinoida
        superorder Gnathostomata
        order Holectypoida
          Clypeasteroida
        superorder Atelostomata

34. May be included in the Elasipoda.

        order Nucleolitoida
              Cassiduloida
              Holasteroida
              Spatangoida

Alternative classification of Echinoidea
class Echinoidea
        order Endocyclica (Echinoida)
              Clypeasteroida ⎱ Exocyclica
              Spatangoida* ⎰
class Asteroidea [35]
        order Phanerozonia
              Spinulosa
              Forcipulata
class Ophiuroidea [35]
        order Ophiurae
              Euryalae
class Auluroidea
class Somasteroidea
phylum Chordata
  subphylum Urochorda(ta) [36]
    class Larvacea (Appendicularia)
        order Copelata
    class Ascidiacea
        order Enterogona
              Pleurogona
    class Thaliacea
        order Pyrosomidae (Luciae)
              Salpidae (Hemimyaria)
              Doliolidae (Cyclomyaria)
  subphylum Cephalochorda(ta) [36]
  subphylum Craniata (Vertebrata)
   Anamniota
    superclass Agnatha
    class Agnatha [37] (Cyclostomata)
      subclass Osteostraci [38]
        order Cephalaspida
              Anaspida
      sublcass Heterostraci [38]
        order Pteraspida
              Coelolepida (Thelodonti)

---

* End of alternative classification.
35. May be considered as subclasses of the class Stellaroidea.
36. See note 33.
37. The Agnatha, Placodermi, Chondrichthyes, and Osteichthyes were formerly considered to be subclasses of the class Pisces.
38. May be placed in the class Ostracodermi.

                    subclass Cyclostomata [39]
                              order Petromyzontia
                                    Myxinoidea
          superclass Gnathostomata
          class Placodermi [37] (Aphetohyoidea)
                              order Arthrodira (Arthrodiriformes)
                                    Rhenanidiformes
                                    Antiarchi (Antiarchiformes)
                                    Stensioelliformes
                                    Palaeospondyliformes
          class Chondrichthyes
            subclass Elasmobranchii
                              order Cladoselachii
                                    Selachii
                                    Batoidea
                                    Xenacanthodii
            subclass Holocephali (Bradyodonti)
          Alternative classification of Chondrichthyes [40]
          class Selachii
                              order Cladoselachiformes
                                    Cladodontiformes
                                    Xenacanthiformes
                                    Heterodontiformes
                                    Hexanchiformes
                                    Lamniformes
                                    Pristiophoriformes
                                    Rajiformes
          class Holocephali
                              order Chimaeriformes
                                    Copodontiformes
                                    Psammodontiformes
                                    Petalodontiformes
                                    Edestiformes
                                    Chondrenchelyiformes *
          class Osteichthyes [40]
            subclass Acanthodii [41]
                              order Climatiiformes
                                    Ischnacanthiformes
                                    Acanthodiformes
            subclass Actinopterygii [41]
              infraclass Chondrostei [42]

39. May be considered to be a class or reduced to an order.
* End of alternative classification.
40. The classes Chondrichthyes and Osteichthyes are not recognized in some classifications.
41. Considered to be classes when the Osteichthyes is not recognized.
42. The number of orders of Chondrostei may be reduced to Palaeonisciformes, Acipenseriformes, and Polypteriformes.

   order Palaeonisciformes
     Tarrasiiformes
     Haplolepiformes
     Perleidiformes
     Redfieldiiformes
     Dorypteriformes
     Bobasatraniiformes
     Pholidopleuriformes
     Peltopleuriformes
     Platysiagiformes
     Cephaloxeniformes
     Luganoiiformes
     Ptycholepiformes
     Saurichthyformes
     Chondrosteiformes
     Acipenseriformes (Acipenseroidea)
     Parasemionotiformes
     Polypteriformes (Polypterini,
     Brachiopterygii)
  infraclass Holostei [43]
   division Holosteans
    order Amiiformes
     Pachycormiformes
     Semionotiformes (Semionotoidea)
     Pycnodontiformes
   division Halecostomi
    order Pholidophoriformes
  infraclass Teleostei
    superorder Leptolepidomorpha
    order Leptolepidiformes
    superorder Elopomorpha
    order Elopiformes
     Anguilliformes (Apodes)
     Notacanthiformes
    superorder Clupeomorpha
    order Clupeiformes
    superorder Osteoglossomorpha
    order Osteoglossiformes
     Mormyriformes
    superorder Protacanthopterygii
    order Salmoniformes
     Ctenothrissiformes
     Gonorhynchiformes
    superorder Ostariophysi
    order Cypriniformes

---

43. The number of orders of Holostei may be reduced to the Semionotiformes and Amiiformes.

order Siluriformes
superorder Scopelomorpha
order Myctophiformes
superorder Paracanthopterygii
order Polymixiiformes
Percopsiformes
Gadiformes
Batrachoidiformes
Lophiiformes
Gobiesociformes
superorder Acanthopterygii
order Atheriniformes
Lampridiformes
Beryciformes
Zeiformes
Gasterosteiformes
Channiformes
Synbranchiformes
Scorpaeniformes
Dactylopteriformes
Pegasiformes
Perciformes
Pleuronectiformes
Tetraodontiformes

Alternative classification of Teleostei I
infraclass Teleostei
order Isospondyli
Ostariophysi
Apodes
Mesichthyes
Acanthopterygii *

Alternative classification of Teleostei II
infraclass Teleostei
order Isospondyli (Clupeiformes)
Scopeliformes (Iniomi)
Giganturiformes
Saccopharyngiformes (Lyomeri)
Ostariophysi (Cypriniformes)
Apodes (Anguilliformes)
Notacanthiformes (Heteromi)
Beloniformes (Synentognathi)
Gadiformes (Anacanthini)
Gasterosteiformes $\Big\}$ Solenichthyes
Syngnathiformes
Lampridiformes (Allotriognathi)

---

* End of alternative classification.

     order Cyprinodontiformes (Microcyprini)
        Percopsiformes (Salmopercae)
        Beryciformes
        Zeiformes
        Mugiliformes
        Ophiocephaliformes
        Symbranchiformes
        Perciformes
        Dactylopteriformes
        Pleuronectiformes (Heterosomata)
        Mastacembeliformes (Opisthomi)
        Echeneiformes (Discocephali)
        Tetraodontiformes (Plectognathi)
        Batrachoidiformes (Haplodoci)
        Lophiiformes (Pediculati)
        Pegasiformes (Hypostomides)*
  class Sarcopterygii[44] (Choanichthyes)
   subclass Crossopterygii[45]
      order Rhipidistia
        Coelacanthina
   subclass Dipnoi[45]
  class Amphibia
   subclass Labyrinthodonta
      order Ichthyostegalia
        Temnospondyli
        Anthracosauria (Batrachosauria)
        Plagiosauria
   subclass Lepospondyli
      order Aistopoda
        Nectridea
        Microsauria
        Lysorophia
   subclass Lissamphibia
      order Anura
        Urodela
        Apoda (Gymnophiona)
Alternative classification of Amphibia I
  class Amphibia
   subclass Apsidospondyli
      superorder Salientia
        order Anura
   subclass Lepospondyli

---

* End of alternative classification.
44. May be considered to be a class or subclass, or may not be recognized, and the Dipnoi and Crossopterygii are then considered to be subgroups of Osteichthyes.
45. May be considered to be orders of Sarcopterygii.

                              order Aistopoda
                                    Nectridea
                                    Microsauria
                                    Urodela
                                    Apoda (Gymnophiona)*
          Alternative classification of Amphibia II
          class Amphibia
                        superorder Ichthyostegalia
                        superorder Temnospondyli
                              order Rhachitomi
                                    Stereospondyli
                                    Plagiosauria
                        superorder Lepospondyli
                              order Aistopoda
                                    Nectridea
                                    Trachysomata
                                    Microsauria
                                    Apoda (Gymnophiona)
                                    Caudata (Urodela)
                        superorder Salientia
                              order Proanura
                                    Anura
                        superorder Anthracosauria
                              order Schizomeri
                                    Embolomeri
                                    Seymouriamorpha[46]
                                    Diplomeri*
          Amniota
              class Reptilia
                  subclass Anapsida
                              order Procolophonia
                                    Captorhinomorpha } Cotylosauria[47]
                                    Chelonia (Testudinata)
                  subclass Lepidosauria[48]
                              order Eosuchia
                                    Rhynchocephalia
                                    Squamata
                  subclass Archosauria[48]
                              order Thecodontia[49]
                                    Crocodilia
                                    Saurischia

---

* End of alternative classification.
46. May be considered to be reptiles and placed in the Cotylosauria.
47. May be subdivided into the Cotylosauria and Mesosauria.
48. May be considered as superorders of the subclass Diapsida.
49. May be placed in the Lepidosauria.

order Ornithischia
    Pterosauria
    Pseudosuchia [50]
    Phytosauria [50]
subclass Sauropterygii [51]
    order Nothosauria
    Plesiosauria
    Placodonti
subclass Ichthyopterygia [51]
    order Ichthyosauria
subclass Synapsida
    order Pelycosauria
    Therapsida
class Aves
subclass Archaeornithes
    order Archaeornithiformes
subclass Neornithes
    superorder Odontognathae
    order Hesperornithiformes
    superorder Palaeognathae [52]
    order Tinamiformes
    Rheiformes
    Struthioniformes
    Casuariiformes
    Aepyornithiformes
    Dinornithiformes
    Apterygiformes
    superorder Neognathae (Carinatae)
    order Colymbiformes (Podicipitiformes)
    Procellariiformes
    Sphenisciformes [53]
    Pelecaniformes
    Anseriformes
    Phoenicopteriformes
    Ciconiiformes
    Falconiformes
    Galliformes
    Gruiformes
    Ichthyornithiformes [54]
    Charadriiformes
    Gaviiformes

50. May be placed in the Thecodontia.
51. May be considered to be superorders of the subclass Parapsida.
52. May be placed in the superorder Neognathae.
53. May be placed in the superorder Impennae.
54. May be placed in the Odontognathae.

order Columbiformes
Psittaciformes
Cuculiformes
Strigiformes
Caprimulgiformes
Apodiformes (Micropodiformes)
Coliiformes
Trogoniformes
Coraciiformes
Piciformes
Passeriformes
class Mammalia
subclass Prototheria
order Monotremata
subclass Allotheria [55]
order Multituberculata
subclass Eotheria
order Morganucondonta
Triconodonta
Docodonta
subclass Theria
infraclass Trituberculata
order Symmetrodonta
Pantotheria
infraclass Metatheria
order Marsupialia
infraclass Eutheria (Placentalia, Monodelphia)
order Insectivora
Macroscelidea
Dermoptera
Tillodonta
Taeniodonta
Chiroptera
Primata (Primates)
Edentata
Pholidota
Tubulidentata
Lagomorpha
Rodentia
Creodonta
Carnivora
Archaeoceti
Odontoceti   Cetacea
Mysticeti
Condylarthra

55. May be considered to be an infraclass of Theria.

order Pyrotheria
      Pantodonta
      Dinocerata
      Proboscidea
      Sirenia
      Desmostylia
      Hyracoidea
      Notoungulata
      Astrapotheria
      Litopterna
      Perissodactyla
      Artiodactyla

C. COMMON CHEMICAL ELEMENTS

Element	symbol	atomic number	atomic mass
aluminium	Al	13	26.98
antimony	Sb	51	121.75
barium	Ba	56	137.34
bismuth	Bi	83	208.98
boron	B	5	10.81
bromine	Br	35	79.90
cadmium	Cd	48	112.40
calcium	Ca	20	40.08
carbon	C	6	12.01
chlorine	Cl	17	35.45
chromium	Cr	24	51.99
cobalt	Co	27	58.93
copper	Cu	29	63.54
fluorine	F	9	19.00
gold	Au	79	196.97
helium	He	2	4.00
hydrogen	H	1	1.01
iodine	I	53	126.90
iron	Fe	26	55.85
lead	Pb	82	207.2
magnesium	Mg	12	24.30
manganese	Mn	25	54.94
mercury	Hg	80	200.59
molybdenum	Mo	42	95.94
neon	Ne	10	20.17
nickel	Ni	28	58.71
nitrogen	N	7	14.01
oxygen	O	8	16.00
phosphorus	P	15	30.97
potassium	K	19	39.10
silicon	Si	14	28.09
silver	Ag	47	107.87
sodium	Na	11	23.00
strontium	Sr	38	87.62
sulphur	S	16	32.06
tin	Sn	50	118.69
titanium	Ti	22	47.90
zinc	Zn	30	65.37